Mastery Learning. Proven

With eBook Integration

ALEKS is a unique, online program that uses artificial intelligence and adaptive questioning proven to raise student proficiency and success rates in math.

ALEKS Delivers a Unique Math Experience:

- **Research-Based, Artificial Intelligence** precisely measures each student's knowledge
- **Individualized Learning** presents the exact topics each student is most **ready to learn**
- **Adaptive, Open-Response Environment** includes comprehensive tutorials and resources
- **Detailed, Automated Reports** track student and class progress toward course mastery
- **Course Management Tools** include textbook integration, custom features, and more

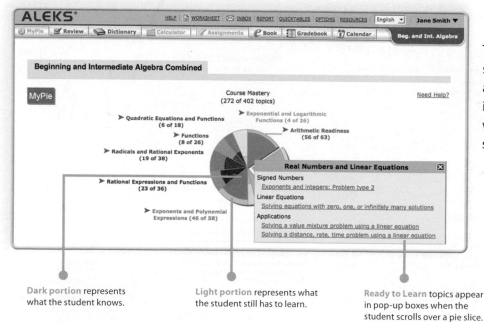

The ALEKS Pie summarizes a student's current knowledge and then delivers an individualized learning path with the exact topics the student is most ready to learn.

Dark portion represents what the student knows.

Light portion represents what the student still has to learn.

Ready to Learn topics appear in pop-up boxes when the student scrolls over a pie slice.

With ALEKS 360, the multimedia eBook is connected to every problem so students can quickly review the exact section they are working on.

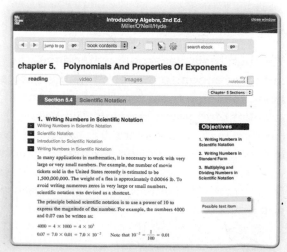

ALEKS is a registered trademark of ALEKS Corporation.

To see ALEKS in action, please visit: **www.SuccessInMath.com**

Visualizing Math Concepts

Dynamic Math Animations

The Miller/O'Neill/Hyde author team has developed a series of Flash animations to illustrate difficult concepts where static images and text fall short. The animations leverage the use of on-screen movement and morphing shapes to enhance conceptual learning.

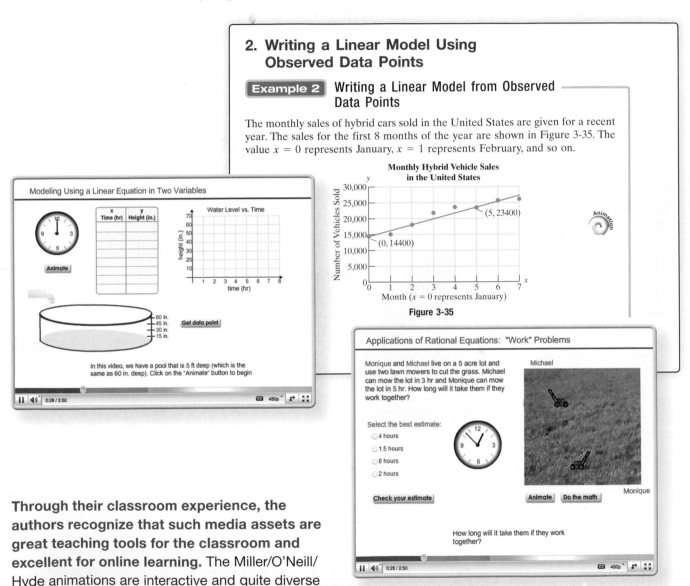

2. Writing a Linear Model Using Observed Data Points

Example 2 Writing a Linear Model from Observed Data Points

The monthly sales of hybrid cars sold in the United States are given for a recent year. The sales for the first 8 months of the year are shown in Figure 3-35. The value $x = 0$ represents January, $x = 1$ represents February, and so on.

Monthly Hybrid Vehicle Sales in the United States

$(5, 23400)$

$(0, 14400)$

Month ($x = 0$ represents January)

Figure 3-35

Modeling Using a Linear Equation in Two Variables

Water Level vs. Time

In this video, we have a pool that is 5 ft deep (which is the same as 60 in. deep). Click on the "Animate" button to begin

Applications of Rational Equations: "Work" Problems

Monique and Michael live on a 5 acre lot and use two lawn mowers to cut the grass. Michael can mow the lot in 3 hr and Monique can mow the lot in 5 hr. How long will it take them if they work together?

Select the best estimate:
- 4 hours
- 1.5 hours
- 8 hours
- 2 hours

Check your estimate

Michael

Monique

How long will it take them if they work together?

Through their classroom experience, the authors recognize that such media assets are great teaching tools for the classroom and excellent for online learning. The Miller/O'Neill/Hyde animations are interactive and quite diverse in their use. Some provide a virtual laboratory for which an application is simulated and where students can collect data points for analysis and modeling. Others provide interactive question-and-answer sessions to test conceptual learning. For word problem applications, the animations ask students to estimate answers and practice "number sense."

Intermediate Algebra

THIRD EDITION

Julie Miller
Daytona State College

Molly O'Neill
Daytona State College

Nancy Hyde
Broward College—
Professor Emeritus

Mc
Graw
Hill
Education

INTERMEDIATE ALGEBRA, THIRD EDITION

Published by McGraw-Hill Education, 2 Penn Plaza, New York, NY 10121. Copyright © 2015 by McGraw-Hill Education. All rights reserved. Printed in the United States of America. Previous editions © 2013, 2010, and 2007. No part of this publication may be reproduced or distributed in any form or by any means, or stored in a database or retrieval system, without the prior written consent of McGraw-Hill Education, including, but not limited to, in any network or other electronic storage or transmission, or broadcast for distance learning.

Some ancillaries, including electronic and print components, may not be available to customers outside the United States.

This book is printed on acid-free paper.

4 5 6 7 8 9 0 DOW/DOW 1 0 9 8 7 6

ISBN 978–0–07–338442–9
MHID 0–07–338442–9

ISBN 978–0–07–734297–5 (Annotated Instructor's Edition)
MHID 0–07–734297–6

Senior Vice President, Products & Markets: *Kurt L. Strand*
Vice President, General Manager, Products & Markets: *Marty Lange*
Vice President, Content Production & Technology Services: *Kimberly Meriwether David*
Managing Director: *Ryan Blankenship*
Brand Manager: *Mary Ellen Rahn*
Director of Marketing: *Alex Gay*
Director of Development: *Rose Koos*
Development Editor: *Emily Williams*
Director of Digital Content: *Nicole Lloyd*
Director, Content Production: *Terri Schiesl*
Content Project Manager: *Peggy Selle*
Buyer: *Nicole Baumgartner*
Senior Designer: *Laurie B. Janssen*
Cover Illustration: *Imagineering Media Services, Inc.*
Lead Content Licensing Specialist: *Carrie K. Burger*
Compositor: *Aptara®, Inc.*
Typeface: *10/12 Times Ten Roman*
Printer: *R. R. Donnelley*

Library of Congress Cataloging-in-Publication Data

Miller, Julie, 1962-
  Intermediate algebra / Julie Miller, Molly O'Neill, Nancy Hyde.–Third edition.
    pages cm
  Includes index.
  ISBN 978–0–07–338442–9—ISBN 0–07–338442–9—ISBN 978–0–07–734297–5—
ISBN 0–07–734297–6 (hard copy : alk. paper) 1. Algebra–Textbooks.
  I. O'Neill, Molly, 1953- II. Hyde, Nancy. III. Title.

QA154.3.M554 2014b
512—dc23

          2013026119

www.mhhe.com

About the Authors

Julie Miller is from Daytona State College, where she has taught developmental and upper-level mathematics courses for 20 years. Prior to her work at Daytona State College, she worked as a software engineer for General Electric in the area of flight and radar simulation. Julie earned a bachelor of science in applied mathematics from Union College in Schenectady, New York, and a master of science in mathematics from the University of Florida. In addition to her textbooks in developmental mathematics, Julie has authored a college algebra textbook and several course supplements for college algebra, trigonometry, and precalculus.

"My father is a medical researcher, and I got hooked on math and science when I was young and would visit his laboratory. I can remember using graph paper to plot data points for his experiments and doing simple calculations. He would then tell me what the peaks and features in the graph meant in the context of his experiment. I think that applications and hands-on experience made math come alive for me and I'd like to see math come alive for my students."

—*Julie Miller*

Molly O'Neill is also from Daytona State College, where she has taught for 22 years in the School of Mathematics. She has taught a variety of courses from developmental mathematics to calculus. Before she came to Florida, Molly taught as an adjunct instructor at the University of Michigan–Dearborn, Eastern Michigan University, Wayne State University, and Oakland Community College. Molly earned a bachelor of science in mathematics and a master of arts and teaching from Western Michigan University in Kalamazoo, Michigan. Besides this textbook, she has authored several course supplements for college algebra, trigonometry, and precalculus and has reviewed texts for developmental mathematics.

"I differ from many of my colleagues in that math was not always easy for me. But in seventh grade I had a teacher who taught me that if I follow the rules of mathematics, even I could solve math problems. Once I understood this, I enjoyed math to the point of choosing it for my career. I now have the greatest job because I get to do math every day and I have the opportunity to influence my students just as I was influenced. Authoring these texts has given me another avenue to reach even more students."

—*Molly O'Neill*

Nancy Hyde served as a full-time faculty member of the Mathematics Department at Broward College for 24 years. During this time she taught the full spectrum of courses from developmental math through differential equations. She received a bachelor of science degree in math education from Florida State University and a master's degree in math education from Florida Atlantic University. She has conducted workshops and seminars for both students and teachers on the use of technology in the classroom. In addition to this textbook, she has authored a graphing calculator supplement for *College Algebra*.

"I grew up in Brevard County, Florida, where my father worked at Cape Canaveral. I was always excited by mathematics and physics in relation to the space program. As I studied higher levels of mathematics I became more intrigued by its abstract nature and infinite possibilities. It is enjoyable and rewarding to convey this perspective to students while helping them to understand mathematics."

—*Nancy Hyde*

Dedications

Contents

Get Better Results

How Will Miller/O'Neill/Hyde Help Your Students *Get Better Results?*

Better Clarity, Quality, and Accuracy!

Julie Miller, Molly O'Neill, and Nancy Hyde know what students need to be successful in mathematics. Better results come from clarity in their exposition, quality of step-by-step worked examples, and accuracy of their exercises sets; but it takes more than just great authors to build a textbook series to help students achieve success in mathematics. Our authors worked with a strong team of mathematics instructors from around the country to ensure that the clarity, quality, and accuracy you expect from the Miller/O'Neill/Hyde series was included in this edition.

> "The most complete text at this level in its thoroughness, accuracy, and pedagogical soundness. The best developmental mathematics text I have seen."
> —Frederick Bakenhus, *Saint Phillips College*

Better Exercise Sets!

Comprehensive sets of exercises are available for every student level. Julie Miller, Molly O'Neill, and Nancy Hyde worked with a board of advisors from across the country to offer the appropriate depth and breadth of exercises for your students. **Problem Recognition Exercises** were created to improve student performance while testing.

Practice exercise sets help students progress from skill development to conceptual understanding. Student tested and instructor approved, the Miller/O'Neill/Hyde exercise sets will help your students *get better results*.

- ▶ **Problem Recognition Exercises**
- ▶ **Skill Practice Exercises**
- ▶ **Study Skills Exercises**
- ▶ **Concept Connections Exercises**
- ▶ **Mixed Exercises**
- ▶ **Expanding Your Skills Exercises**
- ▶ **Vocabulary and Key Concepts Exercises**

> "This series was thoughtfully constructed with students' needs in mind. The Problem Recognition section was extremely well designed to focus on concepts that students often misinterpret."
> —Christine V. Wetzel-Ulrich, *Northampton Community College*

Better Step-By-Step Pedagogy!

Intermediate Algebra provides enhanced step-by-step learning tools to help students *get better results*.

- ▶ **Worked Examples** provide an "easy-to-understand" approach, clearly guiding each student through a step-by-step approach to master each practice exercise for better comprehension.

- ▶ **TIPs** offer students extra cautious direction to help improve understanding through hints and further insight.

- ▶ **Avoiding Mistakes** boxes alert students to common errors and provide practical ways to avoid them. Both of these learning aids will help students get better results by showing how to work through a problem using a clearly defined step-by-step methodology that has been class tested and student approved.

> "The book is designed with both instructors and students in mind. I appreciate that great care was used in the placement of 'Tips' and 'Avoiding Mistakes' as it creates a lot of teachable moments in the classroom."
> —Shannon Vinson, *Wake Tech Community College*

Get Better Results

Formula for Student Success

Step-by-Step Worked Examples

▶ Do you get the feeling that there is a disconnection between your students' class work and homework?

▶ Do your students have trouble finding worked examples that match the practice exercises?

▶ Do you prefer that your students see examples in the textbook that match the ones you use in class?

Miller/O'Neill/Hyde's *Worked Examples* offer a clear, concise methodology that replicates the mathematical processes used in the authors' classroom lectures!

Example 2 Solving a Linear Equation Involving Consecutive Integers

Three times the sum of two consecutive odd integers is 516. Find the integers.

Solution:

Step 1: Read the problem carefully.

Step 2: Label the unknown:
Let x represent the first odd integer.
Then $x + 2$ represents the next odd integer.

Step 3: Write an equation in words.

$3[(\text{first odd integer}) + (\text{second odd integer})] = 516$

Step 4: Write a mathematical equation.

Step 5: Solve for x.

Skill Practice

2. Four times the sum of three consecutive integers is 264. Find the integers.

$\longrightarrow 85$

$x + 2$:

$5 + 2 \longrightarrow 87$

"As always, MOH's Worked Examples are so clear and useful for the students. All steps have wonderfully detailed explanations written with wording that the students can understand. MOH is also excellent with arrows and labels making the Worked Examples extremely clear and understandable."

—Kelli Hammer, *Broward College–South*

"Easy to read step-by-step solutions to sample textbook problems. The 'why' is provided for students, which is invaluable when working exercises without available teacher/tutor assistance."

—Arcola Sullivan, *Copiah-Lincoln Community College*

Classroom Examples

To ensure that the classroom experience also matches the examples in the text and the practice exercises, we have included references to even-numbered exercises to be used as Classroom Examples. These exercises are highlighted in the Practice Exercises at the end of each section.

Skill Practice

Solve the system by graphing.
6. $y = 2x - 3$
   $6x + 2y = 4$

Example 3 Solving a System of Linear Equations by Graphing

Solve the system by the graphing method.

$$x - 2y = -2$$
$$-3x + 2y = 6$$

30. $x + y = -1$
    $2x - y = -5$

Better Learning Tools

Concept Connection Boxes

Concept Connections help students understand the conceptual meaning of the problems they are solving—a vital skill in mathematics.

Concept Connections

Determine if the given value is a solution to the equation.
1. $x + 2 = 8$; -10
2. $3y - 1 = 11$; 4
3. Write a linear equation whose solution is 3. (Answers may vary.)

Equation	Solution	Check
$p + 3 = 11$	8	$p + 3 = 11$ $\downarrow$ $8 + 3 = 11$ ✓	Substitute 8 for p. Right-hand side equals left-hand side.
$-2z = -20$	10	$-2z = -20$ $\downarrow$ $-2(10) = -20$ ✓	Substitute 10 for z. Right-hand side equals left-hand side.

TIP and Avoiding Mistakes Boxes

TIP and **Avoiding Mistakes** boxes have been created based on the authors' classroom experiences—they have also been integrated into the **Worked Examples.** These pedagogical tools will help students get better results by learning how to work through a problem using a clearly defined step-by-step methodology.

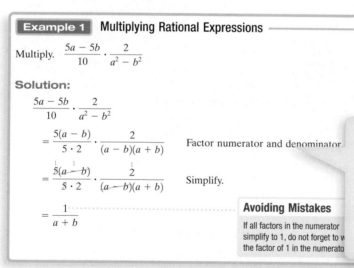

Example 1 Multiplying Rational Expressions

Multiply. $\dfrac{5a - 5b}{10} \cdot \dfrac{2}{a^2 - b^2}$

Solution:

$$\dfrac{5a - 5b}{10} \cdot \dfrac{2}{a^2 - b^2}$$

$$= \dfrac{5(a - b)}{5 \cdot 2} \cdot \dfrac{2}{(a - b)(a + b)} \qquad \text{Factor numerator and denominator.}$$

$$= \dfrac{\overset{1}{5}(a \overset{1}{- b})}{5 \cdot 2} \cdot \dfrac{2}{(a - b)(a + b)} \qquad \text{Simplify.}$$

$$= \dfrac{1}{a + b}$$

Avoiding Mistakes

If all factors in the numerator simplify to 1, do not forget to w... the factor of 1 in the numerato...

Avoiding Mistakes Boxes:

Avoiding Mistakes boxes are integrated throughout the textbook to alert students to common errors and how to avoid them.

TIP Boxes

Teaching tips are usually revealed only in the classroom. Not anymore! TIP boxes offer students helpful hints and extra direction to help improve understanding and provide further insight.

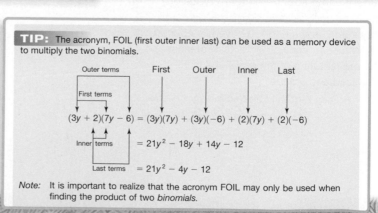

TIP: The acronym, FOIL (first outer inner last) can be used as a memory device to multiply the two binomials.

$$(3y + 2)(7y - 6) = (3y)(7y) + (3y)(-6) + (2)(7y) + (2)(-6)$$

$$= 21y^2 - 18y + 14y - 12$$

$$= 21y^2 - 4y - 12$$

Note: It is important to realize that the acronym FOIL may only be used when finding the product of two *binomials.*

Get Better Results

Better Exercise Sets! Better Practice! Better Results!

▶ Do your students have trouble with problem solving?

▶ Do you want to help students overcome math anxiety?

▶ Do you want to help your students improve performance on math assessments?

Problem Recognition Exercises

Problem Recognition Exercises present a collection of problems that look similar to a student upon first glance, but are actually quite different in the manner of their individual solutions. Students sharpen critical thinking skills and better develop their "solution recall" to help them distinguish the method needed to solve an exercise—an essential skill in developmental mathematics.

Problem Recognition Exercises were tested in the authors' developmental mathematics classes and were created to improve student performance on tests.

> "The PREs are an excellent source of additional mixed problem sets. Frequently students have questions/comments like 'Where do I start?' or 'I know what to do once I get started, but I have trouble getting started.' Perhaps with these PREs, students will be able to overcome this obstacle."
>
> —Erika Blanken, *Daytona State College*

Problem Recognition Exercises

Recognizing Equations and Inequalities

At this point, you have learned how to solve a variety of equations and type of problem being posed is the first step in successfully solving it.

For Exercises 1–20,

a. Identify the problem type. Choose from

- linear equation
- quadratic equation
- rational equation
- absolute value equation
- radical equation
- equation quadratic in form
- polynomial equation
- linear inequality
- polynomial inequality
- rational inequality
- absolute value inequality
- compound inequality

b. Solve the equation or inequality. Write the solution to each inequality in interval notation if possible.

1. $(z^2 - 4)^2 - (z^2 - 4) - 12 = 0$　　**2.** $3 + |4t - 1| < 6$　　**3.** $2y(y - 4) \leq 5 + y$

4. $\sqrt[3]{11x - 3} + 4 = 6$　　**5.** $-5 = -|w - 4|$　　**6.** $\dfrac{5}{x - 2} + \dfrac{3}{x + 2} = 1$

7. $m^3 + 5m^2 - 4m - 20 \geq 0$　　**8.** $-x - 4 > -5$ and $2x - 3 \leq 23$　　**9.** $5 - 2[3 - (x - 4)] \leq 3x + 14$

10. $|2x - 6| = |x + 3|$　　**11.** $\dfrac{3}{x - 2} \leq 1$　　**12.** $9 < |x + 4|$

14. $(4x - 3)^2 = -10$　　**15.** $-4 - x > 2$ or $8 < 2x$

17. $x^2 - 10x \leq -25$　　**18.** $\dfrac{10}{x^2 + 1} < 0$

20. $x^4 - 13x^2 + 36 = 0$

> "These are so important to test whether a student can recognize different types of problems and the method of solving each. They seem very unique—I have not noticed this feature in many other texts or at least your presentation of the problems is very organized and unique."
>
> —Linda Kuroski, *Erie Community College*

Student Centered Applications!

The Miller/O'Neill/Hyde Board of Advisors partnered with our authors to bring the *best applications* from every region in the country! These applications include real data and topics that are more relevant and interesting to today's student.

Mixed Exercises

37. How much pure gold (24K) must be mixed with 60% gold to get 20 grams (g) of 75% gold?

38. A granola mix contains 5% nuts. How many ounces of nuts must be added to get 25 oz of granola with 24% nuts?

39. A rowing team trains on the Halifax River. It can row upstream 10 mi against the current in 2.5 hr and 16 mi downstream with the current in the same amount of time. Find the speed of the boat in still water and the speed of the current.

40. In her kayak, Bonnie can travel 31.5 mi downstream with the current in 7 hr. The return trip against the current takes 9 hr. Find the speed of the kayak in still water and the speed of the current.

41. There are two types of tickets sold at the Canadian Formula One Grand Prix race. The price of 6 grandstand tickets and 2 general admission tickets costs $2330. The price of 4 grandstand tickets and 4 general admission tickets cost $2020. What is the price of each type of ticket?

42. A basketball player scored 19 points by shooting two-point and three-point baskets. If she made a total of eight baskets, how many of each type did she make?

Group Activities!

Each chapter concludes with a Group Activity to promote classroom discussion and collaboration—helping students not only to solve problems but to explain their solutions for better mathematical mastery. Group Activities are great for both full-time and adjunct instructors—bringing a more interactive approach to teaching mathematics! All required materials, activity time, and suggested group sizes are provided in the end-of-chapter material.

Group Activity

Computing the Future Value of an Investment

Materials: Calculator

Estimated time: 15 minutes

Group Size: 3

> "MOH's group activity involves true participation and interaction."
> —Monika Bender, *Central Texas College*

Suppose you are able to save $100 per month. If you invest the money in an account t̶h̶...̶p̶ays 6% annual interest, how much money would you have at the end of 10 yr? This question can be ̶answered by using the following formula.

$$S = R\left[\frac{(1 + i)^n - 1}{i}\right]$$

Where
S is the future value of the investment.
R is the amount saved per period.
i is the interest rate per period.
n is the total number of periods.

In this example, $R = \$100$ (amount invested per month)

$$i = \frac{0.06}{12} = 0.005$$ (annual interest rate divided by 12 months)

$$n = (12)(10) = 120$$ (12 months per year times 10 years)

Therefore, $S = \$100\left[\dfrac{(1 + 0.005)^{120} - 1}{0.005}\right]$

$$S = \$16,387.93$$

... future value of an account if you save $150 per month for 30 yr at an annual interest rate

_____   _____

_____ $S =$ _____

_____

...nvest $5000 in an account on July 1 each year for 20 yr. If the annual growth on your ...8%, how much will the account be worth in 20 yr?

_____

> "This is one part of the book that would have me adopt the MOH book. I am very big on group work and many times it is difficult to think of an activity. I would conclude the chapter doing the group activity in the class. Many books just have problems for this, but the MOH book provides an actual activity."
> —Sharon Giles, *Grossmont College*

Get Better Results

Additional Supplements

Media Suite

NEW Lecture Videos Created by the Authors

Julie Miller began creating these lecture videos for her own students to use when they were absent from class. The student response was overwhelmingly positive, prompting the author team to create the lecture videos for their entire developmental math book series. In these new videos, the authors walk students through the learning objectives using the same language and procedures outlined in the book. Students learn and review right alongside the author! Students can also access the written notes that accompany the videos.

Dynamic Math Animations

The authors have constructed a series of Flash animations to illustrate difficult concepts where static images and text fall short. The animations leverage the use of on-screen movement and morphing shapes to give students an interactive approach to conceptual learning. Some provide a virtual laboratory for which an application is simulated and where students can collect data points for analysis and modeling. Others provide interactive question-and-answer sessions to test conceptual learning.

NEW Exercise Videos

The authors, along with a team of faculty who have used the Miller/O'Neill/Hyde textbooks for many years, have created new exercise videos for designated exercises in the textbook. These videos cover a representative sample of the main objectives in each section of the text. Each presenter works through selected problems, following the solution methodology employed in the text.

The video series is available online as part of Connect Math hosted by ALEKS as well as in ALEKS 360. The videos are closed-captioned for the hearing impaired, and meet the Americans with Disabilities Act Standards for Accessible Design.

Student Resource Manual

The *Student Resource Manual (SRM),* created by the authors, is a printable, electronic supplement available to students through Connect Math hosted by ALEKS. Instructors can also choose to customize this manual and package with their course materials. With increasing demands on faculty schedules, this resource offers a convenient means for both full-time and adjunct faculty to promote active learning and success strategies in the classroom.

This manual supports the series in a variety of different ways:

- NEW Additional Group Activities developed by the authors to supplement what is already available in the text.
- Discovery-based classroom activities written by the authors for each section
- Worksheets for extra practice written by the authors including Problem Recognition Exercise Worksheets
- NEW Lecture Notes designed to help students organize and take notes on key concepts
- Materials for a student portfolio

Annotated Instructor's Edition

In the *Annotated Instructor's Edition* (*AIE*), answers to all exercises appear adjacent to each exercise in a color used *only* for annotations. The *AIE* also contains Instructor Notes that appear in the margin. These notes offer instructors assistance with lecture preparation. In addition, there are Classroom Examples referenced in the text that are highlighted in the Practice Exercises. Also found in the *AIE* are icons within the Practice Exercises that serve to guide instructors in their preparation of homework assignments and lessons.

Powerpoints

The Powerpoints present key concepts and definitions with fully editable slides that follow the textbook. An instructor may project the slides in class or post to a website in an online course.

Instructor's Solutions Manual

The *Instructor's Solutions Manual* provides comprehensive, worked-out solutions to all exercises in the Chapter Openers, the Practice Exercises, the Problem Recognition Exercises, the end-of-chapter Review Exercises, the Chapter Tests, and the Cumulative Review Exercises.

Student's Solutions Manual

The *Student's Solutions Manual* provides comprehensive, worked-out solutions to the odd-numbered exercises in the Practice Exercise sets, the Problem Recognition Exercises, the end-of-chapter Review Exercises, the Chapter Tests, and the Cumulative Review Exercises. Answers to the Chapter Opener Puzzles are also provided.

Instructor's Test Bank

Among the supplements is a computerized test bank using the algorithm-based testing software TestGen® to create customized exams quickly. Hundreds of text-specific, open-ended, and multiple-choice questions are included in the question bank.

With **McGraw-Hill Create™**, you can easily rearrange chapters, combine material from other content sources, and quickly upload content you have written like your course syllabus or teaching notes. Find the content you need in Create by searching through thousands of leading McGraw-Hill textbooks. Arrange your book to fit your teaching style. Create even allows you to personalize your book's appearance by selecting the cover and adding your name, school, and course information. Assemble a Create book and you'll receive a complimentary print review copy in 3—5 business days or a complimentary electronic review copy (eComp) via email in minutes.

To learn more contact your sales rep or visit www.successinmath.com.

Get Better Results

Hosted by ALEKS Corp.

- McGraw-Hill's Connect Math hosted by ALEKS® is a web-based assignment and assessment platform that helps students connect to their coursework and prepares them to succeed in and beyond the course.
- Connect Math hosted by ALEKS enables math and statistics instructors to create and share courses and assignments with colleagues and adjuncts with only a few clicks of the mouse. All exercises, learning objectives, and activities are developed by mathematics instructors to ensure consistency between the textbook and the online resources.
- Connect Math hosted by ALEKS also links students to an interactive eBook with a variety of media assets and a place to study, highlight, and keep track of class notes.

To learn more contact your sales rep or visit www.successinmath.com.

 **ALEKS® Prep Products**

- **ALEKS** is a Web-based program that uses artificial intelligence to assess a student's knowledge and provide personalized instruction on the exact topics the student is most ready to learn. By providing individualized assessment and learning, ALEKS helps students to master course content quickly and easily. ALEKS allows students to easily move between explanations and practice, and provides intuitive feedback to help students correct and analyze errors. ALEKS also includes a powerful instructor module that simplifies course management so instructors spend less time with administrative tasks and more time directing student learning.
- **ALEKS 360** includes a fully integrated, interactive eBook and textbook-specific lecture and exercise videos combined with ALEKS personalized assessment and learning.
- **ALEKS Prep** focuses on prerequisite and introductory material, and can be used during the first six weeks of the term to ensure student success in math courses ranging from Beginning Algebra through Calculus. ALEKS Prep quickly fill gaps in prerequisite knowledge by assessing precisely each student's preparedness and delivering individualized instruction on the exact topics students are most ready to learn.

To learn more contact your sales rep or visit www.successinmath.com.

LEARNSMART®   SMARTBOOK™

- **LearnSmart®** is an adaptive study tool that strengthens student understanding and retention of your course's fundamental concepts. LearnSmart efficiently prepares your students for class so you can review *less* and teach *more*. The LearnSmart methodology is simple: determine the concepts students don't know or understand, and then teach those concepts using personalized plans designed for each student's success.
- **SmartBook®** is the first and only adaptive reading experience available for the higher education market. Powered by the intelligent and adaptive LearnSmart engine, SmartBook facilitates the reading process by identifying what content a student knows and doesn't know. As a student reads, the material continuously adapts to ensure the student is focused on the content he or she needs the most to close specific knowledge gaps.

To learn more contact your sales rep or visit www.successinmath.com.

Faculty Development and Digital Training

McGraw-Hill is excited to partner with our customers to ensure success in the classroom with our course solutions.

Looking for ways to be more effective in the classroom? Interested in the best practices for using available instructor resources?

Workshops are available on these topics for anyone using or considering the Miller, O'Neill, and Hyde Math Series. Led by the authors, contributors, and McGraw-Hill Learning consultants, each workshop is tailored to the needs of individual campuses or programs.

New to McGraw-Hill Digital Solutions? Need help setting up your course, using reports, and building assignments?

No need to wait for that big group training session during faculty development week. The McGraw-Hill Digital Implementation Team is a select group of advisors and experts in Connect Math. The Digital Implementation Team will work one-on-one with each instructor to make sure you are trained on the program and have everything you need to ensure a good experience for you and your students.

Are you redesigning a course or expanding your use of technology? Are you interested in getting ideas from other instructors who have used ALEKS™ or Connect Math in their courses?

Digital Faculty Consultants (DFCs) are instructors who have effectively incorporated technology such as ALEKS and Connect Math hosted by ALEKS in their courses. Discuss goals and best practices and improve outcomes in your course through peer-to-peer interaction and idea sharing.

Contact your local representative for more information about any of the faculty development, training, and support opportunities through McGraw-Hill. http://catalogs.mhhe.com/mhhe/findRep.do

Our Commitment to Market Development and Accuracy

McGraw-Hill's Development Process is an ongoing, never-ending, market-oriented approach to building accurate and innovative print and digital products. We begin developing a series by partnering with authors that desire to make an impact within their discipline to help students succeed. Next, we share these ideas and manuscript with instructors for review for feedback and to ensure that the authors' ideas represent the needs within that discipline. Throughout multiple drafts, we help our authors adapt to incorporate ideas and suggestions from reviewers to ensure that the series carries the same pulse as today's classrooms. With any new edition, we commit to accuracy across the series and its supplements. In addition to involving instructors as we develop our content, we also utilize accuracy checks through our various stages of development and production. The following is a summary of our commitment to market development and accuracy:

1. 2 drafts of author manuscript
2. 2 rounds of manuscript review
3. Multiple focus groups
4. 2 accuracy checks
5. 2 rounds of proofreading and copyediting
6. Toward the final stages of production, we are able to incorporate additional rounds of quality assurance from instructors as they help contribute toward our digital content and print supplements

This process then will start again immediately upon publication in anticipation of the next edition. With our commitment to this process, we are confident that our series has the most developed content the industry has to offer, thus pushing our desire for quality and accurate content that meets the needs of today's students and instructors.

Acknowledgments and Reviewers

Paramount to the development of this series was the invaluable feedback provided by the instructors from around the country who reviewed the manuscript or attended a market development event over the course of the several years the text was in development.

Albert Groccia, *Valencia College–Osceola*

Albert Guerra, *Saint Philips College*

Alexander Kasiukov, *Suffolk County Community College–Brentwood*

Alice Pollock-Cangemi, *Lone Star College*

Amber Smith, *Johnson County Community College*

Amtul Mujeeb Chaudry, *Rio Hondo College*

Anabel Darini, *Suffolk County Community College–Brentwood*

Andrea Blum, *Suffolk County Community College–Brentwood*

Angela Mccombs, *Illinois State University*

Ann Mccormick, *Lone Star College Kingwood*

Anne Prial, *Orange County Community College*

Antonnette Gibbs, *Broward College–North*

Anuradha Vadrevu, *Prince George's Community College*

Arlene Atchison, *South Seattle Community College*

Ashley Fuller, *John Tyler Community College–Chester*

Azzam Shihabi, *Long Beach City College*

Barbara Lott, *Seminole State College*

Barbara Purvis, *Centura College*

Barry Gibson, *Daytona State College–Daytona Beach*

Bashar Zogheib, *Nova Southeastern University*

Becky Schuering, *Blue River Community College–Independence*

Bernadette Turner, *Lincoln University*

Beverly Pepe, *Community College of Rhode Island–Warwick*

Bill Morrow, *Delaware Technical Community College*

Billie Shannon, *Southwestern Oregon Community College*

Brannen Smith, *Central Georgia Technical College*

Brenda Brown, *University of the District of Columbia*

Brent Pohlmann, *California Maritime Academy*

Bruce Legan, *Century Community & Technical College*

Carl Moxey, *Anna Maria College–Paxton*

Carol Curtis, *Fresno City College*

Carol Elias, *John Tyler Community College–Chester*

Carol Marinas, *Barry University*

Carol Mckillip, *Southwestern Oregon Community College*

Carol Rich, *Wallace Community College*

Carol Weideman, *Saint Petersburg College–Gibbs*

Carolyn Chapel, *Western Technical College*

Cassandra Johnson, *Robeson Community College*

Cassie Firth, *Northern Oklahoma College*

Cassondra Thompson, *York Technical College*
Chad Lower, *Pennsylvania College of Technology*
Christina Morian, *Lincoln University*
Cristi Whitfield, *Wallace Community College*
Cylinda Bray, *Yavapai College–Prescott*
Darla Aguilar, *Pima Community College*
Darlene Hatcher, *Metro Community College–South Campus–Omaha*
David Nusbaum, *Cypress College*
Dawn Chapman, *Columbus Technical College*
Deanna Hardy, *Bossier Parish Community College*
Deborah Logan, *Florida State College–South Campus*
Deborah Wolfson, *Suffolk County Community College–Brentwood*
Denise Nunley, *Glendale Community College*
Diana Dwan, *Yavapai College–Prescott*
D'marie Carver, *Portland Community College*
Don Anderson, *Northwest College*
Don Groninger, *Middlesex County College*
Don Solomon, *University of Wisconsin–Milwaukee*
Donald Robertson, *Olympic College*
Dot French, *Community College of Philadelphia*
Ed Thompson, *River Parishes Community College*
Edelma Simes, *Phillips Community College–Helena*
Eden Donahou, *Seminole State College*
Edith Lester, *Volunteer State Community College*
Edward Migliore, *University of California–Santa Cruz*
Eileen Diggle, *Bristol Community College*
Elaine Fitt, *Bucks Community College*
Eldon Baldwin, *Prince Georges Community College*
Elecia Ridley, *Durham Technical Community College*
Elena Litvinova, *Bloomsburg University of Pennsylvania*
Eleni Palmisano, *Centralia College*
Elisha Van Meenan, *Illinois State University*
Emily Simmons, *Centura College*
Eric Bennett, *Michigan State University–East Lansing*
Eric Kaljumagi, *Mt. San Antonio College*
Evon Lisle, *Seminole State College*
Gary Kersting, *North Central Michigan College*
Gene Ponthieux, *River Parishes Community College*
Gerald J. Lepage, *Bristol Community College*
Geri Philley, *Monterey Peninsula College*
Ghytana Goings, *Wallace Community College*
Ginger Eaves, *Bossier Parish Community College*
Gladys Bennett, *Centura College*
Gloria Hernandez, *Louisiana State University*
Greg Longanecker, *Leeward Community College*

Greg Rosik, *Century Community & Technical College*
Hadley Pridgen, *Gulf Coast Community College*
Heather Gallacher, *Cleveland State University*
Heidi Howard, *Florida State College–South Campus*
Heidi Kiley, *Suffolk County Community College–Brentwood*
Heidi Lyman, *South Seattle Community College*
Ignacio Alarcon, *Santa Barbara City College*
Irma Bakenhus, *San Antonio College*
J. Patrick Malone, *Victor Valley Community College*
James Dorn, *Barstow College*
James Miller, *Hillsborough Community College–Dale Mabry*
James Weeks, *Durham Technical Community College*
Jessica Lopez, *Saint Philips College*
Jean-Marie Magnier, *Springfield Technical Community College*
Jennifer Crowley, *Wallace Community College*
Jennifer Lempke, *North Central Michigan College*
Jian Zou, *South Seattle Community College*
Jill Wilsey, *Genesee Community College*
Joanne Strickland, *California Maritime Academy*
Jody Balzer, *Milwaukee Area Technical College*
Joe Jordan, *John Tyler Community College–Chester*
Joe Joyner, *Tidewater Community College–Norfolk*
Jonathan Cornick, *Queensborough Community College*
Joni Dugan, *Johnson County Community College*
Jordan Neus, *Suffolk County Community College–Brentwood*
Joshua Fontenot, *Louisiana State University*
Joyce Davis, *Heart of Georgia Technical College*
Judith Falk, *North Central Michigan College*
Judith Holbrook, *Yavapai College–Prescott*
Justin Dunham, *Johnson County Community College*
Karan Puri, *Queensborough Community College*
Karen Brown, *Wallace Community College*
Khaled Al-Agha, *Wiley College*
Karen Donnelly, *Saint Joseph's College*
Karen Estes, *Saint Petersburg College–Gibbs*
Karl Viehe, *University of the District of Columbia*
Kathleen Kane, *Community College Allegheny County–Pittsburgh*
Ketsia Chapman, *Centura College Online*
Ken Anderson, *Chemeketa Community College*
Kenneth Mead, *Genesee Community College*
Kenneth Williams, *Albany Technical College*

Kim Johnson, *Mesa Community College*
Kristin Good, *Washtenaw Community College*
Lakisha Holmes, *Daytona State College*
Laura Carroll, *Santa Rosa Junior College*
Laura Perez, *Washtenaw Community College*
Laura Stapleton, *Marshall University*
Lee Raubolt, *Yavapai College–Prescott*
Linda Schott, *Ozarks Technical Community College*
Linda Shackelford, *Tidewater Community College–Portsmouth*
Liz Delaney, *Grand Rapids Community College*
Lorena Goebel, *University of Arkansas–Fort Smith*
Loris Zucca, *Lone Star College Kingwood*
Lynette King, *Gadsden State Community College*
Lynn Irons, *College of Southern Idaho*
Mahshid Hassani, *Hillsborough Community College–Brandon*
Marc Campbell, *Daytona State College*
Marcial Echenique, *Broward College–North*
Maria Rodriguez, *Suffolk County Community College–Brentwood*
Marianna Mcclymonds, *Phoenix College*
Marilyn Peacock, *Tidewater Community College–Norfolk*
Marilyn S. Jacobi, *Gateway Community–Technical College*
Mark Anderson, *Durham Technical Community College*
Mark Batell, *Washtenaw Community College*
Mark Billiris, *St. Petersburg College*
Mark Littrell, *Rio Hondo College*
Marwan Abusawwa, *Florida State College–South Campus*
Mary Deas, *Johnson County Community College*
Mary Hito, *Los Angeles Valley College*
Mary Legner, *Riverside Community College*
Mary Wolyniak, *Broome Community College*
Matthew Pitassi, *Rio Hondo College*
Matthew Utz, *University of Arkansas–Fort Smith*
Maureen Loiacano, *Lone Star College*
Mauricio Marroquin, *Los Angeles Valley College*
Michael Cance, *Southeastern Community College*
Michelle Garey, *Delaware Technical & Community College–Dover*
Myrta Groeneveld, *Manchester Community College*
Nancy Eschen, *Florida State College–South Campus*
Natalie Weaver, *Daytona State College–Daytona Beach*
Nataliya Gavryshova, *College of San Mateo*
Nekeith Brown, *Richland College*
Nicole Francis, *Linn-Benton Community College*
Pam Ogaard, *Bismarck State College*
Pat Jones, *Methodist University*
Patricia Arteaga, *Bloomfield College*
Patricia Jones, *Methodist University*

Paula Looney, *Saint Philips College*
Paula Potter, *Yavapai College–Prescott*
Pavel Sikorskii, *Michigan State University–East Lansing*
Penny Marsh, *Johnson County Community College*
Philip Nelson, *Barstow College*
Phillip Taylor, *North Florida Community College*
Ramona Harris, *Gadsden State Community College*
Randey Burnette, *Tallahassee Community College*
Rhoda Oden, *Gadsden State Community College*
Richard Baum, *Santa Barbara City College*
Richard Hobbs, *Mission College*
Richard Pellerin, *Northern Virginia Community College*
Rick Downs, *South Seattle Community College*
Robbert Mckelvy, *Cossatot Community College*
Robert Cohen, *University of District of Columbia*
Robert Evans, *Monterey Peninsula College*
Robert Fusco, *Bergen Community College*
Rodney Oberdick, *Delaware Technical & Community College–Dover*
Roger Mccoach, *County College of Morris*
Roland Trevino, *San Antonio College*

Ron Powers, *Michigan State University–East Lansing*
Rosa Kontos, *Bergen Community College*
Rose Toering, *Kilian Community College*
Ruby Martinez, *San Antonio College*
Ruth Dellinger, *Florida State College Kent Campus*
Ryan Baxter, *Illinois State University*
Sally Jackman, *Richland College*
Sandra Cox, *Kaskaskia College*
Sandra Jovicic, *University of Akron*
Sandra Leabough, *Centura College*
Shanna Goff, *Grand Rapids Community College*
Shannon Miller-Mace, *Marshall University*
Sharon Hudson, *Gulf Coast Community College*
Shawn Krest, *Genesee Community College*
Sherri Kobis, *Erie Community College Northcamp–Williamsville*
Sima Dabir, *Western Iowa Technical Community College*
Spiros Karimbakas, *Fresno City College*
Stanley Hecht, *Santa Monica College*
Stephen Toner, *Victor Valley Community College*
Susan Metzger, *North Central Michigan College*
Susanna Gunther, *Solano Community College*
Suzette Goss, *Lone Star College Kingwood*

Sylvia Brown, *Mountain Empire Community College*
Tammy Potter, *Gadsden State Community College*
Tian Ren, *Queensborough Community College*
Timothy L. Warkentin, *Cloud County Community College*
Toni Houtteman, *Baker College of Clinton Township*
Tonya Michelle Davenport, *Rowan University*
Vernon Bridges, *Durham Technical Community College*
Wayne Barber, *Chemeketa Community College*
William Kirby, *Gadsden State Community College*
Yon Kim, *Passaic County Community College*

The author team most humbly would like to thank all the people who have contributed to this project.

Special thanks to our team of digital authors for their thousands of hours of work: Jody Harris, Linda Schott, Lizette Hernandez Foley, Michael Larkin, Alina Coronel, and to the masters of ceremonies in the digital world: Donna Gerken, Nicole Lloyd, and Steve Toner. We also offer our sincerest appreciation to the video talent: Jody Harris, Alina Coronel, Didi Quesada, Tony Alfonso, and Brianna Kurtz. You folks are the best! To Mitchel Levy our exercise consultant, thank you for your watchful eye fine-tuning the exercise sets and for your ongoing valuable feedback. To Gene Rumann, thank you so much for ensuring accuracy in our manuscripts.

Finally, we greatly appreciate the many people behind the scenes at McGraw-Hill without whom we would still be on page 1. To Emily Williams, the best developmental editor in the business and steady rock that has kept the train on the track. To Mary Ellen Rahn, our executive editor and overall team leader. We're forever grateful for your support and innovative ideas. To our copy editor Pat Steele who has watched over our manuscript for many years and has been a long-time mentor for our writing. You're amazing. To the marketing team Alex Gay, Peter Vanaria, Tim Cote, John Osgood, and Cherie Harshmann, thank you for your creative ideas in making our books come to life in the market. To the director of digital content, Nicole Lloyd, we are most grateful for your long hours of work and innovation in a world that changes day-to-day. And many thanks to the team at ALEKS that oversees quality control in the digital content. To the production team Peggy Selle, Laurie Janssen, and Carrie Burger for making the manuscript beautiful and bringing it all together. And finally to Ryan Blankenship, Marty Lange, and Kurt Strand, thank you for supporting our projects over the years and for the confidence you've always shown in us.

Most importantly, we give special thanks to the students and instructors who use our series in their classes.

Julie Miller   *Molly O'Neill*   *Nancy Hyde*

New and Updated Content for Miller, O'Neill, and Hyde's *Intermediate Algebra*, Third Edition:

Updates throughout the Text:
- New Vocabulary and Key Concept Exercises
- New Tips, Avoiding Mistakes boxes, and Concept Connections
- New and Revised Study Skills Exercises
- Updated Applications and Data in all instances where appropriate
- Writing has been reworked throughout the text in order to improve clarity and understanding

Chapter by Chapter Changes:

Chapter R
- Separated the first three sections of the previous edition Chapter 1 into a new Chapter R
- New Chapter Opener
- New Section on Study Skills
- New Group Activity
- 10 New Exercises

Chapter 1
- Now includes three sections from the previous edition Chapter 8
- Now includes the Problem Recognition Exercise Set from the previous edition Chapter 8
- New Chapter Opener
- New Group Activity

Chapter 2
- Separated previous edition Chapter 2 into current edition Chapter 2 and Chapter 3
- New Chapter Opener
- New Group Activity
- New Cumulative Review with 20 new exercises
- 1 New Example
- 27 New Exercises

Chapter 3
- New Chapter Opener
- New Definition Box of a Quadratic Function
- 12 New Exercises

Chapter 4
- New Section on Linear Inequalities and Systems of Linear Inequalities from previous edition Chapter 8
- New Chapter Opener
- New narrative on the Addition Method
- 32 New Exercises
- 1 New Example

Chapter 5
- New Section on Properties of Integer Exponents and Scientific Notation from previous edition Chapter 1
- New Chapter Opener
- New Problem Recognition Exercise Set
- New Factoring Strategy Procedure Box
- 22 New Exercises

Chapter 6
- New narrative on Restricted Values of a Rational Expression and the Domain of a Rational Function
- 7 New Exercises

Chapter 7
- New Chapter Opener
- New Property Box on Evaluating an nth Root
- 41 New Exercises
- 7 New Examples

Chapter 8
- New Chapter Opener
- New Section on Polynomial and Rational Inequalities from previous edition Chapter 8
- New Problem Recognition Exercise Set
- New Procedure Box on Methods to Solve a Quadratic Equation
- 42 New Exercises
- 4 New Examples

Chapter 9
- New Chapter Opener
- New figures to represent one-to-one functions
- New Definition Box on an Inverse Function
- New Inverse Function Property Box
- Separated Solving Exponential and Logarithmic Equations into two sections
- New Narrative on Solving Exponential Equations
- New Property Box on Equivalence of Logarithmic Expressions
- 67 New Exercises
- 5 New Examples

Chapter 10
- New Chapter Opener
- New figures to illustrate a vertical axis of symmetry for a parabola
- New Procedure Boxes on Graphing an Ellipse and a Hyperbola
- New Problem Recognition Exercise Set
- 9 New Exercises
- 2 New Examples

Additional Topics Appendix
- 14 New Exercises
- 4 New Examples

Vocabulary and Key Concepts

1. **a.** In mathematics, a well-defined collection of elements is called a _____.

    b. The statements $a < b$, $a > b$, and $a \neq b$ are examples of _____.

    c. The statement $a < b$ is read as "_____."

    d. The statement $c \geq d$ is read as "_____."

    e. The statement $5 \neq 6$ is read as "_____."

    f. The symbol ∞ represents _____ and $-\infty$ represents _____.

    g. The set of real numbers greater than 5 can be written in set-builder notation as _____ and in _____ notation as $(5, \infty)$.

    h. The interval $(-2, 5]$ (includes/excludes) the value -2 and (includes/excludes) the value 5.

    i. When expressing interval notation, use a (parenthesis/bracket) with infinity.

Concept 1: The Set of Real Numbers

2. Determine the two consecutive integers between which the given number is located on the number line.

    a. $\dfrac{19}{4}$ **b.** $-\dfrac{2}{3}$ **c.** -4.6 **d.** π

3. Plot the numbers on the number line.

    $$\{1.7, \pi, -5, 4.\overline{2}\}$$

    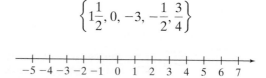

4. Plot the numbers on the number line.

    $$\left\{1\frac{1}{2}, 0, -3, -\frac{1}{2}, \frac{3}{4}\right\}$$

For Exercises 5–10, show that each number is a rational number by finding a ratio of two integers equal to the given number. **(See Example 1.)**

5. -10

6. $7\dfrac{3}{4}$

7. $-\dfrac{3}{5}$

8. -0.1

9. 0

10. 0.35

11. Check the sets to which each number belongs. **(See Example 2.)**

	Real Numbers	Irrational Numbers	Rational Numbers	Integers	Whole Numbers	Natural Numbers
5						
$-\sqrt{9}$						
-1.7						
$\dfrac{1}{2}$						
$\sqrt{7}$						
$\dfrac{0}{4}$						
$0.\overline{2}$						

Animation

Review of Basic Algebraic Concepts

R

Chapter R

In this chapter, we begin our study of algebra by reviewing the sets of numbers used in day-to-day life. We also review how to simplify numerical expressions and algebraic expressions. To prepare for this chapter, practice the following operations on whole numbers, decimals, fractions, and mixed numbers.

Review Your Skills

For Exercises 1–12, simplify each expression. Then find the answer on the right and record the corresponding letter at the bottom of the page. When you are finished, you will have a key definition for this chapter.

Exercises		Answers	
1. $36{,}636 \div 43$	**2.** 0.25×6340	**I.** 389.7842	**B.** 1585
3. $24.0842 + 365.7$	**4.** $\dfrac{4}{7} \times \dfrac{21}{10}$	**R.** $\dfrac{1}{6}$	**D.** $\dfrac{6}{5}$
5. $\dfrac{7}{3} \div 14$	**6.** $\dfrac{5}{3} - \dfrac{1}{2}$	**T.** 852	**S.** 11.95
7. $\dfrac{2}{3} - \dfrac{7}{12}$	**8.** $4\dfrac{5}{9} - 1\dfrac{1}{3}$	**E.** $3\dfrac{2}{9}$	**P.** $\dfrac{1}{12}$
9. $3\dfrac{1}{5} \times 2\dfrac{1}{2}$	**10.** $3.75 + 8\dfrac{1}{5}$	**Y.** $\dfrac{7}{6}$	**O.** 8
11. $582 \div 0.01$	**12.** 582×0.01	**V.** $58{,}200$	**U.** 5.82

In this chapter we will show that $a(b + c) = ab + ac$. This important property is called the:

$$\overline{}_{4} \ \overline{}_{3} \ \overline{}_{10} \ \overline{}_{1} \ \overline{}_{5} \ \overline{}_{3} \ \overline{}_{2} \ \overline{}_{12} \ \overline{}_{1} \ \overline{}_{3} \ \overline{}_{11} \ \overline{}_{8} \quad \overline{}_{7} \ \overline{}_{5} \ \overline{}_{9} \ \overline{}_{7} \ \overline{}_{8} \ \overline{}_{5} \ \overline{}_{1} \ \overline{}_{6} .$$

Section R.1   Study Skills

Concepts

1. Before the Course
2. During the Course
3. Preparation for Exams
4. Where to Go for Help

In taking a course in algebra, you are making a commitment to yourself, your instructor, and your classmates. Following some or all of the study tips below can help you be successful in this endeavor. The features of this text that will assist you are printed in blue.

1. Before the Course

- Purchase the necessary materials for the course before the course begins or on the first day.
- Obtain a three-ring binder to keep and organize your notes, homework, tests, and any other materials acquired in the class. We call this type of notebook a *portfolio*.
- Arrange your schedule so that you have enough time to attend class and to do homework. A common rule is to set aside at least 2 hours for homework for every hour spent in class. That is, if you are taking a 4-credit-hour course, plan on at least 8 hours a week for homework. If you experience difficulty in mathematics, plan for more time. A 4-credit-hour course will then take *at least* 12 hours each week—about the same as a part-time job.
- Communicate with your employer and family members the importance of your success in this course so that they can support you.
- Be sure to find out the type of calculator (if any) that your instructor requires. Also determine if there will be online homework or other computer requirements.

2. During the Course

- To prepare yourself for the next day's class, read the section in the text *before* coming to class. This will help you familiarize yourself with the material and terminology.
- Attend every class and be on time.
- Take notes in class. Write down all of the examples that the instructor presents. Read the notes after class, and add any comments to make your notes clearer to you. Use a tape recorder to record the lecture if the instructor permits the recording of lectures.
- Ask questions in class.
- Read the section in the text *after* the lecture, and pay special attention to the Tip boxes and Avoiding Mistakes boxes.
- After you read an example, try the accompanying Skill Practice problem. The skill practice problem mirrors the example and tests your understanding of what you have read.
- Do homework every night. Even if your class does not meet every day, you should still do some work every night to keep the material fresh in your mind.
- Check your homework with the answers that are supplied in the back of this text. Analyze what you did wrong and correct the exercises that do not match. Circle or star those that you cannot correct yourself. This way you can easily find them and ask your instructor the next day.
- Be sure to do the Vocabulary and Key Concepts exercises found at the beginning of the Practice Exercises.
- The Problem Recognition Exercises are found in Chapters 1–10. These provide additional practice distinguishing among a variety of problem types. Sometimes the most difficult part of learning mathematics is retaining all that you learn. These exercises are excellent tools for retention of material.

- Form a study group with fellow students in your class, and exchange phone numbers. You will be surprised by how much you can learn by talking about mathematics with other students.
- If you use a calculator in your class, read the Calculator Connections boxes to learn how and when to use your calculator.
- Ask your instructor where you might obtain extra help if necessary.

3. Preparation for Exams

- Look over your homework and rework exercises that gave you trouble. Pay special attention to the exercises you have circled or starred to be sure that you have learned that concept.
- Read through the Summary at the end of the chapter. Be sure that you understand each concept and example. If not, go to the section in the text and reread that section.
- Give yourself enough time to take the Chapter Test uninterrupted. Then check the answers. For each problem you answered incorrectly, go to the Review Exercises and do all of the problems that are similar.
- To prepare for the final exam, complete the Cumulative Review Exercises at the end of each chapter, starting with Chapter 2. If you complete the cumulative reviews after finishing each chapter, then you will be preparing for the final exam throughout the course. The Cumulative Review Exercises are another excellent tool for helping you retain material.

4. Where to Go for Help

- At the first sign of trouble, see your instructor. Most instructors have specific office hours set aside to help students. Don't wait until after you have failed an exam to seek assistance.
- Get a tutor. Most colleges and universities have free tutoring available.
- When your instructor and tutor are unavailable, use the Student Solutions Manual for step-by-step solutions to the odd-numbered problems in the exercise sets.
- Work with another student from your class.
- Work on the computer. Many mathematics tutorial programs and websites are available on the Internet, including the website that accompanies this text.

Group Activity

Becoming a Successful Student

Materials: Computer with Internet access (Optional)

Estimated time: 15 minutes

Group Size: 4

Good time management, good study skills, and good organization will help you be successful in this course. Answer the following questions and compare your answers with your group members.

1. To motivate yourself to complete a course, it is helpful to have clear reasons for taking the course. List your goals for taking this course and discuss them with your group.

2. Taking 12 credit-hours is the equivalent of a full-time job. Often students try to work too many hours while taking classes at school.

    a. Write down the number of hours you work per week and the number of credit-hours you are taking this term.

        Number of hours worked per week _____

        Number of credit-hours this term _____

    b. The table gives a recommended limit to the number of hours you should work for the number of credit-hours you are taking at school. (Keep in mind that other responsibilities in your life such as your family might also make it necessary to limit your hours at work even more.) How do your numbers from part (a) compare to those in the table? Are you working too many hours?

Number of Credit-Hours	Maximum Number of Hours of Work per Week
3	40
6	30
9	20
12	10
15	0

    c. It is often suggested that you devote two hours of study and homework time outside of class for each credit-hour you take at school. For example:

            12 credit-hours
        + 24 study hours
            36 total hours ⟵ full-time job!

        Based on the number of credit-hours you are taking, how many study hours should you plan for? _____

        What is the total number of hours (class time plus study time) that you should devote to school? _____

3. For the following week, write down the times each day that you plan to study math.

Monday	Tuesday	Wednesday	Thursday	Friday	Saturday	Sunday

4. Write down the date of your next math test. _____

5. Look through the book in Chapter 2 and find the page number corresponding to each feature in the book. Discuss with your group members how you might use each feature.

    Problem Recognition Exercises: page _____

    Chapter Summary: page _____

    Chapter Review Exercises: page _____

    Chapter Test: page _____ .

    Cumulative Review Exercises: page _____

6. Look at the Skill Practice exercises next to each example (for instance, find Skill Practice exercise 4 in Section 1.2). Where are the answers to these exercises located? Discuss with your group members how you might use the Skill Practice exercises.

7. Discuss with your group members places where you can go for extra help in math. Then write down three of the suggestions.

_____

_____

8. Do you keep an organized notebook for this class? Can you think of any suggestions that you can share with your group members to help them keep their materials organized?

9. Do you think that you have math anxiety? Read the following list for some possible solutions. Check the activities that you can realistically try to help you overcome this problem.

_____ Read a book on math anxiety.

_____ Search the Web for help tips on handling math anxiety.

_____ See a counselor to discuss your anxiety.

_____ Talk with your instructor to discuss strategies to manage math anxiety.

_____ Evaluate your time management to see if you are trying to do too much. Then adjust your schedule accordingly.

10. Some students favor different methods of learning over others. For example, you might prefer:

• Learning through listening and hearing.

• Learning through seeing images, watching demonstrations, and visualizing diagrams and charts.

• Learning by experience through a hands-on approach.

• Learning through reading and writing.

Most experts believe that the most effective learning comes when a student engages in *all* of these activities. However, each individual is different and may benefit from one activity more than another. You can visit a number of different websites to determine your "learning style." Try doing a search on the Internet with the key words "*learning styles assessment.*" Once you have found a suitable website, answer the questionnaire and the site will give you feedback on what method of learning works best for you.

Sets of Numbers and Interval Notation | Section R.2

1. The Set of Real Numbers

Algebra is a powerful mathematical tool that is used to solve real-world problems in science, business, and many other fields. We begin our study of algebra with a review of basic definitions and notations used to express algebraic relationships.

Concepts

1. The Set of Real Numbers
2. Inequalities
3. Interval Notation
4. Translations Involving Inequalities

In mathematics, a collection of items (called elements) is called a **set**, and the set braces { } are used to enclose the elements of the set. For example, the set {a, e, i, o, u} represents the vowels in the English alphabet. The set {1, 3, 5, 7} represents the first four positive odd numbers. Another method to express a set is to *describe* the elements of the set by using **set-builder notation**. Consider the set {a, e, i, o, u} in set-builder notation.

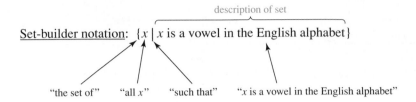

Consider the set {1, 3, 5, 7} in set-builder notation.

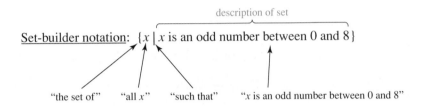

Several sets of numbers are used extensively in algebra. The numbers you are familiar with in day-to-day calculations are elements of the set of **real numbers**. These numbers can be represented graphically on a horizontal number line with a point labeled as 0. Positive real numbers are graphed to the right of 0, and negative real numbers are graphed to the left. Each point on the number line corresponds to exactly one real number, and for this reason, the line is called the **real number line** (Figure R-1).

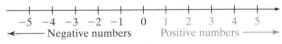

Figure R-1

Several sets of numbers are **subsets** (or part) of the set of real numbers. These are

> The set of natural numbers
>
> The set of whole numbers
>
> The set of integers
>
> The set of rational numbers
>
> The set of irrational numbers

Natural Numbers, Whole Numbers, and Integers

The set of **natural numbers** is {1, 2, 3, . . .}.
The set of **whole numbers** is {0, 1, 2, 3, . . .}.
The set of **integers** is { . . . , −3, −2, −1, 0, 1, 2, 3, . . .}.

The set of rational numbers consists of all the numbers that can be defined as a ratio of two integers.

> ### Rational Numbers
>
> The set of **rational numbers** is $\{\frac{p}{q} \mid p$ and q are integers and q does not equal zero$\}$.

Example 1 — Identifying Rational Numbers

Show that each number is a rational number by finding two integers whose ratio equals the given number.

    a. $\dfrac{-4}{7}$ **b.** 8   **c.** $0.\overline{6}$ **d.** 0.87

Solution:

a. $\frac{-4}{7}$ is a rational number because it can be expressed as the ratio of the integers -4 and 7.

b. 8 is a rational number because it can be expressed as the ratio of the integers 8 and 1 ($8 = \frac{8}{1}$). In this example we see that *an integer is also a rational number.*

c. $0.\overline{6}$ represents the repeating decimal $0.6666666\ldots$ and can be expressed as the ratio of 2 and 3 ($0.\overline{6} = \frac{2}{3}$). In this example we see that *a repeating decimal is a rational number.*

d. 0.87 is the ratio of 87 and 100 ($0.87 = \frac{87}{100}$). In this example we see that *a terminating decimal is a rational number.*

TIP: Any rational number can be represented by a terminating decimal or by a repeating decimal.

Some real numbers such as the number π (pi) cannot be represented by the ratio of two integers. In decimal form, an irrational number is a nonterminating, nonrepeating decimal. The value of π, for example, can be approximated as $\pi \approx 3.1415926535897932$. However, the decimal digits continue indefinitely with no pattern. Other examples of irrational numbers are the square roots of nonperfect squares, such as $\sqrt{3}$ and $\sqrt{10}$.

> ### Irrational Numbers
>
> The set of **irrational numbers** is a subset of the real numbers whose elements cannot be written as a ratio of two integers.
>
> *Note:* An irrational number cannot be written as a terminating decimal or as a repeating decimal.

The set of real numbers consists of both the rational numbers and the irrational numbers. The relationships among the sets of numbers discussed thus far are illustrated in Figure R-2.

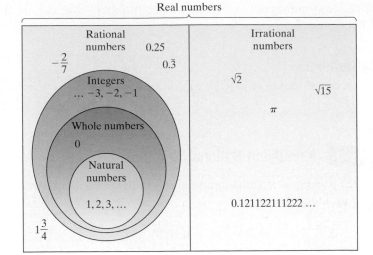

Figure R-2

Skill Practice

8. Check the set(s) to which each number belongs.

	1	0.47	$\sqrt{5}$	$-\frac{1}{2}$
Natural				
Whole				
Integer				
Rational				
Irrational				
Real				

Example 2   Classifying Numbers by Set

Check the set(s) to which each number belongs. The numbers may belong to more than one set.

	Natural Numbers	Whole Numbers	Integers	Rational Numbers	Irrational Numbers	Real Numbers
-6						
$\sqrt{23}$						
$-\frac{2}{7}$						
3						
$2.\overline{3}$						

Solution:

	Natural Numbers	Whole Numbers	Integers	Rational Numbers	Irrational Numbers	Real Numbers
-6			✓	✓		✓
$\sqrt{23}$					✓	✓
$\frac{2}{7}$				✓		✓
3	✓	✓	✓	✓		✓
$2.\overline{3}$				✓		✓

Answers

7. A decimal number that terminates or repeats is rational. A decimal number that is nonterminating and nonrepeating is irrational.

8.

	1	0.47	$\sqrt{5}$	$-\frac{1}{2}$
Natural	✓			
Whole	✓			
Integer	✓			
Rational	✓	✓		✓
Irrational			✓	
Real	✓	✓	✓	✓

2. Inequalities

The relative value of two numbers can be compared by using the real number line. We say that a is less than b (written mathematically as $a < b$) if a lies to the left of b on the number line.

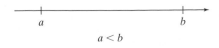

$a < b$

We say that a is greater than b (written mathematically as $a > b$) if a lies to the right of b on the number line.

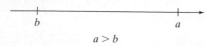

$$a > b$$

From looking at the number line, note that $a > b$ is the same as $b < a$. Table R-1 summarizes the relational operators that compare two real numbers a and b.

Table R-1

Mathematical Expression	Translation	Other Meanings
$a < b$	a is less than b	b exceeds a
		b is greater than a
$a > b$	a is greater than b	a exceeds b
		b is less than a
$a \leq b$	a is less than or equal to b	a is at most b
		a is no more than b
$a \geq b$	a is greater than or equal to b	a is no less than b
		a is at least b
$a = b$	a is equal to b	
$a \neq b$	a is not equal to b	
$a \approx b$	a is approximately equal to b	

The symbols $<$, $>$, $\leq$, $\geq$, and $\neq$ are called inequality signs, and the expressions $a < b, a > b, a \leq b, a \geq b$, and $a \neq b$ are called **inequalities**.

Example 3 **Ordering Real Numbers**

Fill in the blank with the appropriate inequality sign: $<$ or $>$

a. -2 _____ -5 **b.** $\dfrac{4}{7}$ _____ $\dfrac{3}{5}$ **c.** -1.3 _____ $-1.\overline{3}$

Solution:

a. -2 __$>$__ -5

b. To compare $\frac{4}{7}$ and $\frac{3}{5}$, write the fractions as equivalent fractions with a common denominator.

$$\frac{4}{7} \cdot \frac{5}{5} = \frac{20}{35} \quad \text{and} \quad \frac{3}{5} \cdot \frac{7}{7} = \frac{21}{35}$$

Because $\dfrac{20}{35} < \dfrac{21}{35}$, then $\dfrac{4}{7}$ __$<$__ $\dfrac{3}{5}$

c. -1.3 __$\geq$__ $-1.33333\ldots$

Animation

Answers

9. $>$ **10.** $>$ **11.** $<$

3. Interval Notation

The set $\{x \mid x \geq 3\}$ represents all real numbers greater than or equal to 3. This set can be illustrated graphically on the number line.

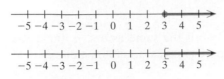

By convention, a closed circle ● or a square bracket [is used to indicate that an "endpoint" ($x = 3$) *is included* in the set.

The set $\{x \mid x > 3\}$ represents all real numbers strictly greater than 3. This set can be illustrated graphically on the number line.

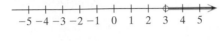

By convention, an open circle ○ or a parenthesis (is used to indicate that an "endpoint" ($x = 3$) is *not* included in the set.

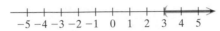

Notice that the sets $\{x \mid x \geq 3\}$ and $\{x \mid x > 3\}$ consist of an infinite number of elements that cannot all be listed. Another method to represent the elements of such sets is by using **interval notation**. To understand interval notation, first consider the real number line, which extends infinitely far to the left and right. The symbol ∞ is used to represent infinity. The symbol $-\infty$ is used to represent negative infinity.

To express a set of real numbers in interval notation, sketch the graph first, using the symbols ( ) or [ ]. Then use these symbols at the endpoints to define the interval.

Example 4   **Expressing Sets by Using Interval Notation**

Graph each set on the number line, and express the set in interval notation.

a. $\{x \mid x \geq -2\}$ **b.** $\{p \mid p > -2\}$

Solution:

a. Set-builder notation: $\{x \mid x \geq -2\}$

Graph:

Interval notation: $[-2, \infty)$

The graph of the set $\{x \mid x \geq -2\}$ "begins" at -2 and extends infinitely far to the right. The corresponding interval notation "begins" at -2 and extends to ∞. Notice that a square bracket [is used at -2 for both the graph and the interval notation to include $x = -2$. *A parenthesis is always used at ∞ and at $-\infty$ because there is no endpoint.*

b. Set-builder notation: $\{p \mid p > -2\}$

Graph:

Interval notation: $(-2, \infty)$

Answers

12.
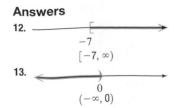
-7
$[-7, \infty)$

13.
0
$(-\infty, 0)$

In general, we use the following guidelines when applying interval notation.

Using Interval Notation

- The endpoints used in interval notation are always written from left to right. That is, the smaller number is written first, followed by a comma, followed by the larger number.
- Parentheses) or (indicate that an endpoint is *excluded* from the set.
- Square brackets] or [indicate that an endpoint is *included* in the set.
- Parentheses are always used with ∞ or $-\infty$.

Example 5 Expressing Sets by Using Interval Notation

Graph each set on the number line, and express the set in interval notation.

a. $\{z \mid z \le -\frac{3}{2}\}$ **b.** $\{x \mid -4 < x \le 2\}$

Solution:

a. Set-builder notation: $\{z \mid z \le -\frac{3}{2}\}$

Graph:

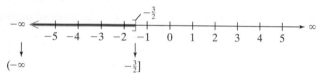

Interval notation: $(-\infty, -\frac{3}{2}]$

The graph of the set $\{z \mid z \le -\frac{3}{2}\}$ extends infinitely far to the left. Interval notation is always written from left to right. Therefore, $-\infty$ is written first, followed by a comma, and then followed by the right-hand endpoint $-\frac{3}{2}$.

b. The inequality $-4 < x \le 2$ means that x is greater than -4 and also less than or equal to 2. More concisely, we can say that x represents the real numbers *between* -4 and 2, including the endpoint, 2.

Set-builder notation: $\{x \mid -4 < x \le 2\}$

Graph:

Interval notation: $(-4, 2]$

Skill Practice

Graph the set on the number line, and express the set in interval notation.

14. $\{w \mid w \ge -\frac{5}{3}\}$

15. $\{y \mid -7 \le y < 4\}$

Concept Connections

Graph and express in set-builder notation.

16. $[4, \infty)$

17. $\left(-\infty, \frac{1}{2}\right)$

Table R-2 summarizes interval notation.

Table R-2

Interval Notation	Graph	Interval Notation	Graph
(a, ∞)		$[a, \infty)$	
$(-\infty, a)$		$(-\infty, a]$	
(a, b)		$[a, b]$	
$(a, b]$		$[a, b)$	

Answers

14.

$\left[-\frac{5}{3}, \infty\right)$

15.

$[-7, 4)$

16.

$\{x \mid x \ge 4\}$

17.

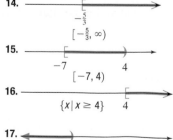

$\{x \mid x < \frac{1}{2}\}$

4. Translations Involving Inequalities

In Table R-1, we learned that phrases such as *at least, at most, no more than, no less than,* and *between* can be translated into mathematical terms by using inequality signs.

Example 6 Translating Inequalities

The intensity of a hurricane is often defined according to its maximum sustained winds, for which wind speed is measured to the nearest mile per hour. Translate the italicized phrases into mathematical inequalities.

a. A tropical storm is updated to hurricane status if the sustained wind speed, *w, is at least 74 mph.*

b. Hurricanes are categorized according to intensity by the Saffir-Simpson scale. On a scale of 1 to 5, a category 5 hurricane is the most destructive. A category 5 hurricane has sustained winds, *w, exceeding 155 mph.*

c. A category 4 hurricane has sustained winds, *w, of at least 131 mph but no more than 155 mph.*

Solution:

a. $w \geq 74$ mph **b.** $w > 155$ mph **c.** 131 mph $\leq w \leq 155$ mph

---

Section R.2 Practice Exercises

Study Skills Exercises

In this text, we will provide skills for you to enhance your learning experience. Many of the Practice Exercises will begin with an activity that focuses on one of the following areas: learning about your course, using your text, taking notes, doing homework, and taking an exam. In subsequent chapters we will insert skills pertaining to the specific material in the chapter. Each activity requires only a few minutes and will help you pass this class and become a better math student.

To begin, write down the following information.

a. Instructor's name

b. Days of the week that the class meets

c. The room number in which the class meets

d. Is there a lab requirement for this course? If so, what is the requirement and what is the location of the lab?

e. Instructor's office number

f. Instructor's telephone number

g. Instructor's e-mail address

h. Instructor's office hours

12. Check the sets to which each number belongs.

	Real Numbers	Irrational Numbers	Rational Numbers	Integers	Whole Numbers	Natural Numbers
$\frac{6}{8}$						
$1\frac{1}{2}$						
π						
0						
$-0.\overline{8}$						
$\frac{8}{2}$						
$4.\overline{2}$						

Concept 2: Inequalities

For Exercises 13–20, fill in the blanks with the appropriate symbol: $<$ or $>$. **(See Example 3.)**

13. -9 ___ -1

14. 0 ___ -6

15. $0.1\overline{5}$ ___ 0.15

16. $-2.\overline{5}$ ___ -2.5

17. $\frac{5}{3}$ ___ $\frac{10}{7}$

18. $-\frac{21}{5}$ ___ $-\frac{17}{4}$

19. $-\frac{5}{8}$ ___ $-\frac{1}{8}$

20. $-\frac{13}{15}$ ___ $-\frac{17}{12}$

Concept 3: Interval Notation

For Exercises 21–28, express the set in interval notation.

21.

at 2

22.

at $\frac{5}{6}$

23.

at 0

24.
at 9

25.
at -5 and 0

26.
at -1 and 15

27.
at -4.7

28.
at 12.8

For Exercises 29–46, graph the sets and express each set in interval notation. **(See Examples 4 and 5.)**

29. $\{x \mid x > -1\}$

30. $\{x \mid x < 3\}$

31. $\{y \mid y \leq -2\}$

32. $\{z \mid z \geq -4\}$

33. $\{w \mid w < \frac{9}{2}\}$

34. $\{p \mid p \geq -\frac{7}{3}\}$

35. $\{x \mid -2.5 < x \leq 4.5\}$

36. $\{x \mid -6 \leq x < 0\}$

37. All real numbers less than -3.

38. All real numbers greater than 2.34.

39. All real numbers greater than $\frac{5}{2}$.

40. All real numbers less than $\frac{4}{7}$.

41. All real numbers not less than 2.

⟶

42. All real numbers no more than 5.

⟶

43. All real numbers between −4 and 4.

⟶

44. All real numbers between −7 and −1.

⟶

45. All real numbers between −3 and 0, inclusive.

⟶

46. All real numbers between −1 and 6, inclusive.

⟶

For Exercises 47–54, write an expression in words that describes the set of numbers given by each interval. (Answers may vary.)

47. $(-\infty, -4)$

48. $[2, \infty)$

49. $(-2, 7]$

50. $(-3.9, 0)$

51. $[-180, 90]$

52. $(3.2, \infty)$

53. $(-\infty, \infty)$

54. $(-\infty, -1]$

Concept 4: Translations Involving Inequalities

For Exercises 55–64, write the expressions as an inequality. **(See Example 6.)**

55. The age, a, to get in to see a certain movie is at least 18 years old.

56. Winston is a cat that was picked up at the Humane Society. His age, a, at the time was no more than 2 years.

57. The cost, c, to have dinner at Jack's Café is at most $25.

58. The number of hours, h, that Katlyn spent studying was no less than 40.

59. The wind speed, s, for an F-5 tornado is no less than 261 mph.

60. The high temperature, t, for a certain December day in Albany is at most 26°F.

61. After a summer drought, the total rainfall, r, for June, July, and August was no more than 4.5 in.

62. Jessica works for a networking firm. Her salary, s, is at least $85,000 per year.

63. To play in a certain division of a tennis tournament, a player's age, a, must be at least 18 years but not more than 25 years.

64. The average age, a, of students at Central Community College is estimated to be between 25 years and 29 years.

The following chart defines the ranges for normal blood pressure, high normal blood pressure, and high blood pressure (*hypertension*). All values are measured in millimeters of mercury (mm Hg). (*Source:* American Heart Association)

Normal	Systolic less than 130	Diastolic less than 85
High normal	Systolic 130–139, inclusive	Diastolic 85–89, inclusive
Hypertension	Systolic 140 or greater	Diastolic 90 or greater

For Exercises 65–68, write an inequality using the variable p that represents each condition.

65. Normal systolic blood pressure

66. Diastolic pressure in hypertension

67. High normal range for systolic pressure

68. Systolic pressure in hypertension

A pH scale determines whether a solution is acidic or alkaline. The pH scale runs from 0 to 14, with 0 being the most acidic and 14 being the most alkaline. A pH of 7 is neutral (distilled water has a pH of 7).

For Exercises 69–72, write the pH ranges as inequalities and label the substances as acidic or alkaline.

69. Lemon juice: 2.2 through 2.4, inclusive

70. Eggs: 7.6 through 8.0, inclusive

71. Carbonated soft drinks: 3.0 through 3.5, inclusive

72. Milk: 6.6 through 6.9, inclusive

Section R.3 Operations on Real Numbers

1. Opposite and Absolute Value

Several key definitions are associated with the set of real numbers and constitute the foundation of algebra. Two important definitions are the opposite of a real number and the absolute value of a real number.

> **Opposite of a Real Number**
>
> Two numbers that are the same distance from 0 but on opposite sides of 0 on the number line are called **opposites** of each other.
>
> Symbolically, we denote the opposite of a real number a as $-a$.

The numbers -4 and 4 are opposites of each other. Similarly, the numbers $\frac{3}{2}$ and $-\frac{3}{2}$ are opposites.

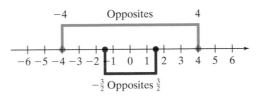

The **absolute value** of a real number a, denoted $|a|$, is the distance between a and 0 on the number line. *Note:* The absolute value of any real number is *nonnegative*.

For example: $|5| = 5$
and
$|-5| = 5$

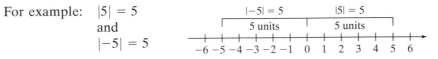

Example 1 Evaluating Absolute Value Expressions

Simplify the expressions.

a. $|-2.5|$ **b.** $\left|\dfrac{5}{4}\right|$ **c.** $-|-4|$

Solution:

a. $|-2.5| = 2.5$

b. $\left|\dfrac{5}{4}\right| = \dfrac{5}{4}$

c. $-|-4| = -(4) = -4$

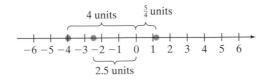

Skill Practice

Simplify.

1. $|-9.2|$
2. $\left|\dfrac{7}{6}\right|$
3. $-|-2|$

Answers

1. 9.2   **2.** $\dfrac{7}{6}$ **3.** -2

Topic: Using the Absolute Value Function

Some calculators have an absolute value function. For example,

```
abs(-2.5)
              2.5
abs(5/4)▶Frac
              5/4
-abs(-4)
              -4
```

The absolute value of a number a is its distance from zero on the number line. The definition of $|a|$ may also be given algebraically depending on whether a is negative or nonnegative.

Absolute Value of a Real Number

Let a be a real number. Then

1. If a is nonnegative (that is, if $a \geq 0$), then $|a| = a$.

2. If a is negative (that is, if $a < 0$), then $|a| = -a$.

This definition states that if a is positive or zero, then $|a|$ equals a itself. If a is a negative number, then $|a|$ equals the opposite of a. For example:

$|9| = 9$ Because 9 is positive, $|9|$ equals the number 9 itself.

$|-7| = 7$ Because -7 is negative, $|-7|$ equals the opposite of -7, which is 7.

2. Addition and Subtraction of Real Numbers

Addition of Real Numbers

- To add two numbers with the *same sign*, add their absolute values and apply the common sign to the sum.
- To add two numbers with *different signs*, subtract the smaller absolute value from the larger absolute value. Then apply the sign of the number having the larger absolute value.

Example 2 **Adding Real Numbers**

Perform the indicated operations.

a. $-2 + (-6)$ **b.** $-10.3 + 13.8$ **c.** $\frac{5}{6} + \left(-1\frac{1}{4}\right)$

Solution:

a. $-2 + (-6)$ First find the absolute value of the addends.

$|-2| = 2$ and $|-6| = 6$

$= -(2 + 6)$ Add their absolute values and apply the common sign. In this case, the common sign is negative.

Common sign is negative.

$= -8$ The sum is -8.

Skill Practice

Perform the indicated operations.
4. $-4 + (-1)$
5. $-2.6 + 1.8$
6. $-1 + \left(-\frac{3}{7}\right)$

Answers
4. -5 **5.** -0.8 **6.** $-\frac{10}{7}$

b. $-10.3 + 13.8$ First find the absolute value of the addends.
$$|-10.3| = 10.3 \quad \text{and} \quad |13.8| = 13.8$$

The absolute value of 13.8 is greater than the absolute value of -10.3. Therefore, the sum is positive.

$= +(13.8 - 10.3)$ Subtract the smaller absolute value from the larger absolute value.

⌐Apply the sign of the number with the larger absolute value.

$= 3.5$

c. $\dfrac{5}{6} + \left(-1\dfrac{1}{4}\right)$

$= \dfrac{5}{6} + \left(-\dfrac{5}{4}\right)$ Write $-1\dfrac{1}{4}$ as a fraction.

$= \dfrac{5 \cdot 2}{6 \cdot 2} + \left(-\dfrac{5 \cdot 3}{4 \cdot 3}\right)$ Write each fraction with the least common denominator (LCD). The LCD is 12.

$= \dfrac{10}{12} + \left(-\dfrac{15}{12}\right)$ Find the absolute value of the addends.
$$\left|\dfrac{10}{12}\right| = \dfrac{10}{12} \quad \text{and} \quad \left|-\dfrac{15}{12}\right| = \dfrac{15}{12}$$

The absolute value of $-\dfrac{15}{12}$ is greater than the absolute value of $\dfrac{10}{12}$. Therefore, the sum is negative.

$= -\left(\dfrac{15}{12} - \dfrac{10}{12}\right)$ Subtract the smaller absolute value from the larger absolute value.

⌐Apply the sign of the number with the larger absolute value.

$= -\dfrac{5}{12}$

Subtraction of real numbers is defined in terms of the addition process. To subtract two real numbers, add the opposite of the second number to the first number.

Subtraction of Real Numbers

If a and b are real numbers, then $a - b = a + (-b)$

Skill Practice

Subtract.

7. $-9 - 8$

8. $1.1 - (-4.2)$

9. $\dfrac{1}{6} - 2\dfrac{1}{4}$

Answers

7. -17 **8.** 5.3 **9.** $-2\dfrac{1}{12}$ or $-\dfrac{25}{12}$

Example 3   **Subtracting Real Numbers**

Perform the indicated operations.

a. $-13 - 5$ **b.** $2.7 - (-3.8)$ **c.** $\dfrac{5}{2} - 4\dfrac{2}{3}$

Solution:

a. $-13 - 5$

$= -13 + (-5)$ Add the opposite of the second number to the first number.

$= -18$ Add.

b. $2.7 - (-3.8)$

$= 2.7 + (3.8)$ Add the opposite of the second number to the first number.

$= 6.5$ Add.

c. $\dfrac{5}{2} - 4\dfrac{2}{3}$

$= \dfrac{5}{2} + \left(-4\dfrac{2}{3}\right)$ Add the opposite of the second number to the first number.

$= \dfrac{5}{2} + \left(-\dfrac{14}{3}\right)$ Write the mixed number as a fraction.

$= \dfrac{5 \cdot 3}{2 \cdot 3} + \left(-\dfrac{14 \cdot 2}{3 \cdot 2}\right)$ The common denominator is 6.

$= \dfrac{15}{6} + \left(-\dfrac{28}{6}\right)$ Get a common denominator and add.

$= -\dfrac{13}{6}$ or $-2\dfrac{1}{6}$

3. Multiplication and Division of Real Numbers

The sign of the product of two real numbers is determined by the signs of the factors.

Multiplication of Real Numbers

1. The product of two real numbers with the *same* sign is *positive*.

2. The product of two real numbers with *different* signs is *negative*.

3. The product of any real number and zero is *zero*.

Example 4   **Multiplying Real Numbers**

Multiply the real numbers.

a. $(2)(-5.1)$ **b.** $-\dfrac{2}{3} \cdot \dfrac{9}{8}$ **c.** $\left(-3\dfrac{1}{3}\right)\left(-\dfrac{3}{10}\right)$

Solution:

a. $(2)(-5.1)$

$= -10.2$ *Different* signs. The product is negative.

b. $-\dfrac{2}{3} \cdot \dfrac{9}{8}$

$= -\dfrac{18}{24}$ *Different* signs. The product is negative.

$= -\dfrac{3}{4}$ Simplify to lowest terms.

Skill Practice

Multiply.
10. $(-5)(2.2)$
11. $\dfrac{5}{7} \cdot \left(-\dfrac{14}{15}\right)$
12. $\left(-5\dfrac{1}{4}\right)\left(-\dfrac{8}{3}\right)$

Answers
10. -11 **11.** $-\dfrac{2}{3}$ **12.** 14

c. $\left(-3\frac{1}{3}\right)\left(-\frac{3}{10}\right)$

$$= \left(-\frac{10}{3}\right)\left(-\frac{3}{10}\right) \qquad \text{Write the mixed number as a fraction.}$$

$$= \frac{30}{30} \qquad \text{\textit{Same} signs. The product is positive.}$$

$$= 1 \qquad \text{Simplify to lowest terms.}$$

TIP: A number and its reciprocal have the same sign. For example:

$$\left(-\frac{10}{3}\right)\left(-\frac{3}{10}\right) = 1$$

and $\quad 3 \cdot \frac{1}{3} = 1$

Notice from Example 4(c) that $\left(-\frac{10}{3}\right)\left(-\frac{3}{10}\right) = 1$. If the product of two numbers is 1, then the numbers are **reciprocals**. That is, the reciprocal of a real *number a* is $\frac{1}{a}$. Furthermore, $a \cdot \frac{1}{a} = 1$.

Recall that subtraction of real numbers was defined in terms of addition. In a similar way, division of real numbers can be defined in terms of multiplication.

Procedure to Divide Real Numbers

To divide two real numbers, multiply the first number by the reciprocal of the second number. For example:

Multiply

$$10 \div 5 = 2 \qquad \text{or equivalently} \qquad 10 \cdot \frac{1}{5} = 2$$

Reciprocal

Concept Connections

13. If $x > 0$ and $y < 0$, which of the following is true about the value of xy?
  a. xy is positive.
  b. xy is negative.
  c. The sign of xy cannot be determined.

14. If $x > 0$ and $y < 0$, which of the following is true about the value of $(x + y)$?
  a. $(x + y)$ is positive.
  b. $(x + y)$ is negative.
  c. The sign of $(x + y)$ cannot be determined.

Because division of real numbers can be expressed in terms of multiplication, the sign rules that apply to multiplication also apply to division.

$$\left. \begin{array}{l} 10 \div 2 = 10 \cdot \dfrac{1}{2} = 5 \\[2mm] -10 \div (-2) = -10 \cdot \left(-\dfrac{1}{2}\right) = 5 \end{array} \right\} \begin{array}{l} \text{Dividing two numbers of the same sign} \\ \text{produces a \textit{positive} quotient.} \end{array}$$

$$\left. \begin{array}{l} 10 \div (-2) = 10 \cdot \left(-\dfrac{1}{2}\right) = -5 \\[2mm] -10 \div 2 = -10 \cdot \dfrac{1}{2} = -5 \end{array} \right\} \begin{array}{l} \text{Dividing two numbers of opposite} \\ \text{signs produces a \textit{negative} quotient.} \end{array}$$

Division of Real Numbers

Assume that a and b are real numbers such that $b \neq 0$.

1. If a and b have the *same* sign, then the quotient $\frac{a}{b}$ is *positive*.

2. If a and b have *different* signs, then the quotient $\frac{a}{b}$ is *negative*.

3. $\dfrac{0}{b} = 0$.

4. $\dfrac{b}{0}$ is undefined.

Answers

13. b   **14.** c

The relationship between multiplication and division can be used to investigate properties 3 and 4 from the preceding box. For example,

$$\frac{0}{6} = 0 \qquad \textit{Because } 6 \cdot 0 = 0 \checkmark$$

$$\frac{6}{0} \textit{ is undefined} \qquad \textit{Because there is no number that when multiplied by 0 will equal 6}$$

Note: The quotient of 0 and 0 *cannot be determined.* Evaluating an expression of the form $\frac{0}{0} = ?$ is equivalent to asking, "What number times zero will equal 0?" That is, $(0)(?) = 0$. Any real number will satisfy this requirement; however, expressions involving $\frac{0}{0}$ are usually discussed in advanced mathematics courses.

Example 5   Dividing Real Numbers

Divide the real numbers. Write the answer as a fraction or whole number.

a. $\dfrac{-42}{7}$ **b.** $\dfrac{-96}{-144}$ **c.** $3\dfrac{1}{10} \div \left(-\dfrac{2}{5}\right)$ **d.** $\dfrac{-8}{-7}$

Solution:

a. $\dfrac{-42}{7} = -6$ *Different* signs. The quotient is negative.

b. $\dfrac{-96}{-144} = \dfrac{2}{3}$ *Same* signs. The quotient is positive. Simplify.

c. $3\dfrac{1}{10} \div \left(-\dfrac{2}{5}\right)$

$= \dfrac{31}{10}\left(-\dfrac{5}{2}\right)$ Write the mixed number as an improper fraction, and multiply by the reciprocal of the second number.

$= \dfrac{31}{\overset{}{\underset{2}{10}}}\left(-\dfrac{\overset{1}{5}}{2}\right)$

$= -\dfrac{31}{4}$ *Different* signs. The quotient is negative.

d. $\dfrac{-8}{-7} = \dfrac{8}{7}$ *Same* signs. The quotient is positive.
Because 7 does not divide into 8 evenly, the answer can be left as a fraction.

Skill Practice

Divide.

15. $\dfrac{42}{-2}$ **16.** $\dfrac{-28}{-4}$

17. $-\dfrac{2}{3} \div 4$ **18.** $\dfrac{-1}{-2}$

TIP: Multiplication may be used to check a division problem.

$$\frac{-42}{7} = -6$$

Check: $(7)(-6) = -42 \checkmark$

Avoiding Mistakes

If the numerator and denominator of a fraction have opposite signs, then the quotient will be negative. Therefore, a fraction has the same value whether the negative sign is written in the numerator, in the denominator, or in front of the fraction.

$$-\frac{31}{4} = \frac{-31}{4} = \frac{31}{-4}$$

4. Exponential Expressions

To simplify the process of repeated multiplication, exponential notation is often used. For example, the quantity $3 \cdot 3 \cdot 3 \cdot 3 \cdot 3$ can be written as 3^5 (3 to the fifth power).

Answers

15. -21 **16.** 7

17. $-\dfrac{1}{6}$ **18.** $\dfrac{1}{2}$

> ## Definition of b^n
>
> Let b represent any real number and n represent a positive integer. Then
>
> $$b^n = \underbrace{b \cdot b \cdot b \cdot b \cdot \ldots b}_{n \text{ factors of } b}$$
>
> b^n is read as "b to the nth power."
> b is called the **base** and n is called the **exponent**, or **power**.
> b^2 is read as "b *squared*," and b^3 is read as "b *cubed*."

Example 6 Evaluating Exponential Expressions

Simplify the expression.

a. 5^3 **b.** $(-2)^4$ **c.** -2^4 **d.** $\left(-\dfrac{1}{3}\right)^3$

Solution:

a. $5^3 = 5 \cdot 5 \cdot 5$ The base is 5, and the exponent is 3.

    $= 125$

b. $(-2)^4 = (-2)(-2)(-2)(-2)$ The base is -2, and the exponent is 4.
The exponent 4 applies to the entire
    $= 16$ contents of the parentheses.

c. $-2^4 = -[2 \cdot 2 \cdot 2 \cdot 2]$ The base is 2, and the exponent is 4.
Because no parentheses enclose the
    $= -16$ negative sign, the exponent applies to only 2.

> **TIP:** The quantity -2^4 can also be interpreted as $-1 \cdot 2^4$.
>
> $$-2^4 = -1 \cdot 2^4 = -1 \cdot (2 \cdot 2 \cdot 2 \cdot 2) = -16$$

d. $\left(-\dfrac{1}{3}\right)^3 = \left(-\dfrac{1}{3}\right)\left(-\dfrac{1}{3}\right)\left(-\dfrac{1}{3}\right)$ The base is $-\frac{1}{3}$, and the exponent is 3.

    $= -\dfrac{1}{27}$

Calculator Connections

Topic: Using the Exponent Keys

On many calculators, the $\boxed{x^2}$ key is used to square a number. The $\boxed{\wedge}$ key is used to raise a base to any power.

```
5^3
            125
( -2)^4
             16
-2^4
            -16
```

5. Square Roots

The inverse operation to squaring a number is to find its square roots. For example, finding a square root of 9 is equivalent to asking, "What number when squared equals 9?" One obvious answer is 3, because $(3)^2 = 9$. However, -3 is also a square root of 9 because $(-3)^2 = 9$. For now, we will focus on the **principal square root**, which is always taken to be nonnegative.

The symbol $\sqrt{}$, called a **radical sign**, is used to denote the principal square root of a number. Therefore, the principal square root of 9 can be written as $\sqrt{9}$. The expression $\sqrt{64}$ represents the principal square root of 64.

Example 7 **Evaluating Square Roots**

Evaluate the expressions, if possible.

  a. $\sqrt{81}$ **b.** $\sqrt{\dfrac{25}{64}}$ **c.** $\sqrt{-16}$ **d.** $-\sqrt{16}$

Solution:

  a. $\sqrt{81} = 9$ because $(9)^2 = 81$

  b. $\sqrt{\dfrac{25}{64}} = \dfrac{5}{8}$ because $\left(\dfrac{5}{8}\right)^2 = \dfrac{25}{64}$

  c. $\sqrt{-16}$ is *not a real number* because no real number when squared will be negative.

  d. $-\sqrt{16} = -4$ because $-\sqrt{16} = -(\sqrt{16}) = -4.$

Example 7(c) illustrates that the square root of a negative number is not a real number because no real number when squared will be negative.

> **Square Root of a Negative Number**
> Let a be a negative real number. Then $\sqrt{a}$ is not a real number.

6. Order of Operations

When algebraic expressions contain numerous operations, it is important to use the proper **order of operations**. Parentheses ( ), brackets [ ], and braces { } are used for grouping numbers and algebraic expressions. It is important to recognize that operations must be done first within parentheses and other grouping symbols.

> **Order of Operations**
> **Step 1** First, simplify expressions within parentheses and other grouping symbols. These include absolute value bars, fraction bars, and radicals. If embedded parentheses are present, start with the innermost parentheses.
> **Step 2** Evaluate expressions involving exponents, radicals, and absolute values.
> **Step 3** Perform multiplication or division in the order in which they occur from left to right.
> **Step 4** Perform addition or subtraction in the order in which they occur from left to right.

Calculator Connections

Topic: Using the Square Root Key

The key is used to find the square root of a non-negative real number.

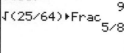

Answers

23. 5   **24.** $\dfrac{7}{10}$
25. Not a real number
26. -3

Skill Practice

Simplify the expression.

27. $36 \div 2^2 \cdot 3 - [(18 - 5) \cdot 2 + 6]$

Avoiding Mistakes

Don't try to perform too many steps at once. Taking a shortcut may result in a careless error. For each step rewrite the entire expression, changing only the operation being evaluated.

Example 8   **Applying the Order of Operations**

Simplify the expression. $10 - [2 - 4(6 - 8)]^2 + \sqrt{16 - 7}$

Solution:

$10 - [2 - 4(6 - 8)]^2 + \sqrt{16 - 7}$

$= 10 - [2 - 4(-2)]^2 + \sqrt{9}$ Simplify inside the innermost parentheses and inside the radical.

$= 10 - [2 + 8]^2 + \sqrt{9}$ Simplify within square brackets. Perform multiplication before addition or subtraction.

$= 10 - [10]^2 + \sqrt{9}$

$= 10 - 100 + 3$ Simplify the exponential expression and the radical.

$= -90 + 3$ Perform addition or subtraction in the order in which they appear from left to right.

$= -87$

Skill Practice

Simplify the expression.

28. $\dfrac{-|5 - 7| + 11}{(-1 - 2)^2}$

Example 9   **Applying the Order of Operations**

Simplify the expression. $\dfrac{|(-3)^3 + (5^2 - 3)|}{-15 \div (-3)(2)}$

Solution:

$\dfrac{|(-3)^3 + (5^2 - 3)|}{-15 \div (-3)(2)}$ Simplify numerator and denominator separately.

$= \dfrac{|(-3)^3 + (25 - 3)|}{5(2)}$ *Numerator:* Simplify within the inner parentheses.
Denominator: Perform division and multiplication (left to right).

$= \dfrac{|(-3)^3 + (22)|}{10}$ *Numerator:* Simplify inner parentheses.
Denominator: Multiply.

$= \dfrac{|-27 + 22|}{10}$ Simplify exponent.

$= \dfrac{|-5|}{10}$ Add within the absolute value.

$= \dfrac{5}{10}$ or $\dfrac{1}{2}$ Evaluate the absolute value and simplify.

Answers

27. -5 **28.** 1

Calculator Connections

Topic: Applying the Order of Operations

To evaluate the expression

$$\frac{|(-3)^3 + (5^2 - 3)|}{-15 \div (-3)(2)}$$

on a graphing calculator, use parentheses to enclose the absolute value expression. Likewise, it is necessary to use parentheses to enclose the entire denominator.

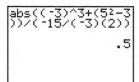

---

7. Evaluating Formulas

An algebraic expression or formula involves operations on numbers and variables. A **variable** is a letter that may represent any numerical value. To evaluate an expression or formula, we substitute known values of the variables into the expression. Then we follow the order of operations.

A list of common geometry formulas is given in the inside back cover of the text. It is important to note that some geometric formulas use Greek letters (such as π) and some use variables with subscripts. A **subscript** is a number or letter written to the right of and below a variable. For example, the area of a trapezoid is given by $A = \frac{1}{2}(b_1 + b_2)h$. The values b_1 and b_2 (read as "b sub 1" and "b sub 2") represent two different bases of the trapezoid (Figure R-3).

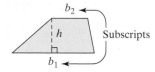

Figure R-3

Example 10 **Evaluating a Formula**

A homeowner in North Carolina wants to buy protective film for a trapezoid-shaped window. The film will adhere to shattered glass in the event that the glass breaks during a bad storm. Find the area of the window whose dimensions are given in Figure R-4.

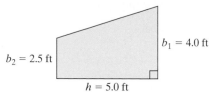

Figure R-4

Solution:

$$A = \frac{1}{2}(b_1 + b_2)h$$

$$= \frac{1}{2}(4.0 \text{ ft} + 2.5 \text{ ft})(5.0 \text{ ft}) \qquad \text{Substitute } b_1 = 4.0 \text{ ft}, b_2 = 2.5 \text{ ft, and } h = 5.0 \text{ ft.}$$

$$= \frac{1}{2}(6.5 \text{ ft})(5.0 \text{ ft}) \qquad \text{Simplify inside parentheses.}$$

$$= 16.25 \text{ ft}^2 \qquad \text{Multiply from left to right.}$$

The area of the window is 16.25 ft².

Skill Practice

29. Use the formula given in Example 10 to find the area of the trapezoid.

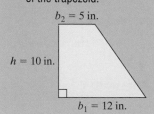

TIP: Subscripts should not be confused with *superscripts*, which are written above a variable. Superscripts are used to denote powers.

$$b_2 \neq b^2$$

Answer
29. The area is 85 in.²

Section R.3 Practice Exercises

Study Skills Exercise

Sometimes you may run into a problem with homework, or you may find that you are having trouble keeping up with the pace of the class. A tutor can be a good resource. Answer the following questions.

a. Does your college offer tutoring? **b.** Is it free?

c. Where should you go to sign up for a tutor? **d.** Is there tutoring available online?

Vocabulary and Key Concepts

1. a. Two numbers that are the same distance from 0 but on opposite sides of 0 on the number line are called _____.

b. The absolute value of a real number, a, is denoted by _____ and is the distance between a and _____ on the number line.

c. Given the expression b^n, the value b is called the _____ and _____ is called the exponent or power.

d. The symbol $\sqrt{}$ is called a _____ sign and is used to find the principal _____ root of a nonnegative real number.

e. If a and b are both negative, then $a + b$ will be (positive/negative) and ab will be (positive/negative).

f. If $a < 0$ and $b > 0$, and if $|a| < |b|$, then the sign of $a + b$ will be (positive/negative) and the sign of $\frac{a}{b}$ will be (positive/negative).

g. The expression $a - b = a +$ _____. If $a > 0$ and $b < 0$, then the sign of $a - b$ is _____.

h. If a is a nonzero real number, then the reciprocal of a is _____. The product of a number and its reciprocal is _____.

i. If either a or b is zero then $ab =$ _____.

j. If $a = 0$ and $b \neq 0$, then $\frac{a}{b} =$ _____ and $\frac{b}{a}$ is _____.

Review Exercises

2. Write the set-builder notation representing the interval $[-5, 3)$.

For Exercises 3–6, express each set in interval notation.

3. $\{z \mid z < 4\}$ **4.** $\{w \mid w \geq -2\}$ **5.** $\{x \mid -3 \leq x < -1\}$ **6.** $\{p \mid 0 < p \leq 8\}$

Concept 1: Opposite and Absolute Value

7. If the absolute value of a number can be thought of as its distance from zero, explain why an absolute value can never be negative.

8. If a number is negative, then its *opposite* will be

    a. Positive **b.** Negative.

9. If a number is negative, then its *reciprocal* will be

    a. Positive **b.** Negative.

10. If a number is negative, then its *absolute value* will be

    a. Positive **b.** Negative.

11. Complete the table. **(See Example 1.)**

Number	Opposite	Reciprocal	Absolute Value
6			
	$-\frac{1}{11}$		
		$-\frac{1}{8}$	
	$\frac{13}{10}$		
0			
		$-0.\overline{3}$	

12. Complete the table.

Number	Opposite	Reciprocal	Absolute Value
-9			
	$\frac{2}{3}$		
		14	
-1			
0			
		$2\frac{1}{9}$	

For Exercises 13–20, fill in the blank with the appropriate symbol ($<$, $>$, $=$). **(See Example 1.)**

13. $-|6|$ _____ $|-6|$

14. $-(-5)$ _____ $-|-5|$

15. $|-4|$ _____ $|4|$

16. $-|2|$ _____ (-2)

17. $-|-1|$ _____ 1

18. -3 _____ $-|-7|$

19. $|2 + (-5)|$ _____ $|2| + |-5|$

20. $|4 + 3|$ _____ $|4| + |3|$

Concept 2: Addition and Subtraction of Real Numbers

For Exercises 21–36, add or subtract as indicated. **(See Examples 2 and 3.)**

21. $-8 + 4$

22. $3 + (-7)$

23. $-12 + (-7)$

24. $-5 + (-11)$

25. $-17 - (-10)$

26. $-14 - (-2)$

27. $5 - (-9)$

28. $8 - (-4)$

29. $-6.3 - 15.8$

30. $-21.9 - 4.7$

31. $1.5 - 9.6$

32. $4.8 - 10$

33. $\frac{2}{3} + \left(-2\frac{1}{3}\right)$

34. $-\frac{4}{7} + \left(1\frac{4}{7}\right)$

35. $-\frac{5}{9} - \frac{14}{15}$

36. $-6 - \frac{2}{9}$

Concept 3: Multiplication and Division of Real Numbers

For Exercises 37–52, perform the indicated operation. **(See Examples 4 and 5.)**

37. $4(-8)$

38. $-21(3)$

39. $\frac{2}{9} \cdot \frac{12}{7}$

40. $\left(-\frac{5}{9}\right) \cdot \left(-1\frac{7}{11}\right)$

41. $\frac{-6}{-10}$

42. $\frac{-15}{-24}$

43. $-2\frac{1}{4} \div \frac{5}{8}$

44. $-\frac{2}{3} \div \left(-1\frac{5}{7}\right)$

45. $7 \div 0$

46. $\frac{1}{16} \div 0$

47. $0 \div (-3)$

48. $0 \div 11$

49. $(-1.2)(-3.1)$

50. $(4.6)(-2.25)$

51. $\frac{-5}{-11}$

52. $\frac{-3}{-13}$

Concept 4: Exponential Expressions

For Exercises 53–60, evaluate the expression. **(See Example 6.)**

53. 4^3

54. -2^3

55. -7^2

56. -3^4

57. $(-7)^2$

58. $(-5)^2$

59. $\left(\frac{5}{3}\right)^3$

60. $\left(\frac{10}{9}\right)^2$

Concept 5: Square Roots

For Exercises 61–68, evaluate the expression, if possible. **(See Example 7.)**

61. $\sqrt{9}$

62. $\sqrt{1}$

63. $\sqrt{-4}$

64. $\sqrt{-36}$

65. $\sqrt{\dfrac{1}{4}}$

66. $\sqrt{\dfrac{9}{4}}$

67. $-\sqrt{49}$

68. $-\sqrt{100}$

Concept 6: Order of Operations

For Exercises 69–96, simplify by using the order of operations. **(See Examples 8 and 9.)**

69. $5 + 3^3$

70. $10 - 2^4$

71. $5 \cdot 2^3$

72. $12 \div 2^2$

73. $(2 + 3)^2$

74. $(4 - 1)^3$

75. $2^2 + 3^2$

76. $4^3 - 1^3$

77. $6 + 10 \div 2 \cdot 3 - 4$

78. $12 \div 3 \cdot 4 - 18$

79. $4^2 - (5 - 2)^2 \cdot 3$

80. $5 - 3(8 \div 4)^2$

81. $2 - 5(9 - 4\sqrt{25})^2$

82. $5^2 - (\sqrt{9} + 4 \div 2)$

83. $\left(-\dfrac{3}{5}\right)^2 - \dfrac{3}{5} \cdot \dfrac{5}{9} + \dfrac{7}{10}$

84. $\dfrac{1}{2} - \left(\dfrac{2}{3} \div \dfrac{5}{9}\right) + \dfrac{5}{6}$

85. $1.75 \div 0.25 - (1.25)^2$

86. $5.4 - (0.3)^2 \div 0.09$

87. $\dfrac{\sqrt{10^2 - 8^2}}{3^2}$

88. $\dfrac{\sqrt{16 - 7} + 3^2}{\sqrt{16} - \sqrt{4}}$

89. $-|-11 + 5| + |7 - 2|$

90. $-|-8 - 3| - (-8 - 3)$

91. $25 - 2[(7 - 3)^2 \div 4] + \sqrt{18 - 2}$

92. $\sqrt{29 - 2^2} + [8 - 3(6 - 2)] \div 4 \cdot 5$

93. $\dfrac{|(10 - 7) - 2^3|}{6 - 16 \div 8 \cdot 3}$

94. $\dfrac{|-12 - (7 - 3^2)^2|}{40 - 6^2 - 8 \div 2}$

95. $\left(\dfrac{1}{2}\right)^2 + \left(\dfrac{6 - 4}{5}\right)^2 + \left(\dfrac{5 + 2}{10}\right)^2$

96. $\left(\dfrac{2^3}{2^3 + 1}\right)^2 \div \left(\dfrac{8 - (-2)}{3^2}\right)^2$

For Exercises 97 and 98, find the average of the set of data values by adding the values and dividing by the number of values.

97. Find the average low temperature for a week in January in St. John's, Newfoundland. Round to the nearest tenth of a degree.

Day	Mon.	Tues.	Wed.	Thur.	Fri.	Sat.	Sun.
Low temperature	−18°C	−16°C	−20°C	−11°C	−4°C	−3°C	1°C

98. Find the average high temperature for a week in January in St. John's, Newfoundland. Round to the nearest tenth of a degree.

Day	Mon.	Tues.	Wed.	Thur.	Fri.	Sat.	Sun.
High temperature	−2°C	−6°C	−7°C	0°C	1°C	8°C	10°C

Concept 7: Evaluating Formulas

99. The formula $C = \frac{5}{9}(F - 32)$ converts temperatures in the Fahrenheit scale to the Celsius scale. Find the equivalent Celsius temperature for each Fahrenheit temperature.

   a. 77°F   **b.** 212°F   **c.** 32°F   **d.** −40°F

100. The formula $F = \frac{9}{5}C + 32$ converts Celsius temperatures to Fahrenheit temperatures. Find the equivalent Fahrenheit temperature for each Celsius temperature.

   a. −5°C   **b.** 0°C   **c.** 37°C   **d.** −40°C

The equation $G_C = \frac{1}{22}c + \frac{1}{30}h$ represents the amount of gasoline used (in gal) for a compact car to drive c miles in the city and h miles on the highway. The equation $G_S = \frac{1}{12}c + \frac{1}{18}h$ represents the amount of gasoline used for a large SUV to make the same trip. Use these formulas for Exercises 101 and 102.

101. Determine the amount of gas used by a compact car that travels 33 mi in the city and 80 mi on the highway.

102. Determine the amount of gas used by a large SUV that travels 33 mi in the city and 80 mi on the highway.

Use the geometry formulas found in the inside back cover of the book to answer Exercises 103–112.

For Exercises 103–106, find the area. **(See Example 10.)**

 103. Trapezoid

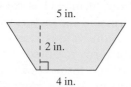

5 in.

2 in.

4 in.

104. Parallelogram

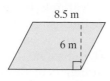

8.5 m

6 m

105. Triangle

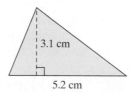

3.1 cm

5.2 cm

106. Rectangle

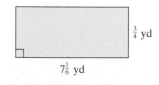

$\frac{3}{4}$ yd

$7\frac{1}{6}$ yd

For Exercises 107–112, find the volume. (Use the π key on your calculator, and round the final answer to one decimal place.)

107. Sphere

$r = 1.5$ ft

108. Sphere

$r = \frac{1}{2}$ yd

109. Right circular cone

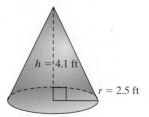

$h = 4.1$ ft

$r = 2.5$ ft

110. Right circular cone

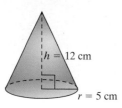

$h = 12$ cm

$r = 5$ cm

111. Right circular cylinder

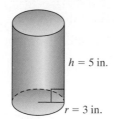

$h = 5$ in.

$r = 3$ in.

112. Right circular cylinder

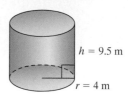

$h = 9.5$ m

$r = 4$ m

Graphing Calculator Exercises

113. Which expression when entered into a graphing calculator will yield the correct value of $\dfrac{12}{6-2}$?

$$12/6 - 2 \quad \text{or} \quad 12/(6-2)$$

114. Which expression when entered into a graphing calculator will yield the correct value of $\dfrac{24-6}{3}$?

$$(24-6)/3 \quad \text{or} \quad 24 - 6/3$$

115. Verify your solution to Exercise 87 by entering the expression into a graphing calculator:

$$(\sqrt{(10^2 - 8^2)})/3^2$$

116. Verify your solution to Exercise 88 by entering the expression into a graphing calculator:

$$(\sqrt{(16-7)} + 3^2)/(\sqrt{(16)} - \sqrt{(4)})$$

| **Section R.4** | **Simplifying Algebraic Expressions** |

1. Recognizing Terms, Factors, and Coefficients

A **term** is a constant or the product of a constant and one or more variables. An algebraic expression is a single term or a sum of two or more terms. For example, the expression

$$-6x^2 + 5xyz - 11 \quad \text{or} \quad -6x^2 + 5xyz + (-11)$$

consists of the terms $-6x^2$, $5xyz$, and -11.

The terms $-6x^2$ and $5xyz$ are **variable terms**, and the term -11 is called a **constant term**. It is important to distinguish between a term and the **factors** within a term. For example, the quantity $5xyz$ is one term, but the values 5, x, y, and z are factors within the term. The constant factor in a term is called the numerical coefficient or simply **coefficient** of the term. In the terms $-6x^2$, $5xyz$, and -11, the coefficients are -6, 5, and -11, respectively. A term containing only variables such as xy has a coefficient of 1.

Terms are called *like* terms if they each have the same variables and the corresponding variables are raised to the same powers. For example:

Concept Connections

1. Write two terms that are *like* terms to $7x^2$. (Answers may vary.)

Like Terms			*Unlike* Terms			
$-6t$	and	$4t$	$-6t$	and	$4s$	(different variables)
$1.8ab$	and	$-3ab$	$1.8xy$	and	$-3x$	(different variables)
$\frac{1}{2}c^2d^3$	and	c^2d^3	$\frac{1}{2}c^2d^3$	and	c^2d	(different powers)
4	and	6	$4p$	and	4	(different variables)

Skill Practice

Given: $-2x^2 + 5x + \dfrac{1}{2} - y^2$

2. List the terms of the expression.
3. Which term is the constant term?
4. Identify the coefficient of the term $-y^2$.

| **Example 1** | **Identifying Terms, Factors, Coefficients, and *Like* Terms** |

a. List the terms of the expression. $\qquad -4x^2 - 7x + \frac{2}{3}$

b. Identify the coefficient of the term. $\qquad yz^3$

c. Identify the pair of *like* terms. $\qquad 16b, 4b^2 \quad$ or $\quad \frac{1}{2}c, -\frac{1}{6}c$

Solution:

a. The terms of the expression $-4x^2 - 7x + \frac{2}{3}$ are $-4x^2$, $-7x$, and $\frac{2}{3}$.

b. The term yz^3 can be written as $1yz^3$; therefore, the coefficient is 1.

c. $\frac{1}{2}c, -\frac{1}{6}c$ are *like* terms because they have the same variable raised to the same power.

2. Properties of Real Numbers

Simplifying algebraic expressions requires several important properties of real numbers that are stated in Table R-3. Assume that a, b, and c represent real numbers or real-valued algebraic expressions.

Answers

1. For example: $-4x^2$ and x^2

2. $-2x^2$, $5x$, $\dfrac{1}{2}$, $-y^2$

3. $\dfrac{1}{2}$ $\qquad$ 4. -1

Table R-3

Property Name	Algebraic Representation	Example	Description/Notes
Commutative property of addition	$a + b = b + a$	$5 + 3 = 3 + 5$	The order in which two real numbers are added or multiplied does not affect the result.
Commutative property of multiplication	$a \cdot b = b \cdot a$	$(5)(3) = (3)(5)$	
Associative property of addition	$(a + b) + c$ $= a + (b + c)$	$(2 + 3) + 7$ $= 2 + (3 + 7)$	The manner in which real numbers are grouped under addition or multiplication does not affect the result.
Associative property of multiplication	$(a \cdot b)c = a(b \cdot c)$	$(2 \cdot 3)7 = 2(3 \cdot 7)$	
Distributive property of multiplication over addition	$a(b + c)$ $= ab + ac$	$3(5 + 2)$ $= 3 \cdot 5 + 3 \cdot 2$	A factor outside the parentheses is multiplied by each term inside the parentheses.
Identity property of addition	0 is the identity element for addition because $a + 0 = 0 + a = a$	$5 + 0 = 0 + 5 = 5$	Any number added to the identity element 0 will remain unchanged.
Identity property of multiplication	1 is the identity element for multiplication because $a \cdot 1 = 1 \cdot a = a$	$5 \cdot 1 = 1 \cdot 5 = 5$	Any number multiplied by the identity element 1 will remain unchanged.
Inverse property of addition	a and $(-a)$ are additive inverses because $a + (-a) = 0$ and $(-a) + a = 0$	$3 + (-3) = 0$	The sum of a number and its additive inverse (opposite) is the identity element 0.
Inverse property of multiplication	a and $\frac{1}{a}$ are multiplicative inverses because $a \cdot \dfrac{1}{a} = 1$ and $\dfrac{1}{a} \cdot a = 1$ (provided $a \neq 0$)	$5 \cdot \frac{1}{5} = 1$	The product of a number and its multiplicative inverse (reciprocal) is the identity element 1.

Animation

The properties of real numbers are used to multiply algebraic expressions. To multiply a term by an algebraic expression containing more than one term, we apply the distributive property of multiplication over addition.

Example 2   **Applying the Distributive Property**

Apply the distributive property.

a. $4(2x + 5)$ **b.** $-(-3.4q + 5.7r)$

c. $-3(a + 2b - 5c)$ **d.** $-\dfrac{2}{3}\left(-9x + \dfrac{3}{8}y - 5\right)$

Solution:

a. $4(2x + 5)$

$= 4(2x) + 4(5)$ Apply the distributive property.

$= 8x + 20$ Simplify, using the associative
                                                property of multiplication.

b. $-(-3.4q + 5.7r)$ The negative sign preceding the
                                                parentheses can be interpreted as a
                                                factor of -1.

$= -1(-3.4q + 5.7r)$

$= -1(-3.4q) + (-1)(5.7r)$ Apply the distributive property.

$= 3.4q - 5.7r$

c. $-3(a + 2b - 5c)$

$= -3(a) + (-3)(2b) + (-3)(-5c)$ Apply the distributive property.

$= -3a - 6b + 15c$ Simplify.

d. $-\dfrac{2}{3}\left(-9x + \dfrac{3}{8}y - 5\right)$

$= -\dfrac{2}{3}(-9x) + \left(-\dfrac{2}{3}\right)\left(\dfrac{3}{8}y\right) + \left(-\dfrac{2}{3}\right)(-5)$ Apply the distributive property.

$= \dfrac{18}{3}x - \dfrac{6}{24}y + \dfrac{10}{3}$ Simplify.

$= 6x - \dfrac{1}{4}y + \dfrac{10}{3}$ Simplify to lowest terms.

TIP: When applying the distributive property, a negative factor preceding the parentheses will change the signs of the terms within the parentheses.

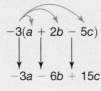

$-3(a + 2b - 5c)$

$-3a - 6b + 15c$

Notice that the parentheses are removed after the distributive property is applied. Sometimes this is referred to as clearing parentheses.

Two terms can be added or subtracted only if they are *like* terms. To add or subtract *like* terms, we use the distributive property, as shown in Example 3.

Example 3 **Using the Distributive Property to Add and Subtract *Like* Terms**

Add and subtract as indicated.

a. $-8x + 3x$ **b.** $4.75y^2 - 9.25y^2 + y^2$

Solution:

a. $-8x + 3x$

$= (-8 + 3)x$ Apply the distributive property.

$= (-5)x$ Simplify.

$= -5x$

b. $4.75y^2 - 9.25y^2 + y^2$

$= 4.75y^2 - 9.25y^2 + 1y^2$ Notice that y^2 is interpreted as $1y^2$.

$= (4.75 - 9.25 + 1)y^2$ Apply the distributive property.

$= (-3.5)y^2$ Simplify.

$= -3.5y^2$

Skill Practice

Combine *like* terms.

9. $-4y + 7y$

10. $a^2 - 6.2a^2 + 2.8a^2$

Although the distributive property is used to add and subtract *like* terms, it is tedious to write each step. Observe that adding or subtracting *like* terms is a matter of combining the coefficients and leaving the variable factors unchanged. This can be shown in one step. This shortcut will be used throughout the text. For example:

$$4w + 7w = 11w \qquad 8ab^2 + 10ab^2 - 5ab^2 = 13ab^2$$

3. Simplifying Expressions

Clearing parentheses and combining *like* terms are important tools to simplifying algebraic expressions. This is demonstrated in Example 4.

Example 4 **Clearing Parentheses and Combining *Like* Terms**

Simplify by clearing parentheses and combining *like* terms.

a. $4 - 3(2x - 8) - 1$ **b.** $-(3s - 11t) - 5(2t + 8s) - 10s$

Solution:

a. $4 - 3(2x - 8) - 1$

$= 4 - 6x + 24 - 1$ Apply the distributive property.

$= -6x + 4 + 24 - 1$ Group *like* terms.

$= -6x + 27$ Combine *like* terms.

b. $-(3s - 11t) - 5(2t + 8s) - 10s$

$= -3s + 11t - 10t - 40s - 10s$ Apply the distributive property.

$= -3s - 40s - 10s + 11t - 10t$ Group *like* terms.

$= -53s + t$ Combine *like* terms.

Skill Practice

Simplify by clearing parentheses and combining *like* terms.

11. $7 - 2(3x - 4) - 5$

12. $-(6z - 10y)$
$\quad - 4(3z + y) - 8y$

Answers

9. $3y$ **10.** $-2.4a^2$

11. $-6x + 10$ **12.** $-2y - 18z$

Example 5 **Clearing Parentheses and Combining *Like* Terms**

Simplify by clearing parentheses and combining *like* terms.

a. $2[1.5x + 4.7(x^2 - 5.2x) - 3x]$ **b.** $-\dfrac{1}{3}(3w - 6) - \left(\dfrac{1}{4}w + 4\right)$

Solution:

a. $2[1.5x + 4.7(x^2 - 5.2x) - 3x]$

$= 2[1.5x + 4.7x^2 - 24.44x - 3x]$ Apply the distributive property to the inner parentheses.

$= 2[1.5x - 24.44x - 3x + 4.7x^2]$ Group *like* terms.

$= 2[-25.94x + 4.7x^2]$ Combine *like* terms.

$= -51.88x + 9.4x^2$ Apply the distributive property.

$= 9.4x^2 - 51.88x$

TIP: By using the commutative property of addition, the expression $-51.88x + 9.4x^2$ can also be written as $9.4x^2 + (-51.88x)$ or simply $9.4x^2 - 51.88x$. Although the expressions are all equal, it is customary to write the terms in descending order of the powers of the variable.

b. $-\dfrac{1}{3}(3w - 6) - \left(\dfrac{1}{4}w + 4\right)$

$= -\dfrac{3}{3}w + \dfrac{6}{3} - \dfrac{1}{4}w - 4$ Apply the distributive property.

$= -w + 2 - \dfrac{1}{4}w - 4$ Simplify fractions.

$= -\dfrac{4}{4}w - \dfrac{1}{4}w + 2 - 4$ Group *like* terms and find a common denominator.

$= -\dfrac{5}{4}w - 2$ Combine *like* terms.

Section R.4 Practice Exercises

Study Skills Exercise

It is very important to attend class every day. Math is cumulative in nature and you must master the material learned in the previous class to understand a new day's lesson. Because this is so important, many instructors tie attendance to the final grade. Write down the attendance policy for your class.

Vocabulary and Key Concepts

1. a. Given the expression $8x + cd - 3y + 90$, the terms $8x$, cd, and $-3y$ are variable terms, whereas 90 is a _____ term.

  b. The constant factor in a term is called the _____.

  c. Given the expression x, the value of the coefficient is _____, and the exponent is _____.

  d. Terms that have the same variables, with corresponding variables raised to the same powers are called _____ terms.

Review Exercises

For Exercises 2–4, simplify.

2. $6 - 2\sqrt{10^2 - 6^2} + 7$

3. $\dfrac{-3^2 - |5 + (-7)|}{26 - 2^2}$

4. $36 \div (-2) \cdot 3 - 4(6 - 8)$

For Exercises 5–8, write the set in interval notation.

5. $\{x \mid x > |-3|\}$

6. $\left\{ x \mid x \le \left| -\dfrac{4}{3} \right| \right\}$

7. $\left\{ w \mid -\dfrac{5}{2} < w \le \sqrt{9} \right\}$

8. $\left\{ z \mid 2 \le z < \dfrac{11}{3} \right\}$

Concept 1: Recognizing Terms, Factors, and Coefficients

For Exercises 9–12:

    a. Determine the number of terms in the expression.

    b. Identify the constant term.

    c. List the coefficients of each term, separated by commas. **(See Example 1.)**

9. $2x^3 - 5xy + 6$

10. $a^2 - 4ab - b^2 + 8$

11. $pq - 7 + q^2 - 4q + p$

12. $7x - 1 + 3xy$

Concept 2: Properties of Real Numbers

For Exercises 13–30, match each expression with the appropriate property.

13. $3 + \dfrac{1}{2} = \dfrac{1}{2} + 3$

14. $7.2(4 + 1) = 7.2(4) + 7.2(1)$

15. $10 + 0 = 10$

16. $7 \cdot 1 = 7$

17. $(6 + 8) + 2 = 6 + (8 + 2)$

18. $(4 + 19) + 7 = (19 + 4) + 7$

19. $6 \cdot \dfrac{1}{6} = 1$

20. $2 + (-2) = 0$

21. $9(4 \cdot 12) = (9 \cdot 4)12$

22. $\left(\dfrac{1}{4} + 2 \right)20 = 5 + 40$

23. $42 \cdot 1 = 42$

24. $4 \cdot \dfrac{1}{4} = 1$

25. $(13 \cdot 41)6 = (41 \cdot 13)6$

26. $6(x + 3) = 6x + 18$

27. $8 + (-8) = 0$

28. $21 + 0 = 21$

29. $3(y + 10) = 3(10 + y)$

30. $5(3 \cdot 7) = (5 \cdot 3)7$

a. Commutative property of addition

b. Associative property of multiplication

c. Distributive property of multiplication over addition

d. Commutative property of multiplication

e. Associative property of addition

f. Identity property of addition

g. Identity property of multiplication

h. Inverse property of addition

i. Inverse property of multiplication

For Exercises 31–42, clear parentheses by applying the distributive property. **(See Example 2.)**

31. $2(x - 3y + 8)$

32. $5(-2a + 4b - 9c)$

33. $-10(4s - 9t - 3)$

34. $-4(-8x + 6y + 3z)$

35. $-(-7w + 5z)$

36. $-(-22a - 17b)$

37. $-\dfrac{1}{5}\left(-\dfrac{5}{2}a + 10b - 8\right)$

38. $-\dfrac{3}{4}\left(6x - 4y + \dfrac{4}{9}\right)$

39. $3(2.6x - 4.1)$

40. $5(-7.2y + 2.3)$

41. $2(7c - 8) - 5(6d - f)$

42. $-2(-3q + r) - 7(5s + 2t)$

Concept 3: Simplifying Expressions

For Exercises 43–80, clear parentheses and combine *like* terms. **(See Examples 3–5.)**

43. $8y - 2x + y + 5y$

44. $-9a + a - b + 5a$

45. $4p^2 - 2p + 3p - 6 + 2p^2$

46. $6q - 9 + 3q^2 - q^2 + 10$

47. $2p - 7p^2 - 5p + 6p^2$

48. $5a^2 - 2a - 7a^2 + 6a + 4$

49. $m - 4n^3 + 3 + 5n^3 - 9$

50. $x + 2y^3 - 2x - 8y^3$

51. $5ab + 2ab + 8a$

52. $-6m^2n - 3mn^2 - 2m^2n$

53. $14xy^2 - 5y^2 + 2xy^2$

54. $9uv + 3u^2 + 5uv + 4u^2$

55. $8(x - 3) + 1$

56. $-4(b + 2) - 3$

57. $-2(c + 3) - 2c$

58. $4(z - 4) - 3z$

59. $-(10w - 1) + 9 + w$

60. $-(2y + 7) - 4 + 3y$

61. $-9 - 4(2 - z) + 1$

62. $3 + 3(4 - w) - 11$

63. $4(2s - 7) - (s - 2)$

64. $2(t - 3) - (t - 7)$

65. $-3(-5 + 2w) - 8w + 2(w - 1)$

66. $5 - (-4t - 7) - t - 9$

67. $8x - 4(x - 2) - 2(2x + 1) - 6$

68. $6(y - 2) - 3(2y - 5) - 3$

69. $\dfrac{1}{2}(4 - 2c) + 5c$

70. $\dfrac{2}{3}(3d + 6) - 4d$

71. $3.1(2x + 2) - 4(1.2x - 1)$

72. $4.5(5 - y) + 3(1.9y + 1)$

73. $2\left[5\left(\dfrac{1}{2}a + 3\right) - (a^2 + a) + 4\right]$

74. $-3\left[3\left(b - \dfrac{2}{3}\right) - 2(b + 4) - 6b^2\right]$

75. $(2y - 5) - 2(y - y^2) - 3y$

76. $-(x + 6) + 3(x^2 + 1) + 2x$

77. $2.2\{4 - 8[6x - 1.5(x + 4) - 6] + 7.5x\}$

78. $-3.2 - \{6.1y - 4[9 - (2y + 2.5)] + 7y\}$

79. $\dfrac{1}{8}(24n - 16m) - \dfrac{2}{3}(3m - 18n - 2) + \dfrac{2}{3}$

80. $\dfrac{1}{5}(25a - 20b) - \dfrac{4}{7}(21a - 14b + 2) + \dfrac{1}{7}$

Expanding Your Skills

81. What is the identity element for addition? Use it in an example.

82. What is the identity element for multiplication? Use it in an example.

83. What is another name for a multiplicative inverse?

84. What is another name for an additive inverse?

85. Is the operation of subtraction commutative? If not, give an example.

86. Is the operation of division commutative? If not, give an example.

87. Given the rectangular regions:

a. Write an expression for the area of region A. (Do not simplify.)

b. Write an expression for the area of region B.

c. Write an expression for the area of region C.

d. Add the expressions for the area of regions B and C.

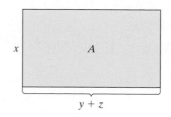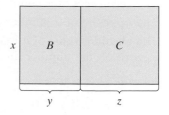

e. Show that the area of region A is equal to the sum of the areas of regions B and C. What property of real numbers does this illustrate?

Chapter R   Summary

Section R.2   Sets of Numbers and Interval Notation

Key Concepts

Natural numbers: $\{1, 2, 3, \ldots\}$
Whole numbers: $\{0, 1, 2, 3, \ldots\}$
Integers: $\{\ldots, -3, -2, -1, 0, 1, 2, 3, \ldots\}$

Rational numbers: $\left\{\dfrac{p}{q} \;\middle|\; p \text{ and } q \text{ are integers and } q \neq 0\right\}$

Rational numbers include all terminating and repeating decimals.

Irrational numbers: A subset of the real numbers whose elements cannot be written as a ratio of two integers.

Irrational numbers cannot be written as repeating or terminating decimals.

Real numbers: $\{x \mid x \text{ is rational or } x \text{ is irrational}\}$

$a < b$ "a is less than b"
$a > b$ "a is greater than b"
$a \leq b$ "a is less than or equal to b"
$a \geq b$ "a is greater than or equal to b"
$a < x < b$ "x is between a and b"

Examples

Example 1

Some rational numbers are:

$\frac{1}{7}, 0.5, 0.\overline{3}$

Some irrational numbers are:

$\sqrt{7}, \sqrt{2}, \pi$

Example 2

Set-Builder Notation	Interval Notation	Graph
$\{x \mid x > 3\}$	$(3, \infty)$	⟶ open at 3
$\{x \mid x \geq 3\}$	$[3, \infty)$	⟶ closed at 3
$\{x \mid x < 3\}$	$(-\infty, 3)$	⟵ open at 3
$\{x \mid x \leq 3\}$	$(-\infty, 3]$	⟵ closed at 3

| Section R.3 | Operations on Real Numbers |

Key Concepts

The **reciprocal** of a number $a \neq 0$ is $\frac{1}{a}$.

The **opposite** of a number a is $-a$.

The **absolute value** of a, denoted $|a|$, is its distance from zero on the number line.

$$|a| = a \qquad \text{if } a \geq 0$$

$$|a| = -a \qquad \text{if } a < 0$$

Addition of Real Numbers

Same Signs: Add the absolute values of the numbers, and apply the common sign to the sum.

Different Signs: Subtract the smaller absolute value from the larger absolute value. Then apply the sign of the number having the larger absolute value.

Subtraction of Real Numbers

Add the opposite of the second number to the first number. $a - b = a + (-b)$

Multiplication and Division of Real Numbers

Same Signs: Product or quotient is positive.

Different Signs: Product or quotient is negative.

The product of any real number and 0 is 0.
The quotient of 0 and a nonzero number is 0.
The quotient of a nonzero number and 0 is undefined.

Exponents and Radicals

$b^4 = b \cdot b \cdot b \cdot b$ (b is the **base**, 4 is the **exponent**)
$\sqrt{b}$ is the **principal square root** of b ($\sqrt{}$ is the radical sign).

Order of Operations

1. Simplify expressions within parentheses and other grouping symbols first.
2. Evaluate expressions involving exponents, radicals, and absolute values.
3. Perform multiplication or division in order from left to right.
4. Perform addition or subtraction in order from left to right.

Examples

Example 1

Given: -5
The reciprocal is $-\frac{1}{5}$. The opposite is 5.

The absolute value is 5.

Example 2

$$-3 + (-4) = -7$$
$$-5 + 7 = 2$$

Example 3

$$7 - (-5) = 7 + (5) = 12$$

Example 4

$(-3)(-4) = 12$ $\qquad$ $\dfrac{-15}{-3} = 5$

$(-2)(5) = -10$ $\qquad$ $\dfrac{6}{-12} = -\dfrac{1}{2}$

$(-7)(0) = 0$ $\qquad$ $0 \div 9 = 0$

$-3 \div 0$ is undefined

Example 5

$$6^3 = 6 \cdot 6 \cdot 6 = 216$$
$$\sqrt{100} = 10$$

Example 6

$$10 - 5(3 - 1)^2 + \sqrt{16}$$
$$= 10 - 5(2)^2 + \sqrt{16}$$
$$= 10 - 5(4) + 4$$
$$= 10 - 20 + 4$$
$$= -10 + 4$$
$$= -6$$

| Section R.4 | Simplifying Algebraic Expressions |

Key Concepts

A **term** is a constant or the product or quotient of a constant and one or more variables.

- A **variable term** contains at least one variable.
- A **constant term** has no variable.

The **coefficient** of a term is the numerical factor of the term.

Like **terms** have the same variables, and the corresponding variables are raised to the same powers.

Distributive Property of Multiplication over Addition

$$a(b + c) = ab + ac$$

Two terms can be added or subtracted if they are *like* terms. First clear parentheses before adding or subtracting *like* terms.

Examples

Example 1

$-2x$	Variable term has coefficient -2.
x^2y	Variable term has coefficient 1.
6	Constant term has coefficient 6.

Example 2

$4ab^3$ and $2ab^3$ are *like* terms.

Example 3

$$2(x + 4y) = 2x + 8y$$
$$-(a + 6b - 5c) = -a - 6b + 5c$$

Example 4

$$-4d + 12d + d$$
$$= 9d$$

Example 5

$$-2[w - 4(w - 2)] + 3$$
$$= -2[w - 4w + 8] + 3$$
$$= -2[-3w + 8] + 3$$
$$= 6w - 16 + 3$$
$$= 6w - 13$$

Chapter R   Review Exercises

Section R.2

1. Find a number that is a whole number but not a natural number.

For Exercises 2 and 3, answers may vary.

2. List three rational numbers that are not integers.

3. List five integers, two of which are not whole numbers.

For Exercises 4–9, write an expression in words that describes the set of numbers given by each interval. (Answers may vary.)

4. $(7, 16)$

5. $(0, 2.6]$

6. $[-6, -3]$

7. $(8, \infty)$

8. $(-\infty, 13]$

9. $(-\infty, \infty)$

For Exercises 10–12, graph each set and write the set in interval notation.

10. $\{x \mid x < 2\}$

11. $\{x \mid x \geq 0\}$

12. $\{x \mid -1 < x < 5\}$

13. True or false? $x < 3$ is equivalent to $3 > x$

Section R.3

For Exercises 14 and 15, find the opposite, reciprocal, and absolute value.

14. -8

15. $\dfrac{4}{9}$

For Exercises 16 and 17, simplify the exponents and the radicals.

16. 4^2, $\sqrt{4}$

17. 25^2, $\sqrt{25}$

For Exercises 18–33, perform the indicated operations.

18. $6 + (-8)$

19. $(-2) - (-5)$

20. $8(-2.7)$

21. $(-1.1)(7.41)$

22. $\dfrac{5}{8} \div \left(-\dfrac{13}{40}\right)$

23. $\left(-\dfrac{1}{4}\right) \div \left(-\dfrac{11}{16}\right)$

24. $2\dfrac{2}{5} - \left(1\dfrac{1}{10}\right)^2$

25. $4\dfrac{2}{3} - 3\left(1\dfrac{1}{6}\right)$

26. $\dfrac{2 - 4(3 - 7)}{-4 - 5(1 - 3)}$

27. $\dfrac{12(2) - 8}{4(-3) + 2(5)}$

28. $24 \div 8 \cdot 2$

29. $40 \div 5 \cdot 6$

30. $3^2 + 2(|-10 + 5| \div 5)$

31. $-91 + \sqrt{4}(\sqrt{25} - 13)^2$

32. $\dfrac{3(3 - 8)^2}{|8 - 3^2|}$

33. $\dfrac{4(5 - 2)^2}{|3 - 7 - 5|}$

34. Given $h = \frac{1}{2}gt^2 + v_0t + h_0$, find h if $g = -32$, $v_0 = 64$, $h_0 = 256$, and $t = 4$.

35. Find the area of a parallelogram with base 42 in. and height 18 in.

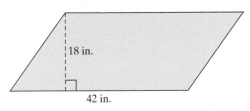

18 in.

42 in.

Section R.4

For Exercises 36–39, apply the distributive property and simplify.

36. $3(x + 5y)$

37. $\dfrac{1}{2}(x + 8y - 5)$

38. $-(-4x + 10y - z)$

39. $-(13a - b - 5c)$

For Exercises 40–43, clear parentheses if necessary, and combine *like* terms.

40. $5 - 6q + 13q - 19$

41. $18p + 3 - 17p + 8p$

42. $7 - 3(y + 4) - 3y$

43. $\dfrac{3}{4}(8x - 4) + \dfrac{1}{2}(6x + 4)$

For Exercises 44 and 45, answers may vary.

44. Write an example of the commutative property of addition.

45. Write an example of the associative property of multiplication.

Chapter R   Test

1. **a.** List the integers between -5 and 2, inclusive.

     b. List three rational numbers between 1 and 2. (Answers may vary.)

2. Write the opposite, reciprocal, and absolute value for each number.

     a. $-\dfrac{1}{2}$ **b.** 4   **c.** 0

3. Explain the difference between the interval $[4, \infty)$ and $(4, \infty)$.

4. Answer true or false: The set $\{x \mid x \geq 5\}$ is the same as $\{x \mid 5 \leq x\}$.

For Exercises 5 and 6, graph the inequality and express the set in interval notation.

5. $\{y \mid y < -\frac{4}{3}\}$ **6.** $\{p \mid 12 \leq p\}$

For Exercises 7 and 8, write each English phrase as an algebraic statement.

7. x is no more than 5.   **8.** p is at least 7.

For Exercises 9–12, simplify the expression.

9. $|-8| - 4(2 - 3)^2 \div \sqrt{4}$ **10.** $\dfrac{-6^2 - 10^2}{-1 + 3^2}$

11. $\left(-\dfrac{1}{6} + \sqrt{\dfrac{4}{9}}\right)^2$ **12.** $-8 \div 3 \cdot 2$

13. Given $z = \dfrac{x - \mu}{\sigma/\sqrt{n}}$, find z when $n = 16$, $x = 18$, $\sigma = 1.8$, and $\mu = 17.5$. (Round the answer to 1 decimal place.)

For Exercises 14–16, simplify the expressions.

14. $5b + 2 - 7b + 6 - 14$

15. $-3(4 - x) + 9(x - 1) - 5(2x - 4)$

16. $\dfrac{1}{2}(2x - 1) - \left(3x - \dfrac{3}{2}\right)$

For Exercises 17–20, answer true or false.

17. $(x + y) + 2 = 2 + (x + y)$ is an example of the associative property of addition.

18. $(2 \cdot 3) \cdot 5 = (3 \cdot 2) \cdot 5$ is an example of the commutative property of multiplication.

19. $\left(\dfrac{1}{2}x + \dfrac{1}{3}\right)6 = 3x + 2$ is an example of the distributive property of multiplication over addition.

20. $(10 + y) + z = 10 + (y + z)$ is an example of the associative property of addition.

Linear Equations and Inequalities in One Variable

Chapter 1

In this chapter, we study linear equations and inequalities and their applications.

Review Your Skills

To prepare yourself, try the crossword puzzle. The clues in the puzzle review formulas from geometry and other important mathematical facts that you will encounter in this chapter as you work through the application problems.

Across

2. What is the next consecutive integer after 1306?

4. Given that distance = rate × time, what is the distance between Atlanta and Los Angeles if it takes 33 hr, traveling 60 mph?

6. What is the next consecutive odd integer after 7803?

7. What is 10% of 64,780?

9. If an angle measures 41°, what is the complement?

10. If an angle measures 70°, what is the supplement?

Down

1. What is 40% of 32,640?

2. What is the sum of the measures of the angles in a triangle?

3. Evaluate $|-7729 + 262|$.

4. What number is 50 more than twice 8707?

5. If the area of a rectangle is 6370 ft², and the width is 65 ft, what is the length?

8. What is the amount of simple interest earned on $4000 at 5% interest for 4 yr?

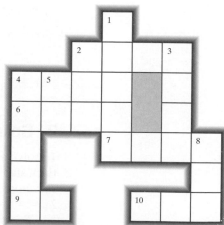

Concepts

1. Linear Equations in One Variable
2. Solving Linear Equations
3. Clearing Fractions and Decimals
4. Conditional Equations, Contradictions, and Identities

1. Linear Equations in One Variable

An **equation** is a statement that indicates that two expressions are equal. The following are equations.

$$x = -4 \qquad p + 3 = 11 \qquad -2z = -20$$

All equations have an equal sign. Furthermore, notice that the equal sign separates the equation into two parts, the left-hand side and the right-hand side. A **solution to an equation** is a value of the variable that makes the equation a true statement. Substituting a solution to an equation for the variable makes the right-hand side equal to the left-hand side.

Equation	Solution	Check	
$p + 3 = 11$	8	$p + 3 = 11$ $\downarrow$ $8 + 3 = 11$ ✓	Substitute 8 for p. Right-hand side equals left-hand side.
$-2z = -20$	10	$-2z = -20$ $\downarrow$ $-2(10) = -20$ ✓	Substitute 10 for z. Right-hand side equals left-hand side.

Concept Connections

Determine if the given value is a solution to the equation.

1. $x + 2 = 8$; -10
2. $3y - 1 = 11$;   4
3. Write a linear equation whose solution is 3. (Answers may vary.)

Throughout this text we will learn to recognize and solve several different types of equations, but in this chapter, we will focus on the specific type of equation called a linear equation in one variable.

Definition of a Linear Equation in One Variable

Let a, b, and c be real numbers such that $a \neq 0$. A **linear equation in one variable** is an equation that can be written in the form

$$ax + b = c$$

Note: A linear equation in one variable is often called a first-degree equation because the variable x has an implied exponent of 1.

Examples	Notes
$3x + 5 = 20$	$a = 3, b = 5, c = 20$
$-2x - 4 = 6$ can be written as $-2x + (-4) = 6$	$a = -2, b = -4, c = 6$
$6x + 7 - 5x = 11$ can be written as $x + 7 = 11$	$a = 1, b = 7, c = 11$

---

2. Solving Linear Equations

To solve a linear equation, the goal is to simplify the equation to isolate the variable. Each step used in simplifying an equation results in an equivalent equation. *Equivalent equations* have the same solution set. For example, the equations $2x + 3 = 7$ and $2x = 4$ are equivalent because 2 is the solution to both equations.

To solve an equation, we may use the addition, subtraction, multiplication, and division properties of equality. These properties state that adding, subtracting, multiplying, or dividing the same quantity on each side of an equation results in an equivalent equation.

Answers

1. No
2. Yes
3. For example: $5x - 6 = 9$

Addition and Subtraction Properties of Equality

Let a, b, and c represent real numbers.

Addition property of equality: If $a = b$, then $a + c = b + c$.

*Subtraction property of equality: If $a = b$, then $a - c = b - c$.

*The subtraction property of equality follows directly from the addition property, because subtraction is defined in terms of addition.

$$\text{If} \quad a + (-c) = b + (-c)$$
$$\text{then,} \quad a - c = b - c$$

Multiplication and Division Properties of Equality

Let a, b, and c represent real numbers with $c \neq 0$.

Multiplication property of equality: If $a = b$, then $a \cdot c = b \cdot c$.

*Division property of equality: If $a = b$, then $\dfrac{a}{c} = \dfrac{b}{c}$.

*The division property of equality follows directly from the multiplication property, because division is defined as multiplication by the reciprocal.

$$\text{If} \quad a \cdot \frac{1}{c} = b \cdot \frac{1}{c}$$
$$\text{then,} \quad \frac{a}{c} = \frac{b}{c}$$

Example 1 Solving a Linear Equation

Solve the equation. $12 + x = 40$

Solution:

$$12 + x = 40$$
$$12 - 12 + x = 40 - 12 \qquad \text{To isolate } x, \text{ subtract 12 from both sides.}$$
$$x = 28 \qquad \text{Simplify.}$$

$$\underline{\text{Check:}} \quad 12 + x = 40$$
$$12 + (28) \stackrel{?}{=} 40$$
$$40 \stackrel{?}{=} 40 \checkmark \quad \text{True.}$$

The solution is 28.

Example 2 Solving Linear Equations

Solve each equation.

a. $-\dfrac{3}{5}p = \dfrac{4}{15}$ **b.** $4 = \dfrac{w}{2.2}$ **c.** $-x = 6$

Answer

4. -6

Solution:

a.
$$-\frac{3}{5}p = \frac{4}{15}$$

$$\left(-\frac{5}{3}\right)\left(-\frac{3}{5}p\right) = \left(-\frac{5}{3}\right)\left(\frac{4}{15}\right)$$ To isolate p, multiply both sides by the reciprocal of $-\dfrac{3}{5}$.

$$p = \left(-\frac{\overset{1}{5}}{3}\right)\left(\frac{4}{\underset{3}{15}}\right)$$ Multiply fractions.

$$p = -\frac{4}{9}$$ The value $-\dfrac{4}{9}$ checks in the original equation.

The solution is $-\dfrac{4}{9}$.

b.
$$4 = \frac{w}{2.2}$$

$$2.2(4) = 2.2 \cdot \left(\frac{w}{2.2}\right)$$ To isolate w, multiply both sides by 2.2.

$$8.8 = w$$ The value 8.8 checks in the original equation.

The solution is 8.8.

c.
$$-x = 6$$

$$-1(-x) = -1(6)$$ To isolate x, multiply both sides by -1.

$$x = -6$$ The value -6 checks in the original equation.

The solution is -6.

For more complicated linear equations, several steps are required to isolate the variable. These steps are listed below.

Solving a Linear Equation in One Variable

Step 1 Simplify both sides of the equation.
- Clear parentheses.
- Consider clearing fractions or decimals (if any are present) by multiplying both sides of the equation by a common denominator of all terms.
- Combine *like* terms.

Step 2 Use the addition or subtraction property of equality to collect the variable terms on one side of the equation.

Step 3 Use the addition or subtraction property of equality to collect the constant terms on the other side of the equation.

Step 4 Use the multiplication or division property of equality to make the coefficient of the variable term equal to 1.

Step 5 Check your answer.

Example 3 Solving a Linear Equation

Solve the linear equation and check the answer. $\qquad 3x + 1 = -7$

Solution:

$$3x + 1 = -7$$

$$3x + 1 - 1 = -7 - 1 \qquad \text{Subtract 1 from both sides.}$$

$$3x = -8 \qquad \text{Combine } like \text{ terms.}$$

$$\frac{3x}{3} = \frac{-8}{3} \qquad \text{To isolate } x, \text{ divide both sides of the equation by 3.}$$

$$x = -\frac{8}{3} \qquad \text{Simplify.}$$

$$\underline{\text{Check:}} \qquad 3x + 1 \overset{?}{=} -7$$

$$3\left(-\frac{8}{3}\right) + 1 \overset{?}{=} -7$$

$$\cancel{3}\left(-\frac{8}{\cancel{3}}\right) + 1 \overset{?}{=} -7$$

$$-8 + 1 \overset{?}{=} -7$$

$$-7 \overset{?}{=} -7 \checkmark \quad \text{True.}$$

The solution is $-\dfrac{8}{3}$.

Skill Practice

Solve the linear equation and check the answer.
8. $5x - 19 = -23$

Example 4 Solving a Linear Equation

Solve the linear equation and check the answer. $\qquad 11z + 2 = 5(z - 2)$

Solution:

$$11z + 2 = 5(z - 2)$$

$$11z + 2 = 5z - 10 \qquad \text{Apply the distributive property to clear parentheses.}$$

$$11z - 5z + 2 = 5z - 5z - 10 \qquad \text{Subtract } 5z \text{ from both sides.}$$

$$6z + 2 = -10 \qquad \text{Combine } like \text{ terms.}$$

$$6z + 2 - 2 = -10 - 2 \qquad \text{Subtract 2 from both sides.}$$

$$6z = -12 \qquad \text{Combine } like \text{ terms.}$$

$$\frac{6z}{6} = \frac{-12}{6} \qquad \text{To isolate } z, \text{ divide both sides of the equation by 6.}$$

$$z = -2 \qquad \text{Simplify.}$$

$$\underline{\text{Check:}} \qquad 11z + 2 = 5(z - 2)$$

$$11(-2) + 2 \overset{?}{=} 5(-2 - 2)$$

$$-22 + 2 \overset{?}{=} 5(-4)$$

$$-20 \overset{?}{=} -20 \checkmark \quad \text{True.}$$

The solution is -2.

Skill Practice

Solve the equations.
9. $7 + 2(y - 3) = 6y + 3$

Answers

8. $-\dfrac{4}{5}$ **9.** $-\dfrac{1}{2}$

Example 5 **Solving a Linear Equation**

Solve the equation. $-3(x - 4) + 2 = 7 - (x + 1)$

Solution:

$$-3(x - 4) + 2 = 7 - (x + 1)$$

$-3x + 12 + 2 = 7 - x - 1$ Clear parentheses.

$-3x + 14 = -x + 6$ Combine *like* terms.

$-3x + x + 14 = -x + x + 6$ Add x to both sides of the equation.

$-2x + 14 = 6$ Combine *like* terms.

$-2x + 14 - 14 = 6 - 14$ Subtract 14 from both sides.

$-2x = -8$ Combine *like* terms.

$\dfrac{-2x}{-2} = \dfrac{-8}{-2}$ To isolate x, divide both sides by -2.

$x = 4$ Simplify. The solution checks in the original equation.

The solution is 4.

Example 6 **Solving a Linear Equation**

Solve the equation. $-4[y - 3(y - 5)] = 2(6 - 5y)$

Solution:

$$-4[y - 3(y - 5)] = 2(6 - 5y)$$

$-4[y - 3y + 15] = 12 - 10y$ Clear parentheses.

$-4[-2y + 15] = 12 - 10y$ Combine *like* terms.

$8y - 60 = 12 - 10y$ Clear parentheses.

$8y + 10y - 60 = 12 - 10y + 10y$ Add $10y$ to both sides of the equation.

$18y - 60 = 12$ Combine *like* terms.

$18y - 60 + 60 = 12 + 60$ Add 60 to both sides of the equation.

$18y = 72$

$\dfrac{18y}{18} = \dfrac{72}{18}$ To isolate y, divide both sides by 18.

$y = 4$ The solution checks.

The solution is 4.

3. Clearing Fractions and Decimals

When an equation contains fractions or decimals, it is sometimes helpful to clear the fractions and decimals. This is accomplished by multiplying both sides of the equation by the least common denominator (LCD) of all terms within the equation. This is demonstrated in Example 7.

Example 7 Solving a Linear Equation by Clearing Fractions

Solve the equation. $\quad \dfrac{1}{4}w + \dfrac{1}{3}w - 1 = \dfrac{1}{2}(w - 4)$

Solution:

$$\dfrac{1}{4}w + \dfrac{1}{3}w - 1 = \dfrac{1}{2}(w - 4)$$

$$\dfrac{1}{4}w + \dfrac{1}{3}w - 1 = \dfrac{1}{2}w - 2 \qquad \text{Clear parentheses.}$$

$$12 \cdot \left(\dfrac{1}{4}w + \dfrac{1}{3}w - 1\right) = 12 \cdot \left(\dfrac{1}{2}w - 2\right) \qquad \begin{array}{l}\text{Multiply both sides of} \\ \text{the equation by the} \\ \text{LCD of all terms. In} \\ \text{this case, the LCD is 12.}\end{array}$$

$$12 \cdot \dfrac{1}{4}w + 12 \cdot \dfrac{1}{3}w + 12 \cdot (-1) = 12 \cdot \dfrac{1}{2}w + 12 \cdot (-2) \qquad \begin{array}{l}\text{Apply the distributive} \\ \text{property.}\end{array}$$

$$3w + 4w - 12 = 6w - 24$$

$$7w - 12 = 6w - 24$$

$$w - 12 = -24 \qquad \text{Subtract } 6w.$$

$$w = -12 \qquad \text{The solution checks.}$$

The solution is -12.

Skill Practice

Solve the equation by first clearing the fractions.

12. $\dfrac{3}{4}a + \dfrac{1}{2} = \dfrac{2}{3}a + \dfrac{1}{3}$

Example 8 Solving a Linear Equation by Clearing Fractions

Solve. $\quad \dfrac{x - 2}{5} - \dfrac{x - 4}{2} = 2 + \dfrac{x + 4}{10}$

Solution:

$$\dfrac{x - 2}{5} - \dfrac{x - 4}{2} = \dfrac{2}{1} + \dfrac{x + 4}{10} \qquad \begin{array}{l}\text{The LCD of} \\ \text{all terms in the} \\ \text{equation is 10.}\end{array}$$

$$10\left(\dfrac{x - 2}{5} - \dfrac{x - 4}{2}\right) = 10\left(\dfrac{2}{1} + \dfrac{x + 4}{10}\right) \qquad \begin{array}{l}\text{Multiply both} \\ \text{sides by 10.}\end{array}$$

$$\dfrac{\overset{2}{\cancel{10}}}{1} \cdot \left(\dfrac{x - 2}{\cancel{5}}\right) - \dfrac{\overset{5}{\cancel{10}}}{1} \cdot \left(\dfrac{x - 4}{\cancel{2}}\right) = \dfrac{10}{1} \cdot \left(\dfrac{2}{1}\right) + \dfrac{\overset{1}{\cancel{10}}}{1} \cdot \left(\dfrac{x + 4}{\cancel{10}}\right) \qquad \begin{array}{l}\text{Apply the} \\ \text{distributive} \\ \text{property.}\end{array}$$

$$2(x - 2) - 5(x - 4) = 20 + 1(x + 4) \qquad \text{Clear fractions.}$$

$$2x - 4 - 5x + 20 = 20 + x + 4 \qquad \begin{array}{l}\text{Apply the distributive} \\ \text{property.}\end{array}$$

$$-3x + 16 = x + 24 \qquad \begin{array}{l}\text{Simplify both sides of the} \\ \text{equation.}\end{array}$$

$$-4x + 16 = 24 \qquad \text{Subtract } x \text{ from both sides.}$$

$$-4x = 8 \qquad \text{Subtract 16 from both sides.}$$

$$x = -2 \qquad \begin{array}{l}\text{The value } -2 \text{ checks in the} \\ \text{original equation.}\end{array}$$

The solution is -2.

Skill Practice

Solve.

13. $\dfrac{1}{8} - \dfrac{x + 3}{4} = \dfrac{3x - 2}{2}$

TIP: Clearing fractions is an application of the multiplication property of equality. We are multiplying both sides of the equation by the same number.

Answers

12. -2 **13.** $\dfrac{3}{14}$

The same procedure used to clear fractions in an equation can be used to clear decimals.

Example 9 Solving a Linear Equation by Clearing Decimals

Solve the equation. $0.55x - 0.6 = 2.05x$

Solution:

Recall that any terminating decimal can be written as a fraction. Therefore, the equation $0.55x - 0.6 = 2.05x$ is equivalent to

$$\frac{55}{100}x - \frac{6}{10} = \frac{205}{100}x$$

A convenient common denominator for all terms in this equation is 100. Multiplying both sides of the equation by 100 will have the effect of "moving" the decimal point 2 places to the right.

$100(0.55x - 0.6) = 100(2.05x)$ Multiply both sides by 100 to clear decimals.

$55x - 60 = 205x$

$-60 = 150x$ Subtract $55x$ from both sides.

$-\dfrac{60}{150} = x$ To isolate x, divide both sides by 150.

$x = -\dfrac{2}{5} = -0.4$ The solution checks.

The solution is -0.4.

4. Conditional Equations, Contradictions, and Identities

The solution to a linear equation is the value of x that makes the equation a true statement. While linear equations have one unique solution, some equations have no solution, and others have infinitely many solutions.

I. Conditional Equations

An equation that is true for some values of the variable but false for other values is called a **conditional equation**. The equation $x + 4 = 6$ is a conditional equation because it is true on the *condition* that $x = 2$. For other values of x, the statement $x + 4 = 6$ is false. The solution is 2.

$$x + 4 = 6$$

$$x + 4 - 4 = 6 - 4$$

$$x = 2 \quad \text{(Conditional equation)} \quad \text{The solution is 2.}$$

II. Contradictions

Some equations have no solution, such as $x + 1 = x + 2$. There is no value of x that when increased by 1 will equal the same value increased by 2. If we tried to solve the equation by subtracting x from both sides, we get the contradiction $1 = 2$.

$$x + 1 = x + 2$$

$$x - x + 1 = x - x + 2$$

$$1 = 2 \quad \text{(Contradiction)}$$

This indicates that the equation has no solution. An equation that has no solution is called a **contradiction**.

III. Identities

An equation that is true for all real numbers is called an **identity**. For example, consider the equation $x + 4 = x + 4$. Because the left- and right-hand sides are *identical*, any real number substituted for x will result in equal quantities on both sides. If we solve the equation, we get the identity $4 = 4$. In such a case, the solution is the set of real numbers.

$$x + 4 = x + 4$$

$$x - x + 4 = x - x + 4$$

$$4 = 4 \quad \text{(Identity)} \qquad \text{The solution is the set of all real numbers.}$$

| **Example 10** | Identifying Conditional Equations, Contradictions, and Identities |

Solve the equations. Identify each equation as a conditional equation, a contradiction, or an identity.

a. $3[x - (x + 1)] = -2$ **b.** $5(3 + c) + 2 = 2c + 3c + 17$

c. $4x - 3 = 17$

Solution:

a. $3[x - (x + 1)] = -2$

$\qquad 3[x - x - 1] = -2 \qquad$ Clear parentheses.

$\qquad\qquad 3[-1] = -2 \qquad$ Combine *like* terms.

$\qquad\qquad\quad -3 = -2 \qquad$ Contradiction

This equation is a contradiction. There is no solution.

b. $5(3 + c) + 2 = 2c + 3c + 17$

$\quad 15 + 5c + 2 = 5c + 17 \qquad$ Clear parentheses and combine *like* terms.

$\qquad\quad 5c + 17 = 5c + 17 \qquad$ Identity

$\qquad\qquad\quad 0 = 0$

This equation is an identity. The solution is the set of all real numbers.

c. $\qquad 4x - 3 = 17$

$\qquad\qquad 4x = 20$

$\qquad\qquad\quad x = 5$

This equation is a conditional equation. The solution is 5.

TIP: Interval notation can also be used to express the set of real numbers, $(-\infty, \infty)$.

Answers

15. Contradiction; there is no solution.
16. Identity; the solution is the set of all real numbers.
17. Conditional equation; 1

Section 1.1 | Practice Exercises

Study Skills Exercise

Some instructors allow the use of calculators. Does your instructor allow the use of a calculator? If so, what kind?

Will you be allowed to use a calculator on tests or just for occasional calculator problems in the text?

Helpful Hint: If you are not permitted to use a calculator on tests, you should do your homework in the same way, without the calculator.

Vocabulary and Key Concepts

1. a. An _____ is a statement that indicates that two expressions are equal.

  b. A _____ to an equation is a value of the variable that makes the equation a true statement.

  c. An equation that can be written in the form $ax + b = c$ $(a \neq 0)$ is called a _____ equation in one variable.

  d. A linear equation is also called a _____ -degree equation because the degree on the variable is 1.

  e. Two equations are equivalent if they have the same _____.

  f. A _____ equation is true for some values of the variable, but false for other values.

  g. An equation that has no solution is called a _____.

  h. An equation that has all real numbers as its solution set is called an _____.

Review Exercises

2. Apply the distributive property. $-\dfrac{1}{2}(6x - 8y + 10)$

For Exercises 3–6, clear parentheses and combine *like* terms.

3. $8x - 3y + 2xy - 5x + 12xy$

4. $5ab + 5a - 13 - 2a + 17$

5. $2(3z - 4) - (z + 12)$

6. $-(6w - 5) + 3(4w - 5)$

Concept 1: Linear Equations in One Variable

For Exercises 7–12, label the equation as linear or nonlinear.

7. $2x + 1 = 5$

8. $10 = x + 6$

9. $x^2 + 7 = 9$

10. $3 + x^3 - x = 4$

11. $-3 = x$

12. $5.2 - 7x = 0$

13. Use substitution to determine which value is the solution to $2x - 1 = 5$.

  a. 2   **b.** 3   **c.** 0   **d.** -1

14. Use substitution to determine which value is the solution to $2y - 3 = -2$.

  a. 1   **b.** $\dfrac{1}{2}$ **c.** 0   **d.** $-\dfrac{1}{2}$

Concept 2: Solving Linear Equations

For Exercises 15–44, solve the equation and check the solution. **(See Examples 1–6.)**

15. $x + 7 = 19$

16. $-3 + y = -28$

17. $-x = 2$

18. $-t = \dfrac{3}{4}$

19. $-\dfrac{7}{8} = -\dfrac{5}{6}z$

20. $-\dfrac{12}{13} = \dfrac{4}{3}b$

21. $\dfrac{a}{5} = -8$

22. $\dfrac{x}{8} = \dfrac{1}{2}$

23. $2.53 = -2.3t$ **24.** $-4.8 = 6.1 + y$ **25.** $p - 2.9 = 3.8$ **26.** $-4.2a = 4.494$

27. $6q - 4 = 62$ **28.** $2w - 15 = 15$ **29.** $4y - 17 = 35$ **30.** $6z - 25 = 83$

31. $-b - 5 = 2$ **32.** $6 = -y + 1$ **33.** $3(x - 6) = 2x - 5$ **34.** $13y + 4 = 5(y - 4)$

35. $6 - (t + 2) = 5(3t - 4)$ **36.** $1 - 5(p + 2) = 2(p + 13)$

37. $6(a + 3) - 10 = -2(a - 4)$ **38.** $8(b - 2) + 3b = -9(b - 1)$

39. $-2[5 - (2z + 1)] - 4 = 2(3 - z)$ **40.** $3[w - (10 - w)] = 7(w + 1)$

41. $6(-y + 4) - 3(2y - 3) = -y + 5 + 5y$ **42.** $13 + 4w = -5(-w - 6) + 2(w + 1)$

43. $14 - 2x + 5x = -4(-2x - 5) - 6$ **44.** $8 - (p + 2) + 6p + 7 = p + 13$

Concept 3: Clearing Fractions and Decimals

For Exercises 45–56, solve the equations. **(See Examples 7–9.)**

45. $\dfrac{2}{3}x - \dfrac{1}{6} = -\dfrac{5}{12}x + \dfrac{3}{2} - \dfrac{1}{6}x$ **46.** $-\dfrac{1}{2}y + 4 = -\dfrac{9}{10}y + \dfrac{2}{5}$ **47.** $\dfrac{1}{5}(p - 5) = \dfrac{3}{5}p + \dfrac{1}{10}p + 1$

48. $\dfrac{5}{6}(q + 2) = -\dfrac{7}{9}q - \dfrac{1}{3} + 2$ **49.** $\dfrac{3x - 7}{2} + \dfrac{3 - 5x}{3} = \dfrac{3 - 6x}{5}$ **50.** $\dfrac{2y - 4}{5} = \dfrac{5y + 13}{4} + \dfrac{y}{2}$

51. $\dfrac{4}{3}(2q + 6) - \dfrac{5q - 6}{6} - \dfrac{q}{3} = 0$ **52.** $\dfrac{-3a + 9}{15} - \dfrac{2a - 5}{5} - \dfrac{a + 2}{10} = 0$ **53.** $6.3w - 1.5 = 4.8$

54. $0.2x + 53.6 = x$ **55.** $0.75(m - 2) + 0.25m = 0.5$ **56.** $0.4(n + 10) + 0.6n = 2$

Concept 4: Conditional Equations, Contradictions, and Identities

57. What is a conditional equation?

58. Explain the difference between a contradiction and an identity.

For Exercises 59–64, solve the following equations. Then label each as a conditional equation, a contradiction, or an identity. **(See Example 10.)**

Animation

59. $4x + 1 = 2(2x + 1) - 1$ **60.** $3x + 6 = 3x$ **61.** $-11x + 4(x - 3) = -2x - 12$

62. $5(x + 2) - 7 = 3$ **63.** $2x - 4 + 8x = 7x - 8 + 3x$ **64.** $-7x + 8 + 4x = -3(x - 3) - 1$

Mixed Exercises

For Exercises 65–96, solve the equations.

65. $-5b + 9 = -71$ **66.** $-3x + 18 = -66$ **67.** $16 = -10 + 13x$

68. $15 = -12 + 9x$ **69.** $10c + 3 = -3 + 12c$ **70.** $2w + 21 = 6w - 7$

71. $12b - 8b - 8 + 13 = 4b + 6 - 1$ **72.** $4z + 2 - 3z + 5 = 3 + z + 4$ **73.** $5(x - 2) - 2x = 3x + 7$

74. $2x + 3(x - 5) = 15$ **75.** $\dfrac{c}{2} - \dfrac{c}{4} + \dfrac{3c}{8} = 1$ **76.** $\dfrac{d}{5} - \dfrac{d}{10} + \dfrac{5d}{20} = \dfrac{7}{10}$

77. $0.75(8x - 4) = \frac{2}{3}(6x - 9)$

78. $-\frac{1}{2}(4z - 3) = -z$

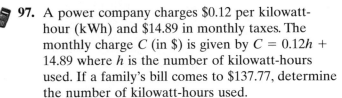 **79.** $7(p + 2) - 4p = 3p + 14$

80. $6(z - 2) = 3z - 8 + 3z$

81. $4[3 + 5(3 - b) + 2b] = 6 - 2b$

82. $\frac{1}{3}(x + 3) - \frac{1}{6} = \frac{1}{6}(2x + 5)$

83. $3 - \frac{3}{4}x = 3\left(3 - \frac{1}{4}x\right)$

84. $\frac{9}{5} - 8w = 8\left(\frac{3}{5} - w\right)$

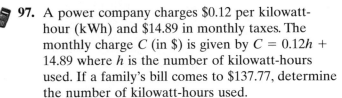 **85.** $\frac{5}{4} + \frac{y - 3}{8} = \frac{2y + 1}{2}$

86. $\frac{2}{3} - \frac{x + 2}{6} = \frac{5x - 2}{2}$

87. $\frac{2y - 9}{10} + \frac{3}{2} = y$

88. $\frac{2}{3}x - \frac{5}{6}x - 3 = \frac{1}{2}x - 5$

89. $0.48x - 0.08x = 0.12(260 - x)$

90. $0.07w + 0.06(140 - w) = 90$

91. $0.5x + 0.25 = \frac{1}{3}x + \frac{5}{4}$

92. $0.2b + \frac{1}{3} = \frac{7}{15}$

93. $0.3b - 1.5 = 0.25(b + 2) + 0.05b$ **94.** $0.7(a - 1) = 0.25 + 0.7a$

95. $-\frac{7}{8}y + \frac{1}{4} = \frac{1}{2}\left(5 - \frac{3}{4}y\right)$

96. $5x - (8 - x) = 2[-4 - (3 + 5x) - 13]$

Expanding Your Skills

97. A power company charges $0.12 per kilowatt-hour (kWh) and $14.89 in monthly taxes. The monthly charge C (in $) is given by $C = 0.12h + 14.89$ where h is the number of kilowatt-hours used. If a family's bill comes to $137.77, determine the number of kilowatt-hours used.

98. For a student's first semester at college, the college charges $105 per credit-hour plus a one-time $50 admissions fee. The cost C (in $) for the first semester is given by $C = 105h + 50$ where h is the number of credit-hours taken. If a student is billed $1415, how many credit-hours is she taking?

99. a. Simplify the expression. $-2(y - 1) + 3(y + 2)$

    b. Solve the equation. $-2(y - 1) + 3(y + 2) = 0$

    c. Explain the difference between simplifying an expression and solving an equation.

100. a. Simplify the expression. $4w - 8(2 + w)$

    b. Solve the equation. $4w - 8(2 + w) = 0$

    c. Explain the difference between simplifying an expression and solving an equation.

Problem Recognition Exercises

Equations Versus Expressions

For Exercises 1–20, identify each exercise as an expression or an equation. Then simplify the expressions and solve the equations.

1. $4x - 2 + 6 - 8x$

2. $-3y - 3 - 4y + 8$

3. $7b - 1 = 2b + 4$

4. $10t + 2 = 2 - 7t$

5. $4(a - 8) - 7(2a + 1)$

6. $10(2x + 3) - 8(5 - x)$

7. $7(2 - w) = 5w + 8$

8. $15(3 - 2y) = 21 + 2y$

9. $2(3x - 4) - 4(5x + 1) = -8x + 7$

10. $\frac{1}{2}v + \frac{3}{5} - \frac{2}{3}v - \frac{7}{10}$

11. $-\frac{7}{8}t - \frac{4}{3}u - \frac{5}{4}t + \frac{11}{6}u$

12. $20x - 8 + 7x + 28 = 27x - 9$

13. $7 + 8w - 12 = 3w - 8 + 5w$ **14.** $\dfrac{5}{6}y - \dfrac{7}{8} = \dfrac{1}{2}y + \dfrac{3}{4}$ **15.** $\dfrac{4}{5} + 3z = \dfrac{1}{2}z + 1$

16. $0.29c + 4.495 - 0.12c$ **17.** $0.45k - 1.67 + 0.89 - 1.456k$ **18.** $0.125(2p - 8) = 0.25(p - 4)$

19. $0.5u + 1.2 - 0.74u = 0.8 - 0.24u + 0.4$

20. $6(2 - 3a) - 2(8a + 3) = -12a - 19$

Applications of Linear Equations in One Variable   Section 1.2

1. Introduction to Problem Solving

One of the important uses of algebra is to develop mathematical models for understanding real-world phenomena. To solve an application problem, relevant information must be extracted from the wording of a problem and then translated into mathematical symbols. This is a skill that requires practice. The key is to stick with it and not to get discouraged.

Problem-Solving Flowchart for Word Problems

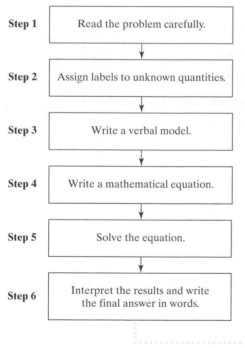

Step 1 Read the problem carefully.
• Familiarize yourself with the problem. Estimate the answer, if possible.

Step 2 Assign labels to unknown quantities.
• Identify the unknown quantity or quantities. Let x represent one of the unknowns. Draw a picture and write down relevant formulas when appropriate.

Step 3 Write a verbal model.
• Write an equation in *words*.

Step 4 Write a mathematical equation.
• Replace the verbal model with a mathematical equation using x or another variable.

Step 5 Solve the equation.
• Solve for the variable, using the steps for solving linear equations.

Step 6 Interpret the results and write the final answer in words.
• Once you've obtained a numerical value for the variable, recall what it represents in the context of the problem. Can this value be used to determine other unknowns in the problem? Write an answer to the word problem in *words*.

Avoiding Mistakes

Once you have reached a solution to a word problem, verify that it is reasonable in the context of the problem.

To write a verbal model as an algebraic expression, review the list of key terms given in Table 1-1.

Table 1-1

Addition: $a + b$	Subtraction: $a - b$
• the **sum** of a and b • a plus b • b added to a • b more than a • a increased by b • the total of a and b	• the **difference** of a and b • a minus b • b subtracted from a • a decreased by b • b less than a
Multiplication: $a \cdot b$	Division: $a \div b, \frac{a}{b}$
• the **product** of a and b • a times b • a multiplied by b	• the **quotient** of a and b • a divided by b • b divided into a • the ratio of a and b • a over b • a per b

Skill Practice

1. One number is 5 more than 3 times another number. The sum of the numbers is 45. Find the numbers.

Example 1 Translating and Solving a Linear Equation

The sum of two numbers is 39. One number is 3 less than twice the other. What are the numbers?

Solution:

Step 1: Read the problem carefully.

Step 2: Let x represent one number.

Let $2x - 3$ represent the other number.

Step 3: (One number) + (other number) = 39

Step 4: Replace the verbal model with a mathematical equation.

$$(\text{One number}) + (\text{other number}) = 39$$
$$x \quad + \quad (2x - 3) \quad = 39$$

Step 5: Solve for x.

$$x + (2x - 3) = 39$$
$$3x - 3 = 39$$
$$3x = 42$$
$$\frac{3x}{3} = \frac{42}{3}$$
$$x = 14$$

Step 6: Interpret your results. Refer back to step 2.

One number is x: $\longrightarrow 14$

The other number is $2x - 3$:

$$2(14) - 3 \longrightarrow 25$$

Answer: The numbers are 14 and 25.

Answer

1. The numbers are 10 and 35.

2. Applications Involving Consecutive Integers

The word *consecutive* means "following one after the other in order."

- The numbers $-2, -1, 0, 1, 2$, and so on are examples of consecutive integers. Notice that two consecutive integers differ by 1. Therefore, if x represents an integer, then $x + 1$ represents the next consecutive integer.
- The numbers $2, 4, 6, 8$, and so on are consecutive *even* integers. Consecutive even integers differ by 2. Therefore, if x represents an even integer, then $x + 2$ represents the next consecutive even integer.
- The numbers $15, 17, 19$, and so on are consecutive *odd* integers. Consecutive odd integers also differ by 2. Therefore, if x represents an odd integer, then $x + 2$ represents the next consecutive odd integer.

Example 2 Solving a Linear Equation Involving Consecutive Integers

Three times the sum of two consecutive odd integers is 516. Find the integers.

Solution:

Step 1: Read the problem carefully.

Step 2: Label the unknown:

Let x represent the first odd integer.

Then $x + 2$ represents the next odd integer.

Step 3: Write an equation in words.

$$3[(\text{first odd integer}) + (\text{second odd integer})] = 516$$

$$3[x + (x + 2)] = 516 \qquad \text{**Step 4:** Write a mathematical equation.}$$

$$3(2x + 2) = 516 \qquad \text{**Step 5:** Solve for } x.$$

$$6x + 6 = 516$$

$$6x = 510$$

$$x = 85$$

Step 6: Interpret your results.

The first odd integer is x: $\longrightarrow$ 85

The second odd integer is $x + 2$:

$$85 + 2 \longrightarrow 87$$

Answer: The integers are 85 and 87.

Avoiding Mistakes

After completing a word problem, it is always a good idea to check that the answer is reasonable. Notice that 85 and 87 are consecutive odd integers, and three times their sum is 3(85 + 87), which equals 516.

Skill Practice

2. Four times the sum of three consecutive integers is 264. Find the integers.

Answer

2. The integers are 21, 22, and 23.

3. Applications Involving Percents and Rates

In many real-world applications, percents are used to represent rates.

- The sales tax rate for a certain county is 6%.
- An ice cream machine is discounted 20%.
- A real estate sales broker receives a $4\frac{1}{2}$% commission on sales.
- A savings account earns 1.5% simple interest.

The following models are used to compute sales tax, commission, and simple interest. In each case the value is found by multiplying the base by the percentage.

$$\textbf{Sales tax} = (\text{cost of merchandise})(\text{tax rate})$$

$$\textbf{Commission} = (\text{dollars in sales})(\text{commission rate})$$

$$\textbf{Simple interest} = (\text{principal})(\text{annual interest rate})(\text{time in years})$$

$$\longrightarrow I = Prt$$

Skill Practice

3. Markos earned $340 in 1 yr on an investment that paid a 4% dividend. Find the amount of money invested.

> **Example 3** Solving a Percent Application
>
> A woman invests $5000 in an account that earns $5\frac{1}{4}$% simple interest. If the money is invested for 3 years (yr), how much money is in the account at the end of the 3-yr period?
>
> **Solution:**
>
> Let x represent the total money in the account.   Label the variables.
>
> $\quad P = \$5000 \qquad$ (principal amount invested)
>
> $\quad r = 0.0525 \qquad$ (interest rate)
>
> $\quad t = 3 \qquad\quad$ (time in years)
>
> The total amount of money includes principal plus interest.
>
> $\quad$ (Total money) = (principal) + (interest)   Verbal model
>
> $\qquad\qquad x = P + Prt$ Mathematical equation
>
> $\qquad\qquad x = \$5000 + (\$5000)(0.0525)(3)$ Substitute for P, r,
>
> $\qquad\qquad x = \$5000 + \$787.50$ and t.
>
> $\qquad\qquad x = \$5787.50$ Solve for x.
>
> The total amount of money in the account is $5787.50.   Interpret the results.

TIP: Remember to use the decimal form of a percent when the number is used in a calculation.

As consumers, we often encounter situations in which merchandise has been marked up or marked down from its original cost. It is important to note that percent increase and percent decrease are based on the original cost. For example, suppose a microwave oven originally priced at $305 is marked down 20%.

The discount is determined by 20% of the original price: $(0.20)(\$305) = \$61.00$. The new price is $\$305.00 - \$61.00 = \$244.00$.

Answer

3. $8500

Example 4 Solving a Percent Increase Application

A college bookstore uses a standard markup of 40% on all books purchased wholesale from the publisher. If the bookstore sells a calculus book for $179.20, what was the original wholesale cost?

Solution:

Let x = original wholesale cost.   Label the variables.

The selling price of the book is based on the original cost of the book plus the bookstore's markup.

(Selling price) = (original cost) + (markup)   Verbal model

(Selling price) = (original cost) + (original cost · markup rate)

$$179.20 = \quad x \quad + (x)(0.40) \qquad \text{Mathematical equation}$$

$$179.20 = x + 0.40x$$

$$179.20 = 1.40x \qquad \text{Combine } like \text{ terms.}$$

$$\frac{179.20}{1.40} = x$$

$$x = 128 \qquad \text{Simplify.}$$

The original wholesale cost of the textbook was $128.00. Interpret the results.

4. Applications Involving Principal and Interest

Example 5 Solving an Investment Growth Application

Miguel had $10,000 to invest in two different mutual funds. One was a relatively safe bond fund that averaged 4% return on his investment at the end of 1 yr. The other fund was a riskier stock fund that averaged 7% return in 1 yr. If at the end of the year Miguel's portfolio grew to $10,625 ($625 above his $10,000 investment), how much money did Miguel invest in each fund?

Solution:

This type of word problem is sometimes categorized as a mixture problem. Miguel is "mixing" his money between two different investments. We have to determine how the money was divided to earn $625.

The information in this problem can be organized in a chart. (*Note:* There are two sources of money: the amount invested and the amount earned.)

	4% Bond Fund	**7% Stock Fund**	**Total**
Amount invested ($)	x	$(10,000 - x)$	10,000
Amount earned ($)	$0.04x$	$0.07(10,000 - x)$	625

Because the amount of principal is unknown for both accounts, we can let x represent the amount invested in the bond fund. If Miguel spends x dollars in the bond fund, then he has $(10,000 - x)$ left over to spend in the stock fund. The return for each fund is found by multiplying the principal and the percent growth rate.

To establish a mathematical model, we know that the total return ($625) must equal the earnings from the bond fund plus the earnings from the stock fund:

(Earnings from bond fund) + (earnings from stock fund) = (total earnings)

$$0.04x \qquad + \qquad 0.07(10{,}000 - x) \qquad = \qquad 625$$

$0.04x + 0.07(10{,}000 - x) = 625$	Mathematical equation
$4x + 7(10{,}000 - x) = 62{,}500$	Multiply by 100 to clear decimals.
$4x + 70{,}000 - 7x = 62{,}500$	
$-3x + 70{,}000 = 62{,}500$	Combine *like* terms.
$-3x = -7500$	Subtract 70,000 from both sides.
$\dfrac{-3x}{-3} = \dfrac{-7500}{-3}$	
$x = 2500$	Solve for x and interpret the results.

The amount invested in the bond fund is $2500.
The amount invested in the stock fund is $10{,}000 - x$, or $7500.

5. Applications Involving Mixtures

Example 6   Solving a Mixture Application

How many liters (L) of a 40% antifreeze solution must be added to 4 L of a 10% antifreeze solution to produce a 35% antifreeze solution?

Solution:

The given information is illustrated in Figure 1-1.

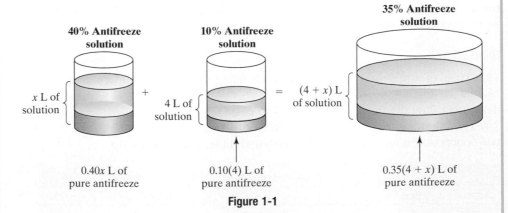

Figure 1-1

The information can also be organized in a table.

	40% Antifreeze	10% Antifreeze	Final Solution: 35% Antifreeze
Number of liters of solution	x	4	$(4 + x)$
Number of liters of pure antifreeze	$0.40x$	$0.10(4)$	$0.35(4 + x)$

TIP: To understand the role of the concentration rate within a mixture problem, consider this example. Suppose you had 30 gal of a 10% antifreeze mixture. How much pure antifreeze is in the mixture?

pure antifreeze = 0.10(30 gal)

= 3 gal

Multiply the concentration rate by the amount of mixture.

Answer

6. 30 oz of the 30% solution is needed.

Notice that an algebraic equation is obtained from the second row of the table relating the number of liters of pure antifreeze in each container.

$$\left(\begin{array}{c}\text{Pure antifreeze}\\\text{from solution 1}\end{array}\right) + \left(\begin{array}{c}\text{pure antifreeze}\\\text{from solution 2}\end{array}\right) = \left(\begin{array}{c}\text{pure antifreeze}\\\text{in the final solution}\end{array}\right)$$

$$0.40x \qquad + \qquad 0.10(4) \qquad = \qquad 0.35(4 + x)$$

$0.40x + 0.10(4) = 0.35(x + 4)$	Mathematical equation
$0.4x + 0.4 = 0.35x + 1.4$	Apply the distributive property.
$0.4x - 0.35x + 0.4 = 0.35x - 0.35x + 1.4$	Subtract $0.35x$ from both sides.
$0.05x + 0.4 = 1.4$	
$0.05x + 0.4 - 0.4 = 1.4 - 0.4$	Subtract 0.4 from both sides.
$0.05x = 1.0$	
$\dfrac{0.05x}{0.05} = \dfrac{1.0}{0.05}$	Divide both sides by 0.05.
$x = 20$	

Therefore, 20 L of a 40% antifreeze solution is needed.

6. Applications Involving Distance, Rate, and Time

The fundamental relationship among the variables distance, rate, and time is given by

$$\text{Distance} = (\text{rate})(\text{time}) \qquad \text{or} \qquad d = rt$$

For example, a motorist traveling 65 mph (miles per hour) for 3 hr (hours) will travel a distance of

$$d = \left(\frac{65 \text{ mi}}{\text{hr}}\right)(3 \text{ hr}) = 195 \text{ mi}$$

Example 7 Solving a Distance, Rate, Time Application

A hiker can hike 1 mph faster downhill to Moose Lake than she can hike uphill back to the campsite. If it takes her 3 hr to hike to the lake and 4.5 hr to hike back, what is her speed hiking back to the campsite?

Skill Practice

7. During a bad rainstorm, Jody drove 15 mph slower on a trip to her mother's house than she normally would when the weather is clear. If a trip to her mother's house takes 3.75 hr in clear weather and 5 hr in a bad storm, what is her normal driving speed during clear weather?

Answer

7. Jody normally drives 60 mph.

Solution:

The information given in the problem can be organized in a table.

	Distance (mi)	Rate (mph)	Time (hr)
Trip to the lake		$x + 1$	3
Return trip		x	4.5

Column 2: Let the rate of the return trip be represented by x.
Then the trip to the lake is 1 mph faster and can be
represented by $x + 1$.

Column 3: The times hiking to and from the lake are given in the
problem.

Column 1: To express the distance, we use the relationship $d = rt$. That is, multiply
the quantities in the second and third columns.

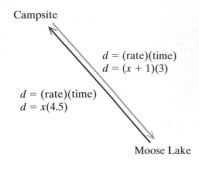

Campsite

$d = (\text{rate})(\text{time})$
$d = (x + 1)(3)$

$d = (\text{rate})(\text{time})$
$d = x(4.5)$

Moose Lake

	Distance (mi)	Rate (mph)	Time (hr)
Trip to the lake	$3(x + 1)$	$x + 1$	3
Return trip	$4.5x$	x	4.5

To create a mathematical model, note that the distances to and from the lake
are equal. Therefore,

(Distance to lake) = (return distance)	Verbal model
$3(x + 1) = 4.5x$	Mathematical equation
$3x + 3 = 4.5x$	Apply the distributive property.
$3x - 3x + 3 = 4.5x - 3x$	Subtract $3x$ from both sides.
$3 = 1.5x$	
$\dfrac{3}{1.5} = \dfrac{1.5x}{1.5}$	Divide by 1.5 to isolate the variable.
$2 = x$	Solve for x.

The hiker's speed on the return trip to the campsite is 2 mph.

Section 1.2 | Practice Exercises

Study Skills Exercise

After doing a section of homework, check the odd-numbered answers in the back of the text. Choose a method
to identify the exercises that gave you trouble (i.e., circle the number or put a star by the number). List some
reasons why it is important to label these problems.

Vocabulary and Key Concepts

1. a. Integers that follow one after the other without "gaps" are called _____ integers.

b. The integers $-2, 0, 2$, and 4 are examples of consecutive _____ integers, and the integers $-3, -1,$ 1, and 3 are examples of consecutive _____ integers.

c. Two consecutive integers differ by _____, two consecutive odd integers differ by _____, and two consecutive even integers differ by _____.

d. If x represents the smaller of two consecutive integers, then the expression _____ represents the next greater consecutive integer.

e. If x represents the smaller of two consecutive odd integers, then the expression _____ represents the next greater consecutive odd integer.

f. If x represents the smallest of three consecutive even integers, then the expressions _____ and _____ represent the next two greater consecutive even integers.

g. Simple interest is interest computed on the principal invested or borrowed. Simple interest is computed as $I =$ _____ where P is the principal, r is the annual _____ rate, and t is the time in years.

h. If \$5000 is borrowed at 6.5% simple interest for 4 yr, then the amount of simple interest is _____.

i. Suppose that 12% of a solution is acid by volume and the remaining 88% is water. Then the amount of acid in a 4-L container of solution is _____. The amount of acid in a container containing $x + 8$ liters of solution is represented by the expression _____.

j. If $d = rt$, then $r = \dfrac{\square}{\square}$ and $t = \dfrac{\square}{\square}$.

Review Exercises

2. Identify the smallest positive integer that could be used to clear fractions in the equation.

$$\frac{2}{3}x - \frac{4}{5} = 1 - \frac{1}{6}x$$

For Exercises 3–8, solve the equations.

3. $7a - 2 = 11$

4. $2z + 6 = -15$

5. $4(x - 3) + 7 = 19$

6. $-3(y - 5) + 4 = 1$

7. $\dfrac{3}{8}p + \dfrac{3}{4} = p - \dfrac{3}{2}$

8. $\dfrac{1}{4} - 2x = 5$

For the remaining exercises, follow the steps outlined in the Problem-Solving Flowchart found on page 55.

Concept 1: Introduction to Problem Solving

9. If x represents a number, write an expression for 5 more than the number.

10. If n represents a number, write an expression for 10 less than the number.

11. If t represents a number, write an expression for 7 less than twice the number.

12. If y represents a number, write an expression for 4 more than 3 times the number.

13. The larger of two numbers is 3 more than twice the smaller. The difference of the larger number and the smaller number is 8. Find the numbers. (See Example 1.)

14. One number is 3 less than another. Their sum is 15. Find the numbers.

15. The sum of 3 times a number and 2 is the same as the difference of the number and 4. Find the number.

16. Twice the sum of a number and 3 is the same as 1 subtracted from the number. Find the number.

Concept 2: Applications Involving Consecutive Integers

17. The sum of two consecutive page numbers in a book is 223. Find the page numbers. **(See Example 2.)**

18. The sum of the numbers on two consecutive raffle tickets is 808,455. Find the numbers on the tickets.

19. The sum of two consecutive odd integers is −148. Find the two integers.

20. The sum of three consecutive integers is −57. Find the integers.

21. Three times the smaller of two consecutive even integers is the same as −146 minus 4 times the larger integer. Find the integers.

22. Four times the smaller of two consecutive odd integers is the same as 73 less than 5 times the larger. Find the integers.

23. Two times the sum of three consecutive odd integers is the same as 23 more than 5 times the largest integer. Find the integers.

24. Five times the smallest of three consecutive even integers is 10 more than twice the largest. Find the integers.

Concept 3: Applications Involving Percents and Rates

25. Belle had the choice of taking out a 4-yr car loan at 8.5% simple interest or a 5-yr car loan at 7.75% simple interest. If she borrows $15,000, how much interest would she pay for each loan? Which option will require less interest? **(See Example 3.)**

26. Robert can take out a 3-yr loan at 8% simple interest or a 2-yr loan at $8\frac{1}{2}$% simple interest. If he borrows $7000, how much interest will he pay for each loan? Which option will require less interest?

27. An account executive earns $600 per month plus a 3% commission on sales. The executive's goal is to earn $2400 this month. How much must she sell to achieve this goal?

28. A salesperson earns $50 a day plus 12% commission on sales over $200. If her daily earnings are $76.88, how much money in merchandise did she sell?

29. J. W. is an artist and sells his pottery each year at a local Renaissance Festival. He keeps track of his sales and the 8.05% sales tax he collects by making notations in a ledger. Every evening he checks his records by counting the total money in his cash drawer. After a day of selling pottery, the cash totaled $1293.38. How much is from the sale of merchandise and how much is sales tax?

30. Wayne County has a sales tax rate of 7%. How much does Mike's used Honda Civic cost before tax if the total cost of the car *plus tax* is $13,888.60?

31. The price of a swimsuit after a 20% markup is $43.08. What was the price before the markup? **(See Example 4.)**

32. The price of a used textbook after a 35% markdown is $29.25. What was the original price?

Concept 4: Applications Involving Principal and Interest

33. Tony has a total of $12,500 in two accounts. One account pays 2% simple interest per year and the other pays 5% simple interest. If he earned $370 in interest in the first year, how much did he invest in each account? **(See Example 5.)**

34. Lillian had $15,000 invested in two accounts, one paying 9% simple interest and one paying 10% simple interest. How much was invested in each account if the interest after 1 yr is $1432?

35. Jason borrowed $18,000 in two loans. One loan charged 11% simple interest and the other charged 6% simple interest. After 1 yr, Jason paid a total of $1380. Find the amount borrowed in each loan.

36. Amanda borrowed $6000 from two sources: her parents and a credit union. Her parents charged 3% simple interest and the credit union charged 8% simple interest. If after 1 yr, Amanda paid $255 in interest, how much did she borrow from her parents, and how much did she borrow from the credit union?

37. Donna invested money in two accounts: one paying 4% simple interest and the other paying 3% simple interest. She invested $4000 more in the 4% account than in the 3% account. If she received $720 in interest at the end of 1 yr how much did she invest in each account?

38. Mr. Hall had some money in his bank earning 4.5% simple interest. He had $5000 more deposited in a credit union earning 6% simple interest. If his total interest for 1 yr was $1140, how much did he deposit in each account?

Concept 5: Applications Involving Mixtures

39. Ahmed mixes two plant fertilizers. How much fertilizer with 15% nitrogen should be mixed with 2 oz of fertilizer with 10% nitrogen to produce a fertilizer that is 14% nitrogen? **(See Example 6.)**

40. How much 8% saline solution should Kent mix with 80 cc (cubic centimeters) of an 18% saline solution to produce a 12% saline solution?

41. Jacque has 3 L of a 50% antifreeze mixture. How much 75% mixture should be added to get a mixture that is 60% antifreeze?

42. One fruit punch has 40% fruit juice and another is 70% fruit juice. How much of the 40% punch should be mixed with 10 gal of the 70% punch to create a fruit punch that is 45% fruit juice?

43. How many liters of an 18% alcohol solution must be added to a 10% alcohol solution to get 20 L of a 15% alcohol solution?

44. How many milliliters of a 2.5% bleach solution must be mixed with a 10% bleach solution to produce 600 mL of a 5% bleach solution?

45. Ronald has a 12% solution of the fertilizer Super Grow. How much pure Super Grow should he add to the mixture to get 32 oz of a 17.5% concentration?

46. How many ounces of water must be added to 20 oz of an 8% salt solution to make a 2% salt solution?

Concept 6: Applications Involving Distance, Rate, and Time

47. An airplane travels 60 mph faster from Atlanta to Fort Lauderdale than it travels on the return trip from Fort Lauderdale to Atlanta. If it takes 2 hr from Atlanta to Fort Lauderdale and 2.5 hr for the return trip, determine the speed of each trip. **(See Example 7.)**

48. A woman can hike 1 mph faster down a trail to Archuletta Lake than she can on the return trip uphill. It takes her 3 hr to get to the lake and 6 hr to return. What is her speed hiking down to the lake?

49. Two cars are 192 mi apart and travel toward each other on the same road. They meet in 2 hr. One car travels 4 mph faster than the other. What is the average speed of each car?

50. Two cars are 190 mi apart and travel toward each other along the same road. They meet in 2 hr. One car travels 5 mph slower than the other car. What is the average speed of each car?

Mixed Exercises

51. The sum of two integers is 30. Ten times one integer is 5 times the other integer. Find the integers. (*Hint:* If one number is x, then the other number is $30 - x$.)

52. The sum of two integers is 10. Three times one integer is 3 less than 8 times the other integer. Find the integers. (*Hint:* If one number is x, then the other number is $10 - x$.)

53. An older model of smart phone is marked down to $89.55 when newer technology comes on the market. If this is the new price after a 55% markdown, determine the original price.

54. A hardcover book is marked down to $15.60 once the book comes out in paperback. If this is the new price after a 35% markdown, determine the original price.

55. Two boats traveling the same direction leave a harbor at noon. After 3 hr they are 60 mi apart. If one boat travels twice as fast as the other, find the average rate of each boat.

56. Two canoes travel down a river, starting at 9:00. One canoe travels twice as fast as the other. After 3.5 hr, the canoes are 5.25 mi apart. Find the average rate of each canoe.

57. Ms. Riley deposited some money in an account paying 5% simple interest and twice that amount in an account paying 6% simple interest. If the total interest from the two accounts is $765 for 1 yr, how much was deposited into each account?

58. Sienna put some money in a certificate of deposit earning 4.2% simple interest. She deposited twice that amount in a money market account paying 4% simple interest. After 1 yr her total interest was $488. How much did Sienna deposit in her money market account?

59. Two different teas are mixed to make a blend that will be sold at a fair. Black tea sells for $2.20 per pound and green tea sells for $3.00 per pound. How much of each should be used to obtain 4 lb of a blend selling for $2.50?

60. A nut mixture consists of almonds and cashews. Almonds are $4.98 per pound, and cashews are $6.98 per pound. How many pounds of each type of nut should be mixed to produce 16 lb selling for $5.73 per pound?

61. After a recent crash of the housing market, the median price of a new home (including land) in the United States was $202,100. This represents a drop of 6% from the previous year. What was the median price the previous year?

62. According to the Bureau of Labor Statistics, the median weekly salary for an individual with a bachelor's degree is $1026 per week. This represents an increase of 35% over the salary of an individual with an associate's degree. What is the median weekly salary of an individual with an associate's degree?

Section 1.3   Applications to Geometry and Literal Equations

Concepts

1. Applications Involving Geometry
2. Literal Equations

1. Applications Involving Geometry

Some word problems involve the use of geometric formulas such as those listed in the inside back cover of this text.

Example 1   Solving an Application Involving Perimeter

The length of a rectangular corral is 2 ft more than 3 times the width. The corral is situated such that one of its shorter sides is adjacent to a barn and does not require fencing. If the total amount of fencing is 774 ft, find the dimensions of the corral.

Solution:

Read the problem and draw a sketch (Figure 1-2).

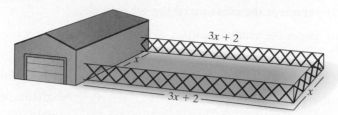

Figure 1-2

Let x represent the width.   Label variables.

Let $3x + 2$ represent the length.

To write a verbal model, we might consider using the formula for the perimeter of a rectangle. However, the formula $P = 2l + 2w$ incorporates all four sides of the rectangle. The formula must be modified to include only one factor of the width.

$$\begin{pmatrix}\text{Distance around} \\ \text{three sides}\end{pmatrix} = \begin{pmatrix}2 \text{ times} \\ \text{the length}\end{pmatrix} + \begin{pmatrix}1 \text{ times} \\ \text{the width}\end{pmatrix} \quad \text{Verbal model}$$

$$774 \qquad = \quad 2(3x + 2) \quad + \qquad x \qquad \text{Mathematical equation}$$

$774 = 2(3x + 2) + x$	Solve for x.
$774 = 6x + 4 + x$	Apply the distributive property.
$774 = 7x + 4$	Combine *like* terms.
$770 = 7x$	Subtract 4 from both sides.
$110 = x$	Divide by 7 on both sides.
$x = 110$	

Because x represents the width, the width of the corral is 110 ft. The length is given by

$$3x + 2 \quad \text{or} \quad 3(110) + 2 = 332 \qquad \text{Interpret the results.}$$

The width of the corral is 110 ft, and the length is 332 ft.

Recall some important facts involving angles.

- Two angles are complementary if the sum of their measures is 90°.
- Two angles are supplementary if the sum of their measures is 180°.
- The sum of the measures of the angles within a triangle is 180°.

Example 2 Solving an Application Involving Angles

Two angles are complementary. One angle measures 10° less than 4 times the other angle. Find the measure of each angle (Figure 1-3).

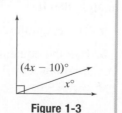

Figure 1-3

Solution:

Let x represent the measure of one angle.

Let $4x - 10$ represent the measure of the other angle.

Recall that two angles are complementary if the sum of their measures is 90°. Therefore, a verbal model is

(One angle) + (the complement of the angle) = 90°	Verbal model
$x + (4x - 10) = 90$	Mathematical equation
$5x - 10 = 90$	Solve for x.
$5x = 100$	
$x = 20$	

If $x = 20$, then $4x - 10 = 4(20) - 10 = 70$. The two angles are 20° and 70°.

2. Literal Equations

Literal equations are equations that contain several variables. A formula is a literal equation with a specific application. For example, the perimeter of a rectangle can be found by the formula $P = 2l + 2w$. In this equation, P is expressed in terms of l and w. However, in science and other branches of applied mathematics, formulas may be more useful in alternative forms.

For example, the formula $P = 2l + 2w$ can be manipulated to solve for either l or w:

Solve for l

$P = 2l + 2w$

$P - 2w = 2l$ Subtract $2w$.

$\dfrac{P - 2w}{2} = l$ Divide by 2.

$l = \dfrac{P - 2w}{2}$

Solve for w

$P = 2l + 2w$

$P - 2l = 2w$ Subtract $2l$.

$\dfrac{P - 2l}{2} = w$ Divide by 2.

$w = \dfrac{P - 2l}{2}$

To solve a literal equation for a specified variable, use the addition, subtraction, multiplication, and division properties of equality.

Skill Practice

The formula to compute the surface area S of a sphere is given by $S = 4\pi r^2$.

3. Solve the equation for π.
4. A sphere has a surface area of 113 in.^{2} and a radius of 3 in. Use the formula found in part (a) to approximate π. Round to two decimal places.

Example 3   Applying a Literal Equation

Buckingham Fountain is one of Chicago's most familiar landmarks. With 133 jets spraying a total of 14,000 gal of water per minute, Buckingham Fountain is one of the world's largest fountains. The circumference of the fountain is approximately 880 ft.

a. The circumference of a circle is given by $C = 2\pi r$. Solve the equation for r.

b. Use the equation from part (a) to find the radius and diameter of the fountain. Use 3.14 for π and round to the nearest foot.

Answers

3. $\pi = \dfrac{S}{4r^2}$ **4.** 3.14

Solution:

a. $C = 2\pi r$

$\dfrac{C}{2\pi} = \dfrac{2\pi r}{2\pi}$

$\dfrac{C}{2\pi} = r$

$r = \dfrac{C}{2\pi}$

b. $r \approx \dfrac{880 \text{ ft}}{2(3.14)}$ Substitute 880 ft for C and 3.14 for π.

$\approx 140 \text{ ft}$

The radius is approximately 140 ft. The diameter is twice the radius ($d = 2r$). Therefore, the diameter is 280 ft.

Example 4 **Solving a Literal Equation**

The formula to find the area of a trapezoid is given by $A = \frac{1}{2}(b_1 + b_2)h$, where b_1 and b_2 are the lengths of the parallel sides and h is the height. (See Figure 1-4.)

Solve this formula for b_1.

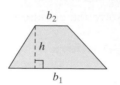

Figure 1-4

Solution:

$A = \frac{1}{2}(b_1 + b_2)h$ The goal is to isolate b_1.

$2A = 2 \cdot \frac{1}{2}(b_1 + b_2)h$ Multiply by 2 to clear fractions.

$2A = (b_1 + b_2)h$ Apply the distributive property.

$2A = b_1 h + b_2 h$

$2A - b_2 h = b_1 h$ Subtract $b_2 h$ from both sides.

$\dfrac{2A - b_2 h}{h} = \dfrac{b_1 h}{h}$ Divide by h.

$\dfrac{2A - b_2 h}{h} = b_1$

Skill Practice

5. The formula for the volume of a right circular cylinder is $V = \pi r^2 h$. Solve for h.

TIP: When solving a literal equation for a specified variable, there is sometimes more than one way to express your final answer. This flexibility often presents difficulty for students. Students may leave their answer in one form, but the answer given in the text looks different. Yet both forms may be correct. To know if your answer is equivalent to the form given in the text you must try to manipulate it to look like the answer in the book, a process called *form fitting*.

The literal equation from Example 4 can be written in several different forms. The quantity $(2A - b_2 h)/h$ can be split into two fractions.

$$b_1 = \frac{2A - b_2 h}{h} = \frac{2A}{h} - \frac{b_2 h}{h} = \frac{2A}{h} - b_2$$

Answer

5. $h = \dfrac{V}{\pi r^2}$

Skill Practice

Solve for *y*.

  6. $5x + 2y = 11$

Example 5   Solving a Literal Equation

Given $-2x + 3y = 5$, solve for *y*.

Solution:

$$-2x + 3y = 5$$

$$3y = 2x + 5 \qquad\qquad \text{Add } 2x \text{ to both sides.}$$

$$\frac{3y}{3} = \frac{2x + 5}{3} \qquad\qquad \text{Divide by 3 on both sides.}$$

$$y = \frac{2x + 5}{3} \qquad \text{or} \qquad y = \frac{2}{3}x + \frac{5}{3}$$

Sometimes the variable we want to isolate may appear in more than one term in a literal equation. In such a case, isolate all terms with that variable on one side of the equation. Then apply the distributive property as demonstrated in Example 6.

Skill Practice

Solve for *t*.

  7. $mt + 4 = nt + 9$

Example 6   Solving a Literal Equation

Solve the equation for *x*. $ax - 3 = cx + 7$

Solution:

$$ax - 3 = cx + 7$$

$$ax - cx = 10 \qquad\qquad \text{Collect the terms containing } x \text{ on one side of the equation. Collect the remaining terms on the other side.}$$

TIP: Applying the distributive property in reverse is called *factoring*. Factoring will be studied in detail in Chapter 5.

The variable *x* appears twice in the equation. To isolate *x*, we want *x* to appear in only one term. To accomplish this, we apply the distributive property in reverse.

$$x(a - c) = 10 \qquad\qquad \text{Apply the distributive property. The variable } x \text{ now appears one time in the equation.}$$

$$\frac{x(a - c)}{(a - c)} = \frac{10}{(a - c)} \qquad \text{Divide both sides by } (a - c).$$

$$x = \frac{10}{a - c}$$

Answers

6. $y = \dfrac{11 - 5x}{2}$ or $y = -\dfrac{5}{2}x + \dfrac{11}{2}$

7. $t = \dfrac{5}{m - n}$

Section 1.3   Practice Exercises

Study Skills Exercise

In your next math class, take notes by drawing a vertical line about three-fourths of the way across the paper, as shown. On the left side, write down what your instructor puts on the board or overhead. On the right side, make your own comments about important words, procedures, or questions that you have.

Vocabulary and Key Concepts

1. **a.** For a rectangle having width w and length l, the perimeter is given by $P =$ _____ .

    b. Two angles are complementary if the sum of their measures is _____ .

    c. Two angles are _____ if the sum of their measures is $180°$.

    d. The sum of the measures of the angles within a triangle is _____ .

Review Exercises

For Exercises 2–6, solve the equations.

2. $7 + 5x - (2x - 6) = 6(x + 1) + 21$

3. $\dfrac{3}{5}y - 3 + 2y = 5$

4. $3[z - (2 - 3z) - 4] = z - 7$

5. $2a - 4 + 8a = 7a - 8 + 3a$

6. $3(t + 6) + t + 2 = 5(t + 4) - t$

Concept 1: Applications Involving Geometry

For Exercises 7–18, use the geometry formulas listed in the inside back cover of the text.

7. A volleyball court is twice as long as it is wide. If the perimeter is 177 ft, find the dimensions of the court. **(See Example 1.)**

8. The length of a rectangular picture frame is 4 in. less than twice the width. The perimeter is 112 in. Find the length and the width.

9. The lengths of the sides of a triangle are given by three consecutive even integers. The perimeter is 24 m. What is the length of each side?

10. A triangular garden has sides that can be represented by three consecutive integers. If the perimeter of the garden is 15 ft, what are the lengths of the sides?

11. Raoul would like to build a rectangular dog run in the rear of his backyard, away from the house. The width of the yard is $12\frac{1}{2}$ yd, and Raoul wants an area of 100 yd^2 for his dog.

    a. Find the dimensions of the dog run.

    b. How much fencing would Raoul need to enclose the dog run?

12. Joanne wants to plant a flower garden in her backyard in the shape of a trapezoid, adjacent to her house (see the figure). She also wants a front yard garden in the same shape, but with sides one-half as long. What should the dimensions be for each garden if Joanne has only a total of 60 ft of fencing?

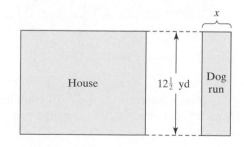

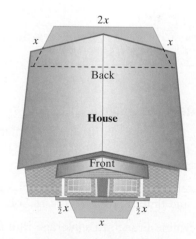

13. George built a rectangular pen for his rabbit such that the length is 7 ft less than twice the width. If the perimeter is 40 ft, what are the dimensions of the pen?

14. Antoine wants to put edging in the form of a square around a tree in his front yard. He has enough money to buy 18 ft of edging. Find the dimensions of the square that will use all the edging.

15. The measures of two angles in a triangle are equal. The third angle measures 2 times the sum of the equal angles. Find the measures of the three angles.

16. The smallest angle in a triangle is one-half the size of the largest. The middle angle measures 25° less than the largest. Find the measures of the three angles.

17. Two angles are complementary. One angle is 5 times as large as the other angle. Find the measure of each angle. **(See Example 2.)**

18. Two angles are supplementary. One angle measures 12° less than 3 times the other. Find the measure of each angle.

In Exercises 19–26, solve for x, and then find the measure of each angle.

19.

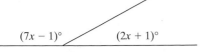

$(7x - 1)°$ $(2x + 1)°$

20.

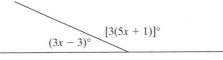

$(10x + 36)°$

$[2(x + 15)]°$

21.

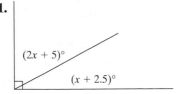

$(2x + 5)°$

$(x + 2.5)°$

22.

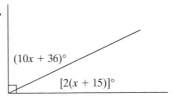

$[3(5x + 1)]°$

$(3x - 3)°$

23.

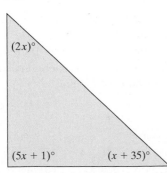

$(2x)°$

$(5x + 1)°$ $(x + 35)°$

24.

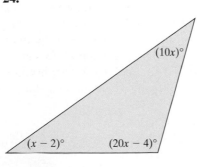

$(10x)°$

$(x - 2)°$ $(20x - 4)°$

25.

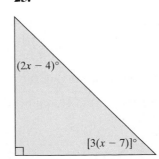

$(2x - 4)°$

$[3(x - 7)]°$

26.

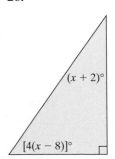

$(x + 2)°$

$[4(x - 8)]°$

Concept 2: Literal Equations

27. In 2010, Dario Franchitti won the Indianapolis 500 car race in 3 hr 5 min 37 sec (≈ 3.094 hr). **(See Example 3.)**

    a. The relationship among the variables distance, rate, and time is given by $d = rt$. Solve for r.

    b. Determine Franchitti's average rate of speed if the total distance for the race is 500 mi. Round to one decimal place.

28. In 2011, Trevor Bayne won the 53rd running of the Daytona 500 car race with an average speed of 130.326 mph.

    a. Solve the formula $d = rt$ for t.

    b. Determine Bayne's total time of the race if the race is 500 mi long. Round to one decimal place.

29. The amount of simple interest earned or borrowed is the product of the principal, the annual interest rate, and the time invested (in years). This is given by $I = Prt$.

    a. Solve $I = Prt$ for t.

    b. Determine the amount of time necessary for the interest on \$5000 invested at 4% to reach \$1400.

30. The force of an object is equal to its mass times the acceleration, or $F = ma$.

    a. Solve $F = ma$ for m.

    b. The force on an object is 24.5 N (newtons), and the acceleration is 9.8 m/sec^2. Find the mass of the object (the answer will be in kilograms).

For Exercises 31–48, solve for the indicated variable. **(See Example 4.)**

31. $A = lw$ for l

32. $C_1 = \frac{5}{2}R$ for R

33. $I = Prt$ for P

34. $a + b + c = P$ for b

35. $W = K_2 - K_1$ for K_1

36. $y = mx + b$ for x

37. $F = \frac{9}{5}C + 32$ for C

38. $C = \frac{5}{9}(F - 32)$ for F

39. $K = \frac{1}{2}mv^2$ for v^2

40. $I = Prt$ for r

41. $v = v_0 + at$ for a

42. $a^2 + b^2 = c^2$ for b^2

43. $w = p(v_2 - v_1)$ for v_2

44. $A = lw$ for w

45. $ax + by = c$ for y

46. $P = 2L + 2W$ for L

47. $V = \frac{1}{3}Bh$ for B

48. $V = \frac{1}{3}\pi r^2 h$ for h

In Chapter 2 it will be necessary to change equations from the form $Ax + By = C$ to $y = mx + b$. For Exercises 49–60, express each equation in the form $y = mx + b$ by solving for y. **(See Example 5.)**

49. $3x + y = 6$

50. $x + y = -4$

51. $5x - 4y = 20$

52. $-4x - 5y = 25$

53. $-6x - 2y = 13$

54. $5x - 7y = 15$

55. $3x - 3y = 6$

56. $2x - 2y = 8$

57. $9x + \frac{4}{3}y = 5$

58. $4x - \frac{1}{3}y = 5$

59. $-x + \frac{2}{3}y = 0$

60. $x - \frac{1}{4}y = 0$

In statistics, the z-score formula $z = \dfrac{x - \mu}{\sigma}$ is used in studying probability. Use this formula for Exercises 61 and 62.

61. a. Solve $z = \dfrac{x - \mu}{\sigma}$ for x.

    b. Find x when $z = 2.5$, $\mu = 100$, and $\sigma = 12$.

62. a. Solve $z = \dfrac{x - \mu}{\sigma}$ for μ.

    b. Find μ when $x = 150$, $z = 2.5$, and $\sigma = 16$.

63. Which expressions are equivalent to $\dfrac{-5}{x - 3}$?

    a. $-\dfrac{5}{x - 3}$ **b.** $\dfrac{5}{3 - x}$ **c.** $\dfrac{5}{-x + 3}$

64. Which expressions are equivalent to $\dfrac{z - 1}{-2}$?

    a. $\dfrac{1 - z}{2}$ **b.** $-\dfrac{z - 1}{2}$ **c.** $\dfrac{-z + 1}{2}$

65. Which expressions are equivalent to $\dfrac{-x - 7}{y}$?

    a. $-\dfrac{x + 7}{y}$ **b.** $\dfrac{x + 7}{-y}$ **c.** $\dfrac{-x - 7}{-y}$

66. Which expressions are equivalent to $\dfrac{-3w}{-x - y}$?

    a. $-\dfrac{3w}{-x - y}$ **b.** $\dfrac{3w}{x + y}$ **c.** $-\dfrac{-3w}{x + y}$

For Exercises 67–75, solve for the indicated variable. **(See Example 6.)**

67. $6t - rt = 12$ for t

68. $5 = 4a + ca$ for a

69. $ax + 5 = 6x + 3$ for x

70. $cx - 4 = dx + 9$ for x

71. $A = P + Prt$ for P

72. $A = P + Prt$ for r

73. $T = mg - mf$ for m

74. $T = mg - mf$ for f

75. $ax + by = cx + z$ for x

Section 1.4   Linear Inequalities in One Variable

Concepts

1. Solving Linear Inequalities
2. Applications of Inequalities

1. Solving Linear Inequalities

In Sections 1.1–1.3, we learned how to solve linear equations and their applications. In this section, we will learn the process of solving linear *inequalities*. A **linear inequality** in one variable, x, is defined as any relationship of the form: $ax + b < c$, $ax + b \le c$, $ax + b > c$, or $ax + b \ge c$, where $a \ne 0$.

The solution to the equation $x = 3$ can be graphed as a single point on the number line.

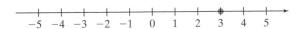

Now consider the related *inequalities*.

Avoiding Mistakes

Recall that parentheses (or) indicate that an endpoint to an interval is not included in the interval. Brackets [or] indicate that an endpoint is included.

Inequality	Solutions	Graph	Interval Notation	Set-Builder Notation
$x \le 3$	All real numbers less than or equal to 3.		$(-\infty, 3]$	$\{x \mid x \le 3\}$
$x < 3$	All real numbers less than 3.		$(-\infty, 3)$	$\{x \mid x < 3\}$
$x \ge 3$	All real numbers greater than or equal to 3.		$[3, \infty)$	$\{x \mid x \ge 3\}$
$x > 3$	All real numbers greater than 3.		$(3, \infty)$	$\{x \mid x > 3\}$

TIP: For a complete review of interval notation and set-builder notation, see Section R.2.

The solution sets to these inequalities have an infinite number of elements that cannot all be listed. Therefore, the solution set can be shown in a graph, or expressed in interval notation or in set-builder notation.

The addition and subtraction properties of equality indicate that a value added to or subtracted from both sides of an equation results in an equivalent equation. The same is true for inequalities.

Addition and Subtraction Properties of Inequality

Let a, b, and c represent real numbers.

*Addition property of inequality: If $a < b$
                                       then $a + c < b + c$

*Subtraction property of inequality: If $a < b$
                                         then $a - c < b - c$

*These properties may also be stated for $a \le b$, $a > b$, and $a \ge b$.

Example 1   Solving a Linear Inequality

Solve the inequality.

$$3x - 7 > 2(x - 4) - 1$$

Solution:

$3x - 7 > 2(x - 4) - 1$	
$3x - 7 > 2x - 8 - 1$	Apply the distributive property.
$3x - 7 > 2x - 9$	Combine *like* terms.
$3x - 2x - 7 > 2x - 2x - 9$	Subtract $2x$ from both sides.
$x - 7 > -9$	
$x - 7 + 7 > -9 + 7$	Add 7 to both sides.
$x > -2$	

Set-builder notation: $\{x \mid x > -2\}$

Interval notation: $(-2, \infty)$

Skill Practice

Solve the inequality.

1. $4(2x - 1) > 7x + 1$

Multiplying both sides of an equation by the same nonzero quantity results in an equivalent equation. However, the same is not always true for an inequality. If you multiply or divide an inequality by a *negative* quantity, the direction of the inequality symbol must be *reversed*.

For example, consider multiplying or dividing the inequality $4 < 5$ by -1.

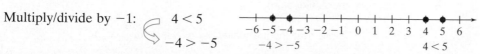

Multiply/divide by -1: $4 < 5$

$-4 > -5$

$-4 > -5$ $4 < 5$

The number 4 lies to the left of 5 on the number line. However, -4 lies to the right of -5. Changing the signs of two numbers changes their relative position on the number line. This is stated formally in the multiplication and division properties of inequality.

Multiplication and Division Properties of Inequality

Let a, b, and c represent real numbers.

*If c is *positive* and $a < b$, then $ac < bc$ and $\dfrac{a}{c} < \dfrac{b}{c}$

*If c is *negative* and $a < b$, then $ac > bc$ and $\dfrac{a}{c} > \dfrac{b}{c}$

The second statement indicates that if both sides of an inequality are multiplied or divided by a negative quantity, the inequality sign must be *reversed*.

*These properties may also be stated for $a \leq b$, $a > b$, and $a \geq b$.

Answer

1. $\{x \mid x > 5\}$
   $(5, \infty)$

Example 2 Solving a Linear Inequality

Solve the inequality. $-2x - 5 < 2$

Solution:

$$-2x - 5 < 2$$

$$-2x - 5 + 5 < 2 + 5 \qquad \text{Add 5 to both sides.}$$

$$-2x < 7$$

$$\frac{-2x}{-2} > \frac{7}{-2} \qquad \text{Divide by } -2 \ (\textit{reverse the inequality sign}).$$

$$x > -\frac{7}{2} \quad \text{or} \quad x > -3.5$$

Set-builder notation: $\{x \mid x > -\frac{7}{2}\}$

Interval notation: $(-\frac{7}{2}, \infty)$

TIP: The inequality $-2x - 5 < 2$ could have been solved by isolating x on the right-hand side of the inequality. This creates a positive coefficient on the x term and eliminates the need to divide by a negative number.

$-2x - 5 < 2$

$\quad -5 < 2x + 2 \qquad$ Add $2x$ to both sides.

$\quad -7 < 2x \qquad\qquad$ Subtract 2 from both sides.

$\quad \dfrac{-7}{2} < \dfrac{2x}{2} \qquad$ Divide by 2 (because 2 is positive, do *not* reverse the inequality sign).

$\quad -\dfrac{7}{2} < x \qquad\qquad$ (Note that the inequality $-\frac{7}{2} < x$ is equivalent to $x > -\frac{7}{2}$.)

Example 3 Solving a Linear Inequality

Solve the inequality. $-6(x - 3) \geq 2 - 2(x - 8)$

Solution:

$$-6(x - 3) \geq 2 - 2(x - 8)$$

$$-6x + 18 \geq 2 - 2x + 16 \qquad \text{Apply the distributive property.}$$

$$-6x + 18 \geq 18 - 2x \qquad\qquad \text{Combine } \textit{like} \text{ terms.}$$

$$-6x + 2x + 18 \geq 18 - 2x + 2x \qquad \text{Add } 2x \text{ to both sides.}$$

$$-4x + 18 \geq 18$$

$$-4x + 18 - 18 \geq 18 - 18 \qquad \text{Subtract 18 from both sides.}$$

$$-4x \geq 0$$

$$\frac{-4x}{-4} \leq \frac{0}{-4} \qquad \text{Divide by } -4 \ (\textit{reverse the inequality sign}).$$

$$x \leq 0$$

Set-builder notation: $\{x \mid x \leq 0\}$

Interval notation: $(-\infty, 0]$

Example 4 Solving a Linear Inequality

Solve the inequality. $\dfrac{-5x + 2}{-3} > x + 2$

Solution:

$$\dfrac{-5x + 2}{-3} > x + 2$$

$$-3\left(\dfrac{-5x + 2}{-3}\right) < -3(x + 2)$$ Multiply by -3 to clear fractions (*reverse* the inequality sign).

$$-5x + 2 < -3x - 6$$

$$-2x + 2 < -6$$ Add $3x$ to both sides.

$$-2x < -8$$ Subtract 2 from both sides.

$$\dfrac{-2x}{-2} > \dfrac{-8}{-2}$$ Divide by -2 (the inequality sign is reversed *again*).

$$x > 4$$ Simplify.

Set-builder notation: $\{x \mid x > 4\}$

Interval notation: $(4, \infty)$

Skill Practice

Solve the inequality.

4. $\dfrac{x + 1}{-3} \geq -x + 1$

In Example 4, the inequality sign was reversed twice: once for multiplying the inequality by -3 and once for dividing by -2. If you are in doubt about whether you have the inequality sign in the correct direction, you can check your final answer by using the **test point method**. That is, pick a point in the proposed solution set, and verify that it makes the original inequality true. Furthermore, any test point picked outside the solution set should make the original inequality false.

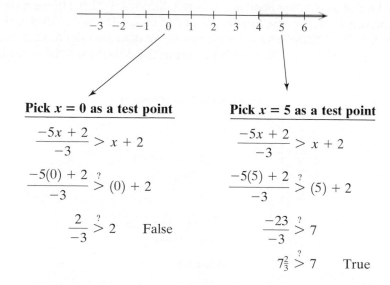

Pick $x = 0$ as a test point

$$\dfrac{-5x + 2}{-3} > x + 2$$

$$\dfrac{-5(0) + 2}{-3} \overset{?}{>} (0) + 2$$

$$\dfrac{2}{-3} \overset{?}{>} 2 \quad \text{False}$$

Pick $x = 5$ as a test point

$$\dfrac{-5x + 2}{-3} > x + 2$$

$$\dfrac{-5(5) + 2}{-3} \overset{?}{>} (5) + 2$$

$$\dfrac{-23}{-3} \overset{?}{>} 7$$

$$7\tfrac{2}{3} \overset{?}{>} 7 \quad \text{True}$$

Concept Connections

5. Check the solution to Skill Practice Exercise 4, using the test point method.

Because a test point to the right of $x = 4$ makes the inequality true, we have shaded the correct part of the number line.

Answers

4. $\{x \mid x \geq 2\}$
$[2, \infty)$

5. Answers may vary.

2. Applications of Inequalities

Example 5   Solving a Linear Inequality Application

Beth received grades of 97, 82, 89, and 99 on her first four algebra tests. To earn an A in the course, she needs an average of 90 or more. What scores can she receive on the fifth test to earn an A?

Solution:

Let x represent the score on the fifth test.

The average of the five tests is given by $\dfrac{97 + 82 + 89 + 99 + x}{5}$

To earn an A, we have:

$$(\text{Average of test scores}) \geq 90 \qquad \text{Verbal model}$$

$$\frac{97 + 82 + 89 + 99 + x}{5} \geq 90 \qquad \text{Mathematical equation}$$

$$\frac{367 + x}{5} \geq 90 \qquad \text{Simplify the numerator.}$$

$$5\left(\frac{367 + x}{5}\right) \geq 5(90) \qquad \text{Clear fractions.}$$

$$367 + x \geq 450 \qquad \text{Simplify.}$$

$$x \geq 83$$

To earn an A, Beth would have to score at least 83 on her fifth test.

Skill Practice

6. Jamie is a salesman who works on commission, so his salary varies from month to month. To qualify for an automobile loan, his monthly salary must average at least $2100 for 6 months. His salaries for the past 5 months have been $1800, $2300, $1500, $2200, and $2800. What amount does he need to earn in the last month to qualify for the loan?

Example 6   Solving a Linear Inequality Application

The water level in a retention pond in northern California is 7.2 ft. During a time of drought, the water level decreases at a rate of 0.05 ft/day. The water level L (in ft) is given by $L = 7.2 - 0.05d$, where d is the number of days after the drought begins (Figure 1-5). For which days after the beginning of the drought will the water level be less than 6 ft?

Skill Practice

7. The population of Alaska has steadily increased since 1950 according to the equation $P = 10t + 117$, where t represents the number of years after 1950 and P represents the population in thousands. For what years since 1950 was the population less than 417 thousand people?

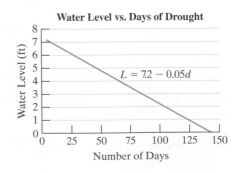

Water Level vs. Days of Drought

$L = 7.2 - 0.05d$

Number of Days

Figure 1-5

Answers

6. Jamie's salary must be at least $2000.
7. The population was less than 417 thousand for $t < 30$. This corresponds to the years before 1980.

Solution:

We require that $L < 6$.

$$\overbrace{7.2 - 0.05d}^{L} < 6 \qquad \text{Substitute } 7.2 - 0.05d \text{ for } L.$$

$$7.2 - 7.2 - 0.05d < 6 - 7.2 \qquad \text{Subtract 7.2 from both sides.}$$

$$-0.05d < -1.2$$

$$\frac{-0.05d}{-0.05} > \frac{-1.2}{-0.05} \qquad \text{Divide both sides by } -0.05 \text{ and reverse the direction of the inequality sign.}$$

$$d > 24$$

The water level in the pond will be less than 6 ft after day 24 of the drought.

Section 1.4   Practice Exercises

Study Skills Exercise

Look over the notes that you took today. Do you understand what you wrote? If there were any rules, definitions, or formulas, highlight them so that they can be easily found when studying for the test. You may want to begin by highlighting the rule indicating when the direction of an inequality sign must be reversed.

Vocabulary and Key Concepts

1. a. A relationship of the form $ax + b > c$ or $ax + b < c \, (a \neq 0)$ is called a _____ _____ in one variable.

  b. When solving an inequality, the direction of the inequality sign must be reversed when multiplying or dividing both sides of the inequality by a _____ number.

  c. Which inequality statement represents the set of real numbers greater than 2? $x > 2$ or $2 < x$

Review Exercises

2. When using interval notation is it proper to use a parenthesis or a bracket with ∞ and $-\infty$?

For Exercises 3–8, complete the table.

	Set-Builder Notation	Interval Notation	Graph
3.	$\{x \mid x > 5\}$		
4.	$\{x \mid x \leq -2\}$		
5.		$(-3, 6]$	
6.		$[0, 4)$	
7.			4
8.			10

Animation

Concept 1: Solving Linear Inequalities

For Exercises 9–46, solve the inequality and graph the solution set. Write the solution set in (a) set-builder notation and (b) interval notation. **(See Examples 1–4.)**

9. $-2x + 4 < 10$

10. $-4x + 2 > -6$

11. $2y + 6 \leq 4$

12. $3y + 11 > 5$

13. $-2x - 5 \leq -25$

14. $-4z - 2 > -22$

15. $6z + 3 > 16$

16. $8w - 2 \leq 13$

17. $-8 > \dfrac{2}{3}t$

18. $-4 \leq \dfrac{1}{5}p$

19. $\dfrac{3}{4}(8y - 9) < 3$

20. $\dfrac{2}{5}(2x - 1) > 10$

21. $0.8a - 0.5 \leq 0.3a - 11$

22. $0.2w - 0.7 < 0.4 - 0.9w$

23. $-5x + 7 < 22$

24. $-3w - 6 > 9$

25. $-\dfrac{5}{6}x \leq -\dfrac{3}{4}$

26. $-\dfrac{3}{2}y > -\dfrac{21}{16}$

27. $\dfrac{3p - 1}{-2} > 5$

28. $\dfrac{3k - 2}{-5} \leq 4$

29. $0.2t + 1 > 2.4t - 10$

30. $20 \leq 8 - \dfrac{1}{3}x$

31. $3 - 4(y + 2) \leq 6 + 4(2y + 1)$

32. $1 + 4(b - 2) < 2(b - 5) + 4$

33. $7.2k - 5.1 \geq 5.7$

34. $6h - 2.92 \leq 16.58$

35. $\dfrac{3}{4}x - 8 \leq 1$

36. $-\dfrac{2}{5}a - 3 > 5$

37. $-1.2b - 0.4 \geq -0.4b$

38. $-0.4t + 1.2 < -2$

39. $-\dfrac{3}{4}c - \dfrac{5}{4} \geq 2c$

40. $-\dfrac{2}{3}q - \dfrac{1}{3} > \dfrac{1}{2}q$

41. $4 - 4(y - 2) < -5y + 6$

42. $6 - 6(k - 3) \geq -4k + 12$

43. $-6(2x + 1) < 5 - (x - 4) - 6x$

44. $2(4p + 3) - p \leq 5 + 3(p - 3)$

45. $6a - (9a + 1) - 3(a - 1) \geq 2$

46. $8(q + 1) - (2q + 1) + 5 > 12$

Concept 2: Applications of Inequalities

47. Nadia received quiz grades of 80, 86, 73, and 91.
(See Example 5.)

  a. What grade would she need to make on the fifth quiz to get a B average, that is, at least 80 but less than 90?

  b. Is it possible for Nadia to get an A average for her quizzes (at least 90)?

48. Ty received test grades of 78, 75, 71, 83, and 73.

  a. What grade would he need to make on the sixth test to get a C if a C is at least 75 but less than 80?

  b. Is it possible for Ty to get a B or better for his test average (at least 80)?

For Exercises 49–52, use the graph that shows the average height for boys based on age. Let a represent a boy's age (in years) and let h represent his height (in inches). **(See Example 6.)**

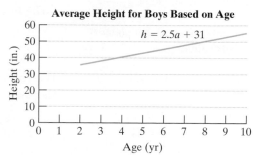

Average Height for Boys Based on Age

$h = 2.5a + 31$

Height (in.) — 60, 50, 40, 30, 20, 10, 0

Age (yr) — 0 1 2 3 4 5 6 7 8 9 10

49. Determine the age range for which the average height of boys is at least 51 in.

50. Determine the age range for which the average height of boys is greater than or equal to 41 in.

51. Determine the age range for which the average height of boys is no more than 46 in.

52. Determine the age range for which the average height of boys is at most 53.5 in.

53. Nolvia sells copy machines, and her salary is $25,000 plus a 4% commission on sales. The equation $S = 25,000 + 0.04x$ represents her salary S in dollars in terms of her total sales x in dollars.

  a. How much money in sales does Nolvia need to earn a salary that exceeds $40,000?

  b. How much money in sales does Nolvia need to earn a salary that exceeds $80,000?

  c. Why is the money in sales required to earn a salary of $80,000 more than twice the money in sales required to earn a salary of $40,000?

54. The amount of money A in a savings account depends on the principal P, the interest rate r, and the time in years t that the money is invested. The equation $A = P + Prt$ shows the relationship among the variables for an account earning simple interest. If an investor deposits $5000 at $6\frac{1}{2}\%$ simple interest, the account will grow according to the formula $A = 5000 + 5000(0.065)t$.

  a. How many years will it take for the investment to exceed $10,000? (Round to the nearest tenth of a year.)

  b. How many years will it take for the investment to exceed $15,000? (Round to the nearest tenth of a year.)

55. The revenue R for selling x fleece jackets is given by the equation $R = 49.95x$. The cost to produce x jackets is $C = 2300 + 18.50x$. Find the number of jackets that the company needs to sell to produce a profit. (*Hint:* A profit occurs when revenue exceeds cost.)

56. The revenue R for selling x mountain bikes is $R = 249.95x$. The cost to produce x bikes is $C = 56,000 + 140x$. Find the number of bikes that the company needs to sell to produce a profit.

Expanding Your Skills

For Exercises 57–60, assume $a > b$. Determine which inequality sign ($>$ or $<$) should be inserted to make a true statement. Assume $a \neq 0$ and $b \neq 0$.

57. $a + c$ _____ $b + c$,   for $c > 0$

58. $a + c$ _____ $b + c$,   for $c < 0$

59. ac _____ bc,   for $c < 0$

60. ac _____ bc,   for $c > 0$

Section 1.5 | Compound Inequalities

1. Union and Intersection of Sets

Two or more sets can be combined by the operations of union and intersection.

> ### Definition of *A* Union *B* and *A* Intersection *B*
>
> The **union** of sets A and B, denoted $A \cup B$, is the set of elements that belong to set A or to set B or to both sets A and B.
>
> The **intersection** of two sets A and B, denoted $A \cap B$, is the set of elements common to both A and B.

The concepts of the union and intersection of two sets are illustrated in Figures 1-6 and 1-7.

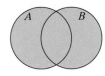

$A \cup B$
A union B
The elements in *A or B or* both

Figure 1-6

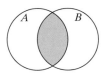

$A \cap B$
A intersection B
The elements in *A and B*

Figure 1-7

Skill Practice

Given: $A = \{r, s, t, u, v, w\}$
$B = \{s, v, w, y, z\}$ $C = \{x, y, z\}$
Find:

1. $B \cup C$
2. $A \cap B$
3. $A \cap C$

TIP: The empty set may be denoted by the symbol { } or by the symbol $\varnothing$.

Example 1   Finding the Union and Intersection of Sets

Given the sets: $A = \{a, b, c, d, e, f\}$ $B = \{a, c, e, g, i, k\}$ $C = \{g, h, i, j, k\}$

Find: **a.** $A \cup B$ **b.** $A \cap B$ **c.** $A \cap C$

Solution:

a. $A \cup B = \{a, b, c, d, e, f, g, i, k\}$ The union of A and B includes all the elements of A along with all the elements of B. Notice that the elements a, c, and e are not listed twice.

b. $A \cap B = \{a, c, e\}$ The intersection of A and B includes only those elements that are common to both sets.

c. $A \cap C = \{ \ \ \}$ (the empty set)   Because A and C share no common elements, the intersection of A and C is the empty set (also called the null set).

Answers
1. $\{s, v, w, x, y, z\}$ 2. $\{s, v, w\}$
3. $\{ \ \}$

Example 2 Finding the Union and Intersection of Sets

Given the sets: $A = \{x \mid x < 3\}$ $\quad B = \{x \mid x \geq -2\}$ $\quad C = \{x \mid x \geq 5\}$

Graph the following sets. Then express each set in interval notation.

a. $A \cap B$ $\quad\quad$ **b.** $A \cup C$

Solution:

It is helpful to visualize the graphs of individual sets on the number line before taking the union or intersection.

a. Graph of $A = \{x \mid x < 3\}$

Graph of $B = \{x \mid x \geq -2\}$

Graph of $A \cap B$ (the "overlap")

Interval notation: $[-2, 3)$

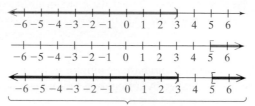

Note that the set $A \cap B$ represents the real numbers greater than or equal to -2 and less than 3. This relationship can be written more concisely as a compound inequality: $-2 \leq x < 3$. We can interpret this inequality as "x is between -2 and 3, including $x = -2$."

b. Graph of $A = \{x \mid x < 3\}$

Graph of $C = \{x \mid x \geq 5\}$

Graph of $A \cup C$

Interval notation: $(-\infty, 3) \cup [5, \infty)$ $\quad$ $A \cup C$ includes all elements from set A along with the elements from set C.

In Example 3, we find the union and intersection of sets expressed in interval notation.

Example 3 Finding the Union and Intersection of Two Intervals

Find the union or intersection as indicated. Write the answer in interval notation.

a. $(-\infty, -2) \cup [-4, 3)$ $\quad\quad$ **b.** $(-\infty, -2) \cap [-4, 3)$

Solution:

a. $(-\infty, -2) \cup [-4, 3)$ $\quad$ To find the union, graph each interval separately. The union is the collection of real numbers that lie in the first interval, the second interval, or both intervals.

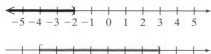

$(-\infty, -2)$

$[-4, 3)$

The union consists of all real numbers in the red interval along with the real numbers in the blue interval: $(-\infty, 3)$

The union is $(-\infty, 3)$.

Animation

b. $(-\infty, -2) \cap [-4, 3)$

$(-\infty, -2)$

$[-4, 3)$

The intersection is the "overlap" of the two intervals: $[-4, -2)$.

The intersection is $[-4, -2)$.

2. Solving Compound Inequalities: And

The solution to two inequalities joined by the word *and* is the intersection of their solution sets.

> ### Solving a Compound Inequality: And
> **Step 1** Solve and graph each inequality separately.
> **Step 2** If the inequalities are joined by the word *and*, find the intersection of the two solution sets.
> **Step 3** Express the solution set in interval notation or in set-builder notation.

As you work through the examples in this section, remember that multiplying or dividing an inequality by a negative factor reverses the direction of the inequality sign.

Skill Practice

Solve the compound inequality.

8. $5x + 2 \geq -8$ and $-4x > -24$

Example 4   **Solving a Compound Inequality: And**

Solve the compound inequality.

$$-2x < 6 \quad \text{and} \quad x + 5 \leq 7$$

Solution:

$-2x < 6 \qquad$ and $\qquad x + 5 \leq 7$ Solve each inequality separately.

$\dfrac{-2x}{-2} > \dfrac{6}{-2} \qquad$ and $\qquad x \leq 2$ Reverse the first inequality sign.

$x > -3 \qquad$ and $\qquad x \leq 2$

$\{x \,|\, x > -3\}$

$\{x \,|\, x \leq 2\}$

Take the intersection of the solution sets: $\{x \,|\, -3 < x \leq 2\}$

The solution is $\{x \,|\, -3 < x \leq 2\}$, or equivalently in interval notation, $(-3, 2]$.

Answer

8. $\{x \,|\, -2 \leq x < 6\}$; $[-2, 6)$

Example 5 Solving a Compound Inequality: And

Solve the compound inequality.

$$4.4a + 3.1 < -12.3 \quad \text{and} \quad -2.8a + 9.1 < -6.3$$

Solution:

$$4.4a + 3.1 < -12.3 \quad \text{and} \quad -2.8a + 9.1 < -6.3$$

$$4.4a < -15.4 \quad \text{and} \quad -2.8a < -15.4 \quad \text{Solve each}$$
inequality
separately.

$$\frac{4.4a}{4.4} < \frac{-15.4}{4.4} \quad \text{and} \quad \frac{-2.8a}{-2.8} > \frac{-15.4}{-2.8} \quad \text{Reverse the}$$
second
inequality sign.

$$a < -3.5 \quad \text{and} \quad a > 5.5$$

$\{a \mid a < -3.5\}$

$\{a \mid a > 5.5\}$

The intersection of the solution
sets is the empty set: $\{\ \}$

There are no real numbers that are simultaneously less than -3.5 and greater than 5.5. There is no solution.

Skill Practice

Solve the compound inequality.

9. $3.2y - 2.4 > 16.8$ and
$-4.1y \geq 8.2$

Example 6 Solving a Compound Inequality: And

Solve the compound inequality.

$$-\frac{2}{3}x \leq 6 \quad \text{and} \quad -\frac{1}{2}x < 1$$

Solution:

$$-\frac{2}{3}x \leq 6 \quad \text{and} \quad -\frac{1}{2}x < 1$$

$$-\frac{3}{2}\left(-\frac{2}{3}x\right) \geq -\frac{3}{2}(6) \quad \text{and} \quad -2\left(-\frac{1}{2}x\right) > -2(1) \quad \text{Solve each}$$
inequality
separately.

$$x \geq -9 \quad \text{and} \quad x > -2$$

$\{x \mid x \geq -9\}$

$\{x \mid x > -2\}$

Take the intersection
of the solution sets:
$\{x \mid x > -2\}$

The solution set is $\{x \mid x > -2\}$, or in interval notation, $(-2, \infty)$.

Skill Practice

Solve the compound inequality.

10. $-\frac{1}{4}z < \frac{5}{8}$ and $\frac{1}{2}z + 1 \geq 3$

Answers

9. No solution
10. $\{z \mid z \geq 4\}$; $[4, \infty)$

3. Solving Inequalities of the Form $a < x < b$

An inequality of the form $a < x < b$ is a type of **compound inequality**, one that defines two simultaneous conditions on x.

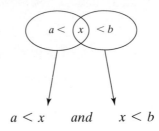

$$a < x \quad and \quad x < b$$

The solution set to the compound inequality $a < x < b$ is the *intersection* of the solution sets to the inequalities $a < x$ and $x < b$.

Skill Practice
Solve the inequality.
11. $-6 \le 2x - 5 < 1$

Example 7 Solving an Inequality of the Form $a < x < b$

Solve the inequality. $\qquad -4 < 3x + 5 \le 10$

Solution:

$$-4 < 3x + 5 \le 10$$

$-4 < 3x + 5$	and	$3x + 5 \le 10$	Set up the intersection of two inequalities.
$-9 < 3x$	and	$3x \le 5$	Solve each inequality.
$\dfrac{-9}{3} < \dfrac{3x}{3}$	and	$\dfrac{3x}{3} \le \dfrac{5}{3}$	
$-3 < x$	and	$x \le \dfrac{5}{3}$	

$$-3 < x \le \frac{5}{3} \qquad \text{Take the intersection of the solution sets.}$$

$$\frac{5}{3}$$

The solution is $\{x \mid -3 < x \le \frac{5}{3}\}$, or equivalently in interval notation, $(-3, \frac{5}{3}]$.

To solve an inequality of the form $a < x < b$, we can also work with the inequality as a "three-part" inequality and isolate x. This is demonstrated in Example 8.

Skill Practice
Solve the inequality.
12. $8 > \dfrac{t + 4}{-2} > -5$

Example 8 Solving an Inequality of the Form $a < x < b$

Solve the inequality. $\qquad 2 \ge \dfrac{p - 2}{-3} \ge -1$

Solution:

$$2 \ge \frac{p - 2}{-3} \ge -1 \qquad \text{Isolate the variable in the middle part.}$$

$$-3(2) \le -3\left(\frac{p - 2}{-3}\right) \le -3(-1) \qquad \text{Multiply all three parts by } -3. \text{ Remember to reverse the inequality signs.}$$

Answers

11. $\{x \mid -\frac{1}{2} \le x < 3\}; [-\frac{1}{2}, 3)$
12. $\{t \mid -20 < t < 6\}; (-20, 6)$

$$-6 \le p - 2 \le 3 \qquad \text{Simplify.}$$

$$-6 + 2 \le p - 2 + 2 \le 3 + 2 \qquad \text{Add 2 to all three parts to isolate } p.$$

$$-4 \le p \le 5$$

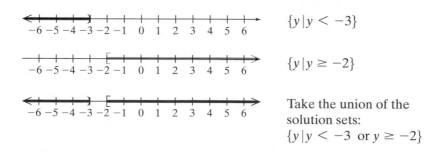

The solution is $\{p \mid -4 \le p \le 5\}$, or equivalently in interval notation $[-4, 5]$.

4. Solving Compound Inequalities: Or

In Examples 9 and 10, we solve compound inequalities that involve inequalities joined by the word *or*. In such a case, the solution to the compound inequality is the union of the solution sets of the individual inequalities.

Solving a Compound Inequality: Or

Step 1 Solve and graph each inequality separately.

Step 2 If the inequalities are joined by the word *or*, find the union of the two solution sets.

Step 3 Express the solution set in interval notation or in set-builder notation.

Example 9 Solving a Compound Inequality: Or

Solve the compound inequality.

$$-3y - 5 > 4 \quad \text{or} \quad 4 - y \le 6$$

Solution:

$$-3y - 5 > 4 \qquad \text{or} \qquad 4 - y \le 6$$

$$-3y > 9 \qquad \text{or} \qquad -y \le 2 \qquad \text{Solve each inequality separately.}$$

$$\frac{-3y}{-3} < \frac{9}{-3} \qquad \text{or} \qquad \frac{-y}{-1} \ge \frac{2}{-1} \qquad \text{Reverse the inequality signs.}$$

$$y < -3 \qquad \text{or} \qquad y \ge -2$$

$\{y \mid y < -3\}$

$\{y \mid y \ge -2\}$

Take the union of the solution sets:
$\{y \mid y < -3 \text{ or } y \ge -2\}$

The solution is $\{y \mid y < -3 \text{ or } y \ge -2\}$ or, equivalently in interval notation, $(-\infty, -3) \cup [-2, \infty)$.

Skill Practice

Solve the compound inequality.
13. $-10t - 8 \ge 12$ or
$3t - 6 > 3$

Answer
13. $\{t \mid t \le -2 \text{ or } t > 3\}$;
$(-\infty, -2] \cup (3, \infty)$

Example 10   **Solving a Compound Inequality: Or**

Solve the compound inequality.

$$4x + 3 < 16 \quad \text{or} \quad -2x < 3$$

Solution:

$$4x + 3 < 16 \quad \text{or} \quad -2x < 3$$

$$4x < 13 \quad \text{or} \quad x > -\frac{3}{2} \qquad \text{Solve each inequality separately.}$$

$$x < \frac{13}{4} \quad \text{or} \quad x > -\frac{3}{2}$$

$\{x \,|\, x < \frac{13}{4}\}$

$\{x \,|\, x > -\frac{3}{2}\}$

Take the union of the solution sets.

The union of the solution sets is $\{x \,|\, x \text{ is a real number}\}$, or equivalently, $(-\infty, \infty)$.

5. Applications of Compound Inequalities

Compound inequalities are used in many applications, as shown in Examples 11 and 12.

Example 11   **Translating Compound Inequalities**

The normal level of thyroid-stimulating hormone (TSH) for adults ranges from 0.4 to 4.8 microunits per milliliter (μU/mL), inclusive. Let x represent the amount of TSH measured in microunits per milliliter.

a. Write an inequality representing the normal range of TSH, inclusive.

b. Write a compound inequality representing abnormal TSH levels.

Solution:

a. $0.4 \le x \le 4.8$ **b.** $x < 0.4 \quad \text{or} \quad x > 4.8$

TIP:

- In mathematics, the word "between" means strictly between two values. That is, the endpoints are *excluded*.

  Example: x is between 4 and 10 $\Rightarrow$ (4, 10).

- If the word "inclusive" is added to the statement, then we *include* the endpoints.

  Example: x is between 4 and 10, inclusive $\Rightarrow$ [4, 10].

Example 12 **Translating and Solving a Compound Inequality**

The sum of a number and 4 is between -5 and 12. Find all such numbers.

Solution:

Let x represent a number.

$$-5 < x + 4 < 12 \qquad \text{Translate the inequality.}$$

$$-5 - 4 < x + 4 - 4 < 12 - 4 \qquad \text{Subtract 4 from all three parts of the inequality.}$$

$$-9 < x < 8$$

The number may be any real number between -9 and 8: $\{x \mid -9 < x < 8\}$.

Section 1.5 Practice Exercises

Study Skills Exercise

Which activities might you try when working in a study group to help you learn and understand the material?

- ☐ Quiz one another by asking one another questions.
- ☐ Practice teaching one another.
- ☐ Share and compare class notes.
- ☐ Support and encourage one another.
- ☐ Work together on exercises and sample problems.

Vocabulary and Key Concepts

1. **a.** The _____ of two sets A and B, denoted by _____, is the set of elements that belong to A or B or both A and B.

b. The _____ of two sets A and B, denoted by _____, is the set of elements common to both A and B.

c. The solution set to the compound inequality $x < c$ and $x > d$ is the (union/intersection) of the solution sets of the individual inequalities.

d. The compound inequality $a < x$ and $x < b$ can be written as the three-part inequality _____.

e. The solution set to the compound inequality $x < a$ or $x > b$ is the (union/intersection) of the solution sets of the individual inequalities.

Review Exercises

2. Write the solution set to the inequality $4 < x + 1$ in set-builder notation and in interval notation.

For Exercises 3–6, solve the linear inequality. Write the solution in interval notation.

3. $-6u + 8 > 2$ **4.** $2 - 3z \geq -4$ **5.** $-12 \leq \frac{3}{4}p$ **6.** $5 > \frac{1}{3}w$

Concept 1: Union and Intersection of Sets

7. Given: $M = \{-3, -1, 1, 3, 5\}$ and $N = \{-4, -3, -2, -1, 0\}$. **(See Example 1.)**

List the elements of the following sets:

a. $M \cap N$ **b.** $M \cup N$

8. Given: $P = \{a, b, c, d, e, f, g, h, i\}$ and $Q = \{a, e, i, o, u\}$.

List the elements of the following sets.

a. $P \cap Q$ **b.** $P \cup Q$

For Exercises 9–20, refer to the sets A, B, C, and D. Determine the union or intersection as indicated. Express the answer in interval notation, if possible. **(See Example 2.)**

$$A = \{x \mid x < -4\}, B = \{x \mid x > 2\}, C = \{x \mid x \geq -7\}, D = \{x \mid 0 \leq x < 5\}$$

9. $A \cap C$ **10.** $B \cap C$ **11.** $A \cup B$

12. $A \cup D$ **13.** $A \cap B$ **14.** $A \cap D$

15. $B \cup C$ **16.** $B \cup D$ **17.** $C \cap D$

18. $B \cap D$ **19.** $C \cup D$ **20.** $A \cup C$

For Exercises 21–26, find the intersection and union of sets as indicated. Write the answers in interval notation. **(See Example 3.)**

21. **a.** $(-2, 5) \cap [-1, \infty)$

  b. $(-2, 5) \cup [-1, \infty)$

22. **a.** $(-\infty, 4) \cap [-1, 5)$

  b. $(-\infty, 4) \cup [-1, 5)$

23. **a.** $\left(-\frac{5}{2}, 3\right) \cap \left(-1, \frac{9}{2}\right)$

  b. $\left(-\frac{5}{2}, 3\right) \cup \left(-1, \frac{9}{2}\right)$

24. **a.** $(-3.4, 1.6) \cap (-2.2, 4.1)$

  b. $(-3.4, 1.6) \cup (-2.2, 4.1)$

25. **a.** $(-4, 5] \cap (0, 2]$

  b. $(-4, 5] \cup (0, 2]$

26. **a.** $[-1, 5) \cap (0, 3)$

  b. $[-1, 5) \cup (0, 3)$

Concept 2: Solving Compound Inequalities: And

For Exercises 27–36, solve the compound inequality and graph the solution. Write the answer in interval notation. **(See Examples 4–6.)**

27. $y - 7 \geq -9$ and $y + 2 \leq 5$

28. $a + 6 > -2$ and $5a < 30$

29. $2t + 7 < 19$ and $5t + 13 > 28$

30. $5p + 2p \geq -21$ and $-9p + 3p \geq -24$

31. $2.1k - 1.1 \leq 0.6k + 1.9$ and
  $0.3k - 1.1 < -0.1k + 0.9$

32. $0.6w + 0.1 > 0.3w - 1.1$ and
  $2.3w + 1.5 \geq 0.3w + 6.5$

33. $\frac{2}{3}(2p - 1) \geq 10$ and $\frac{4}{5}(3p + 4) \geq 20$

34. $\frac{5}{2}(a + 2) < -6$ and $\frac{3}{4}(a - 2) < 1$

35. $-2 < -x - 12$ and $-14 < 5(x - 3) + 6x$

36. $-8 \geq -3y - 2$ and $3(y - 7) + 16 > 4y$

Concept 3: Solving Inequalities of the Form $a < x < b$

37. Write $-4 \leq t < \frac{3}{4}$ as two separate inequalities.

38. Write $-2.8 < y \leq 15$ as two separate inequalities.

39. Explain why $6 < x < 2$ has no solution.

40. Explain why $4 < t < 1$ has no solution.

41. Explain why $-5 > y > -2$ has no solution.

42. Explain why $-3 > w > -1$ has no solution.

For Exercises 43–54, solve the compound inequality and graph the solution set. Write the answer in interval notation. **(See Examples 7 and 8.)**

43. $0 \leq 2b - 5 < 9$

44. $-6 < 3k - 9 \leq 0$

45. $-1 < \frac{a}{6} \leq 1$

46. $-3 \leq \frac{1}{2}x < 0$

47. $-\frac{2}{3} < \frac{y - 4}{-6} < \frac{1}{3}$

48. $\frac{1}{3} > \frac{t - 4}{-3} > -2$

49. $5 \leq -3x - 2 \leq 8$

50. $-1 < -2x + 4 \leq 5$

51. $12 > 6x + 3 \geq 0$

52. $-4 \geq 2x - 5 > -7$

53. $-0.2 < 2.6 + 7t < 4$

54. $-1.5 < 0.1x \leq 8.1$

Concept 4: Solving Compound Inequalities: Or

For Exercises 55–64, solve the compound inequality and graph the solution set. Write the answer in interval notation. **(See Examples 9 and 10.)**

55. $2y - 1 \geq 3$ or $y < -2$

56. $x < 0$ or $3x + 1 \geq 7$

57. $1 > 6z - 8$ or $8z - 6 \leq 10$

58. $22 > 4t - 10$ or $7 > 2t - 5$

59. $5(x - 1) \geq -5$ or $5 - x \leq 11$

60. $-p + 7 \geq 10$ or $3(p - 1) \leq 12$

61. $\frac{5}{3}v \leq 5$ or $-v - 6 < 1$

62. $\frac{3}{8}u + 1 > 0$ or $-2u \geq -4$

63. $0.5w + 5 < 2.5w - 4$ or $0.3w \leq -0.1w - 1.6$

64. $1.25a + 3 \leq 0.5a - 6$ or $2.5a - 1 \geq 9 - 1.5a$

Mixed Exercises

For Exercises 65–74, solve the compound inequality. Write the answer in interval notation.

65. a. $3x - 5 < 19$ and $-2x + 3 < 23$

 b. $3x - 5 < 19$ or $-2x + 3 < 23$

66. a. $0.5(6x + 8) > 0.8x - 7$ and $4(x + 1) < 7.2$

 b. $0.5(6x + 8) > 0.8x - 7$ or $4(x + 1) < 7.2$

67. a. $8x - 4 \geq 6.4$ or $0.3(x + 6) \leq -0.6$

 b. $8x - 4 \geq 6.4$ and $0.3(x + 6) \leq -0.6$

68. a. $-2r + 4 \leq -8$ or $3r + 5 \leq 8$

 b. $-2r + 4 \leq -8$ and $3r + 5 \leq 8$

69. $-4 \leq \dfrac{2 - 4x}{3} < 8$

70. $-1 < \dfrac{3 - x}{2} \leq 0$

71. $5 \geq -4(t - 3) + 3t$ or
$6 < 12t + 8(4 - t)$

72. $3 > -(w - 3) + 4w$ or
$-5 \geq -3(w - 5) + 6w$

73. $\dfrac{-x + 3}{2} > \dfrac{4 + x}{5}$ or $\dfrac{1 - x}{4} > \dfrac{2 - x}{3}$

74. $\dfrac{y - 7}{-3} < \dfrac{1}{4}$ or $\dfrac{y + 1}{-2} > -\dfrac{1}{3}$

Concept 5: Applications of Compound Inequalities

75. The normal number of white blood cells for human blood is between 4800 and 10,800 cells per cubic millimeter, inclusive. Let x represent the number of white blood cells per cubic millimeter. **(See Example 11.)**

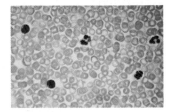

 a. Write an inequality representing the normal range of white blood cells per cubic millimeter.

 b. Write a compound inequality representing abnormal levels of white blood cells per cubic millimeter.

76. Normal hemoglobin levels in human blood for adult males are between 13 and 16 grams per deciliter (g/dL), inclusive. Let x represent the level of hemoglobin measured in grams per deciliter.

 a. Write an inequality representing normal hemoglobin levels for adult males.

 b. Write a compound inequality representing abnormal levels of hemoglobin for adult males.

77. A polling company estimates that a certain candidate running for office will receive between 44% and 48% of the votes. Let x represent the percentage of votes for this candidate.

 a. Write a strict inequality representing the expected percentage of votes for this candidate.

 b. Write a compound inequality representing the percentage of votes that would fall outside the polling company's prediction.

78. A machine is calibrated to cut a piece of wood between 2.4 in. thick and 2.6 in. thick. Let x represent the thickness of the wood after it is cut.

 a. Write a strict inequality representing the expected range of thickness of the wood after it has been cut.

 b. Write a compound inequality representing the thickness of wood that would fall outside the normal range for this machine.

79. Twice a number is between -3 and 12. Find all such numbers. **(See Example 12.)**

80. The difference of a number and 6 is between 0 and 8. Find all such numbers.

81. One plus twice a number is either greater than 5 or less than -1. Find all such numbers.

82. One-third of a number is either less than -2 or greater than 5. Find all such numbers.

83. Amy knows from reading her syllabus in intermediate algebra that the average of her chapter tests accounts for 80% (0.8) of her overall course grade. She also knows that the final exam counts as 20% (0.2) of her grade.

a. Suppose that the average of Amy's chapter tests is 92%. Determine the range of grades that she would need on her final exam to get an "A" in the class. (Assume that a grade of "A" is obtained if Amy's overall average is 90% or better.)

b. Determine the range of grades that Amy would need on her final exam to get a "B" in the class. (Assume that a grade of "B" is obtained if Amy's overall average is at least 80% but less than 90%.)

84. Robert knows from reading his syllabus in intermediate algebra that the average of his chapter tests accounts for 60% (0.6) of his overall course grade. He also knows that the final exam counts as 40% (0.4) of his grade.

a. Suppose that the average of Robert's chapter tests is 89%. Determine the range of grades that he would need on his final exam to get an "A" in the class. (Assume that a grade of "A" is obtained if Robert's overall average is 90% or better.)

b. Determine the range of grades that Robert would need on his final exam to get a "B" in the class. (Assume that a grade of "B" is obtained if Robert's overall average is at least 80% but less than 90%.)

85. The average high and low temperatures for Vancouver, British Columbia, in January are 5.6°C and 0°C, respectively. The formula relating Celsius temperatures to Fahrenheit temperatures is given by $C = \frac{5}{9}(F - 32)$. Convert the inequality $0.0° \le C \le 5.6°$ to an equivalent inequality using Fahrenheit temperatures.

86. For a day in July, the temperatures in Austin, Texas, ranged from 20°C to 29°C. The formula relating Celsius temperatures to Fahrenheit temperatures is given by $C = \frac{5}{9}(F - 32)$. Convert the inequality $20° \le C \le 29°$ to an equivalent inequality using Fahrenheit temperatures.

Absolute Value Equations

Section 1.6

1. Solving Absolute Value Equations

An equation of the form $|x| = a$ is called an **absolute value equation**. For example, consider the equation $|x| = 4$. From the definition of absolute value, the solutions are found by solving the equations $x = 4$ and $-x = 4$. This gives the equivalent equations $x = 4$ and $x = -4$.

Also recall from Section R.3 that the absolute value of a number is its distance from zero on the number line. Therefore, geometrically, the solutions to the equation $|x| = 4$ are the values of x that are 4 units from zero on the number line (Figure 1-8).

Concepts

1. Solving Absolute Value Equations

2. Solving Equations Containing Two Absolute Values

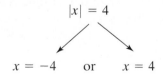

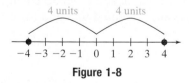

Figure 1-8

Concept Connections

1. Why does the equation $|x| = -2$ have no solution?

Solving Absolute Value Equations of the Form $|x| = a$

If a is a real number, then

- If $a \ge 0$, the solutions to the equation $|x| = a$ are given by $x = a$ and $x = -a$.

- If $a < 0$, there is no solution to the equation $|x| = a$.

Answer

1. The absolute value represents the distance from zero. This cannot be negative.

To solve an absolute value equation of the form $|x| = a$ $(a \geq 0)$, rewrite the equation as $x = a$ or $x = -a$.

Skill Practice

Solve the absolute value equations.
2. $|y| = 7$
3. $|v| + 6 = 10$
4. $|w| = 0$
5. $|z| = -12$

Example 1 **Solving Absolute Value Equations**

Solve the absolute value equations.

a. $|x| = 5$ **b.** $|w| - 2 = 12$ **c.** $|p| = 0$ **d.** $|x| = -6$

Solution:

a. $|x| = 5$ The equation is in the form $|x| = a$, where $a = 5$.

$x = 5$ or $x = -5$ Rewrite the equation as $x = a$ or $x = -a$.

The solutions are 5 and -5.

b. $|w| - 2 = 12$ Isolate the absolute value to write the equation in the form $|x| = a$.

$|w| = 14$

$w = 14$ or $w = -14$ Rewrite the equation as $w = a$ or $w = -a$.

The solutions are 14 and -14.

c. $|p| = 0$

$p = 0$ or $p = -0$ Rewrite as two equations. Notice that the second equation $p = -0$ is the same as the first equation. Intuitively, $p = 0$ is the only number whose absolute value equals 0.

The solution is 0.

d. $|x| = -6$ The equation is of the form $|x| = a$, but a is negative. There is no number whose absolute value is negative.

No solution

We have solved absolute value equations of the form $|x| = a$. Notice that x can represent any algebraic quantity. For example, to solve the equation $|2w - 3| = 5$, we still rewrite the absolute value equation as two equations. In this case, we set the quantity $2w - 3$ equal to 5 and to -5, respectively.

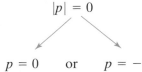

$$|2w - 3| = 5$$

$2w - 3 = 5$ or $2w - 3 = -5$

Solving an Absolute Value Equation

Step 1 Isolate the absolute value. That is, write the equation in the form $|x| = a$, where a is a real number.

Step 2 If $a < 0$, there is no solution.

Step 3 Otherwise, if $a \geq 0$, rewrite the absolute value equation as $x = a$ or $x = -a$.

Step 4 Solve the individual equations from step 3.

Step 5 Check the answers in the original absolute value equation.

Example 2 Solving an Absolute Value Equation

Solve the equation. $|2w - 3| = 5$

Solution:

$|2w - 3| = 5$ The equation is already in the form $|x| = a$, where $x = 2w - 3$.

$2w - 3 = 5$ or $2w - 3 = -5$ Rewrite as two equations.

$2w = 8$ or $2w = -2$ Solve each equation.

$w = 4$ or $w = -1$

Check: $w = 4$ Check: $w = -1$ Check the values in the original equation.

$|2w - 3| = 5$ $|2w - 3| = 5$

$|2(4) - 3| \overset{?}{=} 5$ $|2(-1) - 3| \overset{?}{=} 5$

$|8 - 3| \overset{?}{=} 5$ $|-2 - 3| \overset{?}{=} 5$

$|5| \overset{?}{=} 5$ ✔ $|-5| \overset{?}{=} 5$ ✔

The solutions are 4 and -1.

Skill Practice

Solve the equation.

6. $|4x + 1| = 9$

Example 3 Solving an Absolute Value Equation

Solve the equation. $|2c - 5| + 6 = 2$

Solution:

$|2c - 5| + 6 = 2$

$|2c - 5| = -4$ Isolate the absolute value. The equation is in the form $|x| = a$, where $x = 2c - 5$ and $a = -4$. Because $a < 0$, there is no solution.

No solution   There are no numbers c that will make an absolute value equal to a negative number.

Skill Practice

Solve the equation.

7. $|3z + 10| + 3 = 1$

Avoiding Mistakes

Always isolate the absolute value first. Otherwise you will get answers that do not check.

Answers

6. $2, -\dfrac{5}{2}$ **7.** No solution

Example 4 Solving an Absolute Value Equation

Solve the equation. $-2\left|\dfrac{2}{5}p + 3\right| - 7 = -19$

Solution:

$$-2\left|\frac{2}{5}p + 3\right| - 7 = -19$$

$$-2\left|\frac{2}{5}p + 3\right| = -12 \qquad \text{Isolate the absolute value.}$$

$$\frac{-2\left|\frac{2}{5}p + 3\right|}{-2} = \frac{-12}{-2} \qquad \text{Divide both sides by } -2.$$

$$\left|\frac{2}{5}p + 3\right| = 6$$

$$\frac{2}{5}p + 3 = 6 \qquad \text{or} \qquad \frac{2}{5}p + 3 = -6 \qquad \text{Rewrite as two equations.}$$

$$2p + 15 = 30 \qquad \text{or} \qquad 2p + 15 = -30 \qquad \text{Multiply by 5 to clear fractions.}$$

$$2p = 15 \qquad \text{or} \qquad 2p = -45$$

$$p = \frac{15}{2} \qquad \text{or} \qquad p = -\frac{45}{2} \qquad \begin{array}{l}\text{Both values check in the}\\ \text{original equation.}\end{array}$$

The solutions are $\dfrac{15}{2}$ and $-\dfrac{45}{2}$.

Example 5 Solving an Absolute Value Equation

Solve the equation. $6.9 = |4.1 - p| + 6.9$

Solution:

$$6.9 = |4.1 - p| + 6.9$$

$$|4.1 - p| + 6.9 = 6.9 \qquad \begin{array}{l}\text{First write the absolute value on}\\ \text{the left. Then subtract 6.9 from}\\ \text{both sides to write the equation}\\ \text{in the form } |x| = a.\end{array}$$

$$|4.1 - p| = 0 \qquad \text{Isolate the absolute value.}$$

$$4.1 - p = 0 \qquad \text{or} \qquad 4.1 - p = -0 \qquad \begin{array}{l}\text{Rewrite as two equations. Notice}\\ \text{that the equations are the same.}\end{array}$$

$$-p = -4.1 \qquad \text{Subtract 4.1 from both sides.}$$

$$p = 4.1 \qquad \underline{\text{Check:}}\ p = 4.1$$

$$|4.1 - p| + 6.9 = 6.9$$

$$|4.1 - 4.1| + 6.9 \overset{?}{=} 6.9$$

$$|0| + 6.9 \overset{?}{=} 6.9$$

The solution is 4.1. $\qquad\qquad 6.9 \overset{?}{=} 6.9 \ ✔$

2. Solving Equations Containing Two Absolute Values

Some equations have two absolute values such as $|x| = |y|$. If two quantities have the same absolute value, then the quantities are equal or the quantities are opposites.

Equality of Absolute Values

$$|x| = |y| \text{ implies that } x = y \text{ or } x = -y.$$

Example 6 Solving an Equation Having Two Absolute Values

Solve the equation. $\quad |2w - 3| = |5w + 1|$

Solution:

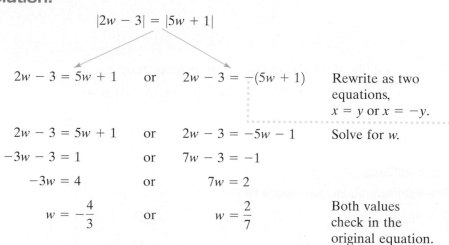

$$|2w - 3| = |5w + 1|$$

$2w - 3 = 5w + 1$	or	$2w - 3 = -(5w + 1)$

Rewrite as two equations, $x = y$ or $x = -y$.

$2w - 3 = 5w + 1$	or	$2w - 3 = -5w - 1$
$-3w - 3 = 1$	or	$7w - 3 = -1$
$-3w = 4$	or	$7w = 2$
$w = -\dfrac{4}{3}$	or	$w = \dfrac{2}{7}$

Solve for w.

Both values check in the original equation.

The solutions are $-\dfrac{4}{3}$ and $\dfrac{2}{7}$.

Skill Practice

Solve the equation.

10. $|3 - 2x| = |3x - 1|$

Avoiding Mistakes

To take the opposite of the quantity $5w + 1$, use parentheses and apply the distributive property.

Example 7 Solving an Equation Having Two Absolute Values

Solve the equation. $\quad |x - 4| = |x + 8|$

Solution:

$$|x - 4| = |x + 8|$$

$x - 4 = x + 8$	or	$x - 4 = -(x + 8)$

Rewrite as two equations, $x = y$ or $x = -y$.

$-4 = 8$	or	$x - 4 = -x - 8$
$\uparrow$		$2x - 4 = -8$
contradiction		$2x = -4$
		$x = -2$

Solve for x.

$x = -2$ checks in the original equation.

The solution is -2.

Skill Practice

Solve the equation.

11. $|4t + 3| = |4t - 5|$

Answers

10. $\dfrac{4}{5}, -2$ **11.** $\dfrac{1}{4}$

Section 1.6   Practice Exercises

Study Skills Exercise

One way to know that you really understand a concept is to explain it to someone else. In your own words, explain how to solve an absolute value equation.

Vocabulary and Key Concepts

1. a. An _____ value equation is an equation of the form $|x| = a$. If a is a positive real number then the solutions are _____.

   b. What is the first step to solve the absolute value equation $|x| + 5 = 7$?

   c. The absolute value equation $|x| = |y|$ implies that $x =$ _____ or $x =$ _____.

   d. There is _____ solution to the equation $|x + 4| = -2$. The solution to the equation $|x + 4| = 0$ is _____.

Review Exercises

For Exercises 2–6, solve the compound inequalities. Write the answers in interval notation.

2. $3(a + 2) - 6 > 2$ and $-2(a - 3) + 14 > -3$ **3.** $3x - 5 \geq 7x + 3$ or $2x - 1 \leq 4x - 5$

4. $-5 < 2 - x \leq 6$ **5.** $5 \geq \dfrac{x - 4}{-2} > -3$ **6.** $4 \leq \dfrac{1}{3}x - 2 < 7$

Concept 1: Solving Absolute Value Equations

For Exercises 7–42, solve the absolute value equations. **(See Examples 1–5.)**

7. $|p| = 7$ **8.** $|q| = 10$ **9.** $|x| + 5 = 11$ **10.** $|x| - 3 = 20$

11. $|y| + 8 = 5$ **12.** $|x| + 12 = 6$ **13.** $|w| - 3 = -1$ **14.** $|z| - 14 = -10$

15. $|y| = \sqrt{2}$ **16.** $|y| = \sqrt{5}$ **17.** $|w| - 3 = -5$ **18.** $|w| + 4 = -8$

19. $|3q| = 0$ **20.** $|4p| = 0$ **21.** $|3x - 4| = 8$ **22.** $|4x + 1| = 6$

23. $5 = |2x - 4|$ **24.** $10 = |3x + 7|$ **25.** $\left|\dfrac{7z}{3} - \dfrac{1}{3}\right| + 3 = 6$ **26.** $\left|\dfrac{w}{2} + \dfrac{3}{2}\right| - 2 = 7$

27. $|0.2x - 3.5| = -5.6$ **28.** $|1.81 + 2x| = -2.2$ **29.** $1 = -4 + \left|2 - \dfrac{1}{4}w\right|$ **30.** $-12 = -6 - |6 - 2x|$

31. $10 = 4 + |2y + 1|$ **32.** $-1 = -|5x + 7|$ **33.** $-2|3b - 7| - 9 = -9$ **34.** $-3|5x + 1| + 4 = 4$

35. $-2|x + 3| = 5$ **36.** $-3|x - 5| = 7$ **37.** $0 = |6x - 9|$ **38.** $7 = |4k - 6| + 7$

39. $\left|-\dfrac{1}{5} - \dfrac{1}{2}k\right| = \dfrac{9}{5}$ **40.** $\left|-\dfrac{1}{6} - \dfrac{2}{9}h\right| = \dfrac{1}{2}$ **41.** $-3|2 - 6x| + 5 = -10$ **42.** $5|1 - 2x| - 7 = 3$

Concept 2: Solving Equations Containing Two Absolute Values

For Exercises 43–60, solve the absolute value equations. **(See Examples 6 and 7.)**

43. $|4x - 2| = |-8|$ **44.** $|3x + 5| = |-5|$ **45.** $|4w + 3| = |2w - 5|$

46. $|3y + 1| = |2y - 7|$ **47.** $|2y + 5| = |7 - 2y|$ **48.** $|9a + 5| = |9a - 1|$

49. $\left|\dfrac{4w - 1}{6}\right| = \left|\dfrac{2w}{3} + \dfrac{1}{4}\right|$

50. $\left|\dfrac{6p + 3}{8}\right| = \left|\dfrac{3}{4}p - 2\right|$

51. $|x + 2| = |-x - 2|$

52. $|2y - 3| = |-2y + 3|$

53. $|3.5m - 1.2| = |8.5m + 6|$

54. $|11.2n + 9| = |7.2n - 2.1|$

55. $|4x - 3| = -|2x - 1|$

56. $-|3 - 6y| = |8 - 2y|$

57. $|8 - 7w| = |7w - 8|$

58. $|4 - 3z| = |3z - 4|$

59. $|x + 2| + |x - 4| = 0$

60. $|t + 6| + |t - 1| = 0$

Expanding Your Skills

61. Write an absolute value equation whose solution is all real numbers 6 units from zero on the number line.

62. Write an absolute value equation whose solution is all real numbers $\frac{7}{2}$ units from zero on the number line.

63. Write an absolute value equation whose solution is all real numbers $\frac{4}{3}$ units from zero on the number line.

64. Write an absolute value equation whose solution is all real numbers 9 units from zero on the number line.

Absolute Value Inequalities

Section 1.7

1. Solving Absolute Value Inequalities by Definition

In Section 1.6, we studied absolute value equations in the form $|x| = a$. In this section, we will solve absolute value *inequalities*. An inequality in any of the forms $|x| < a$, $|x| \leq a$, $|x| > a$, or $|x| \geq a$ is called an **absolute value inequality**.

Recall that an absolute value represents distance from zero on the real number line. Consider the following absolute value equation and inequalities.

Concepts

1. Solving Absolute Value Inequalities by Definition
2. Solving Absolute Value Inequalities by the Test Point Method
3. Translating to an Absolute Value Expression

1. $|x| = 3$
   $x = 3$ or $x = -3$

Solution:

The set of all points 3 units from zero on the number line

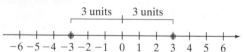

2. $|x| > 3$
   $x < -3$ or $x > 3$

Solution:

The set of all points more than 3 units from zero

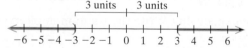

3. $|x| < 3$

   $-3 < x < 3$

Solution:

The set of all points less than 3 units from zero

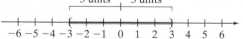

Solving Absolute Value Equations and Inequalities

Let a be a real number such that $a > 0$. Then

Equation/ Inequality	Solution (Equivalent Form)	Graph
$|x| = a$	$x = -a$ or $x = a$	
$|x| > a$	$x < -a$ or $x > a$	
$|x| < a$	$-a < x < a$	

To solve an absolute value inequality, first isolate the absolute value and then rewrite the absolute value inequality in its equivalent form.

Skill Practice

Solve the inequality. Write the solution in interval notation.

1. $|2t + 5| + 2 \le 11$

Example 1 Solving an Absolute Value Inequality

Solve the inequality. Write the solution in interval notation.
$$|3w + 1| - 4 < 7$$

Solution:

$$|3w + 1| - 4 < 7$$

$$|3w + 1| < 11 \longleftarrow \text{Isolate the absolute value first.}$$

The inequality is in the form $|x| < a$, where $x = 3w + 1$.

$$-11 < 3w + 1 < 11 \qquad \text{Rewrite in the equivalent form } -a < x < a.$$

$$-12 < 3w < 10 \qquad \text{Solve for } w.$$

$$-4 < w < \frac{10}{3}$$

The solution is $\left(-4, \frac{10}{3}\right)$.

TIP: Recall that a strict inequality (using the symbols > and <) will have parentheses at the endpoints of the interval form of the solution.

Example 2 Solving an Absolute Value Inequality

Solve the inequality. Write the solution in interval notation.

$$3 \le 1 + \left|\frac{1}{2}t - 5\right|$$

Skill Practice

Solve the inequality. Write the solution in interval notation.

2. $5 < 1 + \left|\frac{1}{3}c - 1\right|$

Solution:

$$3 \le 1 + \left|\frac{1}{2}t - 5\right|$$

$$1 + \left|\frac{1}{2}t - 5\right| \ge 3 \qquad \text{Write the inequality with the absolute value on the left.}$$

$$\left|\frac{1}{2}t - 5\right| \ge 2 \qquad \text{Isolate the absolute value.}$$

The inequality is in the form $|x| \ge a$, where $x = \frac{1}{2}t - 5$.

Answers

1. $[-7, 2]$ **2.** $(-\infty, -9) \cup (15, \infty)$

$\frac{1}{2}t - 5 \leq -2 \qquad$ or $\qquad \frac{1}{2}t - 5 \geq 2 \qquad$ Rewrite in the equivalent form $x \leq -a$ or $x \geq a$.

$\frac{1}{2}t \leq 3 \qquad$ or $\qquad \frac{1}{2}t \geq 7 \qquad$ Solve the compound inequality.

$2\left(\frac{1}{2}t\right) \leq 2(3) \qquad$ or $\qquad 2\left(\frac{1}{2}t\right) \geq 2(7) \qquad$ Clear fractions.

$t \leq 6 \qquad$ or $\qquad t \geq 14$

The solution is $(-\infty, 6] \cup [14, \infty)$.

TIP: It is generally easier to solve an absolute value inequality if the absolute value appears on the left-hand side of the inequality.

By definition, the absolute value of a real number will always be nonnegative. Therefore, the absolute value of any expression will always be greater than a negative number. Similarly, an absolute value can never be less than a negative number. Let a represent a positive real number. Then

- The solution to the inequality $|x| > -a$ is all real numbers, $(-\infty, \infty)$.
- There is no solution to the inequality $|x| < -a$.

Example 3 Solving Absolute Value Inequalities

Solve the inequalities.

  a. $|3d - 5| + 7 < 4$ **b.** $|3d - 5| + 7 > 4$

Solution:

a. $|3d - 5| + 7 < 4$

    $|3d - 5| < -3$ Isolate the absolute value. An absolute value expression cannot be less than a negative number. Therefore, there is no solution.

  No solution

b. $|3d - 5| + 7 > 4$

    $|3d - 5| > -3$ Isolate the absolute value. The inequality is in the form $|x| > a$, where a is negative. An absolute value of any real number is greater than a negative number. Therefore, the

  All real numbers, $(-\infty, \infty)$ solution is all real numbers.

Example 4 Solving Absolute Value Inequalities

Solve the inequalities.

  a. $|4x + 2| \geq 0$ **b.** $|4x + 2| > 0$ **c.** $|4x + 2| \leq 0$

Solution:

a. $|4x + 2| \geq 0 \longleftarrow$ The absolute value is already isolated.

  The absolute value of any real number is nonnegative. Therefore, the solution is all real numbers, $(-\infty, \infty)$.

Skill Practice

Solve the inequalities.

  3. $|4p + 2| + 6 < 2$
  4. $|4p + 2| + 6 > 2$

Skill Practice

Solve the inequalities.

  5. $|3x - 1| \geq 0$
  6. $|3x - 1| > 0$
  7. $|3x - 1| \leq 0$

Answers

3. No solution
4. All real numbers; $(-\infty, \infty)$
5. $(-\infty, \infty)$
6. $\left\{x \mid x \neq \frac{1}{3}\right\}$ or $\left(-\infty, \frac{1}{3}\right) \cup \left(\frac{1}{3}, \infty\right)$
7. $\frac{1}{3}$

b. $|4x + 2| > 0$

An absolute value will be greater than zero at all points *except where it is equal to zero*. That is, the point(s) for which $|4x + 2| = 0$ must be excluded from the solution set.

$|4x + 2| = 0$

$\quad 4x + 2 = 0 \qquad \text{or} \qquad 4x + 2 = -0$ The second equation is the same as the first.

$\qquad 4x = -2$

$\qquad x = -\dfrac{1}{2}$ Therefore, exclude $x = -\frac{1}{2}$ from the solution.

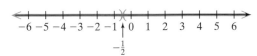

The solution is $\{x \mid x \neq -\frac{1}{2}\}$ or equivalently in interval notation, $(-\infty, -\frac{1}{2}) \cup (-\frac{1}{2}, \infty)$.

c. $|4x + 2| \leq 0$

An absolute value of a number cannot be less than zero. However, it can be *equal* to zero. Therefore, the only solutions to this inequality are the solutions to the related equation:

$|4x + 2| = 0$ From part (b), we see that the solution is $-\dfrac{1}{2}$.

2. Solving Absolute Value Inequalities by the Test Point Method

For the problems in Examples 1 and 2, the absolute value inequality was converted to an equivalent compound inequality. However, sometimes students have difficulty setting up the appropriate compound inequality. To avoid this problem, you may want to use the test point method to solve absolute value inequalities.

Solving Inequalities by Using the Test Point Method

Step 1 Find the boundary points of the inequality. (Boundary points are the real solutions to the related equation and points where the inequality is undefined.)

Step 2 Plot the boundary points on the number line. This divides the number line into intervals.

Step 3 Select a test point from each interval and substitute it into the original inequality.

- If a test point makes the original inequality true, then that interval is part of the solution set.

Step 4 Test the boundary points in the original inequality.

- If the original inequality is strict ($<$ or $>$), do not include the boundary points in the solution set.
- If the original inequality is defined using $\leq$ or $\geq$, then include the boundary points that are well defined within the inequality.

Note: Any boundary point that makes an expression within the inequality undefined must *always* be excluded from the solution set.

To demonstrate the use of the test point method, we will repeat the absolute value inequalities from Examples 1 and 2. Notice that regardless of the method used, the absolute value is always isolated *first* before any further action is taken.

Example 5 Solving an Absolute Value Inequality by the Test Point Method

Solve the inequality by using the test point method. $|3w + 1| - 4 < 7$

Solution:

$|3w + 1| - 4 < 7$

$|3w + 1| < 11$ ⟵ —————————— Isolate the absolute value.

$|3w + 1| = 11$ **Step 1:** Solve the related equation.

$3w + 1 = 11$ or $3w + 1 = -11$ Write as two equations.

$3w = 10$ or $3w = -12$

$w = \dfrac{10}{3}$ or $w = -4$ These are the only boundary points.

Step 2: Plot the boundary points.

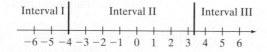

Step 3: Select a test point from each interval.

Test $w = -5$:

$|3(-5) + 1| - 4 \overset{?}{<} 7$

$|-14| - 4 \overset{?}{<} 7$

$14 - 4 \overset{?}{<} 7$

$10 \overset{?}{<} 7$ False

Test $w = 0$:

$|3(0) + 1| - 4 \overset{?}{<} 7$

$|1| - 4 \overset{?}{<} 7$

$1 - 4 \overset{?}{<} 7$

$-3 \overset{?}{<} 7$ True

Test $w = 4$:

$|3(4) + 1| - 4 \overset{?}{<} 7$

$|13| - 4 \overset{?}{<} 7$

$13 - 4 \overset{?}{<} 7$

$9 \overset{?}{<} 7$ False

Step 4: Because the original inequality is a strict inequality, the boundary points (where equality occurs) are not included.

The solution is $\left(-4, \dfrac{10}{3}\right)$.

Skill Practice

Solve the inequality.

9. $\left|\dfrac{1}{2}c + 4\right| + 1 > 6$

Example 6   Solving an Absolute Value Inequality by the Test Point Method

Solve the inequality by using the test point method. $3 \le 1 + \left|\dfrac{1}{2}t - 5\right|$

Solution:

$$3 \le 1 + \left|\dfrac{1}{2}t - 5\right|$$

$$1 + \left|\dfrac{1}{2}t - 5\right| \ge 3 \qquad\qquad \text{Write the inequality with the absolute value on the left.}$$

$$\left|\dfrac{1}{2}t - 5\right| \ge 2 \longleftarrow\qquad\qquad \text{Isolate the absolute value.}$$

$$\left|\dfrac{1}{2}t - 5\right| = 2 \qquad\qquad \textbf{Step 1:} \text{ Solve the related equation.}$$

$$\dfrac{1}{2}t - 5 = 2 \qquad \text{or} \qquad \dfrac{1}{2}t - 5 = -2 \qquad\qquad \text{Write as two equations.}$$

$$\dfrac{1}{2}t = 7 \qquad \text{or} \qquad \dfrac{1}{2}t = 3$$

$$t = 14 \qquad \text{or} \qquad t = 6 \qquad\qquad \text{These are the boundary points.}$$

Interval I | Interval II | Interval III   **Step 2:** Plot the boundary points.

        6   14

Step 3: Select a test point from each interval.

Test $t = 0$:

$$3 \overset{?}{\le} 1 + \left|\dfrac{1}{2}(0) - 5\right|$$

$$3 \overset{?}{\le} 1 + |0 - 5|$$

$$3 \overset{?}{\le} 1 + |-5|$$

$$3 \overset{?}{\le} 6 \text{ True}$$

Test $t = 10$:

$$3 \overset{?}{\le} 1 + \left|\dfrac{1}{2}(10) - 5\right|$$

$$3 \overset{?}{\le} 1 + |5 - 5|$$

$$3 \overset{?}{\le} 1 + |0|$$

$$3 \overset{?}{\le} 1 \text{ False}$$

Test $t = 16$:

$$3 \overset{?}{\le} 1 + \left|\dfrac{1}{2}(16) - 5\right|$$

$$3 \overset{?}{\le} 1 + |8 - 5|$$

$$3 \overset{?}{\le} 1 + |3|$$

$$3 \overset{?}{\le} 4 \text{ True}$$

Step 4: The original inequality uses the sign $\ge$. Therefore, the boundary points (where equality occurs) must be part of the solution set.

   True   False   True

      6   14

The solution is $(-\infty, 6] \cup [14, \infty)$.

Answer

9. $(-\infty, -18) \cup (2, \infty)$

3. Translating to an Absolute Value Expression

Absolute value expressions can be used to describe distances. The distance between c and d is given by $|c - d|$. For example, the distance between -2 and 3 on the number line is $|(-2) - 3| = |-5| = 5$ as expected.

Example 7 Expressing Distances with Absolute Value

Write an absolute value inequality to represent the following phrases.

a. All real numbers x, whose distance from zero is greater than 5 units

b. All real numbers x, whose distance from -7 is less than 3 units

Solution:

a. All real numbers x, whose distance from zero is greater than 5 units

$|x - 0| > 5$ or simply $|x| > 5$

b. All real numbers x, whose distance from -7 is less than 3 units

$|x - (-7)| < 3$ or simply $|x + 7| < 3$

Skill Practice

Write an absolute value inequality to represent the following phrases.

10. All real numbers whose distance from zero is greater than 10 units

11. All real numbers whose distance from 4 is less than 6 units

Absolute value expressions can also be used to describe boundaries for measurement error.

Example 8 Expressing Measurement Error with Absolute Value

Latoya measured a certain compound on a scale in the chemistry lab at school. She measured 8 g of the compound, but the scale is only accurate to ± 0.1 g. Write an absolute value inequality to express an interval for the true mass, x, of the compound she measured.

Solution:

Because the scale is only accurate to ± 0.1 g, the true mass, x, of the compound may deviate by as much as 0.1 g above or below 8 g. This may be expressed as an absolute value inequality:

$|x - 8.0| \leq 0.1$ or equivalently $7.9 \leq x \leq 8.1$

Skill Practice

12. Vonzell molded a piece of metal in her machine shop. She measured the thickness at 12 mm. Her machine is accurate to ± 0.05 mm. Write an absolute value inequality to express an interval for the true measurement of the thickness, t, of the metal.

TIP: The symbol $\pm$ means "plus or minus." The expression ± 0.1 is a short way of writing 0.1 or -0.1.

Answers

10. $|x| > 10$
11. $|x - 4| < 6$
12. $|t - 12| \leq 0.05$

Section 1.7 Practice Exercises

Study Skills Exercise

When you take a test, go through the test and do all the problems that you know first. Then go back and work on the problems that were more difficult. Give yourself a time limit for how much time you spend on each problem (maybe 3 to 5 min the first time through). Circle the importance of each statement.

	Not important	Somewhat important	Very important
a. Read through the entire test first.	1	2	3
b. If time allows, go back and check each problem.	1	2	3
c. Write out all steps instead of doing the work in your head.	1	2	3

Vocabulary and Key Concepts

1. a. If a is a positive real number, then the inequality $|x| < a$ is equivalent to _____ $< x <$ _____.

   b. If a is a positive real number, then the inequality $|x| > a$ is equivalent to $x <$ _____ or x _____ a.

   c. There is _____ solution to the inequality $|x + 2| < -6$, whereas the solution to the inequality $|x + 2| > -6$ is _____.

   d. The solution to the inequality $|x + 4| \leq 0$ (includes/excludes) -4, whereas the solution to the inequality $|x + 4| < 0$ (includes/excludes) -4.

Review Exercises

2. Solve the equation and inequalities. Write the solution sets to the inequalities in interval notation.

   a. $-4x - 3 = 5$ b. $-4x - 3 < 5$ c. $-4x - 3 > 5$

For Exercises 3 and 4, solve the equations.

3. $2 = |5 - 7x| + 1$

4. $|3x - 12| + 4 = 6 - 2$

For Exercises 5–8, solve the inequalities and graph the solution set. Write the solution in interval notation.

5. $-15 < 3w - 6 \leq -9$

6. $5 - 2y \leq 1$ and $3y + 2 \geq 14$

7. $m - 7 \leq -5$ or $m - 7 \geq -10$

8. $3b - 2 < 7$ or $b - 2 > 4$

Concepts 1 and 2: Solving Absolute Value Inequalities

For Exercises 9–20, solve the equations and inequalities. For each inequality, graph the solution set and express the solution in interval notation. **(See Examples 1–6.)**

9. a. $|x| = 5$

   b. $|x| > 5$

   c. $|x| < 5$

10. a. $|a| = 4$

    b. $|a| > 4$

    c. $|a| < 4$

11. a. $|x - 3| = 7$

    b. $|x - 3| > 7$

    c. $|x - 3| < 7$

12. a. $|w + 2| = 6$

    b. $|w + 2| > 6$

    ———————————→

    c. $|w + 2| < 6$

    ———————————→

13. a. $|p| = -2$

    b. $|p| > -2$

    ———————————→

    c. $|p| < -2$

    ———————————→

14. a. $|x| = -14$

    b. $|x| > -14$

    ———————————→

    c. $|x| < -14$

    ———————————→

15. a. $|y + 1| = -6$

    b. $|y + 1| > -6$

    ———————————→

    c. $|y + 1| < -6$

    ———————————→

16. a. $|z - 4| = -3$

    b. $|z - 4| > -3$

    ———————————→

    c. $|z - 4| < -3$

    ———————————→

17. a. $|x| = 0$

    b. $|x| > 0$

    ———————————→

    c. $|x| < 0$

    ———————————→

18. a. $|p + 3| = 0$

    b. $|p + 3| > 0$

    ———————————→

    c. $|p + 3| < 0$

    ———————————→

19. a. $|k - 7| = 0$

    b. $|k - 7| > 0$

    ———————————→

    c. $|k - 7| < 0$

    ———————————→

20. a. $|2x + 4| + 3 = 2$

    b. $|2x + 4| + 3 > 2$

    ———————————→

    c. $|2x + 4| + 3 < 2$

    ———————————→

For Exercises 21–50, solve the absolute value inequalities. Graph the solution set and write the solution in interval notation. **(See Examples 1–6.)**

21. $|x| > 6$

———————————→

22. $|x| \leq 6$

———————————→

23. $|t| \leq 3$

———————————→

24. $|p| > 3$

———————————→

25. $|y + 2| \geq 0$

———————————→

26. $0 \leq |7n + 2|$

———————————→

27. $5 \leq |2x - 1|$

———————————→

28. $7 \leq |x - 2|$

———————————→

29. $|k - 7| < -3$

———————————→

30. $|h + 2| < -9$

———————————→

31. $\left| \dfrac{w - 2}{3} \right| - 3 \leq 1$

———————————→

32. $\left| \dfrac{x + 3}{2} \right| - 2 \geq 4$

———————————→

33. $12 \leq |9 - 4y| - 2$

———————————→

34. $5 > |2m - 7| + 4$

———————————→

35. $4 > -1 + \left| \dfrac{2x + 1}{4} \right|$

———————————→

36. $9 \geq 2 + \left| \dfrac{x - 4}{5} \right|$

———————————→

37. $8 < |4 - 3x| + 12$

———————————→

38. $-16 < |5x - 1| - 1$

———————————→

39. $5 - |2m + 1| > 5$

———————————→

40. $3 - |5x + 3| > 3$

———————————→

41. $|p + 5| \leq 0$

———————————→

42. $|y + 1| - 4 \leq -4$

———————————→

43. $|z - 6| + 5 > 5$

———————————→

44. $|2c - 1| - 4 > -4$

———————————→

45. $5|2y - 6| + 3 \geq 13$

———————————→

46. $7|y + 1| - 3 \geq 11$

———————————→

47. $-3|6 - t| + 1 > -5$

———————————→

48. $-4|8 - x| + 2 > -14$

———————————→

49. $|0.02x + 0.06| - 0.1 < 0.05$

50. $|0.05x - 0.04| - 0.01 < 0.11$

———————————→

Concept 3: Translating to an Absolute Value Expression

For Exercises 51–54, write an absolute value inequality equivalent to the expression given. **(See Example 7.)**

51. All real numbers whose distance from 0 is greater than 7

52. All real numbers whose distance from −3 is less than 4

53. All real numbers whose distance from 2 is at most 13

54. All real numbers whose distance from 0 is at least 6

55. A 32-oz jug of orange juice may not contain exactly 32 oz of juice. The possibility of measurement error exists when the jug is filled in the factory. If the maximum measurement error is ± 0.05 oz, write an absolute value inequality representing the range of volumes, x, in which the orange juice jug may be filled. **(See Example 8.)**

56. The length of a board is measured to be 32.3 in. The maximum measurement error is ± 0.2 in. Write an absolute value inequality that represents the range for the length of the board, x.

57. A bag of potato chips states that its weight is $6\frac{3}{4}$ oz. The maximum measurement error is $\pm \frac{1}{8}$ oz. Write an absolute value inequality that represents the range for the weight, x, of the bag of chips.

58. A $\frac{7}{8}$-in. bolt varies in length by at most $\pm \frac{1}{16}$ in. Write an absolute value inequality that represents the range for the length, x, of the bolt.

59. The width, w, of a bolt is supposed to be 2 cm but may have a 0.01-cm margin of error. Solve $|w - 2| \le 0.01$, and interpret the solution to the inequality in the context of this problem.

60. In a political poll, the front-runner was projected to receive 53% of the votes with a margin of error of 3%. Solve $|p - 0.53| \le 0.03$ and interpret the solution in the context of this problem.

Expanding Your Skills

For Exercises 61–64, match the graph with the inequality.

61.

62.

63.

64.

a. $|x - 2| < 4$ **b.** $|x - 1| > 4$ **c.** $|x - 3| < 2$ **d.** $|x - 5| > 1$

Problem Recognition Exercises

Identifying Equations and Inequalities

For Exercises 1–8, solve each equation or inequality. Express the solution in interval notation where appropriate.

1. a. $3x - 9 = 18$

   b. $|3x - 9| = 18$

   c. $|3x - 9| < 18$

   d. $|3x - 9| \geq 18$

2. a. $5y + 2 = -20$

   b. $|5y + 2| = -20$

   c. $|5y + 2| \leq -20$

   d. $|5y + 2| > -20$

3. a. $-2t - 14 = 0$

   b. $-2t - 14 > 0$

   c. $-2t - 14 \leq 0$

4. a. $\dfrac{x - 2}{3} = 9$

   b. $\dfrac{x - 2}{3} \geq 9$

   c. $\dfrac{x - 2}{3} < 9$

5. a. $|8t - 2| = |-2t + 3|$

   b. $8t - 2 = -2t + 3$

6. a. $-5 < x + 2$ and $x + 2 \leq 8$

   b. $-5 < x + 2 \leq 8$

7. a. $-4x - 9 < 11$ or $2 \leq x + 1$

   b. $-4x - 9 < 11$ and $2 \leq x + 1$

8. a. $4 < 2y$ or $-3(y + 2) > -2y + 1$

   b. $4 < 2y$ and $-3(y + 2) > -2y + 1$

For Exercises 9–28,

   a. Identify the type of equation or inequality. Choose from:
- linear equation
- absolute value equation
- linear inequality
- compound inequality
- absolute value inequality

   b. Solve the equation or inequality. Express the solution in interval notation where appropriate.

9. $-0.5y + 0.7 = 3.7$

10. $3m - 9 = 18$

11. $|2t + 8| \leq 4$

12. $|1 - 3x| < -1$

13. $-11 < 2t + 1 < 19$

14. $2z - 3 \geq 11$ or $3z + 3 < 9$

15. $\left| \dfrac{1}{2}y + 3 \right| = 5$

16. $|4x + 3| = |9 - 2x|$

17. $-\dfrac{3}{4}p \geq -9$

18. $8w + 4 \geq 5w + 1$

19. $\left| \dfrac{2x - 9}{3} \right| \geq 5$

20. $\left| \dfrac{10 - x}{5} \right| < 3$

21. $|2 - c| + 5 = 3$

22. $|10n + 2| + 7 = 7$

23. $\dfrac{w - 4}{5} - \dfrac{w + 1}{3} = 1$

24. $\dfrac{1}{3}y - \dfrac{5}{6} = \dfrac{1}{2}y + 1$

25. $2x - 7 > 9$ and $3x \leq 36$

26. $-3 + x > 2x$ and $2 \geq -\dfrac{1}{3}x$

27. $5(x - 2) + 7 = 2x + 3(x - 1)$

28. $7y - 4 = 3(y + 1) + 4y$

Group Activity

Understanding the Symbolism of Mathematics

Estimated time: 15 minutes

Group Size: 3

As you advance in your study of mathematics, you will notice that mathematics has its own language and syntax. For example, the statement $a > 0$ translates to "a is greater than zero." This also means that a is positive. Read and interpret the following conditions imposed on the variables $a, b, c, d,$ and x. Then determine whether the statements in Exercises 1–12 are true or false. Work three exercises at a time individually. Then stop and check your answers with the other members of the group. Discuss any discrepancies and agree on a final answer.

$$a > 0 \qquad b < 0 \qquad -1 < c < 1 \qquad d > 1 \qquad x = 0$$

1. $ab > 0$ _____

2. $bd < 0$ _____

3. $a + d > 0$ _____

4. $b^2 < 0$ _____

5. $a^2 > 0$ _____

6. $c + d \geq 0$ _____

7. $c^2 < 1$ _____

8. $d^2 > 1$ _____

9. $|c| < 1$ _____

10. $bx < 0$ _____

11. $b + x > 0$ _____

12. $\dfrac{x}{d} = 0$ _____

For Exercises 13–24, suppose that m represents an *odd* integer and n represents an *even* integer. Determine whether the statements are true or false.

13. $m + 1$ is an odd integer. _____

14. $m + 2$ is an odd integer. _____

15. $m - 1$ is an odd integer. _____

16. $m - 2$ is an odd integer. _____

17. $m + n$ is an even integer. _____

18. $m - n$ is an even integer. _____

19. m^2 is an even integer. _____

20. n^2 is an even integer. _____

21. $(m + n)^2$ is an odd integer. _____

22. $(n + n)^2$ is an even integer. _____

23. $(m + n + 1)^3$ is an odd integer. _____

24. $(m + m - 2)^3$ is an odd integer. _____

Chapter 1   Summary

Section 1.1   Linear Equations in One Variable

Key Concepts

A **linear equation in one variable** can be written in the form $ax + b = c \ (a \neq 0)$.

Steps to Solve a Linear Equation in One Variable

1. Simplify both sides of the equation.
   - Clear parentheses.
   - Consider clearing fractions or decimals (if any are present) by multiplying both sides of the equation by a common denominator of all terms.
   - Combine *like* terms.
2. Use the addition or subtraction property of equality to collect the variable terms on one side of the equation.
3. Use the addition or subtraction property of equality to collect the constant terms on the other side.
4. Use the multiplication or division property of equality to make the coefficient on the variable term equal to 1.
5. Check your answer.

An equation that has no solution is called a **contradiction**.

An equation that has all real numbers as its solutions is called an **identity**.

Examples

Example 1

$$\frac{1}{2}(x - 4) - \frac{3}{4}(x + 2) = \frac{1}{4}$$

$$\frac{1}{2}x - 2 - \frac{3}{4}x - \frac{3}{2} = \frac{1}{4}$$

$$4\left(\frac{1}{2}x - 2 - \frac{3}{4}x - \frac{3}{2}\right) = 4\left(\frac{1}{4}\right)$$

$$2x - 8 - 3x - 6 = 1$$

$$-x - 14 = 1$$

$$-x = 15$$

$$x = -15$$

The solution -15 checks in the original equation.

Example 2

$$3x + 6 = 3(x - 5)$$

$$3x + 6 = 3x - 15$$

$$6 = -15 \quad \text{Contradiction}$$

There is no solution.

Example 3

$$-(5x + 12) - 3 = 5(-x - 3)$$

$$-5x - 12 - 3 = -5x - 15$$

$$-5x - 15 = -5x - 15$$

$$-15 = -15 \quad \text{Identity}$$

All real numbers are solutions.

Section 1.2   Applications of Linear Equations in One Variable

Key Concepts

Problem-Solving Steps for Word Problems

1. Read the problem carefully.
2. Assign labels to unknown quantities.
3. Write a verbal model.
4. Write a mathematical equation.
5. Solve the equation.
6. Interpret the results and write the final answer in words.

Sales tax: (cost of merchandise)(tax rate)

Commission: (dollars in sales)(commission rate)

Simple interest: $I = Prt$

Distance = (rate)(time) $d = rt$

Examples

Example 1

1. Estella needs to borrow $8500. She borrows part of the money from a friend and agrees to pay the friend 6% simple interest. She borrows the rest of the money from a bank that charges 10% simple interest. If she pays back the money at the end of 1 yr and also pays $750 in interest, find the amount that Estella borrowed from each source.

2. Let x represent the amount borrowed at 6%. Then $8500 - x$ is the amount borrowed at 10%.

	6% Account	10% Account	Total
Principal	x	$8500 - x$	8500
Interest	$0.06x$	$0.10(8500 - x)$	750

3. $\left(\begin{array}{c} \text{Interest} \\ \text{owed at 6\%} \end{array} \right) + \left(\begin{array}{c} \text{interest} \\ \text{owed at 10\%} \end{array} \right) = \left(\begin{array}{c} \text{total} \\ \text{interest} \end{array} \right)$

4. $0.06x + 0.10(8500 - x) = 750$

5. $6x + 10(8500 - x) = 75{,}000$

   $6x + 85{,}000 - 10x = 75{,}000$

   $\qquad\qquad\qquad -4x = -10{,}000$

   $\qquad\qquad\qquad\quad x = 2500$

6. $x = 2500$

   $8500 - 2500 = 6000$

$2500 was borrowed at 6% and $6000 was borrowed at 10%.

Section 1.3   Applications to Geometry and Literal Equations

Key Concepts

Some useful formulas for word problems:

Perimeter

Rectangle: $P = 2l + 2w$

Area

Rectangle: $A = lw$

Square: $A = s^2$

Triangle: $A = \dfrac{1}{2}bh$

Trapezoid: $A = \dfrac{1}{2}(b_1 + b_2)h$

Angles

Two angles whose measures total 90° are complementary angles.

Two angles whose measures total 180° are supplementary angles.

The sum of the measures of the angles of a triangle is 180°.

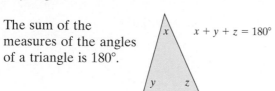 $x + y + z = 180°$

Literal equations (or formulas) are equations with several variables. To solve for a specific variable, follow the steps to solve a linear equation.

Examples

Example 1

A border of marigolds is to enclose a rectangular flower garden. If the length is twice the width and the perimeter is 25.5 ft, what are the dimensions of the garden?

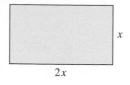

$$P = 2l + 2w$$
$$25.5 = 2(2x) + 2(x)$$
$$25.5 = 4x + 2x$$
$$25.5 = 6x$$
$$4.25 = x$$

The width is 4.25 ft, and the length is 2(4.25) ft or 8.5 ft.

Example 2

Solve for y.

$$4x - 5y = 20$$
$$-5y = -4x + 20$$
$$\frac{-5y}{-5} = \frac{-4x + 20}{-5}$$
$$y = \frac{-4x + 20}{-5} \quad \text{or} \quad y = \frac{4}{5}x - 4$$

Section 1.4 Linear Inequalities in One Variable

Key Concepts

A **linear inequality** is an inequality that can be written in one of the following forms, provided that $a \neq 0$.

$ax + b < c, \quad ax + b > c, \quad ax + b \leq c, \quad$ or
$ax + b \geq c$

Properties of Inequalities

1. If $a < b$, then $a + c < b + c$.
2. If $a < b$, then $a - c < b - c$.
3. If c is positive and $a < b$, then $ac < bc$ and $\dfrac{a}{c} < \dfrac{b}{c}$.
4. If c is negative and $a < b$, then $ac > bc$ and $\dfrac{a}{c} > \dfrac{b}{c}$.

Properties 3 and 4 indicate that if we multiply or divide an inequality by a negative value, the direction of the inequality sign must be reversed.

Examples

Example 1

Solve.

$$\frac{14 - x}{-2} < -3x$$

$$-2\left(\frac{14 - x}{-2}\right) > -2(-3x) \qquad \text{(Reverse the inequality sign.)}$$

$$14 - x > 6x$$

$$-7x > -14$$

$$\frac{-7x}{-7} < \frac{-14}{-7} \qquad \text{(Reverse the inequality sign.)}$$

$$x < 2$$

Set-builder notation: $\{x \mid x < 2\}$
Interval notation: $(-\infty, 2)$

Section 1.5 Compound Inequalities

Key Concepts

$A \cup B$ is the **union** of A and B. This is the set of elements that belong to set A or set B or both sets A and B.

$A \cap B$ is the **intersection** of A and B. This is the set of elements common to both A and B.

- Solve two or more inequalities joined by *and* by finding the intersection of their solution sets.

- Solve two or more inequalities joined by *or* by finding the union of their solution sets.

Examples

Example 1

Union   **Intersection**

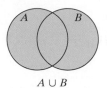

$A \cup B$

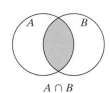

$A \cap B$

Example 2

$$-7x + 3 \geq -11 \qquad \text{and} \qquad 1 - x < 4.5$$
$$-7x \geq -14 \qquad \text{and} \qquad -x < 3.5$$
$$x \leq 2 \qquad \text{and} \qquad x > -3.5$$

$x \leq 2$

$$\overset{\longleftarrow}{\underset{-5\ -4\ -3\ -2\ -1\ \ 0\ \ 1\ \ 2\ \ 3\ \ 4\ \ 5}{\vert\ \vert\ \vert\ \vert\ \vert\ \vert\ \vert\ \vert\ ]\ \vert\ \vert\ \vert}}$$

$x > -3.5$

$$\overset{\longleftarrow}{\underset{-5\ -4\ -3\ -2\ -1\ \ 0\ \ 1\ \ 2\ \ 3\ \ 4\ \ 5}{\vert\ \vert\ (\vert\ \vert\ \vert\ \vert\ \vert\ \vert\ \vert\ \vert\ \vert}}$$

$$\overset{\longleftarrow}{\underset{-5\ -4\ -3\ -2\ -1\ \ 0\ \ 1\ \ 2\ \ 3\ \ 4\ \ 5}{\vert\ \vert\ (\vert\ \vert\ \vert\ \vert\ \vert\ ]\ \vert\ \vert\ \vert}}$$

The solution is $\{x \mid -3.5 < x \leq 2\}$ or equivalently $(-3.5, 2]$.

Inequalities of the form $a < x < b$:

The inequality $a < x < b$ is represented by

$$\overset{\longrightarrow}{\underset{a \qquad b}{(\qquad)}} \quad \text{or, in interval notation, } (a, b).$$

Example 3

$$5y + 1 \geq 6 \qquad \text{or} \qquad 2y - 5 \leq -11$$
$$5y \geq 5 \qquad \text{or} \qquad 2y \leq -6$$
$$y \geq 1 \qquad \text{or} \qquad y \leq -3$$

$y \geq 1$

$$\overset{\longrightarrow}{\underset{-5\ -4\ -3\ -2\ -1\ \ 0\ \ 1\ \ 2\ \ 3\ \ 4\ \ 5}{\vert\ \vert\ \vert\ \vert\ \vert\ \vert\ [\ \vert\ \vert\ \vert\ \vert}}$$

$y \leq -3$

$$\overset{\longleftarrow}{\underset{-5\ -4\ -3\ -2\ -1\ \ 0\ \ 1\ \ 2\ \ 3\ \ 4\ \ 5}{\vert\ \vert\ ]\ \vert\ \vert\ \vert\ \vert\ \vert\ \vert\ \vert\ \vert}}$$

$$\overset{\longleftarrow}{\underset{-5\ -4\ -3\ -2\ -1\ \ 0\ \ 1\ \ 2\ \ 3\ \ 4\ \ 5}{\vert\ \vert\ ]\ \vert\ \vert\ \vert\ [\ \vert\ \vert\ \vert\ \vert}}$$

The solution is $\{y \mid y \geq 1 \text{ or } y \leq -3\}$ or equivalently $(-\infty, -3] \cup [1, \infty)$.

Example 4

Solve.

$$-13 \leq 3x - 1 < 5$$
$$-13 + 1 \leq 3x - 1 + 1 < 5 + 1$$
$$-12 \leq 3x < 6$$
$$\frac{-12}{3} \leq \frac{3x}{3} < \frac{6}{3}$$
$$-4 \leq x < 2$$

$$\overset{\longrightarrow}{\underset{-4 \qquad\qquad 2}{[\qquad\qquad)}}$$

Interval notation: $[-4, 2)$

Section 1.6 Absolute Value Equations

Key Concepts

The equation $|x| = a$ is an **absolute value equation**. For $a \geq 0$, the solution to the equation $|x| = a$ is $x = a$ or $x = -a$.

Steps to Solve an Absolute Value Equation

1. Isolate the absolute value to write the equation in the form $|x| = a$.
2. If $a < 0$, there is no solution.
3. Otherwise, if $a \geq 0$, rewrite the equation $|x| = a$ as $x = a$ or $x = -a$.
4. Solve the equations from step 3.
5. Check answers in the original equation.

The equation $|x| = |y|$ implies $x = y$ or $x = -y$.

Examples

Example 1

$$|2x - 3| + 5 = 10$$
$$|2x - 3| = 5 \qquad \text{Isolate the absolute value.}$$
$$2x - 3 = 5 \quad \text{or} \quad 2x - 3 = -5$$
$$2x = 8 \quad \text{or} \qquad 2x = -2$$
$$x = 4 \quad \text{or} \qquad x = -1$$

The solutions are 4 and -1.

Example 2

$$|x + 2| + 5 = 1$$
$$|x + 2| = -4 \qquad \text{No solution}$$

Example 3

$$|2x - 1| = |x + 4|$$
$$2x - 1 = x + 4 \quad \text{or} \quad 2x - 1 = -(x + 4)$$
$$x = 5 \qquad \text{or} \quad 2x - 1 = -x - 4$$
$$\text{or} \qquad 3x = -3$$
$$\text{or} \qquad x = -1$$

The solutions are 5 and -1.

Section 1.7 Absolute Value Inequalities

Key Concepts

Solutions to Absolute Value Inequalities

For $a > 0$, we have:

$$|x| > a \Rightarrow x < -a \quad \text{or} \quad x > a$$
$$|x| < a \Rightarrow -a < x < a$$

Examples

Example 1

$$|5x - 2| < 12$$
$$-12 < 5x - 2 < 12$$
$$-10 < 5x < 14$$
$$-2 < x < \frac{14}{5}$$

The solution is $\left(-2, \dfrac{14}{5}\right)$.

Test Point Method to Solve Inequalities

1. Find the boundary points of the inequality. (Boundary points are the real solutions to the related equation and points where the inequality is undefined.)
2. Plot the boundary points on the number line. This divides the number line into intervals.
3. Select a test point from each interval and substitute it into the original inequality.
   - If a test point makes the original inequality true, then that interval is part of the solution set.
4. Test the boundary points in the original inequality.
   - If the original inequality is strict ($<$ or $>$), do not include the boundary in the solution set.
   - If the original inequality is defined using $\leq$ or $\geq$, then include the boundary points that are well defined within the inequality.

Note: Any boundary point that makes an expression within the inequality undefined must *always* be excluded from the solution set.

If *a* is *negative* ($a < 0$), then

1. $|x| < a$ has no solution.
2. $|x| > a$ is true for all real numbers.

Example 2

$|x - 3| + 2 \geq 7$

$\qquad |x - 3| \geq 5 \qquad$ Isolate the absolute value.

$\qquad |x - 3| = 5 \qquad$ Solve the related equation.

$x - 3 = 5 \quad$ or $\quad x - 3 = -5$

$\qquad x = 8 \quad$ or $\qquad x = -2 \qquad$ Boundary points

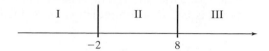

Interval I:

Test $x = -3$: $\qquad |(-3) - 3| + 2 \overset{?}{\geq} 7 \qquad$ True

Interval II:

Test $x = 0$: $\qquad |(0) - 3| + 2 \overset{?}{\geq} 7 \qquad$ False

Interval III:

Test $x = 9$: $\qquad |(9) - 3| + 2 \overset{?}{\geq} 7 \qquad$ True

The solution is $(-\infty, -2] \cup [8, \infty)$.

Example 3

$|x + 5| > -2$

The solution is all real numbers because an absolute value will always be greater than a negative number.

$(-\infty, \infty)$

Example 4

$|x + 5| < -2$

There is no solution because an absolute value cannot be less than a negative number.

There is no solution.

Chapter 1   Review Exercises

Section 1.1

1. Describe the solution set for a contradiction.

2. Describe the solution set for an identity.

For Exercises 3–12, solve the equations and identify each as a conditional equation, a contradiction, or an identity.

3. $x - 27 = -32$

4. $y + \dfrac{7}{8} = 1$

5. $7.23 + 0.6x = 0.2x$

6. $0.1y + 1.122 = 5.2y$

7. $-(4 + 3m) = 9(3 - m)$

8. $-2(5n - 6) = 3(-n - 3)$

9. $\dfrac{x - 3}{5} - \dfrac{2x + 1}{2} = 1$

10. $3(x + 3) - 2 = 3x + 2$

11. $\dfrac{10}{8}m + 18 - \dfrac{7}{8}m = \dfrac{3}{8}m + 25$

12. $\dfrac{2}{3}m + \dfrac{1}{3}(m - 1) = -\dfrac{1}{3}m + \dfrac{1}{3}(4m - 1)$

Section 1.2

13. Explain how you would label three consecutive integers.

14. Explain how you would label two consecutive odd integers.

15. Explain what the formula $d = rt$ means.

16. Explain what the formula $I = Prt$ means.

 17. a. Cory made \$85,200 in taxable income. If he pays an average of 28% in taxes on his income, determine the amount of tax he must pay.

   b. What is his net income (after taxes)?

18. For a recent year, approximately 7.2 million men were in college in the United States. This represents an 8% increase over the number of men in college in 2000. Approximately how many men were in college in 2000? (Round to the nearest tenth of a million.)

19. For a recent year, there were 17,430 deaths due to alcohol-related accidents in the United States. This was a 5% increase over the number of alcohol-related deaths in 1999. How many such deaths were there in 1999?

20. Of three consecutive even integers, the sum of the smallest two integers is equal to 6 less than the largest. Find the integers.

21. To do a rope trick, a magician needs to cut a piece of rope so that one piece is one-third the length of the other piece. If she begins with a $2\frac{2}{3}$-ft rope, what will be the lengths of the two pieces of rope?

22. Sharyn invests \$2000 more in an account that earns 9% simple interest than she invests in an account that earns 6% simple interest. How much did she invest in each account if her total interest is \$405 after 1 yr?

23. How much 10% acid solution should be mixed with 1 L of 25% acid solution to produce a solution that is 15% acid?

24. Two friends plan to meet at a restaurant for lunch. They both leave their homes at 11:30 A.M. and between the two of them, they drive a total of 37.5 mi. Lynn drives in from a neighboring town and averages 15 mph faster than her friend Linda. If they meet at noon, find the average driving speed for each.

Section 1.3

25. The length of a rectangle is 2 ft more than the width. Find the dimensions if the perimeter is 40 ft.

For Exercises 26 and 27, solve for x, and then find the measure of each angle.

26.

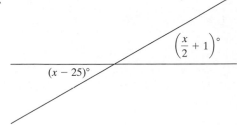

$\left(\dfrac{x}{2} + 1\right)^{\circ}$

$(x - 25)^{\circ}$

27.

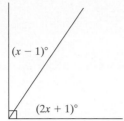

$(x - 1)°$

$(2x + 1)°$

For Exercises 28–31, solve for the indicated variable.

28. $3x - 2y = 4$ for y

29. $-6x + y = 12$ for y

30. $S = 2\pi r + \pi r^2 h$ for h

31. $A = \dfrac{1}{2}bh$ for b

 32. a. The circumference of a circle is given by $C = 2\pi r$. Solve this equation for π.

     b. Tom measures the radius of a circle to be 6 cm and the circumference to be 37.7 cm. Use these values to approximate π. (Round to 2 decimal places.)

Section 1.4

For Exercises 33–38, solve the inequality and graph the solution set. Write the solution set in (a) set-builder notation and (b) interval notation.

33. $-6x - 2 > 6$ ———————————→

34. $-10x \leq 15$ ———————————→

35. $5 - 7(x + 3) > 19x$ ——————→

36. $4 - 3x \geq 10(-x + 5)$ ——————→

37. $\dfrac{5 - 4x}{8} \geq 9$ ———————————→

38. $\dfrac{3 + 2x}{4} \leq 8$ ———————————→

39. Dave earned the following test scores in his biology class: 82, 88, 92, and 93. How high does he have to score on the fifth test to have an average of 90 or more?

Section 1.5

40. Explain the difference between the union and intersection of two sets. You may use the sets C and D in the following diagram to provide an example.

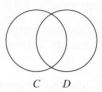

C D

Let $X = \{x \mid x \geq -10\}$, $Y = \{x \mid x < 1\}$, $Z = \{x \mid x > -1\}$, and $W = \{x \mid x \leq -3\}$. For Exercises 41–46, find the intersection or union of the sets X, Y, Z, and W. Write the answers in interval notation.

41. $X \cap Y$ **42.** $X \cup Y$

43. $Y \cup Z$ **44.** $Y \cap Z$

45. $Z \cup W$ **46.** $Z \cap W$

For Exercises 47–56, solve the compound inequalities. Write the solutions in interval notation.

47. $4m > -11$ and $4m - 3 \leq 13$

48. $4n - 7 < 1$ and $7 + 3n \geq -8$

49. $-3y + 1 \geq 10$ and $-2y - 5 \leq -15$

50. $\dfrac{1}{2} - \dfrac{h}{12} \leq -\dfrac{7}{12}$ and $\dfrac{1}{2} - \dfrac{h}{10} > -\dfrac{1}{5}$

51. $\dfrac{2}{3}t - 3 \leq 1$ or $\dfrac{3}{4}t - 2 > 7$

52. $2(3x + 1) < -10$ or $3(2x - 4) \geq 0$

53. $-7 < -7(2w + 3)$ or $-2 < -4(3w - 1)$

54. $5(p + 3) + 4 > p - 1$ or $4(p - 1) + 2 > p + 8$

55. $2 \geq -(b - 2) - 5b \geq -6$

56. $-4 \leq \dfrac{1}{2}(x - 1) < -\dfrac{3}{2}$

57. The product of $\frac{1}{3}$ and the sum of a number and 3 is between -1 and 5. Find all such numbers.

58. Normal levels of total cholesterol vary according to age. For adults between 25 and 40 yr old, the normal range is generally accepted to be between 140 and 225 mg/dL (milligrams per deciliter), inclusive.

   a. Write an inequality representing the normal range for total cholesterol for adults between 25 and 40 yr old.

   b. Write a compound inequality representing abnormal ranges for total cholesterol for adults between 25 and 40 yr old.

59. Normal levels of total cholesterol vary according to age. For adults younger than 25 yr old, the normal range is generally accepted to be between 125 and 200 mg/dL, inclusive.

   a. Write an inequality representing the normal range for total cholesterol for adults younger than 25 yr old.

   b. Write a compound inequality representing abnormal ranges for total cholesterol for adults younger than 25 yr old.

60. One method to approximate your maximum heart rate is to subtract your age from 220. To maintain an aerobic workout, it is recommended that you sustain a heart rate of between 60% and 75% of your maximum heart rate.

   a. If the maximum heart rate h is given by the formula $h = 220 - A$, where A is a person's age, find your own maximum heart rate. (Answers will vary.)

   b. Find the interval for your own heart rate that will sustain an aerobic workout. (Answers will vary.)

Section 1.6

For Exercises 61–74, solve the absolute value equations.

61. $|x| = 10$

62. $|x| = 17$

63. $|8.7 - 2x| = 6.1$

64. $|5.25 - 5x| = 7.45$

65. $16 = |x + 2| + 9$

66. $5 = |x - 2| + 4$

67. $|4x - 1| + 6 = 4$

68. $|3x - 1| + 7 = 3$

69. $\left|\dfrac{7x - 3}{5}\right| + 4 = 4$

70. $\left|\dfrac{4x + 5}{-2}\right| - 3 = -3$

71. $|3x - 5| = |2x + 1|$

72. $|8x + 9| = |8x - 1|$

73. $|2 + 7d| = |-7d - 2|$

74. $-|4y + 6| = |2y - 3|$

75. Which absolute value expression represents the distance between 3 and -2 on the number line?

$$|3 - (-2)| \qquad |-2 - 3|$$

Section 1.7

76. Write the compound inequality $x < -5$ or $x > 5$ as an absolute value inequality.

77. Write the compound inequality $-4 < x < 4$ as an absolute value inequality.

For Exercises 78 and 79, write an absolute value inequality that represents the solution sets graphed here.

78.

79.

For Exercises 80–93, solve the absolute value inequalities. Graph the solution set and write the solution in interval notation.

80. $|x + 6| \geq 8$

81. $|x + 8| \leq 3$

82. $2|7x - 1| + 4 > 4$

83. $4|5x + 1| - 3 > -3$

84. $|3x + 4| - 6 \leq -4$

85. $|5x - 3| + 3 \leq 6$

86. $\left|\dfrac{x}{2} - 6\right| < 5$

87. $\left|\dfrac{x}{3} + 2\right| < 2$

88. $|4 - 2x| + 8 \geq 8$

89. $|9 + 3x| + 1 \geq 1$ ⟶

90. $-2|5.2x - 7.8| < 13$ ⟶

91. $-|2.5x + 15| < 7$ ⟶

92. $|3x - 8| < -1$ ⟶   **93.** $|x + 5| < -4$ ⟶

94. State one possible situation in which an absolute value inequality will have no solution.

95. State one possible situation in which an absolute value inequality will have a solution of all real numbers.

96. The Neilsen ratings estimated that the percent, p, of the television viewing audience watching *American Idol* was 20% with a 3% margin of error. Solve the inequality $|p - 0.20| \leq 0.03$ and interpret the answer in the context of this problem.

97. The length, L, of a screw is supposed to be $3\frac{3}{8}$ in. Due to variation in the production equipment, there is a $\frac{1}{4}$-in. margin of error. Solve the inequality $|L - 3\frac{3}{8}| \leq \frac{1}{4}$ and interpret the answer in the context of this problem.

Chapter 1   Test

For Exercises 1–9, solve the equations.

1. $\dfrac{x}{7} + 1 = 20$

2. $8 - 5(4 - 3z) = 2(4 - z) - 8z$

3. $0.12(x) + 0.08(60{,}000 - x) = 10{,}500$

4. $\dfrac{5 - x}{6} - \dfrac{2x - 3}{2} = \dfrac{x}{3}$

5. $\left|\dfrac{1}{2}x + 3\right| - 4 = 4$

6. $|3x + 4| = |x - 12|$

7. $-5 = -8 + |2y - 3|$

8. $|3.7x - 5| + 7 = 6.2$

9. $|8x + 11| = |8x + 5|$

10. Label each equation as a conditional equation, an identity, or a contradiction.

   a. $(5x - 9) + 19 = 5(x + 2)$

   b. $2a - 2(1 + a) = 5$

   c. $(4w - 3) + 4 = 3(5 - w)$

11. The difference between two numbers is 72. If the larger is 5 times the smaller, find the two numbers.

12. Joëlle is determined to get some exercise and walks to the store at a brisk rate of 4.5 mph. She meets her friend Yun Ling at the store, and together they walk back at a slower rate of 3 mph. Joëlle's total walking time was 1 hr.

   a. How long did it take her to walk to the store?

   b. What is the distance to the store?

13. Shawnna banks at a credit union. Her money is distributed between two accounts: a certificate of deposit (CD) that earns 5% simple interest and a savings account that earns 3.5% simple interest. Shawnna has $100 less in her savings account than in the CD. If after 1 year her total interest is $81.50, how much did she invest in the CD?

14. A yield sign is in the shape of an equilateral triangle (all sides have equal length). Its perimeter is 81 in. Find the length of the sides.

15. The sum of three consecutive odd integers is 41 less than four times the largest. Find the numbers.

16. How many gallons of a 20% acid solution must be mixed with 6 gal of a 30% acid solution to make a 22% solution?

For Exercises 17 and 18, solve the equations for the indicated variable.

   17. $4x + 2y = 6$ for y **18.** $x = \mu + z\sigma$ for z

For Exercises 19–21, solve the inequalities. Graph the solution and write the solution set in interval notation.

19. $x + 8 > 42$

20. $-\dfrac{3}{2}x + 6 \geq x - 3$

21. $-2 < 3x - 1 \leq 5$

For Exercises 22–32, solve the compound and absolute value inequalities. Write the answers in interval notation.

22. $-3 \leq 4x + 5 \leq 7$

23. $-4 \leq \dfrac{6 - 2x}{5} < 2$

24. $-\dfrac{3}{5}x - 1 \leq 8$ or $-\dfrac{2}{3}x \geq 16$

25. $-2x - 3 > -3$ and $x + 3 \geq 0$

26. $5x + 1 \leq 6$ or $2x + 4 > -6$

27. $2x - 3 > 1$ and $x + 4 < -1$

28. $|3 - 2x| + 6 < 2$ **29.** $|3x - 8| \geq 9$

30. $|0.4x + 0.3| - 0.2 < 7$

31. $|7 - 3x| + 1 > -3$ **32.** $6 \geq |2x - 5| - 5$

33. An elevator can accommodate a maximum weight of 2000 lb. If four passengers on the elevator have an average weight of 180 lb each, how many additional passengers of the same average weight can the elevator carry before the maximum weight capacity is exceeded?

34. The normal range in humans of the enzyme adenosine deaminase (ADA), is between 9 and 33 IU (international units), inclusive. Let x represent the ADA level in international units.

a. Write an inequality representing the normal range for ADA.

b. Write a compound inequality representing abnormal ranges for ADA.

35. The mass of a small piece of metal is measured to be 15.41 g. If the measurement error is at most ± 0.01 g, write an absolute value inequality that represents the possible mass, x, of the piece of metal.

Linear Equations in Two Variables

2

Chapter 2

In this chapter, we cover graphing and the applications of graphing. Graphs appear in magazines and newspapers and in other aspects of day-to-day life. In many fields of study such as the sciences and business, graphs are used to display data (information).

Review Your Skills

Answer questions 1–5 below. These problems will refresh your skills at reading and interpreting graphs. Match each answer with the correct letter and record the letter in the space below.

1. Use the graph to approximate the average height of a 5-yr-old girl.
2. Use the graph to estimate a girl's age if her height is 39 in. tall.
3. The equation $h = 2.5a + 31$ can be used to approximate a girl's height, h, in inches according to her age, a, in years. Use the formula $h = 2.5a + 31$ to estimate a girl's height at 6 yr old.
4. Use the formula $h = 2.5a + 31$ to estimate the age of a girl if her height is 50 in.
5. Use the formula $h = 2.5a + 31$ to predict the height of an 11-yr-old girl.

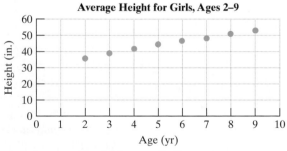

A graph intersects the x-axis at an

__ - __ N __ __ R __ __ P __ .
 1   2   4   3   5   3   4

T. 7.6 yr   **E.** 46 in.   **I.** 3 yr

X. 44 in.   **C.** 58.5 in.

Concepts

1. The Rectangular Coordinate System
2. Linear Equations in Two Variables
3. Graphing Linear Equations in Two Variables
4. *x*- and *y*-Intercepts
5. Horizontal and Vertical Lines

1. The Rectangular Coordinate System

One application of algebra is the graphical representation of numerical information (or data). For example, Table 2-1 shows the percentage of individuals who participate in leisure sports activities according to the age of the individual.

Table 2-1

Age (years)	Percentage of Individuals Participating in Leisure Sports Activities
20	59%
30	52%
40	44%
50	34%
60	21%
70	18%

Source: U.S. National Endowment for the Arts

Information in table form is difficult to picture and interpret. However, when the data are presented in a graph, there appears to be a downward trend in the participation in leisure sports activities as age increases (Figure 2-1).

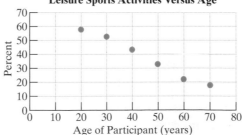

Percentage of Individuals Who Participate in Leisure Sports Activities Versus Age

Figure 2-1

In this example, two variables are related: age and the percentage of individuals who participate in leisure sports activities.

To picture two variables simultaneously, we use a graph with two number lines drawn at right angles to each other (Figure 2-2). This forms a **rectangular coordinate system**. The horizontal line is called the *x*-**axis**, and the vertical line is called the *y*-**axis**. The point where the lines intersect is called the **origin**. On the *x*-axis, the numbers to the right of the origin are positive, and the numbers to the left are negative. On the *y*-axis, the numbers above the origin are positive, and the numbers below are negative. The *x*- and *y*-axes divide the graphing area into four regions called **quadrants**.

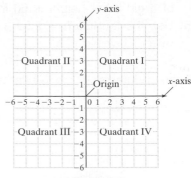

Figure 2-2

Points graphed in a rectangular coordinate system are defined by two numbers as an **ordered pair** (x, y). The first number (called the x-coordinate or abscissa) is the horizontal position from the origin. The second number (called the y-coordinate or ordinate) is the vertical position from the origin. Example 1 shows how points are plotted in a rectangular coordinate system.

Example 1 Plotting Points

Plot each point and state the quadrant or axis where it is located.

a. $(4, 1)$ **b.** $(-3, 4)$ **c.** $(4, -3)$

d. $(-\frac{5}{2}, -2)$ **e.** $(0, 3)$ **f.** $(-4, 0)$

Solution:

a. The point $(4, 1)$ is in quadrant I.

b. The point $(-3, 4)$ is in quadrant II.

c. The point $(4, -3)$ is in quadrant IV.

d. The point $(-\frac{5}{2}, -2)$ can also be written as $(-2.5, -2)$. This point is in quadrant III.

e. The point $(0, 3)$ is on the y-axis.

f. The point $(-4, 0)$ is located on the x-axis.

TIP: Notice that the points $(-3, 4)$ and $(4, -3)$ are in different quadrants. Changing the order of the coordinates changes the location of the point. That is why points are represented by *ordered* pairs (Figure 2-3).

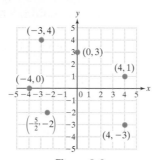

Figure 2-3

Skill Practice

1. Plot the point and state the quadrant or axis where it is located.
   a. $(3, 5)$ **b.** $(-2, 0)$
   c. $(2, -1)$ **d.** $(0, 4)$
   e. $(-2, -2)$ **f.** $(-5, 2)$

Animation

Answers

1. **a.** $(3, 5)$; quadrant I
   b. $(-2, 0)$; x-axis
   c. $(2, -1)$; quadrant IV
   d. $(0, 4)$; y-axis
   e. $(-2, -2)$; quadrant III
   f. $(-5, 2)$; quadrant II

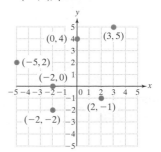

2. Linear Equations in Two Variables

Recall from Section 1.1 that an equation in the form $ax + b = c$ is called a linear equation in one variable. In this section we will study linear equations in *two* variables.

Definition of a Linear Equation in Two Variables

Let A, B, and C be real numbers such that A and B are not both zero. A **linear equation in two variables** is an equation that can be written in the form

$$Ax + By = C \qquad \text{This form is called } standard\ form.$$

A solution to a linear equation in two variables consists of two numbers that when substituted for x and y make the equation a true statement. These solutions are represented as ordered pairs, (x, y). Every linear equation in two variables has infinitely many ordered pairs that are solutions.

Example 2   **Determining Solutions to a Linear Equation**

For the linear equation $-2x + 3y = 8$, determine whether the ordered pair is a solution.

a. $(-4, 0)$ **b.** $(2, -4)$ **c.** $\left(1, \dfrac{10}{3}\right)$

Solution:

a.
$$-2x + 3y = 8$$
The ordered pair $(-4, 0)$ indicates that $x = -4$ and $y = 0$.

$$-2(-4) + 3(0) \stackrel{?}{=} 8$$
Substitute $x = -4$ and $y = 0$ into the equation.

$$8 + 0 \stackrel{?}{=} 8 \checkmark \text{ (true)}$$
The ordered pair $(-4, 0)$ makes the equation a true statement. The ordered pair is a solution to the equation.

b.
$$-2x + 3y = 8$$
Test the point $(2, -4)$.

$$-2(2) + 3(-4) \stackrel{?}{=} 8$$
Substitute $x = 2$ and $y = -4$ into the equation.

$$-4 + (-12) \stackrel{?}{=} 8$$

$$-16 \stackrel{?}{=} 8 \text{ (false)}$$
The ordered pair $(2, -4)$ does not make the equation a true statement. The ordered pair is *not* a solution to the equation.

c.
$$-2x + 3y = 8$$
Test the point $(1, \tfrac{10}{3})$.

$$-2(1) + 3\left(\dfrac{10}{3}\right) \stackrel{?}{=} 8$$
Substitute $x = 1$ and $y = \tfrac{10}{3}$.

$$-2 + 10 \stackrel{?}{=} 8 \checkmark \text{ (true)}$$
The ordered pair $(1, \tfrac{10}{3})$ is a solution to the equation.

3. Graphing Linear Equations in Two Variables

Consider the linear equation $x - y = 3$. The solutions to the equation are ordered pairs such that the difference of x and y is 3. Several solutions are given in the following list:

Solution	Check
(x, y)	$x - y = 3$
$(3, 0)$	$(3) - (0) = 3 \checkmark$
$(4, 1)$	$(4) - (1) = 3 \checkmark$
$(0, -3)$	$(0) - (-3) = 3 \checkmark$
$(-1, -4)$	$(-1) - (-4) = 3 \checkmark$
$(2, -1)$	$(2) - (-1) = 3 \checkmark$

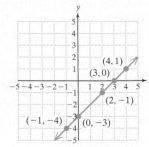

Figure 2-4

By graphing these ordered pairs, we see that the solution points form a line (see Figure 2-4). There are actually an infinite number of solutions to the equation $x - y = 3$. The graph of all solutions to a linear equation forms a line in the xy-plane. Conversely, each ordered pair on the line is a solution to the equation.

The Graph of an Equation in Two Variables

To graph a linear equation in two variables means that we will graph all ordered pair solutions to the equation.

To graph a linear equation, it is sufficient to find two solution points and draw the line through them. We will find three solution points and use the third point as a check point. This is demonstrated in Example 3.

Example 3 Graphing a Linear Equation in Two Variables

Graph the equation $3x + 5y = 15$.

Solution:

We will find three ordered pairs that are solutions to the equation. In the table, we have selected arbitrary values for x or y and must complete the ordered pairs.

x	y
0	
	2
5	

$\longrightarrow$ (0,)
$\longrightarrow$ (, 2)
$\longrightarrow$ (5,)

From the first row, substitute $x = 0$.

$$3x + 5y = 15$$
$$3(0) + 5y = 15$$
$$5y = 15$$
$$y = 3$$

From the second row, substitute $y = 2$.

$$3x + 5y = 15$$
$$3x + 5(2) = 15$$
$$3x + 10 = 15$$
$$3x = 5$$
$$x = \frac{5}{3}$$

From the third row, substitute $x = 5$.

$$3x + 5y = 15$$
$$3(5) + 5y = 15$$
$$15 + 5y = 15$$
$$5y = 0$$
$$y = 0$$

The completed list of ordered pairs is shown. To graph the equation, plot the three solutions and draw the line through the points (Figure 2-5). Arrows on the ends of the line indicate that points on the line extend infinitely in both directions.

x	y
0	3
$\frac{5}{3}$	2
5	0

$\longrightarrow$ (0, 3)
$\longrightarrow$ $\left(\frac{5}{3}, 2\right)$
$\longrightarrow$ (5, 0)

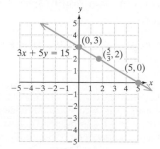

Figure 2-5

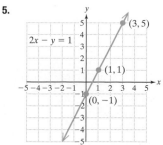

Example 4   Graphing a Linear Equation in Two Variables

Graph the equation $y = \frac{1}{2}x - 2$.

Solution:

Because the *y* variable is isolated in the equation, it is easy to substitute a value for *x* and simplify the right-hand side to find *y*. Since any number can be used for *x*, choose numbers that are multiples of 2 that will simplify easily when multiplied by $\frac{1}{2}$.

x	y
0	
2	
4	

Substitute $x = 0$.

$$y = \frac{1}{2}(0) - 2$$
$$y = 0 - 2$$
$$y = -2$$

Substitute $x = 2$.

$$y = \frac{1}{2}(2) - 2$$
$$y = 1 - 2$$
$$y = -1$$

Substitute $x = 4$.

$$y = \frac{1}{2}(4) - 2$$
$$y = 2 - 2$$
$$y = 0$$

The completed list of ordered pairs is as follows. To graph the equation, plot the three solutions and draw the line through the points (Figure 2-6).

x	y	
0	−2	⟶ (0, −2)
2	−1	⟶ (2, −1)
4	0	⟶ (4, 0)

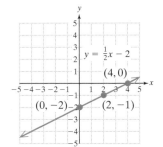

Figure 2-6

4. *x*- and *y*-Intercepts

For many applications of graphing, it is advantageous to know the points where a graph intersects the *x*- or *y*-axis. These points are called the *x*- and *y*-intercepts.

In Figure 2-5, the *x*-intercept is (5, 0). In Figure 2-6, the *x*-intercept is (4, 0). In general, a point on the *x*-axis must have a *y*-coordinate of zero. In Figure 2-5, the *y*-intercept is (0, 3). In Figure 2-6, the *y*-intercept is (0, −2). In general, a point on the *y*-axis must have an *x*-coordinate of zero.

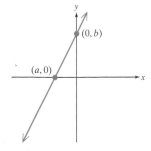

Figure 2-7

Definition of *x*- and *y*-Intercepts

An ***x*-intercept*** is a point $(a, 0)$ where a graph intersects the *x*-axis. (See Figure 2-7.)

A ***y*-intercept** is a point $(0, b)$ where a graph intersects the *y*-axis. (See Figure 2-7.)

*In some applications, an *x*-intercept is defined as the *x-coordinate* of a point of intersection that a graph makes with the *x*-axis. For example, if an *x*-intercept is at the point (3, 0), it is sometimes stated simply as 3 (the *y*-coordinate is understood to be zero). Similarly, a *y*-intercept is sometimes defined as the *y-coordinate* of a point of intersection that a graph makes with the *y*-axis. For example, if a *y*-intercept is at the point (0, 7), it may be stated simply as 7 (the *x*-coordinate is understood to be zero).

Answer

6.

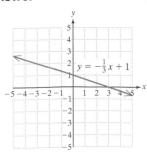

To find the x- and y-intercepts from an equation in x and y, follow these steps:

Determining the x- and y-Intercepts from an Equation

Given an equation in x and y,

- Find the x-intercept(s) by substituting $y = 0$ into the equation and solving for x.
- Find the y-intercept(s) by substituting $x = 0$ into the equation and solving for y.

Example 5 Finding the x- and y-Intercepts of a Line

Given $2x + 4y = 8$, find the x- and y-intercepts. Then graph the equation.

Solution:

To find the x-intercept, substitute $y = 0$.

$$2x + 4y = 8$$
$$2x + 4(0) = 8$$
$$2x = 8$$
$$x = 4$$

The x-intercept is $(4, 0)$.

To find the y-intercept, substitute $x = 0$.

$$2x + 4y = 8$$
$$2(0) + 4y = 8$$
$$4y = 8$$
$$y = 2$$

The y-intercept is $(0, 2)$.

In this case, the intercepts are two distinct points and may be used to graph the line. A third point can be found to verify that the points all fall on the same line (points that lie on the same line are said to be *collinear*). Choose a different value for either x or y, such as $y = 4$.

$$2x + 4y = 8$$

$2x + 4(4) = 8$ Substitute $y = 4$.

$2x + 16 = 8$ Solve for x.

$$2x = -8$$

$x = -4$ The point $(-4, 4)$ lines up with the other two points (Figure 2-8).

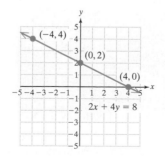

Figure 2-8

Skill Practice

7. Given $-2x + y = -4$, find the x- and y-intercepts. Then graph the equation.

Answer

7.

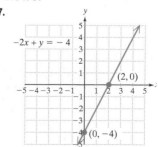

Example 6 Finding the x- and y-Intercepts of a Line

Given $y = \frac{1}{4}x$, find the x- and y-intercepts. Then graph the equation.

Solution:

To find the x-intercept, substitute $y = 0$.

$$y = \frac{1}{4}x$$

$$(0) = \frac{1}{4}x$$

$$0 = x$$

The x-intercept is $(0, 0)$.

To find the y-intercept, substitute $x = 0$.

$$y = \frac{1}{4}x$$

$$y = \frac{1}{4}(0)$$

$$y = 0$$

The y-intercept is $(0, 0)$.

Notice that the x- and y-intercepts are both located at the origin $(0, 0)$. In this case, the intercepts do not yield two distinct points. Therefore, another point is necessary to draw the line. We may pick any value for either x or y. However, for this equation, it would be particularly convenient to pick a value for x that is a multiple of 4 such as $x = 4$.

$$y = \frac{1}{4}x$$

$$y = \frac{1}{4}(4) \qquad \text{Substitute } x = 4.$$

$$y = 1$$

The point $(4, 1)$ is a solution to the equation (Figure 2-9).

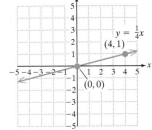

Figure 2-9

Example 7 Interpreting the x- and y-Intercepts of a Line

Companies and corporations are permitted to depreciate assets that have a known useful life span. This accounting practice is called *straight-line depreciation*. In this procedure the useful life span of the asset is determined, and then the asset is depreciated by an equal amount each year until the taxable value of the asset is equal to zero.

The J. M. Gus trucking company purchases a new truck for $65,000. The truck will be depreciated at $13,000 per year. The equation that describes the depreciation line is

$$y = 65{,}000 - 13{,}000x$$

Answer

8.

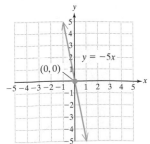

where y represents the value of the truck in dollars and x is the age of the truck in years.

a. Find the x- and y-intercepts. Plot the intercepts on a rectangular coordinate system, and draw the line that represents the straight-line depreciation.

b. What does the x-intercept represent in the context of this problem?

c. What does the y-intercept represent in the context of this problem?

Solution:

a. To find the x-intercept, substitute $y = 0$.

$$0 = 65,000 - 13,000x$$

$$13,000x = 65,000$$

$$x = 5$$

The x-intercept is $(5, 0)$.

To find the y-intercept, substitute $x = 0$.

$$y = 65,000 - 13,000(0)$$

$$y = 65,000$$

The y-intercept is $(0, 65,000)$.

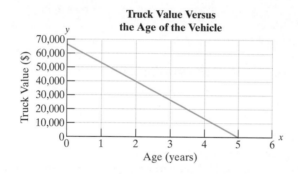

b. The x-intercept $(5, 0)$ indicates that when the truck is 5 yr old, the value of the truck will be $0.

c. The y-intercept $(0, 65,000)$ indicates that when the truck was new (0 yr old), the value was $65,000.

5. Horizontal and Vertical Lines

Recall that a linear equation can be written in the form $Ax + By = C$, where A and B are not both zero. If either A or B is 0, then the resulting line is horizontal or vertical, respectively.

Vertical and Horizontal Lines

1. A **vertical line** is a line that can be written in the form $x = k$, where k is a constant.

2. A **horizontal line** is a line that can be written in the form $y = k$, where k is a constant.

Skill Practice

9. Acme Motor Company tests the engines of its trucks by running the engines in a laboratory. The engines burn 4 gal of fuel per hour. The engines begin the test with 30 gal of fuel. The equation $y = 30 - 4x$ represents the amount of fuel y left in the engine after x hours.
   a. Find the x- and y-intercepts.
   b. Interpret the x-intercept in the context of this problem.
   c. Interpret the y-intercept in the context of this problem.

TIP: In Example 7 we only graphed the line in the first quadrant where both the x- and y-coordinates are positive. (A negative x-coordinate would imply a negative age, and a negative y-coordinate would imply a negative value of a car, neither of which makes sense.)

Answer

9. a. x-intercept: $(7.5, 0)$; y-intercept: $(0, 30)$
   b. The x-intercept $(7.5, 0)$ represents the amount of fuel in the truck after 7.5 hr. After 7.5 hr the tank contains 0 gal. It is empty.
   c. The y-intercept $(0, 30)$ represents the amount of fuel in the truck initially (after 0 hr). After 0 hr, the tank contains 30 gal of fuel.

Skill Practice

10. Graph the equation $x = -4$.

Example 8 **Graphing a Vertical Line**

Graph the equation $x = 6$.

Solution:

Because this equation is in the form $x = k$, the line is vertical and must cross the x-axis at $x = 6$. We can also construct a table of solutions to the equation $x = 6$. The choice for the x-coordinate must be 6, but y can be any real number (Figure 2-10).

x	y
6	-8
6	1
6	5
6	8

→ $(6, -8)$
→ $(6, 1)$
→ $(6, 5)$
→ $(6, 8)$

Figure 2-10

Skill Practice

11. Graph the equation $-2y = 9$.

Example 9 **Graphing a Horizontal Line**

Graph the equation $4y = -7$.

Solution:

The equation $4y = -7$ is equivalent to $y = -\frac{7}{4}$. Because the equation is in the form $y = k$, the line must be horizontal and must pass through the y-axis at $y = -\frac{7}{4}$ (Figure 2-11).

We can also construct a table of solutions to the equation $4y = -7$. The choice for the y-coordinate must be $-\frac{7}{4}$, but x can be any real number.

x	y
0	$-\frac{7}{4}$
-3	$-\frac{7}{4}$
2	$-\frac{7}{4}$

→ $(0, -\frac{7}{4})$
→ $(-3, -\frac{7}{4})$
→ $(2, -\frac{7}{4})$

Figure 2-11

Answers

10.–11.

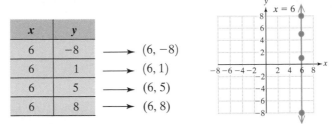

TIP: Notice that horizontal and vertical lines that do not pass through the origin have only one intercept. For instance,

- The vertical line in Example 8 has an x-intercept but no y-intercept.
- The horizontal line in Example 9 has a y-intercept but no x-intercept.

Calculator Connections

Topic: Using the *Table* and *Graph* Features

A viewing window of a graphing calculator shows a portion of a rectangular coordinate system. The *standard viewing window* for most calculators shows both the *x*- and *y*-axes between −10 and 10. Furthermore, the scale defined by the tick marks on both axes is usually set to 1.

The standard viewing window.

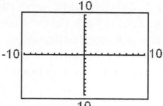

Linear equations can be analyzed with a graphing calculator.

- It is important to isolate the *y* variable in the equation. Then enter the equation in the calculator. For example, to enter the equation from Example 5, we have:

$$2x + 4y = 8 \longrightarrow 4y = -2x + 8$$

$$\frac{4y}{4} = \frac{-2x}{4} + \frac{8}{4}$$

$$y = -\frac{1}{2}x + 2$$

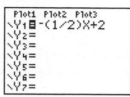

- A *Table* feature can be used to find many solutions to an equation. Several solutions to $y = -\frac{1}{2}x + 2$ are shown here.

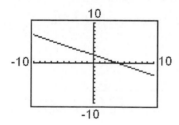

- A *Graph* feature can be used to graph a line.

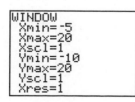

Sometimes the standard viewing window does not provide an adequate display for the graph of an equation. For example, in the standard viewing window, the graph of $y = -x + 15$ is visible only in a small portion of the upper right corner.

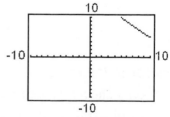

To see the *x*- and *y*-intercepts of this line, we can change the viewing window to accommodate larger values of *x* and *y*. Most calculators have a *Range* or *Window* feature that enables the user to change the minimum and maximum *x* and *y* values. In this case, we changed the values of *x* to range between −5 and 20, and the values of *y* to range between −10 and 20.

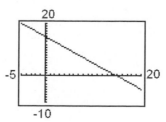

Section 2.1 Practice Exercises

Study Skills Exercise

After getting a test back, it is a good idea to correct the test so that you do not make the same errors again. One recommended approach is to use a clean sheet of paper, and divide the paper down the middle vertically, as shown. For each problem that you missed on the test, rework the problem correctly on the left-hand side of the paper. Then give a written explanation on the right-hand side of the paper. To reinforce the correct procedure, return to the section of text from which the problem was taken and do several more problems.

Take the time this week to make corrections from your last test.

Perform the correct math here.	Explain the process here.
$2 + 4(5)$ $= 2 + 20$ $= 22$	Do multiplication before addition.

Vocabulary and Key Concepts

1. a. In a rectangular coordinate system, two number lines are drawn at right angles to each other. The horizontal line is called the _____-axis, and the vertical line is called the _____.

   b. A point in a rectangular coordinate system is defined by an _____ pair, (x, y).

   c. In a rectangular coordinate system, the point where the x- and y-axes intersect is called the _____ and is represented by the ordered pair _____.

   d. The x- and y-axes divide the coordinate plane into four regions called _____.

   e. A point with a positive x-coordinate and a _____ y-coordinate is located in Quadrant IV.

   f. In Quadrant _____, both the x- and y-coordinates are negative.

   g. A linear equation in two variables is an equation that can be written in the form _____ where A and B are not both zero.

   h. A point where a graph intersects the x-axis is called a(n) _____.

   i. A point where a graph intersects the y-axis is called a(n) _____.

   j. A _____ line can be represented by an equation of the form $x = k$, where k is a constant.

   k. A _____ line can be represented by an equation of the form $y = k$, where k is a constant.

Concept 1: The Rectangular Coordinate System

2. Let a and b represent nonzero real numbers. Then

   a. An ordered pair of the form $(0, b)$ represents a point on which axis?

   b. An ordered pair of the form $(a, 0)$ represents a point on which axis?

3. Given the coordinates of a point, explain how to determine in which quadrant the point is located.

4. What is meant by the word *ordered* in the term *ordered pair*?

5. Plot the points on a rectangular coordinate system. **(See Example 1.)**

   a. $(-2, 1)$ **b.** $(0, 4)$

   c. $(0, 0)$ **d.** $(-3, 0)$

   e. $\left(\dfrac{3}{2}, -\dfrac{7}{3}\right)$ **f.** $(-4.1, -2.7)$

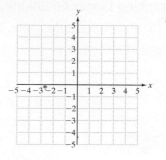

6. Plot the points on a rectangular coordinate system.

   a. $(-2, 5)$ **b.** $\left(\dfrac{5}{2}, 0\right)$

   c. $(4, -3)$ **d.** $(0, -2)$

   e. $(2, 2)$ **f.** $(-3, -3)$

7. A point on the *x*-axis will have what *y*-coordinate?

8. A point on the *y*-axis will have what *x*-coordinate?

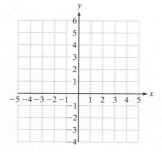

For Exercises 9 and 10, give the coordinates of the labeled points, and state the quadrant or axis where the point is located.

9.

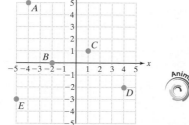

10.

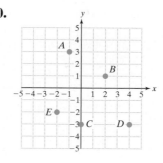

Concept 2: Linear Equations in Two Variables

For Exercises 11–14, determine if the ordered pair is a solution to the linear equation. **(See Example 2.)**

11. $2x - 3y = 9$ **12.** $-5x - 2y = 6$ **13.** $x = \dfrac{1}{3}y + 1$ **14.** $y = -\dfrac{3}{2}x - 4$

   a. $(0, -3)$ **a.** $(0, 3)$ **a.** $(-1, 0)$ **a.** $(0, -4)$

   b. $(-6, 1)$ **b.** $\left(-\dfrac{6}{5}, 0\right)$ **b.** $(2, 3)$ **b.** $(2, -7)$

   c. $\left(1, -\dfrac{7}{3}\right)$ **c.** $(-2, 2)$ **c.** $(-6, 1)$ **c.** $(-4, -2)$

Concept 3: Graphing Linear Equations in Two Variables

For Exercises 15–18, complete the table. Then graph the line defined by the points. **(See Examples 3 and 4.)**

15. $3x - 2y = 4$

x	y
0	
	4
−1	

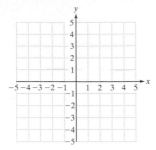

16. $4x + 3y = 6$

x	y
	2
3	
	−1

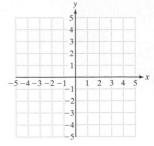

17. $y = -\dfrac{1}{5}x$

x	y
0	
5	
−5	

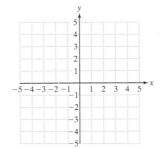

18. $y = \dfrac{1}{3}x$

x	y
0	
3	
6	

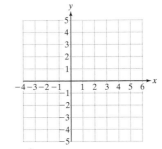

In Exercises 19–30, graph the linear equation by using a table of points. **(See Examples 3 and 4.)**

19. $x + y = 5$

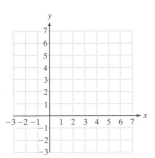

20. $x + y = -8$

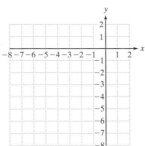

21. $3x - 4y = 12$

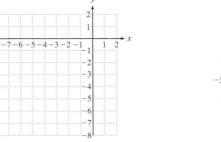

22. $5x + 3y = 15$

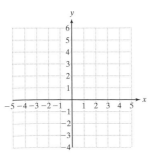

23. $y = -3x + 5$

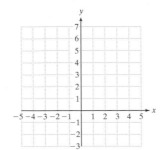

24. $y = -2x + 2$

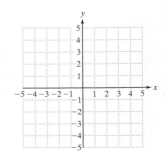

25. $y = \dfrac{2}{5}x - 1$

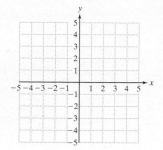

26. $y = \dfrac{5}{3}x + 1$

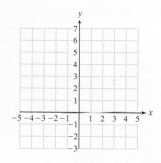

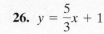 **27.** $x = -5y - 5$

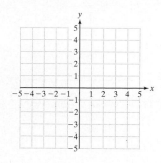

28. $x = 4y + 2$

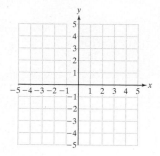

29. $x = 2y$

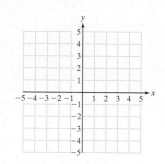

30. $x = -3y$

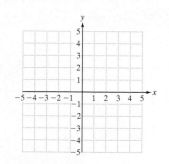

Concept 4: *x*- and *y*-Intercepts

31. Given a linear equation, how do you find an *x*-intercept? How do you find a *y*-intercept?

32. Can the point $(4, -1)$ be an *x*- or *y*-intercept? Why or why not?

For Exercises 33–44, **a.** find the *x*-intercept, **b.** find the *y*-intercept, and **c.** graph the equation. **(See Examples 5 and 6.)**

33. $2x + 3y = 18$

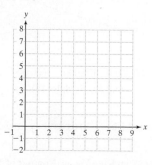

34. $2x - 5y = 10$

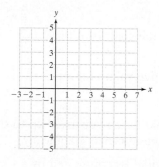

35. $x - 2y = 4$

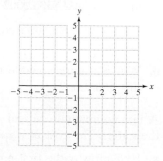

36. $x + y = 8$

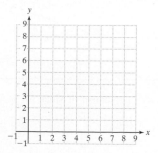

37. $5x = 3y$

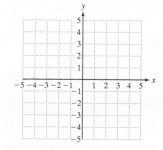

38. $3y = -5x$

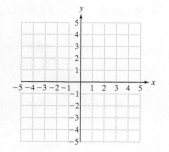

39. $y = 2x + 4$

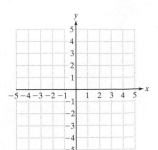

40. $y = -3x - 1$

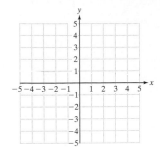

41. $y = -\dfrac{4}{3}x + 2$

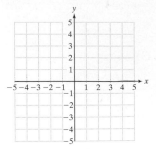

42. $y = -\dfrac{2}{5}x - 1$

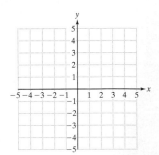

43. $x = \dfrac{1}{4}y$

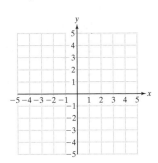

44. $x = \dfrac{2}{3}y$

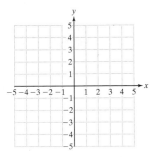

45. A salesperson makes a base salary of $10,000 a year plus a 5% commission on the total sales for the year. The yearly salary can be expressed as a linear equation as

$$y = 10,000 + 0.05x$$

where y represents the yearly salary and x represents the total yearly sales. **(See Example 7.)**

a. What is the salesperson's salary for a year in which his sales total $500,000?

b. What is the salesperson's salary for a year in which his sales total $300,000?

c. What does the y-intercept mean in the context of this problem?

d. Why is it unreasonable to use negative values for x in this equation?

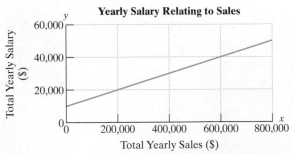

46. A taxi company in Miami charges $2.00 for any distance up to the first mile and $1.10 for every mile thereafter. The cost of a cab ride can be modeled graphically.

a. Explain why the first part of the model is represented by a horizontal line.

b. What does the y-intercept mean in the context of this problem?

c. Explain why the line representing the cost of traveling more than 1 mi is not horizontal.

d. How much would it cost to take a cab $3\frac{1}{2}$ mi?

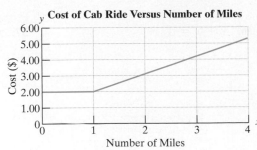

47. A business owner buys several new computers for the office for $1500 each. The accounting office depreciates each computer by $300 per year. The value y (in $) for each computer can be represented by $y = 1500 - 300x$, where x is the number of years after purchase.

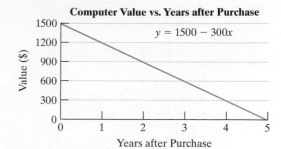

Computer Value vs. Years after Purchase

$y = 1500 - 300x$

a. How much will a computer be worth 1 yr after purchase?

b. After how many years will the computer be worth only $300?

c. Determine the y-intercept and interpret its meaning in the context of this problem.

d. Determine the x-intercept and interpret its meaning in the context of this problem.

48. The equation $y = -3.6x + 59$ can be used to approximate the air temperature y (in °F) at an altitude x (in 1000 ft).

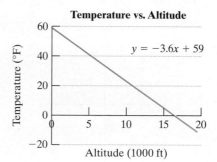

Temperature vs. Altitude

$y = -3.6x + 59$

a. Determine the air temperature at 10,000 ft.

b. At what altitude is the air temperature -5.8°F?

c. Determine the y-intercept and interpret its meaning in the context of this problem.

d. Determine the x-intercept and interpret its meaning in the context of this problem.

Concept 5: Horizontal and Vertical Lines

For Exercises 49–56, identify the line as either vertical or horizontal, and graph the line. **(See Examples 8 and 9.)**

49. $y = -1$

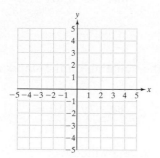

50. $y = 3$

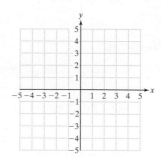

51. $x = 2$

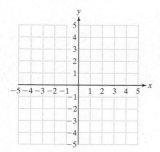

52. $x = -5$

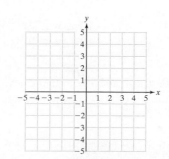

53. $2x + 6 = 5$

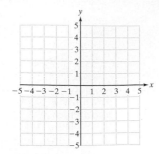

54. $-3x = 12$

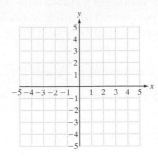

55. $-2y + 1 = 9$

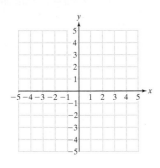

56. $-5y = -10$

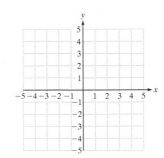

Expanding Your Skills

For Exercises 57–60, find the x- and y-intercepts.

57. $\dfrac{x}{2} + \dfrac{y}{3} = 1$ **58.** $\dfrac{x}{7} + \dfrac{y}{4} = 1$ **59.** $\dfrac{x}{a} + \dfrac{y}{b} = 1$ **60.** $Ax + By = C$

Graphing Calculator Exercises

For Exercises 61–64, solve the equation for y. Use a graphing calculator to graph the equation on the standard viewing window.

61. $2x - 3y = 7$ **62.** $4x + 2y = -2$ **63.** $3y = 9$ **64.** $2y + 10 = 0$

For Exercises 65–68, use a graphing calculator to graph the lines on the suggested viewing window.

65. $y = -\dfrac{1}{2}x - 10$ **66.** $y = -\dfrac{1}{3}x + 12$ **67.** $-2x + 4y = 1$ **68.** $5y = 4x - 1$

$-30 \le x \le 10$ $-10 \le x \le 40$ $-1 \le x \le 1$ $-0.5 \le x \le 0.5$
$-15 \le y \le 5$ $-10 \le y \le 20$ $-1 \le y \le 1$ $-0.5 \le y \le 0.5$

For Exercises 69 and 70, graph the lines in parts (a)–(c) on the same viewing window. Compare the graphs. Are the lines exactly the same?

69. a. $y = x + 3$ **70. a.** $y = 2x + 1$

  b. $y = x + 3.1$ **b.** $y = 1.9x + 1$

  c. $y = x + 2.9$ **c.** $y = 2.1x + 1$

| Slope of a Line and Rate of Change | Section 2.2 |

1. Introduction to the Slope of a Line

In Section 2.1, we learned how to graph a linear equation and to identify its x- and y-intercepts. In this section, we learn about another important feature of a line called the *slope* of a line. Geometrically, slope measures the "steepness" of a line.

Figure 2-12 shows a set of stairs with a wheelchair ramp to the side. Notice that, even though the stairs and ramp both rise the same vertical distance, the stairs are steeper than the ramp. This is because the stairs rise 3 ft over a shorter horizontal distance than the ramp.

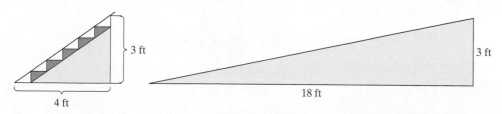

Figure 2-12

To measure the slope of a line quantitatively, consider two points on the line. The slope is the ratio of the vertical change between the two points to the horizontal change. That is, the slope is the ratio of the change in y to the change in x. As a memory device, we might think of the slope of a line as "rise over run."

$$\text{Slope} = \frac{\text{change in } y}{\text{change in } x} = \frac{\text{rise}}{\text{run}}$$

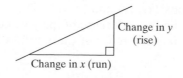

Change in y (rise)

Change in x (run)

To move from point A to point B on the stairs, rise 3 ft and move to the right 4 ft (Figure 2-13).

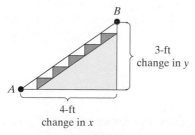

3-ft change in y

4-ft change in x

$$\text{Slope} = \frac{\text{change in } y}{\text{change in } x} = \frac{3 \text{ ft}}{4 \text{ ft}} = \frac{3}{4}$$

Figure 2-13

To move from point A to point B on the wheelchair ramp, rise 3 ft and move to the right 18 ft (Figure 2-14).

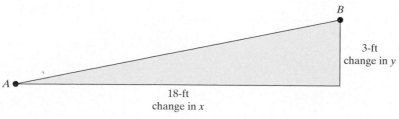

3-ft change in y

18-ft change in x

Figure 2-14

$$\text{Slope} = \frac{\text{change in } y}{\text{change in } x} = \frac{3 \text{ ft}}{18 \text{ ft}} = \frac{1}{6}$$

The slope of the stairs is $\frac{3}{4}$, which is greater than the slope of the ramp, which is $\frac{1}{6}$.

Skill Practice

1. Find the slope of the roof.

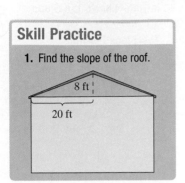

Example 1 Finding the Slope in an Application

Find the slope of the ladder against the wall.

Solution:

$$\text{Slope} = \frac{\text{change in } y}{\text{change in } x}$$

$$= \frac{15 \text{ ft}}{5 \text{ ft}}$$

$$= \frac{3}{1} \text{ or } 3$$

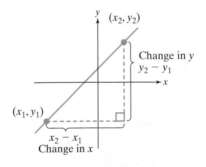

The slope is $\frac{3}{1}$ which indicates that a person climbs 3 ft vertically for every 1 ft traveled horizontally.

2. The Slope Formula

The slope of a line may be found by using *any* two points on the line—call these points (x_1, y_1) and (x_2, y_2). The change in y between the points can be found by taking the difference of the y values: $y_2 - y_1$. The change in x can be found by taking the difference of the x values in the same order: $x_2 - x_1$.

The slope of a line is often symbolized by the letter m and is given by the following formula.

Formula for the Slope of a Line Given Two Points

The **slope** of a line passing through the distinct points (x_1, y_1) and (x_2, y_2) is

$$m = \frac{y_2 - y_1}{x_2 - x_1} \quad \text{provided} \quad x_2 - x_1 \neq 0$$

Answer

1. $\frac{2}{5}$

| Example 2 | **Finding the Slope of a Line Through Two Points** |

Find the slope of the line passing through the points $(1, -1)$ and $(7, 2)$.

Solution:

To use the slope formula, first label the coordinates of each point, and then substitute their values into the slope formula.

$$\underset{(x_1, y_1)}{(1, -1)} \quad \text{and} \quad \underset{(x_2, y_2)}{(7, 2)} \qquad \text{Label the points.}$$

$$m = \frac{y_2 - y_1}{x_2 - x_1} = \frac{2 - (-1)}{7 - 1} \qquad \text{Apply the slope formula.}$$

$$= \frac{3}{6} \qquad \text{Simplify.}$$

$$= \frac{1}{2}$$

The slope of the line can be verified from the graph (Figure 2-15).

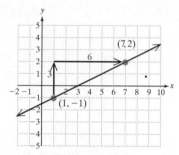

Figure 2-15

TIP: The slope formula does not depend on which point is labeled (x_1, y_1) and which point is labeled (x_2, y_2). For example, reversing the order in which the points are labeled in Example 2 results in the same slope:

$$\underset{(x_2, y_2)}{(1, -1)} \quad \text{and} \quad \underset{(x_1, y_1)}{(7, 2)} \quad \text{then} \quad m = \frac{-1 - 2}{1 - 7} = \frac{-3}{-6} = \frac{1}{2}$$

When you apply the slope formula, you will see that the slope of a line may be positive, negative, zero, or undefined.

- Lines that "increase," or "rise," from left to right have a *positive slope*.
- Lines that "decrease," or "fall," from left to right have a *negative slope*.
- Horizontal lines have a *zero slope*.
- Vertical lines have an *undefined slope*.

Positive slope   Negative slope   Zero slope   Undefined slope

Skill Practice

2. Find the slope of the line that passes through the points $(-4, 5)$ and $(6, 8)$.

Concept Connections

Determine whether the slope of the line is positive, negative, zero, or undefined.

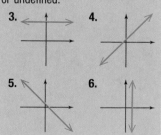

3.   **4.**

5.   **6.**

Answers

2. $\frac{3}{10}$ **3.** Zero   **4.** Positive

5. Negative   **6.** Undefined

Example 3 Finding the Slope of a Line Between Two Points

Find the slope of the line passing through the points $(3, -4)$ and $(-5, -1)$.

Solution:

$(3, -4)$ and $(-5, -1)$

(x_1, y_1) $\qquad$ (x_2, y_2) $\qquad$ Label points.

$$m = \frac{y_2 - y_1}{x_2 - x_1} = \frac{-1 - (-4)}{-5 - 3} \qquad \text{Apply the slope formula.}$$

$$= \frac{3}{-8} = -\frac{3}{8} \qquad \text{Simplify.}$$

The two points can be graphed to verify that $-\frac{3}{8}$ is the correct slope (Figure 2-16).

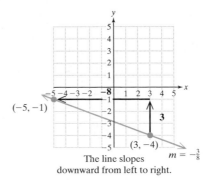

The line slopes
downward from left to right.

Figure 2-16

Example 4 Finding the Slope of a Line Between Two Points

Find the slope of the line passing through the points $(-3, 4)$ and $(-3, -2)$.

Solution:

$(-3, 4)$ and $(-3, -2)$

(x_1, y_1) $\qquad$ (x_2, y_2) $\qquad$ Label points.

$$m = \frac{y_2 - y_1}{x_2 - x_1} = \frac{-2 - 4}{-3 - (-3)} \qquad \text{Apply slope formula.}$$

$$= \frac{-6}{-3 + 3}$$

$$= \frac{-6}{0} \qquad \text{Undefined}$$

The slope is undefined. The points form a vertical line (Figure 2-17).

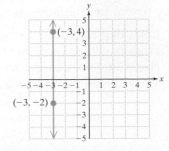

Figure 2-17

Answers

Example 5   **Find the Slope of a Line Between Two Points**

Find the slope of the line passing through the points $(0, 2)$ and $(-4, 2)$.

Solution:

$$(0, 2) \qquad \text{and} \qquad (-4, 2)$$

$(x_1, y_1) \qquad\qquad\qquad (x_2, y_2)$ Label the points.

$$m = \frac{y_2 - y_1}{x_2 - x_1} = \frac{2 - 2}{-4 - 0}$$ Apply the slope formula.

$$= \frac{0}{-4}$$

$$= 0$$ Simplify.

The slope is zero. The line through the two points is a horizontal line (Figure 2-18).

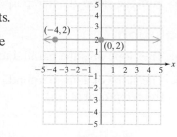

Figure 2-18

Skill Practice

Find the slope of the line that passes through the points.

9. $(1, 6)$ and $(-7, 6)$

3. Parallel and Perpendicular Lines

Lines in the same plane that do not intersect are *parallel*. Nonvertical parallel lines have the same slope and different y-intercepts (Figure 2-19).

Lines that intersect at a right angle are *perpendicular*. If two lines are perpendicular, then the slope of one line is the opposite of the reciprocal of the slope of the other (provided neither line is vertical) (Figure 2-20).

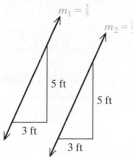

Figure 2-19

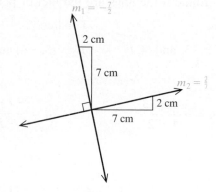

Figure 2-20

Slopes of Parallel Lines

If m_1 and m_2 represent the slopes of two parallel (nonvertical) lines, then

$$m_1 = m_2$$

See Figure 2-19.

Slopes of Perpendicular Lines

If $m_1 \neq 0$ and $m_2 \neq 0$ represent the slopes of two perpendicular lines, then

$$m_1 = -\frac{1}{m_2} \text{ or equivalently, } m_1 \cdot m_2 = -1$$

See Figure 2-20.

Answer

9. 0

Skill Practice

The slope of line L_1 is $-\dfrac{4}{3}$.

10. Find the slope of a line parallel to L_1.

11. Find the slope of a line perpendicular to L_1.

Example 6 Determining the Slope of Parallel and Perpendicular Lines

Suppose a given line has a slope of -5.

  a. Find the slope of a line parallel to the given line.

  b. Find the slope of a line perpendicular to the given line.

Solution:

  a. The slope of a line parallel to the given line is $m = -5$ (same slope).

  b. The slope of a line perpendicular to the given line is $m = \frac{1}{5}$ (the opposite of the reciprocal of -5).

Skill Practice

Two points are given for lines L_1 and L_2. Determine if the lines are parallel, perpendicular, or neither.

12. L_1: $(4, -1)$ and $(-3, 6)$
    L_2: $(-1, 3)$ and $(2, 0)$

Example 7 Determining Whether Two Lines Are Parallel, Perpendicular, or Neither

Two points are given from each of two lines: L_1 and L_2. Without graphing the points, determine if the lines are parallel, perpendicular, or neither.

L_1: $(2, -3)$ and $(4, 1)$

L_2: $(5, -6)$ and $(-3, -2)$

Solution:

First determine the slope of each line. Then compare the values of the slopes to determine if the lines are parallel or perpendicular.

For line 1:	For line 2:	
L_1: $(2, -3)$ and $(4, 1)$	L_2: $(5, -6)$ and $(-3, -2)$	
$(x_1, y_1)\quad\ (x_2, y_2)$	$(x_1, y_1)\quad\ (x_2, y_2)$	Label the points.
$m = \dfrac{1 - (-3)}{4 - 2}$	$m = \dfrac{-2 - (-6)}{-3 - 5}$	Apply the slope formula.
$= \dfrac{4}{2}$	$= \dfrac{4}{-8}$	
$= 2$	$= -\dfrac{1}{2}$	

The slope of L_1 is 2. The slope of L_2 is $-\frac{1}{2}$. The slope of L_1 is the opposite of the reciprocal of L_2. By comparing the slopes, the lines must be perpendicular.

TIP: You can also verify that the lines in Example 7 are perpendicular by noting that the product of their slopes is -1.

$$2 \times \left(-\frac{1}{2}\right) = -1$$

Answers

10. $-\dfrac{4}{3}$ **11.** $\dfrac{3}{4}$ **12.** Parallel

4. Applications and Interpretation of Slope

Example 8 demonstrates that the slope of a line represents a rate of change.

Example 8 **Interpreting the Slope of a Line in an Application**

The number of males 20 years old or older who were employed full time in the United States has grown linearly since 1970. Approximately 43.0 million males 20 years old or older were employed full time in 1970. By 2010, this number had grown to 69.0 million (Figure 2-21).

Number of Males 20 Years or Older Employed Full Time in the United States

Figure 2-21

Source: U.S. Census Bureau

a. Find the slope of the line, using the points $(1970, 43.0)$ and $(2005, 65.4)$.

b. Interpret the meaning of the slope in the context of this problem.

Solution:

a. $(1970, 43.0)$ and $(2005, 65.4)$

$\quad (x_1, y_1) \qquad\qquad (x_2, y_2)$ Label the points.

$$m = \frac{y_2 - y_1}{x_2 - x_1} = \frac{65.4 - 43.0}{2005 - 1970}$$ Apply the slope formula.

$$m = \frac{22.4}{35} \quad \text{or} \quad m = 0.64$$

b. The slope is approximately 0.64, meaning that the full-time workforce has increased at a rate of 0.64 million men (or 640,000 men) per year between 1970 and 2005.

Section 2.2 Practice Exercises

Study Skills Exercise

Go to the online service that accompanies the text. Name two features that this online service offers that can help you in this course.

Vocabulary and Key Concepts

1. a. The ratio of the vertical change to the horizontal change between two distinct points (x_1, y_1) and (x_2, y_2) on a line is called the _____ of the line. The slope can be computed from the formula $m =$ _____.

b. Lines in the same plane that do not intersect are called _____ lines. Such lines, if they are nonvertical, have the _____ slope and different y-intercepts.

c. Two lines are perpendicular if they intersect at a _____ angle.

d. If m_1 and m_2 represent the slopes of two nonvertical perpendicular lines, then $m_1 \cdot m_2 =$ _____.

Review Exercises

2. For each ordered pair, determine the quadrant in which the point lies.

   a. $(-5.73, 102.4)$ b. $(-1500, -2030)$ c. $\left(\dfrac{367}{1013}, \dfrac{41}{512}\right)$ d. $(\pi, -2.1718)$

3. Find the missing coordinate so that the ordered pairs are solutions to the equation $\frac{1}{2}x + y = 4$.

   a. $(0, \ )$ b. $(\ , 0)$ c. $(-4, \ )$

For Exercises 4–6, find the x- and y-intercepts (if possible) for each equation, and sketch the graph.

4. $2x + 8 = 0$ 5. $4 - 2y = 0$ 6. $2x - 2y - 6 = 0$

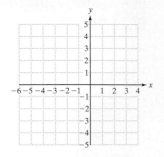   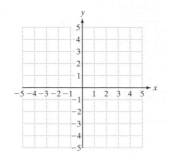   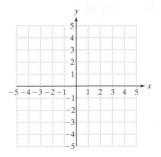

Concept 1: Introduction to the Slope of a Line

7. A 25-ft ladder is leaning against a house, as shown in the diagram. Find the slope of the ladder. **(See Example 1.)**

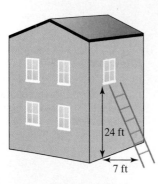

8. Find the pitch (slope) of the roof in the figure.

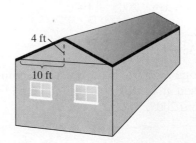

9. Find the slope of the treadmill.

10. Find the average slope of the hill.

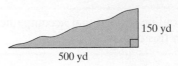

11. The road sign shown in the figure indicates the percent grade of a hill. This gives the slope of the road as the change in elevation per 100 horizontal ft. Given a 4% grade, write this as a slope in fractional form.

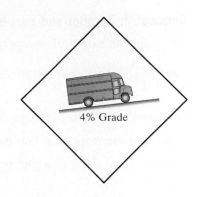

4% Grade

12. If a plane gains 1000 ft in altitude over a distance of 12,000 horizontal ft, what is the slope? Explain what this value means in the context of the problem.

Concept 2: The Slope Formula

For Exercises 13–30, use the slope formula to determine the slope of the line containing the two points. **(See Examples 2–5.)**

13. $(6, 0)$ and $(0, -3)$

14. $(-5, 0)$ and $(0, 4)$

15. $(-2, 3)$ and $(1, -2)$

16. $(4, 5)$ and $(-1, 0)$

17. $(-2, 5)$ and $(-7, 1)$

18. $(4, -2)$ and $(3, -1)$

19. $(0.3, -1.1)$ and $(-0.1, -0.8)$

20. $(0.4, -0.2)$ and $(0.3, -0.1)$

21. $(2, 3)$ and $(2, 7)$

22. $(-1, 5)$ and $(-1, 0)$

23. $(5, -1)$ and $(-3, -1)$

24. $(-8, 4)$ and $(1, 4)$

25. $(-4.6, 4.1)$ and $(0, 6.4)$

26. $(1.1, 4)$ and $(-3.2, -0.3)$

27. $\left(\frac{3}{2}, \frac{4}{3}\right)$ and $\left(\frac{7}{2}, 1\right)$

28. $\left(\frac{2}{3}, -\frac{1}{2}\right)$ and $\left(-\frac{1}{6}, -\frac{3}{2}\right)$

29. $\left(\frac{3}{4}, \frac{7}{3}\right)$ and $\left(\frac{1}{2}, 2\frac{1}{3}\right)$

30. $\left(\frac{9}{4}, \frac{2}{5}\right)$ and $\left(2\frac{1}{4}, \frac{1}{10}\right)$

31. Explain how to use the graph of a line to determine whether the slope of a line is positive, negative, zero, or undefined.

32. If the slope of a line is $\frac{4}{3}$, how many units of change in y will be produced by 6 units of change in x?

For Exercises 33–38, estimate the slope of the line from its graph.

33.
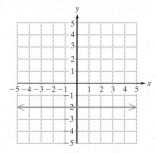

34.
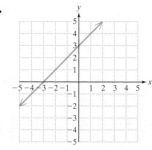

35.
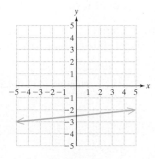

36.
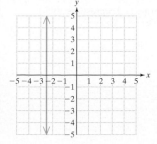

37.
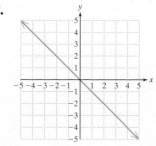

38.
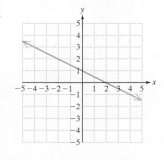

Concept 3: Parallel and Perpendicular Lines

39. Can the slopes of two perpendicular lines both be positive? Explain your answer.

40. Suppose a line is defined by the equation $x = 2$. What is the slope of a line perpendicular to this line?

For Exercises 41–46, the slope of a line is given.

   a. Find the slope of a line parallel to the given line.

   b. Find the slope of a line perpendicular to the given line. **(See Example 6.)**

41. $m = 5$ **42.** $m = 3$ **43.** $m = -\dfrac{4}{7}$

44. $m = -\dfrac{2}{11}$ **45.** $m = 0$ **46.** m is undefined.

In Exercises 47–54, two points are given from each of two lines L_1 and L_2. Without graphing the points, determine if the lines are parallel, perpendicular, or neither. **(See Example 7.)**

47. L_1: $(2, 5)$ and $(4, 9)$
    L_2: $(-1, 4)$ and $(3, 2)$

48. L_1: $(-3, -5)$ and $(-1, 2)$
    L_2: $(0, 4)$ and $(7, 2)$

49. L_1: $(4, -2)$ and $(3, -1)$
    L_2: $(-5, -1)$ and $(-10, -16)$

50. L_1: $(0, 0)$ and $(2, 3)$
    L_2: $(-2, 5)$ and $(0, -2)$

51. L_1: $(5, 3)$ and $(5, 9)$
    L_2: $(4, 2)$ and $(0, 2)$

52. L_1: $(3, 5)$ and $(2, 5)$
    L_2: $(2, 4)$ and $(0, 4)$

53. L_1: $(-3, -2)$ and $(2, 3)$
    L_2: $(-4, 1)$ and $(0, 5)$

54. L_1: $(7, 1)$ and $(0, 0)$
    L_2: $(-10, -8)$ and $(4, -6)$

Concept 4: Applications and Interpretation of Slope

55. The graph shows the number of cellular phone subscriptions (in millions) purchased in the United States for selected years. **(See Example 8.)**

   a. Use the coordinates of the given points to find the slope of the line, and express the answer in decimal form.

   b. Interpret the meaning of the slope in the context of this problem.

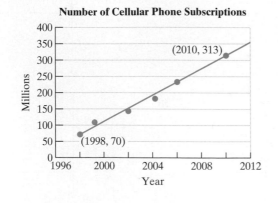

56. The U.S. population (in millions) has grown approximately linearly since the year 2000.

   a. Find the slope of the line defined by the two given points.

   b. Interpret the meaning of the slope in the context of this problem.

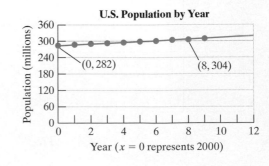

57. The data in the graph show the average weight for boys based on age.

    a. Use the coordinates of the given points to find the slope of the line.

    b. Interpret the meaning of the slope in the context of this problem.

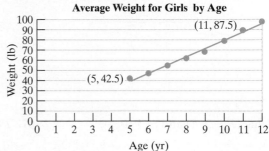

Average Weight for Boys by Age

58. The data in the graph show the average weight for girls based on age.

    a. Use the coordinates of the given points to find the slope of the line, and write the answer in decimal form.

    b. Interpret the meaning of the slope in the context of this problem.

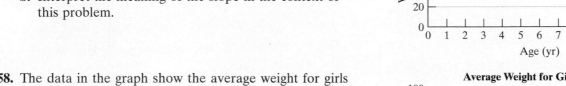

Average Weight for Girls by Age

Mixed Exercises

The slope of a line is constant. This also means that the rate of change of y versus x is constant between any two points on the line. For Exercises 59 and 60, demonstrate this statement by computing the slope of the line using each pair of points on the line.

59. a. $(-1, -4)$ and $(0, -2)$

    b. $(0, -2)$ and $(3, 4)$

    c. $(-1, -4)$ and $(3, 4)$

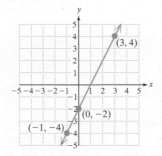

60. a. $(-4, -1)$ and $(0, 1)$

    b. $(0, 1)$ and $(2, 2)$

    c. $(0, 1)$ and $(4, 3)$

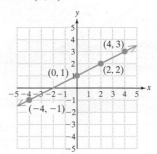

Given a nonlinear graph, the rate of change of y versus x may differ for different pairs of points on the graph. For Exercises 61 and 62, demonstrate this statement by computing the slope formula with each pair of points on the curve.

61. a. $(-2, 0)$ and $(0, -4)$

    b. $(0, -4)$ and $(2, 0)$

    c. $(2, 0)$ and $(3, 5)$

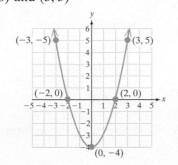

62. a. $(-3, -8)$ and $(-2, -3)$

    b. $(-2, -3)$ and $(0, 1)$

    c. $(0, 1)$ and $(2, -3)$

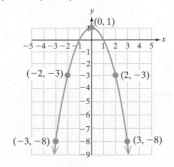

Expanding Your Skills

For Exercises 63–68, given a point P on a line and the slope m of the line, find a second point on the line (answers may vary). *Hint:* Graph the line to help you find the second point.

63. $P(0, 0)$ and $m = 2$

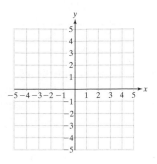

64. $P(-2, 1)$ and $m = -\dfrac{1}{3}$

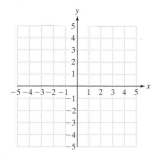

65. $P(2, -3)$ and m is undefined

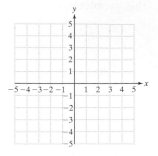

66. $P(-2, 4)$ and $m = 0$

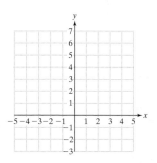

67. $P(-1, 2)$ and $m = -\dfrac{2}{3}$

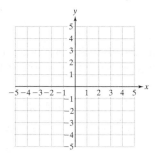

68. $P(-1, -4)$ and $m = \dfrac{4}{5}$

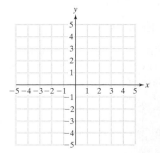

Section 2.3   Equations of a Line

Concepts

1. Slope-Intercept Form
2. The Point-Slope Formula
3. Equations of a Line: A Summary

1. Slope-Intercept Form

In Section 2.1, we learned that an equation of the form $Ax + By = C$ (where A and B are not both zero) represents a line in a rectangular coordinate system. An equation of a line written in this way is in **standard form**. In this section, we will learn a new form, called the **slope-intercept form**, which is useful in determining the slope and y-intercept of a line.

Let $(0, b)$ represent the y-intercept of a line. Let (x, y) represent any other point on the line where $x \neq 0$. Then the slope of the line through the two points is

$$m = \frac{y_2 - y_1}{x_2 - x_1} \quad \rightarrow \quad m = \frac{y - b}{x - 0} \qquad \text{Apply the slope formula.}$$

$$m = \frac{y - b}{x} \qquad \text{Simplify.}$$

$$m \cdot x = \left(\frac{y - b}{x}\right)x \qquad \text{Clear fractions.}$$

$$mx = y - b \qquad \text{Simplify.}$$

$$mx + b = y \quad \text{or} \quad y = mx + b \qquad \text{Solve for } y\text{: slope-intercept form.}$$

Slope-Intercept Form of a Line

$y = mx + b$ is the slope-intercept form of a line.

m is the slope and the point $(0, b)$ is the y-intercept.

The equation $y = -4x + 7$ is written in slope-intercept form. By inspection, we can see that the slope of the line is -4 and the y-intercept is $(0, 7)$.

Example 1 Finding the Slope and y-Intercept of a Line

Given $3x + 4y = 4$, write the equation of the line in slope-intercept form. Then find the slope and y-intercept.

Solution:

Write the equation in slope-intercept form, $y = mx + b$, by solving for y.

$$3x + 4y = 4$$

$$4y = -3x + 4$$

$$\frac{4y}{4} = \frac{-3x}{4} + \frac{4}{4}$$

$$y = -\frac{3}{4}x + 1 \qquad \text{The slope is } -\frac{3}{4} \text{ and the } y\text{-intercept is } (0, 1).$$

The slope-intercept form is a useful tool to graph a line. The y-intercept is a known point on the line, and the slope indicates the "direction" of the line and can be used to find a second point. Using slope-intercept form to graph a line is demonstrated in Example 2.

Example 2 Graphing a Line by Using the Slope and y-Intercept

Graph the equation $y = -\frac{3}{4}x + 1$ by using the slope and y-intercept.

Solution:

First plot the y-intercept $(0, 1)$. The slope $m = -\frac{3}{4}$ can be written as

$$m = \frac{-3}{4}$$

The change in y is -3.

The change in x is 4.

To find a second point on the line, start at the y-intercept and move *down* 3 *units* and to the *right* 4 *units*. Then draw the line through the two points (Figure 2-22).

Similarly, the slope can be written as

$$m = \frac{3}{-4}$$

The change in y is 3.

The change in x is -4.

To find a second point on the line, start at the y-intercept and move *up* 3 *units* and to the *left* 4 *units*. Then draw the line through the two points (see Figure 2-22).

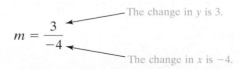

Figure 2-22

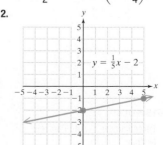

As we have seen earlier, two lines are parallel if they have the same slope and different y-intercepts. Two lines are perpendicular if the slope of one line is the opposite of the reciprocal of the slope of the other line. Otherwise, the lines are neither parallel nor perpendicular.

Example 3   **Determining if Two Lines Are Parallel, Perpendicular, or Neither**

Given the pair of linear equations, determine if the lines are parallel, perpendicular, or neither parallel nor perpendicular.

a. L_1: $y = -2x + 7$ **b.** L_1: $2y = -3x + 2$ **c.** L_1: $x + y = 6$

   L_2: $y = -2x - 1$ L_2: $-4x + 6y = -12$ L_2: $y = 6$

Solution:

a. The equations are written in slope-intercept form.

   L_1: $y = -2x + 7$ The slope is -2 and the y-intercept is $(0, 7)$.

   L_2: $y = -2x - 1$ The slope is -2 and the y-intercept is $(0, -1)$.

   Because the slopes are the same and the y-intercepts are different, the lines are parallel.

b. Write each equation in slope-intercept form by solving for y.

   L_1: $2y = -3x + 2$ L_2: $-4x + 6y = -12$

   $\dfrac{2y}{2} = \dfrac{-3x}{2} + \dfrac{2}{2}$ Divide by 2. $6y = 4x - 12$ Add $4x$ to both sides.

   $y = -\dfrac{3}{2}x + 1$ $\dfrac{6y}{6} = \dfrac{4}{6}x - \dfrac{12}{6}$ Divide by 6.

                                             $y = \dfrac{2}{3}x - 2$

   The slope of L_1 is $-\dfrac{3}{2}$.   The slope of L_2 is $\dfrac{2}{3}$.

   The value $-\frac{3}{2}$ is the opposite of the reciprocal of $\frac{2}{3}$. Therefore, the lines are perpendicular.

c. L_1: $x + y = 6$ is equivalent to $y = -x + 6$. The slope is -1.

   L_2: $y = 6$ is a horizontal line, and the slope is 0.

   The slopes are not the same. Therefore, the lines are not parallel. The slope of one line is not the opposite of the reciprocal of the other slope. Therefore, the lines are not perpendicular. The lines are neither parallel nor perpendicular.

Example 4   **Using Slope-Intercept Form to Find an Equation of a Line**

Use slope-intercept form to find an equation of the line with slope -3 and passing through the point $(1, -4)$.

Solution:

To find an equation of a line in slope-intercept form, $y = mx + b$, it is necessary to find the slope, m, and the y-intercept, b. The slope is given in the problem as $m = -3$. Therefore, the slope-intercept form becomes

$$y = mx + b$$

$$\downarrow$$

$$y = -3x + b$$

Furthermore, because the point $(1, -4)$ is on the line, it is a solution to the equation. Therefore, if we substitute $(1, -4)$ for x and y in the equation, we can solve for b.

$$-4 = -3(1) + b$$
$$-4 = -3 + b$$
$$-1 = b$$

Thus, the slope-intercept form is $y = -3x - 1$.

Skill Practice

6. Use slope-intercept form to find an equation of the line with slope 2 and passing through $(-3, -5)$.

Calculator Connections

Topic: Using the *Value* Feature

The *Value* feature of a graphing calculator prompts the user for a value of x, and then returns the corresponding y value of an equation.

We can check the answer to Example 4 by graphing the equation, $y = -3x - 1$. Using the *Value* feature, we see that the line passes through the point $(1, -4)$ as expected.

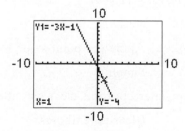

Answer

6. $y = 2x + 1$

2. The Point-Slope Formula

In Example 4, we used the slope-intercept form of a line to construct an equation of a line given its slope and a known point on the line. Here we provide another tool called the *point-slope formula* that (as its name suggests) can accomplish the same result.

Suppose a nonvertical line passes through a given point (x_1, y_1) and has slope m. If (x, y) is any other point on the line, then

$$m = \frac{y - y_1}{x - x_1} \qquad \text{Slope formula}$$

$$m(x - x_1) = \frac{y - y_1}{x - x_1}(x - x_1) \qquad \text{Clear fractions.}$$

$$m(x - x_1) = y - y_1$$

or

$$y - y_1 = m(x - x_1) \qquad \text{Point-slope formula}$$

> ### The Point-Slope Formula
> The **point-slope formula** is given by
> $$y - y_1 = m(x - x_1)$$
> where m is the slope of the line and (x_1, y_1) is a known point on the line.

The point-slope formula is used specifically to find an equation of a line when a point on the line is known and the slope is known. To illustrate the point-slope formula, we will repeat the problem from Example 4.

Example 5 Using the Point-Slope Formula to Find an Equation of a Line

Use the point-slope formula to find an equation of the line having a slope of -3 and passing through the point $(1, -4)$. Write the answer in slope-intercept form.

Solution:

$$m = -3 \qquad \text{and} \qquad (x_1, y_1) = (1, -4)$$

$$y - y_1 = m(x - x_1)$$

$$y - (-4) = -3(x - 1) \qquad \text{Apply the point-slope formula.}$$

$$y + 4 = -3(x - 1) \qquad \text{Simplify.}$$

To write the answer in slope-intercept form, clear parentheses and solve for y.

$$y + 4 = -3x + 3 \qquad \text{Clear parentheses.}$$

$$y = -3x - 1 \qquad \text{Solve for } y. \text{ The answer is written in slope-intercept form. Notice that this is the same equation as in Example 4.}$$

TIP: The solution to Example 5 can also be written in standard form, $Ax + By = C$.

$$y = -3x - 1 \qquad \text{(Slope-intercept form)}$$

$$3x + y = -3x + 3x - 1 \qquad \text{Add } 3x \text{ to both sides.}$$

$$3x + y = -1 \qquad \text{(Standard form)}$$

In general, we will write the solution in slope-intercept form, because the slope and y-intercept can be easily identified.

Example 6 Finding an Equation of a Line Given Two Points

Find an equation of the line passing through the points $(5, -1)$ and $(3, 1)$. Write the answer in slope-intercept form.

Solution:

The slope formula can be used to compute the slope of the line between two points. Once the slope is known, the point-slope formula can be used to find an equation of the line.

First find the slope.

$$m = \frac{y_2 - y_1}{x_2 - x_1} = \frac{1 - (-1)}{3 - 5} = \frac{2}{-2} = -1$$

Next, apply the point-slope formula.

$$y - y_1 = m(x - x_1)$$

$$y - 1 = -1(x - 3) \qquad \text{Substitute } m = -1 \text{ and use } \textit{either} \text{ point for } (x_1, y_1). \text{ We will use } (3, 1) \text{ for } (x_1, y_1).$$

$$y - 1 = -x + 3 \qquad \text{Clear parentheses.}$$

$$y = -x + 4 \qquad \text{Solve for } y. \text{ The final answer is in slope-intercept form.}$$

TIP: In Example 6, the point $(3, 1)$ was used for (x_1, y_1) in the point-slope formula. However, either point could have been used. Using the point $(5, -1)$ for (x_1, y_1) produces the same final equation:

$$y - (-1) = -1(x - 5)$$

$$y + 1 = -x + 5$$

$$y = -x + 4$$

Example 7 Finding an Equation of a Line Parallel to Another Line

Find an equation of the line passing through the point $(-2, -3)$ and parallel to the line $4x + y = 8$. Write the answer in slope-intercept form.

Solution:

To find an equation of a line, we must know a point on the line and the slope. The known point is $(-2, -3)$. Because the line is parallel to $4x + y = 8$, the two lines must have the same slope. Writing the equation $4x + y = 8$ in slope-intercept form, we have $y = -4x + 8$. Therefore, the slope of both lines must be -4.

Skill Practice

8. Use the point-slope formula to write an equation of the line that passes through the points $(-5, 2)$ and $(-1, -1)$. Write the answer in slope-intercept form.

Skill Practice

9. Find an equation of a line containing $(4, -1)$ and parallel to $2x = y - 7$. Write the answer in slope-intercept form.

Answers

8. $y = -\dfrac{3}{4}x - \dfrac{7}{4}$ 9. $y = 2x - 9$

Now find an equation of the line passing through $(-2, -3)$ having a slope of -4.

$$y - y_1 = m(x - x_1) \qquad \text{Apply the point-slope formula.}$$

$$y - (-3) = -4[x - (-2)] \qquad \text{Substitute } m = -4 \text{ and } (-2, -3) \text{ for } (x_1, y_1).$$

$$y + 3 = -4(x + 2)$$

$$y + 3 = -4x - 8 \qquad \text{Clear parentheses.}$$

$$y = -4x - 11 \qquad \text{Write the answer in slope-intercept form.}$$

We can verify the answer to Example 7 by graphing both equations. We see that the line defined by $y = -4x - 11$ passes through the point $(-2, -3)$ and is parallel to the line $y = -4x + 8$. See Figure 2-23.

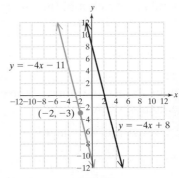

Figure 2-23

Skill Practice

10. Find an equation of the line passing through the point $(1, -6)$ and perpendicular to the line $x + 2y = 8$. Write the answer in slope-intercept form.

Example 8 Finding an Equation of a Line Perpendicular to Another Line

Find an equation of the line passing through the point $(4, 3)$ and perpendicular to the line $2x + 3y = 3$. Write the answer in slope-intercept form.

Solution:

The slope of the given line can be found from its slope-intercept form.

$$2x + 3y = 3$$

$$3y = -2x + 3 \qquad \text{Solve for } y.$$

$$\frac{3y}{3} = \frac{-2x}{3} + \frac{3}{3}$$

$$y = -\frac{2}{3}x + 1 \qquad \text{The slope is } -\frac{2}{3}.$$

The slope of a line *perpendicular* to this line must be the opposite of the reciprocal of $-\frac{2}{3}$; hence, $m = \frac{3}{2}$. Using $m = \frac{3}{2}$ and the known point $(4, 3)$, we can apply the point-slope formula to find an equation of the line.

$$y - y_1 = m(x - x_1) \qquad \text{Apply the point-slope formula.}$$

$$y - 3 = \frac{3}{2}(x - 4) \qquad \text{Substitute } m = \frac{3}{2} \text{ and } (4, 3) \text{ for } (x_1, y_1).$$

$$y - 3 = \frac{3}{2}x - 6 \qquad \text{Clear parentheses.}$$

$$y = \frac{3}{2}x - 3 \qquad \text{Solve for } y.$$

Answer

10. $y = 2x - 8$

Calculator Connections

From Example 8, the line defined by $y = \frac{3}{2}x - 3$ should be perpendicular to the line $y = -\frac{2}{3}x + 1$ and should pass through the point $(4, 3)$.

Note: In this example, we are using a *square window option*, which sets the scale to display distances on the x- and y-axes as equal units of measure.

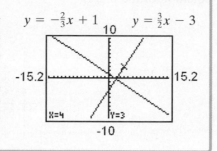

$y = -\frac{2}{3}x + 1 \qquad y = \frac{3}{2}x - 3$

3. Equations of a Line: A Summary

A linear equation can be written in several different forms, as summarized in Table 2-2.

Table 2-2

Form	Example	Comments
Standard Form $Ax + By = C$	$2x + 3y = 6$	A and B must not *both* be zero.
Horizontal Line $y = k$ (k is constant)	$y = 3$	The slope is zero, and the y-intercept is $(0, k)$.
Vertical Line $x = k$ (k is constant)	$x = -2$	The slope is undefined and the x-intercept is $(k, 0)$.
Slope-Intercept Form $y = mx + b$ Slope is m. y-Intercept is $(0, b)$.	$y = -2x + 5$ Slope $= -2$ y-Intercept is $(0, 5)$.	Solving a linear equation for y results in slope-intercept form. The coefficient of the x term is the slope, and the constant defines the location of the y-intercept.
Point-Slope Formula $y - y_1 = m(x - x_1)$ Slope is m and (x_1, y_1) is a point on the line.	$m = -2$ $(x_1, y_1) = (3, 1)$ $y - 1 = -2(x - 3)$	This formula is typically used to build an equation of a line when a point on the line is known and the slope is known.

Although it is important to understand and apply slope-intercept form and the point-slope formula, they are not necessarily applicable to all problems. Example 9 illustrates how a little ingenuity may lead to a simple solution.

Example 9 **Finding an Equation of a Line**

Find an equation of the line passing through the point $(-4, 1)$ and perpendicular to the x-axis.

Solution:

Any line perpendicular to the x-axis must be *vertical*. Recall that all vertical lines can be written in the form $x = k$, where k is constant. A quick sketch can help find the value of the constant (Figure 2-24).

Skill Practice

11. Write an equation of the line through the point $(20, 50)$ and having a slope of 0.

Answer

11. $y = 50$

Because the line must pass through a point whose x-coordinate is -4, the equation of the line is $x = -4$.

Concept Connections

12. Explain why the point-slope formula cannot be used to find an equation for a vertical line.

Answer

12. The slope of a vertical line is undefined. Therefore, there is no value to substitute for the slope in the point-slope formula.

Figure 2-24

Section 2.3 Practice Exercises

Study Skills Exercise

For this chapter, find the page numbers for the Chapter Review Exercises, the Chapter Test, and the Cumulative Review Exercises.

Chapter Review Exercises, page(s)_____

Chapter Test, page(s)_____

Cumulative Review Exercises, page(s)_____

Compare these features and state the advantages of each.

Vocabulary and Key Concepts

1. a. Consider a line with slope m and y-intercept $(0, b)$. The slope-intercept form of an equation of the line is _____ .

   b. An equation of a line written in the form $Ax + By = C$, where A and B are not both zero, is said to be in _____ form.

   c. A line defined by an equation $y = k$, where k is a constant is a (horizontal/vertical) line.

   d. A line defined by an equation $x = k$, where k is a constant is a (horizontal/vertical) line.

   e. Given the slope-intercept form of an equation of a line, $y = mx + b$, the value of m is the _____ and b is the _____ .

   f. Given a point (x_1, y_1) on a line with slope m, the point-slope formula is given by _____ .

Review Exercises

2. Given $\dfrac{x}{2} + \dfrac{y}{3} = 1$

   a. Find the x-intercept.

   b. Find the y-intercept.

   c. Sketch the graph.

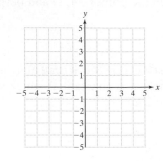

3. a. What is the slope of a horizontal line?

   b. The slope of a vertical line is _____.

4. Using slopes, how do you determine whether two lines are parallel?

5. Using the slopes of two lines, how do you determine whether the lines are perpendicular?

6. Write the formula to find the slope of a line given two points (x_1, y_1) and (x_2, y_2).

Concept 1: Slope-Intercept Form

For Exercises 7–18, determine the slope and the y-intercept of the line. **(See Example 1.)**

7. $y = -\dfrac{2}{3}x - 4$ **8.** $y = \dfrac{3}{7}x - 1$ **9.** $-3x + y = 2$ **10.** $-7x - y = -5$

11. $17x + y = 0$ **12.** $x + y = 0$ **13.** $18 = 2y$ **14.** $-7 = \dfrac{1}{2}y$

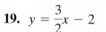 **15.** $8x + 12y = 9$ **16.** $-9x + 10y = -4$ **17.** $y = 0.625x - 1.2$ **18.** $y = -2.5x + 1.8$

In Exercises 19–24, match the equation with the correct graph.

19. $y = \dfrac{3}{2}x - 2$ **20.** $y = -x + 3$ **21.** $y = \dfrac{13}{4}$

22. $y = x + \dfrac{1}{2}$ **23.** $x = -2$ **24.** $y = -\dfrac{1}{2}x + 2$

a.

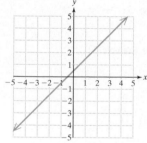

b.

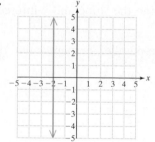

c.

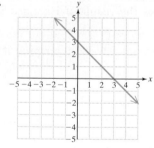

d.

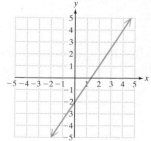

e.

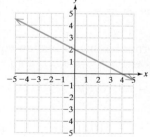

f.

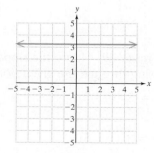

For Exercises 25–30, write the equations in slope-intercept form (if possible). Then graph each line, using the slope and y-intercept. **(See Example 2.)**

25. $y - 2 = 4x$

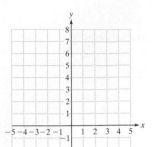

26. $3x = 5 - y$

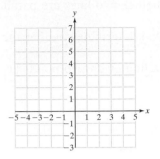

27. $3x + 2y = 6$

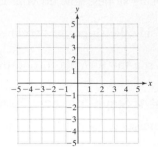

28. $x - 2y = 8$

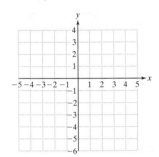

29. $2x - 5y = 0$

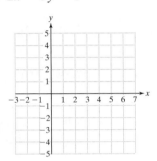

30. $3x - y = 0$

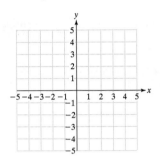

31. Given the standard form of a linear equation $Ax + By = C$, $B \neq 0$, solve for y and write the equation in slope-intercept form. What is the slope of the line? What is the y-intercept?

32. Use the result of Exercise 31 to determine the slope and y-intercept of the line $3x + 7y = 9$.

For Exercises 33–38, determine if the lines are parallel, perpendicular, or neither. **(See Example 3.)**

33. $-3y = 5x - 1$
    $6x = 10y - 12$

34. $x = 6y - 3$
    $3x + \dfrac{1}{2}y = 0$

35. $3x - 4y = 12$
    $\dfrac{1}{2}x - \dfrac{2}{3}y = 1$

36. $4.8x = 1.2y + 3.6$
    $y - 1 = 4x$

37. $3y = 5x + 6$
    $5x + 3y = 9$

38. $-y = 3x - 2$
    $-6x + 2y = 6$

For Exercises 39–44, use the slope-intercept form of a line to find an equation of the line having the given slope and passing through the given point. **(See Example 4.)**

39. $m = 3, (0, 5)$

40. $m = -4, (0, 3)$

41. $m = 2, (4, -3)$

42. $m = 3, (-1, 5)$

43. $m = -\dfrac{4}{5}, (10, 0)$

44. $m = -\dfrac{2}{7}, (-3, 1)$

Concept 2: The Point-Slope Formula

For Exercises 45–72, write an equation of the line satisfying the given conditions. Write the answer in slope-intercept form or standard form.

45. The line passes through the point $(0, -2)$ and has a slope of 3.

46. The line passes through the point $(0, 5)$ and has a slope of $-\frac{1}{2}$.

47. The line passes through the point $(2, 7)$ and has a slope of 2. **(See Example 5.)**

48. The line passes through the point $(3, 10)$ and has a slope of -2.

49. The line passes through the point $(-2, -5)$ and has a slope of -3.

50. The line passes through the point $(-1, -6)$ and has a slope of 4.

51. The line passes through the point $(6, -3)$ and has a slope of $-\frac{4}{5}$.

52. The line passes through the point $(7, -2)$ and has a slope of $\frac{7}{2}$.

53. The line passes through $(0, 4)$ and $(3, 0)$. **(See Example 6.)**

54. The line passes through $(1, 1)$ and $(3, 7)$.

55. The line passes through $(6, 12)$ and $(4, 10)$.

56. The line passes through $(-2, -1)$ and $(3, -4)$.

57. The line passes through $(-5, 2)$ and $(-1, 2)$.

58. The line passes through $(-4, -1)$ and $(2, -1)$.

59. The line contains the point $(3, 2)$ and is parallel to a line with a slope of $-\frac{3}{4}$. **(See Example 7.)**

60. The line contains the point $(-1, 4)$ and is parallel to a line with a slope of $\frac{1}{2}$.

61. The line contains the point $(3, 2)$ and is perpendicular to a line with a slope of $-\frac{3}{4}$. **(See Example 8.)**

62. The line contains the point $(-2, 5)$ and is perpendicular to a line with a slope of $\frac{1}{2}$.

63. The line contains the point $(2, -5)$ and is parallel to $3x - 4y = -7$.

64. The line contains the point $(-6, -1)$ and is parallel to $2x + 3y = -12$.

65. The line contains the point $(-8, -1)$ and is perpendicular to $-15x + 3y = 9$.

66. The line contains the point $(4, -2)$ and is perpendicular to $4x + 3y = -6$.

67. The line contains the point $(4, 0)$ and is parallel to the line defined by $3x = 2y$.

68. The line contains the point $(-3, 0)$ and is parallel to the line defined by $-5x = 6y$.

69. The line is perpendicular to the line defined by $3y + 2x = 21$ and passes through the point $(2, 4)$.

70. The line is perpendicular to $7y - x = -21$ and passes through the point $(-14, 8)$.

71. The line is perpendicular to $\frac{1}{2}y = x$ and passes through $(-3, 5)$.

72. The line is perpendicular to $-\frac{1}{4}y = x$ and passes through $(-1, -5)$.

Concept 3: Equations of a Line: A Summary

For Exercises 73–80, write an equation of the line satisfying the given conditions.

73. The line passes through $(2, -3)$ and has a zero slope.

74. The line contains the point $(\frac{5}{2}, 0)$ and has an undefined slope.

75. The line contains the point $(2, -3)$ and has an undefined slope. **(See Example 9.)**

76. The line contains the point $(\frac{5}{2}, 0)$ and has a zero slope.

77. The line is parallel to the x-axis and passes through $(4, 5)$.

78. The line is perpendicular to the x-axis and passes through $(4, 5)$.

79. The line is parallel to the line $x = 4$ and passes through $(5, 1)$.

80. The line is parallel to the line $y = -2$ and passes through $(-3, 4)$.

Expanding Your Skills

81. Is the equation $x = -2$ in slope-intercept form? Identify the slope and y-intercept.

82. Is the equation $x = 1$ in slope-intercept form? Identify the slope and y-intercept.

83. Is the equation $y = 3$ in slope-intercept form? Identify the slope and y-intercept.

84. Is the equation $y = -5$ in slope-intercept form? Identify the slope and y-intercept.

For Exercises 85–88, write an equation of the given line. Write the answer in slope-intercept form, if possible.

85.

86.
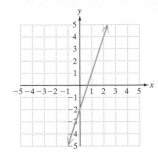

87.
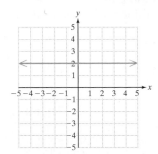

88.
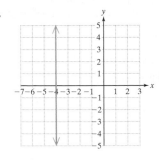

Graphing Calculator Exercises

For Exercises 89–92, use a graphing calculator to graph the lines on the same viewing window. Then explain how the lines are related.

89. $y_1 = \dfrac{1}{2}x + 4$

$y_2 = \dfrac{1}{2}x - 2$

90. $y_1 = -\dfrac{1}{3}x + 5$

$y_2 = -\dfrac{1}{3}x - 3$

91. $y_1 = x - 2$

$y_2 = 2x - 2$

$y_3 = 3x - 2$

92. $y_1 = -2x + 1$

$y_2 = -3x + 1$

$y_3 = -4x + 1$

For Exercises 93–96, use a graphing calculator to graph the lines on a square viewing window. Then explain how the lines are related.

93. $y_1 = 4x - 1$

$y_2 = -\dfrac{1}{4}x - 1$

94. $y_1 = \dfrac{1}{2}x - 3$

$y_2 = -2x - 3$

95. Use a graphing calculator to graph the equation from Exercise 53. Use the *Value* feature to verify that the line passes through the points $(0, 4)$ and $(3, 0)$.

96. Use a graphing calculator to graph the equation from Exercise 54. Use the *Value* feature to verify that the line passes through the points $(1, 1)$ and $(3, 7)$.

Problem Recognition Exercises

Identifying Characteristics of Lines

For Exercises 1–20, choose the equation(s) from column B whose graph satisfies the condition described in column A. Give all possible answers.

Column A		Column B
1. Line whose slope is positive.	**2.** Line whose slope is negative.	**a.** $y = -4x$
3. Line that passes through the origin.	**4.** Line that contains the point $(2, 0)$.	**b.** $2x - 4y = 4$
5. Line whose y-intercept is $(0, -3)$.	**6.** Line whose y-intercept is $(0, 0)$.	
7. Line whose slope is $-\dfrac{1}{3}$.	**8.** Line whose slope is $\dfrac{1}{2}$.	**c.** $y = -\dfrac{1}{3}x - 3$
9. Line whose slope is 0.	**10.** Line whose slope is undefined.	**d.** $3x + 5y = 10$
11. Line that is parallel to the line with equation $x + 3y = 6$.	**12.** Line perpendicular to the line with equation $x - 4y = -4$.	**e.** $3y = -9$
13. Line that is vertical.	**14.** Line that is horizontal.	**f.** $y = 5x - 1$
15. Line whose x-intercept is $(12, 0)$.	**16.** Line whose x-intercept is $\left(\dfrac{1}{5}, 0\right)$.	**g.** $4x + 1 = 9$
17. Line that has no x-intercept.	**18.** Line that is perpendicular to the x-axis.	**h.** $x + 3y = 12$
19. Line with a negative slope and positive y-intercept.	**20.** Line with a positive slope and negative y-intercept.	

Applications of Linear Equations and Modeling   Section 2.4

1. Writing a Linear Model

A **mathematical model** is a formula or equation that represents a relationship between two or more variables in a real-world application. Algebra (or some other field of mathematics) can then be used to solve the problem. The use of mathematical models is found throughout the physical and biological sciences, sports, medicine, economics, business, and many other fields.

For an equation written in slope-intercept form, $y = mx + b$, the term mx is called the *variable term*. The value of this term changes with different values of x. The term b is called the *constant term* and it remains unchanged regardless of the value of x. The slope of the line, m, is called the *rate of change*. A linear equation can be created if the rate of change and the constant are known.

Concepts

1. Writing a Linear Model
2. Interpreting a Linear Model
3. Finding a Linear Model from Observed Data Points

Example 1 Finding a Linear Relationship

Buffalo, New York, had 2 ft (24 in.) of snow on the ground before a snowstorm. During the storm, snow fell at an average rate of $\frac{5}{8}$ in./hr.

a. Write a linear equation to compute the total snow depth y after x hr of the storm.

b. Graph the equation.

c. Use the equation to compute the depth of snow after 8 hr.

d. If the snow depth was 31.5 in. at the end of the storm, determine how long the storm lasted.

Solution:

a. The constant or base amount of snow before the storm began is 24 in. The rate of change is given by $\frac{5}{8}$ in. of snow per hour. If m is replaced by $\frac{5}{8}$ and b is replaced by 24, we have the linear equation

$$y = mx + b$$

$$y = \frac{5}{8}x + 24$$

b. The equation is in slope-intercept form, and the corresponding graph is shown in Figure 2-25.

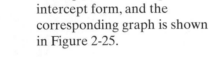

Figure 2-25

c. $y = \frac{5}{8}x + 24$

$y = \frac{5}{8}(8) + 24$ Substitute $x = 8$.

$y = 5 + 24$ Solve for y.

$y = 29$ in.

The snow depth was 29 in. after 8 hr. The corresponding ordered pair is (8, 29) and can be confirmed from the graph.

d. $y = \frac{5}{8}x + 24$

$31.5 = \frac{5}{8}x + 24$ Substitute $y = 31.5$.

$8(31.5) = 8\left(\frac{5}{8}x + 24\right)$ Multiply by 8 to clear fractions.

$252 = 5x + 192$ Clear parentheses.

$60 = 5x$ Solve for x.

$12 = x$

The storm lasted for 12 hr. The corresponding ordered pair is (12, 31.5) and can be confirmed from the graph.

2. Interpreting a Linear Model

Example 2 Interpreting a Linear Model

In 1938, President Franklin D. Roosevelt signed a bill enacting the Fair Labor Standards Act of 1938 (FLSA). In its final form, the act banned oppressive child labor and set the minimum hourly wage at 25 cents and the maximum workweek at 44 hr. Over the years, the minimum hourly wage has been increased by the government to meet the rising cost of living.

The minimum hourly wage y (in dollars per hour) in the United States since 1970 can be approximated by the equation

$$y = 0.14x + 1.60 \qquad x \geq 0$$

where x represents the number of years since 1970 ($x = 0$ corresponds to 1970, $x = 1$ corresponds to 1971, and so on) (Figure 2-26).

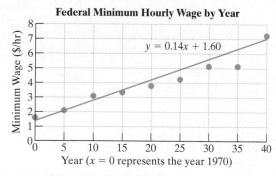

Federal Minimum Hourly Wage by Year

$y = 0.14x + 1.60$

Year ($x = 0$ represents the year 1970)

Figure 2-26

a. Find the slope of the line and interpret the meaning of the slope in the context of this problem.

b. Find the y-intercept of the line and interpret the meaning of the y-intercept in the context of this problem.

c. Use the linear equation to approximate the minimum wage in 1985.

Solution:

a. The equation $y = 0.14x + 1.60$ is written in slope-intercept form. The slope is 0.14 and indicates that minimum hourly wage rose an average of $0.14 per year since 1970.

b. The y-intercept is (0, 1.60). The y-intercept indicates that the minimum wage in the year 1970 ($x = 0$) was approximately $1.60 per hour.

c. The year 1985 is 15 years after the year 1970. Substitute $x = 15$ into the linear equation.

$$y = 0.14x + 1.60$$
$$y = 0.14(15) + 1.60 \qquad \text{Substitute } x = 15.$$
$$y = 2.1 + 1.60$$
$$y = 3.70$$

According to the linear model, the minimum wage in 1985 was approximately $3.70 per hour. (The actual minimum wage in 1985 was $3.35 per hour.)

Answers

4. The slope is 0.40. The customer is charged $0.40 for each minute used beyond 400 min.
5. The y-intercept is (0, 49.99). This means that if 0 min is used beyond 400 min, the customer is charged $49.99.
6. $67.99

3. Finding a Linear Model from Observed Data Points

Graphing a set of data points offers a visual method to determine whether the points follow a linear pattern. If a linear trend exists, we say that there is a linear correlation between the two variables. The better the points "line up," the stronger the correlation.*

When two variables are correlated, it is often desirable to find a mathematical equation (or *model*) to describe the relationship between the variables.

> **Example 3**   Writing a Linear Model from Observed Data

Figure 2-27 represents the winning gold medal times for the women's 100-m freestyle swimming event for selected summer Olympics. Let y represent the winning time in seconds and let x represent the number of years since 1900 ($x = 0$ corresponds to 1900, $x = 1$ corresponds to 1901, and so on).

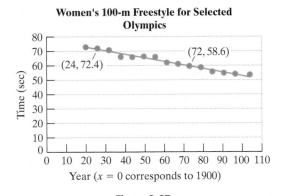

Figure 2-27

*The strength of a linear correlation can be measured mathematically by using techniques often covered in statistics courses.

In 1924, the winning time was 72.4 sec. This corresponds to the ordered pair (24, 72.4). In 1972, the winning time was 58.6 sec, yielding the ordered pair (72, 58.6).

a. Use these ordered pairs to find a linear equation to model the winning time versus the year.

b. What is the slope of the line, and what does it mean in the context of this problem?

c. Use the linear equation to approximate the winning time for the 1964 Olympics.

d. Would it be practical to use the linear model to predict the winning time in the year 2050?

Solution:

a. The slope formula can be used to compute the slope of the line between the two points.

$$\underset{(x_1,\, y_1)}{(24, 72.4)} \quad \text{and} \quad \underset{(x_2,\, y_2)}{(72, 58.6)}$$

$$m = \frac{y_2 - y_1}{x_2 - x_1} = \frac{58.6 - 72.4}{72 - 24} = -0.2875$$

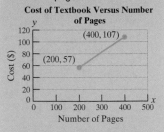

$$y - y_1 = m(x - x_1)$$
Apply the point-slope formula, using $m = -0.2875$ and the point $(24, 72.4)$.

$$y - 72.4 = -0.2875(x - 24)$$

$$y - 72.4 = -0.2875x + 6.9$$
Clear parentheses.

$$y = -0.2875x + 6.9 + 72.4$$
Solve for y.

$$y = -0.2875x + 79.3$$
The answer is in slope-intercept form.

b. The slope is -0.2875 and indicates that the winning time in the women's 100-m Olympic freestyle event has *decreased* on average by 0.2875 sec/yr during this period.

c. Since 1964 is 64 yr after 1900, substitute $x = 64$ into the linear model.

$$y = -0.2875x + 79.3$$

$$y = -0.2875(64) + 79.3$$
Substitute $x = 64$.

$$y = -18.4 + 79.3$$

$$y = 60.9$$

According to the linear model, the winning time in 1964 was approximately 60.9 sec. (The actual winning time in 1964 was set by Dawn Fraser from Australia in 59.5 sec. The linear equation can only be used to *approximate* the winning time.)

d. It would not be practical to use the linear model $y = -0.2875x + 79.3$ to predict the winning time in the year 2050. There is no guarantee that the linear trend will continue beyond the last observed data point in 2004. In fact, the linear trend cannot continue indefinitely; otherwise, the swimmers' times would eventually be negative. The potential for error increases for predictions made beyond the last observed data value.

Section 2.4 Practice Exercises

Study Skills Exercise

On test day, take a look at any formulas or important points that you had to memorize before you enter the classroom. Then when you sit down to take your test, write these formulas on the test or on scrap paper. This is called a memory dump. Write down the formulas from Sections 2.1–2.4.

Vocabulary and Key Concepts

1. A mathematical _____ is a formula or equation that represents a relationship between two or more variables.

Review Exercises

2. Determine the x- and y-intercepts for the line defined by $30x - 40y = 1200$.

For Exercises 3–6,

    a. Find the slope (if possible) of the line passing through the two points.

    b. Find an equation of the line passing through the two points. Write the answer in slope-intercept form (if possible) and in standard form.

    c. Graph the line by using the slope and y-intercept. Verify that the line passes through the two given points.

3. $(-3, 0)$ and $(3, -2)$

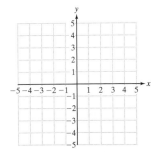

4. $(1, -1)$ and $(3, -5)$

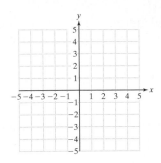

5. $(-4, 3)$ and $(-2, 3)$

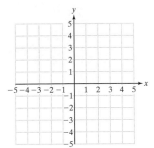

6. $(-2, 4)$ and $(-2, 0)$

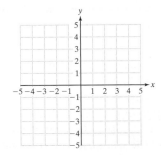

Concept 1: Writing a Linear Model

7. A luxury car rental company charges $120 per day in addition to a flat fee of $65 for insurance. **(See Example 1.)**

    a. Write an equation that represents the cost y (in dollars) to rent the car for x days.

    b. Graph the equation.

    c. What is the y-intercept and what does it mean in the context of this problem?

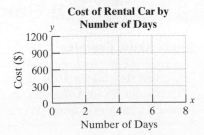

    d. Use the model from part (a) to determine the cost of driving the rental car for 2 days, 5 days, and 7 days.

    e. If the company offers a weekly rate of $799, is this a better deal than paying by the day for 7 days?

    f. Find the total cost of driving the car for 4 days if the sales tax is 6%.

    g. Is it reasonable to use negative values of x in the equation from part (a)? Why or why not?

8. Alex is a sales representative and earns a base salary of $1000 per month plus a 4% commission on his sales for the month.

   a. Write a linear equation that expresses Alex's monthly salary y in terms of his sales x.

   b. Graph the equation.

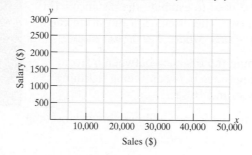

   c. What is the y-intercept and what does it represent in the context of this problem?

   d. What is the slope of the line and what does it represent in the context of this problem?

   e. How much will Alex make if his sales for a given month are $30,000?

9. Ava recently purchased a home in Crescent Beach, Florida. Her property taxes for the first year are $2742. Ava estimates that her taxes will increase at a rate of $52 per year.

   a. Write an equation to compute Ava's yearly property taxes. Let y be the amount she pays in taxes, and let x be the time in years.

   b. Graph the line.

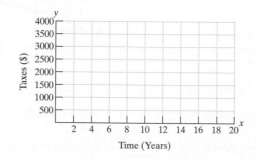

   c. What is the slope of this line? What does the slope of the line represent in the context of this problem?

   d. What is the y-intercept? What does the y-intercept represent in the context of this problem?

   e. What will Ava's yearly property tax be in 10 yr? In 15 yr?

10. Millage rate is the amount per $1000 that is often used to calculate property tax. For example, a home with an $80,000 taxable value in a municipality with a 19 mil tax rate would require $(0.019)(\$80,000) = \$1520$ in property taxes.

   a. In one county, homeowners pay a flat tax of $156 plus a rate of 19 mil on the taxable value of a home. Write a linear equation that represents the total property tax y (in $) for a home with a taxable value of x dollars.

   b. Determine the amount of property tax on a home with a taxable value of $90,000.

   c. If the property tax on a home is $2436, determine the taxable value of the home.

Concept 2: Interpreting a Linear Model

11. Sound travels at approximately one-fifth of a mile per second. Therefore, the difference in time x (in seconds) between seeing lightning and hearing thunder can be used to estimate the distance y (in miles) between a storm and an observer. The distance of the storm can be approximated by the equation $y = 0.2x$, where $x \geq 0$. **(See Example 2.)**

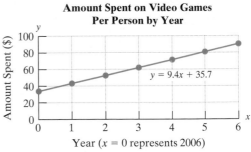

a. Use the linear model to determine the distance between a storm and an observer for the following times between seeing lightning and hearing thunder: 4 sec, 12 sec, and 16 sec.

b. If a storm is 4.2 mi away, how many seconds will pass between seeing lightning and hearing thunder?

12. The force y (in pounds) required to stretch a particular spring x in. beyond its rest (or "equilibrium") position is given by the equation $y = 2.5x$, where $0 \leq x \leq 20$.

a. Use the equation to determine the amount of force necessary to stretch the spring 6 in. from its rest position. How much force is necessary to stretch the spring twice as far?

b. If 45 lb of force is exerted on the spring, how far will the spring be stretched?

13. The equation $y = 9.4x + 35.7$ represents the average amount y (in $) spent on video games per person in the United States where x is the number of years since 2006.

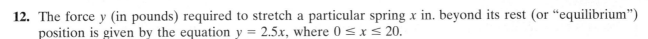

a. Use the equation to approximate the average amount spent per person in the year 2010.

b. Use the equation to approximate the average amount spent per person in the year 2008 and compare it with the actual amount spent of $55.80.

c. What is the slope of the line and what does it mean in the context of this problem?

d. What is the y-intercept and what does it mean in the context of this problem?

14. Let y represent the average number of miles driven per year for passenger cars in the United States since 1980. Let x represent the year where $x = 0$ corresponds to 1980, $x = 1$ corresponds to 1981, and so on. The average yearly mileage for passenger cars can be approximated by the equation $y = 142x + 9060$, where $x \geq 0$.

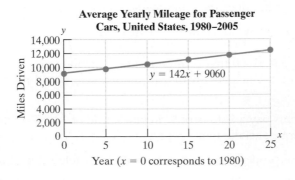

a. Use the linear equation to approximate the average yearly mileage for passenger cars in the United States in the year 2005.

b. Use the linear equation to approximate the average mileage for the year 1985, and compare it with the actual value of 9700 mi.

c. What is the slope of the line and what does it mean in the context of this problem?

d. What is the y-intercept and what does it mean in the context of this problem?

Concept 3: Finding a Linear Model from Observed Data Points

15. The figure represents the winning heights for men's pole vault in selected Olympic games. **(See Example 3.)**

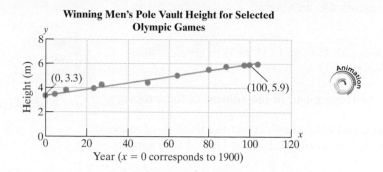

a. Let y represent the winning height (in meters). Let x represent the year, where $x = 0$ corresponds to the year 1900, $x = 4$ represents 1904, and so on. Use the ordered pairs given in the graph (0, 3.3) and (100, 5.9) to find a linear equation to estimate the winning pole vault height versus the year.

b. Use the linear equation from part (a) to approximate the winning vault for the 1920 Olympics.

c. Use the linear equation to approximate the winning vault for 1980.

d. The actual winning vault in 1920 was 4.09 m, and the actual winning vault in 1980 was 5.78 m. Are your answers from parts (b) and (c) different from these? Why?

e. What is the slope of the line? What does the slope of the line mean in the context of this problem?

16. During a drought, the average water level in a retention pond decreased linearly with time.

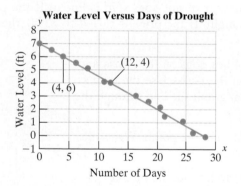

a. Let y represent the water level (in feet) and x represent the number of days since the beginning of the drought. Use the ordered pairs given to write a linear equation to estimate the water level based on the number of days since the drought began.

b. Use the linear equation to approximate the water level after 15 days.

c. Use the linear equation to approximate the water level after 25 days.

d. What is the slope of the line and what does it mean in the context of this problem?

e. Interpret the meaning of the x-intercept. Do you think that this linear trend will continue indefinitely? Explain your answer.

17. The graph displays the number of associate degrees conferred in the United States at the end of selected academic years. The variable x represents the number of years since 1970, and the variable y represents the number of associate degrees in thousands.

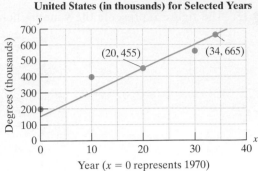

Number of Associate Degrees Awarded in the United States (in thousands) for Selected Years

a. Use the points $(20, 455)$ and $(34, 665)$ to create a linear model of the data.

b. What does the slope mean in the context of this problem?

c. If this linear trend continues, predict the number of associate degrees that will be conferred in the United States in the year 2015.

18. The number of prisoners in federal or state correctional facilities is shown in the figure by year. (*Source:* U.S. Department of Justice)

a. Use the given points to create a linear model of the data.

b. What does the slope mean in the context of this problem?

c. Use the equation in part (a) to determine the number of prisoners in federal or state correctional facilities for the year 2015.

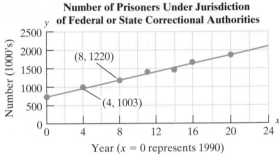

Number of Prisoners Under Jurisdiction of Federal or State Correctional Authorities

19. At a concession stand at a high school football game, the owner notices that the relationship between the price of a hot dog and the number of hot dogs sold is linear. If the price is \$2.50 per hot dog, then approximately 650 are sold each night. If the price is raised to \$3.50, then the number sold drops to 475 per night.

a. Make a graph with the price of hot dogs on the x-axis and the number of hot dogs sold on the y-axis. Use the points $(2.50, 650)$ and $(3.50, 475)$. Then graph the line through the points with $x \geq 0$.

b. Find an equation of the line through the points. Write the equation in slope-intercept form.

c. Use the equation from part (b) to predict the number of hot dogs that would sell if the price were raised to \$4.00. Round to the nearest whole unit.

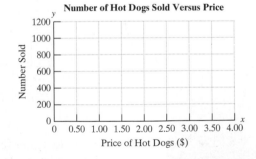

Number of Hot Dogs Sold Versus Price

20. Sales at a concession stand indicate that the relationship between the price of a drink and the number of drinks sold is linear. If drinks are sold at \$1.00 each, then approximately 1020 are sold each night. If the price is raised to \$1.50, then the number sold drops to 820 per night.

a. Make a graph with the price of drinks on the x-axis and the number of drinks sold on the y-axis. Graph the points $(1.00, 1020)$ and $(1.50, 820)$. Then graph the line through the points with $x \geq 0$.

b. Find an equation of the line through the points. Write the equation in slope-intercept form.

c. Use the equation from part (b) to predict the number of drinks that would sell if the price were \$2.00 per drink.

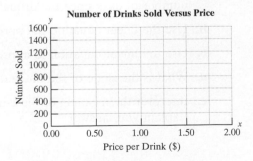

Number of Drinks Sold Versus Price

21. In order to advise students properly, a college advisor is interested in the relationship between a student's GPA and the number of hours the student studies in an average week. The data are shown in the table.

Student	Study Time (in hours)	GPA
1	15	2.5
2	38	3.9
3	10	2.1
4	24	2.8
5	35	3.3
6	15	2.7
7	45	4.0
8	28	3.1
9	35	3.4
10	10	2.2
11	6	1.8

a. Let x represent study time and let y represent GPA. Graph the points.

Grade Point Average vs. Weekly Study Time

b. Does there appear to be a linear trend?

c. Use the data points $(28, 3.1)$ and $(10, 2.2)$ to find an equation of the line through these points.

d. Use the equation in part (c) to estimate the GPA of a student who studies for 30 hours a week.

e. Why would the linear model found in part (c) not be realistic for a student who studies more than 46 hours per week?

22. Loraine is enrolled in an algebra class that meets 5 days per week. Her instructor gives a test every Friday. Loraine has a study plan and keeps a portfolio with notes, homework, test corrections, and vocabulary. She also records the amount of time per day that she studies and does homework. The following data represent the amount of time she studied per day and her weekly test grades.

a. Graph the points on a rectangular coordinate system. Do the data points appear to follow a linear trend?

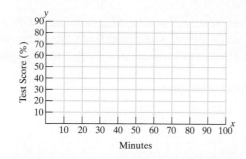

Time Studied per Day (min) x	Weekly Test Grade (percent) y
60	69
70	74
80	79
90	84
100	89

b. Find a linear equation that relates Loraine's weekly test score y to the amount of time she studied per day x. (*Hint:* Pick two ordered pairs from the observed data, and find an equation of the line through the points.)

c. How many minutes should Loraine study per day in order to score at least 90% on her weekly examination? Would the equation used to determine the time Loraine needs to study to get 90% work for other students? Why or why not?

d. If Loraine is only able to spend $\frac{1}{2}$ hr/day studying her math, predict her test score for that week.

Expanding Your Skills

Points are *collinear* if they lie on the same line. For Exercises 23–26, use the slope formula to determine if the points are collinear.

23. $(3, -4)\ (0, -5)\ (9, -2)$

24. $(4, 3)\ (-4, -1)\ (2, 2)$

25. $(0, 2)\ (-2, 12)\ (-1, 6)$

26. $(-2, -2)\ (0, -3)\ (-4, -1)$

Graphing Calculator Exercises

27. Use a *Table* feature to confirm your answers to Exercise 11(a).

28. Use a *Table* feature to confirm your answers to Exercise 12(a).

29. Graph the equation $y = -175x + 1087.5$ on the viewing window $0 \le x \le 5$ and $0 \le y \le 1200$. Use the *Value* feature to support your answer to Exercise 19 by showing that the line passes through the points $(2.5, 650)$ and $(3.5, 475)$.

30. Graph the equation $y = -400x + 1420$ on the viewing window $0 \le x \le 3$ and $0 \le y \le 1600$. Use the *Value* feature to support your answer to Exercise 20 by showing that the line passes through the points $(1, 1020)$ and $(1.5, 820)$.

Group Activity

Using Linear Equations to Construct Images

Estimated Time: 15 minutes

Group Size: 2 or 3

Writing video games is an exciting and creative job for computer programmers, and success in this field requires a fundamental knowledge of the algebraic and graphical representations of lines and curves.

In the example shown here, a schematic drawing of a house is given relative to a coordinate system with origin at the lower left corner of the house. The house is rendered from a series of line segments. Each line segment can be defined mathematically by using an equation of a line with restrictions on the value of x or y. For example,

- The line segment between the points $(0, 10)$ and $(14, 9)$ is represented by

$$y = -\frac{1}{14}x + 10 \text{ for } 0 \le x \le 14$$

- The vertical line segment between the points $(14, -1)$ and $(14, 9)$ is represented by

$$x = 14 \text{ for } -1 \le y \le 9$$

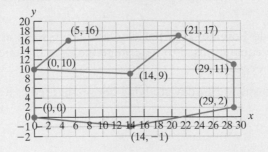

1. For each pair of points given, write a representation of the line segment between the points.

Points	Equation	Restriction on x or y
$(0, 0)$ and $(0, 10)$		
$(0, 10)$ and $(5, 16)$		
$(5, 16)$ and $(21, 17)$		
$(21, 17)$ and $(29, 11)$		
$(29, 11)$ and $(29, 2)$		
$(29, 2)$ and $(14, -1)$		
$(14, -1)$ and $(0, 0)$		
$(0, 10)$ and $(14, 9)$	$y = -\dfrac{1}{14}x + 10$	$0 \le x \le 14$
$(14, 9)$ and $(14, -1)$	$x = 14$	$-1 \le y \le 9$

2. Many graphing utilities are able to graph line segments provided that the slope-intercept form of the line and restrictions on x are known. In the calculator's equation editor, enter the expression $mx + b$ within parentheses. Then enter the restriction(s) on x within parentheses. For example, the line segment defined by $y = -\frac{1}{14}x + 10$ for $0 \le x \le 14$ is entered as

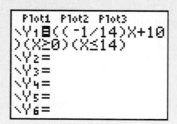

The inequality symbols are found in the TEST menu of many calculators. Use a graphing calculator or online graphing utility to verify your answers from question 1. Discuss with your group members why the calculator cannot be used to graph a vertical line segment using this method.

Chapter 2   Summary

Section 2.1	Linear Equations in Two Variables

Key Concepts

A **linear equation in two variables** can be written in the form $Ax + By = C$, where A, B, and C are real numbers and A and B are not both zero.

    The graph of a linear equation in two variables is a line and can be represented in a rectangular coordinate system.

Examples

Example 1

To graph the equation $3x - 4y = 12$, we can construct a table of points.

x	y
0	-3
4	0
1	$-\dfrac{9}{4}$

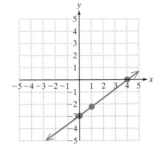

Example 2

Given $2x + 3y = 8$, find the x- and y-intercepts.

x-intercept: $2x + 3(0) = 8$

$$2x = 8$$

$$x = 4 \qquad (4, 0)$$

    An x**-intercept** of an equation is a point $(a, 0)$ where the graph intersects the x-axis. To find an x-intercept, substitute 0 for y and solve for x.

    A y**-intercept** of an equation is a point $(0, b)$ where the graph intersects the y-axis. To find a y-intercept, substitute 0 for x and solve for y.

y-intercept: $2(0) + 3y = 8$

$$3y = 8$$

$$y = \frac{8}{3} \qquad \left(0, \frac{8}{3}\right)$$

A **vertical line** can be written in the form $x = k$.
A **horizontal line** can be written in the form $y = k$.

Example 3

Graph $x = -2$.

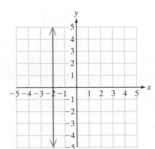

Example 4

Graph $y = 3$.

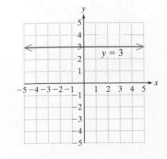

Section 2.2   Slope of a Line and Rate of Change

Key Concepts

The **slope** of a line m between two distinct points (x_1, y_1) and (x_2, y_2) is given by

$$m = \frac{y_2 - y_1}{x_2 - x_1}, \qquad x_2 - x_1 \neq 0$$

The slope of a line may be positive, negative, zero, or undefined.

Two parallel (nonvertical) lines have the same slope: $m_1 = m_2$.

Two lines are perpendicular if the slope of one line is the opposite of the reciprocal of the slope of the other line:

$$m_1 = -\frac{1}{m_2} \text{ or equivalently, } m_1 \cdot m_2 = -1$$

Examples

Example 1

The slope of the line between $(1, -3)$ and $(-3, 7)$ is

$$m = \frac{7 - (-3)}{-3 - 1} = \frac{10}{-4} = -\frac{5}{2}$$

Example 2

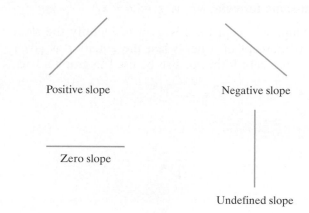

Positive slope   Negative slope

Zero slope

Undefined slope

Example 3

The slopes of two lines are given. Determine whether the lines are parallel, perpendicular, or neither.

a. $m_1 = -7$ and $m_2 = -7$ Parallel

b. $m_1 = -\dfrac{1}{5}$ and $m_2 = 5$ Perpendicular

c. $m_1 = -\dfrac{3}{2}$ and $m_2 = -\dfrac{2}{3}$ Neither

Section 2.3   Equations of a Line

Key Concepts

Standard Form: $Ax + By = C$ (A and B are not both zero)

Horizontal line: $y = k$

Vertical line: $x = k$

Slope-intercept form: $y = mx + b$

Point-slope formula: $y - y_1 = m(x - x_1)$

Slope-intercept form is used to identify the slope and y-intercept of a line when the equation is given. Slope-intercept form can also be used to graph a line.

The point-slope formula can be used to construct an equation of a line, given a point and a slope.

Examples

Example 1

Find the slope and y-intercept. Then graph the equation.

$$7x - 2y = 4 \qquad \text{Solve for } y.$$

$$-2y = -7x + 4$$

$$y = \frac{7}{2}x - 2 \qquad \text{Slope-intercept form}$$

The slope is $\frac{7}{2}$; the y-intercept is $(0, -2)$.

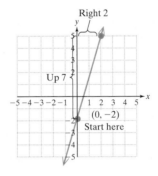

Example 2

Find an equation of the line passing through the point $(2, -3)$ and having slope $m = -4$.

Using the point-slope formula gives

$$y - y_1 = m(x - x_1)$$

$$y - (-3) = -4(x - 2)$$

$$y + 3 = -4x + 8$$

$$y = -4x + 5$$

Section 2.4	Applications of Linear Equations and Modeling

Key Concepts

A linear model can be constructed to describe data for a given situation.

- Given two points from the data, use the point-slope formula to find an equation of the line.

- Interpret the meaning of the slope and y-intercept in the context of the problem.

- Use the equation to predict values.

Examples

Example 1

The per capita income in the United States has been rising linearly since 1980. In the graph, x represents the number of years since 1980, and y represents average income in dollars.

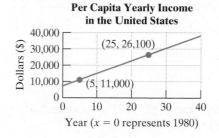

Per Capita Yearly Income in the United States

Dollars ($) vs Year ($x = 0$ represents 1980)

Points shown: (25, 26,100) and (5, 11,000)

Write an equation of the line, using the points (5, 11,000) and (25, 26,100).

Slope: $\dfrac{26,100 - 11,000}{25 - 5} = \dfrac{15,100}{20} = 755$

$y - 11,000 = 755(x - 5)$

$y - 11,000 = 755x - 3775$

$y = 755x + 7225$

- The slope 755 indicates that the average income has increased at a rate of $755 per year.
- The y-intercept $(0, 7225)$ means that the average income in 1980 (year $x = 0$) was $7225.

By substituting different values of x, the equation can be used to approximate the average income for that year. For the year 2020 ($x = 40$), we have:

$y = 755(40) + 7225$

$y = 37,425$

The average per capita income in 2020 would be approximately $37,425.

Chapter 2   Review Exercises

Section 2.1

1. Label the following on the diagram:

   a. Origin

   b. x-Axis

   c. y-Axis

   d. Quadrant I

   e. Quadrant II

   f. Quadrant III

   g. Quadrant IV

2. Determine if $(2, 5)$ is a solution to $-2x + 4y = -16$.

3. Determine if $(-3, 4)$ is a solution to $5x = -15$.

4. Determine the coordinates of the points labeled in the graph.

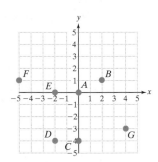

For Exercises 5–7, complete the table and graph the line defined by the points.

5. $3x - 2y = -6$

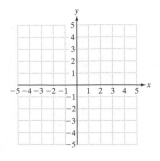

x	y
0	
	0
1	

6. $2y - 3 = 10$

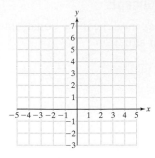

x	y
0	
5	
-4	

7. $6 - x = 2$

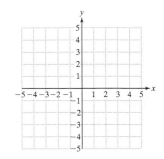

x	y
	0
	1
	-2

For Exercises 8–11, graph the lines. In each case find at least three points and identify the x- and y-intercepts (if possible).

8. $2x = 3y - 6$

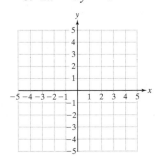

9. $5x - 2y = 0$

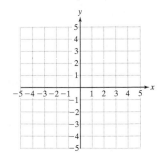

10. $2y = 6$

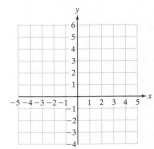

11. $-3x = 6$

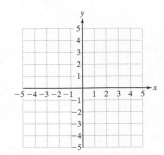

Section 2.2

12. Find the slope of the line.

a.

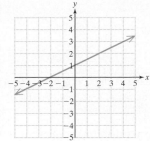

b.

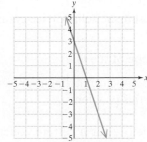

c.

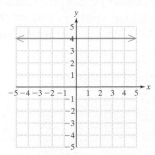

13. Draw a line with slope 2 (answers may vary).

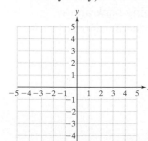

14. Draw a line with slope $-\frac{3}{4}$ (answers may vary).

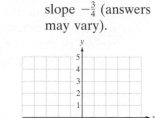

For Exercises 15–18, find the slope of the line that passes through each pair of points.

15. $(2, 6), (-1, 0)$

16. $(7, 2), (-3, -5)$

17. $(8, 2), (3, 2)$

18. $\left(-4, \frac{1}{2}\right), (-4, 1)$

19. Two points for each of two lines are given. Determine if the lines are parallel, perpendicular, or neither.

L_1: $(4, -6)$ and $(3, -2)$

L_2: $(3, -1)$ and $(7, 0)$

For Exercises 20–22, the slopes of two lines are given. Based on the slopes, are the lines parallel, perpendicular, or neither?

20. $m_1 = -\frac{1}{3}, m_2 = 3$ **21.** $m_1 = \frac{5}{4}, m_2 = \frac{4}{5}$

22. $m_1 = 7, m_2 = 7$

23. The graph indicates that the enrollment for a small college has been increasing linearly since 1990.

a. Use the two data points to find the slope of the line.

b. Interpret the meaning of the slope in the context of this problem.

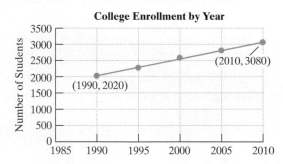

24. Approximate the slope of the stairway pictured here.

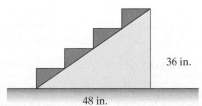

Section 2.3

25. Write a formula.

a. Horizontal line

b. Point-slope formula

c. Standard form

d. Vertical line

e. Slope-intercept form

For Exercises 26–30, write your answer in slope-intercept form or in standard form.

26. Write an equation of the line that has slope $\frac{1}{9}$ and y-intercept $(0, 6)$.

27. Write an equation of the line that has slope $-\frac{2}{3}$ and x-intercept $(3, 0)$.

28. Write an equation of the line that passes through the points $(-8, -1)$ and $(-5, 9)$.

29. Write an equation of the line that passes through the point $(6, -2)$ and is perpendicular to the line $y = -\frac{1}{3}x + 2$.

30. Write an equation of the line that passes through the point $(0, -3)$ and is parallel to the line $4x + 3y = -1$.

31. For each of the given conditions, find an equation of the line

  a. Passing through the point $(-3, -2)$ and parallel to the x-axis.

  b. Passing through the point $(-3, -2)$ and parallel to the y-axis.

  c. Passing through the point $(-3, -2)$ and having an undefined slope.

  d. Passing through the point $(-3, -2)$ and having a zero slope.

32. Are any of the lines in Exercise 31 the same?

Section 2.4

33. Ally loves the beach and decides to spend the summer selling various ice cream products on the beach. From her accounting course, she knows that her total cost is calculated as

  Total cost = fixed cost + variable cost

She estimates that her fixed cost for the summer season is $20 per day. She also knows that each ice cream product costs her $0.25 from her distributor.

  a. Write a relationship for the daily cost y in terms of the number of ice cream products sold per day x.

  b. Graph the equation from part (a) by letting the horizontal axis represent the number of ice cream products sold per day and letting the vertical axis represent the daily cost.

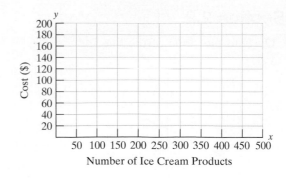

  c. What does the y-intercept represent in the context of this problem?

  d. What is her cost if she sells 450 ice cream products?

  e. What is the slope of the line?

  f. What does the slope of the line represent in the context of this problem?

34. The margin of victory for a certain college football team seems to be linearly related to the number of rushing yards gained by the star running back. The table shows the statistics.

Yards Rushed	Margin of Victory
100	20
60	10
120	24
50	7

  a. Graph the data to determine if a linear trend exists. Let x represent the number of yards rushed by the star running back and y represent the points in the margin of victory.

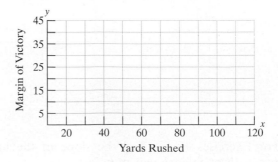

  b. Find an equation for the line through the points $(50, 7)$ and $(100, 20)$.

  c. Based on the equation, what would be the result of the football game if the star running back did not play?

Chapter 2   Test

1. Given the equation $x - \frac{2}{3}y = 6$, complete the ordered pairs and graph the corresponding points. $(0, \ )\ (\ ,0)\ (\ ,-3)$

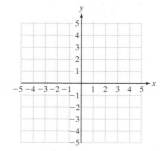

2. Determine whether the statements are true or false and explain your answer.

   a. The product of the x- and y-coordinates is positive only for points in Quadrant I.

   b. The quotient of the x- and y-coordinates is negative only for points in Quadrant IV.

   c. The point $(-2, -3)$ is in Quadrant III.

   d. The point $(0, 0)$ lies on the x-axis.

3. Determine if $(-4, -1)$ is a solution to
   $y = -\frac{1}{2}x - 3.$

4. Explain the process for finding the x- and y-intercepts.

For Exercises 5–8, identify the x- and y-intercepts (if possible) and graph the line.

5. $6x - 8y = 24$

6. $x = -4$

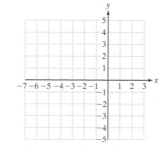

7. $3x = 5y$

8. $2y = -6$

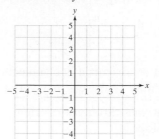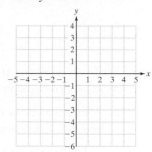

9. Find the slope of the line, given the following information:

   a. The line passes through the points $(7, -3)$ and $(-1, -8)$.

   b. The line is given by $6x - 5y = 1$.

10. Describe the relationship of the slopes of

   a. Two parallel lines

   b. Two perpendicular lines

11. The slope of a line is -7.

   a. Find the slope of a line parallel to the given line.

   b. Find the slope of a line perpendicular to the given line.

For Exercises 12 and 13, two points are given for each of two lines. Determine if the lines are parallel, perpendicular, or neither.

12. L_1: $(4, -4)$ and $(1, -6)$
    L_2: $(-2, 0)$ and $(0, 3)$

13. L_1: $(1, 11)$ and $(-1, 1)$
    L_2: $(0, -2)$ and $(2, 8)$

14. Given the equation $-3x + 4y = 4$,

   a. Write the line in slope-intercept form.

   b. Determine the slope and y-intercept.

   c. Graph the line, using the slope and y-intercept.

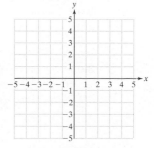

15. Determine if the lines are parallel, perpendicular, or neither.

a. $y = -x + 4$
$y = x - 3$

b. $9x - 3y = 1$
$15x - 5y = 10$

c. $3y = 6$
$x = 0.5$

d. $5x - 3y = 9$
$3x - 5y = 10$

16. Write an equation that represents a line subject to the following conditions. (Answers may vary.)

a. A line that does not pass through the origin and has a positive slope

b. A line with an undefined slope

c. A line perpendicular to the y-axis. What is the slope of such a line?

d. A line that passes through the origin and has a negative slope

17. Write an equation of the line that passes through the point $(8, -\frac{1}{2})$ with slope -2. Write the answer in slope-intercept form.

18. Write an equation of the line containing the points $(2, -3)$ and $(4, 0)$.

19. Write an equation of a line containing $(4, -3)$ and parallel to $6x - 3y = 1$.

20. Write an equation of the line that passes through the point $(-10, -3)$ and is perpendicular to $3x + y = 7$. Write the answer in slope-intercept form.

21. Jack sells used cars. He is paid $800 per month plus $300 commission for each automobile he sells.

a. Write an equation that represents Jack's monthly earnings y in terms of the number of automobiles he sells x.

b. Graph the linear equation you found in part (a).

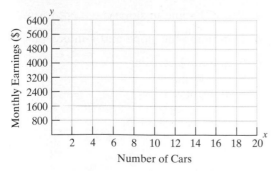

c. What does the y-intercept mean in the context of this problem?

d. How much will Jack earn in a month if he sells 17 automobiles?

 22. The graph represents the life expectancy for females in the United States born from 1940 through 2000.

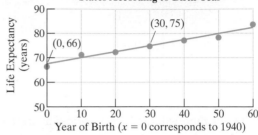

Source: National Center for Health Statistics

a. Determine the y-intercept from the graph. What does the y-intercept represent in the context of this problem?

b. Using the two points $(0, 66)$ and $(30, 75)$, determine the slope of the line. What does the slope of the line represent in the context of this problem?

c. Use the y-intercept and the slope found in parts (a) and (b) to write an equation of the line by letting x represent the year of birth and y represent the corresponding life expectancy.

d. Using the linear equation from part (c), approximate the life expectancy for women born in the United States in 1994. How does your answer compare with the reported life expectancy of 79 yr?

Chapters 1–2 Cumulative Review Exercises

For Exercises 1–3, solve the equations.

1. $10x + 6 = -4(x + 1)$

2. $\dfrac{5}{2} - \dfrac{2}{3}x = x + \dfrac{5}{6}$

3. $-0.5(x + 3) = 0.25(7 - 2x)$

4. A Formula 1 race car travels an average of 48 mph faster than another. It takes the faster car 25 min $\left(\frac{5}{12}\text{hr}\right)$ to cover the same distance as it takes the slower car traveling 35 min $\left(\frac{7}{12}\text{hr}\right)$. What is the speed of the faster car?

5. Two angles are complementary. One angle measures 6° more than twice the other. Find the measure of each angle.

6. Solve the equation for y. $ax = d - by$

For Exercises 7–9, solve the inequality and write the answer in interval notation.

7. $-7x - 2 > 3x + 5$

8. $-7 < 5t + 8 \le 13$

9. $-(y + 5) \le -2(y + 4)$ or $9y + 2 \ge -3$

For Exercises 10–12, solve the absolute value equation or inequality.

10. $\left|\dfrac{x}{3} - 4\right| = 8$

11. $|8x - 4| < 10$

12. $|5x + 7| = |-2x - 1|$

For Exercises 13–15, graph the line.

13. $4y - 3 = 1$

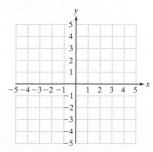

14. $-3x + 2y = 1$

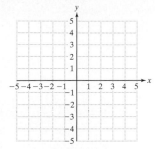

15. $4x + 8 = 0$

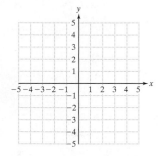

For Exercises 16 and 17, determine if the lines are parallel, perpendicular, or neither.

16. $L_1: 4x - 8y = 10$
    $L_2: 2x + y = 3$

17. $L_1: -2x + 3y = -12$
    $L_2:\ \ \ 3x - 2y = -2$

For Exercises 18–20, write the answer in slope-intercept form.

18. Write an equation of the line that passes through the points $(1, 3)$ and $(-2, 21)$.

19. Write an equation of the line that passes through the point $(-3, -4)$ and is perpendicular to the line $x + 2y = 10$.

20. Write an equation of the line that passes through $(7, -13)$ and is parallel to the y-axis.

Relations and Functions

3

Chapter 3

In Chapter 3 we introduce relations, functions, and graphs of functions.

Review Your Skills

In this chapter, we will be graphing equations that are not linear. That is, the graphs are not straight lines. For this reason, two points are not sufficient to form the graph. More points are needed to determine the general shape. Take a few minutes to graph the equation $y = |x^2 + 2x - 3|$. Use both negative and positive values of x. Finding the x- and y-intercepts will also help.

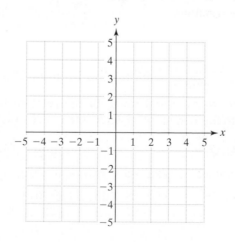

| Section 3.1 | **Relations and Applications** |

1. Definition of a Relation

In many naturally occurring phenomena, two variables may be linked by some type of relationship. Table 3-1 shows a correspondence between the length of a woman's femur and her height. (The femur is the large bone in the thigh attached to the knee and hip.)

Table 3-1

Length of Femur (cm) x	Height (in.) y	Ordered Pair
45.5	65.5	→ (45.5, 65.5)
48.2	68.0	→ (48.2, 68.0)
41.8	62.2	→ (41.8, 62.2)
46.0	66.0	→ (46.0, 66.0)
50.4	70.0	→ (50.4, 70.0)

Each data point from Table 3-1 may be represented as an ordered pair. In this case, the first value represents the length of a woman's femur and the second, the woman's height. The set of ordered pairs {(45.5, 65.5), (48.2, 68.0), (41.8, 62.2), (46.0, 66.0), (50.4, 70.0)} defines a relation between femur length and height.

2. Domain and Range of a Relation

Definition of a Relation in x and y

A set of ordered pairs (x, y) is called a **relation in x and y**. Furthermore,

- The set of first components in the ordered pairs is called the **domain of the relation**.
- The set of second components in the ordered pairs is called the **range of the relation**.

| Example 1 | **Finding the Domain and Range of a Relation** |

Find the domain and range of the relation linking the length of a woman's femur to her height {(45.5, 65.5), (48.2, 68.0), (41.8, 62.2), (46.0, 66.0), (50.4, 70.0)}.

Solution:

Domain:	{45.5, 48.2, 41.8, 46.0, 50.4}	Set of first components
Range:	{65.5, 68.0, 62.2, 66.0, 70.0}	Set of second components

The x- and y-components that constitute the ordered pairs in a relation do not need to be numerical. This is demonstrated in Example 2.

Example 2   **Writing a Relation and Finding Its Domain and Range**

Table 3-2 gives five states in the United States and the corresponding number of representatives in the House of Representatives for a recent year.

a. The data in the table define a relation. Write a list of ordered pairs for this relation.

b. Write the domain and range.

Table 3-2

State x	Number of Representatives y
Alabama	7
California	53
Colorado	7
Florida	25
Kansas	4

Solution:

a. {(Alabama, 7), (California, 53), (Colorado, 7), (Florida, 25), (Kansas, 4)}

b. Domain: {Alabama, California, Colorado, Florida, Kansas}

Range: {7, 53, 25, 4}   (*Note:* The element 7 is not listed twice.)

A relation may consist of a finite number of ordered pairs or an infinite number of ordered pairs. Furthermore, a relation may be defined by several different methods.

• A relation may be defined as a set of ordered pairs.

$$\{(1, 2), (-3, 4), (1, -4), (3, 4)\}$$

• A relation may be defined by a correspondence (Figure 3-1). The corresponding ordered pairs are {(1, 2), (1, −4), (−3, 4), (3, 4)}.

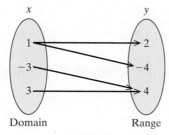

Domain   Range

Figure 3-1

- A relation may be defined by a graph (Figure 3-2). The corresponding ordered pairs are $\{(1, 2), (-3, 4), (1, -4), (3, 4)\}$.

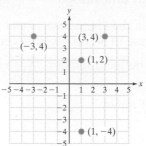

Figure 3-2

- A relation may be expressed by an equation such as $x = y^2$. The solutions to this equation define an infinite set of ordered pairs of the form $\{(x, y)\,|\,x = y^2\}$. The solutions can also be represented by a graph in a rectangular coordinate system (Figure 3-3).

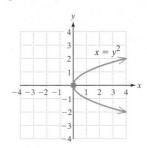

Figure 3-3

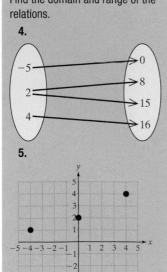
Example 3   **Finding the Domain and Range of a Relation**

Write the relation as a set of ordered pairs. Then find the domain and range.

a.

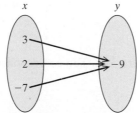

b.

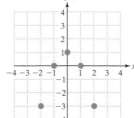

Solution:

a. From the figure, the relation defines the set of ordered pairs:
$\{(3, -9), (2, -9), (-7, -9)\}$

Domain: $\{3, 2, -7\}$

Range: $\{-9\}$

b. The points in the graph make up the set of ordered pairs:
$\{(-2, -3), (-1, 0), (0, 1), (1, 0), (2, -3)\}$

Domain: $\{-2, -1, 0, 1, 2\}$

Range: $\{-3, 0, 1\}$

Example 4   Finding the Domain and Range of a Relation

Use interval notation to express the domain and range of the relation.

a.

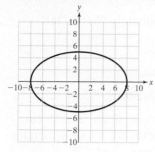

b.

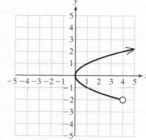

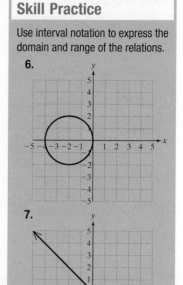
Solution:

a.

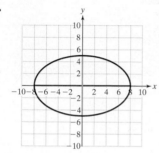

The domain consists of an infinite number of x values extending from -8 to 8 (shown in red). The range consists of all y values from -5 to 5 (shown in blue). Thus, the domain and range must be expressed in set-builder notation or in interval notation.

Domain: $[-8, 8]$

Range: $[-5, 5]$

b.

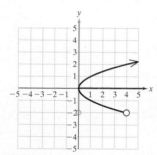

The arrow on the curve indicates that the graph extends infinitely far up and to the right. The open circle means that the graph will end at the point $(4, -2)$, but not include that point.

Domain: $[0, \infty)$.

Range: $(-2, \infty)$.

Animation

3. Applications Involving Relations

Example 5   Analyzing a Relation

The data in Table 3-3 depict the length of a woman's femur and her corresponding height. Based on these data, a forensics specialist can find a linear relationship between height y and femur length x:

$$y = 0.906x + 24.3 \qquad 40 \le x \le 55$$

From this type of relationship, the height of a woman can be inferred based on skeletal remains.

a. Find the height of a woman whose femur is 46.0 cm.

b. Find the height of a woman whose femur is 51.0 cm.

c. Why is the domain restricted to $40 \le x \le 55$?

Table 3-3

Length of Femur (cm) x	Height (in.) y
45.5	65.5
48.2	68.0
41.8	62.2
46.0	66.0
50.4	70.0

Skill Practice

The linear equation,

$y = -0.014x + 64.5$,

$1500 \leq x \leq 4000$ relates the weight of a car, x (in pounds), to its gas mileage, y (in mpg). Use this relationship for the following exercises.

8. Find the gas mileage in miles per gallon for a car weighing 2550 lb.

9. Find the gas mileage for a car weighing 2850 lb.

10. Why is the domain restricted to $1500 \leq x \leq 4000$?

Solution:

a. $y = 0.906x + 24.3$

    $= 0.906(46.0) + 24.3$ Substitute $x = 46.0$ cm.

    $= 65.976$ The woman is approximately 66.0 in. tall.

b. $y = 0.906x + 24.3$

    $= 0.906(51.0) + 24.3$ Substitute $x = 51.0$ cm.

    $= 70.506$ The woman is approximately 70.5 in. tall.

c. The domain restricts femur length to values between 40 cm and 55 cm, inclusive. These values are within the normal lengths for an adult female and are in the proximity of the observed data (Figure 3-4).

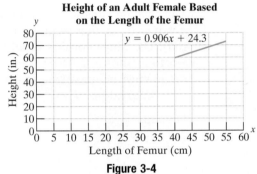

Height of an Adult Female Based on the Length of the Femur

$y = 0.906x + 24.3$

Figure 3-4

Answers

8. 28.8 mpg   9. 24.6 mpg

10. The relation is valid only for cars weighing between 1500 lb and 4000 lb, inclusive.

Section 3.1   Practice Exercises

Study Skills Exercise

Compute your grade at this point. Are you earning the grade that you want? If not, maybe organizing a study group would help.

In a study group, check the activities that you might try to help you learn and understand the material.

_____ Quiz each other by asking each other questions.

_____ Practice teaching each other.

_____ Share and compare class notes.

_____ Support and encourage each other.

_____ Work together on exercises and sample problems.

Vocabulary and Key Concepts

1. a. A set of ordered pairs (x, y) is called a _____ in x and y.

    b. The _____ of a relation is the set of first components in the ordered pairs.

    c. The _____ of a relation is the set of second components in the ordered pairs.

Review Exercises

2. Write an equation of a line passing through the points $(3, -4)$ and $(1, 6)$. Write the answer in slope-intercept form.

For Exercises 3–6,

    a. Determine if the equation represents a horizontal line, a vertical line, or a slanted line.

    b. Determine the slope of the line (if it exists).

    c. Determine the x-intercept (if it exists).

    d. Determine the y-intercept (if it exists).

3. $2x - 3 = 4$ **4.** $2x - 3y = 4$

5. $3x - 2y = 4$ **6.** $2 - 3y = 4$

Concept 2: Domain and Range of a Relation

For Exercises 7–14,

    a. Write the relation as a set of ordered pairs.

    b. Determine the domain and range. **(See Examples 1–3).**

7.

Region	Number Living in Poverty (thousands)
Northeast	54.1
Midwest	65.6
South	110.7
West	70.7

8.

x	y
0	3
-2	$\frac{1}{2}$
-7	1
-2	8
5	1

9.

Country	Year of First Man or Woman in Space
USSR	1961
USA	1962
Poland	1978
Vietnam	1980
Cuba	1980

10.

State, x	Year of Statehood, y
Maine	1820
Nebraska	1823
Utah	1847
Hawaii	1959
Alaska	1959

11.

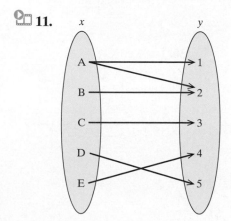

12.

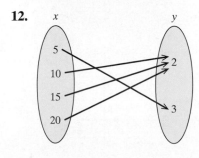

13.

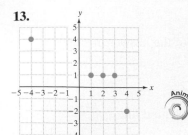

14.

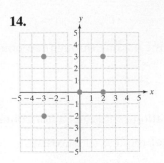

For Exercises 15–30, find the domain and range of the relations. Use interval notation where appropriate. (See Example 4.)

15.

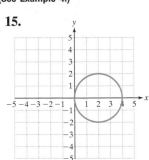

16.

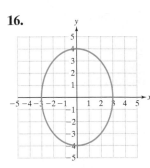

17.

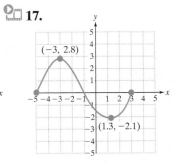

(−3, 2.8), (1.3, −2.1)

18.

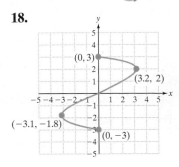

(0, 3), (3.2, 2), (−3.1, −1.8), (0, −3)

19.

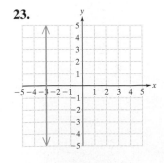

20.

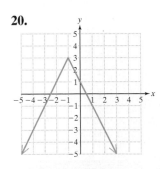

21.

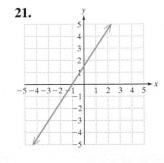

22.

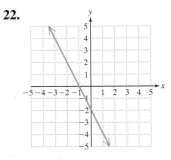

23.

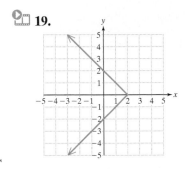

24.

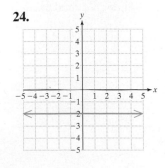

25.

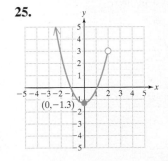

(0, −1.3)

26.

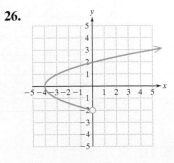

27.

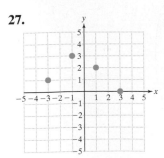

28.

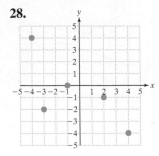

29.

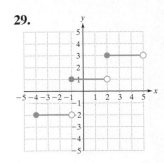

30.

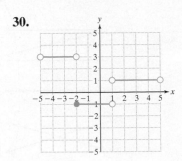

Concept 3: Applications Involving Relations

31. The table gives a relation between the month of the year and the average precipitation for that month for Miami, Florida. **(See Example 5.)**

a. What is the range element corresponding to April?

b. What is the range element corresponding to June?

c. Which element in the domain corresponds to the least value in the range?

d. Complete the ordered pair: (, 2.66)

e. Complete the ordered pair: (Sept.,   )

f. What is the domain of this relation?

Month x	Precipitation (in.) y	Month x	Precipitation (in.) y
Jan.	2.01	July	5.70
Feb.	2.08	Aug.	7.58
Mar.	2.39	Sept.	7.63
Apr.	2.85	Oct.	5.64
May	6.21	Nov.	2.66
June	9.33	Dec.	1.83

Source: U.S. National Oceanic and Atmospheric Administration

32. The table gives a relation between a person's age and the person's maximum recommended heart rate.

a. What is the domain?

b. What is the range?

c. The range element 200 corresponds to what element in the domain?

d. Complete the ordered pair: (50,   )

e. Complete the ordered pair: (, 190)

Age (years) x	Maximum Recommended Heart Rate (Beats per Minute) y
20	200
30	190
40	180
50	170
60	160

33. The population of Canada, y (in millions), can be approximated by the relation $y = 0.146x + 31$, where x represents the number of years since 2000.

a. Approximate the population of Canada in the year 2006.

b. In what year will the population of Canada reach approximately 32,752,000?

34. The world record times for women's track and field events are shown in the table.

The women's world record time y (in seconds) required to run x meters can be approximated by the relation $y = -10.78 + 0.159x$.

a. Predict the time required for a 500-m race.

b. Use this model to predict the time for a 1000-m race. Is this value exactly the same as the data value given in the table? Explain.

Distance (m)	Time (sec)	Winner's Name and Country
100	10.49	Florence Griffith Joyner (United States)
200	21.34	Florence Griffith Joyner (United States)
400	47.60	Marita Koch (East Germany)
800	113.28	Jarmila Kratochvilova (Czechoslovakia)
1000	148.98	Svetlana Masterkova (Russia)
1500	230.46	Qu Yunxia (China)

Expanding Your Skills

35. a. Define a relation with four ordered pairs such that the first element of the ordered pair is the name of a friend and the second element is your friend's place of birth.

b. State the domain and range of this relation.

36. a. Define a relation with four ordered pairs such that the first element is a state and the second element is its capital.

b. State the domain and range of this relation.

37. Use a mathematical equation to define a relation whose second component y is 1 less than 2 times the first component x.

38. Use a mathematical equation to define a relation whose second component y is 3 more than the first component x.

39. Use a mathematical equation to define a relation whose second component is the square of the first component.

40. Use a mathematical equation to define a relation whose second component is one-fourth the first component.

Section 3.2 Introduction to Functions

Concepts

1. Definition of a Function
2. Vertical Line Test
3. Function Notation
4. Finding Function Values from a Graph
5. Domain of a Function

1. Definition of a Function

In this section, we introduce a special type of relation called a function.

> **Definition of a Function**
>
> Given a relation in x and y, we say "y is a **function** of x" if for each element x in the domain, there is exactly one value of y in the range.
>
> *Note:* This means that no two ordered pairs may have the same first coordinate and different second coordinates.

To understand the difference between a relation that is a function and a relation that is not a function, consider Example 1.

Example 1 Determining Whether a Relation Is a Function

Determine which of the relations define y as a function of x.

a.

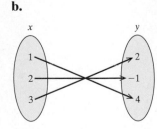

b. **c.**

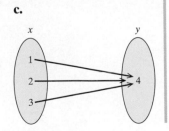

Solution:

a. This relation is defined by the set of ordered pairs

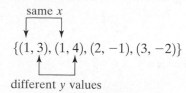

When $x = 1$, there are *two* possible range elements: $y = 3$ and $y = 4$. Therefore, this relation is *not* a function.

b. This relation is defined by the set of ordered pairs $\{(1, 4), (2, -1), (3, 2)\}$.

Notice that no two order pairs have the same value of x but different values of y. Therefore, this relation *is* a function.

c. This relation is defined by the set of ordered pairs $\{(1, 4), (2, 4), (3, 4)\}$.

Notice that no two ordered pairs have the same value of x but different values of y. Therefore, this relation *is* a function.

Skill Practice

Determine if the relations define y as a function of x.

1.

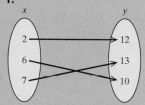

2. $\{(4, 2), (-5, 4), (0, 0), (8, 4)\}$

3. $\{(-1, 6), (8, 9), (-1, 4), (-3, 10)\}$

2. Vertical Line Test

A relation that is not a function has at least one domain element x paired with more than one range value y. For example, the set $\{(4, 2), (4, -2)\}$ does not define a function because two different y values correspond to the same x. These two points are aligned vertically in the xy-plane, and a vertical line drawn through one point also intersects the other point. In general, if a vertical line drawn through a graph of a relation intersects the graph in more than one point, the relation cannot be a function. This idea is stated formally as the **vertical line test**.

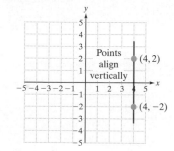

The Vertical Line Test

Consider a relation defined by a set of points (x, y) in a rectangular coordinate system. The graph defines y as a function of x if no vertical line intersects the graph in more than one point.

The vertical line test can be demonstrated by graphing the ordered pairs from the relations in Example 1.

a. $\{(1, 3), (1, 4), (2, -1), (3, -2)\}$ **b.** $\{(1, 4), (2, -1), (3, 2)\}$

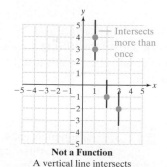

Not a Function
A vertical line intersects
in more than one point.

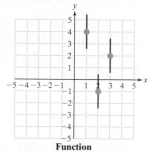

Function
No vertical line
intersects more than once.

Concept Connections

4. What kind of lines are not functions?

Answers

1. Yes   **2.** Yes   **3.** No
4. Vertical lines are not functions.

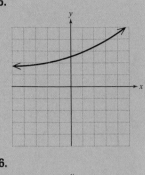
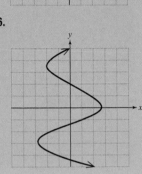
Example 2 Using the Vertical Line Test

Use the vertical line test to determine whether the relations define y as a function of x.

a.

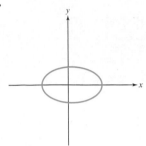

b.

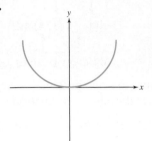

Solution:

a.

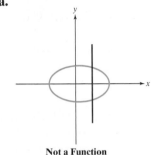

Not a Function
A vertical line intersects in more than one point.

b.

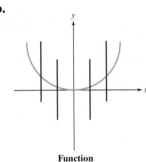

Function
No vertical line intersects in more than one point.

3. Function Notation

A function is defined as a relation with the added restriction that each value in the domain must have only one corresponding y-value in the range. In mathematics, functions are often given by rules or equations to define the relationship between two or more variables. For example, the equation $y = 2x$ defines the set of ordered pairs such that the y value is twice the x value.

When a function is defined by an equation, we often use **function notation**. For example, the equation $y = 2x$ may be written in function notation as

$f(x) = 2x$ where f is the name of the function, x is an input value from the domain of the function, and $f(x)$ is the function value (or y value) corresponding to x.

The notation $f(x)$ is read as "f of x" or "the value of the function f at x."

A function may be evaluated at different values of x by substituting x values from the domain into the function. For example, to evaluate the function defined by $f(x) = 2x$ at $x = 5$, substitute $x = 5$ into the function.

$$f(x) = 2x$$

$$f(5) = 2(5)$$

$$f(5) = 10$$

TIP: The function value $f(5) = 10$ can be written as the ordered pair (5, 10).

Thus, when $x = 5$, the corresponding function value is 10. We say:

- f of 5 is 10.
- f at 5 is 10.
- f evaluated at 5 is 10.

The names of functions are often given by either lowercase or uppercase letters, such as f, g, h, p, K, and M.

Example 3 **Evaluating a Function**

Given the function defined by $g(x) = \frac{1}{2}x - 1$, find the function values.

a. $g(0)$ **b.** $g(2)$ **c.** $g(4)$ **d.** $g(-2)$

Solution:

a. $g(x) = \frac{1}{2}x - 1$

$g(0) = \frac{1}{2}(0) - 1$

$= 0 - 1$

$= -1$ We say, "g of 0 is -1." This is equivalent to the ordered pair $(0, -1)$.

b. $g(x) = \frac{1}{2}x - 1$

$g(2) = \frac{1}{2}(2) - 1$

$= 1 - 1$

$= 0$ We say "g of 2 is 0." This is equivalent to the ordered pair $(2, 0)$.

c. $g(x) = \frac{1}{2}x - 1$

$g(4) = \frac{1}{2}(4) - 1$

$= 2 - 1$

$= 1$ We say "g of 4 is 1." This is equivalent to the ordered pair $(4, 1)$.

d. $g(x) = \frac{1}{2}x - 1$

$g(-2) = \frac{1}{2}(-2) - 1$

$= -1 - 1$

$= -2$ We say "g of -2 is -2." This is equivalent to the ordered pair $(-2, -2)$.

Skill Practice

Given the function defined by $f(x) = -2x - 3$, find the function values.

7. $f(1)$ **8.** $f(0)$

9. $f(-3)$ **10.** $f\left(\frac{1}{2}\right)$

Answers

7. -5 **8.** -3 **9.** 3   **10.** -4

Notice that $g(0)$, $g(2)$, $g(4)$, and $g(-2)$ correspond to the ordered pairs $(0, -1)$, $(2, 0)$, $(4, 1)$, and $(-2, -2)$. In the graph, these points "line up." The graph of *all* ordered pairs defined by this function is a line with a slope of $\frac{1}{2}$ and y-intercept of $(0, -1)$ (Figure 3-5). This should not be surprising because the function defined by $g(x) = \frac{1}{2}x - 1$ is equivalent to $y = \frac{1}{2}x - 1$.

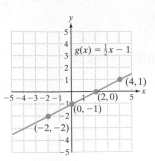

Figure 3-5

Calculator Connections

Topic: Using the *Table* and *Value* Features

The values of $g(x)$ in Example 3 can be found using a *Table* feature.

$$Y_1 = \tfrac{1}{2}x - 1$$

Function values can also be evaluated by using a *Value* feature. The value of $g(4)$ is shown here.

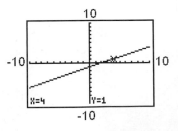

A function may be evaluated at numerical values or at algebraic expressions as shown in Example 4.

Example 4 **Evaluating Functions**

Given the functions defined by $f(x) = x^2 - 2x$ and $g(x) = 3x + 5$, find the function values.

a. $f(t)$ **b.** $g(w + 4)$ **c.** $f(-t)$

Solution:

a. $f(x) = x^2 - 2x$

$\quad f(t) = (t)^2 - 2(t)$ Substitute $x = t$ for all values of x in the function.

$\quad\quad = t^2 - 2t$ Simplify.

b. $g(x) = 3x + 5$

$g(w + 4) = 3(w + 4) + 5$ Substitute $x = w + 4$ for all values of x in
$\qquad = 3w + 12 + 5$ the function.

$\qquad = 3w + 17$ Simplify.

c. $f(x) = x^2 - 2x$ Substitute $-t$ for x.

$f(-t) = (-t)^2 - 2(-t)$

$\qquad = t^2 + 2t$ Simplify.

4. Finding Function Values from a Graph

We can find function values by looking at a graph of the function. The value of $f(a)$ refers to the y-coordinate of a point with x-coordinate a.

Example 5 **Finding Function Values from a Graph**

Consider the function pictured in Figure 3-6.

a. Find $h(-1)$.

b. Find $h(2)$.

c. Find $h(5)$.

d. For what value of x is $h(x) = 3$?

e. For what values of x is $h(x) = 0$?

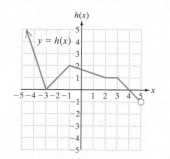

Figure 3-6

Solution:

a. $h(-1) = 2$ This corresponds to the ordered pair $(-1, 2)$.

b. $h(2) = 1$ This corresponds to the ordered pair $(2, 1)$.

c. $h(5)$ is not defined.   The value 5 is not in the domain.

d. $h(x) = 3$ for $x = -4$ This corresponds to the ordered pair $(-4, 3)$.

e. $h(x) = 0$ for $x = -3$ and $x = 4$ These are the ordered pairs $(-3, 0)$ and $(4, 0)$.

Skill Practice

Refer to the function graphed here.

14. Find $f(0)$.

15. Find $f(-2)$.

16. Find $f(5)$.

17. For what value(s) of x is $f(x) = 0$?

18. For what value(s) of x is $f(x) = -4$?

5. Domain of a Function

A function is a relation, and it is often necessary to determine its domain and range. Consider a function defined by the equation $y = f(x)$. The **domain** of f is the set of all x values that when substituted into the function, produce a real number. The **range** of f is the set of all y values corresponding to the values of x in the domain.

To find the domain of a function defined by $y = f(x)$, keep these guidelines in mind.

- Exclude values of x that make the denominator of a fraction zero.
- Exclude values of x that make the expression within a square root negative.

Answers

14. 3   **15.** 1   **16.** not defined
17. $x = -4$ and $x = 3$ **18.** $x = -5$

Example 6 Finding the Domain of a Function

Write the domain in interval notation.

a. $f(x) = \dfrac{x + 7}{2x - 1}$ **b.** $h(x) = \dfrac{x - 4}{x^2 + 9}$

c. $k(t) = \sqrt{t + 4}$ **d.** $g(t) = t^2 - 3t$

Solution:

a. $f(x) = \frac{x + 7}{2x - 1}$ will not be a real number when the denominator is zero, that is, when

$$2x - 1 = 0$$

$$2x = 1$$

$$x = \frac{1}{2} \qquad \text{The value } x = \tfrac{1}{2} \text{ must be } excluded \text{ from the domain.}$$

Interval notation: $\left(-\infty, \dfrac{1}{2}\right) \cup \left(\dfrac{1}{2}, \infty\right)$

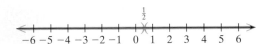

b. For $h(x) = \frac{x - 4}{x^2 + 9}$ the quantity x^2 is greater than or equal to 0 for all real numbers x, and the number 9 is positive. The sum $x^2 + 9$ must be *positive* for all real numbers x. The denominator will never be zero; therefore, the domain is the set of all real numbers.

Interval notation: $(-\infty, \infty)$

c. The function defined by $k(t) = \sqrt{t + 4}$ will not be a real number when the radicand is negative. The domain is the set of all t-values that make the radicand *greater than or equal to zero:*

$$t + 4 \geq 0$$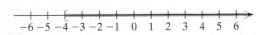

$$t \geq -4$$

Interval notation: $[-4, \infty)$

d. The function defined by $g(t) = t^2 - 3t$ has no restrictions on its domain because any real number substituted for t will produce a real number. The domain is the set of all real numbers.

Interval notation: $(-\infty, \infty)$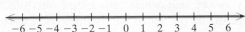

Section 3.2   Practice Exercises

Vocabulary and Key Concepts

1. a. Given a relation in *x* and *y*, we say that *y* is a _____ of *x* if for each element *x* in the domain, there is exactly one value of *y* in the range.

   b. If a _____ line intersects the graph of a relation in more than one point, the relation is not a function.

   c. Function notation for the relation $y = 2x + 1$ is $f(x) = $ _____.

   d. The set of *x* values in a function is called the _____ of the function, and the set of *y* values is called the _____ of the function.

   e. To find the domain of a function defined by $y = f(x)$, exclude all values of *x* that make the _____ of a fraction equal to zero. Also exclude all values of *x* that make the expression within a square root _____.

   f. Given $f(x) = \dfrac{x - 4}{x + 2}$, the domain is restricted so that $x \neq$ _____.

   g. Given $f(x) = \sqrt{x - 3}$ the domain is restricted so that $x \geq$ _____.

   h. Given a function defined by $y = f(x)$, the statement $f(1) = 6$ is equivalent to what ordered pair?

Review Exercises

For Exercises 2–4, **a.** write the relation as a set of ordered pairs, **b.** identify the domain, and **c.** identify the range.

2.

Parent, *x*	Child, *y*
Kevin	Kayla
Kevin	Kira
Kathleen	Katie
Kathleen	Kira

3.

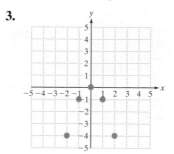

4.

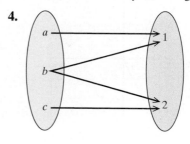

Concept 1: Definition of a Function

For Exercises 5–10, determine if the relation defines *y* as a function of *x*. **(See Example 1.)**

5.

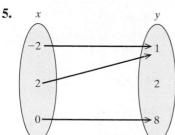

6.

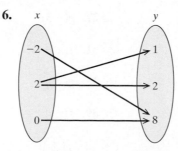

7.

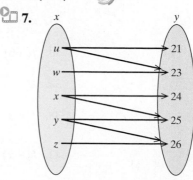

8.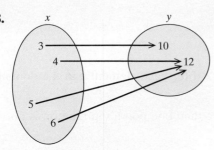

9. $\{(1, 2), (3, 4), (5, 4), (-9, 3)\}$

10. $\left\{(0, -1.1), \left(\frac{1}{2}, 8\right), (1.1, 8), \left(4, \frac{1}{2}\right)\right\}$

Concept 2: Vertical Line Test

For Exercises 11–16, use the vertical line test to determine whether the relation defines y as a function of x.
(See Example 2.)

11.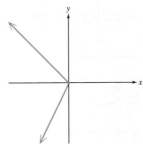

12.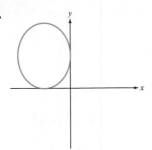

13.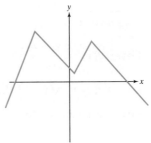

14.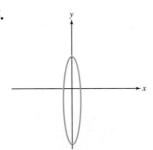

15.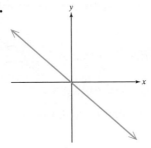

16.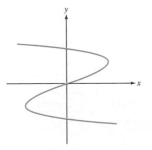

Concept 3: Function Notation

Consider the functions defined by $f(x) = 6x - 2$, $g(x) = -x^2 - 4x + 1$, $h(x) = 7$, and $k(x) = |x - 2|$. For Exercises 17–48, find the following. **(See Examples 3 and 4.)**

17. $g(2)$

18. $k(2)$

19. $g(0)$

20. $h(0)$

21. $k(0)$

22. $f(0)$

23. $f(t)$

24. $g(a)$

25. $h(u)$

26. $k(v)$

27. $g(-3)$

28. $h(-5)$

29. $k(-2)$

30. $f(-6)$

31. $f(x + 1)$

32. $h(x + 1)$

33. $g(2x)$

34. $k(x - 3)$

35. $g(-\pi)$

36. $g(a^2)$

37. $h(a + b)$

38. $f(x + h)$

39. $f(-a)$

40. $g(-b)$

41. $k(-c)$

42. $h(-x)$

43. $f\left(\frac{1}{2}\right)$

44. $g\left(\frac{1}{4}\right)$

45. $h\left(\frac{1}{7}\right)$

46. $k\left(\frac{3}{2}\right)$

47. $f(-2.8)$

48. $k(-5.4)$

Consider the functions $p = \{(\frac{1}{2}, 6), (2, -7), (1, 0), (3, 2\pi)\}$ and $q = \{(6, 4), (2, -5), (\frac{3}{4}, \frac{1}{5}), (0, 9)\}$. For Exercises 49–56, find the function values.

49. $p(2)$

50. $p(1)$

51. $p(3)$

52. $p\left(\dfrac{1}{2}\right)$

53. $q(2)$

54. $q\left(\dfrac{3}{4}\right)$

55. $q(6)$

56. $q(0)$

Concept 4: Finding Function Values from a Graph

57. The graph of $y = f(x)$ is given. **(See Example 5.)**

   a. Find $f(0)$.

   b. Find $f(3)$.

   c. Find $f(-2)$.

   d. For what value(s) of x is $f(x) = -3$?

   e. For what value(s) of x is $f(x) = 3$?

   f. Write the domain of f.

   g. Write the range of f.

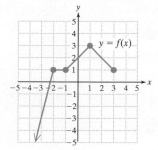

58. The graph of $y = g(x)$ is given.

   a. Find $g(-1)$.

   b. Find $g(1)$.

   c. Find $g(4)$.

   d. For what value(s) of x is $g(x) = 3$?

   e. For what value(s) of x is $g(x) = 0$?

   f. Write the domain of g.

   g. Write the range of g.

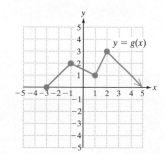

59. The graph of $y = H(x)$ is given.

   a. Find $H(-3)$.

   b. Find $H(4)$.

   c. Find $H(3)$.

   d. For what value(s) of x is $H(x) = 3$?

   e. For what value(s) of x is $H(x) = 2$?

   f. Write the domain of H.

   g. Write the range of H.

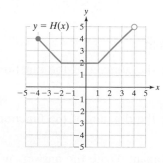

60. The graph of $y = K(x)$ is given.

    a. Find $K(0)$.

    b. Find $K(-5)$.

    c. Find $K(1)$.

    d. For what value(s) of x is $K(x) = 0$?

    e. For what value(s) of x is $K(x) = 3$?

    f. Write the domain of K.

    g. Write the range of K.

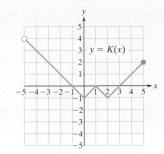

61. The graph of $y = p(x)$ is given.

    a. Find $p(2)$.

    b. Find $p(-1)$.

    c. Find $p(1)$.

    d. For what value(s) of x is $p(x) = 0$?

    e. For what value(s) of x is $p(x) = -2$?

    f. Write the domain of p.

    g. Write the range of p.

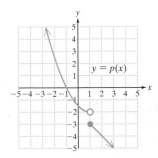

62. The graph of $y = q(x)$ is given.

    a. Find $q(3)$.

    b. Find $q(-1)$.

    c. Find $q(2)$.

    d. For what value(s) of x is $q(x) = -4$?

    e. For what value(s) of x is $q(x) = 3$?

    f. Write the domain of q.

    g. Write the range of q.

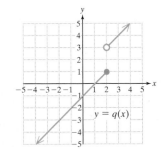

Concept 5: Domain of a Function

63. Explain how to determine the domain of the function defined by $f(x) = \dfrac{x + 6}{x - 2}$.

64. Explain how to determine the domain of the function defined by $g(x) = \sqrt{x - 3}$.

For Exercises 65–80, find the domain. Write the answers in interval notation. **(See Example 6.)**

65. $k(x) = \dfrac{x - 3}{x + 6}$　　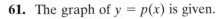 **66.** $m(x) = \dfrac{x - 1}{x - 4}$　　**67.** $f(t) = \dfrac{5}{t}$　　**68.** $g(t) = \dfrac{t - 7}{t}$

69. $h(p) = \dfrac{p - 4}{p^2 + 1}$　　**70.** $n(p) = \dfrac{p + 8}{p^2 + 2}$　　**71.** $h(t) = \sqrt{t + 7}$　　**72.** $k(t) = \sqrt{t - 5}$

73. $f(a) = \sqrt{a - 3}$　　**74.** $g(a) = \sqrt{a + 2}$　　**75.** $m(x) = \sqrt{1 - 2x}$　　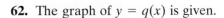 **76.** $n(x) = \sqrt{12 - 6x}$

77. $p(t) = 2t^2 + t - 1$　　**78.** $q(t) = t^3 + t - 1$　　**79.** $f(x) = x + 6$　　**80.** $g(x) = 8x - \pi$

Mixed Exercises

81. The height (in feet) of a ball that is dropped from an 80-ft building is given by $h(t) = -16t^2 + 80$, where t is the time in seconds after the ball is dropped.

    a. Find $h(1)$ and $h(1.5)$.

    b. Interpret the meaning of the function values found in part (a).

82. A ball is dropped from a 50-m building. The height (in meters) after t sec is given by $h(t) = -4.9t^2 + 50$.

    a. Find $h(1)$ and $h(1.5)$.

    b. Interpret the meaning of the function values found in part (a).

83. If Alicia rides a bike at an average of 11.5 mph, the distance that she rides can be represented by $d(t) = 11.5t$, where t is the time in hours.

    a. Find $d(1)$ and $d(1.5)$.

    b. Interpret the meaning of the function values found in part (a).

84. If Miguel walks at an average of 5.9 km/hr, the distance that he walks can be represented by $d(t) = 5.9t$, where t is the time in hours.

    a. Find $d(1)$ and $d(2)$.

    b. Interpret the meaning of the function values found in part (a).

85. Brian's score on an exam is a function of the number of hours he spends studying. The function defined by

$$P(x) = \frac{100x^2}{50 + x^2} \ (x \geq 0)$$ indicates that he will achieve a score of $P\%$ if he studies for x hours.

    a. Evaluate $P(0)$, $P(5)$, $P(10)$, $P(15)$, $P(20)$, and $P(25)$. (Round to 1 decimal place.) Interpret $P(25)$ in the context of this problem.

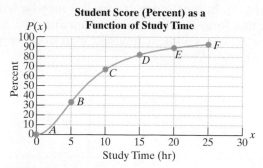

Student Score (Percent) as a Function of Study Time

    b. Match the function values found in part (a) with the points A, B, C, D, E, and F on the graph.

Expanding Your Skills

For Exercises 86 and 87, write the domain in interval notation.

86. $q(x) = \dfrac{2}{\sqrt{x + 2}}$ **87.** $p(x) = \dfrac{8}{\sqrt{x - 4}}$

For Exercises 88–97, refer to the functions $y = f(x)$ and $y = g(x)$, defined as follows:

$$f = \{(-3, 5), (-7, -3), (-\tfrac{3}{2}, 4), (1.2, 5)\} \quad g = \{(0, 6), (2, 6), (6, 0), (1, 0)\}$$

88. Identify the domain of f.   **89.** Identify the range of f.

90. Identify the range of g.

91. Identify the domain of g.

92. For what value(s) of x is $f(x) = 5$?

93. For what value(s) of x is $f(x) = -3$?

94. For what value(s) of x is $g(x) = 0$?

95. For what value(s) of x is $g(x) = 6$?

96. Find $f(-7)$.

97. Find $g(0)$.

Graphing Calculator Exercises

98. Graph $k(t) = \sqrt{t - 5}$. Use the graph to support your answer to Exercise 72.

99. Graph $h(t) = \sqrt{t + 7}$. Use the graph to support your answer to Exercise 71.

100. a. Graph $h(t) = -4.9t^2 + 50$ on a viewing window defined by $0 \le t \le 3$ and $0 \le y \le 60$.

    b. Use the graph to approximate the function at $t = 1$. Use this value to support your answer to Exercise 82.

101. a. Graph $h(t) = -16t^2 + 80$ on a viewing window defined by $0 \le t \le 2$ and $0 \le y \le 100$.

    b. Use the graph to approximate the function at $t = 1$. Use this value to support your answer to Exercise 81.

Section 3.3   Graphs of Basic Functions

Concepts

1. Linear and Constant Functions

2. Graphs of Basic Functions

3. Definition of a Quadratic Function

4. Finding the x- and y-Intercepts of a Function Defined by $y = f(x)$

1. Linear and Constant Functions

A function may be expressed as a mathematical equation that relates two or more variables. In this section, we will look at several elementary functions.

We know from Section 2.1 that an equation in the form $y = k$ where k is a constant is a horizontal line. In function notation, this can be written as $f(x) = k$. For example, the function defined by $f(x) = 3$ is a horizontal line as shown in Figure 3-7.

We say that a function defined by $f(x) = k$ is a constant function because for any value of x, the function value is constant.

An equation of the form $y = mx + b$ is represented graphically by a line with slope m and y-intercept $(0, b)$. In function notation, this may be written as $f(x) = mx + b$. A function in this form is called a linear function. For example, the function defined by $f(x) = 2x - 3$ is a linear function with slope $m = 2$ and y-intercept $(0, -3)$ (Figure 3-8).

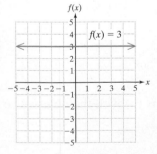

Figure 3-7

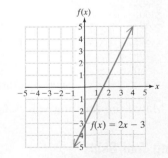

Figure 3-8

Linear Functions and Constant Functions

Let m and b represent real numbers such that $m \ne 0$. Then

A function that can be written in the form $f(x) = mx + b$ is a **linear function**.

A function that can be written in the form $f(x) = b$ is a **constant function**.

Note: The graphs of linear and constant functions are lines.

2. Graphs of Basic Functions

At this point, we are able to recognize the equations and graphs of linear and constant functions. In addition to linear and constant functions, the following equations define six basic functions that will be encountered in the study of algebra:

Equation	Function Notation	Type of Function
$y = x$	$f(x) = x$	Identity function
$y = x^2$	$f(x) = x^2$	Quadratic function
$y = x^3$ equivalent function notation →	$f(x) = x^3$	Cubic function
$y = |x|$	$f(x) = |x|$	Absolute value function
$y = \sqrt{x}$	$f(x) = \sqrt{x}$	Square root function
$y = \dfrac{1}{x}$	$f(x) = \dfrac{1}{x}$	Reciprocal function

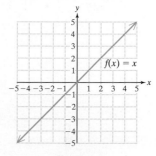

Figure 3-9

The graph of the function defined by $f(x) = x$ is linear, with slope $m = 1$ and y-intercept $(0, 0)$ (Figure 3-9).

To determine the shapes of the other basic functions, we can plot several points to establish the pattern of the graph. Analyzing the equation itself may also provide insight to the domain, range, and shape of the function. To demonstrate this, we will graph $f(x) = x^2$ and $g(x) = \frac{1}{x}$.

Example 1 **Graphing Basic Functions**

Graph the function defined by $f(x) = x^2$.

Solution:

The domain of the function given by $f(x) = x^2$ (or equivalently $y = x^2$) is all real numbers.

To graph the function, choose arbitrary values of x within the domain of the function. Be sure to choose values of x that are positive and values that are negative to determine the behavior of the function to the right and left of the origin (Table 3-4). The graph of $f(x) = x^2$ is shown in Figure 3-10.

The function values are equated to the square of x, so $f(x)$ will always be greater than or equal to zero. Hence, the y-coordinates on the graph will never be negative. The range of the function is $[0, \infty)$. The arrows on each branch of the graph imply that the pattern continues indefinitely.

Table 3-4

x	$f(x) = x^2$
0	0
1	1
2	4
3	9
−1	1
−2	4
−3	9

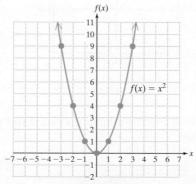

Figure 3-10

Skill Practice

1. Graph $f(x) = -x^2$ by first making a table of points.

Answer

1.

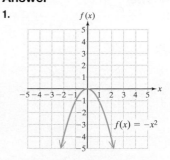

Example 2　**Graphing Basic Functions**

Graph the function defined by $g(x) = \dfrac{1}{x}$.

Solution:

$g(x) = \dfrac{1}{x}$　Notice that $x = 0$ is not in the domain of the function. From the equation $y = \frac{1}{x}$, the y values will be the reciprocal of the x values. The graph defined by $g(x) = \frac{1}{x}$ is shown in Figure 3-11.

x	$g(x) = \dfrac{1}{x}$
1	1
2	$\frac{1}{2}$
3	$\frac{1}{3}$
-1	-1
-2	$-\frac{1}{2}$
-3	$-\frac{1}{3}$

x	$g(x) = \dfrac{1}{x}$
$\frac{1}{2}$	2
$\frac{1}{3}$	3
$\frac{1}{4}$	4
$-\frac{1}{2}$	-2
$-\frac{1}{3}$	-3
$-\frac{1}{4}$	-4

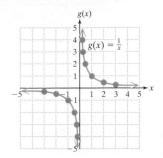

Figure 3-11

　　Notice that as x approaches ∞ and $-\infty$, the y values approach zero, and the graph approaches the x-axis. In this case, the x-axis is called a *horizontal asymptote*. Similarly, the graph of the function approaches the y-axis as x gets close to zero. In this case, the y-axis is called a *vertical asymptote*.

Calculator Connections

Topic: Graphing Nonlinear Functions

To enter a function, let $f(x) = y$. That is, to graph $f(x) = x^2$, enter Y1 $= x^2$.

To graph $g(x) = \dfrac{1}{x}$, enter Y1 $= 1/x$.

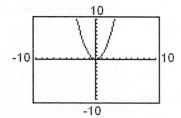
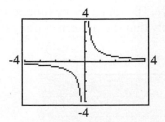

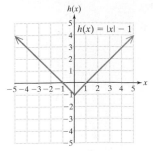

For your reference, we have provided the graphs of six basic functions in the following table.

Summary of Six Basic Functions and Their Graphs

Function	Graph	Domain and Range
1. $f(x) = x$ Identity function		Domain $(-\infty, \infty)$ Range $(-\infty, \infty)$
2. $f(x) = x^2$ Quadratic function		Domain $(-\infty, \infty)$ Range $[0, \infty)$
3. $f(x) = x^3$ Cubic function		Domain $(-\infty, \infty)$ Range $(-\infty, \infty)$
4. $f(x) = |x|$ Absolute value function		Domain $(-\infty, \infty)$ Range $[0, \infty)$
5. $f(x) = \sqrt{x}$ Square root function		Domain $[0, \infty)$ Range $[0, \infty)$
6. $f(x) = \dfrac{1}{x}$ Reciprocal function		Domain $(-\infty, 0) \cup (0, \infty)$ Range $(-\infty, 0) \cup (0, \infty)$

The shapes of these six graphs will be developed in the homework exercises. These functions are used often in the study of algebra. Therefore, we recommend that you associate an equation with its graph and commit each to memory.

3. Definition of a Quadratic Function

In Example 1 we graphed the function defined by $f(x) = x^2$ by plotting points. This function belongs to a special category called quadratic functions.

Definition of a Quadratic Function

A **quadratic function** is a function defined by

$$f(x) = ax^2 + bx + c \qquad \text{where } a, b, \text{ and } c \text{ are real numbers and } a \neq 0.$$

The graph of a quadratic function is in the shape of a **parabola**. The leading coefficient, a, determines the direction of the parabola.

- If $a > 0$, then the parabola opens upward, for example, $f(x) = x^2$. The minimum point on a parabola opening upward is called the vertex (Figure 3-12).
- If $a < 0$, then the parabola opens downward, for example, $f(x) = -x^2$. The maximum point on a parabola opening downward is called the vertex (Figure 3-13).

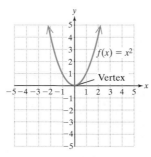

Figure 3-12

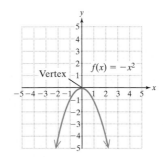

Figure 3-13

Skill Practice

Identify whether the function is constant, linear, quadratic, or none of these.

3. $m(x) = -2x^2 - 3x + 7$

4. $n(x) = -6$

5. $W(x) = \dfrac{4}{3}x - \dfrac{1}{2}$

6. $R(x) = \dfrac{4}{3x} - \dfrac{1}{2}$

Example 3 **Identifying Linear and Constant Functions**

Identify each function as linear, constant, quadratic, or none of these.

a. $f(x) = -4$ **b.** $f(x) = x^2 + 3x + 2$

c. $f(x) = 7 - 2x$ **d.** $f(x) = \dfrac{4x + 8}{8}$

Solution:

a. $f(x) = -4$ is a constant function. It is in the form $f(x) = b$, where $b = -4$.

b. $f(x) = x^2 + 3x + 2$ is a quadratic function. It is in the form $f(x) = ax^2 + bx + c$, where $a \neq 0$.

c. $f(x) = 7 - 2x$ is linear. Writing it in the form $f(x) = mx + b$, we get $f(x) = -2x + 7$, where $m = -2$ and $b = 7$.

d. $f(x) = \dfrac{4x + 8}{8}$ is linear. Writing it in the form $f(x) = mx + b$, we get

$$f(x) = \frac{4x}{8} + \frac{8}{8}$$

$$= \frac{1}{2}x + 1, \text{ where } m = \frac{1}{2} \text{ and } b = 1.$$

Answers

3. Quadratic
4. Constant
5. Linear
6. None of these

4. Finding the *x*- and *y*-Intercepts of a Function Defined by *y* = *f*(*x*)

In Section 2.1, we learned that to find the *x*-intercept, we substitute *y* = 0 and solve the equation for *x*. Using function notation, this is equivalent to finding the real solutions of the equation $f(x) = 0$. To find the *y*-intercept, substitute *x* = 0 and solve the equation for *y*. In function notation, this is equivalent to finding $f(0)$.

Finding the *x*- and *y*-Intercepts of a Function

Given a function defined by $y = f(x)$,

Step 1 The *x*-intercepts are the real solutions to the equation $f(x) = 0$.
Step 2 The *y*-intercept is given by $f(0)$.

Example 4 Finding the *x*- and *y*-Intercepts of a Function

Given the function defined by $f(x) = 2x - 4$:

a. Find the *x*-intercept(s).

b. Find the *y*-intercept.

c. Graph the function.

Solution:

a. To find the *x*-intercept(s), find the real solutions to the equation $f(x) = 0$.

$$f(x) = 2x - 4$$

$$0 = 2x - 4 \qquad \text{Substitute } f(x) = 0.$$

$$4 = 2x$$

$$2 = x \qquad \text{The } x\text{-intercept is } (2, 0).$$

b. To find the *y*-intercept, evaluate $f(0)$.

$$f(0) = 2(0) - 4 \qquad \text{Substitute } x = 0.$$

$$f(0) = -4 \qquad \text{The } y\text{-intercept is } (0, -4).$$

c. This function is linear, with a *y*-intercept of $(0, -4)$, an *x*-intercept of $(2, 0)$, and a slope of 2 (Figure 3-14).

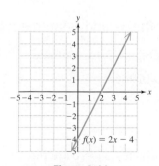

Figure 3-14

Answers
7. $\left(\frac{1}{5}, 0\right)$ **8.** $(0, 1)$
9.

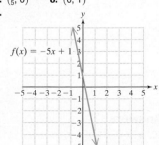

Calculator Connections

Topic: Finding *x*- and *y*-Intercepts

Refer to Example 4 with $f(x) = 2x - 4$. To find the *y*-intercept, let $x = 0$. We can do this using the *Value* key.

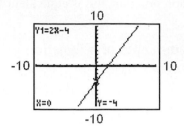

To find the *x*-intercept, use the *Zero* feature. To use this feature you must give bounds for the *x*-intercept. To find a left bound, place the curser to the left of the *x*-intercept and press enter. For the right bound, place the curser to the right of the *x*-intercept.

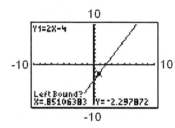 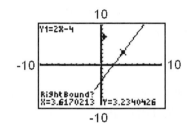

Then make a guess by placing the curser near the *x*-intercept. Press Enter and the *x*-intercept will be displayed.

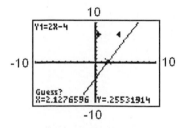 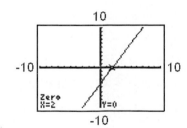

Skill Practice

Use the function pictured below.

10. a. Estimate the real value(s) of *x* for which $f(x) = 0$.

　　b. Estimate the value of $f(0)$.

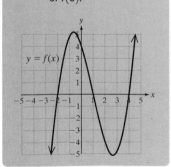

Answers

10. a. $x = -2$, $x = 1$, and $x = 4$

　　b. $f(0) = 4$

Example 5　Finding the *x*- and *y*-Intercepts of a Function

For the function pictured in Figure 3-15, estimate

a. The real values of *x* for which $f(x) = 0$.

b. The value of $f(0)$.

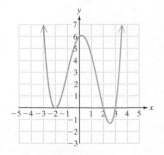

Figure 3-15

Solution:

a. The real values of *x* for which $f(x) = 0$ are the *x*-intercepts of the function. For this graph, the *x*-intercepts are located at $x = -2$, $x = 2$, and $x = 3$.

b. The value of $f(0)$ is the value of *y* at $x = 0$. That is, $f(0)$ is the *y*-intercept, $f(0) = 6$.

Section 3.3 Practice Exercises

Vocabulary and Key Concepts

1. **a.** A function that can be written in the form $f(x) = mx + b, m \neq 0$, is a _____ function.

   b. A function that can be written in the form $f(x) = b$ is a _____ function.

   c. A function that can be written in the form $f(x) = ax^2 + bx + c, a \neq 0$, is a _____ function.

   d. The graph of a quadratic function is in the shape of a _____.

   e. To find the x-intercept(s) of a graph defined by $y = f(x)$, find the real solutions to the equation $f(x) =$ _____.

   f. To find the _____-intercept of a graph defined by $y = f(x)$, evaluate $f(0)$.

Review Exercises

2. Given: $g = \{(6, 1), (5, 2), (4, 3), (3, 4)\}$

   a. Is this relation a function?

   b. List the elements in the domain.

   c. List the elements in the range.

3. Given: $f = \{(7, 3), (2, 3), (-5, 3)\}$

   a. Is this relation a function?

   b. List the elements in the domain.

   c. List the elements in the range.

4. Given: $f(x) = \sqrt{x + 4}$

   a. Evaluate $f(0), f(-3), f(-4)$, and $f(-5)$, if possible.

   b. Write the domain of this function in interval notation.

5. The force (measured in pounds) to stretch a certain spring x inches is given by $f(x) = 3x$. Evaluate $f(3)$ and $f(10)$, and interpret the results in the context of this problem.

6. The velocity (in feet per second) of a falling object is given by $V(t) = -32t$, where t is the time in seconds after the object was released. Evaluate $V(2)$ and $V(5)$, and interpret the results in the context of this problem.

Concept 1: Linear and Constant Functions

7. Fill in the blank with the word *vertical* or *horizontal*. The graph of a constant function is a _____ line.

8. For the linear function $f(x) = mx + b$, identify the slope and y-intercept.

9. Graph the constant function $f(x) = 2$. Then use the graph to identify the domain and range of f.

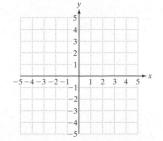

10. Graph the linear function $g(x) = -2x + 1$. Then use the graph to identify the domain and range of g.

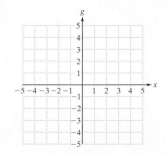

Concept 2: Graphs of Basic Functions

For Exercises 11–16, sketch a graph by completing the table and plotting the points. **(See Examples 1 and 2.)**

11. $f(x) = \dfrac{1}{x}$

x	$f(x)$
-2	
-1	
$-\frac{1}{2}$	
$-\frac{1}{4}$	

x	$f(x)$
$\frac{1}{4}$	
$\frac{1}{2}$	
1	
2	

12. $g(x) = |x|$

x	$g(x)$
-2	
-1	
0	
1	
2	

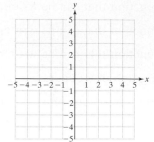

13. $h(x) = x^3$

x	$h(x)$
-2	
-1	
0	
1	
2	

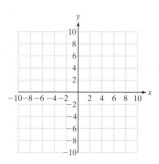

14. $k(x) = x$

x	$k(x)$
-2	
-1	
0	
1	
2	

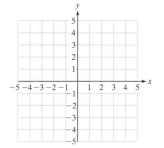

15. $q(x) = x^2$

x	$q(x)$
-2	
-1	
0	
1	
2	

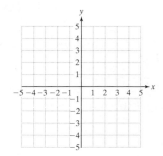

16. $p(x) = \sqrt{x}$

x	$p(x)$
0	
1	
4	
9	
16	

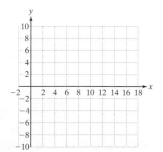

Concept 3: Definition of a Quadratic Function

For Exercises 17–28, determine if the function is constant, linear, quadratic, or none of these. **(See Example 3.)**

17. $f(x) = 2x^2 + 3x + 1$ **18.** $g(x) = -x^2 + 4x + 12$ **19.** $k(x) = -3x - 7$ **20.** $h(x) = -x - 3$

21. $m(x) = \dfrac{4}{3}$ **22.** $n(x) = 0.8$ **23.** $p(x) = \dfrac{2}{3x} + \dfrac{1}{4}$ **24.** $Q(x) = \dfrac{1}{5x} - 3$

25. $t(x) = \dfrac{2}{3}x + \dfrac{1}{4}$ **26.** $r(x) = \dfrac{1}{5}x - 3$ **27.** $w(x) = \sqrt{4 - x}$ **28.** $T(x) = -|x + 10|$

Concept 4: Finding the *x*- and *y*-Intercepts of a Function Defined by *y* = *f*(*x*)

For Exercises 29–36, find the *x*- and *y*-intercepts, and graph the function. **(See Example 4.)**

29. $f(x) = 5x - 10$

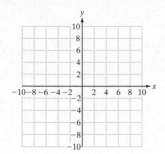

30. $f(x) = -3x + 12$

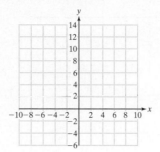

31. $g(x) = -6x + 5$

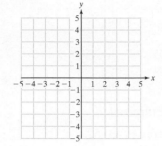

32. $h(x) = 2x + 9$

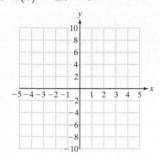

33. $f(x) = 18$

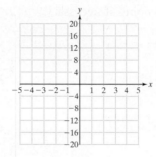

34. $g(x) = -7$

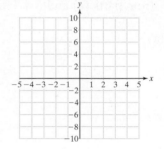

35. $g(x) = \dfrac{2}{3}x + \dfrac{1}{4}$

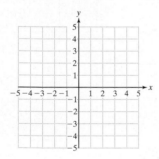

36. $h(x) = -\dfrac{5}{6}x + \dfrac{1}{2}$

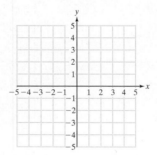

For Exercises 37–42, use the function pictured to estimate

    a. The real values of *x* for which $f(x) = 0$.

    b. The value of $f(0)$. **(See Example 5.)**

37.

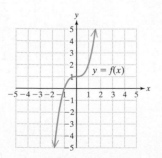

38.

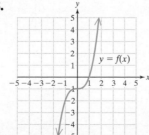

39.

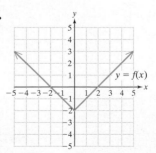

40.

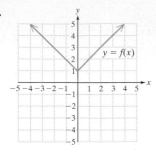

41.

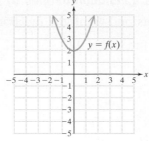

42.

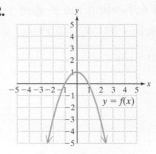

Mixed Exercises

For Exercises 43–52,

   a. Identify the domain of the function.

   b. Identify the *y*-intercept of the function.

   c. Match the function with its graph by recognizing the basic shape of the function and using the results from parts (a) and (b). Plot additional points if necessary.

43. $q(x) = 2x^2$

44. $p(x) = -2x^2 + 1$

45. $h(x) = x^3 + 1$

46. $k(x) = x^3 - 2$

47. $r(x) = \sqrt{x + 1}$

48. $s(x) = \sqrt{x + 4}$

49. $f(x) = \dfrac{1}{x - 3}$

50. $g(x) = \dfrac{1}{x + 1}$

51. $k(x) = |x + 2|$

52. $h(x) = |x - 1| + 2$

i.

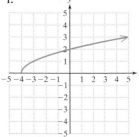

ii.

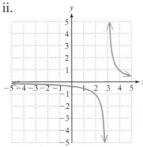

iii.

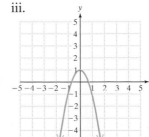

iv.

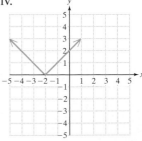

v.

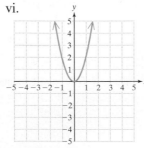

vi.

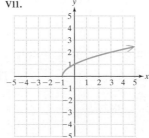

vii.

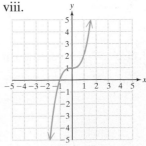

viii.

ix.

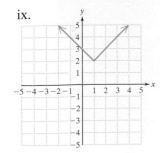

x.

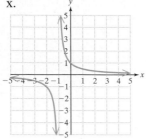

53. Suppose that a student has an 80% average on all of her chapter tests in her Intermediate Algebra class. This counts as $\frac{3}{4}$ of the student's overall grade. The final exam grade counts as the remaining $\frac{1}{4}$ of the student's overall grade. The student's overall course grade, $G(x)$ can be computed by

$$G(x) = \frac{3}{4}(80) + \frac{1}{4}x, \text{ where } x \text{ is the student's grade on the final exam.}$$

a. Is this function linear, quadratic, or neither?

b. Evaluate $G(90)$ and interpret its meaning in the context of the problem.

c. Evaluate $G(50)$ and interpret its meaning in the context of the problem.

54. The median weekly earnings $E(x)$ in dollars, for women 16 years and older working full time can be approximated by $E(x) = 0.14x^2 + 7.8x + 540$. For this function, x represents the number of years since 2000. (*Source:* U.S. Department of Labor)

a. Is this function linear, quadratic, or neither?

b. Evaluate $E(5)$ and interpret its meaning in the context of this problem.

c. Evaluate $E(10)$ and interpret its meaning in the context of this problem.

For Exercises 55–60, write a function defined by $y = f(x)$ under the given conditions.

55. The value of $f(x)$ is two more than the square of x.

56. The value of $f(x)$ is the square of the sum of two and x.

57. The function f is a constant function passing through the point $(4, 3)$.

58. The function f is a constant function with y-intercept $(0, 5)$.

59. The function f is a linear function with slope $\frac{1}{2}$ and y-intercept $(0, -2)$.

60. The function f is a linear function with slope -4 and y-intercept $(0, \frac{1}{3})$.

Graphing Calculator Exercises

For Exercises 61–66, use a graphing calculator to graph the basic functions. Verify your answers from the table on page 213.

61. $f(x) = x$ **62.** $f(x) = x^2$ **63.** $f(x) = x^3$

64. $f(x) = |x|$ **65.** $f(x) = \sqrt{x}$ **66.** $f(x) = \dfrac{1}{x}$

For Exercises 67–70, find the x- and y-intercepts using a graphing calculator and the *Value* and *Zero* features.

67. $y = -\dfrac{1}{8}x + 1$ **68.** $y = -\dfrac{1}{2}x - 3$ **69.** $y = \dfrac{4}{5}x + 4$ **70.** $y = \dfrac{7}{2}x - 7$

Problem Recognition Exercises

Characteristics of Relations

Exercises 1–15 refer to the sets of points (x, y) described in a–h.

a. $\{(0, 8), (1, 4), (\frac{1}{2}, 4), (-3, 5), (2, 1)\}$

b. $\{(-6, 4), (2, 3), (-9, 6), (2, 1), (0, 10)\}$

c. $c(x) = 3x^2 - 2x - 1$

d. $d(x) = 5x - 9$

e.

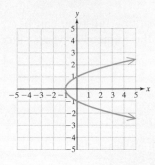

f.

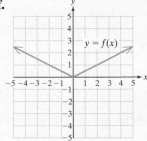

g.

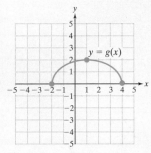

h.

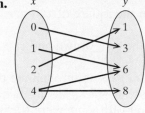

1. Which relations define y as a function of x?

2. Which relations contain the point $(2, 1)$?

3. Use relation (c) to find $c(-1)$.

4. Use relation (f) to find $f(-4)$.

5. Find the domain of relation (a).

6. Find the range of relation (b).

7. Find the domain of relation (g).

8. Find the range of relation (f).

9. Use relation (h) to complete the ordered pair (__, 3).

10. Find the x-intercept(s) of the graph of relation (d).

11. Find the y-intercept(s) of the graph shown in relation (e).

12. Use relation (g) to determine the value of x such that $g(x) = 2$.

13. Which relation describes a quadratic function?

14. Which relation describes a linear function?

15. Use relation (d) to find the value of x such that $d(x) = 6$.

Algebra of Functions, Composition, and Applications

1. Algebra of Functions

Addition, subtraction, multiplication, and division can be used to create a new function from two or more functions. The domain of the new function will be the intersection of the domains of the original functions.

Concepts

1. Algebra of Functions
2. Composition of Functions
3. Operations on Functions

Sum, Difference, Product, and Quotient of Functions

Given two functions f and g, the functions $f + g$, $f - g$, $f \cdot g$, and $\frac{f}{g}$ are defined as

$$(f + g)(x) = f(x) + g(x)$$

$$(f - g)(x) = f(x) - g(x)$$

$$(f \cdot g)(x) = f(x) \cdot g(x)$$

$$\left(\frac{f}{g}\right)(x) = \frac{f(x)}{g(x)} \qquad \text{provided } g(x) \neq 0$$

For example, suppose $f(x) = |x|$ and $g(x) = 3$. Taking the sum of the functions produces a new function denoted by $f + g$. In this case, $(f + g)(x) = |x| + 3$. Graphically, the y values of the function $f + g$ are given by the sum of the corresponding y values of f and g. This is depicted in Figure 3-16. The function $f + g$ appears in red. In particular, notice that $(f + g)(2) = f(2) + g(2) = 2 + 3 = 5$.

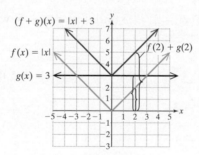

Figure 3-16

Example 1   **Adding, Subtracting, Multiplying and Dividing Functions**

Given: $g(x) = 4x$ $h(x) = x^2 - 3x$ $k(x) = x - 2$

a. Find $(g + h)(x)$.
b. Find $(h - g)(x)$.

c. Find $(g \cdot k)(x)$.
d. Find $\left(\dfrac{k}{h}\right)(x)$.

Solution:

a. $(g + h)(x) = g(x) + h(x)$

$\qquad = (4x) + (x^2 - 3x)$

$\qquad = 4x + x^2 - 3x$

$\qquad = x^2 + x$ Simplify.

Skill Practice

Given $f(x) = x - 1$,
$g(x) = 5x^2 + x$, and $h(x) = x^2$,
find

1. $(f + g)(x)$ **2.** $(g - f)(x)$

3. $(g \cdot h)(x)$ **4.** $\left(\dfrac{f}{h}\right)(x)$

Answers

1. $5x^2 + 2x - 1$ **2.** $5x^2 + 1$

3. $5x^4 + x^3$ **4.** $\dfrac{x - 1}{x^2}, x \neq 0$

b. $(h - g)(x) = h(x) - g(x)$

$$= (x^2 - 3x) - (4x)$$

$$= x^2 - 3x - 4x$$

$$= x^2 - 7x \qquad \text{Simplify.}$$

c. $(g \cdot k)(x) = g(x) \cdot k(x)$

$$= (4x)(x - 2)$$

$$= 4x^2 - 8x \qquad \text{Simplify.}$$

d. $\left(\dfrac{k}{h}\right)(x) = \dfrac{k(x)}{h(x)}$

$$= \frac{x - 2}{x^2 - 3x} \qquad \begin{array}{l} \text{From the denominator we have} \\ x^2 - 3x \neq 0 \text{ or, equivalently,} \\ x(x - 3) \neq 0. \text{ Hence, } x \neq 3 \text{ and } x \neq 0. \end{array}$$

2. Composition of Functions

> **Composition of Functions**
>
> The **composition** of f and g, denoted $f \circ g$, is defined by the rule
>
> $$(f \circ g)(x) = f(g(x)) \qquad \text{provided that } g(x) \text{ is in the domain of } f$$
>
> *Note:* $f \circ g$ is also read as "f compose g."

For example, given $f(x) = 2x - 3$ and $g(x) = x + 5$, we have

$$(f \circ g)(x) = f(g(x))$$

$$= f(x + 5) \qquad \text{Substitute } g(x) = x + 5 \text{ into the function } f.$$

$$= 2(x + 5) - 3$$

$$= 2x + 10 - 3$$

$$= 2x + 7$$

In this composition, the function g is the innermost operation and acts on x first. Then the output value of function g becomes the domain element of the function f, as shown in the figure.

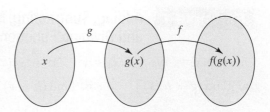

Example 2 **Composing Functions**

Given: $f(x) = x - 5$, $g(x) = x^2$, and $n(x) = \sqrt{x + 2}$, find

a. $(f \circ g)(x)$

b. $(g \circ f)(x)$

c. $(n \circ f)(x)$

Solution:

a. $(f \circ g)(x) = f(g(x))$

$\qquad = f(x^2)$ Evaluate the function f at x^2.

$\qquad = (x^2) - 5$ Replace x by x^2 in the function f.

$\qquad = x^2 - 5$

b. $(g \circ f)(x) = g(f(x))$

$\qquad = g(x - 5)$ Evaluate the function g at $(x - 5)$.

$\qquad = (x - 5)^2$ Replace x by $(x - 5)$ in function g.

$\qquad = x^2 - 10x + 25$ Simplify.

c. $(n \circ f)(x) = n(f(x))$

$\qquad = n(x - 5)$ Evaluate the function n at $x - 5$.

$\qquad = \sqrt{(x - 5) + 2}$ Replace x by the quantity $(x - 5)$ in function n.

$\qquad = \sqrt{x - 3}$

Skill Practice

Given $f(x) = 2x^2$, $g(x) = x + 3$, and $h(x) = \sqrt{x - 1}$, find

5. $(f \circ g)(x)$
6. $(g \circ f)(x)$
7. $(h \circ g)(x)$

TIP: Examples 2(a) and 2(b) illustrate that the order in which two functions are composed may result in different functions. That is, $f \circ g$ does not necessarily equal $g \circ f$.

3. Operations on Functions

Example 3   **Combining Functions**

Given the functions defined by $f(x) = x - 7$ and $h(x) = 2x^3$, find the function values, if possible.

a. $(f \cdot h)(3)$ **b.** $\left(\dfrac{h}{f}\right)(7)$ **c.** $(h \circ f)(2)$

Skill Practice

Given $h(x) = x + 4$ and $k(x) = x^2 - 3$, find

8. $(h \cdot k)(-2)$

9. $\left(\dfrac{h}{k}\right)(4)$

10. $(k \circ h)(1)$

Solution:

a. $(f \cdot h)(3) = f(3) \cdot h(3)$ $(f \cdot h)(3)$ is a product (not a composition).

$\qquad = (3 - 7) \cdot 2(3)^3$

$\qquad = (-4) \cdot 2(27)$

$\qquad = -216$

b. $\left(\dfrac{h}{f}\right)(7) = \dfrac{h(7)}{f(7)} = \dfrac{2(7)^3}{7 - 7}$

The value $x = 7$ will make the denominator equal to 0. Therefore, $\left(\dfrac{h}{f}\right)(7)$ is undefined.

c. $(h \circ f)(2) = h(f(2))$ Evaluate $f(2)$ first. $f(2) = 2 - 7 = -5$.

$\qquad = h(-5)$ Substitute the result into function h.

$\qquad = 2(-5)^3$

$\qquad = 2(-125)$

$\qquad = -250$

Answers

5. $2x^2 + 12x + 18$ **6.** $2x^2 + 3$
7. $\sqrt{x + 2}$ **8.** 2
9. $\dfrac{8}{13}$ **10.** 22

Skill Practice

Find the values from the graph.

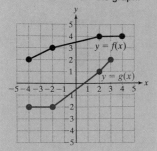

11. $g(3)$

12. $(f + g)(-4)$

13. $\left(\dfrac{f}{g}\right)(2)$

14. $(g \circ f)(-2)$

Answers

11. 2
12. 0
13. 4
14. 2

Example 4 **Finding Function Values from a Graph**

For the functions f and g pictured, find the function values if possible.

a. $g(2)$

b. $(f - g)(-3)$

c. $\left(\dfrac{g}{f}\right)(5)$

d. $(f \circ g)(4)$

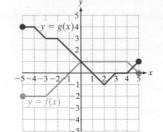

Animation

Solution:

a. $g(2) = -1$ The value $g(2)$ represents the y value of $y = g(x)$ (the red graph) when $x = 2$. Because the point $(2, -1)$ lies on the graph, $g(2) = -1$.

b. $(f - g)(-3) = f(-3) - g(-3)$ Evaluate the difference of $f(-3)$ and $g(-3)$.

$ = -2 - (3)$ Estimate function values from the graph.

$ = -5$

c. $\left(\dfrac{g}{f}\right)(5) = \dfrac{g(5)}{f(5)}$ Evaluate the quotient of $g(5)$ and $f(5)$.

$\phantom{\left(\dfrac{g}{f}\right)(5)} = \dfrac{1}{0}$ (undefined)

The function $\dfrac{g}{f}$ is undefined at 5 because the denominator is zero.

d. $(f \circ g)(4) = f(g(4))$ From the red graph, find the value of $g(4)$ first. We see that $g(4) = 0$.

$ = f(0)$ From the blue graph, find the value of f at $x = 0$.

$ = 1$

Section 3.4   Practice Exercises

Vocabulary and Key Concepts

1. **a.** Given the functions f and g, the function $(f + g)(x) =$ _____ $+$ _____.

    b. Given the functions f and g, the function $\left(\dfrac{f}{g}\right)(x) = \dfrac{f(x)}{\square}$.

    c. The function f composed with g is defined by the rule $(f \circ g)(x) =$ _____.

Concept 1: Algebra of Functions

2. Given $f(x) = x^2$ and $g(x) = 2x - 3$, find

    a. $f(-2)$ **b.** $g(-2)$ **c.** $(f + g)(-2)$

For Exercises 3–14, refer to the functions defined here.

$$f(x) = x + 4 \qquad g(x) = 2x^2 + 4x \qquad h(x) = x^2 + 1 \qquad k(x) = \frac{1}{x}$$

Find the indicated functions. **(See Example 1.)**

3. $(f + g)(x)$

4. $(f - g)(x)$

5. $(g - f)(x)$

6. $(f + h)(x)$

7. $(f \cdot h)(x)$

8. $(h \cdot k)(x)$

9. $(g \cdot f)(x)$

10. $(f \cdot k)(x)$

11. $\left(\dfrac{h}{f}\right)(x)$

12. $\left(\dfrac{g}{f}\right)(x)$

13. $\left(\dfrac{f}{g}\right)(x)$

14. $\left(\dfrac{f}{h}\right)(x)$

Concept 2: Composition of Functions

For Exercises 15–22, find the indicated functions. Use f, g, h, and k as defined in Exercises 3–14. **(See Example 2.)**

15. $(f \circ g)(x)$

16. $(f \circ k)(x)$

17. $(g \circ f)(x)$

18. $(k \circ f)(x)$

19. $(k \circ h)(x)$

20. $(h \circ k)(x)$

21. $(k \circ g)(x)$

22. $(g \circ k)(x)$

23. Based on your answers to Exercises 15 and 17, is it true in general that $(f \circ g)(x) = (g \circ f)(x)$?

24. Based on your answers to Exercises 16 and 18, is it true in general that $(f \circ k)(x) = (k \circ f)(x)$?

For Exercises 25–28, find $(f \circ g)(x)$ and $(g \circ f)(x)$.

25. $f(x) = x^2 - 3x + 1, g(x) = 5x$

26. $f(x) = 3x^2 + 8, g(x) = 2x - 4$

27. $f(x) = |x|, g(x) = x^3 - 1$

28. $f(x) = \dfrac{1}{x + 2}, g(x) = |x + 2|$

29. For $h(x) = 5x - 4$, find $(h \circ h)(x)$.

30. For $k(x) = -x^2 + 1$, find $(k \circ k)(x)$.

Concept 3: Operations on Functions

For Exercises 31–46, refer to the functions defined here.

$$m(x) = x^3 \qquad n(x) = x - 3 \qquad r(x) = \sqrt{x + 4} \qquad p(x) = \frac{1}{x + 2}$$

Find the function values if possible. **(See Example 3.)**

31. $(m \cdot r)(0)$

32. $(n \cdot p)(0)$

33. $(m + r)(-4)$

34. $(n - m)(4)$

35. $(r \circ n)(3)$

36. $(n \circ r)(5)$

37. $(p \circ m)(-1)$

38. $(m \circ n)(5)$

39. $(m \circ p)(2)$

40. $(r \circ m)(2)$

41. $(r + p)(-3)$

42. $(n + p)(-2)$

43. $(m \circ p)(-2)$

44. $(r \circ m)(-2)$

45. $\left(\dfrac{r}{n}\right)(12)$

46. $\left(\dfrac{n}{m}\right)(2)$

For Exercises 47–64, approximate the function values from the graph, if possible. **(See Example 4.)**

47. $f(-4)$ **48.** $f(1)$ **49.** $g(-2)$

50. $g(3)$ 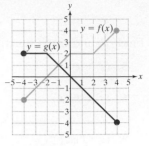 **51.** $(f + g)(2)$ **52.** $(g - f)(3)$

53. $(f \cdot g)(-1)$ **54.** $(g \cdot f)(-4)$ **55.** $\left(\dfrac{g}{f}\right)(0)$

56. $\left(\dfrac{f}{g}\right)(-2)$ **57.** $\left(\dfrac{f}{g}\right)(0)$ **58.** $\left(\dfrac{g}{f}\right)(-2)$

59. $(g \circ f)(-1)$ **60.** $(f \circ g)(0)$ **61.** $(f \circ g)(-4)$

62. $(g \circ f)(-4)$ **63.** $(g \circ g)(2)$ **64.** $(f \circ f)(-2)$

For Exercises 65–80, approximate the function values from the graph, if possible. **(See Example 4.)**

65. $a(-3)$ **66.** $a(1)$ **67.** $b(-1)$

68. $b(3)$ **69.** $(a - b)(-1)$ **70.** $(a + b)(0)$

71. $(b \cdot a)(1)$ **72.** $(a \cdot b)(2)$ **73.** $(b \circ a)(0)$

74. $(a \circ b)(-2)$ **75.** $(a \circ b)(-4)$ **76.** $(b \circ a)(-3)$

77. $\left(\dfrac{b}{a}\right)(3)$ **78.** $\left(\dfrac{a}{b}\right)(4)$ **79.** $(a \circ a)(-2)$ **80.** $(b \circ b)(1)$

 81. The cost in dollars of producing x toy cars is $C(x) = 2.2x + 1$. The revenue received is $R(x) = 5.98x$. To calculate profit, subtract the cost from the revenue.

     a. Write and simplify a function P that represents profit in terms of x.

     b. Find the profit of producing 50 toy cars.

82. The cost in dollars of producing x lawn chairs is $C(x) = 2.5x + 10.1$. The revenue for selling x lawn chairs is $R(x) = 6.99x$. To calculate profit, subtract the cost from the revenue.

     a. Write and simplify a function P that represents profit in terms of x.

     b. Find the profit in producing 100 lawn chairs.

83. The functions defined by $D(t) = 0.925t + 26.958$ and $R(t) = 0.725t + 20.558$ approximate the amount of child support (in billions of dollars) that was due and the amount of child support actually received in the United States since 2000. In each case, $t = 0$ corresponds to the year 2000.

     a. Find the function F defined by $F(t) = D(t) - R(t)$. What does F represent in the context of this problem?

     b. Find $F(4)$. What does this function value represent in the context of this problem?

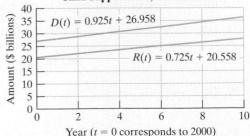

Difference Between Child Support Due and Child Support Paid, United States

Source: U.S. Bureau of the Census

84. The rural and urban populations in the South (in the United States) between the years 1900 and 1970 can be modeled by the following:

$$r(t) = -3.497t^2 + 266.2t + 20{,}220$$

$$u(t) = 0.0566t^3 + 0.952t^2 + 177.8t + 4593$$

The variable t represents the number of years since 1900, $r(t)$ represents the rural population in thousands, and $u(t)$ represents the urban population in thousands.

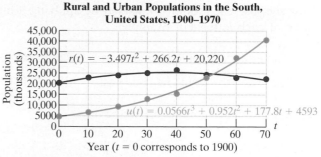

Rural and Urban Populations in the South, United States, 1900–1970

Source: Historical Abstract of the United States

a. Find the function T defined by $T(t) = r(t) + u(t)$. What does the function T represent in the context of this problem?

b. Use the function T to approximate the total population in the South for the year 1940.

85. Joe rides a bicycle and his wheels revolve at 80 revolutions per minute (rpm). Therefore, the total number of revolutions, r, is given by $r(t) = 80t$, where t is the time in minutes. For each revolution of the wheels of the bike, he travels approximately 7 ft. Therefore, the total distance he travels, D, depends on the total number of revolutions, r, according to the function $D(r) = 7r$.

a. Find $(D \circ r)(t)$ and interpret its meaning in the context of this problem.

b. Find Joe's total distance in feet after 10 min.

86. The area of a square is given by the function $a(x) = x^2$, where x is the length of the sides of the square. If carpeting costs \$9.95 per square yard, then the cost, C, to carpet a square room is given by $C(a) = 9.95a$, where a is the area of the room in square yards.

a. Find $(C \circ a)(x)$ and interpret its meaning in the context of this problem.

b. Find the cost to carpet a square room if its floor dimensions are 15 yd by 15 yd.

Group Activity

Deciphering a Coded Message

Materials: A calculator

Estimated Time: 20–25 minutes

Group Size: 4 (two pairs)

Cryptography is the study of coding and decoding messages. One type of coding process assigns a number to each letter of the alphabet and to the space character. For example:

A	B	C	D	E	F	G	H	I	J	K	L	M	N
1	2	3	4	5	6	7	8	9	10	11	12	13	14

O	P	Q	R	S	T	U	V	W	X	Y	Z	space
15	16	17	18	19	20	21	22	23	24	25	26	27

According to the numbers assigned to each letter, the message "*Decimals have a point*" would be coded as follows:

D E C I M A L S _ H A V E _ A _ P O I N T

4 5 3 9 13 1 12 19 27 8 1 22 5 27 1 27 16 15 9 14 20

Now suppose each letter is encoded by applying a function such as $f(x) = 2x + 5$, where x is the numerical value of each letter. For example:

The letter "a" would be coded as: $f(1) = 2(1) + 5 = 7$

The letter "b" would be coded as: $f(2) = 2(2) + 5 = 9$

Using this encoding function, we have

Message:	D E C I M A L S _ H A V E _ A _ P O I N T
Original:	4 5 3 9 13 1 12 19 27 8 1 22 5 27 1 27 16 15 9 14 20
Coded Form:	13 15 11 23 31 7 29 43 59 21 7 49 15 59 7 59 37 35 23 33 45

To decode this message, the receiver would need to reverse the operations assigned by $f(x) = 2x + 5$. Since the function f multiplies x by 2 and then adds 5, we can reverse this process by subtracting 5 and dividing by 2. This is represented by $g(x) = \frac{x-5}{2}$.

1. a. One pair of students will encode the following message according to $f(x) = 4x + 2$.

MATH IS THE KEY TO THE SCIENCES

b. The second pair of students will encode the following message according to $f(x) = 3x - 1$.

MATH IS NOT A SPECTATOR SPORT

2. With each message encoded, the pairs will exchange papers. Each pair will then decode the message.

Chapter 3 Summary

Section 3.1   Relations and Applications

Key Concepts

A set of ordered pairs (x, y) is called a **relation in x and y**.

The **domain** of a relation is the set of first components in the ordered pairs in the relation. The **range** of a relation is the set of second components in the ordered pairs.

Examples

Example 1

Let $A = \{(0, 0), (1, 1), (2, 4), (3, 9), (-1, 1), (-2, 4)\}$.

Domain of A: $\{0, 1, 2, 3, -1, -2\}$

Range of A: $\{0, 1, 4, 9\}$

Example 2

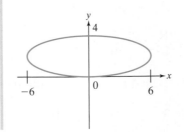

Domain: $[-6, 6]$
Range: $[0, 4]$

Section 3.2   Introduction to Functions

Key Concepts

Given a relation in x and y, we say "y is a **function** of x" if for each element x in the domain, there is exactly one value of y in the range.

Note: This means that no two ordered pairs may have the same first coordinate and different second coordinates.

The Vertical Line Test for Functions

Consider a relation defined by a set of points (x, y) in a rectangular coordinate system. The graph defines y as a function of x if no vertical line intersects the graph in more than one point.

Function Notation

$f(x)$ is the value of the function f at x.

The domain of a function defined by $y = f(x)$ is the set of x values that when substituted into the function produces a real number. In particular,

- Exclude values of x that make the denominator of a fraction zero.
- Exclude values of x that make the expression within a square root negative.

Examples

Example 1

Function $\{(1, 3), (2, 5), (6, 3)\}$

Not a function $\{(1, 3), (2, 5), (1, 4)\}$

Example 2

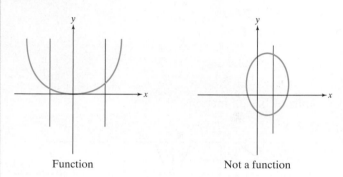

Function   Not a function

Example 3

Given $f(x) = -3x^2 + 5x$, find $f(-2)$.

$f(-2) = -3(-2)^2 + 5(-2)$

$\qquad = -12 - 10$

$\qquad = -22$

Example 4

Find the domain.

1. $f(x) = \dfrac{x + 4}{x - 5}$; Domain: $(-\infty, 5) \cup (5, \infty)$

2. $f(x) = \sqrt{x - 3}$; Domain: $[3, \infty)$

3. $f(x) = 3x^2 - 5$; Domain: $(-\infty, \infty)$

Section 3.3   Graphs of Basic Functions

Key Concepts

A function of the form $f(x) = mx + b \, (m \neq 0)$ is a **linear function**. Its graph is a line with slope m and y-intercept $(0, b)$.

A function of the form $f(x) = k$ is a **constant function**. Its graph is a horizontal line.

A function of the form $f(x) = ax^2 + bx + c \, (a \neq 0)$ is a **quadratic function**. Its graph is a **parabola**.

Graphs of basic functions:

$f(x) = x$ $f(x) = x^2$ $f(x) = x^3$

  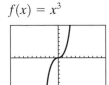

$f(x) = |x|$ $f(x) = \sqrt{x}$ $f(x) = \dfrac{1}{x}$

 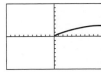 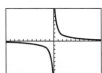

The x-intercepts of a function are determined by finding the real solutions to the equation $f(x) = 0$.

The y-intercept of a function is at $f(0)$.

Examples

Example 1

$f(x) = 2x - 3$ $f(x) = 3$

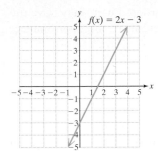   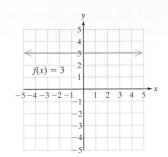

Example 2

Find the x- and y-intercepts for the function pictured.

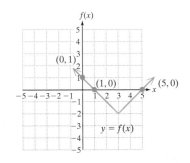

$f(x) = 0$, when $x = 1$ and $x = 5$.

The x-intercepts are $(1, 0)$ and $(5, 0)$.

$f(0) = 1$. The y-intercept is $(0, 1)$.

Section 3.4   Algebra of Functions, Composition, and Applications

Key Concepts

The Algebra of Functions

Given two functions f and g, the functions $f + g, f - g, f \cdot g$, and $\dfrac{f}{g}$ are defined as

$(f + g)(x) = f(x) + g(x)$

$(f - g)(x) = f(x) - g(x)$

$(f \cdot g)(x) = f(x) \cdot g(x)$

$\left(\dfrac{f}{g}\right)(x) = \dfrac{f(x)}{g(x)}$ provided $g(x) \neq 0$

Composition of Functions

The **composition** of f and g, denoted $f \circ g$, is defined by the rule

$(f \circ g)(x) = f(g(x))$ provided that $g(x)$ is in the domain of f

Examples

Example 1

Let $g(x) = 5x + 1$ and $h(x) = x^3$. Find:

1. $(g + h)(3) = g(3) + h(3) = 16 + 27 = 43$

2. $(g \cdot h)(-1) = g(-1) \cdot h(-1) = (-4) \cdot (-1) = 4$

3. $(g - h)(x) = 5x + 1 - x^3$

4. $\left(\dfrac{g}{h}\right)(x) = \dfrac{5x + 1}{x^3}, x \neq 0$

Example 2

Find $(f \circ g)(x)$ and $(g \circ f)(x)$ given the functions defined by $f(x) = 4x + 3$ and $g(x) = 7x$.

$(f \circ g)(x) = f(g(x))$

$= f(7x)$

$= 4(7x) + 3$

$= 28x + 3$

Chapter 3   Review Exercises

Section 3.1

For Exercises 1–4, find the domain and range.

1. $\left\{ \left(\dfrac{1}{3}, 10\right), \left(6, -\dfrac{1}{2}\right), \left(\dfrac{1}{4}, 4\right), \left(7, \dfrac{2}{5}\right) \right\}$

2.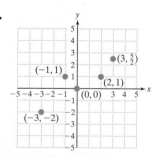

3.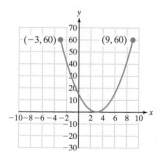

4.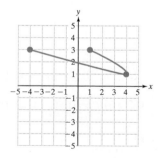

Section 3.2

5. Sketch a relation that is *not* a function. (Answers may vary.)

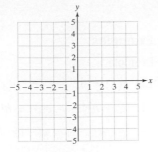

6. Sketch a relation that *is* a function. (Answers may vary.)

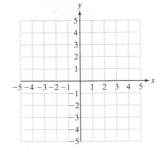

For Exercises 7–12:

a. Determine whether the relation defines y as a function of x.

b. Find the domain.

c. Find the range.

7.

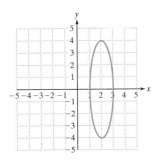

8.

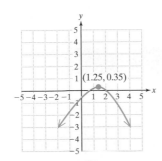

9. $\{(1, 3), (2, 3), (3, 3), (4, 3)\}$

10. $\{(0, 2), (0, 3), (4, 4), (0, 5)\}$

11.

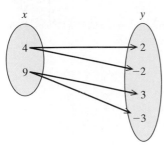

12.

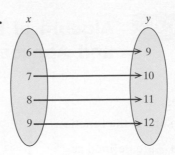

For Exercises 13–20, find the function values given $f(x) = 6x^2 - 4$.

13. $f(0)$ **14.** $f(1)$

15. $f(-1)$ **16.** $f(t)$

17. $f(b)$ **18.** $f(\pi)$

19. $f(\square)$ **20.** $f(-2)$

For Exercises 21–24, write the domain of each function in interval notation.

21. $g(x) = 7x^3 + 1$ **22.** $h(x) = \dfrac{x + 10}{x - 11}$

23. $k(x) = \sqrt{x - 8}$ **24.** $w(x) = \sqrt{x + 2}$

25. Anita is a waitress and makes \$6 per hour plus tips. Her tips average \$5 per table. In one 8-hr shift, Anita's pay can be described by $p(x) = 48 + 5x$, where x represents the number of tables she waits on. Determine how much Anita will earn if she waits on

    a. 10 tables   **b.** 15 tables   **c.** 20 tables

Section 3.3

For Exercises 26–31, sketch the functions from memory.

26. $h(x) = x$ **27.** $f(x) = x^2$

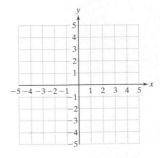
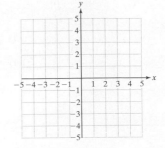

28. $g(x) = x^3$

29. $w(x) = |x|$

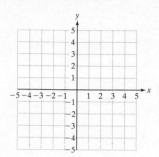

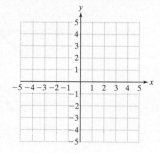

30. $s(x) = \sqrt{x}$

31. $r(x) = \dfrac{1}{x}$

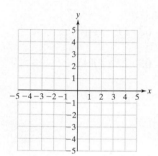

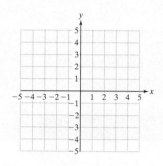

For Exercises 32 and 33, sketch the functions.

32. $q(x) = 3$

33. $k(x) = 2x + 1$

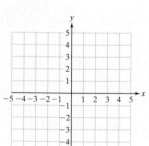

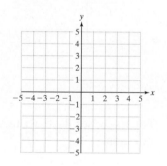

34. Given: $s(x) = (x - 2)^2$

   a. Find $s(4)$, $s(-3)$, $s(2)$, $s(1)$, and $s(0)$.

   b. What is the domain of s?

35. Given: $r(x) = 2\sqrt{x - 4}$

   a. Find $r(2)$, $r(4)$, $r(5)$, and $r(8)$.

   b. What is the domain of r?

36. Given: $h(x) = \dfrac{3}{x - 3}$

   a. Find $h(-3)$, $h(0)$, $h(2)$, and $h(5)$.

   b. What is the domain of h?

37. Given: $k(x) = -|x + 3|$

   a. Find $k(-5)$, $k(-4)$, $k(-3)$, and $k(2)$.

   b. What is the domain of k?

For Exercises 38 and 39, find the x- and y-intercepts.

38. $p(x) = 4x - 7$

39. $q(x) = -2x + 9$

40. The function defined by $b(t) = 0.7t + 4.5$ represents the per capita consumption of bottled water in the United States since 1985. The values of $b(t)$ are measured in gallons, and $t = 0$ corresponds to the year 1985. (*Source:* U.S. Department of Agriculture)

   a. Evaluate $b(0)$ and $b(7)$ and interpret the results in the context of this problem.

   b. What is the slope of this function? Interpret the slope in the context of this problem.

For Exercises 41–46, refer to the graph.

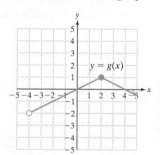

41. Find $g(-2)$.

42. Find $g(4)$.

43. For what value(s) of x is $g(x) = 0$?

44. For what value(s) of x is $g(x) = 4$?

45. Write the domain of g.

46. Write the range of g.

Section 3.4

For Exercises 47–54, refer to the functions defined here.

$$f(x) = x - 7 \qquad g(x) = -2x^3 - 8x$$

$$m(x) = x^2 \qquad n(x) = \dfrac{1}{x - 2}$$

Find the indicated function values.

47. $(f - g)(x)$

48. $(f + g)(x)$

49. $(f \cdot n)(x)$

50. $(f \cdot m)(x)$

51. $\left(\dfrac{f}{g}\right)(x)$

52. $\left(\dfrac{g}{f}\right)(x)$

53. $(m \circ f)(x)$

54. $(n \circ f)(x)$

For Exercises 55–58, refer to the functions defined for Exercises 47–54. Find the function values, if possible.

55. $(m \circ g)(-1)$

56. $(n \circ g)(-1)$

57. $(f \circ g)(4)$

58. $(g \circ f)(8)$

59. Given $f(x) = 2x + 1$ and $g(x) = x^2$

  a. Find $(g \circ f)(x)$.

  b. Find $(f \circ g)(x)$.

  c. Based on your answers to part (a), is $f \circ g$ equal to $g \circ f$?

For Exercises 60–65, refer to the graph. Approximate the function values, if possible.

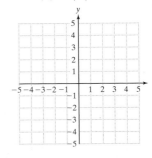

60. $\left(\dfrac{f}{g}\right)(1)$

61. $(f \cdot g)(-2)$

62. $(f + g)(-4)$

63. $(f - g)(2)$

64. $(g \circ f)(-3)$

65. $(f \circ g)(4)$

Chapter 3 Test

For Exercises 1 and 2, **a.** determine if the relation defines y as a function of x, **b.** identify the domain, and **c.** identify the range.

1.

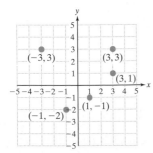

2.

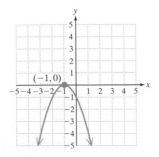

3. Explain how to find the x- and y-intercepts of a function defined by $y = f(x)$.

For Exercises 4–7, graph the functions.

4. $f(x) = -3x - 1$

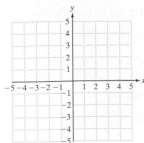

5. $k(x) = -2$

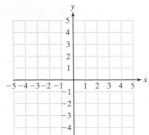

6. $p(x) = x^2$

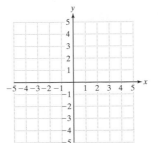

7. $w(x) = |x|$

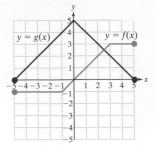

For Exercises 8–10, write the domain in interval notation.

8. $f(x) = \dfrac{x - 5}{x + 7}$

9. $f(x) = \sqrt{x + 7}$

10. $h(x) = (x + 7)(x - 5)$

11. Given: $r(x) = x^2 - 2x + 1$

  a. Find $r(-2)$, $r(-1)$, $r(0)$, $r(2)$, and $r(3)$.

  b. What is the domain of r?

12. The function defined by $s(t) = 1.6t + 36$ approximates the per capita consumption of soft drinks in the United States since 1985. The values of $s(t)$ are measured in gallons, and $t = 0$ corresponds to the year 1985. (*Source:* U.S. Department of Agriculture)

    a. Evaluate $s(0)$ and $s(7)$ and interpret the results in the context of this problem.

    b. What is the slope of the function? Interpret the slope in the context of this problem.

For Exercises 13–20, refer to the graph.

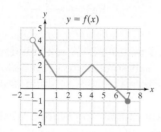

13. Find $f(1)$.

14. Find $f(4)$.

15. Write the domain of f.

16. Write the range of f.

17. True or false? The value $y = 5$ is in the range of f.

18. Find the x-intercept of the function.

19. For what value(s) of x is $f(x) = 0$?

20. For what value(s) of x is $f(x) = 1$?

For Exercises 21–24, determine if the function is constant, linear, quadratic, or none of these.

21. $f(x) = -3x^2$

22. $g(x) = -3x$

23. $h(x) = -3$

24. $k(x) = -\dfrac{3}{x}$

25. Find the x- and y-intercepts for $f(x) = \dfrac{3}{4}x + 9$.

For Exercises 26–34, refer to these functions.

$$f(x) = x - 4 \qquad g(x) = x^2 + 2 \qquad h(x) = \dfrac{1}{x}$$

Find the function values if possible.

26. $\left(\dfrac{f}{g}\right)(x)$

27. $(h \cdot g)(x)$

28. $(g \circ f)(x)$

29. $(h \circ f)(x)$

30. $(f - g)(7)$

31. $(h + f)(2)$

32. $(h \circ g)(4)$

33. $(g \circ f)(0)$

34. $\left(\dfrac{g}{f}\right)(x)$

Chapters 1–3   Cumulative Review Exercises

1. Simplify the expression.
$$\dfrac{5 - 2^3 \div 4 + 7}{-1 - 3(4 - 1)}$$

2. Simplify the expression. $3 + \sqrt{25} - 8(\sqrt{9}) \div 6$

3. Solve. $z - (3 + 2z) + 5 = -2z - 5$

4. Solve. $\dfrac{2b - 3}{6} - \dfrac{b + 1}{4} = -2$

5. A bike rider pedals 10 mph to the top of a mountain and 15 mph on the way down. The total time for the round trip is 10 hr. Find the distance to the top of the hill.

6. Given: $f(x) = \frac{1}{2}x - 1$ and $g(x) = 3x^2 - 2x$
    a. Find $f(4)$.
    b. Find $g(-3)$.

7. Solve the inequalities. Write your answers in interval notation.
    a. $-5x - 4 \le -2(x - 1)$
    b. $-x + 4 > 1$

8. Find the slope of the line that passes through the points $(4, -5)$ and $(-6, -3)$.

9. Determine if $\left(\frac{1}{3}, 7\right)$ is a solution to $y = 6x - 5$.

For Exercises 10 and 11, **a.** find the *x*- and *y*-intercepts, **b.** find the slope, and **c.** graph the line.

10. $3x - 5y = 10$

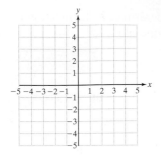

11. $2y + 4 = 10$

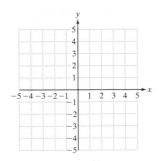

12. Find an equation for the vertical line that passes through the point $(7, 11)$.

13. State the domain and range of the relation. Is the relation a function? $\{(3, -1), (4, -5), (3, -8)\}$

14. Find an equation of the line passing through $(1, -4)$ and parallel to $2x + y = 6$. Write the answer in slope-intercept form.

15. Find an equation of the line passing through $(1, -4)$ and perpendicular to $y = \frac{1}{4}x - 2$. Write the answer in slope-intercept form.

16. For $f(x) = x - 5$ and $g(x) = 4x^2$, find

   a. $(f \circ g)(3)$ **b.** $(g \circ f)(x)$

17. Write the domain of the function in interval notation.

$$f(x) = \frac{1}{x - 15}$$

18. A chemist mixes a 20% salt solution with 15 L of a 50% salt solution to get a 38% salt solution. How much of the 20% solution does she use?

19. The yearly rainfall for Seattle, Washington, is 0.7 in. less than twice the yearly rainfall for Los Angeles, California. If the total yearly rainfall for the two cities is 50 in., how much rain does each city get per year?

20. Simplify. $4[-3x - 5(y - 2x) + 3] - 7(6y + x)$

Systems of Linear Equations

4

CHAPTER OUTLINE

Chapter 4

In this chapter, we solve systems of linear equations in two variables and three variables. Systems of equations occur in a variety of applications when there are two or more unknown quantities. One method to solving systems involves graphing lines.

Review Your Skills

This Sudoku puzzle will help review topics of graphing. Use the clues to fill in additional boxes. Continue to fill in the grid so that every row, every column, and every 2 × 3 box contains the digits 1 through 6.

A. $(A, 0)$ is the x-intercept for the line given by $y = -2x + 4$. Find the value of A.

B. The slope of the line given by $y = 5x - 9$.

C. $(C, 9)$ is a point on the line given by $3x - 2y = -6$. Find the value of C.

D. The slope of the line through the points $(-11, -2)$ and $(-5, 4)$.

E. Quadrant in which the point $(4, -2)$ is located.

F. The slope of a line perpendicular to the line given by $y = -\dfrac{1}{6}x + 7$.

Solving Systems of Linear Equations by the Graphing Method

Concepts

1. Solutions to Systems of Linear Equations
2. Solving Systems of Linear Equations by Graphing

1. Solutions to Systems of Linear Equations

A linear equation in two variables has an infinite number of solutions that form a line in a rectangular coordinate system. Two or more linear equations form a **system of linear equations**. For example:

$$x - 3y = -5$$
$$2x + 4y = 10$$

A **solution to a system of linear equations** is an ordered pair that is a solution to *each* individual linear equation.

Skill Practice

Determine whether the ordered pairs are solutions to the system.

$3x + 2y = -8$
$y = 2x - 18$
  1. $(-2, -1)$
  2. $(4, -10)$

Example 1 Determining Solutions to a System of Linear Equations

Determine whether the ordered pairs are solutions to the system.

$$x + y = -6$$
$$3x - y = -2$$

a. $(-2, -4)$ **b.** $(0, -6)$

Solution:

  a. Substitute the ordered pair $(-2, -4)$ into both equations:

$$x + y = -6 \longrightarrow \quad (-2) + (-4) \overset{?}{=} -6 \; ✔ \text{ True}$$
$$3x - y = -2 \longrightarrow \quad 3(-2) - (-4) \overset{?}{=} -2 \; ✔ \text{ True}$$

  Because the ordered pair $(-2, -4)$ is a solution to each equation, it is a solution to the *system* of equations.

  b. Substitute the ordered pair $(0, -6)$ into both equations:

$$x + y = -6 \longrightarrow \quad (0) + (-6) \overset{?}{=} -6 \; ✔ \text{ True}$$
$$3x - y = -2 \longrightarrow \quad 3(0) - (-6) \overset{?}{=} -2 \qquad \text{ False}$$

  Because the ordered pair $(0, -6)$ is not a solution to the second equation, it is *not* a solution to the system of equations.

A solution to a system of two linear equations may be interpreted graphically as a point of intersection between the two lines. Graphing the lines from Example 1, we have

$$x + y = -6$$
$$3x - y = -2$$

Answers

1. No   **2.** Yes

Notice that the lines intersect at $(-2, -4)$ (Figure 4-1).

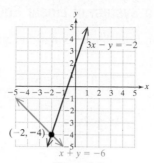

Figure 4-1

When two lines are drawn in a rectangular coordinate system, three geometric relationships are possible:

1. Two lines may intersect at *exactly one point.*

2. Two lines may intersect at *no point.* This occurs if the lines are parallel.

3. Two lines may intersect at *infinitely many points* along the line. This occurs if the equations represent the same line (the lines are coinciding).

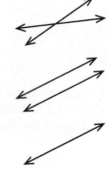

If a system of linear equations has one or more solutions, the system is said to be a **consistent system**. If a linear system has no solution, it is said to be an **inconsistent system**.

If two equations represent the same line, then all points along the line are solutions to the system of equations. In such a case, the equations are said to be **dependent**. If two linear equations represent different lines, then the equations are said to be **independent**. The different possibilities for solutions to systems of linear equations are given in Table 4-1.

Concept Connections

For Exercises 3 and 4, choose the correct response.

3. Two different lines with the same slope and different *y*-intercepts intersect in how many points?
   a. 1   **b.** 2   **c.** 0
   d. Infinitely many

4. Two lines with different slopes intersect in how many points?
   a. 1   **b.** 2   **c.** 0
   d. Infinitely many

Table 4-1   Solutions to Systems of Linear Equations in Two Variables

One Unique Solution	No Solution	Infinitely Many Solutions
One point of intersection	Parallel lines	Coinciding lines
System is consistent.	System is inconsistent.	System is consistent.
Equations are independent.	Equations are independent.	Equations are dependent.

Answers

3. c.   **4.** a.

2. Solving Systems of Linear Equations by Graphing

Example 2   Solving a System of Linear Equations by Graphing

Solve the system by graphing both linear equations and finding the point(s) of intersection.

$$y = \frac{1}{2}x - 2$$

$$4x + 2y = 6$$

Solution:

To graph each equation, write the equation in slope-intercept form $y = mx + b$.

First equation	**Second equation**
$y = \frac{1}{2}x - 2$ Slope: $\frac{1}{2}$	$4x + 2y = 6$
	$2y = -4x + 6$
	$\frac{2y}{2} = \frac{-4x}{2} + \frac{6}{2}$
	$y = -2x + 3$ Slope: -2

From their slope-intercept forms, we see that the lines have different slopes, indicating that the lines must intersect at exactly one point. We can graph the lines using the slope and y-intercept to find the point of intersection (Figure 4-2).

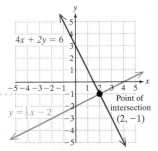

Figure 4-2

The point $(2, -1)$ appears to be the point of intersection. This can be confirmed by substituting $x = 2$ and $y = -1$ into both equations.

$$y = \frac{1}{2}x - 2 \qquad\qquad 4x + 2y = 6$$

$$-1 \stackrel{?}{=} \frac{1}{2}(2) - 2 \qquad\qquad 4(2) + 2(-1) \stackrel{?}{=} 6$$

$$-1 \stackrel{?}{=} 1 - 2 \qquad\qquad 8 - 2 \stackrel{?}{=} 6$$

$$-1 \stackrel{?}{=} -1 \ ✔ \ \text{True} \qquad\qquad 6 \stackrel{?}{=} 6 \ ✔ \ \text{True}$$

The solution is $(2, -1)$.

Answer

5. $(-2, 1)$

> **TIP:** In Example 2, the lines could also have been graphed by using the *x*- and *y*-intercepts or by using a table of points. However, the advantage of writing the equations in slope-intercept form is that we can compare the slope *and* *y*-intercept of each line.
>
> 1. If the slopes differ, the lines are different and nonparallel and must cross in exactly one point.
> 2. If the slopes are the same and the *y*-intercepts are different, the lines are parallel and do not intersect.
> 3. If the slopes are the same and the *y*-intercepts are the same, the two equations represent the same line.

Example 3   Solving a System of Linear Functions by Graphing

Solve the system.

$$f(x) = 3$$

$$g(x) = 2x + 1$$

Solution:

This first function can be written as $y = 3$. This is an equation of a horizontal line. Writing the second equation as $y = 2x + 1$, we have a slope of 2 and a *y*-intercept of $(0, 1)$.

The graphs of the functions are shown in Figure 4-3. The point of intersection is $(1, 3)$. Therefore, the solution to the system is $(1, 3)$.

The equations are independent because they represent different lines. The system is consistent because it has a solution.

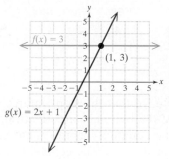

Figure 4-3

Example 4   Solving a System of Linear Equations by Graphing

Solve the system by graphing.

$$-x + 3y = -6$$

$$6y = 2x + 6$$

Solution:

To graph the lines, write each equation in slope-intercept form.

$-x + 3y = -6$	$6y = 2x + 6$
$3y = x - 6$	
$\dfrac{3y}{3} = \dfrac{x}{3} - \dfrac{6}{3}$	$\dfrac{6y}{6} = \dfrac{2x}{6} + \dfrac{6}{6}$
$y = \dfrac{1}{3}x - 2$	$y = \dfrac{1}{3}x + 1$

Because the lines have the same slope but different *y*-intercepts, they are parallel (Figure 4-4). Two parallel lines do not intersect, which implies that the system has no solution. The system is inconsistent.

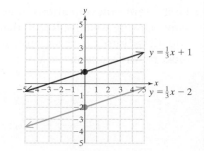

Figure 4-4

Skill Practice

8. Solve the system by graphing.

$$y = \frac{1}{2}x + 1$$

$$x - 2y = -2$$

Example 5   Solving a System of Linear Equations by Graphing

Solve the system by graphing.

$$x + 4y = 8$$

$$y = -\frac{1}{4}x + 2$$

Solution:

Write the first equation in slope-intercept form. The second equation is already in slope-intercept form.

First equation   **Second equation**

$$x + 4y = 8 \qquad\qquad y = -\frac{1}{4}x + 2$$

$$4y = -x + 8$$

$$\frac{4y}{4} = \frac{-x}{4} + \frac{8}{4}$$

$$y = -\frac{1}{4}x + 2$$

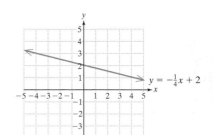

Notice that the slope-intercept forms of the two lines are identical. Therefore, the equations represent the same line (Figure 4-5). The equations are dependent, and the solution to the system of equations is the set of all points on the line.

Figure 4-5

Because the ordered pairs in the solution set cannot all be listed, we can write the solution in set-builder notation. Furthermore, the equations $x + 4y = 8$ and $y = -\frac{1}{4}x + 2$ represent the same line. Therefore, the solution set may be written as $\{(x, y) \mid y = -\frac{1}{4}x + 2\}$ or $\{(x, y) \mid x + 4y = 8\}$.

Calculator Connections

Topic: Using the *Intersect* Feature

The solution to a system of equations can be found by using either a *Trace* feature or an *Intersect* feature on a graphing calculator to find the point of intersection between two curves.

For example, consider the system

$$-2x + y = 6$$

$$5x + y = -1$$

First graph the equations together on the same viewing window. Recall that to enter the equations into the calculator, the equations must be written with the *y* variable isolated.

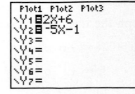

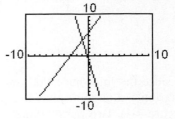

$$\text{Isolate } y.$$

$$-2x + y = 6 \longrightarrow y = 2x + 6$$

$$5x + y = -1 \longrightarrow y = -5x - 1$$

Answer

8. Infinitely many solutions;

$$\{(x, y) \mid y = \frac{1}{2}x + 1\};$$

Dependent equations

By inspection of the graph, it appears that the solution is $(-1, 4)$. The *Trace* option on the calculator may come close to $(-1, 4)$ but may not show the exact solution (Figure 4-6). However, an *Intersect* feature on a graphing calculator may provide the exact solution (Figure 4-7). See your user's manual for further details.

Using *Trace*

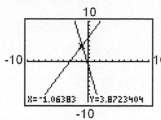

Figure 4-6

Using *Intersect*

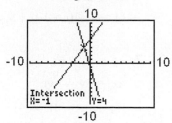

Figure 4-7

Section 4.1   Practice Exercises

Study Skills Exercise

Before you proceed further in Chapter 4, make your test corrections for the Chapter 3 test. See Study Skills Exercise in Section 2.1 for instructions.

Vocabulary and Key Concepts

1. **a.** A _____ of linear equations consists of two or more linear equations.

   b. A _____ to a system of linear equations is an ordered pair that is a solution to each individual equation in the system.

   c. Graphically, a solution to a system of linear equations in two variables is a point where the lines _____.

   d. A system of equations that has one or more solutions is said to be _____.

   e. There is _____ solution to an inconsistent system of equations.

   f. Two equations in a system of linear equations in two variables are said to be _____ if they represent the same line.

   g. Two equations in a system of linear equations in two variables are said to be _____ if they represent different lines.

Concept 1: Solutions to Systems of Linear Equations

2. From the graph shown, determine the solution to the system.

   $x + y = 4$

   $y = 2x + 1$

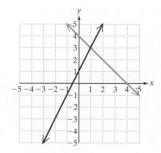

For Exercises 3–8, determine which points are solutions to the given system. **(See Example 1.)**

3. $y = 8x - 5$

$y = 4x + 3$

$(-1, 13), (-1, 1), (2, 11)$

4. $y = -\frac{1}{2}x - 5$

$y = \frac{3}{4}x - 10$

$(4, -7), (0, -10), \left(3, -\frac{9}{2}\right)$

5. $2x - 7y = -30$

$y = 3x + 7$

$(0, -30), \left(\frac{3}{2}, 5\right), (-1, 4)$

6. $x + 2y = 4$

$y = -\frac{1}{2}x + 2$

$(-2, 3), (4, 0), \left(3, \frac{1}{2}\right)$

7. $x - y = 6$

$4x + 3y = -4$

$(4, -2), (6, 0), (2, 4)$

8. $x - 3y = 3$

$2x - 9y = 1$

$(0, 1), (4, -1), (9, 2)$

For Exercises 9–14, the graph of a system of linear equations is given.

a. Identify whether the system is consistent or inconsistent.

b. Identify the equations as dependent or independent.

c. Identify the number of solutions to the system.

9. $y = x + 3$

$3x + y = -1$

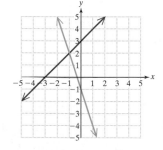

10. $5x - 3y = 6$

$3y = 2x + 3$

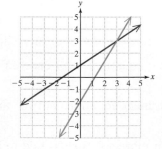

11. $2x = y + 4$

$-4x + 2y = 2$

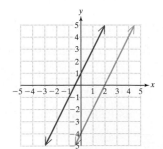

12. $y = -2x - 3$

$-4x - 2y = 0$

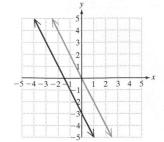

13. $y = \frac{1}{3}x + 2$

$-x + 3y = 6$

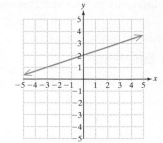

14. $y = -\frac{2}{3}x - 1$

$-4x - 6y = 6$

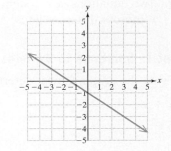

Concept 2: Solving Systems of Linear Equations by Graphing

For Exercises 15–32, solve the systems of equations by graphing. For systems that do not have one unique solution, also state the number of solutions and whether the system is inconsistent or the equations are dependent. **(See Examples 2–5.)**

15. $2x + y = 4$

$\quad x + 2y = -1$

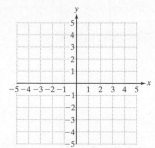

16. $4x - 3y = 12$

$\quad 3x + 4y = -16$

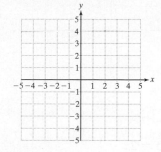

17. $f(x) = -2x + 3$

$\quad g(x) = 5x - 4$

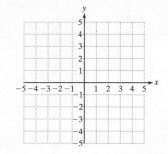

18. $h(x) = 2x + 5$

$\quad g(x) = -x + 2$

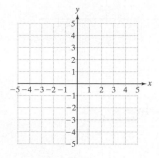

19. $k(x) = \dfrac{1}{3}x - 5$

$\quad f(x) = -\dfrac{2}{3}x - 2$

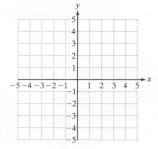

20. $f(x) = \dfrac{1}{2}x + 2$

$\quad g(x) = \dfrac{5}{2}x - 2$

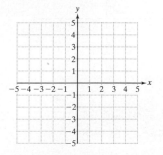

21. $x = 4$

$\quad y = 2x - 3$

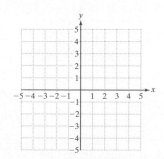

22. $3x + 2y = 6$

$\quad y = -3$

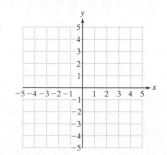

23. $y = -2x + 3$

$\quad -2x = y + 1$

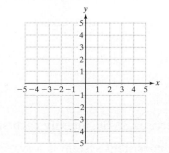

24. $y = \dfrac{1}{3}x - 2$

$x = 3y - 9$

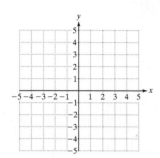

25. $y = \dfrac{2}{3}x - 1$

$2x = 3y + 3$

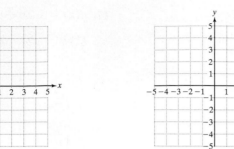

26. $4x = 16 - 8y$

$y = -\dfrac{1}{2}x + 2$

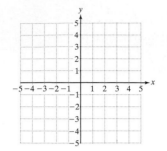

27. $2x = 4$

$\dfrac{1}{2}y = -1$

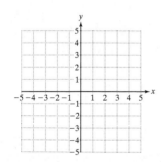

28. $y + 7 = 6$

$-5 = 2x$

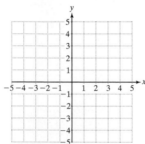

29. $-x + 3y = 6$

$6y = 2x + 12$

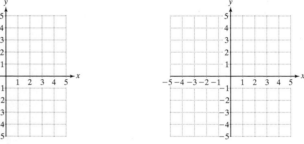

30. $3x = 2y - 4$

$-4y = -6x - 8$

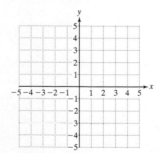

31. $2x - y = 4$

$4x + 2 = 2y$

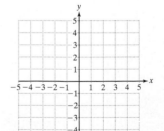

32. $x = 4y + 4$

$-2x + 8y = -16$

For Exercises 33–36, identify each statement as true or false.

33. A consistent system is a system that always has a unique solution.

34. A system of dependent equations has no solution.

35. If two lines coincide, the equations are dependent.

36. If two lines are parallel, the equations are independent.

Expanding Your Skills

37. Write a system of equations with solution $(4, 5)$.

38. Write a system of equations with solution $(-2, 6)$.

39. Find C and D such that the solution to the system is $(1, 3)$.

$$Cx + 2y = 11$$
$$-3x + Dy = 9$$

40. Find M and N such that the solution to the system is $(2, -4)$.

$$3x + My = -22$$
$$Nx + 4y = 6$$

Graphing Calculator Exercises

For Exercises 41–44, use a graphing calculator to graph each linear equation on the same viewing window. Use a *Trace* or *Intersect* feature to find the point(s) of intersection.

41. $y = 5.62x + 15.46$

$y = -1.96x - 11.07$

42. $y = -2.3x - 5.48$

$y = 4.62x + 26.352$

43. $2.4x - 4.8y = -9.36$

$-1.8x + 5.4y = 12.456$

44. $36x - 90y = -36$

$-15.5x - 5y = -80.75$

Solving Systems of Equations by Using the Substitution Method

Section 4.2

1. The Substitution Method

Graphing a system of equations is one method to find the solution of the system. However, sometimes it is difficult to determine the solution using this method because of limitations in the accuracy of the graph. This is particularly true when the coordinates of a solution are not integer values or when the solution is a point not sufficiently close to the origin. Identifying the coordinates of the point $\left(\frac{3}{17}, -\frac{23}{9}\right)$ or $(-251, 8349)$ for example, might be difficult from a graph.

Concepts

1. The Substitution Method
2. Solving Inconsistent Systems and Systems of Dependent Equations

In this section and Section 4.3, we will present two algebraic methods to solve a system of equations. The first is called the *substitution method*. This technique is particularly important because it can be used to solve more advanced problems including nonlinear systems of equations.

The first step in the substitution process is to isolate one of the variables from one of the equations. Consider the system

$$x + y = 16$$
$$x - y = 4$$

Solving the first equation for x yields $x = 16 - y$. Then, because x is equal to $16 - y$, the expression $16 - y$ may replace x in the second equation. This leaves the second equation in terms of y only.

First equation:

$x + y = 16 \xrightarrow{\text{Solve for } x.} x = \underline{16 - y}$

Second equation:

$(16 - y) - y = 4$ Substitute $x = 16 - y$.

$16 - 2y = 4$ Solve for y.

$-2y = -12$

$y = 6$

$x = 16 - y$ To find x, substitute $y = 6$ back into the expression $x = 16 - y$.

$x = 16 - (6)$

$x = 10$ Check the ordered pair $(10, 6)$ in both original equations.

$x + y = 16$ $x - y = 4$

$(10) + (6) \overset{?}{=} 16$ ✓ True $(10) - (6) \overset{?}{=} 4$ ✓ True

The solution is $(10, 6)$.

Solving a System of Equations by the Substitution Method

Step 1 Isolate one of the variables from one equation.

Step 2 Substitute the quantity found in step 1 into the *other* equation.

Step 3 Solve the resulting equation.

Step 4 Substitute the value found in step 3 back into the equation in step 1 to find the value of the remaining variable.

Step 5 Check the solution in both equations, and write the answer as an ordered pair.

Skill Practice

1. Solve by the substitution method.

$3x + \ y = 8$
$x - 2y = 12$

Avoiding Mistakes

Do not substitute $y = -6x + 6$ into the same equation from which it came. This mistake will result in an identity:

$6x + y = 6$

$6x + (-6x + 6) = 6$

$6x - 6x + 6 = 6$

$6 = 6$

Answer

1. $(4, -4)$

> **Example 1** **Using the Substitution Method to Solve a System of Linear Equations**

Solve the system by using the substitution method. $3x - 2y = -7$
                                                          $6x + \ y = 6$

Solution:

The y variable in the second equation is the easiest variable to isolate because its coefficient is 1.

$3x - 2y = -7$

$6x + \ y = 6 \longrightarrow y = \underline{-6x + 6}$ **Step 1:** Solve the second equation for y.

$3x - 2(-6x + 6) = -7$ **Step 2:** Substitute the quantity $-6x + 6$ for y in the *other* equation.

$3x + 12x - 12 = -7$

$15x - 12 = -7$

$15x = 5$ **Step 3:** Solve for x.

$\dfrac{15x}{15} = \dfrac{5}{15}$

$x = \dfrac{1}{3}$

$y = -6x + 6$

$y = -6\left(\dfrac{1}{3}\right) + 6$

$y = -2 + 6$

$y = 4$

Step 4: Substitute $x = \frac{1}{3}$ into the expression $y = -6x + 6$.

$3x - 2y = -7 \qquad\qquad 6x + y = 6$

$3\left(\dfrac{1}{3}\right) - 2(4) \overset{?}{=} -7 \qquad 6\left(\dfrac{1}{3}\right) + 4 \overset{?}{=} 6$

$1 - 8 \overset{?}{=} -7$ ✔ $\qquad\qquad 2 + 4 \overset{?}{=} 6$ ✔

Step 5: Check the ordered pair $(\frac{1}{3}, 4)$ in each original equation.

The solution is $(\frac{1}{3}, 4)$.

Example 2 **Using the Substitution Method to Solve a System of Linear Equations**

Solve the system by using the substitution method.

$$-3x + 4y = 9$$

$$x = -\frac{1}{3}y + 2$$

Solution:

$-3x + 4y = 9$

$x = \underbrace{-\dfrac{1}{3}y + 2}$

Step 1: In the second equation, x is already isolated.

$-3\left(-\dfrac{1}{3}y + 2\right) + 4y = 9$

Step 2: Substitute the quantity $-\frac{1}{3}y + 2$ for x in the *other* equation.

$y - 6 + 4y = 9$

Step 3: Solve for y.

$5y = 15$

$y = 3$

Now use the known value of y to solve for the remaining variable x.

$x = -\dfrac{1}{3}y + 2$

$x = -\dfrac{1}{3}(3) + 2$

Step 4: Substitute $y = 3$ into the equation $x = -\dfrac{1}{3}y + 2$.

$x = -1 + 2$

$x = 1$

Step 5: Check the ordered pair $(1, 3)$ in each original equation.

$-3x + 4y = 9 \qquad\qquad\qquad x = -\dfrac{1}{3}y + 2$

$-3(1) + 4(3) \overset{?}{=} 9$

$-3 + 12 \overset{?}{=} 9$ ✔ True $\qquad\qquad 1 \overset{?}{=} -\dfrac{1}{3}(3) + 2$

$\qquad\qquad\qquad\qquad\qquad\qquad 1 \overset{?}{=} -1 + 2$ ✔ True

The solution is $(1, 3)$.

Skill Practice

2. Solve by using the substitution method.

$x = 2y + 3$

$4x - 2y = 0$

Answer

2. $(-1, -2)$

2. Solving Inconsistent Systems and Systems of Dependent Equations

Example 3 Solving an Inconsistent System

Solve the system by using the substitution method.

$$x = 2y - 4$$
$$-2x + 4y = 6$$

Solution:

$x = 2y - 4$ **Step 1:** The x variable is already isolated.

$-2x + 4y = 6$

$-2(2y - 4) + 4y = 6$ **Step 2:** Substitute the quantity $x = 2y - 4$ into the *other* equation.

$-4y + 8 + 4y = 6$ **Step 3:** Solve for y.

$8 = 6$ The equation reduces to a contradiction, indicating that the system has no solution. The lines never intersect and must be parallel. The system is *inconsistent*.

There is no solution.
The system is inconsistent.

TIP: The answer to Example 3 can be verified by writing each equation in slope-intercept form and graphing the equations.

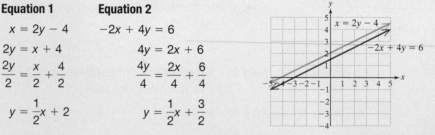

Equation 1	Equation 2
$x = 2y - 4$	$-2x + 4y = 6$
$2y = x + 4$	$4y = 2x + 6$
$\dfrac{2y}{2} = \dfrac{x}{2} + \dfrac{4}{2}$	$\dfrac{4y}{4} = \dfrac{2x}{4} + \dfrac{6}{4}$
$y = \dfrac{1}{2}x + 2$	$y = \dfrac{1}{2}x + \dfrac{3}{2}$

Notice that the equations have the same slope, but different y-intercepts; therefore, the lines must be parallel. There is no solution to this system of equations.

Example 4 Solving a System of Dependent Equations

Solve by using the substitution method.

$$4x - 2y = -6$$
$$y - 3 = 2x$$

Solution:

$4x - 2y = -6$

$y - 3 = 2x \longrightarrow y = \underline{2x + 3}$ **Step 1:** Solve for one of the variables.

$4x - 2(2x + 3) = -6$ **Step 2:** Substitute the quantity $2x + 3$ for y in the *other* equation.

$$4x - 4x - 6 = -6 \qquad \textbf{Step 3:} \quad \text{Solve for } x. \text{ Apply the distributive}$$
property to clear the parentheses.

$$-6 = -6$$

The system reduces to the identity $-6 = -6$. Therefore, the two equations are equivalent. The solution consists of all points on the common line, giving us an infinite number of solutions. The equations are dependent and because the equations $4x - 2y = -6$ and $y - 3 = 2x$ represent the same line, the solution can be written as

$$\{(x, y) \mid 4x - 2y = -6\} \qquad \text{or} \qquad \{(x, y) \mid y - 3 = 2x\}$$

TIP: We can confirm the results of Example 4 by writing each equation in slope-intercept form. The slope-intercept forms are identical, indicating that the lines are the same.

slope-intercept form

$$4x - 2y = -6 \longrightarrow -2y = -4x - 6 \longrightarrow y = 2x + 3$$

$$y - 3 = 2x \longrightarrow y = 2x + 3$$

Section 4.2   Practice Exercises

Study Skills Exercise

Check your progress by answering these questions.

Yes _____ No _____ Did you have sufficient time to study for the test on Chapter 3? If not, what could you have done to create more time for studying?

Yes _____ No _____ Did you work all of the assigned homework problems in Chapter 3?

Yes _____ No _____ If you encountered difficulty, did you see your instructor or tutor for help?

Yes _____ No _____ Have you taken advantage of the textbook supplements such as the *Student Solutions Manual*?

Review Exercises

1. Determine if the ordered pair $(-4, 3)$ is a solution to the system. $-x + 2y = 10$
$$2x - y = 11$$

For Exercises 2–5, use the slope-intercept form of the lines to determine the number of solutions to the system.

2. $y = 8x - 1$
$2x - 16y = 3$

3. $4x + 6y = 1$
$10x + 15y = \dfrac{5}{2}$

4. $2x - 4y = 0$
$x - 2y = 9$

5. $6x + 3y = 8$
$8x + 4y = -1$

For Exercises 6 and 7, solve the system by graphing.

6. $x - y = 4$

$3x + 4y = 12$

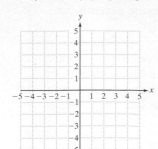

7. $y = 2x + 3$

$6x + 3y = 9$

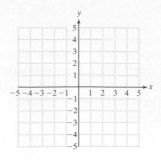

Concept 1: The Substitution Method

For Exercises 8–19, solve by using the substitution method. **(See Examples 1 and 2.)**

8. $4x + 12y = 4$

$y = 5x + 11$

9. $y = -3x - 1$

$2x - 3y = -8$

10. $10y + 34 = x$

$-7x + y = -31$

11. $-3x + 8y = -1$

$4x - 11 = y$

12. $12x - 2y = 0$

$-7x + y = -1$

13. $3x + 12y = 36$

$x - 5y = 12$

14. $x - 3y = -3$

$2x + 3y = -6$

15. $x - y = 8$

$3x + 2y = 9$

16. $5x - 2y = 10$

$y = x - 1$

17. $2x - y = -1$

$y = -2x$

18. $1 + 3y = 10$

$5x + 2y = 6$

19. $2x + 3 = 7$

$3x - 4y = 6$

20. Describe the process of solving a system of linear equations by using substitution.

Concept 2: Solving Inconsistent Systems and Systems of Dependent Equations

For Exercises 21–28, solve the system. For systems that do not have one unique solution, also state the number of solutions and whether the system is inconsistent or the equations are dependent. **(See Examples 3 and 4.)**

21. $2x - 6y = -2$

$x = 3y - 1$

22. $-2x + 4y = 22$

$x = 2y - 11$

23. $y = \dfrac{1}{7}x + 3$

$x - 7y = -4$

24. $x = -\dfrac{3}{2}y + \dfrac{1}{2}$

$4x + 6y = 7$

25. $5x - y = 10$

$2y = 10x - 5$

26. $x + 4y = 8$

$3x = 3 - 12y$

27. $3x - y = 7$

$-14 + 6x = 2y$

28. $x = 4y + 1$

$-12y = -3x + 3$

29. When using the substitution method, explain how to determine whether two linear equations in a system are dependent.

30. When using the substitution method, explain how to determine whether a system of linear equations is inconsistent.

Mixed Exercises

For Exercises 31–54, solve the system by using the substitution method. For systems that do not have one unique solution, also state the number of solutions and whether the system is inconsistent or the equations are dependent.

31. $x = 1.3y + 1.5$
$y = 1.2x - 4.6$

32. $y = 0.8x - 1.8$
$1.1x = -y + 9.6$

33. $y = \dfrac{2}{3}x - \dfrac{1}{3}$
$x = \dfrac{1}{4}y + \dfrac{17}{4}$

34. $x = \dfrac{1}{6}y - \dfrac{5}{3}$
$y = \dfrac{1}{5}x + \dfrac{21}{5}$

35. $-2x + y = 4$
$-\dfrac{1}{4}x + \dfrac{1}{8}y = \dfrac{1}{4}$

36. $8x - y = 8$
$\dfrac{1}{3}x - \dfrac{1}{24}y = \dfrac{1}{2}$

37. $3x + 2y = 6$
$y = x + 3$

38. $-x + 4y = -4$
$y = x - 1$

39. $-300x - 125y = 1350$
$y + 2 = 8$

40. $200y = 150x$
$y - 4 = 1$

41. $2x - y = 6$
$\dfrac{1}{6}x - \dfrac{1}{12}y = \dfrac{1}{2}$

42. $x - 4y = 8$
$\dfrac{1}{16}x - \dfrac{1}{4}y = \dfrac{1}{2}$

43. $y = 200x - 320$
$y = -150x + 1080$

44. $y = -54x + 300$
$y = 20x - 70$

45. $y = -2.7x - 5.1$
$y = 3.1x - 63.1$

46. $y = 6.8x + 2.3$
$y = -4.1x + 56.8$

47. $4x + 4y = 5$
$x - 4y = -\dfrac{5}{2}$

48. $-2x + y = -6$
$6x - 13y = -12$

49. $2(x + 2y) = 12$
$-6x = 5y - 8$

50. $5x - 2y = -25$
$10x = 3(y - 10)$

51. $5(3x - y) = 10$
$4y = 7x - 3$

52. $2x = -3(y + 3)$
$3x - 4y = -22$

53. $2x - 5 = 7$
$4 = 3y + 1$

54. $-2 = 4 - 2y$
$7x - 5 = -5$

55. The **centroid** of a region is the geometric center. For the region shown, the centroid is the point of intersection of the diagonals of the parallelogram.

   a. Find the slope-intercept form of the line through the points $(-4, 1)$ and $(5, 5)$.

   b. Find the slope-intercept form of the line through the points $(-3, 5)$ and $(4, 1)$.

   c. Find the centroid of the region.

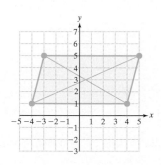

56. a. Find the slope-intercept form of the line through the points $(0, -3)$ and $(-2, 5)$.

b. Find the slope-intercept form of the line through the points $(-4, 4)$ and $(2, -2)$.

c. Find the centroid of the region.

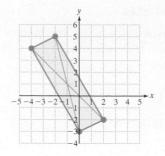

57. The cost to rent an apartment at Glendale Lakes is a $250 pet fee plus $800 per month. An apartment at the Breakers requires a $500 pet fee plus $750 per month.

a. Write the cost y to rent for x months for each apartment.

b. Find the number of months for which the total amount spent for each apartment would be the same.

58. The Surfside Motel charges $159.50 per night plus a $24 parking fee. The Tropical Winds Motel charges $165.50 per night with no parking fee.

a. Write the cost y to stay for x nights for each motel. Assume that the patron needs parking.

b. Find the number of nights for which the cost to stay at each motel would be the same.

Section 4.3 — Solving Systems of Equations by Using the Addition Method

Concepts

1. The Addition Method
2. Solving Inconsistent Systems and Systems of Dependent Equations

1. The Addition Method

The next method we present to solve systems of linear equations is the *addition method* (sometimes called the elimination method). With the addition method, the goal is to manipulate the equations in the system so that one of the variables has opposite coefficients. Then by adding the equations, that variable is eliminated. For example, consider the following system.

$$\begin{array}{rl} x + 2y &= 6 \\ x - 2y &= 14 \\ \hline 2x &= 20 \end{array}$$

The y variables have opposite coefficients 2 and -2. Therefore, by adding the equations, the y terms are eliminated, enabling us to solve for x.

$$x = 10$$

Substituting $x = 10$ in either original equation gives $y = -2$. The solution is $(10, -2)$.

Often, one or both equations must be multiplied by appropriate constants to create opposite coefficients on either the x or y variable.

The steps to solve a system of linear equations in two variables by the addition method is outlined in the following box.

Solving a System of Equations by the Addition Method

Step 1 Write both equations in standard form: $Ax + By = C$.

Step 2 Clear fractions or decimals (optional).

Step 3 Multiply one or both equations by nonzero constants to create opposite coefficients for one of the variables.

Step 4 Add the equations from step 3 to eliminate one variable.

Step 5 Solve for the remaining variable.

Step 6 Substitute the known value found in step 5 into one of the original equations to solve for the other variable.

Step 7 Check the ordered pair in *both* equations.

Example 1 Solving a System by the Addition Method

Solve the system by using the addition method.

$$3x - 4y = 2$$
$$4x + y = 9$$

Solution:

Steps 1 and 2: Both of the equations are written in standard form. There are no fractions or decimals.

$$3x - 4y = 2 \qquad\qquad 3x - 4y = 2$$

$$4x + y = 9 \xrightarrow{\text{Multiply by 4.}} 16x + 4y = 36 \qquad \textbf{Step 3:} \quad \text{Multiply the second equation by 4.}$$

$$\begin{array}{r} 3x - 4y = 2 \\ 16x + 4y = 36 \\ \hline 19x \qquad = 38 \end{array} \qquad \begin{array}{l} \textbf{Step 4:} \quad \text{Add the equations.} \\[4pt] \textbf{Step 5:} \quad \text{Solve for } x. \end{array}$$

$$x = 2$$

$$3x - 4y = 2$$
$$3(2) - 4y = 2 \qquad \textbf{Step 6:} \quad \text{Substitute } x = 2 \text{ back into one of the}$$
$$6 - 4y = 2 \qquad\qquad\quad \text{original equations and solve for } y.$$
$$-4y = -4$$
$$y = 1$$

Step 7: Check the ordered pair $(2, 1)$ in each original equation:

$$3x - 4y = 2 \qquad\qquad 4x + y = 9$$
$$3(2) - 4(1) \stackrel{?}{=} 2 \checkmark \text{ True} \qquad 4(2) + (1) \stackrel{?}{=} 9 \checkmark \text{ True}$$

The solution is $(2, 1)$.

Skill Practice

1. Solve by the addition method.

$$2x - 3y = 13$$
$$x + 2y = 3$$

TIP: Be sure to multiply *both* sides of the equation by 4:

$$4(4x + y) = 4(9)$$

TIP: Substituting $x = 2$ into the other equation, $4x + y = 9$, produces the same value for y.

$$4x + y = 9$$
$$4(2) + y = 9$$
$$8 + y = 9$$
$$y = 1$$

Answer

1. $(5, -1)$

Example 2 **Solving a System by the Addition Method**

Solve the system by using the addition method.

$$4x + 5y = 2$$
$$3x = 1 - 4y$$

Solution:

$4x + 5y = 2 \longrightarrow 4x + 5y = 2$ **Step 1:** Write both equations in standard form. There are no fractions or decimals.

$3x = 1 - 4y \longrightarrow 3x + 4y = 1$

We may choose to eliminate either variable. To eliminate x, change the coefficients to 12 and -12.

$4x + 5y = 2 \xrightarrow{\text{Multiply by 3.}} 12x + 15y = 6$ **Step 3:** Multiply the first equation by 3.

$3x + 4y = 1 \xrightarrow{\text{Multiply by } -4.} -12x - 16y = -4$ Multiply the second equation by -4.

$$\begin{array}{r} 12x + 15y = 6 \\ \underline{-12x - 16y = -4} \\ -y = 2 \\ y = -2 \end{array}$$

Step 4: Add the equations.

Step 5: Solve for y.

$4x + 5y = 2$ **Step 6:** Substitute $y = -2$ back into one of the original equations and solve for x.

$4x + 5(-2) = 2$

$4x - 10 = 2$

$4x = 12$

$x = 3$

The solution is $(3, -2)$.   **Step 7:** Check the ordered pair $(3, -2)$ in both original equations.

TIP: To eliminate the x variable in Example 2, both equations were multiplied by appropriate constants to create $12x$ and $-12x$. We chose 12 because it is the *least common multiple* of 4 and 3.

We could have solved the system by eliminating the y variable. To eliminate y, we would multiply the top equation by 4 and the bottom equation by -5. This would make the coefficients of the y variable 20 and -20, respectively.

$4x + 5y = 2 \xrightarrow{\text{Multiply by 4.}} 16x + 20y = 8$

$3x + 4y = 1 \xrightarrow{\text{Multiply by } -5.} -15x - 20y = -5$

| Example 3 | Solving a System of Equations by the Addition Method |

Solve the system by using the addition method.

$$x - 2y = 6 + y$$
$$0.05y = 0.02x - 0.10$$

Solution:

$x - 2y = 6 + y \longrightarrow x - 3y = 6$
$0.05y = 0.02x - 0.10 \longrightarrow -0.02x + 0.05y = -0.10$

Step 1: Write both equations in standard form.

$x \quad - 3y = 6$
$-0.02x + 0.05y = -0.10 \xrightarrow{\text{Multiply by 100.}} -2x + 5y = -10$

Step 2: Clear decimals.

$x - 3y = 6 \xrightarrow{\text{Multiply by 2.}} 2x - 6y = 12$
$-2x + 5y = -10 \longrightarrow \underline{-2x + 5y = -10}$

Step 3: Create opposite coefficients.

$-y = 2$

Step 4: Add the equations.

$y = -2$

Step 5: Solve for y.

$x - 2y = 6 + y$ **Step 6:** To solve for x, substitute $y = -2$
$x - 2(-2) = 6 + (-2)$ into one of the original equations.
$x + 4 = 4$
$x = 0$

Step 7: The ordered pair $(0, -2)$ checks in each original equation.

The solution is $(0, -2)$.

2. Solving Inconsistent Systems and Systems of Dependent Equations

| Example 4 | Solving a System of Dependent Equations |

Solve the system by using the addition method.

$$\frac{1}{5}x - \frac{1}{2}y = 1$$
$$-4x + 10y = -20$$

Solution:

$\frac{1}{5}x - \frac{1}{2}y = 1$ **Step 1:** Equations are in standard form.

$-4x + 10y = -20$

$$10\left(\frac{1}{5}x - \frac{1}{2}y\right) = 10 \cdot 1 \longrightarrow 2x - 5y = 10$$ **Step 2:** Clear fractions.

$$-4x + 10y = -20$$

$$2x - 5y = 10 \xrightarrow{\text{Multiply by 2.}} 4x - 10y = 20$$ **Step 3:** Multiply the first equation by 2.

$$\underline{-4x + 10y = -20 \longrightarrow -4x + 10y = -20}$$

$$0 = 0$$ **Step 4:** Add the equations.

Notice that both variables were eliminated. The system of equations is reduced to the identity $0 = 0$. Therefore, the two original equations are dependent. The solution consists of an infinite number of ordered pairs (x, y) that fall on the common line of intersection $-4x + 10y = -20$, or equivalently $\frac{1}{5}x - \frac{1}{2}y = 1$. The solution can be written in set notation as

$$\{(x, y) \mid -4x + 10y = -20\} \quad \text{or} \quad \left\{(x, y) \,\middle|\, \frac{1}{5}x - \frac{1}{2}y = 1\right\}$$

Skill Practice

5. Solve by the addition method.

$$18 + 10x = 6y$$
$$5x - 3y = 9$$

| Example 5 | **Solving an Inconsistent System** |

Solve the system by using the addition method.

$$2y = -3x + 4$$
$$20(6x + 5y) = 40 + 20y$$

Solution:

Step 1: Write the equations in standard form.

$$2y = -3x + 4 \longrightarrow 3x + 2y = 4$$
$$20(6x + 5y) = 40 + 20y \longrightarrow 120x + 100y = 40 + 20y \longrightarrow 120x + 80y = 40$$

With both equations now in standard form, we can proceed with the addition method.

Step 2: There are no decimals or fractions.

The goal is to create opposite coefficients on either the x or y terms. The second equation can be divided by -40.

TIP: In Step 2 we see that the division property of equality can also be used to create opposite coefficients on x and y.

$$3x + 2y = 4 \qquad\qquad 3x + 2y = 4$$
$$120x + 80y = 40 \xrightarrow{\text{Divide by } -40.} \underline{-3x - 2y = -1}$$
$$0 = 3$$

Step 3: Divide the second equation by -40.

Step 4: Add the equations.

The equations reduce to a contradiction, indicating that the system has no solution. The system is inconsistent. The two equations represent parallel lines, as shown in Figure 4-8.

There is no solution.

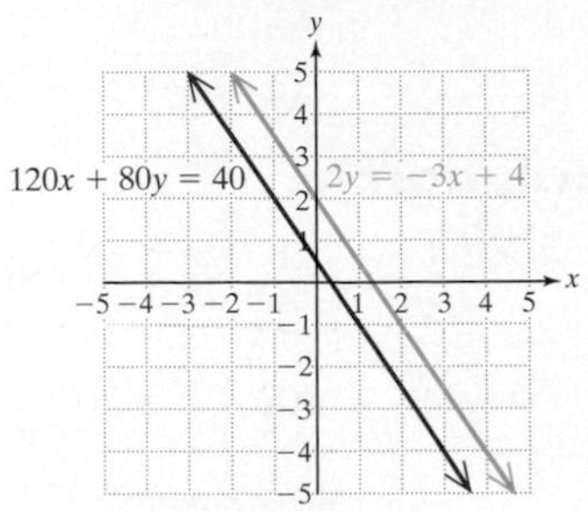

Figure 4-8

Answer

5. No solution; inconsistent system

Section 4.3 Practice Exercises

Study Skills Exercise

Instructors differ in what they emphasize on tests. For example, test material may come from the textbook, notes, handouts, or homework. What does your instructor emphasize?

Vocabulary and Key Concepts

1. **a.** By what nonzero constant would the second equation be multiplied to eliminate the x variable from the system of equations?

$$3x - 4y = 8$$
$$x + 5y = -2$$

b. By what nonzero constant would the first equation be multiplied to eliminate the y variable from the system of equations?

$$3x - y = 8$$
$$-8x + 5y = 12$$

Review Exercises

For Exercises 2–4, use the slope-intercept form of the lines to determine the number of solutions for the system of equations.

2. $y = \dfrac{1}{2}x - 4$

   $y = \dfrac{1}{2}x + 1$

3. $y = 2.32x - 8.1$

   $y = 1.46x - 8.1$

4. $4x = y + 7$

   $-2y = -8x + 14$

Concept 1: The Addition Method

For Exercises 5–16, solve the system by the addition method. **(See Examples 1–3.)**

5. $3x - y = -1$
   $-3x + 4y = -14$

6. $5x - 2y = 15$
   $3x + 2y = -7$

7. $2x + 3y = 3$
   $-10x + 2y = -32$

8. $2x - 5y = 7$
   $3x - 10y = 13$

9. $3x + 7y = -20$
   $-5x + 3y = -84$

10. $6x - 9y = -15$
    $5x - 2y = -40$

11. $3x = 10y + 13$
    $7y = 4x - 11$

12. $-5x = 6y - 4$
    $5y = 1 - 3x$

13. $1.2x - 0.6y = 3$
    $0.8x - 1.4y = 3$

14. $1.8x + 0.8y = 1.4$
    $1.2x + 0.6y = 1.2$

15. $3x + 2 = 4y + 2$
    $7x = 3y$

16. $-4y - 3 = 2x - 3$
    $5y = 3x$

Concept 2: Solving Inconsistent Systems and Systems of Dependent Equations

For Exercises 17–24, solve the system. For systems that do not have one unique solution, also state the number of solutions and whether the system is inconsistent or the equations are dependent. **(See Examples 4 and 5.)**

17. $3x - 2y = 1$
    $-6x + 4y = -2$

18. $3x - y = 4$
    $6x - 2y = 8$

19. $6y = 14 - 4x$
    $2x = -3y - 7$

20. $2x = 4 - y$
    $-y = 2x - 2$

21. $12x - 4y = 2$
    $6x = 1 + 2y$

22. $10x - 15y = 5$
    $3y = 2x - 1$

23. $\dfrac{1}{2}x + y = \dfrac{7}{6}$

    $x + 2y = 4.5$

24. $0.2x - 0.1y = -1.2$

    $x - \dfrac{1}{2}y = 3$

Mixed Exercises

25. Describe a situation in which you would prefer to use the substitution method over the addition method.

26. If you used the addition method to solve the given system, would it be easier to eliminate the x or y variable? Explain.

$$3x - 5y = 4$$
$$7x + 10y = 31$$

For Exercises 27–53, solve by using either the addition method or the substitution method. For systems that do not have one unique solution, also state the number of solutions and whether the system is inconsistent or the equations are dependent.

27. $2x - 4y = 8$
$y = 2x + 1$

28. $8x + 6y = -8$
$x = 6y - 10$

29. $2x + 5y = 9$
$4x - 7y = -16$

30. $0.1x + 0.5y = 0.7$
$0.2x + 0.7y = 0.8$

31. $0.2x - 0.1y = 0.8$
$0.1x - 0.1y = 0.4$

32. $y = \dfrac{1}{2}x - 3$
$4x + y = -3$

33. $4x - 6y = 5$
$2x - 3y = 7$

34. $3x + 6y = 7$
$2x + 4y = 5$

35. $\dfrac{1}{4}x - \dfrac{1}{6}y = -2$
$-\dfrac{1}{6}x + \dfrac{1}{5}y = 4$

36. $\dfrac{1}{3}x + \dfrac{1}{5}y = 7$
$\dfrac{1}{6}x - \dfrac{2}{5}y = -4$

37. $\dfrac{1}{3}x - \dfrac{1}{2}y = 0$
$x = \dfrac{3}{2}y$

38. $\dfrac{2}{5}x - \dfrac{2}{3}y = 0$
$y = \dfrac{3}{5}x$

39. $2(x + 2y) = 20 - y$
$-7(x - y) = 16 + 3y$

40. $-3(x + y) = 10 - 4y$
$4(x + 2y) = 50 + 3y$

41. $-4y = 10$
$4x + 3 = 1$

42. $-9x = 15$
$3y + 2 = 1$

43. $5x - 3y = 18$
$-3x + 5y = 18$

44. $6x - 3y = -3$
$4x + 5y = -9$

45. $3x - 2 = \dfrac{1}{3}(11 + 5y)$
$x + \dfrac{2}{3}(2y - 3) = -2$

46. $2(2y + 3) - 2x = 1 - x$
$x + y = \dfrac{1}{5}(7 + y)$

47. $\dfrac{1}{4}x + \dfrac{1}{2}y = \dfrac{11}{4}$
$\dfrac{2}{3}x + \dfrac{1}{3}y = \dfrac{7}{3}$

48. $\dfrac{1}{10}x - \dfrac{1}{2}y = -\dfrac{8}{5}$
$x + \dfrac{1}{4}y = -\dfrac{11}{2}$

49. $4x + y = -2$
$5x - y = -7$

50. $4y = 8x + 20$
$8x = 24$

51. $4x = 3y$
$y = \dfrac{4}{3}x + 2$

52. $4x - 2y = 6$
$x = \dfrac{1}{2}y + \dfrac{3}{2}$

53. $2x + 7y = -20$
$-6y = 48$

54. A Subaru Forester gets 21 mpg in the city and 27 mpg on the highway. A Jeep Liberty gets 15 mpg in the city and 21 mpg on the highway. Suppose that each vehicle travels c miles in the city and h miles on the highway. The Subaru uses 12 gal of gasoline and the Jeep uses 16 gal. Solve the system of equations to determine the number of city and number of highway miles driven.

Subaru: $\frac{1}{21}c + \frac{1}{27}h = 12$

Jeep: $\frac{1}{15}c + \frac{1}{21}h = 16$

55. A Lincoln Town Car gets 16 mpg in the city and 24 mpg on the highway. A Lincoln Navigator gets 14 mpg in the city and 20 mpg on the highway. Suppose that each vehicle travels c miles in the city and h miles on the highway. The Town Car uses 12 gal of gasoline and the Navigator uses 14 gal. Solve the system of equations to determine the number of city and highway miles driven.

Town Car: $\frac{1}{16}c + \frac{1}{24}h = 12$

Navigator: $\frac{1}{14}c + \frac{1}{20}h = 14$

Problem Recognition Exercises

Solving Systems of Linear Equations

For Exercises 1–4, solve each system by using three different methods.

    a. Use the graphing method.

    b. Use the substitution method.

    c. Use the addition method.

1. $-3x + y = -2$
$\quad 4x - y = 4$

2. $3x - 2y = 4$
$\quad x = \frac{2}{3}y + \frac{4}{3}$

3. $5x = 2y$
$\quad y = \frac{5}{2}x + 1$

4. $2y = 3x + 1$
$\quad 4x = 4$

For Exercises 5–20, solve the system using the method of your choice. For systems that do not have one unique solution, also state the number of solutions and whether the system is inconsistent or the equations are dependent.

5. $y = -4x - 9$
$\quad 8x + 3y = -29$

6. $5x - 2y = -17$
$\quad x + 5y = 2$

7. $5x - 3y = 2$
$\quad 7x + 4y = -30$

8. $\frac{1}{10}x - \frac{2}{5}y = -\frac{3}{5}$
$\quad \frac{3}{4}x + \frac{1}{3}y = \frac{13}{6}$

9. $3x + 2y = 11$
$\quad y = -2x + 4$

10. $x = 8y - 1$
$\quad 3x - 10y = -17$

11. $2x - 3y = -3$
$\quad 4x + 2y = 18$

12. $7x + 15y = -1$
$\quad 6x + 5y = 7$

13. $4x - 8y = 1$
$\quad x = 2y - 6$

14. $y = -3x + 5$
$\quad 6x + 2y = 9$

15. $\frac{1}{2}x - \frac{1}{4}y = 2$
$\quad \frac{1}{5}x + \frac{3}{10}y = 0$

16. $\frac{1}{6}x + \frac{2}{3}y = \frac{4}{3}$
$\quad \frac{1}{2}x - \frac{1}{5}y = \frac{9}{5}$

17. $10x - 6y = 4$
$\quad 15x - 9y = 6$

18. $y = \frac{3}{5}x - 2$
$\quad 6x - 10y = 20$

19. $0.2x - 0.5y = 0.5$
$\quad 0.6x + 0.2y = 3.2$

20. $2.5x + 3.0y = 80$
$\quad 4.0x + 1.5y = 95$

Skill Practice

1. At the movie theater, Tom spent $15.50 on 3 soft drinks and 2 boxes of popcorn. Carly bought 5 soft drinks and 1 box of popcorn for a total of $16.50. Use a system of equations to find the cost of a soft drink and the cost of a box of popcorn.

Section 4.4 Applications of Systems of Linear Equations in Two Variables

1. Applications Involving Cost

In Chapter 1 we solved numerous application problems using equations that contained one variable. However, when an application has more than one unknown, sometimes it is more convenient to use multiple variables. In this section, we will solve applications containing two unknowns. When two variables are present, the goal is to set up a system of two independent equations.

Example 1   Solving a Cost Application

At an amusement park, five hot dogs and one drink cost $22. Two hot dogs and three drinks cost $14. Find the cost per hot dog and the cost per drink.

Solution:

Let h represent the cost per hot dog.   Label the variables.

Let d represent the cost per drink.

$$\left(\begin{array}{c}\text{Cost of 5}\\\text{hot dogs}\end{array}\right) + \left(\begin{array}{c}\text{cost of 1}\\\text{drink}\end{array}\right) = \$22 \longrightarrow 5h + d = 22 \qquad \text{Write two equations.}$$

$$\left(\begin{array}{c}\text{Cost of 2}\\\text{hot dogs}\end{array}\right) + \left(\begin{array}{c}\text{cost of 3}\\\text{drinks}\end{array}\right) = \$14 \longrightarrow 2h + 3d = 14$$

This system can be solved by either the substitution method or the addition method. We will solve by using the substitution method. The d-variable in the first equation is the easiest variable to isolate.

$$5h + d = 22 \longrightarrow d = -5h + 22 \qquad \text{Solve for } d \text{ in the first equation.}$$

$$2h + 3d = 14$$

$$\qquad\qquad\qquad\qquad\qquad\qquad\qquad \text{Substitute the quantity } -5h + 22 \text{ for } d \text{ in the } second \text{ equation.}$$

$$2h + 3(-5h + 22) = 14 \qquad$$

$$2h - 15h + 66 = 14 \qquad \text{Clear parentheses.}$$

$$-13h + 66 = 14 \qquad \text{Solve for } h.$$

$$-13h = -52$$

$$h = 4$$

$$d = -5(4) + 22 \longrightarrow d = 2 \qquad \text{Substitute } h = 4 \text{ in the equation } d = -5h + 22.$$

Because $h = 4$, the cost per hot dog is $4.00.

Because $d = 2$, the cost per drink is $2.00.

Answer

1. Soft drink: $2.50; popcorn: $4.00

TIP: A word problem can be checked by verifying that the solution meets the conditions specified in the problem.

$$5 \text{ hot dogs} + 1 \text{ drink} = 5(\$4.00) + 1(\$2.00) = \$22.00 ✔$$
$$2 \text{ hot dogs} + 3 \text{ drinks} = 2(\$4.00) + 3(\$2.00) = \$14.00 ✔$$

2. Applications Involving Mixtures

Example 2 Solving an Application Involving Chemistry

One brand of cleaner used to etch concrete is 25% acid. A stronger industrial-strength cleaner is 50% acid. How many gallons of each cleaner should be mixed to produce 20 gal of a 40% acid solution?

Solution:

Let x represent the amount of 25% acid cleaner.

Let y represent the amount of 50% acid cleaner.

	25% Acid	50% Acid	40% Acid
Number of gallons of solution	x	y	20
Number of gallons of pure acid	$0.25x$	$0.50y$	0.40(20), or 8

From the first row of the table, we have

$$\begin{pmatrix} \text{Amount of} \\ 25\% \text{ solution} \end{pmatrix} + \begin{pmatrix} \text{amount of} \\ 50\% \text{ solution} \end{pmatrix} = \begin{pmatrix} \text{total amount} \\ \text{of solution} \end{pmatrix} \rightarrow x + y = 20$$

From the second row of the table we have

$$\begin{pmatrix} \text{Amount of} \\ \text{pure acid in} \\ 25\% \text{ solution} \end{pmatrix} + \begin{pmatrix} \text{amount of} \\ \text{pure acid in} \\ 50\% \text{ solution} \end{pmatrix} = \begin{pmatrix} \text{amount of} \\ \text{pure acid in} \\ \text{resulting solution} \end{pmatrix} \rightarrow 0.25x + 0.50y = 8$$

$$\begin{array}{ll} x + \quad y = 20 & \longrightarrow \quad x + \quad y = 20 \\ 0.25x + 0.50y = 8 & \longrightarrow \quad 25x + 50y = 800 \end{array}$$ Multiply by 100 to clear decimals.

$$\begin{array}{ll} x + \quad y = 20 & \xrightarrow{\text{Multiply by } -25.} \quad -25x - 25y = -500 \\ 25x + \quad 50y = 800 & \longrightarrow \quad \underline{25x + 50y = 800} \end{array}$$ Create opposite coefficients of x.

$$25y = 300$$ Add the equations to eliminate x.
$$y = 12$$

$$x + y = 20$$ Substitute $y = 12$ back into one of the original equations.
$$x + (12) = 20$$
$$x = 8$$

Therefore, 8 gal of 25% acid solution must be added to 12 gal of 50% acid solution to create 20 gal of a 40% acid solution.

Skill Practice

2. A pharmacist needs 8 ounces (oz) of a solution that is 50% saline. How many ounces of 60% saline solution and 20% saline solution must be mixed to obtain the mixture needed?

Avoiding Mistakes

It is a common mistake to forget to write the percent as a decimal.

Answer

2. 6 oz of 60% solution and 2 oz of 20% solution

3. Applications Involving Principal and Interest

Example 3 Solving an Application Involving Finance

Serena invested money in two mutual funds. One had a return of 4.5% and the other had a return of 7%. Twice as much was invested at 7% as at 4.5%. If the amount earned on the original principal at the end of 1 yr was $1017.50, determine the amount invested in each fund.

Solution:

Let x represent the amount invested at 4.5%.

Let y represent the amount invested at 7%.

	4.5% Account	7% Account	Total
Principal	x	y	
Amount earned	$0.045x$	$0.07y$	1017.50

Because the amount invested at 7% was twice the amount invested at 4.5%, we have

$$\left(\begin{array}{c}\text{Amount}\\\text{invested}\\\text{at 7\%}\end{array}\right) = 2\left(\begin{array}{c}\text{amount}\\\text{invested}\\\text{at 4.5\%}\end{array}\right) \rightarrow y = 2x$$

From the second row of the table, we have

$$\left(\begin{array}{c}\text{Amount}\\\text{earned from}\\\text{4.5\% account}\end{array}\right) + \left(\begin{array}{c}\text{amount}\\\text{earned from}\\\text{7\% account}\end{array}\right) = \left(\begin{array}{c}\text{total}\\\text{amount}\\\text{earned}\end{array}\right) \rightarrow 0.045x + 0.07y = 1017.50$$

$$y = 2x$$
$$45x + 70y = 1{,}017{,}500 \qquad \text{Multiply by 1000 to clear decimals.}$$

Because the y variable in the first equation is isolated, we will use the substitution method.

$$45x + 70(2x) = 1{,}017{,}500 \qquad \text{Substitute the quantity } 2x \text{ into the second equation.}$$

$$45x + 140x = 1{,}017{,}500 \qquad \text{Solve for } x.$$
$$185x = 1{,}017{,}500$$
$$x = \frac{1{,}017{,}500}{185}$$
$$x = 5500$$

$$y = 2x$$
$$y = 2(5500) \qquad \text{Substitute } x = 5500 \text{ into the equation } y = 2x \text{ to solve for } y.$$
$$y = 11{,}000$$

Because $x = 5500$, the amount invested at 4.5% was $5500.

Because $y = 11{,}000$, the amount invested at 7% was $11,000.

4. Applications Involving Uniform Motion

Example 4 Solving a Distance, Rate, and Time Application

A plane flies 660 mi from Atlanta to Miami in 1.2 hr when traveling with a tailwind. The return flight against the same wind takes 1.5 hr. Find the speed of the plane in still air and the speed of the wind.

Solution:

Let p represent the speed of the plane in still air.

Let w represent the speed of the wind.

The speed of the plane *with* the wind:

$$\text{(Plane's still airspeed)} + \text{(wind speed)}: p + w$$

The speed of the plane *against* the wind:

$$\text{(Plane's still airspeed)} - \text{(wind speed)}: p - w$$

Set up a chart to organize the given information:

	Distance	Rate	Time
With a tailwind	660	$p + w$	1.2
Against a head wind	660	$p - w$	1.5

Two equations can be found by using the relationship $d = rt$.

$$\begin{pmatrix} \text{Distance} \\ \text{with} \\ \text{wind} \end{pmatrix} = \begin{pmatrix} \text{speed} \\ \text{with} \\ \text{wind} \end{pmatrix}\begin{pmatrix} \text{time} \\ \text{with} \\ \text{wind} \end{pmatrix} \longrightarrow 660 = (p + w)(1.2)$$

$$\begin{pmatrix} \text{Distance} \\ \text{against} \\ \text{wind} \end{pmatrix} = \begin{pmatrix} \text{speed} \\ \text{against} \\ \text{wind} \end{pmatrix}\begin{pmatrix} \text{time} \\ \text{against} \\ \text{wind} \end{pmatrix} \longrightarrow 660 = (p - w)(1.5)$$

$660 = (p + w)(1.2)$ Notice that the first equation may be *divided* by 1.2

$660 = (p - w)(1.5)$ and still leave integer coefficients. Similarly, the second equation may be simplified by dividing by 1.5.

$$660 = (p + w)(1.2) \xrightarrow{\text{Divide by 1.2}} \frac{660}{1.2} = \frac{(p + w)1.2}{1.2} \longrightarrow 550 = p + w$$

$$660 = (p - w)(1.5) \xrightarrow{\text{Divide by 1.5}} \frac{660}{1.5} = \frac{(p - w)1.5}{1.5} \longrightarrow 440 = p - w$$

Answer

4. Speed of plane: 540 mph; speed of wind: 60 mph

$$550 = p + w$$

$$\underline{440 = p - w}$$

$$990 = 2p \qquad \text{Add the equations.}$$

$$p = 495$$

$$550 = (495) + w \qquad \text{Substitute } p = 495 \text{ into the equation } 550 = p + w.$$

$$55 = w \qquad \text{Solve for } w.$$

The speed of the plane in still air is 495 mph, and the speed of the wind is 55 mph.

5. Applications Involving Geometry

Example 5 Solving a Geometry Application

The sum of the two acute angles in a right triangle is 90°. The measure of one angle is 6° less than 2 times the measure of the other angle. Find the measure of each angle.

Solution:

Let x represent one acute angle.

Let y represent the other acute angle.

The sum of the two acute angles is 90°: $\qquad x + y = 90$

One angle is 6° less than 2 times the other angle: $\qquad x = 2y - 6$

$$x + y = 90 \qquad \text{Because one variable is already isolated, we}$$
$$x = 2y - 6 \qquad \text{will use the substitution method.}$$

$$(2y - 6) + y = 90 \qquad \text{Substitute } x = 2y - 6 \text{ into the first equation.}$$
$$3y - 6 = 90$$
$$3y = 96$$
$$y = 32$$

$$x = 2y - 6 \qquad \text{To find } x, \text{ substitute } y = 32 \text{ into the equation}$$
$$\qquad\qquad\qquad x = 2y - 6.$$
$$x = 2(32) - 6$$
$$x = 64 - 6$$
$$x = 58$$

The two acute angles in the triangle measure 32° and 58°.

Section 4.4 Practice Exercises

Study Skills Exercise

Make up a practice test for yourself. Use examples or exercises from the text. Be sure to cover each concept that was presented so far in this chapter.

Vocabulary and Key Concepts

1. **a.** If the cost of one T-shirt is $12, how much will five T-shirts cost? How much will x T-shirts cost?

    b. If a solution contains 10% bleach, how much pure bleach is in a 20-L container? How much pure bleach is in a container holding x liters?

    c. If $5000 is in a savings account that earns 4% simple interest, how much interest is earned at the end of 1 yr? If y dollars is invested at 4% simple interest for 1 yr, how much interest is earned?

    d. If a boat travels b km/hr in still water and the current is c km/hr, then what is the speed of the boat traveling against the current? What is the speed of the boat traveling with the current?

    e. The sum of the measures of the angles within a triangle is _____.

    f. If the measure of an angle is x and the measure of its supplement is y, then $x + y =$ _____.

    g. If the measure of an angle is x and the measure of its complement is y, then $x + y =$ _____.

    h. If the measure of one acute angle in a right triangle is a and the other acute angle has measure b, then $a + b =$ _____.

Review Exercises

2. State three methods that can be used to solve a system of linear equations in two variables.

For Exercises 3–6, state which method you would prefer to use to solve the system. Then solve the system.

3. $y = 9 - 2x$
    $3x - y = 16$

4. $7x - y = -25$
    $2x + 5y = 14$

5. $5x + 2y = 6$
    $-2x - y = 3$

6. $x = 5y - 2$
    $-3x + 7y = 14$

Concept 1: Applications Involving Cost

7. The local community college theater put on a production of *Chicago*. There were 186 tickets sold, some for $16 (nonstudent price) and others for $12 (student price). If the receipts for one performance totaled $2640, how many of each type of ticket were sold? **(See Example 1.)**

8. John and Ariana bought school supplies. John spent $10.65 on 4 notebooks and 5 pens. Ariana spent $7.50 on 3 notebooks and 3 pens. What is the cost of 1 notebook and what is the cost of 1 pen?

9. Mickey bought lunch for his fellow office workers on Monday. He spent $24.20 on 3 hamburgers and 2 fish sandwiches. Chloe bought lunch on Tuesday and spent $23.60 for 4 hamburgers and 1 fish sandwich. What is the price of 1 hamburger, and what is the price of 1 fish sandwich?

10. A group of four golfers paid $150 to play a round of golf. Of the golfers, one is a member of the club and three are nonmembers. Another group of golfers consists of two members and one nonmember. They paid a total of $75. What is the cost for a member to play a round of golf, and what is the cost for a nonmember?

11. Jen has 2 scoops of Ben and Jerry's vanilla ice cream and 1 scoop of mud pie ice cream for a total of 40 g of fat. Jim has 1 scoop of vanilla and 2 scoops of mud pie for a total of 44 g of fat. How much fat is in 1 scoop of each ice cream?

12. Roselle has 2 cups of popcorn and 8 oz of soda for a total of 216 calories. Carmel has 1 cup of popcorn and 12 oz of soda for a total of 204 calories. Determine the number of calories per cup of popcorn and per ounce of soda.

Concept 2: Applications Involving Mixtures

13. A jar of one face cream contains 18% moisturizer, and another type contains 24% moisturizer. How many ounces of each should be combined to get 12 oz of a cream that is 22% moisturizer? **(See Example 2.)**

14. Logan wants to mix an 18% acid solution with a 45% acid solution to get 16 L of a 36% acid solution. How many liters of the 18% solution and how many liters of the 45% solution should be mixed?

15. How much fertilizer containing 8% nitrogen should be mixed with a fertilizer containing 12% nitrogen to get 8 L of a fertilizer containing 11% nitrogen?

16. How much 30% acid solution should be added to 10% acid solution to make 100 mL of a 12% acid solution?

17. How much pure bleach should Tim combine with a solution that is 4% bleach to make 12 oz of a 12% bleach solution? (*Hint:* Pure bleach is 100% bleach.)

18. A fruit punch that contains 25% fruit juice is combined with 100% fruit juice. How many ounces of each should be used to make 48 oz of a mixture that is 75% fruit juice? (*Hint:* Pure juice is 100% juice.)

Concept 3: Applications Involving Principal and Interest

19. Mr. Coté invested 3 times as much money in a stock fund that returned 8% interest after one year as he did in a bond fund that earned 5% interest. If his total earnings came to $435 after 1 yr, how much did he invest in each fund? **(See Example 3.)**

20. Aliya deposited half as much money in a savings account earning 2.5% simple interest as she invested in a money market account that earns 3.5% simple interest. If the total interest after one year is $247, how much did she invest in each account?

21. Mr. Levy borrowed money from two lenders. One lender charged 5.5% simple interest and the other charged 3.5% simple interest. Mr. Levy had to borrow $200 more at 5.5% than he did at 3.5%. If the total interest after the first year was $245, how much did he borrow at each rate?

22. Jody invested $5000 less in an account paying 4% simple interest than she did in an account paying 3% simple interest. At the end of the first year, the total interest from both accounts was $675. Find the amount invested in each account.

23. Alina invested $27,000 in two accounts: one that pays 2% simple interest and one that pays 3% simple interest. At the end of 5 yr, her total return was $3425. How much was invested in each account?

24. Didi invested a total of $12,000 in two accounts paying 7.5% and 6% simple interest. If her total return at the end of 2 yr was $1680, how much did she invest in each account?

Concept 4: Applications Involving Uniform Motion

25. It takes a boat 2 hr to travel 16 mi downstream with the current and 4 hr to return against the current. Find the speed of the boat in still water and the speed of the current. **(See Example 4.)**

26. The Gulf Stream is a warm ocean current that extends from the eastern side of the Gulf of Mexico up through the Florida Straits and along the southeastern coast of the United States to Cape Hatteras, North Carolina. A boat travels with the current 100 mi from Miami, Florida, to Freeport, Bahamas, in 2.5 hr. The return trip against the same current takes $3\frac{1}{3}$ hr. Find the speed of the boat in still water and the speed of the current.

27. A plane flew 720 mi in 3 hr with the wind. It would take 4 hr to travel the same distance against the wind. What is the rate of the plane in still air and the rate of wind?

28. A plane flies from Atlanta to Los Angeles against the wind in 5 hr. The return trip back to Atlanta with the wind takes only 4 hr. If the distance between Atlanta and Los Angeles is 3200 km, find the speed of the plane in still air and the speed of the wind.

29. A moving sidewalk in the Atlanta airport moves people between gates. It takes Molly's 8-year-old son Stephen, 20 seconds to travel 100 ft walking with the sidewalk. It takes him 30 seconds to travel 60 ft walking against the moving sidewalk (in the opposite direction). Find the speed of the moving sidewalk and Stephen's walking speed on nonmoving ground.

30. Kim rides a total of 48 km in the bicycle portion of a triathlon. The course is an "out and back" route. It takes her 3 hr on the way out against the wind. The ride back takes her 2 hr with the wind. Find the speed of the wind and Kim's speed riding her bike in still air.

Concept 5: Applications Involving Geometry

For Exercises 31–36, solve the applications involving geometry. If necessary, refer to the geometry formulas listed in the inside back cover of the text.

31. In a right triangle, one acute angle measures 6° more than 3 times the other. If the sum of the measures of the two acute angles must equal 90°, find the measures of the acute angles. **(See Example 5.)**

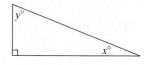

Figure for Exercise 31

32. An isosceles triangle has two angles of the same measure. If the angle represented by y measures 3° less than the angle x, find the measures of all angles of the triangle.

33. Two angles are supplementary. One angle measures 2° less than 3 times the other. What are the measures of the two angles?

34. The measure of one angle is 5 times the measure of another. If the two angles are supplementary, find the measures of the angles.

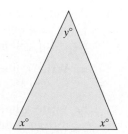

Figure for Exercise 32

35. One angle measures 3° more than twice another. If the two angles are complementary, find the measures of the angles.

36. Two angles are complementary. One angle measures 15° more than 2 times the measure of the other. What are the measures of the two angles?

Mixed Exercises

37. How much pure gold (24K) must be mixed with 60% gold to get 20 grams (g) of 75% gold?

38. A granola mix contains 5% nuts. How many ounces of nuts must be added to get 25 oz of granola with 24% nuts?

39. A rowing team trains on the Halifax River. It can row upstream 10 mi against the current in 2.5 hr and 16 mi downstream with the current in the same amount of time. Find the speed of the boat in still water and the speed of the current.

40. In her kayak, Bonnie can travel 31.5 mi downstream with the current in 7 hr. The return trip against the current takes 9 hr. Find the speed of the kayak in still water and the speed of the current.

41. There are two types of tickets sold at the Canadian Formula One Grand Prix race. The price of 6 grandstand tickets and 2 general admission tickets costs $2330. The price of 4 grandstand tickets and 4 general admission tickets cost $2020. What is the price of each type of ticket?

42. A basketball player scored 19 points by shooting two-point and three-point baskets. If she made a total of eight baskets, how many of each type did she make?

43. A bank offers two accounts, a money market account at 2% simple interest and a regular savings account at 1.3% interest. If Svetlana deposits $3000 between the two accounts and receives $51.25 in total interest in the first year, how much did she invest in each account?

44. Angelo invested $8000 in two accounts: one that pays 3% and one that pays 1.8%. At the end of the first year, his total interest earned was $222. How much did he deposit in the account that pays 3%?

45. The perimeter of a rectangle is 42 m. The length is 1 m longer than the width. Find the dimensions of the rectangle.

46. In a right triangle, the measure of one acute angle is one-fourth the measure of the other. Find the measures of the acute angles.

47. A coin collection consists of 50-cent pieces and $1 coins. If there are 21 coins worth $15.50, how many 50-cent pieces and $1 coins are there?

48. Jacob has a piggy bank consisting of nickels and dimes. If there are 30 coins worth $1.90, how many nickels and dimes are in the bank?

49. A rental car company rents a compact car for $20 a day, plus $0.25 per mile. A midsize car rents for $30 a day, plus $0.20 per mile.

    a. Write a linear function representing the cost to rent the compact car for x miles.

    b. Write a linear function representing the cost to rent a midsize car for x miles.

    c. Find the number of miles at which the cost to rent either car would be the same.

50. One phone plan charges $0.15 per text message. A second plan charges $0.10 per text message but adds a monthly fee of $4.95.

    a. Write a linear function representing the cost for the first plan for x text messages.

    b. Write a linear function representing the cost for the second plan for x text messages.

    c. Find the number of text messages for which the cost of either plan would be the same.

Linear Inequalities and Systems of Linear Inequalities in Two Variables

1. Graphing Linear Inequalities in Two Variables

A **linear inequality in two variables** x and y is an inequality that can be written in one of the following forms: $Ax + By < C$, $Ax + By > C$, $Ax + By \le C$, or $Ax + By \ge C$, provided A and B are not both zero.

A solution to a linear inequality in two variables is an ordered pair that makes the inequality true. For example, solutions to the inequality $x + y < 6$ are ordered pairs (x, y) such that the sum of the x- and y-coordinates is less than 6. This inequality has an infinite number of solutions, and therefore it is convenient to express the solution set as a graph.

To graph a linear inequality in two variables, we will follow these steps.

Concepts

1. Graphing Linear Inequalities in Two Variables
2. Systems of Linear Inequalities in Two Variables
3. Graphing a Feasible Region

Graphing a Linear Inequality in Two Variables

Step 1 Write the inequality with the y variable isolated, if possible.

Step 2 Graph the related equation. Draw a dashed line if the inequality is strict, $<$ or $>$. Otherwise, draw a solid line.

Step 3 Shade above or below the line as follows:

- Shade *above* the line if the inequality is of the form $y > mx + b$ or $y \ge mx + b$.
- Shade *below* the line if the inequality is of the form $y < mx + b$ or $y \le mx + b$.

Concept Connections

1. Determine whether each ordered pair is a solution to the inequality $x + y > 6$.
   a. $(5, 3)$
   b. $(1, 2)$

This process is demonstrated in Example 1.

Example 1 Graphing a Linear Inequality in Two Variables

Graph the solution set. $-3x + y \le 1$

Solution:

$-3x + y \le 1$

$\qquad y \le 3x + 1$ \qquad\qquad Solve for y.

Next graph the line defined by the related equation $y = 3x + 1$.

Because the inequality is of the form $y \le mx + b$, the solution to the inequality is the set of points in the region *below* the line $y = 3x + 1$. To indicate the solution set, shade the region below the line. (See Figure 4-9.)

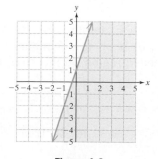

Figure 4-9

Skill Practice

Graph the solution set.
2. $2x + y \ge -4$

Answers
1a. Yes   **b.** No
2.

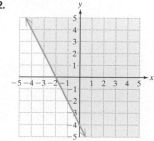

After graphing the solution to a linear inequality, we can verify that we have shaded the correct side of the line by using test points. In Example 1, we can pick an arbitrary ordered pair within the shaded region. Then substitute the x- and y-coordinates in the original inequality. If the result is a true statement, then that ordered pair is a solution to the inequality and suggests that other points from the same region are also solutions.

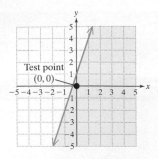

Figure 4-10

Skill Practice

Graph the solution set.

3. $-3y < x$

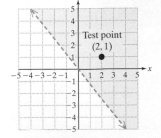

Figure 4-12

Answer

3.

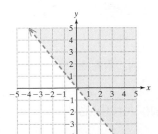

For example, the point $(0, 0)$ lies within the shaded region (Figure 4-10).

$$-3x + y \leq 1 \qquad \text{Substitute } (0, 0) \text{ in the original inequality.}$$

$$-3(0) + (0) \overset{?}{\leq} 1$$

$$0 + 0 \overset{?}{\leq} 1 \quad ✔ \text{ True} \qquad \text{The point } (0, 0) \text{ from the shaded region is a solution.}$$

In Example 2, we will graph the solution set to a strict inequality. A strict inequality does not include equality and therefore, is expressed with the symbols $<$ or $>$. In such a case, the boundary line will be drawn as a dashed line. This indicates that the boundary itself is *not* part of the solution set.

Example 2 **Graphing a Linear Inequality in Two Variables**

Graph the solution set. $-4y < 5x$

Solution:

$$-4y < 5x$$

$$\frac{-4y}{-4} > \frac{5x}{-4} \qquad \text{Solve for } y. \text{ Reverse the inequality sign.}$$

$$y > -\frac{5}{4}x$$

Graph the line defined by the related equation, $y = -\frac{5}{4}x$. The boundary line is drawn as a dashed line because the inequality is strict. Also note that the line passes through the origin.

Because the inequality is of the form $y > mx + b$, the solution to the inequality is the set of points in the region *above* the line. (See Figure 4-11.)

Figure 4-11

In Example 2, we cannot use the origin as a test point, because the point $(0, 0)$ is on the boundary line. Be sure to select a test point strictly within the shaded region. In this case, we choose $(2, 1)$. (See Figure 4-12.)

$$-4y < 5x$$

$$-4(1) \overset{?}{<} 5(2) \qquad \text{Substitute } (2, 1) \text{ in the original inequality.}$$

$$-4 \overset{?}{<} 10 \quad ✔ \text{ True} \qquad \text{The point } (2, 1) \text{ from the shaded region is a solution to the original inequality.}$$

In Example 3, we encounter a situation in which we cannot solve for the y variable.

| Example 3 | **Graphing a Linear Inequality in Two Variables**

Graph the solution set. $4x \geq -12$

Solution:

$4x \geq -12$

$x \geq -3$

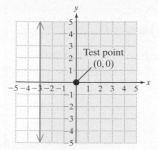

Figure 4-13

In this inequality, there is no y variable. However, we can simplify the inequality by solving for x.

Graph the related equation $x = -3$. This is a vertical line. The boundary is drawn as a solid line, because the inequality is not strict, $\geq$.

To shade the appropriate region, refer to the inequality, $x \geq -3$. The points for which x is greater than -3 are to the right of $x = -3$. Therefore, shade the region to the *right* of the line (Figure 4-13).

Selecting a test point such as $(0, 0)$ from the shaded region indicates that we have shaded the correct side of the line.

$$4x \geq -12 \qquad \text{Substitute } x = 0.$$
$$4(0) \overset{?}{\geq} -12 \checkmark \qquad \text{True}$$

Skill Practice

Graph the solution set.

4. $-2x \geq 2$

2. Systems of Linear Inequalities in Two Variables

Some applications require us to find the solutions to a system of linear inequalities.

| Example 4 | **Graphing a System of Linear Inequalities**

Graph the solution set.

$$y > \tfrac{1}{2}x + 1$$
$$x + y < 1$$

Solution:

Solve each inequality for y.

<u>First inequality</u>

$$y > \tfrac{1}{2}x + 1$$

The inequality $y > \tfrac{1}{2}x + 1$ is of the form $y > mx + b$. Graph *above* the boundary line. (See Figure 4-14.)

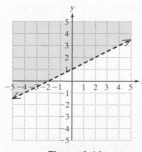

Figure 4-14

Skill Practice

Graph the solution set.

5. $x - 3y > 3$
  $y < -2x + 4$

Answers

4.

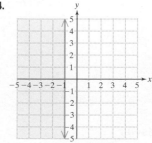

5.

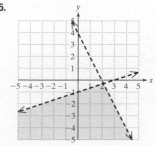

Second inequality

$x + y < 1$

$\qquad y < -x + 1$

The inequality $y < -x + 1$ is of the form $y < mx + b$. Graph *below* the boundary line. (See Figure 4-15.)

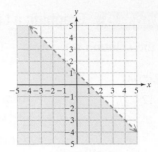

Figure 4-15

The region bounded by the inequalities is the region above the line $y = \frac{1}{2}x + 1$ and below the line $y = -x + 1$. This is the intersection or "overlap" of the two shaded regions (shown in purple in Figure 4-16).

The intersection is the solution set to the system of inequalities. (See Figure 4-17.)

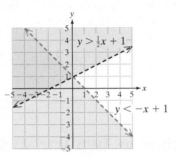

Figure 4-16

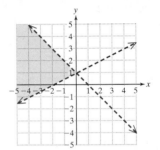

Figure 4-17

Example 5 Graphing a System of Linear Inequalities

Graph the solution set.

$$3y \leq 6$$
$$y - x \leq 0$$

Solution:

First inequality

$3y \leq 6$

$\qquad y \leq 2$

The graph of $y \leq 2$ is the region on and below the horizontal line $y = 2$. (See Figure 4-18.)

Second inequality

$y - x \leq 0$

$\qquad y \leq x$

The inequality $y \leq x$ is of the form $y \leq mx + b$. Graph a solid line and shade the region below the line. (See Figure 4-19.)

Skill Practice

Graph the solution set.

6. $2y \leq 4$
$\qquad y \leq x + 1$

Answer

6.

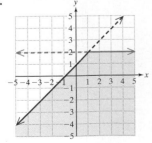

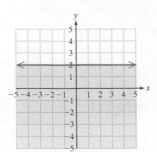

Figure 4-18

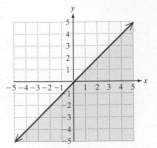

Figure 4-19

The solution to the system of inequalities is the intersection of the shaded regions. Notice that the portions of the lines not bounding the solution are dashed. (See Figure 4-20.)

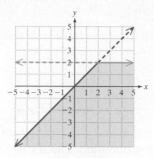

Figure 4-20

Example 6   Graphing a System of Linear Inequalities

Describe the region of the plane defined by the system of inequalities.

$$x \leq 0$$

$$y \geq 0$$

Solution:

$x \leq 0$ $x \leq 0$ for points on the y-axis and in the second and third quadrants.

$y \geq 0$ $y \geq 0$ for points on the x-axis and in the first and second quadrants.

The intersection of these regions is the set of points in the second quadrant (with the boundaries included).

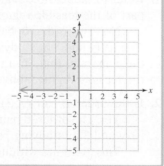

Skill Practice

Graph the region defined by the system of inequalities.

7. $x \leq 0$
     $y \leq 0$

---

3. Graphing a Feasible Region

When two variables are related under certain constraints, a system of linear inequalities can be used to show a region of feasible values for the variables.

Example 7   Graphing a Feasible Region

Susan has two tests on Friday: one in chemistry and one in psychology. Because the two classes meet in consecutive hours, she has no study time between tests. Susan estimates that she has a maximum of 12 hr of study time before the tests, and she must divide her time between chemistry and psychology.

Let x represent the number of hours Susan spends studying chemistry.

Let y represent the number of hours Susan spends studying psychology.

  a. Find a set of inequalities to describe the constraints on Susan's study time.

  b. Graph the constraints to find the feasible region defining Susan's study time.

Answer

7.

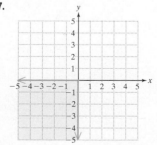

8. A local pet rescue group has a total of 30 cages that can be used to hold cats and dogs. Let x represent the number of cages used for cats, and let y represent the number used for dogs.

   a. Write a set of inequalities to express the fact that the number of cat and dog cages cannot be negative.

   b. Write an inequality to describe the constraint on the total number of cages for cats and dogs.

   c. Graph the system of inequalities to find the feasible region describing the available cages.

Solution:

a. Because Susan cannot study chemistry or psychology for a negative period of time, we have $x \geq 0$ and $y \geq 0$. Furthermore, her total time studying cannot exceed 12 hr: $x + y \leq 12$.

A system of inequalities that defines the constraints on Susan's study time is:

$$x \geq 0$$
$$y \geq 0$$
$$x + y \leq 12$$

b. The first two conditions $x \geq 0$ and $y \geq 0$ represent the set of points in the first quadrant. The third condition $x + y \leq 12$ represents the set of points below and including the line $x + y = 12$ (Figure 4-21).

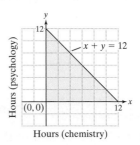

Figure 4-21

Discussion:

1. Refer to the feasible region drawn in Example 7(b). Is the ordered pair $(8, 5)$ part of the feasible region?

   No. The ordered pair $(8, 5)$ indicates that Susan spent 8 hr studying chemistry and 5 hr studying psychology. This is a total of 13 hr, which exceeds the constraint that Susan only had 12 hr to study. The point $(8, 5)$ lies outside the feasible region, above the line $x + y = 12$ (Figure 4-22).

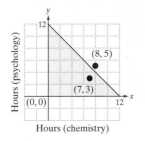

Figure 4-22

2. Is the ordered pair $(7, 3)$ part of the feasible region?

   Yes. The ordered pair $(7, 3)$ indicates that Susan spent 7 hr studying chemistry and 3 hr studying psychology.

   This point lies within the feasible region and satisfies all three constraints.

$$x \geq 0 \longrightarrow \qquad 7 \geq 0 \qquad \text{True}$$
$$y \geq 0 \longrightarrow \qquad 3 \geq 0 \qquad \text{True}$$
$$x + y \leq 12 \longrightarrow (7) + (3) \leq 12 \qquad \text{True}$$

   Notice that the ordered pair $(7, 3)$ corresponds to a point where Susan is not making full use of the 12 hr of study time.

3. Suppose there was one additional constraint imposed on Susan's study time. She knows she needs to spend at least twice as much time studying chemistry as she does studying psychology. Graph the feasible region with this additional constraint.

   Because the time studying chemistry must be at least twice the time studying psychology, we have $x \geq 2y$.

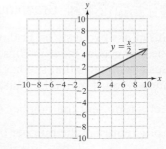

This inequality may also be written as $y \leq \dfrac{x}{2}$.

Figure 4-23

Figure 4-23 shows the first quadrant with the constraint $y \leq \dfrac{x}{2}$.

Answer

8a. $x \geq 0$ and $y \geq 0$
b. $x + y \leq 30$
c.

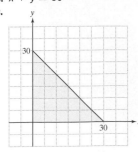

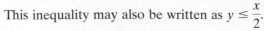

4. At what point in the feasible region is Susan making the most efficient use of her time for both classes?

First and foremost, Susan must make use of *all* 12 hr. This occurs for points along the line $x + y = 12$. Susan will also want to study for both classes with approximately twice as much time devoted to chemistry. Therefore, Susan will be deriving the maximum benefit at the point of intersection of the line $x + y = 12$ and the line $y = \dfrac{x}{2}$.

Using the substitution method, replace $y = \dfrac{x}{2}$ into the equation $x + y = 12$.

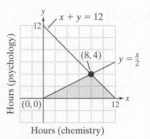

$$x + \dfrac{x}{2} = 12$$

$2x + x = 24$ Clear fractions.

$3x = 24$

$x = 8$ Solve for x.

$y = \dfrac{(8)}{2}$ To solve for y, substitute $x = 8$ into the equation $y = \dfrac{x}{2}$.

$y = 4$

Therefore, Susan should spend 8 hr studying chemistry and 4 hr studying psychology.

Section 4.5 Practice Exercises

Vocabulary and Key Concepts

1. **a.** An inequality that can be written in the form $Ax + By > C$ is called a _____ inequality in two variables.

 b. Given the inequality $3x - 4y < 6$, the boundary line $3x - 4y = 6$ (is/is not) included in the solution set. However, given $3x - 4y \leq 6$, then the boundary line (is/is not) included in the solution set.

 c. Given the inequality $5x + y > 5$, the boundary line $5x + y = 5$ should be drawn as a (solid/dashed) line to indicate that the line (is/is not) part of the solution set.

 d. Given the inequality $y \geq -x + 4$, the boundary line $y = -x + 4$ should be drawn as a (solid/dashed) line to indicate that the line (is/is not) part of the solution set.

Review Exercises

For Exercises 2–5, solve the inequalities.

2. $5 < x + 1$ and $-2x + 6 \geq -6$

3. $5 - x \leq 4$ and $6 > 3x - 3$

4. $4 - y < 3y + 12$ or $-2(y + 3) \geq 12$

5. $-2x < 4$ or $3x - 1 \leq -13$

Concept 1: Graphing Linear Inequalities in Two Variables

For Exercises 6–9, determine if the given point is a solution to the inequality.

6. $2x - y > 8$

   a. $(3, -5)$ **c.** $(4, -2)$

   b. $(-1, -10)$ **d.** $(0, 0)$

7. $3y + x < 5$

   a. $(-1, 7)$ **c.** $(0, 0)$

   b. $(5, 0)$ **d.** $(2, -3)$

8. $y \leq -2$

   a. $(5, -3)$ **c.** $(0, 0)$

   b. $(-4, -2)$ **d.** $(3, 2)$

9. $x \geq 5$

   a. $(4, 5)$ **c.** $(8, 8)$

   b. $(5, -1)$ **d.** $(0, 0)$

10. When should you use a dashed line to graph the solution to a linear inequality?

For Exercises 11–16, decide which inequality symbol should be used $(<, >, \geq, \leq)$ by looking at the graph.

11.

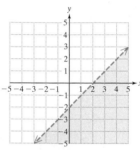

$x - y \underline{\hspace{1cm}} 2$

12.

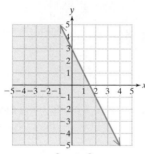

$y \underline{\hspace{1cm}} -2x + 3$

13.

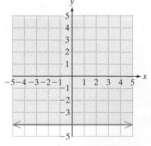

$y \underline{\hspace{1cm}} -4$

14.

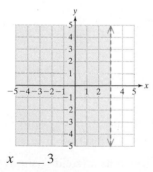

$x \underline{\hspace{1cm}} 3$

15.

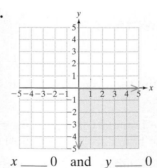

$x \underline{\hspace{1cm}} 0 \quad$ and $\quad y \underline{\hspace{1cm}} 0$

16.

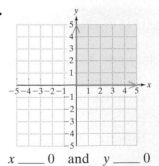

$x \underline{\hspace{1cm}} 0 \quad$ and $\quad y \underline{\hspace{1cm}} 0$

For Exercises 17–40, graph the solution set. **(See Examples 1–3.)**

17. $x - 2y > 4$

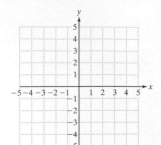

18. $x - 3y \geq 6$

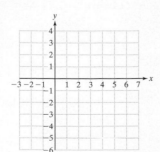

19. $5x - 2y < 10$

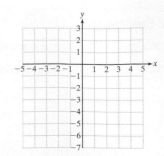

20. $x - 3y < 8$

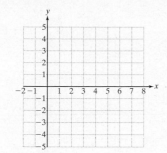

21. $2x \le -6y + 12$

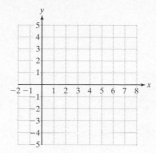

22. $4x < 3y + 12$

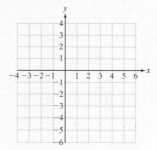

23. $2y \le 4x$

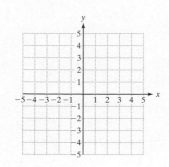

24. $-6x < 2y$

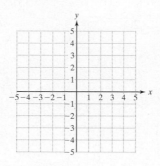

25. $y \ge -2$

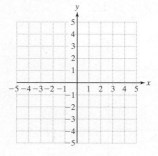

26. $y \ge 5$

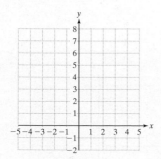

27. $4x < 5$

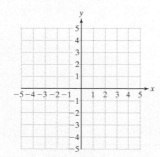

28. $x + 6 < 7$

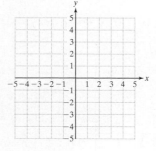

29. $y \ge \dfrac{2}{5}x - 4$

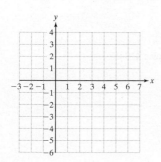

30. $y \ge -\dfrac{5}{2}x - 4$

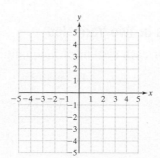

31. $y \le \dfrac{1}{3}x + 6$

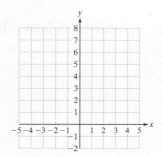

32. $y \leq -\dfrac{1}{4}x + 2$

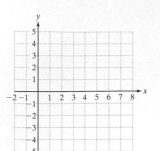

33. $y - 5x > 0$

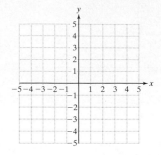

34. $y - \dfrac{1}{2}x > 0$

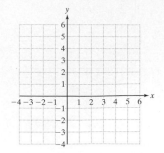

35. $\dfrac{x}{5} + \dfrac{y}{4} < 1$

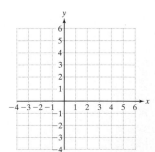

36. $x + \dfrac{y}{2} \geq 2$

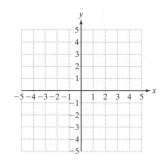

37. $0.1x + 0.2y \leq 0.6$

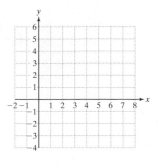

38. $0.3x - 0.2y < 0.6$

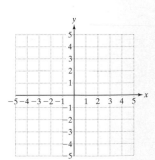

39. $x \leq -\dfrac{2}{3}y$

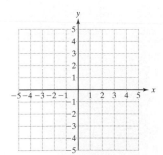

40. $x \geq -\dfrac{5}{4}y$

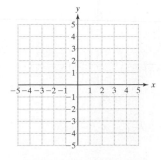

Concept 2: Systems of Linear Inequalities in Two Variables

For Exercises 41–56, graph the solution set of each system of inequalities. **(See Examples 4–6.)**

41. $y < 4$
$\quad\;\; y > -x + 2$

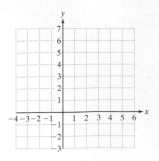

42. $y < 3$
$\quad\;\; x + 2y < 6$

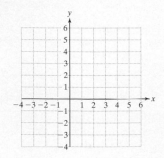

43. $2x + y \leq 5$

  $x \geq 3$

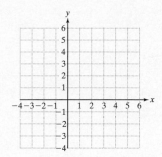

44. $x + 3y \geq 3$

  $x \leq -2$

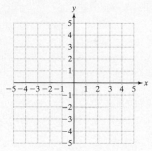

45. $x + y < 3$

  $4x + y < 6$

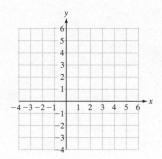

46. $x + y < 4$

  $3x + y < 9$

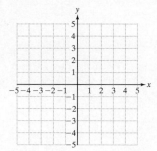

47. $2x - y \leq 2$

  $2x + 3y \geq 6$

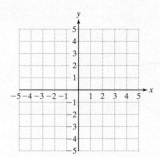

48. $3x + 2y \geq 4$

  $x - y \leq 3$

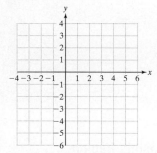

49. $x > 4$

  $y < 2$

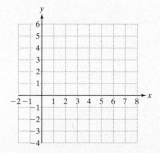

50. $x < 3$

  $y > 4$

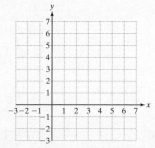

51. $x \le -2$
$y \le 0$

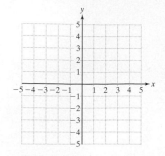

52. $x \ge 0$
$y \ge -3$

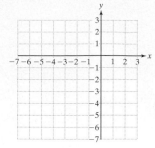

53. $x > 0$
$x + y < 6$

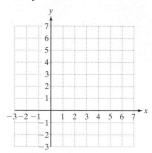

54. $x < 0$
$x + y < 2$

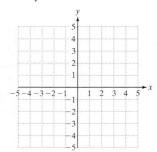

55. $y \ge 0$
$-2x + y \le -4$

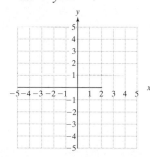

56. $y \ge 0$
$x - y \ge -3$

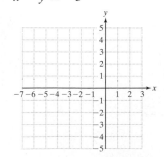

Concept 3: Graphing a Feasible Region

For Exercises 57–62, graph the feasible regions.

57. $x + y \le 3$
$x \ge 0$
$y \ge 0$

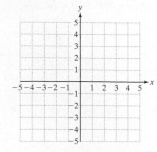

58. $x - y \le 2$
$x \ge 0$
$y \ge 0$

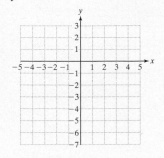

59. $y < \dfrac{1}{2}x - 3$
$x \le 0$
$y \ge -5$

60. $y > \dfrac{1}{2}x - 3$

$x \geq -2$

$y \leq 0$

61. $x \geq 0, y \geq 0$

$x + y \leq 8$

$3x + 5y \leq 30$

62. $x \geq 0, y \geq 0$

$x + y \leq 5$

$x + 2y \leq 6$

63. Suppose Sue has 50 ft of fencing with which she can build a rectangular dog run. Let x represent the length of the dog run and let y represent the width.

   a. Write an inequality representing the fact that the total perimeter of the dog run is at most 50 ft.

   b. Sketch part of the solution set for this inequality that represents all possible values for the length and width of the dog run. (*Hint:* Note that both the length and the width must be positive.)

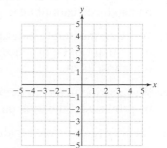

64. Suppose Rick has 40 ft of fencing with which he can build a rectangular garden. Let x represent the length of the garden and let y represent the width.

   a. Write an inequality representing the fact that the total perimeter of the garden is at most 40 ft.

   b. Sketch part of the solution set for this inequality that represents all possible values for the length and width of the garden. (*Hint:* Note that both the length and the width must be positive.)

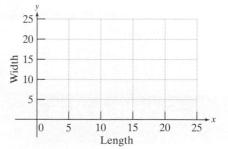

65. In scheduling two drivers for delivering pizza, James needs to have at least 65 hr scheduled this week. His two drivers, Karen and Todd, are not allowed to get overtime, so each one can work at most 40 hr. Let x represent the number of hours that Karen can be scheduled, and let y represent the number of hours Todd can be scheduled. **(See Example 7.)**

   a. Write two inequalities that express the fact that Karen and Todd cannot work a negative number of hours.

   b. Write two inequalities that express the fact that neither Karen nor Todd is allowed overtime (i.e., each driver can have at most 40 hr).

   c. Write an inequality that expresses the fact that the total number of hours from both Karen and Todd needs to be at least 65 hr.

   d. Graph the feasible region formed by graphing the inequalities.

   e. Is the point $(35, 40)$ in the feasible region? What does the point $(35, 40)$ represent in the context of this problem?

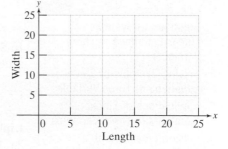

   f. Is the point $(20, 40)$ in the feasible region? What does the point $(20, 40)$ represent in the context of this problem?

66. A manufacturer produces two models of desks. Model A requires 4 hr to stain and finish and 3 hr to assemble. Model B requires 3 hr to stain and finish and 1 hr to assemble. The total amount of time available for staining and finishing is 24 hr and for assembling is 12 hr. Let x represent the number of Model A desks produced, and let y represent the number of Model B desks produced.

   a. Write two inequalities that express the fact that the number of desks to be produced cannot be negative.

   b. Write an inequality in terms of the number of Model A and Model B desks that can be produced if the total time for staining and finishing is at most 24 hr.

   c. Write an inequality in terms of the number of Model A and Model B desks that can be produced if the total time for assembly is no more than 12 hr.

   d. Identify the feasible region formed by graphing the preceding inequalities.

   e. Is the point $(3, 1)$ in the feasible region? What does the point $(3, 1)$ represent in the context of this problem?

   f. Is the point $(5, 4)$ in the feasible region? What does the point $(5, 4)$ represent in the context of this problem?

Section 4.6	Systems of Linear Equations in Three Variables and Applications

Concepts

1. Solutions to Systems of Linear Equations in Three Variables
2. Solving Systems of Linear Equations in Three Variables
3. Applications of Linear Equations in Three Variables
4. Solving Inconsistent Systems and Systems of Dependent Equations

1. Solutions to Systems of Linear Equations in Three Variables

In Sections 4.1–4.3, we solved systems of linear equations in two variables. In this section, we will expand the discussion to solving systems involving three variables.

A **linear equation in three variables** can be written in the form $Ax + By + Cz = D$, where A, B, and C are not all zero. For example, the equation $2x + 3y + z = 6$ is a linear equation in three variables. Solutions to this equation are **ordered triples** of the form (x, y, z) that satisfy the equation. Some solutions to the equation $2x + 3y + z = 6$ are

Solution:	Check:
$(1, 1, 1) \longrightarrow$	$2(1) + 3(1) + (1) \overset{?}{=} 6$ ✔ True
$(2, 0, 2) \longrightarrow$	$2(2) + 3(0) + (2) \overset{?}{=} 6$ ✔ True
$(0, 1, 3) \longrightarrow$	$2(0) + 3(1) + (3) \overset{?}{=} 6$ ✔ True

Infinitely many ordered triples serve as solutions to the equation $2x + 3y + z = 6$.

The set of all ordered triples that are solutions to a linear equation in three variables may be represented graphically by a plane in space. Figure 4-24 shows a portion of the plane $2x + 3y + z = 6$ in a 3-dimensional coordinate system.

An example of a system of three linear equations in three variables is shown here.

$$2x + y - 3z = -7$$
$$3x - 2y + z = 11$$
$$-2x - 3y - 2z = 3$$

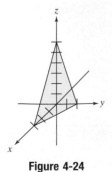

Figure 4-24

A solution to a system of linear equations in three variables is an ordered triple that satisfies *each* equation. Geometrically, a solution is a point of intersection of the planes represented by the equations in the system.

A system of linear equations in three variables may have *one unique solution, infinitely many solutions,* or *no solution* (Table 4-2, Table 4-3, and Table 4-4).

Table 4-2

One unique solution (planes intersect at one point)

- The system is consistent.
- The equations are independent.

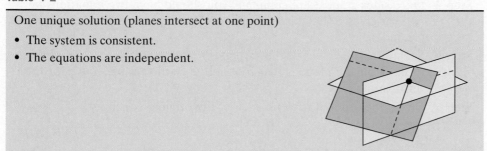

Table 4-3

No solution (the three planes do not all intersect)

- The system is inconsistent.
- The equations are independent.

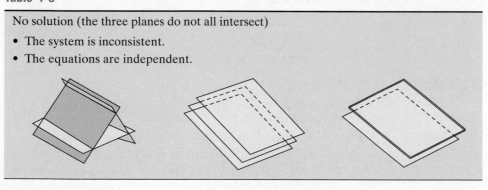

Table 4-4

Infinitely many solutions (planes intersect at infinitely many points)

- The system is consistent.
- The equations are dependent.

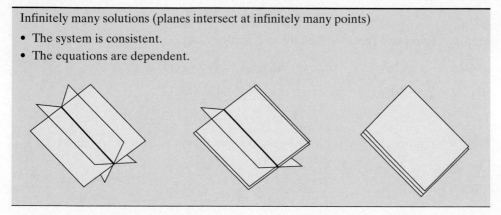

2. Solving Systems of Linear Equations in Three Variables

To solve a system involving three variables, the goal is to eliminate one variable. This reduces the system to two equations in two variables. One strategy for eliminating a variable is to pair up the original equations two at a time.

> ## Solving a System of Three Linear Equations in Three Variables
> **Step 1** Write each equation in standard form $Ax + By + Cz = D$.
> **Step 2** Choose a pair of equations, and eliminate one of the variables by using the addition method.
> **Step 3** Choose a different pair of equations and eliminate the *same* variable.
> **Step 4** Once steps 2 and 3 are complete, you should have two equations in two variables. Solve this system by using the methods from Sections 4.2 and 4.3.
> **Step 5** Substitute the values of the variables found in step 4 into any of the three original equations that contain the third variable. Solve for the third variable.
> **Step 6** Check the ordered triple in each of the original equations.

Skill Practice

1. Solve the system.
$$x + 2y + z = 1$$
$$3x - y + 2z = 13$$
$$2x + 3y - z = -8$$

Example 1 **Solving a System of Linear Equations in Three Variables**

Solve the system.
$$2x + y - 3z = -7$$
$$3x - 2y + z = 11$$
$$-2x - 3y - 2z = 3$$

Solution:

$\boxed{A}$ $2x + y - 3z = -7$ **Step 1:** The equations are already in standard form.

$\boxed{B}$ $3x - 2y + z = 11$ • It is often helpful to label the equations.

$\boxed{C}$ $-2x - 3y - 2z = 3$ • The y variable can be easily eliminated from equations $\boxed{A}$ and $\boxed{B}$ and from equations $\boxed{A}$ and $\boxed{C}$. This is accomplished by creating opposite coefficients for the y terms and then adding the equations.

Step 2: Eliminate the y variable from equations $\boxed{A}$ and $\boxed{B}$.

$\boxed{A}$ $2x + y - 3z = -7$ $\xrightarrow{\text{Multiply by 2.}}$ $4x + 2y - 6z = -14$

$\boxed{B}$ $3x - 2y + z = 11$ $\xrightarrow{\hspace{1cm}}$ $\underline{3x - 2y + z = 11}$

$$7x \qquad - 5z = -3 \;\boxed{D}$$

TIP: It is important to note that in steps 2 and 3, the *same* variable is eliminated.

Step 3: Eliminate the y variable again, this time from equations $\boxed{A}$ and $\boxed{C}$.

$\boxed{A}$ $2x + y - 3z = -7$ $\xrightarrow{\text{Multiply by 3.}}$ $6x + 3y - 9z = -21$

$\boxed{C}$ $-2x - 3y - 2z = 3$ $\xrightarrow{\hspace{1cm}}$ $\underline{-2x - 3y - 2z = 3}$

$$4x \qquad - 11z = -18 \;\boxed{E}$$

Step 4: Now equations $\boxed{D}$ and $\boxed{E}$ can be paired up to form a linear system in two variables. Solve this system.

$\boxed{D}$ $7x - 5z = -3$ $\xrightarrow{\text{Multiply by } -4.}$ $-28x + 20z = 12$

$\boxed{E}$ $4x - 11z = -18$ $\xrightarrow{\text{Multiply by 7.}}$ $\underline{28x - 77z = -126}$

$$-57z = -114$$

$$z = 2$$

Answer

1. $(1, -2, 4)$

Once one variable has been found, substitute this value into either equation in the two-variable system, that is, either equation $\boxed{D}$ or $\boxed{E}$.

$\boxed{D}$ $7x - 5z = -3$

$7x - 5(2) = -3$ Substitute $z = 2$ into equation $\boxed{D}$.

$7x - 10 = -3$

$7x = 7$

$x = 1$

$\boxed{A}$ $2x + y - 3z = -7$

$2(1) + y - 3(2) = -7$

$2 + y - 6 = -7$

$y - 4 = -7$

$y = -3$

Step 5: Now that two variables are known, substitute these values (x and z) into any of the original three equations to find the remaining variable y.

Substitute $x = 1$ and $z = 2$ into equation $\boxed{A}$.

The solution is $(1, -3, 2)$.

Step 6: Check the ordered triple in the three original equations.

Check: $2x + y - 3z = -7 \rightarrow 2(1) + (-3) - 3(2) \overset{?}{=} -7$ ✔ True

$3x - 2y + z = 11 \rightarrow 3(1) - 2(-3) + (2) \overset{?}{=} 11$ ✔ True

$-2x - 3y - 2z = 3 \rightarrow -2(1) - 3(-3) - 2(2) \overset{?}{=} 3$ ✔ True

3. Applications of Linear Equations in Three Variables

Example 2 Applying Systems of Linear Equations in Three Variables

In a triangle, the smallest angle measures $10°$ more than one-half the measure of the largest angle. The middle angle measures $12°$ more than the measure of the smallest angle. Find the measure of each angle.

Solution:

Let x represent the measure of the smallest angle.

Let y represent the measure of the middle angle.

Let z represent the measure of the largest angle.

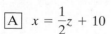

To solve for three variables, we need to establish three independent relationships among x, y, and z.

$\boxed{A}$ $x = \dfrac{1}{2}z + 10$

The smallest angle measures $10°$ more than one-half the measure of the largest angle.

$\boxed{B}$ $y = x + 12$

The middle angle measures $12°$ more than the measure of the smallest angle.

$\boxed{C}$ $x + y + z = 180$

The sum of the angles inscribed in a triangle is $180°$.

Clear fractions and write each equation in standard form.

Skill Practice

2. The perimeter of a triangle is 30 in. The shortest side is 4 in. shorter than the longest side. The longest side is 6 in. less than the sum of the other two sides. Find the length of each side.

Answer

2. 8 in., 10 in., and 12 in.

$$\boxed{A}\quad x \qquad\qquad = \frac{1}{2}z + 10 \xrightarrow{\text{Multiply by 2.}} 2x = z + 20 \longrightarrow \begin{array}{r} \textbf{Standard Form} \\ 2x \qquad - z = \quad 20 \end{array}$$

$$\boxed{B}\quad\quad y \;=\; x + 12 \longrightarrow\qquad\qquad -x + y \qquad = \;\; 12$$

$$\boxed{C}\quad x + y + z = \qquad 180 \longrightarrow\qquad\qquad x + y + z = 180$$

Notice equation $\boxed{B}$ is missing the z-variable. Therefore, we can eliminate z again by pairing up equations $\boxed{A}$ and $\boxed{C}$.

$$\boxed{A}\;\; 2x \qquad - z = \;\; 20$$
$$\boxed{C}\;\; \underline{x + y + z = 180}$$
$$\quad 3x + y \qquad = 200 \quad \boxed{D}$$

$$\boxed{B}\;\; -x + y = 12 \xrightarrow{\text{Multiply by }-1.} x - y = -12 \quad \text{Pair up equations } \boxed{B} \text{ and}$$
$$\boxed{D}\;\; 3x + y = 200 \longrightarrow \underline{3x + y = \;\; 200} \quad \boxed{D} \text{ to form a system of two}$$
$$\qquad\qquad\qquad\qquad\qquad\qquad 4x \quad = \;\; 188 \quad \text{variables.}$$
$$\qquad\qquad\qquad\qquad\qquad\qquad\qquad x = 47 \quad \text{Solve for } x.$$

From equation $\boxed{B}$ we have $-x + y = 12 \longrightarrow -47 + y = 12 \rightarrow y = 59$

From equation $\boxed{C}$ we have $x + y + z = 180 \rightarrow 47 + 59 + z = 180 \rightarrow z = 74$

The smallest angle measures $47°$, the middle angle measures $59°$, and the largest angle measures $74°$.

Example 3 **Applying Systems of Linear Equations to Nutrition**

Doctors have become increasingly concerned about the sodium intake in the U.S. diet. Recommendations by the American Medical Association indicate that most individuals should not exceed 2400 mg of sodium per day.

Torie ate 1 slice of pizza, 1 serving of ice cream, and 1 glass of soda for a total of 1030 mg of sodium. David ate 3 slices of pizza, no ice cream, and 2 glasses of soda for a total of 2420 mg of sodium. Emilie ate 2 slices of pizza, 1 serving of ice cream, and 2 glasses of soda for a total of 1910 mg of sodium. How much sodium is in one serving of each item?

Solution:

Let x represent the sodium content of 1 slice of pizza.
Let y represent the sodium content of 1 serving of ice cream.
Let z represent the sodium content of 1 glass of soda.

From Torie's meal we have: $\boxed{A}\;\; x + y + z = 1030$

From David's meal we have: $\boxed{B}\;\; 3x \quad + 2z = 2420$

From Emilie's meal we have: $\boxed{C}\;\; 2x + y + 2z = 1910$

Equation $\boxed{B}$ is missing the y variable. Eliminating y from equations $\boxed{A}$ and $\boxed{C}$, we have

$$\boxed{A}\;\; x + y + z = 1030 \xrightarrow{\text{Multiply by }-1.} -x - y - z = -1030$$
$$\boxed{C}\;\; 2x + y + 2z = 1910 \longrightarrow \underline{2x + y + 2z = \;\; 1910}$$
$$\boxed{D}\;\; x \qquad + z = \qquad 880$$

Solve the system formed by equations $\boxed{B}$ and $\boxed{D}$.

$\boxed{B}$ $3x + 2z = 2420$ $\longrightarrow$ $3x + 2z = 2420$

$\boxed{D}$ $x + z = 880$ $\longrightarrow$ $-2x - 2z = -1760$

$\phantom{\boxed{D}}$ $\underset{\text{Multiply by } -2.}{}$ $\overline{x = 660}$

From equation $\boxed{D}$ we have $x + z = 880 \longrightarrow 660 + z = 880 \longrightarrow z = 220$

From equation $\boxed{A}$ we have $x + y + z = 1030 \rightarrow 660 + y + 220 = 1030 \rightarrow y = 150$

Therefore, 1 slice of pizza has 660 mg of sodium, 1 serving of ice cream has 150 mg of sodium, and 1 glass of soda has 220 mg of sodium.

4. Solving Inconsistent Systems and Systems of Dependent Equations

| Example 4 | Solving a Dependent System of Linear Equations

Solve the system. If the system does not have one unique solution, state the number of solutions and whether the system is inconsistent or the equations are dependent.

$$\boxed{A} \quad 3x + y - z = 8$$
$$\boxed{B} \quad 2x - y + 2z = 3$$
$$\boxed{C} \quad x + 2y - 3z = 5$$

Solution:

The first step is to make a decision regarding the variable to eliminate. The y variable is particularly easy to eliminate because the coefficients of y in equations $\boxed{A}$ and $\boxed{B}$ are already opposites. The y variable can be eliminated from equations $\boxed{B}$ and $\boxed{C}$ by multiplying equation $\boxed{B}$ by 2.

$\boxed{A}$ $3x + y - z = 8$

$\boxed{B}$ $\underline{2x - y + 2z = 3}$

$\phantom{\boxed{B}}$ $5x + z = 11$ $\boxed{D}$

Pair up equations $\boxed{A}$ and $\boxed{B}$ to eliminate y.

$\boxed{B}$ $2x - y + 2z = 3$ $\xrightarrow{\text{Multiply by 2.}}$ $4x - 2y + 4z = 6$

$\boxed{C}$ $x + 2y - 3z = 5$ $\longrightarrow$ $\underline{x + 2y - 3z = 5}$

$\phantom{\boxed{C}}$ $5x + z = 11$ $\boxed{E}$

Pair up equations $\boxed{B}$ and $\boxed{C}$ to eliminate y.

Because equations $\boxed{D}$ and $\boxed{E}$ are equivalent equations, it appears that this is a dependent system. By eliminating variables we obtain the identity $0 = 0$.

$\boxed{D}$ $5x + z = 11$ $\xrightarrow{\text{Multiply by } -1.}$ $-5x - z = -11$

$\boxed{E}$ $5x + z = 11$ $\longrightarrow$ $\underline{5x + z = 11}$

$\phantom{\boxed{E}}$ $0 = 0$

The result $0 = 0$ indicates that there are infinitely many solutions and that the equations are dependent.

Skill Practice

4. Solve the system. If the system does not have one unique solution, state the number of solutions and whether the system is inconsistent or the equations are dependent.

$$x + y + z = 8$$
$$2x - y + z = 6$$
$$-5x - 2y - 4z = -30$$

Answer

4. Infinitely many solutions; dependent equations

5. Solve the system. If the system does not have one unique solution, state the number of solutions and whether the system is inconsistent or the equations are dependent.

$$x - 2y + z = 5$$
$$x - 3y + 2z = -7$$
$$-2x + 4y - 2z = 6$$

Example 5 Solving an Inconsistent System of Linear Equations

Solve the system. If the system does not have one unique solution, state the number of solutions and whether the system is inconsistent or the equations are dependent.

$$2x + 3y - 7z = 4$$
$$-4x - 6y + 14z = 1$$
$$5x + y - 3z = 6$$

Solution:

We will eliminate the x variable from equations $\boxed{A}$ and $\boxed{B}$.

$\boxed{A}$ $\quad 2x + 3y - 7z = 4 \quad \xrightarrow{\text{Multiply by 2.}} \quad 4x + 6y - 14z = 8$

$\boxed{B}$ $\quad -4x - 6y + 14z = 1 \quad \xrightarrow{\hspace{2cm}} \quad \underline{-4x - 6y + 14z = 1}$

$$\hspace{6cm} 0 = 9 \quad \text{(contradiction)}$$

The result $0 = 9$ is a contradiction, indicating that the system has no solution. The system is inconsistent.

Answer

5. No solution; inconsistent system

Section 4.6 Practice Exercises

Study Skills Exercise

Look back over your notes for this chapter. Have you highlighted the important topics? Have you underlined the key terms? Have you indicated the places where you are having trouble? If you find that you have problems with a particular topic, write a question that you can ask your instructor either in class or in the office.

Vocabulary and Key Concepts

1. **a.** An equation written in the form $Ax + By + Cz = D$ where A, B, and C are not all zero is called a _____ equation in three variables.

   b. Solutions to a linear equation in three variables are of the form (x, y, z) and are called _____ _____.

Review Exercises

For Exercises 2–4, solve the systems by using two methods: **(a)** the substitution method and **(b)** the addition method.

2. $4y = x + 8$
   $5x = 7y + 12$

3. $3x + y = 4$
   $4x + y = 5$

4. $2x - 5y = 3$
   $-4x + 10y = 3$

5. Marge can ride her bike 24 mi in $1\frac{1}{3}$ hr riding with the wind. Riding against the wind she can ride 24 mi in 2 hr. Find the speed at which Marge can ride in still air and the speed of the wind.

Concept 1: Solutions to Systems of Linear Equations in Three Variables

6. How many solutions are possible when solving a system of three equations with three variables?

7. Which of the following points are solutions to the system? $(2, 1, 7), (3, -10, -6), (4, 0, 2)$

$$2x - y + z = 10$$
$$4x + 2y - 3z = 10$$
$$x - 3y + 2z = 8$$

8. Which of the following points are solutions to the system? $(1, 1, 3), (0, 0, 4), (4, 2, 1)$

$$-3x - 3y - 6z = -24$$
$$-9x - 6y + 3z = -45$$
$$9x + 3y - 9z = 33$$

9. Which of the following points are solutions to the system? $(12, 2, -2), (4, 2, 1), (1, 1, 1)$

$$-x - y - 4z = -6$$
$$x - 3y + z = -1$$
$$4x + y - z = 4$$

10. Which of the following points are solutions to the system? $(0, 4, 3), (3, 6, 10), (3, 3, 1)$

$$x + 2y - z = 5$$
$$x - 3y + z = -5$$
$$-2x + y - z = -4$$

Concept 2: Solving Systems of Linear Equations in Three Variables

For Exercises 11–22, solve the system of equations. **(See Example 1.)**

11.
$$2x + y - 3z = -12$$
$$3x - 2y - z = 3$$
$$-x + 5y + 2z = -3$$

12.
$$-3x - 2y + 4z = -15$$
$$2x + 5y - 3z = 3$$
$$4x - y + 7z = 15$$

13.
$$x - 3y - 4z = -7$$
$$5x + 2y + 2z = -1$$
$$4x - y - 5z = -6$$

14.
$$6x - 5y + z = 7$$
$$5x + 3y + 2z = 0$$
$$-2x + y - 3z = 11$$

15.
$$4x + 2z = 12 + 3y$$
$$2y = 3x + 3z - 5$$
$$y = 2x + 7z + 8$$

16.
$$y = 2x + z + 1$$
$$-3x - 1 = -2y + 2z$$
$$5x + 3z = 16 - 3y$$

17.
$$x + y + z = 6$$
$$-x + y - z = -2$$
$$2x + 3y + z = 11$$

18.
$$x - y - z = -11$$
$$x + y - z = 15$$
$$2x - y + z = -9$$

19.
$$2x - 3y + 2z = -1$$
$$x + 2y = -4$$
$$x + z = 1$$

20.
$$x + y + z = 2$$
$$2x - z = 5$$
$$3y + z = 2$$

21.
$$4x + 9y = 8$$
$$8x + 6z = -1$$
$$6y + 6z = -1$$

22.
$$3x + 2z = 11$$
$$y - 7z = 4$$
$$x - 6y = 1$$

Concept 3: Applications of Linear Equations in Three Variables

23. A triangle has one angle that measures 5° more than twice the smallest angle, and the third angle measures 11° less than 3 times the measure of the smallest angle. Find the measures of the three angles. **(See Example 2.)**

24. The largest angle of a triangle measures 4° less than 5 times the measure of the smallest angle. The middle angle measures twice that of the smallest angle. Find the measures of the three angles.

25. The perimeter of a triangle is 55 cm. The measure of the shortest side is 8 cm less than the middle side. The measure of the longest side is 1 cm less than the sum of the other two sides. Find the lengths of the sides.

26. The perimeter of a triangle is 5 ft. The longest side of the triangle measures 20 in. more than the shortest side. The middle side is 3 times the measure of the shortest side. Find the lengths of the three sides in *inches*.

27. Sean kept track of his fiber intake from three sources for 3 weeks. The first week he had 3 servings of a fiber supplement, 1 serving of oatmeal, and 4 servings of cereal, which totaled 19 g of fiber. The second week he had 2 servings of the fiber supplement, 4 servings of oatmeal, and 2 servings of cereal totaling 25 g. The third week he had 5 servings of the fiber supplement, 3 servings of oatmeal, and 2 servings of cereal for a total of 30 g. Find the amount of fiber in one serving of each of the following: the fiber supplement, the oatmeal, and the cereal. **(See Example 3.)**

28. Natalie kept track of her calcium intake from three sources for 3 days. The first day she had 1 glass of milk, 1 serving of ice cream, and 1 calcium supplement in pill form which totaled 1180 mg of calcium. The second day she had 2 glasses of milk, 1 serving of ice cream, and 1 calcium supplement totaling 1680 mg. The third day she had 1 glass of milk, 2 servings of ice cream, and 1 calcium supplement for a total of 1260 mg. Find the amount of calcium in one glass of milk, in one serving of ice cream, and in one calcium supplement.

29. Goofie Golf has 18 holes that are par 3, par 4, or par 5. Most of the holes are par 4. In fact, there are 3 times as many par 4s as par 3s. There are 3 more par 5s than par 3s. How many of each type are there?

30. Combining peanuts, pecans, and cashews makes a party mixture of nuts. If the amount of peanuts equals the amount of pecans and cashews combined, and if there are twice as many cashews as pecans, how many ounces of each nut is used to make 48 oz of party mixture?

31. Souvenir hats, T-shirts, and jackets are sold at a rock concert. Three hats, two T-shirts, and one jacket cost $140. Two hats, two T-shirts, and two jackets cost $170. One hat, three T-shirts, and two jackets cost $180. Find the prices of the individual items.

32. Annie and Maria traveled overseas for 7 days and stayed in three different hotels in three different cities: Stockholm, Sweden; Oslo, Norway; and Paris, France.

The total bill for all seven nights (not including tax) was $1040. The total tax was $106. The nightly cost (excluding tax) to stay at the hotel in Paris was $80 more than the nightly cost (excluding tax) to stay in Oslo. Find the cost per night for each hotel excluding tax.

City	Number of Nights	Cost/Night ($)	Tax Rate
Paris, France	1	x	8%
Stockholm, Sweden	4	y	11%
Oslo, Norway	2	z	10%

33. Walter had $25,000 to invest. He split the money into three types of investment: small caps earning 6%, global market investments earning 10%, and a balanced fund earning 9%. He put twice as much money in the global account as he did in the balanced fund. If his earnings for the first year totaled $2160, how much did he invest in each account?

34. Raeann deposited $8000 into three accounts at her credit union: a checking account that pays 1.2% interest, a savings account that pays 2.5% interest, and a money market account that pays 3% interest. If she put three times more money in the 3% account than she did in the 1.2% account, and her total interest for 1 year was $202, how much did she deposit into each account?

Concept 4: Solving Inconsistent Systems and Systems of Dependent Equations (Mixed Exercises)

For Exercises 35–46, solve the system. For systems that do not have one unique solution, state whether the system is inconsistent or the equations are dependent. (See Examples 1, 4, and 5.)

35.
$$2x + y + 3z = 2$$
$$x - y + 2z = -4$$
$$-2x + 2y - 4z = 8$$

36.
$$x + y = z$$
$$2x + 4y - 2z = 6$$
$$3x + 6y - 3z = 9$$

37.
$$6x - 2y + 2z = 2$$
$$4x + 8y - 2z = 5$$
$$-2x - 4y + z = -2$$

38.
$$3x + 2y + z = 3$$
$$x - 3y + z = 4$$
$$-6x - 4y - 2z = 1$$

39.
$$\frac{1}{2}x + \frac{2}{3}y = \frac{5}{2}$$
$$\frac{1}{5}x - \frac{1}{2}z = -\frac{3}{10}$$
$$\frac{1}{3}y - \frac{1}{4}z = \frac{3}{4}$$

40.
$$\frac{1}{2}x + \frac{1}{4}y + z = 3$$
$$\frac{1}{8}x + \frac{1}{4}y + \frac{1}{4}z = \frac{9}{8}$$
$$x - y - \frac{2}{3}z = \frac{1}{3}$$

41.
$$-3x + y - z = 8$$
$$-4x + 2y + 3z = -3$$
$$2x + 3y - 2z = -1$$

42.
$$2x + 3y + 3z = 15$$
$$3x - 6y - 6z = -23$$
$$-9x - 3y + 6z = 8$$

43.
$$2x + y = 3(z - 1)$$
$$3x - 2(y - 2z) = 1$$
$$2(2x - 3z) = -6 - 2y$$

44.
$$2x + y = -3$$
$$2y + 16z = -10$$
$$-7x - 3y + 4z = 8$$

45.
$$-0.1y + 0.2z = 0.2$$
$$0.1x + 0.1y + 0.1z = 0.2$$
$$-0.1x - 0.3z = 0.2$$

46.
$$-0.4x - 0.3y = 0$$
$$0.3y + 0.1z = -0.1$$
$$0.4x - 0.1z = 1.2$$

Expanding Your Skills

The systems in Exercises 47–50 are called homogeneous systems because each system has $(0, 0, 0)$ as a solution. However, if the equations are dependent, the system will have infinitely many more solutions. For each system determine whether $(0, 0, 0)$ is the only solution or if the equations are dependent.

47.
$$2x - 4y + 8z = 0$$
$$-x - 3y + z = 0$$
$$x - 2y + 5z = 0$$

48.
$$2x - 4y + z = 0$$
$$x - 3y - z = 0$$
$$3x - y + 2z = 0$$

49.
$$4x - 2y - 3z = 0$$
$$-8x - y + z = 0$$
$$2x - y - \frac{3}{2}z = 0$$

50.
$$5x + y = 0$$
$$4y - z = 0$$
$$5x + 5y - z = 0$$

Solving Systems of Linear Equations by Using Matrices

Section 4.7

1. Introduction to Matrices

In Sections 4.2, 4.3, and 4.6, we solved systems of linear equations by using the substitution method and the addition method. We now present a third method called the Gauss-Jordan method that uses matrices to solve a linear system.

A **matrix** is a rectangular array of numbers (the plural of *matrix* is *matrices*). The rows of a matrix are read horizontally, and the columns of a matrix are read vertically. Every number or entry within a matrix is called an element of the matrix.

The **order of a matrix** is determined by the number of rows and number of columns. A matrix with m rows and n columns is an $m \times n$ (read as "m by n") matrix. Notice that with the order of a matrix, the number of rows is given first, followed by the number of columns.

Concepts

1. Introduction to Matrices
2. Solving Systems of Linear Equations by Using the Gauss-Jordan Method

Example 1 **Determining the Order of a Matrix**

Determine the order of each matrix.

a. $\begin{bmatrix} 2 & -4 & 1 \\ 5 & \pi & \sqrt{7} \end{bmatrix}$ **b.** $\begin{bmatrix} 1.9 \\ 0 \\ 7.2 \\ -6.1 \end{bmatrix}$ **c.** $\begin{bmatrix} 1 & 0 & 0 \\ 0 & 1 & 0 \\ 0 & 0 & 1 \end{bmatrix}$ **d.** $\begin{bmatrix} a & b & c \end{bmatrix}$

Solution:

a. This matrix has two rows and three columns. Therefore, it is a 2 × 3 matrix.

b. This matrix has four rows and one column. Therefore, it is a 4 × 1 matrix. A matrix with one column is called a **column matrix**.

c. This matrix has three rows and three columns. Therefore, it is a 3 × 3 matrix. A matrix with the same number of rows and columns is called a **square matrix**.

d. This matrix has one row and three columns. Therefore, it is a 1 × 3 matrix. A matrix with one row is called a **row matrix**.

A matrix can be used to represent a system of linear equations written in standard form. To do so, we extract the coefficients of the variable terms and the constants within the equation. For example, consider the system

$$2x - y = 5$$
$$x + 2y = -5$$

The matrix **A** is called the **coefficient matrix**.

$$\mathbf{A} = \begin{bmatrix} 2 & -1 \\ 1 & 2 \end{bmatrix}$$

If we extract both the coefficients and the constants from the equations, we can construct the **augmented matrix** of the system:

$$\left[\begin{array}{cc|c} 2 & -1 & 5 \\ 1 & 2 & -5 \end{array} \right]$$

A vertical bar is inserted into an augmented matrix to designate the position of the equal signs.

Example 2 **Writing the Augmented Matrix of a System of Linear Equations**

Write the augmented matrix for each linear system.

a. $-3x - 4y = 3$
$2x + 4y = 2$

b. $2x - 3z = 14$
$2y + z = 2$
$x + y = 4$

Solution:

a. $\left[\begin{array}{cc|c} -3 & -4 & 3 \\ 2 & 4 & 2 \end{array} \right]$ **b.** $\left[\begin{array}{ccc|c} 2 & 0 & -3 & 14 \\ 0 & 2 & 1 & 2 \\ 1 & 1 & 0 & 4 \end{array} \right]$

TIP: Notice that zeros are inserted to denote the coefficient of each missing term.

Example 3 Writing a Linear System from an Augmented Matrix

Write a system of linear equations represented by each augmented matrix.

a. $\begin{bmatrix} 2 & -5 & | & -8 \\ 4 & 1 & | & 6 \end{bmatrix}$ **b.** $\begin{bmatrix} 2 & -1 & 3 & | & 14 \\ 1 & 1 & -2 & | & -5 \\ 3 & 1 & -1 & | & 2 \end{bmatrix}$ **c.** $\begin{bmatrix} 1 & 0 & 0 & | & 4 \\ 0 & 1 & 0 & | & -1 \\ 0 & 0 & 1 & | & 0 \end{bmatrix}$

Solution:

a. $2x - 5y = -8$
$4x + y = 6$

b. $2x - y + 3z = 14$
$x + y - 2z = -5$
$3x + y - z = 2$

c. $x + 0y + 0z = 4$
$0x + y + 0z = -1 \quad$ or $\quad y = -1$
$0x + 0y + z = 0$
$ x = 4$
$ z = 0$

Skill Practice

Write a system of linear equations represented by each augmented matrix.

7. $\begin{bmatrix} 2 & 3 & | & 5 \\ -1 & 8 & | & 1 \end{bmatrix}$

8. $\begin{bmatrix} -3 & 2 & 1 & | & 4 \\ 14 & 1 & 0 & | & 20 \\ -8 & 3 & 5 & | & 6 \end{bmatrix}$

9. $\begin{bmatrix} 1 & 0 & 0 & | & -8 \\ 0 & 1 & 0 & | & 2 \\ 0 & 0 & 1 & | & 15 \end{bmatrix}$

2. Solving Systems of Linear Equations by Using the Gauss-Jordan Method

We know that interchanging two equations results in an equivalent system of linear equations. Interchanging two rows in an augmented matrix results in an equivalent augmented matrix. Similarly, because each row in an augmented matrix represents a linear equation, we can perform the following elementary row operations that result in an equivalent augmented matrix.

Elementary Row Operations

The following *elementary row operations* performed on an augmented matrix produce an equivalent augmented matrix:

- Interchange two rows.
- Multiply every element in a row by a nonzero real number.
- Add a multiple of one row to another row.

When we are solving a system of linear equations by any method, the goal is to write a series of simpler but equivalent systems of equations until the solution is obvious. The *Gauss-Jordan method* uses a series of elementary row operations performed on the augmented matrix to produce a simpler augmented matrix. In particular, we want to produce an augmented matrix that has 1's along the diagonal of the matrix of coefficients and 0's for the remaining entries in the matrix of coefficients. A matrix written in this way is said to be written in **reduced row echelon form**. For example, the augmented matrix from Example 3(c) is written in reduced row echelon form.

$$\begin{bmatrix} 1 & 0 & 0 & | & 4 \\ 0 & 1 & 0 & | & -1 \\ 0 & 0 & 1 & | & 0 \end{bmatrix}$$

Answers

7. $2x + 3y = 5$
$-x + 8y = 1$

8. $-3x + 2y + z = 4$
$14x + y = 20$
$-8x + 3y + 5z = 6$

9. $x = -8, y = 2, z = 15$

The solution to the corresponding system of equations is easily recognized as $x = 4$, $y = -1$, and $z = 0$.

Similarly, matrix **B** represents a solution of $x = a$ and $y = b$.

$$\mathbf{B} = \begin{bmatrix} 1 & 0 & | & a \\ 0 & 1 & | & b \end{bmatrix}$$

Skill Practice

10. Solve by using the Gauss-Jordan method.
$$x - 2y = -21$$
$$2x + y = -2$$

Example 4 Solving a System of Linear Equations by Using the Gauss-Jordan Method

Solve by using the Gauss-Jordan method.

$$2x - y = 5$$
$$x + 2y = -5$$

Solution:

$$\begin{bmatrix} 2 & -1 & | & 5 \\ 1 & 2 & | & -5 \end{bmatrix}$$
Set up the augmented matrix.

$$\xrightarrow{R_1 \Leftrightarrow R_2} \begin{bmatrix} 1 & 2 & | & -5 \\ 2 & -1 & | & 5 \end{bmatrix}$$
Switch row 1 and row 2 to get a 1 in the upper left position.

$$\xrightarrow{-2R_1 + R_2 \Rightarrow R_2} \begin{bmatrix} 1 & 2 & | & -5 \\ 0 & -5 & | & 15 \end{bmatrix}$$
Multiply row 1 by -2 and add the result to row 2. This produces an entry of 0 below the upper left position.

$$\xrightarrow{-\frac{1}{5}R_2 \Rightarrow R_2} \begin{bmatrix} 1 & 2 & | & -5 \\ 0 & 1 & | & -3 \end{bmatrix}$$
Multiply row 2 by $-\frac{1}{5}$ to produce a 1 along the diagonal in the second row.

$$\xrightarrow{-2R_2 + R_1 \Rightarrow R_1} \begin{bmatrix} 1 & 0 & | & 1 \\ 0 & 1 & | & -3 \end{bmatrix}$$
Multiply row 2 by -2 and add the result to row 1. This produces a 0 in the first row, second column.

$$\mathbf{C} = \begin{bmatrix} 1 & 0 & | & 1 \\ 0 & 1 & | & -3 \end{bmatrix}$$

The matrix **C** is in reduced row echelon form. From the augmented matrix, we have $x = 1$ and $y = -3$. The solution to the system is $(1, -3)$.

The order in which we manipulate the elements of an augmented matrix to produce reduced row echelon form was demonstrated in Example 4. In general, the order is as follows.

- First produce a 1 in the first row, first column. Then use the first row to obtain 0's in the first column below this element.
- Next, if possible, produce a 1 in the second row, second column. Use the second row to obtain 0's above and below this element.
- Next, if possible, produce a 1 in the third row, third column. Use the third row to obtain 0's above and below this element.
- The process continues until reduced row echelon form is obtained.

Answer

10. $(-5, 8)$

Example 5 **Solving a System of Linear Equations by Using the Gauss-Jordan Method**

Solve by using the Gauss-Jordan method.

$$x \qquad = -y + 5$$
$$-2x \qquad + 2z = y - 10$$
$$3x + 6y + 7z = 14$$

Skill Practice

Solve by using the Gauss-Jordan method.

11. $x + y + z = 2$
$x - y + z = 4$
$x + 4y + 2z = 1$

Solution:

First write each equation in the system in standard form.

$$x \qquad = -y + 5 \longrightarrow \quad x + y \qquad = \quad 5$$
$$-2x + \qquad 2z = y - 10 \longrightarrow \quad -2x - y + 2z = -10$$
$$3x + 6y + 7z = 14 \qquad \longrightarrow \quad 3x + 6y + 7z = \quad 14$$

$$\begin{bmatrix} 1 & 1 & 0 & | & 5 \\ -2 & -1 & 2 & | & -10 \\ 3 & 6 & 7 & | & 14 \end{bmatrix}$$
Set up the augmented matrix.

$$2R_1 + R_2 \Rightarrow R_2 \longrightarrow$$
$$-3R_1 + R_3 \Rightarrow R_3 \longrightarrow$$
$$\begin{bmatrix} 1 & 1 & 0 & | & 5 \\ 0 & 1 & 2 & | & 0 \\ 0 & 3 & 7 & | & -1 \end{bmatrix}$$
Multiply row 1 by 2 and add the result to row 2. Multiply row 1 by -3 and add the result to row 3.

$$-1R_2 + R_1 \Rightarrow R_1 \longrightarrow$$
$$-3R_2 + R_3 \Rightarrow R_3 \longrightarrow$$
$$\begin{bmatrix} 1 & 0 & -2 & | & 5 \\ 0 & 1 & 2 & | & 0 \\ 0 & 0 & 1 & | & -1 \end{bmatrix}$$
Multiply row 2 by -1 and add the result to row 1. Multiply row 2 by -3 and add the result to row 3.

$$2R_3 + R_1 \Rightarrow R_1 \longrightarrow$$
$$-2R_3 + R_2 \Rightarrow R_2 \longrightarrow$$
$$\begin{bmatrix} 1 & 0 & 0 & | & 3 \\ 0 & 1 & 0 & | & 2 \\ 0 & 0 & 1 & | & -1 \end{bmatrix}$$
Multiply row 3 by 2 and add the result to row 1. Multiply row 3 by -2 and add the result to row 2.

From the reduced row echelon form of the matrix, we have $x = 3$, $y = 2$, and $z = -1$. The solution to the system is $(3, 2, -1)$.

Answer

11. $(1, -1, 2)$

It is particularly easy to recognize a system of dependent equations or an inconsistent system of equations from the reduced row echelon form of an augmented matrix. This is demonstrated in Examples 6 and 7.

Skill Practice

Solve by using the Gauss-Jordan method.
12. $4x - 6y = 16$
      $6x - 9y = 24$

Example 6 Solving a System of Dependent Equations by Using the Gauss-Jordan Method

Solve by using the Gauss-Jordan method.

$$x - 3y = 4$$
$$\frac{1}{2}x - \frac{3}{2}y = 2$$

Solution:

$$\begin{bmatrix} 1 & -3 & \Big| & 4 \\ \frac{1}{2} & -\frac{3}{2} & \Big| & 2 \end{bmatrix}$$ Set up the augmented matrix.

$$\xrightarrow{-\frac{1}{2}R_1 + R_2 \Rightarrow R_2} \begin{bmatrix} 1 & -3 & \Big| & 4 \\ 0 & 0 & \Big| & 0 \end{bmatrix}$$ Multiply row 1 by $-\frac{1}{2}$ and add the result to row 2.

The second row of the augmented matrix represents the equation $0 = 0$. The equations are dependent. The solution is $\{(x, y) \,|\, x - 3y = 4\}$.

Skill Practice

Solve by using the Gauss-Jordan method.
13. $6x + 10y = 1$
      $15x + 25y = 3$

Example 7 Solving an Inconsistent System of Equations by Using the Gauss-Jordan Method

Solve by using the Gauss-Jordan method.

$$x + 3y = 2$$
$$-3x - 9y = 1$$

Solution:

$$\begin{bmatrix} 1 & 3 & \Big| & 2 \\ -3 & -9 & \Big| & 1 \end{bmatrix}$$ Set up the augmented matrix.

$$\xrightarrow{3R_1 + R_2 \Rightarrow R_2} \begin{bmatrix} 1 & 3 & \Big| & 2 \\ 0 & 0 & \Big| & 7 \end{bmatrix}$$ Multiply row 1 by 3 and add the result to row 2.

The second row of the augmented matrix represents the contradiction $0 = 7$. The system is inconsistent. There is no solution.

Answers
12. Infinitely many solutions;
     $\{(x, y) \,|\, 4x - 6y = 16\}$;
     dependent equations
13. No solution; inconsistent system

Calculator Connections

Topic: Entering a Matrix into a Calculator

Many graphing calculators have a matrix editor in which the user defines the order of the matrix and then enters the elements of the matrix. For example, the 2×3 matrix

$$\mathbf{D} = \begin{bmatrix} 2 & -3 & | & -13 \\ 3 & 1 & | & 8 \end{bmatrix}$$

is entered as shown.

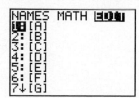

Once an augmented matrix has been entered into a graphing calculator, a *rref* function can be used to transform the matrix into reduced row echelon form.

```
rref([A])
        [[1 0 1]
         [0 1 5]]
```

Section 4.7 Practice Exercises

Study Skills Exercise

Prepare a one-page summary sheet with the most important information that you need for the next test. On the day of the test, look at this sheet several times to refresh your memory, instead of trying to memorize new information.

Vocabulary and Key Concepts

1. **a.** A _____ is a rectangular array of numbers. The order of a matrix is $m \times n$, where m is the number of _____ and n is the number of _____.

   b. A matrix that has exactly one column is called a _____ matrix. A matrix that has exactly _____ row is called a row matrix, and a matrix with the same number of rows and columns is called a _____ matrix.

   c. Given the system of equations shown, matrix **A** is called the _____ matrix. The matrix **B** is called the _____ matrix.

   $$\begin{aligned} 3x - 2y &= 6 \\ 4x + 5y &= 9 \end{aligned} \qquad \mathbf{A} = \begin{bmatrix} 3 & -2 \\ 4 & 5 \end{bmatrix} \qquad \mathbf{B} = \begin{bmatrix} 3 & -2 & | & 6 \\ 4 & 5 & | & 9 \end{bmatrix}$$

   d. The matrix **A** is said to be written in reduced _____ _____ form.

   $$\mathbf{A} = \begin{bmatrix} 1 & 0 & | & 8 \\ 0 & 1 & | & 2 \end{bmatrix}$$

Review Exercises

2. How much 50% acid solution should be mixed with pure acid to obtain 20 L of a mixture that is 70% acid?

For Exercises 3–5, solve the system by using any method.

3. $x - 6y = 9$
$x + 2y = 13$

4. $x + y - z = 8$
$x - 2y + z = 3$
$x + 3y + 2z = 7$

5. $2x - y + z = -4$
$-x + y + 3z = -7$
$x + 3y - 4z = 22$

Concept 1: Introduction to Matrices

For Exercises 6–14, **(a)** determine the order of each matrix and **(b)** determine if the matrix is a row matrix, a column matrix, a square matrix, or none of these. **(See Example 1.)**

6. $\begin{bmatrix} 4 \\ 5 \\ -3 \\ 0 \end{bmatrix}$

7. $\begin{bmatrix} 5 \\ -1 \\ 2 \end{bmatrix}$

8. $\begin{bmatrix} -9 & 4 & 3 \\ -1 & -8 & 4 \\ 5 & 8 & 7 \end{bmatrix}$

9. $\begin{bmatrix} 3 & -9 \\ -1 & -3 \end{bmatrix}$

10. $\begin{bmatrix} 4 & -7 \end{bmatrix}$

11. $\begin{bmatrix} 0 & -8 & 11 & 5 \end{bmatrix}$

12. $\begin{bmatrix} 5 & -8.1 & 4.2 & 0 \\ 4.3 & -9 & 18 & 3 \end{bmatrix}$

13. $\begin{bmatrix} \frac{1}{3} & \frac{3}{4} & 6 \\ -2 & 1 & -\frac{7}{8} \end{bmatrix}$

14. $\begin{bmatrix} 5 & 1 \\ -1 & 2 \\ 0 & 7 \end{bmatrix}$

For Exercises 15–18, set up the augmented matrix. **(See Example 2.)**

15. $x - 2y = -1$
$2x + y = -7$

16. $x - 3y = 3$
$2x - 5y = 4$

17. $x - 2y = 5 - z$
$2x + 6y + 3z = -2$
$3x - y - 2z = 1$

18. $5x - 17 = -2z$
$8x + 6z = 26 + y$
$8x + 3y - 12z = 24$

For Exercises 19–22, write a system of linear equations represented by the augmented matrix. **(See Example 3.)**

19. $\left[\begin{array}{cc|c} 4 & 3 & 6 \\ 12 & 5 & -6 \end{array}\right]$

20. $\left[\begin{array}{cc|c} -2 & 5 & -15 \\ -7 & 15 & -45 \end{array}\right]$

21. $\left[\begin{array}{ccc|c} 1 & 0 & 0 & 4 \\ 0 & 1 & 0 & -1 \\ 0 & 0 & 1 & 7 \end{array}\right]$

22. $\left[\begin{array}{ccc|c} 1 & 0 & 0 & 0.5 \\ 0 & 1 & 0 & 6.1 \\ 0 & 0 & 1 & 3.9 \end{array}\right]$

Concept 2: Solving Systems of Linear Equations by Using the Gauss-Jordan Method

23. Given the matrix **E**

$$\mathbf{E} = \left[\begin{array}{cc|c} 3 & -2 & 8 \\ 9 & -1 & 7 \end{array}\right]$$

a. What is the element in the second row and third column?

b. What is the element in the first row and second column?

24. Given the matrix **F**

$$\mathbf{F} = \left[\begin{array}{cc|c} 1 & 8 & 0 \\ 12 & -13 & -2 \end{array}\right]$$

a. What is the element in the second row and second column?

b. What is the element in the first row and third column?

25. Given the matrix **Z**

$$\mathbf{Z} = \left[\begin{array}{cc|c} 2 & 1 & 11 \\ 2 & -1 & 1 \end{array}\right]$$

write the matrix obtained by multiplying the elements in the first row by $\frac{1}{2}$.

26. Given the matrix **J**

$$\mathbf{J} = \left[\begin{array}{cc|c} 1 & 1 & 7 \\ 0 & 3 & -6 \end{array}\right]$$

write the matrix obtained by multiplying the elements in the second row by $\frac{1}{3}$.

27. Given the matrix **K**

$$\mathbf{K} = \left[\begin{array}{cc|c} 5 & 2 & 1 \\ 1 & -4 & 3 \end{array}\right]$$

write the matrix obtained by interchanging rows 1 and 2.

28. Given the matrix **L**

$$\mathbf{L} = \left[\begin{array}{cc|c} 9 & 6 & 13 \\ -7 & 2 & 19 \end{array}\right]$$

write the matrix obtained by interchanging rows 1 and 2.

29. Given the matrix **M**

$$\mathbf{M} = \left[\begin{array}{cc|c} 1 & 5 & 2 \\ -3 & -4 & -1 \end{array}\right]$$

write the matrix obtained by multiplying the first row by 3 and adding the result to row 2.

30. Given the matrix **N**

$$\mathbf{N} = \left[\begin{array}{cc|c} 1 & 3 & -5 \\ -2 & 2 & 12 \end{array}\right]$$

write the matrix obtained by multiplying the first row by 2 and adding the result to row 2.

31. Given the matrix **R**

$$\mathbf{R} = \left[\begin{array}{ccc|c} 1 & 3 & 0 & -1 \\ 4 & 1 & -5 & 6 \\ -2 & 0 & -3 & 10 \end{array}\right]$$

a. Write the matrix obtained by multiplying the first row by -4 and adding the result to row 2.

b. Using the matrix obtained from part (a), write the matrix obtained by multiplying the first row by 2 and adding the result to row 3.

32. Given the matrix **S**

$$\mathbf{S} = \left[\begin{array}{ccc|c} 1 & 2 & 0 & 10 \\ 5 & 1 & -4 & 3 \\ -3 & 4 & 5 & 2 \end{array}\right]$$

a. Write the matrix obtained by multiplying the first row by -5 and adding the result to row 2.

b. Using the matrix obtained from part (a), write the matrix obtained by multiplying the first row by 3 and adding the result to row 3.

For Exercises 33–36, use the augmented matrices **A**, **B**, **C**, and **D** to answer true or false.

$$\mathbf{A} = \left[\begin{array}{cc|c} 6 & -4 & 2 \\ 5 & -2 & 7 \end{array}\right] \quad \mathbf{B} = \left[\begin{array}{cc|c} 5 & -2 & 7 \\ 6 & -4 & 2 \end{array}\right] \quad \mathbf{C} = \left[\begin{array}{cc|c} 1 & -\frac{2}{3} & \frac{1}{3} \\ 5 & -2 & 7 \end{array}\right] \quad \mathbf{D} = \left[\begin{array}{cc|c} 5 & -2 & 7 \\ -12 & 8 & -4 \end{array}\right]$$

33. The matrix **A** is a 2×3 matrix.

34. Matrix **B** is equivalent to matrix **A**.

35. Matrix **A** is equivalent to matrix **C**.

36. Matrix **B** is equivalent to matrix **D**.

37. What does the notation $R_2 \Leftrightarrow R_1$ mean when one is performing the Gauss-Jordan method?

38. What does the notation $2R_3 \Rightarrow R_3$ mean when one is performing the Gauss-Jordan method?

39. What does the notation $-3R_1 + R_2 \Rightarrow R_2$ mean when one is performing the Gauss-Jordan method?

40. What does the notation $4R_2 + R_3 \Rightarrow R_3$ mean when one is performing the Gauss-Jordan method?

For Exercises 41–56, solve the system by using the Gauss-Jordon method. For systems that do not have one unique solution, also state the number of solutions and whether the system is inconsistent or the equations are dependent. (See Examples 4–7.)

41. $\quad x - 2y = -1$
$\quad\quad 2x + y = -7$

42. $\quad x - 3y = 3$
$\quad\quad 2x - 5y = 4$

43. $\quad\quad x + 3y = 6$
$\quad\quad -4x - 9y = 3$

44. $2x - 3y = -2$
$\quad\quad x + 2y = 13$

45. $\quad x + 3y = 3$
$\quad\quad 4x + 12y = 12$

46. $\quad 2x + 5y = 1$
$\quad\quad -4x - 10y = -2$

47. $\quad x - y = 4$
$\quad\quad 2x + y = 5$

48. $2x - y = 0$
$\quad\quad x + y = 3$

49. $\quad\quad x + 3y = -1$
$\quad\quad -3x - 6y = 12$

50. $\quad x + y = 4$
$\quad\quad 2x - 4y = -4$

51. $\quad 3x + y = -4$
$\quad\quad -6x - 2y = 3$

52. $2x + y = 4$
$\quad\quad 6x + 3y = -1$

53. $x + y + z = 6$
$x - y + z = 2$
$x + y - z = 0$

54. $2x - 3y - 2z = 11$
$x + 3y + 8z = 1$
$3x - y + 14z = -2$

55. $x - 2y = 5 - z$
$2x + 6y + 3z = -10$
$3x - y - 2z = 5$

56. $5x = 10z + 15$
$x - y + 6z = 23$
$x + 3y - 12z = 13$

Graphing Calculator Exercises

For Exercises 57–62, use the matrix features on a graphing calculator to express each augmented matrix in reduced row echelon form. Compare your results to the solution you obtained in the indicated exercise.

57. $\begin{bmatrix} 1 & -2 & | & -1 \\ 2 & 1 & | & -7 \end{bmatrix}$

Compare with Exercise 41.

58. $\begin{bmatrix} 1 & -3 & | & 3 \\ 2 & -5 & | & 4 \end{bmatrix}$

Compare with Exercise 42.

59. $\begin{bmatrix} 1 & 3 & | & 6 \\ -4 & -9 & | & 3 \end{bmatrix}$

Compare with Exercise 43.

60. $\begin{bmatrix} 2 & -3 & | & -2 \\ 1 & 2 & | & 13 \end{bmatrix}$

Compare with Exercise 44.

61. $\begin{bmatrix} 1 & 1 & 1 & | & 6 \\ 1 & -1 & 1 & | & 2 \\ 1 & 1 & -1 & | & 0 \end{bmatrix}$

Compare with Exercise 53.

62. $\begin{bmatrix} 2 & -3 & -2 & | & 11 \\ 1 & 3 & 8 & | & 1 \\ 3 & -1 & 14 & | & -2 \end{bmatrix}$

Compare with Exercise 54.

Group Activity

Creating a Quadratic Model of the Form $y = at^2 + bt + c$

Estimated time: 20 minutes

Group Size: 3

Natalie Dalton was a player on the Daytona State College women's fast pitch softball team. She threw a ball 120 ft from right field to make a play at third base. A photographer uses strobe photography to follow the path of the ball. The height of the ball (in ft) at 0.2-sec intervals was recorded in the table.

Time (sec) t	0	0.2	0.4	0.6	0.8	1.0	1.2	1.4	1.6	1.8	2.0
Height (ft) y	5	11	16	19	21	22	21	19	16	12	6

1. Make a scatter diagram of the data.

2. Select three data points and label them (t_1, y_1), (t_2, y_2), and (t_3, y_3).

$(t_1, y_1) = $ _____

$(t_2, y_2) = $ _____

$(t_3, y_3) = $ _____

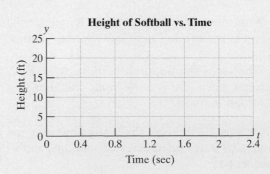

Height of Softball vs. Time

3. Substitute the values of t_1 and y_1 in for t and y in the quadratic model $y = at^2 + bt + c$. The equation you are left with should only have the variables a, b, and c. Call this equation 1. Then repeat this step two more times, using the points (t_2, y_2) and (t_3, y_3). You should then be left with three equations involving a, b, and c only.

Equation 1: _____

Equation 2: _____

Equation 3: _____

4. Solve the system of equations from step 3 for the variables *a*, *b*, and *c*.

5. Replace the values of *a*, *b*, and *c* into the quadratic function $y = at^2 + bt + c$. What does this equation represent?

6. Use the function found in step 5 to approximate the height of the ball after $\frac{1}{2}$ sec.

7. Use the function found in step 5 to approximate the height of the ball after 1.4 seconds. How well does this value match the observed value of 19 ft?

Chapter 4   Summary

Section 4.1   Solving Systems of Linear Equations by the Graphing Method

Key Concepts

A **system of linear equations** in two variables can be solved by graphing.

A **solution to a system of linear equations** is an ordered pair that satisfies each equation in the system. Graphically, this represents a point of intersection of the lines.

There may be one solution, infinitely many solutions, or no solution.

One solution	Infinitely many	No solution
Consistent	solutions	Inconsistent
Independent	Consistent	Independent
	Dependent	

A system of equations is **consistent** if there is at least one solution. A system is **inconsistent** if there is no solution.

Two linear equations in *x* and *y* are **dependent** if the equations represent the same line. The solution to the system is the set of all points on the line.

If two linear equations represent different lines, then the equations are **independent**.

Examples

Example 1

Solve by graphing. $x + y = 3$
$$2x - y = 0$$

Write each equation in slope-intercept form $(y = mx + b)$ to graph the lines.

$$y = -x + 3$$

$$y = 2x$$

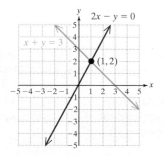

The solution is the point of intersection $(1, 2)$.

Section 4.2   Solving Systems of Equations by Using the Substitution Method

Key Concepts

Substitution Method

1. Isolate one of the variables.
2. Substitute the quantity found in step 1 into the *other* equation.
3. Solve the resulting equation.
4. Substitute the value from step 3 back into the equation from step 1 to solve for the remaining variable.
5. Check the ordered pair in both equations, and write the answer as an ordered pair.

Examples

Example 1

$$2y = -6x + 14$$

$$2x + y = 5 \xrightarrow{\text{Isolate a variable.}} y = \underbrace{-2x + 5}$$

Substitute

$$2(-2x + 5) = -6x + 14$$

$$-4x + 10 = -6x + 14$$

$$2x + 10 = 14$$

$$2x = 4$$

$$x = 2$$

$$y = -2x + 5 \qquad \text{Now solve for } y.$$

$$y = -2(2) + 5$$

$$y = 1$$

The solution is $(2, 1)$ and checks in both equations.

A system is consistent if there is at least one solution. A system is inconsistent if there is no solution. An inconsistent system is detected by a contradiction (such as $0 = 5$).

Example 2

$$y = -2x + 3$$

$$-4x - 2y = 1 \longrightarrow -4x - 2(-2x + 3) = 1$$

$$-4x + 4x - 6 = 1$$

$$-6 = 1$$

Contradiction. The system is inconsistent. There is no solution.

Two linear equations are independent if the equations represent different lines. The equations are dependent if they represent the same line. This produces infinitely many solutions. Two equations in a system of equations are dependent if the system reduces to an identity (such as $0 = 0$).

Example 3

$$x = -3y + 1$$

$$2x + 6y = 2 \longrightarrow 2(-3y + 1) + 6y = 2$$

$$-6y + 2 + 6y = 2$$

$$2 = 2$$

Identity. The equations are dependent. There are infinitely many solutions.

$$\{(x, y) | x = -3y + 1\}$$

Section 4.3   Solving Systems of Equations by Using the Addition Method

Key Concepts

Addition Method

1. Write both equations in standard form $Ax + By = C$.
2. Clear fractions or decimals (optional).
3. Multiply one or both equations by nonzero constants to create opposite coefficients for one of the variables.
4. Add the equations from step 3 to eliminate one variable.
5. Solve for the remaining variable.
6. Substitute the known value from step 5 back into one of the original equations to solve for the other variable.
7. Check the ordered pair in both equations.

Examples

Example 1

$$3x - 4y = 18 \xrightarrow{\text{Mult. by 3.}} 9x - 12y = 54$$
$$-5x - 3y = -1 \xrightarrow{\text{Mult. by } -4.} 20x + 12y = 4$$
$$\overline{ 29x = 58}$$
$$x = 2$$

$$3(2) - 4y = 18$$
$$6 - 4y = 18$$
$$-4y = 12$$
$$y = -3$$

The solution is $(2, -3)$ and checks in both equations.

Section 4.4   Applications of Systems of Linear Equations in Two Variables

Key Concepts

Solve application problems by using systems of linear equations in two variables.

- Cost applications
- Mixture applications
- Applications involving principal and interest
- Applications involving uniform motion
- Geometry applications

Steps to Solve Applications:

1. Label two variables.
2. Construct two equations in words.
3. Write two equations.
4. Solve the system.
5. Write the answer.

Examples

Example 1

Mercedes borrowed $1500 more from a lender that charges 6.5% interest than she did from a lender that charges 4% interest. If the total interest owed at the end of 1 yr is $622.50, find the amount she borrowed at 6.5%.

Let x represent the amount borrowed at 6.5%.
Let y represent the amount borrowed at 4%.

$$\left(\begin{array}{c}\text{Amount borrowed} \\ \text{at } 6.5\%\end{array}\right) = \left(\begin{array}{c}\text{amount borrowed} \\ \text{at } 4\%\end{array}\right) + \$1500$$

$$\left(\begin{array}{c}\text{Interest owed} \\ \text{from } 6.5\% \\ \text{account}\end{array}\right) + \left(\begin{array}{c}\text{interest owed} \\ \text{from } 4\% \\ \text{account}\end{array}\right) = \$622.50$$

$$x = y + 1500$$
$$0.065x + 0.04y = 622.50$$

Using substitution gives

$$0.065(y + 1500) + 0.04y = 622.50$$
$$0.065y + 97.5 + 0.04y = 622.50$$
$$0.105y = 525$$
$$y = 5000$$
$$x = (5000) + 1500 = 6500$$

Mercedes borrowed $6500 at 6.5%.

Section 4.5   Linear Inequalities and Systems of Linear Inequalities in Two Variables

Key Concepts

A **linear inequality in two variables** is an inequality of the form $Ax + By < C, Ax + By > C, Ax + By \leq C$, or $Ax + By \geq C$.

Graphing a Linear Inequality in Two Variables

1. Write the inequality with the y variable isolated, if possible.
2. Graph the related equation. Draw a dashed line if the inequality is strict, < or >. Otherwise, draw a solid line.
3. Shade above or below the line according to the following convention.
   - Shade *above* the line if the inequality is of the form $y > mx + b$ or $y \geq mx + b$.
   - Shade *below* the line if the inequality is of the form $y < mx + b$ or $y \leq mx + b$.

   You can use test points to check that you have shaded the correct region. Select an ordered pair from the proposed solution set and substitute the values of x and y in the original inequality. If the test point produces a true statement, then you have shaded the correct region.

   The solution to a system of linear inequalities is the intersection of the solution sets.

Examples

Example 1

Graph the solution to the inequality $2x - y < 4$. Graph the related equation, $y = 2x - 4$, with a dashed line.

Solve for y: $2x - y < 4$

$$-y < -2x + 4$$
$$y > 2x - 4$$

Shade above the line.

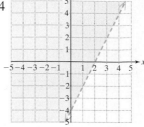

Example 2

Graph.

$x < 0$ and $y > 2$

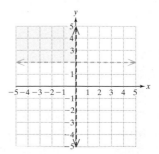

Section 4.6   Systems of Linear Equations in Three Variables and Applications

Key Concepts

A **linear equation in three variables** can be written in the form $Ax + By + Cz = D$, where A, B, and C are not all zero. The graph of a linear equation in three variables is a plane in space.

A solution to a system of linear equations in three variables is an **ordered triple** that satisfies each equation. Graphically, a solution is a point of intersection among three planes.

A system of linear equations in three variables may have one unique solution, infinitely many solutions (dependent equations), or no solution (inconsistent system).

Examples

Example 1

$\boxed{A}$ $x + 2y - z = 4$

$\boxed{B}$ $3x - y + z = 5$

$\boxed{C}$ $2x + 3y + 2z = 7$

$\boxed{A}$ $x + 2y - z = 4$

$\boxed{B}$ $\underline{3x - y + z = 5}$

     $4x + y \quad\ = 9$ $\boxed{D}$

$2 \cdot \boxed{A} \ 2x + 4y - 2z = 8$

     $\boxed{C} \ \underline{2x + 3y + 2z = 7}$

      $4x + 7y \quad\ = 15$ $\boxed{E}$

$\boxed{D}$ $4x + y = 9 \ \rightarrow \ -4x - y = -9$

$\boxed{E}$ $4x + 7y = 15 \ \rightarrow \ \underline{4x + 7y = \ 15}$

                        $6y = \ 6$

                         $y = \ 1$

Substitute $y = 1$ into either equation $\boxed{D}$ or $\boxed{E}$.

$\boxed{D}$ $4x + (1) = 9$

         $4x = 8$

         $x = 2$

Substitute $x = 2$ and $y = 1$ into equation $\boxed{A}$, $\boxed{B}$, or $\boxed{C}$.

$\boxed{A}$ $(2) + 2(1) - z = 4$

                $z = 0$

The solution is $(2, 1, 0)$.

Key Concepts

A **matrix** is a rectangular array of numbers displayed in rows and columns. Every number or entry within a matrix is called an element of the matrix.

The **order of a matrix** is determined by the number of rows and number of columns. A matrix with m rows and n columns is an $m \times n$ matrix.

A system of equations written in standard form can be represented by an **augmented matrix** consisting of the coefficients of the terms of each equation in the system.

The Gauss-Jordan method can be used to solve a system of equations by using the following elementary row operations on an augmented matrix.

1. Interchange two rows.
2. Multiply every element in a row by a nonzero real number.
3. Add a multiple of one row to another row.

These operations are used to write the matrix in **reduced row echelon form**.

$$\begin{bmatrix} 1 & 0 & | & a \\ 0 & 1 & | & b \end{bmatrix}$$

This represents the solution, $x = a$ and $y = b$.

Examples

Example 1

$[1 \quad 2 \quad 5]$ is a 1×3 matrix (a row matrix).

$\begin{bmatrix} -1 & 8 \\ 1 & 5 \end{bmatrix}$ is a 2×2 matrix (a square matrix).

$\begin{bmatrix} 4 \\ 1 \end{bmatrix}$ is a 2×1 matrix (a column matrix).

Example 2

The augmented matrix for

$$\begin{aligned} 4x + y &= -12 \\ x - 2y &= 6 \end{aligned}$$

is $\begin{bmatrix} 4 & 1 & | & -12 \\ 1 & -2 & | & 6 \end{bmatrix}$

Example 3

Solve the system from Example 2 by using the Gauss-Jordan method.

$R_1 \Leftrightarrow R_2$ $\quad \begin{bmatrix} 1 & -2 & | & 6 \\ 4 & 1 & | & -12 \end{bmatrix}$

$\xrightarrow{-4R_1 + R_2 \Rightarrow R_2} \begin{bmatrix} 1 & -2 & | & 6 \\ 0 & 9 & | & -36 \end{bmatrix}$

$\xrightarrow{\frac{1}{9}R_2 \Rightarrow R_2} \begin{bmatrix} 1 & -2 & | & 6 \\ 0 & 1 & | & -4 \end{bmatrix}$

$\xrightarrow{2R_2 + R_1 \Rightarrow R_1} \begin{bmatrix} 1 & 0 & | & -2 \\ 0 & 1 & | & -4 \end{bmatrix}$

From the reduced row echelon form of the matrix we have $x = -2$ and $y = -4$. The solution is $(-2, -4)$.

Chapter 4   Review Exercises

Section 4.1

1. Determine if the ordered pair is a solution to the system.

$$-5x - 7y = 4$$

$$y = -\frac{1}{2}x - 1$$

    a. $(2, 2)$ **b.** $(2, -2)$

For Exercises 2–4, answer true or false.

2. An inconsistent system has one solution.

3. Parallel lines form an inconsistent system.

4. Lines with different slopes intersect in one point.

For Exercises 5–8, solve the system by graphing. For systems that do not have one unique solution, also state the number of solutions and whether the system is inconsistent or the equations are dependent.

5. $f(x) = x - 1$
   $g(x) = 2x - 4$

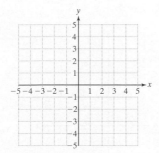

6. $y = 2x + 7$
   $y = -x - 5$

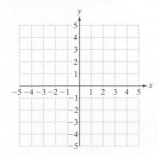

7. $6x + 2y = 4$
   $3x = -y + 2$

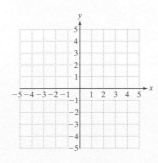

8. $y = \frac{1}{2}x - 2$
   $-4x + 8y = -8$

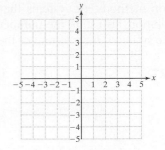

Section 4.2

For Exercises 9–14, solve the system by using the substitution method. For systems that do not have one unique solution, also state the number of solutions and whether the system is inconsistent or the equations are dependent.

9. $y = \frac{3}{4}x - 4$
   $-x + 2y = -6$

10. $3x = 11y - 9$
    $y = \frac{3}{11}x + \frac{6}{11}$

11. $4x + y = 7$
    $x + \frac{1}{4}y = \frac{7}{4}$

12. $6x + y = 5$
    $5x + y = 3$

13. $2(x + y) = 10 - 3y$
    $1.5x + y = 2$

14. $60(5x - y) = 30$
    $10x = 3y - 1$

15. The cost y (in $) to rent a 5-ft by 5-ft storage space for x months is given for two companies.

Company A: $y = 105 + 45x$

Company B: $y = 48.50x$

Determine the number of months for which the cost would be the same to rent from either company.

16. The cost y (in $) to rent a car for x days is given for two rental car companies.

Company A: $y = 44 + 81.50x$

Company B: $y = 87x$

Determine the number of days for which the cost would be the same from each company.

Section 4.3

For Exercises 17–26, solve the system by using the addition method. For systems that do not have one unique solution, also state the number of solutions and whether the system is inconsistent or the equations are dependent.

17. $\frac{2}{5}x + \frac{3}{5}y = 1$

$x - \frac{2}{3}y = \frac{1}{3}$

18. $4x + 3y = 5$

$3x - 4y = 10$

19. $3x + 4y = 2$

$2x + 5y = -1$

20. $3x + y = 1$

$-x - \frac{1}{3}y = -\frac{1}{3}$

21. $2y = 3x - 8$

$-6x = -4y + 4$

22. $3x + y = 16$

$3(x + y) = y + 2x + 2$

23. $-(y + 4x) = 2x - 9$

$-2x + 2y = -10$

24. $-(4x - 35) = 3y$

$-(x - 15) = y$

25. $-0.4x + 0.3y = 1.8$

$0.6x - 0.2y = -1.2$

26. $0.02x - 0.01y = -0.11$

$0.01x + 0.04y = 0.26$

Section 4.4

27. Melinda invested twice as much money in an account paying 5% simple interest as she did in an account paying 3.5% simple interest. If her total interest at the end of 1 yr is $303.75, find the amount she invested in the 5% account.

28. A school carnival sold tickets to ride on a Ferris wheel. The charge was $1.50 for adults and $1.00 for students. If 54 tickets were sold for a total of $70.50, how many of each type of ticket were sold?

29. How many liters of 20% saline solution must be mixed with 50% saline solution to produce 16 L of a 31.25% saline solution?

30. It takes a pilot $1\frac{3}{4}$ hr to travel with the wind to get from Jacksonville, Florida, to Myrtle Beach, South Carolina. Her return trip takes 2 hr flying against the wind. What is the speed of the wind and the speed of the plane in still air if the distance between Jacksonville and Myrtle Beach is 280 mi?

31. Two phone companies offer discount rates to students.

Company 1: $9.95 per month, plus $0.10 per minute for long-distance calls

Company 2: $12.95 per month, plus $0.08 per minute for long-distance calls

a. Write a linear function describing the total cost for x min of long-distance calls from Company 1.

b. Write a linear function describing the total cost for x min of long-distance calls from Company 2.

c. How many minutes of long-distance calls would result in equal cost for both offers?

32. Two angles are complementary. One angle measures 6° more than 5 times the measure of the other. What are the measures of the two angles?

Section 4.5

For Exercises 33–40, solve the inequalities by graphing.

33. $2x > -y + 5$

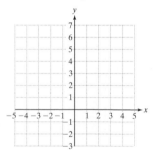

34. $2x \leq -8 - 3y$

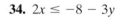
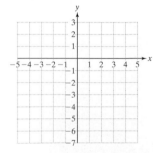

35. $y \geq -\frac{2}{3}x + 3$

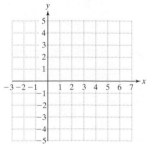

36. $y > \frac{3}{4}x - 2$

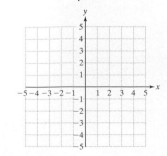

37. $x > -3$

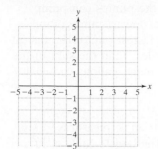

38. $x \leq 2$

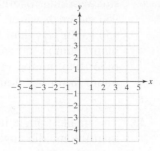

39. $x \geq \dfrac{1}{2}y$

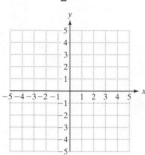

40. $x < \dfrac{2}{5}y$

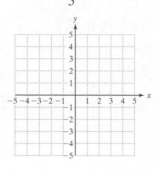

For Exercises 41–44, graph the system of inequalities.

41. $2x - y > -2$

$2x - y \leq 2$

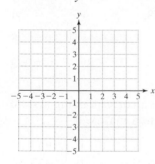

42. $3x + y < 6$

$-3x + y < -2$

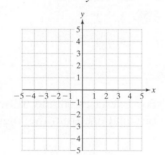

43. $x \geq 0$

$y \geq 0$

$y \geq -\dfrac{3}{2}x + 4$

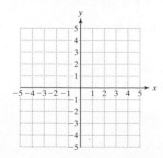

44. $x \geq 0$

$y \geq 0$

$y \leq -\dfrac{2}{3}x + 4$

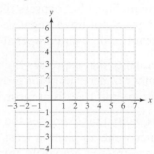

45. A pirate's treasure is buried on a small, uninhabited island in the eastern Caribbean. A shipwrecked sailor finds a treasure map at the base of a coconut palm tree. The treasure is buried within the intersection of three linear inequalities. The palm tree is located at the origin, and the positive y-axis is oriented due north. The scaling on the map is in 1-yd increments. Find the region where the sailor should dig for the treasure.

$$-2x + y \leq 4$$

$$y \leq -x + 6$$

$$y \geq 0$$

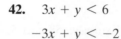

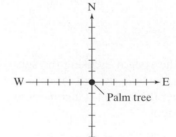

46. Suppose a farmer has 100 acres of land on which to grow oranges and grapefruit. Furthermore, because of demand from his customers, he wants to plant at least 4 times as many acres of orange trees as grapefruit trees.

Let x represent the number of acres of orange trees.

Let y represent the number of acres of grape-fruit trees.

a. Write two inequalities that express the fact that the farmer cannot use a negative number of acres to plant orange and grapefruit trees.

b. Write an inequality that expresses the fact that the total number of acres used for growing orange and grapefruit trees is at most 100.

c. Write an inequality that expresses the fact that the farmer wants to plant at least 4 times as many orange trees as grapefruit trees.

d. Sketch the inequalities in parts (a)–(c) to find the feasible region for the farmer's distribution of orange and grapefruit trees.

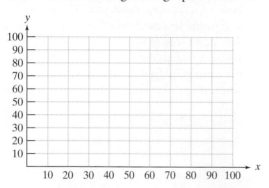

Section 4.6

For Exercises 47–50, solve the system by using the substitution method. For systems that do not have one unique solution, state whether the system is inconsistent or the equations are dependent.

47. $5x + 5y + 5z = 30$
 $-x + y + z = 2$
 $10x + 6y - 2z = 4$

48. $5x + 3y - z = 5$
 $x + 2y + z = 6$
 $-x - 2y - z = 8$

49. $x + y + z = 4$
 $-x - 2y - 3z = -6$
 $2x + 4y + 6z = 12$

50. $3x \qquad + 4z = 5$
 $2y + 3z = 2$
 $2x - 5y \qquad = 8$

51. The perimeter of a right triangle is 30 ft. One leg is 2 ft longer than twice the shortest leg. The hypotenuse is 2 ft less than 3 times the shortest leg. Find the lengths of the sides of this triangle.

52. Three pumps are working to drain a construction site. Working together, the pumps can pump 950 gal/hr of water. The slowest pump pumps 150 gal/hr less than the fastest pump. The fastest pump pumps 150 gal/hr less than the sum of the other two pumps. How many gallons can each pump drain per hour?

53. The smallest angle in a triangle measures 9° less than the middle angle. The largest angle is 26° more than 3 times the measure of the smallest angle. Find the measure of the three angles.

Section 4.7

For Exercises 54–57, determine the order of each matrix.

54. $\begin{bmatrix} 2 & 4 & -1 \\ 5 & 0 & -3 \\ -1 & 6 & 10 \end{bmatrix}$

55. $\begin{bmatrix} -5 & 6 \\ 9 & 2 \\ 0 & -3 \end{bmatrix}$

56. $\begin{bmatrix} 0 & 13 & -4 & 16 \end{bmatrix}$

57. $\begin{bmatrix} 7 \\ 12 \\ -4 \end{bmatrix}$

For Exercises 58 and 59, set up the augmented matrix.

58. $x + y = 3$
 $x - y = -1$

59. $x - y + z = 4$
 $2x - y + 3z = 8$
 $-2x + 2y - z = -9$

For Exercises 60 and 61, write a corresponding system of equations from the augmented matrix.

60. $\left[\begin{array}{cc|c} 1 & 0 & 9 \\ 0 & 1 & -3 \end{array}\right]$

61. $\left[\begin{array}{ccc|c} 1 & 0 & 0 & -5 \\ 0 & 1 & 0 & 2 \\ 0 & 0 & 1 & -8 \end{array}\right]$

62. Given the matrix **C**

$$\mathbf{C} = \left[\begin{array}{cc|c} 1 & 3 & 1 \\ 4 & -1 & 6 \end{array}\right]$$

a. What is the element in the second row and first column?

b. Write the matrix obtained by multiplying the first row by -4 and adding the result to row 2.

63. Given the matrix **D**

$$\mathbf{D} = \begin{bmatrix} 1 & 2 & 0 & | & -3 \\ 4 & -1 & 1 & | & 0 \\ -3 & 2 & 2 & | & 5 \end{bmatrix}$$

a. Write the matrix obtained by multiplying the first row by -4 and adding the result to row 2.

b. Using the matrix obtained in part (a), write the matrix obtained by multiplying the first row by 3 and adding the result to row 3.

For Exercises 64–67, solve the system by using the Gauss-Jordan method.

64. $x + y = 3$
    $x - y = -1$

65. $4x + 3y = 6$
    $12x + 5y = -6$

66. $x - y + z = -4$
    $2x + y - 2z = 9$
    $x + 2y + z = 5$

67. $x - y + z = 4$
    $2x - y + 3z = 8$
    $-2x + 2y - z = -9$

Chapter 4   Test

1. Determine if the ordered pair $\left(\frac{1}{4}, 2\right)$ is a solution to the system.

$$4x - 3y = -5$$
$$12x + 2y = 7$$

For Exercises 2–4, match each figure with the appropriate description, a, b, or c.

2.

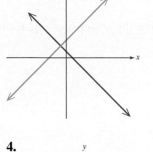

3.

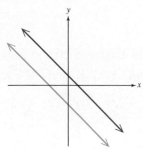

4.

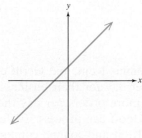

a. The system is consistent and the equations are dependent. There are infinitely many solutions.

b. The system is consistent and the equations are independent. There is one solution.

c. The system is inconsistent and the equations are independent. There are no solutions.

5. Solve the system by graphing.

$$4x - 2y = -4$$
$$3x + y = 7$$

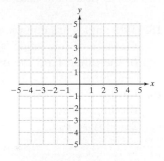

6. Solve the system by graphing.

$$f(x) = x + 3$$
$$g(x) = -\frac{3}{2}x - 2$$

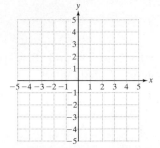

7. Solve the system by using the substitution method.

$$3x + 5y = 13$$
$$y = x + 9$$

8. Solve the system by using the addition method.

$$6x + 8y = 5$$
$$3x - 2y = 1$$

For Exercises 9–13, solve the system by using the substitution method. For systems that do not have one unique solution, also state the number of solutions and whether the system is inconsistent or the equations are dependent.

9. $7y = 5x - 21$

$9y + 2x = -27$

10. $3x - 5y = -7$

$-18x + 30y = 42$

11. $\dfrac{1}{5}x = \dfrac{1}{2}y + \dfrac{17}{5}$

$\dfrac{1}{4}(x + 2) = -\dfrac{1}{6}y$

12. $4x = 5 - 2y$

$y = -2x + 4$

13. $-0.03y + 0.06x = 0.3$

$0.4x - 2 = -0.5y$

14. How many liters of a 20% acid solution should be mixed with a 60% acid solution to produce 200 L of a 44% acid solution?

15. Two angles are complementary. Two times the measure of one angle is 60° less than the measure of the other. Find the measure of each angle.

16. Graph the solution to the inequality $2x - 5y \geq 10$.

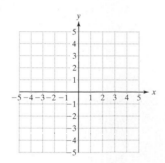

For Exercises 17 and 18, graph the solution to the systems of inequalities.

17. $x + y < 3$
$3x - 2y > -6$

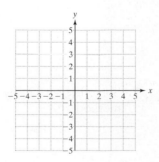

18. $5x \leq 5$
$x + y \leq 0$

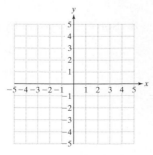

19. After menopause, women are at higher risk for hip fractures as a result of low calcium. As early as their teen years, women need at least 1200 mg of calcium per day (the USDA recommended daily allowance). One 8-oz glass of skim milk contains 300 mg of calcium, and one Antacid (regular strength) contains 400 mg of calcium. Let x represent the number of 8-oz glasses of milk that a woman drinks per day. Let y represent the number of Antacid tablets (regular strength) that a woman takes per day.

a. Write two inequalities that express the fact that the number of glasses of milk and the number of Antacid taken each day cannot be negative.

b. Write a linear inequality in terms of x and y for which the daily calcium intake is at least 1200 mg.

c. Graph the inequalities.

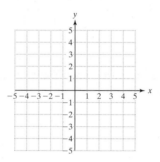

For Exercises 20 and 21, solve the systems of equations.

20. $2x + 2y + 4z = -6$

$3x + y + 2z = 29$

$x - y - z = 44$

21. $2(x + z) = 6 + x - 3y$

$2x = 11 + y - z$

$x + 2(y + z) = 8$

22. Working together, Joanne, Kent, and Geoff can process 504 orders per day for their business. Kent can process 20 more orders per day than Joanne can process. Geoff can process 104 fewer orders per day than Kent and Joanne combined. Find the number of orders that each person can process per day.

23. Write an example of a 3×2 matrix.

24. Given the matrix **A**

$$A = \begin{bmatrix} 1 & 2 & 1 & | & -3 \\ 4 & 0 & 1 & | & -2 \\ -5 & -6 & 3 & | & 0 \end{bmatrix}$$

a. Write the matrix obtained by multiplying the first row by -4 and adding the result to row 2.

b. Using the matrix obtained in part (a), write the matrix obtained by multiplying the first row by 5 and adding the result to row 3.

For Exercises 25 and 26, solve by using the Gauss-Jordan method.

25. $5x - 4y = 34$
$x - 2y = 8$

26. $x + y + z = 1$
$2x + y = 0$
$-2y - z = 5$

Chapters 1–4 Cumulative Review Exercises

1. Simplify. $2[3^2 - 8(7 - 5)]$

2. Simplify. $7[4 - 2(w - 5) - 3(2w + 1)] + 20$

For Exercises 3 and 4, solve the equation.

3. $-5(2x - 1) - 2(3x + 1) = 7 - 2(8x + 1)$

4. $\frac{1}{2}(a - 2) - \frac{3}{4}(2a + 1) = -\frac{1}{6}$

5. Simplify the expression. $\dfrac{(6a^2b)^{-2}}{a^{-5}b^0}$

6. Solve the inequality. Write the answer in interval notation.

$$-3y - 2(y + 1) < 5$$

7. Identify the slope and the x- and y-intercepts of the line $5x - 2y = 15$.

For Exercises 8 and 9, graph the equations.

8. $y = -\dfrac{1}{3}x - 4$

9. $x = -2$

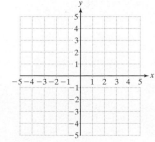

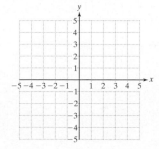

10. Find the slope of the line passing through the points $(4, -10)$ and $(6, -10)$.

11. Find an equation for the line that passes through the points $(3, -8)$ and $(2, -4)$. Write the answer in slope-intercept form.

12. Solve the system by using the addition method.

$$2x - 3y = 6$$
$$\frac{1}{2}x - \frac{3}{4}y = 1$$

13. Solve the system by using the substitution method.

$$2x + y = 4$$
$$y = 3x - 1$$

14. Two video clubs rent DVDs according to the following fee schedules:

Club 1: $25 initiation fee plus $2.50 per DVD
Club 2: $10 initiation fee plus $3.00 per DVD

a. Write a linear function describing the total cost of renting x DVDs from club 1.

b. Write a linear function describing the total cost of renting x DVDs from club 2.

c. How many DVDs would have to be rented to make the cost for club 1 the same as the cost for club 2?

15. Graph the inequality. $4x - y > 12$

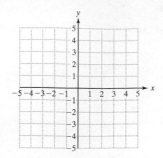

16. Solve the system.

$$3x = 3 - 2y - 3z$$
$$4x - 5y + 7z = 1$$
$$2x + 3y - 2z = 6$$

17. Determine the order of the matrix.

$$\begin{bmatrix} 4 & 5 & 1 \\ -2 & 6 & 0 \end{bmatrix}$$

18. Write an example of a 2×4 matrix.

19. List at least two different row operations.

20. Solve the system by using the Gauss-Jordan method.

$$2x - 4y = -2$$
$$4x + y = 5$$

Polynomials

CHAPTER OUTLINE

Chapter 5

In this chapter, we study addition, subtraction, multiplication, and division of polynomials, along with an important operation called factoring.

Review Your Skills

To prepare for this chapter, use the crossword puzzle to review the properties of real numbers and exponents from Chapter R.

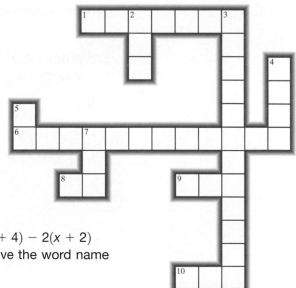

Across

1. Simplify and give the word name for $(-2)^4$.

6. $a(b + c) = ab + ac$ illustrates the _____ property.

8. Simplify. $\left(\dfrac{1}{6}\right)^{-1} + \left(\dfrac{1}{3}\right)^{-2}$

9. Simplify and give the word name for y^0.

10. Simplify. $\left(\dfrac{3r^4}{r^5}\right)^2 \cdot (2r)^3$

Down

2. Simplify. $3(x + 4) - 2(x + 2)$

3. Simplify and give the word name for -2^2.

4. True or false: $(x + 2)(x + 3) = (x + 3)(x + 2)$

5. Simplify. $\dfrac{16d^4}{2d^3}$

7. Simplify. $-4(5 + t) + 5(t + 3) + 10$

Section 5.1 Properties of Integer Exponents and Scientific Notation

1. Simplifying Expressions with Exponents

In Section R.3, we learned that exponents are used to represent repeated multiplication. The following properties of exponents (Table 5-1) are often used to simplify algebraic expressions.

Table 5-1 Properties of Exponents*

Description	Property	Example	Expanded Form
Multiplication of like bases	$b^m \cdot b^n = b^{m+n}$	$b^2 \cdot b^4 = b^{2+4}$ $= b^6$	$b^2 \cdot b^4 = (b \cdot b)(b \cdot b \cdot b \cdot b)$ $= b^6$
Division of like bases	$\dfrac{b^m}{b^n} = b^{m-n}$	$\dfrac{b^5}{b^2} = b^{5-2}$ $= b^3$	$\dfrac{b^5}{b^2} = \dfrac{b \cdot b \cdot b \cdot b \cdot b}{b \cdot b}$ $= b^3$
Power rule	$(b^m)^n = b^{m \cdot n}$	$(b^4)^2 = b^{4 \cdot 2}$ $= b^8$	$(b^4)^2 = (b \cdot b \cdot b \cdot b)(b \cdot b \cdot b \cdot b)$ $= b^8$
Power of a product	$(ab)^m = a^m b^m$	$(ab)^3 = a^3 b^3$	$(ab)^3 = (ab)(ab)(ab)$ $= (a \cdot a \cdot a)(b \cdot b \cdot b) = a^3 b^3$
Power of a quotient	$\left(\dfrac{a}{b}\right)^m = \dfrac{a^m}{b^m}$	$\left(\dfrac{a}{b}\right)^3 = \dfrac{a^3}{b^3}$	$\left(\dfrac{a}{b}\right)^3 = \left(\dfrac{a}{b}\right)\left(\dfrac{a}{b}\right)\left(\dfrac{a}{b}\right)$ $= \dfrac{a \cdot a \cdot a}{b \cdot b \cdot b} = \dfrac{a^3}{b^3}$

*Assume that a and b are real numbers ($b \neq 0$) and that m and n represent integers.

In addition to the properties of exponents, two definitions are used to simplify algebraic expressions.

Definition of b^0 and b^{-n}

Let n be an integer, and b be a real number such that $b \neq 0$.

*1. $b^0 = 1$ <u>Example:</u> $5^0 = 1$

2. $b^{-n} = \left(\dfrac{1}{b}\right)^n = \dfrac{1}{b^n}$ <u>Example:</u> $4^{-3} = \left(\dfrac{1}{4}\right)^3 = \dfrac{1}{4^3}$ or $\dfrac{1}{64}$

3. From the definition of b^{-n} we also have:

$\left(\dfrac{a}{b}\right)^{-n} = \left(\dfrac{b}{a}\right)^n = \dfrac{b^n}{a^n}$ for $a \neq 0$, $b \neq 0$.

<u>Example:</u> $\left(\dfrac{3}{7}\right)^{-2} = \left(\dfrac{7}{3}\right)^2 = \dfrac{7^2}{3^2}$ or $\dfrac{49}{9}$

*Note: The value of 0^0 is not defined by definition 1 because the base, b, must not equal 0.

The definition of b^0 is consistent with the properties of exponents. For example, if b is a nonzero real number and n is an integer, then

$$\dfrac{b^n}{b^n} = 1 \longleftarrow$$

The expression $b^0 = 1$

$$\dfrac{b^n}{b^n} = b^{n-n} = b^0 \longrightarrow$$

The definition of b^{-n} is also consistent with the properties of exponents. If b is a nonzero real number, then

$$\frac{b^3}{b^5} = \frac{b \cdot b \cdot b}{b \cdot b \cdot b \cdot b \cdot b} = \frac{1}{b^2}$$

The expression $b^{-2} = \frac{1}{b^2}$

$$\frac{b^3}{b^5} = b^{3-5} = b^{-2}$$

Example 1 **Simplifying Expressions with Exponents**

Simplify the expressions.

a. $(-2)^4$ **b.** -2^4 **c.** -2^{-4} **d.** $(-7x)^0$ **e.** $-7x^0$

Solution:

a. $(-2)^4 = (-2)(-2)(-2)(-2)$

$\qquad = 16$

b. $-2^4 = -1 \cdot 2^4$

$\qquad = -1 \cdot (2 \cdot 2 \cdot 2 \cdot 2)$

$\qquad = -16$

c. $-2^{-4} = -1 \cdot (2^{-4})$

$\qquad = -1 \cdot \left(\frac{1}{2^4}\right)$

$\qquad = -1 \cdot \frac{1}{2 \cdot 2 \cdot 2 \cdot 2}$

$\qquad = -\frac{1}{16}$

d. $(-7x)^0 = 1$ because $b^0 = 1$

e. $-7x^0 = -7 \cdot x^0$

$\qquad = -7 \cdot 1$

$\qquad = -7$

Example 2 **Using Properties of Exponents**

Simplify the expressions. Write the final answer with positive exponents only.

a. $x^3 x^5 x^{-2}$ **b.** $\dfrac{y^7}{y^4}$ **c.** $(b^2)^{-5}$

Solution:

a. $x^3 x^5 x^{-2}$

$\qquad = x^{3+5+(-2)}$ Multiply like bases by adding exponents.

$\qquad = x^6$ Simplify.

b. $\dfrac{y^7}{y^4}$

$\qquad = y^{7-4}$ 　　　　Divide like bases by subtracting exponents.

$\qquad = y^3$ 　　　　Simplify.

c. $(b^2)^{-5}$

$\qquad = b^{2(-5)}$ 　　　　Apply the power rule.

$\qquad = b^{-10}$ 　　　　Multiply the exponents.

$\qquad = \dfrac{1}{b^{10}}$ 　　　　Write the answer with positive exponents.

Example 3　**Simplifying an Expression with Exponents**

Simplify the expression. 　　$\left(\dfrac{1}{5}\right)^{-3} - (2)^{-2} + 3^0$

Solution:

$\left(\dfrac{1}{5}\right)^{-3} - (2)^{-2} + 3^0$

$\qquad = 5^3 - \left(\dfrac{1}{2}\right)^2 + 3^0$ 　　　Simplify negative exponents.

$\qquad = 125 - \dfrac{1}{4} + 1$ 　　　Evaluate expressions with exponents.

$\qquad = \dfrac{500}{4} - \dfrac{1}{4} + \dfrac{4}{4}$ 　　　Write the expressions with a common denominator.

$\qquad = \dfrac{503}{4}$ 　　　Simplify.

Example 4　**Simplifying an Expression with Exponents**

Simplify the expression. Write the answer with positive exponents only.

$$\dfrac{(2a^7b^{-4})^3}{(4a^3b^{-2})^2}$$

Solution:

$\dfrac{(2a^7b^{-4})^3}{(4a^3b^{-2})^2}$

$\qquad = \dfrac{2^3a^{21}b^{-12}}{4^2a^6b^{-4}}$ 　　　Apply the power rule.

$\qquad = \dfrac{8a^{21}b^{-12}}{16a^6b^{-4}}$ 　　　Simplify the coefficients.

$$= \frac{8a^{21-6}b^{-12-(-4)}}{16} \qquad \text{Divide like bases by subtracting exponents.}$$

$$= \frac{\overset{1}{8}a^{15}b^{-8}}{\underset{2}{16}} \qquad \text{Simplify.}$$

$$= \frac{a^{15}}{2b^8} \qquad \text{Simplify negative exponents.}$$

Example 5 Simplifying an Expression with Exponents

Simplify the expression. Write the answer with positive exponents only.

$$4xy^{-3}\left(\frac{8x^2}{3x^5y^2}\right)^{-2}$$

Solution:

$$4xy^{-3}\left(\frac{8x^2}{3x^5y^2}\right)^{-2} = 4xy^{-3} \cdot \left(\frac{8}{3x^3y^2}\right)^{-2} \qquad \text{Simplify within parentheses.}$$

$$= \frac{4xy^{-3}}{1} \cdot \frac{8^{-2}}{3^{-2}x^{-6}y^{-4}} \qquad \begin{array}{l}\text{Raise the expression in} \\ \text{parentheses to the } -2 \text{ power.}\end{array}$$

$$= \frac{4x}{y^3} \cdot \frac{3^2x^6y^4}{8^2} \qquad \text{Simplify negative exponents.}$$

$$= \frac{4 \cdot 9 \cdot x \cdot x^6 \cdot y^4}{64y^3} \qquad \begin{array}{l}\text{Simplify the expressions } 3^2 \text{ and} \\ 8^2 \text{ and multiply the fractions.}\end{array}$$

$$= \frac{\overset{1}{4} \cdot 9 \cdot x^{1+6}y^{4-3}}{\underset{16}{64}} \qquad \begin{array}{l}\text{Add the exponents on } x. \\ \text{Subtract the exponents on } y.\end{array}$$

$$= \frac{9x^7y}{16} \qquad \text{Simplify.}$$

2. Scientific Notation

Scientists in a variety of fields often work with very large or very small numbers. For instance, the distance between the Earth and the Sun is approximately 93,000,000 mi. The mass of an electron is 0.000 000 000 000 000 000 000 000 000 911 kg.

Scientific notation was devised as a shortcut method of expressing very large and very small numbers. The principle behind scientific notation is to use a power of 10 to express the magnitude of the number. For example, a number such as 50,000 can be written as $5 \times 10,000$ or equivalently as 5×10^4. Similarly, the number 0.0035 is equal to $3.5 \times \frac{1}{1000}$ or, equivalently, 3.5×10^{-3}.

Definition of Scientific Notation

A number expressed in the form $a \times 10^n$, where $1 \leq |a| < 10$ and n is an integer, is said to be written in **scientific notation**.

Consider the following numbers in scientific notation:

The distance between the Sun and the Earth: $93{,}000{,}000$ mi $= 9.3 \times 10^7$ mi

7 places

The mass of an electron: $0.000\,000\,000\,000\,000\,000\,000\,000\,000\,000\,911$ kg

$= 9.11 \times 10^{-31}$ kg   31 places

In each case, the power of 10 corresponds to the number of place positions that the decimal point is moved. The power of 10 is sometimes called the order of magnitude (or simply the magnitude) of the number. The order of magnitude of the distance between the Earth and Sun is 10^7 mi (tens of millions). The mass of an electron has an order of magnitude of 10^{-31} kg.

Skill Practice

Rewrite each number in either scientific notation or standard notation.

13. 2,600,000 **14.** 0.00088

15. -5.7×10^{-8} **16.** 1.9×10^5

Example 6 ── **Writing Numbers in Scientific Notation**

Fill in the table by writing the numbers in scientific notation or standard notation as indicated.

Quantity	Standard Notation	Scientific Notation
Number of NASCAR fans	75,000,000 people	
Width of an influenza virus	0.000000001 m	
Cost of hurricane Katrina		$\$1.25 \times 10^{11}$
Probability of winning the Florida state lottery		$4.35587878 \times 10^{-8}$

Solution:

Quantity	Standard Notation	Scientific Notation
Number of NASCAR fans	75,000,000 people	7.5×10^7 people
Width of an influenza virus	0.000000001 m	1×10^{-9} m
Cost of hurricane Katrina	$125,000,000,000	$\$1.25 \times 10^{11}$
Probability of winning the Florida state lottery	0.0000000435587878	$4.35587878 \times 10^{-8}$

Calculator Connections

Topic: Using Scientific Notation on a Calculator

Calculators use scientific notation to display very large or very small numbers. To enter scientific notation in a calculator, try using the **EE** key or the **EXP** key to express the power of 10.

Notice that E in the display of the calculator introduces the power of 10.

```
9.3E7
              93000000
7.25E-2
                 .0725
```

Example 7 **Applying Scientific Notation**

a. During a recent economic crisis, the U.S. government lent money to troubled financial institutions that had a large number of mortgage-related assets. The government committed an estimated $750,000,000,000. How much money does this represent per person if the U.S. population was 300,000,000 at that time?

b. The mean distance between the Earth and the Andromeda Galaxy is approximately 1.8×10^6 light-years. Assuming that 1 light-year is 6×10^{12} mi, what is the distance in miles to the Andromeda Galaxy?

Solution:

a. Express each value in scientific notation. Then divide the total amount to be paid off by the number of people.

$$\frac{7.5 \times 10^{11}}{3 \times 10^8}$$

$$= \left(\frac{7.5}{3}\right) \times \left(\frac{10^{11}}{10^8}\right) \quad \text{Divide 7.5 by 3 and subtract the powers of 10.}$$

$$= 2.5 \times 10^3$$

In standard notation, this amounts to $2500 per person.

b. Multiply the number of light-years by the number of miles per light-year.

$$(1.8 \times 10^6)(6 \times 10^{12})$$

$$= (1.8)(6) \times (10^6)(10^{12})$$

$$= 10.8 \times 10^{18} \qquad \text{Multiply 1.8 and 6 and add the powers of 10.}$$

The number 10.8×10^{18} is not in "proper" scientific notation because 10.8 is not between 1 and 10.

$$= (1.08 \times 10^1) \times 10^{18} \qquad \text{Rewrite 10.8 as } 1.08 \times 10^1.$$

$$= 1.08 \times (10^1 \times 10^{18}) \qquad \text{Apply the associative property of multiplication.}$$

$$= 1.08 \times 10^{19}$$

The distance between the Earth and the Andromeda Galaxy is 1.08×10^{19} mi.

Calculator Connections

Topic: Multiplying and Dividing with Scientific Notation

Use a calculator to check the solutions to Example 7.

```
(7.5E11)/(3.0E8)
              2500
(1.8E6)*(6.0E12)
           1.08E19
```

Answers

17. Approximately 2.5×10^5 pennies
18. 4.56×10^{13} mi

Section 5.1   Practice Exercises

Vocabulary and Key Concepts

1. **a.** A(n) _____ is used to show repeated multiplication of the base.

   b. For $b \neq 0$, the expression b^0 is defined to be _____.

   c. For $b \neq 0$, the expression b^{-n} is defined as _____.

   d. A number expressed in the form $a \times 10^n$, where $1 \leq |a| < 10$ and n is an integer is said to be written in

   _____ _____.

Concept 1: Simplifying Expressions with Exponents

2. Write the expressions in expanded form and simplify.

   $$b^4 \cdot b^3 \text{ and } (b^4)^3$$

3. Write the expressions in expanded form and simplify.

   $$ab^3 \text{ and } (ab)^3$$

For Exercises 4–9, write an example of each property. (Answers may vary.)

4. $b^n \cdot b^m = b^{n+m}$

5. $(ab)^n = a^n b^n$

6. $(b^n)^m = b^{nm}$

7. $\dfrac{b^n}{b^m} = b^{n-m} \quad (b \neq 0)$

8. $\left(\dfrac{a}{b}\right)^n = \dfrac{a^n}{b^n} \quad (b \neq 0)$

9. $b^0 = 1 \quad (b \neq 0)$

For Exercises 10–28, simplify. **(See Example 1.)**

10. $\left(\dfrac{2}{3}\right)^{-1}$

11. $\left(\dfrac{1}{3}\right)^{-1}$

12. 3^{-1}

13. 5^{-2}

14. 8^{-2}

15. -5^{-2}

16. -8^{-2}

17. $(-5)^{-2}$

18. $(-8)^{-2}$

19. $\left(-\dfrac{1}{4}\right)^{-3}$

20. $\left(-\dfrac{3}{8}\right)^{-1}$

21. $\left(-\dfrac{3}{2}\right)^{-4}$

22. $\left(-\dfrac{1}{9}\right)^{-2}$

23. $-\left(\dfrac{2}{5}\right)^{-3}$

24. $-\left(\dfrac{1}{2}\right)^{-5}$

25. $(10ab)^0$

26. $(13x)^0$

27. $10ab^0$

28. $13x^0$

For Exercises 29–80, simplify and write the answer with positive exponents only. **(See Examples 2–5.)**

29. $y^3 \cdot y^5$

30. $x^4 \cdot x^8$

31. $\dfrac{13^8}{13^6}$

32. $\dfrac{5^7}{5^3}$

33. $(y^2)^4$

34. $(z^3)^4$

35. $(3x^2)^4$

36. $(2y^5)^3$

37. p^{-3}

38. q^{-5}

39. $7^{10} \cdot 7^{-13}$

40. $11^{-9} \cdot 11^7$

41. $\dfrac{w^3}{w^5}$

42. $\dfrac{t^4}{t^8}$

43. $a^{-2}a^{-5}$

44. $b^{-1}b^{-8}$

45. $\dfrac{r}{r^{-1}}$

46. $\dfrac{s^{-1}}{s}$

47. $\dfrac{z^{-6}}{z^{-2}}$

48. $\dfrac{w^{-8}}{w^{-3}}$

49. $\dfrac{a^3}{b^{-2}}$

50. $\dfrac{c^4}{d^{-1}}$

51. $(6xyz^2)^0$

52. $(-7ab^3)^0$

53. $2^4 + 2^{-2}$

54. $3^2 + 3^{-1}$

55. $1^{-2} + 5^{-2}$

56. $4^{-2} + 2^{-2}$

57. $\left(\dfrac{2}{3}\right)^{-2} - \left(\dfrac{1}{2}\right)^2 + \left(\dfrac{1}{3}\right)^0$ **58.** $\left(\dfrac{1}{6}\right)^{-1} + \left(\dfrac{2}{3}\right)^0 - \left(\dfrac{1}{4}\right)^{-2}$ **59.** $\left(\dfrac{4}{5}\right)^{-1} + \left(\dfrac{3}{2}\right)^2 - \left(\dfrac{2}{7}\right)^0$ **60.** $\left(\dfrac{4}{5}\right)^0 - \left(\dfrac{2}{3}\right)^2 + \left(\dfrac{9}{5}\right)^{-1}$

61. $\dfrac{p^2 q}{p^5 q^{-1}}$

62. $\dfrac{m^{-1} n^3}{m^4 n^{-2}}$

63. $\dfrac{-48ab^{10}}{32a^4 b^3}$

64. $\dfrac{25x^2 y^{12}}{10x^5 y^7}$

65. $(-3x^{-4} y^5 z^2)^{-4}$

66. $(-6a^{-2} b^3 c)^{-2}$

67. $(4m^{-2} n)(-m^6 n^{-3})$

68. $(-6pq^{-3})(2p^4 q)$

 69. $(p^{-2} q)^3 (2pq^4)^2$

70. $(mn^3)^2 (5m^{-2} n^2)$

71. $\left(\dfrac{x^2}{y}\right)^3 (5x^2 y)$

72. $\left(\dfrac{a}{b^2}\right)^2 (3a^2 b^3)$

73. $\dfrac{(-8a^2 b^2)^4}{(16a^3 b^7)^2}$

74. $\dfrac{(-3x^2 y^3)^2}{(-2xy^4)^3}$

75. $\left(\dfrac{-2x^6 y^{-5}}{3x^{-2} y^4}\right)^{-3}$

76. $\left(\dfrac{-6a^2 b^{-3}}{5a^{-1} b}\right)^{-2}$

77. $\left(\dfrac{2x^{-3} y^0}{4x^6 y^{-5}}\right)^{-2}$

78. $\left(\dfrac{a^3 b^2 c^0}{a^{-1} b^{-2} c^{-3}}\right)^{-2}$

79. $3xy^5 \left(\dfrac{2x^4 y}{6x^5 y^3}\right)^{-2}$

80. $7x^{-3} y^{-4} \left(\dfrac{3x^{-1} y^5}{9x^3 y^{-2}}\right)^{-3}$

Concept 2: Scientific Notation

81. The European Organization for Nuclear Research (known by the acronym CERN) has built the world's largest high-energy particle accelerator, called the Large Hadron Collider (LHC). Scientists hope the LHC will answer many open questions in physics. Write the following numbers in scientific notation. **(See Example 6.)**

   a. The LHC cost $8,000,000,000 to build.

   b. 3,000,000 DVDs worth of data will be produced each year.

   c. 14,000,000,000,000 electron volts (eV) of energy will be produced to smash the protons together.

   d. 1 eV is equivalent to 0.000 000 000 000 000 000 1602 joules (J).

82. Write the numbers in scientific notation.

   a. The approximate population of the United States in 2013 was 313,000,000.

   b. The size of the smallest visible object in an optical microscope is 0.0000002 m.

   c. A trillion is defined as 1,000,000,000,000.

83. Write the numbers in standard notation.

   a. The Andromeda Galaxy contains at least 2×10^{11} stars.

   b. The diameter of a capillary is 4×10^{-6} m.

   c. The mean distance of Venus from the Sun is 1.082×10^{11} m.

84. Write the numbers in standard notation.

   a. At the end of a recent year, the Department of Energy's inventory of high-level radioactive waste was approximately 3.784×10^5 m³.

   b. The diameter of a water molecule is 3×10^{-10} m.

   c. The distance a bullet will travel in 1 sec when fired from a 0.22 caliber gun is 4.1×10^2 m.

For Exercises 85–90, determine which numbers are in "proper" scientific notation. If the number is not in "proper" scientific notation, correct it.

85. 35×10^4

86. 0.469×10^{-7}

87. 7×10^0

88. 8.12×10^1

89. 9×10^1

90. 6.9×10^0

For Exercises 91–98, perform the indicated operations and write the answer in scientific notation.

91. $(6.5 \times 10^3)(5.2 \times 10^{-8})$

92. $(3.26 \times 10^{-6})(8.2 \times 10^9)$

93. $(0.0000024)(6,700,000,000)$

94. $(3,400,000,000)(70,000,000,000,000)$

95. $(8.5 \times 10^{-2}) \div (2.5 \times 10^{-15})$

96. $(3 \times 10^9) \div (1.5 \times 10^{13})$

97. $(900,000,000) \div (360,000)$

98. $(0.0000000002) \div (8,000,000)$

99. If one H_2O molecule contains 2 hydrogen atoms and 1 oxygen atom, and 10 H_2O molecules contain 20 hydrogen atoms and 10 oxygen atoms, how many hydrogen atoms and oxygen atoms are contained in 6.02×10^{23} H_2O molecules? **(See Example 7.)**

100. The star named Alpha Centauri is 4.3 light-years from the Earth. If 1 light-year is approximately 6×10^9 mi, how far is Alpha Centauri?

101. If the county of Queens, New York, has a population of approximately 2,200,000 and the area is 110 mi^2, how many people are there per square mile? **(See Example 7.)**

102. If the county of Catawba, North Carolina, has a population of approximately 150,000 and the area is 400 mi^2, how many people are there per square mile?

103. According to the Federal Emergency Management Agency (FEMA), the annual loss due to earthquakes in California is approximately $\$3.5 \times 10^9$. If this is representative as a yearly average, find the loss over 15 yr.

104. Avogadro's number $N_a = 6.02 \times 10^{23}$ is the number of atoms in 1 *mole* of an element.

    a. How many atoms are in 5 moles of carbon-12?

    b. If 75 g of carbon-12 has 4.515×10^{25} atoms, how many moles is this?

Expanding Your Skills

105. A 20-yr-old starts a savings plan for her retirement. She will put \$20 per month into a mutual fund that she hopes will average 6% growth annually.

    a. If she plans to retire at age 65, for how many months will she be depositing money?

    b. By age 65, how much money will she have deposited?

    c. The value of an account built in this fashion is given by

$$A = P \cdot \left[\left(1 + \frac{r}{12} \right)^N - 1 \right] \cdot \left(1 + \frac{12}{r} \right)$$

where A is the final amount of money in the account, P is the amount of the monthly deposit, and N is the number of months. Use a calculator to find the total amount in the woman's retirement account at age 65.

For Exercises 106–111, simplify each expression. Assume that a and b represent positive integers and x and y are nonzero real numbers.

106. $x^{a+1}x^{a+5}$

107. $y^{a-5}y^{a+7}$

108. $\dfrac{y^{2a+1}}{y^{a-1}}$

109. $\dfrac{x^{3a-3}}{x^{a+1}}$

110. $\dfrac{x^{3b-2}y^{b+1}}{x^{2b+1}y^{2b+2}}$

111. $\dfrac{x^{2a-2}y^{a+3}}{x^{a+4}y^{a-3}}$

	Section 5.2
Addition and Subtraction of Polynomials and Polynomial Functions	

1. Polynomials: Basic Definitions

One commonly used algebraic expression is called a polynomial. A **polynomial** in x is defined as a finite sum of terms of the form ax^n, where a is a real number and the exponent n is a whole number. For each term, a is called the **coefficient**, and n is called the **degree of the term**. For example:

Term (Expressed in the Form ax^n)	Coefficient	Degree
$3x^5$	3	5
x^{14} → rewrite as $1x^{14}$	1	14
7 → rewrite as $7x^0$	7	0
$\frac{1}{2}p$ → rewrite as $\frac{1}{2}p^1$	$\frac{1}{2}$	1

If a polynomial has exactly one term, it is called a **monomial**. A two-term polynomial is called a **binomial**, and a three-term polynomial is called a **trinomial**. Usually the terms of a polynomial are written in descending order according to degree. In descending order, the highest-degree term is written first and is called the **leading term**. Its coefficient is called the **leading coefficient**. The **degree of a polynomial** is the greatest degree of all its terms. Thus, the leading term determines the degree of the polynomial.

	Expression	Descending Order	Leading Coefficient	Degree of Polynomial
Monomials	$2x^9$	$2x^9$	2	9
	-49	-49	-49	0
Binomials	$10y - 7y^2$	$-7y^2 + 10y$	-7	2
	$6 - \frac{2}{3}b$	$-\frac{2}{3}b + 6$	$-\frac{2}{3}$	1
Trinomials	$w + 2w^3 + 9w^6$	$9w^6 + 2w^3 + w$	9	6
	$2.5a^4 - a^8 + 1.3a^3$	$-a^8 + 2.5a^4 + 1.3a^3$	-1	8

Polynomials may have more than one variable. In such a case, the degree of a term is the sum of the exponents of the variables contained in the term. For example, the term $2x^3y^4z$ has degree 8 because the exponents applied to x, y, and z are 3, 4, and 1, respectively.

The following polynomial has a degree of 12 because the highest degree of its terms is 12.

$$11x^4y^3z \quad - \quad 5x^3y^2z^7 \quad + \quad 2x^2y \quad + \quad 7$$

$$\begin{array}{cccc} \text{degree} & \text{degree} & \text{degree} & \text{degree} \\ 8 & 12 & 3 & 0 \end{array}$$

Concepts

1. Polynomials: Basic Definitions
2. Addition of Polynomials
3. Subtraction of Polynomials
4. Polynomial Functions

Concept Connections

Give the coefficient and the degree of each term.

1. $-5y$ **2.** $\frac{3}{4}c^8$

Concept Connections

Write the polynomial in descending order. Then give the degree and the leading coefficient.

3. $12 + 6t^3 - t^4$
4. $6x^3y^4$
5. Write a trinomial whose degree is 5. (Answers may vary.)

Answers

1. Degree: 1; coefficient: -5
2. Degree: 8; coefficient: $\frac{3}{4}$
3. $-t^4 + 6t^3 + 12$
   Degree: 4; leading coefficient: -1
4. $6x^3y^4$
   Degree: 7; leading coefficient: 6
5. For example: $2x^5 - 4x^2 + 1$

2. Addition of Polynomials

To add or subtract two polynomials, we combine *like* terms. Recall that two terms are **like terms** if they each have the same variables and the corresponding variables are raised to the same powers.

Example 1   Adding Polynomials

Add the polynomials.

a. $(3t^3 + 2t^2 - 5t) + (t^3 - 6t)$ **b.** $\left(\dfrac{2}{3}w^2 - w + \dfrac{1}{8}\right) + \left(\dfrac{4}{3}w^2 + 8w - \dfrac{1}{4}\right)$

Solution:

a. $(3t^3 + 2t^2 - 5t) + (t^3 - 6t)$

$\qquad = 3t^3 + t^3 + 2t^2 + (-5t) + (-6t)$ Group *like* terms.

$\qquad = 4t^3 + 2t^2 - 11t$ Add *like* terms.

b. $\left(\dfrac{2}{3}w^2 - w + \dfrac{1}{8}\right) + \left(\dfrac{4}{3}w^2 + 8w - \dfrac{1}{4}\right)$

$\qquad = \dfrac{2}{3}w^2 + \dfrac{4}{3}w^2 + (-w) + 8w + \dfrac{1}{8} + \left(-\dfrac{1}{4}\right)$ Group *like* terms.

$\qquad = \dfrac{6}{3}w^2 + 7w + \left(\dfrac{1}{8} - \dfrac{2}{8}\right)$ Add fractions with common denominators.

$\qquad = 2w^2 + 7w - \dfrac{1}{8}$ Simplify.

Example 2   Adding Polynomials

Add the polynomials. $(a^2b + 7ab + 6) + (5a^2b - 2ab - 7)$

Solution:

Polynomials can be added vertically. Be sure to line up the *like* terms.

$$
\begin{array}{r}
a^2b + 7ab + 6 \\
+\ 5a^2b - 2ab - 7 \\
\hline
6a^2b + 5ab - 1
\end{array}
$$
  Add *like* terms.

3. Subtraction of Polynomials

Subtraction of two polynomials is similar to subtracting real numbers. Add the opposite of the second polynomial to the first polynomial.

The opposite (or additive inverse) of a real number a is $-a$. Similarly, if A is a polynomial, then $-A$ is its opposite.

Example 3 Finding the Opposite of a Polynomial

Find the opposite of the polynomials.

a. $5a - 2b - c$ **b.** $-5.5y^4 - 2.4y^3 + 1.1y$

Solution:

Expression	Opposite	Simplified Form
a. $5a - 2b - c$	$-(5a - 2b - c)$	$-5a + 2b + c$
b. $-5.5y^4 - 2.4y^3 + 1.1y$	$-(-5.5y^4 - 2.4y^3 + 1.1y)$	$5.5y^4 + 2.4y^3 - 1.1y$

Skill Practice

Find the opposite of the polynomials.

9. $-7z + 6w$

10. $2p - 3q + r + 1$

TIP: Notice that the sign of each term is changed when finding the opposite of a polynomial.

Subtraction of Polynomials

If A and B are polynomials, then $A - B = A + (-B)$.

Example 4 Subtracting Polynomials

Subtract the polynomials. $(3x^2 + 2x - 5) - (4x^2 - 7x + 2)$

Solution:

$$(3x^2 + 2x - 5) - (4x^2 - 7x + 2)$$

$$= (3x^2 + 2x - 5) + (-4x^2 + 7x - 2) \qquad \text{Add the opposite of the second polynomial.}$$

$$= 3x^2 + (-4x^2) + 2x + 7x + (-5) + (-2) \qquad \text{Group } like \text{ terms.}$$

$$= -x^2 + 9x - 7 \qquad \text{Combine } like \text{ terms.}$$

Skill Practice

Subtract the polynomials.

11. $(6a^2 - 2a)$
$\quad - (-3a^2 + 2a + 3)$

Example 5 Subtracting Polynomials

Subtract the polynomials. $(6x^2y - 2xy + 5) - (x^2y - 3)$

Solution:

$$(6x^2y - 2xy + 5) - (x^2y - 3)$$

Subtraction of polynomials can be performed vertically by vertically aligning *like* terms. Then add the opposite of the second polynomial. "Placeholders" (shown in red) may be used to help line up *like* terms.

$$
\begin{array}{r}
6x^2y - 2xy + 5 \\
-(x^2y + 0xy - 3) \\
\end{array}
\quad \xrightarrow[\text{opposite.}]{\text{Add the}} \quad
\begin{array}{r}
6x^2y - 2xy + 5 \\
+ \ -x^2y - 0xy + 3 \\
\hline
5x^2y - 2xy + 8 \\
\end{array}
$$

Skill Practice

Subtract the polynomials.

12. $(7p^2q - 6)$
$\quad - (2p^2q + 4pq + 4)$

Answers

9. $7z - 6w$

10. $-2p + 3q - r - 1$

11. $9a^2 - 4a - 3$

12. $5p^2q - 4pq - 10$

Example 6 **Subtracting Polynomials**

Subtract $\frac{1}{2}x^4 - \frac{3}{4}x^2 + \frac{1}{5}$ from $\frac{3}{2}x^4 + \frac{1}{2}x^2 - 4x$.

Solution:

In general, to subtract a from b, we write $b - a$. Therefore, to subtract

$$\frac{1}{2}x^4 - \frac{3}{4}x^2 + \frac{1}{5} \qquad \text{from} \qquad \frac{3}{2}x^4 + \frac{1}{2}x^2 - 4x$$

we have

$$\left(\frac{3}{2}x^4 + \frac{1}{2}x^2 - 4x\right) - \left(\frac{1}{2}x^4 - \frac{3}{4}x^2 + \frac{1}{5}\right)$$

$$= \frac{3}{2}x^4 + \frac{1}{2}x^2 - 4x - \frac{1}{2}x^4 + \frac{3}{4}x^2 - \frac{1}{5} \qquad \begin{array}{l}\text{Subtract the polynomials by}\\\text{adding the opposite of the}\\\text{second polynomial.}\end{array}$$

$$= \frac{3}{2}x^4 - \frac{1}{2}x^4 + \frac{1}{2}x^2 + \frac{3}{4}x^2 - 4x - \frac{1}{5} \qquad \text{Group } like \text{ terms.}$$

$$= \frac{3}{2}x^4 - \frac{1}{2}x^4 + \frac{2}{4}x^2 + \frac{3}{4}x^2 - 4x - \frac{1}{5} \qquad \begin{array}{l}\text{Write } like \text{ terms with a}\\\text{common denominator.}\end{array}$$

$$= \frac{2}{2}x^4 + \frac{5}{4}x^2 - 4x - \frac{1}{5} \qquad \text{Combine } like \text{ terms.}$$

$$= x^4 + \frac{5}{4}x^2 - 4x - \frac{1}{5} \qquad \text{Simplify.}$$

4. Polynomial Functions

A **polynomial function** is a function defined by a finite sum of terms of the form ax^n, where a is a real number and n is a whole number. For example, the functions defined here are polynomial functions:

$$f(x) = 3x - 8$$

$$g(x) = 4x^5 - 2x^3 + 5x - 3$$

$$h(x) = -\frac{1}{2}x^4 + \frac{3}{5}x^3 - 4x^2 + \frac{5}{9}x - 1$$

$$k(x) = 7 \qquad (7 = 7x^0, \text{ which is of the form } ax^n, \text{ where } n = 0 \text{ is a whole number})$$

The following functions are *not* polynomial functions:

$$m(x) = \frac{1}{x} - 8 \qquad \left(\frac{1}{x} = x^{-1}, \text{ the exponent } -1 \text{ is not a whole number}\right)$$

$$q(x) = |x| \qquad (|x| \text{ is not of the form } ax^n)$$

Answer

13. $-\frac{1}{6}p^3 + \frac{5}{12}p^2 - \frac{3}{2}p$

Example 7 Evaluating a Polynomial Function

Given $P(x) = x^3 + 2x^2 - x - 2$, find the function values.

   a. $P(-3)$ **b.** $P(-1)$ **c.** $P(0)$ **d.** $P(2)$

Solution:

a. $P(x) = x^3 + 2x^2 - x - 2$

    $P(-3) = (-3)^3 + 2(-3)^2 - (-3) - 2$

          $= -27 + 2(9) + 3 - 2$

          $= -27 + 18 + 3 - 2$

          $= -8$

b. $P(-1) = (-1)^3 + 2(-1)^2 - (-1) - 2$

          $= -1 + 2(1) + 1 - 2$

          $= -1 + 2 + 1 - 2$

          $= 0$

c. $P(0) = (0)^3 + 2(0)^2 - (0) - 2$

       $= -2$

d. $P(2) = (2)^3 + 2(2)^2 - (2) - 2$

       $= 8 + 2(4) - 2 - 2$

       $= 8 + 8 - 2 - 2$

       $= 12$

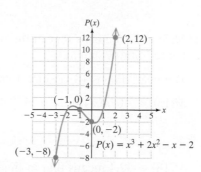

The function values can be confirmed from the graph of $y = P(x)$ (Figure 5-1).

Figure 5-1

Skill Practice

Given: $P(x) = -2x^3 - 4x + 6$

14. Find $P(0)$.

15. Find $P(-2)$.

16. Find $P(-1)$.

17. Find $P(2)$.

Example 8 Using Polynomial Functions in an Application

The length of a rectangle is 4 m less than 3 times the width. Let x represent the width. Write a polynomial function P that represents the perimeter of the rectangle and simplify the result.

Solution:

Let x represent the width. Then $3x - 4$ is the length. The perimeter of a rectangle is given by $P = 2L + 2W$. Thus,

        $P(x) = 2(3x - 4) + 2(x)$

             $= 6x - 8 + 2x$

             $= 8x - 8$

x

$3x - 4$

Skill Practice

18. The longest side of a triangle is 2 ft less than 4 times the shortest side. The middle side is 3 ft more than twice the shortest side. Let x represent the shortest side. Find a polynomial function P that represents the perimeter of the triangle, and simplify the result.

Answers

14. $P(0) = 6$ **15.** $P(-2) = 30$

16. $P(-1) = 12$ **17.** $P(2) = -18$

18. $P(x) = x + (4x - 2) + (2x + 3)$

       $= 7x + 1$

| **Example 9** | **Using a Polynomial Function in an Application** |

The percent of male smokers in the United States has varied by year since 1990. The function defined by $M(x) = -0.003x^2 - 0.257x + 28.0$ approximates the percent of male smokers, $M(x)$, where x is the number of years since 1990. See Figure 5-2.

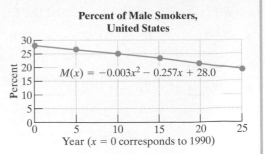

Percent of Male Smokers, United States

$M(x) = -0.003x^2 - 0.257x + 28.0$

Year ($x = 0$ corresponds to 1990)

Source: U.S. National Center for Health Statistics

Figure 5-2

a. Evaluate $M(5)$ and interpret the meaning in the context of this problem.

b. Determine the percent of male smokers in the year 2008.

Solution:

a. $M(5) = -0.003(5)^2 - 0.257(5) + 28.0$ Substitute $x = 5$ into the function.

≈ 26.6

$M(5) \approx 26.6$ means that in the year 1995, 26.6% of males in the United States smoked.

b. The year 2008 is 18 yr since 1990. Therefore, evaluate $M(18)$.

$M(18) = -0.003(18)^2 - 0.257(18) + 28.0$ Substitute $x = 18$ into the function.

≈ 22.4

$M(18) \approx 22.4$ means that in the year 2008, 22.4% of males in the United States smoked.

Section 5.2   Practice Exercises

Vocabulary and Key Concepts

1. a. A _____ in the variable, x, is a single term or a sum of terms of the form ax^n, where a is a real number and n is a nonnegative integer.

b. Given the term ax^n, a is called the _____, and _____ is called the degree of the term.

c. Given the term x, the coefficient of the term is _____ and the degree is _____.

d. A monomial is a polynomial with exactly _____ term(s).

e. A _____ is a polynomial with exactly two terms.

f. A _____ is a polynomial with exactly three terms.

g. The term with the highest degree is called the _____ term and its coefficient is called the _____ _____.

h. The degree of a polynomial is the _____ degree of all of its terms.

  i. The degree of a nonzero constant such as 7 is _____.

  j. If a term of a polynomial has more than one variable, then the degree of the term is the sum of the _____ of the variables contained in the term.

  k. A _____ function is a function defined by a finite sum of terms of the form ax^n, where a is a real number and n is a whole number.

Review Exercises

For Exercises 2–6, simplify the expression.

2. $\left(\dfrac{6tv^2}{15t^3}\right)^{-1}$

3. $(2ac^{-2})(5a^{-1}c^4)$

4. $\left(\dfrac{1}{4}\right)^{-2} - \left(\dfrac{2}{3}\right)^0$

5. $(3.4 \times 10^5)(5.0 \times 10^{-2})$

6. $\dfrac{6.2 \times 10^3}{3.1 \times 10^5}$

Concept 1: Polynomials: Basic Definitions

For Exercises 7–12, write the polynomial in descending order. Then identify the leading coefficient and the degree.

7. $a^2 - 6a^3 - a$

8. $2b - b^4 + 5b^2$

9. $6x^2 - x + 3x^4 - 1$

10. $8 - 4y + y^5 - y^2$

11. $100 - t^2$

12. $-51 + s^2$

For Exercises 13–18, write a polynomial in one variable that is described by the following. (Answers may vary.)

13. A monomial of degree 5

14. A monomial of degree 4

15. A trinomial of degree 2

16. A trinomial of degree 3

17. A binomial of degree 4

18. A binomial of degree 2

Concept 2: Addition of Polynomials

For Exercises 19–30, add the polynomials and simplify. (See Examples 1 and 2.)

19. $(-4m^2 + 4m) + (5m^2 + 6m)$

20. $(3n^3 + 5n) + (2n^3 - 2n)$

21. $(3x^4 - x^3 - x^2) + (3x^3 - 7x^2 + 2x)$

22. $(6x^3 - 2x^2 - 12) + (x^2 + 3x + 9)$

23. $\left(\dfrac{1}{2}w^3 + \dfrac{2}{9}w^2 - 1.8w\right) + \left(\dfrac{3}{2}w^3 - \dfrac{1}{9}w^2 + 2.7w\right)$

24. $\left(2.9t^4 - \dfrac{7}{8}t + \dfrac{5}{3}\right) + \left(-8.1t^4 - \dfrac{1}{8}t - \dfrac{1}{3}\right)$

25. Add $(9x^2y - 5xy + 1)$ to $(8x^2y + xy - 15)$.

26. Add $(-x^3y^2 + 5xy)$ to $(10x^3y^2 + x^2y - 10)$.

27. Add $(-7a + 6a^2 + 1)$ to $(-8 - 4a - 2a^2)$.

28. Add $(1 - 12p + 8p^3)$ to $(6p^2 + p^3 - 14)$.

29. $\begin{array}{r} 12x^3 \qquad\ + 6x - 8 \\ + (-3x^3 - 5x^2 - 4x \qquad\ ) \\ \hline \end{array}$

30. $\begin{array}{r} -8y^4 - 8y^3 - 6y^2 \qquad\qquad - 9 \\ + \ (4y^4 + 5y^3 \qquad\ - 10y - 3) \\ \hline \end{array}$

Concept 3: Subtraction of Polynomials

For Exercises 31–36, write the opposite of the given polynomial. (See Example 3.)

31. $-30y^3$

32. $-2x^2$

33. $4p^3 + 2p - 12$

34. $8t^2 - 4t - 3$

35. $-11ab^2 + a^2b$

36. $-23rs - 4r + 9s$

For Exercises 37–46, subtract the polynomials and simplify. **(See Examples 4 and 5.)**

37. $(13z^5 - z^2) - (7z^5 + 5z^2)$

38. $(8w^4 + 3w^2) - (12w^4 - w^2)$

39. $(-3x^3 + 3x^2 - x + 6) - (1 - x - x^2 - x^3)$

40. $(-8x^3 + 6x + 7) - (-4 - 2x - 5x^3)$

41. $(-3xy^3 + 3x^2y - x + 6) - (-xy^3 - xy - x + 1)$

42. $(-8x^2y^2 + 6xy^2 + 7xy) - (5xy^2 - 2xy - 4)$

43. $\begin{array}{r} 4t^3 - 6t^2 \qquad - 18 \\ - (3t^3 + 7t^2 + 9t - 5) \end{array}$

44. $\begin{array}{r} 5w^3 - 9w^2 + 6w + 13 \\ - (7w^3 \qquad - 10w - 8) \end{array}$

45. $\left(\dfrac{1}{5}a^2 - \dfrac{1}{2}ab + \dfrac{1}{10}b^2 + 3\right) - \left(-\dfrac{3}{10}a^2 + \dfrac{2}{5}ab - \dfrac{1}{2}b^2 - 5\right)$

46. $\left(\dfrac{4}{7}a^2 - \dfrac{1}{7}ab + \dfrac{1}{14}b^2 - 7\right) - \left(\dfrac{1}{2}a^2 - \dfrac{2}{7}ab - \dfrac{9}{14}b^2 + 1\right)$

47. Subtract $(9x^2 - 5x + 1)$ from $(8x^2 + x - 15)$. **(See Example 6.)**

48. Subtract $(-x^3 + 5x)$ from $(10x^3 + x^2 - 10)$.

49. Find the difference of $(3x^5 - 2x^3 + 4)$ and $(x^4 + 2x^3 - 7)$.

50. Find the difference of $(7x^{10} - 2x^4 - 3x)$ and $(-4x^3 - 5x^4 + x + 5)$.

Mixed Exercises

For Exercises 51–74, add or subtract as indicated. Write the answers in descending order, if possible.

51. $(8y^2 - 4y^3) - (3y^2 - 8y^3)$

52. $(-9y^2 - 8) - (4y^2 + 3)$

53. $(-2r - 6r^4) + (-r^4 - 9r)$

54. $(-8s^9 + 7s^2) + (7s^9 - s^2)$

55. $(5xy + 13x^2 + 3y) - (4x^2 - 8y)$

56. $(6p^2q - 2q) - (-2p^2q + 13)$

57. $(11ab - 23b^2) + (7ab - 19b^2)$

58. $(-4x^2y + 9) + (8x^2y - 12)$

59. $[2p - (3p + 5)] + (4p - 6) + 2$

60. $-(q - 2) - [4 - (2q - 3) + 5]$

61. $5 - [2m^2 - (4m^2 + 1)]$

62. $[4n^3 - (n^3 + 4)] + 3n^3$

63. $(6x^3 - 5) - (-3x^3 + 2x) - (2x^3 - 6x)$

64. $(9p^4 - 2) + (7p^4 + 1) - (8p^4 - 10)$

65. $(-ab + 5a^2b) - [7ab^2 - 2ab - (7a^2b + 2ab^2)]$

66. $(m^3n^2 + 4m^2n) - [-5m^3n^2 - 4mn - (7m^2n - 6mn)]$

67. $(8x^3 - x^2 + 3) - [5x^2 + x - (4x^3 + x - 2)]$

68. $(y^2 + 6y - 6) - [(2y^3 - 4y) - (3y^2 + y + 1)]$

69. $\begin{array}{r} 12a^2b - 4ab^2 - ab \\ -(4a^2b + ab^2 - 5ab) \end{array}$

70. $\begin{array}{r} 2x^2 - 7xy + 3y^2 \\ - (9x^2 - 10xy - y^2) \end{array}$

71. $\begin{array}{r} -5x^4 \qquad - 11x^2 \qquad + 6 \\ - (-5x^4 + 3x^3 + 5x^2 - 10x + 5) \end{array}$

72. $\begin{array}{r} 9z^4 \qquad + 2z^2 \qquad + 11 \\ - (9z^4 - 4z^3 + 8z^2 - 9z - 4) \end{array}$

73. $\begin{array}{r} -2.2p^5 - 9.1p^4 \qquad + 5.3p^2 - 7.9p \\ + \qquad - 6.4p^4 - 8.5p^3 - 10.3p^2 \end{array}$

74. $\begin{array}{r} 5.5w^4 \qquad + 4.6w^2 - 9.3w - 8.3 \\ + 0.4w^4 - 7.3w^3 \qquad - 5.8w + 4.6 \end{array}$

For Exercises 75 and 76, find the perimeter.

75.

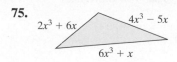

76.

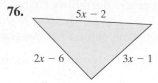

Concept 4: Polynomial Functions

For Exercises 77–84, determine whether the given function is a polynomial function. If it is a polynomial function, state the degree. If not, state the reason why.

77. $h(x) = \frac{2}{3}x^2 - 5$

78. $k(x) = -7x^4 - 0.3x + x^3$

79. $p(x) = 8x^3 + 2x^2 - \frac{3}{x}$

80. $q(x) = x^2 - 4x^{-3}$

81. $g(x) = -7$

82. $g(x) = 4x$

83. $M(x) = |x| + 5x$

84. $N(x) = x^2 + |x|$

85. Given $P(x) = -x^4 + 2x - 5$, find the function values. **(See Example 7.)**

   a. $P(2)$ **b.** $P(-1)$ **c.** $P(0)$ **d.** $P(1)$

86. Given $N(x) = -x^2 + 5x$, find the function values.

   a. $N(1)$ **b.** $N(-1)$ **c.** $N(2)$ **d.** $N(0)$

87. Given $H(x) = \frac{1}{2}x^3 - x + \frac{1}{4}$, find the function values.

   a. $H(0)$ **b.** $H(2)$ **c.** $H(-2)$ **d.** $H(-1)$

88. Given $K(x) = \frac{2}{3}x^2 + \frac{1}{9}$, find the function values.

   a. $K(0)$ **b.** $K(3)$ **c.** $K(-3)$ **d.** $K(-1)$

89. A rectangular garden is designed to be 3 ft longer than it is wide. Let x represent the width of the garden. Find a function P that represents the perimeter in terms of x. **(See Example 8.)**

90. Pauline measures a rectangular conference room and finds that the length is 4 yd greater than twice the width. Let x represent the width. Find a function P that represents the perimeter in terms of x.

91. The cost in dollars of producing x calendars is $C(x) = 5.40x + 99$. The revenue for selling x calendars is $R(x) = 12x$. To calculate profit, subtract the cost from the revenue.

   a. Write and simplify a function P that represents profit in terms of x.

   b. Find the profit of producing and selling 50 calendars.

92. The cost in dollars of producing x lawn chairs is $C(x) = 4.5x + 10.1$. The revenue for selling x chairs is $R(x) = 12.99x$. To calculate profit, subtract the cost from the revenue.

   a. Write and simplify a function P that represents profit in terms of x.

   b. Find the profit of producing and selling 100 lawn chairs.

93. The function defined by $D(x) = 5.2x^2 + 40.4x + 1636$ approximates the average yearly dormitory charge for 4-yr universities x years since 1990. $D(x)$ is the cost in dollars, and x represents the number of years since 1990. **(See Example 9.)**

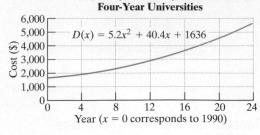

Yearly Dormitory Cost, Four-Year Universities

a. Evaluate $D(0)$ and $D(18)$ and interpret their meaning in the context of this problem.

b. If this trend continues, what will the annual dormitory charge be in the year 2015?

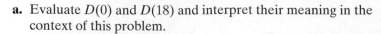

Source: U.S. National Center for Education Statistics

94. The population of bacteria in a culture can be modeled by $P(t) = -0.01t^3 + 12.96t + 10$, where t is the time in hours after the culture was started and $P(t)$ is the population in thousands.

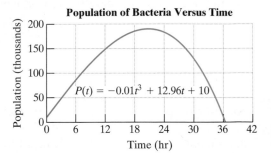

Population of Bacteria Versus Time

a. Evaluate $P(0)$ and $P(14)$ and interpret their meaning in the context of this problem.

b. Predict the population of bacteria 24 hr after the culture was started.

95. The number of women, W, to be paid child support in the United States can be approximated by

$$W(t) = 143t + 6580$$

where t is the number of years since 2000, and $W(t)$ is the yearly total measured in thousands. (*Source:* U.S. Bureau of the Census)

a. Evaluate $W(0)$, $W(5)$, and $W(10)$.

b. Interpret the meaning of the function value $W(10)$.

96. The total yearly amount of child support due (in billions of dollars) in the United States can be approximated by

$$D(t) = 0.925t + 4.625$$

where t is the number of years since 2000, and $D(t)$ is the amount due (in billions of dollars).

a. Evaluate $D(0)$, $D(4)$, and $D(8)$.

b. Interpret the meaning of the function value of $D(8)$.

Expanding Your Skills

97. A toy rocket is shot from ground level at an angle of 60° from the horizontal. See the figure. The x- and y-positions of the rocket (measured in feet) vary with time t according to

$$x(t) = 25t$$
$$y(t) = -16t^2 + 43.3t$$

a. Evaluate $x(0)$ and $y(0)$, and write the values as an ordered pair. Interpret the meaning of these function values in the context of this problem. Match the ordered pair with a point on the graph.

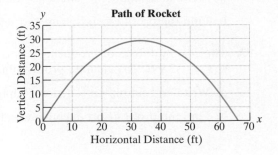

Path of Rocket

b. Evaluate $x(1)$ and $y(1)$ and write the values as an ordered pair. Interpret the meaning of these function values in the context of this problem. Match the ordered pair with a point on the graph.

c. Evaluate $x(2)$ and $y(2)$, and write the values as an ordered pair. Match the ordered pair with a point on the graph.

Multiplication of Polynomials	Section 5.3

1. Multiplying Polynomials

The properties of exponents covered in Section 5.1 can be used to simplify many algebraic expressions including the multiplication of monomials. To multiply monomials, first use the associative and commutative properties of multiplication to group coefficients and like bases. Then simplify the result by using the properties of exponents.

Concepts

1. Multiplying Polynomials
2. Special Case Products: Difference of Squares and Perfect Square Trinomials
3. Applications Involving a Product of Polynomials

Example 1 Multiplying Monomials

Multiply the monomials. $(3x^2y^7)(5x^3y)$

Solution:

$(3x^2y^7)(5x^3y)$

$= (3 \cdot 5)(x^2 \cdot x^3)(y^7 \cdot y)$ Group coefficients and like bases.

$= 15x^5y^8$ Add exponents and simplify.

Skill Practice

Multiply the monomials.
1. $(-8r^3s)(-4r^4s^4)$

The distributive property is used to multiply polynomials: $a(b + c) = ab + ac$.

Example 2 Multiplying a Polynomial by a Monomial

Multiply the polynomials.

a. $5y^3(2y^2 - 7y + 6)$ **b.** $-4a^3b^7c\left(2ab^2c^4 - \dfrac{1}{2}a^5b\right)$

Solution:

a. $5y^3(2y^2 - 7y + 6)$

$= (5y^3)(2y^2) + (5y^3)(-7y) + (5y^3)(6)$ Apply the distributive property.

$= 10y^5 - 35y^4 + 30y^3$ Simplify each term.

b. $-4a^3b^7c\left(2ab^2c^4 - \dfrac{1}{2}a^5b\right)$

$= (-4a^3b^7c)(2ab^2c^4) + (-4a^3b^7c)\left(-\dfrac{1}{2}a^5b\right)$ Apply the distributive property.

$= -8a^4b^9c^5 + 2a^8b^8c$ Simplify each term.

Skill Practice

Multiply the polynomials.
2. $-6b^2(2b^2 + 3b - 8)$
3. $8st^3\left(\dfrac{1}{2}s^2 - \dfrac{1}{4}st\right)$

Answers

1. $32r^7s^5$
2. $-12b^4 - 18b^3 + 48b^2$
3. $4s^3t^3 - 2s^2t^4$

Thus far, we have illustrated polynomial multiplication involving monomials. Next, the distributive property will be used to multiply polynomials with more than one term. For example:

$$(x + 3)(x + 5) = (x + 3)x + (x + 3)5$$ Apply the distributive property.

$$= (x + 3)x + (x + 3)5$$ Apply the distributive property again.

$$= x^2 + 3x + 5x + 15$$

$$= x^2 + 8x + 15$$ Combine *like* terms.

Note: Using the distributive property results in multiplying each term of the first polynomial by each term of the second polynomial:

$$(x + 3)(x + 5) = x^2 + 5x + 3x + 15$$

$$= x^2 + 8x + 15$$

Skill Practice

Multiply the polynomials.

4. $(2y - 1)(3y^2 - 2y - 1)$

Example 3 Multiplying Polynomials

Multiply the polynomials. $(2x^2 + 4)(3x^2 - x + 5)$

Solution:

$$(2x^2 + 4)(3x^2 - x + 5)$$ Multiply each term in the first polynomial by each term in the second.

$$= (2x^2)(3x^2) + (2x^2)(-x) + (2x^2)(5)$$
$$+ (4)(3x^2) + (4)(-x) + (4)(5)$$ Apply the distributive property.

$$= 6x^4 - 2x^3 + 10x^2 + 12x^2 - 4x + 20$$ Simplify each term.

$$= 6x^4 - 2x^3 + 22x^2 - 4x + 20$$ Combine *like* terms.

> **TIP:** Multiplication of polynomials can be performed vertically by a process similar to column multiplication of real numbers.
>
> $$(2x^2 + 4)(3x^2 - x + 5) \longrightarrow$$
>
> $$\begin{array}{r} 3x^2 - x + 5 \\ \times\ 2x^2\qquad\ + 4 \\ \hline 12x^2 - 4x + 20 \\ 6x^4 - 2x^3 + 10x^2 \\ \hline 6x^4 - 2x^3 + 22x^2 - 4x + 20 \end{array}$$
>
> *Note:* When multiplying by the column method, it is important to align *like* terms vertically before adding terms.

Answer

4. $6y^3 - 7y^2 + 1$

Example 4   **Multiplying Polynomials**

Multiply the polynomials. $(3y + 2)(7y - 6)$

Solution:

$(3y + 2)(7y - 6)$ Multiply each term in the first polynomial by each term in the second.

$= (3y)(7y) + (3y)(-6) + (2)(7y) + (2)(-6)$ Apply the distributive property.

$= 21y^2 - 18y + 14y - 12$ Simplify each term.

$= 21y^2 - 4y - 12$ Combine *like* terms.

Skill Practice

Multiply the polynomials.
5. $(4t + 5)(2t + 3)$

TIP: The acronym, FOIL (first outer inner last) can be used as a memory device to multiply the two binomials.

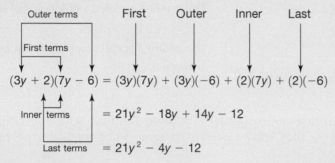

Note: It is important to realize that the acronym FOIL may only be used when finding the product of two *binomials*.

2. Special Case Products: Difference of Squares and Perfect Square Trinomials

In some cases, the product of two binomials takes on a special pattern.

I.   The first special case occurs when multiplying the sum and difference of the same two terms. For example:

$(2x + 3)(2x - 3)$

$= 4x^2 - 6x + 6x - 9$

$= 4x^2 - 9$

Notice that the "middle terms" are opposites. This leaves only the difference between the square of the first term and the square of the second term. For this reason, the product is called a *difference of squares*.

Note: The binomials $2x + 3$ and $2x - 3$ are called **conjugates**. In one expression, $2x$ and 3 are added, and in the other, $2x$ and 3 are subtracted. In general, $a + b$ and $a - b$ are conjugates of each other.

Answer
5. $8t^2 + 22t + 15$

II. The second special case involves the square of a binomial. For example:

$$(3x + 7)^2$$

$$= (3x + 7)(3x + 7)$$

$$= 9x^2 + 21x + 21x + 49$$

$$= 9x^2 + 42x + 49$$

$$= (3x)^2 + 2(3x)(7) + (7)^2$$

When squaring a binomial, the product will be a trinomial called a *perfect square trinomial*. The first and third terms are formed by squaring the terms of the binomial. The middle term is twice the product of the terms in the binomial.

Note: The expression $(3x - 7)^2$ also results in a perfect square trinomial, but the middle term is negative.

$$(3x - 7)(3x - 7) = 9x^2 - 21x - 21x + 49 = 9x^2 - 42x + 49$$

The following summarizes these special case products.

Special Case Product Formulas

1. $(a + b)(a - b) = a^2 - b^2$ The product is called a **difference of squares**.

2. $(a + b)^2 = a^2 + 2ab + b^2$
$(a - b)^2 = a^2 - 2ab + b^2$ The product is called a **perfect square trinomial**.

It is advantageous for you to become familiar with these special case products because they will be presented again when we factor polynomials.

Example 5   Using Special Products

Use the special product formulas to multiply the polynomials.

a. $(6c - 7d)(6c + 7d)$ **b.** $(5x - 2)^2$ **c.** $(4x^3 + 3y^2)^2$

Solution:

a. $(6c - 7d)(6c + 7d)$ $a = 6c, b = 7d$

$\quad = (6c)^2 - (7d)^2$ Apply the formula $a^2 - b^2$.

$\quad = 36c^2 - 49d^2$ Simplify each term.

b. $(5x - 2)^2$ $a = 5x, b = 2$

$\quad = (5x)^2 - 2(5x)(2) + (2)^2$ Apply the formula $a^2 - 2ab + b^2$.

$\quad = 25x^2 - 20x + 4$ Simplify each term.

c. $(4x^3 + 3y^2)^2$ $a = 4x^3, b = 3y^2$

$\quad = (4x^3)^2 + 2(4x^3)(3y^2) + (3y^2)^2$ Apply the formula $a^2 + 2ab + b^2$.

$\quad = 16x^6 + 24x^3y^2 + 9y^4$ Simplify each term.

The special case products can be used to simplify more complicated algebraic expressions.

Skill Practice

Multiply the polynomials.

6. $(5x - 4y)(5x + 4y)$

7. $(c - 3)^2$

8. $(7s^2 + 2t)^2$

Answers

6. $25x^2 - 16y^2$

7. $c^2 - 6c + 9$

8. $49s^4 + 28s^2t + 4t^2$

Example 6 Using Special Products

Multiply. $[x + (y + z)][x - (y + z)]$

Solution:

$[x + (y + z)][x - (y + z)]$ This product is in the form $(a + b)(a - b)$, where $a = x$ and $b = (y + z)$.

$= (x)^2 - (y + z)^2$ Apply the formula $a^2 - b^2$.

$= (x)^2 - (y^2 + 2yz + z^2)$ Expand $(y + z)^2$ by using the special case product formula.

$= x^2 - y^2 - 2yz - z^2$ Apply the distributive property.

Example 7 Using Special Products

Multiply. $(x + y)^3$

Solution:

$(x + y)^3$

$= (x + y)^2(x + y)$ Rewrite as the square of a binomial and another factor.

$= (x^2 + 2xy + y^2)(x + y)$ Expand $(x + y)^2$ by using the special case product formula.

$= (x^2)(x) + (x^2)(y) + (2xy)(x)$
$\quad + (2xy)(y) + (y^2)(x) + (y^2)(y)$ Apply the distributive property.

$= x^3 + x^2y + 2x^2y + 2xy^2 + xy^2 + y^3$ Simplify each term.

$= x^3 + 3x^2y + 3xy^2 + y^3$ Combine *like* terms.

3. Applications Involving a Product of Polynomials

Example 8 Translating Between English Form and Algebraic Form

Complete the table.

English Form	Algebraic Form
The square of the sum of x and y	
	$x^2 + y^2$
The square of the product of 3 and x	

Solution:

English Form	Algebraic Form	Notes
The square of the sum of x and y	$(x + y)^2$	The *sum* is squared, not the individual terms.
The sum of the squares of x and y	$x^2 + y^2$	The individual terms x and y are squared first. Then the sum is taken.
The square of the product of 3 and x	$(3x)^2$	The product of 3 and x is taken. Then the result is squared.

Skill Practice

A rectangular photograph is mounted on a square piece of cardboard whose sides have length x. The border that surrounds the photo is 3 in. on each side and 4 in. on both top and bottom.

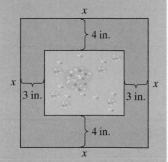

14. Write a function A for the area of the photograph in terms of x.

15. Determine the area of the photograph if x is 12 in.

Example 9   Applying a Product of Polynomials

A box is created from a sheet of cardboard 20 in. on a side by cutting a square from each corner and folding up the sides (Figures 5-3 and 5-4). Let x represent the length of the sides of the squares removed from each corner.

a. Write a function V that represents the volume of the box in terms of x.

b. Find the volume if a 4-in. square is removed.

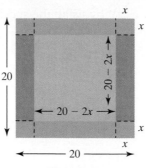

Figure 5-3

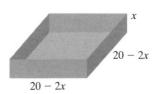

Figure 5-4

Solution:

a. The volume of a rectangular box is given by the formula $V = lwh$. The length and width can both be expressed as $20 - 2x$. The height of the box is x. The volume is given by

$$V = l \cdot w \cdot h$$
$$V(x) = (20 - 2x)(20 - 2x)x$$
$$= (20 - 2x)^2 x$$
$$= (400 - 80x + 4x^2)x$$
$$= 400x - 80x^2 + 4x^3$$
$$= 4x^3 - 80x^2 + 400x$$

b. If a 4-in. square is removed from the corners of the box, we have $x = 4$. Thus, the volume is

$$V(x) = 4x^3 - 80x^2 + 400x$$
$$V(4) = 4(4)^3 - 80(4)^2 + 400(4)$$
$$= 4(64) - 80(16) + 400(4)$$
$$= 256 - 1280 + 1600$$
$$= 576$$

The volume is 576 in.^{3}

Answers

14. $A(x) = (x - 8)(x - 6)$;
      $A(x) = x^2 - 14x + 48$
15. 24 in.^{2}

Section 5.3 Practice Exercises

Vocabulary and Key Concepts

1. **a.** To multiply $2(4x - 5)$, apply the _____ property.

   b. The conjugate of $4x + 7$ is _____.

   c. When two conjugates are multiplied the resulting binomial is a difference of _____. This is given by the formula $(a + b)(a - b) = $ _____.

   d. When a binomial is squared, the resulting trinomial is a _____ square trinomial. This is given by the formula $(a + b)^2 = $ _____.

Review Exercises

2. Simplify. $(-4x^2y - 2xy + 3xy^2) - (2x^2y - 4xy^2) + (6x^2y + 5xy)$

3. Simplify. $(-2 - 3x) - [5 - (6x^2 + 4x + 1)]$

4. Given $f(x) = 4x^3 - 5$, find the function values.

   a. $f(3)$ **b.** $f(0)$ **c.** $f(-2)$

5. Given $g(x) = x^4 - x^2 - 3$, find the function values.

   a. $g(-1)$ **b.** $g(2)$ **c.** $g(0)$

6. Write the distributive property of multiplication over addition. Give an example of the distributive property. (Answers may vary.)

Concept 1: Multiplying Polynomials

For Exercises 7–40, multiply the polynomials. **(See Examples 1–4.)**

7. $(7x^4y)(-6xy^5)$

8. $(-4a^3b^7)(-2ab^3)$

9. $(2.2a^6b^4c^7)(5ab^4c^3)$

10. $(8.5c^4d^5e)(6cd^2e)$

11. $\dfrac{1}{5}(2a - 3)$

12. $\dfrac{1}{3}(6b + 4)$

13. $2m^3n^2(m^2n^3 - 3mn^2 + 4n)$

14. $3p^2q(p^3q^3 - pq^2 - 4p)$

15. $6xy^2\left(\dfrac{1}{2}x - \dfrac{2}{3}xy\right)$

16. $12ab\left(\dfrac{5}{6}a + \dfrac{1}{4}ab^2\right)$

17. $(x + y)(x - 2y)$

18. $(3a + 5)(a - 2)$

19. $(6x - 1)(5 + 2x)$

20. $(7 + 3x)(x - 8)$

21. $(y^2 - 12)(2y^2 + 3)$

22. $(4p^2 - 1)(2p^2 + 5)$

23. $(5s + 3t)(5s - 2t)$

24. $(4a + 3b)(4a - b)$

25. $(n^2 + 10)(5n + 3)$

26. $(m^2 + 8)(3m + 7)$

27. $(1.3a - 4b)(2.5a + 7b)$

28. $(2.1x - 3.5y)(4.7x + 2y)$

29. $(2x + y)(3x^2 + 2xy + y^2)$

30. $(h - 5k)(h^2 - 2hk + 3k^2)$

31. $(x - 7)(x^2 + 7x + 49)$

32. $(x + 3)(x^2 - 3x + 9)$

33. $(4a - b)(a^3 - 4a^2b + ab^2 - b^3)$

34. $(3m + 2n)(m^3 + 2m^2n - mn^2 + 2n^3)$

35. $\left(\dfrac{1}{2}a - 2b + c\right)(a + 6b - c)$

36. $\left(\dfrac{1}{2}a^2 - 2ab + b^2\right)(2a + b)$

37. $(-x^2 + 2x + 1)(3x - 5)$

38. $(x + y - 2z)(5x - y + z)$

39. $\left(\dfrac{1}{5}y - 10\right)\left(\dfrac{1}{2}y - 15\right)$

40. $\left(\dfrac{2}{3}x + 6\right)\left(\dfrac{1}{2}x - 9\right)$

Concept 2: Special Case Products: Difference of Squares and Perfect Square Trinomials

For Exercises 41–60, multiply by using the special case products. **(See Example 5.)**

41. $(a - 8)(a + 8)$ **42.** $(b + 2)(b - 2)$ **43.** $(3p + 1)(3p - 1)$ **44.** $(5q - 3)(5q + 3)$

45. $\left(x - \dfrac{1}{3}\right)\left(x + \dfrac{1}{3}\right)$ **46.** $\left(\dfrac{1}{2}x + \dfrac{1}{3}\right)\left(\dfrac{1}{2}x - \dfrac{1}{3}\right)$ **47.** $(3h - k)(3h + k)$ **48.** $(x - 7y)(x + 7y)$

49. $(3h - k)^2$ **50.** $(x - 7y)^2$ **51.** $(t - 7)^2$ **52.** $(w + 9)^2$

53. $(u + 3v)^2$ **54.** $(a - 4b)^2$ **55.** $\left(h + \dfrac{1}{6}k\right)^2$ **56.** $\left(\dfrac{2}{5}x + 1\right)^2$

57. $(2z^2 - w^3)(2z^2 + w^3)$ **58.** $(a^4 - 2b^3)(a^4 + 2b^3)$ **59.** $(5x^2 - 3y)^2$ **60.** $(4p^3 - 2m)^2$

For Exercises 61 and 62, the product of two binomials is shown. Determine if the binomials are conjugates.

61. a. $(-5x + 4)(5x + 4)$

    b. $(-5x + 4)(5x - 4)$

62. a. $(-3 - 7x)(3 + 7x)$

    b. $(-3 + 7x)(3 + 7x)$

63. Multiply the expressions. Explain their similarities.

    a. $(A - B)(A + B)$

    b. $[(x + y) - B][(x + y) + B]$

64. Multiply the expressions. Explain their similarities.

    a. $(A + B)(A - B)$

    b. $[A + (3h + k)][A - (3h + k)]$

For Exercises 65–70, multiply the expressions. **(See Example 6.)**

65. $[(w + v) - 2][(w + v) + 2]$ **66.** $[(x + y) - 6][(x + y) + 6]$ **67.** $[2 - (x + y)][2 + (x + y)]$

68. $[a - (b + 1)][a + (b + 1)]$ **69.** $[(3a - 4) + b][(3a - 4) - b]$ **70.** $[(5p - 7) - q][(5p - 7) + q]$

71. Explain how to multiply $(x + y)^3$.

72. Explain how to multiply $(a - b)^3$.

For Exercises 73–76, multiply the expressions. **(See Example 7.)**

73. $(2x + y)^3$

74. $(x - 5y)^3$

75. $(4a - b)^3$

76. $(3a + 4b)^3$

77. Explain how you would multiply the binomials

$$(x - 2)(x + 6)(2x + 1)$$

78. Explain how you would multiply the binomials

$$(a + b)(a - b)(2a + b)(2a - b)$$

For Exercises 79–86, simplify the expressions.

79. $2a^2(a + 5)(3a + 1)$

80. $-5y(2y - 3)(y + 3)$

81. $(x + 3)(x - 3)(x + 5)$

82. $(t + 2)(t - 3)(t + 1)$

83. $-3(2x + 7) - (4x - 1)^2$

84. $(p + 10)^2 - 4(p + 6)^2$

85. $(y + 1)^2 - (2y + 3)^2$

86. $(b - 3)^2 - (3b - 1)^2$

Concept 3: Applications Involving a Product of Polynomials

For Exercises 87–90, translate from English form to algebraic form. **(See Example 8.)**

87. The square of the sum of r and t

88. The square of a plus the cube of b

89. The difference of x squared and y cubed

90. The square of the product of 3 and a

For Exercises 91–94, translate from algebraic form to English form. **(See Example 8.)**

91. $p^3 + q^2$

92. $a^3 - b^3$

93. xy^2

94. $(c + d)^3$

95. A rectangular garden has a walk around it of width x. The garden is 20 ft by 15 ft. Write a function representing the combined area $A(x)$ of the garden and walk. Simplify the result.

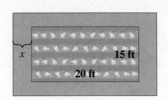

96. An 8-in. by 10-in. photograph is in a frame of width x. Write a function that represents the area $A(x)$ of the frame alone. Simplify the result.

97. A box is created from a square piece of cardboard 8 in. on a side by cutting a square from each corner and folding up the sides. Let x represent the length of the sides of the squares removed from each corner. **(See Example 9.)**

    a. Write a function representing the volume of the box.

    b. Find the volume if 1-in. squares are removed from the corners.

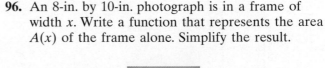
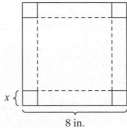

98. A box is created from a rectangular piece of metal with dimensions 12 in. by 9 in. by removing a square from each corner of the metal sheet and folding up the sides. Let x represent the length of the sides of the squares removed from each corner.

    a. Write a function representing the volume of the box.

    b. Find the volume if 2-in. squares are removed from the corners.

For Exercises 99–104, write an expression for the area and simplify your answer.

99. Square

$x - 2$

100. Square

$x + 3$

101. Rectangle

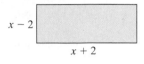

$x - 2$

$x + 2$

102. Rectangle

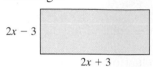

$2x - 3$

$2x + 3$

103. Triangle

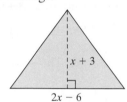

$x + 3$

$2x - 6$

104. Triangle

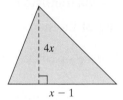

$4x$

$x - 1$

For Exercises 105 and 106, write an expression for the volume and simplify your answer.

105.

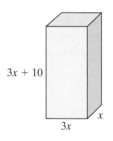

$3x + 10$

x

$3x$

106.

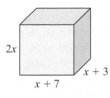

$2x$

$x + 3$

$x + 7$

Expanding Your Skills

For Exercises 107 and 108, simplify completely.

107. $\dfrac{[(x + h)^2 - 3(x + h) - 5] - (x^2 - 3x - 5)}{h}$

108. $\dfrac{[(x + h)^2 - 4(x + h) + 2] - (x^2 - 4x + 2)}{h}$

109. Explain how to multiply $(x + 2)^4$.

110. Explain how to multiply $(y - 3)^4$.

111. $(2x - 3)$ multiplied by what binomial will result in the trinomial $10x^2 - 27x + 18$? Check your answer by multiplying the binomials.

112. $(4x + 1)$ multiplied by what binomial will result in the trinomial $12x^2 - 5x - 2$? Check your answer by multiplying the binomials.

113. $(4y + 3)$ multiplied by what binomial will result in the trinomial $8y^2 + 2y - 3$? Check your answer by multiplying the binomials.

114. $(3y - 2)$ multiplied by what binomial will result in the trinomial $3y^2 - 17y + 10$? Check your answer by multiplying the binomials.

Division of Polynomials

1. Division by a Monomial

Concepts

1. **Division by a Monomial**
2. **Long Division**
3. **Synthetic Division**

Division of polynomials is presented in this section as two separate cases. The first case illustrates division by a monomial divisor. The second case illustrates division by a polynomial with two or more terms.

To divide a polynomial by a monomial, divide each individual term in the polynomial by the divisor and simplify the result.

Dividing a Polynomial by a Monomial

If a, b, and c are polynomials such that $c \neq 0$, then

$$\frac{a + b}{c} = \frac{a}{c} + \frac{b}{c} \qquad \text{Similarly,} \qquad \frac{a - b}{c} = \frac{a}{c} - \frac{b}{c}$$

Example 1   Dividing a Polynomial by a Monomial

Divide the polynomials.

a. $\dfrac{3x^4 - 6x^3 + 9x}{3x}$ **b.** $(10c^3d - 15c^2d^2 + 2cd^3) \div (-5c^2d^2)$

Solution:

a. $\dfrac{3x^4 - 6x^3 + 9x}{3x}$

$= \dfrac{3x^4}{3x} - \dfrac{6x^3}{3x} + \dfrac{9x}{3x}$ Divide each term in the numerator by $3x$.

$= x^3 - 2x^2 + 3$ Simplify each term, using the properties of exponents.

b. $(10c^3d - 15c^2d^2 + 2cd^3) \div (-5c^2d^2)$

$= \dfrac{10c^3d - 15c^2d^2 + 2cd^3}{-5c^2d^2}$

$= \dfrac{10c^3d}{-5c^2d^2} - \dfrac{15c^2d^2}{-5c^2d^2} + \dfrac{2cd^3}{-5c^2d^2}$ Divide each term in the numerator by $-5c^2d^2$.

$= -\dfrac{2c}{d} + 3 - \dfrac{2d}{5c}$ Simplify each term.

Skill Practice

Divide.

1. $\dfrac{18y^3 - 6y^2 - 12y}{6y}$

2. $(-24a^3b^2 - 16a^2b^3 + 8ab)$
$\div (-8ab)$

---

2. Long Division

If the divisor has two or more terms, a long division process similar to the division of whole numbers is used.

Answers

1. $3y^2 - y - 2$
2. $3a^2b + 2ab^2 - 1$

Example 2 Using Long Division to Divide Polynomials

Divide the polynomials by using long division.

$$(3x^2 - 14x - 10) \div (x - 2)$$

Solution:

$$x - 2 \overline{)3x^2 - 14x - 10}$$

Divide the leading term in the dividend by the leading term in the divisor.

$\dfrac{3x^2}{x} = 3x$. This is the first term in the quotient.

The terms in the quotient are aligned with the *like* terms in the dividend.

$$\begin{array}{r} 3x \\ x - 2 \overline{)3x^2 - 14x - 10} \\ -(3x^2 - 6x) \end{array}$$

Multiply $3x$ by the divisor and record the result: $3x(x - 2) = 3x^2 - 6x$.

TIP: Take a minute to compare the long division process with numbers to the process of dividing polynomials.

$$\begin{array}{r} 148 \\ 31 \overline{)4602} \\ -31 \\ \hline 150 \\ -124 \\ \hline 262 \\ -248 \\ \hline 14 \end{array}$$

$$\begin{array}{r} 3x \\ x - 2 \overline{)3x^2 - 14x - 10} \\ -3x^2 + 6x \\ \hline -8x \end{array}$$

Next, subtract the quantity $3x^2 - 6x$. To do this, add its opposite.

$$\begin{array}{r} 3x - 8 \\ x - 2 \overline{)3x^2 - 14x - 10} \\ -3x^2 + 6x \\ \hline -8x - 10 \\ -(-8x + 16) \end{array}$$

Bring down the next column and repeat the process.

Divide the leading term by x: $\dfrac{-8x}{x} = -8$

Multiply the divisor by -8 and record the result: $-8(x - 2) = -8x + 16$.

$$\begin{array}{r} 3x - 8 \\ x - 2 \overline{)3x^2 - 14x - 10} \\ -3x^2 + 6x \\ \hline -8x - 10 \\ +8x - 16 \\ \hline -26 \end{array}$$

Subtract the quantity $-8x + 16$ by adding its opposite.

The remainder is -26. We do not continue because the degree of the remainder is less than the degree of the divisor.

Summary:

The quotient is $\quad 3x - 8$

The remainder is $\quad -26$

The divisor is $\quad x - 2$

The dividend is $\quad 3x^2 - 14x - 10$

The solution to a long division problem is often written in the form:

$$\text{Quotient} + \frac{\text{remainder}}{\text{divisor}}.$$

This answer is $3x - 8 + \dfrac{-26}{x - 2}$.

Answers

3. $4x - 6 + \dfrac{10}{x + 3}$

4. Long division is required when the divisor has two or more terms.

The division of polynomials can be checked in the same fashion as the division of real numbers. To check we use the division algorithm.

Division Algorithm

$$\text{Dividend} = (\text{divisor})(\text{quotient}) + \text{remainder}$$

$$3x^2 - 14x - 10 \stackrel{?}{=} (x - 2)(3x - 8) + (-26)$$
$$\stackrel{?}{=} 3x^2 - 8x - 6x + 16 + (-26)$$
$$\stackrel{?}{=} 3x^2 - 14x - 10 \checkmark \quad \text{True}$$

Example 3   Using Long Division to Divide Polynomials

Divide the polynomials by using long division: $(-2x^3 - 10x^2 + 56) \div (2x - 4)$

Solution:

First note that the dividend has a missing power of x and can be written as $-2x^3 - 10x^2 + 0x + 56$. The term $0x$ is a placeholder for the missing term. It is helpful to use the placeholder to keep the powers of x lined up.

$$
\begin{array}{r}
-x^2 \\
2x - 4 \overline{) -2x^3 - 10x^2 + 0x + 56} \\
-(-2x^3 + 4x^2)
\end{array}
$$

Leave space for the missing power of x.

Divide $\dfrac{-2x^3}{2x} = -x^2$ to get the first term of the quotient.

$$
\begin{array}{r}
-x^2 - 7x \\
2x - 4 \overline{) -2x^3 - 10x^2 + 0x + 56} \\
2x^3 - 4x^2 \\
\hline
-14x^2 + 0x
\end{array}
$$

Subtract by adding the opposite.
Bring down the next column.

Divide $\dfrac{-14x^2}{2x} = -7x$ to get the next term in the quotient.

$$
\begin{array}{r}
-x^2 - 7x - 14 \\
2x - 4 \overline{) -2x^3 - 10x^2 + 0x + 56} \\
2x^3 - 4x^2 \\
\hline
-14x^2 + 0x \\
14x^2 - 28x \\
\hline
-28x + 56
\end{array}
$$

Subtract by adding the opposite.
Bring down the next column.

Divide $\dfrac{-28x}{2x} = -14$ to get the next term in the quotient.

$$
\begin{array}{r}
-x^2 - 7x - 14 \\
2x - 4 \overline{) -2x^3 - 10x^2 + 0x + 56} \\
2x^3 - 4x^2 \\
\hline
-14x^2 + 0x \\
14x^2 - 28x \\
\hline
-28x + 56 \\
28x - 56 \\
\hline
0
\end{array}
$$

Subtract by adding the opposite.

$\leftarrow$ The remainder is 0.

The answer is $-x^2 - 7x - 14$.

In Example 3, the quotient is $-x^2 - 7x - 14$ and the remainder is 0. Because the remainder is zero, $2x - 4$ divides *evenly* into $-2x^3 - 10x^2 + 56$. For this reason, the divisor and quotient are *factors* of $-2x^3 - 10x^2 + 56$. To check, we have

$$\text{Dividend} = (\text{divisor})(\text{quotient}) + \text{remainder}$$

$$-2x^3 - 10x^2 + 56 \overset{?}{=} (2x - 4)(-x^2 - 7x - 14) + 0$$
$$\overset{?}{=} -2x^3 - 14x^2 - 28x + 4x^2 + 28x + 56$$
$$\overset{?}{=} -2x^3 - 10x^2 + 56 \quad ✔ \quad \text{True}$$

Skill Practice

Divide.

6. $(x^3 + 1 + 2x^2) \div (x^2 + 1)$

TIP: Both the divisor and dividend must be written in descending order before performing polynomial division.

Example 4 Using Long Division to Divide Polynomials

Divide. $(15x^3 - 4x + 9x^4 - 5) \div (3x^2 - 4)$

Solution:

Write the dividend in descending powers of x. $\quad 9x^4 + 15x^3 - 4x - 5$
The dividend has a missing power of x^2. $\quad 9x^4 + 15x^3 + 0x^2 - 4x - 5$

The divisor has a missing power of x and can be written as $3x^2 + 0x - 4$.

$$
\require{enclose}
\begin{array}{r}
3x^2 + 5x + 4 \\
3x^2 + 0x - 4 \enclose{longdiv}{9x^4 + 15x^3 + 0x^2 - 4x - 5} \\
\end{array}
$$

$\quad\quad\quad\quad -(9x^4 + 0x^3 - 12x^2)$
$\quad\quad\quad\quad\quad\quad\quad 15x^3 + 12x^2 - 4x$
$\quad\quad\quad\quad\quad\quad -(15x^3 + 0x^2 - 20x)$
$\quad\quad\quad\quad\quad\quad\quad\quad\quad 12x^2 + 16x - 5$
$\quad\quad\quad\quad\quad\quad\quad\quad -(12x^2 + 0x - 16)$
$\quad\quad\quad\quad\quad\quad\quad\quad\quad\quad\quad 16x + 11$

The remainder is $16x + 11$. The degree of the remainder is less than the degree of the divisor.

The answer is $3x^2 + 5x + 4 + \dfrac{16x + 11}{3x^2 - 4}$.

3. Synthetic Division

In this section, we introduced the process of long division to divide two polynomials. Next, we will learn another technique, called **synthetic division**, to divide two polynomials. Synthetic division can be used when dividing a polynomial by a first-degree divisor of the form $x - r$, where r is a constant. Synthetic division is considered a "shortcut" because it uses the coefficients of the divisor and dividend without writing the variables.

Answer

6. $x + 2 + \dfrac{-x - 1}{x^2 + 1}$

Consider dividing the polynomials $(3x^2 - 14x - 10) \div (x - 2)$.

$$
\begin{array}{r}
3x - 8 \\
x - 2 \overline{)\, 3x^2 - 14x - 10} \\
-(3x^2 - 6x) \\
\hline
-8x - 10 \\
-(-8x + 16) \\
\hline
-26
\end{array}
$$

First note that the divisor $x - 2$ is in the form $x - r$, where $r = 2$. Therefore, synthetic division can be used to find the quotient and remainder.

Step 1: Write the value of r in a box. $\longrightarrow$ $\underline{2}\rfloor$ 3   −14   −10 $\longleftarrow$ **Step 2:** Write the coefficients of the dividend to the right of the box.

     3

Step 3: Skip a line and draw a horizontal line below the list of coefficients.

Step 4: Bring down the leading coefficient from the dividend and write it below the line.

$\underline{2}\rfloor$ 3   −14   −10   **Step 6:** Add the numbers in the column above the line $(-14 + 6)$, and write the result below the line.

Step 5: Multiply the value of r by the number below the line $(2 \times 3 = 6)$. Write the result in the next column above the line.

     6

   3   −8

Repeat steps 5 and 6 until all columns have been completed.

Step 7: To get the final result, we use the numbers below the line. The number in the last column is the remainder. The other numbers are the coefficients of the quotient.

$\underline{2}\rfloor$ 3   −14   −10

        6   −16

   3   −8 $\;\lfloor$−26$\rfloor$

A box is usually drawn around the remainder.

Quotient: $3x - 8$,   remainder: −26

Avoiding Mistakes

It is important to check that the divisor is in the form $(x - r)$ before applying synthetic division. The variable x in the divisor must be of first degree, and its coefficient must be 1.

Concept Connections

Determine whether synthetic division can be used directly to divide the polynomials. Explain your answer.

7. $(x^3 + 6x^2 + 2x - 1)$
$\div (2x + 3)$

8. $(y^2 - 5y + 8) \div (y^2 + 1)$

9. $(a^3 + 5a^2 + a - 2)$
$\div (a - 5)$

The degree of the quotient will always be 1 less than that of the dividend. Because the dividend is a second-degree polynomial, the quotient will be a first-degree polynomial. In this case, the quotient is $3x - 8$ and the remainder is −26.

Answers

7. No. The leading coefficient of the divisor is not equal to 1.

8. No. The divisor is not first-degree.

9. Yes. The divisor is in the form $(x - r)$, where $r = 5$.

Skill Practice

Divide the polynomials by using synthetic division.

10. $(5y^2 - 4y + 2y^3 - 5)$
$\div (y + 3)$

Example 5 Using Synthetic Division to Divide Polynomials

Divide the polynomials $(5x + 4x^3 - 6 + x^4) \div (x + 3)$ by using synthetic division.

Solution:

As with long division, the terms of the dividend and divisor should be written in descending order. Furthermore, missing powers must be accounted for by using placeholders (shown here in red).

$$5x + 4x^3 - 6 + x^4$$
$$= x^4 + 4x^3 + 0x^2 + 5x - 6$$

To use synthetic division, the divisor must be in the form $(x - r)$. The divisor $x + 3$ can be written as $x - (-3)$. Hence, $r = -3$.

Step 1: Write the value of r in a box.

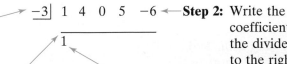

Step 2: Write the coefficients of the dividend to the right of the box.

$$-3 \, | \; 1 \;\; 4 \;\; 0 \;\; 5 \;\; -6$$
$$\overline{}$$
$$1$$

Step 3: Skip a line and draw a horizontal line below the list of coefficients.

Step 4: Bring down the leading coefficient from the dividend and write it below the line.

Step 5: Multiply the value of r by the number below the line $(-3 \times 1 = -3)$. Write the result in the next column above the line.

$$-3 \, | \; 1 \;\; 4 \;\; 0 \;\; 5 \;\; -6$$
$$ -3$$
$$\overline{}$$
$$1 \;\; 1$$

Step 6: Add the numbers in the column above the line: $4 + (-3) = 1$.

Repeat steps 5 and 6:

$$-3 \, | \; 1 \;\;\; 4 \;\;\; 0 \;\;\; 5 \;\;\; -6$$
$$ \;\; -3 \;\; -3 \;\; 9 \;\; -42$$
$$\overline{}$$
$$1 \;\; 1 \;\; -3 \;\; 14 \;\; | -48 \;\;\;\leftarrow \text{remainder}$$

The quotient is

$x^3 + x^2 - 3x + 14.$

The remainder is -48.

constant
x term coefficient
x^2 term coefficient
x^3 term coefficient

The answer is $x^3 + x^2 - 3x + 14 + \dfrac{-48}{x + 3}$.

Answer

10. $2y^2 - y - 1 + \dfrac{-2}{y + 3}$

TIP: It is interesting to compare the long division process to the synthetic division process. For Example 5, long division is shown on the left, and synthetic division is shown on the right. Notice that the same pattern of coefficients used in long division appears in the synthetic division process.

$$\begin{array}{r} x^3 + x^2 - 3x + 14 \\ x + 3 \overline{)x^4 + 4x^3 + 0x^2 + 5x - 6} \\ -\underline{(x^4 + 3x^3)} \\ x^3 + 0x^2 \\ -\underline{(x^3 + 3x^2)} \\ -3x^2 + 5x \\ -\underline{(-3x^2 - 9x)} \\ 14x - 6 \\ -\underline{(14x + 42)} \\ -48 \end{array}$$

$$\begin{array}{r|rrrrr} -3 & 1 & 4 & 0 & 5 & -6 \\ & & -3 & -3 & 9 & -42 \\ \hline & 1 & 1 & -3 & 14 & \boxed{-48} \end{array}$$

x^3 x^2 x constant   remainder

Quotient: $x^3 + x^2 - 3x + 14$

Remainder: -48

Example 6 **Using Synthetic Division to Divide Polynomials**

Divide the polynomials by using synthetic division. $(p^4 - 81) \div (p - 3)$

Solution:

$(p^4 - 81) \div (p - 3)$

$(p^4 + 0p^3 + 0p^2 + 0p - 81) \div (p - 3)$ Insert placeholders (red) for missing powers of p.

$$\begin{array}{r|rrrrr} 3 & 1 & 0 & 0 & 0 & -81 \\ & & 3 & 9 & 27 & 81 \\ \hline & 1 & 3 & 9 & 27 & \boxed{0} \end{array}$$

Quotient: $p^3 + 3p^2 + 9p + 27$

Remainder: 0

The answer is $p^3 + 3p^2 + 9p + 27$.

> **Skill Practice**
>
> Divide the polynomials by using synthetic division.
>
> **11.** $(x^3 + 1) \div (x + 1)$

Answer

11. $x^2 - x + 1$

Section 5.4   Practice Exercises

Vocabulary and Key Concepts

1. a. The _____ algorithm states that: Dividend = (divisor)(_____) + (_____).

   b. _____ division or long division can be used when dividing a polynomial by a divisor of the form $x - r$, where r is a constant.

Review Exercises

2. a. Add $(3x + 1) + (2x - 5)$.

   b. Multiply $(3x + 1)(2x - 5)$.

3. a. Subtract $(a - 10b) - (5a + b)$.

   b. Multiply $(a - 10b)(5a + b)$.

4. a. Subtract $(2y^2 + 1) - (y^2 - 5y + 1)$.

   b. Multiply $(2y^2 + 1)(y^2 - 5y + 1)$.

5. a. Add $(x^2 - x) + (6x^2 + x + 2)$.

   b. Multiply $(x^2 - x)(6x^2 + x + 2)$.

356 **Chapter 5** Polynomials

For Exercises 6–8, answers may vary.

6. Write an example of a product of two binomials and simplify.

7. Write an example of the square of a binomial and simplify.

8. Write an example of the product of conjugates and simplify.

Concept 1: Division by a Monomial

For Exercises 9–24, divide the polynomials. Check your answer by multiplication. **(See Example 1.)**

9. $\dfrac{16t^4 - 4t^2 + 20t}{-4t}$

10. $\dfrac{2x^3 + 8x^2 - 2x}{-2x}$

11. $(36y + 24y^2 + 6y^3) \div (3y)$

12. $(6p^2 - 18p^4 + 30p^5) \div (6p)$

13. $(4x^3y + 12x^2y^2 - 4xy^3) \div (4xy)$

14. $(25m^5n - 10m^4n + m^3n) \div (5m^3n)$

15. $(-8y^4 - 12y^3 + 32y^2) \div (-4y^2)$

16. $(12y^5 - 8y^6 + 16y^4 - 10y^3) \div (2y^3)$

17. $(3p^4 - 6p^3 + 2p^2 - p) \div (-6p)$

18. $(-4q^3 + 8q^2 - q) \div (-12q)$

19. $(a^3 + 5a^2 + a - 5) \div (a)$

20. $(2m^5 - 3m^4 + m^3 - m^2 + 9m) \div (m^2)$

21. $\dfrac{6s^3t^5 - 8s^2t^4 + 10st^2}{-2st^4}$

22. $\dfrac{-8r^4w^2 - 4r^3w + 2w^3}{-4r^3w}$

23. $(8p^4q^7 - 9p^5q^6 - 11p^3q - 4) \div (p^2q)$

24. $(20a^5b^5 - 20a^3b^2 + 5a^2b + 6) \div (a^2b)$

Concept 2: Long Division

25. **a.** Divide $(2x^3 - 7x^2 + 5x - 1) \div (x - 2)$, and identify the divisor, quotient, and remainder.

    b. Explain how to check by using multiplication.

26. **a.** Divide $(x^3 + 4x^2 + 7x - 3) \div (x + 3)$, and identify the divisor, quotient, and remainder.

    b. Explain how to check by using multiplication.

For Exercises 27–46, divide the polynomials by using long division. Check your answer by multiplication. **(See Examples 2–4.)**

27. $(x^2 + 11x + 19) \div (x + 4)$

28. $(x^3 - 7x^2 - 13x + 3) \div (x + 2)$

29. $(3y^3 - 7y^2 - 4y + 3) \div (y - 3)$

30. $(z^3 - 2z^2 + 2z - 5) \div (z - 4)$

31. $(-12a^2 + 77a - 121) \div (3a - 11)$

32. $(28x^2 - 29x + 6) \div (4x - 3)$

33. $(9y + 18y^2 - 20) \div (3y + 4)$

34. $(-2y + 3y^2 - 1) \div (y - 1)$

35. $(18x^3 + 7x + 12) \div (3x - 2)$

36. $(8x^3 - 6x + 22) \div (2x - 1)$

37. $(8a^3 + 1) \div (2a + 1)$

38. $(81x^4 - 1) \div (3x + 1)$

39. $(x^4 - x^3 - x^2 + 4x - 2) \div (x^2 + x - 1)$

40. $(2a^5 - 7a^4 + 11a^3 - 22a^2 + 29a - 10) \div (2a^2 - 5a + 2)$

41. $(2x^3 - 10x + x^4 - 25) \div (x^2 - 5)$

42. $(-5x^3 + x^4 - 4 - 10x) \div (x^2 + 2)$

43. $(x^4 - 3x^2 + 10) \div (x^2 - 2)$

44. $(3y^4 - 25y^2 - 18) \div (y^2 - 3)$

45. $(n^4 - 16) \div (n - 2)$

46. $(m^3 + 27) \div (m + 3)$

Concept 3: Synthetic Division

47. Explain the conditions under which you may use synthetic division to divide polynomials.

48. Can synthetic division be used directly to divide $(4x^4 + 3x^3 - 7x + 9)$ by $(2x + 5)$? Explain why or why not.

49. Can synthetic division be used to divide $(6x^5 - 3x^2 + 2x - 14)$ by $(x^2 - 3)$? Explain why or why not.

50. Can synthetic division be used to divide $(3x^4 - x + 1)$ by $(x - 5)$? Explain why or why not.

51. The following table represents the result of a synthetic division.

$$
\begin{array}{r|rrrr}
5 & 1 & -2 & -4 & 3 \\
  &   & 5 & 15 & 55 \\
\hline
  & 1 & 3 & 11 & \underline{58}
\end{array}
$$

Use x as the variable.

 a. Identify the divisor.

 b. Identify the quotient.

 c. Identify the remainder.

52. The following table represents the result of a synthetic division.

$$
\begin{array}{r|rrrrr}
-2 & 2 & 3 & 0 & -1 & 6 \\
   &   & -4 & 2 & -4 & 10 \\
\hline
   & 2 & -1 & 2 & -5 & \underline{16}
\end{array}
$$

Use x as the variable.

 a. Identify the divisor.

 b. Identify the quotient.

 c. Identify the remainder.

For Exercises 53–70, divide by using synthetic division. Check your answer by multiplication. (See Examples 5 and 6.)

53. $(x^2 - 2x - 48) \div (x - 8)$

54. $(x^2 - 4x - 12) \div (x - 6)$

55. $(t^2 - 3t - 4) \div (t + 1)$

56. $(h^2 + 7h + 12) \div (h + 3)$

57. $(5y^2 + 5y + 1) \div (y - 1)$

58. $(3w^2 + w - 5) \div (w + 2)$

59. $(3 + 7y^2 - 4y + 3y^3) \div (y + 3)$

60. $(2z - 2z^2 + z^3 - 5) \div (z + 3)$

61. $(x^3 - 3x^2 + 4) \div (x - 2)$

62. $(3y^4 - 25y^2 - 18) \div (y - 3)$

63. $(a^5 - 32) \div (a - 2)$

64. $(b^3 + 27) \div (b + 3)$

65. $(x^3 - 216) \div (x - 6)$

66. $(y^4 - 16) \div (y + 2)$

67. $(6t^3 + 7t^2 - t + 3) \div \left(t + \dfrac{2}{3}\right)$

68. $(5z^3 - 6z^2 - 4z + 4) \div \left(z - \dfrac{1}{5}\right)$

69. $(4w^4 - w^2 + 6w - 3) \div \left(w - \dfrac{1}{2}\right)$

70. $(-12y^4 - 5y^3 - y^2 + y + 3) \div \left(y + \dfrac{3}{4}\right)$

Mixed Exercises

For Exercises 71–82, divide the polynomials by using an appropriate method.

71. $(-x^3 - 8x^2 - 3x - 2) \div (x + 4)$

72. $(8xy^2 - 9x^2y + 6x^2y^2) \div (x^2y^2)$

73. $(22x^2 - 11x + 33) \div (11x)$

74. $(2m^3 - 4m^2 + 5m - 33) \div (m - 3)$

75. $(12y^3 - 17y^2 + 30y - 10) \div (3y^2 - 2y + 5)$

76. $(90h^{12} - 63h^9 + 45h^8 - 36h^7) \div (9h^9)$

77. $(4x^4 + 6x^3 + 3x - 1) \div (2x^2 + 1)$

78. $(y^4 - 3y^3 - 5y^2 - 2y + 5) \div (y + 2)$

79. $(16k^{11} - 32k^{10} + 8k^8 - 40k^4) \div (8k^8)$

80. $(4m^3 - 18m^2 + 22m - 10) \div (2m^2 - 4m + 3)$

81. $(5x^3 + 9x^2 + 10x) \div (5x^2)$

82. $(15k^4 + 3k^3 + 4k^2 + 4) \div (3k^2 - 1)$

Expanding Your Skills

83. Given $P(x) = 4x^3 + 10x^2 - 8x - 20$,

    a. Evaluate $P(-4)$.

    b. Divide. $(4x^3 + 10x^2 - 8x - 20) \div (x + 4)$

    c. Compare the value found in part (a) to the remainder found in part (b).

84. Given $P(x) = -3x^3 - 12x^2 + 5x - 8$,

    a. Evaluate $P(-6)$.

    b. Divide. $(-3x^3 - 12x^2 + 5x - 8) \div (x + 6)$

    c. Compare the value found in part (a) to the remainder found in part (b).

85. Based on your solutions to Exercises 81 and 82, make a conjecture about the relationship between the value of a polynomial function $P(x)$ at $x = r$, and the value of the remainder of $P(x) \div (x - r)$.

86. **a.** Use synthetic division to divide. $(7x^2 - 16x + 9) \div (x - 1)$

    b. Based on your solution to part (a), is $x - 1$ a *factor* of $7x^2 - 16x + 9$?

87. **a.** Use synthetic division to divide. $(8x^2 + 13x + 5) \div (x + 1)$

    b. Based on your solution to part (a), is $x + 1$ a *factor* of $8x^2 + 13x + 5$?

Problem Recognition Exercises

Operations on Polynomials

Perform the indicated operations.

1. **a.** $(3x + 1)^2$

    b. $(3x + 1)(3x - 1)$

    c. $(3x + 1) - (3x - 1)$

3. **a.** $\dfrac{4x^2 + 8x - 10}{2x}$

    b. $\dfrac{4x^2 + 8x - 10}{2x - 1}$

    c. $(4x^2 + 8x - 10) \div (x - 1)$

2. **a.** $(9m - 5) - (9m + 5)$

    b. $(9m - 5)(9m + 5)$

    c. $(9m - 5)^2$

4. **a.** $\dfrac{3y^2 - 15y + 4}{3y}$

    b. $\dfrac{3y^2 - 15y + 4}{3y + 6}$

    c. $(3y^2 - 15y + 4) \div (y + 6)$

5. a. $(p - 5)(p + 5) - (p^2 + 5)$

   b. $(p - 5)(p + 5) - (p + 5)^2$

   c. $(p - 5)(p + 5) - (p^2 - 25)$

6. a. $(x + 4)(x - 4) - (x + 4)^2$

   b. $(x + 4)(x - 4) - (x^2 + 4)$

   c. $(x + 4)(x - 4) - (x^2 - 16)$

7. $(5t^2 - 6t + 2) - (3t^2 - 7t + 3)$

8. $-5x^2(3x^2 + x - 2)$

9. $(6z + 5)(6z - 5)$

10. $(6y^3 + 2y^2 + y - 2) + (3y^3 - 4y + 3)$

11. $(3b - 4)(2b - 1)$

12. $(5a + 2)(2a^2 + 3a + 1)$

13. $(t^3 - 4t^2 + t - 9) + (t + 12) - (2t^2 - 6t)$

14. $(2b^3 - 3b - 10) \div (b - 2)$

15. $(k + 4)^2 + (-4k + 9)$

16. $(3x^4 - 11x^3 - 4x^2 - 5x + 20) \div (x - 4)$

17. $-2t(t^2 + 6t - 3) + t(3t + 2)(3t - 2)$

18. $\dfrac{7x^2y^3 - 14xy^2 - x^2}{-7xy}$

19. $\left(\dfrac{1}{4}p^3 - \dfrac{1}{6}p^2 + 5\right) - \left(-\dfrac{2}{3}p^3 + \dfrac{1}{3}p^2 - \dfrac{1}{5}p\right)$

20. $-6w^3(1.2w - 2.6w^2 + 5.1w^3)$

21. $(6a^2 - 4b)^2$

22. $\left(\dfrac{1}{2}z^2 - \dfrac{1}{3}\right)\left(\dfrac{1}{2}z^2 + \dfrac{1}{3}\right)$

23. $(m - 3)^2 - 2(m + 8)$

24. $(2x - 5)(x + 1) - (x - 3)^2$

25. $(m^2 - 6m + 7)(2m^2 + 4m - 3)$

26. $(x^3 - 64) \div (x - 4)$

27. $[5 - (a + b)]^2$

28. $[a - (x - y)][a + (x - y)]$

29. $(x + y)^2 - (x - y)^2$

30. $(a - 4)^3$

31. $\left(-\dfrac{1}{2}x + \dfrac{1}{3}\right)\left(\dfrac{1}{4}x - \dfrac{1}{2}\right)$

32. $-3x^2y^3z^4\left(\dfrac{1}{6}x^4yzw^3\right)$

Greatest Common Factor and Factoring by Grouping

1. Factoring Out the Greatest Common Factor

Sections 5.5 through 5.7 are devoted to a mathematical operation called factoring. To factor an integer means to write the integer as a product of two or more integers. To factor a polynomial means to express the polynomial as a product of two or more polynomials. For example, in the product $5 \cdot 7 = 35$, the numbers 5 and 7 are factors of 35. In the product $(2x + 1)(x - 6) = 2x^2 - 11x - 6$, the quantities $(2x + 1)$ and $(x - 6)$ are factors of $2x^2 - 11x - 6$.

Concepts

1. Factoring Out the Greatest Common Factor

2. Factoring Out a Negative Factor

3. Factoring Out a Binomial Factor

4. Factoring by Grouping

The **greatest common factor (GCF)** of a polynomial is the greatest factor that divides each term of the polynomial evenly. For example, the greatest common factor of $9x^4 + 18x^3 - 6x^2$ is $3x^2$. To factor out the greatest common factor from a polynomial, follow these steps:

Factoring Out the Greatest Common Factor

Step 1   Identify the greatest common factor of all terms of the polynomial.

Step 2   Write each term as the product of the GCF and another factor.

Step 3   Use the distributive property to factor out the greatest common factor.

Note: To check the factorization, multiply the polynomials.

Example 1   Factoring Out the Greatest Common Factor

Factor out the greatest common factor.

a. $12x^3 + 30x^2$ **b.** $12c^2d^3 - 30c^3d^2 - 3cd$

Solution:

a. $12x^3 + 30x^2$ The GCF is $6x^2$.

$\quad = 6x^2(2x) + 6x^2(5)$ Write each term as the product of the GCF and another factor.

$\quad = 6x^2(2x + 5)$ Factor out $6x^2$ by using the distributive property.

TIP: Any factoring problem can be checked by multiplying the factors.

Check: $6x^2(2x + 5) = 12x^3 + 30x^2$ ✔

---

b. $12c^2d^3 - 30c^3d^2 - 3cd$ The GCF is $3cd$.

$\quad = 3cd(4cd^2) - 3cd(10c^2d) - 3cd(1)$ Write each term as the product of the GCF and another factor.

$\quad = 3cd(4cd^2 - 10c^2d - 1)$ Factor out $3cd$ by using the distributive property.

Check: $3cd(4cd^2 - 10c^2d - 1) = 12c^2d^3 - 30c^3d^2 - 3cd$ ✔

---

2. Factoring Out a Negative Factor

Sometimes it is advantageous to factor out the *opposite* of the GCF, particularly when the leading coefficient of the polynomial is negative. This is demonstrated in Example 2. Notice that this *changes the signs* of the remaining terms inside the parentheses.

Example 2 Factoring Out a Negative Factor

Factor out the quantity $-5a^2b$ from the polynomial $-5a^4b - 10a^3b^2 + 15a^2b^3$.

Solution:

$-5a^4b - 10a^3b^2 + 15a^2b^3$ The GCF is $5a^2b$. However, in this case we will factor out the opposite of the GCF, $-5a^2b$.

$= -5a^2b(a^2) + -5a^2b(2ab) + -5a^2b(-3b^2)$ Write each term as the product of $-5a^2b$ and another factor.

$= -5a^2b(a^2 + 2ab - 3b^2)$ Factor out $-5a^2b$ by using the distributive property.

3. Factoring Out a Binomial Factor

The distributive property may also be used to factor out a common factor that consists of more than one term. This is shown in Example 3.

Example 3 Factoring Out a Binomial Factor

Factor out the greatest common factor.

$$x^3(x + 2) - x(x + 2) - 9(x + 2)$$

Solution:

$x^3(x + 2) - x(x + 2) - 9(x + 2)$ The GCF is the quantity $(x + 2)$.

$= (x + 2)(x^3) - (x + 2)(x) - (x + 2)(9)$ Write each term as the product of $(x + 2)$ and another factor.

$= (x + 2)(x^3 - x - 9)$ Factor out $(x + 2)$ by using the distributive property.

4. Factoring by Grouping

When two binomials are multiplied, the product before simplifying contains four terms. For example:

$$(3a + 2)(2b - 7) = (3a + 2)(2b) + (3a + 2)(-7)$$

$$= (3a + 2)(2b) + (3a + 2)(-7)$$

$$= 6ab + 4b - 21a - 14$$

In Example 4, we learn how to reverse this process. That is, given a four-term polynomial, we will factor it as a product of two binomials. The process is called **factoring by grouping**.

> **Factoring by Grouping**
>
> To factor a four-term polynomial by grouping:
>
> **Step 1** Identify and factor out the GCF from all four terms.
>
> **Step 2** Factor out the GCF from the first pair of terms. Factor out the GCF from the second pair of terms. (Sometimes it is necessary to factor out the *opposite* of the GCF.)
>
> **Step 3** If the two terms share a common binomial factor, factor out the binomial factor.

Skill Practice

Factor by grouping.

5. $7c^2 + cd + 14c + 2d$

Avoiding Mistakes

In step 2, the expression $3a(2b - 7) + 2(2b - 7)$ is not yet factored because it is a *sum*, not a product. To factor the expression, you must carry it one step further.

$$3a\,(2b - 7) + 2(2b - 7)$$

$$= (2b - 7)(3a + 2)$$

The factored form must be represented as a product.

Example 4 Factoring by Grouping

Factor by grouping.

$$6ab - 21a + 4b - 14$$

Solution:

$6ab - 21a + 4b - 14$	**Step 1:** Identify and factor out the GCF from all four terms. In this case the GCF is 1.
$= 6ab - 21a \mid\ + 4b - 14$	Group the first pair of terms and the second pair of terms.
$= 3a(2b - 7) + 2(2b - 7)$	**Step 2:** Factor out the GCF from each pair of terms.
	Note: The two terms now share a common binomial factor of $(2b - 7)$.
$= (2b - 7)(3a + 2)$	**Step 3:** Factor out the common binomial factor.

Check: $(2b - 7)(3a + 2) = 2b(3a) + 2b(2) - 7(3a) - 7(2)$

$$= 6ab + 4b - 21a - 14 \checkmark$$

Skill Practice

Factor by grouping.

6. $a^3 - 4a^2 - 3a + 12$

Example 5 Factoring by Grouping

Factor by grouping.

$$x^3 + 3x^2 - 3x - 9$$

Solution:

$x^3 + 3x^2 - 3x - 9$	**Step 1:** Identify and factor out the GCF from all four terms. In this case the GCF is 1.
$= x^3 + 3x^2 \mid\ - 3x - 9$	Group the first pair of terms and the second pair of terms.

Answers

5. $(7c + d)(c + 2)$
6. $(a^2 - 3)(a - 4)$

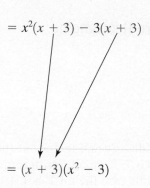

$= x^2(x + 3) - 3(x + 3)$ **Step 2:** Factor out x^2 from the first pair of terms.

Factor out -3 from the second pair of terms (this causes the signs to change in the second parentheses). The terms now contain a common binomial factor.

$= (x + 3)(x^2 - 3)$ **Step 3:** Factor out the common binomial $(x + 3)$.

TIP: One frequent question is, can the order be switched between factors? The answer is yes. Because multiplication is commutative, the order in which two or more factors are written does not matter. Thus, the following factorizations are equivalent:

$$(x + 3)(x^2 - 3) = (x^2 - 3)(x + 3)$$

Example 6 **Factoring by Grouping**

Factor by grouping.

$$24p^2q^2 - 18p^2q + 60pq^2 - 45pq$$

Solution:

$24p^2q^2 - 18p^2q + 60pq^2 - 45pq$

$= 3pq(8pq - 6p + 20q - 15)$ **Step 1:** Remove the GCF $3pq$ from all four terms.

$= 3pq(8pq - 6p \;\vdots\; + 20q - 15)$ Group the first pair of terms and the second pair of terms.

$= 3pq[2p(4q - 3) + 5(4q - 3)]$ **Step 2:** Factor out the GCF from each pair of terms. The terms share the binomial factor $(4q - 3)$.

$= 3pq(4q - 3)(2p + 5)$ **Step 3:** Factor out the common binomial $(4q - 3)$.

Skill Practice

Factor the polynomial.

7. $24x^2y - 12x^2 + 20xy - 10x$

Notice that in step 3 of factoring by grouping, a common binomial is factored from the two terms. These binomials must be *exactly* the same in each term. If the two binomial factors differ, try rearranging the original four terms.

Answer

7. $2x(6x + 5)(2y - 1)$

Skill Practice

Factor the polynomial.

8. $3ry + 2s + sy + 6r$

Avoiding Mistakes

Remember that when factoring by grouping, the binomial factors must be *exactly* the same.

Answer

8. $(3r + s)(2 + y)$

> **Example 7** **Factoring by Grouping Where Rearranging Terms Is Necessary**

Factor the polynomial.

$$4x + 6pa - 8a - 3px$$

Solution:

$$4x + 6pa - 8a - 3px$$

Step 1: Identify and factor out the GCF from all four terms. In this case the GCF is 1.

$$= 4x + 6pa \quad \vert \quad - 8a - 3px$$

$$= 2(2x + 3pa) - 1(8a + 3px)$$

Step 2: The binomial factors in each term are different.

$$= 4x - 8a \quad \vert \quad - 3px + 6pa$$

Try rearranging the original four terms in such a way that the first pair of coefficients is in the same ratio as the second pair of coefficients. Notice that the ratio 4 to −8 is the same as the ratio −3 to 6.

$$= 4(x - 2a) - 3p(x - 2a)$$

Step 2: Factor out 4 from the first pair of terms.

Factor out $-3p$ from the second pair of terms.

$$= (x - 2a)(4 - 3p)$$

Step 3: Factor out the common binomial factor.

Section 5.5 Practice Exercises

Vocabulary and Key Concepts

1. a. Factoring a polynomial means to write it as a _____ of two or more polynomials.

   b. The _____ _____ _____ (GCF) of a polynomial is the greatest factor that divides each term of the polynomial evenly.

   c. The first step toward factoring a polynomial is to factor out the _____ _____ _____.

   d. To factor a four-term polynomial, we try the process of factoring by _____.

Review Exercises

For Exercises 2–8, perform the indicated operation.

2. $(-4a^3b^5c)(-2a^7c^2)$

3. $(7t^4 + 5t^3 - 9t) - (-2t^4 + 6t^2 - 3t)$

4. $(5x^3 - 9x + 5) + (4x^3 + 3x^2 - 2x + 1) - (6x^3 - 3x^2 + x + 1)$

5. $(5y^2 - 3)(y^2 + y + 2)$

6. $(a + 6b)^2$

7. $\dfrac{6v^3 - 12v^2 + 2v}{-2v}$

8. $\dfrac{3x^3 + 2x^2 - 4}{x + 2}$

Concept 1: Factoring Out the Greatest Common Factor

For Exercises 9–24, factor out the greatest common factor. **(See Example 1.)**

9. $3x + 12$

10. $15x - 10$

11. $6z^2 + 4z$

12. $49y^3 - 35y^2$

13. $4p^6 - 4p$

14. $5q^2 - 5q$

15. $12x^4 - 36x^2$

16. $51w^4 - 34w^3$

17. $9st^2 + 27t$

18. $8a^2b^3 + 12a^2b$

19. $9a^4b^3 + 27a^3b^4 - 18a^2b^5$ **20.** $3x^5y^4 - 15x^4y^5 + 9x^2y^7$

21. $10x^2y + 15xy^2 - 5xy$ **22.** $12c^3d - 15c^2d + 3cd$ **23.** $13b^2 - 11a^2b - 12ab$ **24.** $6a^3 - 2a^2b + 5a^2$

Concept 2: Factoring Out a Negative Factor

For Exercises 25–30, factor out the indicated quantity. **(See Example 2.)**

25. $-x^2 - 10x + 7$: Factor out -1.

26. $-5y^2 + 10y + 3$: Factor out -1.

27. $-12x^3y - 6x^2y - 3xy$: Factor out $-3xy$.

28. $-32a^4b^2 + 24a^3b + 16a^2b$: Factor out $-8a^2b$.

29. $-2t^3 + 11t^2 - 3t$: Factor out $-t$.

30. $-7y^2z - 5yz - z$: Factor out $-z$.

Concept 3: Factoring Out a Binomial Factor

For Exercises 31–38, factor out the GCF. **(See Example 3.)**

31. $2a(3z - 2b) - 5(3z - 2b)$

32. $5x(3x + 4) + 2(3x + 4)$

33. $2x^2(2x - 3) + (2x - 3)$

34. $z(w - 9) + (w - 9)$

35. $y(2x + 1)^2 - 3(2x + 1)^2$

36. $a(b - 7)^2 + 5(b - 7)^2$

37. $3y(x - 2)^2 + 6(x - 2)^2$

38. $10z(z + 3)^2 - 2(z + 3)^2$

39. Construct a polynomial that has a greatest common factor of $3x^2$. (Answers may vary.)

40. Construct two different trinomials that have a greatest common factor of $5x^2y^3$. (Answers may vary.)

41. Construct a binomial that has a greatest common factor of $(c + d)$. (Answers may vary.)

Concept 4: Factoring by Grouping

42. If a polynomial has four terms, what technique would you use to factor it?

43. Factor the polynomials by grouping.

a. $2ax - ay + 6bx - 3by$

b. $10w^2 - 5w - 6bw + 3b$

c. Explain why you factored out $3b$ from the second pair of terms in part (a) but factored out the quantity $-3b$ from the second pair of terms in part (b).

44. Factor the polynomials by grouping.

    a. $3xy + 2bx + 6by + 4b^2$

    b. $15ac + 10ab - 6bc - 4b^2$

    c. Explain why you factored out $2b$ from the second pair of terms in part (a) but factored out the quantity $-2b$ from the second pair of terms in part (b).

For Exercises 45–64, factor each polynomial by grouping (if possible). **(See Examples 4–7.)**

45. $y^3 + 4y^2 + 3y + 12$ **46.** $ab + b + 2a + 2$

47. $6p - 42 + pq - 7q$ **48.** $2t - 8 + st - 4s$

49. $2mx + 2nx + 3my + 3ny$ **50.** $4x^2 + 6xy - 2xy - 3y^2$

51. $10ax - 15ay - 8bx + 12by$ **52.** $35a^2 - 15a + 14a - 6$

53. $x^3 - x^2 - 3x + 3$ **54.** $2rs + 4s - r - 2$

55. $6p^2q + 18pq - 30p^2 - 90p$ **56.** $5s^2t + 20st - 15s^2 - 60s$

57. $100x^3 - 300x^2 + 200x - 600$ **58.** $2x^5 - 10x^4 + 6x^3 - 30x^2$

59. $6ax - by + 2bx - 3ay$ **60.** $5pq - 12 - 4q + 15p$

61. $4a - 3b - ab + 12$ **62.** $x^2y + 6x - 3x^3 - 2y$

63. $7y^3 - 21y^2 + 5y - 10$ **64.** $5ax + 10bx - 2ac + 4bc$

65. Explain why the grouping method failed for Exercise 63.

66. Explain why the grouping method failed for Exercise 64.

Mixed Exercises

67. Solve the equation $U = Av + Acw$ for A by first factoring out A.

68. Solve the equation $S = rt + wt$ for t by first factoring out t.

69. Solve the equation $ay + bx = cy$ for y.

70. Solve the equation $cd + 2x = ac$ for c.

71. The area of a rectangle of width w is given by $A = 2w^2 + w$. Factor the right-hand side of the equation to find an expression for the length of the rectangle.

72. The amount in a savings account bearing simple interest at an annual interest rate r for t years is given by

$A = P + Prt$ where P is the principal amount invested.

    a. Solve the equation for P.

    b. Compute the amount of principal originally invested if the account is worth $12,705 after 3 yr at a 7% interest rate.

Expanding Your Skills

For Exercises 73–80, factor out the greatest common factor and simplify.

73. $(a + 3)^4 + 6(a + 3)^5$

74. $(4 - b)^4 - 2(4 - b)^3$

75. $24(3x + 5)^3 - 30(3x + 5)^2$

76. $10(2y + 3)^2 + 15(2y + 3)^3$

77. $(t + 4)^2 - (t + 4)$

78. $(p + 6)^2 - (p + 6)$

79. $15w^2(2w - 1)^3 + 5w^3(2w - 1)^2$

80. $8z^4(3z - 2)^2 + 12z^3(3z - 2)^3$

Factoring Trinomials and Perfect Square Trinomials

Section 5.6

1. Factoring Trinomials: AC-Method

In Section 5.5, we learned how to factor out the greatest common factor from a polynomial and how to factor a four-term polynomial by grouping. In this section, we present two methods to factor trinomials. The first method is called the ac-method. The second method is called the trial-and-error method.

The product of two binomials results in a four-term expression that can sometimes be simplified to a trinomial. To factor the trinomial, we want to reverse the process.

Multiply: $(2x + 3)(x + 2) = \xrightarrow{\text{Multiply the binomials.}} 2x^2 + 4x + 3x + 6$

$= \xrightarrow{\text{Add the middle terms.}} 2x^2 + 7x + 6$

Factor: $2x^2 + 7x + 6 = \xrightarrow{\substack{\text{Rewrite the middle term as} \\ \text{a sum or difference of terms.}}} 2x^2 + 4x + 3x + 6$

$= \xrightarrow{\text{Factor by grouping.}} (2x + 3)(x + 2)$

To factor a trinomial $ax^2 + bx + c$ by the ac-method, we rewrite the middle term bx as a sum or difference of terms. The goal is to produce a four-term polynomial that can be factored by grouping. The process is outlined as follows.

The AC-Method to Factor $ax^2 + bx + c$ ($a \neq 0$)

Step 1   After factoring out the GCF, multiply the coefficients of the first and last terms, ac.

Step 2   Find two integers whose product is ac and whose sum is b. (If no pair of integers can be found, then the trinomial cannot be factored further and is called a **prime polynomial**.)

Step 3   Rewrite the middle term bx as the sum of two terms whose coefficients are the integers found in step 2.

Step 4   Factor by grouping.

The ac-method for factoring trinomials is illustrated in Example 1. Before we begin, however, keep these two important guidelines in mind.

- For any factoring problem you encounter, always factor out the GCF from all terms first.
- To factor a trinomial, write the trinomial in the form $ax^2 + bx + c$.

Concepts

1. Factoring Trinomials: AC-Method
2. Factoring Trinomials: Trial-and-Error Method
3. Factoring Perfect Square Trinomials
4. Factoring by Using Substitution

Concept Connections

Given: the trinomial
$3x^2 + 8x + 4$

1. Identify a, b, c.
2. What is the value of ac?
3. Find the two factors of ac whose sum is equal to b.

Answers

1. $a = 3$; $b = 8$; $c = 4$ **2.** 12
3. 6 and 2

Skill Practice

4. Factor by using the ac-method. $10x^2 + x - 3$.

Example 1 Factoring a Trinomial by the AC-Method

Factor by using the ac-method. $12x^2 - 5x - 2$

Solution:

$12x^2 - 5x - 2$ The GCF is 1.

$a = 12 \quad b = -5 \quad c = -2$ **Step 1:** The expression is written in the form $ax^2 + bx + c$. Find the product $ac = 12(-2) = -24$.

Factors of –24	Factors of –24
$(1)(-24)$	$(-1)(24)$
$(2)(-12)$	$(-2)(12)$
$(3)(-8)$	$(-3)(8)$
$(4)(-6)$	$(-4)(6)$

Step 2: List all the factors of -24, and find the pair whose sum equals -5.

The numbers 3 and -8 produce a product of -24 and a sum of -5.

$12x^2 - 5x - 2$

$= 12x^2 + 3x - 8x - 2$

Step 3: Write the middle term of the trinomial as two terms whose coefficients are the selected numbers 3 and -8.

$= 12x^2 + 3x \ \vdots \ - 8x - 2$ **Step 4:** Factor by grouping.

$= 3x(4x + 1) - 2(4x + 1)$

$= (4x + 1)(3x - 2)$ The check is left for the reader.

Skill Practice

5. Factor by using the ac-method.
$-4wz^3 - 2w^2z^2 + 20w^3z$

Example 2 Factoring a Trinomial by the AC-Method

Factor the trinomial by using the ac-method. $-20c^3 + 34c^2d - 6cd^2$

Solution:

$-20c^3 + 34c^2d - 6cd^2$

$= -2c(10c^2 - 17cd + 3d^2)$ Factor out $-2c$.

Step 1: Find the product
$a \cdot c = (10)(3) = 30$

Factors of 30	Factors of 30
$1 \cdot 30$	$(-1)(-30)$
$2 \cdot 15$	$(-2)(-15)$
$3 \cdot 10$	$(-3)(-10)$
$5 \cdot 6$	$(-5)(-6)$

Step 2: The numbers -2 and -15 form a product of 30 and a sum of -17.

$= -2c(10c^2 - 17cd + 3d^2)$

Answers

4. $(5x + 3)(2x - 1)$
5. $-2wz(2z + 5w)(z - 2w)$

$$= -2c(10c^2 - 2cd \mid - 15cd + 3d^2)$$

Step 3: Write the middle term of the trinomial as two terms whose coefficients are -2 and -15.

$$= -2c[2c(5c - d) - 3d(5c - d)]$$

Step 4: Factor by grouping.

$$= -2c(5c - d)(2c - 3d)$$

TIP: In Example 2, removing the GCF from the original trinomial produced a new trinomial with smaller coefficients. This makes the factoring process simpler because the product ac is smaller.

Original trinomial	**With the GCF factored out**
$-20c^3 + 34c^2d - 6cd^2$	$-2c(10c^2 - 17cd + 3d^2)$
$ac = (-20)(-6) = 120$	$ac = (10)(3) = 30$

2. Factoring Trinomials: Trial-and-Error Method

Another method that is widely used to factor trinomials of the form $ax^2 + bx + c$ is the trial-and-error method. To understand how the trial-and-error method works, first consider the multiplication of two binomials:

Product of $2 \cdot 1$ Product of $3 \cdot 2$

$$(2x + 3)(1x + 2) = 2x^2 + 4x + 3x + 6 = 2x^2 + 7x + 6$$

sum of products
of inner terms
and outer terms

In Example 3, we will factor this trinomial by reversing this process.

Example 3 Factoring a Trinomial by the Trial-and-Error Method

Factor by the trial-and-error method. $2x^2 + 7x + 6$

Solution:

To factor by the trial-and-error method, we must fill in the blanks to create the correct product.

Factors of 2

$$2x^2 + 7x + 6 = (\Box x \quad \Box)(\Box x \quad \Box)$$

Factors of 6

Skill Practice

Factor by the trial-and-error method.

6. $5y^2 - 9y + 4$

Answer
6. $(5y - 4)(y - 1)$

Two binomials are multiplied, and the result is shown.

$$(3x + 5)(2x + 3)$$
$$6x^2 + 9x + 10x + 15$$
$$6x^2 + 19x + 15$$

7. Use arrows to show which two terms of the binomials were multiplied to obtain the first term of the product.

$$(3x + 5)(2x + 3)$$

8. Use arrows to show which two terms of the binomials were multiplied to obtain the last term of the product.

$$(3x + 5)(2x + 3)$$

9. Use arrows to show which two terms were multiplied and then added to obtain the middle term of the product.

$$(3x + 5)(2x + 3)$$

- The first terms in the binomials must be $2x$ and x. This creates a product of $2x^2$, which is the first term in the trinomial.
- The second terms in the binomials must form a product of 6. This means that the factors must both be positive or both be negative. Because the middle term of the trinomial is positive, we will consider only *positive* factors of 6. The options are $1 \cdot 6$, $2 \cdot 3$, $6 \cdot 1$, and $3 \cdot 2$.
- Test each combination of factors until the correct product of binomials is found.

$$(2x + 1)(x + 6) = 2x^2 + 12x + 1x + 6$$
$$= 2x^2 + 13x + 6 \qquad \textit{Incorrect.} \quad \text{Wrong middle term.}$$

$$(2x + 2)(x + 3) = 2x^2 + 6x + 2x + 6$$
$$= 2x^2 + 8x + 6 \qquad \textit{Incorrect.} \quad \text{Wrong middle term.}$$

$$(2x + 6)(x + 1) = 2x^2 + 2x + 6x + 6$$
$$= 2x^2 + 8x + 6 \qquad \textit{Incorrect.} \quad \text{Wrong middle term.}$$

$$(2x + 3)(x + 2) = 2x^2 + 4x + 3x + 6$$
$$= 2x^2 + 7x + 6 \qquad \textit{Correct.}$$

The factored form of $2x^2 + 7x + 6$ is $(2x + 3)(x + 2)$.

When applying the trial-and-error method, sometimes it is not necessary to test all possible combinations of factors. For the trinomial, $2x^2 + 7x + 6$, the GCF is 1. Therefore, any binomial factor that shares a common factor greater than 1 will not work and does not need to be tested. For example, the following binomials cannot work:

$$(2x + 2)(x + 3) \qquad\qquad (2x + 6)(x + 1)$$
Common factor of 2 $\qquad\qquad$ Common factor of 2

Although the trial-and-error method is tedious, its principle is generally easy to remember. We reverse the process of multiplying binomials.

The Trial-and-Error Method to Factor $ax^2 + bx + c$

Step 1 Factor out the greatest common factor.

Step 2 List all pairs of positive factors of a and pairs of positive factors of c. Consider the reverse order for either list of factors.

Step 3 Construct two binomials of the form

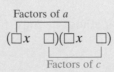

Factors of a

$$(\square x \quad \square)(\square x \quad \square)$$

Factors of c

Step 4 Test each combination of factors and signs until the correct product is found.

Step 5 If no combination of factors produces the correct product, the trinomial cannot be factored further and is a **prime polynomial**.

Answers

7. $(3x + 5)(2x + 3)$

8. $(3x + 5)(2x + 3)$

9. $(3x + 5)(2x + 3)$

Example 4   Factoring a Trinomial by the Trial-and-Error Method

Factor by the trial-and-error method. $13y - 6 + 8y^2$

Solution:

$8y^2 + 13y - 6$ Write in the form $ax^2 + bx + c$.

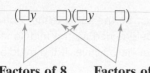
$(\square y \quad \square)(\square y \quad \square)$ **Step 1:** The GCF is 1.

Factors of 8	**Factors of 6**
$1 \cdot 8$	$1 \cdot 6$
$2 \cdot 4$	$2 \cdot 3$
	$\left.\begin{array}{c} 3 \cdot 2 \\ 6 \cdot 1 \end{array}\right\}$ (reverse order)

Step 2: List the positive factors of 8 and positive factors of 6. Consider the reverse order in one list of factors.

$\left.\begin{array}{l} (2y \quad 1)(4y \quad 6) \\ (2y \quad 2)(4y \quad 3) \\ (2y \quad 3)(4y \quad 2) \\ (2y \quad 6)(4y \quad 1) \\ (1y \quad 1)(8y \quad 6) \\ (1y \quad 3)(8y \quad 2) \end{array}\right\}$

Step 3: Construct all possible binomial factors by using different combinations of the factors of 8 and 6.

Without regard to signs, these factorizations cannot work because the terms in the binomial share a common factor greater than 1.

Test the remaining factorizations. Keep in mind that to produce a product of -6, the signs within the parentheses must be opposite (one positive and one negative). Also, the sum of the products of the inner terms and outer terms must be combined to form $13y$.

$(1y \quad 6)(8y \quad 1)$ *Incorrect.*   Wrong middle term.

Regardless of signs, the product of inner terms $48y$ and the product of outer terms $1y$ cannot be combined to form the middle term $13y$.

$(1y \quad 2)(8y \quad 3)$ *Correct.*   The terms $16y$ and $3y$ can be combined to form the middle term $13y$, provided the signs are applied correctly. We require $+16y$ and $-3y$.

The correct factorization of $8y^2 + 13y - 6$ is $(y + 2)(8y - 3)$.

Answer

10. $(4t - 3)(t + 2)$

In Example 4, the factors of -6 must have opposite signs to produce a negative product. Therefore, one binomial factor is a sum and one is a difference. Determining the correct signs is an important aspect of factoring trinomials. We suggest the following guidelines:

TIP: Given the trinomial $ax^2 + bx + c\,(a > 0)$, the signs can be determined as follows:

1. If c is *positive*, then the signs in the binomials must be the same (either both positive or both negative). The correct choice is determined by the middle term. If the middle term is positive, then both signs must be positive. If the middle term is negative, then both signs must be negative.

c is positive.

Example: $20x^2 + 43x + 21$
$(4x + 3)(5x + 7)$
same signs

c is positive.

Example: $20x^2 - 43x + 21$
$(4x - 3)(5x - 7)$
same signs

2. If c is *negative*, then the signs in the binomials must be different. The middle term in the trinomial determines which factor gets the positive sign and which factor gets the negative sign.

c is negative.

Example: $x^2 + 3x - 28$
$(x + 7)(x - 4)$
different signs

c is negative.

Example: $x^2 - 3x - 28$
$(x - 7)(x + 4)$
different signs

Skill Practice

Factor by the trial-and-error method.

11. $-4z^3 - 22z^2 - 30z$

Example 5 **Factoring a Trinomial by the Trial-and-Error Method**

Factor by the trial-and-error method.

$$-80x^3y + 208x^2y^2 - 20xy^3$$

Solution:

$-80x^3y + 208x^2y^2 - 20xy^3$

$= -4xy(20x^2 - 52xy + 5y^2)$ **Step 1:** Factor out $-4xy$.

$= -4xy(\Box x\ \Box y)(\Box x\ \Box y)$

Factors of 20	**Factors of 5**
$1 \cdot 20$	$1 \cdot 5$
$2 \cdot 10$	$5 \cdot 1$
$4 \cdot 5$	

Step 2: List the positive factors of 20 and positive factors of 5. Consider the reverse order in one list of factors.

Step 3: Construct all possible binomial factors by using different combinations of the factors of 20 and factors of 5. The signs in the parentheses must both be negative.

Answer

11. $-2z(2z + 5)(z + 3)$

$$-4xy(1x - 1y)(20x - 5y)$$
$$-4xy(2x - 1y)(10x - 5y) \Bigg\} \qquad \textit{Incorrect.} \quad \text{These binomials contain a}$$
$$-4xy(4x - 1y)(5x - 5y) \qquad\qquad\qquad\quad \text{common factor.}$$

$$-4xy(1x - 5y)(20x - 1y) \qquad \textit{Incorrect.} \quad \text{Wrong middle term.}$$
$$-4xy(x - 5y)(20x - 1y)$$
$$= -4xy(20x^2 - 101xy + 5y^2)$$

$$-4xy(4x - 5y)(5x - 1y) \qquad \textit{Incorrect.} \quad \text{Wrong middle term.}$$
$$-4xy(4x - 5y)(5x - 1y)$$
$$= -4xy(20x^2 - 29x + 5y^2)$$

$$-4xy(2x - 5y)(10x - 1y) \qquad \textbf{\textit{Correct.}} \quad -4xy(2x - 5y)(10x - 1y)$$
$$= \mathbf{-4xy(20x^2 - 52xy + 5y^2)}$$
$$= -80x^3y + 208x^2y^2 - 20xy^3$$

The correct factorization of $-80x^3y + 208x^2y^2 - 20xy^3$ is $-4xy(2x - 5y)(10x - y)$.

Example 6 **Factoring a Trinomial by the Trial-and-Error Method**

Factor completely.

$$2x^2 + 9x + 14$$

Solution:

$2x^2 + 9x + 14$	The GCF is 1 and the trinomial is written in the form $ax^2 + bx + c$.
$(2x + 14)(x + 1)$	*Incorrect.* $(2x + 14)$ contains a common factor of 2.
$(2x + 2)(x + 7)$	*Incorrect.* $(2x + 2)$ contains a common factor of 2.
$(2x + 1)(x + 14) = 2x^2 + 15x + 14$	*Incorrect.* Wrong middle term.
$(2x + 7)(x + 2) = 2x^2 + 11x + 14$	*Incorrect.* Wrong middle term.

No combination of factors results in the correct product. Therefore, the trinomial is prime (cannot be factored).

Skill Practice

Factor completely.

12. $6r^2 - 13r + 10$

If a trinomial has a leading coefficient of 1, the factoring process simplifies significantly. Consider the trinomial $x^2 + bx + c$. To produce a leading term of x^2, we can construct binomials of the form $(x + \square)(x + \square)$. The remaining terms may be satisfied by two numbers p and q whose product is c and whose sum is b:

Factors of c

$$(x + p)(x + q) = x^2 + qx + px + pq = x^2 + \underbrace{(p + q)}x + \underbrace{pq}$$
$$\text{Sum} = b \qquad \text{Product} = c$$

This process is demonstrated in Example 7.

Answer

12. Prime

Skill Practice

Factor completely.

13. $c^2 + 6c - 27$

Example 7 Factoring a Trinomial with a Leading Coefficient of 1

Factor completely.

$$x^2 - 10x + 16$$

Solution:

$x^2 - 10x + 16$ Factor out the GCF from all terms. In this case, the GCF is 1.

$= (x \;\;\square)(x \;\;\square)$ The trinomial is written in the form $x^2 + bx + c$. To form the product x^2, use the factors x and x.

Next, look for two numbers whose product is 16 and whose sum is -10. Because the middle term is negative, we will consider only the negative factors of 16.

Factors of 16	Sum
$-1(-16)$	$-1 + (-16) = -17$
$-2(-8)$	$-2 + (-8) = -10$
$-4(-4)$	$-4 + (-4) = -8$

The numbers are -2 and -8.

Therefore, $x^2 - 10x + 16 = (x - 2)(x - 8)$.

3. Factoring Perfect Square Trinomials

Recall from Section 5.3 that the square of a binomial always results in a **perfect square trinomial**.

$$(a + b)^2 = (a + b)(a + b) = a^2 + ab + ab + b^2 = a^2 + 2ab + b^2$$

$$(a - b)^2 = (a - b)(a - b) = a^2 - ab - ab + b^2 = a^2 - 2ab + b^2$$

For example, $(2x + 7)^2 = (2x)^2 + 2(2x)(7) + (7)^2 = 4x^2 + 28x + 49$

$a = 2x \quad b = 7$ $a^2 + 2ab + b^2$

TIP: The following are perfect squares.

$1^2 = 1$	$(x^1)^2 = x^2$
$2^2 = 4$	$(x^2)^2 = x^4$
$3^2 = 9$	$(x^3)^2 = x^6$
$4^2 = 16$	$(x^4)^2 = x^8$
$\vdots$	$\vdots$

Any expression raised to an even power (multiple of 2) is a perfect square.

To factor the trinomial $4x^2 + 28x + 49$, the ac-method or the trial-and-error method can be used. However, recognizing that the trinomial is a perfect square trinomial, we can use one of the following patterns to reach a quick solution.

Factored Form of a Perfect Square Trinomial

$$a^2 + 2ab + b^2 = (a + b)^2$$

$$a^2 - 2ab + b^2 = (a - b)^2$$

Answer

13. $(c + 9)(c - 3)$

TIP: To determine if a trinomial is a perfect square trinomial, follow these steps:

1. Check if the first and third terms are both perfect squares with positive coefficients.
2. If this is the case, identify a and b, and determine if the middle term equals $2ab$ or $-2ab$.

Example 8 **Factoring a Perfect Square Trinomial**

Factor completely. $x^2 + 12x + 36$

Solution:

$x^2 + 12x + 36$ The GCF is 1.

- The first and third terms are positive.
- The first term is a perfect square:
$$x^2 = (x)^2$$
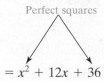
Perfect squares

$= x^2 + 12x + 36$
- The third term is a perfect square:
$$36 = (6)^2$$
- The middle term is twice the product of x and 6:

$$12x = 2(x)(6)$$

$= (x)^2 + 2(x)(6) + (6)^2$ The trinomial is in the form $a^2 + 2ab + b^2$, where $a = x$ and $b = 6$.

$= (x + 6)^2$ Factor as $(a + b)^2$.

Skill Practice

Factor completely.

14. $x^2 + 2x + 1$

Example 9 **Factoring a Perfect Square Trinomial**

Factor completely. $4x^2 - 36xy + 81y^2$

Solution:

$4x^2 - 36xy + 81y^2$ The GCF is 1.

- The first and third terms are positive.
- The first term is a perfect square:
$$4x^2 = (2x)^2.$$

Perfect squares

$= 4x^2 - 36xy + 81y^2$
- The third term is a perfect square:
$$81y^2 = (9y)^2.$$
- The middle term:

$$-36xy = -2(2x)(9y)$$

$= (2x)^2 - 2(2x)(9y) + (9y)^2$ The trinomial is in the form $a^2 - 2ab + b^2$, where $a = 2x$ and $b = 9y$.

$= (2x - 9y)^2$ Factor as $(a - b)^2$.

Skill Practice

Factor completely.

15. $9y^2 - 12yz + 4z^2$

Answers

14. $(x + 1)^2$
15. $(3y - 2z)^2$

4. Factoring by Using Substitution

Sometimes it is convenient to use substitution to convert a polynomial into a simpler form before factoring.

Example 10 Using Substitution to Factor a Polynomial

Factor by using substitution. $(2x - 7)^2 - 3(2x - 7) - 40$

Solution:

$(2x - 7)^2 - 3(2x - 7) - 40$

$\quad = u^2 - 3u - 40$ Substitute $u = 2x - 7$. The trinomial is simpler in form.

$\quad = (u - 8)(u + 5)$ Factor the trinomial.

$\quad = [(2x - 7) - 8][(2x - 7) + 5]$ Reverse substitute. Replace u by $2x - 7$.

$\quad = (2x - 7 - 8)(2x - 7 + 5)$ Simplify.

$\quad = (2x - 15)(2x - 2)$ The second binomial has a GCF of 2.

$\quad = (2x - 15)(2)(x - 1)$ Factor out the GCF from the second binomial.

$\quad = 2(2x - 15)(x - 1)$

Example 11 Using Substitution to Factor a Polynomial

Factor by using substitution. $6y^6 - 5y^3 - 4$

Solution:

$6y^6 - 5y^3 - 4$

$\quad = 6(y^3)^2 - 5(y^3) - 4$ Let $u = y^3$.

$\quad = 6u^2 - 5u - 4$ Substitute u for y^3 in the trinomial.

$\quad = (2u + 1)(3u - 4)$ Factor the trinomial.

$\quad = (2y^3 + 1)(3y^3 - 4)$ Reverse substitute. Replace u with y^3.

TIP: The ac-method or trial-and-error method can also be used for Example 11 without using substitution.

As you work through the exercises in this section, keep these guidelines in mind to factor trinomials.

Factoring Trinomials of the Form $ax^2 + bx + c$ ($a \neq 0$)

When factoring trinomials, the following guidelines should be considered:

Step 1 Factor out the greatest common factor.

Step 2 Check to see if the trinomial is a perfect square trinomial. If so, factor it as either $(a + b)^2$ or $(a - b)^2$. (With a perfect square trinomial, you do not need to use the ac-method or trial-and-error method.)

Step 3 If the trinomial is not a perfect square, use either the ac-method or the trial-and-error method to factor.

Step 4 Check the factorization by multiplication.

Note: Consider using substitution if a trinomial is in the form $au^2 + bu + c$, where u is an algebraic expression.

Section 5.6 Practice Exercises

Vocabulary and Key Concepts

1. **a.** Given a trinomial $x^2 + bx + c$, if c is positive, then the signs in the binomial factors are either both _____ or both negative.

   b. Given a trinomial $x^2 + bx + c$, if c is negative, then the signs in the binomial factors are (choose one: both positive, both negative, opposite).

   c. Which is the correct factored form of $2x^2 - 5x - 12$, the product $(2x + 3)(x - 4)$ or $(x - 4)(2x + 3)$?

   d. Which is the complete factorization of $6x^2 - 4x - 10$, the product $(3x - 5)(2x + 2)$ or $2(3x - 5)(x + 1)$?

   e. A perfect square trinomial $a^2 + 2ab + b^2$ factors as _____.
   Likewise $a^2 - 2ab + b^2$ factors as _____.

Review Exercises

2. Explain how to check a factoring problem.

For Exercises 3–8, factor the polynomial completely.

3. $36c^2d^7e^{11} + 12c^3d^5e^{15} - 6c^2d^4e^7$

4. $5x^3y^3 + 15x^4y^2 - 35x^2y^4$

5. $2x(3a - b) - (3a - b)$

6. $6(v - 8) - 3u(v - 8)$

7. $wz^2 + 2wz - 33az - 66a$

8. $3a^2x + 9ab - abx - 3b^2$

Concepts 1–2: Factoring Trinomials

In Exercises 9–46, factor the trinomial completely by using any method. Remember to look for a common factor first. **(See Examples 1–7.)**

9. $b^2 - 12b + 32$

10. $a^2 - 12a + 27$

11. $y^2 + 10y - 24$

12. $w^2 + 3w - 54$

13. $x^2 + 13x + 30$

14. $t^2 + 9t + 8$

15. $c^2 - 6c - 16$

16. $z^2 - 3z - 28$

17. $2x^2 - 7x - 15$

18. $2y^2 - 13y + 15$

19. $a + 6a^2 - 5$

20. $10b^2 - 3 - 29b$

21. $s^2 + st - 6t^2$

22. $p^2 - pq - 20q^2$

23. $3x^2 - 60x + 108$

24. $4c^2 + 12c - 72$

25. $2c^2 - 2c - 24$

26. $3x^2 + 12x - 15$

27. $2x^2 + 8xy - 10y^2$

28. $20z^2 + 26zw - 28w^2$

29. $33t^2 - 18t + 2$

30. $5p^2 - 10p + 7$

31. $3x^2 + 14xy + 15y^2$

32. $2a^2 + 15ab - 27b^2$

33. $5u^3v - 30u^2v^2 + 45uv^3$

34. $3a^3 + 30a^2b + 75ab^2$

35. $x^3 - 5x^2 - 14x$

36. $p^3 + 2p^2 - 24p$

37. $-23z - 5 + 10z^2$

38. $3 + 16y^2 + 14y$

39. $b^2 + 2b + 15$

40. $x^2 - x - 1$

41. $-2t^2 + 12t + 80$

42. $-3c^2 + 33c - 72$ **43.** $14a^2 + 13a - 12$ **44.** $12x^2 - 16x + 5$

45. $6a^2b + 22ab + 12b$ **46.** $6cd^2 + 9cd - 42c$

Concept 3: Factoring Perfect Square Trinomials

47. a. Multiply the binomials $(x + 5)(x + 5)$.
 b. Factor $x^2 + 10x + 25$.

48. a. Multiply the binomials $(2w - 5)(2w - 5)$.
 b. Factor $4w^2 - 20w + 25$.

49. a. Multiply the binomials $(3x - 2y)^2$.
 b. Factor $9x^2 - 12xy + 4y^2$.

50. a. Multiply the binomials $(x + 7y)^2$.
 b. Factor $x^2 + 14xy + 49y^2$.

For Exercises 51–54, fill in the blank to make the trinomial a perfect square trinomial.

51. $9x^2 + (\underline{}) + 25$ **52.** $16x^4 - (\underline{}) + 1$

53. $64z^4 + (\underline{}) + t^2$ **54.** $9m^4 - (\underline{}) + 49n^2$

For Exercises 55–66, factor out the greatest common factor, if necessary. Then determine if the polynomial is a perfect square trinomial. If it is, factor it. **(See Examples 8 and 9.)**

55. $y^2 - 8y + 16$ **56.** $x^2 + 10x + 25$ **57.** $64m^2 + 80m + 25$

58. $100c^2 - 140c + 49$ **59.** $w^2 - 5w + 9$ **60.** $2a^2 + 14a + 98$

61. $9a^2 - 30ab + 25b^2$ **62.** $16x^4 - 48x^2y + 9y^2$ **63.** $16t^2 - 80tv + 20v^2$

64. $12x^2 - 12xy + 3y^2$ **65.** $5b^4 - 20b^2 + 20$ **66.** $a^4 + 12a^2 + 36$

Concept 4: Factoring by Using Substitution

For Exercises 67–70, factor the polynomial in part (a). Then use substitution to help factor the polynomials in parts (b) and (c).

67. a. $u^2 - 10u + 25$
 b. $x^4 - 10x^2 + 25$
 c. $(a + 1)^2 - 10(a + 1) + 25$

68. a. $u^2 + 12u + 36$
 b. $y^4 + 12y^2 + 36$
 c. $(b - 2)^2 + 12(b - 2) + 36$

69. a. $u^2 + 11u - 26$
 b. $w^6 + 11w^3 - 26$
 c. $(y - 4)^2 + 11(y - 4) - 26$

70. a. $u^2 + 17u + 30$
 b. $z^6 + 17z^3 + 30$
 c. $(x + 3)^2 + 17(x + 3) + 30$

For Exercises 71–82, factor by using substitution. **(See Examples 10 and 11.)**

71. $(3x - 1)^2 - (3x - 1) - 6$ **72.** $(2x + 5)^2 - (2x + 5) - 12$ **73.** $2(x - 5)^2 + 9(x - 5) + 4$

74. $4(x - 3)^2 + 7(x - 3) + 3$ **75.** $3(y + 4)^2 + 5(y + 4) - 2$ **76.** $(3t - 2)^2 - (3t - 2) - 20$

77. $3y^6 + 11y^3 + 6$ **78.** $3x^4 - 5x^2 - 12$ **79.** $4p^4 + 5p^2 + 1$

80. $t^4 + 3t^2 + 2$ **81.** $x^4 + 15x^2 + 36$ **82.** $t^6 - 16t^3 + 63$

Mixed Exercises

83. A student factored $4y^2 - 10y + 4$ as $(2y - 1)(2y - 4)$ on her factoring test. Why did her professor deduct several points, even though $(2y - 1)(2y - 4)$ does multiply out to $4y^2 - 10y + 4$?

84. A student factored $9w^2 + 36w + 36$ as $(3w + 6)^2$ on his factoring test. Why did his instructor deduct several points, even though $(3w + 6)^2$ does multiply out to $9w^2 + 36w + 36$?

For Exercises 85–105, factor completely by using an appropriate method. (Be sure to note the number of terms in the polynomial.)

85. $w^4 + 12w^2 + 36$

86. $9 - 6t^2 + t^4$

87. $81w^2 + 90w + 25$

88. $49a^2 - 28ab + 4b^2$

89. $3x(a + b) - 6(a + b)$

90. $4p(t - 8) + 2(t - 8)$

91. $12a^2bc^2 + 4ab^2c^2 - 6abc^3$

92. $18x^2z - 6xyz + 30xz^2$

93. $-20x^3 + 74x^2 - 60x$

94. $-24y^3 + 90y^2 - 75y$

95. $2y^2 - 9y - 4$

96. $3w^2 - 12w + 4$

97. $2(w^2 - 5)^2 + (w^2 - 5) - 15$

98. $5(t^2 + 3)^2 + 21(t^2 + 3) + 4$

99. $1 - 4d + 3d^2$

100. $2 - 5a + 2a^2$

101. $ax - 5a^2 + 2bx - 10ab$

102. $my + y^2 - 3xm - 3xy$

103. $8z^2 + 24zw - 224w^2$

104. $9x^2 - 18xy - 135y^2$

105. $ay + ax - 5cy - 5cx$

For Exercises 106–114, factor the expression that defines each function.

106. $f(x) = 2x^2 + 13x - 7$

107. $g(x) = 3x^2 + 14x + 8$

108. $m(t) = t^2 - 22t + 121$

109. $n(t) = t^2 + 20t + 100$

110. $P(x) = x^3 + 4x^2 + 3x$

111. $Q(x) = x^4 + 6x^3 + 8x^2$

112. $h(a) = a^3 + 5a^2 - 6a - 30$

113. $k(a) = a^3 - 4a^2 + 2a - 8$

114. $f(x) = 3x^3 - 9x^2 + 5x - 15$

Factoring Binomials Including Sum and Difference of Cubes

Section 5.7

1. Difference of Squares

Up to this point we have learned how to

- Factor out the greatest common factor from a polynomial.
- Factor a four-term polynomial by grouping.
- Recognize and factor perfect square trinomials.
- Factor trinomials by the ac-method and by the trial-and-error method.

   Next, we will learn how to factor binomials that fit the pattern of a difference of squares. Recall from Section 5.3 that the product of two conjugates results in a **difference of squares**

$$(a + b)(a - b) = a^2 - b^2$$

Therefore, to factor a difference of squares, the process is reversed. Identify a and b and construct the conjugate factors.

Concepts

1. Difference of Squares
2. Using a Difference of Squares in Grouping
3. Sum and Difference of Cubes
4. Summary of Factoring Binomials
5. Factoring Binomials of the Form $x^6 - y^6$

> **Factored Form of a Difference of Squares**
>
> $$a^2 - b^2 = (a + b)(a - b)$$

Example 1 Factoring a Difference of Squares

Factor the binomial completely. $16x^2 - 9$

Solution:

$16x^2 - 9$ — The GCF is 1. The binomial is a difference of squares.

$= (4x)^2 - (3)^2$ — Write in the form $a^2 - b^2$, where $a = 4x$ and $b = 3$.

$= (4x + 3)(4x - 3)$ — Factor as $(a + b)(a - b)$.

Example 2 Factoring a Difference of Squares

Factor the binomial completely. $98c^2d - 50d^3$

Solution:

$98c^2d - 50d^3$

$= 2d(49c^2 - 25d^2)$ — The GCF is $2d$. The resulting binomial is a difference of squares.

$= 2d[(7c)^2 - (5d)^2]$ — Write in the form $a^2 - b^2$, where $a = 7c$ and $b = 5d$.

$= 2d(7c + 5d)(7c - 5d)$ — Factor as $(a + b)(a - b)$.

Example 3 Factoring a Difference of Squares

Factor the binomial completely. $z^4 - 81$

Solution:

$z^4 - 81$ — The GCF is 1. The binomial is a difference of squares.

$= (z^2)^2 - (9)^2$ — Write in the form $a^2 - b^2$, where $a = z^2$ and $b = 9$.

$= (z^2 + 9)(z^2 - 9)$ — Factor as $(a + b)(a - b)$.

$z^2 - 9$ is also a difference of squares.

$= (z^2 + 9)(z + 3)(z - 3)$

The difference of squares $a^2 - b^2$ factors as $(a - b)(a + b)$. However, the *sum* of squares is not factorable.

Sum of Squares

Suppose a and b have no common factors. Then the **sum of squares** $a^2 + b^2$ is *not* factorable over the real numbers.

That is, $a^2 + b^2$ is prime over the real numbers.

To see why $a^2 + b^2$ is not factorable, consider the product of binomials:

$$(a \quad b)(a \quad b) \stackrel{?}{=} a^2 + b^2$$

If all possible combinations of signs are considered, none produces the correct product.

$$(a + b)(a - b) = a^2 - b^2 \qquad \text{Wrong sign}$$
$$(a + b)(a + b) = a^2 + 2ab + b^2 \qquad \text{Wrong middle term}$$
$$(a - b)(a - b) = a^2 - 2ab + b^2 \qquad \text{Wrong middle term}$$

After exhausting all possibilities, we see that if a and b share no common factors, then the sum of squares $a^2 + b^2$ is a prime polynomial.

2. Using a Difference of Squares in Grouping

Sometimes a difference of squares can be used along with other factoring techniques.

Example 4 **Using a Difference of Squares in Grouping**

Factor completely. $\quad y^3 - 6y^2 - 4y + 24$

Solution:

$\quad y^3 - 6y^2 - 4y + 24 \qquad\qquad$ The GCF is 1.

$\quad = y^3 - 6y^2 \ \vdots\ - 4y + 24 \qquad$ The polynomial has four terms.
$\qquad\qquad\qquad\qquad\qquad\qquad$ Factor by grouping.

$\quad = y^2(y - 6) - 4(y - 6)$

$\quad = (y - 6)(y^2 - 4) \qquad\qquad y^2 - 4$ is a difference of squares.

$\quad = (y - 6)(y + 2)(y - 2)$

Skill Practice

Factor completely.

4. $a^3 + 5a^2 - 9a - 45$

Answer

4. $(a + 5)(a - 3)(a + 3)$

Example 5 Factoring a Four-Term Polynomial by Grouping Three Terms

Factor completely. $x^2 - y^2 - 6y - 9$

Solution:

Grouping "2 by 2" will not work to factor this polynomial. However, if we factor out -1 from the last three terms, the resulting trinomial will be a perfect square trinomial.

$$x^2 \,\vdots\, - y^2 - 6y - 9 \qquad \text{Group the last three terms.}$$

$$= x^2 - 1(y^2 + 6y + 9) \qquad \text{Factor out } -1 \text{ from the last three terms.}$$

$$= x^2 - (y + 3)^2 \qquad \text{Factor the perfect square trinomial } y^2 + 6y + 9 \text{ as } (y + 3)^2.$$

The quantity $x^2 - (y + 3)^2$ is a difference of squares, $a^2 - b^2$, where $a = x$ and $b = (y + 3)$.

$$= [x - (y + 3)][x + (y + 3)] \qquad \text{Factor as } a^2 - b^2 = (a + b)(a - b).$$

$$= (x - y - 3)(x + y + 3) \qquad \text{Apply the distributive property to clear the inner parentheses.}$$

Avoiding Mistakes

When factoring the expression $x^2 - (y + 3)^2$ as a difference of squares, be sure to use parentheses around the quantity $(y + 3)$. This will help you remember to "distribute the negative" in the expression $[x - (y + 3)]$.

$$[x - (y + 3)] = (x - y - 3)$$

TIP: From Example 5, the expression $x^2 - (y + 3)^2$ can also be factored by using substitution. Let $u = y + 3$.

$$x^2 - (y + 3)^2$$

$$= x^2 - u^2 \qquad \text{Substitution } u = y + 3.$$

$$= (x - u)(x + u) \qquad \text{Factor as a difference of squares.}$$

$$= [x - (y + 3)][x + (y + 3)] \qquad \text{Substitute back.}$$

$$= (x - y - 3)(x + y + 3) \qquad \text{Apply the distributive property.}$$

3. Sum and Difference of Cubes

For binomials that represent the sum or difference of cubes, factor by using the following formulas.

TIP: The following are perfect cubes.

$$1^3 = 1 \qquad (x^1)^3 = x^3$$
$$2^3 = 8 \qquad (x^2)^3 = x^6$$
$$3^3 = 27 \qquad (x^3)^3 = x^9$$
$$4^3 = 64 \qquad (x^4)^3 = x^{12}$$
$$\vdots \qquad\qquad \vdots$$

Any expression raised to a multiple of 3 is a perfect cube.

Factored Form of a Sum and Difference of Cubes

Sum of cubes: $\qquad a^3 + b^3 = (a + b)(a^2 - ab + b^2)$

Difference of cubes: $\quad a^3 - b^3 = (a - b)(a^2 + ab + b^2)$

Multiplication can be used to confirm the formulas for factoring a sum or difference of cubes.

$$(a + b)(a^2 - ab + b^2) = a^3 - a^2b + ab^2 + a^2b - ab^2 + b^3 = a^3 + b^3 ✔$$

$$(a - b)(a^2 + ab + b^2) = a^3 + a^2b + ab^2 - a^2b - ab^2 - b^3 = a^3 - b^3 ✔$$

To help you remember the formulas for factoring a sum or difference of cubes, keep the following guidelines in mind.

- The factored form is the product of a binomial and a trinomial.
- The first and third terms in the trinomial are the squares of the terms within the binomial factor. Therefore, these terms are always positive.
- Without regard to sign, the middle term in the trinomial is the product of terms in the binomial factor.

Square the first term of the binomial.   Product of terms in the binomial

$$x^3 + 8 = (x)^3 + (2)^3 = (x + 2)[(x)^2 - (x)(2) + (2)^2]$$

Square the last term of the binomial.

- The sign within the binomial factor is the same as the sign of the original binomial.
- The first and third terms in the trinomial are always positive.
- The sign of the middle term in the trinomial is opposite the sign within the binomial.

Same sign   Positive

$$x^3 + 8 = (x)^3 + (2)^3 = (x + 2)[(x)^2 - (x)(2) + (2)^2]$$

Opposite signs

> **TIP:** To help remember the placement of the signs in factoring the sum or difference of cubes, remember SOAP: Same sign, Opposite signs, Always Positive.

Example 6   Factoring a Difference of Cubes

Factor. $8x^3 - 27$

Solution:

$8x^3 - 27$	$8x^3$ and 27 are perfect cubes.
$= (2x)^3 - (3)^3$	Write as $a^3 - b^3$, where $a = 2x$ and $b = 3$.
$a^3 - b^3 = (a - b)(a^2 + ab + b^2)$	
$(2x)^3 - (3)^3 = (2x - 3)[(2x)^2 + (2x)(3) + (3)^2]$	Apply the difference of cubes formula.
$= (2x - 3)(4x^2 + 6x + 9)$	Simplify.

> **Skill Practice**
>
> Factor completely.
>
> **6.** $125p^3 - 8$

Answer

6. $(5p - 2)(25p^2 + 10p + 4)$

Example 7 Factoring a Sum of Cubes

Factor. $125t^3 + 64z^6$

Solution:

$$125t^3 + 64z^6$$ $125t^3$ and $64z^6$ are perfect cubes.

$$= (5t)^3 + (4z^2)^3$$ Write as $a^3 + b^3$, where $a = 5t$ and $b = 4z^2$.

$$a^3 + b^3 = (a + b)(a^2 - ab + b^2)$$ Apply the sum of cubes formula.

$$(5t)^3 + (4z^2)^3 = [(5t) + (4z^2)][(5t)^2 - (5t)(4z^2) + (4z^2)^2]$$

$$= (5t + 4z^2)(25t^2 - 20tz^2 + 16z^4)$$ Simplify.

4. Summary of Factoring Binomials

After factoring out the greatest common factor, the next step in any factoring problem is to recognize what type of pattern it follows. Exponents that are divisible by 2 are perfect squares, and those divisible by 3 are perfect cubes. The formulas for factoring binomials are summarized here.

Summary of Factoring Binomials
- Difference of squares: $a^2 - b^2 = (a + b)(a - b)$
- Difference of cubes: $a^3 - b^3 = (a - b)(a^2 + ab + b^2)$
- Sum of cubes: $a^3 + b^3 = (a + b)(a^2 - ab + b^2)$

Example 8 Review of Factoring Binomials

Factor the binomials.

a. $m^3 - \dfrac{1}{8}$ **b.** $9k^2 + 24m^2$ **c.** $128y^6 + 54x^3$ **d.** $50y^6 - 8x^2$

Solution:

a. $m^3 - \dfrac{1}{8}$ m^3 is a perfect cube: $m^3 = (m)^3$. $\frac{1}{8}$ is a perfect cube: $\frac{1}{8} = \left(\frac{1}{2}\right)^3$.

$$= (m)^3 - \left(\frac{1}{2}\right)^3$$ This is a difference of cubes, where $a = m$ and $b = \frac{1}{2}$:

$$= \left(m - \frac{1}{2}\right)\left(m^2 + \frac{1}{2}m + \frac{1}{4}\right)$$ $a^3 - b^3 = (a - b)(a^2 + ab + b^2)$.

b. $9k^2 + 24m^2$ | Factor out the GCF.

$= 3(3k^2 + 8m^2)$ | The resulting binomial is not a difference of squares or a sum or difference of cubes. It cannot be factored further over the real numbers.

c. $128y^6 + 54x^3$ | Factor out the GCF.

$= 2(64y^6 + 27x^3)$ | Both 64 and 27 are perfect cubes, and the exponents of both x and y are multiples of 3. This is a sum of cubes, where $a = 4y^2$ and

$= 2[(4y^2)^3 + (3x)^3]$ | $b = 3x$.

$= 2(4y^2 + 3x)(16y^4 - 12xy^2 + 9x^2)$ | $a^3 + b^3 = (a + b)(a^2 - ab + b^2)$.

d. $50y^6 - 8x^2$ | Factor out the GCF.

$= 2(25y^6 - 4x^2)$ | Both 25 and 4 are perfect squares. The exponents of both x and y are multiples of 2. This is a difference of squares, where

$= 2[(5y^3)^2 - (2x)^2]$ | $a = 5y^3$ and $b = 2x$.

$= 2(5y^3 + 2x)(5y^3 - 2x)$ | $a^2 - b^2 = (a + b)(a - b)$.

5. Factoring Binomials of the Form $x^6 - y^6$

Example 9 Factoring Binomials

Factor the binomial $x^6 - y^6$ as

a. A difference of cubes

b. A difference of squares

Skill Practice

Factor completely.

12. $a^6 - 64$

Solution:

Notice that the expressions x^6 and y^6 are both perfect squares and perfect cubes because the exponents are both multiples of 2 and of 3. Consequently, $x^6 - y^6$ can be interpreted initially as either a difference of cubes or a difference of squares.

a. $x^6 - y^6$

Difference
of cubes

$= (x^2)^3 - (y^2)^3$ | Write as $a^3 - b^3$, where $a = x^2$ and $b = y^2$.

$= (x^2 - y^2)[(x^2)^2 + (x^2)(y^2) + (y^2)^2]$ | Apply the formula $a^3 - b^3 = (a - b)(a^2 + ab + b^2)$.

$= (x^2 - y^2)(x^4 + x^2y^2 + y^4)$ | Factor $x^2 - y^2$ as a difference of squares.

$= (x + y)(x - y)(x^4 + x^2y^2 + y^4)$ | The expression $x^4 + x^2y^2 + y^4$ cannot be factored by using the skills learned thus far.

Answer

12. $(a - 2)(a + 2)(a^2 + 2a + 4)$
$(a^2 - 2a + 4)$

b. $x^6 - y^6$

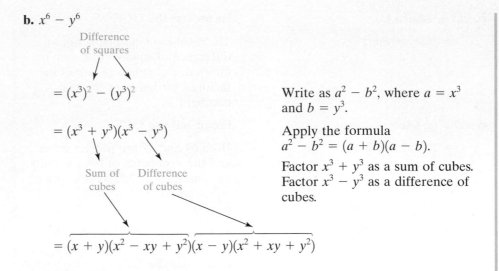

Difference
of squares

$= (x^3)^2 - (y^3)^2$

Write as $a^2 - b^2$, where $a = x^3$
and $b = y^3$.

$= (x^3 + y^3)(x^3 - y^3)$

Apply the formula
$a^2 - b^2 = (a + b)(a - b)$.

Sum of   Difference
cubes   of cubes

Factor $x^3 + y^3$ as a sum of cubes.
Factor $x^3 - y^3$ as a difference of
cubes.

$= (x + y)(x^2 - xy + y^2)(x - y)(x^2 + xy + y^2)$

TIP: Notice that the expressions x^6 and y^6 are both perfect squares and perfect cubes because both exponents are multiples of 2 and of 3. Consequently, $x^6 - y^6$ can be factored initially as either the difference of squares or as the difference of cubes. In such a case, it is recommended that you factor the expression as a difference of squares first because it factors more completely into polynomials of lower degree.

$$x^6 - y^6 = (x + y)(x^2 - xy + y^2)(x - y)(x^2 + xy + y^2)$$

Section 5.7   Practice Exercises

Study Skills Exercise

Multiplying polynomials and factoring polynomials are inverse operations. That is, to check a factoring problem you can multiply, and to check a multiplication problem you can factor. To practice both operations, write a factored polynomial on one side of a 3×5 card with the directions, *Multiply*. On the other side of the card, write the expanded form of the polynomial with the directions, *Factor*. Now you can mix up the cards and get a good sense of what is meant by the directions: *Factor* and *Multiply*.

Vocabulary and Key Concepts

1. a. The binomial $x^2 - 36$ is an example of a _____ of squares. A difference of squares $a^2 - b^2$ factors as _____.

b. The binomial $y^2 + 9$ is an example of a _____ of squares.

c. A sum of squares with greatest common factor 1 (is/is not) factorable over the real numbers.

d. The square of a binomial always results in a perfect _____ trinomial.

e. The binomial $x^3 + 64$ is an example of a _____ of _____.

f. The binomial $c^3 - 27$ is an example of a _____ of _____.

g. A difference of cubes $a^3 - b^3$ factors as (   )(   ).

h. A sum of cubes $a^3 + b^3$ factors as (   )(   ).

Review Exercises

2. a. Identify which expressions represent perfect squares.
$2,\ 4,\ 8,\ 16,\ 25,\ 64,\ x^2,\ x^3,\ x^4,\ x^5,\ x^9$

b. Identify which expressions represent perfect cubes.
$2,\ 4,\ 8,\ 16,\ 25,\ 64,\ x^2,\ x^3,\ x^4,\ x^5,\ x^9$

For Exercises 3–8, factor completely.

3. $4x^2 - 20x + 25$ **4.** $9t^2 - 42t + 49$ **5.** $10x + 6xy + 5 + 3y$

6. $21a + 7ab - 3b - b^2$ **7.** $32p^2 - 28p - 4$ **8.** $6q^2 + 37q - 35$

Concept 1: Difference of Squares

9. Explain how to identify and factor a difference of squares.

10. Can you factor $25x^2 + 4$?

For Exercises 11–22, factor the binomials. Identify the binomials that are prime. **(See Examples 1–3.)**

11. $x^2 - 9$ **12.** $y^2 - 25$ **13.** $16 - 49w^2$ **14.** $81 - 64b^2$

15. $8a^2 - 162b^2$ **16.** $50c^2 - 72d^2$ **17.** $25u^2 + 1$ **18.** $w^2 + 4$

19. $2a^4 - 32$ **20.** $5y^4 - 5$ **21.** $49 - k^6$ **22.** $4 - h^6$

Concept 2: Using a Difference of Squares in Grouping

For Exercises 23–36, use the difference of squares along with factoring by grouping. **(See Examples 4 and 5.)**

23. $x^3 - x^2 - 16x + 16$ **24.** $x^3 + 5x^2 - x - 5$ **25.** $4x^3 + 12x^2 - x - 3$

26. $5x^3 - x^2 - 45x + 9$ **27.** $9y^3 + 7y^2 - 36y - 28$ **28.** $9z^3 - 5z^2 - 36z + 20$

29. $49x^2 + 28x + 4 - y^2$ **30.** $100y^2 + 140y + 49 - z^2$ **31.** $w^2 - 9n^2 + 6n - 1$

32. $m^2 - 25c^2 + 20c - 4$ **33.** $p^4 - 10p^2 + 25 - t^4$ **34.** $m^4 - 14m^2 + 49 - z^4$

35. $9u^4 - 4v^4 + 20v^2 - 25$ **36.** $x^4 - 9y^4 - 42y^2 - 49$

Concept 3: Sum and Difference of Cubes

37. Explain how to identify and factor a sum of cubes.

38. Explain how to identify and factor a difference of cubes.

For Exercises 39–52, factor the sum or difference of cubes. **(See Examples 6 and 7.)**

39. $8x^3 - 1$ (Check by multiplying.)   **40.** $y^3 + 64$ (Check by multiplying.)

41. $125c^3 + 27$ **42.** $216u^3 - v^3$ **43.** $x^3 - 1000$ **44.** $y^3 - 27$

45. $64t^6 + 1$ **46.** $125r^6 + 1$ **47.** $2000y^6 + 2x^3$ **48.** $3a^6 + 24b^3$

49. $16z^4 - 54z$ **50.** $x^5 - 64x^2$ **51.** $p^{12} - 125$ **52.** $t^9 - 8$

Concept 4: Summary of Factoring Binomials

For Exercises 53–80, factor completely. **(See Example 8.)**

53. $36y^2 - \dfrac{1}{25}$ **54.** $16p^2 - \dfrac{1}{9}$ **55.** $18d^{12} - 32$ **56.** $3z^8 - 12$

57. $242v^2 + 32$ **58.** $8p^2 + 200$ **59.** $4x^2 - 16$ **60.** $9m^2 - 81n^2$

61. $25 - 49q^2$ **62.** $1 - 25p^2$ **63.** $(t + 2s)^2 - 36$ **64.** $(5x + 4)^2 - y^2$

65. $27 - t^3$ **66.** $8 + y^3$ **67.** $27a^3 + \dfrac{1}{8}$ **68.** $b^3 + \dfrac{27}{125}$

69. $2m^3 + 16$ **70.** $3x^3 - 375$ **71.** $x^4 - y^4$ **72.** $81u^4 - 16v^4$

73. $a^9 + b^9$ **74.** $27m^9 - 8n^9$ **75.** $\dfrac{1}{8}p^3 - \dfrac{1}{125}$ **76.** $1 - \dfrac{1}{27}d^3$

77. $4w^2 + 25$ **78.** $64 + a^2$ **79.** $\dfrac{1}{25}x^2 - \dfrac{1}{4}y^2$ **80.** $\dfrac{1}{100}a^2 - \dfrac{4}{49}b^2$

Concept 5: Factoring Binomials of the Form $x^6 - y^6$

For Exercises 81–88, factor completely. **(See Example 9.)**

81. $a^6 - b^6$ (*Hint:* First factor as a difference of squares.)

82. $64x^6 - y^6$ **83.** $64 - y^6$ **84.** $1 - p^6$ **85.** $h^6 + k^6$ (*Hint:* Factor as a sum of cubes.)

86. $27q^6 + 125p^6$ **87.** $8x^6 + 125$ **88.** $t^6 + 1$

Mixed Exercises

89. Find a difference of squares that has $(2x + 3)$ as one of its factors.

90. Find a difference of squares that has $(4 - p)$ as one of its factors.

91. Find a difference of cubes that has $(4a^2 + 6a + 9)$ as its trinomial factor.

92. Find a sum of cubes that has $(25c^2 - 10cd + 4d^2)$ as its trinomial factor.

93. Find a sum of cubes that has $(4x^2 + y)$ as its binomial factor.

94. Find a difference of cubes that has $(3t - r^2)$ as its binomial factor.

95. Consider the shaded region:

   a. Find an expression that represents the area of the shaded region.

   b. Factor the expression found in part (a).

   c. Find the area of the shaded region if $x = 6$ in. and $y = 4$ in.

 96. A manufacturer needs to know the area of a metal washer. The outer radius of the washer is R and the inner radius is r.

   a. Find an expression that represents the area of the washer.

   b. Factor the expression found in part (a).

   c. Find the area of the washer if $R = \frac{1}{2}$ in. and $r = \frac{1}{4}$ in. (Round to the nearest 0.01 in.^{2})

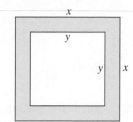

Expanding Your Skills

For Exercises 97–102, factor the polynomials by using the difference of squares, sum of cubes, or difference of cubes with grouping.

97. $x^2 - y^2 + x + y$

98. $64m^2 - 25n^2 + 8m + 5n$

99. $x^3 + y^3 + x + y$

100. $4pu^3 - 4pv^3 - 7yu^3 + 7yv^3$

101. $576a^5 - 9a^2 - 64a^3c^2 + c^2$

102. $32t^5 - 108t^2 - 72t^3v^2 + 243v^2$

Problem Recognition Exercises

Factoring Summary

We now review the techniques of factoring presented thus far along with a general strategy for factoring polynomials.

Step 1   Factor out the greatest common factor (Section 5.5).

Step 2   Identify whether the remaining polynomial has two terms, three terms, or more than three terms.

- If the polynomial has more than three terms, try factoring by grouping (Section 5.5 and Section 5.8).
- If the polynomial has three terms, check first for a perfect square trinomial. Otherwise, factor the trinomial with the ac-method or the trial-and-error method (Section 5.6).
- If the polynomial has two terms, determine if it fits the pattern for a difference of squares, difference of cubes, or sum of cubes. Remember, a sum of squares is not factorable over the real numbers (Section 5.7).

Step 3   For the polynomial in each factor, return to step 2 and factor further if necessary.

Step 4   Check by multiplying.

1. What is meant by a prime polynomial?

2. What is the first step in factoring any polynomial?

3. When factoring a binomial, what patterns do you look for?

4. When factoring a trinomial, what pattern do you look for first?

5. What do you look for when factoring a four-term polynomial?

6. How would you use substitution to factor $3(4x^2 + 1)^2 + 20(4x^2 + 1) + 12$.

For Exercises 7–66,

    a. Factor out the GCF from each polynomial and identify the category in which the remaining polynomial best fits. Choose from

        • difference of squares   • sum of squares

        • difference of cubes   • sum of cubes

        • perfect square trinomial   • trinomial (ac-method or trial-and-error)

        • four terms—grouping   • none of these

    b. Factor the polynomial completely.

7. $6x^2 - 21x - 45$

8. $8m^3 - 10m^2 - 3m$

9. $8a^2 - 50$

10. $ab + ay - b^2 - by$

11. $14u^2 - 11uv + 2v^2$

12. $9p^2 - 12pq + 4q^2$

13. $16x^3 - 2$

14. $9m^2 + 16n^2$

15. $27y^3 + 125$

16. $3x^2 - 16$

17. $128p^6 + 54q^3$

18. $5b^2 - 30b + 45$

19. $16a^4 - 1$

20. $81u^2 - 90uv + 25v^2$

21. $p^2 - 12p + 36 - c^2$

22. $4x^2 + 16$

23. $12ax - 6ay + 4bx - 2by$

24. $125y^3 - 8$

25. $5y^2 + 14y - 3$

26. $2m^4 - 128$

27. $t^2 - 100$

28. $4m^2 - 49n^2$

29. $y^3 + 27$

30. $x^3 + 1$

31. $d^2 + 3d - 28$

32. $c^2 + 5c - 24$

33. $x^2 - 12x + 36$

34. $p^2 + 16p + 64$

35. $2ax^2 - 5ax + 2bx - 5b$

36. $8x^2 - 4bx + 2ax - ab$

37. $10y^2 + 3y - 4$

38. $12z^2 + 11z + 2$

39. $10p^2 - 640$

40. $50a^2 - 72$

41. $z^4 - 64z$

42. $t^4 - 8t$

43. $b^3 - 4b^2 - 45b$

44. $y^3 - 14y^2 + 40y$

45. $9w^2 + 24wx + 16x^2$

46. $4k^2 - 20kp + 25p^2$

47. $60x^2 - 20x + 30ax - 10a$

48. $50x^2 - 200x + 10cx - 40c$

49. $w^4 - 16$

50. $k^4 - 81$

51. $t^6 - 8$

52. $p^6 + 27$

53. $8p^2 - 22p + 5$

54. $9m^2 - 3m - 20$

55. $36y^2 - 12y + 1$

56. $9a^2 + 42a + 49$

57. $2x^2 + 50$

58. $4y^2 + 64$

59. $12r^2s^2 + 7rs^2 - 10s^2$

60. $7z^2w^2 - 10zw^2 - 8w^2$

61. $x^2 + 8xy - 33y^2$

62. $s^2 - 9st - 36t^2$

63. $m^6 + n^3$

64. $a^3 - b^6$

65. $x^2 - 4x$

66. $y^2 - 9y$

For Exercises 67–101, factor completely using the strategy found on page 389.

67. $x^2(x + y) - y^2(x + y)$

68. $u^2(u - v) - v^2(u - v)$

69. $(a + 3)^4 + 6(a + 3)^5$

70. $(4 - b)^4 - 2(4 - b)^3$

71. $24(3x + 5)^3 - 30(3x + 5)^2$

72. $10(2y + 3)^2 + 15(2y + 3)^3$

73. $\dfrac{1}{100}x^2 + \dfrac{1}{35}x + \dfrac{1}{49}$

74. $\dfrac{1}{25}a^2 + \dfrac{1}{15}a + \dfrac{1}{36}$

75. $(5x^2 - 1)^2 - 4(5x^2 - 1) - 5$

76. $(x^3 + 4)^2 - 10(x^3 + 4) + 24$

77. $16p^4 - q^4$

78. $s^4t^4 - 81$

79. $y^3 + \dfrac{1}{64}$

80. $z^3 + \dfrac{1}{125}$

81. $6a^3 + a^2b - 6ab^2 - b^3$

82. $4p^3 + 12p^2q - pq^2 - 3q^3$

83. $\dfrac{1}{9}t^2 + \dfrac{1}{6}t + \dfrac{1}{16}$

84. $\dfrac{1}{25}y^2 + \dfrac{1}{5}y + \dfrac{1}{4}$

85. $x^2 + 12x + 36 - a^2$

86. $a^2 + 10a + 25 - b^2$

87. $p^2 + 2pq + q^2 - 81$

88. $m^2 - 2mn + n^2 - 9$

89. $b^2 - (x^2 + 4x + 4)$

90. $p^2 - (y^2 - 6y + 9)$

91. $4 - u^2 + 2uv - v^2$

92. $25 - a^2 - 2ab - b^2$

93. $6ax - by + 2bx - 3ay$

94. $5pq - 12 - 4q + 15p$

95. $u^6 - 64$ [*Hint:* Factor first as a difference of squares, $(u^3)^2 - (8)^2$.]

96. $1 - v^6$

97. $x^8 - 1$

98. $y^8 - 256$

99. $a^2 - b^2 + a + b$

100. $25c^2 - 9d^2 + 5c - 3d$

101. $5wx^3 + 5wy^3 - 2zx^3 - 2zy^3$

Solving Equations and Applications by Factoring

1. Solving Equations by Using the Zero Product Rule

In Section 1.1, we defined a linear equation in one variable as an equation of the form $ax + b = c$ $(a \neq 0)$. A linear equation in one variable is sometimes called a first-degree polynomial equation because the highest degree of all its terms is 1. A second-degree polynomial equation is called a quadratic equation.

Concepts

1. Solving Equations by Using the Zero Product Rule
2. Applications of Quadratic Equations
3. Definition of a Quadratic Function
4. Applications of Quadratic Functions

Definition of a Quadratic Equation in One Variable

If $a, b,$ and c are real numbers such that $a \neq 0$, then a **quadratic equation** is an equation that can be written in the form

$$ax^2 + bx + c = 0$$

The following equations are quadratic because they can each be written in the form $ax^2 + bx + c = 0$ $(a \neq 0)$.

$$-4x^2 + 4x = 1 \qquad x(x - 2) = 3 \qquad (x - 4)(x + 4) = 9$$
$$-4x^2 + 4x - 1 = 0 \qquad x^2 - 2x = 3 \qquad x^2 - 16 = 9$$
$$x^2 - 2x - 3 = 0 \qquad x^2 - 25 = 0$$
$$x^2 + 0x - 25 = 0$$

One method to solve a quadratic equation is to factor and apply the zero product rule. The **zero product rule** states that if the product of two factors is zero, then one or both of its factors is equal to zero.

The Zero Product Rule

If $ab = 0$, then $a = 0$ or $b = 0$.

For example, the quadratic equation $x^2 - x - 12 = 0$ can be written in factored form as $(x - 4)(x + 3) = 0$. By the zero product rule, one or both factors must be zero: $x - 4 = 0$ or $x + 3 = 0$. Therefore, to solve the quadratic equation, set each factor to zero and solve for x.

$$(x - 4)(x + 3) = 0 \qquad \text{Apply the zero product rule.}$$

$$x - 4 = 0 \quad \text{or} \quad x + 3 = 0 \qquad \text{Set each factor to zero.}$$

$$x = 4 \quad \text{or} \quad x = -3 \qquad \text{Solve each equation for } x.$$

Quadratic equations, like linear equations, arise in many applications of mathematics, science, and business. The following steps summarize the factoring method to solve a quadratic equation.

Solving a Quadratic Equation by Factoring

Step 1   Write the equation in the form $ax^2 + bx + c = 0$.

Step 2   Factor completely.

Step 3   Apply the zero product rule. That is, set each factor equal to zero and solve the resulting equations.*

*The solution(s) found in step 3 may be checked by substitution in the original equation.

Example 1 Solving a Quadratic Equation

Solve. $2x^2 - 5x = 12$

Solution:

$2x^2 - 5x = 12$

$2x^2 - 5x - 12 = 0$ Write the equation in the form $ax^2 + bx + c = 0$.

$(2x + 3)(x - 4) = 0$ Factor completely.

$2x + 3 = 0$ or $x - 4 = 0$ Set each factor equal to zero.

$2x = -3$ or $x = 4$ Solve each equation.

$x = -\dfrac{3}{2}$ or $x = 4$

Check: $x = -\dfrac{3}{2}$ Check: $x = 4$

$2x^2 - 5x = 12$ $2x^2 - 5x = 12$

$2\left(-\dfrac{3}{2}\right)^2 - 5\left(-\dfrac{3}{2}\right) \stackrel{?}{=} 12$ $2(4)^2 - 5(4) \stackrel{?}{=} 12$

$2\left(\dfrac{9}{4}\right) + \dfrac{15}{2} \stackrel{?}{=} 12$ $2(16) - 20 \stackrel{?}{=} 12$

$\dfrac{9}{2} + \dfrac{15}{2} \stackrel{?}{=} 12$ $32 - 20 \stackrel{?}{=} 12$ ✔

$\dfrac{24}{2} \stackrel{?}{=} 12$ ✔

The solutions are $-\dfrac{3}{2}$ and 4.

Skill Practice

Solve.
 1. $y^2 - 2y = 35$

Avoiding Mistakes

The zero product rule tells us that if $ab = 0$, then $a = 0$ or $b = 0$. This property does not hold for other numbers. For example if $ab = 12$, then it is not necessary that a or b must equal 12.

Example 2 Solving a Quadratic Equation

Solve. $6x^2 + 8x = 0$

Solution:

$6x^2 + 8x = 0$

$2x(3x + 4) = 0$ Factor completely.

$2x = 0$ or $3x + 4 = 0$ Set each factor equal to zero.

$x = 0$ $3x = -4$ Solve each equation for x.

$x = -\dfrac{4}{3}$

The solutions are 0 and $-\dfrac{4}{3}$.   The solutions check.

Skill Practice

Solve.
 2. $9x^2 = 21x$

Answers
 1. 7, −5   **2.** 0, $\dfrac{7}{3}$

Skill Practice

Solve.
3. $5a(2a - 3) + 4(a + 1)$
$= 3a(3a - 2)$

Example 3 Solving a Quadratic Equation

Solve. $9x(4x + 2) - 10x = 8x + 25$

Solution:

$$9x(4x + 2) - 10x = 8x + 25$$

$$36x^2 + 18x - 10x = 8x + 25 \qquad \text{Clear parentheses.}$$

$$36x^2 + 8x = 8x + 25 \qquad \text{Combine } like \text{ terms.}$$

$$36x^2 - 25 = 0 \qquad \begin{array}{l} \text{Make one side of the equation} \\ \text{equal to zero. The equation is in} \\ \text{the form } ax^2 + bx + c = 0. \\ (\textit{Note: } b = 0.) \end{array}$$

$$(6x - 5)(6x + 5) = 0 \qquad \text{Factor completely.}$$

$$6x - 5 = 0 \qquad \text{or} \qquad 6x + 5 = 0 \qquad \text{Set each factor equal to zero.}$$

$$6x = 5 \qquad \text{or} \qquad 6x = -5 \qquad \text{Solve each equation.}$$

$$x = \frac{5}{6} \qquad \text{or} \qquad x = -\frac{5}{6} \qquad \text{The check is left to the reader.}$$

The solutions are $\dfrac{5}{6}$ and $-\dfrac{5}{6}$.

Skill Practice

Solve.
4. $t^2 - 3t + 1$
$= t^2 + 2t + 11$

Example 4 Solving an Equation

Solve. $2x(x + 5) + 3 = 2x^2 - 5x + 1$

Solution:

$$2x(x + 5) + 3 = 2x^2 - 5x + 1$$

$$2x^2 + 10x + 3 = 2x^2 - 5x + 1 \qquad \text{Clear parentheses.}$$

$$15x + 2 = 0 \qquad \begin{array}{l} \text{Make one side of the equation equal to} \\ \text{zero. The equation is not quadratic. It is} \\ \text{in the form } ax + b = 0, \text{ which is linear.} \\ \text{Solve by using the method for linear} \\ \text{equations.} \end{array}$$

$$15x = -2$$

$$x = -\frac{2}{15}$$

The solution is $-\dfrac{2}{15}$. \qquad The check is left to the reader.

The zero product rule can be used to solve higher-degree polynomial equations provided one side of the equation is zero and the other is written in factored form.

Answers
3. 4, 1 **4.** −2

Example 5 Solving a Higher-Degree Polynomial Equation

Solve. $-2(y + 7)(y - 1)(10y + 3) = 0$

Solution:

$-2(y + 7)(y - 1)(10y + 3) = 0$

One side of the equation is zero, and the other side is already factored.

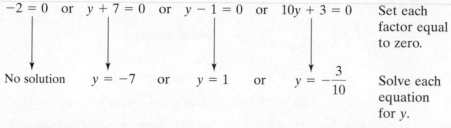

$-2 = 0$ or $y + 7 = 0$ or $y - 1 = 0$ or $10y + 3 = 0$ Set each factor equal to zero.

No solution $y = -7$ or $y = 1$ or $y = -\dfrac{3}{10}$ Solve each equation for y.

Notice that when the constant factor is set to zero, the result is the contradiction $-2 = 0$. The constant factor does not produce a solution to the equation. Therefore, the only solutions are -7, 1, and $-\frac{3}{10}$. Each solution can be checked in the original equation.

The solutions are -7, 1, and $-\dfrac{3}{10}$.

Example 6 Solving a Higher-Degree Polynomial Equation

Solve. $z^3 + 3z^2 - 4z - 12 = 0$

Solution:

$z^3 + 3z^2 - 4z - 12 = 0$ This is a higher-degree polynomial equation.

$z^3 + 3z^2 \; \vdots \; - 4z - 12 = 0$ One side of the equation is zero. Now factor. Because there are four terms, try factoring by grouping.

$z^2(z + 3) - 4(z + 3) = 0$

$(z + 3)(z^2 - 4) = 0$ $z^2 - 4$ can be factored further as a difference of squares.

$(z + 3)(z - 2)(z + 2) = 0$

$z + 3 = 0$ or $z - 2 = 0$ or $z + 2 = 0$ Set each factor equal to zero.

$z = -3$ or $z = 2$ or $z = -2$ Solve each equation.

The solutions are -3, 2, and -2.

Answers

5. $-2, -\dfrac{1}{2}, 8$ **6.** $-1, 3, -3$

2. Applications of Quadratic Equations

Skill Practice

7. The product of two consecutive even integers is 48. Find the integers.

Example 7 Solving an Application of a Quadratic Equation

The product of two consecutive odd integers is 35. Find the integers.

Solution:

Let x represent the smaller odd integer and $x + 2$ represent the next consecutive odd integer.

$$\begin{pmatrix}\text{First odd}\\ \text{integer}\end{pmatrix} \cdot \begin{pmatrix}\text{next odd}\\ \text{integer}\end{pmatrix} = 35 \qquad \text{Verbal model}$$

$$x \cdot (x + 2) = 35 \qquad \text{Mathematical equation}$$

$$x^2 + 2x = 35 \qquad \text{Clear parentheses.}$$

$$x^2 + 2x - 35 = 0 \qquad \text{Set the equation equal to zero.}$$

$$(x + 7)(x - 5) = 0 \qquad \text{Factor.}$$

$$x + 7 = 0 \quad \text{or} \quad x - 5 = 0 \qquad \text{Set each factor equal to zero.}$$

$$x = -7 \quad \text{or} \quad x = 5 \qquad \text{Solve each equation.}$$

If $x = -7$ then the next odd integer is $x + 2 = -5$.
If $x = 5$ then the next odd integer is $x + 2 = 7$.

There are two pairs of odd integers that are solutions, $-7, -5$ and $5, 7$.

Skill Practice

8. The width of a rectangle is 5 in. less than 3 times the length. The area is 2 in.^{2} Find the length and width.

Example 8 Solving an Application of a Quadratic Equation

The length of a basketball court is 6 ft less than 2 times the width. If the total area is 4700 ft^2, find the dimensions of the court.

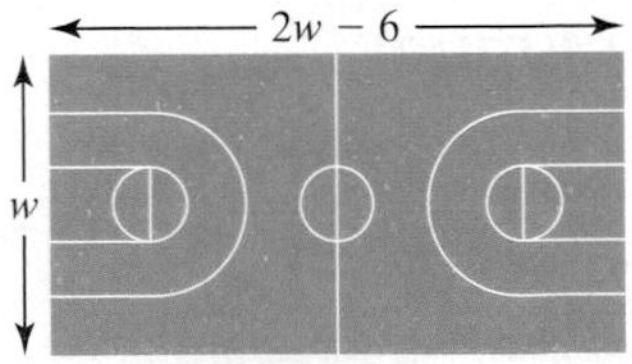

Figure 5-5

Solution:

If the width of the court is represented by w, then the length can be represented by $2w - 6$ (Figure 5-5).

$$A = (\text{length})(\text{width}) \qquad \text{Area of a rectangle}$$

$$4700 = (2w - 6)w \qquad \text{Mathematical equation}$$

$$4700 = 2w^2 - 6w$$

$$2w^2 - 6w - 4700 = 0 \qquad \text{Set the equation equal to zero and factor.}$$

$$2(w^2 - 3w - 2350) = 0 \qquad \text{Factor out the GCF.}$$

$$2(w - 50)(w + 47) = 0 \qquad \text{Factor the trinomial.}$$

$$2 \not= 0 \quad \text{or} \quad w - 50 = 0 \quad \text{or} \quad w + 47 = 0 \qquad \text{Set each factor equal to zero.}$$

contradiction

$$w = 50 \quad \text{or} \quad w \not= -47 \qquad \text{A negative width is not possible.}$$

The width is 50 ft.

The length is $2w - 6 = 2(50) - 6 = 94$ ft.

Answers

7. The integers are -8 and -6 or 6 and 8.

8. The width is 1 in., and the length is 2 in.

A right triangle is a triangle that contains a 90° angle. Furthermore, the sum of the squares of the two legs (the shorter sides) of a right triangle equals the square of the hypotenuse (the longest side). This important fact is known as the Pythagorean theorem. For the right triangle shown in Figure 5-6, the Pythagorean theorem is stated as

$$a^2 + b^2 = c^2$$

In this formula, a and b are the legs and c is the hypotenuse. Notice that the hypotenuse is the longest side and is opposite the right angle.

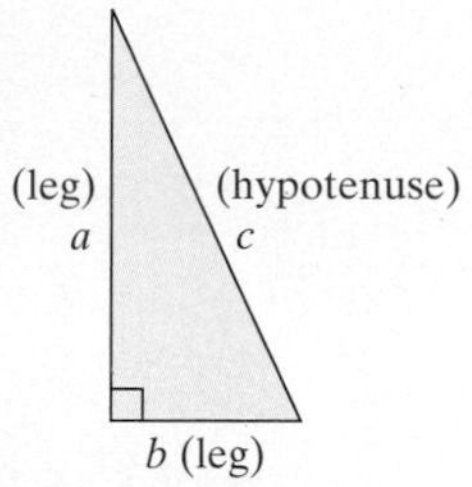

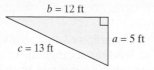

Figure 5-6

The triangle given in Figure 5-7 is a right triangle. We have

$$a^2 + b^2 = c^2$$
$$(5 \text{ ft})^2 + (12 \text{ ft})^2 = (13 \text{ ft})^2$$
$$25 \text{ ft}^2 + 144 \text{ ft}^2 = 169 \text{ ft}^2$$
$$169 \text{ ft}^2 = 169 \text{ ft}^2 \ ✔$$

Figure 5-7

> **TIP:** When applying the Pythagorean theorem, it does not matter which leg you label a and which you label b. Since the lengths of the legs are interchangeable you can also write the Pythagorean theorem as $\text{leg}^2 + \text{leg}^2 = \text{hyp}^2$.

Example 9   **Application of a Quadratic Equation**

A region of coastline off Biscayne Bay is approximately in the shape of a right angle. The corresponding triangular area has sandbars and is marked off on navigational charts as being shallow water. If one leg of the triangle is 0.5 mi shorter than the other leg, and the hypotenuse is 2.5 mi, find the lengths of the legs of the triangle (Figure 5-8).

Figure 5-8

Skill Practice

9. The longer leg of a right triangle measures 7 ft more than the shorter leg. The hypotenuse is 8 ft longer than the shorter leg. Find the lengths of the sides of the triangle.

Solution:

Let x represent the longer leg.

Then $x - 0.5$ represents the shorter leg.

$$a^2 + b^2 = c^2 \qquad \text{Pythagorean theorem}$$
$$x^2 + (x - 0.5)^2 = (2.5)^2$$

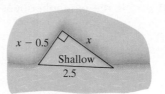

$$x^2 + (x)^2 - 2(x)(0.5) + (0.5)^2 = 6.25$$
$$x^2 + x^2 - x + 0.25 = 6.25$$

> **TIP:** Recall that the square of a binomial results in a perfect square trinomial.
> $$(a - b)^2 = a^2 - 2ab + b^2$$
> $$(x - 0.5)^2 = (x)^2 - 2(x)(0.5) + (0.5)^2$$
> $$= x^2 - x + 0.25$$

$$2x^2 - x - 6 = 0 \qquad \text{Write the equation in the form } ax^2 + bx + c = 0.$$

$$(2x + 3)(x - 2) = 0 \qquad \text{Factor.}$$

$$2x + 3 = 0 \quad \text{or} \quad x - 2 = 0 \qquad \text{Set both factors to zero.}$$

$$x = -\frac{3}{2} \quad \text{or} \quad x = 2 \qquad \text{Solve both equations for } x.$$

The side of a triangle cannot be negative, so we reject the solution $x = -\frac{3}{2}$.

Therefore, one leg of the triangle is 2 mi.

The other leg is $x - 0.5 = 2 - 0.5 = 1.5$ mi.

Answer

9. The sides are 5 ft, 12 ft, and 13 ft.

3. Definition of a Quadratic Function

In Section 3.3, we graphed several basic functions by plotting points, including $f(x) = x^2$. This function is called a quadratic function, and its graph is in the shape of a **parabola**. In general, any second-degree polynomial function is a quadratic function.

> ### Definition of a Quadratic Function
> Let a, b, and c represent real numbers such that $a \neq 0$. Then a function defined by $f(x) = ax^2 + bx + c$ is called a **quadratic function**.

The graph of a quadratic function is a parabola that opens upward or downward. The leading coefficient a determines the direction of the parabola. For the quadratic function defined by $f(x) = ax^2 + bx + c$:

If $a > 0$, the parabola opens upward. For example, $f(x) = x^2$

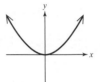

If $a < 0$, the parabola opens downward. For example, $g(x) = -x^2$

Recall from Section 3.3 that the x-intercepts of a function $y = f(x)$ are the real solutions to the equation $f(x) = 0$. The y-intercept is found by evaluating $f(0)$.

Skill Practice

10. Find the x- and y-intercepts of the function defined by $f(x) = x^2 + 8x + 12$.

Example 10   Finding the x- and y-Intercepts of a Quadratic Function

Find the x- and y-intercepts. $f(x) = x^2 - x - 12$

Solution:

To find the x-intercept, substitute $f(x) = 0$.

$$f(x) = x^2 - x - 12$$

$$0 = x^2 - x - 12 \qquad \text{Substitute 0 for } f(x). \text{ The result is a quadratic equation.}$$

$$0 = (x - 4)(x + 3) \qquad \text{Factor.}$$

$$x - 4 = 0 \quad \text{or} \quad x + 3 = 0 \qquad \text{Set each factor equal to zero.}$$

$$x = 4 \quad \text{or} \quad x = -3 \qquad \text{Solve each equation.}$$

The x-intercepts are $(4, 0)$ and $(-3, 0)$.

To find the y-intercept, find $f(0)$.

$$f(x) = x^2 - x - 12$$

$$f(0) = (0)^2 - (0) - 12 \qquad \text{Substitute } x = 0.$$

$$= -12$$

The y-intercept is $(0, -12)$.

Answer

10. x-intercepts: $(-6, 0)$ and $(-2, 0)$; y-intercept: $(0, 12)$

Topic: Using the Zero Feature

The graph of $f(x) = x^2 - x - 12$ supports the solution to Example 10. Use the *zero* feature to confirm that the graph crosses the x-axis at -3 and 4. The y-intercept is given as $(0, -12)$.

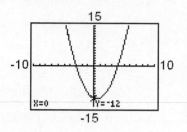

4. Applications of Quadratic Functions

Example 11 Application of a Quadratic Function

A model rocket is shot vertically upward with an initial velocity of 288 ft/sec. The function given by $h(t) = -16t^2 + 288t$ relates the rocket's height $h(t)$ (in feet) to the time t after launch (in seconds).

a. Find $h(0)$, $h(5)$, $h(10)$, and $h(15)$, and interpret the meaning of these function values in the context of the rocket's height and time after launch.

b. Find the t-intercepts of the function, and interpret their meaning in the context of the rocket's height and time after launch.

c. Find the time(s) at which the rocket is at a height of 1152 ft.

Solution:

a. $h(t) = -16t^2 + 288t$

$h(0) = -16(0)^2 + 288(0) = 0$

$h(5) = -16(5)^2 + 288(5) = 1040$

$h(10) = -16(10)^2 + 288(10) = 1280$

$h(15) = -16(15)^2 + 288(15) = 720$

$h(0) = 0$ means that at $t = 0$ sec, the height of the rocket is 0 ft.

$h(5) = 1040$ means that 5 sec after launch, the height is 1040 ft.

$h(10) = 1280$ means that 10 sec after launch, the height is 1280 ft.

$h(15) = 720$ means that 15 sec after launch, the height is 720 ft.

b. The t-intercepts of the function are represented by the real solutions of the equation $h(t) = 0$.

$-16t^2 + 288t = 0$ Set $h(t) = 0$.

$-16t(t - 18) = 0$ Factor.

$-16t = 0$ or $t - 18 = 0$ Apply the zero product rule.

$t = 0$ or $t = 18$

The rocket is at ground level initially (at $t = 0$ sec) and then again after 18 sec when it hits the ground.

Skill Practice

An object is dropped from the top of a building that is 144 ft high. The function given by $h(t) = -16t^2 + 144$ relates the height $h(t)$ of the object (in feet) to the time t in seconds after it is dropped.

11. a. Find $h(0)$ and interpret the meaning of the function value in the context of this problem.
    b. Find the t-intercept(s) and interpret the meaning in the context of this problem.
    c. When will the object be 128 ft above ground level?

Answer

11a. $h(0) = 144$, which is the initial height of the object (after 0 sec).
  b. The t-intercept is $(3, 0)$, which means the object is at ground level (0 ft high) after 3 sec. The intercept $(-3, 0)$ does not make sense for this problem since time cannot be negative.
  c. One second after release, the object will be 128 ft above ground level.

c. Set $h(t) = 1152$ and solve for t.

$$h(t) = -16t^2 + 288t$$

$$1152 = -16t^2 + 288t \qquad \text{Substitute 1152 for } h(t).$$

$$16t^2 - 288t + 1152 = 0 \qquad \text{Set the equation equal to zero.}$$

$$16(t^2 - 18t + 72) = 0 \qquad \text{Factor out the GCF.}$$

$$16(t - 6)(t - 12) = 0 \qquad \text{Factor.}$$

$$t = 6 \qquad \text{or} \qquad t = 12$$

The rocket will reach a height of 1152 ft after 6 sec (on the way up) and after 12 sec (on the way down). (See Figure 5-9.)

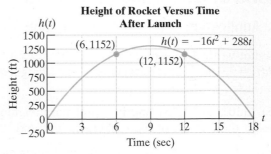

Height of Rocket Versus Time After Launch

$h(t) = -16t^2 + 288t$

(6, 1152) (12, 1152)

Figure 5-9

Section 5.8 Practice Exercises

Vocabulary and Key Concepts

1. a. An equation that can be written in the form $ax^2 + bx + c = 0$, $a \neq 0$, is called a _____ equation.

b. The zero product rule states that if $ab = 0$, then $a =$ _____ or $b =$ _____ .

c. The _____ theorem states that given a right triangle with legs a and b and hypotenuse c, then $a^2 + b^2 =$ _____ .

d. A function defined by $f(x) = ax^2 + bx + c$, $a \neq 0$, is called a _____ function.

e. Given a quadratic function $f(x) = ax^2 + bx + c$, $a \neq 0$, find the x-intercept(s) by solving the equation _____ . Find the _____ -intercept by evaluating $f(0)$.

f. If x is an integer, then _____ represents the next greater integer. If x is an odd integer, then _____ represents the next greater odd integer. Likewise if x is an even integer, then _____ represents the next greater even integer.

g. The area of a rectangle of length l and width w is given by $A =$ _____ .

h. The area of a triangle with base b and height h is given by the formula $A =$ _____ .

Review Exercises

2. Write the factored form for each binomial, if possible.

a. $x^2 - y^2$ **b.** $x^2 + y^2$ **c.** $x^3 - y^3$ **d.** $x^3 + y^3$

For Exercises 3–8, factor completely.

3. $10x^2 + 3x$ **4.** $7x^2 - 28$ **5.** $2p^2 - 9p - 5$

6. $3q^2 - 4q - 4$ **7.** $t^3 - 1$ **8.** $z^2 - 11z + 30$

Concept 1: Solving Equations by Using the Zero Product Rule

9. What conditions are necessary to solve an equation by using the zero product rule?

10. State the zero product rule.

For Exercises 11–16, determine which of the equations are written in the correct form to apply the zero product rule directly. If an equation is not in the correct form, explain what is wrong.

11. $2x(x - 3) = 0$ **12.** $(u + 1)(u - 3) = 10$ **13.** $3p^2 - 7p + 4 = 0$

14. $t^2 - t - 12 = 0$ **15.** $a(a + 3)^2 = 5$ **16.** $\left(\dfrac{2}{3}x - 5\right)\left(x + \dfrac{1}{2}\right) = 0$

For Exercises 17–20, factor the polynomial or solve the equation as indicated.

17. a. Factor. $w^2 - 81$
     b. Solve. $w^2 - 81 = 0$

18. a. Factor. $p^2 - 25$
     b. Solve. $p^2 - 25 = 0$

19. a. Factor. $3x^2 + 14x - 5$
     b. Solve. $3x^2 + 14x - 5 = 0$

20. a. Factor. $2y^2 - y - 3$
     b. Solve. $2y^2 - y - 3 = 0$

For Exercises 21–56, solve the equation. **(See Examples 1–6.)**

21. $(x + 3)(x + 5) = 0$ **22.** $(x + 7)(x - 4) = 0$ **23.** $(2w + 9)(5w - 1) = 0$

24. $(3a + 1)(4a - 5) = 0$ **25.** $x(x + 4)(10x - 3) = 0$ **26.** $t(t - 6)(3t - 11) = 0$

27. $0 = 5(y - 0.4)(y + 2.1)$ **28.** $0 = -4(z - 7.5)(z - 9.3)$ **29.** $x^2 + 6x - 27 = 0$

30. $2x^2 + x - 15 = 0$ **31.** $2x^2 + 5x = 3$ **32.** $-11x = 3x^2 - 4$

33. $10x^2 = 15x$ **34.** $5x^2 = 7x$ **35.** $6(y - 2) - 3(y + 1) = 8$

36. $4x + 3(x - 9) = 6x + 1$ **37.** $-9 = y(y + 6)$ **38.** $-62 = t(t - 16) + 2$

39. $9p^2 - 15p - 6 = 0$ **40.** $6y^2 + 2y = 48$ **41.** $(x + 1)(2x - 1)(x - 3) = 0$

42. $2x(x - 4)^2(4x + 3) = 0$ **43.** $(y - 3)(y + 4) = 8$ **44.** $(t + 10)(t + 5) = 6$

45. $(2a - 1)(a - 1) = 6$ **46.** $w(6w + 1) = 2$ **47.** $p^2 + (p + 7)^2 = 169$

48. $x^2 + (x + 2)^2 = 100$ **49.** $3t(t + 5) - t^2 = 2t^2 + 4t - 1$ **50.** $a^2 - 4a - 2 = (a + 3)(a - 5)$

51. $2x^3 - 8x^2 - 24x = 0$ **52.** $2p^3 + 20p^2 + 42p = 0$ **53.** $w^3 = 16w$

54. $12x^3 = 27x$ **55.** $0 = 2x^3 + 5x^2 - 18x - 45$ **56.** $0 = 3y^3 + y^2 - 48y - 16$

Concept 2: Applications of Quadratic Equations

57. If 5 is added to the square of a number, the result is 30. Find all such numbers.

58. Four less than the square of a number is 77. Find all such numbers.

59. The square of a number is equal to 12 more than the number. Find all such numbers.

60. The square of a number is equal to 20 more than the number. Find all such numbers.

61. The product of two consecutive integers is 42. Find the integers.

62. The product of two consecutive integers is 110. Find the integers.

63. The product of two consecutive odd integers is 63. Find the integers. **(See Example 7.)**

64. The product of two consecutive even integers is 120. Find the integers.

65. A rectangular pen is to contain 35 ft^2 of area. If the width is 2 ft less than the length, find the dimensions of the pen. **(See Example 8.)**

66. The length of a rectangular photograph is 7 in. more than the width. If the area is 78 in.^{2}, what are the dimensions of the photograph?

67. The length of a rectangular room is 5 yd more than the width. If the area is 300 yd^2, find the length and the width of the room.

68. The top of a rectangular dining room table is twice as long as it is wide. Find the dimensions of the table if the area is 18 ft^2.

69. The height of a triangle is 1 in. more than the base. If the height is increased by 2 in. while the base remains the same, the new area becomes 20 in.^{2}

    a. Find the base and height of the original triangle.

    b. Find the area of the original triangle.

70. The base of a triangle is 2 cm more than the height. If the base is increased by 4 cm while the height remains the same, the new area is 56 cm^2.

    a. Find the base and height of the original triangle.

    b. Find the area of the original triangle.

71. The area of a triangular garden is 25 ft^2. The base is twice the height. Find the base and height of the triangle.

72. The height of a triangle is 1 in. more than twice the base. If the area is 18 in.^{2}, find the base and height of the triangle.

73. The sum of the squares of two consecutive positive integers is 41. Find the integers.

74. The sum of the squares of two consecutive, positive even integers is 164. Find the integers.

75. Justin must travel from Summersville to Clayton. He can drive 10 mi through the mountains at 40 mph. Or he can drive east and then north on superhighways at 60 mph. The alternative route forms a right angle as shown in the diagram. The eastern leg is 2 mi less than the northern leg. **(See Example 9.)**

    a. Find the total distance Justin would travel in going the alternative route.

    b. If Justin wants to minimize the time of the trip, which route should he take?

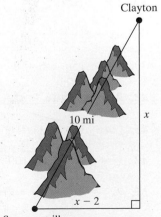

76. A 17-ft ladder is standing up against a wall. The distance between the base of the ladder and the wall is 7 ft less than the distance between the top of the ladder and the base of the wall. Find the distance between the base of the ladder and the wall.

77. A right triangle has side lengths represented by three consecutive even integers. Find the lengths of the three sides, measured in meters.

78. The hypotenuse of a right triangle is 3 m more than twice the short leg. The longer leg is 2 m more than twice the shorter leg. Find the lengths of the sides.

79. Determine the length of the radius of a circle whose area is numerically equal to its circumference.

80. Determine the length of the radius of a circle whose area is numerically twice its circumference.

Concept 3: Definition of a Quadratic Function

For Exercises 81–84, **a.** Find the values of x for which $f(x) = 0$. **b.** Find $f(0)$.

81. $f(x) = x^2 - 3x$

82. $f(x) = 4x^2 + 2x$

83. $f(x) = x^2 - 6x - 7$

84. $f(x) = 2x^2 + 11x + 5$

For Exercises 85–88, find the x- and y-intercepts for the functions defined by $y = f(x)$. **(See Example 10.)**

85. $f(x) = \dfrac{1}{2}(x - 2)(x + 1)(2x)$

86. $f(x) = (x + 1)(x - 2)(x + 3)^2$

87. $f(x) = x^2 - 2x + 1$

88. $f(x) = x^2 + 4x + 4$

For Exercises 89–92, find the x-intercepts of each function and use that information to match the function with its graph.

89. $g(x) = x^2 - 9$

90. $h(x) = x(x - 2)(x + 4)$

91. $f(x) = 4(x + 1)$

92. $k(x) = (x + 1)(x + 3)(x - 2)(x - 1)$

a.

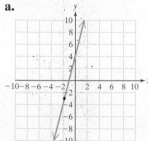

b.

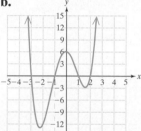

c.

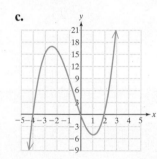

d.

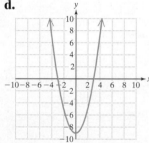

Concept 4: Applications of Quadratic Functions

 93. A rocket is fired upward from ground level with an initial velocity of 490 m/sec. The height of the rocket $s(t)$ in meters is a function of the time t in seconds after launch. **(See Example 11.)**

$$s(t) = -4.9t^2 + 490t$$

a. What characteristics of s indicate that it is a quadratic function?

b. Find the t-intercepts of the function.

c. What do the t-intercepts mean in the context of this problem?

d. At what times is the rocket at a height of 485.1 m?

94. A certain company makes water purification systems. The factory can produce x water systems per year. The profit $P(x)$ the company makes is a function of the number of systems x it produces.

$$P(x) = -2x^2 + 1000x$$

a. Is this function linear or quadratic?

b. Find the number of water systems x that would produce a zero profit.

c. What points on the graph do the answers in part (b) represent?

d. Find the number of systems for which the profit is $80,000.

For Exercises 95–100, factor the expressions represented by $f(x)$. Explain how the factored form relates to the graph of the function. Can the graph of the function help you determine the factors?

95. $f(x) = x^2 - 7x + 10$

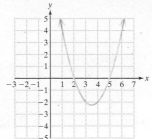

96. $f(x) = x^2 - 2x - 3$

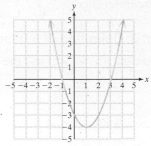

97. $f(x) = x^2 + 2x + 1$

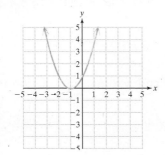

98. $f(x) = x^2 - 8x + 16$

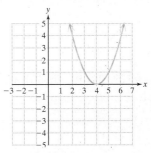

99. $f(x) = -x^2 - 6x - 5$

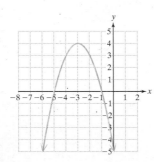

100. $f(x) = -x^2 + 6x - 8$

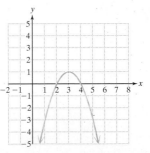

Expanding Your Skills

101. The surface area of a right circular cylinder is represented by SA $= 2\pi r^2 + 2\pi rh$. If the surface area is 156π ft^2 and the height is 7 ft, determine the radius of the cylinder.

102. Determine the length and width of a rectangle with a perimeter of 20 yd and an area of 16 yd^2.

103. Determine the length and width of a rectangle with a perimeter of 28 ft and an area of 48 ft^2.

For Exercises 104–107, find an equation that has the given solutions. For example, 2 and -1 are solutions to $(x - 2)(x + 1) = 0$ or $x^2 - x - 2 = 0$. In general, x_1 and x_2 are solutions to the equation $a(x - x_1)(x - x_2) = 0$, where a can be any nonzero real number. For each problem, there is more than one correct answer depending on your choice of a.

104. $x = -3$ and $x = 1$

105. $x = 2$ and $x = -2$

106. $x = 0$ and $x = -5$

107. $x = 0$ and $x = -3$

Graphing Calculator Exercises

For Exercises 108–111, graph Y_1. Use the *Zero* feature to approximate the x-intercepts. Then solve $Y_1 = 0$ and compare the solutions to the x-intercepts.

108. $Y_1 = -x^2 + x + 2$

109. $Y_1 = -x^2 - x + 20$

110. $Y_1 = x^2 - 6x + 9$

111. $Y_1 = x^2 + 4x + 4$

Group Activity

Investigating Pascal's Triangle

Estimated time: 15 minutes

Group Size: 2 or 3

1. Determine the value of $(a + b)^0$. Write your answer here: _____

2. Write the binomials in expanded form. [For example, for $(a + b)^2$, multiply the binomials $(a + b)(a + b)$ and write the result in the space provided.] Check your answers with another group.

   a. $(a + b)^1$ _____

   b. $(a + b)^2$ _____

   c. $(a + b)^3$ _____

3. **a.** How many terms were in the expansion of $(a + b)^2$? _____

   b. How many terms were in the expansion of $(a + b)^3$? _____

   c. How many terms would you expect in the expansion of $(a + b)^4$? _____

   d. How many terms would you expect in the expansion of $(a + b)^5$? _____

   e. How many terms would you expect in the expansion of $(a + b)^n$? _____

4. Based on your answers from question 2, do you see a pattern regarding the exponents on the factors of a and b in each term? Explain.

5. The coefficients of each expansion also follow a pattern. The triangle of numbers shown here is called **Pascal's triangle**. Pascal's triangle gives the coefficients of a binomial expansion of the form $(a + b)^n$. Work with your group members to see if you can figure out the pattern for the coefficients of $(a + b)^4$, $(a + b)^5$, $(a + b)^6$, and $(a + b)^7$.

$(a + b)^n$	Coefficients of the Expansion
$(a + b)^0$	1
$(a + b)^1$	1 1
$(a + b)^2$	1 2 1
$(a + b)^3$	1 3 3 1
$(a + b)^4$	
$(a + b)^5$	
$(a + b)^6$	

6. Using the expected number of terms found in question 3, the pattern of the exponents found in question 4, and the pattern for the coefficients in question 5, write the expansions for the following.

   a. $(a + b)^4 =$ _____

   b. $(a + b)^5 =$ _____

   c. $(a + b)^6 =$ _____

Chapter 5   Summary

Section 5.1   Properties of Integer Exponents and Scientific Notation

Key Concepts

Let a and b ($b \neq 0$) represent real numbers and m and n represent positive integers.

$$b^m \cdot b^n = b^{m+n} \qquad \frac{b^m}{b^n} = b^{m-n}$$

$$(b^m)^n = b^{m \cdot n} \qquad (ab)^m = a^m b^m$$

$$\left(\frac{a}{b}\right)^m = \frac{a^m}{b^m} \qquad b^0 = 1$$

$$b^{-n} = \left(\frac{1}{b}\right)^n = \frac{1}{b^n}$$

A number expressed in the form $a \times 10^n$, where $1 \leq |a| < 10$ and n is an integer, is written in **scientific notation**.

Examples

Example 1

$$\left(\frac{2x^2 y}{z^{-1}}\right)^{-3} (x^{-4} y^0)$$

$$= \left(\frac{2^{-3} x^{-6} y^{-3}}{z^3}\right)(x^{-4} \cdot 1)$$

$$= \frac{2^{-3} x^{-10} y^{-3}}{z^3}$$

$$= \frac{1}{2^3 x^{10} y^3 z^3} \quad \text{or} \quad \frac{1}{8 x^{10} y^3 z^3}$$

Example 2

$$0.0000002 \times 35{,}000$$

$$= (2 \times 10^{-7})(3.5 \times 10^4)$$

$$= 7 \times 10^{-3} \text{ or } 0.007$$

Section 5.2   Addition and Subtraction of Polynomials and Polynomial Functions

Key Concepts

A **polynomial** in x is defined by a sum of terms of the form ax^n, where a is a real number and n is a whole number.

- a is the **coefficient** of the term.
- n is the **degree of the term**.

The **degree of a polynomial** is the greatest degree of its terms.

The term of a polynomial with the greatest degree is the **leading term**. Its coefficient is the **leading coefficient**.

A one-term polynomial is a **monomial**.

A two-term polynomial is a **binomial**.

A three-term polynomial is a **trinomial**.

To add or subtract polynomials, add or subtract *like* terms.

Examples

Example 1

$$7y^4 - 2y^2 + 3y + 8$$

is a polynomial with leading coefficient 7 and degree 4.

Example 2

$$f(x) = 4x^3 - 6x - 11$$

f is a polynomial function with leading term $4x^3$ and leading coefficient 4. The degree of f is 3.

Example 3

For $f(x) = 4x^3 - 6x - 11$, find $f(-1)$.

$$f(-1) = 4(-1)^3 - 6(-1) - 11$$

$$= -9$$

Example 4

$$(-4x^3 y + 3x^2 y^2) - (7x^3 y - 5x^2 y^2)$$

$$= -4x^3 y + 3x^2 y^2 - 7x^3 y + 5x^2 y^2$$

$$= -11x^3 y + 8x^2 y^2$$

Section 5.3   Multiplication of Polynomials

Key Concepts

To multiply polynomials, multiply each term in the first polynomial by each term in the second polynomial.

Special Products

1. Multiplication of **conjugates**

   $(a + b)(a - b) = a^2 - b^2$

   The product is called a **difference of squares**.

2. Square of a binomial

   $(a + b)^2 = a^2 + 2ab + b^2$

   $(a - b)^2 = a^2 - 2ab + b^2$

   The product is called a **perfect square trinomial**.

Examples

Example 1

$(x - 2)(3x^2 - 4x + 11)$
$= 3x^3 - 4x^2 + 11x - 6x^2 + 8x - 22$
$= 3x^3 - 10x^2 + 19x - 22$

Example 2

$(3x + 5)(3x - 5)$
$= (3x)^2 - (5)^2$
$= 9x^2 - 25$

Example 3

$(4y + 3)^2$
$= (4y)^2 + (2)(4y)(3) + (3)^2$
$= 16y^2 + 24y + 9$

Section 5.4   Division of Polynomials

Key Concepts

Division of polynomials:

1. For division by a monomial, use the properties

   $\dfrac{a + b}{c} = \dfrac{a}{c} + \dfrac{b}{c}$ and $\dfrac{a - b}{c} = \dfrac{a}{c} - \dfrac{b}{c}$

   for $c \neq 0$.

2. If the divisor has more than one term, use long division.
   - First write each polynomial in descending order.
   - Insert placeholders for missing powers.

Examples

Example 1

$\dfrac{-12a^2 - 6a + 9}{-3a}$

$= \dfrac{-12a^2}{-3a} - \dfrac{6a}{-3a} + \dfrac{9}{-3a}$

$= 4a + 2 - \dfrac{3}{a}$

Example 2

$(3x^2 - 5x + 1) \div (x + 2)$

$$
\begin{array}{r}
3x - 11 \\
x + 2 \overline{\smash{)}\, 3x^2 - 5x + 1} \\
-(3x^2 + 6x) \\
\hline
-11x + 1 \\
-(-11x - 22) \\
\hline
23
\end{array}
$$

Answer: $3x - 11 + \dfrac{23}{x + 2}$

3. **Synthetic division** may be used to divide a polynomial by a binomial in the form $x - r$, where r is a constant.

Example 3

$(3x^2 - 5x + 1) \div (x + 2)$

$$\begin{array}{r|rr} -2 & 3 & -5 & 1 \\ & & -6 & 22 \\ \hline & 3 & -11 & \underline{|23} \end{array}$$

Answer: $3x - 11 + \dfrac{23}{x + 2}$

Section 5.5   Greatest Common Factor and Factoring by Grouping

Key Concepts

The **greatest common factor (GCF)** is the largest factor common to all terms of a polynomial. To factor out the GCF from a polynomial, use the distributive property.

A four-term polynomial may be **factored by grouping**.

Steps to Factor by Grouping

1. Identify and factor out the GCF from all four terms.
2. Factor out the GCF from the first pair of terms. Factor out the GCF from the second pair of terms. (Sometimes it is necessary to factor out the *opposite* of the GCF.)
3. If the two pairs of terms share a common binomial factor, factor out the binomial factor.

Examples

Example 1

$3x^2(a + b) - 6x(a + b)$

$= 3x(a + b)x - 3x(a + b)(2)$

$= 3x(a + b)(x - 2)$

Example 2

$60xa - 30xb - 80ya + 40yb$

$= 10[6xa - 3xb \mathbin{\vdots} - 8ya + 4yb]$

$= 10[3x(2a - b) - 4y(2a - b)]$

$= 10(2a - b)(3x - 4y)$

Section 5.6   Factoring Trinomials and Perfect Square Trinomials

Key Concepts

AC-Method

To factor trinomials of the form $ax^2 + bx + c$:

1. Factor out the GCF. Find the product ac.
2. Find two integers whose product is ac and whose sum is b. (If no pair of numbers can be found, then the trinomial is prime.)
3. Rewrite the middle term bx as the sum of two terms whose coefficients are the numbers found in step 2.
4. Factor the polynomial by grouping.

Examples

Example 1

$10y^2 + 35y - 20 = 5(2y^2 + 7y - 4)$

$$ac = (2)(-4) = -8$$

Find two integers whose product is -8 and whose sum is 7. The numbers are 8 and -1.

$5[2y^2 + 8y \mathbin{\vdots} - 1y - 4]$

$= 5[2y(y + 4) - 1(y + 4)]$

$= 5(y + 4)(2y - 1)$

Trial-and-Error Method

To factor trinomials in the form $ax^2 + bx + c$:

1. Factor out the GCF.
2. List the pairs of factors of a and the pairs of factors of c. Consider the reverse order in either list.
3. Construct two binomials of the form

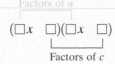

4. Test each combination of factors and signs until the correct product is found.
5. If no combination of factors works, the polynomial is prime.

The factored form of a **perfect square trinomial** is the square of a binomial:

$a^2 + 2ab + b^2 = (a + b)^2$

$a^2 - 2ab + b^2 = (a - b)^2$

Sometimes it is easier to factor a polynomial after making a substitution.

Example 2

$10y^2 + 35y - 20 = 5(2y^2 + 7y - 4)$

The pairs of factors of 2 are $2 \cdot 1$.
The pairs of factors of -4 are

$$\begin{array}{cc} -1 \cdot 4 & 1 \cdot (-4) \\ -2 \cdot 2 & 2 \cdot (-2) \\ -4 \cdot 1 & 4 \cdot (-1) \end{array}$$

$(2y - 2)(y + 2) = 2y^2 + 2y - 4$ No

$(2y - 4)(y + 1) = 2y^2 - 2y - 4$ No

$(2y + 1)(y - 4) = 2y^2 - 7y - 4$ No

$(2y + 2)(y - 2) = 2y^2 - 2y - 4$ No

$(2y + 4)(y - 1) = 2y^2 + 2y - 4$ No

$(2y - 1)(y + 4) = 2y^2 + 7y - 4$ Yes

Therefore, $10y^2 + 35y - 20$ factors as $5(2y - 1)(y + 4)$.

Example 3

$9w^2 - 30wz + 25z^2$

$= (3w)^2 - 2(3w)(5z) + (5z)^2$

$= (3w - 5z)^2$

Example 4

$(7v^2 - 1)^2 - (7v^2 - 1) - 12$ Let $u = (7v^2 - 1)$.

$= u^2 - u - 12$ Substitute.

$= (u + 3)(u - 4)$ Factor.

$= (7v^2 - 1 + 3)(7v^2 - 1 - 4)$ Back substitute.

$= (7v^2 + 2)(7v^2 - 5)$ Simplify.

Section 5.7 Factoring Binomials Including Sum and Difference of Cubes

Key Concepts

Factoring Binomials: Summary

Difference of squares:

$a^2 - b^2 = (a + b)(a - b)$

Sum of squares:

If a and b share no common factors, then $a^2 + b^2$ is prime.

Examples

Example 1

$25u^2 - 9v^4 = (5u)^2 - (3v^2)^2$

$= (5u + 3v^2)(5u - 3v^2)$

Example 2

$32 + 2w^2 = 2(16 + w^2)$ cannot be factored further because $16 + w^2$ is a sum of squares.

Difference of cubes:

$a^3 - b^3 = (a - b)(a^2 + ab + b^2)$

Sum of cubes:

$a^3 + b^3 = (a + b)(a^2 - ab + b^2)$

Sometimes it is necessary to group three terms with one term.

Example 3

$8c^3 - d^6 = (2c - d^2)(4c^2 + 2cd^2 + d^4)$

Example 4

$27w^9 + 64x^3$

$\quad = (3w^3 + 4x)(9w^6 - 12w^3x + 16x^2)$

Example 5

$4a^2 - 12ab + 9b^2 - c^2$

$= 4a^2 - 12ab + 9b^2 \mid - c^2$ Group 3 by 1.

$= (2a - 3b)^2 - c^2$ Perfect square trinomial.

$= (2a - 3b - c)(2a - 3b + c)$ Difference of squares.

Section 5.8   Solving Equations and Applications by Factoring

Key Concepts

An equation of the form $ax^2 + bx + c = 0$, where $a \neq 0$, is a **quadratic equation**.

    The **zero product rule** states that if $ab = 0$, then $a = 0$ or $b = 0$. The zero product rule can be used to solve a quadratic equation or higher-degree polynomial equation that is factored and equal to zero.

$f(x) = ax^2 + bx + c \, (a \neq 0)$ defines a **quadratic function**. The x-intercepts of a function defined by $y = f(x)$ are determined by finding the real solutions to the equation $f(x) = 0$. The y-intercept of $y = f(x)$ is at $f(0)$.

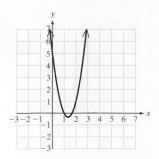

Examples

Example 1

$0 = x(2x - 3)(x + 4)$

$x = 0$ or $2x - 3 = 0$ or $x + 4 = 0$

$\quad\quad\quad\quad\quad\quad x = \dfrac{3}{2}$ or $x = -4$

The solutions are $0, \dfrac{3}{2},$ and -4.

Example 2

Find the x-intercepts.

$f(x) = 3x^2 - 8x + 5$

$0 = 3x^2 - 8x + 5$

$0 = (3x - 5)(x - 1)$

$3x - 5 = 0$ or $x - 1 = 0$

$\quad x = \dfrac{5}{3}$ or $x = 1$

The x-intercepts are $\left(\frac{5}{3}, 0\right)$ and $(1, 0)$.

Find the y-intercept.

$f(x) = 3x^2 - 8x + 5$

$f(0) = 3(0)^2 - 8(0) + 5$

$f(0) = 5$

The y-intercept is $(0, 5)$.

Chapter 5   Review Exercises

Section 5.1

For Exercises 1–8, simplify the expression and write the answer with positive exponents.

1. $(3x)^3(3x)^2$

2. $(-6x^{-4})(3x^{-8})$

3. $\dfrac{24x^5y^3}{-8x^4y}$

4. $\dfrac{-18x^{-2}y^3}{-12x^{-5}y^5}$

5. $(-2a^2b^{-5})^{-3}$

6. $(-4a^{-2}b^3)^{-2}$

7. $\left(\dfrac{-4x^4y^{-2}}{5x^{-1}y^4}\right)^{-4}$

8. $\left(\dfrac{25x^2y^{-3}}{5x^4y^{-2}}\right)^{-5}$

9. Write the numbers in scientific notation.

   a. The population of Asia was 3,686,600,000 in 2000.

   b. A nanometer is 0.000001 of a millimeter.

10. Write the numbers in scientific notation.

   a. A millimeter is 0.001 of a meter.

   b. The population of Asia is predicted to be 5,155,700,000 by 2040.

11. Write the numbers in standard form.

   a. A micrometer is 1×10^{-3} of a millimeter.

   b. A nanometer is 1×10^{-9} of a meter.

12. Write the numbers in standard form.

   a. The total square footage of shopping centers in the United States is approximately 5.23×10^9 ft^2.

   b. The total yearly sales of those shopping centers is $\$1.091 \times 10^{12}$. (*Source:* International Council of Shopping Centers)

For Exercises 13–16, perform the indicated operations. Write the answer in scientific notation.

13. $\dfrac{2,500,000}{0.0004}$

14. $\dfrac{0.0005}{25,000}$

15. $(3.6 \times 10^8)(9 \times 10^{-2})$

16. $(7 \times 10^{-12})(5.2 \times 10^3)$

Section 5.2

For Exercises 17 and 18, identify the polynomial as a monomial, binomial, or trinomial; then give the degree of the polynomial.

17. $6x^4 + 10x - 1$ **18.** 18

19. Given the polynomial function defined by $g(x) = 4x - 7$, find the function values.

   a. $g(0)$ **b.** $g(-4)$ **c.** $g(3)$

20. Given the polynomial function defined by $p(x) = -x^4 - x + 12$, find the function values.

   a. $p(0)$ **b.** $p(1)$ **c.** $p(-2)$

21. The amount, $A(x)$, of bottled water consumed per capita in the United States can be approximated by

$$A(x) = 0.047x^2 + 1.46x + 16.8$$

where x is the number of years since the year 2000 and $A(x)$ is measured in gallons.

   a. Evaluate $A(5)$ and interpret the meaning in the context of the problem.

   b. Interpret the meaning of $A(15)$.

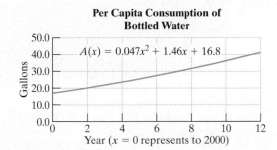

Per Capita Consumption of Bottled Water

For Exercises 22–33, add or subtract the polynomials as indicated.

22. $(x^2 - 2x - 3xy - 7) + (-3x^2 - x + 2xy + 6)$

23. $7xy - 3xz + 5yz$
   $+13xy - 15xz - 8yz$

24. $-4a^3 + 8a^2 - 3a$
   $-(-7a^3 + 3a^2 - 9a)$

25. $(3a^2 - 2a - a^3) - (5a^2 - a^3 - 8a)$

26. $\left(\dfrac{5}{8}x^4 - \dfrac{1}{4}x^2 - \dfrac{1}{2}\right) - \left(-\dfrac{3}{8}x^4 + \dfrac{3}{4}x^2 + \dfrac{1}{2}\right)$

27. $\left(\dfrac{5}{6}x^4 + \dfrac{1}{2}x^2 - \dfrac{1}{3}\right) - \left(-\dfrac{1}{6}x^4 - \dfrac{1}{4}x^2 - \dfrac{1}{3}\right)$

28. $(7x - y) - [-(2x + y) - (-3x - 6y)]$

29. $-(4x - 4y) - [(4x + 2y) - (3x + 7y)]$

30. Add $-4x + 6$ to $-7x - 5$.

31. Add $2x^2 - 4x$ to $2x^2 - 7x$.

32. Subtract $-4x + 6$ from $-7x - 5$.

33. Subtract $2x^2 - 4x$ from $2x^2 - 7x$.

Section 5.3

For Exercises 34–51, multiply the polynomials.

34. $2x(x^2 - 7x - 4)$ **35.** $-3x(6x^2 - 5x + 4)$

36. $(x + 6)(x - 7)$ **37.** $(x - 2)(x - 9)$

38. $\left(\dfrac{1}{2}x + 1\right)\left(\dfrac{1}{2}x - 5\right)$ **39.** $\left(-\dfrac{1}{5} + 2y\right)\left(\dfrac{1}{5} + y\right)$

40. $(3x + 5)(9x^2 - 15x + 25)$

41. $(x - y)(x^2 + xy + y^2)$

42. $(2x - 5)^2$ **43.** $\left(\dfrac{1}{2}x + 4\right)^2$

44. $(3y - 11)(3y + 11)$ **45.** $(6w - 1)(6w + 1)$

46. $\left(\dfrac{2}{3}t + 4\right)\left(\dfrac{2}{3}t - 4\right)$ **47.** $\left(z + \dfrac{1}{4}\right)\left(z - \dfrac{1}{4}\right)$

48. $[(x + 2) - b][(x + 2) + b]$

49. $[c - (w + 3)][c + (w + 3)]$

50. $(2x + 1)^3$ **51.** $(y^2 - 3)^3$

52. A square garden is surrounded by a walkway of uniform width x. If the sides of the garden are given by the expression $2x + 3$, find and simplify a polynomial for the following.

a. The area of the garden.

b. The area of the walkway and garden.

c. The area of the walkway only.

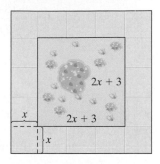

53. The length of a rectangle is 2 ft more than 3 times the width. Let x represent the width of the rectangle.

a. Write a function P that represents the perimeter of the rectangle.

b. Write a function A that represents the area of the rectangle.

54. In parts (a) and (b), one of the statements is true and the other is false. Identify the true statement and explain why the false statement is incorrect.

a. $2x^2 + 5x = 7x^3$ $(2x^2)(5x) = 10x^3$

b. $4x - 7x = -3x$ $4x - 7x = -3$

Section 5.4

For Exercises 55 and 56, divide the polynomials.

55. $(6x^3y + 12x^2y^2 - 9xy^3) \div (3xy)$

56. $(10x^4 + 15x^3 - 20x^2) \div (-5x^2)$

57. a. Divide $(9y^4 + 14y^2 - 8) \div (3y + 2)$.

b. Identify the quotient and the remainder.

c. Explain how you can check your answer.

For Exercises 58–61, divide the polynomials by using long division.

58. $(x^2 + 7x + 10) \div (x + 5)$

59. $(x^2 + 8x - 16) \div (x + 4)$

60. $(2x^5 - 4x^4 + 2x^3 - 4) \div (x^2 - 3x)$

61. $(2x^5 + 3x^3 + x^2 - 4) \div (x^2 + x)$

62. Explain the conditions under which you may use synthetic division.

63. The following table is the result of a synthetic division.

$$\begin{array}{r|rrrrr} 3 & 2 & 5 & -2 & 6 & 1 \\ & & 6 & 33 & 93 & 297 \\ \hline & 2 & 11 & 31 & 99 & \underline{|298} \end{array}$$

Use x as the variable.

  a. Identify the divisor.

  b. Identify the quotient.

  c. Identify the remainder.

For Exercises 64–68, divide the polynomials by using synthetic division.

64. $(t^3 - 3t^2 + 8t - 12) \div (t - 2)$

65. $(x^2 + 7x + 14) \div (x + 5)$

66. $(x^2 + 8x + 20) \div (x + 4)$

67. $(w^3 - 6w^2 + 8) \div (w - 3)$

68. $(p^4 - 16) \div (p - 2)$

Section 5.5

For Exercises 69–72, factor by removing the greatest common factor.

69. $-x^3 - 4x^2 + 11x$

70. $21w^3 - 7w + 14$

71. $5x(x - 7) - 2(x - 7)$

72. $3t(t + 4) + 5(t + 4)$

For Exercises 73–76, factor by grouping (remember to take out the GCF first).

73. $m^3 - 8m^2 + m - 8$

74. $24x^3 - 36x^2 + 72x - 108$

75. $4ax^2 + 2bx^2 - 6ax - 3xb$

76. $y^3 - 6y^2 + y - 6$

Section 5.6

77. What characteristics determine a perfect square trinomial?

For Exercises 78–87, factor the polynomials by using any method.

78. $18x^2 + 27xy + 10y^2$ **79.** $3m^2 + mt - 10t^2$

80. $60a^2 + 65a^3 - 20a^4$ **81.** $2k^2 + 7k^3 + 6k^4$

82. $49x^2 + 36 - 84x$ **83.** $80z + 32 + 50z^2$

84. $(9w + 2)^2 + 4(9w + 2) - 5$

85. $(4x + 3)^2 - 12(4x + 3) + 36$

86. $18a^4 + 39a^2 - 15$ **87.** $3w^4 - 2w^2 - 5$

Section 5.7

For Exercises 88–94, factor the binomials.

88. $25 - y^2$ **89.** $x^3 - \dfrac{1}{27}$

90. $b^2 + 64$

91. $h^3 + 9h$ **92.** $a^3 + 64$

93. $k^4 - 16$ **94.** $9y^3 - 4y$

For Exercises 95–98, factor by grouping and by using the difference of squares.

95. $x^2 - 8xy + 16y^2 - 9$ (*Hint:* Group three terms that make up a perfect square trinomial, then factor as a difference of squares.)

96. $a^2 + 12a + 36 - b^2$

97. $t^2 + 16t + 64 - 25c^2$ **98.** $y^2 - 6y + 9 - 16x^2$

Section 5.8

99. How do you determine if an equation is quadratic?

100. What shape is the graph of a quadratic function?

For Exercises 101–104, label the equation as quadratic or linear.

101. $x^2 + 6x = 7$ **102.** $(x - 3)(x + 4) = 9$

103. $2x - 5 = 3$ **104.** $x + 3 = 5x^2$

105. a. Factor. $5x^2 + 6x - 8$

     b. Solve. $5x^2 + 6x - 8 = 0$

106. a. Factor. $3x^2 - 19x + 28$

     b. Solve. $3x^2 - 19x + 28 = 0$

For Exercises 107–110, use the zero product rule to solve the equations.

107. $x^2 - 2x - 15 = 0$

108. $8x^2 = 59x - 21$

109. $2t(t + 5) + 1 = 3t - 3 - t^2$

110. $3(x - 1)(x + 5)(2x - 9) = 0$

For Exercises 111–114, find the x- and y-intercepts of the function. Then match the function with its graph.

111. $f(x) = -4x^2 + 4$

112. $g(x) = 2x^2 - 2$

113. $h(x) = 5x^3 - 10x^2 - 20x + 40$

114. $k(x) = -\dfrac{1}{8}x^2 + \dfrac{1}{2}$

a.

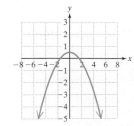

b.

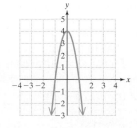

c.

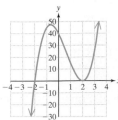

d.

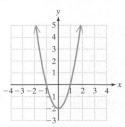

115. A moving van has the capacity to hold 1200 ft³ in volume. If the van is 10 ft high and the length is 1 ft less than twice the width, find the dimensions of the van.

 116. A missile is shot upward from a submarine 1280 ft below sea level. The initial velocity of the missile is 672 ft/sec. A function that approximates the height of the missile (relative to sea level) is given by

$$h(t) = -16t^2 + 672t - 1280$$

where $h(t)$ is the height in feet and t is the time in seconds.

a. Complete the table to determine the height of the missile for the given values of t.

Time t (sec)	Height $h(t)$ (ft)
0	
1	
3	
10	
20	
30	
42	

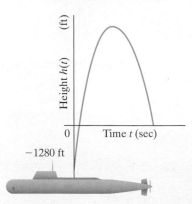

b. Interpret the meaning of a negative value of $h(t)$.

c. Factor the function to find the time required for the missile to emerge from the water and the time required for the missile to reenter the water. (*Hint:* The height of the missile will be zero at sea level.)

Chapter 5 Test

For Exercises 1–4, simplify the expression, and write the answer with positive exponents only.

1. $\dfrac{20a^7}{4a^{-6}}$

2. $\dfrac{x^6 x^3}{x^{-2}}$

3. $\left(\dfrac{-3x^6}{5y^7}\right)^2$

4. $\dfrac{(2^{-1}xy^{-2})^{-3}(x^{-4}y)}{(x^0 y^5)^{-1}}$

5. Multiply. $(8 \times 10^{-6})(7.1 \times 10^5)$

6. Divide. Write the answer in scientific notation.
$9{,}200{,}000 \div 0.004$

7. For the function defined by $F(x) = 5x^3 - 2x^2 + 8$, find the function values $F(-1)$, $F(2)$, and $F(0)$.

8. Perform the indicated operations. Write the answer in descending order.
$$(5x^2 - 7x + 3) - (x^2 + 5x - 25)$$
$$+ (4x^2 + 4x - 20)$$

For Exercises 9–11, multiply the polynomials.

9. $(2a - 5)(a^2 - 4a - 9)$

10. $\left(\dfrac{1}{3}x - \dfrac{3}{2}\right)(6x + 4)$

11. $(5x - 4y^2)(5x + 4y^2)$

12. Explain why $(5x + 7)^2 \neq 25x^2 + 49$.

13. Write and simplify an expression that describes the area of the square.

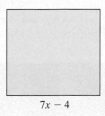

$7x - 4$

14. Divide the polynomials.
$$(2x^3y^4 + 5x^2y^2 - 6xy^3 - xy) \div (2xy)$$

15. Divide the polynomials.
$$(10p^3 + 13p^2 - p + 3) \div (2p + 3)$$

16. Divide the polynomials by using synthetic division. $(y^4 - 2y + 5) \div (y - 2)$

17. Explain the strategy for factoring a polynomial expression.

18. Explain the process to solve a quadratic equation by factoring.

For Exercises 19–32, factor completely.

19. $3a^2 + 27ab + 54b^2$

20. $c^4 - 1$

21. $xy - 7x + 3y - 21$

22. $49 + p^2$

23. $-10u^2 + 30u - 20$

24. $12t^2 - 75$

25. $5y^2 - 50y + 125$

26. $21q^2 + 14q$

27. $2x^3 + x^2 - 8x - 4$

28. $y^3 - 125$

29. $x^2 + 8x + 16 - y^2$

30. $r^6 - 256r^2$

31. $(x^2 + 1)^2 + 3(x^2 + 1) + 2$

32. $12a - 6ac + 2b - bc$

For Exercises 33–38, solve the equation.

33. $(2x - 3)(x + 5) = 0$

34. $w^2 - 7w = 0$

35. $y^2 - 6y = 16$

36. $x(5x + 4) = 1$

37. $4p - 64p^3 = 0$

38. $t^2 + \dfrac{1}{2}t + \dfrac{1}{16} = 0$

39. A child launches a toy rocket from the ground. The height, $h(x)$, of the rocket can be determined by its horizontal distance, x, from the launch pad by

$$h(x) = -\dfrac{x^2}{256} + x$$

where x and h are in feet and $x \geq 0$ and $h \geq 0$.

How many feet from the launch pad will the rocket hit the ground?

40. The population of Japan, $P(t)$ (in millions), can be approximated by:

$$P(t) = -0.01t^2 - 0.062t + 127.7,$$

where $t = 0$ represents the year 2000.

     a. Evaluate $p(4)$ and interpret in the context of the problem.

     b. Approximate the number of people in Japan in the year 2006.

     c. If the trend continues, predict the population of Japan in the year 2015.

For Exercises 41–44, find the x- and y-intercepts of the function. Then match the function with its graph.

41. $f(x) = x^2 - 6x + 8$

42. $k(x) = x^3 + 4x^2 - 9x - 36$

43. $p(x) = -2x^2 - 8x - 6$

44. $q(x) = x^3 - x^2 - 12x$

a.

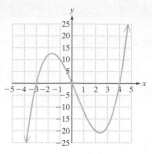

b.

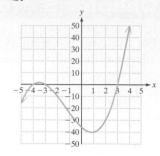

c.

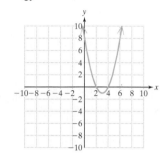

d.

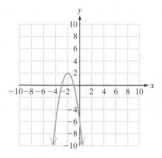

Chapters 1–5   Cumulative Review Exercises

1. Graph the inequality and express the set in interval notation: All real numbers at least 5, but not more than 12

2. Simplify the expression $3x^2 - 5x + 2 - 4(x^2 + 3)$.

3. Graph from memory.

     a. $y = x^2$ **b.** $y = |x|$

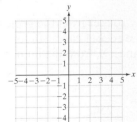

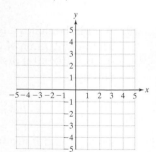

4. Simplify the expression $\left(\dfrac{1}{3}\right)^{-2} - \left(\dfrac{1}{2}\right)^3$.

5. The recent population of Mexico was approximately 9.85×10^7. At the current growth rate of 1.7%, this number is expected to double after 42 yr. Determine the population after 42 yr. Express the answer in scientific notation.

6. In the 2006 Orange Bowl football championship, Penn State scored 3 points more than Florida State in a three-overtime thriller. The total number of points scored was 49. Find the number of points scored by each team.

7. Find the value of each angle in the triangle.

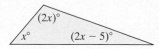

8. Divide $(x^3 + 64) \div (x + 4)$.

9. Determine the slope and y-intercept of the line $4x - 3y = -9$, and graph the line.

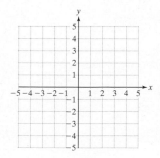

10. Write an equation of the line that is perpendicular to the y-axis and passes through the point $(-2, -5)$.

11. Simplify the expression.

$$\left(\frac{36a^{-2}b^4}{18b^{-6}}\right)^{-3}$$

12. Solve the system.

$$2x - y + 2z = 1$$
$$-3x + 5y - 2z = 11$$
$$x + y - 2z = -1$$

13. Determine whether the relation is a function.

a. $\{(2, 1), (3, 1), (-8, 1), (5, 1)\}$

b.

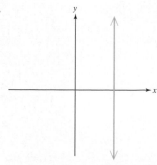

14. A telephone pole is leaning after a storm (see figure). What is the slope of the pole?

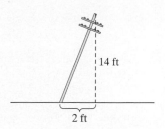

15. Given $P(x) = \frac{1}{6}x^2 + x - 5$, find the function value $P(6)$.

16. Solve. $\dfrac{1}{3}x - \dfrac{1}{6} = \dfrac{1}{2}(x - 3)$

17. Given $3x - 2y = 5$, solve for y.

18. A student scores 76, 85, and 92 on her first three algebra tests.

a. Is it possible for her to score high enough on the fourth test to bring her test average up to 90? Assume that each test is weighted equally and that the maximum score on a test is 100 points.

b. What is the range of values required for the fourth test so that the student's test average will be between 80 and 89, inclusive?

19. How many liters of a 40% acid solution and how many liters of a 15% acid solution must be mixed to obtain 25 L of a 30% acid solution?

20. Multiply the polynomials $(4b - 3)(2b^2 + 1)$.

21. Add the polynomials.

$$(5a^2 + 3a - 1) + (3a^3 - 5a + 6)$$

22. Divide the polynomials $(6w^3 - 5w^2 - 2w) \div (2w^2)$.

For Exercises 23 and 24, solve the equations.

23. $y^2 - 5y = 14$

24. $a^3 + 9a^2 + 20a = 0$

25. Solve the inequality and write the solution set in interval notation. $-5 < -8 + |2x - 5|$

Rational Expressions and Rational Equations

6

Chapter 6

In this chapter we define a rational expression as a ratio of two polynomials. First we focus on simplifying, adding, subtracting, multiplying, and dividing rational expressions. Then the chapter concludes with solving rational equations and showing how they are used in applications.

Review Your Skills

A magic square is shown here. The numbers in each row and each column add to 6. Positive and negative integers from −30 to 30 are used. The hints on the left are important concepts from this chapter and will help you fill in some additional squares.

A is the LCD for $\dfrac{3}{4}, \dfrac{5}{8},$ *and* $-\dfrac{5}{3}$.

B is the numerator of the sum of $\dfrac{3}{4} + \dfrac{5}{8} + \left(-\dfrac{5}{3}\right)$.

C is the numerator of the sum of $\dfrac{3}{x} + \dfrac{4}{5x}$.

D equals $\dfrac{a-5}{5-a}$.

E is the smallest positive number used to clear fractions in the equation $\dfrac{2}{3}y - \dfrac{1}{5} = \dfrac{1}{2}y + \dfrac{3}{10}$.

F is the solution to $\dfrac{2}{3}y - \dfrac{1}{5} = \dfrac{1}{2}y + \dfrac{3}{10}$.

		C	
			−2
	E		B
D		A	
	−24		7
F			
		−30	

419

Section 6.1 | Rational Expressions and Rational Functions

1. Evaluating Rational Expressions

In Chapter 5, we introduced polynomials and polynomial functions. In this chapter, polynomials will be used to define rational expressions and rational functions.

> **Definition of a Rational Expression**
>
> An expression is a **rational expression** if it can be written in the form $\frac{p}{q}$, where p and q are polynomials and $q \neq 0$.

A rational expression is the quotient of two polynomials. The following expressions are examples of rational expressions.

$$\frac{5}{x}, \qquad \frac{2y+1}{y^2-4y+3}, \qquad \frac{c^3+5}{c-4}, \qquad \frac{a}{a+10}$$

Skill Practice

Evaluate the expression for the given values of y.

$$\frac{3y}{y-4}$$

1. $y = -1$ 2. $y = 10$
3. $y = 0$ 4. $y = 4$
5. $y = -2$

Example 1   Evaluating a Rational Expression

Evaluate the expression for the given values of x. $\dfrac{x-2}{x+3}$

a. 1   **b.** 2   **c.** 0   **d.** −1   **e.** −3

Solution:

a. $\dfrac{(1)-2}{(1)+3} = -\dfrac{1}{4}$ **b.** $\dfrac{(2)-2}{(2)+3} = \dfrac{0}{5} = 0$ **c.** $\dfrac{(0)-2}{(0)+3} = -\dfrac{2}{3}$

d. $\dfrac{(-1)-2}{(-1)+3} = -\dfrac{3}{2}$ **e.** $\dfrac{(-3)-2}{(-3)+3} = \dfrac{-5}{0}$ (undefined)

2. Restricted Values of a Rational Expression

From Example 1 we see that not all values of x can be substituted into a rational expression. The values that make the denominator zero must be restricted. The expression $\dfrac{x-2}{x+3}$ is undefined for $x = -3$, so we call $x = -3$ a restricted value. **Restricted values of a rational expression** are all values that make the expression undefined, that is, make the denominator equal to zero.

> **Finding the Restricted Values of a Rational Expression**
> - Set the denominator equal to zero and solve the resulting equation.
> - The restricted values are the solutions to the equation.

Answers

1. $\dfrac{3}{5}$ **2.** 5   **3.** 0
4. Undefined   **5.** 1

Example 2 **Finding the Restricted Values of a Rational Expression**

Identify the restricted values.

a. $\dfrac{5}{4t - 1}$ **b.** $\dfrac{x + 4}{2x^2 - 11x + 5}$ **c.** $\dfrac{x}{x^2 + 4}$

Solution:

a. $\dfrac{5}{4t - 1}$ The expression is undefined when the denominator is zero.

$4t - 1 = 0$ Set the denominator equal to zero and solve for t.

$4t = 1$

$t = \dfrac{1}{4}$ The restricted value is $t = \dfrac{1}{4}$.

b. $\dfrac{x + 4}{2x^2 - 11x + 5}$

$2x^2 - 11x + 5 = 0$ Set the denominator equal to zero and solve for x.

$(2x - 1)(x - 5) = 0$ This is a factorable quadratic equation.

$2x - 1 = 0$ or $x - 5 = 0$

$x = \dfrac{1}{2}$ or $x = 5$

The restricted values are $x = \frac{1}{2}$ and $x = 5$.

c. $\dfrac{x}{x^2 + 4}$

Because the quantity x^2 is nonnegative for any real number x, the denominator $x^2 + 4$ cannot equal zero; therefore, there are no restricted values.

3. Simplifying Rational Expressions to Lowest Terms

A rational expression is an expression in the form $\frac{p}{q}$, where p and q are polynomials and $q \neq 0$. As with fractions, it is often advantageous to simplify rational expressions to lowest terms.

The method for simplifying rational expressions to lowest terms mirrors the process to simplify fractions. In each case, factor the numerator and denominator. Common factors in the numerator and denominator form a ratio of 1 and can be simplified.

Simplifying a fraction: $\dfrac{15}{35} \xrightarrow{\text{factor}} \dfrac{3 \cdot \overset{1}{\cancel{5}}}{7 \cdot \cancel{5}} = \dfrac{3}{7}(1) = \dfrac{3}{7}$

Simplifying a rational expression: $\dfrac{x^2 - x - 12}{x^2 - 16} \xrightarrow{\text{factor}} \dfrac{(x + 3)\overset{1}{\cancel{(x - 4)}}}{(x + 4)\cancel{(x - 4)}} = \dfrac{(x + 3)}{(x + 4)}(1) = \dfrac{x + 3}{x + 4}$

This process is stated formally as the fundamental principle of rational expressions.

Answers

6. $v = 9$
7. $x = -\frac{1}{2}, x = 1$
8. There are no restricted values.

> ### Fundamental Principle of Rational Expressions
> Let p, q, and r represent polynomials. Then
> $$\frac{pr}{qr} = \frac{p}{q} \cdot \frac{r}{r} = \frac{p}{q} \cdot 1 = \frac{p}{q} \qquad \text{for } q \neq 0 \text{ and } r \neq 0$$

Example 3 Simplifying Rational Expressions to Lowest Terms

$$\frac{2x^3 + 12x^2 + 16x}{6x + 24}$$

a. Factor the numerator and denominator.

b. Identify the restricted values.

c. Simplify the expression.

Solution:

a. $\dfrac{2x^3 + 12x^2 + 16x}{6x + 24}$

$= \dfrac{2x(x^2 + 6x + 8)}{6(x + 4)}$ Factor the numerator and denominator.

$= \dfrac{2x(x + 4)(x + 2)}{6(x + 4)}$

Avoiding Mistakes

The restricted values of a rational expression are always determined *before* simplifying the expression.

b. Set the denominator equal to zero and solve for x.

$6(x + 4) = 0$

$x = -4$

The restricted value is $x = -4$.

c. $\dfrac{\overset{1}{2}}{\underset{1}{2}} \cdot \dfrac{\overset{1}{(x + 4)}}{\underset{1}{(x + 4)}} \cdot \dfrac{x(x + 2)}{3}$ Simplify the ratio of common factors to 1.

$= \dfrac{x(x + 2)}{3}$ provided $x \neq -4$

It is important to note that the expressions

$$\frac{2x^3 + 12x^2 + 16x}{6x + 24} \qquad \text{and} \qquad \frac{x(x + 2)}{3}$$

are equal for all values of x that make each expression a real number. Therefore,

$$\frac{2x^3 + 12x^2 + 16x}{6x + 24} = \frac{x(x + 2)}{3}$$

for all values of x *except* $x = -4$. (At $x = -4$, the original expression is undefined.) This is why the *restricted values of a rational expression are always determined before the expression is simplified.*

Answers

9. a. $x = -7$

   b. $\dfrac{x - 4}{2}$

The objective to simplifying a rational expression is to create an equivalent expression that is simpler to evaluate. Consider the expression from Example 3 in its original form and in its simplified form. If we substitute an arbitrary value of x into the expression (such as $x = 3$), we see that the simplified form is easier to evaluate.

Original Expression	**Simplified Form**
$\dfrac{2x^3 + 12x^2 + 16x}{6x + 24}$	$\dfrac{x(x + 2)}{3}$

Substitute $x = 3$ ⟶

$$\frac{2(3)^3 + 12(3)^2 + 16(3)}{6(3) + 24} \qquad \frac{(3)[(3) + 2]}{3}$$

$$= \frac{2(27) + 12(9) + 48}{18 + 24} \qquad = \frac{3(5)}{3}$$

$$= \frac{54 + 108 + 48}{42} \qquad = 5$$

$$= \frac{210}{42}$$

$$= 5$$

Example 4 involves reducing a quotient of two monomials. This was also covered in Section 5.1.

Example 4 **Simplifying a Rational Expression to Lowest Terms**

Simplify the rational expression. $\dfrac{2x^2y^5}{8x^4y^3}$

Solution:

$$\frac{2x^2y^5}{8x^4y^3}$$

This expression has the restriction that $x \neq 0$ and $y \neq 0$.

$$= \frac{2x^2y^5}{2^3x^4y^3}$$

Factor the denominator.

$$= \frac{2x^{\overset{1}{2}}y^3}{2x^2y^3} \cdot \frac{y^2}{2^2x^2}$$

Reduce the common factors whose ratio is 1.

$$= \frac{y^2}{4x^2}$$

Answer

10. $\dfrac{b^2}{2a^3}$

Skill Practice

Simplify to lowest terms.

11. $\dfrac{8y^2 - 14y + 3}{2y^2 - y - 3}$

Example 5 **Simplifying a Rational Expression to Lowest Terms**

Simplify to lowest terms. $\dfrac{t^3 + 8}{t^2 + 6t + 8}$

Solution:

$\dfrac{t^3 + 8}{t^2 + 6t + 8}$

> **TIP:** $t^3 + 8$ is a sum of cubes.
>
> Recall: $a^3 + b^3 = (a + b)(a^2 - ab + b^2)$
>
> $t^3 + 8 = (t + 2)(t^2 - 2t + 4)$

$= \dfrac{(t + 2)(t^2 - 2t + 4)}{(t + 2)(t + 4)}$ Factor the numerator and denominator.

$= \dfrac{\cancel{(t + 2)}}{\cancel{(t + 2)}} \cdot \dfrac{(t^2 - 2t + 4)}{(t + 4)}$ Reduce common factors whose ratio is 1.

$= \dfrac{t^2 - 2t + 4}{t + 4}$

Avoiding Mistakes

Because the fundamental property of rational expressions is based on the identity property of multiplication, reducing applies only to factors (remember that factors are multiplied). Therefore, terms that are added or subtracted cannot be reduced. For example:

$$\dfrac{3x}{3y} = \dfrac{\cancel{3}}{\cancel{3}} \cdot \dfrac{x}{y} = (1) \cdot \dfrac{x}{y} = \dfrac{x}{y} \qquad \text{However, } \dfrac{x + 3}{y + 3} \text{ cannot be simplified.}$$

     ↑   ↑
   Reduce common   Cannot reduce
    factor.   common terms.

---

4. Simplifying Ratios of −1

When two factors are identical in the numerator and denominator, they form a ratio of 1 and can be simplified. Sometimes we encounter two factors that are *opposites* and form a ratio of −1. For example:

Concept Connections

12. Which of the following expressions simplifies to −1?

a. $\dfrac{x + y}{y + x}$ **b.** $\dfrac{x - y}{y - x}$

c. $\dfrac{x - y}{x + y}$

Simplified Form	Details/Notes
$\dfrac{-5}{5} = -1$	The ratio of a number and its opposite is −1.
$\dfrac{100}{-100} = -1$	The ratio of a number and its opposite is −1.
$\dfrac{x + 7}{-x - 7} = -1$	$\dfrac{x + 7}{-x - 7} = \dfrac{x + 7}{-1(x + 7)} = \dfrac{\cancel{x + 7}}{-1\cancel{(x + 7)}} = \dfrac{1}{-1} = -1$
	Factor out −1.
$\dfrac{2 - x}{x - 2} = -1$	$\dfrac{2 - x}{x - 2} = \dfrac{-1(-2 + x)}{x - 2} = \dfrac{-1\cancel{(x - 2)}}{\cancel{x - 2}} = \dfrac{-1}{1} = -1$

Answers

11. $\dfrac{4y - 1}{y + 1}$

12. b

Recognizing factors that are opposites is useful when simplifying rational expressions. For example, $a - b$ and $b - a$ are opposites because the opposite of $a - b$ can be written $-(a - b) = -a + b = b - a$. Therefore, in general, $\frac{a-b}{b-a} = -1$.

Example 6 **Simplifying a Rational Expression**

Simplify the rational expression to lowest terms. $\dfrac{x-5}{25-x^2}$

Solution:

$\dfrac{x-5}{25-x^2}$

$= \dfrac{x-5}{(5+x)(5-x)}$ Factor.

               Notice that $x - 5$ and $5 - x$ are opposites and form a ratio of -1.

$= \dfrac{1}{(5+x)} \cdot \dfrac{\overset{-1}{(x-5)}}{(5-x)}$ In general, $\dfrac{a-b}{b-a} = -1$.

$= \dfrac{1}{5+x}(-1)$

$= -\dfrac{1}{5+x}$ or $-\dfrac{1}{x+5}$

> **TIP:** The factor of -1 may be applied in front of the rational expression, or it may be applied to the numerator or to the denominator. Therefore, the final answer may be written in different forms.
>
> $$-\dfrac{1}{x+5} \quad \text{or} \quad \dfrac{-1}{x+5} \quad \text{or} \quad \dfrac{1}{-(x+5)}$$

Skill Practice

Simplify the expression.

13. $\dfrac{20-5x}{x^2-x-12}$

5. Rational Functions

Thus far in the text, we have introduced several types of functions including constant, linear, and quadratic functions. Now we will introduce another category of functions called rational functions.

> ### Definition of a Rational Function
>
> A function f is a **rational function** if it can be written in the form $f(x) = \dfrac{p(x)}{q(x)}$, where $p(x)$ and $q(x)$ are polynomial functions and $q(x) \neq 0$.

For example, the functions f, g, h, and k are rational functions.

$$f(x) = \frac{1}{x}, \qquad g(x) = \frac{2}{x-3}, \qquad h(a) = \frac{a+6}{a^2-5}, \qquad k(x) = \frac{x+4}{2x^2-11x+5}$$

Answer

13. $-\dfrac{5}{x+3}$

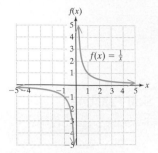

Figure 6-1

In Section 3.3, we introduced the rational function defined by $f(x) = \frac{1}{x}$. Recall that $f(x) = \frac{1}{x}$ has a restriction on its domain that $x \neq 0$ and the graph of $f(x) = \frac{1}{x}$ has a vertical asymptote at $x = 0$ (Figure 6-1).

In this course, we restrict our discussion of functions defined by $y = f(x)$ to those whose domain and range are subsets of the set of real numbers. Therefore, the domain of a function is the set of real numbers that when substituted into the function produce a real number. For a rational function, we must exclude values that make the denominator zero. To find the domain of a rational function, we offer these guidelines.

Finding the Domain of a Rational Function

Step 1 Set the denominator equal to zero and solve the resulting equation, that is, identify the restricted values.

Step 2 The domain is the set of all real numbers *excluding* the values found in step 1.

Example 7   **Evaluating a Rational Function**

Given $g(x) = \dfrac{2}{x - 3}$

a. Find the function values (if they exist).

$g(0), g(1), g(2), g(2.5), g(2.9), g(3), g(3.1), g(3.5), g(4)$, and $g(5)$.

b. Write the domain of the function in interval notation.

Solution:

a. $g(0) = \dfrac{2}{(0) - 3} = -\dfrac{2}{3}$ $g(3) = \dfrac{2}{(3) - 3} = \dfrac{2}{0}$ (undefined)

$g(1) = \dfrac{2}{(1) - 3} = -1$ $g(3.1) = \dfrac{2}{(3.1) - 3} = \dfrac{2}{0.1} = 20$

$g(2) = \dfrac{2}{(2) - 3} = -2$ $g(3.5) = \dfrac{2}{(3.5) - 3} = \dfrac{2}{0.5} = 4$

$g(2.5) = \dfrac{2}{(2.5) - 3} = \dfrac{2}{-0.5} = -4$ $g(4) = \dfrac{2}{(4) - 3} = 2$

$g(2.9) = \dfrac{2}{(2.9) - 3} = \dfrac{2}{-0.1} = -20$ $g(5) = \dfrac{2}{(5) - 3} = 1$

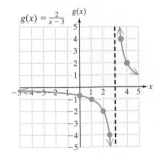

$g(x) = \frac{2}{x-3}$

Figure 6-2

b. $g(x) = \dfrac{2}{x - 3}$ The value of the function is undefined when the denominator equals zero.

$x - 3 = 0$ Set the denominator equal to zero and solve for x to find the restricted values.

$x = 3$ The value $x = 3$ must be excluded from the domain.

Domain: $(-\infty, 3) \cup (3, \infty)$

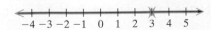

The graph of $g(x) = \frac{2}{x-3}$ is shown in Figure 6-2. Notice that the function has a vertical asymptote at $x = 3$ where the function is undefined (Figure 6-2).

Calculator Connections

Topic: Using *Dot* Mode on a Graphing Calculator

A *Table* feature can be used to check the function values in Example 7, $g(x) = \frac{2}{x-3}$

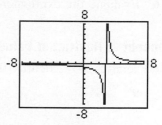

The graph of a rational function may be misleading on a graphing calculator. For example, $g(x) = 2/(x-3)$ has a vertical asymptote at $x = 3$, but the vertical asymptote is *not* part of the graph (Figure 6-3). A graphing calculator may try to "connect the dots" between two consecutive points, one to the left of $x = 3$ and one to the right of $x = 3$. This creates a line that is nearly vertical and appears to be part of the graph.

To show that this line is not part of the graph, we can graph the function in a *Dot Mode* (see the owner's manual for your calculator). The graph of $g(x) = 2/(x-3)$ in dot mode indicates that the line $x = 3$ is not part of the function (Figure 6-4).

Figure 6-3

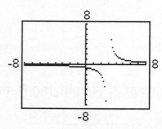

Figure 6-4

Section 6.1 Practice Exercises

Vocabulary and Key Concepts

1. **a.** A _____ expression is the ratio of two polynomials, $\frac{p}{q}$, where $q \neq 0$.

   b. Restricted values of a rational expression are all values that make the _____ equal to _____.

   c. For polynomials p, q, and r, where ($q \neq 0$ and $r \neq 0$), $\frac{pr}{qr} = $ _____.

   d. The ratio $\frac{a-b}{a-b} = $ _____, whereas the ratio $\frac{a-b}{b-a} = $ _____ provided that $a \neq b$.

   e. A _____ function is a function that can be written as $f(x) = \frac{p(x)}{q(x)}$, where $p(x)$ and $q(x)$ are polynomial functions and $q(x) \neq 0$.

   f. The domain of a rational function is the set of all real numbers excluding the values of the variable that make the _____ equal to zero.

Concept 1: Evaluating Rational Expressions

2. Evaluate the expression $\frac{5}{x+1}$ for $x = 0$, $x = 2$, $x = -1$, $x = -6$.

3. Evaluate the expression $\frac{x-4}{x+6}$ for $x = -6$, $x = -4$, $x = 0$, and $x = 4$. **(See Example 1.)**

4. Evaluate the expression $\frac{y-3}{y-2}$ for $y = 0$, $y = 2$, $y = 3$, $y = -4$.

5. Evaluate the expression $\dfrac{3a + 1}{a^2 + 1}$ for $a = 1$, $a = 0$, $a = -\dfrac{1}{3}$, $a = -1$.

6. Evaluate the expression $\dfrac{2z - 8}{z^2 + 9}$ for $z = 4$, $z = -4$, $z = 3$, $z = -3$.

Concept 2: Restricted Values of a Rational Expression

For Exercises 7–18, identify the restricted values of the rational expression. **(See Example 2.)**

7. $\dfrac{9}{y}$

8. $\dfrac{-10}{a}$

9. $\dfrac{v + 1}{v - 8}$

10. $\dfrac{t + 9}{t + 3}$

11. $\dfrac{3x - 1}{2x - 5}$

12. $\dfrac{6t + 5}{3t + 8}$

13. $\dfrac{q + 1}{q^2 + 6q - 27}$

14. $\dfrac{b^2}{2b^2 + 3b - 5}$

15. $\dfrac{c}{c^2 + 25}$

16. $\dfrac{d}{d^2 + 16}$

17. $\dfrac{x + 5}{x^2 - 25}$

18. $\dfrac{t + 4}{t^2 - 16}$

Concept 3: Simplifying Rational Expressions to Lowest Terms

19. Given: $\dfrac{x^2 + 6x + 8}{x^2 + 3x - 4}$,

    a. Factor the numerator and denominator.

    b. Identify the restricted values.

    c. Simplify the expression. **(See Example 3.)**

20. Given: $\dfrac{x^2 - 6x}{2x^2 - 11x - 6}$

    a. Factor the numerator and denominator.

    b. Identify the restricted values.

    c. Simplify the expression.

21. Given: $\dfrac{x^2 - 18x + 81}{x^2 - 81}$

    a. Factor the numerator and denominator.

    b. Identify the restricted values.

    c. Simplify the expression.

22. Given: $\dfrac{x^2 + 14x + 49}{x^2 - 49}$

    a. Factor the numerator and denominator.

    b. Identify the restricted values.

    c. Simplify the expression.

For Exercises 23–44, simplify the rational expressions. **(See Examples 4 and 5.)**

23. $\dfrac{100x^3y^5}{36xy^8}$

24. $\dfrac{48ab^3c^2}{6a^7bc^0}$

25. $\dfrac{7w^{11}z^6}{14w^3z^3}$

26. $\dfrac{12r^9s^3}{24r^8s^4}$

27. $\dfrac{-3m^4n}{12m^6n^4}$

28. $\dfrac{-5x^3y^2}{20x^4y^2}$

29. $\dfrac{6a + 18}{9a + 27}$

30. $\dfrac{5y - 15}{3y - 9}$

31. $\dfrac{x - 5}{x^2 - 25}$

32. $\dfrac{3z - 6}{3z^2 - 12}$

33. $\dfrac{-7c}{21c^2 - 35c}$

34. $\dfrac{2p + 3}{2p^2 + 7p + 6}$

35. $\dfrac{2t^2 + 7t - 4}{-2t^2 - 5t + 3}$

36. $\dfrac{y^2 + 8y - 9}{y^2 - 5y + 4}$

37. $\dfrac{(p + 1)(2p - 1)^4}{(p + 1)^2(2p - 1)^2}$

38. $\dfrac{r(r - 3)^5}{r^3(r - 3)^2}$

39. $\dfrac{9 - z^2}{2z^2 + z - 15}$

40. $\dfrac{2c^2 + 2c - 12}{-8 + 2c + c^2}$

41. $\dfrac{2z^3 + 16}{-10 - 3z + z^2}$

42. $\dfrac{5p^2 - p - 4}{p^3 - 1}$

43. $\dfrac{10x^3 - 25x^2 + 4x - 10}{-4 - 10x^2}$

44. $\dfrac{8x^3 - 12x^2 + 6x - 9}{16x^4 - 9}$

Concept 4: Simplifying Ratios of −1

For Exercises 45–56, simplify the rational expressions. **(See Example 6.)**

45. $\dfrac{r + 6}{6 + r}$

46. $\dfrac{a + 2}{2 + a}$

47. $\dfrac{b + 8}{-b - 8}$

48. $\dfrac{7 + w}{-7 - w}$

49. $\dfrac{10 - x}{x - 10}$

50. $\dfrac{y - 14}{14 - y}$

51. $\dfrac{2t - 2}{1 - t}$

52. $\dfrac{5p - 10}{2 - p}$

53. $\dfrac{c + 4}{c - 4}$

54. $\dfrac{b + 2}{b - 2}$

55. $\dfrac{y - x}{12x^2 - 12y^2}$

56. $\dfrac{4w^2 - 49z^2}{14z - 4w}$

Mixed Exercises

For Exercises 57–76, simplify the rational expression.

57. $\dfrac{t^2 - 1}{t^2 + 7t + 6}$

58. $\dfrac{x^2 + 4x + 4}{x^2 - 4}$

59. $\dfrac{8p + 8}{2p^2 - 4p - 6}$

60. $\dfrac{15y - 15}{3y^2 + 9y - 12}$

61. $\dfrac{-16a^2bc^4}{8ab^2c^4}$

62. $\dfrac{-9x^3yz^2}{27x^4yz}$

63. $\dfrac{x^2 - y^2}{8y - 8x}$

64. $\dfrac{p^2 - 49}{14 - 2p}$

65. $\dfrac{b + 4}{2b^2 + 5b - 12}$

66. $\dfrac{c - 6}{3c^2 - 17c - 6}$

67. $\dfrac{-2x + 34}{-4x + 6}$

68. $\dfrac{-9w - 3}{3w + 12}$

69. $\dfrac{(a - 2)^2(a - 5)^3}{(a - 2)^3(a - 5)}$

70. $\dfrac{t^2(t - 11)^4}{t^5(t - 11)^2}$

71. $\dfrac{4x - 2x^2}{5x - 10}$

72. $\dfrac{2y - 6}{3y^2 - y^3}$

73. $\dfrac{x^3 - 2x^2 - 25x + 50}{x^3 + 5x^2 - 4x - 20}$

74. $\dfrac{4y^3 + 12y^2 - y - 3}{2y^3 + y^2 - 18y - 9}$

75. $\dfrac{t^3 + 8}{3t^2 + t - 10}$

76. $\dfrac{w^3 - 27}{4w^2 - 5w - 21}$

Concept 5: Rational Functions

77. Let $h(x) = \dfrac{-3}{x - 1}$. Find the function values $h(0), h(1), h(-3), h(-1), h(\frac{1}{2})$. **(See Example 7.)**

78. Let $k(x) = \dfrac{2}{x + 1}$. Find the function values $k(0), k(1), k(-3), k(-1), k(\frac{1}{2})$.

For Exercises 79–82, write the domain of each function in interval notation and use that information to match the function with its graph.

79. $m(x) = \dfrac{1}{x + 4}$

80. $n(x) = \dfrac{1}{x + 1}$

81. $q(x) = \dfrac{1}{x - 4}$

82. $p(x) = \dfrac{1}{x - 1}$

a.
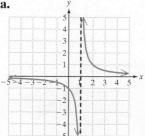

b.
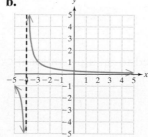

c.
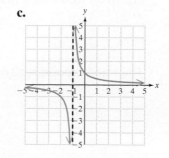

d.
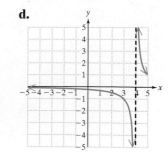

For Exercises 83–91, write the domain of the function in interval notation.

83. $f(x) = \dfrac{1}{3x - 1}$

84. $r(t) = \dfrac{1}{2t - 1}$

85. $q(t) = \dfrac{t + 2}{8}$

86. $p(x) = \dfrac{x - 5}{3}$

87. $w(x) = \dfrac{x - 2}{3x - 6}$

88. $s(t) = \dfrac{-2}{4t + 6}$

89. $m(x) = \dfrac{3}{6x^2 - 7x - 10}$

90. $n(x) = \dfrac{1}{2x^2 + 11x + 12}$

91. $r(x) = \dfrac{x + 1}{6x^3 - x^2 - 15x}$

Expanding Your Skills

92. Write a rational expression whose domain is $(-\infty, 2) \cup (2, \infty)$. (Answers may vary.)

93. Write a rational expression whose domain is $(-\infty, 3) \cup (3, \infty)$. (Answers may vary.)

94. Write a rational function whose domain is $(-\infty, -5) \cup (-5, \infty)$. (Answers may vary.)

95. Write a rational function whose domain is $(-\infty, -6) \cup (-6, \infty)$. (Answers may vary.)

Section 6.2 Multiplication and Division of Rational Expressions

Concepts

1. Multiplication of Rational Expressions
2. Division of Rational Expressions

1. Multiplication of Rational Expressions

Recall that to multiply fractions, we multiply the numerators and multiply the denominators. The same is true for multiplying rational expressions.

> **Multiplication Property of Rational Expressions**
> Let p, q, r, and s represent polynomials, such that $q \neq 0$ and $s \neq 0$. Then
>
> $$\frac{p}{q} \cdot \frac{r}{s} = \frac{pr}{qs}$$

For example:

Multiply the Fractions	Multiply the Rational Expressions
$\dfrac{2}{3} \cdot \dfrac{5}{7} = \dfrac{10}{21}$	$\dfrac{2x}{3y} \cdot \dfrac{5z}{7} = \dfrac{10xz}{21y}$

Sometimes it is possible to simplify a ratio of common factors to 1 *before* multiplying. To do so, we must first factor the numerators and denominators of each fraction.

$$\frac{7}{10} \cdot \frac{15}{21} \xrightarrow{\text{Factor.}} \frac{\overset{1}{\cancel{7}}}{2 \cdot \cancel{5}} \cdot \frac{\overset{1}{\cancel{5}} \cdot \overset{1}{\cancel{3}}}{3 \cdot \cancel{7}} = \frac{1}{2}$$

The same process is also used to multiply rational expressions.

> **Multiplying Rational Expressions**
>
> **Step 1** Factor the numerators and denominators of all rational expressions.
>
> **Step 2** Simplify the ratios of common factors to 1.
>
> **Step 3** Multiply the remaining factors in the numerator, and multiply the remaining factors in the denominator.

Example 1 Multiplying Rational Expressions

Multiply. $\dfrac{5a - 5b}{10} \cdot \dfrac{2}{a^2 - b^2}$

Solution:

$\dfrac{5a - 5b}{10} \cdot \dfrac{2}{a^2 - b^2}$

$= \dfrac{5(a - b)}{5 \cdot 2} \cdot \dfrac{2}{(a - b)(a + b)}$ Factor numerator and denominator.

$= \dfrac{\overset{1}{\cancel{5}}(\overset{1}{\cancel{a - b}})}{\cancel{5} \cdot 2} \cdot \dfrac{2}{(\cancel{a - b})(a + b)}$ Simplify.

$= \dfrac{1}{a + b}$

> **Avoiding Mistakes**
>
> If all factors in the numerator simplify to 1, do not forget to write the factor of 1 in the numerator.

Example 2 Multiplying Rational Expressions

Multiply. $\dfrac{4w - 20p}{2w^2 - 50p^2} \cdot \dfrac{2w^2 + 7wp - 15p^2}{3w + 9p}$

Solution:

$\dfrac{4w - 20p}{2w^2 - 50p^2} \cdot \dfrac{2w^2 + 7wp - 15p^2}{3w + 9p}$

$= \dfrac{4(w - 5p)}{2(w^2 - 25p^2)} \cdot \dfrac{(2w - 3p)(w + 5p)}{3(w + 3p)}$ Factor numerator and denominator.

$= \dfrac{2 \cdot 2(w - 5p)}{2(w - 5p)(w + 5p)} \cdot \dfrac{(2w - 3p)(w + 5p)}{3(w + 3p)}$ Factor further.

$= \dfrac{2 \cdot 2(\overset{1}{\cancel{w - 5p}})}{2(\cancel{w - 5p})(w + 5p)} \cdot \dfrac{(2w - 3p)(\overset{1}{\cancel{w + 5p}})}{3(w + 3p)}$ Simplify common factors.

$= \dfrac{2(2w - 3p)}{3(w + 3p)}$

Notice that the expression is left in factored form to show that it has been simplified to lowest terms.

2. Division of Rational Expressions

Recall that to divide fractions, multiply the first fraction by the reciprocal of the second fraction.

Divide: $\dfrac{15}{14} \div \dfrac{10}{49}$ $\xrightarrow[\text{of the second fraction.}]{\text{Multiply by the reciprocal}}$ $\dfrac{15}{14} \cdot \dfrac{49}{10} = \dfrac{3 \cdot \overset{1}{\cancel{5}}}{2 \cdot \cancel{7}} \cdot \dfrac{\overset{1}{\cancel{7}} \cdot 7}{2 \cdot \cancel{5}} = \dfrac{21}{4}$

The same process is used for dividing rational expressions.

> ### Division Property of Rational Expressions
> Let p, q, r, and s represent polynomials, such that $q \ne 0, r \ne 0, s \ne 0$. Then
> $$\frac{p}{q} \div \frac{r}{s} = \frac{p}{q} \cdot \frac{s}{r} = \frac{ps}{qr}$$

Skill Practice

Divide.

3. $\dfrac{x^2 + x}{5x^3 - x^2} \div \dfrac{10x^2 + 12x + 2}{25x^2 - 1}$

Example 3 Dividing Rational Expressions

Divide. $\dfrac{8t^3 + 27}{9 - 4t^2} \div \dfrac{4t^2 - 6t + 9}{2t^2 - t - 3}$

Solution:

$\dfrac{8t^3 + 27}{9 - 4t^2} \div \dfrac{4t^2 - 6t + 9}{2t^2 - t - 3}$

$= \dfrac{8t^3 + 27}{9 - 4t^2} \cdot \dfrac{2t^2 - t - 3}{4t^2 - 6t + 9}$

Multiply the first fraction by the reciprocal of the second.

$= \dfrac{(2t + 3)(4t^2 - 6t + 9)}{(3 - 2t)(3 + 2t)} \cdot \dfrac{(2t - 3)(t + 1)}{4t^2 - 6t + 9}$

Factor numerator and denominator. Notice $8t^3 + 27$ is a sum of cubes. Furthermore, $4t^2 - 6t + 9$ does not factor over the real numbers.

$= \dfrac{(\cancel{2t + 3})(\overset{1}{\cancel{4t^2 - 6t + 9}})}{(\cancel{3 - 2t})(\cancel{3 + 2t})} \cdot \dfrac{(\overset{-1}{\cancel{2t - 3}})(t + 1)}{\cancel{4t^2 - 6t + 9}}$

Simplify to lowest terms.

$= (-1)\dfrac{(t + 1)}{1}$

$= -(t + 1) \qquad \text{or} \qquad -t - 1$

> **TIP:** In Example 3, the factors $(2t - 3)$ and $(3 - 2t)$ are opposites and form a ratio of -1. The factors $(2t + 3)$ and $(3 + 2t)$ are equal and form a ratio of 1.
>
> $$\frac{2t - 3}{3 - 2t} = -1 \quad \text{whereas} \quad \frac{2t + 3}{3 + 2t} = 1$$

Answer

3. $\dfrac{1}{2x}$

Section 6.2   Practice Exercises

Study Skills Exercise

Write an example of how to multiply two fractions, divide two fractions, add two fractions, and subtract two fractions. For example: $\dfrac{5}{12} \cdot \dfrac{9}{10}$, $\dfrac{3}{4} \div \dfrac{15}{8}$, $\dfrac{5}{6} + \dfrac{7}{9}$, and $\dfrac{2}{7} - \dfrac{2}{7}$. Then as you learn about rational expressions, compare the operations on rational expressions to those on fractions. This is a great place to use 3×5 cards again. Write an example of an operation with fractions on one side and the same operation with rational expressions on the other side.

Vocabulary and Key Concepts

1. a. Given polynomials p, q, r, and s such that $q \neq 0$ and $s \neq 0$, $\dfrac{p}{q} \cdot \dfrac{r}{s} = \dfrac{\square}{\square}$.

   b. Given polynomials p, q, r, and s such that $q \neq 0$, $r \neq 0$, and $s \neq 0$, $\dfrac{p}{q} \div \dfrac{r}{s} = \dfrac{\square}{\square}$.

Review Exercises

For Exercises 2–9, simplify the rational expression to lowest terms.

2. $\dfrac{3w^2 - 12}{w^2 + 5w + 6}$

3. $\dfrac{t^2 - 5t - 6}{t^2 - 7t + 6}$

4. $\dfrac{5y + 15}{5y^2 + 16y + 3}$

5. $\dfrac{2 - p}{p^2 - p - 2}$

6. $\dfrac{5x^2yz^3}{20xyz}$

7. $\dfrac{7x + 14}{7x^2 - 7x - 42}$

8. $\dfrac{25 - x^2}{x^2 - 10x + 25}$

9. $\dfrac{a^3b^2c^5}{2a^3bc^2}$

Concept 1: Multiplication of Rational Expressions

For Exercises 10–21, multiply the rational expressions. **(See Examples 1 and 2.)**

10. $\dfrac{8w^2}{9} \cdot \dfrac{3}{2w^4}$

11. $\dfrac{16}{z^7} \cdot \dfrac{z^4}{8}$

12. $\dfrac{5p^2q^4}{12pq^3} \cdot \dfrac{6p^2}{20q^2}$

13. $\dfrac{27r^5}{7s} \cdot \dfrac{28rs^3}{9r^3s^2}$

14. $\dfrac{3z + 12}{8z^3} \cdot \dfrac{16z^3}{9z + 36}$

15. $\dfrac{x^2y}{x^2 - 4x - 5} \cdot \dfrac{2x^2 - 13x + 15}{xy^3}$

16. $\dfrac{3y^2 + 18y + 15}{6y + 6} \cdot \dfrac{y - 5}{y^2 - 25}$

17. $\dfrac{10w - 8}{w + 2} \cdot \dfrac{3w^2 - w - 14}{25w^2 - 16}$

18. $\dfrac{x - 5y}{x^2 + xy} \cdot \dfrac{y^2 - x^2}{10y - 2x}$

19. $\dfrac{3x - 15}{4x^2 - 2x} \cdot \dfrac{10x - 20x^2}{5 - x}$

20. $x(x + 5)^2 \cdot \dfrac{2}{x^2 - 25}$

21. $y(y^2 - 4) \cdot \dfrac{y}{y + 2}$

Concept 2: Division of Rational Expressions

For Exercises 22–33, divide the rational expressions. **(See Example 3.)**

22. $\dfrac{6x^2y^2}{(x - 2)} \div \dfrac{3xy^2}{(x - 2)^2}$

23. $\dfrac{(r + 3)^2}{4r^3s} \div \dfrac{r + 3}{rs}$

24. $\dfrac{t^2 + 5t}{t + 1} \div (t + 5)$

25. $\dfrac{6p + 7}{p + 2} \div (36p^2 - 49)$

26. $\dfrac{a}{a - 10} \div \dfrac{a^3 + 6a^2 - 40a}{a^2 - 100}$

27. $\dfrac{b^2 - 6b + 9}{b^2 - b - 6} \div \dfrac{b^2 - 9}{4}$

28. $\dfrac{2x^2 + 5xy + 2y^2}{4x^2 - y^2} \div \dfrac{x^2 + xy - 2y^2}{2x^2 + xy - y^2}$

29. $\dfrac{6s^2 + st - 2t^2}{6s^2 - 5st + t^2} \div \dfrac{3s^2 + 17st + 10t^2}{6s^2 + 13st - 5t^2}$

30. $\dfrac{x^4 - x^3 + x^2 - x}{2x^3 + 2x^2 + x + 1} \div \dfrac{x^3 - 4x^2 + x - 4}{2x^3 - 8x^2 + x - 4}$

31. $\dfrac{a^3 + a + a^2 + 1}{a^3 + a^2 + ab^2 + b^2} \div \dfrac{a^3 + a + a^2b + b}{2a^2 + 2ab + ab^2 + b^3}$

32. $\dfrac{3y - y^2}{y^3 - 27} \div \dfrac{y}{y^2 + 3y + 9}$

33. $\dfrac{8x - 4x^2}{xy - 2y + 3x - 6} \div \dfrac{3x + 6}{y + 3}$

Mixed Exercises

For Exercises 34–49, perform the indicated operations.

34. $\dfrac{8a^4b^3}{3c} \div \dfrac{a^7b^2}{9c}$

35. $\dfrac{3x^5}{2x^2y^7} \div \dfrac{4x^3y}{6y^6}$

36. $\dfrac{2}{25x^2} \cdot \dfrac{5x}{12} \div \dfrac{2}{15x}$

37. $\dfrac{4y}{7} \div \dfrac{y^2}{14} \cdot \dfrac{3}{y}$

38. $\dfrac{10x^2 - 13xy - 3y^2}{8x^2 - 10xy - 3y^2} \cdot \dfrac{2y + 8x}{2x^2 + 2y^2}$

39. $\dfrac{6a^2 + ab - b^2}{10a^2 + 5ab} \cdot \dfrac{2a^3 + 4a^2b}{3a^2 + 5ab - 2b^2}$

40. $\dfrac{(a + b)^2}{a - b} \cdot \dfrac{a^3 - b^3}{a^2 - b^2} \div \dfrac{a^2 + ab + b^2}{(a - b)^2}$

41. $\dfrac{m^2 - n^2}{(m - n)^2} \div \dfrac{m^2 - 2mn + n^2}{m^2 - mn + n^2} \cdot \dfrac{(m - n)^4}{m^3 + n^3}$

42. $\dfrac{x^2 - 4y^2}{x + 2y} \div (x + 2y) \cdot \dfrac{2y}{x - 2y}$

43. $\dfrac{x^2 - 6xy + 9y^2}{x^2 - 4y^2} \cdot \dfrac{x^2 - 5xy + 6y^2}{3y - x} \div \dfrac{x^2 - 9y^2}{x + 2y}$

44. $\dfrac{8x^3 - 27y^3}{4x^2 - 9y^2} \div \dfrac{8x^2 + 12xy + 18y^2}{2x + 3y}$

45. $\dfrac{25m^2 - 1}{125m^3 - 1} \div \dfrac{5m + 1}{25m^2 + 5m + 1}$

46. $\dfrac{m^3 + 2m^2 - mn^2 - 2n^2}{m^3 - m^2 - 20m} \cdot \dfrac{m^3 - 25m}{m^3 + m^2n - 4m - 4n}$

47. $\dfrac{2a^2 + ab - 8a - 4b}{2a^2 - 6a + ab - 3b} \cdot \dfrac{a^2 - 6a + 9}{a^2 - 16}$

48. $\dfrac{12y + 3}{6y^2 - y - 12} \div \dfrac{4y^2 - 19y - 5}{2y^2 - y - 3}$

49. $\dfrac{2x^2 - 11x - 6}{3x - 2} \div \dfrac{2x^2 - 5x - 3}{3x^2 - 7x - 6}$

Expanding Your Skills

For Exercises 50–53, write an expression for the area of the figure and simplify.

50.

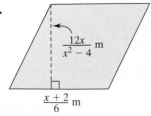

$\dfrac{b^2}{5a}$ in.

$\dfrac{4a^2}{b}$ in.

51.

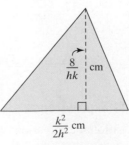

$\dfrac{8}{hk}$ cm

$\dfrac{k^2}{2h^2}$ cm

52.

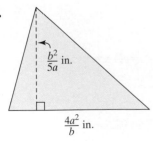

$\dfrac{12x}{x^2 - 4}$ m

$\dfrac{x + 2}{6}$ m

53.

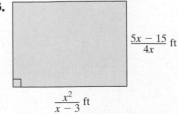

$\dfrac{5x - 15}{4x}$ ft

$\dfrac{x^2}{x - 3}$ ft

Addition and Subtraction of Rational Expressions	**Section 6.3**

1. Addition and Subtraction of Rational Expressions with Like Denominators

To add or subtract rational expressions, the expressions must have the same denominator. As with fractions, we add or subtract rational expressions with the same denominator by combining the terms in the numerator and then writing the result over the common denominator. Then, if possible, we simplify the expression to lowest terms.

> ### Addition and Subtraction Properties of Rational Expressions
>
> Let p, q, and r represent polynomials where $q \neq 0$. Then
>
> **1.** $\dfrac{p}{q} + \dfrac{r}{q} = \dfrac{p + r}{q}$ **2.** $\dfrac{p}{q} - \dfrac{r}{q} = \dfrac{p - r}{q}$

Concepts

1. Addition and Subtraction of Rational Expressions with Like Denominators
2. Identifying the Least Common Denominator
3. Writing Equivalent Rational Expressions
4. Addition and Subtraction of Rational Expressions with Unlike Denominators

---

Example 1   **Adding and Subtracting Rational Expressions with Like Denominators**

Add or subtract as indicated.

a. $\dfrac{1}{8} + \dfrac{3}{8}$ **b.** $\dfrac{5x}{2x - 1} + \dfrac{3}{2x - 1}$ **c.** $\dfrac{x^2}{x - 4} - \dfrac{x + 12}{x - 4}$

Solution:

a. $\dfrac{1}{8} + \dfrac{3}{8} = \dfrac{1 + 3}{8}$ Add the terms in the numerator.

$= \dfrac{4}{8}$

$= \dfrac{1}{2}$ Simplify the fraction.

b. $\dfrac{5x}{2x - 1} + \dfrac{3}{2x - 1} = \dfrac{5x + 3}{2x - 1}$ Add the terms in the numerator. The answer is already in lowest terms.

c. $\dfrac{x^2}{x - 4} - \dfrac{x + 12}{x - 4}$ Combine the terms in the numerator. Use parentheses to group the terms in the numerator that follow the subtraction sign. This will help you remember to apply the distributive property.

$= \dfrac{x^2 - (x + 12)}{x - 4}$

$= \dfrac{x^2 - x - 12}{x - 4}$ Apply the distributive property.

$= \dfrac{(x - 4)(x + 3)}{(x - 4)}$ Factor the numerator and denominator.

$= \dfrac{(x \overset{1}{-\!\!\!\diagup} 4)(x + 3)}{(x \overset{}{-\!\!\!\diagup} 4)}$ Simplify the rational expression.

$= x + 3$

Skill Practice

Add or subtract as indicated.

1. $\dfrac{5}{12} - \dfrac{1}{12}$

2. $\dfrac{b^2 + 1}{b - 2} + \dfrac{b - 7}{b - 2}$

3. $\dfrac{t^2}{t - 7} - \dfrac{5t + 14}{t - 7}$

Answers

1. $\dfrac{1}{3}$ **2.** $b + 3$ **3.** $t + 2$

2. Identifying the Least Common Denominator

If two rational expressions have different denominators, each expression must be rewritten with a common denominator before adding or subtracting the expressions. The **least common denominator (LCD)** of two or more rational expressions is defined as the least common multiple of the denominators.

For example, consider the fractions $\frac{1}{20}$ and $\frac{1}{8}$. By inspection, you can probably see that the least common denominator is 40. To understand why, find the prime factorization of both denominators.

$$20 = 2^2 \cdot 5 \quad \text{and} \quad 8 = 2^3$$

A common multiple of 20 and 8 must be a multiple of 5, a multiple of 2^2, and a multiple of 2^3. However, any number that is a multiple of $2^3 = 8$ is automatically a multiple of $2^2 = 4$. Therefore, it is sufficient to construct the least common denominator as the product of unique prime factors, where each factor is raised to its highest power.

The LCD of $\dfrac{1}{2^2 \cdot 5}$ and $\dfrac{1}{2^3}$ is $2^3 \cdot 5 = 40$.

Finding the LCD of Two or More Rational Expressions

Step 1 Factor all denominators completely.

Step 2 The LCD is the product of unique prime factors from the denominators, where each factor is raised to the highest power to which it appears in any denominator.

Example 2 Finding the LCD of Rational Expressions

Find the LCD of the rational expressions.

a. $\dfrac{1}{12}, \dfrac{5}{18}, \dfrac{7}{30}$ **b.** $\dfrac{1}{2x^3y}, \dfrac{5}{16xy^2z}$

Solution:

a. $\dfrac{1}{12}, \dfrac{5}{18}, \dfrac{7}{30}$

$\dfrac{1}{2^2 \cdot 3}, \dfrac{5}{2 \cdot 3^2}, \dfrac{7}{2 \cdot 3 \cdot 5}$ Factor the denominators completely.

$\text{LCD} = 2^2 \cdot 3^2 \cdot 5 = 180$ The LCD is the product of the factors 2, 3, and 5. Each factor is raised to its highest power.

b. $\dfrac{1}{2x^3y}, \dfrac{5}{16xy^2z}$

$\dfrac{1}{2x^3y}, \dfrac{5}{2^4xy^2z}$ Factor the denominators completely.

$\text{LCD} = 2^4x^3y^2z$ The LCD is the product of the factors 2, x, y, and z. Each factor is raised to its highest

$= 16x^3y^2z$ power.

Example 3 Finding the LCD of Rational Expressions

Find the LCD of the rational expressions.

a. $\dfrac{x^2 + 3}{x^2 + 9x + 20}, \dfrac{6}{x^2 + 8x + 16}$

b. $\dfrac{x + 4}{x - 3}, \dfrac{1}{3 - x}$

Solution:

a. $\dfrac{x^2 + 3}{x^2 + 9x + 20}, \dfrac{6}{x^2 + 8x + 16}$

$\dfrac{x^2 + 3}{(x + 4)(x + 5)}, \dfrac{6}{(x + 4)^2}$ Factor the denominators completely.

$\text{LCD} = (x + 5)(x + 4)^2$ The LCD is the product of the factors $(x + 5)$ and $(x + 4)$. Each factor is raised to its highest power.

b. $\dfrac{x + 4}{x - 3}, \dfrac{1}{3 - x}$ The denominators are already factored.

Notice that $x - 3$ and $3 - x$ are opposites. If -1 is factored from either expression, the binomial factors will be the same.

$\dfrac{x + 4}{x - 3}, \dfrac{1}{-1(-3 + x)}$

Factor out -1.

same binomial factors

$\dfrac{x + 4}{-1(-x + 3)}, \dfrac{1}{3 - x}$

Factor out -1.

same binomial factors

$\text{LCD} = (x - 3)(-1)$

$= -x + 3$

$= 3 - x$

$\text{LCD} = (-1)(3 - x)$

$= -3 + x$

$= x - 3$

The LCD can be expressed as either $(3 - x)$ or $(x - 3)$.

---

3. Writing Equivalent Rational Expressions

Rational expressions can be added if they have common denominators. Once the LCD has been determined, each rational expression must be converted to an **equivalent rational expression** with the indicated denominator.

Using the identity property of multiplication, we know that for $q \neq 0$ and $r \neq 0$,

$$\frac{p}{q} = \frac{p}{q} \cdot 1 = \frac{p}{q} \cdot \frac{r}{r} = \frac{pr}{qr}$$

This principle is used to convert a rational expression to an equivalent expression with a different denominator. For example, $\frac{1}{2}$ can be converted to an equivalent expression with a denominator of 12 as follows:

$$\frac{1}{2} = \frac{1}{2} \cdot 1 = \frac{1}{2} \cdot \frac{6}{6} = \frac{1 \cdot 6}{2 \cdot 6} = \frac{6}{12}$$

In this example, we multiplied $\frac{1}{2}$ by a convenient form of 1. The ratio $\frac{6}{6}$ was chosen so that the product produced a new denominator of 12. Notice that multiplying $\frac{1}{2}$ by $\frac{6}{6}$ is equivalent to multiplying the numerator and denominator of the original expression by 6.

Skill Practice

Fill in the blank to make an equivalent fraction with the given denominator.

8. $\dfrac{1}{8xy} = \dfrac{}{16x^3y^2}$

9. $\dfrac{5b}{b-3} = \dfrac{}{b^2-9}$

Example 4 **Creating Equivalent Rational Expressions**

Convert each expression to an equivalent rational expression with the indicated denominator.

a. $\dfrac{7}{5p^2} = \dfrac{}{20p^6}$ **b.** $\dfrac{w}{w+5} = \dfrac{}{w^2+3w-10}$

Solution:

a. $\dfrac{7}{5p^2} = \dfrac{}{20p^6}$

$= \dfrac{}{5p^2 \cdot 4p^4}$ $20p^6$ can be written as $5p^2 \cdot 4p^4$.

$\dfrac{7}{5p^2} = \dfrac{7}{5p^2} \cdot \dfrac{4p^4}{4p^4}$ Multiply the numerator and denominator by the missing factor, $4p^4$.

$= \dfrac{28p^4}{20p^6}$

TIP: Notice that in Example 4 we multiplied the polynomials in the numerator but left the denominator in factored form. This convention is followed because when we add and subtract rational expressions, the terms in the numerators must be combined.

b. $\dfrac{w}{w+5} = \dfrac{}{w^2+3w-10}$

$= \dfrac{}{(w+5)(w-2)}$ Factor the denominator.

$\dfrac{w}{w+5} = \dfrac{w}{w+5} \cdot \dfrac{w-2}{w-2}$ Multiply the numerator and denominator by the missing factor $(w-2)$.

$= \dfrac{w^2-2w}{(w+5)(w-2)}$

4. Addition and Subtraction of Rational Expressions with Unlike Denominators

To add or subtract rational expressions with unlike denominators, we must convert each expression to an equivalent expression with the same denominator. For example, consider adding the expressions $\frac{3}{x-2} + \frac{5}{x+1}$. The LCD is $(x-2)(x+1)$. For each expression, identify the factors from the LCD that are missing in the denominator. Then multiply the numerator and denominator of the expression by the missing factor(s):

$\dfrac{(3)}{(x-2)} \cdot \dfrac{(x+1)}{(x+1)} + \dfrac{(5)}{(x+1)} \cdot \dfrac{(x-2)}{(x-2)}$ The rational expressions now have the same denominator and can be added.

$= \dfrac{3(x+1) + 5(x-2)}{(x-2)(x+1)}$ Combine terms in the numerator.

$= \dfrac{3x + 3 + 5x - 10}{(x-2)(x+1)}$ Clear parentheses and simplify.

$= \dfrac{8x - 7}{(x-2)(x+1)}$

Answers

8. $2x^2y$ **9.** $5b^2 + 15b$

Adding and Subtracting Rational Expressions

Step 1 Factor the denominator of each rational expression.

Step 2 Identify the LCD.

Step 3 Rewrite each rational expression as an equivalent expression with the LCD as its denominator.

Step 4 Add or subtract the numerators, and write the result over the common denominator.

Step 5 Simplify, if possible.

Example 5 Adding Rational Expressions with Unlike Denominators

Add. $\dfrac{3}{7b} + \dfrac{4}{b^2}$

Solution:

$\dfrac{3}{7b} + \dfrac{4}{b^2}$ **Step 1:** The denominators are already factored.

           Step 2: The LCD is $7b^2$.

$= \dfrac{3}{7b} \cdot \dfrac{b}{b} + \dfrac{4}{b^2} \cdot \dfrac{7}{7}$ **Step 3:** Write each expression with the LCD.

$= \dfrac{3b}{7b^2} + \dfrac{28}{7b^2}$ **Step 4:** Add the numerators, and write the result over the LCD.

$= \dfrac{3b + 28}{7b^2}$ **Step 5:** Simplify.

Skill Practice

Add.

10. $\dfrac{4}{5y} + \dfrac{1}{3y^3}$

Example 6 Subtracting Rational Expressions with Unlike Denominators

Subtract. $\dfrac{3t - 2}{t^2 + 4t - 12} - \dfrac{5}{2t + 12}$

Solution:

$\dfrac{3t - 2}{t^2 + 4t - 12} - \dfrac{5}{2t + 12}$

$= \dfrac{3t - 2}{(t + 6)(t - 2)} - \dfrac{5}{2(t + 6)}$ **Step 1:** Factor the denominators.

           Step 2: The LCD is $2(t + 6)(t - 2)$.

$= \dfrac{(2)}{(2)} \cdot \dfrac{(3t - 2)}{(t + 6)(t - 2)} - \dfrac{5}{2(t + 6)} \cdot \dfrac{(t - 2)}{(t - 2)}$ **Step 3:** Write each expression with the LCD.

$= \dfrac{2(3t - 2) - 5(t - 2)}{2(t + 6)(t - 2)}$ **Step 4:** Add the numerators and write the result over the LCD.

Skill Practice

Subtract.

11. $\dfrac{2x + 3}{x^2 + x - 2} - \dfrac{5}{3x - 3}$

Answers

10. $\dfrac{12y^2 + 5}{15y^3}$ **11.** $\dfrac{1}{3(x + 2)}$

$$= \frac{6t - 4 - 5t + 10}{2(t + 6)(t - 2)}$$ **Step 5:** Simplify.

$$= \frac{t + 6}{2(t + 6)(t - 2)}$$ Combine *like* terms.

$$= \frac{\overset{1}{\cancel{t + 6}}}{2\cancel{(t + 6)}(t - 2)}$$ Simplify.

$$= \frac{1}{2(t - 2)}$$

Skill Practice

Add.

12. $\dfrac{a^2 + a + 24}{a^2 - 9} + \dfrac{5}{a + 3}$

Example 7 **Adding and Subtracting Rational Expressions with Unlike Denominators**

Add and subtract as indicated. $\dfrac{2}{x} + \dfrac{x}{x + 3} - \dfrac{3x + 18}{x^2 + 3x}$

Solution:

$$\frac{2}{x} + \frac{x}{x + 3} - \frac{3x + 18}{x^2 + 3x}$$

$$= \frac{2}{x} + \frac{x}{x + 3} - \frac{3x + 18}{x(x + 3)}$$ **Step 1:** Factor the denominators.

Step 2: The LCD is $x(x + 3)$.

$$= \frac{2}{x} \cdot \frac{(x + 3)}{(x + 3)} + \frac{x}{(x + 3)} \cdot \frac{x}{x} - \frac{3x + 18}{x(x + 3)}$$ **Step 3:** Write each expression with the LCD.

Avoiding Mistakes

It is important to insert parentheses around the quantity being subtracted.

$$= \frac{2(x + 3) + x^2 - (3x + 18)}{x(x + 3)}$$ **Step 4:** Add the numerators, and write the result over the LCD.

$$= \frac{2x + 6 + x^2 - 3x - 18}{x(x + 3)}$$ **Step 5:** Simplify.

$$= \frac{x^2 - x - 12}{x(x + 3)}$$ Combine *like* terms.

$$= \frac{(x - 4)(x + 3)}{x(x + 3)}$$ Factor the numerator.

$$= \frac{(x - 4)\overset{1}{\cancel{(x + 3)}}}{x\cancel{(x + 3)}}$$ Simplify.

$$= \frac{x - 4}{x}$$

Answer

12. $\dfrac{a + 3}{a - 3}$

Example 8 Subtracting Rational Expressions with Unlike Denominators

Subtract. $\dfrac{6}{w} - \dfrac{4}{-w}$

Skill Practice

Subtract.

13. $\dfrac{3}{-y} - \dfrac{5}{y}$

Solution:

$\dfrac{6}{w} - \dfrac{4}{-w}$

Step 1: The denominators are already factored.

Step 2: The denominators are opposites and differ by a factor of -1. The LCD can either be taken as w or $-w$. We will use an LCD of w.

$= \dfrac{6}{w} - \dfrac{4}{-w} \cdot \dfrac{(-1)}{(-1)}$

Step 3: Write each expression with the LCD. Note that $(-w)(-1) = w$.

$= \dfrac{6}{w} - \dfrac{-4}{w}$

$= \dfrac{6 - (-4)}{w}$

Step 4: Subtract the numerators, and write the result over the LCD.

$= \dfrac{10}{w}$

Step 5: Simplify.

Example 9 Adding Rational Expressions with Unlike Denominators

Add. $\dfrac{x^2}{x - y} + \dfrac{y^2}{y - x}$

Skill Practice

Add.

14. $\dfrac{3a}{a - 5} + \dfrac{15}{5 - a}$

Solution:

$\dfrac{x^2}{x - y} + \dfrac{y^2}{y - x}$

Step 1: The denominators are already factored.

Step 2: The denominators are opposites and differ by a factor of -1. The LCD can be taken as either $(x - y)$ or $(y - x)$. We will use an LCD of $(x - y)$.

$= \dfrac{x^2}{(x - y)} + \dfrac{y^2}{(y - x)} \cdot \dfrac{(-1)}{(-1)}$

Step 3: Write each expression with the LCD. Note that $(y - x)(-1) = -y + x = x - y$.

$= \dfrac{x^2}{x - y} + \dfrac{-y^2}{x - y}$

$= \dfrac{x^2 - y^2}{x - y}$

Step 4: Combine the numerators, and write the result over the LCD.

$= \dfrac{(x + y)(\overset{1}{\cancel{x - y}})}{\cancel{x - y}}$

Step 5: Factor and simplify to lowest terms.

$= x + y$

Answers

13. $-\dfrac{8}{y}$ 14. 3

Section 6.3   Practice Exercises

Vocabulary and Key Concepts

1. a. Given polynomials p, q, and r such that $q \neq 0$, $\dfrac{p}{q} + \dfrac{r}{q} = \dfrac{\square}{\square}$ and $\dfrac{p}{q} - \dfrac{r}{q} = \dfrac{\square}{\square}$

   b. The _____ _____ _____ (LCD) of two rational expressions is defined as the least common multiple of their denominators.

Review Exercises

For Exercises 2–6, perform the indicated operation.

2. $\dfrac{x}{x-y} \div \dfrac{x^2}{y-x}$

3. $\dfrac{9b+9}{4b+8} \cdot \dfrac{2b+4}{3b-3}$

4. $\dfrac{8a^2}{7b^3} \div \dfrac{4a}{21b}$

5. $\dfrac{(5-a)^2}{10a-2} \cdot \dfrac{25a^2-1}{a^2-10a+25}$

6. $\dfrac{x^2-z^2}{14x^2z^4} \div \dfrac{x^2+2xz+z^2}{3xz^3}$

Concept 1: Addition and Subtraction of Rational Expressions with Like Denominators

For Exercises 7–14, add or subtract as indicated and simplify if possible. **(See Example 1.)**

7. $\dfrac{3}{5x} + \dfrac{7}{5x}$

8. $\dfrac{1}{2x^2} - \dfrac{5}{2x^2}$

9. $\dfrac{x}{x^2-2x-3} - \dfrac{3}{x^2-2x-3}$

10. $\dfrac{x}{x^2+4x-12} + \dfrac{6}{x^2+4x-12}$

11. $\dfrac{5x-1}{(2x+9)(x-6)} - \dfrac{3x-6}{(2x+9)(x-6)}$

12. $\dfrac{4-x}{8x+1} - \dfrac{5x-6}{8x+1}$

13. $\dfrac{x+2}{x-5} + \dfrac{x-12}{x-5}$

14. $\dfrac{2x-1}{x-2} + \dfrac{x-5}{x-2}$

Concept 2: Identifying the Least Common Denominator

For Exercises 15–26, find the least common denominator (LCD). **(See Examples 2 and 3.)**

15. $\dfrac{5}{8}$; $\dfrac{3}{20x}$

16. $\dfrac{y}{15a}$; $\dfrac{y^2}{35}$

17. $\dfrac{-5}{6m^4}$; $\dfrac{1}{15mn^7}$

18. $\dfrac{13}{12cd^5}$; $\dfrac{9}{8c^3}$

19. $\dfrac{6}{(x-4)(x+2)}$; $\dfrac{-8}{(x-4)(x-6)}$

20. $\dfrac{x}{(2x-1)(x-7)}$; $\dfrac{2}{(2x-1)(x+1)}$

21. $\dfrac{3}{x(x-1)(x+7)^2}$; $\dfrac{-1}{x^2(x+7)}$

22. $\dfrac{14}{(x-2)^2(x+9)}$; $\dfrac{41}{x(x-2)(x+9)}$

23. $\dfrac{5}{x-6}$; $\dfrac{x-5}{x^2-8x+12}$

24. $\dfrac{7a}{a+4}$; $\dfrac{a+12}{a^2-16}$

25. $\dfrac{3a}{a-4}$; $\dfrac{5}{4-a}$

26. $\dfrac{10}{x-6}$; $\dfrac{x+1}{6-x}$

Concept 3: Writing Equivalent Rational Expressions

For Exercises 27–32, fill in the blank to make an equivalent fraction with the given denominator. **(See Example 4.)**

27. $\dfrac{5}{3x} = \dfrac{}{9x^2y}$

28. $\dfrac{-5}{xy} = \dfrac{}{4x^2y^3}$

29. $\dfrac{2x}{x-1} = \dfrac{}{x(x-1)(x+2)}$

30. $\dfrac{5x}{2x-5} = \dfrac{}{(2x-5)(x+8)}$

31. $\dfrac{y}{y+6} = \dfrac{}{y^2+5y-6}$

32. $\dfrac{t^2}{t-8} = \dfrac{}{t^2-6t-16}$

Concept 4: Addition and Subtraction of Rational Expressions with Unlike Denominators

For Exercises 33–56, add or subtract as indicated. **(See Examples 5–9.)**

33. $\dfrac{4}{3p} - \dfrac{5}{2p^2}$

34. $\dfrac{6}{5a^2b} - \dfrac{1}{10ab}$

35. $\dfrac{s-1}{s} - \dfrac{t+1}{t}$

36. $\dfrac{x+2}{x} - \dfrac{y-2}{y}$

37. $\dfrac{4a-2}{3a+12} - \dfrac{a-2}{a+4}$

38. $\dfrac{6y+5}{5y-25} - \dfrac{y+2}{y-5}$

39. $\dfrac{10}{b(b+5)} + \dfrac{2}{b}$

40. $\dfrac{6}{w(w-2)} + \dfrac{3}{w}$

41. $\dfrac{x-2}{x-6} - \dfrac{x+2}{6-x}$

42. $\dfrac{x-10}{x-8} - \dfrac{x+10}{8-x}$

43. $\dfrac{6b}{b-4} - \dfrac{1}{b+1}$

44. $\dfrac{a}{a-3} - \dfrac{5}{a+6}$

45. $\dfrac{2}{2x+1} + \dfrac{4}{x-2}$

46. $\dfrac{3}{y+6} + \dfrac{1}{3y+1}$

47. $\dfrac{y-2}{y-4} + \dfrac{2y^2-15y+12}{y^2-16}$

48. $\dfrac{x^2+13x+18}{x^2-9} + \dfrac{x+1}{x+3}$

49. $\dfrac{x+2}{x^2-36} - \dfrac{x}{x^2+9x+18}$

50. $\dfrac{7}{x^2-x-2} + \dfrac{x}{x^2+4x+3}$

51. $\dfrac{5}{w} + \dfrac{8}{-w}$

52. $\dfrac{4}{y} + \dfrac{5}{-y}$

53. $\dfrac{n}{5-n} + \dfrac{2n-5}{n-5}$

54. $\dfrac{c}{7-c} + \dfrac{2c-7}{c-7}$

55. $\dfrac{2}{3x-15} + \dfrac{x}{25-x^2}$

56. $\dfrac{5}{9-x^2} - \dfrac{4}{x^2+4x+3}$

Mixed Exercises

For Exercises 57–74, simplify.

57. $\dfrac{x+3}{x^2} + \dfrac{x+5}{2x}$

58. $\dfrac{x+2}{5x^2} + \dfrac{x+4}{15x}$

59. $w+2+\dfrac{1}{w-2}$

60. $h-3+\dfrac{1}{h+3}$

61. $\dfrac{9}{x^2-2x+1} - \dfrac{x-3}{x^2-x}$

62. $\dfrac{2}{4z^2-12z+9} - \dfrac{z+1}{2z^2-3z}$

63. $\dfrac{t+1}{t+3} - \dfrac{t-2}{t-3} + \dfrac{6}{t^2-9}$

64. $\dfrac{y-3}{y-2} - \dfrac{y+1}{2y-5} + \dfrac{-4y+7}{2y^2-9y+10}$

65. $(x-1) \cdot \left[\dfrac{3}{x^2-1} + \dfrac{x}{2x-2}\right]$

66. $(3x - 2) \cdot \left[\dfrac{x}{3x^2 + x - 2} + \dfrac{2}{x + 1} \right]$

67. $\dfrac{3z}{z - 3} - \dfrac{z}{z + 4}$

68. $\dfrac{2p}{p - 5} - \dfrac{p}{p + 6}$

69. $\dfrac{2x}{x^2 - y^2} - \dfrac{1}{x - y} + \dfrac{1}{y - x}$

70. $\dfrac{3w - 1}{2w^2 + w - 3} - \dfrac{2 - w}{w - 1} - \dfrac{w}{1 - w}$

71. $(2p + 1) \cdot \left[\dfrac{2p}{6p + 3} - \dfrac{1}{p + 4} \right]$

72. $(y + 8) \cdot \left[\dfrac{4}{2y + 1} - \dfrac{y}{2y^2 + 17y + 8} \right]$

73. $\dfrac{3}{y} + \dfrac{2}{y - 6}$

74. $\dfrac{-8}{p} + \dfrac{p}{p + 5}$

For Exercises 75–78, write an expression that represents the perimeter of the figure and simplify.

75.

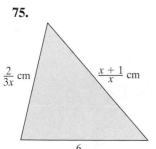

$\frac{2}{3x}$ cm $\quad$ $\frac{x + 1}{x}$ cm $\quad$ $\frac{6}{x^2}$ cm

76.

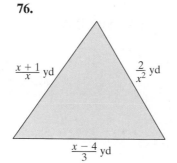

$\frac{x + 1}{x}$ yd $\quad$ $\frac{2}{x^2}$ yd $\quad$ $\frac{x - 4}{3}$ yd

77.

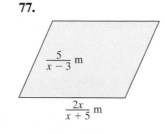

$\frac{5}{x - 3}$ m $\quad$ $\frac{2x}{x + 5}$ m

78.

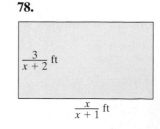

$\frac{3}{x + 2}$ ft $\quad$ $\frac{x}{x + 1}$ ft

Section 6.4 | Complex Fractions

Concepts

1. Simplifying Complex Fractions by Method I
2. Simplifying Complex Fractions by Method II

1. Simplifying Complex Fractions by Method I

A **complex fraction** is an expression containing one or more fractional expressions in the numerator, denominator, or both. For example:

$$\dfrac{\dfrac{5x^2}{y}}{\dfrac{10x}{y^2}} \quad \text{and} \quad \dfrac{2 + \dfrac{1}{2} - \dfrac{1}{3}}{\dfrac{3}{4} + \dfrac{1}{6}}$$

are complex fractions.

Two methods will be presented to simplify complex fractions. The first method (Method I) follows the order of operations to simplify the numerator and denominator separately before dividing. The process is summarized as follows.

> **Simplifying a Complex Fraction—Method I**
>
> **Step 1** Add or subtract expressions in the numerator to form a single fraction. Add or subtract expressions in the denominator to form a single fraction.
>
> **Step 2** Divide the rational expressions from step 1 by multiplying the numerator of the complex fraction by the reciprocal of the denominator of the complex fraction.
>
> **Step 3** Simplify to lowest terms, if possible.

Example 1 Simplifying a Complex Fraction by Method I

Simplify the expression. $\dfrac{\dfrac{5x^2}{y}}{\dfrac{10x}{y^2}}$

Solution:

$\dfrac{\dfrac{5x^2}{y}}{\dfrac{10x}{y^2}}$ **Step 1:** The numerator and denominator of the complex fraction are already single fractions.

$= \dfrac{5x^2}{y} \div \dfrac{10x}{y^2}$

$= \dfrac{5x^2}{y} \cdot \dfrac{y^2}{10x}$ **Step 2:** Multiply the numerator of the complex fraction by the reciprocal of the denominator.

$= \dfrac{5x^2 y^2}{10xy}$

$= \dfrac{1}{2} x^{2-1} y^{2-1}$ **Step 3:** Simplify.

$= \dfrac{1}{2}xy$ or $\dfrac{xy}{2}$

Sometimes it is necessary to simplify the numerator and denominator of a complex fraction before the division is performed. This is illustrated in Example 2.

Example 2 Simplifying a Complex Fraction by Method I

Simplify the expression. $\dfrac{2 + \dfrac{1}{2} - \dfrac{1}{3}}{\dfrac{3}{4} + \dfrac{1}{6}}$

Solution:

$\dfrac{2 + \dfrac{1}{2} - \dfrac{1}{3}}{\dfrac{3}{4} + \dfrac{1}{6}}$ **Step 1:** Combine fractions in numerator and denominator separately.

$= \dfrac{\dfrac{12}{6} + \dfrac{3}{6} - \dfrac{2}{6}}{\dfrac{9}{12} + \dfrac{2}{12}}$ The LCD in the numerator is 6. The LCD in the denominator is 12.

$= \dfrac{\dfrac{13}{6}}{\dfrac{11}{12}}$ Form single fractions in the numerator and denominator.

$$= \frac{13}{6} \cdot \frac{12}{11}$$ **Step 2:** Multiply by the reciprocal of $\frac{11}{12}$, which is $\frac{12}{11}$.

$$= \frac{13}{\cancel{6}} \cdot \frac{\cancel{12}^{2}}{11}$$

$$= \frac{26}{11}$$ **Step 3:** Simplify.

2. Simplifying Complex Fractions by Method II

We will now use a second method to simplify complex fractions—Method II. Recall that multiplying the numerator and denominator of a rational expression by the same quantity does not change the value of the expression. This is the basis for Method II.

Simplifying a Complex Fraction—Method II

Step 1 Multiply the numerator and denominator of the complex fraction by the LCD of *all* individual fractions within the expression.

Step 2 Apply the distributive property, and simplify the numerator and denominator.

Step 3 Simplify to lowest terms, if possible.

Skill Practice

Simplify the expression.

3. $\dfrac{y - \dfrac{1}{y}}{1 - \dfrac{1}{y^2}}$

Example 3 Simplifying Complex Fractions by Method II

Simplify by using Method II. $\dfrac{4 - \dfrac{6}{x}}{2 - \dfrac{3}{x}}$

Solution:

$$\dfrac{4 - \dfrac{6}{x}}{2 - \dfrac{3}{x}}$$ The LCD of all individual terms is x.

$$= \dfrac{x \cdot \left(4 - \dfrac{6}{x}\right)}{x \cdot \left(2 - \dfrac{3}{x}\right)}$$ **Step 1:** Multiply the numerator and denominator of the complex fraction by the LCD, which is x.

$$= \dfrac{x \cdot (4) - x \cdot \left(\dfrac{6}{x}\right)}{x \cdot (2) - x \cdot \left(\dfrac{3}{x}\right)}$$ **Step 2:** Apply the distributive property.

Answer

3. y

$$= \frac{4x - 6}{2x - 3}$$ **Step 3:** Simplify numerator and denominator.

$$= \frac{2(2x - 3)}{2x - 3}$$ Factor and simplify to lowest terms.

$$= \frac{2(\overset{1}{\cancel{2x - 3}})}{\cancel{2x - 3}}$$

$$= 2$$

Example 4   **Simplifying Complex Fractions by Method II**

Simplify by using Method II. $\dfrac{x^{-1} - x^{-2}}{1 + 2x^{-1} - 3x^{-2}}$

Solution:

$$\frac{x^{-1} - x^{-2}}{1 + 2x^{-1} - 3x^{-2}}$$

$$= \frac{\dfrac{1}{x} - \dfrac{1}{x^2}}{1 + \dfrac{2}{x} - \dfrac{3}{x^2}}$$ Rewrite the expression with positive exponents. The LCD of all individual terms is x^2.

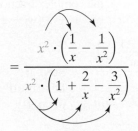

$$= \frac{x^2 \cdot \left(\dfrac{1}{x} - \dfrac{1}{x^2}\right)}{x^2 \cdot \left(1 + \dfrac{2}{x} - \dfrac{3}{x^2}\right)}$$ **Step 1:** Multiply the numerator and denominator of the complex fraction by the LCD x^2.

$$= \frac{x^2\left(\dfrac{1}{x}\right) - x^2\left(\dfrac{1}{x^2}\right)}{x^2(1) + x^2\left(\dfrac{2}{x}\right) - x^2\left(\dfrac{3}{x^2}\right)}$$ **Step 2:** Apply the distributive property.

$$= \frac{x - 1}{x^2 + 2x - 3}$$

$$= \frac{x - 1}{(x + 3)(x - 1)}$$ **Step 3:** Factor and simplify to lowest terms.

$$= \frac{\overset{1}{\cancel{x - 1}}}{(x + 3)(\cancel{x - 1})}$$

$$= \frac{1}{x + 3}$$

TIP: When writing $2x^{-1}$ with positive exponents recall that

$$2x^{-1} = 2 \cdot \frac{1}{x}$$

$$= \frac{2}{1} \cdot \frac{1}{x} = \frac{2}{x}$$

Answer

4. $b - c$

Skill Practice

Simplify the expression.

5. $\dfrac{\dfrac{2}{x+1} - \dfrac{1}{x-1}}{\dfrac{x}{x-1} - \dfrac{1}{x+1}}$

Example 5 **Simplifying Complex Fractions by Method II**

Simplify the expression by Method II. $\dfrac{\dfrac{1}{w+3} - \dfrac{1}{w-3}}{1 + \dfrac{9}{w^2-9}}$

Solution:

$$\dfrac{\dfrac{1}{w+3} - \dfrac{1}{w-3}}{1 + \dfrac{9}{w^2-9}}$$

$$= \dfrac{\dfrac{1}{w+3} - \dfrac{1}{w-3}}{1 + \dfrac{9}{(w+3)(w-3)}} \qquad \text{Factor all denominators to find the LCD.}$$

The LCD of $\dfrac{1}{1}, \dfrac{1}{w+3}, \dfrac{1}{w-3}$, and $\dfrac{9}{(w+3)(w-3)}$ is $(w+3)(w-3)$.

$$= \dfrac{(w+3)(w-3)\left(\dfrac{1}{w+3} - \dfrac{1}{w-3}\right)}{(w+3)(w-3)\left[1 + \dfrac{9}{(w+3)(w-3)}\right]} \qquad \begin{array}{l} \textbf{Step 1:} \;\; \text{Multiply numerator} \\ \text{and denominator of} \\ \text{the complex fraction} \\ \text{by } (w+3)(w-3). \end{array}$$

$$= \dfrac{(w+3)(w-3)\left(\dfrac{1}{w+3}\right) - (w+3)(w-3)\left(\dfrac{1}{w-3}\right)}{(w+3)(w-3)1 + (w+3)(w-3)\left[\dfrac{9}{(w+3)(w-3)}\right]} \qquad \begin{array}{l} \textbf{Step 2:} \\ \text{Distributive} \\ \text{property.} \end{array}$$

$$= \dfrac{(w-3) - (w+3)}{(w+3)(w-3) + 9} \qquad \textbf{Step 3:} \;\; \text{Simplify.}$$

$$= \dfrac{w-3-w-3}{w^2-9+9} \qquad \text{Apply the distributive property.}$$

$$= \dfrac{-6}{w^2}$$

$$= -\dfrac{6}{w^2}$$

Answer

5. $\dfrac{x-3}{x^2+1}$

Section 6.4 Practice Exercises

Vocabulary and Key Concepts

1. A _____ fraction is an expression containing one or more fractional expressions in the numerator, denominator, or both.

Review Exercises

For Exercises 2–4, simplify to lowest terms.

2. $\dfrac{x^3 + y^3}{5x + 5y}$

3. $\dfrac{25a^3b^3c}{15a^4bc}$

4. $\dfrac{6t^2 - 27t + 30}{12t - 30}$

For Exercises 5–8, perform the indicated operations.

5. $\dfrac{5}{x^2} + \dfrac{3}{2x}$

6. $\dfrac{2y - 4}{y + 1} \cdot \dfrac{y^2 + 3y + 2}{y^2 - 4}$

7. $\dfrac{3}{a - 5} - \dfrac{1}{a + 1}$

8. $\dfrac{7}{12 - 6b} \div \dfrac{14b}{b^2 + b - 6}$

Concept 1: Simplifying Complex Fractions by Method I

For Exercises 9–16, simplify the complex fractions by using Method I. **(See Examples 1 and 2.)**

9. $\dfrac{\dfrac{5x^2}{9y^2}}{\dfrac{3x}{y^2x}}$

10. $\dfrac{\dfrac{3w^2}{4rs}}{\dfrac{15wr}{s^2}}$

11. $\dfrac{\dfrac{x - 6}{3x}}{\dfrac{3x - 18}{9}}$

12. $\dfrac{\dfrac{a + 4}{6}}{\dfrac{16 - a^2}{3}}$

13. $\dfrac{\dfrac{2}{3} + \dfrac{1}{6}}{\dfrac{1}{2} - \dfrac{1}{4}}$

14. $\dfrac{\dfrac{7}{8} + \dfrac{3}{4}}{\dfrac{1}{3} - \dfrac{5}{6}}$

15. $\dfrac{2 - \dfrac{1}{y}}{4 + \dfrac{1}{y}}$

16. $\dfrac{\dfrac{1}{x} - 3}{\dfrac{1}{x} + 3}$

Concept 2: Simplifying Complex Fractions by Method II

For Exercises 17–44, simplify the complex fractions by using Method II. **(See Examples 3–5.)**

17. $\dfrac{\dfrac{7y}{y + 3}}{\dfrac{1}{4y + 12}}$

18. $\dfrac{\dfrac{6x}{x - 5}}{\dfrac{1}{4x - 20}}$

19. $\dfrac{1 + \dfrac{1}{3}}{\dfrac{5}{6} - 1}$

20. $\dfrac{2 + \dfrac{4}{5}}{-1 + \dfrac{3}{10}}$

21. $\dfrac{\dfrac{3q}{p} - q}{q - \dfrac{q}{p}}$

22. $\dfrac{\dfrac{b}{a} + 3b}{b + \dfrac{2b}{a}}$

23. $\dfrac{\dfrac{2}{a} + \dfrac{3}{a^2}}{\dfrac{4}{a^2} - \dfrac{9}{a}}$

24. $\dfrac{\dfrac{2}{y^2} + \dfrac{1}{y}}{\dfrac{4}{y^2} - \dfrac{1}{y}}$

25. $\dfrac{t^{-1} - 1}{1 - t^{-2}}$

26. $\dfrac{d^{-2} - c^{-2}}{c^{-1} - d^{-1}}$

27. $\dfrac{-8}{\dfrac{6w}{w - 1} - 4}$

28. $\dfrac{6}{2z - \dfrac{10}{z - 4}}$

29. $\dfrac{\dfrac{y}{y+3}}{\dfrac{y}{y+3}+y}$

30. $\dfrac{\dfrac{4}{w-4}}{\dfrac{4}{w-4}-1}$

31. $\dfrac{1-\dfrac{1}{x}-\dfrac{6}{x^2}}{1-\dfrac{4}{x}+\dfrac{3}{x^2}}$

32. $\dfrac{1+\dfrac{1}{x}-\dfrac{12}{x^2}}{\dfrac{9}{x^2}+\dfrac{3}{x}-2}$

33. $\dfrac{2-\dfrac{2}{t+1}}{2+\dfrac{2}{t}}$

34. $\dfrac{3+\dfrac{3}{p-1}}{3-\dfrac{3}{p}}$

35. $\dfrac{\dfrac{2}{a}-\dfrac{3}{a+1}}{\dfrac{2}{a+1}-\dfrac{3}{a}}$

36. $\dfrac{\dfrac{5}{b}+\dfrac{4}{b+1}}{\dfrac{4}{b}-\dfrac{5}{b+1}}$

37. $\dfrac{\dfrac{1}{y+2}+\dfrac{4}{y-3}}{\dfrac{2}{y-3}-\dfrac{7}{y+2}}$

38. $\dfrac{\dfrac{1}{t-4}+\dfrac{1}{t+5}}{\dfrac{6}{t+5}+\dfrac{2}{t-4}}$

39. $\dfrac{\dfrac{2}{x+h}-\dfrac{2}{x}}{h}$

40. $\dfrac{\dfrac{1}{2x+2h}-\dfrac{1}{2x}}{h}$

41. $\dfrac{x^{-2}}{x+3x^{-1}}$

42. $\dfrac{x^{-1}+x^{-2}}{5x^{-2}}$

43. $\dfrac{2a^{-1}+3b^{-2}}{a^{-1}-b^{-1}}$

44. $\dfrac{2m^{-1}+n^{-1}}{m^{-2}-4n^{-1}}$

Expanding Your Skills

45. The slope formula is used to find the slope of the line passing through the points (x_1, y_1) and (x_2, y_2). Write the slope formula from memory.

For Exercises 46–49, find the slope of the line that passes through the given points.

46. $\left(1\dfrac{1}{2},\dfrac{2}{5}\right),\left(\dfrac{1}{4},-2\right)$

47. $\left(-\dfrac{3}{7},\dfrac{3}{5}\right),(-1,-3)$

48. $\left(\dfrac{5}{8},\dfrac{9}{10}\right),\left(-\dfrac{1}{16},-\dfrac{1}{5}\right)$

49. $\left(\dfrac{1}{4},\dfrac{1}{3}\right),\left(\dfrac{1}{8},\dfrac{1}{6}\right)$

50. Show that $(x+x^{-1})^{-1}=\dfrac{x}{x^2+1}$ by writing the expression on the left without negative exponents and simplifying.

51. Show that $(x^{-1}+y^{-1})^{-1}=\dfrac{xy}{x+y}$ by writing the expression on the left without negative exponents and simplifying.

52. Simplify. $\dfrac{x}{1-\left(1+\dfrac{1}{x}\right)^{-1}}$

53. Simplify. $\dfrac{x}{1-\left(1-\dfrac{1}{x}\right)^{-1}}$

Problem Recognition Exercises

Operations on Rational Expressions

For Exercises 1–20, perform the indicated operations and simplify.

1. $\dfrac{2}{2y - 3} - \dfrac{3}{2y} + 1$

2. $(x + 5) + \left(\dfrac{7}{x - 4}\right)$

3. $\dfrac{5x^2 - 6x + 1}{x^2 - 1} \div \dfrac{16x^2 - 9}{4x^2 + 7x + 3} - \dfrac{x}{4x - 3}$

4. $\dfrac{a^2 - 25}{3a^2 + 3ab} \cdot \dfrac{a^2 + 4a + ab + 4b}{a^2 + 9a + 20}$

5. $\dfrac{4}{y + 1} + \dfrac{y + 2}{y^2 - 1} - \dfrac{3}{y - 1}$

6. $\dfrac{8w^2}{w^3 - 16w} - \dfrac{4w}{w^2 - 4w}$

7. $\dfrac{a^2 - 16}{2x + 6} \cdot \dfrac{x + 3}{a - 4}$

8. $\dfrac{t^2 - 9}{t} \div \dfrac{t + 3}{t + 2}$

9. $\dfrac{2 + \dfrac{1}{a}}{4 - \dfrac{1}{a^2}}$

10. $\dfrac{\dfrac{6x^2y}{5}}{\dfrac{3x}{y}}$

11. $\dfrac{6xy}{x^2 - y^2} + \dfrac{x + y}{y - x}$

12. $(x^2 - 6x + 8) \cdot \left(\dfrac{3}{x - 2}\right)$

13. $\dfrac{1}{w - 1} - \dfrac{w + 2}{3w - 3}$

14. $\dfrac{3y + 6}{y^2 - 3y - 10} \div \dfrac{27}{y - 5}$

15. $\dfrac{y + \dfrac{2}{y} - 3}{1 - \dfrac{2}{y}}$

16. $\dfrac{2}{t - 3} - \dfrac{3}{t + 2} + 5$

17. $\dfrac{4x^2 + 22x + 24}{4x + 4} \cdot \dfrac{6x + 6}{4x^2 - 9}$

18. $\dfrac{12x^3y^5z}{5x^4} \div \dfrac{16xy^7}{10z^2}$

19. $(y + 2) \cdot \dfrac{2y + 1}{y^2 - 4} - \dfrac{y - 2}{y + 3}$

20. $\dfrac{a^2}{a - 10} - \dfrac{20a - 100}{a - 10}$

Solving Rational Equations

1. Solving Rational Equations

Thus far, we have studied two types of equations in one variable: linear equations and quadratic equations. In this section, we will study another type of equation called a rational equation.

Definition of a Rational Equation

An equation with one or more rational expressions is called a **rational equation**.

The following equations are rational equations:

$$\dfrac{1}{2}x + \dfrac{1}{3} = \dfrac{1}{4}x \qquad \dfrac{3}{5} + \dfrac{1}{x} = \dfrac{2}{3} \qquad 3 - \dfrac{6w}{w + 1} = \dfrac{6}{w + 1}$$

To understand the process of solving a rational equation, first review the procedure of clearing fractions from Section 1.1.

Example 1 **Solving an Equation with Fractions**

Solve the equation. $\frac{1}{2}x + \frac{1}{3} = \frac{1}{4}x$

Solution:

$$\frac{1}{2}x + \frac{1}{3} = \frac{1}{4}x \qquad \text{The LCD of all terms in the equation is 12.}$$

$$12\left(\frac{1}{2}x + \frac{1}{3}\right) = 12\left(\frac{1}{4}x\right) \qquad \text{Multiply both sides by 12 to clear fractions.}$$

$$12 \cdot \frac{1}{2}x + 12 \cdot \frac{1}{3} = 12 \cdot \frac{1}{4}x \qquad \text{Apply the distributive property.}$$

$$6x + 4 = 3x \qquad \text{Solve the resulting equation.}$$

$$3x = -4$$

$$x = -\frac{4}{3}$$

$$\underline{\text{Check:}} \qquad \frac{1}{2}x + \frac{1}{3} = \frac{1}{4}x$$

$$\frac{1}{2}\left(-\frac{4}{3}\right) + \frac{1}{3} \overset{?}{=} \frac{1}{4}\left(-\frac{4}{3}\right)$$

$$-\frac{2}{3} + \frac{1}{3} \overset{?}{=} -\frac{1}{3}$$

$$-\frac{1}{3} \overset{?}{=} -\frac{1}{3} \quad \checkmark$$

The solution is $-\frac{4}{3}$.

The same process of clearing fractions is used to solve rational equations when variables are present in the denominator.

Example 2 **Solving a Rational Equation**

Solve the equation. $\frac{3}{5} + \frac{1}{x} = \frac{2}{3}$

Solution:

$$\frac{3}{5} + \frac{1}{x} = \frac{2}{3} \qquad \begin{array}{l}\text{The LCD of all terms in the equation} \\ \text{is } 15x. \text{ Note that in this equation there is a} \\ \text{restriction that } x \neq 0.\end{array}$$

$$15x\left(\frac{3}{5} + \frac{1}{x}\right) = 15x\left(\frac{2}{3}\right) \qquad \text{Multiply by } 15x \text{ to clear fractions.}$$

$$15x \cdot \frac{3}{5} + 15x \cdot \frac{1}{x} = 15x \cdot \frac{2}{3} \qquad \text{Apply the distributive property.}$$

$$9x + 15 = 10x \qquad \text{Solve the resulting equation.}$$

$$15 = x$$

$$\underline{\text{Check: }} x = 15 \qquad \frac{3}{5} + \frac{1}{x} = \frac{2}{3}$$

$$\frac{3}{5} + \frac{1}{(15)} \overset{?}{=} \frac{2}{3}$$

$$\frac{9}{15} + \frac{1}{15} \overset{?}{=} \frac{2}{3}$$

$$\frac{10}{15} \overset{?}{=} \frac{2}{3} \checkmark$$

The solution is 15.

| Example 3 | **Solving a Rational Equation**

Solve the equation. $\quad 3 - \dfrac{6w}{w+1} = \dfrac{6}{w+1}$

Solution:

Skill Practice

Solve the equation.

3. $\dfrac{4}{b+2} + \dfrac{2}{b-2} = \dfrac{8}{b^2-4}$

$$3 - \frac{6w}{w+1} = \frac{6}{w+1}$$
The LCD of all terms in the equation is $w + 1$. Note that in this equation there is a restriction that $w \neq -1$.

$$(w+1)(3) - (w+1)\left(\frac{6w}{w+1}\right) = (w+1)\left(\frac{6}{w+1}\right)$$
Multiply by $(w + 1)$ on both sides to clear fractions.

$$(w+1)(3) - (w+1)\left(\frac{6w}{w+1}\right) = (w+1)\left(\frac{6}{w+1}\right)$$
Apply the distributive property.

$$3w + 3 - 6w = 6$$
Solve the resulting equation.

$$-3w = 3$$

$$w = -1$$

$$\underline{\text{Check:}} \qquad 3 - \frac{6w}{w+1} = \frac{6}{w+1}$$

No solution

$$3 - \frac{6(-1)}{(-1)+1} \overset{?}{=} \frac{6}{(-1)+1}$$

The denominator is 0 for the value of $w = -1$.

Because the value -1 makes the denominator zero in one (or more) of the rational expressions within the equation, the equation is *undefined* for $w = -1$. No other potential solutions exist, so the equation has no solution.

Examples 1–3 show that the steps for solving a rational equation mirror the process of clearing fractions from Section 1.1. However, there is one significant difference. The solutions of a rational equation must be defined in each rational expression in the equation. When $w = -1$ is substituted into the expression $\dfrac{6w}{w+1}$

Answer

3. No solution (The value 2 does not check.)

or $\dfrac{6}{w+1}$, the denominator is zero and the expression is undefined. Therefore, $w = -1$ cannot be a solution to the equation

$$3 - \frac{6w}{w+1} = \frac{6}{w+1}$$

The steps for solving a rational equation are summarized as follows.

Solving a Rational Equation

Step 1 Factor the denominators of all rational expressions. Identify any values of the variable for which any expression is undefined.

Step 2 Identify the LCD of all terms in the equation.

Step 3 Multiply both sides of the equation by the LCD.

Step 4 Solve the resulting equation.

Step 5 Check the potential solutions in the original equation. Note that any value from step 1 for which the equation is undefined cannot be a solution to the equation.

Skill Practice

Solve.

4. $\dfrac{y}{4} - \dfrac{1}{2} = \dfrac{2}{y}$

Example 4 Solving a Rational Equation

Solve the equation. $\quad 1 + \dfrac{3}{x} = \dfrac{28}{x^2}$

Solution:

$$1 + \frac{3}{x} = \frac{28}{x^2} \qquad \text{The LCD of all terms in the equation is } x^2.$$
$$\text{Expressions will be undefined for } x = 0.$$

$$x^2\left(1 + \frac{3}{x}\right) = x^2\left(\frac{28}{x^2}\right) \qquad \text{Multiply both sides by } x^2 \text{ to clear fractions.}$$

$$x^2 \cdot 1 + x^2 \cdot \frac{3}{x} = x^2 \cdot \frac{28}{x^2} \qquad \text{Apply the distributive property.}$$

$$x^2 + 3x = 28 \qquad \text{The resulting equation is quadratic.}$$

$$x^2 + 3x - 28 = 0 \qquad \text{Set the equation equal to zero and factor.}$$

$$(x + 7)(x - 4) = 0$$

$$x = -7 \quad \text{or} \quad x = 4$$

$$\underline{\text{Check: } x = -7} \qquad\qquad \underline{\text{Check: } x = 4}$$

$$1 + \frac{3}{x} = \frac{28}{x^2} \qquad\qquad 1 + \frac{3}{x} = \frac{28}{x^2}$$

$$1 + \frac{3}{-7} \overset{?}{=} \frac{28}{(-7)^2} \qquad\qquad 1 + \frac{3}{4} \overset{?}{=} \frac{28}{(4)^2}$$

$$\frac{49}{49} - \frac{21}{49} \overset{?}{=} \frac{28}{49} \qquad\qquad \frac{16}{16} + \frac{12}{16} \overset{?}{=} \frac{28}{16}$$

$$\frac{28}{49} \overset{?}{=} \frac{28}{49} \; \checkmark \qquad\qquad \frac{28}{16} \overset{?}{=} \frac{28}{16} \; \checkmark \quad \text{(True)}$$

Answer

4. $4, -2$

The solutions are -7 and 4.

Example 5 Solving a Rational Equation

Solve. $\dfrac{36}{p^2 - 9} = \dfrac{2p}{p + 3} - 1$

Solution:

$$\dfrac{36}{p^2 - 9} = \dfrac{2p}{p + 3} - 1$$

$$\dfrac{36}{(p + 3)(p - 3)} = \dfrac{2p}{p + 3} - 1$$

The LCD is $(p + 3)(p - 3)$.
Expressions will be undefined for $p = 3$ and $p = -3$.

Multiply both sides by the LCD to clear fractions.

$$(p + 3)(p - 3)\left[\dfrac{36}{(p + 3)(p - 3)}\right] = (p + 3)(p - 3)\left(\dfrac{2p}{p + 3}\right) - (p + 3)(p - 3)1$$

$$\cancel{(p + 3)(p - 3)}\left[\dfrac{36}{\cancel{(p + 3)(p - 3)}}\right] = \cancel{(p + 3)}(p - 3)\left(\dfrac{2p}{\cancel{p + 3}}\right) - (p + 3)(p - 3)1$$

$$36 = 2p(p - 3) - (p + 3)(p - 3)$$ Solve the resulting equation.

$$36 = 2p^2 - 6p - (p^2 - 9)$$ The equation is quadratic.

$$36 = 2p^2 - 6p - p^2 + 9$$

$$36 = p^2 - 6p + 9$$

$$0 = p^2 - 6p - 27$$ Set the equation equal to zero and factor.

$$0 = (p - 9)(p + 3)$$

$$p = 9 \quad \text{or} \quad p = -3$$

<u>Check:</u> $p = 9$

$$\dfrac{36}{p^2 - 9} = \dfrac{2p}{p + 3} - 1$$

$$\dfrac{36}{(9)^2 - 9} \stackrel{?}{=} \dfrac{2(9)}{(9) + 3} - 1$$

$$\dfrac{36}{72} \stackrel{?}{=} \dfrac{18}{12} - 1$$

$$\dfrac{1}{2} \stackrel{?}{=} \dfrac{3}{2} - 1$$

$$\dfrac{1}{2} \stackrel{?}{=} \dfrac{1}{2} \ \checkmark \ \text{(True)}$$

<u>Check:</u> $p = -3$

$$\dfrac{36}{p^2 - 9} = \dfrac{2p}{p + 3} - 1$$

$$\dfrac{36}{(-3)^2 - 9} \stackrel{?}{=} \dfrac{2(-3)}{(-3) + 3} - 1$$

Denominator is zero.

Here the value -3 is *not* a solution to the original equation because it is undefined in the original equation. However, 9 checks in the original equation.

The solution is 9.

Answer

5. -9, (The value -2 does not check.)

2. Formulas Involving Rational Equations

Example 6 Solving a Literal Equation Involving Rational Expressions

Solve for the indicated variable. $V = \dfrac{mv}{m + M}$ for *m*

Solution:

$$V = \frac{mv}{m + M} \qquad \text{for } m$$

$$V(m + M) = \left(\frac{mv}{m + M}\right)(m + M) \qquad$$ Multiply by the LCD and clear fractions.

$$V(m + M) = mv$$

$$Vm + VM = mv \qquad$$ Use the distributive property to clear parentheses.

$$Vm - mv = -VM \qquad$$ Collect all *m* terms on one side.

$$m(V - v) = -VM \qquad$$ Factor out *m*.

$$\frac{m(V - v)}{(V - v)} = \frac{-VM}{(V - v)} \qquad$$ Divide by $(V - v)$.

$$m = \frac{-VM}{V - v}$$

TIP: The factor of −1 that appears in the numerator may be written in the denominator or out in front of the expression. The following expressions are equivalent:

$$m = \frac{-VM}{V - v} \qquad \text{or}$$

$$\frac{VM}{-(V - v)} = \frac{VM}{v - V} \qquad \text{or} \qquad -\frac{VM}{V - v}$$

Section 6.5 Practice Exercises

Vocabulary and Key Concepts

1. a. The equation $\dfrac{5}{x + 2} + \dfrac{1}{2} = \dfrac{4}{5}$ is an example of a _____ equation.

b. After solving a rational equation, check each potential solution to determine if it makes the _____ equal to zero in one or more of the rational expressions. If so, that value is not part of the solution.

c. Given $\dfrac{3}{2x + 1} + \dfrac{36}{2x^2 - 7x - 4} = \dfrac{4}{x - 4}$, is it possible for 4 to be a solution to the equation?

Review Exercises

For Exercises 2–7, perform the indicated operations.

2. $\dfrac{1}{x^2 - 16} + \dfrac{1}{x^2 + 8x + 16}$

3. $\dfrac{3}{y^2 - 1} - \dfrac{2}{y^2 - 2y + 1}$

4. $\dfrac{m^2 - 9}{m^2 - 3m} \div (m^2 - m - 12)$

5. $\dfrac{2t^2 + 7t + 3}{4t^2 - 1} \div (t + 3)$

6. $\dfrac{1 + x^{-1}}{1 - x^{-2}}$

7. $\dfrac{x + y}{x^{-1} + y^{-1}}$

Concept 1: Solving Rational Equations

8. Why is it important to check your answer when solving a rational equation?

For Exercises 9–40, solve the rational equations. **(See Examples 1–5.)**

9. $\dfrac{x + 2}{3} - \dfrac{x - 4}{4} = \dfrac{1}{2}$

10. $\dfrac{x + 6}{3} - \dfrac{x + 8}{5} = 0$

11. $\dfrac{3y}{4} - 2 = \dfrac{5y}{6}$

12. $\dfrac{2w}{5} - 8 = \dfrac{4w}{2}$

13. $\dfrac{5}{4p} - \dfrac{7}{6} + 3 = 0$

14. $\dfrac{7}{15w} - \dfrac{3}{10} - 2 = 0$

15. $\dfrac{1}{2} - \dfrac{3}{2x} = \dfrac{4}{x} - \dfrac{5}{12}$

16. $\dfrac{2}{3x} + \dfrac{1}{4} = \dfrac{11}{6x} - \dfrac{1}{3}$

17. $\dfrac{3}{x - 4} + 2 = \dfrac{5}{x - 4}$

18. $\dfrac{5}{x + 3} - 2 = \dfrac{7}{x + 3}$

19. $\dfrac{1}{3} + \dfrac{2}{w - 3} = 1$

20. $\dfrac{3}{5} + \dfrac{7}{p + 2} = 2$

21. $\dfrac{12}{x} - \dfrac{12}{x - 5} = \dfrac{2}{x}$

22. $\dfrac{25}{y} - \dfrac{25}{y - 2} = \dfrac{2}{y}$

23. $\dfrac{3}{a^2} - \dfrac{4}{a} = -1$

24. $\dfrac{3}{w^2} = 2 + \dfrac{1}{w}$

25. $\dfrac{1}{4}a - 4a^{-1} = 0$

26. $\dfrac{1}{3}t - 12t^{-1} = 0$

27. $\dfrac{y}{y + 3} + \dfrac{2}{y^2 + 3y} = \dfrac{6}{y}$

28. $\dfrac{-8}{t^2 - 6t} + \dfrac{t}{t - 6} = \dfrac{1}{t}$

29. $\dfrac{4}{t - 2} - \dfrac{8}{t^2 - 2t} = -2$

30. $\dfrac{x}{x + 6} = \dfrac{72}{x^2 - 36} + 4$

31. $\dfrac{6}{5y + 10} - \dfrac{1}{y - 5} = \dfrac{4}{y^2 - 3y - 10}$

32. $\dfrac{-3}{x^2 - 7x + 12} - \dfrac{2}{x^2 + x - 12} = \dfrac{10}{x^2 - 16}$

33. $\dfrac{x}{x - 5} + \dfrac{1}{5} = \dfrac{5}{x - 5}$

34. $\dfrac{x}{x - 2} + \dfrac{2}{3} = \dfrac{2}{x - 2}$

35. $\dfrac{6}{x^2 - 4x + 3} - \dfrac{1}{x - 3} = \dfrac{1}{4x - 4}$

36. $\dfrac{1}{4x^2 - 36} - \dfrac{5}{x + 3} + \dfrac{2}{x - 3} = 0$

37. $\dfrac{1}{k + 2} - \dfrac{4}{k - 2} - \dfrac{k^2}{4 - k^2} = 0$

38. $\dfrac{h}{2} - \dfrac{h}{h - 4} = \dfrac{4}{4 - h}$

39. $\dfrac{5}{x^2 - 7x + 12} = \dfrac{2}{x - 3} + \dfrac{5}{x - 4}$

40. $\dfrac{9}{x^2 + 7x + 10} = \dfrac{5}{x + 2} - \dfrac{3}{x + 5}$

Concept 2: Formulas Involving Rational Equations

For Exercises 41–58, solve the formula for the indicated variable. **(See Example 6.)**

41. $K = \dfrac{ma}{F}$ for m

42. $K = \dfrac{ma}{F}$ for a

43. $K = \dfrac{IR}{E}$ for E

44. $K = \dfrac{IR}{E}$ for R

45. $I = \dfrac{E}{R + r}$ for R

46. $I = \dfrac{E}{R + r}$ for r

47. $h = \dfrac{2A}{B + b}$ for B

48. $\dfrac{V}{\pi h} = r^2$ for h

49. $x = \dfrac{at + b}{t}$ for t

50. $\dfrac{T + mf}{m} = g$ for m

51. $\dfrac{x - y}{xy} = z$ for x

52. $\dfrac{w - n}{wn} = P$ for w

53. $a + b = \dfrac{2A}{h}$ for h

54. $1 + rt = \dfrac{A}{P}$ for P

55. $\dfrac{1}{R} = \dfrac{1}{R_1} + \dfrac{1}{R_2}$ for R

56. $\dfrac{b + a}{ab} = \dfrac{1}{f}$ for b

57. $v = \dfrac{s_2 - s_1}{t_2 - t_1}$ for t_2

58. $a = \dfrac{v_2 - v_1}{t_2 - t_1}$ for v_1

Mixed Exercises

For Exercises 59–72, solve the equations.

59. $\dfrac{3}{x + 2} + \dfrac{2}{x} = \dfrac{-4}{x^2 + 2x}$

60. $\dfrac{1}{y^2 - 3y} + \dfrac{8}{y} = \dfrac{2}{y - 3}$

61. $4c(c + 1) = 3(c^2 + 4)$

62. $3t(2t - 2) = 5(t^2 - 1)$

63. $\dfrac{2}{v - 1} - \dfrac{4}{v + 5} = \dfrac{3}{v^2 + 4v - 5}$

64. $\dfrac{-2}{a + 4} - \dfrac{3}{a - 5} = \dfrac{6}{a^2 - a - 20}$

65. $5(x - 9) = 3(x + 4) - 2(4x + 1)$

66. $4z - 3(5z - 3) = z - 12$

67. $\dfrac{3y}{10} - \dfrac{5}{2y} = \dfrac{y}{5}$

68. $\dfrac{2h}{3} - \dfrac{8}{3h} = \dfrac{h}{2}$

69. $\dfrac{1}{2}(4d - 1) + \dfrac{2}{3}(2d + 2) = \dfrac{5}{6}(4d + 1)$

70. $\dfrac{2}{5}(2b + 5) + \dfrac{1}{10}(7b - 10) = \dfrac{1}{2}(3b + 2)$

71. $8t^{-1} + 2 = 3t^{-1}$

72. $6z^{-2} - 5z^{-1} = 0$

Expanding Your Skills

73. Find the value of y so that the slope of the line between the points $(3, 1)$ and $(11, y)$ is $\frac{1}{2}$.

74. Find the value of x so that the slope of the line between the points $(-2, -5)$ and $(x, 10)$ is 3.

75. Find the value of x so that the slope of the line between the points $(4, -2)$ and $(x, 2)$ is 4.

76. Find the value of y so that the slope of the line between the points $(3, 2)$ and $(-1, y)$ is $-\frac{3}{4}$.

Problem Recognition Exercises

Rational Equations Versus Expressions

1. a. Simplify. $\dfrac{3}{w-5} + \dfrac{10}{w^2-25} - \dfrac{1}{w+5}$

  b. Solve. $\dfrac{3}{w-5} + \dfrac{10}{w^2-25} - \dfrac{1}{w+5} = 0$

  c. Identify each problem in parts (a) and (b) as either an equation or an expression.

2. a. Simplify. $\dfrac{x}{2x+4} + \dfrac{2}{3x+6} - 1$

  b. Solve. $\dfrac{x}{2x+4} + \dfrac{2}{3x+6} = 1$

  c. Identify each problem in parts (a) and (b) as either an equation or an expression.

For Exercises 3–20, first ask yourself whether the problem is an expression to simplify or an equation to solve. Then simplify or solve as indicated.

3. $\dfrac{2}{a^2+4a+3} + \dfrac{1}{a+3}$

4. $\dfrac{1}{c+6} + \dfrac{4}{c^2+8c+12}$

5. $\dfrac{7}{y^2-y-2} + \dfrac{1}{y+1} - \dfrac{3}{y-2} = 0$

6. $\dfrac{3}{b+2} - \dfrac{1}{b-1} - \dfrac{5}{b^2+b-2} = 0$

7. $\dfrac{x}{x-1} - \dfrac{12}{x^2-x}$

8. $\dfrac{3}{5t-20} + \dfrac{4}{t-4}$

9. $\dfrac{3}{w} - 5 = \dfrac{7}{w} - 1$

10. $\dfrac{-3}{y^2} - \dfrac{1}{y} = -2$

11. $\dfrac{4p+1}{8p-12} + \dfrac{p-3}{2p-3}$

12. $\dfrac{x+1}{2x+4} - \dfrac{x^2}{x+2}$

13. $\dfrac{1}{2x^2} + \dfrac{1}{6x}$

14. $\dfrac{5}{4a} + \dfrac{1}{6a^2}$

15. $\dfrac{3}{2t} + \dfrac{2}{3t^2} = \dfrac{-1}{t}$

16. $\dfrac{-3}{b^2} + \dfrac{1}{5b} = \dfrac{1}{2b}$

17. $\dfrac{3}{c^2+4c+3} - \dfrac{2}{c^2+6c+9}$

18. $\dfrac{1}{y^2-10y+25} - \dfrac{3}{y^2-7y+10}$

19. $\dfrac{4}{w-4} - \dfrac{36}{2w^2-7w-4} = \dfrac{3}{2w+1}$

20. $\dfrac{2}{x-3} - \dfrac{5}{x+2} = \dfrac{25}{x^2-x-6}$

Applications of Rational Equations and Proportions

Concepts

1. Solving Proportions
2. Applications of Proportions
3. Similar Triangles
4. Applications of Rational Equations

1. Solving Proportions

A proportion is a rational equation that equates two ratios.

Definition of Ratio and Proportion

1. The **ratio** of a to b is $\dfrac{a}{b}$ $(b \neq 0)$ and can also be expressed as $a:b$ or $a \div b$.

2. An equation that equates two ratios or rates is called a **proportion**. Therefore, if $b \neq 0$ and $d \neq 0$, then $\dfrac{a}{b} = \dfrac{c}{d}$ is a proportion.

The process for solving rational equations can be used to solve proportions.

Skill Practice

Solve the proportion.

1. $\dfrac{8}{5} = \dfrac{12}{x}$

Example 1 Solving a Proportion

Solve the proportion. $\quad \dfrac{5}{19} = \dfrac{95}{y}$

Solution:

$$\dfrac{5}{19} = \dfrac{95}{y} \qquad \text{The LCD is } 19y. \text{ Note that } y \neq 0.$$

$$19y\left(\dfrac{5}{19}\right) = 19y\left(\dfrac{95}{y}\right) \qquad \text{Multiply both sides by the LCD.}$$

$$\cancel{19}y\left(\dfrac{5}{\cancel{19}}\right) = 19\cancel{y}\left(\dfrac{95}{\cancel{y}}\right) \qquad \text{Clear fractions.}$$

$$5y = 1805 \qquad \text{Solve the resulting equation.}$$

$$\dfrac{5y}{5} = \dfrac{1805}{5}$$

$$y = 361 \qquad \text{The solution 361 checks in the original equation.}$$

TIP: For any proportion

$$\dfrac{a}{b} = \dfrac{c}{d} \qquad b \neq 0, d \neq 0$$

the cross products of terms are equal: $ad = bc$. Finding the cross product is a quick way to clear fractions in a proportion.* Consider Example 1:

$$\dfrac{5}{19} \diagup\!\!\!\!\diagdown \dfrac{95}{y}$$

$$5y = (19)(95) \qquad \text{Equate the cross products.}$$

$$5y = 1805$$

$$y = 361$$

*It is important to realize that this method is only valid for proportions.

Answer

1. $\dfrac{15}{2}$ or 7.5

2. Applications of Proportions

Example 2 Solving a Proportion

The recommended ratio of total cholesterol to HDL cholesterol is 7 to 2. If Rich's blood test revealed that he has a total cholesterol level of 210 mg/dL (milligrams per deciliter), what should his HDL level be to fit within the recommendations?

Solution:

One method of solving this problem is to set up a proportion. Write two equivalent ratios depicting the amount of total cholesterol to HDL cholesterol. Let x represent the unknown amount of HDL cholesterol.

$$\boxed{\begin{array}{c}\text{Given}\\\text{ratio}\end{array}} \longrightarrow \frac{7}{2} = \frac{210}{x} \longleftarrow \boxed{\begin{array}{l}\text{Amount of total cholesterol}\\\text{Amount of HDL cholesterol}\end{array}}$$

$$2x\left(\frac{7}{2}\right) = 2x\left(\frac{210}{x}\right) \qquad \text{Multiply both sides by the LCD } 2x.$$

$$7x = 420 \qquad \text{Clear fractions.}$$

$$x = 60$$

Rich's HDL cholesterol level should be 60 mg/dL to fit within the recommended level.

Skill Practice

2. The ratio of cats to dogs at an animal rescue facility is 8 to 5. How many dogs are in the facility if there are 400 cats?

Example 3 Solving a Proportion

The ratio of male to female police officers in a certain town is 11:3. If the total number of officers is 112, how many are men and how many are women?

Solution:

Let x represent the number of male police officers.

Then $112 - x$ represents the number of female police officers.

$$\boxed{\frac{\text{Male}}{\text{Female}}} \begin{array}{c}\longrightarrow\\\longrightarrow\end{array} \frac{11}{3} = \frac{x}{112 - x} \begin{array}{c}\longleftarrow\\\longleftarrow\end{array} \boxed{\begin{array}{l}\text{Number of males}\\\text{Number of females}\end{array}}$$

$$3(112 - x)\left(\frac{11}{3}\right) = 3(112 - x)\left(\frac{x}{112 - x}\right) \qquad \begin{array}{l}\text{Multiply both sides by}\\3(112 - x).\end{array}$$

$$11(112 - x) = 3x \qquad \begin{array}{l}\text{The resulting equation is}\\\text{linear.}\end{array}$$

$$1232 - 11x = 3x$$

$$1232 = 14x$$

$$\frac{1232}{14} = \frac{14x}{14}$$

$$x = 88$$

The number of male police officers is $x = 88$.

The number of female officers is $112 - x = 112 - 88 = 24$.

Skill Practice

3. Professor Wolfe has a ratio of passing students to failing students of 5 to 4. One semester he had a total of 207 students. How many students passed and how many failed?

Answers

2. 250 dogs
3. 115 passed and 92 failed.

3. Similar Triangles

Proportions are used in geometry with **similar triangles**. Two triangles are similar if their corresponding angles are equal. In such a case, the lengths of the corresponding sides are proportional. In Figure 6-5, triangle ABC is similar to triangle XYZ. Therefore, the following ratios are equivalent.

$$\frac{a}{x} = \frac{b}{y} = \frac{c}{z}$$

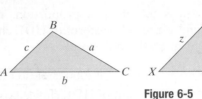

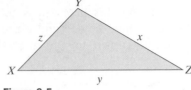

Figure 6-5

Skill Practice

4. Triangle *XYZ* is similar to triangle *ABC*. Solve for *x*.

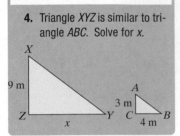

Example 4 **Using Similar Triangles in an Application**

The shadow cast by a yardstick is 2 ft long. The shadow cast by a tree is 11 ft long. Find the height of the tree.

Solution:

Let x represent the height of the tree.

We will assume that the measurements were taken at the same time of day. Therefore, the angle of the sun is the same on both objects, and we can set up similar triangles (Figure 6-6).

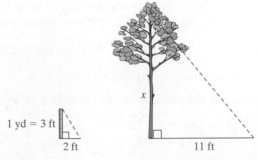

Figure 6-6

Height of yardstick	→	$\dfrac{3 \text{ ft}}{x \text{ ft}} = \dfrac{2}{11 \text{ ft}}$	←	Length of yardstick's shadow
Height of tree	→		←	Length of tree's shadow

$$\frac{3}{x} = \frac{2}{11} \qquad \text{Write an equation.}$$

$$11x \cdot \left(\frac{3}{x}\right) = \cancel{11}x \cdot \left(\frac{2}{\cancel{11}}\right) \qquad \text{Multiply by the LCD.}$$

$$33 = 2x \qquad \text{Solve the equation.}$$

$$16.5 = x \qquad \text{Interpret the results.}$$

The tree is 16.5 ft high.

Answer

4. $x = 12$ m

4. Applications of Rational Equations

Example 5 Solving an Application Involving Distance, Rate, and Time

An athlete's average speed on her bike is 14 mph faster than her average speed running. She can bike 31.5 mi in the same time that it takes her to run 10.5 mi. Find her speed running and her speed biking.

Solution:

Because the speed biking is given in terms of the speed running, let x represent the running speed.

Let x represent the speed running.

Then $x + 14$ represents the speed biking.

Organize the given information in a chart.

	Distance	Rate	Time
Running	10.5	x	$\dfrac{10.5}{x}$
Biking	31.5	$x + 14$	$\dfrac{31.5}{x + 14}$

Because $d = rt$, then $t = \dfrac{d}{r}$

The time required to run 10.5 mi is the same as the time required to bike 31.5 mi, so we can equate the two expressions for time:

$$\frac{10.5}{x} = \frac{31.5}{x + 14}$$ The LCD is $x(x + 14)$.

$$x(x + 14)\left(\frac{10.5}{x}\right) = x(x + 14)\left(\frac{31.5}{x + 14}\right)$$ Multiply by $x(x + 14)$ to clear fractions.

$$10.5(x + 14) = 31.5x$$ The resulting equation is linear.

$$10.5x + 147 = 31.5x$$ Solve for x.

$$-21x = -147$$

$$x = 7$$

The athlete's speed running is 7 mph.

The speed biking is $x + 14$ or $7 + 14 = 21$ mph.

6. Jason drives 50 mi to a train station and then continues his trip with a 210-mi train ride. The car travels 20 mph slower than the train. If the total travel time is 4 hr, find the average speed of the car and the average speed of the train.

Example 6 Solving an Application Involving Distance, Rate, and Time

Valentina travels 70 mi to Rome by train, and then takes a bus 30 mi to the Coliseum. The bus travels 24 mi per hour slower than the train. If the total time traveling on the bus and train is 2 hr, find the average speed of the train and the average speed of the bus.

Solution:

Because the speed of the bus is given in terms of the speed of the train, let x represent the speed of the train.

Let x represent the speed of the train.

Let $x - 24$ represent the speed of the bus.

Organize the given information in a chart.

	Distance	Rate	Time
Train	70	x	$\dfrac{70}{x}$
Bus	30	$x - 24$	$\dfrac{30}{x - 24}$

Fill in the last column with $t = \dfrac{d}{r}$

In this problem, we are given that the total time is 2 hr. So we add the two times to equal 2.

$$\frac{70}{x} + \frac{30}{x - 24} = 2$$

The LCD is $x(x - 24)$.

Avoiding Mistakes

Remember to multiply all terms on both sides of the equation by the LCD.

$$x(x - 24)\left(\frac{70}{x}\right) + x(x - 24)\frac{30}{x - 24} = x(x - 24)(2)$$

Multiply by the LCD to clear the fractions.

$$70(x - 24) + 30x = 2x(x - 24)$$

$$70x - 1680 + 30x = 2x^2 - 48x$$

The resulting equation is quadratic.

$$0 = 2x^2 - 148x + 1680$$

Set the equation equal to 0.

$$0 = 2(x^2 - 74x + 840)$$

Factor.

$$0 = 2(x - 60)(x - 14)$$

Solve for x.

$$x - 60 = 0 \quad \text{or} \quad x - 14 = 0$$

Set each factor equal to 0.

$$x = 60 \quad \text{or} \quad x = 14$$

If $x = 14$, then the rate of the bus would be $14 - 24 = -10$. Because this is not reasonable, the solution is $x = 60$. That is, the average rate of the train is 60 mph and the average rate of the bus is $60 - 24$ or 36 mph.

Answer

6. The car's average speed is 50 mph and the train's average speed is 70 mph.

Example 7 **Solving an Application Involving "Work"**

Skill Practice

7. Antonio can install a new roof in 4 days. Carlos can install the same size roof in 6 days. How long will it take them to install a roof if they work together?

JoAn can wallpaper a bathroom in 3 hr. Bonnie can wallpaper the same bathroom in 5 hr. How long would it take them if they worked together?

Solution:

Let x represent the time required for both people working together to complete the job.

One method to approach this problem is to determine the portion of the job that each person can complete in 1 hr and extend that rate to the portion of the job completed in x hr.

- JoAn can perform the job in 3 hr. Therefore, she completes $\frac{1}{3}$ of the job in 1 hr and $\frac{1}{3}x$ jobs in x hr.
- Bonnie can perform the job in 5 hr. Therefore, she completes $\frac{1}{5}$ of the job in 1 hr and $\frac{1}{5}x$ jobs in x hr.

	Work Rate	Time	Portion of Job Completed
JoAn	$\frac{1}{3}$ job/hr	x hr	$\frac{1}{3}x$ jobs
Bonnie	$\frac{1}{5}$ job/hr	x hr	$\frac{1}{5}x$ jobs

The sum of the portions of the job completed by each person must equal one whole job:

$$\begin{pmatrix} \text{Portion of job} \\ \text{completed} \\ \text{by JoAn} \end{pmatrix} + \begin{pmatrix} \text{portion of job} \\ \text{completed} \\ \text{by Bonnie} \end{pmatrix} = \begin{pmatrix} 1 \\ \text{whole} \\ \text{job} \end{pmatrix}$$

$$\frac{1}{3}x + \frac{1}{5}x = 1 \qquad \text{The LCD is 15.}$$

$$15\left(\frac{1}{3}x + \frac{1}{5}x\right) = 15(1) \qquad \text{Multiply by the LCD.}$$

$$15 \cdot \frac{1}{3}x + 15 \cdot \frac{1}{5}x = 15 \cdot 1 \qquad \text{Apply the distributive property.}$$

$$5x + 3x = 15 \qquad \text{Solve the resulting linear equation.}$$

$$8x = 15$$

$$x = \frac{15}{8} \qquad \text{or} \qquad x = 1\frac{7}{8}$$

Together JoAn and Bonnie can wallpaper the bathroom in $1\frac{7}{8}$ hr.

Answer

7. $\frac{12}{5}$ days or $2\frac{2}{5}$ days

TIP: An alternative approach to solving a "work" problem is to add rates of speed. In Example 7, we could have set up an equation as follows.

$$\begin{pmatrix} \text{JoAn's} \\ \text{speed} \end{pmatrix} + \begin{pmatrix} \text{Bonnie's} \\ \text{speed} \end{pmatrix} = \begin{pmatrix} \text{Speed working} \\ \text{together} \end{pmatrix}$$

$$\frac{1 \text{ job}}{3 \text{ hr}} + \frac{1 \text{ job}}{5 \text{ hr}} = \frac{1 \text{ job}}{x \text{ hr}}$$

$$\frac{1}{3} + \frac{1}{5} = \frac{1}{x}$$

$$15x\left(\frac{1}{3} + \frac{1}{5}\right) = 15x\left(\frac{1}{x}\right) \qquad \text{Multiply by the LCD, } 15x.$$

$$5x + 3x = 15$$

$$8x = 15$$

$$x = \frac{15}{8} \qquad \text{The time required for both JoAn and Bonnie working together is } 1\frac{7}{8} \text{ hr.}$$

Section 6.6 Practice Exercises

Vocabulary and Key Concepts

1. **a.** An equation that equates two rates or ratios is called a _____.

    b. Given similar triangles, the lengths of corresponding sides are _____.

Review Exercises

For Exercises 2–10, perform the indicated operation and simplify, or solve the equation.

2. $3 - \dfrac{6}{x} = x + 8$

3. $2 + \dfrac{6}{x} = x + 7$

4. $\dfrac{5}{3x - 6} - \dfrac{3}{4x - 8}$

5. $\dfrac{4}{5t - 1} + \dfrac{1}{10t - 2}$

6. $\dfrac{2}{y - 1} - \dfrac{5}{4} = \dfrac{-1}{y + 1}$

7. $\dfrac{5}{w - 2} = 7 - \dfrac{10}{w + 2}$

8. $\dfrac{ab}{6} \div \dfrac{a^2}{12} \cdot \dfrac{a + 1}{ab + b}$

9. $\dfrac{8p^2 - 32}{p^2 - 4p + 4} \cdot \dfrac{3p^2 - 3p - 6}{2p^2 + 20p + 32}$

10. $\dfrac{3t}{t + 6} = t + \dfrac{2t}{t + 6}$

Concept 1: Solving Proportions

For Exercises 11–26, solve the proportions. **(See Example 1.)**

11. $\dfrac{y}{6} = \dfrac{20}{15}$

12. $\dfrac{12}{18} = \dfrac{14}{x}$

13. $\dfrac{9}{75} = \dfrac{m}{50}$

14. $\dfrac{n}{15} = \dfrac{12}{45}$

15. $\dfrac{p - 1}{4} = \dfrac{p + 3}{3}$

16. $\dfrac{q - 5}{2} = \dfrac{q + 2}{3}$

17. $\dfrac{x + 1}{5} = \dfrac{4}{15}$

18. $\dfrac{t - 1}{7} = \dfrac{2}{21}$

19. $\dfrac{5 - 2x}{x} = \dfrac{1}{4}$ **20.** $\dfrac{2y + 3}{y} = \dfrac{3}{2}$ 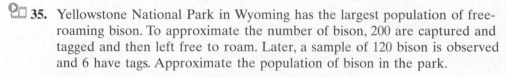 **21.** $\dfrac{2}{y - 1} = \dfrac{y - 3}{4}$ **22.** $\dfrac{1}{x - 5} = \dfrac{x - 3}{3}$

23. $\dfrac{1}{49w} = \dfrac{w}{9}$ **24.** $\dfrac{1}{4z} = \dfrac{z}{25}$ **25.** $\dfrac{x + 3}{5x + 26} = \dfrac{2}{x + 4}$ **26.** $\dfrac{-2}{x - 2} = \dfrac{x - 3}{8x + 11}$

Concept 2: Applications of Proportions

27. A preschool advertises that it has a 3-to-1 ratio of children to adults. If 18 children are enrolled, how many adults must be on the staff? (See Example 2.)

28. An after-school care facility tries to maintain a 4-to-1 ratio of children to adults. If the facility hired five adults, what is the maximum number of children that can enroll?

29. A 3.5-oz box of candy has a total of 21.0 g of fat. How many grams of fat would a 14-oz box of candy contain?

30. A 6-oz box of candy has 350 calories. How many calories would a 10-oz box contain?

31. A fisherman in the North Atlantic catches eight swordfish for a total of 1840 lb. How many swordfish were caught if a commercial fishing boat arrives in port with 230,000 lb of swordfish?

32. If a 64-fl-oz bottle of laundry detergent costs $4.00, how much would an 80-fl-oz bottle cost?

33. Pam drives her Toyota Prius 243 mi in city driving on 4.5 gal of gas. At this rate how many gallons of gas are required to drive 621 mi?

34. On a map, the distance from Sacramento, California, to San Francisco, California, is 8 cm. The legend gives the actual distance as 96 mi. On the same map, Fatima measured 7 cm from Sacramento to Modesto, California. What is the actual distance?

35. Yellowstone National Park in Wyoming has the largest population of free-roaming bison. To approximate the number of bison, 200 are captured and tagged and then left free to roam. Later, a sample of 120 bison is observed and 6 have tags. Approximate the population of bison in the park.

36. Laws have been instituted in Florida to help save the manatee. To establish the number of manatees in Florida, 150 manatees were tagged. A new sample was taken later, and among the 40 manatees in the sample, 3 were tagged. Approximate the number of manatees in Florida.

37. The ratio of men to women accountants in a large accounting firm is 2 to 1. If the total number of accountants is 81, how many are men and how many are women? (See Example 3.)

38. The ratio of Hank's income spent on rent to his income spent on car payments is 3 to 1. If he spends a total of $1640 per month on the rent and car payment, how much does he spend on each item?

39. The ratio of single men in their 20s to single women in their 20s is 119 to 100. (*Source:* U.S. Census) In a random group of 1095 single college students in their 20s, how many are men and how many are women?

40. A chemist mixes water and alcohol in a 7 to 8 ratio. If she makes a 450-L solution, how much is water and how much is alcohol?

Concept 3: Similar Triangles

For Exercises 41–44, triangle ABC is similar to triangle XYZ. Find the lengths of the missing sides. **(See Example 4.)**

41.

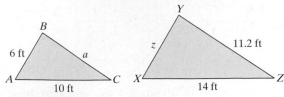

42.

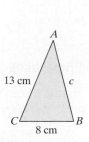

43.

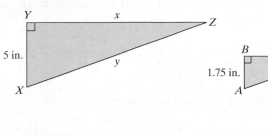

44.

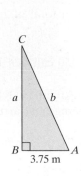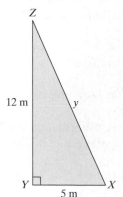

Concept 4: Applications of Rational Equations

45. If 5 is added to the reciprocal of a number, the result is $\frac{16}{3}$. Find the number.

46. If $\frac{2}{3}$ is added to the reciprocal of a number, the result is $\frac{17}{3}$. Find the number.

47. If 7 is decreased by the reciprocal of a number, the result is $\frac{9}{2}$. Find the number.

48. If a number is added to its reciprocal, the result is $\frac{13}{6}$. Find the number.

For Exercises 49 and 50, use the fact that distance = (rate)(time).

49. A truck travels 7 mph faster than a car. Let x represent the speed of the car.

   a. Write an expression for the speed of the truck.

   b. Write an expression for the time it takes the car to travel 48 mi.

   c. Write an expression for the time it takes the truck to travel 83 mi.

50. A car travels 4 mph slower than a motorcycle. Let x represent the speed of the motorcycle.

   a. Write an expression for the speed of the car.

   b. Write an expression for the time it takes the motorcycle to travel 50 mi.

   c. Write an expression for the time it takes the car to travel 145 mi.

51. A motorist travels 80 mi while driving in a bad rainstorm. In sunny weather, the motorist drives 20 mph faster and covers 120 mi in the same amount of time. Find the speed of the motorist in the rainstorm and the speed in sunny weather. **(See Example 5.)**

52. Brooke walks 2 km/hr slower than her older sister Adrianna. If Brooke can walk 12 km in the same amount of time that Adrianna can walk 18 km, find their speeds.

53. The current in a stream is 2 mph. Find the speed of a boat in still water if it goes 26 mi downstream (with the current) in the same amount of time it takes to go 18 mi upstream (against the current).

54. Kathy can run 3 mi to the beach in the same time Dennis can ride his bike 7 mi to work. Kathy runs 8 mph slower than Dennis rides his bike. Find their speeds.

55. A bicyclist rides 30 mi against a wind and returns 30 mi with the wind. His average speed for the return trip is 5 mph faster. How fast did the cyclist ride against the wind if the total time of the trip was 5 hr? **(See Example 6.)**

56. A boat travels 60 mi to an island and 60 mi back again. Changes in the wind and tide made the average speed on the return trip 3 mph slower than the speed on the way out. If the total time of the trip took 9 hr, find the speed going to the island and the speed of the return trip.

57. Celeste walked 140 ft on a moving walkway at the airport. Then she walked on the ground for 100 ft. She travels 2 ft/sec faster on the walkway than she does on the ground. If the time it takes her to travel the total distance of 240 ft is 40 sec, how fast does Celeste travel on and off the moving walkway?

58. Julio rides his bike for 6 mi and gets a flat tire. Then he has to walk with the bike for another mile. His speed walking is 6 mph less than his speed riding the bike. If the total time is 1 hr, find his speed riding the bike and his speed walking.

59. Beatrice does professional triathlons. She runs 2 mph faster than her friend Joe, a weekend athlete. If they each run 12 mi, Beatrice finishes 30 min ($\frac{1}{2}$ hr) ahead of Joe. Determine how fast each person runs.

60. A bus leaves a terminal at 9:00. A car leaves 3 hr later and averages a speed 21 mph faster than that of the bus. If the car overtakes the bus after 196 mi, find the average speed of the bus and the average speed of the car.

61. One painter can paint a room in 6 hr. Another painter can paint the same room in 8 hr. How long would it take them working together? **(See Example 7.)**

62. Karen can wax her SUV in 2 hr. Clarann can wax the same SUV in 3 hr. If they work together, how long will it take them to wax the SUV?

63. A new housing development offers fenced-in yards that all have the same dimensions. Joel can fence a yard in 12 hr, and Michael can fence a yard in 15 hr. How long will it take if they work together?

64. Ted can change an advertisement on a billboard in 4 hr. Marie can do the same job in 5 hr. How long would it take them if they worked together?

65. One carpenter can complete a kitchen in 8 days. With the help of another carpenter, they can do the job together in 4 days. How long would it take the second carpenter if he worked alone?

66. It takes 5 min to fill the tub of a washing machine with hot water only. If it takes 3 min to fill the tub with both hot and cold water, how long would it take to fill the tub with just cold water?

67. Gus works twice as fast as Sid. Together they can dig a garden in 4 hr. How long would it take each person working alone?

68. It takes a child 3 times longer to vacuum a house than an adult. If it takes 1 hr for one adult and one child working together to vacuum a house, how long would it take each person working alone?

Section 6.7   Variation

Concepts

1. Definition of Direct and Inverse Variation
2. Translations Involving Variation
3. Applications of Variation

1. Definition of Direct and Inverse Variation

In this section, we introduce the concept of variation. Direct and inverse variation models can show how one quantity varies in proportion to another.

Direct and Inverse Variation

Let k be a nonzero constant real number. Then,

1. y varies **directly** as x. $\left.\right\}$ $y = kx$
   y is directly proportional to x.

2. y varies **inversely** as x. $\left.\right\}$ $y = \dfrac{k}{x}$
   y is inversely proportional to x.

Note: The value of k is called the constant of variation.

For a car traveling at 30 mph, the equation $d = 30t$ indicates that the distance traveled is *directly proportional* to the time of travel. For positive values of k, when two values are directly related, as one value increases, the other value will also increase. Likewise, if one value decreases, the other will decrease. In the equation $d = 30t$, the longer the time of the trip, the greater the distance traveled. The shorter the time of the trip, the shorter the distance traveled.

For positive values of k, when two positive values are *inversely related*, as one value increases, the other will decrease, and vice versa. Consider a car traveling between Toronto and Montreal, a distance of 500 km. The time required to make the trip is inversely proportional to the speed of travel: $t = 500/r$. As the rate of speed r increases, the quotient $500/r$ will decrease. Hence, the time will decrease. Similarly, as the rate of speed decreases, the trip will take longer.

2. Translations Involving Variation

The first step in using a variation model is to translate an English phrase into an equivalent mathematical equation.

Example 1   Translating to a Variation Model

Translate each expression into an equivalent mathematical model.

a. The circumference of a circle varies directly as the radius.

b. At a constant temperature, the volume of a gas varies inversely as the pressure.

c. The length of time of a meeting is directly proportional to the *square* of the number of people present.

Solution:

a. Let C represent circumference and r represent radius. The variables are directly related, so use the model $C = kr$.

b. Let V represent volume and P represent pressure. Because the variables are inversely related, use the model $V = \dfrac{k}{P}$.

c. Let t represent time and let N be the number of people present at a meeting. Because t is directly related to N^2, use the model $t = kN^2$.

Skill Practice

Translate to a variation model.
1. The time t it takes to drive a particular distance is inversely proportional to the speed s.
2. The amount of your paycheck P varies directly with the number of hours h that you work.
3. q varies inversely as the square of t.

Sometimes a variable varies directly as the product of two or more other variables. In this case, we have joint variation.

Joint Variation

Let k be a nonzero constant real number. Then,

$$\left.\begin{array}{l} y \text{ varies } \textbf{jointly} \text{ as } w \text{ and } z. \\ y \text{ is jointly proportional to } w \text{ and } z. \end{array}\right\} \quad y = kwz$$

Example 2   Translating to a Variation Model

Translate each expression into an equivalent mathematical model.

a. y varies jointly as u and the square root of v.

b. The gravitational force of attraction between two planets varies jointly as the product of their masses and inversely as the square of the distance between them.

Solution:

a. $y = ku\sqrt{v}$

b. Let m_1 and m_2 represent the masses of the two planets. Let F represent the gravitational force of attraction and d represent the distance between the planets. The variation model is $F = \dfrac{km_1 m_2}{d^2}$.

Skill Practice

Translate to a variation model.
4. a varies jointly as b and c.
5. x varies directly as the square root of y and inversely as z.

Answers

1. $t = \dfrac{k}{s}$ 2. $P = kh$

3. $q = \dfrac{k}{t^2}$ 4. $a = kbc$

5. $x = \dfrac{k\sqrt{y}}{z}$

3. Applications of Variation

Consider the variation models $y = kx$ and $y = k/x$. In either case, if values for x and y are known, we can solve for k. Once k is known, we can use the variation equation to find y if x is known, or to find x if y is known. This concept is the basis for solving many problems involving variation.

Finding a Variation Model

Step 1 Write a general variation model that relates the variables given in the problem. Let k represent the constant of variation.

Step 2 Solve for k by substituting known values of the variables into the model from step 1.

Step 3 Substitute the value of k into the original variation model from step 1.

Example 3 Solving an Application Involving Direct Variation

The variable z varies directly as w. When w is 16, z is 56.

a. Write a variation model for this situation. Use k as the constant of variation.

b. Solve for the constant of variation.

c. Find the value of z when w is 84.

Solution:

a. $z = kw$

b. $z = kw$

$56 = k(16)$ Substitute known values for z and w. Then solve for the unknown value of k.

$\dfrac{56}{16} = \dfrac{k(\cancel{16})}{\cancel{16}}$ To isolate k, divide both sides by 16.

$\dfrac{7}{2} = k$ Simplify $\dfrac{56}{16}$ to $\dfrac{7}{2}$.

c. With the value of k known, the variation model can now be written as $z = \dfrac{7}{2}w$.

$z = \dfrac{7}{2}(84)$ To find z when $w = 84$, substitute $w = 84$ into the equation.

$z = 294$

Example 4   Solving an Application Involving Direct Variation

The speed of a racing canoe in still water varies directly as the square root of the length of the canoe.

a. If a 16-ft canoe can travel 6.2 mph in still water, find a variation model that relates the speed of a canoe to its length.

b. Find the speed of a 25-ft canoe.

Solution:

a. Let s represent the speed of the canoe and L represent the length. The general variation model is $s = k\sqrt{L}$. To solve for k, substitute the known values for s and L.

$$s = k\sqrt{L}$$

$$6.2 = k\sqrt{16} \qquad \text{Substitute } s = 6.2 \text{ mph and } L = 16 \text{ ft.}$$

$$6.2 = k \cdot 4$$

$$\frac{6.2}{4} = \frac{4k}{4} \qquad \text{Solve for } k.$$

$$k = 1.55$$

$$s = 1.55\sqrt{L} \qquad \text{Substitute } k = 1.55 \text{ into the model } s = k\sqrt{L}.$$

b. $s = 1.55\sqrt{L}$

$$= 1.55\sqrt{25} \qquad \text{Find the speed when } L = 25 \text{ ft.}$$

$$= 7.75 \text{ mph}$$

Skill Practice

The amount of water needed by a mountain hiker varies directly as the time spent hiking. The hiker needs 2.4 L for a 3-hr hike.

9. Write a model that relates the amount of water needed to the time of the hike.

10. How much water will be needed for a 5-hr hike?

Animation

Example 5   Solving an Application Involving Inverse Variation

The loudness of sound measured in decibels (dB) varies inversely as the square of the distance between the listener and the source of the sound. If the loudness of sound is 17.92 dB at a distance of 10 ft from a stereo speaker, what is the decibel level 20 ft from the speaker?

Solution:

Let L represent the loudness of sound in decibels and d represent the distance in feet. The inverse relationship between decibel level and the square of the distance is modeled by

$$L = \frac{k}{d^2}$$

$$17.92 = \frac{k}{(10)^2} \qquad \text{Substitute } L = 17.92 \text{ dB and } d = 10 \text{ ft.}$$

$$17.92 = \frac{k}{100}$$

$$(17.92)100 = \frac{k}{100} \cdot 100 \qquad \text{Solve for } k \text{ (clear fractions).}$$

$$k = 1792$$

$$L = \frac{1792}{d^2} \qquad \text{Substitute } k = 1792 \text{ into the original model } L = \frac{k}{d^2}.$$

Skill Practice

11. The yield on a bond varies inversely as the price. The yield on a particular bond is 4% when the price is $100. Find the yield when the price is $80.

Answers
9. $w = 0.8t$ **10.** 4 L   **11.** 5%

With the value of k known, we can find L for any value of d.

$$L = \frac{1792}{(20)^2}$$ Find the loudness when $d = 20$ ft.

$$= 4.48 \text{ dB}$$

Notice that the loudness of sound is 17.92 dB at a distance 10 ft from the speaker. When the distance from the speaker is increased to 20 ft, the decibel level decreases to 4.48 dB. This is consistent with an inverse relationship. For $k > 0$, as one variable is increased, the other is decreased. It also seems reasonable that the farther one moves away from the source of a sound, the softer the sound becomes.

Example 6   Solving an Application Involving Joint Variation

In the early morning hours of August 29, 2005, Hurricane Katrina plowed into the Gulf Coast of the United States, bringing unprecedented destruction to southern Louisiana, Mississippi, and Alabama. The winds of a hurricane are strong enough to send a piece of plywood through a tree.

The kinetic energy of an object varies jointly as the weight of the object at sea level and as the square of its velocity. During a hurricane, a 0.5-lb stone traveling at 60 mph has 81 joules (J) of kinetic energy. Suppose the wind speed doubles to 120 mph. Find the kinetic energy.

Solution:

Let E represent the kinetic energy, let w represent the weight, and let v represent the velocity of the stone. The variation model is

$$E = kwv^2$$

$$81 = k(0.5)(60)^2$$ Substitute $E = 81$ J, $w = 0.5$ lb, and $v = 60$ mph.

$$81 = k(0.5)(3600)$$ Simplify exponents.

$$81 = k(1800)$$

$$\frac{81}{1800} = \frac{k(\cancel{1800})}{\cancel{1800}}$$ Divide by 1800.

$$0.045 = k$$ Solve for k.

With the value of k known, the model $E = kwv^2$ can be written as $E = 0.045wv^2$. We now find the kinetic energy of a 0.5-lb stone traveling at 120 mph.

$$E = 0.045(0.5)(120)^2$$

$$= 324$$

The kinetic energy of a 0.5-lb stone traveling at 120 mph is 324 J.

In Example 6, when the velocity increased by 2 times, the kinetic energy increased by 4 times (note that 324 J $= 4 \cdot 81$ J). This factor of 4 occurs because the kinetic energy is proportional to the *square* of the velocity. When the velocity increased by 2 times, the kinetic energy increased by 2^2 times.

Section 6.7 Practice Exercises

Study Skills Exercise

It is not too early to think about your final exam. Write the page number of the cumulative review for Chapters 1–6. Make this exercise set part of your homework this week.

Vocabulary and Key Concepts

1. a. Let k be a nonzero constant. If y varies directly as x, then $y =$ _____, where k is the constant of variation.

b. Let k be a nonzero constant. If y varies inversely as x, then $y =$ _____, where k is the constant of variation.

c. Let k be a nonzero constant. If y varies jointly as x and w, then $y =$ _____, where k is the constant of variation.

d. If y varies directly as x and the constant of variation is positive, then as x increases y (increases/decreases).

e. For $x > 0, y > 0$, if y varies inversely as x and the constant of variation is positive, then as x increases y (increases/decreases).

Review Exercises

For Exercises 2–7, solve the proportion.

2. $\dfrac{x}{4} = \dfrac{13}{10}$

3. $\dfrac{8}{y} = \dfrac{6}{11}$

4. $\dfrac{3}{8} = \dfrac{w + 2}{6}$

5. $\dfrac{2}{3} = \dfrac{x - 4}{2}$

6. $\dfrac{p - 5}{p} = \dfrac{2}{7}$

7. $\dfrac{k}{k + 1} = \dfrac{1}{9}$

Concept 1: Definition of Direct and Inverse Variation

8. In the equation $r = kt$, does r vary directly or inversely with t?

9. In the equation $w = \frac{k}{v}$, does w vary directly or inversely with v?

10. a. Given $y = 12x$, if x increases, then y will (increase/decrease).

b. Given $y = \frac{12}{x}$, if $x > 0$ increases, then y will (increase/decrease).

Concept 2: Translations Involving Variation

For Exercises 11–22, write a variation model. Use k as the constant of variation. **(See Examples 1 and 2.)**

11. T varies directly as q.

12. W varies directly as z.

13. b varies inversely as c.

14. m varies inversely as t.

15. Q is directly proportional to x and inversely proportional to y.

16. d is directly proportional to p and inversely proportional to n.

17. c varies jointly as s and t.

18. w varies jointly as p and f.

19. L varies jointly as w and the square root of v.

20. q varies jointly as v and the square root of w.

21. x varies directly as the square of y and inversely as z.

22. a varies directly as n and inversely as the square of d.

Concept 3: Applications of Variation

For Exercises 23–28, find the constant of variation k. **(See Example 3.)**

23. y varies directly as x, and when x is 4, y is 18.

24. m varies directly as x, and when x is 8, m is 22.

25. p is inversely proportional to q, and when q is 16, p is 32.

26. T is inversely proportional to x, and when x is 40, T is 200.

27. y varies jointly as w and v. When w is 50 and v is 0.1, y is 8.75.

28. N varies jointly as t and p. When t is 1 and p is 7.5, N is 330.

Solve Exercises 29–40 by using the steps found on page 472. **(See Examples 3 and 4.)**

29. x varies directly as p. If $x = 50$ when $p = 10$, find x when p is 14.

30. y is directly proportional to z. If $y = 12$ when $z = 36$, find y when z is 21.

31. b is inversely proportional to c. If b is 4 when c is 3, find b when $c = 2$.

32. q varies inversely as w. If q is 8 when w is 50, find q when w is 125.

33. Z varies directly as the square of w, and $Z = 14$ when $w = 4$. Find Z when $w = 8$.

34. m varies directly as the square of x. If $m = 200$ when $x = 20$, find m when x is 32.

35. Q varies inversely as the square of p, and $Q = 4$ when $p = 3$. Find Q when $p = 2$.

36. z is inversely proportional to the square of t. If $z = 15$ when $t = 4$, find z when $t = 10$.

37. L varies jointly as a and the square root of b, and $L = 72$ when $a = 8$ and $b = 9$. Find L when $a = \frac{1}{2}$ and $b = 36$.

38. Y varies jointly as the cube of x and the square root of w, and $Y = 128$ when $x = 2$ and $w = 16$. Find Y when $x = \frac{1}{2}$ and $w = 64$.

39. B varies directly as m and inversely as n, and $B = 20$ when $m = 10$ and $n = 3$. Find B when $m = 15$ and $n = 12$.

40. R varies directly as s and inversely as t, and $R = 14$ when $s = 2$ and $t = 9$. Find R when $s = 4$ and $t = 3$.

For Exercises 41–58, use a variation model to solve for the unknown value.

41. The amount of medicine that a physician prescribes for a patient varies directly as the weight of the patient. A physician prescribes 3 grams (g) of a medicine for a 150-lb person.

   a. How many grams should be prescribed for a 180-lb person?

   b. How many grams should be prescribed for a 225-lb person?

   c. How many grams should be prescribed for a 120-lb person?

42. The number of people that a turkey can serve varies directly as the weight of the turkey. A 15-lb turkey can serve 10 people.

   a. How many people will a 12-lb turkey serve?

   b. How many people will an 18-lb turkey serve?

   c. How many people will a 21-lb turkey serve?

43. The unit cost of producing CDs is inversely proportional to the number of CDs produced. If 5000 CDs are produced, the cost per CD is $0.48.

   a. What would be the unit cost if 6000 CDs were produced?

   b. What would be the unit cost if 8000 CDs were produced?

   c. What would be the unit cost if 2400 CDs were produced?

44. An author self-publishes a book and finds that the number of books she can sell per month varies inversely as the price of the book. The author can sell 1500 books per month when the price is set at $8 per book.

   a. How many books would she expect to sell if the price were $12?

   b. How many books would she expect to sell if the price were $15?

   c. How many books would she expect to sell if the price were $6?

45. The amount of pollution entering the atmosphere varies directly as the number of people living in an area. If 80,000 people create 56,800 tons of pollutants, how many tons enter the atmosphere in a city with a population of 500,000?

46. The area of a picture projected on a wall varies directly as the square of the distance from the projector to the wall. If a 10-ft distance produces a 16-ft^2 picture, what is the area of a picture produced when the projection unit is moved to a distance 20 ft from the wall?

47. The intensity of light from a light source varies inversely as the square of the distance from the source. If the intensity of a light bulb is 400 lumens/m^2 (lux) at a distance of 5 m, determine the intensity at 8 m. **(See Example 5.)**

48. The frequency of a vibrating string is inversely proportional to its length. A 24-in. piano string vibrates at 252 cycles/sec. What is the frequency of an 18-in. piano string?

49. The current in a wire varies directly as the voltage and inversely as the resistance. If the current is 9 amperes (A) when the voltage is 90 volts (V) and the resistance is 10 ohms (Ω), find the current when the voltage is 185 V and the resistance is 10 Ω.

50. The resistance of a wire varies directly as its length and inversely as the square of its diameter. A 40-ft wire with 0.1-in. diameter has a resistance of 4 Ω. What is the resistance of a 50-ft wire with a diameter of 0.20 in.?

51. The amount of simple interest earned in an account varies jointly as the amount of principal invested and the amount of time the money is invested. If $2500 in principal earns $500 in interest after 4 yr, then how much interest will be earned on $7000 invested for 10 yr? **(See Example 6.)**

52. The amount of simple interest earned in an account varies jointly as the amount of principal invested and the amount of time the money is invested. If $6000 in principal earns $840 in interest after 2 yr, then how much interest will be earned on $4500 invested for 8 yr?

53. The stopping distance of a car is directly proportional to the square of the speed of the car. If a car traveling at 40 mph has a stopping distance of 109 ft, find the stopping distance of a car that is traveling at 25 mph. (Round your answer to one decimal place.)

54. The weight of a medicine ball varies directly as the cube of its radius. A ball with a radius of 3 in. weighs 4.32 lb. How much would a medicine ball weigh if its radius were 5 in.?

55. The surface area of a cube varies directly as the square of the length of an edge. The surface area is 24 ft^2 when the length of an edge is 2 ft. Find the surface area of a cube with an edge that is 5 ft.

56. The period of a pendulum is the length of time required to complete one swing back and forth. The period varies directly as the square root of the length of the pendulum. If it takes 1.8 sec for a 0.81-m pendulum to complete one period, what is the period of a 1-m pendulum?

57. The power in an electric circuit varies jointly as the resistance and the square of the current. If the power is 96 watts (W) when the current is 4 A and the resistance is 6 Ω, find the power when the current is 3 A and the resistance is 10 Ω.

58. The strength of a wooden beam varies jointly as the width of the beam and the square of the thickness of the beam and inversely as the length of the beam. A beam that is 48 in. long, 6 in. wide, and 2 in. thick can support a load of 417 lb. Find the maximum load that can be safely supported by a board that is 12 in. wide, 72 in. long, and 4 in. thick.

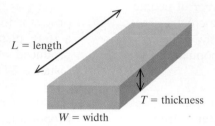

Expanding Your Skills

59. The area A of a square varies directly as the square of the length l of its sides.

   a. Write a general variation model with k as the constant of variation.

   b. If the length of the sides is doubled, what effect will that have on the area?

   c. If the length of the sides is tripled, what effect will that have on the area?

60. In a physics laboratory, a spring is fixed to the ceiling. With no weight attached to the end of the spring, the spring is said to be in its equilibrium position. As weights are applied to the end of the spring, the force stretches the spring a distance d from its equilibrium position. A student in the laboratory collects the following data:

Force F (lb)	2	4	6	8	10
Distance d (cm)	2.5	5.0	7.5	10.0	12.5

   a. Based on the data, do you suspect a direct relationship between force and distance or an inverse relationship?

   b. Find a variation model that describes the relationship between force and distance.

Group Activity

Computing the Future Value of an Investment

Materials: Calculator

Estimated time: 15 minutes

Group Size: 3

Suppose you are able to save $100 per month. If you invest the money in an account that pays 6% annual interest, how much money would you have at the end of 10 yr? This question can be answered by using the following formula.

$$S = R\left[\frac{(1 + i)^n - 1}{i}\right]$$ 　Where　

S is the future value of the investment.
R is the amount saved per period.
i is the interest rate per period.
n is the total number of periods.

In this example, $R = \$100$ 　　　　(amount invested per month)

$$i = \frac{0.06}{12} = 0.005$$ 　　　　(annual interest rate divided by 12 months)

$$n = (12)(10) = 120$$ 　　　　(12 months per year times 10 years)

Therefore, $S = \$100\left[\dfrac{(1 + 0.005)^{120} - 1}{0.005}\right]$

$$S = \$16,387.93$$

1. Compute the future value of an account if you save $150 per month for 30 yr at an annual interest rate of 6%.

   $R =$ _____

   $i =$ _____　　　　　$S =$ _____

   $n =$ _____

2. Suppose you invest $5000 in an account on July 1 each year for 20 yr. If the annual growth on your investment is 8%, how much will the account be worth in 20 yr?

   $R =$ _____

   $i =$ _____　　　　　$S =$ _____

   $n =$ _____

3. Suppose you need to accumulate $50,000 in 15 yr to pay for your child's college tuition. You decide to invest the money in a mutual fund and the average rate of return is 9% per year. How much should you save each month to reach your goal? Round up to the nearest dollar.

Chapter 6 Summary

Section 6.1 Rational Expressions and Rational Functions

Key Concepts

A **rational expression** is in the form $\dfrac{p}{q}$, where p and q are polynomials and $q \neq 0$.

Restricted values of a rational expression are all values that make the expression undefined, that is, make the denominator equal to zero.

To simplify a rational expression to lowest terms, factor the numerator and denominator completely. Then simplify the factors whose ratio is 1 or -1. A rational expression will still have the same restricted values as the original expression.

A **rational function** is a function of the form $\dfrac{p(x)}{q(x)}$, where $p(x)$ and $q(x)$ are polynomial functions and $q(x) \neq 0$.

The **domain** of a rational function is the set of all real numbers excluding the restricted values.

Examples

Example 1

Identify the restricted values. $\dfrac{x - 3}{(x + 4)(2x - 1)}$

$(x + 4)(2x - 1) = 0$

$x + 4 = 0$ or $2x - 1 = 0$

The restricted values are $x = -4$ and $x = \dfrac{1}{2}$.

Example 2

Simplify to lowest terms.

$$\dfrac{t^2 - 6t - 16}{5t + 10}$$

$$= \dfrac{(t - 8)}{5} \cdot \dfrac{\overset{1}{\cancel{(t + 2)}}}{\cancel{(t + 2)}}$$

$$= \dfrac{(t - 8)}{5} \cdot (1) = \dfrac{t - 8}{5} \qquad \text{provided } t \neq -2$$

Example 3

Find the domain of the function.

$$f(x) = \dfrac{x - 5}{(x + 2)(x - 8)} \, .$$

Domain: $(-\infty, -2) \cup (-2, 8) \cup (8, \infty)$

Section 6.2	**Multiplication and Division of Rational Expressions**

Key Concepts

To multiply rational expressions, factor the numerators and denominators completely. Then simplify factors whose ratio is 1 or -1. Multiply the remaining factors in the numerator, and multiply the remaining factors in the denominator.

To divide rational expressions, multiply by the reciprocal of the divisor.

Examples

Example 1

$$\frac{b^2 - a^2}{a^2 - 2ab + b^2} \cdot \frac{a^2 - 3ab + 2b^2}{2a + 2b}$$

$$= \frac{\overset{-1}{(b - a)}\overset{1}{(b + a)}}{(a - b)(a - b)} \cdot \frac{(a - 2b)\overset{1}{(a - b)}}{2(a + b)} \qquad \text{Factor.}$$

$$= -\frac{a - 2b}{2} \qquad \text{or} \qquad \frac{2b - a}{2} \qquad \text{Simplify.}$$

Example 2

$$\frac{9x + 3}{x^2 - 4} \div \frac{3x + 1}{4x + 8}$$

$$= \frac{3\overset{1}{(3x + 1)}}{(x - 2)(x + 2)} \cdot \frac{4\overset{1}{(x + 2)}}{3x + 1} \qquad \begin{array}{l}\text{Factor. Multiply by} \\ \text{the reciprocal.}\end{array}$$

$$= \frac{12}{x - 2} \qquad \text{Simplify.}$$

Section 6.3   Addition and Subtraction of Rational Expressions

Key Concepts

To add or subtract rational expressions, the expressions must have the same denominator.

   The **least common denominator (LCD)** is the product of unique factors from the denominators, in which each factor is raised to its highest power.

Steps to Add or Subtract Rational Expressions

1. Factor the denominator of each rational expression.
2. Identify the LCD.
3. Rewrite each rational expression as an equivalent expression with the LCD as its denominator. [This is accomplished by multiplying the numerator and denominator of each rational expression by the missing factor(s) from the LCD.]
4. Add or subtract the numerators, and write the result over the common denominator.
5. Simplify, if possible.

Examples

Example 1

For $\dfrac{1}{3(x-1)^3(x+2)}$ and $\dfrac{-5}{6(x-1)(x+7)^2}$

LCD $= 6(x-1)^3(x+2)(x+7)^2$

Example 2

$\dfrac{c}{c^2-c-12} - \dfrac{1}{2c-8}$

$= \dfrac{c}{(c-4)(c+3)} - \dfrac{1}{2(c-4)}$ Factor the denominators.

The LCD is $2(c-4)(c+3)$

$\dfrac{2}{2} \cdot \dfrac{c}{(c-4)(c+3)} - \dfrac{1}{2(c-4)} \cdot \dfrac{(c+3)}{(c+3)}$ Write equivalent fractions with LCD.

$= \dfrac{2c-(c+3)}{2(c-4)(c+3)}$ Subtract.

$= \dfrac{2c-c-3}{2(c-4)(c+3)}$ Simplify.

$= \dfrac{c-3}{2(c-4)(c+3)}$

Section 6.4   Complex Fractions

Key Concepts

Complex fractions can be simplified by using Method I or Method II.

Method I uses the order of operations to simplify the numerator and denominator separately before multiplying by the reciprocal of the denominator of the complex fraction.

Examples

Example 1

Simplify by using Method I.

$$\dfrac{\dfrac{5x+15}{7}}{\dfrac{x+3}{2}}$$

$$= \dfrac{5x+15}{7} \div \dfrac{x+3}{2}$$

$$= \dfrac{5x+15}{7} \cdot \dfrac{2}{x+3}$$

$$= \dfrac{5(\overset{1}{\cancel{x+3}})}{7} \cdot \dfrac{2}{\cancel{x+3}}$$

$$= \dfrac{10}{7}$$

To use Method II, multiply the numerator and denominator of the complex fraction by the LCD of all the individual fractions. Then simplify the result.

Example 2

Simplify by using Method II.

$$\dfrac{1 - \dfrac{4}{w^2}}{1 - \dfrac{1}{w} - \dfrac{6}{w^2}} \qquad \text{The LCD is } w^2.$$

$$= \dfrac{w^2\left(1 - \dfrac{4}{w^2}\right)}{w^2\left(1 - \dfrac{1}{w} - \dfrac{6}{w^2}\right)}$$

$$= \dfrac{w^2 - 4}{w^2 - w - 6}$$

$$= \dfrac{(w-2)(\overset{1}{\cancel{w+2}})}{(w-3)(\cancel{w+2})}$$

$$= \dfrac{w-2}{w-3}$$

| Section 6.5 | **Solving Rational Equations** |

Key Concepts

Steps to Solve a Rational Equation

1. Factor the denominators of all rational expressions. Identify any restrictions on the variable.
2. Identify the LCD of all expressions in the equation.
3. Multiply both sides of the equation by the LCD.
4. Solve the resulting equation.
5. Check each potential solution.

Examples

Example 1

$$\frac{1}{w} - \frac{1}{2w - 1} = \frac{-2w}{2w - 1}$$

The LCD is $w(2w - 1)$.

$$w(2w - 1)\frac{1}{w} - w(2w - 1) \cdot \frac{1}{2w - 1}$$

$$= w(2w - 1) \cdot \frac{-2w}{2w - 1}$$

$$(2w - 1)1 - w(1) = w(-2w)$$

$$2w - 1 - w = -2w^2 \quad \text{(quadratic equation)}$$

$$2w^2 + w - 1 = 0$$

$$(2w - 1)(w + 1) = 0$$

$$w = \frac{1}{2} \quad \text{or} \quad w = -1$$

The solution is -1. (The value $\frac{1}{2}$ does not check.)

| Section 6.6 | **Applications of Rational Equations and Proportions** |

Key Concepts

An equation that equates two **ratios** is called a **proportion**:

$$\frac{a}{b} = \frac{c}{d} \quad \text{provided } b \neq 0, d \neq 0$$

Examples

Example 1

A sample of 85 g of a particular ice cream contains 17 g of fat. How much fat does 324 g of the same ice cream contain?

$$\frac{17 \text{ g fat}}{85 \text{ g ice cream}} = \frac{x \text{ g fat}}{324 \text{ g ice cream}}$$

$$\frac{17}{85} = \frac{x}{324} \quad \text{Multiply by the LCD.}$$

$$(\overset{1}{85} \cdot 324) \cdot \frac{17}{85} = (85 \cdot \overset{1}{324}) \cdot \frac{x}{324}$$

$$5508 = 85x$$

$$x = 64.8 \text{ g}$$

Section 6.7   Variation

Key Concepts

Direct Variation

y varies directly as x.
y is directly proportional to x. $\left.\right\}$ $y = kx$

Inverse Variation

y varies inversely as x.
y is inversely proportional to x. $\left.\right\}$ $y = \dfrac{k}{x}$

Joint Variation

y varies jointly as w and z.
y is jointly proportional to w and z. $\left.\right\}$ $y = kwz$

Steps to Find a Variation Model

1. Write a general variation model that relates the variables given in the problem. Let k represent the constant of variation.

2. Solve for k by substituting known values of the variables into the model from step 1.

3. Substitute the value of k into the original variation model from step 1.

Examples

Example 1

t varies directly as the square root of x.

$t = k\sqrt{x}$

Example 2

W is inversely proportional to the cube of x.

$W = \dfrac{k}{x^3}$

Example 3

y is jointly proportional to x and to the square of z.

$y = kxz^2$

Example 4

C varies directly as the square root of d and inversely as t. If $C = 12$ when d is 9 and t is 6, find C if d is 16 and t is 12.

Step 1: $C = \dfrac{k\sqrt{d}}{t}$

Step 2: $12 = \dfrac{k\sqrt{9}}{6} \Rightarrow 12 = \dfrac{k \cdot 3}{6} \Rightarrow k = 24$

Step 3: $C = \dfrac{24\sqrt{d}}{t} \Rightarrow C = \dfrac{24\sqrt{16}}{12} \Rightarrow C = 8$

Chapter 6   Review Exercises

Section 6.1

1. For the rational expression $\dfrac{t-2}{t+9}$

   a. Evaluate the expression (if possible) for $t = 0, 1, 2, -3, -9$

   b. Identify the restricted values.

2. For the rational expression $\dfrac{k+1}{k-5}$

   a. Evaluate the expression (if possible) for $k = 0, 1, 5, -1, -2$

   b. Identify the restricted values.

3. Let $k(y) = \dfrac{y}{y^2 - 1}$.

  a. Find the function values (if they exist): $k(2)$, $k(0)$, $k(1)$, $k(-1)$, $k\left(\dfrac{1}{2}\right)$.

  b. Identify the domain for k. Write the answer in interval notation.

4. Let $h(x) = \dfrac{x}{x^2 + 1}$.

  a. Find the function values (if they exist): $h(1)$, $h(0)$, $h(-1)$, $h(-3)$, $h\left(\dfrac{1}{2}\right)$.

  b. Identify the domain for h. Write the answer in interval notation.

For Exercises 5–12, simplify the rational expressions.

5. $\dfrac{28a^3b^3}{14a^2b^3}$

6. $\dfrac{25x^2yz^3}{125xyz}$

7. $\dfrac{x^2 - 4x + 3}{x - 3}$

8. $\dfrac{k^2 + 3k - 10}{k^2 - 5k + 6}$

9. $\dfrac{x^3 - 27}{9 - x^2}$

10. $\dfrac{a^4 - 81}{3 - a}$

11. $\dfrac{2t^2 + 3t - 5}{7 - 6t - t^2}$

12. $\dfrac{y^3 - 4y}{y^2 - 5y + 6}$

For Exercises 13–16, write the domain of each function in interval notation. Use that information to match the function with its graph.

13. $f(x) = \dfrac{1}{x - 3}$

14. $m(x) = \dfrac{1}{x + 2}$

15. $k(x) = \dfrac{6}{x^2 - 3x}$

16. $p(x) = \dfrac{-2}{x^2 + 4}$

a.

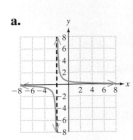

b.

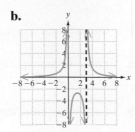

c.

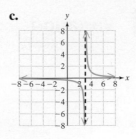

d.

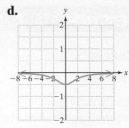

Section 6.2

For Exercises 17–28, multiply or divide as indicated.

17. $\dfrac{3a + 9}{a^2} \cdot \dfrac{a^3}{6a + 18}$

18. $\dfrac{4 - y}{5} \div \dfrac{2y - 8}{15}$

19. $\dfrac{x - 4y}{x^2 + xy} \div \dfrac{20y - 5x}{x^2 - y^2}$

20. $(x^2 + 5x - 24)\left(\dfrac{x + 8}{x - 3}\right)$

21. $\dfrac{7k + 28}{2k + 4} \cdot \dfrac{k^2 - 2k - 8}{k^2 + 2k - 8}$

22. $\dfrac{ab + 2a + b + 2}{ab - 3b + 2a - 6} \cdot \dfrac{ab - 3b + 4a - 12}{ab - b + 4a - 4}$

23. $\dfrac{x^2 + 8x - 20}{x^2 + 6x - 16} \div \dfrac{x^2 + 6x - 40}{x^2 + 3x - 40}$

24. $\dfrac{2b - b^2}{b^3 - 8} \cdot \dfrac{b^2 + 2b + 4}{b^2}$

25. $\dfrac{2w}{21} \div \dfrac{3w^2}{7} \cdot \dfrac{4}{w}$

26. $\dfrac{5y^2 - 20}{y^3 + 8} \div \dfrac{7y^2 - 14y}{y^3 + y}$

27. $\dfrac{x^2 + x - 20}{x^2 - 4x + 4} \cdot \dfrac{x^2 + x - 6}{12 + x - x^2} \div \dfrac{2x + 10}{10 - 5x}$

28. $(9k^2 - 25) \cdot \left(\dfrac{k + 5}{3k - 5}\right)$

Section 6.3

For Exercises 29–40, add or subtract as indicated.

29. $\dfrac{1}{x} + \dfrac{1}{x^2} - \dfrac{1}{x^3}$

30. $\dfrac{1}{x + 2} + \dfrac{5}{x - 2}$

31. $\dfrac{y}{2y - 1} + \dfrac{3}{1 - 2y}$

32. $\dfrac{a + 2}{2a + 6} - \dfrac{3}{a + 3}$

33. $\dfrac{4k}{k^2 + 2k + 1} + \dfrac{3}{k^2 - 1}$

34. $4x + 3 - \dfrac{2x + 1}{x + 4}$

35. $\dfrac{2}{a + 3} + \dfrac{2a^2 - 2a}{a^2 - 2a - 15}$

36. $\dfrac{6}{x^2 + 4x + 3} + \dfrac{7}{x^2 + 5x + 6}$

37. $\dfrac{2}{3x - 5} - 8$

38. $\dfrac{7}{4k^2 - k - 3} + \dfrac{1}{4k^2 - 7k + 3}$

39. $\dfrac{6a}{3a^2 - 7a + 2} + \dfrac{2}{1 - 3a} + \dfrac{3a}{a - 2}$

40. $\dfrac{y}{3 - y} + \dfrac{2y - 5}{y + 2} + 4$

Section 6.4

For Exercises 41–48, simplify the complex fraction.

41. $\dfrac{\dfrac{2x}{3x^2 - 3}}{\dfrac{4x}{6x - 6}}$

42. $\dfrac{\dfrac{k + 2}{3}}{\dfrac{5}{k - 2}}$

43. $\dfrac{\dfrac{2}{x} + \dfrac{1}{xy}}{\dfrac{4}{x^2}}$

44. $\dfrac{\dfrac{4}{y} - 1}{\dfrac{1}{y} - \dfrac{4}{y^2}}$

45. $\dfrac{\dfrac{1}{a - 1} + 1}{\dfrac{1}{a + 1} - 1}$

46. $\dfrac{\dfrac{3}{x - 1} - \dfrac{1}{1 - x}}{\dfrac{2}{x - 1} - \dfrac{2}{x}}$

47. $\dfrac{1 + xy^{-1}}{x^2y^{-2} - 1}$

48. $\dfrac{5a^{-1} + (ab)^{-1}}{3a^{-2}}$

For Exercises 49 and 50, find the slope of the line containing the two points.

49. $\left(\dfrac{2}{3}, -\dfrac{7}{4}\right)$ and $\left(\dfrac{13}{6}, -\dfrac{5}{3}\right)$

50. $\left(\dfrac{8}{15}, -\dfrac{1}{3}\right)$ and $\left(\dfrac{13}{10}, \dfrac{9}{5}\right)$

Section 6.5

For Exercises 51–56, solve the equation.

51. $\dfrac{x + 3}{x^2 - x} - \dfrac{8}{x^2 - 1} = 0$

52. $\dfrac{y}{y + 3} + \dfrac{3}{3 - y} = \dfrac{18}{y^2 - 9}$

53. $x - 9 = \dfrac{72}{x - 8}$ **54.** $\dfrac{3x + 1}{x + 5} = \dfrac{x - 1}{x + 1} + 2$

55. $5y^{-2} + 1 = 6y^{-1}$ **56.** $1 + \dfrac{7}{6}m^{-1} = \dfrac{13}{6}m^{-1}$

57. Solve for x. $c = \dfrac{ax + b}{x}$

58. Solve for P. $\dfrac{A}{rt} = P + \dfrac{P}{rt}$

Section 6.6

For Exercises 59–62, solve the proportions.

59. $\dfrac{5}{4} = \dfrac{x}{6}$ **60.** $\dfrac{x}{36} = \dfrac{6}{7}$

61. $\dfrac{x + 2}{3} = \dfrac{5(x + 1)}{4}$ **62.** $\dfrac{x}{x + 2} = \dfrac{-3}{5}$

63. In one game Peyton Manning completed 34 passes for 357 yd. At this rate how many yards would be gained for 22 passes?

64. Erik bought $108 Canadian with $100 American. At this rate, how many Canadian dollars can he buy with $235 American?

65. Stephen drove his car 45 mi. He ran out of gas and had to walk 3 mi to a gas station. His speed driving is 15 times his speed walking. If the total time for the drive and walk was $1\frac{1}{2}$ hr, what was his speed driving?

66. Two pipes can fill a tank in 6 hr. The larger pipe works twice as fast as the smaller pipe. How long would it take each pipe to fill the tank if they worked separately?

Section 6.7

67. The force applied to a spring varies directly with the distance that the spring is stretched.

  a. Write a variation model using k as the constant of variation.

  b. When 6 lb of force is applied, the spring stretches 2 ft. Find k.

  c. How much force is required to stretch the spring 4.2 ft?

68. Suppose y varies inversely with x and $y = 32$ when $x = 2$. Find y when $x = 4$.

69. Suppose y varies jointly with x and the square root of z, and $y = 3$ when $x = 3$ and $z = 4$. Find y when $x = 8$ and $z = 9$.

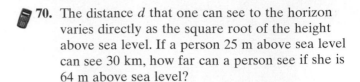 **70.** The distance d that one can see to the horizon varies directly as the square root of the height above sea level. If a person 25 m above sea level can see 30 km, how far can a person see if she is 64 m above sea level?

Chapter 6   Test

1. For the expression $\dfrac{2x + 6}{x^2 - x - 12}$,

  a. Identify the restricted values.

  b. Simplify the expression to lowest terms.

2. For the function $h(x) = \dfrac{2x - 14}{x^2 - 49}$,

  a. Evaluate $h(0)$, $h(5)$, $h(7)$, and $h(-7)$, if possible.

  b. Write the domain of h in interval notation.

3. Write the domain of $k(x) = \dfrac{5x - 3}{7}$ in interval notation.

For Exercises 4 and 5, simplify to lowest terms.

4. $\dfrac{12m^3n^7}{18mn^8}$

5. $\dfrac{9x^2 - 9}{3x^2 + 2x - 5}$

6. Find the slope of the line containing the points $\left(\dfrac{1}{12}, -\dfrac{3}{4}\right)$ and $\left(\dfrac{5}{6}, -\dfrac{8}{3}\right)$.

For Exercises 7–13, simplify.

7. $\dfrac{2x - 5}{25 - 4x^2} \cdot (2x^2 - x - 15)$

8. $\dfrac{x^2}{x - 4} - \dfrac{8x - 16}{x - 4}$

9. $\dfrac{4x}{x + 1} + x + \dfrac{2}{x + 1}$

10. $\dfrac{3 + \dfrac{3}{k}}{4 + \dfrac{4}{k}}$

11. $\dfrac{2u^{-1} + 2v^{-1}}{4u^{-3} + 4v^{-3}}$

12. $\dfrac{ax + bx + 2a + 2b}{ax - 3a + bx - 3b} \cdot \dfrac{x - 3}{5 - x} \div \dfrac{x + 2}{ax - 5a}$

13. $\dfrac{3}{x^2 + 8x + 15} - \dfrac{1}{x^2 + 7x + 12} - \dfrac{1}{x^2 + 9x + 20}$

For Exercises 14–16, solve the equation.

14. $\dfrac{7}{z+1} - \dfrac{z-5}{z^2-1} = \dfrac{6}{z}$ **15.** $\dfrac{3}{y^2-9} + \dfrac{4}{y+3} = 1$

16. $\dfrac{4x}{x-4} = 3 + \dfrac{16}{x-4}$

17. Solve for T. $\dfrac{1+Tv}{T} = p$

18. Solve for m_1. $F = \dfrac{Gm_1m_2}{r^2}$

19. If the reciprocal of a number is added to 3 times the number, the result is $\frac{13}{2}$. Find the number.

20. Triangle ABC is similar to triangle XYZ. Find the lengths of the missing sides.

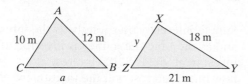

21. On a certain map, the distance between New York and Los Angeles is 8.2 in., and the actual distance is 2820 mi. What is the distance between two cities that are 5.7 in. apart on the same map? Round to the nearest mile.

22. Lance can ride 48 mi on his bike against the wind. With the wind at his back, he rides 4 mph faster and can ride 60 mi in the same amount of time. Find his speed riding against the wind and his speed riding with the wind.

23. Barbara can type a chapter in a book in 4 hr. Jack can type a chapter in a book in 10 hr. How long would it take them to type a chapter if they worked together?

24. Write a variation model using k as the constant of variation. The variable x varies directly as y and inversely as the square of t.

25. The period of a pendulum varies directly as the square root of the length of the pendulum. If the period of the pendulum is 2.2 sec when the length is 4 ft, find the period when the length is 9 ft.

Chapters 1–6   Cumulative Review Exercises

1. Check the sets to which each number belongs.

Set \ Number	-22	π	6	$-\sqrt{2}$
Real numbers				
Irrational numbers				
Rational numbers				
Integers				
Whole numbers				
Natural numbers				

2. Solve the system of equations.

$$3x + 4y = -4$$
$$2x - 8y = -8$$

3. Perform the indicated operations.
$(2x - 3)(x - 4) - (x - 5)^2$

4. The area of a trapezoid is given by
$A = \frac{1}{2}h(b_1 + b_2)$.

  a. Solve for b_1.

  b. Find b_1 when $h = 4$ cm, $b_2 = 6$ cm, and $A = 32$ cm^2.

5. The dimensions of a rectangular swimming pool are such that the length is 10 m less than twice the width. If the perimeter is 160 m, find the length and width.

6. Solve the system of equations.

$$3x - 3y + z = -13$$
$$2x - y - 2z = 4$$
$$x + 2y - 3z = 15$$

7. Find an equation of the line through $(-3, 5)$ that is perpendicular to the line $y = 3x$. Write the answer in slope-intercept form.

8. Graph the line $2x - y = 3$.

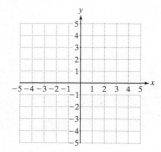

9. The speed of a car varies inversely as the time to travel a fixed distance. A car traveling the speed limit of 60 mph travels between two points in 10 sec. How fast is a car moving if it takes only 8 sec to cover the same distance?

10. Find the x-intercepts. $f(x) = -12x^3 + 17x^2 - 6x$

11. Factor $64y^3 - 8z^6$ completely over the real numbers.

12. Factor. $10x^2 - x - 2$

13. Perform the indicated operations.

$$\frac{2x^2 + 11x - 21}{4x^2 - 10x + 6} \div \frac{2x^2 - 98}{x^2 - x + xa - a}$$

14. Perform the indicated operations.

$$\frac{x}{x^2 + 5x - 50} - \frac{1}{x^2 - 7x + 10} + \frac{1}{x^2 + 8x - 20}$$

15. Simplify the complex fraction.

$$\frac{1 - \dfrac{49}{c^2}}{\dfrac{7}{c} + 1}$$

16. Solve. $y^3 - 5y^2 - y = -5$

17. Solve the equation.

$$\frac{4y}{y + 2} - \frac{y}{y - 1} = \frac{9}{y^2 + y - 2}$$

18. Max knows that the distance between Roanoke, Virginia, and Washington, D.C., is 195 mi. On a certain map, the distance between the two cities is 6.5 in. On the same map, the distance between Roanoke and Cincinnati, Ohio, is 9.25 in. Find the distance in miles between Roanoke and Cincinnati. Round to the nearest mile.

19. Determine whether the equation represents a horizontal or vertical line. Then identify the slope of the line.

    a. $x = -5$ b. $2y = 8$

20. Simple interest varies jointly as the interest rate and as the time the money is invested. If an investment yields $1120 interest at 8% for 2 yr, how much interest will the investment yield at 10% for 5 yr?

Radicals and Complex Numbers

7

Chapter 7

In this chapter, we study radical expressions. This includes operations on square roots, cube roots, fourth roots, and so on. We also revisit the Pythagorean theorem and its applications as well as introduce the set of complex numbers.

Review Your Skills

Take a minute to review your knowledge of square roots and operations on polynomials. These skills will help you be successful in this chapter.

For Exercises 1–10, perform the indicated operations. Then write the letter corresponding to each answer in the space at the bottom of the page.

1. $(4x - 3)(x + 5)$ **2.** $4x^2(3x^2 + 5x + 1)$ **3.** $(4x - 3)(4x + 3)$ **4.** $(4x - 3)^2$

5. $\sqrt{4}$ **6.** 4^2 **7.** $\sqrt{-4}$ **8.** $\sqrt{10^2 - 6^2}$

9. $(-3x^4 + 9x^2 - 15x) - (-3x^4 - 7x^2 + 9x + 9)$ **10.** $(2x)(6x^3)$

R 16	N $16x^2 - 9$	I $16x^2 - 24x + 9$	M 8
Y $16x^2 - 24x - 9$	A not a real number	E $12x^4 + 20x^3 + 4x^2$	U 2
G $4x^2 + 17x - 15$	B $12x^4$		

In this chapter, we also introduce a new type of number called an

$\overline{}\ \overline{}\ \overline{}\ \overline{}\ \overline{}\ \overline{}\ \overline{}\ \overline{}\ \overline{}\ \ \ \overline{}\ \overline{}\ \overline{}\ \overline{}\ \overline{}\ \overline{}$.

4   8   7   1   4   3   7   6   9   3   5   8   10   2   6

Section 7.1 Definition of an *n*th Root

1. Definition of a Square Root

The inverse operation to squaring a number is to find its square roots. For example, finding a square root of 36 is equivalent to asking, "what number when squared equals 36?"

One obvious answer to this question is 6 because $(6)^2 = 36$, but -6 will also work, because $(-6)^2 = 36$.

> **Definition of a Square Root**
>
> b is a **square root** of a if $b^2 = a$.

Skill Practice

Identify the square roots of the real numbers.

1. 64　　2. 16
3. 1　　4. -100

Example 1　Identifying Square Roots

Identify the square roots of the real numbers.

　a. 25　　　**b.** 49　　　**c.** 0　　　**d.** -9

Solution:

　a. 5 is a square root of 25 because $(5)^2 = 25$.

　　　-5 is a square root of 25 because $(-5)^2 = 25$.

　b. 7 is a square root of 49 because $(7)^2 = 49$.

　　　-7 is a square root of 49 because $(-7)^2 = 49$.

　c. 0 is a square root of 0 because $(0)^2 = 0$.

　d. There are no real numbers that when squared will equal a negative number; therefore, there are no real-valued square roots of -9.

> **TIP:** All positive real numbers have two real-valued square roots: one positive and one negative. Zero has only one square root, which is 0 itself. Finally, for any negative real number, there are no real-valued square roots.

Recall from Section R.3 that the positive square root of a real number can be denoted with a **radical sign** $\sqrt{}$.

> **Positive and Negative Square Roots**
>
> Let a represent a positive real number. Then
>
> 　**1.** $\sqrt{a}$ is the *positive* square root of a. The positive square root is also called the **principal square root**.
> 　**2.** $-\sqrt{a}$ is the *negative* square root of a.
> 　**3.** $\sqrt{0} = 0$

Answers

1. -8 and 8
2. -4 and 4
3. -1 and 1
4. No real-valued square roots

Example 2   **Simplifying a Square Root**

Simplify the square roots.

    a. $\sqrt{36}$ **b.** $\sqrt{\dfrac{4}{9}}$ **c.** $\sqrt{0.04}$

Solution:

    a. $\sqrt{36}$ denotes the positive square root of 36.

        $\sqrt{36} = 6$ because $(6)^2 = 36$

    b. $\sqrt{\dfrac{4}{9}}$ denotes the positive square root of $\dfrac{4}{9}$.

        $\sqrt{\dfrac{4}{9}} = \dfrac{2}{3}$ because $\left(\dfrac{2}{3}\right)^2 = \dfrac{4}{9}$

    c. $\sqrt{0.04}$ denotes the positive square root of 0.04.

        $\sqrt{0.04} = 0.2$ because $(0.2)^2 = 0.04$

The numbers 36, $\frac{4}{9}$, and 0.04 are **perfect squares** because their square roots are rational numbers.

Radicals that cannot be simplified to rational numbers are irrational numbers. Recall that an irrational number cannot be written as a terminating or repeating decimal. For example, the symbol $\sqrt{13}$ is used to represent the *exact* value of the square root of 13. The symbol $\sqrt{42}$ is used to represent the *exact* value of the square root of 42. These values can be approximated by a rational number by using a calculator.

$$\sqrt{13} \approx 3.605551275 \qquad \sqrt{42} \approx 6.480740698$$

TIP: Before using a calculator to evaluate a square root, try estimating the value first.

$\sqrt{13}$ must be a number between 3 and 4 because $\sqrt{9} < \sqrt{13} < \sqrt{16}$.
$\sqrt{42}$ must be a number between 6 and 7 because $\sqrt{36} < \sqrt{42} < \sqrt{49}$.

Calculator Connections

Topic: Approximating Square Roots

Use a calculator to approximate the values of $\sqrt{13}$ and $\sqrt{42}$.

```
√(13)
          3.605551275
√(42)
          6.480740698
```

A negative number cannot have a real number as a square root because no real number when squared is negative. For example, $\sqrt{-25}$ is *not* a real number because there is no real number b for which $(b)^2 = -25$.

Example 3 **Evaluating Square Roots**

Simplify the square roots, if possible.

a. $\sqrt{-144}$ **b.** $-\sqrt{144}$ **c.** $\sqrt{-0.01}$ **d.** $-\sqrt{\dfrac{1}{9}}$

Solution:

a. $\sqrt{-144}$ is *not* a real number. No real number when squared equals -144.

b. $-\sqrt{144}$

$\qquad = -1 \cdot \sqrt{144}$

$\qquad \quad\;\; \downarrow \qquad\;\; \downarrow$

$\qquad = -1 \cdot \quad 12$

$\qquad = -12$

> **TIP:** For the expression $-\sqrt{144}$, the factor of -1 is *outside* the radical.

c. $\sqrt{-0.01}$ is *not* a real number. No real number when squared equals -0.01.

d. $-\sqrt{\dfrac{1}{9}}$

$\qquad = -1 \cdot \sqrt{\dfrac{1}{9}}$

$\qquad = -1 \cdot \dfrac{1}{3}$

$\qquad = -\dfrac{1}{3}$

2. Definition of an *n*th Root

Finding a square root of a number is the reverse process of squaring a number. This concept can be extended to finding a third root (called a cube root), a fourth root, and in general an ***n*th root**.

> ### Definition of an *n*th Root
>
> b is an nth root of a if $b^n = a$.
>
> Example: 2 is a square root of 4 because 2^2 is 4.
> Example: 2 is a third root of 8 because 2^3 is 8.
> Example: 2 is a fourth root of 16 because 2^4 is 16.

The radical sign $\sqrt{}$ is used to denote the principal square root of a number. The symbol $\sqrt[n]{}$ is used to denote the principal nth root of a number. In the expression $\sqrt[n]{a}$, n is called the **index** of the radical, and a is called the **radicand**. For a square root, the index is 2, but it is usually not written ($\sqrt[2]{a}$ is denoted simply as $\sqrt{a}$). A radical with an index of 3 is called a **cube root**, denoted by $\sqrt[3]{a}$.

> ### Evaluating $\sqrt[n]{a}$
>
> **1.** If $n > 1$ is an *even* integer and $a > 0$, then $\sqrt[n]{a}$ is the principal (positive) nth root of a. Example: $\sqrt{9} = 3$
> **2.** If $n > 1$ is an *odd* integer, then $\sqrt[n]{a}$ is the nth root of a.
>   Example: $\sqrt[3]{8} = 2$, $\sqrt[3]{-8} = -2$
> **3.** If $n > 1$ is an integer, then $\sqrt[n]{0} = 0$.

For the purpose of simplifying radicals, it is helpful to know the following powers.

Perfect Cubes	Perfect Fourth Powers	Perfect Fifth Powers
$1^3 = 1$	$1^4 = 1$	$1^5 = 1$
$2^3 = 8$	$2^4 = 16$	$2^5 = 32$
$3^3 = 27$	$3^4 = 81$	$3^5 = 243$
$4^3 = 64$	$4^4 = 256$	$4^5 = 1024$
$5^3 = 125$	$5^4 = 625$	$5^5 = 3125$

Example 4 Identifying the *n*th Root of a Real Number

Simplify the expressions, if possible.

a. $\sqrt{4}$ **b.** $\sqrt[3]{64}$ **c.** $\sqrt[5]{-32}$ **d.** $\sqrt[4]{81}$

e. $\sqrt[6]{1,000,000}$ **f.** $\sqrt{-100}$ **g.** $\sqrt[4]{-16}$

Solution:

a. $\sqrt{4} = 2$ because $(2)^2 = 4$

b. $\sqrt[3]{64} = 4$ because $(4)^3 = 64$

c. $\sqrt[5]{-32} = -2$ because $(-2)^5 = -32$

d. $\sqrt[4]{81} = 3$ because $(3)^4 = 81$

e. $\sqrt[6]{1,000,000} = 10$ because $(10)^6 = 1,000,000$

f. $\sqrt{-100}$ is not a real number.   No real number when squared equals -100.

g. $\sqrt[4]{-16}$ is not a real number.   No real number when raised to the fourth power equals -16.

Skill Practice

Simplify if possible.

14. $\sqrt[4]{16}$ **15.** $\sqrt[3]{1000}$

16. $\sqrt[5]{-1}$ **17.** $\sqrt[5]{32}$

18. $\sqrt[5]{100,000}$

19. $\sqrt{-36}$ **20.** $\sqrt[3]{-27}$

Examples 4(f) and 4(g) illustrate that an *n*th root of a negative quantity is not a real number if the index is even. This is because no real number raised to an even power is negative.

Calculator Connections

Topic: Approximating *n*th Roots

A calculator can be used to approximate *n*th roots by using the $\sqrt[x]{}$ function. On most calculators, the index is entered first.

```
3 ˣ√(64)
                    4
5 ˣ√(-32)
                   -2
6 ˣ√(1000000)
                   10
```

3. Roots of Variable Expressions

Finding an *n*th root of a variable expression is similar to finding an *n*th root of a numerical expression. For roots with an even index, however, particular care must be taken to obtain a nonnegative result.

Answers

14. 2   **15.** 10   **16.** -1
17. 2   **18.** 10
19. Not a real number   **20.** -3

> **Evaluating $\sqrt[n]{a^n}$**
>
> 1. If n is a positive *odd* integer, then $\sqrt[n]{a^n} = a$.
> 2. If n is a positive *even* integer, then $\sqrt[n]{a^n} = |a|$.

The absolute value bars are necessary for roots with an even index because the variable a may represent a positive quantity or a negative quantity. By using absolute value bars, $\sqrt[n]{a^n} = |a|$ is nonnegative and represents the principal nth root of a^n.

Example 5   Simplifying Expressions of the Form $\sqrt[n]{a^n}$

Simplify the expressions.

a. $\sqrt[4]{(-3)^4}$ b. $\sqrt[5]{(-3)^5}$ c. $\sqrt{(x + 2)^2}$ d. $\sqrt[3]{(a + b)^3}$ e. $\sqrt{y^4}$

Solution:

a. $\sqrt[4]{(-3)^4} = |-3| = 3$ Because this is an *even*-indexed root, absolute value bars are necessary to make the answer positive.

b. $\sqrt[5]{(-3)^5} = -3$ This is an *odd*-indexed root, so absolute value bars are not necessary.

c. $\sqrt{(x + 2)^2} = |x + 2|$ Because this is an *even*-indexed root, absolute value bars are necessary. The sign of the quantity $x + 2$ is unknown; however, $|x + 2| \geq 0$ regardless of the value of x.

d. $\sqrt[3]{(a + b)^3} = a + b$ This is an *odd*-indexed root, so absolute value bars are not necessary.

e. $\sqrt{y^4} = \sqrt{(y^2)^2}$

$\quad = |y^2|$ Because this is an even-indexed root, use absolute value bars.

$\quad = y^2$ However, because y^2 is nonnegative, the absolute value bars are not necessary.

If n is an even integer, then $\sqrt[n]{a^n} = |a|$; however, if the variable a is assumed to be *nonnegative,* then the absolute value bars may be dropped. That is, $\sqrt[n]{a^n} = a$ provided $a \geq 0$. In many examples and exercises, we will make the assumption that the variables within a radical expression are positive real numbers. In such a case, the absolute value bars are not needed to evaluate $\sqrt[n]{a^n}$.

In Chapter 5 you became familiar with the patterns associated with perfect squares and perfect cubes. In general, any expression raised to an even power is a perfect square. An expression raised to a power that is a multiple of three is a perfect cube.

Perfect Squares	Perfect Cubes
$(x^1)^2 = x^2$	$(x^1)^3 = x^3$
$(x^2)^2 = x^4$	$(x^2)^3 = x^6$
$(x^3)^2 = x^6$	$(x^3)^3 = x^9$
$(x^4)^2 = x^8$	$(x^4)^3 = x^{12}$

These patterns may be extended to higher powers.

| Example 6 | Simplifying *n*th Roots |

Simplify the expressions. Assume that all variables are positive real numbers.

a. $\sqrt{y^8}$ **b.** $\sqrt[3]{27a^3}$ **c.** $\sqrt[5]{\dfrac{a^5}{b^5}}$ **d.** $-\sqrt[4]{\dfrac{81x^4y^8}{16}}$

Solution:

a. $\sqrt{y^8} = \sqrt{(y^4)^2} = y^4$

b. $\sqrt[3]{27a^3} = \sqrt[3]{(3a)^3} = 3a$

c. $\sqrt[5]{\dfrac{a^5}{b^5}} = \sqrt[5]{\left(\dfrac{a}{b}\right)^5} = \dfrac{a}{b}$

d. $-\sqrt[4]{\dfrac{81x^4y^8}{16}} = -\sqrt[4]{\left(\dfrac{3xy^2}{2}\right)^4} = -\dfrac{3xy^2}{2}$

> **Skill Practice**
>
> Simplify the expressions. Assume all variables represent positive real numbers.
>
> **26.** $\sqrt{t^6}$
>
> **27.** $\sqrt[3]{64y^{12}}$
>
> **28.** $\sqrt[4]{\dfrac{x^4}{y^4}}$
>
> **29.** $-\sqrt[5]{\dfrac{32a^5}{b^{10}}}$

4. Pythagorean Theorem

In Section 5.8, we used the Pythagorean theorem in several applications. For the triangle shown in Figure 7-1, the **Pythagorean theorem** may be stated as $a^2 + b^2 = c^2$. In this formula, a and b are the legs of the right triangle, and c is the hypotenuse.

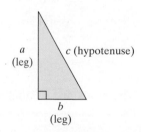

a (leg) c (hypotenuse)

b (leg)

Figure 7-1

| Example 7 | Applying the Pythagorean Theorem |

Use the Pythagorean theorem and the definition of the principal square root to find the length of the unknown side.

Solution:

Label the sides of the triangle.

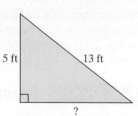

5 ft   13 ft   ?

$a = 5$ ft $c = 13$ ft $b = ?$

$a^2 + b^2 = c^2$ Apply the Pythagorean theorem.

$(5)^2 + b^2 = (13)^2$

$25 + b^2 = 169$ Simplify.

$b^2 = 169 - 25$ Isolate b^2.

$b^2 = 144$

By definition, b must be one of the square roots of 144. Because b represents the length of a side of a triangle, choose the positive square root of 144.

$b = 12$

The third side is 12 ft long.

> **Skill Practice**
>
> **30.** Use the Pythagorean theorem and the definition of the principal square root to find the length of the unknown side of the right triangle.
>
>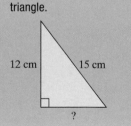
>
> 12 cm   15 cm   ?

Answers

26. t^3 **27.** $4y^4$ **28.** $\dfrac{x}{y}$

29. $-\dfrac{2a}{b^2}$ **30.** 9 cm

Example 8   Applying the Pythagorean Theorem

Two boats leave a dock at 12:00 noon. One travels due north at 6 mph, and the other travels due east at 8 mph (Figure 7-2). How far apart are the two boats after 2 hr?

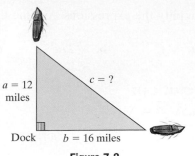

$a = 12$ miles

$c = ?$

Dock $b = 16$ miles

Figure 7-2

Solution:

The boat traveling north travels a distance of $(6 \text{ mph})(2 \text{ hr}) = 12 \text{ mi}$. The boat traveling east travels a distance of $(8 \text{ mph})(2 \text{ hr}) = 16 \text{ mi}$. The course of the boats forms a right triangle where the hypotenuse represents the distance between them.

$$a^2 + b^2 = c^2$$

$$(12)^2 + (16)^2 = c^2 \qquad \text{Apply the Pythagorean theorem.}$$

$$144 + 256 = c^2 \qquad \text{Simplify.}$$

$$400 = c^2$$

$$\sqrt{400} = c \qquad \text{By definition, } c \text{ must be one of the square roots of}$$
400. Choose the positive square root of 400 to represent distance between the two boats.

$$20 = c$$

The boats are 20 mi apart.

5. Radical Functions

If n is an integer greater than 1, then a function written in the form $f(x) = \sqrt[n]{x}$ is called a **radical function**. Note that if n is an even integer, then the function will be a real number only if the radicand is nonnegative. Therefore, the domain is restricted to nonnegative real numbers, or equivalently, $[0, \infty)$. If n is an odd integer, then the domain is all real numbers.

Example 9   Determining the Domain of Radical Functions

For each function, write the domain in interval notation.

a. $g(t) = \sqrt[4]{t - 2}$ **b.** $h(a) = \sqrt[3]{a - 3}$ **c.** $k(x) = \sqrt{3 - 5x}$

Solution:

a. $g(t) = \sqrt[4]{t - 2}$ The index is even. The radicand must be nonnegative.

$t - 2 \geq 0$ Set the radicand greater than or equal to zero.

$t \geq 2$ Solve for t.

The domain is $[2, \infty)$.

b. $h(a) = \sqrt[3]{a - 3}$ The index is odd; therefore, the domain is all real numbers.

The domain is $(-\infty, \infty)$.

c. $k(x) = \sqrt{3 - 5x}$ The index is even; therefore, the radicand must be nonnegative.

$3 - 5x \geq 0$ Set the radicand greater than or equal to zero.

$-5x \geq -3$ Solve for x.

$\dfrac{-5x}{-5} \leq \dfrac{-3}{-5}$ Reverse the inequality sign.

$x \leq \dfrac{3}{5}$

The domain is $\left(-\infty, \frac{3}{5}\right]$.

Calculator Connections

Topic: Graphing Radical Functions

We can graph the functions defined in Example 9. The graphs support the answers we obtained for the domain of each function.

$g(t) = \sqrt[4]{t - 2}$ $h(a) = \sqrt[3]{a - 3}$ $k(x) = \sqrt{3 - 5x}$

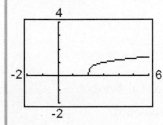 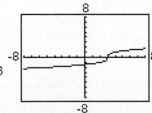 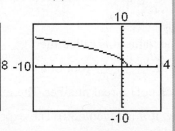

Example 10   Graphing a Radical Function

Given $f(x) = \sqrt{3 - x}$

a. Write the domain of f in interval notation.

b. Graph f by making a table of ordered pairs.

Solution:

a. $f(x) = \sqrt{3 - x}$

$3 - x \geq 0$ The index is even. The radicand must be greater than or equal to zero.

$-x \geq -3$

$x \leq 3$ Reverse the inequality sign.

The domain is $(-\infty, 3]$.

Skill Practice

35. Given $f(x) = \sqrt{x + 4}$

  a. Write the domain of f in interval notation.

  b. Graph f by making a table of ordered pairs.

Answers

35. a. $[-4, \infty)$

  b.

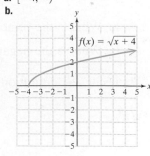

b. Create a table of ordered pairs where x values are taken to be less than or equal to 3.

x	$f(x)$
3	0
2	1
−1	2
−6	3

$f(3) = \sqrt{3 - 3} = \sqrt{0} = 0$

$f(2) = \sqrt{3 - 2} = \sqrt{1} = 1$

$f(-1) = \sqrt{3 - (-1)} = \sqrt{4} = 2$

$f(-6) = \sqrt{3 - (-6)} = \sqrt{9} = 3$

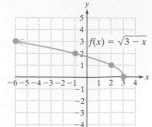

Section 7.1 Practice Exercises

Vocabulary and Key Concepts

1. a. If $b^2 = a$, then _____ is a square root of _____.

 b. The symbol $\sqrt{a}$ denotes the positive or _____ square root of a.

 c. b is an nth root of a if _____ = _____.

 d. Given the symbol $\sqrt[n]{a}$, the value n is called the _____ and a is called the _____.

 e. The symbol $\sqrt[3]{a}$ denotes the _____ root of a.

 f. The expression $\sqrt{-4}$ (is/is not) a real number. The expression $-\sqrt{4}$ (is/is not) a real number.

 g. The expression $\sqrt[n]{a^n} = |a|$ if n is (even/odd). The expression $\sqrt[n]{a^n} = a$ if n is (even/odd).

 h. Given a right triangle with legs a and b and hypotenuse c, the _____ theorem is stated as $a^2 + b^2 =$ _____.

 i. In interval notation, the domain of $f(x) = \sqrt{x}$ is _____, whereas the domain of $g(x) = \sqrt[3]{x}$ is _____.

 j. Which of the following values of x are *not* in the domain of $h(x) = \sqrt{x + 3}$: $x = -5, x = -4, x = -3, x = -2, x = -1, x = 0$?

Concept 1: Definition of a Square Root

2. Simplify the expression $\sqrt[3]{8}$. Explain how you can check your answer.

3. a. Find the square roots of 64. **(See Example 1.)**

 b. Find $\sqrt{64}$.

 c. Explain the difference between the answers in part (a) and part (b).

4. a. Find the square roots of 121.

 b. Find $\sqrt{121}$.

 c. Explain the difference between the answers in part (a) and part (b).

5. a. What is the principal square root of 81?

 b. What is the negative square root of 81?

6. a. What is the principal square root of 100?

 b. What is the negative square root of 100?

7. Using the definition of a square root, explain why $\sqrt{-36}$ is not a real number.

For Exercises 8–19, evaluate the roots without using a calculator. Identify those that are not real numbers. (See Examples 2 and 3.)

8. $\sqrt{25}$ **9.** $\sqrt{49}$ **10.** $-\sqrt{25}$ **11.** $-\sqrt{49}$

12. $\sqrt{-25}$ **13.** $\sqrt{-49}$ **14.** $\sqrt{\dfrac{100}{121}}$ **15.** $\sqrt{\dfrac{64}{9}}$

16. $\sqrt{0.64}$ **17.** $\sqrt{0.81}$ **18.** $-\sqrt{0.0144}$ **19.** $-\sqrt{0.16}$

Concept 2: Definition of an *n*th Root

20. Using the definition of an *n*th root, explain why $\sqrt[4]{-16}$ is not a real number.

For Exercises 21–38, evaluate the roots without using a calculator. Identify those that are not real numbers. (See Example 4.)

21. a. $\sqrt{64}$ **b.** $\sqrt[3]{64}$ **c.** $-\sqrt{64}$
   d. $-\sqrt[3]{64}$ **e.** $\sqrt{-64}$ **f.** $\sqrt[3]{-64}$

22. a. $\sqrt{16}$ **b.** $\sqrt[4]{16}$ **c.** $-\sqrt{16}$
   d. $-\sqrt[4]{16}$ **e.** $\sqrt{-16}$ **f.** $\sqrt[4]{-16}$

23. $\sqrt[3]{-27}$ **24.** $\sqrt[3]{-125}$ **25.** $\sqrt[3]{\dfrac{1}{8}}$ **26.** $\sqrt[5]{\dfrac{1}{32}}$

27. $\sqrt[5]{32}$ **28.** $\sqrt[4]{1}$ **29.** $\sqrt[3]{-\dfrac{125}{64}}$ **30.** $\sqrt[3]{-\dfrac{8}{27}}$

31. $\sqrt[4]{-1}$ **32.** $\sqrt[6]{-1}$ **33.** $\sqrt[6]{1,000,000}$ **34.** $\sqrt[4]{10,000}$

35. $-\sqrt[3]{0.008}$ **36.** $-\sqrt[4]{0.0016}$ **37.** $\sqrt[4]{0.0625}$ **38.** $\sqrt[3]{0.064}$

Concept 3: Roots of Variable Expressions

For Exercises 39–58, simplify the radical expressions. (See Example 5.)

39. $\sqrt{a^2}$ **40.** $\sqrt[4]{a^4}$ **41.** $\sqrt[3]{a^3}$ **42.** $\sqrt[5]{a^5}$

43. $\sqrt[6]{a^6}$ **44.** $\sqrt[7]{a^7}$ **45.** $\sqrt{(x+1)^2}$ **46.** $\sqrt[3]{(y+3)^3}$

47. $\sqrt{x^2y^4}$ **48.** $\sqrt[3]{(u+v)^3}$ **49.** $-\sqrt[3]{\dfrac{x^3}{y^3}}$, $y \neq 0$ **50.** $\sqrt[4]{\dfrac{a^4}{b^8}}$, $b \neq 0$

51. $\dfrac{2}{\sqrt[4]{x^4}}$, $x \neq 0$ **52.** $\sqrt{(-5)^2}$ **53.** $\sqrt[3]{(-92)^3}$ **54.** $\sqrt[6]{(50)^6}$

55. $\sqrt[10]{(-2)^{10}}$ **56.** $\sqrt[5]{(-2)^5}$ **57.** $\sqrt[7]{(-923)^7}$ **58.** $\sqrt[6]{(-417)^6}$

For Exercises 59–74, simplify the expressions. Assume all variables are positive real numbers. (See Example 6.)

59. $\sqrt{y^8}$ **60.** $\sqrt{x^4}$ **61.** $\sqrt{\dfrac{a^6}{b^2}}$ **62.** $\sqrt{\dfrac{w^2}{z^4}}$

63. $-\sqrt{\dfrac{25}{q^2}}$ **64.** $-\sqrt{\dfrac{p^6}{81}}$ **65.** $\sqrt{9x^2y^4z^2}$ **66.** $\sqrt{4a^4b^2c^6}$

67. $\sqrt{\dfrac{h^2k^4}{16}}$ **68.** $\sqrt{\dfrac{4x^2}{y^8}}$ **69.** $-\sqrt[3]{\dfrac{t^3}{27}}$ **70.** $\sqrt[4]{\dfrac{16}{w^4}}$

71. $\sqrt[5]{32y^{10}}$ **72.** $\sqrt[3]{64x^6y^3}$ **73.** $\sqrt[6]{64p^{12}q^{18}}$ **74.** $\sqrt[4]{16r^{12}s^8}$

Concept 4: Pythagorean Theorem

For Exercises 75–78, find the length of the third side of each triangle by using the Pythagorean theorem. **(See Example 7.)**

75.

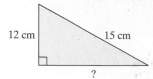

76.

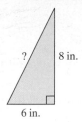

77.

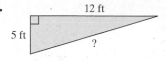

78.

For Exercises 79–82, use the Pythagorean theorem.

79. Roberto and Sherona began running from the same place at the same time. They ran along two different paths that formed right angles with each other. Roberto ran 4 mi and stopped, while Sherona ran 3 mi and stopped. How far apart were they when they stopped? **(See Example 8.)**

80. Leine and Laura began hiking from their campground. Laura headed south while Leine headed east. Laura walked 12 mi and Leine walked 5 mi. How far apart were they when they stopped walking?

81. Two mountain bikers take off from the same place at the same time. One travels north at 4 mph, and the other travels east at 3 mph. How far apart are they after 5 hr?

82. Professor Ortiz leaves campus on her bike, heading west at 6 mph. Professor Wilson leaves campus at the same time and walks south at 2.5 mph. How far apart are they after 4 hr?

Concept 5: Radical Functions

For Exercises 83–86, evaluate the function for the given values of x. Then write the domain of the function in interval notation. **(See Example 9.)**

83. $h(x) = \sqrt{x-2}$ **84.** $k(x) = \sqrt{x+1}$ **85.** $g(x) = \sqrt[3]{x-2}$ **86.** $f(x) = \sqrt[3]{x+1}$

a. $h(0)$ **a.** $k(-3)$ **a.** $g(-6)$ **a.** $f(-9)$

b. $h(1)$ **b.** $k(-2)$ **b.** $g(1)$ **b.** $f(-2)$

c. $h(2)$ **c.** $k(-1)$ **c.** $g(2)$ **c.** $f(0)$

d. $h(3)$ **d.** $k(0)$ **d.** $g(3)$ **d.** $f(7)$

e. $h(6)$ **e.** $k(3)$

For each function defined in Exercises 87–94, write the domain in interval notation. **(See Example 9.)**

87. $f(x) = \sqrt{5-2x}$ **88.** $g(x) = \sqrt{3-4x}$ **89.** $k(x) = \sqrt[3]{4x-7}$ **90.** $R(x) = \sqrt[3]{x+1}$

91. $M(x) = \sqrt{x-5}+3$ **92.** $N(x) = \sqrt{x+3}-1$ **93.** $F(x) = \sqrt[3]{x+7}-2$ **94.** $G(x) = \sqrt[3]{x-10}+4$

For Exercises 95–102,

   a. Write the domain of *f* in interval notation.

   b. Graph *f* by making a table of ordered pairs. **(See Example 10.)**

95. $f(x) = \sqrt{1 - x}$ **96.** $f(x) = \sqrt{2 - x}$ **97.** $f(x) = \sqrt{x + 3}$ **98.** $f(x) = \sqrt{x + 1}$

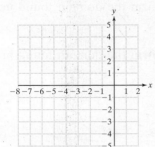 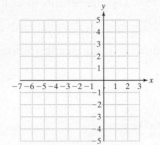 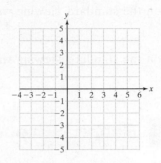 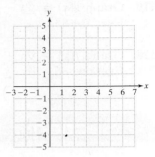

99. $f(x) = \sqrt{x} + 2$ **100.** $f(x) = \sqrt{x} - 1$ **101.** $f(x) = \sqrt[3]{x - 1}$ **102.** $f(x) = \sqrt[3]{x + 2}$

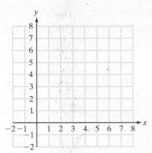 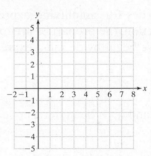 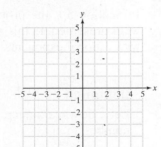 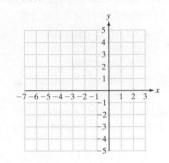

Mixed Exercises

For Exercises 103–106, translate the English phrase to an algebraic expression.

103. The sum of *q* and the square of *p*

104. The product of 11 and the cube root of *x*

105. The quotient of 6 and the cube root of *x*

106. The difference of *y* and the principal square root of *x*

107. If a square has an area of 64 in.², then what are the lengths of the sides?

108. If a square has an area of 121 m², then what are the lengths of the sides?

$s = ?$

$A = 64 \text{ in.}^2$ $s = ?$

$s = ?$

$A = 121 \text{ m}^2$ $s = ?$

Graphing Calculator Exercises

For Exercises 109–116, use a calculator to evaluate the expressions to four decimal places.

109. $\sqrt{69}$ **110.** $\sqrt{5798}$ **111.** $2 + \sqrt[3]{5}$ **112.** $3 - 2\sqrt[4]{10}$

113. $7\sqrt[4]{25}$ **114.** $-3\sqrt[3]{9}$ **115.** $\dfrac{3 - \sqrt{19}}{11}$ **116.** $\dfrac{5 + 2\sqrt{15}}{12}$

117. Graph $h(x) = \sqrt{x - 2}$ on the standard viewing window. Use the graph to confirm the domain found in Exercise 83.

118. Graph $k(x) = \sqrt{x + 1}$ on the standard viewing window. Use the graph to confirm the domain found in Exercise 84.

119. Graph $g(x) = \sqrt[3]{x - 2}$ on the standard viewing window. Use the graph to confirm the domain found in Exercise 85.

120. Graph $f(x) = \sqrt[3]{x + 1}$ on the standard viewing window. Use the graph to confirm the domain found in Exercise 86.

Section 7.2 Rational Exponents

Concepts

1. Definition of $a^{1/n}$ and $a^{m/n}$
2. Converting Between Rational Exponents and Radical Notation
3. Properties of Rational Exponents
4. Applications Involving Rational Exponents

1. Definition of $a^{1/n}$ and $a^{m/n}$

In Section 5.1, the properties for simplifying expressions with integer exponents were presented. In this section, the properties are expanded to include expressions with rational exponents. We begin by defining expressions of the form $a^{1/n}$.

> **Definition of $a^{1/n}$**
>
> Let a be a real number, and let n be an integer such that $n > 1$. If $\sqrt[n]{a}$ is a real number, then
>
> $$a^{1/n} = \sqrt[n]{a}$$

Skill Practice

Convert the expression to radical form and simplify, if possible.

1. $(-64)^{1/3}$
2. $16^{1/4}$
3. $-36^{1/2}$
4. $(-36)^{1/2}$
5. $64^{-1/3}$

Example 1 Evaluating Expressions of the Form $a^{1/n}$

Convert the expression to radical form and simplify, if possible.

a. $(-8)^{1/3}$ **b.** $81^{1/4}$ **c.** $-100^{1/2}$ **d.** $(-100)^{1/2}$ **e.** $16^{-1/2}$

Solution:

a. $(-8)^{1/3} = \sqrt[3]{-8} = -2$

b. $81^{1/4} = \sqrt[4]{81} = 3$

c. $-100^{1/2} = -1 \cdot 100^{1/2}$ The exponent applies only to the base of 100.

$\qquad = -1\sqrt{100}$

$\qquad = -10$

d. $(-100)^{1/2}$ is not a real number because $\sqrt{-100}$ is not a real number.

e. $16^{-1/2} = \dfrac{1}{16^{1/2}}$ Write the expression with a positive exponent.

$\qquad\qquad\qquad$ Recall that $b^{-n} = \dfrac{1}{b^n}$.

$\qquad = \dfrac{1}{\sqrt{16}}$

$\qquad = \dfrac{1}{4}$

Answers

1. -4 2. 2   3. -6
4. Not a real number   5. $\dfrac{1}{4}$

If $\sqrt[n]{a}$ is a real number, then we can define an expression of the form $a^{m/n}$ in such a way that the multiplication property of exponents still holds true. For example:

$$16^{3/4} \nearrow (16^{1/4})^3 = (\sqrt[4]{16})^3 = (2)^3 = 8$$
$$\searrow (16^3)^{1/4} = \sqrt[4]{16^3} = \sqrt[4]{4096} = 8$$

Definition of $a^{m/n}$

Let a be a real number, and let m and n be positive integers such that m and n share no common factors and $n > 1$. If $\sqrt[n]{a}$ is a real number, then

$$a^{m/n} = (a^{1/n})^m = (\sqrt[n]{a})^m \qquad \text{and} \qquad a^{m/n} = (a^m)^{1/n} = \sqrt[n]{a^m}$$

The rational exponent in the expression $a^{m/n}$ is essentially performing two operations. The numerator of the exponent raises the base to the mth power. The denominator takes the nth root.

Example 2 Evaluating Expressions of the Form $a^{m/n}$

Convert each expression to radical form and simplify.

a. $8^{2/3}$　　　**b.** $100^{5/2}$　　　**c.** $\left(\dfrac{1}{25}\right)^{3/2}$　　　**d.** $4^{-3/2}$　　　**e.** $(-81)^{3/4}$

Solution:

a. $8^{2/3} = (\sqrt[3]{8})^2$　　　Take the cube root of 8 and square the result.

　　$= (2)^2$　　　Simplify.

　　$= 4$

b. $100^{5/2} = (\sqrt{100})^5$　　　Take the square root of 100 and raise the result to the fifth power.

　　$= (10)^5$　　　Simplify.

　　$= 100{,}000$

c. $\left(\dfrac{1}{25}\right)^{3/2} = \left(\sqrt{\dfrac{1}{25}}\right)^3$　　　Take the square root of $\dfrac{1}{25}$ and cube the result.

　　$= \left(\dfrac{1}{5}\right)^3$　　　Simplify.

　　$= \dfrac{1}{125}$

Skill Practice

Convert each expression to radical form and simplify.

6. $9^{3/2}$　　　**7.** $8^{5/3}$

8. $32^{-4/5}$　　　**9.** $\left(\dfrac{1}{27}\right)^{4/3}$

10. $(-4)^{3/2}$

Calculator Connections

Topic: Evaluating Expressions with Rational Exponents

A calculator can be used to confirm the results of Example 2(a)–2(c).

```
(8)^(2/3)
                 4
(100)^(5/2)
            100000
(1/25)^(3/2)
              .008
```

Answers

6. 27　　**7.** 32

8. $\dfrac{1}{16}$　**9.** $\dfrac{1}{81}$　**10.** Not a real number

d. $4^{-3/2} = \dfrac{1}{4^{3/2}}$ Write the expression with positive exponents.

$= \dfrac{1}{(\sqrt{4})^3}$ Take the square root of 4 and cube the result.

$= \dfrac{1}{2^3}$ Simplify.

$= \dfrac{1}{8}$

e. $(-81)^{3/4}$ is not a real number because $\sqrt[4]{-81}$ is not a real number.

2. Converting Between Rational Exponents and Radical Notation

Example 3 Using Radical Notation and Rational Exponents

Convert each expression to radical notation. Assume all variables represent positive real numbers.

a. $a^{3/5}$ **b.** $5^{1/3}x^{2/3}$ **c.** $3y^{1/4}$ **d.** $9z^{-3/4}$

Solution:

a. $a^{3/5} = \sqrt[5]{a^3}$ or $\left(\sqrt[5]{a}\right)^3$

b. $5^{1/3}x^{2/3} = (5x^2)^{1/3} = \sqrt[3]{5x^2}$

c. $3y^{1/4} = 3\sqrt[4]{y}$ Note that the coefficient 3 is not raised to the $\frac{1}{4}$ power.

d. $9z^{-3/4} = 9 \cdot \dfrac{1}{z^{3/4}} = \dfrac{9}{z^{3/4}} = \dfrac{9}{\sqrt[4]{z^3}}$ Note that the coefficient 9 has an implied exponent of 1, not $-\frac{3}{4}$.

Example 4 Using Radical Notation and Rational Exponents

Convert each expression to an equivalent expression by using rational exponents. Assume that all variables represent positive real numbers.

a. $\sqrt[4]{b^3}$ **b.** $\sqrt{7a}$ **c.** $7\sqrt{a}$

Solution:

a. $\sqrt[4]{b^3} = b^{3/4}$ **b.** $\sqrt{7a} = (7a)^{1/2}$ **c.** $7\sqrt{a} = 7a^{1/2}$

3. Properties of Rational Exponents

In Section 5.1, several properties and definitions were introduced to simplify expressions with integer exponents. These properties also apply to rational exponents.

Properties of Exponents and Definitions

Let a and b be nonzero real numbers. Let m and n be rational numbers such that a^m, a^n, and b^m are real numbers.

Description	Property	Example
1. Multiplying like bases	$a^m a^n = a^{m+n}$	$x^{1/3} x^{4/3} = x^{5/3}$
2. Dividing like bases	$\dfrac{a^m}{a^n} = a^{m-n}$	$\dfrac{x^{3/5}}{x^{1/5}} = x^{2/5}$
3. The power rule	$(a^m)^n = a^{mn}$	$(2^{1/3})^{1/2} = 2^{1/6}$
4. Power of a product	$(ab)^m = a^m b^m$	$(xy)^{1/2} = x^{1/2} y^{1/2}$
5. Power of a quotient	$\left(\dfrac{a}{b}\right)^m = \dfrac{a^m}{b^m}$	$\left(\dfrac{4}{25}\right)^{1/2} = \dfrac{4^{1/2}}{25^{1/2}} = \dfrac{2}{5}$

Description	Definition	Example
1. Negative exponents	$a^{-m} = \left(\dfrac{1}{a}\right)^m = \dfrac{1}{a^m}$	$(8)^{-1/3} = \left(\dfrac{1}{8}\right)^{1/3} = \dfrac{1}{2}$
2. Zero exponent	$a^0 = 1$	$5^0 = 1$

Example 5 **Simplifying Expressions with Rational Exponents**

Use the properties of exponents to simplify the expressions. Assume all variables represent positive real numbers.

a. $y^{2/5} y^{3/5}$ **b.** $\dfrac{6a^{-1/2}}{a^{3/2}}$ **c.** $\left(\dfrac{s^{1/2} t^{1/3}}{w^{3/4}}\right)^4$

Solution:

a. $y^{2/5} y^{3/5} = y^{(2/5)+(3/5)}$ Multiply like bases by adding exponents.

   $= y^{5/5}$ Simplify.

   $= y$

b. $\dfrac{6a^{-1/2}}{a^{3/2}}$ Divide like bases by subtracting exponents.

   $= 6a^{(-1/2)-(3/2)}$ Simplify: $-\dfrac{4}{2} = -2$

   $= 6a^{-2}$

   $= \dfrac{6}{a^2}$ Simplify the negative exponent.

c. $\left(\dfrac{s^{1/2} t^{1/3}}{w^{3/4}}\right)^4 = \dfrac{s^{(1/2)\cdot 4} t^{(1/3)\cdot 4}}{w^{(3/4)\cdot 4}}$ Apply the power rule. Multiply exponents.

   $= \dfrac{s^2 t^{4/3}}{w^3}$ Simplify.

Skill Practice

Use the properties of exponents to simplify the expressions. Assume all variables represent positive real numbers.

18. $x^{1/2} \cdot x^{3/4}$

19. $\dfrac{4k^{-2/3}}{k^{1/3}}$

20. $\left(\dfrac{a^{1/3} b^{1/2}}{c^{5/8}}\right)^6$

Answers

18. $x^{5/4}$ **19.** $\dfrac{4}{k}$

20. $\dfrac{a^2 b^3}{c^{15/4}}$

4. Applications Involving Rational Exponents

21. The formula for the radius of a sphere is

$$r = \left(\frac{3V}{4\pi}\right)^{1/3}$$

where V is the volume. Find the radius of a sphere whose volume is 113.04 in^3. (Use 3.14 for π.)

> **Example 6** Applying Rational Exponents
>
> Suppose P dollars in principal is invested in an account that earns interest annually. If after t years the investment grows to A dollars, then the annual rate of return r on the investment is given by
>
> $$r = \left(\frac{A}{P}\right)^{1/t} - 1$$
>
> Find the annual rate of return on \$5000, which grew to \$12,500 after 6 yr.
>
> **Solution:**
>
> $$r = \left(\frac{A}{P}\right)^{1/t} - 1$$
>
> $$= \left(\frac{12{,}500}{5000}\right)^{1/6} - 1 \qquad \text{where } A = \$12{,}500, \ P = \$5000, \text{ and } t = 6$$
>
> $$= (2.5)^{1/6} - 1$$
>
> $$\approx 1.165 - 1$$
>
> $$\approx 0.165 \text{ or } 16.5\%$$
>
> The annual rate of return is 16.5%.

Answer

21. The radius is 3 in.

Section 7.2 Practice Exercises

For the exercises in this set, assume that all variables represent positive real numbers unless otherwise stated.

Study Skills Exercise

Before you do your homework for this section, go back to Section 5.1 and review the properties of exponents. Do several problems from the Section 5.1 exercises. This will help you with the concepts in Section 7.2.

Vocabulary and Key Concepts

1. a. If n is an integer greater than 1, then radical notation for $a^{1/n}$ is _____.

  b. If m and n are positive integers that share no common factors and $n > 1$, then in radical notation $a^{m/n} = $ _____.

  c. The radical notation for $x^{-1/2}$ is _____.

  d. $8^{1/3} = $ _____ and $8^{-1/3} = $ _____.

Review Exercises

For Exercises 2–4, identify the index and radicand.

2. $\sqrt[4]{16x}$ **3.** $\sqrt[3]{27}$ **4.** $\sqrt{18}$

For Exercises 5–8, evaluate the radicals.

5. $\sqrt{25}$ **6.** $\sqrt[3]{8}$ **7.** $\sqrt[4]{81}$ **8.** $(\sqrt[4]{16})^3$

Concept 1: Definition of $a^{1/n}$ and $a^{m/n}$

For Exercises 9–20, convert the expressions to radical form and simplify. **(See Example 1.)**

9. $144^{1/2}$ **10.** $16^{1/4}$ **11.** $-144^{1/2}$ **12.** $-16^{1/4}$

13. $(-144)^{1/2}$ **14.** $(-16)^{1/4}$ **15.** $(-64)^{1/3}$ **16.** $(-32)^{1/5}$

17. $25^{-1/2}$ **18.** $27^{-1/3}$ **19.** $-49^{-1/2}$ **20.** $-64^{-1/2}$

21. Explain how to interpret the expression $a^{m/n}$ as a radical.

22. Explain why $(\sqrt[3]{8})^4$ is easier to evaluate than $\sqrt[3]{8^4}$.

For Exercises 23–26, simplify the expression, if possible. **(See Example 2.)**

23. a. $16^{3/4}$ **24. a.** $81^{3/4}$ **25. a.** $25^{3/2}$ **26. a.** $4^{3/2}$

       b. $-16^{3/4}$ **b.** $-81^{3/4}$ **b.** $-25^{3/2}$ **b.** $-4^{3/2}$

       c. $(-16)^{3/4}$ **c.** $(-81)^{3/4}$ **c.** $(-25)^{3/2}$ **c.** $(-4)^{3/2}$

       d. $16^{-3/4}$ **d.** $81^{-3/4}$ **d.** $25^{-3/2}$ **d.** $4^{-3/2}$

       e. $-16^{-3/4}$ **e.** $-81^{-3/4}$ **e.** $-25^{-3/2}$ **e.** $-4^{-3/2}$

       f. $(-16)^{-3/4}$ **f.** $(-81)^{-3/4}$ **f.** $(-25)^{-3/2}$ **f.** $(-4)^{-3/2}$

For Exercises 27–50, simplify the expression. **(See Example 2.)**

27. $64^{-3/2}$ **28.** $81^{-3/2}$ **29.** $243^{3/5}$ **30.** $1^{5/3}$

31. $-27^{-4/3}$ **32.** $-16^{-5/4}$ **33.** $\left(\dfrac{100}{9}\right)^{-3/2}$ **34.** $\left(\dfrac{49}{100}\right)^{-1/2}$

35. $(-4)^{-3/2}$ **36.** $(-49)^{-3/2}$ **37.** $(-8)^{1/3}$ **38.** $(-9)^{1/2}$

39. $-8^{1/3}$ **40.** $-9^{1/2}$ **41.** $\dfrac{1}{36^{-1/2}}$ **42.** $\dfrac{1}{16^{-1/2}}$

43. $\dfrac{1}{1000^{-1/3}}$ **44.** $\dfrac{1}{81^{-3/4}}$ **45.** $\left(\dfrac{1}{8}\right)^{2/3} + \left(\dfrac{1}{4}\right)^{1/2}$ **46.** $\left(\dfrac{1}{8}\right)^{-2/3} + \left(\dfrac{1}{4}\right)^{-1/2}$

47. $\left(\dfrac{1}{16}\right)^{-3/4} - \left(\dfrac{1}{49}\right)^{-1/2}$ **48.** $\left(\dfrac{1}{16}\right)^{1/4} - \left(\dfrac{1}{49}\right)^{1/2}$ **49.** $\left(\dfrac{1}{4}\right)^{1/2} + \left(\dfrac{1}{64}\right)^{-1/3}$ **50.** $\left(\dfrac{1}{36}\right)^{1/2} + \left(\dfrac{1}{64}\right)^{-5/6}$

Concept 2: Converting Between Rational Exponents and Radical Notation

For Exercises 51–58, convert each expression to radical notation. **(See Example 3.)**

51. $q^{2/3}$ **52.** $t^{3/5}$ **53.** $6y^{3/4}$ **54.** $8b^{4/9}$

55. $x^{2/3}y^{1/3}$ **56.** $c^{2/5}d^{3/5}$ **57.** $6r^{-2/5}$ **58.** $7x^{-3/4}$

For Exercises 59–66, write each expression by using rational exponents rather than radical notation. **(See Example 4.)**

59. $\sqrt[3]{x}$ **60.** $\sqrt[4]{a}$ **61.** $10\sqrt{b}$ **62.** $-2\sqrt[3]{t}$

63. $\sqrt[3]{y^2}$ **64.** $\sqrt[6]{z^5}$ **65.** $\sqrt[4]{a^2b^3}$ **66.** $\sqrt{abc}$

Concept 3: Properties of Rational Exponents

For Exercises 67–90, simplify the expressions by using the properties of rational exponents. Write the final answer using positive exponents only. **(See Example 5.)**

67. $x^{1/4}x^{-5/4}$ **68.** $2^{2/3}2^{-5/3}$ **69.** $\dfrac{p^{5/3}}{p^{2/3}}$ **70.** $\dfrac{q^{5/4}}{q^{1/4}}$

71. $(y^{1/5})^{10}$ **72.** $(x^{1/2})^{8}$ **73.** $6^{-1/5}6^{3/5}$ **74.** $a^{-1/3}a^{2/3}$

75. $\dfrac{4t^{-1/3}}{t^{4/3}}$ **76.** $\dfrac{5s^{-1/3}}{s^{5/3}}$ **77.** $(a^{1/3}a^{1/4})^{12}$ **78.** $(x^{2/3}x^{1/2})^{6}$

79. $(5a^2c^{-1/2}d^{1/2})^2$ **80.** $(2x^{-1/3}y^2z^{5/3})^3$ **81.** $\left(\dfrac{x^{-2/3}}{y^{-3/4}}\right)^{12}$ **82.** $\left(\dfrac{m^{-1/4}}{n^{-1/2}}\right)^{-4}$

83. $\left(\dfrac{16w^{-2}z}{2wz^{-8}}\right)^{1/3}$ **84.** $\left(\dfrac{50p^{-1}q}{2pq^{-3}}\right)^{1/2}$ **85.** $(25x^2y^4z^6)^{1/2}$ **86.** $(8a^6b^3c^9)^{2/3}$

87. $(x^2y^{-1/3})^6(x^{1/2}yz^{2/3})^2$ **88.** $(a^{-1/3}b^{1/2})^4(a^{-1/2}b^{3/5})^{10}$ **89.** $\left(\dfrac{x^{3m}y^{2m}}{z^{5m}}\right)^{1/m}$ **90.** $\left(\dfrac{a^{4n}b^{3n}}{c^n}\right)^{1/n}$

Concept 4: Applications Involving Rational Exponents

91. If P dollars in principal grows to A dollars after t years with annual interest, then the interest rate is given by $r = \left(\dfrac{A}{P}\right)^{1/t} - 1$. **(See Example 6.)**

     a. In one account, \$10,000 grows to \$16,802 after 5 yr. Compute the interest rate. Round your answer to a tenth of a percent.

     b. In another account \$10,000 grows to \$18,000 after 7 yr. Compute the interest rate. Round your answer to a tenth of a percent.

     c. Which account produced a higher average yearly return?

92. If the area A of a square is known, then the length of its sides, s, can be computed by the formula $s = A^{1/2}$.

     a. Compute the length of the sides of a square having an area of 100 in.^{2}

     b. Compute the length of the sides of a square having an area of 72 in.^{2} Round your answer to the nearest 0.1 in.

93. The radius r of a sphere of volume V is given by $r = \left(\dfrac{3V}{4\pi}\right)^{1/3}$. Find the radius of a sphere having a volume of 85 in.^{3} Round your answer to the nearest 0.1 in.

94. Is $(a + b)^{1/2}$ the same as $a^{1/2} + b^{1/2}$? If not, give a counterexample.

Expanding Your Skills

For Exercises 95–102, write the expression using rational exponents. Then simplify and convert back to radical notation. Assume all variables represent nonnegative real numbers.

Example: $\sqrt[15]{x^{10}}$ $\xrightarrow[\text{exponents}]{\text{Rational}}$ $x^{10/15}$ $\xrightarrow{\text{Simplify}}$ $x^{2/3}$ $\xrightarrow[\text{notation}]{\text{Radical}}$ $\sqrt[3]{x^2}$

95. $\sqrt[6]{y^3}$ **96.** $\sqrt[4]{w^2}$ **97.** $\sqrt[12]{z^3}$ **98.** $\sqrt[18]{t^3}$

99. $\sqrt[9]{x^6}$ **100.** $\sqrt[12]{p^9}$ **101.** $\sqrt[8]{x^{16}y^2}$ **102.** $\sqrt[6]{a^{18}b^2}$

For Exercises 103–106, write the expression as a single radical.

103. $\sqrt{\sqrt[3]{x}}$ **104.** $\sqrt[3]{\sqrt{x}}$ **105.** $\sqrt[5]{\sqrt[3]{w}}$ **106.** $\sqrt[3]{\sqrt[4]{w}}$

For Exercises 107–114, use a calculator to approximate the expressions. Round to 4 decimal places, if necessary.

107. $9^{1/2}$ **108.** $125^{-1/3}$ **109.** $50^{-1/4}$ **110.** $(172)^{3/5}$

111. $\sqrt[3]{5^2}$ **112.** $\sqrt[4]{6^3}$ **113.** $\sqrt{10^3}$ **114.** $\sqrt[3]{16}$

Simplifying Radical Expressions

1. Multiplication Property of Radicals

You may have already noticed certain properties of radicals involving a product or quotient.

> **Multiplication Property of Radicals**
>
> Let a and b represent real numbers such that $\sqrt[n]{a}$ and $\sqrt[n]{b}$ are both real. Then
>
> $$\sqrt[n]{ab} = \sqrt[n]{a} \cdot \sqrt[n]{b}$$

Concepts

1. Multiplication Property of Radicals
2. Simplifying Radicals by Using the Multiplication Property of Radicals
3. Simplifying Radicals by Using the Order of Operations

The multiplication property of radicals follows from the property of rational exponents.

$$\sqrt[n]{ab} = (ab)^{1/n} = a^{1/n}b^{1/n} = \sqrt[n]{a} \cdot \sqrt[n]{b}$$

The multiplication property of radicals indicates that a product within a radicand can be written as a product of radicals, provided the roots are real numbers. For example:

$$\sqrt{144} = \sqrt{16} \cdot \sqrt{9}$$

The reverse process is also true. A product of radicals can be written as a single radical provided the roots are real numbers and they have the same indices.

$$\sqrt{3} \cdot \sqrt{12} = \sqrt{36}$$

In algebra, it is customary to simplify radical expressions as much as possible.

> ### Simplified Form of a Radical
>
> Consider any radical expression where the radicand is written as a product of prime factors. The expression is in *simplified form* if all the following conditions are met:
>
> 1. The radicand has no factor raised to a power greater than or equal to the index.
> 2. The radicand does not contain a fraction.
> 3. There are no radicals in the denominator of a fraction.

For example, the following radicals are not simplified.

1. The expression $\sqrt[3]{x^5}$ fails rule 1.

2. The expression $\sqrt{\dfrac{1}{4}}$ fails rule 2.

3. The expression $\dfrac{1}{\sqrt[3]{8}}$ fails rule 3.

2. Simplifying Radicals by Using the Multiplication Property of Radicals

The expression $\sqrt{x^2}$ is not simplified because it fails condition 1. Because x^2 is a perfect square, $\sqrt{x^2}$ is easily simplified:

$$\sqrt{x^2} = x \qquad \text{for } x \geq 0$$

However, how is an expression such as $\sqrt{x^9}$ simplified? This and many other radical expressions are simplified by using the multiplication property of radicals. The following examples illustrate how nth powers can be removed from the radicands of nth roots.

Skill Practice

Simplify the expression. Assume $a \geq 0$.

1. $\sqrt{a^{11}}$

Example 1 **Using the Multiplication Property to Simplify a Radical Expression**

Simplify the expression assuming that $x \geq 0$. $\sqrt{x^9}$

Solution:

The expression $\sqrt{x^9}$ is equivalent to $\sqrt{x^8 \cdot x}$.

$$\sqrt{x^9} = \sqrt{x^8 \cdot x}$$

$$= \sqrt{x^8} \cdot \sqrt{x} \qquad \text{Apply the multiplication property of radicals.}$$

$$\qquad\qquad\qquad \text{Note that } x^8 \text{ is a perfect square because } x^8 = (x^4)^2.$$

$$= x^4 \sqrt{x} \qquad \text{Simplify.}$$

In Example 1, the expression x^9 is not a perfect square. Therefore, to simplify $\sqrt{x^9}$, it was necessary to write the expression as the product of the largest perfect square and a remaining or "left-over" factor: $\sqrt{x^9} = \sqrt{x^8 \cdot x}$. This process also applies to simplifying nth roots, as shown in Example 2.

Answer

1. $a^5 \sqrt{a}$

Example 2 Using the Multiplication Property to Simplify a Radical Expression

Simplify each expression. Assume all variables represent positive real numbers.

a. $\sqrt[4]{b^7}$ **b.** $\sqrt[3]{w^7 z^9}$

Solution:

The goal is to rewrite each radicand as the product of the greatest perfect square (perfect cube, perfect fourth power, and so on) and a left-over factor.

a. $\sqrt[4]{b^7} = \sqrt[4]{b^4 \cdot b^3}$ b^4 is the greatest perfect fourth power in the radicand.

$= \sqrt[4]{b^4} \cdot \sqrt[4]{b^3}$ Apply the multiplication property of radicals.

$= b\sqrt[4]{b^3}$ Simplify.

b. $\sqrt[3]{w^7 z^9} = \sqrt[3]{(w^6 z^9) \cdot (w)}$ $w^6 z^9$ is the greatest perfect cube in the radicand.

$= \sqrt[3]{w^6 z^9} \cdot \sqrt[3]{w}$ Apply the multiplication property of radicals.

$= w^2 z^3 \sqrt[3]{w}$ Simplify.

Skill Practice

Simplify the expressions. Assume all variables represent positive real numbers.

2. $\sqrt[4]{v^{25}}$

3. $\sqrt[3]{p^8 q^{12}}$

Each expression in Example 2 involves a radicand that is a product of variable factors. If a numerical factor is present, sometimes it is necessary to factor the coefficient before simplifying the radical.

Example 3 Using the Multiplication Property to Simplify a Radical

Simplify the expression. $\sqrt{56}$

Solution:

$\sqrt{56} = \sqrt{2^3 \cdot 7}$ Factor the radicand.

$= \sqrt{(2^2) \cdot (2 \cdot 7)}$ 2^2 is the greatest perfect square in the radicand.

$= \sqrt{2^2} \cdot \sqrt{2 \cdot 7}$ Apply the multiplication property of radicals.

$= 2\sqrt{14}$ Simplify.

$$\begin{array}{r|l} 2 & 56 \\ 2 & 28 \\ 2 & 14 \\ & 7 \end{array}$$

Skill Practice

Simplify.

4. $\sqrt{24}$

TIP: It may be easier to visualize the greatest perfect square factor within the radicand as follows:

$\sqrt{56} = \sqrt{4 \cdot 14}$

$= \sqrt{4} \cdot \sqrt{14}$

$= 2\sqrt{14}$

Calculator Connections

Topic: Verifying a Simplified Radical

A calculator can be used to support the solution to Example 3. The decimal approximation for $\sqrt{56}$ and $2\sqrt{14}$ agree for the first 10 digits. This in itself does not make $\sqrt{56} = 2\sqrt{14}$. It is the multiplication property of radicals that guarantees that the expressions are equal.

```
√(56)
            7.483314774
2*√(14)
            7.483314774
```

Answers

2. $v^6 \sqrt[4]{v}$

3. $p^2 q^4 \sqrt[3]{p^2}$

4. $2\sqrt{6}$

Example 4 **Using the Multiplication Property to Simplify Radicals**

Simplify. $\quad 6\sqrt{50}$

Solution:

$$6\sqrt{50} = 6\sqrt{5^2 \cdot 2} \qquad \text{Factor the radicand.}$$

$$= 6 \cdot \sqrt{5^2} \cdot \sqrt{2} \qquad \text{Apply the multiplication property of radicals.}$$

$$= 6 \cdot 5 \cdot \sqrt{2} \qquad \text{Simplify.}$$

$$= 30\sqrt{2} \qquad \text{Simplify.}$$

TIP: The radical can also be simplified as:

$$6\sqrt{50} = 6\sqrt{25 \cdot 2}$$

$$= 6\sqrt{25} \cdot \sqrt{2}$$

$$= 6 \cdot 5\sqrt{2}$$

$$= 30\sqrt{2}$$

Example 5 **Using the Multiplication Property to Simplify Radicals**

Simplify the expression. Assume that x, y, and z represent positive real numbers.

$$\sqrt[3]{40x^3y^5z^7}$$

Solution:

$$\sqrt[3]{40x^3y^5z^7}$$

$$= \sqrt[3]{2^3 5 x^3 y^5 z^7} \qquad \text{Factor the radicand.}$$

$$\begin{array}{r} 2\underline{|40} \\ 2\underline{|20} \\ 2\underline{|10} \\ 5 \end{array}$$

$$= \sqrt[3]{(2^3 x^3 y^3 z^6) \cdot (5y^2z)} \qquad 2^3 x^3 y^3 z^6 \text{ is the greatest perfect cube.}$$

$$= \sqrt[3]{2^3 x^3 y^3 z^6} \cdot \sqrt[3]{5y^2z} \qquad \text{Apply the multiplication property of radicals.}$$

$$= \quad 2xyz^2 \sqrt[3]{5y^2z} \qquad \text{Simplify.}$$

TIP: In Example 5, the numerical coefficient within the radicand can be written $8 \cdot 5$ because 8 is the greatest perfect cube factor of 40:

$$\sqrt[3]{40x^3y^5z^7} = \sqrt[3]{8 \cdot 5x^3y^5z^7}$$

$$= \sqrt[3]{(8x^3y^3z^6)(5y^2z)}$$

$$= \sqrt[3]{8x^3y^3z^6} \cdot \sqrt[3]{5y^2z}$$

$$= 2xyz^2 \sqrt[3]{5y^2z}$$

3. Simplifying Radicals by Using the Order of Operations

Often a radical can be simplified by applying the order of operations. In Examples 6 and 7, the first step will be to simplify the expression within the radicand.

Example 6 Using the Order of Operations to Simplify Radicals

Use the order of operations to simplify the expressions. Assume $a > 0$.

a. $\sqrt{\dfrac{a^7}{a^3}}$ **b.** $\sqrt[3]{\dfrac{3}{81}}$

Solution:

a. $\sqrt{\dfrac{a^7}{a^3}}$ The radicand contains a fraction. However, the fraction can be reduced to lowest terms.

$= \sqrt{a^4}$

$= a^2$ Simplify the radical.

b. $\sqrt[3]{\dfrac{3}{81}}$ The radical contains a fraction that can be simplified.

$= \sqrt[3]{\dfrac{1}{27}}$ Reduce to lowest terms.

$= \dfrac{1}{3}$ Simplify.

Example 7 Using the Order of Operations

Use the order of operations to simplify. $\dfrac{7\sqrt{50}}{15}$

Solution:

$\dfrac{7\sqrt{50}}{15}$ Following the order of operations, first we simplify the radical.

$= \dfrac{7\sqrt{25 \cdot 2}}{15}$ 25 is the greatest perfect square in the radicand.

$= \dfrac{7\sqrt{25} \cdot \sqrt{2}}{15}$ Apply the multiplication property of radicals.

$= \dfrac{7 \cdot 5\sqrt{2}}{15}$ Simplify the radicals.

$= \dfrac{7 \cdot \overset{1}{5}\sqrt{2}}{\underset{3}{15}}$ Reduce to lowest terms.

$= \dfrac{7\sqrt{2}}{3}$

Section 7.3 Practice Exercises

For the exercises in this set, assume that all variables represent positive real numbers unless otherwise stated.

Study Skills Exercise

> The final exam is just around the corner. Your old tests and quizzes provide good material to study for the final exam. Use your old tests to make a list of the concepts on which you need to concentrate. Ask your professor for help if there are still concepts that you do not understand.

Vocabulary and Key Concepts

1. **a.** The multiplication property of radicals indicates that if $\sqrt[n]{a}$ and $\sqrt[n]{b}$ are real numbers, then $\sqrt[n]{ab} =$ _____ · _____ .

   b. Explain why the following radical is not in simplified form. $\sqrt{x^3}$

   c. The radical expression $\sqrt[3]{x^{10}}$ (is/is not) in simplified form.

   d. The radical expression $\sqrt{18}$ simplifies to __$\sqrt{2}$.

   e. To simplify the radical expression $\sqrt[3]{t^{14}}$ the radicand is rewritten as $\sqrt[3]{\underline{} \cdot t^2}$.

   f. On a calculator, $\sqrt{2}$ is given as 1.414213562. Is this decimal number the exact value of $\sqrt{2}$?

Review Exercises

For Exercises 2–4, simplify the expression. Write the answer with positive exponents only.

2. $(a^2 b^{-4})^{1/2}\left(\dfrac{a}{b^{-3}}\right)$

3. $\left(\dfrac{p^4}{q^{-6}}\right)^{-1/2}(p^3 q^{-2})$

4. $(x^{1/3} y^{5/6})^{-6}$

5. Write $x^{4/7}$ in radical notation.

6. Write $y^{2/5}$ in radical notation.

7. Write $\sqrt{y^9}$ by using rational exponents.

8. Write $\sqrt[3]{x^2}$ by using rational exponents.

Concept 2: Simplifying Radicals by Using the Multiplication Property of Radicals

For Exercises 9–28, simplify the radicals. **(See Examples 1–5.)**

9. $\sqrt{x^5}$

10. $\sqrt{p^{15}}$

11. $\sqrt[3]{q^7}$

12. $\sqrt[3]{r^{17}}$

13. $\sqrt{a^5 b^4}$

14. $\sqrt{c^9 d^6}$

15. $-\sqrt[4]{x^8 y^{13}}$

16. $-\sqrt[4]{p^{16} q^{17}}$

17. $\sqrt{28}$

18. $\sqrt{63}$

19. $5\sqrt{18}$

20. $2\sqrt{24}$

21. $\sqrt[3]{54}$

22. $\sqrt[3]{250}$

23. $\sqrt{25ab^3}$

24. $\sqrt{64m^5 n^{20}}$

25. $\sqrt[3]{-16x^6 yz^3}$

26. $\sqrt[3]{-192a^6 bc^2}$

27. $\sqrt[4]{80w^4 z^7}$

28. $\sqrt[4]{32p^8 qr^5}$

Concept 3: Simplifying Radicals by Using the Order of Operations

For Exercises 29–40, simplify the radicals. **(See Examples 6 and 7.)**

29. $\sqrt{\dfrac{x^3}{x}}$

30. $\sqrt{\dfrac{y^5}{y}}$

31. $\sqrt{\dfrac{p^7}{p^3}}$

32. $\sqrt{\dfrac{q^{11}}{q^5}}$

33. $\sqrt{\dfrac{50}{2}}$

34. $\sqrt{\dfrac{98}{2}}$

35. $\sqrt[3]{\dfrac{3}{24}}$

36. $\sqrt[3]{\dfrac{2}{250}}$

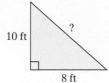 **37.** $\dfrac{5\sqrt[3]{16}}{6}$

38. $\dfrac{7\sqrt{18}}{9}$

39. $\dfrac{5\sqrt[3]{72}}{12}$

40. $\dfrac{3\sqrt[3]{250}}{10}$

Mixed Exercises

For Exercises 41–68, simplify the radicals.

41. $\sqrt{80}$

42. $\sqrt{108}$

43. $-6\sqrt{75}$

44. $-8\sqrt{8}$

45. $\sqrt{25x^4y^3}$

46. $\sqrt{125p^3q^2}$

47. $\sqrt[3]{27x^2y^3z^4}$

48. $\sqrt[3]{108a^3bc^2}$

49. $\sqrt{\dfrac{12w^5}{3w}}$

50. $\sqrt{\dfrac{64x^9}{4x^3}}$

51. $\sqrt{\dfrac{3y^3}{300y^{15}}}$

52. $\sqrt{\dfrac{4h}{100h^5}}$

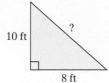 **53.** $\sqrt[3]{\dfrac{16a^2b}{2a^2b^4}}$

54. $\sqrt[3]{\dfrac{-27a^4}{8a}}$

55. $\sqrt{2^3a^{14}b^8c^{31}d^{22}}$

56. $\sqrt{7^5u^{12}v^{20}w^{65}x^{80}}$

57. $\sqrt[3]{54a^6b^4}$

58. $\sqrt[3]{72m^5n^3}$

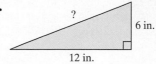 **59.** $-5a\sqrt{12a^3b^4c}$

60. $-7y\sqrt{75xy^5z^6}$

61. $\sqrt[4]{7x^5y}$

62. $\sqrt[4]{10cd^7}$

63. $\sqrt{54a^4b^2}$

64. $\sqrt{48r^6s^2}$

65. $\dfrac{2\sqrt{27}}{3}$

66. $\dfrac{7\sqrt{24}}{2}$

67. $\dfrac{3\sqrt{125}}{20}$

68. $\dfrac{10\sqrt{63}}{12}$

For Exercises 69–72, write a mathematical expression for the English phrase and simplify.

69. The quotient of 1 and the cube root of w^6

70. The principal square root of the quotient of h^2 and 49

71. The principal square root of the quantity k raised to the third power

72. The cube root of $2x^4$

For Exercises 73–76, determine the length of the third side of the right triangle. Write the answer as a simplified radical.

73.

10 ft ?

8 ft

74.

? 6 in.

12 in.

75.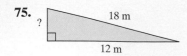

76.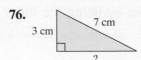

77. On a baseball diamond, the bases are 90 ft apart. Find the exact distance from home plate to second base. Then round to the nearest tenth of a foot.

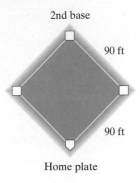

78. Linda is at the beach flying a kite. The kite is directly over a sand castle 60 ft away from Linda. If 100 ft of kite string is out (ignoring any sag in the string), how high is the kite? (Assume that Linda is 5 ft tall.) See figure.

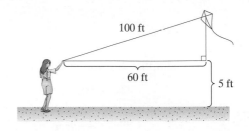

Expanding Your Skills

79. Tom has to travel from town A to town C across a small mountain range. He can travel one of two routes. He can travel on a four-lane highway from A to B and then from B to C at an average speed of 55 mph. Or he can travel on a two-lane road directly from town A to town C, but his average speed will be only 35 mph. If Tom is in a hurry, which route will take him to town C faster?

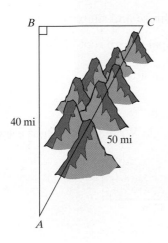

80. One side of a rectangular pasture is 80 ft in length. The diagonal distance is 110 yd. If fencing costs $3.29 per foot, how much will it cost to fence the pasture?

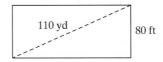

Addition and Subtraction of Radicals

1. Definition of *Like* Radicals

Definition of *Like* Radicals

Two radical terms are ***like* radicals** if they have the same index and the same radicand.

The following are pairs of *like* radicals:

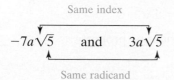

Indices and radicands are the same.

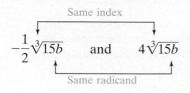

Indices and radicands are the same.

These pairs are not *like* radicals:

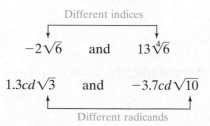

Radicals have different indices.

$1.3cd\sqrt{3}$ and $-3.7cd\sqrt{10}$ Radicals have different radicands.

Different radicands

2. Addition and Subtraction of Radicals

To add or subtract *like* radicals, use the distributive property. For example:

$$2\sqrt{5} + 6\sqrt{5} = (2 + 6)\sqrt{5}$$
$$= 8\sqrt{5}$$
$$9\sqrt[3]{2y} - 4\sqrt[3]{2y} = (9 - 4)\sqrt[3]{2y}$$
$$= 5\sqrt[3]{2y}$$

Example 1 Adding and Subtracting Radicals

Add or subtract as indicated.

a. $6\sqrt{11} - 2\sqrt{11}$ **b.** $\sqrt{3} + \sqrt{3}$

Solution:

a. $6\sqrt{11} - 2\sqrt{11}$

$= (6 - 2)\sqrt{11}$ Apply the distributive property.

$= 4\sqrt{11}$ Simplify.

b. $\sqrt{3} + \sqrt{3}$

$= 1\sqrt{3} + 1\sqrt{3}$ Note that $\sqrt{3} = 1\sqrt{3}$.

$= (1 + 1)\sqrt{3}$ Apply the distributive property.

$= 2\sqrt{3}$ Simplify.

Avoiding Mistakes

The process of adding *like* radicals with the distributive property is similar to adding *like* terms. The end result is that the numerical coefficients are added and the radical factor is unchanged.

$$\sqrt{3} + \sqrt{3} = 1\sqrt{3} + 1\sqrt{3} = 2\sqrt{3}$$

Be careful: $\sqrt{3} + \sqrt{3} \neq \sqrt{6}$

In general: $\sqrt{x} + \sqrt{y} \neq \sqrt{x + y}$

Example 2 Adding and Subtracting Radicals

Add or subtract as indicated.

a. $-2\sqrt[3]{ab} + 7\sqrt[3]{ab} - \sqrt[3]{ab}$ **b.** $\frac{1}{4}x\sqrt{3y} - \frac{3}{2}x\sqrt{3y}$

Solution:

a. $-2\sqrt[3]{ab} + 7\sqrt[3]{ab} - \sqrt[3]{ab}$

$= (-2 + 7 - 1)\sqrt[3]{ab}$ Apply the distributive property.

$= 4\sqrt[3]{ab}$ Simplify.

b. $\frac{1}{4}x\sqrt{3y} - \frac{3}{2}x\sqrt{3y}$

$= \left(\frac{1}{4} - \frac{3}{2}\right)x\sqrt{3y}$ Apply the distributive property.

$= \left(\frac{1}{4} - \frac{6}{4}\right)x\sqrt{3y}$ Get a common denominator.

$= -\frac{5}{4}x\sqrt{3y}$ Simplify.

Example 3 shows that it is often necessary to simplify radicals before adding or subtracting.

| Example 3 | Adding and Subtracting Radicals |

Simplify the radicals and add or subtract as indicated. Assume all variables represent positive real numbers.

a. $3\sqrt{8} + \sqrt{2}$ **b.** $8\sqrt{x^3y^2} - 3y\sqrt{x^3}$

c. $\sqrt{50x^2y^5} - 13y\sqrt{2x^2y^3} + xy\sqrt{98y^3}$

Solution:

a. $3\sqrt{8} + \sqrt{2}$ The radicands are different. Try simplifying the radicals first.

$= 3 \cdot 2\sqrt{2} + \sqrt{2}$ Simplify: $\sqrt{8} = \sqrt{4 \cdot 2} = 2\sqrt{2}$

$= 6\sqrt{2} + \sqrt{2}$

$= (6 + 1)\sqrt{2}$ Apply the distributive property.

$= 7\sqrt{2}$ Simplify.

b. $8\sqrt{x^3y^2} - 3y\sqrt{x^3}$ The radicands are different. Simplify the radicals first.

$= 8xy\sqrt{x} - 3xy\sqrt{x}$ Simplify $\sqrt{x^3y^2} = xy\sqrt{x}$ and $\sqrt{x^3} = x\sqrt{x}$.

$= (8 - 3)xy\sqrt{x}$ Apply the distributive property.

$= 5xy\sqrt{x}$ Simplify.

c. $\sqrt{50x^2y^5} - 13y\sqrt{2x^2y^3} + xy\sqrt{98y^3}$ Simplify each radical.

$$\begin{cases} \sqrt{50x^2y^5} = \sqrt{25 \cdot 2x^2y^5} \\ \qquad = 5xy^2\sqrt{2y} \\ -13y\sqrt{2x^2y^3} = -13xy^2\sqrt{2y} \\ xy\sqrt{98y^3} = xy\sqrt{49 \cdot 2y^3} \\ \qquad = 7xy^2\sqrt{2y} \end{cases}$$

$= 5xy^2\sqrt{2y} - 13xy^2\sqrt{2y} + 7xy^2\sqrt{2y}$

$= (5 - 13 + 7)xy^2\sqrt{2y}$ Apply the distributive property.

$= -xy^2\sqrt{2y}$

| Section 7.4 | **Practice Exercises** |

For the exercises in this set, assume that all variables represent positive real numbers, unless otherwise stated.

Vocabulary and Key Concepts

1. a. Two radical terms are called *like* radicals if they have the same _____ and the same _____.

    b. The expression $\sqrt{3x} + \sqrt{3x}$ simplifies to _____.

    c. The expression $\sqrt{2} + \sqrt{3}$ (can/cannot) be simplified further whereas the expression $\sqrt{2} \cdot \sqrt{3}$ (can/cannot) be simplified further.

    d. The expression $\sqrt{2} + \sqrt{18}$ simplifies to _____.

Review Exercises

For Exercises 2–5, simplify the radicals.

2. $\sqrt[3]{-16s^4t^9}$ **3.** $-\sqrt[4]{x^7y^4}$ **4.** $\sqrt{36a^2b^3}$ **5.** $\sqrt[3]{\dfrac{7b^8}{56b^2}}$

6. Write $(4x^2)^{1/3}$ in radical notation.   **7.** Write $\sqrt[4]{x^3y}$ using rational exponents.

8. Simplify. $32^{-1/5}$

For Exercises 9 and 10, simplify the expressions. Write the answer with positive exponents only.

9. $y^{2/3}y^{1/4}$ **10.** $(x^{1/2}y^{-3/4})^{-4}$

Concept 1: Definition of *Like* Radicals

For Exercises 11 and 12, determine if the radical terms are *like*.

11. a. $\sqrt{2}$ and $\sqrt[3]{2}$ **12. a.** $7\sqrt[3]{x}$ and $\sqrt[3]{x}$

     b. $\sqrt{2}$ and $3\sqrt{2}$ **b.** $\sqrt[3]{x}$ and $\sqrt[4]{x}$

     c. $\sqrt{2}$ and $\sqrt{5}$ **c.** $2\sqrt[4]{x}$ and $x\sqrt[4]{2}$

13. Explain the similarities between the pairs of expressions.

     a. $7\sqrt{5} + 4\sqrt{5}$ and $7x + 4x$

     b. $-2\sqrt{6} - 9\sqrt{3}$ and $-2x - 9y$

14. Explain the similarities between the pairs of expressions.

     a. $-4\sqrt{3} + 5\sqrt{3}$ and $-4z + 5z$

     b. $13\sqrt{7} - 18$ and $13a - 18$

Concept 2: Addition and Subtraction of Radicals

For Exercises 15–35, add or subtract the radical expressions, if possible. **(See Examples 1 and 2.)**

15. $3\sqrt{5} + 6\sqrt{5}$ **16.** $5\sqrt{a} + 3\sqrt{a}$ **17.** $3\sqrt[3]{tw} - 2\sqrt[3]{tw} + \sqrt[3]{tw}$

18. $6\sqrt[3]{7} - 2\sqrt[3]{7} + \sqrt[3]{7}$ **19.** $6\sqrt{10} - \sqrt{10}$ **20.** $13\sqrt{11} - \sqrt{11}$

21. $\sqrt[4]{3} + 7\sqrt[4]{3} - \sqrt[4]{14}$ **22.** $2\sqrt{11} + 3\sqrt{13} + 5\sqrt{11}$ **23.** $8\sqrt{x} + 2\sqrt{y} - 6\sqrt{x}$

24. $10\sqrt{10} - 8\sqrt{10} + \sqrt{2}$ **25.** $\sqrt[3]{ab} + a\sqrt[3]{b}$ **26.** $x\sqrt[4]{y} - y\sqrt[4]{x}$

27. $\sqrt{2t} + \sqrt[3]{2t}$ **28.** $\sqrt[4]{5c} + \sqrt[3]{5c}$ **29.** $\dfrac{5}{6}z\sqrt[3]{6} + \dfrac{7}{9}z\sqrt[3]{6}$

30. $\dfrac{3}{4}a\sqrt[4]{b} + \dfrac{1}{6}a\sqrt[4]{b}$ **31.** $0.81x\sqrt{y} - 0.11x\sqrt{y}$ **32.** $7.5\sqrt{pq} - 6.3\sqrt{pq}$

33. $5x\sqrt{x} + 6\sqrt{x}$ **34.** $9y^2\sqrt{2} + 4\sqrt{2}$ **35.** $14a\sqrt{2a} - 3\sqrt{2a}$

36. Explain the process for adding the two radicals. Then find the sum. $3\sqrt{2} + 7\sqrt{50}$

For Exercises 37–62, add or subtract the radical expressions as indicated. **(See Example 3.)**

37. $\sqrt{36} + \sqrt{81}$

38. $3\sqrt{80} - 5\sqrt{45}$

39. $2\sqrt{12} + \sqrt{48}$

40. $5\sqrt{32} + 2\sqrt{50}$

41. $4\sqrt{7} + \sqrt{63} - 2\sqrt{28}$

42. $8\sqrt{3} - 2\sqrt{27} + \sqrt{75}$

43. $5\sqrt{18} + \sqrt{27} - 4\sqrt{50}$

44. $7\sqrt{40} - \sqrt{12} + 2\sqrt{75}$

45. $\sqrt[3]{81} - \sqrt[3]{24}$

46. $17\sqrt[3]{81} - 2\sqrt[3]{24}$

47. $3\sqrt{2a} - \sqrt{8a} - \sqrt{72a}$

48. $\sqrt{12t} - \sqrt{27t} + 5\sqrt{3t}$

49. $2s^2\sqrt[3]{s^2t^6} + 3t^2\sqrt[3]{8s^8}$

50. $4\sqrt[3]{x^4} - 2x\sqrt[3]{x}$

51. $7\sqrt[3]{x^4} - x\sqrt[3]{x}$

52. $6\sqrt[3]{y^{10}} - 3y^2\sqrt[3]{y^4}$

53. $5p\sqrt{20p^2} + p^2\sqrt{80}$

54. $2q\sqrt{48q^2} - \sqrt{27q^4}$

55. $\sqrt[3]{a^2b} - \sqrt[3]{8a^2b}$

56. $w\sqrt{80} - 3\sqrt{125w^2}$

57. $11\sqrt[3]{54cd^3} - 2\sqrt[3]{2cd^3} + d\sqrt[3]{16c}$

58. $x\sqrt[3]{64x^5y^2} - x^2\sqrt[3]{x^2y^2} + 5\sqrt[3]{x^8y^2}$

59. $\dfrac{3}{2}ab\sqrt{24a^3} + \dfrac{4}{3}\sqrt{54a^5b^2} - a^2b\sqrt{150a}$

60. $mn\sqrt{72n} + \dfrac{2}{3}n\sqrt{8m^2n} - \dfrac{5}{6}\sqrt{50m^2n^3}$

61. $x\sqrt[3]{16} - 2\sqrt[3]{27x} + \sqrt[3]{54x^3}$

62. $5\sqrt[4]{y^5} - 2y\sqrt[4]{y} + \sqrt[4]{16y^7}$

Mixed Exercises

For Exercises 63–68, answer true or false. If an answer is false, explain why or give a counterexample.

63. $\sqrt{x} + \sqrt{y} = \sqrt{x + y}$

64. $\sqrt{x} + \sqrt{x} = 2\sqrt{x}$

65. $5\sqrt[3]{x} + 2\sqrt[3]{x} = 7\sqrt[3]{x}$

66. $6\sqrt{x} + 5\sqrt[3]{x} = 11\sqrt{x}$

67. $\sqrt{y} + \sqrt{y} = \sqrt{2y}$

68. $\sqrt{c^2 + d^2} = c + d$

For Exercises 69–72, translate the English phrase to an algebraic expression. Simplify each expression.

69. The sum of the principal square root of 48 and the principal square root of 12

70. The sum of the cube root of 16 and the cube root of 2

71. The difference of 5 times the cube root of x^6 and the square of x

72. The sum of the cube of y and the principal fourth root of y^{12}

For Exercises 73–76, write an English phrase that translates the mathematical expression. (Answers may vary.)

73. $\sqrt{18} - 5^2$

74. $4^3 - \sqrt[3]{4}$

75. $\sqrt[4]{x} + y^3$

76. $a^4 + \sqrt{a}$

For Exercises 77–80, find the exact value of the perimeter, and then approximate the value to 1 decimal place.

77.

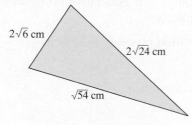

$2\sqrt{6}$ cm

$2\sqrt{24}$ cm

$\sqrt{54}$ cm

78.

$\sqrt{75}$ ft

$\sqrt{3}$ ft

$\sqrt{27}$ ft

79. The figure has perimeter $14\sqrt{2}$ ft. Find the value of x.

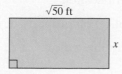

$\sqrt{50}$ ft

x

80. The figure has perimeter $12\sqrt{7}$ cm. Find the value of x.

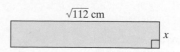

$\sqrt{112}$ cm

x

Expanding Your Skills

 81. a. An irregularly shaped garden is shown in the figure. All distances are expressed in yards. Find the perimeter. *Hint:* Use the Pythagorean theorem to find the length of each side. Write the final answer in radical form.

    b. Approximate your answer to two decimal places.

    c. If edging costs \$1.49 per foot and sales tax is 6%, find the total cost of edging the garden.

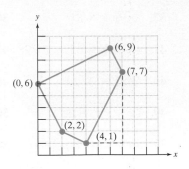

Section 7.5	**Multiplication of Radicals**

Concepts

1. Multiplication Property of Radicals
2. Expressions of the Form $(\sqrt[n]{a})^n$
3. Special Case Products
4. Multiplying Radicals with Different Indices

1. Multiplication Property of Radicals

In this section, we will learn how to multiply radicals by using the multiplication property of radicals first introduced in Section 7.3.

The Multiplication Property of Radicals

Let a and b represent real numbers such that $\sqrt[n]{a}$ and $\sqrt[n]{b}$ are both real. Then

$$\sqrt[n]{ab} = \sqrt[n]{a} \cdot \sqrt[n]{b}$$

To multiply two radical expressions, we use the multiplication property of radicals along with the commutative and associative properties of multiplication.

Example 1 Multiplying Radical Expressions

Multiply each expression and simplify the result. Assume all variables represent positive real numbers.

a. $(3\sqrt{2})(5\sqrt{6})$ **b.** $(2x\sqrt{y})(-7\sqrt{xy})$ **c.** $(15c\sqrt[3]{cd})\left(\dfrac{1}{3}\sqrt[3]{cd^2}\right)$

Solution:

a. $(3\sqrt{2})(5\sqrt{6})$

$= (3 \cdot 5)(\sqrt{2} \cdot \sqrt{6})$ Commutative and associative properties of multiplication

$= 15\sqrt{12}$ Multiplication property of radicals

$= 15\sqrt{4 \cdot 3}$

$= 15 \cdot 2\sqrt{3}$ Simplify the radical.

$= 30\sqrt{3}$

b. $(2x\sqrt{y})(-7\sqrt{xy})$

$= (2x)(-7)(\sqrt{y} \cdot \sqrt{xy})$ Commutative and associative properties of multiplication

$= -14x\sqrt{xy^2}$ Multiplication property of radicals

$= -14xy\sqrt{x}$ Simplify the radical.

Skill Practice

Multiply the expressions and simplify the results. Assume all variables represent positive real numbers.

1. $(4\sqrt{6})(-2\sqrt{6})$
2. $(3ab\sqrt{b})(-2\sqrt{ab})$
3. $(2\sqrt[3]{4ab})(5\sqrt[3]{2a^2b})$

Concept Connections

4. Explain why the multiplication property of radicals cannot be used to multiply $\sqrt{-2} \cdot \sqrt{-8}$

Answers

1. -48
2. $-6ab^2\sqrt{a}$
3. $20a\sqrt[3]{b^2}$
4. The multiplication property of radicals cannot be used because the radicals $\sqrt{-2}$ and $\sqrt{-8}$ are not real numbers.

c. $\left(15c\sqrt[3]{cd}\right)\left(\dfrac{1}{3}\sqrt[3]{cd^2}\right)$

$$= \left(15c \cdot \dfrac{1}{3}\right)\left(\sqrt[3]{cd} \cdot \sqrt[3]{cd^2}\right) \qquad \text{Commutative and associative properties of multiplication}$$

$$= 5c\sqrt[3]{c^2d^3} \qquad\qquad\qquad \text{Multiplication property of radicals}$$

$$= 5cd\sqrt[3]{c^2} \qquad\qquad\qquad\; \text{Simplify the radical.}$$

When multiplying radical expressions with more than one term, we use the distributive property.

Example 2 **Multiplying Radical Expressions**

Multiply. $3\sqrt{11}(2 + \sqrt{11})$

Solution:

$3\sqrt{11}(2 + \sqrt{11})$

$$= 3\sqrt{11} \cdot (2) + 3\sqrt{11} \cdot \sqrt{11} \qquad \text{Apply the distributive property.}$$

$$= 6\sqrt{11} + 3\sqrt{11^2} \qquad\qquad\;\; \text{Multiplication property of radicals}$$

$$= 6\sqrt{11} + 3 \cdot 11 \qquad\qquad\;\;\; \text{Simplify the radical.}$$

$$= 6\sqrt{11} + 33$$

Skill Practice

Multiply.

5. $5\sqrt{5}(2\sqrt{5} - 2)$

Example 3 **Multiplying Radical Expressions**

Multiply.

a. $(\sqrt{5} + 3\sqrt{2})(2\sqrt{5} - \sqrt{2})$ **b.** $(-10a\sqrt{b} + 7b)(a\sqrt{b} + 2b)$

Solution:

a. $(\sqrt{5} + 3\sqrt{2})(2\sqrt{5} - \sqrt{2})$

$$= 2\sqrt{5^2} - \sqrt{10} + 6\sqrt{10} - 3\sqrt{2^2} \qquad \text{Apply the distributive property.}$$

$$= 2 \cdot 5 + 5\sqrt{10} - 3 \cdot 2 \qquad\qquad\quad \text{Simplify radicals and combine } like \text{ radicals.}$$

$$= 10 + 5\sqrt{10} - 6$$

$$= 4 + 5\sqrt{10} \qquad\qquad\qquad\qquad\quad \text{Combine } like \text{ terms.}$$

b. $(-10a\sqrt{b} + 7b)(a\sqrt{b} + 2b)$

$$= -10a^2\sqrt{b^2} - 20ab\sqrt{b} + 7ab\sqrt{b} + 14b^2 \qquad \text{Apply the distributive property.}$$

$$= -10a^2b - 13ab\sqrt{b} + 14b^2 \qquad\qquad\qquad \text{Simplify and combine } like \text{ terms.}$$

Skill Practice

Multiply.

6. $(2\sqrt{3} - 3\sqrt{10}) \cdot (\sqrt{3} + 2\sqrt{10})$

7. $(x\sqrt{y} + y)(3x\sqrt{y} - 2y)$

Answers

5. $50 - 10\sqrt{5}$ **6.** $-54 + \sqrt{30}$
7. $3x^2y + xy\sqrt{y} - 2y^2$

Example 4 Multiplying Radical Expressions

Multiply. $(2\sqrt{x} + \sqrt{y})(6 - \sqrt{x} + 8\sqrt{y})$

Solution:

$(2\sqrt{x} + \sqrt{y})(6 - \sqrt{x} + 8\sqrt{y})$

$= 12\sqrt{x} - 2\sqrt{x^2} + 16\sqrt{xy} + 6\sqrt{y} - \sqrt{xy} + 8\sqrt{y^2}$ Apply the distributive property.

$= 12\sqrt{x} - 2x + 16\sqrt{xy} + 6\sqrt{y} - \sqrt{xy} + 8y$ Simplify the radicals.

$= 12\sqrt{x} - 2x + 15\sqrt{xy} + 6\sqrt{y} + 8y$ Combine *like* terms.

2. Expressions of the Form $(\sqrt[n]{a})^n$

The multiplication property of radicals can be used to simplify an expression of the form $(\sqrt{a})^2$, where $a \geq 0$.

$$(\sqrt{a})^2 = \sqrt{a} \cdot \sqrt{a} = \sqrt{a^2} = a \qquad \text{where } a \geq 0$$

This logic can be applied to nth roots.

- If $\sqrt[n]{a}$ is a real number, then $(\sqrt[n]{a})^n = a$.

Example 5 Simplifying Radical Expressions

Simplify the expressions. Assume all variables represent positive real numbers.

a. $(\sqrt{11})^2$ **b.** $(\sqrt[5]{z})^5$ **c.** $(\sqrt[3]{pq})^3$

Solution:

a. $(\sqrt{11})^2 = 11$ **b.** $(\sqrt[5]{z})^5 = z$ **c.** $(\sqrt[3]{pq})^3 = pq$

3. Special Case Products

From Examples 2–4, you may have noticed a similarity between multiplying radical expressions and multiplying polynomials.

In Section 5.3 we learned that the square of a binomial results in a perfect square trinomial:

$$(a + b)^2 = a^2 + 2ab + b^2$$
$$(a - b)^2 = a^2 - 2ab + b^2$$

The same patterns occur when squaring a radical expression with two terms.

Example 6 **Squaring a Two-Term Radical Expression**

Square the radical expressions. Assume all variables represent positive real numbers.

a. $(\sqrt{d} + 3)^2$ **b.** $(5\sqrt{y} - \sqrt{2})^2$

Solution:

a. $(\sqrt{d} + 3)^2$

$\quad$ This expression is in the form $(a + b)^2$, where $a = \sqrt{d}$ and $b = 3$.

$\overset{a^2 + 2ab + b^2}{}$

$= (\sqrt{d})^2 + 2(\sqrt{d})(3) + (3)^2$ $\quad$ Apply the formula $(a + b)^2 = a^2 + 2ab + b^2$.

$= d + 6\sqrt{d} + 9$ $\quad$ Simplify.

> **TIP:** The product $(\sqrt{d} + 3)^2$ can also be found by using the distributive property.
>
> $(\sqrt{d} + 3)^2 = (\sqrt{d} + 3)(\sqrt{d} + 3) = \sqrt{d} \cdot \sqrt{d} + \sqrt{d} \cdot 3 + 3 \cdot \sqrt{d} + 3 \cdot 3$
> $= \sqrt{d^2} + 3\sqrt{d} + 3\sqrt{d} + 9$
> $= d + 6\sqrt{d} + 9$

b. $(5\sqrt{y} - \sqrt{2})^2$

$\quad$ This expression is in the form $(a - b)^2$, where $a = 5\sqrt{y}$ and $b = \sqrt{2}$.

$\overset{a^2 - 2ab + b^2}{}$

$= (5\sqrt{y})^2 - 2(5\sqrt{y})(\sqrt{2}) + (\sqrt{2})^2$ $\quad$ Apply the formula $(a - b)^2 = a^2 - 2ab + b^2$.

$= 25y - 10\sqrt{2y} + 2$ $\quad$ Simplify.

Recall from Section 5.3 that the product of two conjugate binomials results in a difference of squares.

$$(a + b)(a - b) = a^2 - b^2$$

The same pattern occurs when multiplying two conjugate radical expressions.

Skill Practice

Square the radical expressions. Assume all variables represent positive real numbers.

12. $(\sqrt{a} - 5)^2$

13. $(4\sqrt{x} + \sqrt{3})^2$

Concept Connections

14. Give the conjugate of $(\sqrt{a} + \sqrt{b})$.

Answers

12. $a - 10\sqrt{a} + 25$
13. $16x + 8\sqrt{3x} + 3$
14. $(\sqrt{a} - \sqrt{b})$

Example 7 **Multiplying Conjugate Radical Expressions**

Multiply the radical expressions. Assume all variables represent positive real numbers.

a. $(\sqrt{3} + 2)(\sqrt{3} - 2)$ **b.** $\left(\dfrac{1}{3}\sqrt{s} - \dfrac{3}{4}\sqrt{t}\right)\left(\dfrac{1}{3}\sqrt{s} + \dfrac{3}{4}\sqrt{t}\right)$

Solution:

a. $(\sqrt{3} + 2)(\sqrt{3} - 2)$ The expression is in the form $(a + b)(a - b)$, where $a = \sqrt{3}$ and $b = 2$.

$$\overset{a^2 - b^2}{}$$

$$= (\sqrt{3})^2 - (2)^2 \qquad \text{Apply the formula } (a + b)(a - b) = a^2 - b^2.$$

$$= 3 - 4 \qquad \text{Simplify.}$$

$$= -1$$

TIP: The product $(\sqrt{3} + 2)(\sqrt{3} - 2)$ can also be found by using the distributive property.

$$(\sqrt{3} + 2)(\sqrt{3} - 2) = \sqrt{3} \cdot \sqrt{3} + \sqrt{3} \cdot (-2) + 2 \cdot \sqrt{3} + 2 \cdot (-2)$$

$$= 3 - 2\sqrt{3} + 2\sqrt{3} - 4$$

$$= 3 - 4$$

$$= -1$$

b. $\left(\dfrac{1}{3}\sqrt{s} - \dfrac{3}{4}\sqrt{t}\right)\left(\dfrac{1}{3}\sqrt{s} + \dfrac{3}{4}\sqrt{t}\right)$ This expression is in the form $(a - b)(a + b)$, where $a = \frac{1}{3}\sqrt{s}$ and $b = \frac{3}{4}\sqrt{t}$.

$$\overset{a^2 - b^2}{}$$

$$= \left(\dfrac{1}{3}\sqrt{s}\right)^2 - \left(\dfrac{3}{4}\sqrt{t}\right)^2 \qquad \begin{array}{l}\text{Apply the formula}\\ (a + b)(a - b) = a^2 - b^2.\end{array}$$

$$= \dfrac{1}{9}s - \dfrac{9}{16}t \qquad \text{Simplify.}$$

4. Multiplying Radicals with Different Indices

The product of two radicals can be simplified provided the radicals have the same index. If the radicals have different indices, then we can use the properties of rational exponents to obtain a common index.

Example 8 **Multiplying Radicals with Different Indices**

Multiply the expressions. Write the answers in radical form.

a. $\sqrt[3]{5} \cdot \sqrt[4]{5}$　　b. $\sqrt[3]{7} \cdot \sqrt{2}$

Solution:

a. $\sqrt[3]{5} \cdot \sqrt[4]{5}$

$= 5^{1/3}5^{1/4}$　　Rewrite each expression with rational exponents.

$= 5^{(1/3)+(1/4)}$　　Because the bases are equal, we can add the exponents.

$= 5^{(4/12)+(3/12)}$　　Write the fractions with a common denominator.

$= 5^{7/12}$　　Simplify the exponent.

$= \sqrt[12]{5^7}$　　Rewrite the expression as a radical.

b. $\sqrt[3]{7} \cdot \sqrt{2}$

$= 7^{1/3}2^{1/2}$　　Rewrite each expression with rational exponents.

$= 7^{2/6}2^{3/6}$　　Write the rational exponents with a common denominator.

$= (7^2 2^3)^{1/6}$　　Apply the power rule of exponents.

$= \sqrt[6]{7^2 2^3}$　　Rewrite the expression as a single radical.

$= \sqrt[6]{392}$　　Simplify.

Answers

17. $\sqrt[6]{x^5}$　　18. $\sqrt[12]{a^9 b^4}$

Section 7.5 Practice Exercises

For the exercises in this set, assume that all variables represent positive real numbers, unless otherwise stated.

Vocabulary and Key Concepts

1. **a.** If $\sqrt[n]{a}$ and $\sqrt[n]{b}$ are real numbers, then $\sqrt[n]{a} \cdot \sqrt[n]{b} =$ _____.

   b. If $x \geq 0$, then $\sqrt{x} \cdot \sqrt{x} =$ _____.

   c. If $\sqrt[n]{a}$ is a real number, then $(\sqrt[n]{a})^n =$ _____.

   d. Two binomials $(x + \sqrt{2})$ and $(x - \sqrt{2})$ are called _____ of each other, and their product is $(x)^2 - (\sqrt{2})^2$.

   e. If $m \geq 0$ and $n \geq 0$, then $(\sqrt{m} + \sqrt{n})(\sqrt{m} - \sqrt{n}) =$ _____.

   f. Which is the correct simplification of $(\sqrt{c} + 4)^2$?

   $c + 16$　or　$c + 8\sqrt{c} + 16$

Review Exercises

For Exercises 2–4, simplify the radicals.

2. $\sqrt[3]{-16x^5 y^6 z^7}$　　　3. $-\sqrt{20a^2 b^3 c}$　　　4. $\sqrt{\dfrac{8y^3 z^5}{y}}$

For Exercises 5 and 6, simplify the expressions. Write the answer with positive exponents only.

5. $x^{1/3}y^{1/4}x^{-1/6}y^{1/3}$

6. $\dfrac{b^{1/4}}{b^{3/2}}$

For Exercises 7 and 8, add or subtract as indicated.

7. $-2\sqrt[3]{7} + 4\sqrt[3]{7}$

8. $4\sqrt{8x^3} - x\sqrt{50x}$

Concept 1: Multiplication Property of Radicals

For Exercises 9–44, multiply the radical expressions. **(See Examples 1–4.)**

9. $\sqrt[3]{7} \cdot \sqrt[3]{3}$

10. $\sqrt[4]{6} \cdot \sqrt[4]{2}$

11. $\sqrt{2} \cdot \sqrt{10}$

12. $\sqrt[3]{4} \cdot \sqrt[3]{12}$

13. $\sqrt[4]{16} \cdot \sqrt[4]{64}$

14. $\sqrt{5x^3} \cdot \sqrt{10x^4}$

15. $(4\sqrt[3]{4})(2\sqrt[3]{5})$

16. $(2\sqrt{5})(3\sqrt{7})$

17. $(8a\sqrt{b})(-3\sqrt{ab})$

18. $(p\sqrt[4]{q^3})(\sqrt[4]{pq})$

19. $\sqrt{30} \cdot \sqrt{12}$

20. $\sqrt{20} \cdot \sqrt{54}$

21. $\sqrt{6x} \cdot \sqrt{12x}$

22. $(\sqrt{3ab^2})(\sqrt{21a^2b})$

23. $\sqrt{5a^3b^2} \cdot \sqrt{20a^3b^3}$

24. $\sqrt[3]{m^2n^2} \cdot \sqrt[3]{48m^4n^2}$

25. $(4\sqrt{3xy^3})(-2\sqrt{6x^3y^2})$

26. $(2\sqrt[4]{3x})(4\sqrt[4]{27x^6})$

27. $(\sqrt[3]{4a^2b})(\sqrt[3]{2ab^3})(\sqrt[3]{54a^2b})$

28. $(\sqrt[3]{9x^3y})(\sqrt[3]{6xy})(\sqrt[3]{8x^2y^5})$

29. $\sqrt{3}(4\sqrt{3} - 6)$

30. $3\sqrt{5}(2\sqrt{5} + 4)$

31. $\sqrt{2}(\sqrt{6} - \sqrt{3})$

32. $\sqrt{5}(\sqrt{3} + \sqrt{7})$

33. $-\dfrac{1}{3}\sqrt{x}(6\sqrt{x} + 7)$

34. $-\dfrac{1}{2}\sqrt{y}(8 - 3\sqrt{y})$

35. $(\sqrt{3} + 2\sqrt{10})(4\sqrt{3} - \sqrt{10})$

36. $(8\sqrt{7} - \sqrt{5})(\sqrt{7} + 3\sqrt{5})$

37. $(\sqrt{x} + 4)(\sqrt{x} - 9)$

38. $(\sqrt{w} - 2)(\sqrt{w} - 9)$

39. $(\sqrt[3]{y} + 2)(\sqrt[3]{y} - 3)$

40. $(4 + \sqrt[5]{p})(5 + \sqrt[5]{p})$

41. $(\sqrt{a} - 3\sqrt{b})(9\sqrt{a} - \sqrt{b})$

42. $(11\sqrt{m} + 4\sqrt{n})(\sqrt{m} + \sqrt{n})$

43. $(\sqrt{p} + 2\sqrt{q})(8 + 3\sqrt{p} - \sqrt{q})$

44. $(5\sqrt{s} - \sqrt{t})(\sqrt{s} + 5 + 6\sqrt{t})$

Concept 2: Expressions of the Form $(\sqrt[n]{a})^n$

For Exercises 45–52, simplify the expressions. Assume all variables represent positive real numbers. **(See Example 5.)**

45. $(\sqrt{15})^2$

46. $(\sqrt{58})^2$

47. $(\sqrt{3y})^2$

48. $(\sqrt{19yz})^2$

49. $(\sqrt[3]{6})^3$

50. $(\sqrt[5]{24})^5$

51. $\sqrt{709} \cdot \sqrt{709}$

52. $\sqrt{401} \cdot \sqrt{401}$

Concept 3: Special Case Products

53. a. Write the formula for the product of two conjugates. $(x + y)(x - y) =$

b. Multiply $(x + 5)(x - 5)$.

54. a. Write the formula for squaring a binomial. $(x + y)^2 =$

b. Multiply $(x + 5)^2$.

For Exercises 55–66, multiply the radical expressions. **(See Examples 6 and 7.)**

55. $(\sqrt{13} + 4)^2$

56. $(6 - \sqrt{11})^2$

57. $(\sqrt{p} - \sqrt{7})^2$

58. $(\sqrt{q} + \sqrt{2})^2$

59. $(\sqrt{2a} - 3\sqrt{b})^2$

60. $(\sqrt{3w} + 4\sqrt{z})^2$

61. $(\sqrt{3} + x)(\sqrt{3} - x)$

62. $(y + \sqrt{6})(y - \sqrt{6})$

63. $(\sqrt{6} + \sqrt{2})(\sqrt{6} - \sqrt{2})$

64. $(\sqrt{15} + \sqrt{5})(\sqrt{15} - \sqrt{5})$

65. $\left(\frac{2}{3}\sqrt{x} + \frac{1}{2}\sqrt{y}\right)\left(\frac{2}{3}\sqrt{x} - \frac{1}{2}\sqrt{y}\right)$

66. $\left(\frac{1}{4}\sqrt{s} + \frac{1}{5}\sqrt{t}\right)\left(\frac{1}{4}\sqrt{s} - \frac{1}{5}\sqrt{t}\right)$

For Exercises 67 and 68, multiply the expressions.

67. a. $(\sqrt{3} + \sqrt{x})(\sqrt{3} - \sqrt{x})$

    b. $(\sqrt{3} + \sqrt{x})(\sqrt{3} + \sqrt{x})$

    c. $(\sqrt{3} - \sqrt{x})(\sqrt{3} - \sqrt{x})$

68. a. $(\sqrt{5} + \sqrt{y})(\sqrt{5} - \sqrt{y})$

    b. $(\sqrt{5} + \sqrt{y})(\sqrt{5} + \sqrt{y})$

    c. $(\sqrt{5} - \sqrt{y})(\sqrt{5} - \sqrt{y})$

Mixed Exercises

For Exercises 69–76, identify each statement as true or false. If an answer is false, explain why.

69. $\sqrt{3} \cdot \sqrt{2} = \sqrt{6}$

70. $\sqrt{5} \cdot \sqrt[3]{2} = \sqrt{10}$

71. $(x - \sqrt{5})^2 = x - 5$

72. $3(2\sqrt{5x}) = 6\sqrt{5x}$

73. $5(3\sqrt{4x}) = 15\sqrt{20x}$

74. $\frac{\sqrt{5x}}{5} = \sqrt{x}$

75. $\frac{3\sqrt{x}}{3} = \sqrt{x}$

76. $(\sqrt{t} - 1)(\sqrt{t} + 1) = t - 1$

For Exercises 77–84, perform the indicated operations.

77. $(-\sqrt{6x})^2$

78. $(-\sqrt{8a})^2$

79. $(\sqrt{3x} + 1)^2$

80. $(\sqrt{x} - 1)^2$

81. $(\sqrt{x + 3} - 4)^2$

82. $(\sqrt{x + 1} + 3)^2$

83. $(\sqrt{2t - 3} + 5)^2$

84. $(\sqrt{3w - 2} - 4)^2$

For Exercises 85–88, find the exact area.

85.

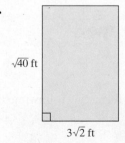

$\sqrt{40}$ ft

$3\sqrt{2}$ ft

86.

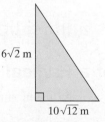

$6\sqrt{2}$ m

$10\sqrt{12}$ m

87.

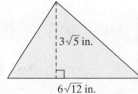

$3\sqrt{5}$ in.

$6\sqrt{12}$ in.

88.

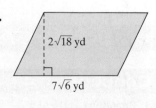

$2\sqrt{18}$ yd

$7\sqrt{6}$ yd

Concept 4: Multiplying Radicals with Different Indices

For Exercises 89–100, multiply or divide the radicals with different indices. Write the answers in radical form and simplify. **(See Example 8.)**

89. $\sqrt{x} \cdot \sqrt[4]{x}$

90. $\sqrt[3]{y} \cdot \sqrt{y}$

91. $\sqrt[5]{2z} \cdot \sqrt[3]{2z}$

92. $\sqrt[3]{5w} \cdot \sqrt[4]{5w}$

93. $\sqrt[3]{p^2} \cdot \sqrt{p^3}$

94. $\sqrt[4]{q^3} \cdot \sqrt[3]{q^2}$

95. $\frac{\sqrt{u^3}}{\sqrt[3]{u}}$

96. $\frac{\sqrt{v^5}}{\sqrt[4]{v}}$

97. $\sqrt[3]{x} \cdot \sqrt[6]{y}$ **98.** $\sqrt{a} \cdot \sqrt[6]{b}$ **99.** $\sqrt[4]{8} \cdot \sqrt{3}$ **100.** $\sqrt{11} \cdot \sqrt[6]{2}$

Expanding Your Skills

101. Multiply $(\sqrt[3]{a} + \sqrt[3]{b})(\sqrt[3]{a^2} - \sqrt[3]{ab} + \sqrt[3]{b^2})$.   **102.** Multiply $(\sqrt[3]{a} - \sqrt[3]{b})(\sqrt[3]{a^2} + \sqrt[3]{ab} + \sqrt[3]{b^2})$.

Problem Recognition Exercises

Simplifying Radical Expressions

For Exercises 1–20, simplify the expressions.

1. a. $\sqrt{24}$
   b. $\sqrt[3]{24}$

2. a. $\sqrt{54}$
   b. $\sqrt[3]{54}$

3. a. $\sqrt{200y^6}$
   b. $\sqrt[3]{200y^6}$

4. a. $\sqrt{32z^{15}}$
   b. $\sqrt[3]{32z^{15}}$

5. a. $\sqrt{80}$
   b. $\sqrt[3]{80}$
   c. $\sqrt[4]{80}$

6. a. $\sqrt{48}$
   b. $\sqrt[3]{48}$
   c. $\sqrt[4]{48}$

7. a. $\sqrt{x^5y^6}$
   b. $\sqrt[3]{x^5y^6}$
   c. $\sqrt[4]{x^5y^6}$

8. a. $\sqrt{a^{10}b^9}$
   b. $\sqrt[3]{a^{10}b^9}$
   c. $\sqrt[4]{a^{10}b^9}$

9. a. $\sqrt[3]{32s^5t^6}$
   b. $\sqrt[4]{32s^5t^6}$
   c. $\sqrt[5]{32s^5t^6}$

10. a. $\sqrt[3]{96v^7w^{20}}$
   b. $\sqrt[4]{96v^7w^{20}}$
   c. $\sqrt[5]{96v^7w^{20}}$

11. a. $\sqrt{5} + \sqrt{5}$
   b. $\sqrt{5} \cdot \sqrt{5}$

12. a. $\sqrt{10} + \sqrt{10}$
   b. $\sqrt{10} \cdot \sqrt{10}$

13. a. $2\sqrt{6} - 5\sqrt{6}$
   b. $2\sqrt{6} \cdot 5\sqrt{6}$

14. a. $3\sqrt{7} - 10\sqrt{7}$
   b. $3\sqrt{7} \cdot 10\sqrt{7}$

15. a. $\sqrt{8} + \sqrt{2}$
   b. $\sqrt{8} \cdot \sqrt{2}$

16. a. $\sqrt{12} + \sqrt{3}$
   b. $\sqrt{12} \cdot \sqrt{3}$

17. a. $5\sqrt{18} - 4\sqrt{8}$
   b. $5\sqrt{18} \cdot 4\sqrt{8}$

18. a. $\sqrt{50} - \sqrt{72}$
   b. $\sqrt{50} \cdot \sqrt{72}$

19. a. $4\sqrt[3]{24} + 6\sqrt[3]{3}$
   b. $4\sqrt[3]{24} \cdot 6\sqrt[3]{3}$

20. a. $2\sqrt[3]{54} - 5\sqrt[3]{2}$
   b. $2\sqrt[3]{54} \cdot 5\sqrt[3]{2}$

Section 7.6   Division of Radicals and Rationalization

Concepts

1. Simplified Form of a Radical
2. Division Property of Radicals
3. Rationalizing the Denominator—One Term
4. Rationalizing the Denominator—Two Terms

1. Simplified Form of a Radical

Recall that for a radical expression to be in simplified form the following three conditions must be met.

Simplified Form of a Radical

Consider any radical expression in which the radicand is written as a product of prime factors. The expression is in simplified form if all the following conditions are met:

1. The radicand has no factor raised to a power greater than or equal to the index.
2. The radicand does not contain a fraction.
3. No radicals are in the denominator of a fraction.

In the previous sections, we have concentrated on the first condition in the simplification process. Next, we will demonstrate how to satisfy the second and third conditions involving radicals and fractions.

2. Division Property of Radicals

The multiplication property of radicals enables a product within a radical to be separated and written as a product of radicals. We now state a similar property for radicals involving quotients.

> **Division Property of Radicals**
>
> Let a and b represent real numbers such that $\sqrt[n]{a}$ and $\sqrt[n]{b}$ are both real. Then,
>
> $$\sqrt[n]{\dfrac{a}{b}} = \dfrac{\sqrt[n]{a}}{\sqrt[n]{b}} \quad b \neq 0$$

The division property of radicals indicates that a quotient within a radicand can be written as a quotient of radicals provided the roots are real numbers. For example:

$$\sqrt{\dfrac{25}{36}} = \dfrac{\sqrt{25}}{\sqrt{36}}$$

The reverse process is also true. A quotient of radicals can be written as a single radical provided that the roots are real numbers and that they have the same index.

$$\text{Same index} \longrightarrow \dfrac{\sqrt[3]{125}}{\sqrt[3]{8}} = \sqrt[3]{\dfrac{125}{8}}$$

In Examples 1 and 2, we will apply the division property of radicals to simplify radical expressions.

Example 1 Using the Division Property to Simplify Radicals

Use the division property of radicals to simplify the expressions. Assume the variables represent positive real numbers.

a. $\sqrt{\dfrac{a^6}{b^4}}$ **b.** $\sqrt[3]{\dfrac{81y^5}{x^3}}$

Solution:

a. $\sqrt{\dfrac{a^6}{b^4}}$ The radicand contains an irreducible fraction.

$= \dfrac{\sqrt{a^6}}{\sqrt{b^4}}$ Apply the division property to rewrite as a quotient of radicals.

$= \dfrac{a^3}{b^2}$ Simplify the radicals.

b. $\sqrt[3]{\dfrac{81y^5}{x^3}}$ The radicand contains an irreducible fraction.

$= \dfrac{\sqrt[3]{81y^5}}{\sqrt[3]{x^3}}$ Apply the division property to rewrite as a quotient of radicals.

$= \dfrac{\sqrt[3]{3^4 \cdot y^5}}{\sqrt[3]{x^3}}$ Factor the radicand in the numerator to simplify the radical.

$= \dfrac{3y\sqrt[3]{3y^2}}{x}$ Simplify the radicals in the numerator and the denominator. The expression is simplified because it now satisfies all conditions.

Example 2   **Using the Division Property to Simplify Radicals**

Use the division property of radicals to simplify the expressions. Assume the variables represent positive real numbers.

$$\frac{\sqrt[4]{8p^7}}{\sqrt[4]{p^3}}$$

Solution:

$\dfrac{\sqrt[4]{8p^7}}{\sqrt[4]{p^3}}$ There is a radical in the denominator of the fraction.

$= \sqrt[4]{\dfrac{8p^7}{p^3}}$ Apply the division property to write the quotient under a single radical.

$= \sqrt[4]{8p^4}$ Simplify the fraction.

$= p\sqrt[4]{8}$ Simplify the radical.

---

3. Rationalizing the Denominator—One Term

The third condition restricts radicals from the denominator of an expression. The process to remove a radical from the denominator is called **rationalizing the denominator**. In many cases, rationalizing the denominator creates an expression that is computationally simpler. In Examples 5 and 8 we will show that

$$\frac{6}{\sqrt{3}} = 2\sqrt{3} \quad \text{and} \quad \frac{-2}{2 + \sqrt{6}} = 2 - \sqrt{6}$$

We will demonstrate the process to rationalize the denominator as two separate cases:

- Rationalizing the denominator (one term)
- Rationalizing the denominator (two terms involving square roots)

To begin the first case, recall that the nth root of a perfect nth power simplifies completely.

$$\sqrt{x^2} = x \qquad x \geq 0$$
$$\sqrt[3]{x^3} = x$$
$$\sqrt[4]{x^4} = x \qquad x \geq 0$$
$$\sqrt[5]{x^5} = x$$

$\cdots$

Therefore, to rationalize a radical expression, use the multiplication property of radicals to create an nth root of an nth power.

| **Example 3** | **Rationalizing Radical Expressions** |

Fill in the missing radicand to rationalize the radical expressions. Assume all variables represent positive real numbers.

a. $\sqrt{a} \cdot \sqrt{?} = \sqrt{a^2} = a$ **b.** $\sqrt[3]{y} \cdot \sqrt[3]{?} = \sqrt[3]{y^3} = y$

c. $\sqrt[4]{2z^3} \cdot \sqrt[4]{?} - \sqrt[4]{2^4z^4} = 2z$

Solution:

a. $\sqrt{a} \cdot \sqrt{?} = \sqrt{a^2} = a$ What multiplied by $\sqrt{a}$ will equal $\sqrt{a^2}$?

$\sqrt{a} \cdot \sqrt{a} = \sqrt{a^2} = a$

b. $\sqrt[3]{y} \cdot \sqrt[3]{?} = \sqrt[3]{y^3} = y$ What multiplied by $\sqrt[3]{y}$ will equal $\sqrt[3]{y^3}$?

$\sqrt[3]{y} \cdot \sqrt[3]{y^2} = \sqrt[3]{y^3} = y$

c. $\sqrt[4]{2z^3} \cdot \sqrt[4]{?} = \sqrt[4]{2^4z^4} = 2z$ What multiplied by $\sqrt[4]{2z^3}$ will equal $\sqrt[4]{2^4z^4}$?

$\sqrt[4]{2z^3} \cdot \sqrt[4]{2^3z} = \sqrt[4]{2^4z^4} = 2z$

To rationalize the denominator of a radical expression, multiply the numerator and denominator by an appropriate expression to create an nth root of an nth power in the denominator.

| **Example 4** | **Rationalizing the Denominator—One Term** |

Simplify the expression. $\dfrac{5}{\sqrt[3]{a}}$ $a \neq 0$

Solution:

To remove the radical from the denominator, a cube root of a perfect cube is needed in the denominator. Multiply numerator and denominator by $\sqrt[3]{a^2}$ because $\sqrt[3]{a} \cdot \sqrt[3]{a^2} = \sqrt[3]{a^3} = a$.

$$\frac{5}{\sqrt[3]{a}} = \frac{5}{\sqrt[3]{a}} \cdot \frac{\sqrt[3]{a^2}}{\sqrt[3]{a^2}}$$

$$= \frac{5\sqrt[3]{a^2}}{\sqrt[3]{a^3}} \qquad \text{Multiply the radicals.}$$

$$= \frac{5\sqrt[3]{a^2}}{a} \qquad \text{Simplify.}$$

Note that for $a \neq 0$, the expression $\dfrac{\sqrt[3]{a^2}}{\sqrt[3]{a^2}} = 1$. In Example 4, multiplying the

expression $\dfrac{5}{\sqrt[3]{a}}$ by this ratio does not change its value.

Skill Practice

Simplify the expression.

12. $\dfrac{12}{\sqrt{2}}$

Example 5 Rationalizing the Denominator—One Term

Simplify the expression. $\dfrac{6}{\sqrt{3}}$

Solution:

To rationalize the denominator, a square root of a perfect square is needed. Multiply numerator and denominator by $\sqrt{3}$ because $\sqrt{3} \cdot \sqrt{3} = \sqrt{3^2} = 3$.

$$\dfrac{6}{\sqrt{3}} = \dfrac{6}{\sqrt{3}} \cdot \dfrac{\sqrt{3}}{\sqrt{3}} \qquad \text{Rationalize the denominator.}$$

$$= \dfrac{6\sqrt{3}}{\sqrt{3^2}} \qquad \text{Multiply the radicals.}$$

$$= \dfrac{6\sqrt{3}}{3} \qquad \text{Simplify.}$$

$$= 2\sqrt{3} \qquad \text{Reduce to lowest terms.}$$

Skill Practice

Simplify the expression.

13. $\sqrt{\dfrac{8}{3}}$

Example 6 Rationalizing the Denominator—One Term

Simplify the expression. $\sqrt{\dfrac{y^5}{7}}$

Solution:

$$\sqrt{\dfrac{y^5}{7}} \qquad \text{The radical contains an irreducible fraction.}$$

$$= \dfrac{\sqrt{y^5}}{\sqrt{7}} \qquad \text{Apply the division property of radicals.}$$

$$= \dfrac{y^2\sqrt{y}}{\sqrt{7}} \qquad \text{Simplify the radical in the numerator.}$$

$$= \dfrac{y^2\sqrt{y}}{\sqrt{7}} \cdot \dfrac{\sqrt{7}}{\sqrt{7}} \qquad \text{Rationalize the denominator.}$$

Note: $\sqrt{7} \cdot \sqrt{7} = \sqrt{7^2} = 7$.

$$= \dfrac{y^2\sqrt{7y}}{\sqrt{7^2}}$$

$$= \dfrac{y^2\sqrt{7y}}{7} \qquad \text{Simplify.}$$

> **Avoiding Mistakes**
>
> A factor within a radicand cannot be simplified with a factor outside the radicand. For example, $\dfrac{\sqrt{7y}}{7}$ cannot be simplified.

Answers

12. $6\sqrt{2}$ **13.** $\dfrac{2\sqrt{6}}{3}$

Example 7 Rationalizing the Denominator—One Term

Simplify the expression. $\dfrac{15}{\sqrt[3]{25s}}$

Solution:

$\dfrac{15}{\sqrt[3]{25s}}$

$= \dfrac{15}{\sqrt[3]{5^2 s}} \cdot \dfrac{\sqrt[3]{5s^2}}{\sqrt[3]{5s^2}}$ Because $25 = 5^2$, one additional factor of 5 is needed to form a perfect cube. Two additional factors of s are needed to make a perfect cube. Multiply numerator and denominator by $\sqrt[3]{5s^2}$.

$= \dfrac{15\sqrt[3]{5s^2}}{\sqrt[3]{5^3 s^3}}$

$= \dfrac{15\sqrt[3]{5s^2}}{5s}$ Simplify the perfect cube.

$= \dfrac{\overset{3}{\cancel{15}}\sqrt[3]{5s^2}}{\underset{1}{\cancel{5}}s}$ Reduce to lowest terms.

$= \dfrac{3\sqrt[3]{5s^2}}{s}$

TIP: In the expression $\dfrac{15\sqrt[3]{5s^2}}{5s}$, the factor of 15 and the factor of 5 may be reduced because both are outside the radical.

$$\dfrac{15\sqrt[3]{5s^2}}{5s} = \dfrac{15}{5} \cdot \dfrac{\sqrt[3]{5s^2}}{s} = \dfrac{3\sqrt[3]{5s^2}}{s}$$

4. Rationalizing the Denominator—Two Terms

Example 8 demonstrates how to rationalize a two-term denominator involving square roots.

First recall from the multiplication of polynomials that the product of two conjugates results in a difference of squares.

$$(a + b)(a - b) = a^2 - b^2$$

If either a or b has a square root factor, the expression will simplify without a radical. That is, the expression is *rationalized*. For example:

$$(2 + \sqrt{6})(2 - \sqrt{6}) = (2)^2 - (\sqrt{6})^2$$

$$= 4 - 6$$

$$= -2$$

Skill Practice

Simplify the expression.

14. $\dfrac{18}{\sqrt[3]{3y^2}}$

Concept Connections

15. Fill in the blank to rationalize the expression.

$(5 - \sqrt{x})(\underline{\quad})$

Answers

14. $\dfrac{6\sqrt[3]{9y}}{y}$ **15.** $(5 + \sqrt{x})$

Example 8 **Rationalizing the Denominator—Two Terms**

Simplify the expression by rationalizing the denominator.

$$\frac{-2}{2 + \sqrt{6}}$$

Solution:

$$\frac{-2}{2 + \sqrt{6}}$$

$$= \frac{(-2)}{(2 + \sqrt{6})} \cdot \frac{(2 - \sqrt{6})}{(2 - \sqrt{6})} \qquad \text{Multiply the numerator and denominator by the conjugate of the denominator.}$$

$$\underset{\text{conjugates}}{}$$

$$= \frac{-2(2 - \sqrt{6})}{(2)^2 - (\sqrt{6})^2} \qquad \text{In the denominator, apply the formula } (a + b)(a - b) = a^2 - b^2.$$

$$= \frac{-2(2 - \sqrt{6})}{4 - 6} \qquad \text{Simplify.}$$

$$= \frac{-2(2 - \sqrt{6})}{-2}$$

$$= \frac{-\cancel{2}(2 - \sqrt{6})}{-\cancel{2}}$$

$$= 2 - \sqrt{6}$$

Example 9 **Rationalizing the Denominator—Two Terms**

Rationalize the denominator of the expression. Assume all variables represent positive real numbers and $c \neq d$.

$$\frac{\sqrt{c} + \sqrt{d}}{\sqrt{c} - \sqrt{d}}$$

Solution:

$$\frac{\sqrt{c} + \sqrt{d}}{\sqrt{c} - \sqrt{d}}$$

$$= \frac{(\sqrt{c} + \sqrt{d})}{(\sqrt{c} - \sqrt{d})} \cdot \frac{(\sqrt{c} + \sqrt{d})}{(\sqrt{c} + \sqrt{d})} \qquad \text{Multiply numerator and denominator by the conjugate of the denominator.}$$

$$\underset{\text{conjugates}}{}$$

$$= \frac{(\sqrt{c} + \sqrt{d})^2}{(\sqrt{c})^2 - (\sqrt{d})^2} \qquad \text{In the denominator apply the formula } (a + b)(a - b) = a^2 - b^2.$$

$$= \frac{(\sqrt{c})^2 + 2\sqrt{c}\sqrt{d} + (\sqrt{d})^2}{c - d} \qquad \text{In the numerator apply the formula } (a + b)^2 = a^2 + 2ab + b^2.$$

$$= \frac{c + 2\sqrt{cd} + d}{c - d}$$

Answers

16. $6 - 2\sqrt{5}$

17. $\dfrac{3 - 2\sqrt{3y} + y}{3 - y}$

Section 7.6 Practice Exercises

For the exercises in this set, assume that all variables represent positive real numbers unless otherwise stated.

Vocabulary and Key Concepts

1. a. In the simplified form of a radical expression, no _____ may appear in the denominator of a fraction.

b. The division property of radicals indicates that if $\sqrt[n]{a}$ and $\sqrt[n]{b}$ are real numbers, then $\sqrt[n]{\dfrac{a}{b}} = \dfrac{\square}{\square}$ provided that $b \neq 0$.

c. The simplified form of the expression $\sqrt[3]{\dfrac{64}{x^6}}$ is _____.

d. The process of removing a radical from the denominator of a fraction is called _____ the denominator.

e. The expression $\dfrac{\sqrt{3}}{3}$ (is/is not) in simplified form whereas $\dfrac{3}{\sqrt{3}}$ (is/is not) in simplified form.

f. To rationalize the denominator for the expression $\dfrac{\sqrt{x}+3}{\sqrt{x}-2}$, multiply numerator and denominator by the conjugate of the _____.

Review Exercises

2. Multiply and simplify. $\sqrt{x} \cdot \sqrt{x}$

For Exercises 3–10, perform the indicated operations.

3. $2y\sqrt{45} + 3\sqrt{20y^2}$ **4.** $3x\sqrt{72x} - 9\sqrt{50x^3}$ **5.** $(-6\sqrt{y}+3)(3\sqrt{y}+1)$ **6.** $(\sqrt{w}+12)(2\sqrt{w}-4)$

7. $(8 - \sqrt{t})^2$ **8.** $(\sqrt{p}+4)^2$ **9.** $(\sqrt{2}+\sqrt{7})(\sqrt{2}-\sqrt{7})$ **10.** $(\sqrt{3}+5)(\sqrt{3}-5)$

Concept 2: Division Property of Radicals

For Exercises 11–26, simplify using the division property of radicals. Assume all variables represent positive real numbers. **(See Examples 1 and 2.)**

11. $\sqrt{\dfrac{49x^4}{y^6}}$ **12.** $\sqrt{\dfrac{100p^2}{q^8}}$ **13.** $\sqrt{\dfrac{8a^2}{x^6}}$ **14.** $\sqrt{\dfrac{4w^3}{25y^4}}$

15. $\sqrt[3]{\dfrac{-16j^3}{k^3}}$ **16.** $\sqrt[5]{\dfrac{32x}{y^{10}}}$ **17.** $\sqrt[4]{\dfrac{3s^2t^4}{10,000}}$ **18.** $\sqrt[4]{\dfrac{x^7y}{z^8}}$

19. $\dfrac{\sqrt{72ab^5}}{\sqrt{8ab}}$ **20.** $\dfrac{\sqrt{6x^3}}{\sqrt{24x}}$ **21.** $\dfrac{\sqrt[4]{3b^3}}{\sqrt[4]{48b^{11}}}$ **22.** $\dfrac{\sqrt[3]{128wz^8}}{\sqrt[3]{2wz^2}}$

23. $\dfrac{\sqrt{54x^5}}{\sqrt{3x}}$ **24.** $\dfrac{\sqrt{48z^3}}{\sqrt{6z}}$ **25.** $\dfrac{\sqrt{3yz^2}}{\sqrt{w^4}}$ **26.** $\dfrac{\sqrt{50x^3z}}{\sqrt{9y^4}}$

Concept 3: Rationalizing the Denominator—One Term

The radical expressions in Exercises 27–34 have radicals in the denominator. Fill in the missing radicands to rationalize the denominators. **(See Example 3.)**

27. $\dfrac{x}{\sqrt{5}} = \dfrac{x}{\sqrt{5}} \cdot \dfrac{\sqrt{?}}{\sqrt{?}}$

28. $\dfrac{2}{\sqrt{x}} = \dfrac{2}{\sqrt{x}} \cdot \dfrac{\sqrt{?}}{\sqrt{?}}$

29. $\dfrac{7}{\sqrt[3]{x}} = \dfrac{7}{\sqrt[3]{x}} \cdot \dfrac{\sqrt[3]{?}}{\sqrt[3]{?}}$

30. $\dfrac{5}{\sqrt[4]{y}} = \dfrac{5}{\sqrt[4]{y}} \cdot \dfrac{\sqrt[4]{?}}{\sqrt[4]{?}}$

31. $\dfrac{8}{\sqrt{3z}} = \dfrac{8}{\sqrt{3z}} \cdot \dfrac{\sqrt{??}}{\sqrt{??}}$

32. $\dfrac{10}{\sqrt{7w}} = \dfrac{10}{\sqrt{7w}} \cdot \dfrac{\sqrt{??}}{\sqrt{??}}$

33. $\dfrac{1}{\sqrt[4]{8a^2}} = \dfrac{1}{\sqrt[4]{8a^2}} \cdot \dfrac{\sqrt[4]{??}}{\sqrt[4]{??}}$

34. $\dfrac{1}{\sqrt[3]{9b^2}} = \dfrac{1}{\sqrt[3]{9b^2}} \cdot \dfrac{\sqrt[3]{??}}{\sqrt[3]{??}}$

For Exercises 35–62, rationalize the denominator. **(See Examples 4–7.)**

35. $\dfrac{1}{\sqrt{3}}$

36. $\dfrac{1}{\sqrt{7}}$

37. $\sqrt{\dfrac{1}{x}}$

38. $\sqrt{\dfrac{1}{z}}$

39. $\dfrac{6}{\sqrt{2y}}$

40. $\dfrac{9}{\sqrt{3t}}$

41. $\sqrt{\dfrac{a^3}{2}}$

42. $\sqrt{\dfrac{b^3}{3}}$

43. $\dfrac{6}{\sqrt{8}}$

44. $\dfrac{2}{\sqrt{48}}$

45. $\dfrac{3}{\sqrt[3]{2}}$

46. $\dfrac{2}{\sqrt[3]{7}}$

47. $\dfrac{-6}{\sqrt[4]{x}}$

48. $\dfrac{-2}{\sqrt[5]{y}}$

49. $\dfrac{7}{\sqrt[3]{4}}$

50. $\dfrac{1}{\sqrt[3]{9}}$

51. $\sqrt[3]{\dfrac{4}{w^2}}$

52. $\sqrt[3]{\dfrac{5}{z^2}}$

53. $\sqrt[4]{\dfrac{16}{3}}$

54. $\sqrt[4]{\dfrac{81}{8}}$

55. $\dfrac{2}{\sqrt[3]{4x^2}}$

56. $\dfrac{6}{\sqrt[3]{3y^2}}$

57. $\dfrac{8}{7\sqrt{24}}$

58. $\dfrac{5}{3\sqrt{50}}$

59. $\dfrac{1}{\sqrt{x^7}}$

60. $\dfrac{1}{\sqrt{y^5}}$

61. $\dfrac{2}{\sqrt{8x^5}}$

62. $\dfrac{6}{\sqrt{27t^7}}$

Concept 4: Rationalizing the Denominator—Two Terms

63. What is the conjugate of $\sqrt{2} - \sqrt{6}$?

64. What is the conjugate of $\sqrt{11} + \sqrt{5}$?

65. What is the conjugate of $\sqrt{x} + 23$?

66. What is the conjugate of $17 - \sqrt{y}$?

For Exercises 67–86, rationalize the denominators. **(See Examples 8 and 9.)**

67. $\dfrac{4}{\sqrt{2} + 3}$

68. $\dfrac{6}{4 - \sqrt{3}}$

69. $\dfrac{8}{\sqrt{6} - 2}$

70. $\dfrac{-12}{\sqrt{5} - 3}$

71. $\dfrac{\sqrt{7}}{\sqrt{3} + 2}$

72. $\dfrac{\sqrt{8}}{\sqrt{3} + 1}$

73. $\dfrac{-1}{\sqrt{p} + \sqrt{q}}$

74. $\dfrac{6}{\sqrt{a} - \sqrt{b}}$

75. $\dfrac{x-5}{\sqrt{x}+\sqrt{5}}$

76. $\dfrac{y-2}{\sqrt{y}-\sqrt{2}}$

77. $\dfrac{-7}{2\sqrt{a}-5\sqrt{b}}$

78. $\dfrac{-4}{3\sqrt{w}-2\sqrt{z}}$

79. $\dfrac{3\sqrt{x}-\sqrt{y}}{\sqrt{y}+\sqrt{x}}$

80. $\dfrac{2\sqrt{a}+\sqrt{b}}{\sqrt{b}-\sqrt{a}}$

81. $\dfrac{3\sqrt[3]{10}}{2+\sqrt{10}}$

82. $\dfrac{4\sqrt{7}}{3+\sqrt{7}}$

83. $\dfrac{2\sqrt{3}+\sqrt{7}}{3\sqrt{3}-\sqrt{7}}$

84. $\dfrac{5\sqrt{2}-\sqrt{5}}{5\sqrt{2}+\sqrt{5}}$

85. $\dfrac{\sqrt{5}+4}{2-\sqrt{5}}$

86. $\dfrac{3+\sqrt{2}}{\sqrt{2}-5}$

Mixed Exercises

For Exercises 87–90, translate the English phrase to an algebraic expression. Then simplify the expression.

87. 16 divided by the cube root of 4

88. 21 divided by the principal fourth root of 27

89. 4 divided by the difference of x and the principal square root of 2

90. 8 divided by the sum of y and the principal square root of 3

91. The time $T(x)$ (in seconds) for a pendulum to make one complete swing back and forth is approximated by

$$T(x) = 2\pi\sqrt{\dfrac{x}{32}}$$

where x is the length of the pendulum in feet.

Determine the exact time required for one swing for a pendulum that is 1 ft long. Then approximate the time to the nearest hundredth of a second.

92. An object is dropped off a building x meters tall. The time $T(x)$ (in seconds) required for the object to hit the ground is given by

$$T(x) = \sqrt{\dfrac{10x}{49}}$$

Find the exact time required for the object to hit the ground if it is dropped off the First National Plaza Building in Chicago, a height of 230 m. Then round the time to the nearest hundredth of a second.

For Exercises 93–96, rationalize the denominator.

93. a. $\dfrac{1}{\sqrt{2}}$

b. $\dfrac{1}{\sqrt[3]{2}}$

94. a. $\dfrac{1}{\sqrt[3]{x}}$

b. $\dfrac{1}{\sqrt[3]{x^2}}$

95. a. $\dfrac{1}{\sqrt{5a}}$

b. $\dfrac{1}{\sqrt{5}+a}$

96. a. $\dfrac{4}{\sqrt{2x}}$

b. $\dfrac{4}{2-\sqrt{x}}$

Expanding Your Skills

For Exercises 97–102, simplify each term of the expression. Then add or subtract as indicated.

97. $\dfrac{\sqrt{6}}{2} + \dfrac{1}{\sqrt{6}}$

98. $\dfrac{1}{\sqrt{7}} + \sqrt{7}$

99. $\sqrt{15} - \sqrt{\dfrac{3}{5}} + \sqrt{\dfrac{5}{3}}$

100. $\sqrt{\dfrac{6}{2}} - \sqrt{12} + \sqrt{\dfrac{2}{6}}$

101. $\sqrt[3]{25} + \dfrac{3}{\sqrt[3]{5}}$

102. $\dfrac{1}{\sqrt[3]{4}} + \sqrt[3]{54}$

For Exercises 103–106, rationalize the numerator by multiplying both numerator and denominator by the conjugate of the numerator.

103. $\dfrac{\sqrt{3} + 6}{2}$

104. $\dfrac{\sqrt{7} - 2}{5}$

105. $\dfrac{\sqrt{a} - \sqrt{b}}{\sqrt{a} + \sqrt{b}}$

106. $\dfrac{\sqrt{p} + \sqrt{q}}{\sqrt{p} - \sqrt{q}}$

Section 7.7   Radical Equations and Applications

Concepts

1. Solutions to Radical Equations

2. Solving Radical Equations Involving One Radical

3. Solving Radical Equations Involving More than One Radical

4. Applications of Radical Equations and Functions

1. Solutions to Radical Equations

An equation with one or more radicals containing a variable is called a **radical equation**. For example, $\sqrt[3]{x} = 5$ is a radical equation. Recall that $(\sqrt[n]{a})^n = a$, provided $\sqrt[n]{a}$ is a real number. The basis of solving a radical equation is to eliminate the radical by raising both sides of the equation to a power equal to the index of the radical.

To solve the equation $\sqrt[3]{x} = 5$, cube both sides of the equation.

$$\sqrt[3]{x} = 5$$
$$(\sqrt[3]{x})^3 = (5)^3$$
$$x = 125$$

By raising each side of a radical equation to a power equal to the index of the radical, a new equation is produced. Note, however, that some of or all the solutions to the new equation may *not* be solutions to the original radical equation. For this reason, it is necessary to *check all potential solutions* in the original equation.

For example, consider the equation $\sqrt{x} = -7$. This equation has no solution because by definition, the principal square root of x must be a nonnegative number. However, if we square both sides of the equation, it appears as though a solution exists.

$$(\sqrt{x})^2 = (-7)^2$$
$$x = 49 \qquad \text{The value 49 does not check in the original equation}$$
$$\sqrt{x} = -7. \text{ Therefore, 49 is an } \textbf{extraneous solution}.$$

Concept Connections

1. The equation shown must be raised to what power to clear the radical?
$$\sqrt[4]{a} = 2$$

Solving a Radical Equation

Step 1 Isolate the radical. If an equation has more than one radical, choose one of the radicals to isolate.

Step 2 Raise each side of the equation to a power equal to the index of the radical.

Step 3 Solve the resulting equation. If the equation still has a radical, repeat steps 1 and 2.

*Step 4 Check the potential solutions in the original equation.

*In solving a radical equation, extraneous solutions *potentially* occur only when each side of the equation is raised to an even power.

Answer

1. Fourth power

2. Solving Radical Equations Involving One Radical

Example 1 Solving an Equation Containing One Radical

Solve the equation. $\sqrt{p} + 5 = 9$

Solution:

$$\sqrt{p} + 5 = 9$$

$\sqrt{p} = 4$ Isolate the radical.

$(\sqrt{p})^2 = 4^2$ Because the index is 2, square both sides.

$p = 16$

Check: $p = 16$ Check $p = 16$ as a potential solution.

$\sqrt{p} + 5 = 9$

$\sqrt{16} + 5 \stackrel{?}{=} 9$

$4 + 5 \stackrel{?}{=} 9 \checkmark$ True, 16 is a solution to the original equation.

Skill Practice

Solve the equation.
2. $\sqrt{x} - 3 = 2$

Example 2 Solving an Equation Containing One Radical

Solve the equation. $\sqrt{3x - 2} + 4 = 5$

Solution:

$$\sqrt{3x - 2} + 4 = 5$$

$\sqrt{3x - 2} = 1$ Isolate the radical.

$(\sqrt{3x - 2})^2 = (1)^2$ Because the index is 2, square both sides.

$3x - 2 = 1$ Simplify.

$3x = 3$ Solve the resulting equation.

$x = 1$

Check: $x = 1$ Check $x = 1$ as a potential solution.

$\sqrt{3x - 2} + 4 = 5$

$\sqrt{3(1) - 2} + 4 \stackrel{?}{=} 5$

$\sqrt{1} + 4 \stackrel{?}{=} 5$

$5 \stackrel{?}{=} 5 \checkmark$ True, 1 is a solution to the original equation.

Skill Practice

Solve.
3. $\sqrt{5y + 1} - 2 = 4$

Answers
2. 25   3. 7

Skill Practice

Solve.

4. $(t + 2)^{1/3} + 5 = 3$

Example 3 **Solving an Equation Containing One Radical**

Solve the equation. $(w - 1)^{1/3} - 2 = 2$

Solution:

$(w - 1)^{1/3} - 2 = 2$ Note that $(w - 1)^{1/3} = \sqrt[3]{w - 1}$.

$\sqrt[3]{w - 1} - 2 = 2$

$\sqrt[3]{w - 1} = 4$ Isolate the radical.

$(\sqrt[3]{w - 1})^3 = (4)^3$ Because the index is 3, cube both sides.

$w - 1 = 64$ Simplify.

$w = 65$

Check: $w = 65$

$(w - 1)^{1/3} - 2 = 2$ Check $w = 65$ as a potential solution.

$\sqrt[3]{65 - 1} - 2 \stackrel{?}{=} 2$

$\sqrt[3]{64} - 2 \stackrel{?}{=} 2$

$4 - 2 \stackrel{?}{=} 2$

$2 \stackrel{?}{=} 2$ ✔   True, 65 is a solution to the original equation.

Skill Practice

Solve.

5. $3 = 6 + \sqrt[4]{b - 1}$

TIP: After isolating the radical in Example 4, the equation shows a fourth root equated to a negative number:

$-2 = \sqrt[4]{x + 3}$

By definition, a principal fourth root of any real number must be nonnegative. Therefore, there can be no real solution to this equation.

Example 4 **Solving an Equation Containing One Radical**

Solve the equation. $7 = \sqrt[4]{x + 3} + 9$

Solution:

$7 = \sqrt[4]{x + 3} + 9$

$-2 = \sqrt[4]{x + 3}$ Isolate the radical.

$(-2)^4 = (\sqrt[4]{x + 3})^4$ Because the index is 4, raise both sides to the fourth power.

$16 = x + 3$

$x = 13$ Solve for x.

Check: $x = 13$

$7 = \sqrt[4]{x + 3} + 9$

$7 \stackrel{?}{=} \sqrt[4]{(13) + 3} + 9$

$7 \stackrel{?}{=} \sqrt[4]{16} + 9$

$7 \stackrel{?}{=} 2 + 9$ (false)   13 is *not* a solution to the original equation.

The equation $7 = \sqrt[4]{x + 3} + 9$ has no solution.

Answers

4. -10

5. No solution (The value 82 does not check.)

Example 5 Solving an Equation Containing One Radical

Solve the equation. $y + \sqrt{y - 2} = 8$

Solution:

$$y + \sqrt{y - 2} = 8$$

$$\sqrt{y - 2} = 8 - y \qquad \text{Isolate the radical.}$$

$$(\sqrt{y - 2})^2 = (8 - y)^2 \qquad \text{Because the index is 2, square both sides.}$$

Note that $(8 - y)^2 = (8 - y)(8 - y) = 64 - 16y + y^2$.

$$y - 2 = 64 - 16y + y^2 \qquad \text{Simplify.}$$

$$0 = y^2 - 17y + 66 \qquad \begin{array}{l}\text{The equation is quadratic. Set one} \\ \text{side equal to zero. Write the other} \\ \text{side in descending order.}\end{array}$$

$$0 = (y - 11)(y - 6) \qquad \text{Factor.}$$

$$y - 11 = 0 \quad \text{or} \quad y - 6 = 0 \qquad \text{Set each factor equal to zero.}$$

$$y = 11 \quad \text{or} \quad y = 6 \qquad \text{Solve.}$$

Check: $y = 11$ Check: $y = 6$

$$y + \sqrt{y - 2} = 8 \qquad\qquad y + \sqrt{y - 2} = 8$$

$$11 + \sqrt{11 - 2} \overset{?}{=} 8 \qquad\qquad 6 + \sqrt{6 - 2} \overset{?}{=} 8$$

$$11 + \sqrt{9} \overset{?}{=} 8 \qquad\qquad 6 + \sqrt{4} \overset{?}{=} 8$$

$$11 + 3 \overset{?}{=} 8 \qquad\qquad 6 + 2 \overset{?}{=} 8$$

$$14 \overset{?}{=} 8 \text{ (False)} \qquad\qquad 8 \overset{?}{=} 8 ✔$$

11 is not a solution to the original equation.

True, 6 is a solution to the original equation.

Skill Practice

Solve.

6. $\sqrt{x + 1} + 5 = x$

Avoiding Mistakes

Be sure to square the binomial, not the individual terms.

3. Solving Radical Equations Involving More than One Radical

Example 6 Solving Equations with Two Radicals

Solve the equation.

$$\sqrt[3]{2x - 4} = \sqrt[3]{1 - 8x}$$

Solution:

$$\sqrt[3]{2x - 4} = \sqrt[3]{1 - 8x}$$

$$(\sqrt[3]{2x - 4})^3 = (\sqrt[3]{1 - 8x})^3 \qquad \text{Because the index is 3, cube both sides.}$$

$$2x - 4 = 1 - 8x \qquad \text{Simplify.}$$

$$10x - 4 = 1 \qquad \text{Solve the resulting equation.}$$

$$10x = 5$$

$$x = \frac{1}{2} \qquad \text{Solve for } x.$$

Skill Practice

Solve the equation.

7. $\sqrt[5]{2y - 1} = \sqrt[5]{10y + 3}$

Answers

6. 8 (The value 3 does not check.)

7. $-\dfrac{1}{2}$

Check: $x = \frac{1}{2}$

$$\sqrt[3]{2x - 4} = \sqrt[3]{1 - 8x}$$

$$\sqrt[3]{2\left(\frac{1}{2}\right) - 4} \stackrel{?}{=} \sqrt[3]{1 - 8\left(\frac{1}{2}\right)}$$

$$\sqrt[3]{1 - 4} \stackrel{?}{=} \sqrt[3]{1 - 4}$$

$$\sqrt[3]{-3} \stackrel{?}{=} \sqrt[3]{-3} \ \checkmark \ \text{(True)}$$

Therefore, $\frac{1}{2}$ is a solution to the original equation.

Calculator Connections

Topic: Verifying a Solution to an Equation

In Example 6, the expressions on the right- and left-hand sides of the equation $\sqrt[3]{2x - 4} = \sqrt[3]{1 - 8x}$ are each functions of x. Consider the graphs of the functions:

$$\text{Y}_1 = \sqrt[3]{2x - 4} \qquad \text{and} \qquad \text{Y}_2 = \sqrt[3]{1 - 8x}$$

The x-coordinate of the point of intersection of the two functions is the solution to the equation $\sqrt[3]{2x - 4} = \sqrt[3]{1 - 8x}$. The point of intersection can be approximated by using an *Intersect* function.

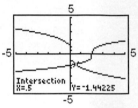

Skill Practice

Solve the equation.

8. $\sqrt{3c + 1} - \sqrt{c - 1} = 2$

Example 7   Solving Equations with Two Radicals

Solve the equation. $\sqrt{3m + 1} - \sqrt{m + 4} = 1$

Solution:

$$\sqrt{3m + 1} - \sqrt{m + 4} = 1$$

$$\sqrt{3m + 1} = \sqrt{m + 4} + 1 \qquad \text{Isolate one of the radicals.}$$

$$(\sqrt{3m + 1})^2 = (\sqrt{m + 4} + 1)^2 \qquad \text{Square both sides.}$$

$$3m + 1 = m + 4 + 2\sqrt{m + 4} + 1$$

$$\begin{aligned} \textit{Note: } (\sqrt{m + 4} + 1)^2 \\ = (\sqrt{m + 4})^2 + 2(1)\sqrt{m + 4} + (1)^2 \\ = m + 4 + 2\sqrt{m + 4} + 1 \end{aligned}$$

$$3m + 1 = m + 5 + 2\sqrt{m + 4} \qquad \text{Combine } \textit{like } \text{terms.}$$

$$2m - 4 = 2\sqrt{m + 4} \qquad \text{Isolate the radical again.}$$

$$m - 2 = \sqrt{m + 4} \qquad \text{Divide both sides by 2.}$$

$$(m - 2)^2 = (\sqrt{m + 4})^2 \qquad \text{Square both sides again.}$$

$$m^2 - 4m + 4 = m + 4 \qquad \begin{array}{l} \text{The resulting equation is} \\ \text{quadratic.} \end{array}$$

$$m^2 - 5m = 0 \qquad \text{Set one side equal to zero.}$$

$$m(m - 5) = 0 \qquad \text{Factor.}$$

$$m = 0 \qquad \text{or} \qquad m = 5$$

Answer

8. 1, 5

Check: $m = 0$ Check: $m = 5$

$\sqrt{3(0) + 1} - \sqrt{(0) + 4} \overset{?}{=} 1$ $\sqrt{3(5) + 1} - \sqrt{(5) + 4} = 1$

$\sqrt{1} - \sqrt{4} \overset{?}{=} 1$ $\sqrt{16} - \sqrt{9} \overset{?}{=} 1$

$1 - 2 \overset{?}{=} 1$ (False) $4 - 3 \overset{?}{=} 1$ ✔

The solution is 5. The value 0 does not check.

4. Applications of Radical Equations and Functions

Example 8 **Applying a Radical Equation in Geometry**

For a pyramid with a square base, the length of a side of the base b is given by

$$b = \sqrt{\frac{3V}{h}}$$

where V is the volume and h is the height.

The Pyramid of the Pharoah Khufu (known as the Great Pyramid) at Giza has a square base (Figure 7-3). If the distance around the bottom of the pyramid is 921.6 m and the height is 146.6 m, what is the volume of the pyramid?

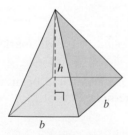

Figure 7-3

Solution:

$$b = \sqrt{\frac{3V}{h}}$$

$$b^2 = \left(\sqrt{\frac{3V}{h}}\right)^2 \qquad \text{Because the index is 2, square both sides.}$$

$$b^2 = \frac{3V}{h} \qquad \text{Simplify.}$$

$$b^2 \cdot h = \frac{3V}{h} \cdot h \qquad \text{Multiply both sides by } h.$$

$$b^2 h = 3V$$

$$\frac{b^2 h}{3} = \frac{3V}{3} \qquad \text{Divide both sides by 3.}$$

$$\frac{b^2 h}{3} = V$$

The length of a side b (in meters) is given by $\dfrac{921.6}{4} = 230.4$ m.

Answer

9. 40.5 in.²

$$\frac{(230.4)^2(146.6)}{3} = V \qquad \text{Substitute } b = 230.4 \text{ and } h = 146.6.$$

$$2{,}594{,}046 \approx V$$

The volume of the Great Pyramid at Giza is approximately 2,594,046 m³.

Skill Practice

When an object is dropped from a height of 64 ft, the time $t(x)$ (in seconds) it takes to reach a height x (in feet) is given by

$$t(x) = \frac{1}{4}\sqrt{64 - x}$$

10. Find the time to reach a height of 28 ft from the ground.

11. What is the height after 1 sec?

Example 9 **Applying a Radical Function**

On a certain surface, the speed $s(x)$ (in miles per hour) of a car before the brakes were applied can be approximated from the length of its skid marks x (in feet) by

$$s(x) = 3.8\sqrt{x} \qquad x \geq 0 \quad \text{See Figure 7-4.}$$

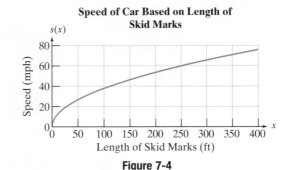

Speed of Car Based on Length of Skid Marks

Figure 7-4

a. Find the speed of a car before the brakes were applied if its skid marks are 361 ft long.

b. How long would you expect the skid marks to be if the car had been traveling the speed limit of 50 mph? (Round to the nearest foot.)

Solution:

a. $\quad s(x) = 3.8\sqrt{x}$

$\quad s(361) = 3.8\sqrt{361} \qquad$ Substitute $x = 361$.

$\qquad = 3.8(19)$

$\qquad = 72.2$

If the skid marks are 361 ft, the car was traveling approximately 72.2 mph before the brakes were applied.

b. $\quad s(x) = 3.8\sqrt{x}$

$\quad 50 = 3.8\sqrt{x} \qquad$ Substitute $s(x) = 50$ and solve for x.

$\quad \dfrac{50}{3.8} = \sqrt{x} \qquad$ Isolate the radical.

$\quad \left(\dfrac{50}{3.8}\right)^2 = x$

$\qquad x \approx 173$

If the car had been going the speed limit (50 mph), then the length of the skid marks would have been approximately 173 ft.

Answers

10. $\frac{3}{2}$ sec **11.** 48 ft

Section 7.7   Practice Exercises

Vocabulary and Key Concepts

1. **a.** The equation $\sqrt{x + 5} + 7 = 11$ is an example of a _____ equation.

   b. The first step to solve the equation $\sqrt{x + 5} + 7 = 11$ is to _____ the radical by subtracting _____ from both sides of the equation.

   c. When solving a radical equation some potential solutions may not check in the original equation. These are called _____ solutions.

   d. To solve the equation $\sqrt[3]{w - 1} = 5$, raise both sides of the equation to the _____ power.

Review Exercises

2. Identify the equation as linear or quadratic. Then solve the equation.

   a. $2x + 3 = 23$ **b.** $2x^2 - 9x = 5$

For Exercises 3–6, simplify the radical expressions. Assume all variables represent positive real numbers.

3. $\sqrt{\dfrac{9w^3}{16}}$ 4. $\sqrt{\dfrac{a^2}{3}}$ 5. $\sqrt[3]{54c^4}$ 6. $\sqrt{\dfrac{49}{5t^3}}$

For Exercises 7–10, simplify each expression. Assume all radicands represent positive real numbers.

7. $(\sqrt{4x - 6})^2$ 8. $(\sqrt{5y + 2})^2$ 9. $(\sqrt[3]{9p + 7})^3$ 10. $(\sqrt[3]{4t + 13})^3$

Concept 2: Solving Radical Equations Involving One Radical

For Exercises 11–26, solve the equations. **(See Examples 1–4.)**

11. $\sqrt{x} + 4 = 6$ 12. $\sqrt{x} + 2 = 8$ 13. $\sqrt{5y + 1} = 4$ 14. $\sqrt{9z - 5} - 2 = 9$

15. $(2z - 3)^{1/2} - 3 = 6$ 16. $(8 + 3a)^{1/2} - 1 = 4$ 17. $\sqrt[3]{x - 2} - 1 = 2$ 18. $\sqrt[3]{2x - 5} - 1 = 1$

19. $(15 - w)^{1/3} + 7 = 2$ 20. $(k + 18)^{1/3} + 5 = 3$ 21. $3 + \sqrt{x - 16} = 0$ 22. $12 + \sqrt{2x + 1} = 0$

23. $2\sqrt{6a + 7} - 2a = 0$ 24. $2\sqrt{3 - w} - w = 0$ 25. $\sqrt[4]{2x - 5} = -1$ 26. $\sqrt[4]{x + 16} = -4$

For Exercises 27–30, assume all variables represent positive real numbers.

27. Solve for V: $r = \sqrt[3]{\dfrac{3V}{4\pi}}$ 28. Solve for V: $r = \sqrt{\dfrac{V}{h\pi}}$

29. Solve for h^2: $r = \pi\sqrt{r^2 + h^2}$ 30. Solve for d: $s = 1.3\sqrt{d}$

For Exercises 31–36, square the expression as indicated.

31. $(a + 5)^2$ 32. $(b + 7)^2$ 33. $(\sqrt{5a} - 3)^2$

34. $(2 + \sqrt{b})^2$ 35. $(\sqrt{r - 3} + 5)^2$ 36. $(2 - \sqrt{2t - 4})^2$

For Exercises 37–42, solve the radical equations, if possible. **(See Example 5.)**

37. $\sqrt{a^2 + 2a + 1} = a + 5$ 38. $\sqrt{b^2 - 5b - 8} = b + 7$ 39. $\sqrt{25w^2 - 2w - 3} = 5w - 4$

40. $\sqrt{4p^2 - 2p + 1} = 2p - 3$ 41. $\sqrt{5y + 1} + 2 = y + 3$ 42. $\sqrt{2x - 2} + 3 = x + 2$

Concept 3: Solving Radical Equations Involving More than One Radical

For Exercises 43–66, solve the radical equations, if possible. **(See Examples 6 and 7.)**

43. $\sqrt[4]{h+4} = \sqrt[4]{2h-5}$

44. $\sqrt[4]{3b+6} = \sqrt[4]{7b-6}$

45. $\sqrt[3]{5a+3} - \sqrt[3]{a-13} = 0$

46. $\sqrt[3]{k-8} - \sqrt[3]{4k+1} = 0$

47. $\sqrt{5a-9} = \sqrt{5a} - 3$

48. $\sqrt{8+b} = 2 + \sqrt{b}$

49. $\sqrt{2h+5} - \sqrt{2h} = 1$

50. $\sqrt{3k-5} - \sqrt{3k} = -1$

51. $(t-9)^{1/2} - t^{1/2} = 3$

52. $(y-16)^{1/2} - y^{1/2} = 4$

53. $6 = \sqrt{x^2+3} - x$

54. $2 = \sqrt{y^2+5} - y$

55. $\sqrt{3t-7} = 2 - \sqrt{3t+1}$

56. $\sqrt{p-6} = \sqrt{p+2} - 4$

57. $\sqrt{z+1} + \sqrt{2z+3} = 1$

58. $\sqrt{2y+6} = \sqrt{7-2y} + 1$

59. $\sqrt{6m+7} - \sqrt{3m+3} = 1$

60. $\sqrt{5w+1} - \sqrt{3w} = 1$

61. $2 + 2\sqrt{2t+3} + 2\sqrt{3t-5} = 0$

62. $6 + 3\sqrt{3x+1} + 3\sqrt{x-1} = 0$

63. $3\sqrt{y-3} = \sqrt{4y+3}$

64. $\sqrt{5x-8} = 2\sqrt{x-1}$

65. $\sqrt{p+7} = \sqrt{2p} + 1$

66. $\sqrt{t} = \sqrt{t-12} + 2$

Concept 4: Applications of Radical Equations and Functions

67. If an object is dropped from an initial height h, its velocity at impact with the ground is given by

$$v = \sqrt{2gh}$$

where g is the acceleration due to gravity and h is the initial height. **(See Example 8.)**

a. Find the initial height (in feet) of an object if its velocity at impact is 44 ft/sec. (Assume that the acceleration due to gravity is $g = 32$ ft/sec^2.)

b. Find the initial height (in meters) of an object if its velocity at impact is 26 m/sec. (Assume that the acceleration due to gravity is $g = 9.8$ m/sec^2.) Round to the nearest tenth of a meter.

68. The time T (in seconds) required for a pendulum to make one complete swing back and forth is approximated by

$$T = 2\pi\sqrt{\frac{L}{g}}$$

where g is the acceleration due to gravity and L is the length of the pendulum (in feet).

a. Find the length of a pendulum that requires 1.36 sec to make one complete swing back and forth. (Assume that the acceleration due to gravity is $g = 32$ ft/sec^2.) Round to the nearest tenth of a foot.

b. Find the time required for a pendulum to complete one swing back and forth if the length of the pendulum is 4 ft. (Assume that the acceleration due to gravity is $g = 32$ ft/sec^2.) Round to the nearest tenth of a second.

69. The airline cost for x thousand passengers to travel round trip from New York to Atlanta is given by

$$C(x) = \sqrt{0.3x + 1}$$

where $C(x)$ is measured in millions of dollars and $x \geq 0$. **(See Example 9.)**

a. Find the airline's cost for 10,000 passengers ($x = 10$) to travel from New York to Atlanta.

b. If the airline charges $320 per passenger, find the profit made by the airline for flying 10,000 passengers from New York to Atlanta.

c. Approximate the number of passengers who traveled from New York to Atlanta if the total cost for the airline was $4 million.

70. The time $t(d)$ in seconds it takes an object to drop d meters is given by

$$t(d) = \sqrt{\frac{d}{4.9}}$$

a. Approximate the height of the Texas Commerce Tower in Houston if it takes an object 7.89 sec to drop from the top. Round to the nearest meter.

b. Approximate the height of the Willis Tower in Chicago if it takes an object 9.51 sec to drop from the top. Round to the nearest meter.

Expanding Your Skills

 71. The number of hours needed to cook a turkey that weighs x lb can be approximated by

$$t(x) = 0.90\sqrt[5]{x^3}$$

where $t(x)$ is the time in hours and x is the weight of the turkey in pounds.

a. Find the weight of a turkey that cooked for 4 hr. Round to the nearest pound.

b. Find $t(18)$ and interpret the result. Round to the nearest tenth of an hour.

For Exercises 72–75, use the Pythagorean theorem to find a, b, or c.

72. Find b when $a = 2$ and $c = y$.

73. Find b when $a = h$ and $c = 5$.

74. Find a when $b = x$ and $c = 8$.

75. Find a when $b = 14$ and $c = k$.

Graphing Calculator Exercises

76. Graph Y_1 and Y_2 on a viewing window defined by $-10 \le x \le 40$ and $-5 \le y \le 10$.

$$Y_1 = \sqrt{2x} \quad \text{and} \quad Y_2 = 8$$

Use an *Intersect* feature to approximate the x-coordinate of the point of intersection of the two graphs.

77. Graph Y_1 and Y_2 on a viewing window defined by $-10 \le x \le 20$ and $-5 \le y \le 10$.

$$Y_1 = \sqrt{4x} \quad \text{and} \quad Y_2 = 6$$

Use an *Intersect* feature to approximate the x-coordinate of the point of intersection of the two graphs.

78. Refer to Exercise 44. Graph Y_1 and Y_2 on a viewing window defined by $-5 \le x \le 20$ and $-1 \le y \le 4$.

$$Y_1 = \sqrt[4]{3x + 6} \quad \text{and} \quad Y_2 = \sqrt[4]{7x - 6}$$

Use an *Intersect* feature to approximate the x-coordinate of the point of intersection of the two graphs to support your solution to Exercise 44.

79. Refer to Exercise 43. Graph Y_1 and Y_2 on a viewing window defined by $-5 \le x \le 20$ and $-1 \le y \le 4$.

$$Y_1 = \sqrt[4]{x + 4} \quad \text{and} \quad Y_2 = \sqrt[4]{2x - 5}$$

Use an *Intersect* feature to approximate the x-coordinate of the point of intersection of the two graphs to support your solution to Exercise 43.

| **Section 7.8** | **Complex Numbers** |

Concepts

1. Definition of i
2. Powers of i
3. Definition of a Complex Number
4. Addition, Subtraction, and Multiplication of Complex Numbers
5. Division and Simplification of Complex Numbers

Concept Connections

1. $i = ?$
2. $i^2 = ?$

1. Definition of i

In Section 7.1, we learned that there are no real-valued square roots of a negative number. For example, $\sqrt{-9}$ is not a real number because no real number when squared equals -9. However, the square roots of a negative number are defined over another set of numbers called the **imaginary numbers**. The foundation of the set of imaginary numbers is the definition of the imaginary number i.

> **Definition of the Imaginary Number i**
>
> $$i = \sqrt{-1}$$
>
> *Note:* From the definition of i, it follows that $i^2 = -1$.

Using the imaginary number i, we can define the square root of any negative real number.

> **Definition of $\sqrt{-b}$ for $b > 0$**
>
> Let b be a positive real number. Then $\sqrt{-b} = i\sqrt{b}$.

Skill Practice

Simplify the expressions in terms of i.

3. $\sqrt{-81}$ 4. $\sqrt{-20}$
5. $-\sqrt{-36}$ 6. $\sqrt{-7}$

| **Example 1** | **Simplifying Expressions in Terms of i** |

Simplify the expressions in terms of i.

a. $\sqrt{-64}$ **b.** $\sqrt{-50}$ **c.** $-\sqrt{-4}$ **d.** $\sqrt{-29}$

Solution:

a. $\sqrt{-64} = i\sqrt{64}$
$$= 8i$$

b. $\sqrt{-50} = i\sqrt{50}$
$$= i\sqrt{25 \cdot 2}$$
$$= 5i\sqrt{2}$$

c. $-\sqrt{-4} = -1 \cdot \sqrt{-4}$
$$= -1 \cdot 2i$$
$$= -2i$$

d. $\sqrt{-29} = i\sqrt{29}$

Avoiding Mistakes

In an expression such as $i\sqrt{29}$, the i is often written in front of the square root. The expression $\sqrt{29}\,i$ is also correct, but may be misinterpreted as $\sqrt{29i}$ (with i incorrectly placed under the radical).

The multiplication and division properties of radicals were presented in Sections 7.3 and 7.5 as follows:

If a and b represent real numbers such that $\sqrt[n]{a}$ and $\sqrt[n]{b}$ are both real, then

$$\sqrt[n]{ab} = \sqrt[n]{a} \cdot \sqrt[n]{b} \quad \text{and} \quad \sqrt[n]{\frac{a}{b}} = \frac{\sqrt[n]{a}}{\sqrt[n]{b}} \quad b \neq 0$$

Answers

1. $\sqrt{-1}$ 2. -1 3. $9i$
4. $2i\sqrt{5}$ 5. $-6i$ 6. $i\sqrt{7}$

The conditions that $\sqrt[n]{a}$ and $\sqrt[n]{b}$ must both be real numbers prevent us from applying the multiplication and division properties of radicals for square roots with a negative radicand. Therefore, to multiply or divide radicals with a negative radicand, first write the radical in terms of the imaginary number i. This is demonstrated in Example 2.

Example 2 Simplifying a Product or Quotient in Terms of i

Simplify the expressions.

a. $\dfrac{\sqrt{-100}}{\sqrt{-25}}$ b. $\sqrt{-25} \cdot \sqrt{-9}$ c. $\sqrt{-5} \cdot \sqrt{-5}$

Solution:

a. $\dfrac{\sqrt{-100}}{\sqrt{-25}}$

$= \dfrac{10i}{5i}$ Simplify each radical in terms of i *before* dividing.

$= 2$

b. $\sqrt{-25} \cdot \sqrt{-9}$

$= 5i \cdot 3i$ Simplify each radical in terms of i first *before* multiplying.

$= 15i^2$ Multiply.

$= 15(-1)$ Recall that $i^2 = -1$.

$= -15$ Simplify.

c. $\sqrt{-5} \cdot \sqrt{-5}$

$= i\sqrt{5} \cdot i\sqrt{5}$

$= i^2 \cdot (\sqrt{5})^2$

$= -1 \cdot 5$

$= -5$

Avoiding Mistakes

In Example 2, we wrote the radical expressions in terms of i first, before multiplying or dividing. If we had mistakenly applied the multiplication or division property first, we would have obtained an incorrect answer.

Correct: $\sqrt{-25} \cdot \sqrt{-9}$

$= (5i)(3i) = 15i^2$

$= 15(-1) = -15$
        ↑ correct

Be careful: $\sqrt{-25} \cdot \sqrt{-9}$ $\sqrt{-25}$ and $\sqrt{-9}$ are not real numbers. Therefore, the multiplication property of radicals cannot be applied.

$\neq \sqrt{225} = 15$
    ↑ (incorrect answer)

Answers

7. 2   **8.** −28   **9.** −2

2. Powers of i

From the definition of $i = \sqrt{-1}$, it follows that

$$i = i$$

$$i^2 = -1$$

$$i^3 = -i \qquad \text{because } i^3 = i^2 \cdot i = (-1)i = -i$$

$$i^4 = 1 \qquad \text{because } i^4 = i^2 \cdot i^2 = (-1)(-1) = 1$$

$$i^5 = i \qquad \text{because } i^5 = i^4 \cdot i = (1)i = i$$

$$i^6 = -1 \qquad \text{because } i^6 = i^4 \cdot i^2 = (1)(-1) = -1$$

This pattern of values $i, -1, -i, 1, i, -1, -i, 1, \ldots$ continues for all subsequent powers of i. Table 7-1 lists several powers of i.

Table 7-1   Powers of i

$i^1 = i$	$i^5 = i$	$i^9 = i$
$i^2 = -1$	$i^6 = -1$	$i^{10} = -1$
$i^3 = -i$	$i^7 = -i$	$i^{11} = -i$
$i^4 = 1$	$i^8 = 1$	$i^{12} = 1$

To simplify higher powers of i, we can decompose the expression into multiples of i^4 ($i^4 = 1$) and write the remaining factors as i, i^2, or i^3.

Skill Practice

Simplify the powers of i.

10. i^{45} **11.** i^{22}

12. i^{31} **13.** i^{80}

Example 3   Simplifying Powers of i

Simplify the powers of i.

a. i^{13} **b.** i^{18} **c.** i^{107} **d.** i^{32}

Solution:

a. $i^{13} = (i^{12}) \cdot (i)$

$\qquad = (i^4)^3 \cdot (i)$

$\qquad = (1)^3(i)$ Recall that $i^4 = 1$.

$\qquad = i$ Simplify.

b. $i^{18} = (i^{16}) \cdot (i^2)$

$\qquad = (i^4)^4 \cdot (i^2)$

$\qquad = (1)^4 \cdot (-1)$ $i^4 = 1$ and $i^2 = -1$

$\qquad = -1$ Simplify.

c. $i^{107} = (i^{104}) \cdot (i^3)$

$\qquad = (i^4)^{26}(i^3)$

$\qquad = (1)^{26}(-i)$ $i^4 = 1$ and $i^3 = -i$

$\qquad = -i$ Simplify.

Answers

10. i **11.** -1

12. $-i$ **13.** 1

d. $i^{32} = (i^4)^8$

$\quad\quad = (1)^8 \quad\quad i^4 = 1$

$\quad\quad = 1 \quad\quad\quad$ Simplify.

3. Definition of a Complex Number

We have already learned the definitions of the integers, rational numbers, irrational numbers, and real numbers. In this section, we define the complex numbers.

> **Definition of a Complex Number**
>
> A **complex number** is a number of the form $a + bi$, where a and b are real numbers and $i = \sqrt{-1}$.
>
> *Notes:*
>
> - If $b = 0$, then the complex number $a + bi$ is a real number.
> - If $b \neq 0$, then we say that $a + bi$ is an imaginary number.
> - The complex number $a + bi$ is said to be written in standard form. The quantities a and b are called the real and imaginary parts (respectively) of the complex number.
> - The complex numbers $a - bi$ and $a + bi$ are called **complex conjugates**.

Concept Connections

14. Give an example of a complex number whose imaginary part is zero. (Answers may vary.)

15. Give an example of a complex number whose real part is zero. (Answers may vary.)

16. Give the complex conjugate of $-7 + 2i$.

From the definition of a complex number, it follows that all real numbers are complex numbers and all imaginary numbers are complex numbers. Figure 7-5 illustrates the relationship among the sets of numbers we have learned so far.

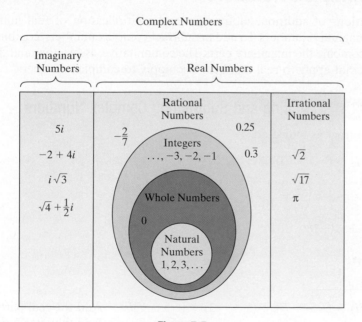

Figure 7-5

Answers

14. For example: -4

15. For example: $2i$

16. $-7 - 2i$

Skill Practice

Identify the real and imaginary parts of the complex numbers.

17. $22 - 14i$

18. -50

19. $15i$

Example 4 **Identifying the Real and Imaginary Parts of a Complex Number**

Identify the real and imaginary parts of the complex numbers.

a. $-8 + 2i$ **b.** $\dfrac{3}{2}$ **c.** $-1.75i$

Solution:

a. $-8 + 2i$ -8 is the real part, and 2 is the imaginary part.

b. $\dfrac{3}{2} = \dfrac{3}{2} + 0i$ Rewrite $\frac{3}{2}$ in the form $a + bi$.
$\frac{3}{2}$ is the real part, and 0 is the imaginary part.

c. $-1.75i$

$= 0 + -1.75i$ Rewrite $-1.75i$ in the form $a + bi$.
0 is the real part, and -1.75 is the imaginary part.

TIP: Example 4(b) illustrates that a real number is also a complex number.

Example 4(c) illustrates that an imaginary number is also a complex number.

$$\dfrac{3}{2} = \dfrac{3}{2} + 0i$$

$$-1.75i = 0 + -1.75i$$

4. Addition, Subtraction, and Multiplication of Complex Numbers

The operations of addition, subtraction, and multiplication of real numbers also apply to imaginary numbers. To add or subtract complex numbers, combine the real parts and combine the imaginary parts. The commutative, associative, and distributive properties that apply to real numbers also apply to complex numbers.

Skill Practice

Perform the indicated operations.

20. $\left(\dfrac{1}{2} - \dfrac{1}{4}i\right) + \left(\dfrac{3}{5} + \dfrac{2}{3}i\right)$

21. $(-6 + 11i) - (-9 - 12i)$

22. $\sqrt{-45} - \sqrt{-20}$

Example 5 **Adding and Subtracting Complex Numbers**

Add or subtract as indicated.

a. $(1 - 5i) + (-3 + 7i)$

b. $\left(-\dfrac{1}{4} + \dfrac{3}{5}i\right) - \left(\dfrac{1}{2} - \dfrac{1}{10}i\right)$

c. $\sqrt{-8} + \sqrt{-18}$

Solution:

real parts

a. $(1 - 5i) + (-3 + 7i) = (1 + -3) + (-5 + 7)i$ Add the real parts.
Add the imaginary parts.

imaginary parts

$= -2 + 2i$ Simplify.

Answers

17. real: 22; imaginary: -14

18. real: -50; imaginary: 0

19. real: 0; imaginary: 15

20. $\dfrac{11}{10} + \dfrac{5}{12}i$ **21.** $3 + 23i$

22. $i\sqrt{5}$

b. $\left(-\dfrac{1}{4} + \dfrac{3}{5}i\right) - \left(\dfrac{1}{2} - \dfrac{1}{10}i\right) = -\dfrac{1}{4} + \dfrac{3}{5}i - \dfrac{1}{2} + \dfrac{1}{10}i$ Apply the distributive property.

$$= \left(-\dfrac{1}{4} - \dfrac{1}{2}\right) + \left(\dfrac{3}{5} + \dfrac{1}{10}\right)i \quad \begin{array}{l}\text{Add real parts.}\\ \text{Add imaginary}\\ \text{parts.}\end{array}$$

$$= \left(-\dfrac{1}{4} - \dfrac{2}{4}\right) + \left(\dfrac{6}{10} + \dfrac{1}{10}\right)i \quad \begin{array}{l}\text{Get common}\\ \text{denominators.}\end{array}$$

$$= -\dfrac{3}{4} + \dfrac{7}{10}i \qquad\qquad\qquad \text{Simplify.}$$

c. $\sqrt{-8} + \sqrt{-18} = 2i\sqrt{2} + 3i\sqrt{2}$ Simplify each radical in terms of i.

$$= 5i\sqrt{2} \qquad\qquad \text{Combine } like \text{ radicals.}$$

Example 6   Multiplying Complex Numbers

Multiply.

a. $(10 - 5i)(2 + 3i)$ **b.** $(1.2 + 0.5i)(1.2 - 0.5i)$

Solution:

a. $(10 - 5i)(2 + 3i)$

$$= (10)(2) + (10)(3i) + (-5i)(2) + (-5i)(3i) \qquad \begin{array}{l}\text{Apply the distributive}\\ \text{property.}\end{array}$$

$$= 20 + 30i - 10i - 15i^2$$

$$= 20 + 20i - (15)(-1) \qquad \text{Recall } i^2 = -1.$$

$$= 20 + 20i + 15$$

$$= 35 + 20i \qquad\qquad\qquad \text{Write in the form } a + bi.$$

b. $(1.2 + 0.5i)(1.2 - 0.5i)$

The expressions $(1.2 + 0.5i)$ and $(1.2 - 0.5i)$ are complex conjugates. The product is a difference of squares.

$$(a + b)(a - b) = a^2 - b^2$$

$(1.2 + 0.5i)(1.2 - 0.5i) = (1.2)^2 - (0.5i)^2$ Apply the formula, where $a = 1.2$ and $b = 0.5i$.

$$= 1.44 - 0.25i^2$$

$$= 1.44 - 0.25(-1) \qquad \text{Recall } i^2 = -1.$$

$$= 1.44 + 0.25$$

$$= 1.69 + 0i$$

Skill Practice

Multiply.

23. $(4 - 6i)(2 - 3i)$

24. $(1.5 + 0.8i)(1.5 - 0.8i)$

Answers

23. $-10 - 24i$ **24.** 2.89

TIP: In Example 6(b), the complex numbers $(1.2 + 0.5i)$ and $(1.2 - 0.5i)$ can also be multiplied by using the distributive property.

$$(1.2 + 0.5i)(1.2 - 0.5i) = 1.44 - 0.6i + 0.6i - 0.25i^2$$
$$= 1.44 - 0.25(-1)$$
$$= 1.69 + 0i$$

5. Division and Simplification of Complex Numbers

The product of a complex number and its complex conjugate is a real number. For example:

$$(5 + 3i)(5 - 3i) = 25 - 9i^2$$
$$= 25 - 9(-1)$$
$$= 25 + 9$$
$$= 34$$

To divide by a complex number, multiply the numerator and denominator by the complex conjugate of the denominator. This produces a real number in the denominator so that the resulting expression can be written in the form $a + bi$.

Skill Practice

Divide the complex numbers. Write the answer in the form $a + bi$.

25. $\dfrac{2 + i}{3 - 2i}$

Example 7 **Dividing by a Complex Number**

Divide the complex numbers and write the answer in the form $a + bi$.

$$\frac{4 - 3i}{5 + 2i}$$

Solution:

$$\frac{4 - 3i}{5 + 2i}$$ Multiply the numerator and denominator by the complex conjugate of the denominator:

$$\frac{(4 - 3i)}{(5 + 2i)} \cdot \frac{(5 - 2i)}{(5 - 2i)} = \frac{(4)(5) + (4)(-2i) + (-3i)(5) + (-3i)(-2i)}{(5)^2 - (2i)^2}$$

$$= \frac{20 - 8i - 15i + 6i^2}{25 - 4i^2}$$ Simplify numerator and denominator.

$$= \frac{20 - 23i + 6(-1)}{25 - 4(-1)}$$ Recall $i^2 = -1$.

$$= \frac{20 - 23i - 6}{25 + 4}$$

$$= \frac{14 - 23i}{29}$$ Simplify.

$$= \frac{14}{29} - \frac{23}{29}i$$ Write in the form $a + bi$.

Answer

25. $\dfrac{4}{13} + \dfrac{7}{13}i$

Example 8   **Simplifying Complex Numbers**

Simplify the complex numbers. $\dfrac{6 + \sqrt{-18}}{9}$

Solution:

$$\dfrac{6 + \sqrt{-18}}{9} = \dfrac{6 + i\sqrt{18}}{9}$$ Write the radical in terms of i.

$$= \dfrac{6 + 3i\sqrt{2}}{9}$$ Simplify $\sqrt{18} = 3\sqrt{2}$.

$$= \dfrac{3(2 + i\sqrt{2})}{9}$$ Factor the numerator.

$$= \dfrac{\overset{1}{3}(2 + i\sqrt{2})}{\underset{3}{9}}$$ Simplify.

$$= \dfrac{2 + i\sqrt{2}}{3}   \quad \text{or} \quad \dfrac{2}{3} + \dfrac{\sqrt{2}}{3}i$$

Skill Practice

Simplify the complex numbers.

26. $\dfrac{8 - \sqrt{-24}}{6}$

Answer

26. $\dfrac{4 - i\sqrt{6}}{3}$ or $\dfrac{4}{3} - \dfrac{\sqrt{6}}{3}i$

Section 7.8   Practice Exercises

Vocabulary and Key Concepts

1. a. A square root of a negative number is not a real number, but rather is an _____ number.

   b. $i = $ _____, and $i^2 = $ _____.

   c. For a positive number b, the value $\sqrt{-b} = $ _____.

   d. A complex number is a number in the form _____, where a and b are real numbers and $i = $ _____.

   e. Given a complex number $a + bi$, the value a is called the _____ part, and _____ is called the imaginary part.

   f. The complex conjugate of $a - bi$ is _____.

   g. Answer true or false. All real numbers are complex numbers.

   h. Answer true or false. All imaginary numbers are complex numbers.

Review Exercises

For Exercises 2–4, perform the indicated operations.

2. $-2\sqrt{5} - 3\sqrt{50} + \sqrt{125}$ 　　　　**3.** $(3 - \sqrt{x})(3 + \sqrt{x})$ 　　　　**4.** $(\sqrt{5} + \sqrt{2})^2$

For Exercises 5–7, solve the equations.

5. $\sqrt[3]{3p + 7} - \sqrt[3]{2p - 1} = 0$ 　　**6.** $\sqrt[3]{t - 5} - \sqrt[3]{2t + 1} = 0$ 　　**7.** $\sqrt{4a + 29} = 2\sqrt{a} + 5$

8. Rationalize the denominator. $\dfrac{2}{\sqrt{x} - 3}$

Concept 1: Definition of *i*

9. Simplify the expressions $\sqrt{-1}$ and $-\sqrt{1}$. 　　　　**10.** Simplify i^2.

For Exercises 11–30, simplify the expressions. **(See Examples 1 and 2.)**

11. $\sqrt{-144}$ **12.** $\sqrt{-81}$ **13.** $\sqrt{-3}$ **14.** $\sqrt{-17}$

15. $-\sqrt{-20}$ **16.** $-\sqrt{-75}$ **17.** $(2\sqrt{-25})(3\sqrt{-4})$ **18.** $(-4\sqrt{-9})(-3\sqrt{-1})$

19. $7\sqrt{-63} - 4\sqrt{-28}$ **20.** $7\sqrt{-3} - 4\sqrt{-27}$ **21.** $\sqrt{-7} \cdot \sqrt{-7}$ **22.** $\sqrt{-11} \cdot \sqrt{-11}$

23. $\sqrt{-9} \cdot \sqrt{-16}$ **24.** $\sqrt{-25} \cdot \sqrt{-36}$ **25.** $\sqrt{-15} \cdot \sqrt{-6}$ **26.** $\sqrt{-12} \cdot \sqrt{-50}$

27. $\dfrac{\sqrt{-50}}{\sqrt{25}}$ **28.** $\dfrac{\sqrt{-27}}{\sqrt{9}}$ **29.** $\dfrac{\sqrt{-90}}{\sqrt{10}}$ **30.** $\dfrac{\sqrt{-125}}{\sqrt{45}}$

Concept 2: Powers of *i*

For Exercises 31–42, simplify the powers of *i*. **(See Example 3.)**

31. i^7 **32.** i^{38} **33.** i^{64} **34.** i^{75}

35. i^{41} **36.** i^{25} **37.** i^{52} **38.** i^0

39. i^{23} **40.** i^{103} **41.** i^6 **42.** i^{82}

Concept 3: Definition of a Complex Number

43. What is the conjugate of a complex number $a + bi$?

44. True or false?

     a. Every real number is a complex number.   **b.** Every complex number is a real number.

For Exercises 45–52, identify the real and imaginary parts of the complex number. **(See Example 4.)**

45. $-5 + 12i$ **46.** $22 - 16i$ **47.** $-6i$ **48.** $10i$

49. 35 **50.** -1 **51.** $\dfrac{3}{5} + i$ **52.** $-\dfrac{1}{2} - \dfrac{1}{4}i$

Concept 4: Addition, Subtraction, and Multiplication of Complex Numbers

For Exercises 53–76, perform the indicated operations. Write the answer in the form $a + bi$. **(See Examples 5 and 6.)**

53. $(2 - i) + (5 + 7i)$ **54.** $(5 - 2i) + (3 + 4i)$ **55.** $\left(\dfrac{1}{2} + \dfrac{2}{3}i\right) - \left(\dfrac{1}{5} - \dfrac{5}{6}i\right)$

56. $\left(\dfrac{11}{10} - \dfrac{7}{5}i\right) - \left(-\dfrac{2}{5} + \dfrac{3}{5}i\right)$ **57.** $\sqrt{-98} - \sqrt{-8}$ **58.** $\sqrt{-75} + \sqrt{-12}$

59. $(2 + 3i) - (1 - 4i) + (-2 + 3i)$ **60.** $(2 + 5i) - (7 - 2i) + (-3 + 4i)$

61. $(8i)(3i)$ **62.** $(2i)(4i)$ **63.** $6i(1 - 3i)$ **64.** $-i(3 + 4i)$

65. $(2 - 10i)(3 + 2i)$ **66.** $(4 + 7i)(2 - 3i)$ **67.** $(-5 + 2i)(5 + 2i)$ **68.** $(4 - 11i)(4 + 11i)$

69. $(4 + 5i)^2$ **70.** $(3 - 2i)^2$ **71.** $(2 + i)(3 - 2i)(4 + 3i)$ **72.** $(3 - i)(3 + i)(4 - i)$

73. $(-4 - 6i)^2$ **74.** $(-3 - 5i)^2$ **75.** $\left(-\dfrac{1}{2} - \dfrac{3}{4}i\right)\left(-\dfrac{1}{2} + \dfrac{3}{4}i\right)$ **76.** $\left(-\dfrac{2}{3} + \dfrac{1}{6}i\right)\left(-\dfrac{2}{3} - \dfrac{1}{6}i\right)$

Concept 5: Division and Simplification of Complex Numbers

For Exercises 77–90, divide the complex numbers. Write the answer in the form $a + bi$. **(See Example 7.)**

77. $\dfrac{2}{1 + 3i}$

78. $\dfrac{-2}{3 + i}$

79. $\dfrac{-i}{4 - 3i}$

80. $\dfrac{3 - 3i}{1 - i}$

81. $\dfrac{5 + 2i}{5 - 2i}$

82. $\dfrac{7 + 3i}{4 - 2i}$

83. $\dfrac{3 + 7i}{-2 - 4i}$

84. $\dfrac{-2 + 9i}{-1 - 4i}$

85. $\dfrac{13i}{-5 - i}$

86. $\dfrac{15i}{-2 - i}$

87. $\dfrac{2 + 3i}{6i}$

(*Hint:* Consider multiplying numerator and denominator by i or $-i$. This will make the denominator a real number.)

88. $\dfrac{4 - i}{2i}$

89. $\dfrac{-10 + i}{i}$

90. $\dfrac{-6 - i}{-i}$

For Exercises 91–98, simplify the complex numbers. Write the answer in the form $a + bi$. **(See Example 8.)**

91. $\dfrac{2 + \sqrt{-16}}{8}$

92. $\dfrac{6 - \sqrt{-4}}{4}$

93. $\dfrac{-6 + \sqrt{-72}}{6}$

94. $\dfrac{-20 + \sqrt{-500}}{10}$

95. $\dfrac{-8 - \sqrt{-48}}{4}$

96. $\dfrac{-18 - \sqrt{-72}}{3}$

97. $\dfrac{-5 + \sqrt{-50}}{10}$

98. $\dfrac{14 + \sqrt{-98}}{7}$

Expanding Your Skills

For Exercises 99–102, determine if the complex number is a solution to the equation.

99. $x^2 - 4x + 5 = 0$; $2 + i$

100. $x^2 - 6x + 25 = 0$; $3 - 4i$

101. $x^2 + 12 = 0$; $-2i\sqrt{3}$

102. $x^2 + 18 = 0$; $3i\sqrt{2}$

Group Activity

Margin of Error of Survey Results

Materials: Calculator

Estimated time: 20 minutes

Group Size: 3

The members of the class will conduct a survey to estimate the percentage of the college population that answers "yes" to the several survey questions asked.

1. The members of the class will decide on three "Yes or No" questions that they will use to perform a survey. For example, here are some possible questions.

   "Do you study more than 10 hr a week?"

   "Do you work more than 25 hr a week?"

   "Are you taking more than 10 credit hours?"

List the questions here:

a. _____

b. _____

c. _____

2. Each student in the class will survey 20 people and record the number of "yes" responses in the table.

Question	Number of "Yes" Responses
1	
2	
3	

3. The instructor will then pool the results obtained from each student to get the total number of people surveyed by the whole class. Record the number of people surveyed (sample size) and the number of "yes" responses for each question in the table. Once the data are entered into the table, compute the percent, p, of "yes" responses in the fourth column. Write this value in decimal form. Round the values of p to three decimal places. (*Note:* For now, leave the last column blank.)

Question	Sample Size	Number of "Yes" Responses	Percent "Yes" Responses, p	Margin of Error, E
1				
2				
3				

4. The value of p represents the percent (written in decimal form) of people in the *sample* who answered "yes." However, the percentage of people in the entire college population who would answer "yes" is unknown. The value of p is only an *estimate*. Statisticians often compute a margin of error associated with such an estimate by using the following formula.

$$E = 1.96 \sqrt{\frac{p(1-p)}{n}}$$ where E is the margin of error.

n is the sample size.

p is the percent (in decimal form) of "yes" responses.

Compute the margin of error for each of the three questions. Round the values of E to three decimal places. Record the results in the table.

5. The margin of error computed in step 4 is associated with what statisticians call a 95% level of confidence. To interpret the results, a statistician would say "With 95% confidence, the researcher estimates the percentage of "yes" responses from the entire college population to be between $p \pm E$."

Interpret the results from each of the three questions.

Chapter 7 Summary

Section 7.1 Definition of an *n*th root

Key Concepts

b is an **nth root** of a if $b^n = a$.

The expression $\sqrt{a}$ represents the **principal square root** of a.

The expression $\sqrt[n]{a}$ represents the nth root of a.

$\sqrt[n]{a^n} = |a|$ if n is even.

$\sqrt[n]{a^n} = a$ if n is odd.

$\sqrt[n]{a}$ is not a real number if $a < 0$ and n is even.

$f(x) = \sqrt[n]{x}$ defines a **radical function**.

The Pythagorean Theorem

$a^2 + b^2 = c^2$

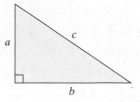

Examples

Example 1

2 is a square root of 4.

-2 is a square root of 4.

-3 is a cube root of -27.

Example 2

$\sqrt{36} = 6 \qquad \sqrt[3]{-64} = -4$

Example 3

$\sqrt[4]{(x + 3)^4} = |x + 3| \qquad \sqrt[5]{(x + 3)^5} = x + 3$

Example 4

$\sqrt[4]{-16}$ is not a real number.

Example 5

For $g(x) = \sqrt{x}$ the domain is $[0, \infty)$.

For $h(x) = \sqrt[3]{x}$ the domain is $(-\infty, \infty)$.

Example 6

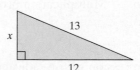

$x^2 + 12^2 = 13^2$

$x^2 + 144 = 169$

$x^2 = 25$

$x = \sqrt{25}$

$x = 5$

Section 7.2 Rational Exponents

Key Concepts

Let a be a real number and n be an integer such that $n > 1$. If $\sqrt[n]{a}$ exists, then

$$a^{1/n} = \sqrt[n]{a}$$

$$a^{m/n} = \left(\sqrt[n]{a}\right)^m = \sqrt[n]{a^m}$$

All properties of integer exponents hold for rational exponents, provided the roots are real-valued.

Examples

Example 1

$$121^{1/2} = \sqrt{121} = 11$$

Example 2

$$27^{2/3} = \left(\sqrt[3]{27}\right)^2 = (3)^2 = 9$$

Example 3

$$p^{1/3} \cdot p^{1/4}$$

$$= p^{1/3 + 1/4}$$

$$= p^{4/12 + 3/12}$$

$$= p^{7/12} \quad \text{or} \quad \sqrt[12]{p^7}$$

Section 7.3 Simplifying Radical Expressions

Key Concepts

Let a and b represent real numbers such that $\sqrt[n]{a}$ and $\sqrt[n]{b}$ are both real. Then

$$\sqrt[n]{ab} = \sqrt[n]{a} \cdot \sqrt[n]{b} \quad \text{Multiplication property}$$

A radical expression whose radicand is written as a product of prime factors is in simplified form if all the following conditions are met:

1. The radicand has no factor raised to a power greater than or equal to the index.
2. The radicand does not contain a fraction.
3. No radicals are in the denominator of a fraction.

Examples

Example 1

$$\sqrt{12} = \sqrt{4 \cdot 3}$$

$$= \sqrt{4} \cdot \sqrt{3}$$

$$= 2\sqrt{3}$$

Example 2

$$\sqrt[3]{16x^5 y^7}$$

$$= \sqrt[3]{2^4 x^5 y^7}$$

$$= \sqrt[3]{2^3 x^3 y^6 \cdot 2x^2 y}$$

$$= \sqrt[3]{2^3 x^3 y^6} \cdot \sqrt[3]{2x^2 y}$$

$$= 2xy^2 \sqrt[3]{2x^2 y}$$

Section 7.4   Addition and Subtraction of Radicals

Key Concepts

Like **radicals** have radical factors with the same index and the same radicand.

Use the distributive property to add and subtract *like* radicals.

Examples

Example 1

$3x\sqrt{7} - 5x\sqrt{7} + x\sqrt{7}$

$= (3 - 5 + 1)x\sqrt{7}$

$= -x\sqrt{7}$

Example 2

$x\sqrt[4]{16x} - 3\sqrt[4]{x^5}$

$= 2x\sqrt[4]{x} - 3x\sqrt[4]{x}$

$= (2 - 3)x\sqrt[4]{x}$

$= -x\sqrt[4]{x}$

Section 7.5   Multiplication of Radicals

Key Concepts

The Multiplication Property of Radicals

If $\sqrt[n]{a}$ and $\sqrt[n]{b}$ are real numbers, then

$\sqrt[n]{a} \cdot \sqrt[n]{b} = \sqrt[n]{ab}$

To multiply or divide radicals with different indices, convert to rational exponents and use the properties of exponents.

Examples

Example 1

$3\sqrt{2}(\sqrt{2} + 5\sqrt{7} - \sqrt{6})$

$= 3\sqrt{4} + 15\sqrt{14} - 3\sqrt{12}$

$= 3 \cdot 2 + 15\sqrt{14} - 3 \cdot 2\sqrt{3}$

$= 6 + 15\sqrt{14} - 6\sqrt{3}$

Example 2

$\sqrt{p} \cdot \sqrt[5]{p^2}$

$= p^{1/2} \cdot p^{2/5}$

$= p^{5/10} \cdot p^{4/10}$

$= p^{9/10}$

$= \sqrt[10]{p^9}$

Section 7.6   Division of Radicals and Rationalization

Key Concepts	**Examples**
The Division Property of Radicals	**Example 1**

The Division Property of Radicals

If $\sqrt[n]{a}$ and $\sqrt[n]{b}$ are real numbers, then

$$\sqrt[n]{\frac{a}{b}} = \frac{\sqrt[n]{a}}{\sqrt[n]{b}} \qquad b \neq 0$$

The process of removing a radical from the denominator of an expression is called **rationalizing the denominator**.

Example 1

Simplify.

$$\sqrt{\frac{4x^5}{y^4}}$$

$$= \frac{\sqrt{4x^5}}{\sqrt{y^4}}$$

$$= \frac{2x^2\sqrt{x}}{y^2}$$

Rationalizing a denominator with one term

Example 2

Rationalize the denominator.

$$\frac{4}{\sqrt[3]{t}}$$

$$= \frac{4}{\sqrt[3]{t}} \cdot \frac{\sqrt[3]{t^2}}{\sqrt[3]{t^2}}$$

$$= \frac{4\sqrt[3]{t^2}}{\sqrt[3]{t^3}} = \frac{4\sqrt[3]{t^2}}{t}$$

Rationalizing a denominator with two terms involving square roots

Example 3

Rationalize the denominator.

$$\frac{\sqrt{2}}{\sqrt{x} - \sqrt{3}}$$

$$= \frac{\sqrt{2}}{(\sqrt{x} - \sqrt{3})} \cdot \frac{(\sqrt{x} + \sqrt{3})}{(\sqrt{x} + \sqrt{3})}$$

$$= \frac{\sqrt{2x} + \sqrt{6}}{x - 3}$$

Section 7.7 Radical Equations and Applications

Key Concepts

Steps to Solve a Radical Equation

1. Isolate the radical. If an equation has more than one radical, choose one of the radicals to isolate.
2. Raise each side of the equation to a power equal to the index of the radical.
3. Solve the resulting equation. If the equation still has a radical, repeat steps 1 and 2.
4. Check the potential solutions in the original equation.

Examples

Example 1

Solve:

$$\sqrt{b - 5} - \sqrt{b + 3} = 2$$
$$\sqrt{b - 5} = \sqrt{b + 3} + 2$$
$$(\sqrt{b - 5})^2 = (\sqrt{b + 3} + 2)^2$$
$$b - 5 = b + 3 + 4\sqrt{b + 3} + 4$$
$$b - 5 = b + 7 + 4\sqrt{b + 3}$$
$$-12 = 4\sqrt{b + 3}$$
$$-3 = \sqrt{b + 3}$$
$$(-3)^2 = (\sqrt{b + 3})^2$$
$$9 = b + 3$$
$$6 = b$$

<u>Check</u>:

$$\sqrt{6 - 5} - \sqrt{6 + 3} \stackrel{?}{=} 2$$
$$\sqrt{1} - \sqrt{9} \stackrel{?}{=} 2$$
$$1 - 3 \stackrel{?}{=} 2 \qquad \text{(false)}$$

There is no solution.

Section 7.8 Complex Numbers

Key Concepts

$i = \sqrt{-1}$ and $i^2 = -1$

For a real number $b > 0$, $\sqrt{-b} = i\sqrt{b}$

A **complex number** is in the form $a + bi$, where a and b are real numbers. The value a is called the real part, and b is called the imaginary part.

To add or subtract complex numbers, combine the real parts and combine the imaginary parts.

Multiply complex numbers by using the distributive property.

Divide complex numbers by multiplying the numerator and denominator by the **complex conjugate** of the denominator.

Examples

Example 1

$\sqrt{-4} \cdot \sqrt{-9}$

$= (2i)(3i)$

$= 6i^2$

$= -6$

Example 2

$(3 - 5i) - (2 + i) + (3 - 2i)$

$= 3 - 5i - 2 - i + 3 - 2i$

$= 4 - 8i$

Example 3

$(1 + 6i)(2 + 4i)$

$= 2 + 4i + 12i + 24i^2$

$= 2 + 16i + 24(-1)$

$= -22 + 16i$

Example 4

$\dfrac{3}{2 - 5i}$

$= \dfrac{3}{(2 - 5i)} \cdot \dfrac{(2 + 5i)}{(2 + 5i)}$

$= \dfrac{6 + 15i}{4 - 25i^2}$

$= \dfrac{6 + 15i}{4 + 25}$

$= \dfrac{6 + 15i}{29}$

$= \dfrac{6}{29} + \dfrac{15}{29}i$

Chapter 7   Review Exercises

For the exercises in this set, assume that all variables represent positive real numbers unless otherwise stated.

Section 7.1

1. True or false?

   a. The principal nth root of an even-indexed root is always positive.

   b. The principal nth root of an odd-indexed root is always positive.

2. Explain why $\sqrt{(-3)^2} \neq -3$.

3. For $a > 0$ and $b > 0$, are the statements true or false?

   a. $\sqrt{a^2 + b^2} = a + b$

   b. $\sqrt{(a + b)^2} = a + b$

For Exercises 4–6, simplify the radicals.

4. $\sqrt{\dfrac{50}{32}}$ **5.** $\sqrt[4]{625}$ **6.** $\sqrt{(-6)^2}$

7. Evaluate the function values for $f(x) = \sqrt{x - 1}$.

   a. $f(10)$ **b.** $f(1)$ **c.** $f(8)$

   d. Write the domain of f in interval notation.

8. Evaluate the function values for $g(t) = \sqrt{5 + t}$.

   a. $g(-5)$ **b.** $g(-4)$ **c.** $g(4)$

   d. Write the domain of g in interval notation.

9. Translate the English expression to an algebraic expression: Four more than the quotient of the cube root of $2x$ and the principal fourth root of $2x$.

For Exercises 10 and 11, simplify the expression. Assume that x and y represent *any* real number.

10. a. $\sqrt{x^2}$ **b.** $\sqrt[3]{x^3}$

   c. $\sqrt[4]{x^4}$ **d.** $\sqrt[5]{(x + 1)^5}$

11. a. $\sqrt{4y^2}$ **b.** $\sqrt[3]{27y^3}$

   c. $\sqrt[100]{y^{100}}$ **d.** $\sqrt[101]{y^{101}}$

12. Use the Pythagorean theorem to find the length of the third side of the triangle.

17 cm

15 cm

Section 7.2

13. Are the properties of exponents the same for rational exponents and integer exponents? Give an example. (Answers may vary.)

14. In the expression $x^{m/n}$ what does n represent?

15. Explain the process of eliminating a negative exponent from an algebraic expression.

For Exercises 16–21, simplify the expressions. Write the answer with positive exponents only.

16. $(-125)^{1/3}$ **17.** $16^{-1/4}$

18. $\left(\dfrac{1}{16}\right)^{-3/4} - \left(\dfrac{1}{8}\right)^{-2/3}$ **19.** $(b^{1/2} \cdot b^{1/3})^{12}$

20. $\left(\dfrac{x^{-1/4}y^{-1/3}z^{3/4}}{2^{1/3}x^{-1/3}y^{2/3}}\right)^{-12}$ **21.** $\left(\dfrac{a^{12}b^{-4}c^7}{a^3b^2c^4}\right)^{1/3}$

For Exercises 22 and 23, rewrite the expressions by using rational exponents.

22. $\sqrt[4]{x^3}$ **23.** $\sqrt[3]{2y^2}$

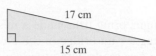 For Exercises 24–26, use a calculator to approximate the expressions to 4 decimal places.

24. $10^{1/3}$ **25.** $17.8^{2/3}$ **26.** $\sqrt[5]{147^4}$

Section 7.3

27. List the criteria for a radical expression to be simplified.

For Exercises 28–31, simplify the radicals.

28. $\sqrt{108}$ **29.** $\sqrt[4]{x^5yz^4}$

30. $-2\sqrt[3]{250a^3b^{10}}$ **31.** $\sqrt[3]{\dfrac{-16a^4}{2ab^3}}$

32. Write an English phrase that describes the following mathematical expressions: (Answers may vary.)

a. $\sqrt{\dfrac{2}{x}}$ **b.** $(x + 1)^3$

33. An engineering firm made a mistake when building a $\frac{1}{4}$-mi bridge in the Florida Keys. The bridge was made without adequate expansion joints to prevent buckling during the heat of summer. During mid-June, the bridge expanded 1.5 ft, causing a vertical bulge in the middle. Calculate the height of the bulge h in feet. (*Note:* 1 mi = 5280 ft.) Round to the nearest foot.

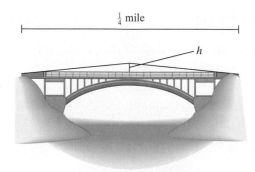

Section 7.4

34. Complete the statement: Radicals may be added or subtracted if . . .

For Excrcises 35–38, determine whether the radicals may be combined, and explain your answer.

35. $\sqrt[3]{2x} - 2\sqrt{2x}$ **36.** $2 + \sqrt{x}$

37. $\sqrt[4]{3xy} + 2\sqrt[4]{3xy}$ **38.** $-4\sqrt{32} + 7\sqrt{50}$

For Exercises 39–42, add or subtract as indicated.

39. $4\sqrt{7} - 2\sqrt{7} + 3\sqrt{7}$

40. $2\sqrt[3]{64} + 3\sqrt[3]{54} - 16$

41. $\sqrt{50} + 7\sqrt{2} - \sqrt{8}$

42. $x\sqrt[3]{16x^2} - 4\sqrt[3]{2x^5} + 5x\sqrt[3]{54x^2}$

For Exercises 43 and 44, answer true or false. If an answer is false, explain why. Assume all variables represent positive real numbers.

43. $5 + 3\sqrt{x} = 8\sqrt{x}$

44. $\sqrt{y} + \sqrt{y} = \sqrt{2y}$

Section 7.5

For Exercises 45–56, multiply the radicals and simplify the answer.

45. $\sqrt{3} \cdot \sqrt{12}$ **46.** $\sqrt[4]{4} \cdot \sqrt[4]{8}$

47. $-2\sqrt{3}(\sqrt{7} - 3\sqrt{11})$ **48.** $-3\sqrt{5}(2\sqrt{3} - \sqrt{5})$

49. $(2\sqrt{x} - 3)(2\sqrt{x} + 3)$ **50.** $(\sqrt{y} + 4)(\sqrt{y} - 4)$

51. $(\sqrt{7y} - \sqrt{3x})^2$ **52.** $(2\sqrt{3w} + 5)^2$

53. $(-\sqrt{z} - \sqrt{6})(2\sqrt{z} + 7\sqrt{6})$

54. $(3\sqrt{a} - \sqrt{5})(\sqrt{a} + 2\sqrt{5})$

55. $\sqrt[3]{u} \cdot \sqrt{u^5}$ **56.** $\sqrt{2} \cdot \sqrt[4]{w^3}$

Section 7.6

For Exercises 57–60, simplify the radicals.

57. $\sqrt{\dfrac{3y^5}{25x^6}}$ **58.** $\sqrt[3]{\dfrac{-16x^7y^6}{z^9}}$

59. $\dfrac{\sqrt{324w^7}}{\sqrt{4w^3}}$ **60.** $\dfrac{\sqrt[3]{3t^{14}}}{\sqrt[3]{192t^2}}$

For Exercises 61–68, rationalize the denominator.

61. $\sqrt{\dfrac{7}{2y}}$ **62.** $\sqrt{\dfrac{5}{3w}}$

63. $\dfrac{4}{\sqrt[3]{9p^2}}$ **64.** $\dfrac{-2}{\sqrt[3]{2x}}$

65. $\dfrac{-5}{\sqrt{15} - \sqrt{10}}$ **66.** $\dfrac{-6}{\sqrt{7} - \sqrt{5}}$

67. $\dfrac{t - 3}{\sqrt{t} - \sqrt{3}}$ **68.** $\dfrac{w - 7}{\sqrt{w} - \sqrt{7}}$

69. Translate the mathematical expression to an English phrase. (Answers may vary.)

$$\dfrac{\sqrt{2}}{x^2}$$

Section 7.7

Solve the radical equations in Exercises 70–77, if possible.

70. $\sqrt{2y} = 7$ **71.** $\sqrt{a - 6} - 5 = 0$

72. $\sqrt[3]{2w - 3} + 5 = 2$

73. $\sqrt[4]{p + 12} - \sqrt[4]{5p - 16} = 0$

74. $\sqrt{t} + \sqrt{t - 5} = 5$

75. $\sqrt{8x + 1} = -\sqrt{x - 13}$

76. $\sqrt{2m^2 + 4} - \sqrt{9m} = 0$

77. $\sqrt{x + 2} = 1 - \sqrt{2x + 5}$

78. A tower is supported by stabilizing wires. Find the exact length of each wire, and then round to the nearest tenth of a meter.

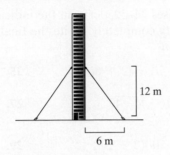

12 m

6 m

79. The velocity, $v(d)$, of an ocean wave depends on the water depth d as the wave approaches land.

$$v(d) = \sqrt{32d}$$

where $v(d)$ is in feet per second and d is in feet.

a. Find $v(20)$ and interpret its value. Round to 1 decimal place.

b. Find the depth of the water at a point where a wave is traveling at 16 ft/sec.

Section 7.8

80. Define a complex number.

81. Define an imaginary number.

82. Describe the first step in the process to simplify the expression. $\dfrac{3}{4 + 6i}$

For Exercises 83–86, rewrite the expressions in terms of i.

83. $\sqrt{-16}$

84. $-\sqrt{-5}$

85. $\sqrt{-75} \cdot \sqrt{-3}$

86. $\dfrac{-\sqrt{-24}}{\sqrt{6}}$

For Exercises 87–90, simplify the powers of i.

87. i^{38}

88. i^{101}

89. i^{19}

90. $i^{1000} + i^{1002}$

For Exercises 91–94, perform the indicated operations. Write the final answer in the form $a + bi$.

91. $(-3 + i) - (2 - 4i)$

92. $(1 + 6i)(3 - i)$

93. $(4 - 3i)(4 + 3i)$

94. $(5 - i)^2$

For Exercises 95 and 96, write the expressions in the form $a + bi$, and determine the real and imaginary parts.

95. $\dfrac{17 - 4i}{-4}$

96. $\dfrac{-16 - 8i}{8}$

For Exercises 97–100, divide and simplify. Write the final answer in the form $a + bi$.

97. $\dfrac{2 - i}{3 + 2i}$

98. $\dfrac{10 + 5i}{2 - i}$

99. $\dfrac{5 + 3i}{-2i}$

100. $\dfrac{4i}{4 - i}$

For Exercises 101 and 102, simplify the expression.

101. $\dfrac{-8 + \sqrt{-40}}{12}$

102. $\dfrac{6 - \sqrt{-144}}{3}$

Chapter 7   Test

1. a. What is the principal square root of 36?

b. What is the negative square root of 36?

2. Which of the following are real numbers?

a. $-\sqrt{100}$ **b.** $\sqrt{-100}$

c. $-\sqrt[3]{1000}$ **d.** $\sqrt[3]{-1000}$

3. Simplify.

a. $\sqrt[3]{y^3}$ **b.** $\sqrt[4]{y^4}$

For Exercises 4–11, simplify the radicals. Assume that all variables represent positive numbers.

4. $\sqrt[4]{81}$ **5.** $\sqrt{\dfrac{16}{9}}$

6. $\sqrt[3]{32}$ **7.** $\sqrt{a^4b^3c^5}$

8. $\sqrt{18x^5y^3z^4}$ **9.** $\sqrt{\dfrac{32w^6}{2w}}$

10. $\sqrt[3]{\dfrac{x^6}{125y^3}}$ **11.** $\dfrac{2\sqrt{72}}{8}$

12. a. Evaluate the function values $f(-8), f(-6),$ $f(-4),$ and $f(-2)$ for $f(x) = \sqrt{-2x - 4}$.

b. Write the domain of f in interval notation.

13. Use a calculator to evaluate $\dfrac{-3 - \sqrt{5}}{17}$ to 4 decimal places.

For Exercises 14 and 15, simplify the expressions. Assume that all variables represent positive numbers.

14. $-27^{1/3}$ **15.** $\dfrac{t^{-1} \cdot t^{1/2}}{t^{1/4}}$

For Exercises 16 and 17, use rational exponents to multiply or divide. Write the final answer in radical form.

16. $\sqrt[6]{7} \cdot \sqrt{y}$ **17.** $\dfrac{\sqrt[3]{10}}{\sqrt[4]{10}}$

18. Add or subtract as indicated.
$$3\sqrt{5} + 4\sqrt{5} - 2\sqrt{20}$$

For Exercises 19 and 20, multiply the radicals.

19. $3\sqrt{x}(\sqrt{2} - \sqrt{5})$

20. $(2\sqrt{5} - 3\sqrt{x})(4\sqrt{5} + \sqrt{x})$

For Exercises 21 and 22, rationalize the denominator. Assume $x > 0$.

21. $\dfrac{-2}{\sqrt[3]{x}}$ **22.** $\dfrac{\sqrt{x} + 2}{3 - \sqrt{x}}$

23. Rewrite the expressions in terms of i.

a. $\sqrt{-8}$ **b.** $2\sqrt{-16}$ **c.** $\dfrac{2 + \sqrt{-8}}{4}$

For Exercises 24–30, perform the indicated operations and simplify completely. Write the final answer in the form $a + bi$.

24. $(3 - 5i) - (2 + 6i)$ **25.** $(4 + i)(8 + 2i)$

26. $\sqrt{-16} \cdot \sqrt{-49}$ **27.** $(4 - 7i)^2$

28. $(2 - 10i)(2 + 10i)$ **29.** $\dfrac{3 - 2i}{3 - 4i}$

30. $(10 + 3i)[(-5i + 8) - (5 - 3i)]$

31. If the volume V of a sphere is known, the radius of the sphere can be computed by
$$r(V) = \sqrt[3]{\dfrac{3V}{4\pi}}.$$
Find $r(10)$ to 2 decimal places. Interpret the meaning in the context of the problem.

32. A patio 20 ft wide has a slanted roof, as shown in the picture. Find the length of the roof if there is an 8-in. overhang. Round the answer to the nearest foot.

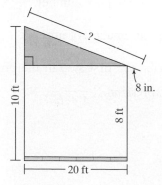

For Exercises 33–35, solve the radical equation.

33. $\sqrt[3]{2x + 5} = -3$

34. $\sqrt{5x + 8} = \sqrt{5x - 1} + 1$

35. $\sqrt{t + 7} - \sqrt{2t - 3} = 2$

Chapters 1–7 Cumulative Review Exercises

1. Simplify the expression.
$6^2 - 2[5 - 8(3 - 1) + 4 \div 2]$

2. Simplify the expression.
$3x - 3(-2x + 5) - 4y + 2(3x + 5) - y$

3. Solve. $9(2y + 8) = 20 - (y + 5)$

4. Solve the inequality. Write the answer in interval notation.

$$2a - 4 < -14$$

5. Write an equation of the line that is parallel to the line $2x + y = 9$ and passes through the point $(3, -1)$. Write the answer in slope-intercept form.

6. On the same coordinate system, graph the line $2x + y = 9$ and the line that you derived in Exercise 5. Verify that these two lines are indeed parallel.

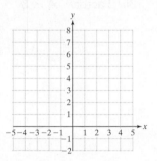

7. Solve the system of equations by using the addition method.

$$2x - 3y = 0$$

$$-4x + 3y = -1$$

8. Determine if $(2, -2, \frac{1}{2})$ is a solution to the system.

$$2x + y - 4z = 0$$

$$x - y + 2z = 5$$

$$3x + 2y + 2z = 4$$

9. Write a system of linear equations from the augmented matrix. Use x, y, and z for the variables.

$$\begin{bmatrix} 1 & 0 & 0 & | & 6 \\ 0 & 1 & 0 & | & 3 \\ 0 & 0 & 1 & | & 8 \end{bmatrix}$$

10. Given the function defined by $f(x) = 4x - 2$.

   a. Find $f(-2), f(0), f(4)$, and $f(\frac{1}{2})$.

   b. Write the ordered pairs that correspond to the function values in part (a).

   c. Graph $y = f(x)$.

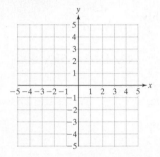

11. Determine if the graph defines y as a function of x.

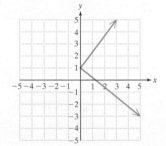

12. Simplify the expression. Write the final answer with positive exponents only.

$$\left(\frac{a^3 b^{-1} c^3}{ab^{-5} c^2} \right)^2$$

13. Simplify the expression. Write the final answer with positive exponents only.

$$\left(\frac{a^{3/2} b^{-1/4} c^{1/3}}{ab^{-5/4} c^0} \right)^{12}$$

14. Multiply or divide as indicated, and write the answer in scientific notation.

   a. $(3.5 \times 10^7)(4 \times 10^{-12})$

   b. $\dfrac{6.28 \times 10^5}{2 \times 10^{-4}}$

15. Multiply the polynomials $(2x + 5)(x - 3)$. What is the degree of the product?

16. Perform the indicated operations and simplify.
$\sqrt{3}(\sqrt{5} + \sqrt{6} + \sqrt{3})$

17. Divide. $(x^2 - x - 12) \div (x + 3)$

18. Simplify and subtract. $\sqrt[4]{\dfrac{1}{16}} - \sqrt[3]{\dfrac{8}{27}}$

19. Simplify. $\sqrt[3]{\dfrac{54c^4}{cd^3}}$

20. Add. $4\sqrt{45b^3} + 5b\sqrt{80b}$

21. Divide. $\dfrac{13i}{3 + 2i}$ Write the answer in the form $a + bi$.

22. Solve. $\dfrac{5}{y - 2} - \dfrac{3}{y - 4} = \dfrac{6}{y^2 - 6y + 8}$

23. Add. $\dfrac{3}{x^2 + 5x} + \dfrac{-2}{x^2 - 25}$

24. Divide. $\dfrac{a + 10}{2a^2 - 11a - 6} \div \dfrac{a^2 + 12a + 20}{6 - a}$

25. Perform the indicated operations.
$(-5x^2 - 4x + 8) - (3x - 5)^2$

26. Simplify. $\dfrac{-4}{\sqrt{3} - \sqrt{5}}$

27. Divide. $\dfrac{4}{3 - 5i}$

28. Solve. $12x^2 + 4x - 21 = 0$

29. Factor. $x^2 + 6x + 9 - y^2$

30. Factor. $x^6 + 8$

Quadratic Equations, Functions, and Inequalities

8

CHAPTER OUTLINE

Chapter 8

In Chapter 8, we revisit quadratic equations. We have already learned how to solve quadratic equations by factoring and applying the zero product rule. Now we present two additional techniques that can be used to solve a quadratic equation even if the quadratic polynomial does not factor.

Review Your Skills

To prepare for this chapter, we recommend that you review how to solve quadratic equations by using the zero product rule and how to simplify radical expressions. Match the answers to the exercises below to the letter on the right. Then write the letter in the appropriate space at the bottom of the page to define an important vocabulary word.

Solve.

1. $x^2 - 9 = 0$

2. $x^2 - 6x + 8 = 0$

3. $3x(x + 1) = 60$

Simplify.

4. $\dfrac{-5 \pm \sqrt{50}}{10}$ **5.** $\dfrac{7 \pm 21i}{7}$ **6.** $\dfrac{2 \pm \sqrt{-12}}{4}$

7. Determine the x-intercept(s). $f(x) = x^2 - 6x + 5$

8. Determine the y-intercept. $f(x) = x^2 - 6x + 5$

A function defined by $f(x) = ax^2 + bx + c$ $(a \neq 0)$ is called a

___ ___ ___ ___ ___ ___ ___ ___ ___ function.
 2   1   6   5   8   6   4   7   3

Answers

D $1 \pm 3i$

A $\dfrac{1}{2} \pm \dfrac{\sqrt{3}}{2}i$

C $4, -5$

U $-3, 3$

T $\dfrac{-1 \pm \sqrt{2}}{2}$

Q $2, 4$

R $(0, 5)$

I $(1, 0)$ and $(5, 0)$

Square Root Property and Completing the Square

1. Solving Quadratic Equations by Using the Square Root Property

In Section 5.8 we learned how to solve a quadratic equation by factoring and applying the zero product rule. For example:

$$x^2 = 81$$

$$x^2 - 81 = 0 \qquad \text{Set one side equal to zero.}$$

$$(x - 9)(x + 9) = 0 \qquad \text{Factor.}$$

$$x - 9 = 0 \quad \text{or} \quad x + 9 = 0 \qquad \text{Set each factor equal to zero.}$$

$$x = 9 \quad \text{or} \qquad x = -9$$

The solutions are 9 and -9.

It is important to note that the zero product rule can only be used if the equation is factorable. In this section and Section 8.2, we will learn how to solve quadratic equations, factorable and nonfactorable.

Consider a quadratic equation of the form $x^2 = k$. The solutions are all numbers (real or imaginary) that when squared equal k, so for any nonzero constant k, there will be two solutions, $\sqrt{k}$ or $-\sqrt{k}$. For example:

$$x^2 = 81 \qquad \text{The solutions are 9 and } -9.$$

$$x^2 = -81 \qquad \text{The solutions are } 9i \text{ and } -9i.$$

The first technique utilizes the **square root property**.

The Square Root Property

For any real number, k, if $x^2 = k$, then $x = \sqrt{k}$ or $x = -\sqrt{k}$.

Note: The solution may also be written as $\pm\sqrt{k}$, read "plus or minus the square root of k."

Example 1 Solving a Quadratic Equation by Using the Square Root Property

Use the square root property to solve the equation. $\qquad 4p^2 = 9$

Solution:

$$4p^2 = 9$$

$$p^2 = \frac{9}{4} \qquad \text{Isolate } p^2 \text{ by dividing both sides by 4.}$$

$$p = \pm\sqrt{\frac{9}{4}} \qquad \text{Apply the square root property.}$$

$$p = \pm\frac{3}{2} \qquad \text{Simplify the radical.}$$

The solutions are $\dfrac{3}{2}$ and $-\dfrac{3}{2}$.

Answer

1. $\dfrac{4}{5}, -\dfrac{4}{5}$

For a quadratic equation, $ax^2 + bx + c = 0$, if $b = 0$, then the equation is easily solved by using the square root property. This is demonstrated in Example 2.

Example 2 Solving a Quadratic Equation by Using the Square Root Property

Use the square root property to solve the equation.

$$3x^2 + 75 = 0$$

Solution:

$$3x^2 + 75 = 0 \qquad \text{Rewrite the equation to fit the form } x^2 = k.$$

$$3x^2 = -75$$

$$x^2 = -25 \qquad \text{The equation is now in the form } x^2 = k.$$

$$x = \pm\sqrt{-25} \qquad \text{Apply the square root property.}$$

$$= \pm 5i$$

Check: $x = 5i$ Check: $x = -5i$

$3x^2 + 75 = 0$ $3x^2 + 75 = 0$

$3(5i)^2 + 75 \overset{?}{=} 0$ $3(-5i)^2 + 75 \overset{?}{=} 0$

$3(25i^2) + 75 \overset{?}{=} 0$ $3(25i^2) + 75 \overset{?}{=} 0$

$3(-25) + 75 \overset{?}{=} 0$ $3(-25) + 75 \overset{?}{=} 0$

$-75 + 75 \overset{?}{=} 0$ ✔ $-75 + 75 \overset{?}{=} 0$ ✔

The solutions are $5i$ and $-5i$.

Skill Practice

Solve using the square root property.

2. $8x^2 + 72 = 0$

Avoiding Mistakes

A common mistake is to forget the $\pm$ symbol when solving the equation $x^2 = k$:

$$x = \pm\sqrt{k}$$

Example 3 Solving a Quadratic Equation by Using the Square Root Property

Use the square root property to solve the equation.

$$(w + 3)^2 = 20$$

Solution:

$$(w + 3)^2 = 20 \qquad \text{The equation is in the form } x^2 = k, \text{ where } x = (w + 3).$$

$$w + 3 = \pm\sqrt{20} \qquad \text{Apply the square root property.}$$

$$w + 3 = \pm\sqrt{4 \cdot 5} \qquad \text{Simplify the radical.}$$

$$w + 3 = \pm 2\sqrt{5}$$

$$w = -3 \pm 2\sqrt{5} \qquad \text{Solve for } w.$$

The solutions are $-3 + 2\sqrt{5}$ and $-3 - 2\sqrt{5}$.

Skill Practice

Solve using the square root property.

3. $(t - 5)^2 = 18$

Answers

2. $\pm 3i$ **3.** $5 \pm 3\sqrt{2}$

2. Solving Quadratic Equations by Completing the Square

In Example 3 we used the square root property to solve an equation where the square of a binomial was equal to a constant.

$$(w + 3)^2 = 20$$

$\underbrace{}_{\substack{\text{Square of a} \\ \text{binomial}}}$ $\underset{\text{Constant}}{\uparrow}$

The square of a binomial is the factored form of a perfect square trinomial. For example:

Perfect Square Trinomial		Factored Form
$x^2 + 10x + 25$	$\longrightarrow$	$(x + 5)^2$
$t^2 - 6t + 9$	$\longrightarrow$	$(t - 3)^2$
$p^2 - 14p + 49$	$\longrightarrow$	$(p - 7)^2$

For a perfect square trinomial with a leading coefficient of 1, the constant term is the square of one-half the linear term coefficient. For example:

$$x^2 + 10x + 25$$
$$[\tfrac{1}{2}(10)]^2$$

In general an expression of the form $x^2 + bx + n$ is a perfect square trinomial if $n = (\tfrac{1}{2}b)^2$. The process to create a perfect square trinomial is called **completing the square**.

| Example 4 | **Completing the Square** |

Determine the value of n that makes the polynomial a perfect square trinomial. Then factor the expression as the square of a binomial.

a. $x^2 + 12x + n$ b. $x^2 - 26x + n$

c. $x^2 + 11x + n$ d. $x^2 - \dfrac{4}{7}x + n$

Solution:

The expressions are in the form $x^2 + bx + n$. The value of n equals the square of one-half the linear term coefficient $(\tfrac{1}{2}b)^2$.

a. $x^2 + 12x + n$

     $x^2 + 12x + 36$ $n = [\tfrac{1}{2}(12)]^2 = (6)^2 = 36.$

     $(x + 6)^2$ Factored form

b. $x^2 - 26x + n$

     $x^2 - 26x + 169$ $n = [\tfrac{1}{2}(-26)]^2 = (-13)^2 = 169.$

     $(x - 13)^2$ Factored form

c. $x^2 + 11x + n$

     $x^2 + 11x + \dfrac{121}{4}$ $n = [\tfrac{1}{2}(11)]^2 = (\tfrac{11}{2})^2 = \tfrac{121}{4}.$

     $\left(x + \dfrac{11}{2}\right)^2$ Factored form

d. $x^2 - \dfrac{4}{7}x + n$

$x^2 - \dfrac{4}{7}x + \dfrac{4}{49}$ $n = [\frac{1}{2}(-\frac{4}{7})]^2 = (-\frac{2}{7})^2 = \frac{4}{49}$

$\left(x - \dfrac{2}{7}\right)^2$ Factored form

The process of completing the square can be used to write a quadratic equation $ax^2 + bx + c = 0$ $(a \neq 0)$ in the form $(x - h)^2 = k$. Then the square root property can be used to solve the equation. The following steps outline the procedure.

Solving a Quadratic Equation $ax^2 + bx + c = 0$ by Completing the Square and Applying the Square Root Property

Step 1 Divide both sides by a to make the leading coefficient 1.

Step 2 Isolate the variable terms on one side of the equation.

Step 3 Complete the square. (Add the square of one-half the linear term coefficient to both sides of the equation. Then factor the resulting perfect square trinomial.)

Step 4 Apply the square root property and solve for x.

Example 5   **Solving a Quadratic Equation by Completing the Square and Applying the Square Root Property**

Solve by completing the square and applying the square root property.

$$x^2 - 6x + 13 = 0$$

Solution:

$x^2 - 6x + 13 = 0$	**Step 1:** Since the leading coefficient a is equal to 1, we do not have to divide by a. We can proceed to step 2.
$x^2 - 6x = -13$	**Step 2:** Isolate the variable terms on one side.
$x^2 - 6x + 9 = -13 + 9$	**Step 3:** To complete the square, add $[\frac{1}{2}(-6)]^2 = 9$ to both sides of the equation.
$(x - 3)^2 = -4$	Factor the perfect square trinomial.
$x - 3 = \pm\sqrt{-4}$	**Step 4:** Apply the square root property.
$x - 3 = \pm 2i$	Simplify the radical.
$x = 3 \pm 2i$	Solve for x.

The solutions are imaginary numbers and can be written as $3 + 2i$ and $3 - 2i$.

Skill Practice

Solve by completing the square and applying the square root property.

9. $z^2 - 4z - 2 = 0$

Answer

9. $2 \pm \sqrt{6}$

Example 6 **Solving a Quadratic Equation by Completing the Square and Applying the Square Root Property**

Solve by completing the square and applying the square root property.

$$2x(2x - 10) = -30 + 6x$$

Solution:

$$2x(2x - 10) = -30 + 6x$$

$$4x^2 - 20x = -30 + 6x \qquad \text{Clear parentheses.}$$

$$4x^2 - 26x + 30 = 0 \qquad \text{Write the equation in the form } ax^2 + bx + c = 0.$$

$$\frac{4x^2}{4} - \frac{26x}{4} + \frac{30}{4} = \frac{0}{4} \qquad \textbf{Step 1:} \quad \text{Divide both sides by the leading coefficient 4.}$$

$$x^2 - \frac{13}{2}x + \frac{15}{2} = 0$$

$$x^2 - \frac{13}{2}x = -\frac{15}{2} \qquad \textbf{Step 2:} \quad \text{Isolate the variable terms on one side.}$$

$$x^2 - \frac{13}{2}x + \frac{169}{16} = -\frac{15}{2} + \frac{169}{16} \qquad \textbf{Step 3:} \quad \text{Add } [\tfrac{1}{2}(-\tfrac{13}{2})]^2 = (-\tfrac{13}{4})^2 = \tfrac{169}{16} \text{ to both sides.}$$

$$\left(x - \frac{13}{4}\right)^2 = -\frac{120}{16} + \frac{169}{16} \qquad \text{Factor the perfect square trinomial. Rewrite the right-hand side with a common denominator.}$$

$$\left(x - \frac{13}{4}\right)^2 = \frac{49}{16}$$

$$x - \frac{13}{4} = \pm\sqrt{\frac{49}{16}} \qquad \textbf{Step 4:} \quad \text{Apply the square root property.}$$

$$x - \frac{13}{4} = \pm\frac{7}{4} \qquad \text{Simplify the radical.}$$

$$x = \frac{13}{4} \pm \frac{7}{4}$$

$$x = \frac{13}{4} + \frac{7}{4} = \frac{20}{4} = 5$$

$$x = \frac{13}{4} - \frac{7}{4} = \frac{6}{4} = \frac{3}{2}$$

The solutions are rational numbers: $\frac{3}{2}$ and 5.

TIP: In general, if the solutions to a quadratic equation are rational numbers, the equation can be solved by factoring and using the zero product rule. Consider the equation from Example 6.

$$2x(2x - 10) = -30 + 6x$$

$$4x^2 - 20x = -30 + 6x$$

$$4x^2 - 26x + 30 = 0$$

$$2(2x^2 - 13x + 15) = 0$$

$$2(x - 5)(2x - 3) = 0$$

$$x = 5 \quad \text{or} \quad x = \frac{3}{2}$$

Example 7   **Solving a Quadratic Equation by Completing the Square and Applying the Square Root Property**

Solve by completing the square and applying the square root property.

$$p^2 + \frac{4}{5}p - \frac{7}{25} = 0$$

Solution:

$$p^2 + \frac{4}{5}p - \frac{7}{25} = 0$$ **Step 1:**   The leading coefficient is 1.

$$p^2 + \frac{4}{5}p = \frac{7}{25}$$ **Step 2:**   Isolate the variable term on one side.

$$p^2 + \frac{4}{5}p + \frac{4}{25} = \frac{7}{25} + \frac{4}{25}$$ **Step 3:**   Add $\left[\frac{1}{2}\left(\frac{4}{5}\right)\right]^2 = \left(\frac{2}{5}\right)^2 = \frac{4}{25}$.

$$\left(p + \frac{2}{5}\right)^2 = \frac{11}{25}$$ Factor the perfect square trinomial. Add the fractions.

$$p + \frac{2}{5} = \pm\sqrt{\frac{11}{25}}$$ **Step 4:**   Apply the square root property.

$$p + \frac{2}{5} = \pm\frac{\sqrt{11}}{5}$$ Simplify the radical.

$$p = -\frac{2}{5} \pm \frac{\sqrt{11}}{5}$$ Solve for p.

The solutions are $-\frac{2}{5} + \frac{\sqrt{11}}{5}$ and $-\frac{2}{5} - \frac{\sqrt{11}}{5}$.

TIP: The solutions to Example 7 can also be written as:

$$\frac{-2 \pm \sqrt{11}}{5}$$

3. Literal Equations

Example 8   **Solving a Literal Equation**

Ignoring air resistance, the distance d (in meters) that an object falls in t sec is given by the equation

$$d = 4.9t^2 \qquad \text{where } t \geq 0$$

a. Solve the equation for t. Do not rationalize the denominator.

b. Using the equation from part (a), determine the amount of time required for an object to fall 500 m. Round to the nearest second.

Solution:

a. $d = 4.9t^2$

$$\frac{d}{4.9} = t^2$$ Isolate the quadratic term. The equation is in the form $t^2 = k$.

$$t = \pm\sqrt{\frac{d}{4.9}}$$ Apply the square root property.

$$= \sqrt{\frac{d}{4.9}}$$ Because $t \geq 0$, reject the negative solution.

b. $t = \sqrt{\dfrac{d}{4.9}}$

$= \sqrt{\dfrac{500}{4.9}}$ Substitute $d = 500$.

$t \approx 10.1$

The object will require approximately 10.1 sec to fall 500 m.

Section 8.1   Practice Exercises

Vocabulary and Key Concepts

1. a. The zero product rule states that if $ab = 0$, then $a =$ _____ or $b =$ _____.

  b. To apply the zero product rule, one side of the equation must be equal to _____ and the other side must be written in factored form.

  c. The square root property states that for any real number k, if $x^2 = k$, then $x =$ _____ or $x =$ _____.

  d. To apply the square root property to the equation $t^2 + 2 = 11$, first subtract _____ from both sides. The solutions are _____.

  e. The process to create a perfect square trinomial is called _____ the square.

  f. Fill in the blank to complete the square for the trinomial $x^2 + 20x +$ _____.

  g. To use completing the square to solve the equation $4x^2 + 3x + 5 = 0$, the first step is to divide by _____ so that the coefficient of the x^2 term is _____.

  h. Given the trinomial $y^2 + 8y + 16$, the coefficient of the linear term is _____.

Concept 1: Solving Quadratic Equations by Using the Square Root Property

For Exercises 2–17, solve the equations by using the square root property. **(See Examples 1–3.)**

2. $x^2 = 100$ **3.** $y^2 = 4$ **4.** $4t^2 = 81$ **5.** $36u^2 = 121$

6. $3v^2 + 33 = 0$ **7.** $-2m^2 = 50$ **8.** $(p - 5)^2 = 9$ **9.** $(q + 3)^2 = 4$

10. $(3x - 2)^2 - 5 = 0$ **11.** $(2y + 3)^2 - 7 = 0$ **12.** $(h - 4)^2 = -8$ **13.** $(t + 5)^2 = -18$

14. $6p^2 - 3 = 2$ **15.** $15 = 4 + 3w^2$ **16.** $\left(x - \dfrac{3}{2}\right)^2 + \dfrac{7}{4} = 0$ **17.** $\left(m + \dfrac{4}{5}\right)^2 - \dfrac{3}{25} = 0$

18. Given the equation $x^2 = k$, match the following statements.

  a. If $k > 0$, then _____   **i.** there will be one real solution.

  b. If $k < 0$, then _____   **ii.** there will be two real solutions.

  c. If $k = 0$, then _____   **iii.** there will be two imaginary solutions.

19. State two methods that can be used to solve the equation $x^2 - 36 = 0$. Then solve the equation by using both methods.

20. Explain the difference between solving the equations: $x = \sqrt{16}$ and $x^2 = 16$.

Concept 2: Solving Quadratic Equations by Completing the Square

For Exercises 21–32, find the value of n so that the expression is a perfect square trinomial. Then factor the trinomial. **(See Example 4.)**

21. $x^2 - 6x + n$

22. $x^2 + 12x + n$

23. $t^2 + 8t + n$

24. $v^2 - 18v + n$

25. $c^2 - c + n$

26. $x^2 + 9x + n$

27. $y^2 + 5y + n$

28. $a^2 - 7a + n$

29. $b^2 + \dfrac{2}{5}b + n$

30. $m^2 - \dfrac{2}{7}m + n$

31. $p^2 - \dfrac{2}{3}p + n$

32. $w^2 + \dfrac{3}{4}w + n$

33. Summarize the steps used in solving a quadratic equation by completing the square and applying the square root property.

34. What types of quadratic equations can be solved by completing the square and applying the square root property?

For Exercises 35–54, solve the quadratic equation by completing the square and applying the square root property. Write imaginary solutions in the form $a + bi$. **(See Examples 5–7.)**

35. $t^2 + 8t + 15 = 0$

36. $m^2 + 6m + 8 = 0$

37. $x^2 + 6x = -16$

38. $x^2 - 4x = 3$

39. $p^2 + 4p + 6 = 0$

40. $q^2 + 2q + 2 = 0$

41. $-3y - 10 = -y^2$

42. $-24 = -2y^2 + 2y$

43. $2a^2 + 4a + 5 = 0$

44. $3a^2 + 6a - 7 = 0$

45. $9x^2 - 36x + 40 = 0$

46. $9y^2 - 12y + 5 = 0$

47. $p^2 - \dfrac{2}{5}p = \dfrac{2}{25}$

48. $n^2 - \dfrac{2}{3}n = \dfrac{1}{9}$

49. $(2w + 5)(w - 1) = 2$

50. $(3p - 5)(p + 1) = -3$

51. $n(n - 4) = 7$

52. $m(m + 10) = 2$

53. $2x(x + 6) = 14$

54. $3x(x - 2) = 24$

Concept 3: Literal Equations

55. The distance (in feet) that an object falls in t sec is given by the equation $d = 16t^2$, where $t \geq 0$.

    a. Solve the equation for t. **(See Example 8.)**

    b. Using the equation from part (a), determine the amount of time required for an object to fall 1024 ft.

56. The volume of a can that is 4 in. tall is given by the equation $V = 4\pi r^2$, where r is the radius of the can, measured in inches.

    a. Solve the equation for r. Do not rationalize the denominator.

    b. Using the equation from part (a), determine the radius of a can with volume of 12.56 in.^{3} Use 3.14 for π.

For Exercises 57–62, solve for the indicated variable.

57. $A = \pi r^2$ for r $(r > 0)$

58. $E = mc^2$ for c $(c > 0)$

59. $a^2 + b^2 + c^2 = d^2$ for a $(a > 0)$

60. $a^2 + b^2 = c^2$ for b $(b > 0)$

61. $V = \dfrac{1}{3}\pi r^2 h$ for r $(r > 0)$

62. $V = \dfrac{1}{3}s^2 h$ for s $(s > 0)$

63. A corner shelf is to be made from a triangular piece of plywood, as shown in the diagram. Find the distance x that the shelf will extend along the walls. Assume that the walls are at right angles. Round the answer to a tenth of a foot.

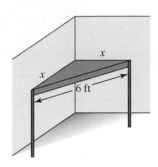

64. The volume of a box with a square bottom and a height of 4 in. is given by $V(x) = 4x^2$, where x is the length (in inches) of the sides of the bottom of the box.

  a. If the volume of the box is 289 in.^{3}, find the dimensions of the box.

  b. Are there two possible answers to part (a)? Why or why not?

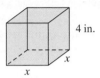

65. A square has an area of 50 in.^{2} What are the lengths of the sides? (Round to 1 decimal place.)

66. The amount of money A in an account with an interest rate r compounded annually is given by

$$A = P(1 + r)^t$$

where P is the initial principal and t is the number of years the money is invested.

  a. If a \$10,000 investment grows to \$11,664 after 2 yr, find the interest rate.

  b. If a \$6000 investment grows to \$7392.60 after 2 yr, find the interest rate.

  c. Jamal wants to invest \$5000. He wants the money to grow to at least \$6500 in 2 yr to cover the cost of his son's first year at college. What interest rate does Jamal need for his investment to grow to \$6500 in 2 yr? Round to the nearest hundredth of a percent.

67. A textbook company has discovered that the profit for selling its books is given by

$$P(x) = -\frac{1}{8}x^2 + 5x$$

where x is the number of textbooks produced (in thousands) and $P(x)$ is the corresponding profit (in thousands of dollars). The graph of the function is shown at right.

  a. Approximate the number of books required to make a profit of \$20,000. [*Hint:* Let $P(x) = 20$. Then complete the square to solve for x.] Round to one decimal place.

  b. Why are there two answers to part (a)?

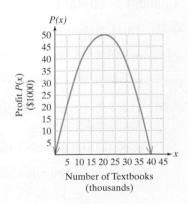

68. If we ignore air resistance, the distance (in feet) that an object travels in free fall can be approximated by $d(t) = 16t^2$, where t is the time in seconds after the object was dropped.

    a. If the CN Tower in Toronto is 1815 ft high, how long will it take an object to fall from the top of the building? Round to one decimal place.

    b. If the Renaissance Tower in Dallas is 886 ft high, how long will it take an object to fall from the top of the building? Round to one decimal place.

Quadratic Formula and Applications

Section 8.2

1. Derivation of the Quadratic Formula

If we solve a quadratic equation in standard form $ax^2 + bx + c = 0\ (a > 0)$ by completing the square and using the square root property, the result is a formula that gives the solutions for x in terms of a, b, and c.

$ax^2 + bx + c = 0$	Begin with a quadratic equation in standard form.
$\dfrac{ax^2}{a} + \dfrac{bx}{a} + \dfrac{c}{a} = \dfrac{0}{a}$	Divide by the leading coefficient.
$x^2 + \dfrac{b}{a}x + \dfrac{c}{a} = 0$	
$x^2 + \dfrac{b}{a}x = -\dfrac{c}{a}$	Isolate the terms containing x.
$x^2 + \dfrac{b}{a}x + \left(\dfrac{1}{2} \cdot \dfrac{b}{a}\right)^2 = \left(\dfrac{1}{2} \cdot \dfrac{b}{a}\right)^2 - \dfrac{c}{a}$	Add the square of $\frac{1}{2}$ the linear term coefficient to both sides of the equation.
$\left(x + \dfrac{b}{2a}\right)^2 = \dfrac{b^2}{4a^2} - \dfrac{c}{a}$	Factor the left side as a perfect square.
$\left(x + \dfrac{b}{2a}\right)^2 = \dfrac{b^2 - 4ac}{4a^2}$	Combine fractions on the right side by getting a common denominator.
$x + \dfrac{b}{2a} = \pm\sqrt{\dfrac{b^2 - 4ac}{4a^2}}$	Apply the square root property.
$x + \dfrac{b}{2a} = \dfrac{\pm\sqrt{b^2 - 4ac}}{2a}$	Simplify the denominator.
$x = -\dfrac{b}{2a} \pm \dfrac{\sqrt{b^2 - 4ac}}{2a}$	Subtract $\dfrac{b}{2a}$ from both sides.
$\quad = \dfrac{-b \pm \sqrt{b^2 - 4ac}}{2a}$	Combine fractions.

The solutions to the equation $ax^2 + bx + c = 0$ for x in terms of the coefficients a, b, and c are given by the **quadratic formula**.

Concepts

1. Derivation of the Quadratic Formula
2. Solving Quadratic Equations by Using the Quadratic Formula
3. Using the Quadratic Formula in Applications
4. Discriminant
5. Mixed Review: Methods to Solve a Quadratic Equation

The Quadratic Formula

For a quadratic equation of the form $ax^2 + bx + c = 0$ $(a \neq 0)$ the solutions are

$$x = \frac{-b \pm \sqrt{b^2 - 4ac}}{2a}$$

The quadratic formula gives us another technique to solve a quadratic equation. This method will work regardless of whether the equation is factorable or not factorable.

2. Solving Quadratic Equations by Using the Quadratic Formula

Example 1 Solving a Quadratic Equation by Using the Quadratic Formula

Solve the quadratic equation by using the quadratic formula.

$$2x^2 - 3x = 5$$

Solution:

$$2x^2 - 3x = 5$$

$$2x^2 - 3x - 5 = 0 \qquad \text{Write the equation in the form } ax^2 + bx + c = 0.$$

$$a = 2, \qquad b = -3, \qquad c = -5 \qquad \text{Identify } a, b, \text{ and } c.$$

$$x = \frac{-b \pm \sqrt{b^2 - 4ac}}{2a} \qquad \text{Apply the quadratic formula.}$$

$$x = \frac{-(-3) \pm \sqrt{(-3)^2 - 4(2)(-5)}}{2(2)} \qquad \text{Substitute } a = 2, b = -3, \text{ and } c = -5.$$

$$= \frac{3 \pm \sqrt{9 + 40}}{4} \qquad \text{Simplify.}$$

$$= \frac{3 \pm \sqrt{49}}{4}$$

$$= \frac{3 \pm 7}{4} \qquad\qquad x = \frac{3 + 7}{4} = \frac{10}{4} = \frac{5}{2}$$

$$x = \frac{3 - 7}{4} = \frac{-4}{4} = -1$$

There are two rational solutions, $\frac{5}{2}$ and -1. Both solutions check in the original equation.

TIP: Always remember to write the *entire* numerator over 2a.

Example 2   Solving a Quadratic Equation by Using the Quadratic Formula

Solve the quadratic equation by using the quadratic formula.

$$-x(x - 6) = 11$$

Solution:

$$-x(x - 6) = 11$$

$$-x^2 + 6x - 11 = 0$$ Write the equation in the form $ax^2 + bx + c = 0$.

$$-1(-x^2 + 6x - 11) = -1(0)$$ If the leading coefficient of the quadratic polynomial is negative, we suggest multiplying both sides of the equation by -1. Although this is not mandatory, it is generally easier to simplify the quadratic formula when the value of a is positive.

$$x^2 - 6x + 11 = 0$$

$$a = 1, b = -6, \text{ and } c = 11$$ Identify a, b, and c.

$$x = \frac{-b \pm \sqrt{b^2 - 4ac}}{2a}$$ Apply the quadratic formula.

$$x = \frac{-(-6) \pm \sqrt{(-6)^2 - 4(1)(11)}}{2(1)}$$ Substitute $a = 1, b = -6,$ and $c = 11$.

$$= \frac{6 \pm \sqrt{36 - 44}}{2}$$ Simplify.

$$= \frac{6 \pm \sqrt{-8}}{2}$$

$$= \frac{6 \pm 2i\sqrt{2}}{2}$$ Simplify the radical.

$$= \frac{2(3 \pm i\sqrt{2})}{2}$$ Factor the numerator.

$$= \frac{2(3 \pm i\sqrt{2})}{2}$$ Simplify the fraction to lowest terms.

$$= 3 \pm i\sqrt{2} \quad \longrightarrow \quad x = 3 + i\sqrt{2}$$
$$\quad \longrightarrow \quad x = 3 - i\sqrt{2}$$

There are two imaginary solutions, $3 + i\sqrt{2}$ and $3 - i\sqrt{2}$.

Skill Practice

Solve the equation by using the quadratic formula.

3. $y(y + 3) = 2$

Avoiding Mistakes

When identifying a, b, and c, use the coefficients only, not the variable.

Answer

3. $\dfrac{-3 \pm \sqrt{17}}{2}$

3. Using the Quadratic Formula in Applications

Example 3 Using the Quadratic Formula in an Application

A delivery truck travels south from Hartselle, Alabama, to Birmingham, Alabama, along Interstate 65. The truck then heads east to Atlanta, Georgia, along Interstate 20. The distance from Birmingham to Atlanta is 8 mi less than twice the distance from Hartselle to Birmingham. If the straight-line distance from Hartselle to Atlanta is 165 mi, find the distance from Hartselle to Birmingham and from Birmingham to Atlanta. (Round the answers to the nearest mile.)

Solution:

The motorist travels due south and then due east. Therefore, the three cities form the vertices of a right triangle (Figure 8-1).

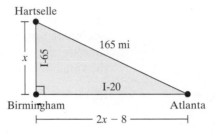

Figure 8-1

Let x represent the distance between Hartselle and Birmingham.

Then $2x - 8$ represents the distance between Birmingham and Atlanta.

Use the Pythagorean theorem to establish a relationship among the three sides of the triangle.

$$(x)^2 + (2x - 8)^2 = (165)^2$$

$$x^2 + 4x^2 - 32x + 64 = 27{,}225$$

$$5x^2 - 32x - 27{,}161 = 0 \qquad \text{Write the equation in the form } ax^2 + bx + c = 0.$$

$$a = 5 \qquad b = -32 \qquad c = -27{,}161 \qquad \text{Identify } a, b, \text{ and } c.$$

$$x = \frac{-(-32) \pm \sqrt{(-32)^2 - 4(5)(-27{,}161)}}{2(5)} \qquad \text{Apply the quadratic formula.}$$

$$x = \frac{32 \pm \sqrt{1024 + 543{,}220}}{10} \qquad \text{Simplify.}$$

$$x = \frac{32 \pm \sqrt{544{,}244}}{10} \quad \begin{cases} \dfrac{32 + \sqrt{544{,}244}}{10} \approx 76.97 \\[2mm] \dfrac{32 - \sqrt{544{,}244}}{10} \approx -70.57 \end{cases}$$

We reject the negative solution because distance cannot be negative.

Recall that x represents the distance from Hartselle to Birmingham; therefore, to the nearest mile, the distance between Hartselle and Birmingham is 77 mi.

The distance between Birmingham and Atlanta is $2x - 8 = 2(77) - 8 = 146$ mi.

Answer

4. Tammy hiked 2.8 mi, and Steve hiked 8.6 mi.

Example 4 **Analyzing a Quadratic Function**

A model rocket is launched straight upward from the side of a 144-ft cliff (Figure 8-2). The initial velocity is 112 ft/sec. The height of the rocket $h(t)$ is given by

$$h(t) = -16t^2 + 112t + 144$$

where $h(t)$ is measured in feet and t is the time in seconds. Find the time(s) at which the rocket is 208 ft above the ground.

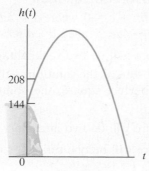

Figure 8-2

Solution:

$$h(t) = -16t^2 + 112t + 144$$

$$208 = -16t^2 + 112t + 144 \qquad \text{Substitute 208 for } h(t).$$

$$16t^2 - 112t + 64 = 0 \qquad \text{Write the equation in the form } at^2 + bt + c = 0.$$

$$\frac{16t^2}{16} - \frac{112t}{16} + \frac{64}{16} = \frac{0}{16} \qquad \text{Divide by 16. This makes the coefficients smaller, and it is less cumbersome to apply the quadratic formula.}$$

$$t^2 - 7t + 4 = 0 \qquad \text{The equation is not factorable. Apply the quadratic formula.}$$

$$t = \frac{-(-7) \pm \sqrt{(-7)^2 - 4(1)(4)}}{2(1)} \qquad \text{Let } a = 1, b = -7, \text{ and } c = 4.$$

$$= \frac{7 \pm \sqrt{33}}{2}$$

$$t = \frac{7 + \sqrt{33}}{2} \approx 6.37$$

$$t = \frac{7 - \sqrt{33}}{2} \approx 0.63$$

The rocket will reach a height of 208 ft after approximately 0.63 sec (on the way up) and after 6.37 sec (on the way down).

Skill Practice

5. A rocket is launched from the top of a 96-ft building with an initial velocity of 64 ft/sec. The height $h(t)$ of the rocket is given by $h(t) = -16t^2 + 64t + 96$. Find the time it takes for the rocket to hit the ground. [*Hint: $h(t) = 0$ when the rocket hits the ground.*]

Answer

5. The time it takes to hit the ground is approximately $2 + \sqrt{10} \approx 5.16$ sec.

4. Discriminant

The radicand within the quadratic formula is the expression $b^2 - 4ac$. This is called the **discriminant**. The discriminant can be used to determine the number of solutions to a quadratic equation as well as whether the solutions are rational, irrational, or imaginary numbers.

Using the Discriminant to Determine the Number and Type of Solutions to a Quadratic Equation

Consider the equation $ax^2 + bx + c = 0$, where a, b, and c are rational numbers and $a \neq 0$. The expression $b^2 - 4ac$ is called the *discriminant*. Furthermore,

- If $b^2 - 4ac > 0$, then there will be two real solutions.
  - **a.** If $b^2 - 4ac$ is a perfect square, the solutions will be rational numbers.
  - **b.** If $b^2 - 4ac$ is not a perfect square, the solutions will be irrational numbers.
- If $b^2 - 4ac < 0$, then there will be two imaginary solutions.
- If $b^2 - 4ac = 0$, then there will be one rational solution.

<table>
<tr><td width="30%" valign="top">

Skill Practice

Determine the discriminant. Use the discriminant to determine the type and number of solutions for the equation.

6. $3y^2 + y + 3 = 0$

7. $4t^2 = 6t - 2$

8. $3t(t + 1) = 9$

9. $\dfrac{2}{3}x^2 - \dfrac{2}{3}x + \dfrac{1}{6} = 0$

</td><td valign="top">

Example 5   **Using the Discriminant**

Use the discriminant to determine the type and number of solutions for each equation.

a. $2x^2 - 5x + 9 = 0$ **b.** $3x^2 = -x + 2$

c. $-2x(2x - 3) = -1$ **d.** $3.6x^2 = -1.2x - 0.1$

Solution:

For each equation, first write the equation in standard form $ax^2 + bx + c = 0$. Then determine the discriminant.

</td></tr>
</table>

Equation	Discriminant	Solution Type and Number
a. $2x^2 - 5x + 9 = 0$	$b^2 - 4ac$ $= (-5)^2 - 4(2)(9)$ $= 25 - 72$ $= -47$	Because $-47 < 0$, there will be two imaginary solutions.
b. $3x^2 = -x + 2$ $\quad 3x^2 + x - 2 = 0$	$b^2 - 4ac$ $= (1)^2 - 4(3)(-2)$ $= 1 - (-24)$ $= 25$	$25 > 0$ and 25 is a perfect square. There will be two rational solutions.
c. $\quad -2x(2x - 3) = -1$ $\quad\quad -4x^2 + 6x = -1$ $-4x^2 + 6x + 1 = 0$	$b^2 - 4ac$ $= (6)^2 - 4(-4)(1)$ $= 36 - (-16)$ $= 52$	$52 > 0$, but 52 is *not* a perfect square. There will be two irrational solutions.

Answers

6. -35; two imaginary solutions
7. 4; two rational solutions
8. 117; two irrational solutions
9. 0; one rational solution

d. $3.6x^2 = -1.2x - 0.1$

$3.6x^2 + 1.2x + 0.1 = 0$ $b^2 - 4ac$

$$= (1.2)^2 - 4(3.6)(0.1)$$

$$= 1.44 - 1.44$$

$$= 0$$

> Because the discriminant equals 0, there will be only one rational solution.

With the discriminant we can determine the number of real-valued solutions to the equation $ax^2 + bx + c = 0$, and thus the number of x-intercepts to the function $f(x) = ax^2 + bx + c$. The following illustrations show the graphical interpretation of the three cases of the discriminant.

$f(x) = x^2 - 4x + 3$

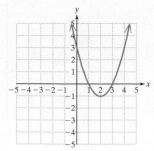

Use $x^2 - 4x + 3 = 0$ to find the value of the discriminant.

$$b^2 - 4ac = (-4)^2 - 4(1)(3)$$
$$= 4$$

Since the discriminant is positive, there are two real solutions to the quadratic equation. Therefore, there are two x-intercepts to the corresponding quadratic function, $(1, 0)$ and $(3, 0)$.

$f(x) = x^2 - x + 1$

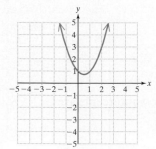

Use $x^2 - x + 1 = 0$ to find the value of the discriminant.

$$b^2 - 4ac = (-1)^2 - 4(1)(1)$$
$$= -3$$

Since the discriminant is negative, there are no real solutions to the quadratic equation. Therefore, there are no x-intercepts to the corresponding quadratic function.

$f(x) = x^2 - 2x + 1$

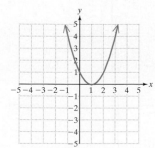

Use $x^2 - 2x + 1 = 0$ to find the value of the discriminant.

$$b^2 - 4ac = (-2)^2 - 4(1)(1)$$
$$= 0$$

Since the discriminant is zero, there is one real solution to the quadratic equation. Therefore, there is one x-intercept to the corresponding quadratic function, $(1, 0)$.

Example 6 Finding x- and y-Intercepts of a Quadratic Function

Given $f(x) = x^2 - 3x + 1$

a. Find the discriminant and use it to determine if there are any x-intercepts.

b. Find the x-intercept(s), if they exist.

c. Find the y-intercept.

Solution:

a. $a = 1$, $b = -3$, and $c = 1$.

The discriminant is $b^2 - 4ac = (-3)^2 - 4(1)(1)$

$$= 9 - 4$$

$$= 5$$

Since $5 > 0$, there are two x-intercepts.

b. The x-intercepts are given by the real solutions to the equation $f(x) = 0$. In this case, we have

$$f(x) = x^2 - 3x + 1 = 0$$

$$x^2 - 3x + 1 = 0 \qquad \text{The equation is in the form } ax^2 + bx + c = 0.$$

$$x = \frac{3 \pm \sqrt{(-3)^2 - (4)(1)(1)}}{2(1)} \qquad \text{Apply the quadratic formula.}$$

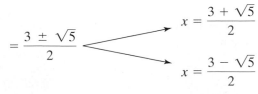

$$= \frac{3 \pm \sqrt{5}}{2} \qquad \qquad x = \frac{3 + \sqrt{5}}{2}$$

$$x = \frac{3 - \sqrt{5}}{2}$$

The solutions are $\dfrac{3 + \sqrt{5}}{2}$ and $\dfrac{3 - \sqrt{5}}{2}$. Therefore, the x-intercepts are

$\left(\dfrac{3 + \sqrt{5}}{2}, 0\right)$ and $\left(\dfrac{3 - \sqrt{5}}{2}, 0\right)$.

TIP: Recall that an x-intercept is a point $(a, 0)$ where the graph of a function intersects the x-axis. A y-intercept is a point $(0, b)$ where the graph intersects the y-axis.

c. To find the y-intercept, evaluate $f(0)$.

$$f(0) = (0)^2 - 3(0) + 1 = 1$$

The y-intercept is located at $(0, 1)$.

The parabola is shown in the graph with the x- and y-intercepts labeled.

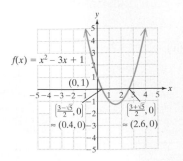

Example 7 Finding *x*- and *y*-Intercepts of a Quadratic Function

Given $f(x) = x^2 + x + 2$,

a. Find the discriminant and use it to determine if there are any *x*-intercepts.

b. Find the *x*-intercept(s), if they exist.

c. Find the *y*-intercept.

Solution:

a. $a = 1$, $b = 1$, and $c = 2$.

The discriminant is $b^2 - 4ac = (1)^2 - 4(1)(2)$

$$= 1 - 8$$

$$= -7$$

Since $-7 < 0$, there are no *x*-intercepts.

b. There are no *x*-intercepts.

c. $f(0) = (0)^2 + (0) + 2$

$$= 2$$

The *y*-intercept is located at $(0, 2)$.

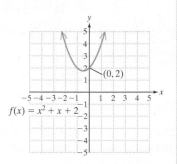

The parabola is shown. *Note:* the graph does not intersect the *x*-axis.

5. Mixed Review: Methods to Solve a Quadratic Equation

Three methods have been presented to solve quadratic equations.

Methods to Solve a Quadratic Equation	
Factor and use the zero product rule (Section 5.8). • This method works well if you can factor the equation easily.	Example: $x^2 + 8x + 15 = 0$ factors as $(x + 3)(x + 5) = 0$
Use the square root property. Complete the square if necessary (Section 8.1). This method is particularly good if • the equation is of the form $ax^2 + c = 0$ or • the equation is of the form $(x - a)^2 = k$ or • the equation is of the form $x^2 + bx + c = 0$, where b is even.	Examples: $4x^2 + 9 = 0$ $x^2 + 10x - 3 = 0$
Quadratic Formula (Section 8.2) • This method works in all cases—just be sure to write the equation in the form $ax^2 + bx + c = 0$ before applying the formula.	Example: $7x^2 - 3x + 11 = 0$

Before solving a quadratic equation, take a minute to analyze it first. Each problem must be evaluated individually before choosing the most efficient method to find its solutions.

Example 8   Solving a Quadratic Equation by Using Any Method

Solve the equation by using any method.

$$(x + 3)^2 + x^2 - 9x = 8$$

Solution:

$$(x + 3)^2 + x^2 - 9x = 8$$

$$x^2 + 6x + 9 + x^2 - 9x - 8 = 0 \qquad \text{Clear parentheses and write the equation in the form } ax^2 + bx + c = 0.$$

$$2x^2 - 3x + 1 = 0 \qquad \text{This equation is factorable.}$$

$$(2x - 1)(x - 1) = 0 \qquad \text{Factor.}$$

$$2x - 1 = 0 \quad \text{or} \quad x - 1 = 0 \qquad \text{Apply the zero product rule.}$$

$$x = \tfrac{1}{2} \quad \text{or} \quad x = 1 \qquad \text{Solve for } x.$$

The solutions are $\frac{1}{2}$ and 1.

This equation could have been solved by using any of the three methods, but factoring was the most efficient method.

Example 9   Solving a Quadratic Equation by Using Any Method

Solve the equation by using any method.

$$x^2 + 5 = -2x$$

Solution:

$$x^2 + 2x + 5 = 0 \qquad \text{The equation does not factor.}$$

$$x^2 + 2x = -5 \qquad \text{Because } a = 1 \text{ and } b \text{ is even, we can easily complete the square.}$$

$$x^2 + 2x + 1 = -5 + 1 \qquad \text{Add } [\tfrac{1}{2}(2)]^2 = 1^2 = 1 \text{ to both sides.}$$

$$(x + 1)^2 = -4$$

$$x + 1 = \pm\sqrt{-4} \qquad \text{Apply the square root property.}$$

$$x = -1 \pm 2i \qquad \text{Solve for } x.$$

The solutions are $-1 + 2i$ and $-1 - 2i$.

This problem could also have been solved by using the quadratic formula.

Example 10 Solving a Quadratic Equation by Using Any Method

Solve the equation by using any method. $\dfrac{1}{3}x^2 - \dfrac{5}{9}x + \dfrac{1}{9} = 0$

Solution:

$9\left(\dfrac{1}{3}x^2 - \dfrac{5}{9}x + \dfrac{1}{9}\right) = 9(0)$ Clear the fractions and write the equation in the form $ax^2 + bx + c = 0$.

$3x^2 - 5x + 1 = 0$ This is not factorable. Because $a \neq 1$, we will use the quadratic formula.

$a = 3, b = -5,$ and $c = 1$ Identify $a, b,$ and c.

$x = \dfrac{5 \pm \sqrt{(-5)^2 - 4(3)(1)}}{2(3)}$ Apply the quadratic formula.

$= \dfrac{5 \pm \sqrt{13}}{6}$ Simplify.

The solutions are $\dfrac{5 + \sqrt{13}}{6}$ and $\dfrac{5 - \sqrt{13}}{6}$.

Skill Practice

Solve using any method.

14. $\dfrac{1}{2}x^2 - \dfrac{5}{2}x + \dfrac{1}{4} = 0$

Example 11 Solving a Quadratic Equation by Using Any Method

Solve the equation by using any method. $9p^2 - 11 = 0$

Solution:

$9p^2 - 11 = 0$ Because $b = 0$, use the square root property.

$9p^2 = 11$ Isolate the variable term.

$p^2 = \dfrac{11}{9}$ The equation is in the form $x^2 = k$.

$p = \pm\sqrt{\dfrac{11}{9}}$ Apply the square root property.

$p = \pm\dfrac{\sqrt{11}}{3}$ Simplify the radical.

The solutions are $\dfrac{\sqrt{11}}{3}$ and $-\dfrac{\sqrt{11}}{3}$.

Skill Practice

Solve using any method.
15. $4y^2 - 13 = 0$

Answers

14. $\dfrac{5 \pm \sqrt{23}}{2}$

15. $\pm\dfrac{\sqrt{13}}{2}$

Section 8.2 Practice Exercises

Vocabulary and Key Concepts

1. **a.** For the equation $ax^2 + bx + c = 0$ $(a \neq 0)$, the _____ formula gives the solutions as
   $x =$ _____.

   b. To apply the quadratic formula, a quadratic equation must be written in the form _____ where $a \neq 0$.

   c. To apply the quadratic formula to solve the equation $8x^2 - 42x - 27 = 0$, the value of a is _____, the value of b is _____, and the value of c is _____.

d. To apply the quadratic formula to solve the equation $3x^2 - 7x - 4 = 0$, the value of $-b$ is _____ and the value of the radicand is _____.

e. The radicand within the quadratic formula is _____ and is called the _____.

f. If the discriminant is negative, then the solutions to a quadratic equation will be (real/imaginary) numbers.

g. If the discriminant is positive, then the solutions to a quadratic equation will be (real/imaginary) numbers.

h. Given a quadratic function $f(x) = ax^2 + bx + c = 0$, the function will have no x-intercepts if the discriminant is (less than, greater than, equal to) zero.

Review Exercises

2. Use substitution to determine if $x = -3 + \sqrt{5}$ is a solution to $x^2 + 6x + 4 = 0$.

For Exercises 3–6, solve by completing the square and using the square root property.

3. $(x + 5)^2 = 49$ **4.** $16 = (2x - 3)^2$ **5.** $x^2 - 2x + 10 = 0$ **6.** $x^2 - 4x + 15 = 0$

For Exercises 7–10, simplify the expressions.

7. $\dfrac{16 - \sqrt{320}}{4}$ **8.** $\dfrac{18 + \sqrt{180}}{3}$ **9.** $\dfrac{14 - \sqrt{-147}}{7}$ **10.** $\dfrac{10 - \sqrt{-175}}{5}$

For Exercises 11 and 12, determine whether the equation is linear or quadratic.

11. $2(x - 5) + x^2 = 3x(x - 4) - 2x^2$ **12.** $5x(x + 3) - 9 = 4x^2 - 3(x + 1)$

Concept 2: Solving Quadratic Equations by Using the Quadratic Formula

For Exercises 13–38, solve the equation by using the quadratic formula. Write imaginary solutions in the form $a + bi$.
(See Examples 1 and 2.)

13. $x^2 + 11x - 12 = 0$ **14.** $5x^2 - 14x - 3 = 0$ **15.** $9y^2 - 2y + 5 = 0$

16. $2t^2 + 3t - 7 = 0$ **17.** $12p^2 - 4p + 5 = 0$ **18.** $-5n^2 + 4n - 6 = 0$

19. $-z^2 = -2z - 35$ **20.** $12x^2 - 5x = 2$ **21.** $y^2 + 3y = 8$

22. $k^2 + 4 = 6k$ **23.** $25x^2 - 20x + 4 = 0$ **24.** $9y^2 = -12y - 4$

25. $w(w - 6) = -14$ **26.** $m(m + 6) = -11$ **27.** $(x + 2)(x - 3) = 1$

28. $3y(y + 1) - 7y(y + 2) = 6$ **29.** $-4a^2 - 2a + 3 = 0$ **30.** $-2m^2 - 5m + 3 = 0$

31. $\dfrac{1}{2}y^2 + \dfrac{2}{3} = -\dfrac{2}{3}y$
(*Hint:* Clear the fractions first.)   **32.** $\dfrac{2}{3}p^2 - \dfrac{1}{6}p + \dfrac{1}{2} = 0$ **33.** $\dfrac{1}{5}h^2 + h + \dfrac{3}{5} = 0$

34. $\dfrac{1}{4}w^2 + \dfrac{7}{4}w + 1 = 0$ **35.** $0.01x^2 + 0.06x + 0.08 = 0$
(*Hint:* Clear the decimals first.)   **36.** $0.5y^2 - 0.7y + 0.2 = 0$

37. $0.3t^2 + 0.7t - 0.5 = 0$ **38.** $0.01x^2 + 0.04x - 0.07 = 0$

For Exercises 39–42, factor the expression. Then use the zero product rule and the quadratic formula to solve the equation. There should be three solutions to each equation. Write imaginary solutions in the form $a + bi$.

39. a. Factor. $x^3 - 27$

    b. Solve. $x^3 - 27 = 0$

40. a. Factor. $64x^3 + 1$

    b. Solve. $64x^3 + 1 = 0$

41. a. Factor. $3x^3 - 6x^2 + 6x$

    b. Solve. $3x^3 - 6x^2 + 6x = 0$

42. a. Factor. $5x^3 + 5x^2 + 10x$

    b. Solve. $5x^3 + 5x^2 + 10x = 0$

Concept 3: Using the Quadratic Formula in Applications

43. The volume of a cube is 27 ft^3. Find the lengths of the sides.

44. The volume of a rectangular box is 64 ft^3. If the width is 3 times longer than the height, and the length is 9 times longer than the height, find the dimensions of the box.

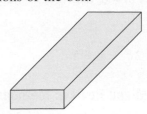

45. The hypotenuse of a right triangle measures 4 in. One leg of the triangle is 2 in. longer than the other leg. Find the lengths of the legs of the triangle. Round to one decimal place. **(See Example 3.)**

46. The length of one leg of a right triangle is 1 cm more than twice the length of the other leg. The hypotenuse measures 6 cm. Find the lengths of the legs. Round to one decimal place.

47. The hypotenuse of a right triangle is 10.2 m long. One leg is 2.1 m shorter than the other leg. Find the lengths of the legs. Round to one decimal place.

48. The hypotenuse of a right triangle is 17 ft long. One leg is 3.4 ft longer than the other leg. Find the lengths of the legs.

49. The fatality rate (in fatalities per 100 million vehicle miles driven) can be approximated for drivers x years old according to the function, $F(x) = 0.0036x^2 - 0.35x + 9.2$. (*Source:* U.S. Department of Transportation)

    a. Approximate the fatality rate for drivers 16 yr old.

    b. Approximate the fatality rate for drivers 40 yr old.

    c. Approximate the fatality rate for drivers 80 yr old.

    d. For what age(s) is the fatality rate approximately 2.5?

50. The braking distance (in feet) of a car going v mph is given by

$$d(v) = \frac{v^2}{20} + v \qquad v \geq 0$$

    a. How fast would a car have been traveling for a braking distance of 150 ft? Round to the nearest mile per hour.

    b. How fast would a car have been traveling for a braking distance of 100 ft? Round to the nearest mile per hour.

51. Mitch throws a baseball straight up in the air from a cliff that is 48 ft high. The initial velocity is 48 ft/sec. The height (in feet) of the object after t sec is given by $h(t) = -16t^2 + 48t + 48$. Find the time at which the height of the object is 64 ft. **(See Example 4.)**

52. An astronaut on the moon throws a rock into the air from the deck of a spacecraft that is 8 m high. The initial velocity of the rock is 2.4 m/sec. The height (in meters) of the rock after t sec is given by $h(t) = -0.8t^2 + 2.4t + 8$. Find the time at which the height of the rock is 6 m.

Concept 4: Discriminant

For Exercises 53–60,

   a. Write the equation in the form $ax^2 + bx + c = 0, a > 0$.

   b. Find the value of the discriminant.

   c. Use the discriminant to determine the number and type of solutions. **(See Example 5.)**

53. $x^2 + 2x = -1$

54. $12y - 9 = 4y^2$

55. $19m^2 = 8m$

56. $2n - 5n^2 = 0$

57. $5p^2 - 21 = 0$

58. $3k^2 = 7$

59. $4n(n - 2) - 5n(n - 1) = 4$

60. $(2x + 1)(x - 3) = -9$

For Exercises 61–66, find the x- and y-intercepts of the quadratic function. **(See Examples 6 and 7.)**

61. $f(x) = x^2 - 5x + 3$

62. $g(x) = 2x^2 + 7x + 2$

63. $g(x) = -x^2 + x - 1$

64. $f(x) = 2x^2 + x + 5$

65. $p(x) = 2x^2 + 5x - 2$

66. $h(x) = 3x^2 + 2x - 2$

Concept 5: Mixed Review: Methods to Solve a Quadratic Equation

For Exercises 67–84, solve the quadratic equations by using any method. Write imaginary solutions in the form $a + bi$. **(See Examples 8–11.)**

67. $a^2 + 3a + 4 = 0$

68. $4z^2 + 7z = 0$

69. $(x - 2)^2 + 2x^2 - 13x = 10$

70. $(x - 3)^2 + 3x^2 - 5x = 12$

71. $4y^2 + 8y - 5 = 0$

72. $k^2 - 4k = -13$

73. $\left(x + \dfrac{1}{2} \right)^2 + 4 = 0$

74. $(2y + 3)^2 = 9$

75. $2y(y - 3) = -1$

76. $w(w - 5) = 4$

77. $(2t + 5)(t - 1) = (t - 3)(t + 8)$

78. $(b - 1)(b + 4) = (3b + 2)(b + 1)$

79. $\dfrac{1}{8}x^2 - \dfrac{1}{2}x + \dfrac{1}{4} = 0$

80. $\dfrac{1}{6}x^2 - \dfrac{1}{2}x + \dfrac{1}{4} = 0$

81. $32z^2 - 20z - 3 = 0$

82. $8k^2 - 14k + 3 = 0$

83. $3p^2 - 27 = 0$

84. $5h^2 - 120 = 0$

Sometimes students shy away from completing the square and using the square root property to solve a quadratic equation. However, sometimes this process leads to a simple solution. For Exercises 85 and 86, solve the equations two ways.

   a. Solve the equation by completing the square and applying the square root property.

   b. Solve the equation by applying the quadratic formula.

   c. Which technique was easier for you?

85. $x^2 + 6x = 5$

86. $x^2 - 10x = -27$

Expanding Your Skills

87. An artist has been commissioned to make a stained glass window in the shape of a regular octagon. The octagon must fit inside an 18-in. square space. See the figure.

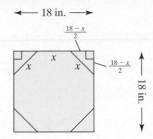

a. Let x represent the length of each side of the octagon. Verify that the legs of the small triangles formed by the corners of the square can be expressed as $\dfrac{18 - x}{2}$.

b. Use the Pythagorean theorem to set up an equation in terms of x that represents the relationship between the legs of the triangle and the hypotenuse.

c. Simplify the equation by clearing parentheses and clearing fractions.

d. Solve the resulting quadratic equation by using the quadratic formula. Use a calculator and round your answers to the nearest tenth of an inch.

e. There are two solutions for x. Which one is appropriate and why?

Graphing Calculator Exercises

88. Graph $Y_1 = 64x^3 + 1$. Compare the x-intercepts with the solutions to the equation $64x^3 + 1 = 0$ found in Exercise 40.

89. Graph $Y_1 = x^3 - 27$. Compare the x-intercepts with the solutions to the equation $x^3 - 27 = 0$ found in Exercise 39.

90. Graph $Y_1 = 5x^3 + 5x^2 + 10x$. Compare the x-intercepts with the solutions to the equation $5x^3 + 5x^2 + 10x = 0$ found in Exercise 42.

91. Graph $Y_1 = 3x^3 - 6x^2 + 6x$. Compare the x-intercepts with the solutions to the equation $3x^3 - 6x^2 + 6x = 0$ found in Exercise 41.

Equations in Quadratic Form

Section 8.3

1. Solving Equations by Using Substitution

Concepts

1. Solving Equations by Using Substitution

2. Solving Equations Reducible to a Quadratic

We have learned to solve a variety of different types of equations, including linear, radical, rational, and polynomial equations. Sometimes, however, it is necessary to use a quadratic equation as a tool to solve other types of equations.

In Example 1, we will solve the equation $(2x^2 - 5)^2 - 16(2x^2 - 5) + 39 = 0$. Notice that the terms in the equation are written in descending order by degree. Furthermore, the first two terms have the same base, $2x^2 - 5$, and the exponent on the first term is exactly double the exponent on the second term. The third term is a constant. An equation in this pattern is called **quadratic in form**.

Exponent is double.   Third term is constant.

$$(2x^2 - 5)^2 - 16(2x^2 - 5)^1 + 39 = 0.$$

To solve this equation we will use substitution as demonstrated in Example 1.

Example 1 Solving an Equation in Quadratic Form

Solve the equation.

$$(2x^2 - 5)^2 - 16(2x^2 - 5) + 39 = 0$$

Solution:

$(2x^2 - 5)^2 - 16(2x^2 - 5) + 39 = 0$ If the substitution $u = (2x^2 - 5)$ is made, the equation becomes quadratic in the variable u.

Substitute $u = (2x^2 - 5)$.

$u^2 \quad - \quad 16u \quad + 39 = 0$ The equation is in the form $au^2 + bu + c = 0$.

$(u - 13)(u - 3) = 0$ Factor.

$u = 13 \quad$ or $\quad u = 3$ Apply the zero product rule.

Reverse substitute.

$2x^2 - 5 = 13 \quad$ or $\quad 2x^2 - 5 = 3$

$2x^2 = 18 \quad$ or $\quad 2x^2 = 8$

$x^2 = 9 \quad$ or $\quad x^2 = 4$ Write the equations in the form $x^2 = k$.

$x = \pm\sqrt{9} \quad$ or $\quad x = \pm\sqrt{4}$ Apply the square root property.

$= \pm 3 \quad$ or $\quad = \pm 2$

The solutions are $3, -3, 2,$ and -2.   Substituting these values in the original equation verifies that these are all valid solutions.

For an equation written in descending order, notice that u was set equal to the variable factor on the middle term. This is generally the case.

Example 2 Solving an Equation in Quadratic Form

Solve the equation. $p^{2/3} - 2p^{1/3} = 8$

Solution:

$$p^{2/3} - 2p^{1/3} = 8$$

$p^{2/3} - 2p^{1/3} - 8 = 0$ Set the equation equal to zero.

$(p^{1/3})^2 - 2(p^{1/3})^1 - 8 = 0$ Make the substitution $u = p^{1/3}$.

Substitute $u = p^{1/3}$.   Then the equation is in the form $au^2 + bu + c = 0$.

$u^2 \quad - \quad 2u \quad - 8 = 0$

$$(u - 4)(u + 2) = 0 \qquad \text{Factor.}$$

$$u = 4 \qquad \text{or} \qquad u = -2 \qquad \text{Apply the zero product rule.}$$

Reverse substitute.

$$p^{1/3} = 4 \qquad \text{or} \qquad p^{1/3} = -2$$

$$\sqrt[3]{p} = 4 \qquad \text{or} \qquad \sqrt[3]{p} = -2 \qquad \begin{array}{l}\text{The equations are radical} \\ \text{equations.}\end{array}$$

$$(\sqrt[3]{p})^3 = (4)^3 \qquad \text{or} \qquad (\sqrt[3]{p})^3 = (-2)^3 \qquad \text{Cube both sides.}$$

$$p = 64 \qquad \text{or} \qquad p = -8$$

Check: $p = 64$ Check: $p = -8$

$$p^{2/3} - 2p^{1/3} = 8 \qquad\qquad p^{2/3} - 2p^{1/3} = 8$$

$$(64)^{2/3} - 2(64)^{1/3} \stackrel{?}{=} 8 \qquad (-8)^{2/3} - 2(-8)^{1/3} \stackrel{?}{=} 8$$

$$16 - 2(4) \stackrel{?}{=} 8 \qquad\qquad 4 - 2(-2) \stackrel{?}{=} 8$$

$$8 \stackrel{?}{=} 8 \checkmark \qquad\qquad\qquad 4 + 4 \stackrel{?}{=} 8 \checkmark$$

The solutions are 64 and -8.

Example 3   **Solving an Equation in Quadratic Form**

Solve the equation. $x - \sqrt{x} - 12 = 0$

Solution:

The equation can be solved by first isolating the radical and then squaring both sides (this is left as an exercise—see Exercise 24). However, this equation is also quadratic in form. By writing $\sqrt{x}$ as $x^{1/2}$, we see that the exponent on the first term is exactly double the exponent on the middle term.

$$x^1 - x^{1/2} - 12 = 0$$

$$(x^{1/2})^2 - (x^{1/2})^1 - 12 = 0 \qquad \text{Let } u = x^{1/2}.$$

$$u^2 - u - 12 = 0$$

$$(u - 4)(u + 3) = 0 \qquad \text{Factor.}$$

$$u = 4 \quad \text{or} \quad u = -3 \qquad \text{Solve for } u.$$

$$x^{1/2} = 4 \quad \text{or} \quad x^{1/2} = -3 \qquad \text{Back substitute.}$$

$$\sqrt{x} = 4 \quad \text{or} \quad \sqrt{x} = -3 \qquad \begin{array}{l}\text{Solve each equation for } x.\text{ Notice that the} \\ \text{principal square root of a number cannot} \\ \text{equal } -3.\end{array}$$

$$x = 16$$

The solution 16 checks in the original equation.

Skill Practice

Solve the equation.

3. $z - \sqrt{z} - 2 = 0$

Answer

3. 4 (the value 1 does not check).

2. Solving Equations Reducible to a Quadratic

Some equations are reducible to a quadratic equation. In Example 4, we solve a polynomial equation by factoring. The resulting factors are quadratic.

Example 4 Solving a Polynomial Equation

Solve the equation. $\quad 4x^4 + 7x^2 - 2 = 0$

Solution:

$4x^4 + 7x^2 - 2 = 0$	This is a polynomial equation.
$(4x^2 - 1)(x^2 + 2) = 0$	Factor.
$4x^2 - 1 = 0 \text{ or } x^2 + 2 = 0$	Set each factor equal to zero. Notice that the two equations are quadratic.
$x^2 = \dfrac{1}{4} \qquad \text{or} \quad x^2 = -2$	Each can be solved by the square root property.
$x = \pm\sqrt{\dfrac{1}{4}} \quad \text{or} \quad x = \pm\sqrt{-2}$	Apply the square root property.
$x = \pm\dfrac{1}{2} \qquad \text{or} \quad x = \pm i\sqrt{2}$	Simplify the radicals.

The equation has four solutions: $\frac{1}{2}$, $-\frac{1}{2}$, $i\sqrt{2}$, and $-i\sqrt{2}$.

Example 5 Solving a Rational Equation

Solve the equation. $\quad \dfrac{3y}{y + 2} - \dfrac{2}{y - 1} = 1$

Solution:

$\dfrac{3y}{y + 2} - \dfrac{2}{y - 1} = 1 \qquad$ This is a rational equation. The LCD is $(y + 2)(y - 1)$.

$\left(\dfrac{3y}{y + 2} - \dfrac{2}{y - 1}\right) \cdot (y + 2)(y - 1) = 1 \cdot (y + 2)(y - 1) \qquad$ Multiply both sides by the LCD.

$\dfrac{3y}{y+2} \cdot (y+2)(y - 1) - \dfrac{2}{y-1} \cdot (y + 2)(y-1) = 1 \cdot (y + 2)(y - 1)$

$3y(y - 1) - 2(y + 2) = (y + 2)(y - 1)$	Clear fractions.
$3y^2 - 3y - 2y - 4 = y^2 - y + 2y - 2$	Apply the distributive property.
$3y^2 - 5y - 4 = y^2 + y - 2$	The equation is quadratic.
$2y^2 - 6y - 2 = 0$	Write the equation in descending order.
$\dfrac{2y^2}{2} - \dfrac{6y}{2} - \dfrac{2}{2} = \dfrac{0}{2}$	Each coefficient in the equation is divisible by 2. Therefore, if we divide both sides by 2, the coefficients in the equation are smaller. This will make it easier to apply the quadratic formula.
$y^2 - 3y - 1 = 0$	

$$y = \frac{-(-3) \pm \sqrt{(-3)^2 - 4(1)(-1)}}{2(1)}$$ Apply the quadratic formula.

$$= \frac{3 \pm \sqrt{9 + 4}}{2}$$

$$= \frac{3 \pm \sqrt{13}}{2} \begin{cases} y = \dfrac{3 + \sqrt{13}}{2} \\ y = \dfrac{3 - \sqrt{13}}{2} \end{cases}$$

The values $\dfrac{3 + \sqrt{13}}{2}$ and $\dfrac{3 - \sqrt{13}}{2}$ both check in the original equation.

Section 8.3   Practice Exercises

Vocabulary and Key Concepts

1. **a.** An equation that can be written in the form $au^2 + bu + c = 0$, where u represents an algebraic expression, is said to be in _____ form.

   b. To use the method of substitution to solve the equation $(3x - 1)^2 + 2(3x - 1) - 8 = 0$, let $u =$ _____.

   c. To use the method of substitution to solve the equation $p^{2/3} - 2p^{1/3} - 15 = 0$, let $u =$ _____.

Review Exercises

For Exercises 2–7, solve the quadratic equations.

2. $16 = (2x - 3)^2$

3. $\left(x - \dfrac{3}{2}\right)^2 = \dfrac{7}{4}$

4. $n(n - 6) = -13$

5. $x(x + 8) = -16$

6. $6k^2 + 7k = 6$

7. $2x^2 - 8x - 44 = 0$

Concept 1: Solving Equations by Using Substitution

8. **a.** Solve the quadratic equation by factoring. $u^2 + 10u + 24 = 0$

   b. Solve the equation by using substitution. $(y^2 + 5y)^2 + 10(y^2 + 5y) + 24 = 0$

9. **a.** Solve the quadratic equation by factoring. $u^2 - 2u - 35 = 0$

   b. Solve the equation by using substitution. $(w^2 - 6w)^2 - 2(w^2 - 6w) - 35 = 0$

10. **a.** Solve the quadratic equation by factoring. $u^2 - 2u - 24 = 0$

    b. Solve the equation by using substitution. $(x^2 - 5x)^2 - 2(x^2 - 5x) - 24 = 0$

11. **a.** Solve the quadratic equation by factoring. $u^2 - 4u + 3 = 0$

    b. Solve the equation by using substitution. $(2p^2 + p)^2 - 4(2p^2 + p) + 3 = 0$

For Exercises 12–23, solve the equation by using substitution. **(See Examples 1–3.)**

12. $(x^2 - 2x)^2 + 2(x^2 - 2x) = 3$

13. $(x^2 + x)^2 - 8(x^2 + x) = -12$

14. $m^{2/3} - m^{1/3} - 6 = 0$

15. $2n^{2/3} + 7n^{1/3} - 15 = 0$

16. $2t^{2/5} + 7t^{1/5} + 3 = 0$

17. $p^{2/5} + p^{1/5} - 2 = 0$

18. $y + 6\sqrt{y} = 16$

19. $p - 8\sqrt{p} = -15$

20. $2x + 3\sqrt{x} - 2 = 0$

21. $3t + 5\sqrt{t} - 2 = 0$

22. $16\left(\dfrac{x+6}{4}\right)^2 + 8\left(\dfrac{x+6}{4}\right) + 1 = 0$

23. $9\left(\dfrac{x+3}{2}\right)^2 - 6\left(\dfrac{x+3}{2}\right) + 1 = 0$

24. In Example 3, we solved the equation $x - \sqrt{x} - 12 = 0$ by using substitution. Now solve this equation by first isolating the radical and then squaring both sides. Don't forget to check the potential solutions in the original equation. Do you obtain the same solution as in Example 3?

Concept 2: Solving Equations Reducible to a Quadratic

For Exercises 25–34, solve the equations. **(See Examples 4 and 5.)**

25. $t^4 + t^2 - 12 = 0$

26. $w^4 + 4w^2 - 45 = 0$

27. $x^2(9x^2 + 7) = 2$

28. $y^2(4y^2 + 17) = 15$

29. $\dfrac{y}{10} - 1 = -\dfrac{12}{5y}$

30. $\dfrac{x+5}{x} + \dfrac{x}{2} = \dfrac{x+19}{4x}$

31. $\dfrac{3x}{x+1} - \dfrac{2}{x-3} = 1$

32. $\dfrac{2t}{t-3} - \dfrac{1}{t+4} = 1$

33. $\dfrac{x}{2x-1} = \dfrac{1}{x-2}$

34. $\dfrac{z}{3z+2} = \dfrac{2}{z+1}$

Mixed Exercises

For Exercises 35–54, solve the equations.

35. $x^4 - 16 = 0$

36. $t^4 - 625 = 0$

37. $(4x + 5)^2 + 3(4x + 5) + 2 = 0$

38. $2(5x + 3)^2 - (5x + 3) - 28 = 0$

39. $4m^4 - 9m^2 + 2 = 0$

40. $x^4 - 7x^2 + 12 = 0$

41. $x^6 - 9x^3 + 8 = 0$

42. $x^6 - 26x^3 - 27 = 0$

43. $\sqrt{x^2 + 20} = 3\sqrt{x}$

44. $\sqrt{4t + 1} = t + 1$

45. $2\left(\dfrac{t-4}{3}\right)^2 - \left(\dfrac{t-4}{3}\right) - 3 = 0$

46. $\left(\dfrac{x+1}{5}\right)^2 - 3\left(\dfrac{x+1}{5}\right) - 10 = 0$

47. $x^{2/3} + x^{1/3} = 20$

48. $x^{2/5} - 3x^{1/5} = -2$

49. $m^4 + 2m^2 - 8 = 0$

50. $2c^4 + c^2 - 1 = 0$

51. $a^3 + 16a - a^2 - 16 = 0$ (*Hint:* Factor by grouping first.)

52. $b^3 + 9b - b^2 - 9 = 0$

53. $x^3 + 5x - 4x^2 - 20 = 0$

54. $y^3 + 8y - 3y^2 - 24 = 0$

Graphing Calculator Exercises

55. a. Solve the equation $x^4 + 4x^2 + 4 = 0$.

b. How many solutions are real and how many solutions are imaginary?

c. How many x-intercepts do you anticipate for the function defined by $y = x^4 + 4x^2 + 4$?

d. Graph $Y_1 = x^4 + 4x^2 + 4$ on a standard viewing window.

56. a. Solve the equation $x^4 - 2x^2 + 1 = 0$.

  b. How many solutions are real and how many solutions are imaginary?

  c. How many x-intercepts do you anticipate for the function defined by $y = x^4 - 2x^2 + 1$?

  d. Graph $Y_1 = x^4 - 2x^2 + 1$ on a standard viewing window.

57. a. Solve the equation $x^4 - x^3 - 6x^2 = 0$.

  b. How many solutions are real and how many solutions are imaginary?

  c. How many x-intercepts do you anticipate for the function defined by $y = x^4 - x^3 - 6x^2$?

  d. Graph $Y_1 = x^4 - x^3 - 6x^2$ on a standard viewing window.

58. a. Solve the equation $x^4 - 10x^2 + 9 = 0$.

  b. How many solutions are real and how many solutions are imaginary?

  c. How many x-intercepts do you anticipate for the function defined by $y = x^4 - 10x^2 + 9$?

  d. Graph $Y_1 = x^4 - 10x^2 + 9$ on a standard viewing window.

Problem Recognition Exercises

Identifying and Solving Equations

For Exercises 1–4, solve each equation by

  a. Completing the square and applying the square root property.

  b. Using the quadratic formula.

  1. $x^2 + 10x + 3 = 0$ **2.** $v^2 - 16v + 5 = 0$ **3.** $3t^2 + t + 4 = 0$ **4.** $4y^2 + 3y + 5 = 0$

In Exercises 5–24, we have presented all types of equations that you have learned up to this point. For each equation,

  a. First determine the type of equation that is presented. Choose from: linear equation, quadratic equation, quadratic in form, rational equation, or radical equation.

  b. Solve the equation by using a suitable method.

  5. $t^2 + 5t - 14 = 0$ **6.** $a^2 - 9a + 20 = 0$ **7.** $a^4 - 10a^2 + 9 = 0$

  8. $s^4 - 29s^2 + 100 = 0$ **9.** $y^2 + 7y + 4 = 0$ **10.** $x^2 - 9x - 2 = 0$

  11. $8b(b + 1) + 2(3b - 4) = 4b(2b + 3)$ **12.** $6x(x + 1) - 3(x + 4) = 3x(2x + 5)$

  13. $5a(a + 6) = 10(3a - 1)$ **14.** $4x(x + 3) = 6(2x - 4)$ **15.** $\dfrac{t}{t + 5} + \dfrac{3}{t - 4} = \dfrac{17}{t^2 + t - 20}$

  16. $\dfrac{v}{v + 4} + \dfrac{12}{v^2 + 7v + 12} = \dfrac{5}{v + 3}$ **17.** $c^2 - 20c - 1 = 0$ **18.** $d^2 + 18d + 4 = 0$

  19. $2u(u - 3) = 4(2 - u)$ **20.** $3y(y + 2) = 9(y + 1)$ **21.** $\sqrt{2b + 3} = b$

  22. $\sqrt{5t + 6} = t$ **23.** $x^{2/3} + 2x^{1/3} - 15 = 0$ **24.** $y^{2/3} + 5y^{1/3} + 4 = 0$

Section 8.4   Graphs of Quadratic Functions

Concepts

1. Quadratic Functions of the Form $f(x) = x^2 + k$
2. Quadratic Functions of the Form $f(x) = (x - h)^2$
3. Quadratic Functions of the Form $f(x) = ax^2$
4. Quadratic Functions of the Form $f(x) = a(x - h)^2 + k$

In Section 5.8, we defined a quadratic function as a function of the form $f(x) = ax^2 + bx + c$ $(a \neq 0)$. We also learned that the graph of a quadratic function is a **parabola**. In this section we will learn how to graph parabolas.

A parabola opens upward if $a > 0$ (Figure 8-3) and opens downward if $a < 0$ (Figure 8-4). If a parabola opens upward, the **vertex** is the lowest point on the graph. If a parabola opens downward, the **vertex** is the highest point on the graph. The **axis of symmetry** is the vertical line that passes through the vertex.

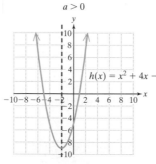

Figure 8-3

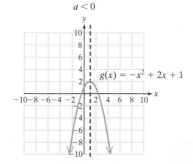

Figure 8-4

1. Quadratic Functions of the Form $f(x) = x^2 + k$

One technique for graphing a function is to plot a sufficient number of points on the graph until the general shape and defining characteristics can be determined. Then sketch a curve through the points.

Skill Practice

1. Refer to the graph of $f(x) = x^2 + k$ to determine the value of k.

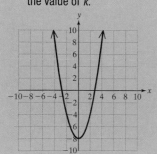

Answer

1. $k = -8$

Example 1   Graphing Quadratic Functions of the Form $f(x) = x^2 + k$

Graph the functions f, g, and h on the same coordinate system.

$$f(x) = x^2 \qquad g(x) = x^2 + 1 \qquad h(x) = x^2 - 2$$

Solution:

Several function values for f, g, and h are shown in Table 8-1 for selected values of x. The corresponding graphs are pictured in Figure 8-5.

Table 8-1

x	$f(x) = x^2$	$g(x) = x^2 + 1$	$h(x) = x^2 - 2$
-3	9	10	7
-2	4	5	2
-1	1	2	-1
0	0	1	-2
1	1	2	-1
2	4	5	2
3	9	10	7

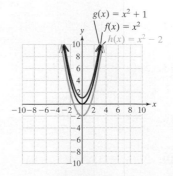

Figure 8-5

Notice that the graphs of $g(x) = x^2 + 1$ and $h(x) = x^2 - 2$ take on the same shape as $f(x) = x^2$. However, the y values of g are 1 greater than the y values of f. Hence, the graph of $g(x) = x^2 + 1$ is the same as the graph of $f(x) = x^2$ shifted *up* 1 unit. Likewise the y values of h are 2 less than those of f. The graph of $h(x) = x^2 - 2$ is the same as the graph of $f(x) = x^2$ shifted *down* 2 units.

A function of the form $f(x) = x^2 + k$ represents a vertical shift of the function $f(x) = x^2$. The functions in Example 1 illustrate the following properties of quadratic functions of the form $f(x) = x^2 + k$.

Graphs of $f(x) = x^2 + k$

A function of the form $f(x) = x^2 + k$ represents a vertical shift of the function $f(x) = x^2$.

If $k > 0$, then the graph of $f(x) = x^2 + k$ is the same as the graph of $y = x^2$ shifted *up* k units.

If $k < 0$, then the graph of $f(x) = x^2 + k$ is the same as the graph of $y = x^2$ shifted *down* $|k|$ units.

Calculator Connections

Topic: Investigating Vertical Shifts of a Function

Try experimenting with a graphing calculator by graphing functions of the form $y = x^2 + k$ for several values of k.

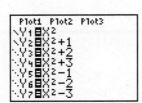
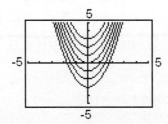

Example 2   ### Graphing Quadratic Functions of the Form $f(x) = x^2 + k$

Sketch the functions defined by

a. $m(x) = x^2 - 4$ **b.** $n(x) = x^2 + \dfrac{7}{2}$

Solution:

a. $m(x) = x^2 - 4$

$m(x) = x^2 + (-4)$

Because $k = -4$, the graph is obtained by shifting the graph of $y = x^2$ down $|-4|$ units (Figure 8-6).

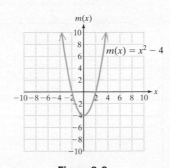

Figure 8-6

Skill Practice

2. Graph the functions f, g, and h on the same coordinate system.

$f(x) = x^2$

$g(x) = x^2 + 3$

$h(x) = x^2 - 5$

Answer

2.

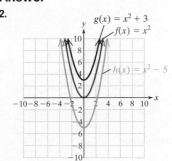

b. $n(x) = x^2 + \dfrac{7}{2}$

Because $k = \frac{7}{2}$, the graph is obtained by shifting the graph of $y = x^2$ up $\frac{7}{2}$ units (Figure 8-7).

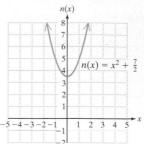

Figure 8-7

2. Quadratic Functions of the Form $f(x) = (x - h)^2$

The graph of $f(x) = x^2 + k$ represents a vertical shift (up or down) of the function $y = x^2$. Example 3 shows that functions of the form $f(x) = (x - h)^2$ represent a horizontal shift (left or right) of the function $y = x^2$.

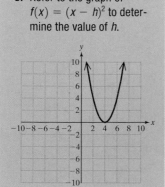

Example 3 **Graphing Quadratic Functions of the Form** $f(x) = (x - h)^2$

Graph the functions f, g, and h on the same coordinate system.

$$f(x) = x^2 \qquad g(x) = (x + 1)^2 \qquad h(x) = (x - 2)^2$$

Solution:

Several function values for f, g, and h are shown in Table 8-2 for selected values of x. The corresponding graphs are pictured in Figure 8-8.

Table 8-2

x	$f(x) = x^2$	$g(x) = (x + 1)^2$	$h(x) = (x - 2)^2$
-4	16	9	36
-3	9	4	25
-2	4	1	16
-1	1	0	9
0	0	1	4
1	1	4	1
2	4	9	0
3	9	16	1
4	16	25	4
5	25	36	9

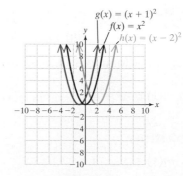

Figure 8-8

Example 3 illustrates the following properties of quadratic functions of the form $f(x) = (x - h)^2$.

Graphs of $f(x) = (x - h)^2$

A function of the form $f(x) = (x - h)^2$ represents a horizontal shift of the function $f(x) = x^2$.

If $h > 0$, then the graph of $f(x) = (x - h)^2$ is the same as the graph of $y = x^2$ shifted h units to the *right*.

If $h < 0$, then the graph of $f(x) = (x - h)^2$ is the same as the graph of $y = x^2$ shifted $|h|$ units to the *left*.

From Example 3 we have

$$h(x) = (x - 2)^2 \qquad \text{and} \qquad g(x) = (x + 1)^2$$

$$g(x) = [x - (-1)]^2$$

$y = x^2$ shifted 2 units
to the right

$y = x^2$ shifted $|-1|$ unit
to the left

Example 4   **Graphing Functions of the Form $f(x) = (x - h)^2$**

Sketch the functions p and q.

   a. $p(x) = (x - 7)^2$ **b.** $q(x) = (x + 1.6)^2$

Solution:

   a. $p(x) = (x - 7)^2$

   Because $h = 7 > 0$, shift the graph of $y = x^2$ to the *right* 7 units (Figure 8-9).

   b. $q(x) = (x + 1.6)^2$

   $q(x) = [x - (-1.6)]^2$

   Because $h = -1.6 < 0$, shift the graph of $y = x^2$ to the *left* 1.6 units (Figure 8-10).

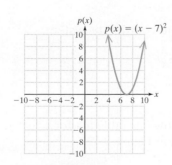

Figure 8-9

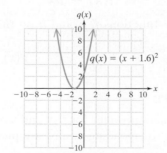

Figure 8-10

Answer

4.

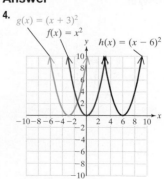

3. Quadratic Functions of the Form $f(x) = ax^2$

Examples 5 and 6 investigate functions of the form $f(x) = ax^2$ $(a \neq 0)$.

> **Example 5** **Graphing Functions of the Form** $f(x) = ax^2$
>
> Graph the functions f, g, and h on the same coordinate system.
>
> $$f(x) = x^2 \qquad g(x) = 2x^2 \qquad h(x) = \frac{1}{2}x^2$$
>
> **Solution:**
>
> Several function values for f, g, and h are shown in Table 8-3 for selected values of x. The corresponding graphs are pictured in Figure 8-11.
>
> **Table 8-3**
>
> | x | $f(x) = x^2$ | $g(x) = 2x^2$ | $h(x) = \frac{1}{2}x^2$ |
> |---|---|---|---|
> | -3 | 9 | 18 | $\frac{9}{2}$ |
> | -2 | 4 | 8 | 2 |
> | -1 | 1 | 2 | $\frac{1}{2}$ |
> | 0 | 0 | 0 | 0 |
> | 1 | 1 | 2 | $\frac{1}{2}$ |
> | 2 | 4 | 8 | 2 |
> | 3 | 9 | 18 | $\frac{9}{2}$ |
>
>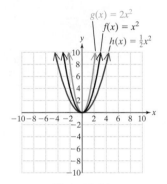
>
> **Figure 8-11**

In Example 5, the function values defined by $g(x) = 2x^2$ are twice those of $f(x) = x^2$. The graph of $g(x) = 2x^2$ is the same as the graph of $f(x) = x^2$ *stretched vertically* by a factor of 2 (the graph appears narrower than $f(x) = x^2$).

In Example 5, the function values defined by $h(x) = \frac{1}{2}x^2$ are one-half those of $f(x) = x^2$. The graph of $h(x) = \frac{1}{2}x^2$ is the same as the graph of $f(x) = x^2$ *shrunk vertically* by a factor of $\frac{1}{2}$ (the graph appears wider than $f(x) = x^2$).

> **Example 6** **Graphing Functions of the Form** $f(x) = ax^2$
>
> Graph the functions f, g, and h on the same coordinate system.
>
> $$f(x) = -x^2 \qquad g(x) = -3x^2 \qquad h(x) = -\frac{1}{3}x^2$$
>
> **Solution:**
>
> Several function values for f, g, and h are shown in Table 8-4 for selected values of x. The corresponding graphs are pictured in Figure 8-12.

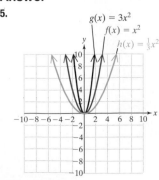

Table 8-4

x	$f(x) = -x^2$	$g(x) = -3x^2$	$h(x) = -\frac{1}{3}x^2$
-3	-9	-27	-3
-2	-4	-12	$-\frac{4}{3}$
-1	-1	-3	$-\frac{1}{3}$
0	0	0	0
1	-1	-3	$-\frac{1}{3}$
2	-4	-12	$-\frac{4}{3}$
3	-9	-27	-3

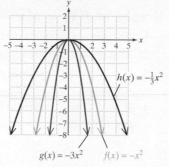

Figure 8-12

Example 6 illustrates that if the coefficient of the square term is negative, the parabola opens downward. The graph of $g(x) = -3x^2$ is the same as the graph of $f(x) = -x^2$ with a *vertical stretch* by a factor of $|-3|$. The graph of $h(x) = -\frac{1}{3}x^2$ is the same as the graph of $f(x) = -x^2$ with a *vertical shrink* by a factor of $|-\frac{1}{3}|$.

Graphs of $f(x) = ax^2$

1. If $a > 0$, the parabola opens upward. Furthermore,
   - If $0 < a < 1$, then the graph of $f(x) = ax^2$ is the same as the graph of $y = x^2$ with a *vertical shrink* by a factor of a.
   - If $a > 1$, then the graph of $f(x) = ax^2$ is the same as the graph of $y = x^2$ with a *vertical stretch* by a factor of a.

2. If $a < 0$, the parabola opens downward. Furthermore,
   - If $0 < |a| < 1$, then the graph of $f(x) = ax^2$ is the same as the graph of $y = -x^2$ with a *vertical shrink* by a factor of $|a|$.
   - If $|a| > 1$, then the graph of $f(x) = ax^2$ is the same as the graph of $y = -x^2$ with a *vertical stretch* by a factor of $|a|$.

4. Quadratic Functions of the Form $f(x) = a(x - h)^2 + k$

We can summarize our findings from Examples 1–6 by graphing functions of the form $f(x) = a(x - h)^2 + k \quad (a \neq 0)$.

The graph of $y = x^2$ has its vertex at the origin $(0, 0)$. The graph of $f(x) = a(x - h)^2 + k$ is the same as the graph of $y = x^2$ shifted to the right or left h units and shifted up or down k units. Therefore, the vertex is shifted from $(0, 0)$ to (h, k). The axis of symmetry is the vertical line through the vertex. Therefore, the axis of symmetry must be the line $x = h$.

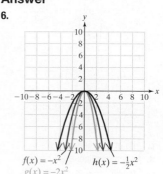

Graphs of $f(x) = a(x - h)^2 + k$

1. The vertex is located at (h, k).
2. The axis of symmetry is the line $x = h$.
3. If $a > 0$, the parabola opens upward, and k is the **minimum value** of the function.
4. If $a < 0$, the parabola opens downward, and k is the **maximum value** of the function.

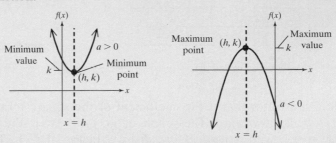

Skill Practice

7. Given the function defined by $g(x) = 3(x + 1)^2 - 3$

a. Identify the vertex.

b. Sketch the graph.

c. Identify the axis of symmetry.

d. Identify the maximum or minimum value of the function.

e. Determine the domain and range in interval notation.

Example 7 **Graphing a Function of the Form** $f(x) = a(x - h)^2 + k$

Given the function defined by

$$f(x) = 2(x - 3)^2 + 4$$

a. Identify the vertex.

b. Sketch the function.

c. Identify the axis of symmetry.

d. Identify the maximum or minimum value of the function.

e. Use the graph to determine the domain and range of f in interval notation.

Solution:

a. The function is in the form $f(x) = a(x - h)^2 + k$, where $a = 2$, $h = 3$, and $k = 4$. Therefore, the vertex is $(3, 4)$.

b. The graph of f is the same as the graph of $f(x) = x^2$ shifted to the right 3 units, shifted up 4 units, and stretched vertically by a factor of 2 (Figure 8-13).

c. The axis of symmetry is the line $x = 3$.

Answers

7. a. Vertex: $(-1, -3)$

b.

d. Because $a > 0$, the function opens upward. Therefore, the minimum function value is 4. Notice that the minimum value is the minimum y value on the graph.

e. The domain is all real numbers: $(-\infty, \infty)$.

The vertex is the lowest point on the parabola, and its y-coordinate is 4. Therefore, the range is $[4, \infty)$.

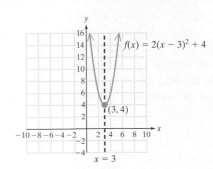

Figure 8-13

c. Axis of symmetry: $x = -1$

d. Minimum value: -3

e. Domain: $(-\infty, \infty)$; range: $[-3, \infty)$

TIP: We can find additional points to the right or left of the vertex and use the symmetry of the parabola to refine the sketch of the curve.

x	f(x)
4	6
5	12

Calculator Connections

Topic: Using the Maximum and Minimum Features

Some graphing calculators have *Minimum* and *Maximum* features that enable the user to approximate the minimum and maximum values of a function.

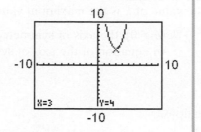

Example 8 **Graphing a Function of the Form**
$$f(x) = a(x - h)^2 + k$$

Given the function defined by $g(x) = -(x + 2)^2 - \dfrac{7}{4}$

a. Identify the vertex.

b. Sketch the function.

c. Identify the axis of symmetry.

d. Identify the maximum or minimum value of the function.

e. Use the graph to determine the domain and range of f in interval notation.

Solution:

a. $g(x) = -(x + 2)^2 - \dfrac{7}{4}$

$= -1[x - (-2)]^2 + \left(-\dfrac{7}{4}\right)$

The function is in the form $g(x) = a(x - h)^2 + k$, where $a = -1$, $h = -2$, and $k = -\dfrac{7}{4}$. Therefore, the vertex is $\left(-2, -\dfrac{7}{4}\right)$.

b. The graph of g is the same as the graph of $f(x) = x^2$ shifted to the left 2 units, shifted down $\dfrac{7}{4}$ units, and opening downward (Figure 8-14).

Figure 8-14

c. The axis of symmetry is the line $x = -2$.

d. The parabola opens downward, so the maximum function value is $-\dfrac{7}{4}$.

e. The domain is all real numbers: $(-\infty, \infty)$.

The vertex is the highest point on the parabola, and its y-coordinate is $-\dfrac{7}{4}$. Therefore, the range is $\left(-\infty, -\dfrac{7}{4}\right]$.

Skill Practice

8. Given the function defined by

$$h(x) = -\dfrac{1}{2}(x - 4)^2 + 2$$

a. Identify the vertex.

b. Sketch the graph.

c. Identify the axis of symmetry.

d. Identify the maximum or minimum value of the function.

e. Determine the domain and range in interval notation.

Answers

8. a. Vertex: (4, 2)

b.

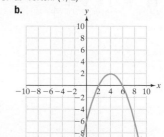

c. Axis of symmetry: $x = 4$

d. Maximum value: 2

e. Domain: $(-\infty, \infty)$; range: $(-\infty, 2]$

Section 8.4 Practice Exercises

Vocabulary and Key Concepts

1. **a.** The graph of a quadratic function, $f(x) = ax^2 + bx + c$, is a _____.

    b. The parabola defined by $f(x) = ax^2 + bx + c$ $(a \neq 0)$ will open upward if a _____ 0 and will open downward if a _____ 0.

    c. If a parabola opens upward, the vertex is the (highest/lowest) point on the graph. If a parabola opens downward, the vertex is the (highest/lowest) point on the graph.

    d. Given $f(x) = a(x - h)^2 + k$ $(a \neq 0)$, the vertex of the parabola is given by the ordered pair _____. If the parabola opens _____, the value of k is the minimum value of the function. If the parabola opens _____, the value of k is the maximum value of the function.

    e. Given $f(x) = a(x - h)^2 + k$ $(a \neq 0)$, the axis of symmetry of the parabola is the vertical line that passes through the _____. An equation of the axis of symmetry is _____.

Review Exercises

For Exercises 2–8, solve the equations.

2. $x^2 + x - 5 = 0$ 3. $(y - 3)^2 = -4$ 4. $\sqrt{2a + 2} = a + 1$ 5. $5t(t - 2) = -3$

6. $2z^2 - 3z - 9 = 0$ 7. $x^{2/3} + 5x^{1/3} + 6 = 0$ 8. $m^2(m^2 + 6) = 27$

Concept 1: Quadratic Functions of the Form $f(x) = x^2 + k$

9. Describe how the value of k affects the graph of a function defined by $f(x) = x^2 + k$.

For Exercises 10–17, graph the functions. **(See Examples 1 and 2.)**

10. $g(x) = x^2 + 1$ 11. $f(x) = x^2 + 2$ 12. $p(x) = x^2 - 3$ 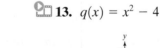 13. $q(x) = x^2 - 4$

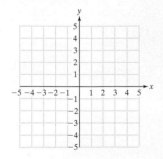 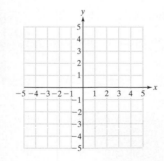 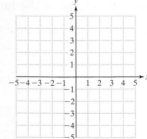 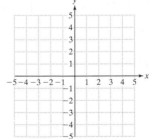

14. $T(x) = x^2 + \dfrac{3}{4}$ 15. $S(x) = x^2 + \dfrac{3}{2}$ 16. $M(x) = x^2 - \dfrac{5}{4}$ 17. $n(x) = x^2 - \dfrac{1}{3}$

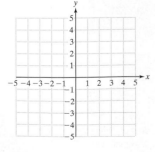 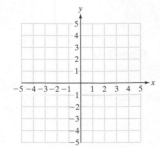 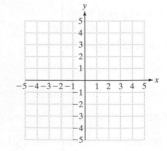 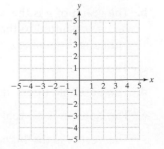

Concept 2: Quadratic Functions of the Form $f(x) = (x - h)^2$

18. Describe how the value of h affects the graph of a function defined by $f(x) = (x - h)^2$.

For Exercises 19–26, graph the functions. **(See Examples 3 and 4.)**

19. $r(x) = (x + 1)^2$ **20.** $h(x) = (x + 2)^2$ **21.** $k(x) = (x - 3)^2$ **22.** $L(x) = (x - 4)^2$

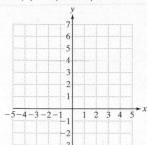   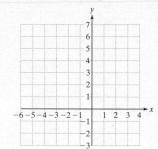   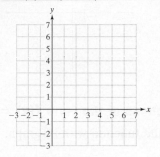   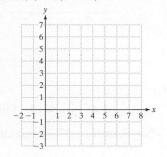

23. $A(x) = \left(x + \dfrac{3}{4}\right)^2$ **24.** $r(x) = \left(x + \dfrac{3}{2}\right)^2$ **25.** $W(x) = (x - 1.25)^2$ **26.** $V(x) = (x - 2.5)^2$

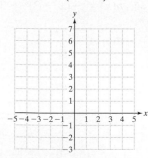   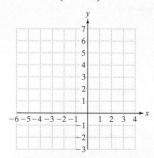   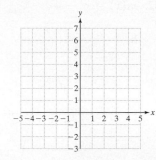   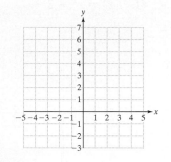

Concept 3: Quadratic Functions of the Form $f(x) = ax^2$

27. Describe how the value of a affects the graph of a function defined by $f(x) = ax^2$, where $a \neq 0$.

28. How do you determine whether the graph of a function defined by $h(x) = ax^2 + bx + c$ $(a \neq 0)$ opens upward or downward?

For Exercises 29–36, graph the functions. **(See Examples 5 and 6.)**

29. $f(x) = 2x^2$ **30.** $g(x) = 3x^2$ **31.** $h(x) = \dfrac{1}{2}x^2$ **32.** $f(x) = \dfrac{1}{3}x^2$

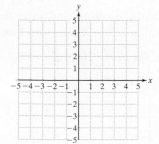   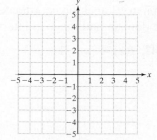   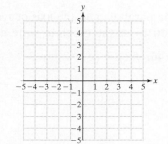   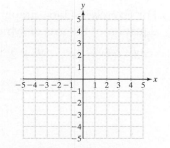

33. $c(x) = -x^2$

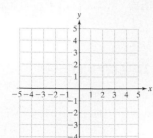

34. $g(x) = -2x^2$

35. $v(x) = -\dfrac{1}{3}x^2$

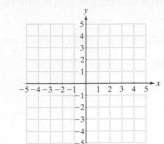

36. $f(x) = -\dfrac{1}{2}x^2$

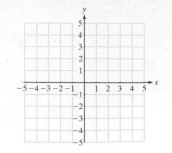

Concept 4: Quadratic Functions of the Form $f(x) = a(x - h)^2 + k$

For Exercises 37–44, match the function with its graph.

37. $f(x) = -\dfrac{1}{4}x^2$

38. $g(x) = (x + 3)^2$

39. $k(x) = (x - 3)^2$

40. $h(x) = \dfrac{1}{4}x^2$

41. $t(x) = x^2 + 2$

42. $m(x) = x^2 - 4$

43. $p(x) = (x + 1)^2 - 3$

44. $n(x) = -(x - 2)^2 + 3$

a.

b.

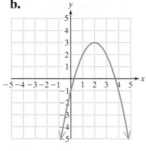

c.

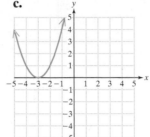

d.

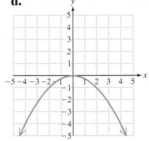

e.

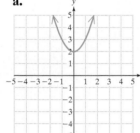

f.

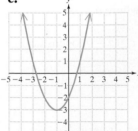

g.

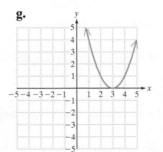

h.

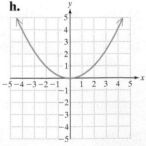

For Exercises 45–64, graph the parabola and the axis of symmetry. Label the coordinates of the vertex, and write the equation of the axis of symmetry. Use the graph to write the domain and range in interval notation. **(See Examples 7 and 8.)**

45. $f(x) = (x - 3)^2 + 2$

46. $f(x) = (x - 2)^2 + 3$

47. $f(x) = (x + 1)^2 - 3$

48. $f(x) = (x + 3)^2 - 1$

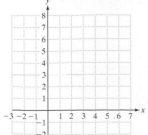

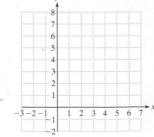

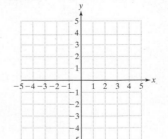

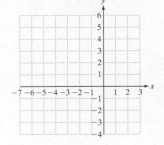

49. $f(x) = -(x - 4)^2 - 2$ **50.** $f(x) = -(x - 2)^2 - 4$ **51.** $f(x) = -(x + 3)^2 + 3$ **52.** $f(x) = -(x + 2)^2 + 2$

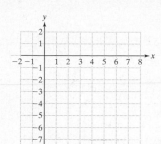 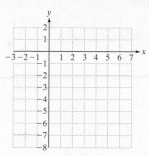 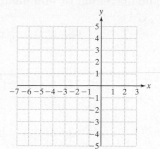 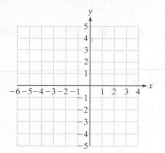

53. $f(x) = (x + 1)^2 + 1$ **54.** $f(x) = (x - 4)^2 - 4$ **55.** $f(x) = 3(x - 1)^2$ **56.** $f(x) = -3(x - 1)^2$

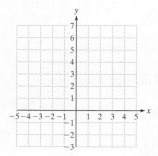 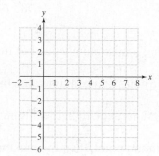 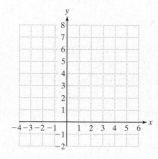 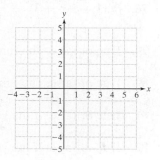

57. $f(x) = -4x^2 + 3$ **58.** $f(x) = 4x^2 + 3$ **59.** $f(x) = 2(x + 3)^2 - 1$ **60.** $f(x) = -2(x + 3)^2 - 1$

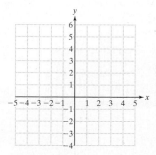 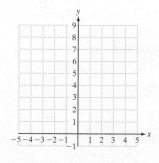 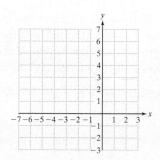 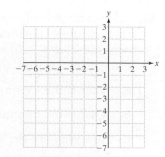

61. $f(x) = -\frac{1}{4}(x - 1)^2 + 2$ **62.** $f(x) = \frac{1}{4}(x - 1)^2 + 2$ **63.** $f(x) = \frac{1}{3}(x - 2)^2 + 1$ **64.** $f(x) = -\frac{1}{3}(x - 2)^2 + 1$

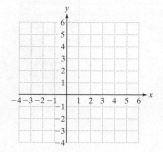 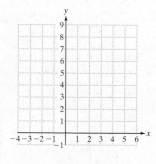 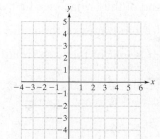 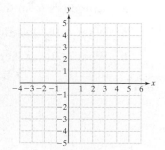

65. Compare the graphs of the following equations to the graph of $y = x^2$.

   a. $y = x^2 + 3$

   b. $y = (x + 3)^2$

   c. $y = 3x^2$

66. Compare the graphs of the following equations to the graph of $y = x^2$.

   a. $y = (x - 2)^2$

   b. $y = 2x^2$

   c. $y = x^2 - 2$

For Exercises 67–78, write the coordinates of the vertex and determine if the vertex is a maximum point or a minimum point. Then write the maximum or minimum value.

67. $f(x) = 4(x - 6)^2 - 9$

68. $g(x) = 3(x - 4)^2 - 7$

69. $p(x) = -\dfrac{2}{5}(x - 2)^2 + 5$

70. $h(x) = -\dfrac{3}{7}(x - 5)^2 + 10$

71. $k(x) = \dfrac{1}{2}(x + 8)^2$

72. $m(x) = \dfrac{2}{9}(x + 11)^2$

73. $n(x) = -6x^2 + \dfrac{21}{4}$

74. $q(x) = -4x^2 + \dfrac{1}{6}$

75. $A(x) = 2(x - 7)^2 - \dfrac{3}{2}$

76. $B(x) = 5(x - 3)^2 - \dfrac{1}{4}$

77. $F(x) = 7x^2$

78. $G(x) = -7x^2$

79. True or false: The function defined by $g(x) = -5x^2$ has a maximum value but no minimum value.

80. True or false: The function defined by $f(x) = 2(x - 5)^2$ has a maximum value but no minimum value.

81. True or false: If the vertex $(-2, 8)$ represents a minimum point, then the minimum value is -2.

82. True or false: If the vertex $(-2, 8)$ represents a maximum point, then the maximum value is 8.

Expanding Your Skills

83. A suspension bridge is 120 ft long. Its supporting cable hangs in a shape that resembles a parabola. The function defined by $H(x) = \frac{1}{90}(x - 60)^2 + 30$ (where $0 \le x \le 120$) approximates the height of the supporting cable a distance of x ft from the end of the bridge (see figure).

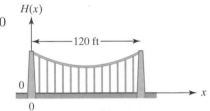

   a. What is the location of the vertex of the parabolic cable?

   b. What is the minimum height of the cable?

   c. How high are the towers at either end of the supporting cable?

84. A 50-m bridge over a crevasse is supported by a parabolic arch. The function defined by $f(x) = -0.16(x - 25)^2 + 100$ (where $0 \le x \le 50$) approximates the height of the supporting arch x meters from the end of the bridge (see figure).

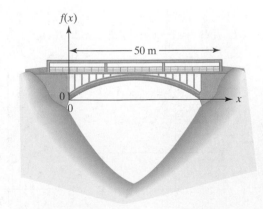

   a. What is the location of the vertex of the arch?

   b. What is the maximum height of the arch (relative to the origin)?

Graphing Calculator Exercises

For Exercises 85–88, verify the maximum and minimum points found in Exercises 67–70, by graphing each function on the calculator.

85. $Y_1 = 4(x - 6)^2 - 9$ (Exercise 67)

86. $Y_1 = 3(x - 4)^2 - 7$ (Exercise 68)

87. $Y_1 = -\dfrac{2}{5}(x - 2)^2 + 5$ (Exercise 69)

88. $Y_1 = -\dfrac{3}{7}(x - 5)^2 + 10$ (Exercise 70)

Vertex of a Parabola: Applications and Modeling | Section 8.5

1. Writing a Quadratic Function in the Form $f(x) = a(x - h)^2 + k$

The graph of a quadratic function is a parabola, and if the function is written in the form $f(x) = a(x - h)^2 + k$ $(a \neq 0)$, then the vertex is at (h, k). A quadratic function can be written in the form $f(x) = a(x - h)^2 + k$ $(a \neq 0)$ by completing the square. The process is similar to the steps outlined in Section 8.1 except that all algebraic manipulations are performed on the right-hand side of the equation.

> **Example 1** Writing a Quadratic Function in the Form $f(x) = a(x - h)^2 + k$ $(a \neq 0)$
>
> Given: $f(x) = x^2 + 8x + 13$
>
> **a.** Write the function in the form $f(x) = a(x - h)^2 + k$.
>
> **b.** Identify the vertex, axis of symmetry, and minimum function value.

Solution:

a. $f(x) = x^2 + 8x + 13$

$\quad = 1(x^2 + 8x) + 13$

Rather than dividing by the leading coefficient on both sides, we will factor out the leading coefficient from the variable terms on the right-hand side.

$\quad = 1(x^2 + 8x \quad\quad) + 13$

Next, complete the square on the expression within the parentheses: $[\frac{1}{2}(8)]^2 = 16$.

$\quad = 1(x^2 + 8x + 16 - 16) + 13$

Rather than add 16 to both sides of the function, we *add and subtract* 16 within the parentheses on the right-hand side. This has the effect of adding 0 to the right-hand side.

$\quad = 1(x^2 + 8x + 16) - 16 + 13$

Use the associative property of addition to regroup terms and isolate the perfect square trinomial within the parentheses.

$\quad = (x + 4)^2 - 3$

Factor and simplify.

b. $f(x) = (x + 4)^2 - 3$

The vertex is $(-4, -3)$.

The axis of symmetry is $x = -4$.

Because $a > 0$, the parabola opens upward.

The minimum value is -3 (Figure 8-15).

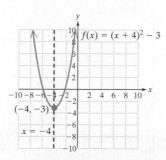

Figure 8-15

Example 2 Analyzing a Quadratic Function

Given: $f(x) = -2x^2 + 12x - 16$

a. Write the function in the form $f(x) = a(x - h)^2 + k.$

b. Find the vertex, axis of symmetry, and maximum function value.

c. Find the x- and y-intercepts.

d. Sketch the graph of the function.

Solution:

a. $f(x) = -2x^2 + 12x - 16$ To find the vertex, write the function in the form $f(x) = a(x - h)^2 + k.$

$= -2(x^2 - 6x \quad) - 16$ If the leading coefficient is not 1, factor the coefficient from the variable terms.

$= -2(x^2 - 6x + 9 - 9) - 16$ Add and subtract the quantity $[\frac{1}{2}(-6)]^2 = 9$ within the parentheses.

$= -2(x^2 - 6x + 9) + (-2)(-9) - 16$ To remove the term -9 from the parentheses, we must first apply the distributive property. When -9 is removed from the parentheses, it carries with it a factor of -2.

$= -2(x - 3)^2 + 18 - 16$ Factor and simplify.

$= -2(x - 3)^2 + 2$

b. $f(x) = -2(x - 3)^2 + 2$

The vertex is $(3, 2)$. The axis of symmetry is $x = 3$. Because $a < 0$, the parabola opens downward and the maximum value is 2.

c. The y-intercept is given by $f(0) = -2(0)^2 + 12(0) - 16 = -16$.

The y-intercept is $(0, -16)$.

To find the x-intercept(s), find the real solutions to the equation $f(x) = 0$.

$f(x) = -2x^2 + 12x - 16$

$0 = -2x^2 + 12x - 16$ Substitute $f(x) = 0$.

$0 = -2(x^2 - 6x + 8)$ Factor.

$0 = -2(x - 4)(x - 2)$

$x = 4$ or $x = 2$

The x-intercepts are $(4, 0)$ and $(2, 0)$.

d. Using the information from parts (a)–(c), sketch the graph (Figure 8-16).

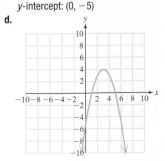

Figure 8-16

2. Vertex Formula

Completing the square and writing a quadratic function in the form $f(x) = a(x - h)^2 + k$ $(a \neq 0)$ is one method to find the vertex of a parabola. Another method is to use the vertex formula. The **vertex formula** can be derived by completing the square on $f(x) = ax^2 + bx + c$ $(a \neq 0)$.

$f(x) = ax^2 + bx + c$ $\quad (a \neq 0)$

$= a\left(x^2 + \dfrac{b}{a}x \qquad\right) + c$ $\qquad$ Factor a from the variable terms.

$= a\left(x^2 + \dfrac{b}{a}x + \dfrac{b^2}{4a^2} - \dfrac{b^2}{4a^2}\right) + c$ $\qquad$ Add and subtract $[\frac{1}{2}(b/a)]^2 = b^2/(4a^2)$ within the parentheses.

$= a\left(x^2 + \dfrac{b}{a}x + \dfrac{b^2}{4a^2}\right) + (a)\left(-\dfrac{b^2}{4a^2}\right) + c$ $\qquad$ Apply the distributive property and remove the term $-b^2/(4a^2)$ from the parentheses.

$= a\left(x + \dfrac{b}{2a}\right)^2 - \dfrac{b^2}{4a} + c$ $\qquad$ Factor the trinomial and simplify.

$= a\left(x + \dfrac{b}{2a}\right)^2 + c - \dfrac{b^2}{4a}$ $\qquad$ Apply the commutative property of addition to reverse the last two terms.

$= a\left(x + \dfrac{b}{2a}\right)^2 + \dfrac{4ac}{4a} - \dfrac{b^2}{4a}$ $\qquad$ Obtain a common denominator.

$= a\left(x + \dfrac{b}{2a}\right)^2 + \dfrac{4ac - b^2}{4a}$

$= a\left[x - \left(-\dfrac{b}{2a}\right)\right]^2 + \dfrac{4ac - b^2}{4a}$

$f(x) = a(x \quad - \quad h)^2 \quad + \quad k$

The function is in the form $f(x) = a(x - h)^2 + k$, where

$$h = \frac{-b}{2a} \quad \text{and} \quad k = \frac{4ac - b^2}{4a}$$

Therefore, the vertex is $\left(\dfrac{-b}{2a}, \dfrac{4ac - b^2}{4a}\right)$

Although the y-coordinate of the vertex is given as $(4ac - b^2)/(4a)$, it is usually easier to determine the x-coordinate of the vertex first and then find y by evaluating the function at $x = -b/(2a)$.

The Vertex Formula

For $f(x) = ax^2 + bx + c$ $\quad (a \neq 0)$, the vertex is given by

$$\left(\frac{-b}{2a}, \frac{4ac - b^2}{4a}\right) \quad \text{or} \quad \left(\frac{-b}{2a}, f\left(\frac{-b}{2a}\right)\right)$$

3. Determining the Vertex and Intercepts of a Quadratic Function

Example 3 Determining the Vertex and Intercepts of a Quadratic Function

Given: $h(x) = x^2 - 2x + 5$

a. Use the vertex formula to find the vertex.

b. Find the x- and y-intercepts.

c. Sketch the function.

Solution:

a. $h(x) = x^2 - 2x + 5$

$a = 1$ $b = -2$ $c = 5$ Identify a, b, and c.

The x-coordinate of the vertex is $\dfrac{-b}{2a} = \dfrac{-(-2)}{2(1)} = 1$.

The y-coordinate of the vertex is $h(1) = (1)^2 - 2(1) + 5 = 4$.

The vertex is $(1, 4)$.

b. The y-intercept is given by $h(0) = (0)^2 - 2(0) + 5 = 5$.

The y-intercept is $(0, 5)$.

To find the x-intercept(s), find the real solutions to the equation $h(x) = 0$.

$h(x) = x^2 - 2x + 5$

$0 = x^2 - 2x + 5$ This quadratic equation is not factorable. Apply the quadratic formula: $a = 1, b = -2, c = 5$.

$x = \dfrac{-(-2) \pm \sqrt{(-2)^2 - 4(1)(5)}}{2(1)}$

$= \dfrac{2 \pm \sqrt{4 - 20}}{2(1)}$

$= \dfrac{2 \pm \sqrt{-16}}{2}$

$= \dfrac{2 \pm 4i}{2}$

$= 1 \pm 2i$

The solutions to the equation $h(x) = 0$ are not real numbers. Therefore, there are no x-intercepts.

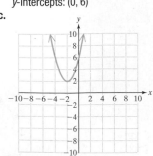

c.

TIP: The location of the vertex and the direction that the parabola opens can be used to determine whether the function has any x-intercepts.
  Given $h(x) = x^2 - 2x + 5$, the vertex $(1, 4)$ is *above* the x-axis. Furthermore, because $a > 0$, the parabola opens upward. Therefore, it is not possible for the graph to cross the x-axis (Figure 8-17).

Because $a > 0$, the parabola opens up.

Figure 8-17

4. Applications and Modeling of Quadratic Functions

| Example 4 | Applying a Quadratic Function

The crew from Extravaganza Entertainment launches fireworks at an angle of 60° from the horizontal. The height of one particular type of display can be approximated by:

$$h(t) = -16t^2 + 128\sqrt{3}t$$

where $h(t)$ is measured in feet and t is measured in seconds.

a. How long will it take the fireworks to reach their maximum height? Round to the nearest second.

b. Find the maximum height. Round to the nearest foot.

Solution:

$$h(t) = -16t^2 + 128\sqrt{3}t$$

This parabola opens downward; therefore, the maximum height of the fireworks will occur at the vertex of the parabola.

$$a = -16 \qquad b = 128\sqrt{3} \qquad c = 0$$

Identify a, b, and c, and apply the vertex formula.

The x-coordinate of the vertex is

$$\frac{-b}{2a} = \frac{-128\sqrt{3}}{2(-16)} = \frac{-128\sqrt{3}}{-32} \approx 6.9$$

The y-coordinate of the vertex is approximately

$$h(6.9) = -16(6.9)^2 + 128\sqrt{3}(6.9) \approx 768$$

The vertex is $(6.9, 768)$.

a. The fireworks will reach their maximum height in 6.9 sec.

b. The maximum height is 768 ft.

Skill Practice

4. An object is launched into the air with an initial velocity of 48 ft/sec from the top of a building 288 ft high. The height $h(t)$ of the object after t seconds is given by
$$h(t) = -16t^2 + 48t + 288$$
a. Find the time it takes for the object to reach its maximum height.
b. Find the maximum height.

In Section 2.3 we wrote a linear model given two points by using the point-slope formula: $y - y_1 = m(x - x_1)$. Now we will learn how to write a quadratic model of a parabola given three points by using systems of equations and the standard form of a parabola: $y = ax^2 + bx + c$.

This process involves substituting the x- and y-coordinates of the three given points into the quadratic model. Then we solve the resulting system of three equations (Section 4.6).

Answers

4. a. 1.5 sec   **b.** 324 ft

Example 5 **Writing a Quadratic Model**

Write an equation of a parabola that passes through the points $(1, -1)$, $(-1, -5)$, and $(2, 4)$.

Solution:

Substitute $(1, -1)$ into the equation

$y = ax^2 + bx + c$ $\longrightarrow$ $(-1) = a(1)^2 + b(1) + c$
$-1 = a + b + c$
(standard form) $\quad a + b + c = -1$

Substitute $(-1, -5)$ into the equation

$y = ax^2 + bx + c$ $\longrightarrow$ $(-5) = a(-1)^2 + b(-1) + c$
$-5 = a - b + c$
(standard form) $\quad a - b + c = -5$

Substitute $(2, 4)$ into the equation

$y = ax^2 + bx + c$ $\longrightarrow$ $(4) = a(2)^2 + b(2) + c$
$4 = 4a + 2b + c$
(standard form) $\quad 4a + 2b + c = 4$

Solve the system:
$\boxed{A} \quad a + \ b + c = -1$
$\boxed{B} \quad a - \ b + c = -5$
$\boxed{C} \quad 4a + 2b + c = \ \ 4$

Notice that the c-variables all have a coefficient of 1. Therefore, we choose to eliminate the c-variable.

$\boxed{A} \quad a + b + c = -1 \longrightarrow \qquad a + b + c = -1$
$\boxed{B} \quad a - b + c = -5 \xrightarrow[\text{Multiply by } -1.]{} \underline{-a + b - c = \ \ 5}$
$\qquad\qquad\qquad\qquad\qquad\qquad\qquad 2b \qquad = \ \ 4$
$\qquad\qquad\qquad\qquad\qquad\qquad\qquad\quad b = \ \ 2 \quad \boxed{D}$

$\boxed{B} \quad a - b + c = -5 \longrightarrow \qquad a - b + c = -5$
$\boxed{C} \quad 4a + 2b + c = \ \ 4 \xrightarrow[\text{Multiply by } -1.]{} \underline{-4a - 2b - c = -4}$
$\qquad\qquad\qquad\qquad\qquad\qquad\qquad -3a - 3b \quad = -9 \quad \boxed{E}$

Substitute b with 2 in equation $\boxed{E}$: $\quad -3a - 3(2) = -9$
$\qquad\qquad\qquad\qquad\qquad\qquad\qquad\quad -3a - 6 = -9$
$\qquad\qquad\qquad\qquad\qquad\qquad\qquad\qquad -3a = -3$
$\qquad\qquad\qquad\qquad\qquad\qquad\qquad\qquad\quad a = 1$

Substitute a and b in equation $\boxed{A}$ to solve for c: $\quad (1) + (2) + c = -1$
$\qquad\qquad\qquad\qquad\qquad\qquad\qquad\qquad\qquad\qquad 3 + c = -1$
$\qquad\qquad\qquad\qquad\qquad\qquad\qquad\qquad\qquad\qquad\qquad c = -4$

Substitute $a = 1$, $b = 2$, and $c = -4$ in the standard form of the parabola for the final answer.

$y = ax^2 + bx + c$

$y = (1)x^2 + (2)x + (-4) \longrightarrow y = x^2 + 2x - 4$

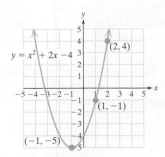

A graph $y = x^2 + 2x - 4$ is shown in Figure 8-18. Notice that the graph of the function passes through the points $(1, -1)$, $(-1, -5)$, and $(2, 4)$.

Figure 8-18

Section 8.5   Practice Exercises

Vocabulary and Key Concepts

1. a. Given $f(x) = ax^2 + bx + c$ $(a \neq 0)$, the vertex formula gives the x-coordinate of the vertex as _____.
The y-coordinate can be found by evaluating the function at $x =$ _____.

b. Answer true or false. A parabola may have no x-intercepts.

c. Answer true or false. A parabola may have one x-intercept.

d. Answer true or false. A parabola may have two x-intercepts.

e. Answer true or false. A parabola may have three x-intercepts.

Review Exercises

2. How does the graph of $f(x) = -2x^2$ compare with the graph of $y = x^2$?

3. How does the graph of $p(x) = \frac{1}{4}x^2$ compare with the graph of $y = x^2$?

4. How does the graph of $Q(x) = x^2 - \frac{8}{3}$ compare with the graph of $y = x^2$?

5. How does the graph of $r(x) = x^2 + 7$ compare with the graph of $y = x^2$?

6. How does the graph of $s(x) = (x - 4)^2$ compare with the graph of $y = x^2$?

7. How does the graph of $t(x) = (x + 10)^2$ compare with the graph of $y = x^2$?

8. Find the coordinates of the vertex of the parabola defined by $g(x) = 2(x + 3)^2 - 4$.

For Exercises 9–16, find the value of n to complete the square. Factor the resulting trinomial.

9. $x^2 - 8x + n$ **10.** $x^2 + 4x + n$ **11.** $y^2 + 7y + n$ **12.** $a^2 - a + n$

13. $b^2 + \frac{2}{9}b + n$ **14.** $m^2 - \frac{2}{7}m + n$ **15.** $t^2 - \frac{1}{3}t + n$ **16.** $p^2 + \frac{1}{4}p + n$

Concept 1: Writing a Quadratic Function in the Form $f(x) = a(x - h)^2 + k$

For Exercises 17–28, write the function in the form $f(x) = a(x - h)^2 + k$ by completing the square. Then identify the vertex. **(See Examples 1 and 2.)**

17. $g(x) = x^2 - 8x + 5$ **18.** $h(x) = x^2 + 4x + 5$ **19.** $n(x) = 2x^2 + 12x + 13$

20. $f(x) = 4x^2 + 16x + 19$ **21.** $p(x) = -3x^2 + 6x - 5$ **22.** $q(x) = -2x^2 + 12x - 11$

23. $k(x) = x^2 + 7x - 10$ **24.** $m(x) = x^2 - x - 8$ **25.** $F(x) = 5x^2 + 10x + 1$

26. $G(x) = 4x^2 + 4x - 7$ **27.** $P(x) = -2x^2 + x$ **28.** $Q(x) = -3x^2 + 12x$

Concept 2: Vertex Formula

For Exercises 29–40, find the vertex by using the vertex formula. **(See Example 3.)**

29. $Q(x) = x^2 - 4x + 7$ **30.** $T(x) = x^2 - 8x + 17$ **31.** $r(x) = -3x^2 - 6x - 5$

32. $s(x) = -2x^2 - 12x - 19$ **33.** $N(x) = x^2 + 8x + 1$ **34.** $M(x) = x^2 + 6x - 5$

35. $m(x) = \dfrac{1}{2}x^2 + x + \dfrac{5}{2}$

36. $n(x) = \dfrac{1}{2}x^2 + 2x + 3$

37. $k(x) = -x^2 + 2x + 2$

38. $h(x) = -x^2 + 4x - 3$

39. $A(x) = -\dfrac{1}{3}x^2 + x$

40. $B(x) = -\dfrac{2}{3}x^2 - 2x$

For Exercises 41–44, find the vertex two ways:

   a. by completing the square and writing in the form $f(x) = a(x - h)^2 + k$.

   b. by using the vertex formula.

41. $p(x) = x^2 + 8x + 1$ **42.** $F(x) = x^2 + 4x - 2$ **43.** $f(x) = 2x^2 + 4x + 6$ **44.** $g(x) = 3x^2 + 12x + 9$

Concept 3: Determining the Vertex and Intercepts of a Quadratic Function

For Exercises 45–52

   a. Find the vertex.

   b. Find the y-intercept.

   c. Find the x-intercept(s), if they exist.

   d. Use this information to graph the function. **(See Examples 2 and 3.)**

45. $f(x) = x^2 + 2x - 3$ **46.** $f(x) = x^2 + 4x + 3$ **47.** $f(x) = 2x^2 - 2x + 4$ **48.** $f(x) = 2x^2 - 12x + 19$

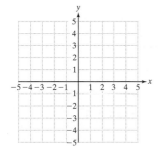 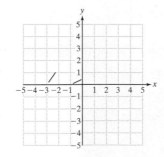 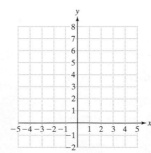 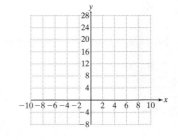

49. $f(x) = -x^2 + 3x - \dfrac{9}{4}$ **50.** $f(x) = -x^2 - \dfrac{3}{2}x - \dfrac{9}{16}$ **51.** $f(x) = -x^2 - 2x + 3$ **52.** $f(x) = -x^2 - 4x$

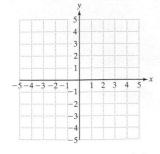 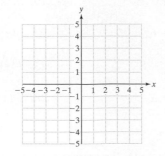 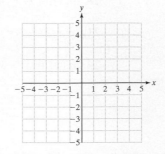 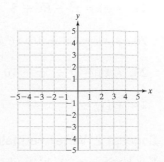

Concept 4: Applications and Modeling of Quadratic Functions

53. Mia sells used MP3 players. The average cost to package MP3 players is given by the equation $C(x) = 2x^2 - 40x + 2200$, where x is the number of MP3 players she packages in a week. How many players must she package to minimize her average cost?

54. Ben sells used iPods. The average cost to package iPods is given by the equation $C(x) = 3x^2 - 120x + 1300$, where x is the number of iPods packaged per month. Determine the number of iPods that Ben needs to package to minimize the average cost.

55. The pressure x in an automobile tire can affect its wear. Both overinflated and underinflated tires can lead to poor performance and poor mileage. For one particular tire, the function P represents the number of miles that a tire lasts (in thousands) for a given pressure x.

$$P(x) = -0.857x^2 + 56.1x - 880$$

where x is the tire pressure in pounds per square inch (psi). **(See Example 4.)**

a. Find the tire pressure that will yield the maximum mileage. Round to the nearest pound per square inch.

b. What is the maximum number of miles that a tire can last? Round to the nearest thousand.

56. A baseball player throws a ball, and the height of the ball (in feet) can be approximated by

$$y(x) = -0.011x^2 + 0.577x + 5$$

where x is the horizontal position of the ball measured in feet from the origin.

a. For what value of x will the ball reach its highest point? Round to the nearest foot.

b. What is the maximum height of the ball? Round to the nearest tenth of a foot.

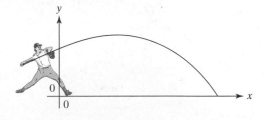

57. Gas mileage depends in part on the speed of the car. The gas mileage of a subcompact car is given by $m(x) = -0.04x^2 + 3.6x - 49$, where $20 \leq x \leq 70$ represents the speed in miles per hour and $m(x)$ is given in miles per gallon. At what speed will the car get the maximum gas mileage?

58. Gas mileage depends in part on the speed of the car. The gas mileage of a luxury car is given by $L(x) = -0.015x^2 + 1.44x - 21$, where $25 \leq x \leq 70$ represents the speed in miles per hour and $L(x)$ is given in miles per gallon. At what speed will the car get the maximum gas mileage?

59. Tetanus bacillus bacterium is cultured to produce tetanus toxin used in an inactive form for the tetanus vaccine. The amount of toxin produced per batch increases with time and then becomes unstable. The amount of toxin (in grams) as a function of time t (in hours) can be approximated by:

$$b(t) = -\frac{1}{1152}t^2 + \frac{1}{12}t$$

How many hours will it take to produce the maximum yield?

60. The bacterium *Pseudomonas aeruginosa* is cultured with an initial population of 10^4 active organisms. The population of active bacteria increases up to a point, and then due to a limited food supply and an increase of waste products, the population of living organisms decreases. Over the first 48 hr, the population can be approximated by:

$$P(t) = -1718.75t^2 + 82{,}500t + 10{,}000$$

where $0 \leq t \leq 48$

Find the time required for the population to reach its maximum value.

For Exercises 61–66, use the standard form of a parabola given by $y = ax^2 + bx + c$ to write an equation of a parabola that passes through the given points. **(See Example 5.)**

61. $(0, 4), (1, 0),$ and $(-1, -10)$ **62.** $(0, 3), (3, 0), (-1, 8)$ **63.** $(2, 1), (-2, 5),$ and $(1, -4)$

64. $(1, 2), (-1, -6),$ and $(2, -3)$ **65.** $(2, -4), (1, 1),$ and $(-1, -7)$ **66.** $(1, 4), (-1, 6),$ and $(2, 18)$

Expanding Your Skills

67. A farmer wants to fence a rectangular corral adjacent to the side of a barn; however, she has only 200 ft of fencing and wants to enclose the largest possible area. See the figure.

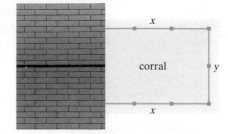

  a. If x represents the length of the corral and y represents the width, explain why the dimensions of the corral are subject to the constraint $2x + y = 200$.

  b. The area of the corral is given by $A = xy$. Use the constraint equation from part (a) to express A as a function of x, where $0 < x < 100$.

  c. Use the function from part (b) to find the dimensions of the corral that will yield the maximum area. [*Hint:* Find the vertex of the function from part (b).]

68. A veterinarian wants to construct two equal-sized pens of maximum area out of 240 ft of fencing. See the figure.

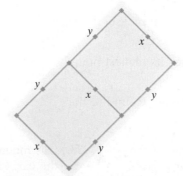

  a. If x represents the length of each pen and y represents the width of each pen, explain why the dimensions of the pens are subject to the constraint $3x + 4y = 240$.

  b. The area of each individual pen is given by $A = xy$. Use the constraint equation from part (a) to express A as a function of x, where $0 < x < 80$.

  c. Use the function from part (b) to find the dimensions of an individual pen that will yield the maximum area. [*Hint:* Find the vertex of the function from part (b).]

Graphing Calculator Exercises

For Exercises 69–74, graph the functions in Exercises 45–50 on a graphing calculator. Use the *Max* or *Min* feature to approximate the vertex.

69. $Y_1 = x^2 + 2x - 3$ (Exercise 45)   **70.** $Y_1 = x^2 + 4x + 3$ (Exercise 46)

71. $Y_1 = 2x^2 - 2x + 4$ (Exercise 47)   **72.** $Y_1 = 2x^2 - 12x + 19$ (Exercise 48)

73. $Y_1 = -x^2 + 3x - \dfrac{9}{4}$ (Exercise 49)   **74.** $Y_1 = -x^2 - \dfrac{3}{2}x - \dfrac{9}{16}$ (Exercise 50)

Polynomial and Rational Inequalities

1. Solving Polynomial Inequalities

In Sections 1.4 and 1.5, we solved simple and compound linear inequalities. In this section we will solve polynomial and rational inequalities. We begin by defining a quadratic inequality.

Quadratic inequalities are inequalities that can be written in any of the following forms:

$$ax^2 + bx + c \geq 0 \qquad ax^2 + bx + c \leq 0$$

$$ax^2 + bx + c > 0 \qquad ax^2 + bx + c < 0 \qquad \text{where } a \neq 0$$

Recall from Section 8.4 that the graph of a quadratic function defined by $f(x) = ax^2 + bx + c \, (a \neq 0)$ is a parabola that opens upward or downward. The quadratic inequality $f(x) > 0$ or equivalently $ax^2 + bx + c > 0$ is asking the question, "For what values of x is the value of the function positive (above the x-axis)?" The inequality $f(x) < 0$ or equivalently $ax^2 + bx + c < 0$ is asking, "For what values of x is the value of the function negative (below the x-axis)?" The graph of a quadratic function can be used to answer these questions.

Concepts

1. Solving Polynomial Inequalities
2. Solving Rational Inequalities
3. Inequalities with "Special Case" Solution Sets

| Example 1 | Using a Graph to Solve a Quadratic Inequality |

Use the graph of $f(x) = x^2 - 6x + 8$ in Figure 8-19 to solve the inequalities.

a. $x^2 - 6x + 8 < 0$ **b.** $x^2 - 6x + 8 > 0$

Solution:

From Figure 8-19, we see that the graph of $f(x) = x^2 - 6x + 8$ is a parabola opening upward. The function factors as $f(x) = (x - 2)(x - 4)$. The x-intercepts are at $x = 2$ and $x = 4$, and the y-intercept is $(0, 8)$.

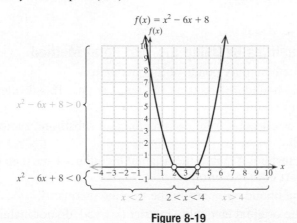

$f(x) = x^2 - 6x + 8$

Figure 8-19

a. The solution to $x^2 - 6x + 8 < 0$ is the set of real numbers x for which $f(x) < 0$. Graphically, this is the set of all x-values corresponding to the points where the parabola is below the x-axis (shown in red).

$$x^2 - 6x + 8 < 0 \qquad \text{for } \{x \,|\, 2 < x < 4\}, \text{ or equivalently, } (2, 4).$$

Skill Practice

Refer to the graph of $f(x) = x^2 + 3x - 4$ to solve the inequalities.

1. $x^2 + 3x - 4 > 0$
2. $x^2 + 3x - 4 < 0$

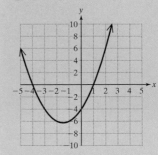

Answers

1. $\{x \,|\, x < -4 \text{ or } x > 1\}$; $(-\infty, -4) \cup (1, \infty)$
2. $\{x \,|\, -4 < x < 1\}$; $(-4, 1)$

b. The solution to $x^2 - 6x + 8 > 0$ is the set of x-values for which $f(x) > 0$. This is the set of x-values where the parabola is above the x-axis (shown in blue).

$$x^2 - 6x + 8 > 0 \qquad \text{for } \{x \mid x < 2 \quad \text{or} \quad x > 4\}, \text{ or equivalently in interval notation, } (-\infty, 2) \cup (4, \infty).$$

TIP: The inequalities in Example 1 are strict inequalities. Therefore, $x = 2$ and $x = 4$ (where $f(x) = 0$) are not included in the solution set. However, the corresponding inequalities using the symbols $\leq$ and $\geq$ do include the values where $f(x) = 0$.

The solution to $x^2 - 6x + 8 \leq 0$ is $\{x \mid 2 \leq x \leq 4\}$, or equivalently, $[2, 4]$

The solution to $x^2 - 6x + 8 \geq 0$ is $\{x \mid x \leq 2 \text{ or } x \geq 4\}$, or equivalently, $(-\infty, 2] \cup [4, \infty)$

Notice that $x = 2$ and $x = 4$ define the boundaries of the solutions sets to the inequalities given in Example 1. These values are the solutions to the related equation $x^2 - 6x + 8 = 0$. In general, to solve a polynomial inequality, we first need to find the boundary points.

Definition of Boundary Points

The **boundary points** of an inequality consist of the real solutions to the related equation and the points where the inequality is undefined.

Testing points in regions bounded by these points is the basis of the **test point method** to solve inequalities.

Solving Inequalities by Using the Test Point Method

Step 1   Find the boundary points of the inequality.

Step 2   Plot the boundary points on the number line. This divides the number line into regions.

Step 3   Select a test point from each region and substitute it into the original inequality.

 • If a test point makes the original inequality true, then that region is part of the solution set.

Step 4   Test the boundary points in the original inequality.

 • If the original inequality is strict ($<$ or $>$), do not include the boundary points in the solution set.

 • If the original inequality is defined using $\leq$ or $\geq$, then include the boundary points that are well defined within the inequality.

Note: Any boundary point that makes an expression within the inequality undefined must *always* be excluded from the solution set.

Example 2 Solving a Quadratic Inequality by Using the Test Point Method

Solve the inequality by using the test point method. $2x^2 + 5x < 12$

Solution:

$2x^2 + 5x < 12$ **Step 1:** Find the boundary points. Because polynomials are defined for all values of x, the only boundary points are the real solutions to the related equation.

$2x^2 + 5x = 12$ Solve the related equation.

$2x^2 + 5x - 12 = 0$

$(2x - 3)(x + 4) = 0$

$x = \frac{3}{2}$ $x = -4$ The boundary points are $\frac{3}{2}$ and -4.

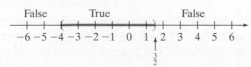

| Region I | Region II | Region III |

Step 2: Plot the boundary points.

Step 3: Select a test point from each region.

Test $x = -5$

$2x^2 + 5x < 12$

$2(-5)^2 + 5(-5) \overset{?}{<} 12$

$50 - 25 \overset{?}{<} 12$

$25 \overset{?}{<} 12$ False

Test $x = 0$

$2x^2 + 5x < 12$

$2(0)^2 + 5(0) \overset{?}{<} 12$

$0 \overset{?}{<} 12$

$0 \overset{?}{<} 12$ True

Test $x = 2$

$2x^2 + 5x < 12$

$2(2)^2 + 5(2) \overset{?}{<} 12$

$8 + 10 \overset{?}{<} 12$

$18 \overset{?}{<} 12$ False

The strict inequality, $<$, excludes values of x for which $2x^2 + 5x = 12$. This implies that the boundary points are not included in the solution set.

Step 4: Test the boundary points.

The solution is $\{x \mid -4 < x < \frac{3}{2}\}$ or equivalently in interval notation $(-4, \frac{3}{2})$.

Calculator Connections

Topic: Analyzing a Quadratic Inequality Graphically

Graph $Y_1 = 2x^2 + 5x$ and $Y_2 = 12$.
Notice that $Y_1 < Y_2$ for $\{x \mid -4 < x < \frac{3}{2}\}$.

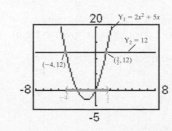

Skill Practice

Solve the inequality.

3. $x^2 + x > 6$

Answer

3. $(-\infty, -3) \cup (2, \infty)$

Example 3   Solving a Polynomial Inequality by Using the Test Point Method

Solve the inequality by using the test point method.

$$x(x + 4)^2(x - 4) \geq 0$$

Solution:

$$x(x + 4)^2(x - 4) \geq 0$$

$$x(x + 4)^2(x - 4) = 0$$ **Step 1:** Find the boundary points.

$$x = 0 \qquad x = -4 \qquad x = 4$$

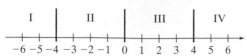

Step 2: Plot the boundary points.

Step 3: Select a test point from each interval.

Test $x = -5$: $-5(-5 + 4)^2(-5 - 4) \overset{?}{\geq} 0$ $45 \overset{?}{\geq} 0$ True

Test $x = -1$: $-1(-1 + 4)^2(-1 - 4) \overset{?}{\geq} 0$ $45 \overset{?}{\geq} 0$ True

Test $x = 1$: $1(1 + 4)^2(1 - 4) \overset{?}{\geq} 0$ $-75 \overset{?}{\geq} 0$ False

Test $x = 5$: $5(5 + 4)^2(5 - 4) \overset{?}{\geq} 0$ $405 \overset{?}{\geq} 0$ True

Step 4: The inequality symbol, $\geq$, includes equality. Therefore, include the boundary points in the solution set.

The solution is $\{x \mid x \leq 0 \text{ and } x \geq 4\}$, or equivalently in interval notation, $(-\infty, 0] \cup [4, \infty)$.

TIP: In Example 3, one side of the inequality is factored, and the other side is zero. For inequalities written in this form, we can use a sign chart to determine the sign of each factor. Then the sign of the product (bottom row) is easily determined.

Sign of x	$-$	$-$	$+$	$+$
Sign of $(x + 4)^2$	$+$	$+$	$+$	$+$
Sign of $(x - 4)$	$-$	$-$	$-$	$+$
Sign of $x(x + 4)^2(x - 4)$	$+$	$+$	$-$	$+$

$$-4 \qquad 0 \qquad 4$$

The solution to the inequality $x(x + 4)^2(x - 4) \geq 0$ includes the intervals for which the product is positive (shown in blue).

The solution is $(-\infty, 0] \cup [4, \infty)$.

Example 4	Solving a Polynomial Inequality Using the Test Point Method

Solve the inequality by using the test point method. $x^2 + x - 4 \geq 0$

Solution:

$x^2 + x - 4 \geq 0$

$x^2 + x - 4 = 0$

$x = \dfrac{-1 \pm \sqrt{(1)^2 - 4(1)(-4)}}{2(1)}$

Step 1: Find the boundary points in the related equation. Since this equation is not factorable, use the quadratic formula to find the solutions.

$x = \dfrac{-1 \pm \sqrt{17}}{2}$

$$x = \dfrac{-b \pm \sqrt{b^2 - 4ac}}{2a}$$

$x = \dfrac{-1 - \sqrt{17}}{2} \approx -2.56 \text{ and } x = \dfrac{-1 + \sqrt{17}}{2} \approx 1.56$

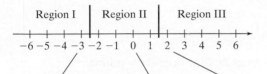

Region I   Region II   Region III

-6 -5 -4 -3 -2 -1 0 1 2 3 4 5 6

Step 2: Plot the boundary points.

Step 3: Select a test point from each region.

Test $x = -3$

$x^2 + x - 4 \geq 0$

$(-3)^2 + (-3) - 4 \overset{?}{\geq} 0$

$2 \overset{?}{\geq} 0$ True

Test $x = 0$

$x^2 + x - 4 \geq 0$

$(0)^2 + (0) - 4 \overset{?}{\geq} 0$

$-4 \overset{?}{\geq} 0$ False

Test $x = 2$

$x^2 + x - 4 \geq 0$

$(2)^2 + (2) - 4 \overset{?}{\geq} 0$

$2 \overset{?}{\geq} 0$ True

  True   False   True

-6 -5 -4 -3 -2 -1 0 1 2 3 4 5 6

Step 4: Test the boundary points. Both boundary points make the inequality true. Therefore, both boundary points are included in the solution set.

The solution is $\left\{ x \mid x \leq \dfrac{-1 - \sqrt{17}}{2} \text{ or } x \geq \dfrac{-1 + \sqrt{17}}{2} \right\}$ or equivalently

in interval notation: $\left(-\infty, \dfrac{-1 - \sqrt{17}}{2} \right] \cup \left[\dfrac{-1 + \sqrt{17}}{2}, \infty \right)$.

2. Solving Rational Inequalities

The test point method can be used to solve rational inequalities. A **rational inequality** is an inequality in which one or more terms is a rational expression. The solution set to a rational inequality must exclude all values of the variable that make the inequality undefined. That is, exclude all values that make the denominator equal to zero for any rational expression in the inequality.

Answer

5. $\left(\dfrac{3 - \sqrt{13}}{2}, \dfrac{3 + \sqrt{13}}{2} \right)$

Example 5 Solving a Rational Inequality by Using the Test Point Method

Solve the inequality. $\quad \dfrac{3}{x-1} > 0$

Solution:

$$\dfrac{3}{x-1} > 0$$

Step 1: Find the boundary points. Note that the inequality is undefined for $x = 1$, so $x = 1$ is a boundary point. To find any other boundary points, solve the related equation.

$$\dfrac{3}{x-1} = 0$$

 $(x-1) \cdot \left(\dfrac{3}{x-1} \right) = (x-1) \cdot 0$ Clear fractions.

$$3 = 0$$

There is no solution to the related equation.

The only boundary point is $x = 1$.

Region I   Region II

$$-5 \;-4 \;-3 \;-2 \;-1 \;\; 0 \;\; 1 \;\; 2 \;\; 3 \;\; 4 \;\; 5$$

Step 2: Plot boundary points.

Test $x = 0$:   **Test $x = 2$:**   **Step 3:** Select test points.

$$\dfrac{3}{(0)-1} \overset{?}{>} 0 \qquad \dfrac{3}{(2)-1} \overset{?}{>} 0$$

$$\dfrac{3}{-1} \overset{?}{>} 0 \text{ False} \qquad \dfrac{3}{1} > 0 \text{ True}$$

Step 4: The boundary point $x = 1$ cannot be included in the solution set because it is undefined in the original inequality.

False   True

$$-5 \;-4 \;-3 \;-2 \;-1 \;\; 0 \;\; 1 \;\; 2 \;\; 3 \;\; 4 \;\; 5$$

The solution is $\{x \mid x > 1\}$ or equivalently in interval notation, $(1, \infty)$.

TIP: Using a sign chart we see that the quotient of the factors 3 and $(x-1)$ is positive on the interval $(1, \infty)$.

Therefore, the solution to the inequality $\dfrac{3}{x-1} > 0$ is $(1, \infty)$.

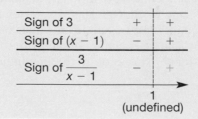

Sign of 3	$+$	$+$
Sign of $(x-1)$	$-$	$+$
Sign of $\dfrac{3}{x-1}$	$-$	$+$

1
(undefined)

Answer

6. $(-2, \infty)$

The solution to the inequality $\dfrac{3}{x-1} > 0$ can be confirmed from the graph of the

related rational function, $f(x) = \dfrac{3}{x-1}$ (see Figure 8-20).

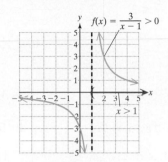

Figure 8-20

- The graph is above the x-axis where $f(x) = \dfrac{3}{x-1} > 0$ for $x > 1$ (shaded red).

- Also note that $x = 1$ cannot be included in the solution set because 1 is not in the domain of f.

Example 6 **Solving a Rational Inequality by Using the Test Point Method**

Solve the inequality by using the test point method. $\dfrac{x+2}{x-4} \le 3$

Solution:

$\dfrac{x+2}{x-4} \le 3$ **Step 1:** Find the boundary points. Note that the inequality is undefined for $x = 4$. Therefore, $x = 4$ is automatically a boundary point. To find any other boundary points, solve the related equation.

$$\dfrac{x+2}{x-4} = 3$$

$$(x-4)\left(\dfrac{x+2}{x-4}\right) = (x-4)(3) \qquad \text{Clear fractions.}$$

$$x + 2 = 3(x-4) \qquad \text{Solve for } x.$$

$$x + 2 = 3x - 12$$

$$-2x = -14$$

$$x = 7$$

The solution to the related equation is $x = 7$, and the inequality is undefined for $x = 4$. Therefore, the boundary points are $x = 4$ and $x = 7$.

Region I   Region II   Region III   **Step 2:** Plot boundary points.

$-3\ -2\ -1\ \ 0\ \ 1\ \ 2\ \ 3\ \ 4\ \ 5\ \ 6\ \ 7\ \ 8$ **Step 3:** Select test points.

Answer

7. $\left(-4, \dfrac{1}{2}\right]$

Test $x = 0$

$$\dfrac{x + 2}{x - 4} \leq 3$$

$$\dfrac{0 + 2}{0 - 4} \overset{?}{\leq} 3$$

$$-\dfrac{1}{2} \overset{?}{\leq} 3 \quad \text{True}$$

Test $x = 5$

$$\dfrac{x + 2}{x - 4} \leq 3$$

$$\dfrac{5 + 2}{5 - 4} \overset{?}{\leq} 3$$

$$\dfrac{7}{1} \overset{?}{\leq} 3 \quad \text{False}$$

Test $x = 8$

$$\dfrac{x + 2}{x - 4} \leq 3$$

$$\dfrac{8 + 2}{8 - 4} \overset{?}{\leq} 3$$

$$\dfrac{10}{4} \overset{?}{\leq} 3$$

$$\dfrac{5}{2} \overset{?}{\leq} 3 \quad \text{True}$$

Test $x = 4$:

$$\dfrac{x + 2}{x - 4} \leq 3$$

$$\dfrac{4 + 2}{4 - 4} \overset{?}{\leq} 3$$

$$\dfrac{6}{0} \overset{?}{\leq} 3 \quad \text{Undefined}$$

Test $x = 7$:

$$\dfrac{x + 2}{x - 4} \leq 3$$

$$\dfrac{7 + 2}{7 - 4} \overset{?}{\leq} 3$$

$$\dfrac{9}{3} \overset{?}{\leq} 3 \quad \text{True}$$

Step 4: Test the boundary points.

The boundary point $x = 4$ cannot be included in the solution set, because it is undefined in the inequality. The boundary point $x = 7$ makes the original inequality true and must be included in the solution set.

$$\begin{array}{c}\xleftarrow{\hspace{0.5em}} \overset{}{|} \ \overset{}{|} \ \overset{}{|} \ \overset{}{|} \ \overset{}{|} \ \overset{}{|} \ \overset{)}{|} \ \overset{}{|} \ \overset{}{|} \ \overset{[}{|} \ \xrightarrow{\hspace{0.5em}} \\ -3\ -2\ -1\ \ 0\ \ 1\ \ 2\ \ 3\ \ 4\ \ 5\ \ 6\ \ 7\ \ 8 \end{array}$$

The solution is $\{x \mid x < 4 \text{ or } x \geq 7\}$, or equivalently in interval notation, $(-\infty, 4) \cup [7, \infty)$.

3. Inequalities with "Special Case" Solution Sets

The solution to an inequality is often one or more regions on the real number line. Sometimes, however, the solution to an inequality may be a single point on the number line, the empty set, or the set of all real numbers.

Skill Practice

Solve the inequalities.
8. $x^2 - 4x + 4 \geq 0$
9. $x^2 - 4x + 4 > 0$
10. $x^2 - 4x + 4 \leq 0$
11. $x^2 - 4x + 4 < 0$

Example 7   Solving Inequalities

Solve the inequalities.

a. $x^2 + 6x + 9 \geq 0$ **b.** $x^2 + 6x + 9 > 0$

c. $x^2 + 6x + 9 \leq 0$ **d.** $x^2 + 6x + 9 < 0$

Solution:

a. $x^2 + 6x + 9 \geq 0$ Notice that $x^2 + 6x + 9$ is a perfect square trinomial.

      $(x + 3)^2 \geq 0$ Factor $x^2 + 6x + 9 = (x + 3)^2$.

The quantity $(x + 3)^2$ is a perfect square and is greater than or equal to zero for all real numbers, x. The solution is all real numbers, $(-\infty, \infty)$.

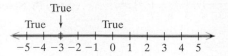

Answers

8. All real numbers; $(-\infty, \infty)$
9. $(-\infty, 2) \cup (2, \infty)$ **10.** $\{2\}$
11. No solution

b. $x^2 + 6x + 9 > 0$

$(x + 3)^2 > 0$ This is the same inequality as in part (a) with the exception that the inequality is strict. The solution set does not include the point where $x^2 + 6x + 9 = 0$. Therefore, the boundary point $x = -3$ is *not* included in the solution set.

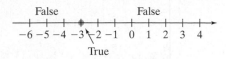

The solution set is $\{x \mid x \neq -3\}$ or equivalently $(-\infty, -3) \cup (-3, \infty)$.

c. $x^2 + 6x + 9 \leq 0$

$(x + 3)^2 \leq 0$

A perfect square cannot be less than zero. However, $(x + 3)^2$ is equal to zero at $x = -3$. Therefore, the solution set is $\{-3\}$.

False   False

$-6\ -5\ -4\ -3\ -2\ -1\ \ 0\ \ 1\ \ 2\ \ 3\ \ 4$

True

d. $x^2 + 6x + 9 < 0$

$(x + 3)^2 < 0$

A perfect square cannot be negative; therefore, there are no real numbers x such that $(x + 3)^2 < 0$. There is no solution.

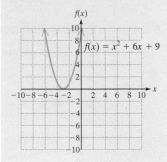

TIP: The graph of $f(x) = x^2 + 6x + 9$, or equivalently $f(x) = (x + 3)^2$, is equal to zero at $x = -3$ and positive (above the x-axis) for all other values of x in its domain.

Section 8.6   Practice Exercises

Vocabulary and Key Concepts

1. a. An inequality of the form $ax^2 + bx + c > 0$ $(a \neq 0)$ is an example of a _____ inequality.

b. The boundary points of an inequality consist of the real _____ to the related equation and the points where the inequality is _____.

c. In solving an inequality by using the _____ _____ method, a point is selected from each interval formed by the boundary points and substituted into the original inequality.

d. If a test point makes the original inequality (true/false) then that interval is part of the solution set.

e. The inequality $\dfrac{4}{x + 7} > 0$ is an example of a _____ inequality.

f. The solution set to a rational inequality must exclude all values that make the _____ equal to zero for any rational expression in the original inequality.

g. True or False: The inequality $(x + 3)^2 \leq -4$ has no solution.

h. True or False: The solution to the inequality $(x + 3)^2 \geq -4$ is $(-\infty, \infty)$.

i. The solution set to the inequality $(x + 3)^2 > 0$ (includes/excludes) -3, whereas the solution set to the inequality $(x + 3)^2 \leq 0$ (includes/excludes) -3.

Review Exercises

For Exercises 2–8, solve the compound inequalities. Write the solutions in interval notation.

2. $6x - 10 > 8$　or　$8x + 2 < 5$

3. $3(a - 1) + 2 > 0$　or　$2a > 5a + 12$

4. $5(k - 2) > -25$　and　$7(1 - k) > 7$

5. $2y + 4 \geq 10$　and　$5y - 3 \leq 13$

6. $0 < 3(x + 1) \leq 4$

7. $6 \geq 4 - 2x \geq -2$

8. $-4 > 5 - x > -6$

Concept 1: Solving Polynomial Inequalities

For Exercises 9–12, estimate from the graph the intervals for which the inequality is true. **(See Example 1.)**

 9.

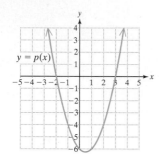

a. $p(x) > 0$　　**b.** $p(x) < 0$

c. $p(x) \leq 0$　　**d.** $p(x) \geq 0$

10.

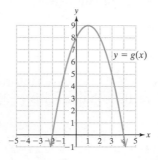

a. $g(x) > 0$　　**b.** $g(x) < 0$

c. $g(x) \leq 0$　　**d.** $g(x) \geq 0$

11.

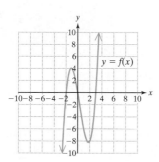

a. $f(x) > 0$　　**b.** $f(x) < 0$

c. $f(x) \leq 0$　　**d.** $f(x) \geq 0$

12.

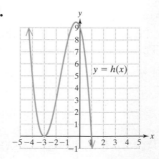

a. $h(x) > 0$　　**b.** $h(x) < 0$

c. $h(x) \leq 0$　　**d.** $h(x) \geq 0$

For Exercises 13–18, solve the equation and related inequalities. **(See Examples 2–4.)**

13. a. $3(4 - x)(2x + 1) = 0$

b. $3(4 - x)(2x + 1) < 0$

c. $3(4 - x)(2x + 1) > 0$

14. a. $5(y + 6)(3 - 5y) = 0$

b. $5(y + 6)(3 - 5y) < 0$

c. $5(y + 6)(3 - 5y) > 0$

15. a. $x^2 + 7x = 30$

b. $x^2 + 7x < 30$

c. $x^2 + 7x > 30$

16. a. $q^2 - 4q = 5$

b. $q^2 - 4q \leq 5$

c. $q^2 - 4q \geq 5$

17. a. $2p(p - 2) = p + 3$

b. $2p(p - 2) \leq p + 3$

c. $2p(p - 2) \geq p + 3$

18. a. $3w(w + 4) = 10 - w$

b. $3w(w + 4) < 10 - w$

c. $3w(w + 4) > 10 - w$

For Exercises 19–38, solve the polynomial inequality. Write the answer in interval notation. **(See Examples 2–4.)**

19. $(t - 7)(t + 1) < 0$

20. $(p - 4)(p + 2) > 0$

21. $-6(4 + 2x)(5 - x) > 0$

22. $-8(2t + 5)(6 - t) < 0$

23. $m(m + 1)^2(m + 5) \leq 0$

24. $w^2(3 - w)(w + 2) \geq 0$

25. $a^2 - 12a \leq -32$

26. $w^2 + 20w \geq -64$

27. $5x^2 - 2x - 1 > 0$

28. $3x^2 + 2x - 4 < 0$

29. $x^2 + 3x \leq 6$

30. $x^2 - 8 \leq 3x$

31. $b^2 - 121 < 0$

32. $c^2 - 25 < 0$

33. $3p(p - 2) - 3 \geq 2p$

34. $2t(t + 3) - t \leq 12$

35. $x^3 - x^2 \leq 12x$

36. $x^3 + 36 > 4x^2 + 9x$

37. $w^3 + w^2 > 4w + 4$

38. $2p^3 - 5p^2 \leq 3p$

Concept 2: Solving Rational Inequalities

For Exercises 39–42, solve the equation and related inequalities. **(See Examples 5 and 6.)**

39. a. $\dfrac{10}{x - 5} = 5$

40. a. $\dfrac{8}{a + 1} = 4$

41. a. $\dfrac{z + 2}{z - 6} = -3$

42. a. $\dfrac{w - 8}{w + 6} = 2$

b. $\dfrac{10}{x - 5} < 5$

b. $\dfrac{8}{a + 1} > 4$

b. $\dfrac{z + 2}{z - 6} \leq -3$

b. $\dfrac{w - 8}{w + 6} \leq 2$

c. $\dfrac{10}{x - 5} > 5$

c. $\dfrac{8}{a + 1} < 4$

c. $\dfrac{z + 2}{z - 6} \geq -3$

c. $\dfrac{w - 8}{w + 6} \geq 2$

For Exercises 43–54, solve the rational inequalities. Write the answer in interval notation. **(See Examples 5 and 6.)**

43. $\dfrac{2}{x - 1} \geq 0$

44. $\dfrac{-3}{x + 2} \leq 0$

45. $\dfrac{b + 4}{b - 4} > 0$

46. $\dfrac{a + 1}{a - 3} < 0$

47. $\dfrac{3}{2x - 7} < -1$

48. $\dfrac{8}{4x + 9} > 1$

49. $\dfrac{x + 1}{x - 5} \geq 4$

50. $\dfrac{x - 2}{x + 6} \leq 5$

51. $\dfrac{1}{x} \leq 2$

52. $\dfrac{1}{x} \geq 3$

53. $\dfrac{(x + 2)^2}{x} > 0$

54. $\dfrac{(x - 3)^2}{x} < 0$

Concept 3: Inequalities with "Special Case" Solution Sets

For Exercises 55–66, solve the inequalities. **(See Example 7.)**

55. $x^2 + 10x + 25 \geq 0$

56. $x^2 + 6x + 9 < 0$

57. $x^2 + 2x + 1 < 0$

58. $x^2 + 8x + 16 \geq 0$

59. $x^4 + 3x^2 \leq 0$

60. $x^4 + 2x^2 \leq 0$

61. $x^2 + 12x + 36 < 0$

62. $x^2 + 12x + 36 \geq 0$

63. $x^2 + 3x + 5 < 0$

64. $2x^2 + 3x + 3 > 0$

65. $-5x^2 + x < 1$

66. $-3x^2 - x > 6$

Mixed Exercises

For Exercises 67–90, identify the inequality as one of the following types: linear, quadratic, rational, or polynomial (degree > 2). Then solve the inequality and write the answer in interval notation.

67. $2y^2 - 8 \leq 24$

68. $8p^2 - 18 > 0$

69. $(5x + 2)^2 > -4$

70. $(3 - 7x)^2 < -1$

71. $4(x - 2) < 6x - 3$

72. $-7(3 - y) > 4 + 2y$

73. $\dfrac{2x + 3}{x + 1} \le 2$

74. $\dfrac{5x - 1}{x + 3} \ge 5$

75. $4x^3 - 40x^2 + 100x > 0$

76. $2y^3 - 12y^2 + 18y < 0$

77. $2p^3 > 4p^2$

78. $w^3 \le 5w^2$

79. $-3(x + 4)^2(x - 5) \ge 0$

80. $5x(x - 2)(x - 6)^2 \ge 0$

81. $x^2 - 2 < 0$

82. $y^2 - 3 > 0$

83. $x^2 + 5x - 2 \ge 0$

84. $t^2 + 7t + 3 \le 0$

85. $\dfrac{a + 2}{a - 5} \ge 0$

86. $\dfrac{t + 1}{t - 2} \le 0$

87. $2 \ge t - 3$

88. $-5p + 8 < p$

89. $4x^2 + 2 \ge -x$

90. $5x - 4 \ge 2x^2$

Graphing Calculator Exercises

91. To solve the inequality $\dfrac{x}{x - 2} > 0$ enter Y_1 as $x/(x - 2)$ and determine where the graph is above the x-axis. Write the solution in interval notation.

92. To solve the inequality $\dfrac{x}{x - 2} < 0$ enter Y_1 as $x/(x - 2)$ and determine where the graph is below the x-axis. Write the solution in interval notation.

93. To solve the inequality $x^2 - 1 < 0$, enter Y_1 as $x^2 - 1$ and determine where the graph is below the x-axis. Write the solution in interval notation.

94. To solve the inequality $x^2 - 1 > 0$, enter Y_1 as $x^2 - 1$ and determine where the graph is above the x-axis. Write the solution in interval notation.

For Exercises 95–98, determine the solution by graphing the inequalities.

95. $x^2 + 10x + 25 \le 0$

96. $-x^2 + 10x - 25 \ge 0$

97. $\dfrac{8}{x^2 + 2} < 0$

98. $\dfrac{-6}{x^2 + 3} > 0$

Problem Recognition Exercises

Recognizing Equations and Inequalities

At this point, you have learned how to solve a variety of equations and inequalities. Being able to distinguish the type of problem being posed is the first step in successfully solving it.

For Exercises 1–20,

a. Identify the problem type. Choose from

- linear equation
- quadratic equation
- rational equation
- absolute value equation
- radical equation
- equation quadratic in form

- polynomial equation
- linear inequality
- polynomial inequality
- rational inequality
- absolute value inequality
- compound inequality

b. Solve the equation or inequality. Write the solution to each inequality in interval notation if possible.

1. $(z^2 - 4)^2 - (z^2 - 4) - 12 = 0$

2. $3 + |4t - 1| < 6$

3. $2y(y - 4) \le 5 + y$

4. $\sqrt[3]{11x - 3} + 4 = 6$

5. $-5 = -|w - 4|$

6. $\dfrac{5}{x - 2} + \dfrac{3}{x + 2} = 1$

7. $m^3 + 5m^2 - 4m - 20 \ge 0$

8. $-x - 4 > -5$ and $2x - 3 \le 23$

9. $5 - 2[3 - (x - 4)] \le 3x + 14$

10. $|2x - 6| = |x + 3|$

11. $\dfrac{3}{x - 2} \le 1$

12. $9 < |x + 4|$

13. $\sqrt{t + 8} - 6 = t$

14. $(4x - 3)^2 = -10$

15. $-4 - x > 2$ or $8 < 2x$

16. $\dfrac{1}{3}x - 2 = \dfrac{3}{4} + \dfrac{5}{6}x$

17. $x^2 - 10x \le -25$

18. $\dfrac{10}{x^2 + 1} < 0$

19. $x - 13\sqrt{x} + 36 = 0$

20. $x^4 - 13x^2 + 36 = 0$

Group Activity

Creating a Quadratic Model of the Form $y = a(x - h)^2 + k$

Estimated time: 20 minutes

Group Size: 3

In the Group Activity in Chapter 4, we modeled the path of a softball that was thrown from right field to third base. The data points are given in the table. The values of t represent the time in seconds after the ball was released, and y represents the height of the ball in feet.

Time (sec) t	0	0.2	0.4	0.6	0.8	1.0	1.2	1.4	1.6	1.8	2.0
Height (ft) y	5	11	16	19	21	22	21	19	16	12	6

1. Graph the points defined by (t, y) in the table.

2. Select a point that you think best represents the vertex of the parabola. Label this point (h, k).

$(h, k) = $ _____

3. Substitute the values of h and k into the formula $y = a(t - h)^2 + k$, and write the equation here.

$y = a(t - $ _____$)^2 + $ _____

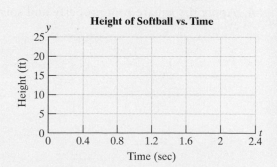

Height of Softball vs. Time

4. Choose a different point (t, y) from the graph. Substitute these values into the equation in step 3 and then solve for a.

5. Substitute the values of h, k, and a that you found in steps 2 and 4 into the equation $y = a(t - h)^2 + k$.

6. Use the function found in step 5 to approximate the height of the ball after 0.7 sec.

7. Use the function found in step 5 to approximate the height of the ball after 1.8 sec. How well does this value match the observed value of 12 ft?

Chapter 8   Summary

Section 8.1   Square Root Property and Completing the Square

Key Concepts	Examples

Key Concepts

The **square root property** states that

If $x^2 = k$ then $x = \pm\sqrt{k}$

Follow these steps to solve a quadratic equation in the form $ax^2 + bx + c = 0 \, (a \neq 0)$ by completing the square and applying the square root property:

1. Divide both sides by a to make the leading coefficient 1.
2. Isolate the variable terms on one side of the equation.
3. Complete the square: Add the square of one-half the linear term coefficient to both sides of the equation. Then factor the resulting perfect square trinomial.
4. Apply the square root property and solve for x.

Examples

Example 1

$$(x - 5)^2 = -13$$
$$x - 5 = \pm\sqrt{-13} \qquad \text{(square root property)}$$
$$x = 5 \pm i\sqrt{13}$$

Example 2

$$2x^2 - 6x - 5 = 0$$
$$\frac{2x^2}{2} - \frac{6x}{2} - \frac{5}{2} = \frac{0}{2}$$
$$x^2 - 3x = \frac{5}{2}$$
$$\text{Note: } \left[\frac{1}{2} \cdot (-3)\right]^2 = \frac{9}{4}$$
$$x^2 - 3x + \frac{9}{4} = \frac{5}{2} + \frac{9}{4}$$
$$x^2 - 3x + \frac{9}{4} = \frac{10}{4} + \frac{9}{4}$$
$$\left(x - \frac{3}{2}\right)^2 = \frac{19}{4}$$
$$x - \frac{3}{2} = \pm\sqrt{\frac{19}{4}}$$
$$x = \frac{3}{2} \pm \frac{\sqrt{19}}{2} \qquad \text{or} \qquad x = \frac{3 \pm \sqrt{19}}{2}$$

Section 8.2   Quadratic Formula and Applications

Key Concepts

The solutions to a quadratic equation $ax^2 + bx + c = 0 \ (a \neq 0)$ are given by the **quadratic formula**

$$x = \frac{-b \pm \sqrt{b^2 - 4ac}}{2a}$$

The **discriminant** of a quadratic equation $ax^2 + bx + c = 0$ is $b^2 - 4ac$. If a, b, and c are rational numbers, then

1. If $b^2 - 4ac > 0$, then there will be two real solutions. Moreover,
   a. If $b^2 - 4ac$ is a perfect square, the solutions will be rational numbers.
   b. If $b^2 - 4ac$ is not a perfect square, the solutions will be irrational numbers.
2. If $b^2 - 4ac < 0$, then there will be two imaginary solutions.
3. If $b^2 - 4ac = 0$, then there will be one rational solution.

Three methods to solve a quadratic equation are:

1. Factoring and applying the zero product rule.
2. Completing the square and applying the square root property.
3. Using the quadratic formula.

Example

Example 1

$$3x^2 - 2x + 4 = 0$$

$$a = 3 \qquad b = -2 \qquad c = 4$$

$$x = \frac{-(-2) \pm \sqrt{(-2)^2 - 4(3)(4)}}{2(3)}$$

$$= \frac{2 \pm \sqrt{4 - 48}}{6}$$

$$= \frac{2 \pm \sqrt{-44}}{6} \qquad \text{The discriminant is } -44. \text{ Therefore, there will be two imaginary solutions.}$$

$$= \frac{2 \pm 2i\sqrt{11}}{6}$$

$$= \frac{1 \pm i\sqrt{11}}{3}$$

$$= \frac{1}{3} \pm \frac{\sqrt{11}}{3}i$$

Section 8.3   Equations in Quadratic Form

Key Concepts

Substitution can be used to solve equations that are in quadratic form.

Examples

Example 1

$$x^{2/3} - x^{1/3} - 12 = 0 \quad \text{Let } u = x^{1/3}.$$

$$u^2 - u - 12 = 0$$

$$(u - 4)(u + 3) = 0$$

$$u = 4 \qquad \text{or} \qquad u = -3$$

$$x^{1/3} = 4 \qquad \text{or} \qquad x^{1/3} = -3$$

$$x = 64 \qquad \text{or} \qquad x = -27 \quad \text{Cube both sides.}$$

Section 8.4	**Graphs of Quadratic Functions**

Key Concepts

A quadratic function of the form $f(x) = x^2 + k$ shifts the graph of $y = x^2$ up k units if $k > 0$ and down $|k|$ units if $k < 0$.

Examples

Example 1

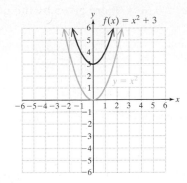

A quadratic function of the form $f(x) = (x - h)^2$ shifts the graph of $y = x^2$ right h units if $h > 0$ and left $|h|$ units if $h < 0$.

Example 2

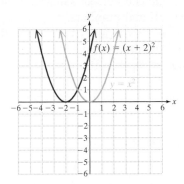

The graph of a quadratic function of the form $f(x) = ax^2$ is a parabola that opens upward when $a > 0$ and opens downward when $a < 0$. If $|a| > 1$, the graph of $y = x^2$ is stretched vertically by a factor of $|a|$. If $0 < |a| < 1$, the graph of $y = x^2$ is shrunk vertically by a factor of $|a|$.

Example 3

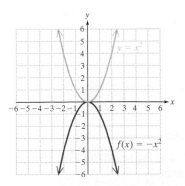

A quadratic function of the form $f(x) = a(x - h)^2 + k$ has vertex (h, k). If $a > 0$, the vertex represents the minimum point. If $a < 0$, the vertex represents the maximum point.

Example 4

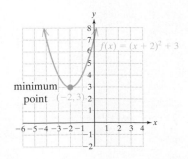

Section 8.5 Vertex of a Parabola: Applications and Modeling

Key Concepts

Completing the square is a technique used to write a quadratic function $f(x) = ax^2 + bx + c$ $(a \neq 0)$ in the form $f(x) = a(x - h)^2 + k$ for the purpose of identifying the vertex (h, k).

The **vertex formula** finds the vertex of a quadratic function $f(x) = ax^2 + bx + c$ $(a \neq 0)$.

The vertex is

$$\left(\frac{-b}{2a}, \frac{4ac - b^2}{4a}\right) \quad \text{or} \quad \left(\frac{-b}{2a}, f\left(\frac{-b}{2a}\right)\right)$$

Examples

Example 1

$$f(x) = 3x^2 + 6x + 11$$
$$= 3(x^2 + 2x \qquad) + 11$$
$$= 3(x^2 + 2x + 1 - 1) + 11$$
$$= 3(x^2 + 2x + 1) - 3 + 11$$
$$= 3(x + 1)^2 + 8$$
$$= 3[x - (-1)]^2 + 8$$

The vertex is $(-1, 8)$. Because $a = 3 > 0$, the parabola opens upward and the vertex $(-1, 8)$ is a minimum point.

Example 2

$$f(x) = 3x^2 + 6x + 11$$
$$a = 3 \qquad b = 6 \qquad c = 11$$
$$x = \frac{-6}{2(3)} = -1$$
$$f(-1) = 3(-1)^2 + 6(-1) + 11 = 8$$

The vertex is $(-1, 8)$.

| Section 8.6 | Polynomial and Rational Inequalities |

Key Concepts

The Test Point Method to Solve Polynomial and Rational Inequalities

1. Find the boundary points of the inequality. (Boundary points are the real solutions to the related equation and points where the inequality is undefined.)
2. Plot the boundary points on the number line. This divides the number line into regions.
3. Select a test point from each region and substitute it into the original inequality.
   - If a test point makes the original inequality true, then that region is part of the solution set.
4. Test the boundary points in the original inequality.
   - If the original inequality is strict ($<$ or $>$), do not include the boundary points in the solution set.
   - If the original inequality is defined using $\leq$ or $\geq$, then include the boundary points that are well defined within the inequality.

Note: Any boundary point that makes an expression within the inequality undefined must *always* be excluded from the solution set.

Examples

Example 1

$$\frac{28}{2x - 3} \leq 4$$ The inequality is undefined for $x = \frac{3}{2}$. Find other possible boundary points by solving the related equation.

$$\frac{28}{2x - 3} = 4$$ Related equation

$$(2x - 3) \cdot \frac{28}{2x - 3} = (2x - 3) \cdot 4$$

$$28 = 8x - 12$$

$$40 = 8x$$

$$x = 5$$

The boundaries are $x = \dfrac{3}{2}$ and $x = 5$

Region I: Test $x = 1$: $\dfrac{28}{2(1) - 3} \overset{?}{\leq} 4$ True

Region II: Test $x = 2$: $\dfrac{28}{2(2) - 3} \overset{?}{\leq} 4$ False

Region III: Test $x = 6$: $\dfrac{28}{2(6) - 3} \overset{?}{\leq} 4$ True

The boundary point $x = \frac{3}{2}$ is not included because $\dfrac{28}{2x - 3}$ is undefined there. The boundary $x = 5$ does check in the original inequality.

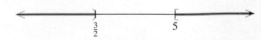

The solution is $\left(-\infty, \frac{3}{2}\right) \cup [5, \infty)$.

Chapter 8   Review Exercises

Section 8.1

For Exercises 1–8, solve the equations by using the square root property.

1. $x^2 = 5$ **2.** $2y^2 = -8$

3. $a^2 = 81$ **4.** $3b^2 = -19$

5. $(x - 2)^2 = 72$ **6.** $(2x - 5)^2 = -9$

7. $(3y - 1)^2 = 3$ **8.** $3(m - 4)^2 = 15$

 9. The length of each side of an equilateral triangle is 10 in. Find the height of the triangle. Round the answer to the nearest tenth of an inch.

10 in. 10 in. h ← 10 in. →

10. Use the square root property to find the length of the sides of a square whose area is 81 in^2.

 11. Use the square root property to find the length of the sides of a square whose area is 150 in^2. Round the answer to the nearest tenth of an inch.

For Exercises 12–15, find the value of n so that the expression is a perfect square trinomial. Then factor the trinomial.

12. $x^2 + 16x + n$ **13.** $x^2 - 9x + n$

14. $y^2 + \dfrac{1}{2}y + n$ **15.** $z^2 - \dfrac{2}{5}z + n$

For Exercises 16–21, solve the equation by completing the square and applying the square root property.

16. $w^2 + 4w + 13 = 0$ **17.** $4y^2 - 12y + 13 = 0$

18. $3x^2 + 2x = 1$ **19.** $b^2 + \dfrac{7}{2}b = 2$

20. $2x^2 = 12x + 6$ **21.** $-t^2 + 8t - 25 = 0$

22. Solve for r. $V = \pi r^2 h$ $(r > 0)$

23. Solve for s. $A = 6s^2$ $(s > 0)$

Section 8.2

24. Explain how the discriminant can determine the type and number of solutions to a quadratic equation with rational coefficients.

For Exercises 25–30, determine the type (rational, irrational, or imaginary) and number of solutions for the equations by using the discriminant.

25. $x^2 - 5x = -6$ **26.** $2y^2 = -3y$

27. $z^2 + 23 = 17z$ **28.** $a^2 + a + 1 = 0$

29. $10b + 1 = -25b^2$ **30.** $3x^2 + 15 = 0$

For Exercises 31–38, solve the equations by using the quadratic formula.

31. $y^2 - 4y + 1 = 0$ **32.** $m^2 - 5m + 25 = 0$

33. $6a(a - 1) = 10 + a$ **34.** $3x(x - 3) = x - 8$

35. $b^2 - \dfrac{4}{25} = \dfrac{3}{5}b$ **36.** $k^2 + 0.4k = 0.05$

37. $-32 + 4x - x^2 = 0$ **38.** $8y - y^2 = 10$

For Exercises 39–42, solve using any method.

39. $3x^2 - 4x = 6$ **40.** $\dfrac{w}{8} - \dfrac{2}{w} = \dfrac{3}{4}$

41. $y^2 + 14y = -46$ **42.** $(a + 1)^2 = 11$

43. The landing distance that a certain plane will travel on a runway is determined by the initial landing speed at the instant the plane touches down. The function D relates landing distance in feet to initial landing speed s:

$$D(s) = \frac{1}{10}s^2 - 3s + 22 \text{ for } s \geq 50$$

where s is in feet per second.

a. Find the landing distance for a plane traveling 150 ft/sec at touchdown.

b. If the landing speed is too fast, the pilot may run out of runway. If the speed is too slow, the plane may stall. Find the maximum initial landing speed of a plane for a runway that is 1000 ft long. Round to one decimal place.

44. The recent population of Kenya (in thousands) can be approximated by $P(t) = 4.62t^2 + 564.6t + 13,128$, where t is the number of years since 1974.

    a. If this trend continues, approximate the number of people in Kenya for the year 2025.

    b. In what year after 1974 will the population of Kenya reach 50 million? (*Hint:* 50 million equals 50,000 thousand.)

Section 8.3

For Exercises 45–54, solve the equations.

45. $x - 4\sqrt{x} - 21 = 0$

46. $n - 6\sqrt{n} + 8 = 0$

47. $y^4 - 11y^2 + 18 = 0$

48. $2m^4 - m^2 - 3 = 0$

49. $t^{2/5} + t^{1/5} - 6 = 0$

50. $p^{2/5} - 3p^{1/5} + 2 = 0$

51. $\dfrac{2t}{t + 1} + \dfrac{-3}{t - 2} = 1$

52. $\dfrac{1}{m - 2} - \dfrac{m}{m + 3} = 2$

53. $(x^2 + 5)^2 + 2(x^2 + 5) - 8 = 0$

54. $(x^2 - 3)^2 - 5(x^2 - 3) + 4 = 0$

Section 8.4

For Exercises 55–62, graph the functions and write the domain and range in interval notation.

55. $g(x) = x^2 - 5$

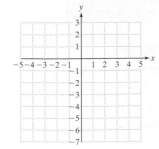

56. $f(x) = x^2 + 3$

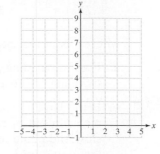

57. $h(x) = (x - 5)^2$

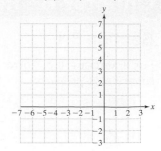

58. $k(x) = (x + 3)^2$

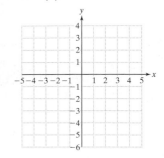

59. $m(x) = -2x^2$

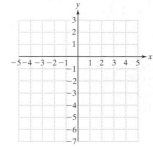

60. $n(x) = -4x^2$

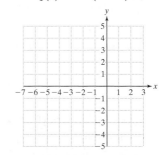

61. $p(x) = -2(x - 5)^2 - 5$

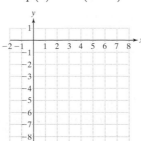

62. $q(x) = -4(x + 3)^2 + 3$

For Exercises 63 and 64, write the coordinates of the vertex of the parabola and determine if the vertex is a maximum point or a minimum point. Then write the maximum or the minimum value.

63. $t(x) = \dfrac{1}{3}(x - 4)^2 + \dfrac{5}{3}$

64. $s(x) = -\dfrac{5}{7}(x - 1)^2 - \dfrac{1}{7}$

For Exercises 65 and 66, write the equation of the axis of symmetry of the parabola.

65. $a(x) = -\dfrac{3}{2}\left(x + \dfrac{2}{11}\right)^2 - \dfrac{4}{13}$

66. $w(x) = -\dfrac{4}{3}\left(x - \dfrac{3}{16}\right)^2 + \dfrac{2}{9}$

Section 8.5

For Exercises 67–70, write the function in the form $f(x) = a(x - h)^2 + k$ by completing the square. Then write the coordinates of the vertex.

67. $z(x) = x^2 - 6x + 7$ **68.** $b(x) = x^2 - 4x - 44$

69. $p(x) = -5x^2 - 10x - 13$

70. $q(x) = -3x^2 - 24x - 54$

For Exercises 71–74, find the coordinates of the vertex of each function by using the vertex formula.

71. $f(x) = -2x^2 + 4x - 17$

72. $g(x) = -4x^2 - 8x + 3$

73. $m(x) = 3x^2 - 3x + 11$

74. $n(x) = 3x^2 + 2x - 7$

75. For the quadratic equation $y = \dfrac{3}{4}x^2 - 3x$,

    a. Write the coordinates of the vertex.

    b. Find the x- and y-intercepts.

    c. Use this information to sketch a graph of the parabola.

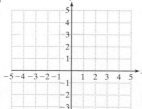

76. For the quadratic equation $y = -(x + 2)^2 + 4$,

    a. Write the coordinates of the vertex.

    b. Find the x- and y-intercepts.

    c. Use this information to sketch a graph of the parabola.

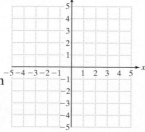

77. The height $h(t)$ (in feet) of a projectile fired vertically into the air from the ground is given by the equation $h(t) = -16t^2 + 96t$, where t represents the number of seconds that the projectile has been in the air. How long will it take the projectile to reach its maximum height?

78. The weekly profit, $P(x)$ (in dollars), for a catering service is given by $P(x) = -0.053x^2 + 15.9x + 7.5$. In this context, x is the number of meals prepared.

    a. Find the number of meals that should be prepared to obtain the maximum profit.

    b. What is the maximum profit?

79. Write an equation of a parabola that passes through the points $(-3, -4)$, $(-2, -5)$, and $(1, 4)$

Section 8.6

80. Solve the equation and inequalities. How do your answers to parts (a), (b), and (c) relate to the graph of $g(x) = x^2 - 4$?

    a. $x^2 - 4 = 0$

    b. $x^2 - 4 < 0$

    c. $x^2 - 4 > 0$

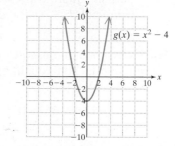

81. Solve the equation and inequalities. How do your answers to parts (a), (b), (c), and (d) relate to the graph of $k(x) = \dfrac{4x}{(x - 2)}$?

    a. $\dfrac{4x}{x - 2} = 0$

    b. For which values is $k(x)$ undefined?

    c. $\dfrac{4x}{x - 2} \geq 0$

    d. $\dfrac{4x}{x - 2} \leq 0$

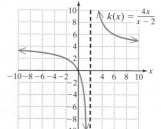

For Exercises 82–93, solve the inequalities. Write the answers in interval notation.

82. $w^2 - 4w - 12 < 0$ **83.** $t^2 + 6t + 9 \geq 0$

84. $\dfrac{12}{x + 2} \leq 6$ **85.** $\dfrac{8}{p - 1} \geq -4$

86. $3y(y - 5)(y + 2) > 0$

90. $\dfrac{w + 1}{w - 3} > 1$

91. $\dfrac{2a}{a + 3} \le 2$

87. $-3c(c + 2)(2c - 5) < 0$

92. $t^2 + 10t + 25 \le 0$

88. $-x^2 - 4x \le 1$

89. $y^2 + 4y > 5$

93. $-x^2 - 4x < 4$

Chapter 8   Test

For Exercises 1–3, solve the equation by using the square root property.

1. $(x + 3)^2 = 25$

2. $(p - 2)^2 = 12$

3. $(m + 1)^2 = -1$

4. Find the value of n so that the expression is a perfect square trinomial. Then factor the trinomial $d^2 + 11d + n$.

For Exercises 5 and 6, solve the equation by completing the square and applying the square root property.

5. $2x^2 + 12x - 36 = 0$

6. $2x^2 = 3x - 7$

For Exercises 7 and 8,

   a. Write the equation in standard form $ax^2 + bx + c = 0$.

   b. Identify a, b, and c.

   c. Find the discriminant.

   d. Determine the number and type (rational, irrational, or imaginary) of solutions.

7. $x^2 - 3x = -12$

8. $y(y - 2) = -1$

For Exercises 9 and 10, solve the equation by using the quadratic formula.

9. $3x^2 - 4x + 1 = 0$

10. $x(x + 6) = -11 - x$

11. The base of a triangle is 3 ft less than twice the height. The area of the triangle is 14 ft². Find the base and the height. Round the answers to the nearest tenth of a foot.

12. A circular garden has an area of approximately 450 ft². Find the radius. Round the answer to the nearest tenth of a foot.

For Exercises 13–17, solve the equation.

13. $x - \sqrt{x} - 6 = 0$

14. $y^{2/3} + 2y^{1/3} = 8$

15. $(3y - 8)^2 - 13(3y - 8) + 30 = 0$

16. $p^4 - 15p^2 = -54$

17. $3 = \dfrac{y}{2} - \dfrac{1}{y + 1}$

18. Find the vertex of $y = x^2 - 6x - 8$ two ways

   a. by completing the square.

   b. by using the vertex formula.

For Exercises 19–21, graph the functions. Use the graph to write the domain and range in interval notation.

19. $h(x) = x^2 - 4$

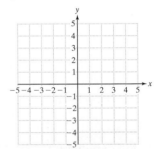

20. $f(x) = -(x - 4)^2$

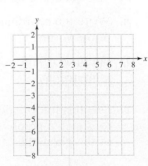

21. $g(x) = \dfrac{1}{2}(x + 2)^2 - 3$

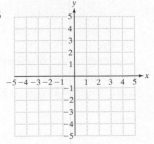

22. A child launches a toy rocket from the ground. The height of the rocket can be determined by its horizontal distance from the launch pad x by

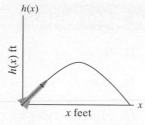

$$h(x) = -\frac{x^2}{256} + x$$

where x and $h(x)$ are in feet. How many feet from the launch pad will the rocket hit the ground?

23. The recent population of India (in millions) can be approximated by $P(t) = 0.135t^2 + 12.6t + 600$, where $t = 0$ corresponds to the year 1974.

   a. If this trend continues, approximate the number of people in India in the year 2014.

   b. Approximate the year in which the population of India reached 1 billion (1000 million). (Round to the nearest year.)

24. Explain the relationship between the graphs of $y = x^2$ and $y = x^2 - 2$.

25. Explain the relationship between the graphs of $y = x^2$ and $y = (x + 3)^2$.

26. Explain the relationship between the graphs of $y = 4x^2$ and $y = -4x^2$.

27. Given the function defined by

$$f(x) = -(x - 4)^2 + 2$$

   a. Identify the vertex of the parabola.

   b. Does this parabola open upward or downward?

   c. Does the vertex represent the maximum or minimum point of the function?

   d. What is the maximum or minimum value of the function f?

   e. What is the axis of symmetry for this parabola?

28. For the function defined by $g(x) = 2x^2 - 20x + 51$, find the vertex by using two methods.

   a. Complete the square to write $g(x)$ in the form $g(x) = a(x - h)^2 + k$. Identify the vertex.

   b. Use the vertex formula to find the vertex.

29. Given: $f(x) = x^2 + 4x - 12$

   a. Write the equation for the function in the form $f(x) = a(x - h)^2 + k$.

   b. Determine the vertex.

   c. Find the x- and y-intercepts.

   d. Determine the maximum or minimum value.

   e. Determine the axis of symmetry.

30. A farmer has 400 ft of fencing with which to enclose a rectangular field. The field is situated such that one of its sides is adjacent to a river and requires no fencing. The area of the field (in square feet) can be modeled by

$$A(x) = -\frac{x^2}{2} + 200x$$

where x is the length of the side parallel to the river (measured in feet).

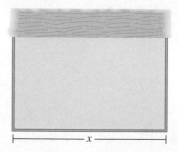

   a. Determine the value of x that maximizes the area of the field.

   b. Determine the maximum area that can be enclosed.

For Exercises 31–36, solve the polynomial and rational inequalities.

31. $\dfrac{2x - 1}{x - 6} \le 0$

32. $50 - 2a^2 > 0$

33. $y^3 + 3y^2 - 4y - 12 < 0$

34. $\dfrac{3}{w + 3} > 2$

35. $5x^2 - 2x + 2 < 0$

36. $t^2 + 22t + 121 \le 0$

Chapters 1–8 Cumulative Review Exercises

1. Given: $A = \{2, 4, 6, 8, 10\}$ and $B = \{2, 8, 12, 16\}$

    a. Find $A \cup B$.   **b.** Find $A \cap B$.

2. Perform the indicated operations and simplify.

$$(2x^2 - 5) - (x + 3)(5x - 2)$$

3. Simplify completely. $4^0 - \left(\dfrac{1}{2}\right)^{-3} - 81^{1/2}$

4. Perform the indicated operations. Write the answer in scientific notation:

$$(3 \times 10^{12})(6 \times 10^{-3})$$

5. a. Factor completely. $x^3 + 2x^2 - 9x - 18$

    b. Divide by using long division. Identify the quotient and remainder.

$$(x^3 + 2x^2 - 9x - 18) \div (x - 3)$$

6. Multiply. $(\sqrt{x} - \sqrt{2})(\sqrt{x} + \sqrt{2})$

7. Simplify. $\dfrac{4}{\sqrt{2x}}$

8. Jacques invests a total of $10,000 in two mutual funds. After 1 yr, one fund produced 12% growth, and the other lost 3%. Find the amount invested in each fund if the total investment grew by $900.

9. Solve the system of equations.

$$\frac{1}{9}x - \frac{1}{3}y = -\frac{13}{9}$$

$$x - \frac{1}{2}y = \frac{9}{2}$$

10. An object is fired straight up into the air from an initial height of 384 ft with an initial velocity of 160 ft/sec. The height in feet is given by

$$h(t) = -16t^2 + 160t + 384$$

where t is the time in seconds after launch.

    a. Find the height of the object after 3 sec.

    b. Find the height of the object after 7 sec.

    c. Find the time required for the object to hit the ground.

11. Solve. $(x - 3)^2 + 16 = 0$

12. Solve. $2x^2 + 5x - 1 = 0$

13. Find the value of n so that the expression is a perfect square trinomial. Then factor the trinomial $x^2 + 10x + n$.

14. Factor completely. $2x^3 + 250$

15. Graph the equation. $3x - 5y = 10$

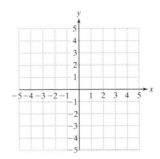

16. a. Find the x-intercepts of the function defined by $g(x) = 2x^2 - 9x + 10$.

    b. What is the y-intercept?

17. Michael Jordan was the NBA leading scorer for 10 of 12 seasons between 1987 and 1998. In his 1998 season, he scored a total of 2357 points consisting of 1-point free throws, 2-point field goals, and 3-point field goals. He scored 286 more 2-point shots than he did free throws. The number of 3-point shots was 821 less than the number of 2-point shots. Determine the number of free throws, 2-point shots, and 3-point shots scored by Michael Jordan during his 1998 season.

18. Explain why this relation is *not* a function.

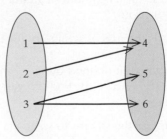

19. Graph the function defined by $f(x) = \dfrac{1}{x}$.

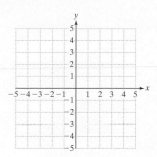

20. The quantity y varies directly as x and inversely as z. If $y = 15$ when $x = 50$ and $z = 10$, find y when $x = 65$ and $z = 5$.

21. The total number of flights (including passenger flights and cargo flights) at a large airport can be approximated by $F(x) = 300{,}000 + 0.008x$, where x is the number of passengers.

    a. Is this function linear, quadratic, constant, or other?

    b. Find the y-intercept and interpret its meaning in the context of this problem.

    c. What is the slope of the function and what does the slope mean in the context of this problem?

22. Given the function defined by $g(x) = \sqrt{2 - x}$, find the function values (if they exist) over the set of real numbers.

    a. $g(-7)$ **b.** $g(0)$ **c.** $g(3)$

23. Let $m(x) = \sqrt{x + 4}$ and $n(x) = x^2 + 2$. Find

    a. The domain of m **b.** The domain of n

24. Consider the function $y = f(x)$ graphed here. Find

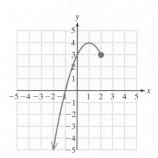

    a. The domain   **b.** The range

    c. $f(1)$ **d.** $f(0)$

    e. For what value(s) of x is $f(x) = 0$?

25. Solve. $\sqrt{8x + 5} = \sqrt{2x + 2}$

26. Solve for f. $\dfrac{1}{p} + \dfrac{1}{q} = \dfrac{1}{f}$

27. Solve. $\dfrac{15}{t^2 - 2t - 8} = \dfrac{1}{t - 4} + \dfrac{2}{t + 2}$

28. Simplify. $\dfrac{y - \dfrac{4}{y - 3}}{y - 4}$

29. Given the function defined by $f(x) = 2(x - 3)^2 + 1$

    a. Write the coordinates of the vertex.

    b. Does the graph of the function open upward or downward?

    c. Write the coordinates of the y-intercept.

    d. Find the x-intercepts, if possible.

    e. Sketch the function.

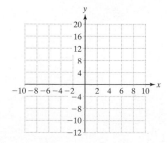

30. Solve the inequality. Write the answer in interval notation.

$$\dfrac{5x}{x - 2} \geq 0$$

Exponential and Logarithmic Functions

Chapter 9

This chapter is devoted to the study of exponential and logarithmic functions. These functions are used to study many naturally occurring phenomena such as population growth, decay of radioactive matter, and growth of investments.

Review Your Skills

To prepare yourself for this chapter, practice evaluating the expressions with exponents. Use the clues to fill in the boxes labeled A–H. Then fill in the remaining part of the grid so that every row, every column, and every 2 × 3 box contains the digits 1 through 6.

A. Value of x in the equation $5^x = 25$.

B. Value of x in the equation $3^x = 81$.

C. Value of x in the equation $8^{1/x} = 2$.

D. Real value of x in the equation $x^4 = 16$ (assume that $x > 0$).

E. Absolute value of x in the equation $2^x = \dfrac{1}{16}$.

F. Value of x in the equation $2^x = 64$.

G. Value of x in the equation $5^0 = x$.

H. Value of x in the equation $e^0 = x$.

Section 9.1 Inverse Functions

1. Introduction to Inverse Functions

In Section 3.2, we defined a function as a set of ordered pairs (x, y) such that for every element x in the domain, there corresponds exactly one element y in the range. For example, the function f relates the weight of a package of deli meat, x, to its cost, y.

$$f = \{(1, 4), (2.5, 10), (4, 16)\}$$

That is, 1 lb of meat sells for $4, 2.5 lb sells for $10, and 4 lb sells for $16. Now suppose we create a new function in which the values of x and y are interchanged. The new function, called the **inverse of f**, denoted f^{-1}, relates the price of meat, x, to its weight, y.

$$f^{-1} = \{(4, 1), (10, 2.5), (16, 4)\}$$

Notice that interchanging the x- and y-values has the following outcome. The domain of f is the same as the range of f^{-1}, and the range of f is the domain of f^{-1}.

> **Avoiding Mistakes**
>
> f^{-1} denotes the inverse of a function. The -1 does not represent an exponent.

2. Definition of a One-to-One Function

A necessary condition for a function f to have an inverse function is that no two ordered pairs in f have different x-coordinates and the same y-coordinate. A function that satisfies this condition is called a **one-to-one function**. The function relating the weight of a package of meat to its price is a one-to-one function. However, consider the function g defined by

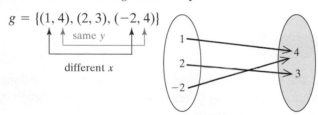

$$g = \{(1, 4), (2, 3), (-2, 4)\}$$

Not a one-to-one function because two different x values map to the same y value.

This function is not one-to-one because the range element 4 has two different x-coordinates, 1 and -2. Interchanging the x- and y-values produces a relation that violates the definition of a function.

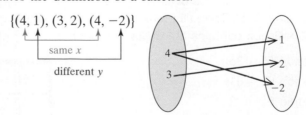

$$\{(4, 1), (3, 2), (4, -2)\}$$

This relation is not a function because for $x = 4$, there are two different y-values.

> **Concept Connections**
>
> Determine whether the relation is a one-to-one function. If it is not, explain why.
>
> 1. $\{(4, 2), (1, 2), (3, 0), (-4, 4)\}$
> 2. $\{(0, 0), (5, -1), (2, 3), (5, 6)\}$
> 3. $\{(3, 4), (2, 2), (0, 8), (-1, 7)\}$

In Section 3.2, you learned the vertical line test to determine visually if a graph represents a function. Similarly, we use a **horizontal line test** to determine whether a function is one-to-one.

> **Answers**
>
> 1. Not one-to-one. The range element 2 has two different x-coordinates.
> 2. Not a function. The domain element 5 is paired with two different y-coordinates.
> 3. One-to-one function.

Using the Horizontal Line Test

Consider a function defined by a set of points (x, y) in a rectangular coordinate system. The graph of the ordered pairs defines y as a *one-to-one* function of x if no horizontal line intersects the graph in more than one point.

To understand the horizontal line test, consider the functions f and g.

$$f = \{(1, 4), (2.5, 10), (4, 16)\} \qquad g = \{(1, 4), (2, 3), (-2, 4)\}$$

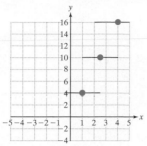 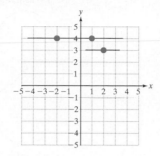

This function is one-to-one. No horizontal line intersects more than once.

This function is *not* one-to-one. A horizontal line intersects more than once.

Example 1 Identifying One-to-One Functions

Determine whether the function is one-to-one.

a.
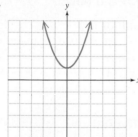

b.
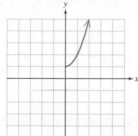

Solution:

a. Function is not one-to-one. A horizontal line intersects in more than one point.

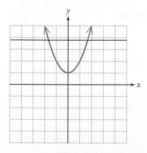

b. Function is one-to-one. No horizontal line intersects more than once.

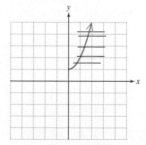

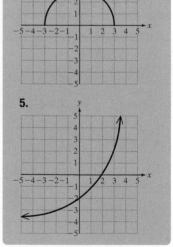
3. Definition of the Inverse of a Function

All one-to-one functions have an inverse function.

Definition of an Inverse Function

If f is a one-to-one function represented by ordered pairs of the form (x, y), then the inverse function, denoted by f^{-1}, is the set of ordered pairs given by (y, x).

Answers
4. Not one-to-one
5. One-to-one

Because the values of x and y are interchanged between a function, f, and its inverse, f^{-1}, we have the following important relationship (Figure 9-1).

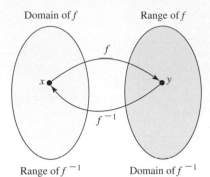

Figure 9-1

From Figure 9-1, we see that the operations performed by f are reversed by f^{-1}. This observation leads to the following important property.

Inverse Function Property

If f is a one-to-one function, then g is the inverse of f if and only if

$$(f \circ g)(x) = x \qquad \text{for all } x \text{ in the domain of } g$$

and

$$(g \circ f)(x) = x \qquad \text{for all } x \text{ in the domain of } f.$$

Example 2 **Composing a Function with Its Inverse**

Show that the functions are inverses.

$$f(x) = 5x + 4 \qquad g(x) = \frac{x - 4}{5}$$

Solution:

To show that the functions f and g are inverses, confirm that $(f \circ g)(x) = x$ and $(g \circ f)(x) = x$.

$$(f \circ g)(x) = f(g(x)) \qquad\qquad (g \circ f)(x) = g(f(x))$$

$$= f\left(\frac{x - 4}{5}\right) \qquad\qquad\qquad = g(5x + 4)$$

$$= 5\left(\frac{x - 4}{5}\right) + 4 \qquad\qquad = \frac{(5x + 4) - 4}{5}$$

$$= x - 4 + 4 \qquad\qquad\qquad = \frac{5x}{5}$$

$$= x \; ✔ \qquad\qquad\qquad\qquad = x \; ✔$$

Answer

6. $(f \circ g)(x) = f(g(x))$

$\qquad = 3\left(\dfrac{x + 2}{3}\right) - 2 = x$

$(g \circ f)(x) = g(f(x))$

$\qquad = \dfrac{3x - 2 + 2}{3} = x$

4. Finding an Equation of the Inverse of a Function

Another way to view the construction of the inverse of a function is to find a function that performs the inverse operations in the reverse order. For example, the function defined by $f(x) = 2x + 1$ multiplies x by 2 and then adds 1. Therefore, the inverse function must *subtract* 1 from x and *divide* by 2. We have

$$f^{-1}(x) = \frac{x - 1}{2}$$

To facilitate the process of finding an equation of the inverse of a one-to-one function, we offer the following steps.

> ### Finding an Equation for the Inverse of a Function
>
> For a one-to-one function defined by $y = f(x)$, the equation of the inverse can be found as follows:
>
> **Step 1** Replace $f(x)$ by y.
> **Step 2** Interchange x and y.
> **Step 3** Solve for y.
> **Step 4** Replace y by $f^{-1}(x)$.

Example 3 Finding an Equation of the Inverse of a Function

Find the inverse. $f(x) = 2x + 1$

Solution:

Foremost, we know the graph of f is a nonvertical line. Therefore, $f(x) = 2x + 1$ defines a one-to-one function. To find the inverse we have

$y = 2x + 1$	**Step 1:** Replace $f(x)$ by y.
$x = 2y + 1$	**Step 2:** Interchange x and y.
$x - 1 = 2y$	**Step 3:** Solve for y. Subtract 1 from both sides.
$\dfrac{x - 1}{2} = y$	Divide both sides by 2.
$f^{-1}(x) = \dfrac{x - 1}{2}$	**Step 4:** Replace y by $f^{-1}(x)$.

Skill Practice

Find the inverse.
7. $f(x) = 4x + 6$

Avoiding Mistakes

It is important to check that a function is one-to-one before finding the inverse function.

In Example 3, we can verify that $f(x) = 2x + 1$ and $f^{-1}(x) = \dfrac{x - 1}{2}$ are inverses by using function composition.

$$(f \circ f^{-1})(x) = f\left(f^{-1}(x)\right) = 2\left(\frac{x - 1}{2}\right) + 1 = x \quad \checkmark \quad \text{and}$$

$$(f^{-1} \circ f)(x) = f^{-1}(f(x)) = \frac{(2x + 1) - 1}{2} = x \quad \checkmark$$

The key step in determining the equation of the inverse of a function is to interchange x and y. By so doing, a point (a, b) on f corresponds to a point (b, a) on f^{-1}. For this reason, the graphs of f and f^{-1} are symmetric with respect to the line $y = x$ (Figure 9-2). Notice that the point $(-3, -5)$ of the function f corresponds to the point $(-5, -3)$ of f^{-1}. Likewise, $(1, 3)$ of f corresponds to $(3, 1)$ of f^{-1}.

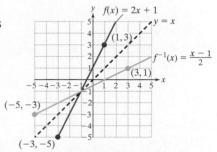

Figure 9-2

Answer
7. $f^{-1}(x) = \dfrac{x - 6}{4}$

Example 4 **Finding an Equation of the Inverse of a Function**

Find the inverse of the one-to-one function. $\quad g(x) = \sqrt[3]{5x} - 4$

Solution:

$y = \sqrt[3]{5x} - 4$	**Step 1:** Replace $g(x)$ by y.
$x = \sqrt[3]{5y} - 4$	**Step 2:** Interchange x and y.
$x + 4 = \sqrt[3]{5y}$	**Step 3:** Solve for y. Add 4 to both sides.
$(x + 4)^3 = (\sqrt[3]{5y})^3$	To eliminate the cube root, cube both sides.
$(x + 4)^3 = 5y$	Simplify the right side.
$\dfrac{(x + 4)^3}{5} = y$	Divide both sides by 5.
$g^{-1}(x) = \dfrac{(x + 4)^3}{5}$	**Step 4:** Replace y by $g^{-1}(x)$.

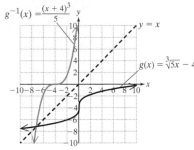

Figure 9-3

The graphs of g and g^{-1} are shown in Figure 9-3. Once again we see that the graphs of a function and its inverse are symmetric with respect to the line $y = x$.

For a function that is not one-to-one, sometimes we can restrict its domain to create a new function that is one-to-one. This is demonstrated in Example 5.

Example 5 **Finding the Equation of an Inverse of a Function with a Restricted Domain**

Given the function defined by $m(x) = x^2 + 4$ for $x \geq 0$, find an equation defining m^{-1}.

Solution:

From Section 8.4, we know that $y = x^2 + 4$ is a parabola with vertex at $(0, 4)$ (Figure 9-4). The graph represents a function that is not one-to-one. However, with the restriction on the domain $x \geq 0$, the graph of $m(x) = x^2 + 4$, $x \geq 0$, consists of only the "right" branch of the parabola (Figure 9-5). This *is* a one-to-one function.

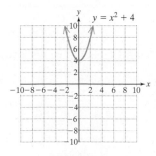

Figure 9-4

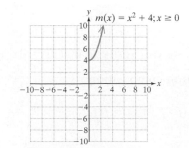

Figure 9-5

To find the inverse, we have

$y = x^2 + 4 \qquad x \geq 0$ **Step 1:** Replace $m(x)$ by y.

$x = y^2 + 4 \qquad y \geq 0$ **Step 2:** Interchange x and y. Notice that the restriction $x \geq 0$ becomes $y \geq 0$.

$x - 4 = y^2 \qquad y \geq 0$ **Step 3:** Solve for y. Subtract 4 from both sides.

$\sqrt{x - 4} = y \qquad y \geq 0$ Apply the square root property. Notice that we obtain the *positive* square root of $x - 4$ because of the restriction $y \geq 0$.

$m^{-1}(x) = \sqrt{x - 4}$ **Step 4:** Replace y by $m^{-1}(x)$. Notice that the domain of m^{-1} has the same values as the range of m.

Figure 9-6 shows the graphs of m and m^{-1}.

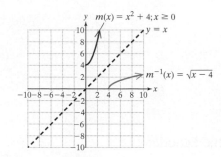

Figure 9-6

Section 9.1 Practice Exercises

Vocabulary and Key Concepts

1. **a.** Given the function $f = \{(1, 2), (2, 3), (3, 4)\}$, write the set of ordered pairs representing f^{-1}.

   b. A necessary condition for a function f to be a _____-_____-_____ function is that no two ordered pairs in f have different x-coordinates and the same _____-coordinate.

   c. The function $f = \{(1, 5), (-2, 3), (-4, 2), (2, 5)\}$ (is/is not) a one-to-one function.

   d. A function defined by $y = f(x)$ (is/is not) a one-to-one function if no horizontal line intersects the graph of f in more than one point.

   e. The graph of a function and its inverse are symmetric with respect to the line _____.

   f. Let f be a one-to-one function and let g be the inverse of f. Then $(f \circ g)(x) =$ _____ and $(g \circ f)(x) =$ _____.

   g. The notation _____ is often used to represent the inverse of a function f and not the reciprocal of f.

   h. If (a, b) is a point on the graph of a one-to-one function f, then the corresponding ordered pair _____ is a point on the graph of f^{-1}.

Review Exercises

2. Write the domain and range of the relation $\{(3, 4), (5, -2), (6, 1), (3, 0)\}$.

For Exercises 3–8, determine if the relation is a function by using the vertical line test. **(See Section 3.2.)**

3.

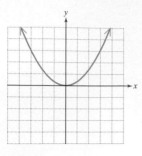

4.

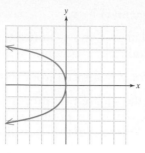

5.

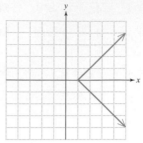

6.

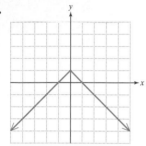

7.

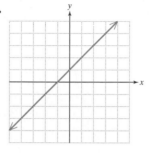

8.

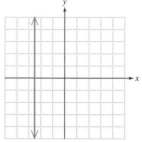

Concept 1: Introduction to Inverse Functions

For Exercises 9–12, write the inverse function for each function.

9. $g = \{(3, 5), (8, 1), (-3, 9), (0, 2)\}$

10. $f = \{(-6, 2), (-9, 0), (-2, -1), (3, 4)\}$

11. $r = \{(a, 3), (b, 6), (c, 9)\}$

12. $s = \{(-1, x), (-2, y), (-3, z)\}$

Concept 2: Definition of a One-to-One Function

13. The table relates a state, x, to the number of representatives in the House of Representatives, y, for a recent year. Does this relation define a one-to-one function? If so, write a function defining the inverse as a set of ordered pairs.

State x	Number of Representatives y
Colorado	7
California	53
Texas	32
Louisiana	7
Pennsylvania	19

14. The table relates a city, x, to its average January temperature, y. Does this relation define a one-to-one function? If so, write a function defining the inverse as a set of ordered pairs.

City x	Temperature y (°C)
Gainesville, Florida	13.6
Keene, New Hampshire	−6.0
Wooster, Ohio	−4.0
Rock Springs, Wyoming	−6.0
Lafayette, Louisiana	10.9

For Exercises 15–20, determine if the function is one-to-one by using the horizontal line test. **(See Example 1.)**

15.

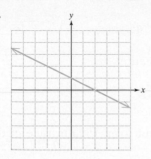

16.

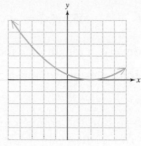

17.

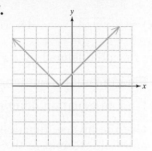

18.

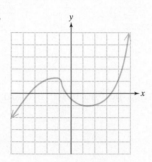

19.

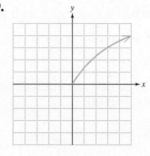

20.

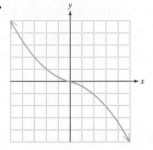

Concept 3: Definition of the Inverse of a Function

For Exercises 21–26, verify that f and g are inverse functions by showing that

    a. $(f \circ g)(x) = x$ **b.** $(g \circ f)(x) = x$ **(See Example 2.)**

21. $f(x) = 6x + 1$ and $g(x) = \dfrac{x - 1}{6}$

22. $f(x) = 5x - 2$ and $g(x) = \dfrac{x + 2}{5}$

23. $f(x) = \dfrac{\sqrt[3]{x}}{2}$ and $g(x) = 8x^3$

24. $f(x) = \dfrac{\sqrt[3]{x}}{3}$ and $g(x) = 27x^3$

25. $f(x) = x^2 + 1$, $x \geq 0$, and $g(x) = \sqrt{x - 1}$, $x \geq 1$ **26.** $f(x) = x^2 - 3$, $x \geq 0$, and $g(x) = \sqrt{x + 3}$, $x \geq -3$

Concept 4: Finding an Equation of the Inverse of a Function

For Exercises 27–46, write an equation of the inverse for each one-to-one function as defined. **(See Examples 3–5.)**

27. $h(x) = x + 4$ **28.** $k(x) = x - 3$ **29.** $m(x) = \dfrac{1}{3}x - 2$ **30.** $n(x) = 4x + 2$

31. $p(x) = -x + 10$ **32.** $q(x) = -x - \dfrac{2}{3}$ **33.** $n(x) = \dfrac{3x + 2}{5}$ **34.** $p(x) = \dfrac{2x - 7}{4}$

35. $h(x) = \dfrac{4x - 1}{3}$ **36.** $f(x) = \dfrac{6x + 3}{2}$ **37.** $f(x) = x^3 + 1$ **38.** $g(x) = \sqrt[3]{x} + 2$

39. $g(x) = \sqrt[3]{2x - 1}$ **40.** $f(x) = x^3 - 4$ **41.** $g(x) = x^2 + 9 \quad x \geq 0$ **42.** $m(x) = x^2 - 1 \quad x \geq 0$

43. $q(x) = \sqrt{x + 4}$ **44.** $v(x) = \sqrt{x + 16}$ **45.** $z(x) = -\sqrt{x + 4}$ **46.** $u(x) = -\sqrt{x + 16}$

47. The function defined by $f(x) = 0.3048x$ converts a length of x feet into $f(x)$ meters.

   a. Find the equivalent length in meters for a 4-ft board and a 50-ft wire.

   b. Find an equation defining $y = f^{-1}(x)$.

   c. Use the inverse function from part (b) to find the equivalent length in feet for a 1500-m race in track and field. Round to the nearest tenth of a foot.

48. The function defined by $s(x) = 1.47x$ converts a speed of x mph to $s(x)$ ft/sec.

   a. Find the equivalent speed in feet per second for a car traveling 60 mph.

   b. Find an equation defining $y = s^{-1}(x)$.

   c. Use the inverse function from part (b) to find the equivalent speed in miles per hour for a train traveling 132 ft/sec. Round to the nearest tenth.

For Exercises 49–55, answer true or false.

49. The function defined by $y = 2$ has an inverse function defined by $x = 2$.

50. The domain of any one-to-one function is the same as the domain of its inverse.

51. All linear functions with a nonzero slope have an inverse function.

52. The function defined by $g(x) = |x|$ is one-to-one.

53. The function defined by $k(x) = x^2$ is one-to-one.

54. The function defined by $h(x) = |x|$ for $x \geq 0$ is one-to-one.

55. The function defined by $L(x) = x^2$ for $x \geq 0$ is one-to-one.

56. Explain how the domain and range of a one-to-one function and its inverse are related.

57. If $(0, b)$ is the y-intercept of a one-to-one function, what is the x-intercept of its inverse?

58. If $(a, 0)$ is the x-intercept of a one-to-one function, what is the y-intercept of its inverse?

59. **a.** Find the domain and range of the function defined by $f(x) = \sqrt{x - 1}$.

   b. Find the domain and range of the function defined by $f^{-1}(x) = x^2 + 1, x \geq 0$.

60. **a.** Find the domain and range of the function defined by $g(x) = x^2 - 4, x \leq 0$.

   b. Find the domain and range of the function defined by $g^{-1}(x) = -\sqrt{x + 4}$.

For Exercises 61–64, the graph of $y = f(x)$ is given.

    a. State the domain of f.   **b.** State the range of f.

    c. State the domain of f^{-1}.   **d.** State the range of f^{-1}.

    e. Graph the function defined by $y = f^{-1}(x)$. The line $y = x$ is provided for your reference.

61.
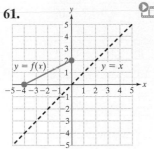

62.
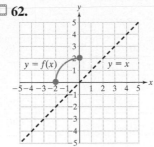

63.
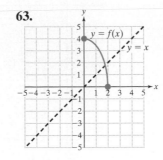

64.
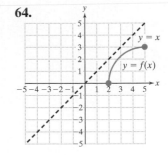

Expanding Your Skills

For Exercises 65–70, write an equation of the inverse of the one-to-one function.

65. $f(x) = \dfrac{x - 1}{x + 1}$

66. $p(x) = \dfrac{3 - x}{x + 3}$

67. $t(x) = \dfrac{2}{x - 1}$

68. $w(x) = \dfrac{4}{x + 2}$

69. $n(x) = x^2 + 9 \quad x \le 0$

70. $g(x) = x^2 - 1 \quad x \le 0$

Graphing Calculator Exercises

For Exercises 71–74, use a graphing calculator to graph each function on the standard viewing window. Use the graph of the function to determine if the function is one-to-one on the interval $-10 \le x \le 10$. If the function is one-to-one, find its inverse and graph both functions on the standard viewing window.

71. $f(x) = \sqrt[3]{x + 5}$

72. $k(x) = x^3 - 4$

73. $g(x) = 0.5x^3 - 2$

74. $m(x) = 3x - 4$

Exponential Functions   Section 9.2

1. Definition of an Exponential Function

The concept of a function was first introduced in Section 3.2. Since then we have learned to recognize several categories of functions, including constant, linear, rational, and quadratic functions. In this section and the next, we will define two new types of functions called exponential and logarithmic functions.

    To introduce the concept of an exponential function, consider the following salary plans for a new job. Plan A pays $1 million for a month's work. Plan B starts with 2¢ on the first day, and every day thereafter the salary is doubled.

    At first glance, the million-dollar plan appears to be more favorable. Look, however, at Table 9-1, which shows the daily payments for 30 days under plan B.

Concepts

1. Definition of an Exponential Function

2. Approximating Exponential Expressions with a Calculator

3. Graphs of Exponential Functions

4. Applications of Exponential Functions

Table 9-1

Day	Payment	Day	Payment	Day	Payment
1	2¢	11	$20.48	21	$20,971.52
2	4¢	12	$40.96	22	$41,943.04
3	8¢	13	$81.92	23	$83,886.08
4	16¢	14	$163.84	24	$167,772.16
5	32¢	15	$327.68	25	$335,544.32
6	64¢	16	$655.36	26	$671,088.64
7	$1.28	17	$1310.72	27	$1,342,177.28
8	$2.56	18	$2621.44	28	$2,684,354.56
9	$5.12	19	$5242.88	29	$5,368,709.12
10	$10.24	20	$10,485.76	30	$10,737,418.24

Notice that the salary on the 30th day for plan B is over $10 million. Taking the sum of the payments, we see the total salary for the 30-day period is $21,474,836.46.

The daily salary for plan B can be represented by the function $y = 2^x$, where x is the number of days on the job and y is the salary (in cents) for that day. An interesting feature of this function is that for every positive 1-unit change in x, the y-value doubles. The function $y = 2^x$ is called an exponential function.

Concept Connections

Determine if the function is exponential.

1. $y = x^3$

2. $y = 3^x$

3. $y = \left(\dfrac{1}{2}\right)^x$

Definition of an Exponential Function

Let b be any real number such that $b > 0$ and $b \neq 1$. Then for any real number x, a function of the form $f(x) = b^x$ is called an **exponential function**.

An exponential function is easily recognized as a function with a constant base and exponent, x. Notice that the base of an exponential function must be a positive real number not equal to 1.

2. Approximating Exponential Expressions with a Calculator

Up to this point, we have evaluated exponential expressions with integer exponents and rational exponents, for example, $4^3 = 64$ and $4^{1/2} = \sqrt{4} = 2$. However, how do we evaluate an exponential expression with an irrational exponent such as 4^π? In such a case, the exponent is a nonterminating and nonrepeating decimal. The value of 4^π can be thought of as the limiting value of a sequence of approximations using rational exponents:

$$4^{3.14} \approx 77.7084726$$

$$4^{3.141} \approx 77.81627412$$

$$4^{3.1415} \approx 77.87023095$$

$$\vdots$$

$$4^\pi \approx 77.88023365$$

An exponential expression can be evaluated at all rational numbers and at all irrational numbers. Therefore, the domain of an exponential function is all real numbers.

Answers

1. No   **2.** Yes   **3.** Yes

Example 1 Approximating Exponential Expressions with a Calculator

Approximate the expressions. Round the answers to four decimal places.

a. $8^{\sqrt{3}}$ **b.** $5^{-\sqrt{17}}$ **c.** $\sqrt{10}^{\sqrt{2}}$

Solution:

Calculator Connections

Topic: Investigating Exponential Expressions

On a graphing calculator, use the [^] key to approximate an expression with an irrational exponent.

```
8^√(3)
          36.66044576
5^-√(17)
          .00131242
(√(10))^√(2)
          5.09456117
```

a. $8^{\sqrt{3}} \approx 36.6604$ **b.** $5^{-\sqrt{17}} \approx 0.0013$

c. $\sqrt{10}^{\sqrt{2}} \approx 5.0946$

3. Graphs of Exponential Functions

The functions defined by $f(x) = 2^x$, $g(x) = (\frac{1}{2})^x$, $h(x) = 3^x$, and $k(x) = 5^x$ are all examples of exponential functions. Example 2 illustrates the two general shapes of exponential functions.

Example 2 Graphing Exponential Functions

Graph the functions f and g.

a. $f(x) = 2^x$ **b.** $g(x) = (\frac{1}{2})^x$

Solution:

Table 9-2 shows several function values $f(x)$ and $g(x)$ for both positive and negative values of x. The graphs are shown in Figure 9-7.

Table 9-2

x	$f(x) = 2^x$	$g(x) = (\frac{1}{2})^x$
-4	$\frac{1}{16}$	16
-3	$\frac{1}{8}$	8
-2	$\frac{1}{4}$	4
-1	$\frac{1}{2}$	2
0	1	1
1	2	$\frac{1}{2}$
2	4	$\frac{1}{4}$
3	8	$\frac{1}{8}$
4	16	$\frac{1}{16}$

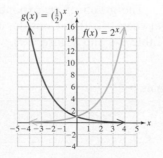

Figure 9-7

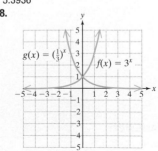

The graphs in Figure 9-7 illustrate several important features of exponential functions.

> ## Graphs of $f(x) = b^x$
>
> The graph of an exponential function defined by $f(x) = b^x$ ($b > 0$ and $b \neq 1$) has the following properties.
>
> **1.** If $b > 1$, f is an *increasing* exponential function, sometimes called an **exponential growth function**.
>
>   If $0 < b < 1$, f is a *decreasing* exponential function, sometimes called an **exponential decay function**.
>
> **2.** The domain is the set of all real numbers, $(-\infty, \infty)$.
>
> **3.** The range is $(0, \infty)$.
>
> **4.** The line $y = 0$ (x-axis) is a horizontal asymptote.
>
> **5.** The function passes through the point $(0, 1)$ because $f(0) = b^0 = 1$.

These properties indicate that the graph of an exponential function is an increasing function if the base is greater than 1. Furthermore, the base affects its "steepness." Consider the graphs of $f(x) = 2^x$, $h(x) = 3^x$, and $k(x) = 5^x$ (Figure 9-8). For every positive 1-unit change in x, $f(x) = 2^x$ increases by 2 times, $h(x) = 3^x$ increases by 3 times, and $k(x) = 5^x$ increases by 5 times (Table 9-3).

Concept Connections

9.

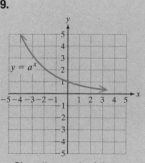

Given the graph of the exponential function $y = a^x$, which is true?

a. $a > 1$

b. $0 < a < 1$

c. It cannot be determined.

Table 9-3

x	$f(x) = 2^x$	$h(x) = 3^x$	$k(x) = 5^x$
-3	$\frac{1}{8}$	$\frac{1}{27}$	$\frac{1}{125}$
-2	$\frac{1}{4}$	$\frac{1}{9}$	$\frac{1}{25}$
-1	$\frac{1}{2}$	$\frac{1}{3}$	$\frac{1}{5}$
0	1	1	1
1	2	3	5
2	4	9	25
3	8	27	125

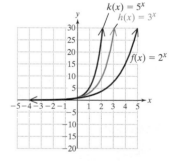

Figure 9-8

The graph of an exponential function is a *decreasing function* if the base is between 0 and 1. Consider the graphs of $g(x) = (\frac{1}{2})^x$, $m(x) = (\frac{1}{3})^x$, and $n(x) = (\frac{1}{5})^x$ (Table 9-4 and Figure 9-9).

Table 9-4

x	$g(x) = (\frac{1}{2})^x$	$m(x) = (\frac{1}{3})^x$	$n(x) = (\frac{1}{5})^x$
-3	8	27	125
-2	4	9	25
-1	2	3	5
0	1	1	1
1	$\frac{1}{2}$	$\frac{1}{3}$	$\frac{1}{5}$
2	$\frac{1}{4}$	$\frac{1}{9}$	$\frac{1}{25}$
3	$\frac{1}{8}$	$\frac{1}{27}$	$\frac{1}{125}$

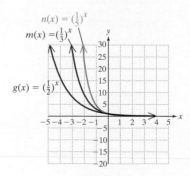

Figure 9-9

Answer

9. b

4. Applications of Exponential Functions

Exponential growth and decay can be found in a variety of real-world phenomena; for example,

- Population growth can often be modeled by an exponential function.
- The growth of an investment under compound interest increases exponentially.
- The mass of a radioactive substance decreases exponentially with time.
- The temperature of a cup of coffee decreases exponentially as it approaches room temperature.

A substance that undergoes radioactive decay is said to be radioactive. The *half-life* of a radioactive substance is the amount of time it takes for one-half of the original amount of the substance to change into something else. That is, after each half-life the amount of the original substance decreases by one-half.

In 1898, Marie Curie discovered the highly radioactive element radium. She shared the 1903 Nobel Prize in physics for her research on radio-activity and was awarded the 1911 Nobel Prize in chemistry for her discovery of radium and polonium. Radium 226 (an isotope of radium) has a half-life of 1620 years and decays into radon 222 (a radio-active gas).

Marie and Pierre Curie

Example 3   Applying an Exponential Decay Function

In a sample originally having 1 g of radium 226, the amount of radium 226 present after t years is given by

$$A(t) = \left(\tfrac{1}{2}\right)^{t/1620}$$

where A is the amount of radium in grams and t is the time in years.

a. How much radium 226 will be present after 1620 yr?

b. How much radium 226 will be present after 3240 yr?

c. How much radium 226 will be present after 4860 yr?

Solution:

$$A(t) = \left(\frac{1}{2}\right)^{t/1620}$$

a. $A(1620) = \left(\dfrac{1}{2}\right)^{1620/1620}$ Substitute $t = 1620$.

$$= \left(\frac{1}{2}\right)^{1}$$

$$= 0.5$$

After 1620 yr (1 half-life), 0.5 g remains.

Answers
10. 0.5 g
11. 0.125 g

b. $A(3240) = \left(\dfrac{1}{2}\right)^{3240/1620}$ Substitute $t = 3240$.

$$= \left(\dfrac{1}{2}\right)^2$$

$$= 0.25$$

After 3240 yr (2 half-lives), the amount of the original substance is reduced by one-half, 2 times: 0.25 g remains.

c. $A(4860) = \left(\dfrac{1}{2}\right)^{4860/1620}$ Substitute $t = 4860$.

$$= \left(\dfrac{1}{2}\right)^3$$

$$= 0.125$$

After 4860 yr (3 half-lives), the amount of the original substance is reduced by one-half, 3 times: 0.125 g remains.

Exponential functions are often used to model population growth. Suppose the initial value of a population at some time $t = 0$ is P_0. If the rate of increase of a population is r, then after 1, 2, and 3 yr, the new population can be found as follows:

After 1 yr: $\left(\begin{array}{c}\text{Total} \\ \text{population}\end{array}\right) = \left(\begin{array}{c}\text{initial} \\ \text{population}\end{array}\right) + \left(\begin{array}{c}\text{increase in} \\ \text{population}\end{array}\right)$

$$= P_0 + P_0 r$$

$$= P_0(1 + r) \qquad \text{Factor out } P_0.$$

After 2 yr: $\left(\begin{array}{c}\text{Total} \\ \text{population}\end{array}\right) = \left(\begin{array}{c}\text{population} \\ \text{after 1 yr}\end{array}\right) + \left(\begin{array}{c}\text{increase in} \\ \text{population}\end{array}\right)$

$$= P_0(1 + r) + P_0(1 + r)r$$

$$= P_0(1 + r)1 + P_0(1 + r)r$$

$$= P_0(1 + r)(1 + r) \qquad \text{Factor out } P_0(1 + r).$$

$$= P_0(1 + r)^2$$

After 3 yr: $\left(\begin{array}{c}\text{Total} \\ \text{population}\end{array}\right) = \left(\begin{array}{c}\text{population} \\ \text{after 2 yr}\end{array}\right) + \left(\begin{array}{c}\text{increase in} \\ \text{population}\end{array}\right)$

$$= P_0(1 + r)^2 + P_0(1 + r)^2 r$$

$$= P_0(1 + r)^2 1 + P_0(1 + r)^2 r$$

$$= P_0(1 + r)^2(1 + r) \qquad \text{Factor out } P_0(1 + r)^2.$$

$$= P_0(1 + r)^3$$

This pattern continues, and after t years, the population $P(t)$ is given by

$$P(t) = P_0(1 + r)^t$$

| Example 4 | Applying an Exponential Growth Function |

The population of the Bahamas in 2008 was estimated at 321,000 with an annual rate of increase of 1.39%.

a. Find a mathematical model that relates the population of the Bahamas as a function of the number of years since 2008.

b. If the annual rate of increase remains the same, use this model to predict the population of the Bahamas in the year 2016. Round to the nearest thousand.

Solution:

a. The initial population is $P_0 = 321,000$, and the rate of increase is 13.9%.

$$P(t) = P_0(1 + r)^t \qquad \text{Substitute } P_0 = 321,000 \text{ and } r = 0.0139.$$

$$= 321,000(1 + 0.0139)^t$$

$$= 321,000(1.0139)^t \qquad \text{Here } t = 0 \text{ corresponds to the year 2008.}$$

b. Because the initial population ($t = 0$) corresponds to the year 2008, we use $t = 8$ to find the population in the year 2016.

$$P(8) = 321,000(1.0139)^8$$

$$\approx 358,000$$

Section 9.2 Practice Exercises

Vocabulary and Key Concepts

1. a. Given a real number b, where $b > 0$ and $b \neq 1$, then a function of the form $f(x) =$ _____ is called an exponential function.

b. The function defined by $f(x) = x^4$ (is/is not) an exponential function, whereas the function defined by $f(x) = 4^x$ (is/is not) an exponential function.

c. The graph of $f(x) = \left(\dfrac{7}{2}\right)^x$ is (increasing/decreasing) over its domain.

d. The graph of $f(x) = \left(\dfrac{2}{7}\right)^x$ is (increasing/decreasing) over its domain.

e. In interval notation, the domain of an exponential function $f(x) = b^x$ is _____ and the range is _____.

f. All exponential functions $f(x) = b^x$ pass through the point _____.

g. The horizontal asymptote of an exponential function $f(x) = b^x$ is the line _____.

h. The function defined by $f(x) = 1^x$ (is/is not) an exponential function.

Review Exercises

For Exercises 2–7, find the functions, using f and g as given. $f(x) = 2x^2 + x + 2 \qquad g(x) = 3x - 1$

2. $(f + g)(x)$ **3.** $(g - f)(x)$ **4.** $(f \cdot g)(x)$

5. $\left(\dfrac{g}{f}\right)(x)$ **6.** $(f \circ g)(x)$ **7.** $(g \circ f)(x)$

8. Find the inverse. $\{(2, 3), (0, 0), (-8, 4), (10, 12)\}$

For Exercises 9–16, evaluate the expression without the use of a calculator.

9. 5^2 **10.** 2^{-3} **11.** 10^{-3} **12.** 3^4

13. $36^{1/2}$ **14.** $27^{1/3}$ **15.** $16^{3/4}$ **16.** $8^{2/3}$

Concept 2: Approximating Exponential Expressions with a Calculator

For Exercises 17–24, evaluate the expression by using a calculator. Round to four decimal places. **(See Example 1.)**

17. $5^{1.1}$ **18.** $2^{\sqrt{3}}$ **19.** 10^{π} **20.** $3^{4.8}$

21. $36^{-\sqrt{2}}$ **22.** $27^{-0.5126}$ **23.** $16^{-0.04}$ **24.** $8^{-0.61}$

25. Solve for x.

     a. $3^x = 9$ **b.** $3^x = 27$ **c.** Between what two consecutive integers must the solution to $3^x = 11$ lie?

26. Solve for x.

     a. $5^x = 125$ **b.** $5^x = 625$ **c.** Between what two consecutive integers must the solution to $5^x = 130$ lie?

27. Solve for x.

     a. $2^x = 16$ **b.** $2^x = 32$ **c.** Between what two consecutive integers must the solution to $2^x = 30$ lie?

28. Solve for x.

     a. $4^x = 16$ **b.** $4^x = 64$ **c.** Between what two consecutive integers must the solution to $4^x = 20$ lie?

29. For $f(x) = \left(\frac{1}{5}\right)^x$ find $f(0), f(1), f(2), f(-1),$ and $f(-2)$.

30. For $g(x) = \left(\frac{2}{3}\right)^x$ find $g(0), g(1), g(2), g(-1),$ and $g(-2)$.

31. For $h(x) = 3^x$ use a calculator to find $h(0), h(1), h(-1), h\left(\sqrt{2}\right),$ and $h(\pi)$. Round to two decimal places.

32. For $k(x) = 5^x$ use a calculator to find $k(0), k(1), k(-1), k(\pi),$ and $k\left(\sqrt{2}\right)$. Round to two decimal places.

Concept 3: Graphs of Exponential Functions

33. How do you determine whether the graph of $f(x) = b^x$ $(b > 0, b \neq 1)$ is increasing or decreasing?

34. For $f(x) = b^x$ $(b > 0, b \neq 1)$, find $f(0)$.

Graph the functions defined in Exercises 35–42. Plot at least three points for each function. **(See Example 2.)**

35. $f(x) = 4^x$

36. $g(x) = 6^x$

37. $m(x) = \left(\dfrac{1}{8}\right)^x$

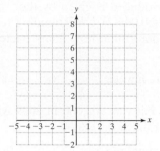
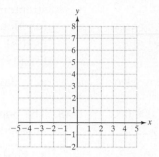
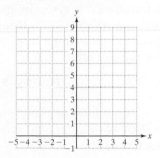

38. $n(x) = \left(\dfrac{1}{3}\right)^x$

 39. $h(x) = 2^{x+1}$

40. $k(x) = 5^{x-1}$

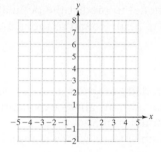
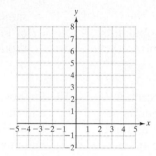
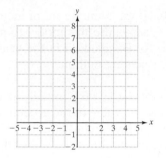

41. $g(x) = 5^{-x}$

42. $f(x) = 2^{-x}$

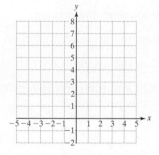
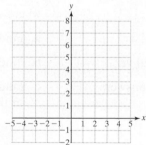

Concept 4: Applications of Exponential Functions

43. The half-life of the element radon (Rn 86) is 3.8 days. In a sample originally containing 1 g of radon, the amount left after t days is given by $A(t) = (0.5)^{t/3.8}$. (Round to two decimal places, if necessary.) **(See Example 3.)**

   a. How much radon will be present after 7.6 days?

   b. How much radon will be present after 10 days?

44. Nobelium, an element discovered in 1958, has a half-life of 10 min under certain conditions. In a sample containing 1 g of nobelium, the amount left after t min is given by $A(t) = (0.5)^{t/10}$. (Round to three decimal places.)

   a. How much nobelium is left after 5 min?

   b. How much nobelium is left after 1 hr?

45. Once an antibiotic is introduced to bacteria, the number of bacteria decreases exponentially. For example, beginning with 1 million bacteria, the amount present t days from the time penicillin is introduced is given by the function $A(t) = 1,000,000(2)^{-t/5}$. Rounding to the nearest thousand, determine how many bacteria are present after

a. 2 days   **b.** 1 week   **c.** 2 weeks

46. Once an antibiotic is introduced to bacteria, the number of bacteria decreases exponentially. For example, beginning with 1 million bacteria, the amount present t days from the time streptomycin is introduced is given by the function $A(t) = 1,000,000(2)^{-t/10}$. Rounding to the nearest thousand, determine how many bacteria are present after

a. 5 days   **b.** 1 week   **c.** 2 weeks

47. The population of Bangladesh was 153,000,000 in 2009 with an annual growth rate of 1.25%. (See Example 4.)

a. Find a mathematical model that relates the population of Bangladesh as a function of the number of years after 2009.

b. If the annual rate of increase remains the same, use this model to predict the population of Bangladesh in the year 2050. Round to the nearest million.

48. The population of Fiji was 908,000 in 2009 with an annual growth rate of 0.07%.

a. Find a mathematical model that relates the population of Fiji as a function of the number of years after 2009.

b. If the annual rate of increase remains the same, use this model to predict the population of Fiji in the year 2050. Round to the nearest thousand.

49. Suppose $1000 is initially invested in an account and the value of the account grows exponentially. If the investment doubles in 7 yr, then the amount in the account t years after the initial investment is given by

$$A(t) = 1000(2)^{t/7}$$

where $A(t)$ is the amount in the account.

a. Find the amount in the account after 5 yr.

b. Find the amount in the account after 10 yr.

c. Find $A(0)$ and $A(7)$ and interpret the answers in the context of this problem.

50. Suppose $1500 is initially invested in an account and the value of the account grows exponentially. If the investment doubles in 8 yr, then the amount in the account t years after the initial investment is given by

$$A(t) = 1500(2)^{t/8}$$

where $A(t)$ is the amount in the account.

a. Find the amount in the account after 5 yr.

b. Find the amount in the account after 10 yr.

c. Find $A(0)$ and $A(8)$ and interpret the answers in the context of this problem.

Graphing Calculator Exercises

For Exercises 51–58, graph the functions on your calculator to support your solutions to the indicated exercises.

51. $f(x) = 4^x$ (see Exercise 35)

52. $g(x) = 6^x$ (see Exercise 36)

53. $m(x) = \left(\frac{1}{8}\right)^x$ (see Exercise 37)

54. $n(x) = \left(\frac{1}{3}\right)^x$ (see Exercise 38)

55. $h(x) = 2^{x+1}$ (see Exercise 39)

56. $k(x) = 5^{x-1}$ (see Exercise 40)

57. $g(x) = 5^{-x}$ (see Exercise 41)

58. $f(x) = 2^{-x}$ (see Exercise 42)

Logarithmic Functions

1. Definition of a Logarithmic Function

Consider the following equations in which the variable is located in the exponent of an expression. In some cases the solution can be found by inspection because the constant on the right-hand side of the equation is a perfect power of the base.

Equation	Solution
$5^x = 5$	$x = 1$
$5^x = 20$	$x = ?$
$5^x = 25$	$x = 2$
$5^x = 60$	$x = ?$
$5^x = 125$	$x = 3$

The equation $5^x = 20$ cannot be solved by inspection. However, we might suspect that x is between 1 and 2. Similarly, the solution to the equation $5^x = 60$ is between 2 and 3. To solve an exponential equation for an unknown exponent, we must use a new type of function called a logarithmic function.

> ### Definition of a Logarithmic Function
>
> If x and b are positive real numbers such that $b \neq 1$, then $y = \log_b x$ is called the **logarithmic function** with base b and
>
> $$y = \log_b x \quad \text{is equivalent to} \quad b^y = x$$
>
> *Note:* In the expression $y = \log_b x$, y is called the **logarithm**, b is called the **base**, and x is called the **argument**.

The expression $y = \log_b x$ is equivalent to $b^y = x$ and indicates that *the logarithm y is the exponent to which b must be raised to obtain x.* The expression $y = \log_b x$ is called the logarithmic form of the equation, and the expression $b^y = x$ is called the exponential form of the equation.

The definition of a logarithmic function suggests a close relationship with an exponential function of the same base. In fact, a logarithmic function is the inverse of the corresponding exponential function. For example, the following steps find the inverse of the exponential function defined by $f(x) = b^x$.

$f(x) = b^x$	
$y = b^x$	Replace $f(x)$ by y.
$x = b^y$	Interchange x and y.
$y = \log_b x$	Solve for y using the definition of a logarithmic function.
$f^{-1}(x) = \log_b x$	Replace y by $f^{-1}(x)$.

Example 1 **Converting from Logarithmic Form to Exponential Form**

Rewrite the logarithmic equations in exponential form.

a. $\log_2 32 = 5$ **b.** $\log_{10}\left(\dfrac{1}{1000}\right) = -3$ **c.** $\log_5 1 = 0$

Concepts

1. Definition of a Logarithmic Function
2. Evaluating Logarithmic Expressions
3. The Common Logarithmic Function
4. Graphs of Logarithmic Functions
5. Applications of the Common Logarithmic Function

Concept Connections

1. Write the logarithmic equation in exponential form.
$a = \log_r w$

Answer

1. $r^a = w$

Skill Practice

Rewrite the logarithmic equations in exponential form.

2. $\log_3 9 = 2$

3. $\log_{10}\left(\dfrac{1}{100}\right) = -2$

4. $\log_8 1 = 0$

Solution:

Because the concept of a logarithm is new and unfamiliar, it may be advantageous to rewrite a logarithm in its equivalent exponential form.

Logarithmic Form		**Exponential Form**
a. $\log_2 32 = 5$	$\Leftrightarrow$	$2^5 = 32$
b. $\log_{10}\left(\dfrac{1}{1000}\right) = -3$	$\Leftrightarrow$	$10^{-3} = \dfrac{1}{1000}$
c. $\log_5 1 = 0$	$\Leftrightarrow$	$5^0 = 1$

2. Evaluating Logarithmic Expressions

Example 2 Evaluating Logarithmic Expressions

Evaluate the logarithmic expressions.

a. $\log_{10} 10{,}000$ **b.** $\log_5\left(\dfrac{1}{125}\right)$ **c.** $\log_{1/2}\left(\dfrac{1}{8}\right)$

Skill Practice

Evaluate the logarithmic expressions.

5. $\log_{10} 1000$

6. $\log_4\left(\dfrac{1}{16}\right)$

7. $\log_{1/3} 3$

Solution:

a. $\log_{10} 10{,}000$ is the exponent to which 10 must be raised to obtain 10,000.

$\quad y = \log_{10} 10{,}000$ Let y represent the value of the logarithm.

$\quad 10^y = 10{,}000$ Rewrite the expression in exponential form.

$\quad 10^y = 10^4$

$\quad y = 4$ Therefore, $\log_{10} 10{,}000 = 4$.

b. $\log_5\left(\frac{1}{125}\right)$ is the exponent to which 5 must be raised to obtain $\frac{1}{125}$.

$\quad y = \log_5\left(\dfrac{1}{125}\right)$ Let y represent the value of the logarithm.

$\quad 5^y = \dfrac{1}{125}$ Rewrite the expression in exponential form.

$\quad 5^y = \dfrac{1}{5^3} = 5^{-3}$

$\quad y = -3$ Therefore, $\log_5\left(\frac{1}{125}\right) = -3$.

c. $\log_{1/2}\left(\frac{1}{8}\right)$ is the exponent to which $\frac{1}{2}$ must be raised to obtain $\frac{1}{8}$.

$\quad y = \log_{1/2}\left(\dfrac{1}{8}\right)$ Let y represent the value of the logarithm.

$\quad \left(\dfrac{1}{2}\right)^y = \dfrac{1}{8}$ Rewrite the expression in exponential form.

$\quad \left(\dfrac{1}{2}\right)^y = \left(\dfrac{1}{2}\right)^3$

$\quad y = 3$ Therefore, $\log_{1/2}\left(\frac{1}{8}\right) = 3$.

Answers

2. $3^2 = 9$ **3.** $10^{-2} = \dfrac{1}{100}$

4. $8^0 = 1$ **5.** 3

6. -2 **7.** -1

Example 3 Evaluating Logarithmic Expressions

Evaluate the logarithmic expressions.

a. $\log_b b$ **b.** $\log_c c^7$ **c.** $\log_3 \sqrt[4]{3}$

Solution:

a. $\log_b b$ is the exponent to which b must be raised to obtain b.

$y = \log_b b$	Let y represent the value of the logarithm.
$b^y = b$	Rewrite the expression in exponential form.
$y = 1$	Therefore, $\log_b b = 1$.

b. $\log_c c^7$ is the exponent to which c must be raised to obtain c^7.

$y = \log_c c^7$	Let y represent the value of the logarithm.
$c^y = c^7$	Rewrite the expression in exponential form.
$y = 7$	Therefore, $\log_c c^7 = 7$.

c. $\log_3 \sqrt[4]{3} = \log_3 3^{1/4}$ is the exponent to which 3 must be raised to obtain $3^{1/4}$.

$y = \log_3 3^{1/4}$	Let y represent the value of the logarithm.
$3^y = 3^{1/4}$	Rewrite the expression in exponential form.
$y = \dfrac{1}{4}$	Therefore, $\log_3 \sqrt[4]{3} = \frac{1}{4}$.

Skill Practice

Evaluate the logarithmic expressions.

8. $\log_x x$

9. $\log_b b^{10}$

10. $\log_5 \sqrt[3]{5}$

3. The Common Logarithmic Function

The logarithmic function with base 10 is called the **common logarithmic function** and is denoted by $y = \log x$. Notice that the base is not explicitly written but is understood to be 10. That is, $y = \log_{10} x$ is written simply as $y = \log x$.

Example 4 Evaluating Common Logarithms

Evaluate the logarithmic expressions.

a. $\log 100{,}000$ **b.** $\log 0.01$

Solution:

a. $\log 100{,}000$ is the exponent to which 10 must be raised to obtain 100,000.

$y = \log 100{,}000$	Let y represent the value of the logarithm.
$10^y = 100{,}000$	Rewrite the expression in exponential form.
$10^y = 10^5$	
$y = 5$	Therefore, $\log 100{,}000 = 5$.

b. $\log 0.01$ is the exponent to which 10 must be raised to obtain 0.01 or $\frac{1}{100}$.

$y = \log 0.01$	Let y represent the value of the logarithm.
$10^y = 0.01$	Rewrite the expression in exponential form.
$10^y = 10^{-2}$	Note that $0.01 = \frac{1}{100}$ or 10^{-2}.
$y = -2$	Therefore, $\log 0.01 = -2$.

Skill Practice

Evaluate the logarithmic expressions.

11. $\log 1{,}000{,}000$

12. $\log 0.0001$

Answers

8. 1   **9.** 10

10. $\dfrac{1}{3}$ **11.** 6

12. -4

On most calculators the LOG key is used to compute logarithms with base 10. This is demonstrated in Example 5.

Example 5 Evaluating Common Logarithms on a Calculator

Evaluate the common logarithms. Round the answers to four decimal places.

a. $\log 420$ **b.** $\log (8.2 \times 10^9)$ **c.** $\log 0.0002$

Solution:

a. $\log 420 \approx 2.6232$

b. $\log (8.2 \times 10^9) \approx 9.9138$

c. $\log 0.0002 \approx -3.6990$

Calculator Connections

Topic: Evaluating Common Logarithms

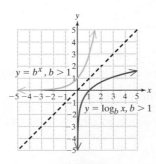

```
log(420)
         2.62324929
log(8.2E9)
         9.913813852
log(0.0002)
        -3.698970004
```

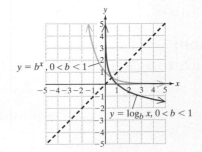

TIP: On a scientific calculator, you may need to enter the logarithm and argument in reverse order. For example: 420 log .

4. Graphs of Logarithmic Functions

In Section 9.2 we studied the graphs of exponential functions. In this section, we will graph logarithmic functions. First, recall that $f(x) = \log_b x$ is the inverse of $g(x) = b^x$. Therefore, the graph of $y = f(x)$ is symmetric to the graph of $y = g(x)$ about the line $y = x$, as shown in Figures 9-10 and 9-11.

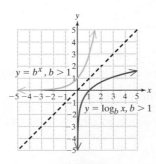

Figure 9-10   **Figure 9-11**

From Figures 9-10 and 9-11, we see that the range of $y = b^x$ is the set of positive real numbers. As expected, the domain of its inverse, the logarithmic function $y = \log_b x$, is also the set of positive real numbers. Therefore, the **domain of the logarithmic function** $y = \log_b x$ is the set of positive real numbers.

Example 6 Graphing Logarithmic Functions

Graph the functions and compare the graphs to examine the effect of the base on the shape of the graph.

a. $y = \log_2 x$ **b.** $y = \log x$

Solution:

We can write each equation in its equivalent exponential form and create a table of values (Table 9-5). To simplify the calculations, choose integer values of y and then solve for x.

$$y = \log_2 x \quad \text{or} \quad 2^y = x \qquad\qquad y = \log x \quad \text{or} \quad 10^y = x$$

The graphs of $y = \log_2 x$ and $y = \log x$ are shown in Figure 9-12. Both graphs exhibit the same general behavior, and the steepness of the curve is affected by the base. The function $y = \log x$ requires a 10-fold increase in x to increase the y-value by 1 unit. The function $y = \log_2 x$ requires a 2-fold increase in x to increase the y-value by 1 unit. In addition, the following characteristics are true for both graphs.

- The domain is $(0, \infty)$.
- The range is $(-\infty, \infty)$.
- The line $x = 0$ (y-axis) is a vertical asymptote.
- Both graphs pass through the point $(1, 0)$.

Choose values for y.

Table 9-5

$x = 2^y$	$x = 10^y$	y
$\frac{1}{8}$	$\frac{1}{1000}$	-3
$\frac{1}{4}$	$\frac{1}{100}$	-2
$\frac{1}{2}$	$\frac{1}{10}$	-1
1	1	0
2	10	1
4	100	2
8	1000	3

Solve for x.

Skill Practice

Graph the function.

16. $y = \log_3 x$

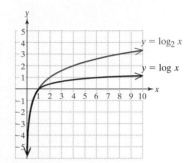

Figure 9-12

Example 6 illustrates that a logarithmic function with base $b > 1$ is an increasing function. In Example 7, we see that if the base b is between 0 and 1, the function decreases over its entire domain.

Skill Practice

17. Graph $y = \log_{1/2} x$.

Example 7 Graphing a Logarithmic Function

Graph $y = \log_{1/4} x$.

Solution:

The equation $y = \log_{1/4} x$ can be written in exponential form as $\left(\frac{1}{4}\right)^y = x$. By choosing several values for y, we can solve for x and plot the corresponding points (Table 9-6).

The expression $y = \log_{1/4} x$ defines a decreasing logarithmic function (Figure 9-13). Notice that the vertical asymptote, domain, and range are the same for both increasing and decreasing logarithmic functions.

Answers

16.

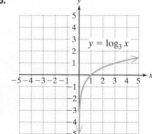

17.

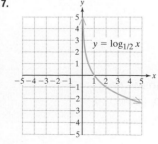

Choose values for y.

Table 9-6

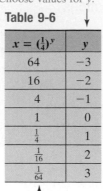

$x = \left(\frac{1}{4}\right)^y$	y
64	-3
16	-2
4	-1
1	0
$\frac{1}{4}$	1
$\frac{1}{16}$	2
$\frac{1}{64}$	3

Solve for x.

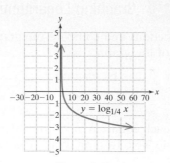

Figure 9-13

When graphing a logarithmic equation, it is helpful to know its domain.

Skill Practice

Find the domain of the functions.

18. $f(x) = \log_3 (x + 7)$

19. $g(x) = \log (4 - 8x)$

Example 8   Identifying the Domain of a Logarithmic Function

Find the domain of the functions.

a. $f(x) = \log (4 - x)$ **b.** $g(x) = \log (2x + 6)$

Solution:

The domain of the function $y = \log_b x$ is the set of all positive real numbers. That is, the argument x must be greater than zero: $x > 0$.

a. $f(x) = \log (4 - x)$ The argument is $4 - x$.

$4 - x > 0$ The argument of the logarithm must be greater than zero.

$-x > -4$ Solve for x.

$x < 4$ Divide by -1 and reverse the inequality sign.

The domain is $(-\infty, 4)$.

b. $g(x) = \log (2x + 6)$ The argument is $2x + 6$.

$2x + 6 > 0$ The argument of the logarithm must be greater than zero.

$2x > -6$

$x > -3$

The domain is $(-3, \infty)$.

Answers

18. Domain: $(-7, \infty)$

19. Domain: $\left(-\infty, \frac{1}{2}\right)$

Topic: Graphing Common Logarithmic Functions

The graphs of $Y_1 = \log(4 - x)$ and $Y_2 = \log(2x + 6)$ are shown here and can be used to confirm the solutions to Example 8. Notice that each function has a vertical asymptote at the value of x where the argument equals zero.

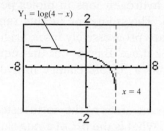
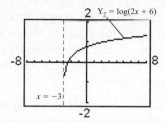

The general shape and important features of exponential and logarithmic functions are summarized as follows.

Graphs of Exponential and Logarithmic Functions—A Summary

Exponential Functions

$y = b^x$

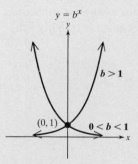

Logarithmic Functions

$y = \log_b x$

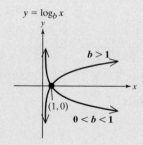

Domain: $(-\infty, \infty)$

Range: $(0, \infty)$

Horizontal asymptote: $y = 0$

Passes through $(0, 1)$

If $b > 1$, the function is increasing.

If $0 < b < 1$, the function is decreasing.

Domain: $(0, \infty)$

Range: $(-\infty, \infty)$

Vertical asymptote: $x = 0$

Passes through $(1, 0)$

If $b > 1$, the function is increasing.

If $0 < b < 1$, the function is decreasing.

Notice that the roles of x and y are interchanged for the functions $y = b^x$ and $b^y = x$. Therefore, it is not surprising that the domain and range are reversed between exponential and logarithmic functions. Moreover, an exponential function passes through $(0, 1)$, whereas a logarithmic function passes through $(1, 0)$. An exponential function has a horizontal asymptote at $y = 0$, whereas a logarithmic function has a vertical asymptote at $x = 0$.

5. Applications of the Common Logarithmic Function

Example 9 Applying a Common Logarithm to Compute pH

The pH (hydrogen potential) of a solution is defined as

$$pH = -\log [H^+]$$

where $[H^+]$ represents the concentration of hydrogen ions in moles per liter (mol/L). The pH scale ranges from 0 to 14. The midpoint of this range, 7, represents a neutral solution. Values below 7 are progressively more acidic, and values above 7 are progressively more alkaline. Based on the equation $pH = -\log [H^+]$, a 1-unit change in pH means a 10-fold change in hydrogen ion concentration.

a. Normal rain has a pH of 5.6. However, in some areas of the northeastern United States the rainwater is more acidic. What is the pH of a rain sample for which the concentration of hydrogen ions is 0.0002 mol/L?

b. Find the pH of household ammonia if the concentration of hydrogen ions is 1.0×10^{-11} mol/L.

Solution:

a. $pH = -\log [H^+]$

$\quad = -\log (0.0002) \qquad$ Substitute $[H^+] = 0.0002$.

$\quad \approx 3.7 \qquad$ (To compare this value with a familiar substance, note that the pH of orange juice is roughly 3.5.)

b. $pH = -\log [H^+]$

$\quad = -\log (1.0 \times 10^{-11}) \qquad$ Substitute $[H^+] = 1.0 \times 10^{-11}$.

$\quad = -\log (10^{-11})$

$\quad = -(-11)$

$\quad = 11$

The pH of household ammonia is 11.

Example 10 Applying Logarithmic Functions to a Memory Model

One method of measuring a student's retention of material after taking a course is to retest the student at specified time intervals after the course has been completed. A student's score on a calculus test t months after completing a course in calculus is approximated by

$$S(t) = 85 - 25 \log (t + 1)$$

where t is the time in months after completing the course and $S(t)$ is the student's score as a percent.

a. What was the student's score at $t = 0$?

b. What was the student's score after 2 months?

c. What was the student's score after 1 year?

Solution:

a. $S(t) = 85 - 25 \log (t + 1)$

$S(0) = 85 - 25 \log (0 + 1)$ Substitute $t = 0$.

     $= 85 - 25 \log 1$ $\log 1 = 0$ because $10^0 = 1$.

     $= 85 - 25(0)$

     $= 85 - 0$

     $= 85$ The student's score at the time the course was completed was 85%.

b. $S(t) = 85 - 25 \log (t + 1)$

$S(2) = 85 - 25 \log (2 + 1)$

     $= 85 - 25 \log 3$ Use a calculator to approximate $\log 3$.

     ≈ 73.1 The student's score dropped to 73.1%.

c. $S(t) = 85 - 25 \log (t + 1)$

$S(12) = 85 - 25 \log (12 + 1)$

     $= 85 - 25 \log 13$ Use a calculator to approximate $\log 13$.

     ≈ 57.2 The student's score dropped to 57.2%.

Section 9.3   Practice Exercises

Vocabulary and Key Concepts

1. a. The function defined by $y = \log_b x$ is called the _____ function with base _____. The values of x and b are positive, and $b \neq 1$.

b. Given $y = \log_b x$, we call y the _____, we call b the _____, and we call x the _____.

c. In interval notation, the domain of $y = \log_b x$ is _____ and the range is _____.

d. A logarithmic function base b is the inverse of the _____ function base b.

e. The logarithmic function base 10 is called the _____ logarithmic function and is denoted by $y = \log x$.

f. The graph of a logarithmic function is (increasing/decreasing) if the base $b > 1$, whereas the graph is (increasing/decreasing) if the base is between 0 and 1.

g. Which values of x are *not* in the domain of $y = \log(x - 4)$? $x = 2, x = 3, x = 4, x = 5, x = 6$

h. The graphs of a logarithmic function base b and an exponential function base b are symmetric with respect to the line _____.

Review Exercises

2. For which graph of $y = b^x$ is $0 < b < 1$?

3. For which graph of $y = b^x$ is $b > 1$?

i.

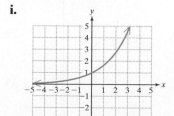

ii.

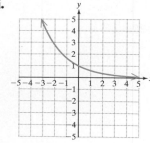

4. Let $f(x) = 6^x$.

  a. Find $f(-2), f(-1), f(0), f(1)$, and $f(2)$.

  b. Graph $y = f(x)$.

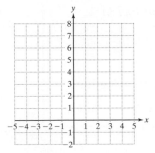

5. Let $g(x) = 3^x$.

  a. Find $g(-2), g(-1), g(0), g(1)$, and $g(2)$.

  b. Graph $y = g(x)$.

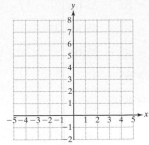

6. Let $r(x) = \left(\frac{3}{4}\right)^x$.

  a. Find $r(-2), r(-1), r(0), r(1)$, and $r(2)$.

  b. Graph $y = r(x)$.

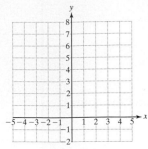

7. Let $s(x) = \left(\frac{2}{5}\right)^x$.

  a. Find $s(-2), s(-1), s(0), s(1)$, and $s(2)$.

  b. Graph $y = s(x)$.

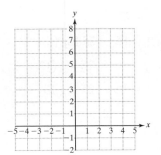

Concept 1: Definition of a Logarithmic Function

8. For the equation $y = \log_b x$, identify the base, the argument, and the logarithm.

9. Rewrite the equation in exponential form. $y = \log_b x$

10. Write the equation in logarithmic form. $b^y = x$

For Exercises 11–22, write the equation in exponential form. **(See Example 1.)**

11. $\log_5 625 = 4$

12. $\log_{125} 25 = \dfrac{2}{3}$

13. $\log_{10}(0.0001) = -4$

14. $\log_{25}\left(\dfrac{1}{5}\right) = -\dfrac{1}{2}$

15. $\log_6 36 = 2$

16. $\log_2 128 = 7$

17. $\log_b 15 = x$

18. $\log_b 82 = y$

19. $\log_3 5 = x$

20. $\log_2 7 = x$

21. $\log_{1/4} x = 10$

22. $\log_{1/2} x = 6$

For Exercises 23–34, write the equation in logarithmic form.

23. $3^x = 81$

24. $10^3 = 1000$

25. $5^2 = 25$

26. $8^{1/3} = 2$

27. $7^{-1} = \dfrac{1}{7}$

28. $8^{-2} = \dfrac{1}{64}$

29. $b^x = y$

30. $b^y = x$

31. $e^x = y$

32. $e^y = x$

33. $\left(\dfrac{1}{3}\right)^{-2} = 9$

34. $\left(\dfrac{5}{2}\right)^{-1} = \dfrac{2}{5}$

Concept 2: Evaluating Logarithmic Expressions

For Exercises 35–50, evaluate the logarithms without the use of a calculator. **(See Examples 2 and 3.)**

35. $\log_7 49$

36. $\log_3 81$

37. $\log_{10} 0.1$

 38. $\log_2 \left(\dfrac{1}{16} \right)$

39. $\log_{16} 4$

40. $\log_8 2$

41. $\log_{7/2} 1$

42. $\log_{1/2} 2$

43. $\log_3 3^5$

44. $\log_9 9^3$

45. $\log_{10} 10$

46. $\log_7 1$

47. $\log_a a^3$

48. $\log_r r^4$

49. $\log_x \sqrt{x}$

50. $\log_y \sqrt[3]{y}$

Concept 3: The Common Logarithmic Function

For Exercises 51–58, evaluate the common logarithm without the use of a calculator. **(See Example 4.)**

51. $\log 10$

52. $\log 100$

 53. $\log 1000$

54. $\log 10,000$

55. $\log (1.0 \times 10^6)$

56. $\log 0.1$

57. $\log 0.01$

58. $\log 0.001$

For Exercises 59–70, use a calculator to approximate the logarithms. Round to four decimal places. **(See Example 5.)**

59. $\log 6$

60. $\log 18$

61. $\log \pi$

62. $\log \left(\dfrac{1}{8} \right)$

63. $\log \left(\dfrac{1}{32} \right)$

64. $\log \sqrt{5}$

65. $\log 0.0054$

66. $\log 0.0000062$

67. $\log (3.4 \times 10^5)$

68. $\log (4.78 \times 10^9)$

69. $\log (3.8 \times 10^{-8})$

70. $\log (2.77 \times 10^{-4})$

71. Given: $\log 10 = 1$ and $\log 100 = 2$

    a. Estimate $\log 93$.

    b. Estimate $\log 12$.

    c. Evaluate the logarithms in parts (a) and (b) on a calculator and compare to your estimates.

72. Given: $\log \left(\frac{1}{10} \right) = -1$ and $\log 1 = 0$

    a. Estimate $\log \left(\frac{9}{10} \right)$.

    b. Estimate $\log \left(\frac{1}{5} \right)$.

    c. Evaluate the logarithms in parts (a) and (b) on a calculator and compare to your estimates.

Concept 4: Graphs of Logarithmic Functions

73. Let $f(x) = \log_4 x$.

    a. Find the values of $f(\frac{1}{64}), f(\frac{1}{16}), f(\frac{1}{4}), f(1), f(4),$ $f(16),$ and $f(64)$.

    b. Graph $y = f(x)$. **(See Example 6.)**

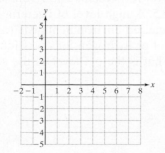

74. Let $g(x) = \log_2 x$.

    a. Find the values of $g(\frac{1}{8}), g(\frac{1}{4}), g(\frac{1}{2}), g(1), g(2),$ $g(4),$ and $g(8)$.

    b. Graph $y = g(x)$.

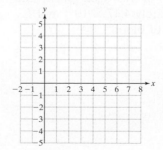

Graph the logarithmic functions in Exercises 75–78 by writing the function in exponential form and making a table of points.

75. $y = \log_3 x$

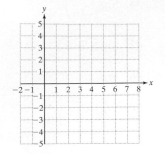

x	y
	−2
	−1
	0
	1
	2

76. $y = \log_5 x$

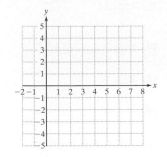

x	y
	−2
	−1
	0
	1
	2

77. $y = \log_{1/2} x$ **(See Example 7.)**

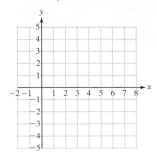

x	y
	−2
	−1
	0
	1
	2

78. $y = \log_{1/3} x$

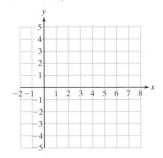

x	y
	−2
	−1
	0
	1
	2

For Exercises 79–90, find the domain of the function and express the domain in interval notation. **(See Example 8.)**

79. $f(x) = \log_7 (x - 5)$

80. $h(x) = \log (x - 4)$

81. $g(x) = \log (2 - x)$

82. $g(x) = \log (1 - x)$

83. $h(x) = \log (3x - 1)$

84. $f(x) = \log_3 (2x + 1)$

85. $f(x) = \log_3 (x + 1.2)$

86. $f(x) = \log \left(x - \dfrac{1}{2} \right)$

87. $g(x) = \log (4 - 2x)$

88. $h(x) = \log (6 - 2x)$

89. $f(x) = \log x^2$

90. $f(x) = \log (x^2 + 1)$

Concept 5: Applications of the Common Logarithmic Function

For Exercises 91 and 92, use the formula $pH = -\log[H^+]$, where $[H^+]$ represents the concentration of hydrogen ions in moles per liter. Round to two decimal places. **(See Example 9.)**

91. Normally, the level of hydrogen ions in the blood is approximately 4.47×10^{-8} mol/L. Find the pH level of blood.

92. The normal pH level for streams and rivers is between 6.5 and 8.5. A high level of bacteria in a particular stream caused environmentalists to test the water. The level of hydrogen ions was found to be 0.006 mol/L. What is the pH level of the stream?

93. A graduate student in education is doing research to compare the effectiveness of two different teaching techniques designed to teach vocabulary to sixth-graders. The first group of students (group 1) was taught with method I, in which the students worked individually to complete the assignments in a workbook. The second group (group 2) was taught with method II, in which the students worked cooperatively in groups of four to complete the assignments in the same workbook.

None of the students knew any of the vocabulary words before the study began. After completing the assignments in the workbook, the students were then tested on the vocabulary at 1-month intervals to assess how much material they had retained over time. The students' average score t months after completing the assignments are given by the following functions:

Method I: $S_1(t) = 91 - 30 \log (t + 1)$, where t is the time in months and $S_1(t)$ is the average score of students in group 1.

Method II: $S_2(t) = 88 - 15 \log (t + 1)$, where t is the time in months and $S_2(t)$ is the average score of students in group 2.

a. Complete the table to find the average scores for each group of students after the indicated number of months. Round to one decimal place. **(See Example 10.)**

b. Based on the table of values, what were the average scores for each group immediately after completion of the assigned material $(t = 0)$?

t (months)	0	1	2	6	12	24
$S_1(t)$						
$S_2(t)$						

c. Based on the table of values, which teaching method helped students retain the material better for a long time?

94. Generally, the more money a company spends on advertising, the higher the sales. Let a represent the amount of money spent on advertising (in \$100s). Then the amount of money in sales $S(a)$ (in \$1000s) is given by

$$S(a) = 10 + 20 \log (a + 1) \text{ where } a \geq 0.$$

a. The value of $S(1) \approx 16.0$ means that if \$100 is spent on advertising, \$16,000 is returned in sales. Find the values of $S(11)$, $S(21)$, and $S(31)$. Round to one decimal place. Interpret the meaning of each function value in the context of this problem.

b. The graph of $y = S(a)$ is shown here. Use the graph and your answers from part (a) to explain why the money spent in advertising becomes less "efficient" as it is used in larger amounts.

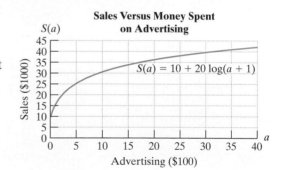

Graphing Calculator Exercises

For Exercises 95–100, graph the function on an appropriate viewing window. From the graph, identify the domain of the function and the location of the vertical asymptote.

95. $y = \log (x + 6)$

96. $y = \log (2x + 4)$

97. $y = \log (0.5x - 1)$

98. $y = \log (x + 8)$

99. $y = \log (2 - x)$

100. $y = \log (3 - x)$

Problem Recognition Exercises

Identifying Graphs of Functions

Match the function with the appropriate graph. Do not use a calculator.

1. $g(x) = 3^x$ **2.** $f(x) = \log x$ **3.** $h(x) = x^2$ **4.** $k(x) = -2x - 3$

5. $L(x) = |x|$ **6.** $m(x) = \sqrt{x}$ **7.** $B(x) = 3$ **8.** $A(x) = -x^2$

9. $n(x) = \sqrt[3]{x}$ **10.** $p(x) = x^3$ **11.** $q(x) = \dfrac{1}{x}$ **12.** $r(x) = x$

a.
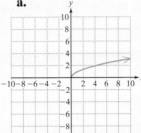

b.
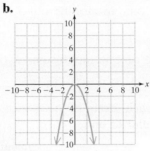

c.
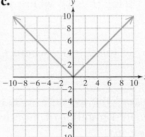

d.
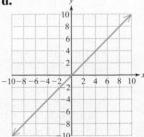

e.
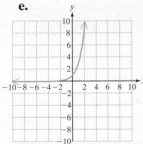

f.
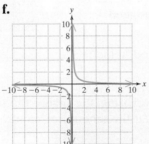

g.
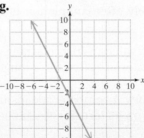

h.
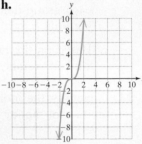

i.
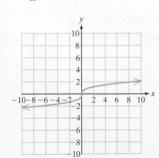

j.
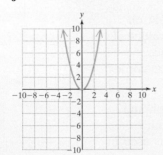

k.
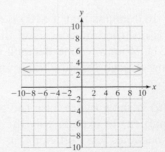

l.
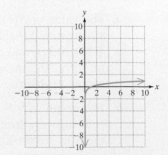

Properties of Logarithms

1. Properties of Logarithms

You have already been exposed to certain properties of logarithms that follow directly from the definition. Recall

$$y = \log_b x \quad \text{is equivalent to} \quad b^y = x \quad \text{for } x > 0, b > 0, \text{ and } b \neq 1$$

The following properties follow directly from the definition.

Property	Explanation
1. $\log_b 1 = 0$	Exponential form: $b^0 = 1$
2. $\log_b b = 1$	Exponential form: $b^1 = b$
3. $\log_b b^p = p$	Exponential form: $b^p = b^p$
4. $b^{\log_b x} = x$	Logarithmic form: $\log_b x = \log_b x$

Concepts

1. Properties of Logarithms
2. Expanded Logarithmic Expressions
3. Single Logarithmic Expressions

Example 1 **Applying the Properties of Logarithms to Simplify Expressions**

Use the properties of logarithms to simplify the expressions. Assume that all variable expressions within the logarithms represent positive real numbers.

a. $\log_8 8 + \log_8 1$ **b.** $10^{\log (x+2)}$ **c.** $\log_{1/2}\left(\dfrac{1}{2}\right)^x$

Solution:

a. $\log_8 8 + \log_8 1 = 1 + 0 = 1$ Properties 2 and 1

b. $10^{\log (x+2)} = x + 2$ Property 4

c. $\log_{1/2}\left(\dfrac{1}{2}\right)^x = x$ Property 3

Skill Practice

Use the properties of logarithms to simplify the expressions.

1. $\log_5 1 + \log_5 5$
2. $15^{\log_{15} 7}$
3. $\log_{1/3}\left(\dfrac{1}{3}\right)^c$

Three additional properties are useful when simplifying logarithmic expressions. The first is the product property for logarithms.

Product Property of Logarithms

Let b, x, and y be positive real numbers where $b \neq 1$. Then

$$\log_b (xy) = \log_b x + \log_b y$$

The logarithm of a product equals the sum of the logarithms of the factors.

Proof:

Let $M = \log_b x$, which implies $b^M = x$.

Let $N = \log_b y$, which implies $b^N = y$.

Then $xy = b^M b^N = b^{M+N}$.

Answers
1. 1   **2.** 7   **3.** c

Writing the expression $xy = b^{M+N}$ in logarithmic form, we have,

$$\log_b (xy) = M + N$$

$$\log_b (xy) = \log_b x + \log_b y \checkmark$$

To demonstrate the product property for logarithms, simplify the following expressions by using the order of operations.

$$\log_3 (3 \cdot 9) \overset{?}{=} \log_3 3 + \log_3 9$$

$$\log_3 27 \overset{?}{=} 1 + 2$$

$$3 \overset{?}{=} 3 \checkmark \quad \text{True}$$

Quotient Property of Logarithms

Let b, x, and y be positive real numbers where $b \neq 1$. Then

$$\log_b \left(\frac{x}{y} \right) = \log_b x - \log_b y$$

The logarithm of a quotient equals the difference of the logarithms of the numerator and denominator.

The proof of the quotient property for logarithms is similar to the proof of the product property and is omitted here. To demonstrate the quotient property for logarithms, simplify the following expressions by using the order of operations.

$$\log \left(\frac{1{,}000{,}000}{100} \right) \overset{?}{=} \log (1{,}000{,}000) - \log (100)$$

$$\log (10{,}000) \overset{?}{=} 6 - 2$$

$$4 \overset{?}{=} 4 \checkmark \quad \text{True}$$

Concept Connections

4. Use the product property of logarithms to simplify $\log_6 3 + \log_6 12$.
5. Use the quotient property for logarithms to simplify $\log_4 24 - \log_4 6$.
6. Use the power property for logarithms to simplify $\log_7 49^{12}$.

Power Property of Logarithms

Let b and x be positive real numbers where $b \neq 1$. Let p be any real number. Then

$$\log_b x^p = p \log_b x$$

To demonstrate the power property for logarithms, simplify the following expressions by using the order of operations.

$$\log_4 4^2 \overset{?}{=} 2 \log_4 4$$

$$2 \overset{?}{=} 2 \cdot 1$$

$$2 \overset{?}{=} 2 \checkmark \quad \text{True}$$

Answers
4. 2   **5.** 1   **6.** 24

The properties of logarithms are summarized in the box.

Summary of the Properties of Logarithms

Let b, x, and y be positive real numbers where $b \neq 1$, and let p be a real number. Then the following properties of logarithms are true.

1. $\log_b 1 = 0$	**5.** $\log_b(xy) = \log_b x + \log_b y$	**Product property for logarithms**
2. $\log_b b = 1$	**6.** $\log_b\left(\dfrac{x}{y}\right) = \log_b x - \log_b y$	**Quotient property for logarithms**
3. $\log_b b^p = p$	**7.** $\log_b x^p = p \log_b x$	**Power property for logarithms**
4. $b^{\log_b x} = x$		

---

2. Expanded Logarithmic Expressions

In many applications, it is advantageous to expand a logarithm into a sum or difference of simpler logarithms.

Example 2 **Writing a Logarithmic Expression in Expanded Form**

Write the expression as the sum or difference of logarithms of x, y, and z. Assume all variables represent positive real numbers.

$$\log_3\left(\frac{xy^3}{z^2}\right)$$

Solution:

$\log_3\left(\dfrac{xy^3}{z^2}\right)$

$= \log_3(xy^3) - \log_3 z^2$ Quotient property for logarithms (property 6)

$= [\log_3 x + \log_3 y^3] - \log_3 z^2$ Product property for logarithms (property 5)

$= \log_3 x + 3\log_3 y - 2\log_3 z$ Power property for logarithms (property 7)

Skill Practice

Write the expression as the sum or difference of logarithms of a, b, and c. Assume all variables represent positive real numbers.

7. $\log_5\left(\dfrac{a^2 b^3}{c}\right)$

Answer

7. $2\log_5 a + 3\log_5 b - \log_5 c$

Example 3 Writing a Logarithmic Expression in Expanded Form

Write the expression as the sum or difference of logarithms of x and y. Assume all variables represent positive real numbers.

$$\log\left(\frac{\sqrt{x + y}}{10}\right)$$

Solution:

$$\log\left(\frac{\sqrt{x + y}}{10}\right)$$

$$= \log\left(\sqrt{x + y}\right) - \log 10 \qquad \text{Quotient property for logarithms (property 6)}$$

$$= \log\left(x + y\right)^{1/2} - 1 \qquad \text{Write } \sqrt{x + y} \text{ as } (x + y)^{1/2} \text{ and simplify } \log 10 = 1.$$

$$= \frac{1}{2}\log\left(x + y\right) - 1 \qquad \text{Power property for logarithms (property 7)}$$

Example 4 Writing a Logarithmic Expression in Expanded Form

Write the expression as the sum or difference of logarithms of x, y, and z. Assume all variables represent positive real numbers.

$$\log_b \sqrt[5]{\frac{x^4}{yz^3}}$$

Solution:

$$\log_b \sqrt[5]{\frac{x^4}{yz^3}}$$

$$= \log_b\left(\frac{x^4}{yz^3}\right)^{1/5} \qquad \text{Write } \sqrt[5]{\frac{x^4}{yz^3}} \text{ as } \left(\frac{x^4}{yz^3}\right)^{1/5}.$$

$$= \frac{1}{5}\log_b\left(\frac{x^4}{yz^3}\right) \qquad \text{Power property for logarithms (property 7)}$$

$$= \frac{1}{5}\left[\log_b x^4 - \log_b\left(yz^3\right)\right] \qquad \text{Quotient property for logarithms (property 6)}$$

$$= \frac{1}{5}\left[\log_b x^4 - \left(\log_b y + \log_b z^3\right)\right] \qquad \text{Product property for logarithms (property 5)}$$

$$= \frac{1}{5}\left[\log_b x^4 - \log_b y - \log_b z^3\right] \qquad \text{Distributive property}$$

$$= \frac{1}{5}\left[4\log_b x - \log_b y - 3\log_b z\right] \qquad \text{Power property for logarithms (property 7)}$$

$$\text{or} \qquad \frac{4}{5}\log_b x - \frac{1}{5}\log_b y - \frac{3}{5}\log_b z$$

3. Single Logarithmic Expressions

In some applications, it is necessary to write a sum or difference of logarithms as a single logarithm.

Example 5 **Writing a Sum or Difference of Logarithms as a Single Logarithm**

Rewrite the expression as a single logarithm, and simplify the result, if possible.

$$\log_2 560 - \log_2 7 - \log_2 5$$

Solution:

$\log_2 560 - \log_2 7 - \log_2 5$

$= \log_2 560 - (\log_2 7 + \log_2 5)$	Factor out -1 from the last two terms.
$= \log_2 560 - \log_2 (7 \cdot 5)$	Product property for logarithms (property 5)
$= \log_2 \left(\dfrac{560}{7 \cdot 5} \right)$	Quotient property for logarithms (property 6)
$= \log_2 (16)$	Simplify inside parentheses.
$= \log_2 2^4$	Write 16 as 2^4.
$= 4$	Property 3

Skill Practice

Write the expression as a single logarithm, and simplify the result, if possible.

10. $\log_3 54 + \log_3 10 - \log_3 20$

Example 6 **Writing a Sum or Difference of Logarithms as a Single Logarithm**

Rewrite the expression as a single logarithm, and simplify the result, if possible. Assume all variable expressions within the logarithms represent positive real numbers.

$$2 \log x - \frac{1}{2} \log y + 3 \log z$$

Solution:

$2 \log x - \dfrac{1}{2} \log y + 3 \log z$

$= \log x^2 - \log y^{1/2} + \log z^3$	Power property for logarithms (property 7)
$= \log x^2 + \log z^3 - \log y^{1/2}$	Group terms with positive coefficients.
$= \log (x^2 z^3) - \log y^{1/2}$	Product property for logarithms (property 5)
$= \log \left(\dfrac{x^2 z^3}{y^{1/2}} \right)$ or $\log \left(\dfrac{x^2 z^3}{\sqrt{y}} \right)$	Quotient property for logarithms (property 6)

Skill Practice

Write the expression as a single logarithm, and simplify, if possible.

11. $3 \log x + \dfrac{1}{3} \log y - 2 \log z$

Answers

10. 3   **11.** $\log \left(\dfrac{x^3 \sqrt[3]{y}}{z^2} \right)$

It is important to note that the properties of logarithms may be used to write a single logarithm as a sum or difference of logarithms. Furthermore, the properties may be used to write a sum or difference of logarithms as a single logarithm. In either case, these operations may change the domain.

For example, consider the function $y = \log_b x^2$. Using the power property for logarithms, we have $y = 2 \log_b x$. Consider the domain of each function:

$y = \log_b x^2$ Domain: $(-\infty, 0) \cup (0, \infty)$

$y = 2 \log_b x$ Domain: $(0, \infty)$

These two functions are equivalent only for values of x in the intersection of the two domains, that is, for $(0, \infty)$.

Section 9.4 Practice Exercises

Vocabulary and Key Concepts

1. a. Fill in the blanks to complete the basic properties of logarithms.
   $\log_b b = $ _____, $\log_b 1 = $ _____, $\log_b b^x = $ _____, and $b^{\log_b x} = $ _____.

b. If b, x, and y are positive real numbers and $b \neq 1$, then $\log_b(xy) = $ _____ and $\log_b\left(\dfrac{x}{y}\right) = $ _____.

c. If b and x are positive real numbers and $b \neq 1$, then for any real number p, $\log_b x^p$ can be written as

   _____.

d. Determine if the statement is true or false: $\log_b(xy) = (\log_b x)(\log_b y)$
   Use the expression $\log_2 (4 \cdot 8)$ to help you answer.

e. Determine if the statement is true or false: $\log_b\left(\dfrac{x}{y}\right) = \dfrac{\log_b x}{\log_b y}$

   Use the expression $\log_3\left(\dfrac{27}{9}\right)$ to help you answer.

f. Determine if the statement is true or false: $\log_b (x)^p = (\log_b x)^p$
   Use the expression $\log(1000)^2$ to help you answer.

Review Exercises

For Exercises 2–5, find the values of the logarithmic and exponential expressions without using a calculator.

2. 8^{-2} **3.** $\log 10,000$ **4.** $\log_2 32$ **5.** 6^{-1}

For Exercises 6–9, approximate the values of the logarithmic and exponential expressions by using a calculator. Round to four decimal places.

6. $(\sqrt{2})^{\pi}$ **7.** $\log 8$ **8.** $\log 27$ **9.** $\pi^{\sqrt{2}}$

For Exercises 10–13, match the function with the appropriate graph.

10. $f(x) = 4^x$ **11.** $q(x) = \left(\dfrac{1}{5}\right)^x$ **12.** $h(x) = \log_5 x$ **13.** $k(x) = \log_{1/3} x$

a.

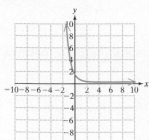

b.

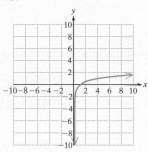

c.

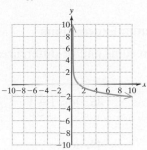

d.

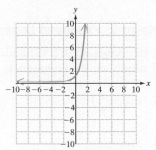

Concept 1: Properties of Logarithms

14. Select the values that are equivalent to $\log_5 5^2$.

   a. $\log_5 25$

   b. $2 \log_5 5$

   c. $\log_5 5 + \log_5 5$

15. Select the values that are equivalent to $\log_2 2^3$.

   a. $3 \log_2 2$

   b. $\log_2 8$

   c. 3

16. Select the values that are equivalent to $\log 10^4$.

   a. 4

   b. $4 \log 10$

   c. $\log 10{,}000$

For Exercises 17–40, evaluate each expression. **(See Example 1.)**

17. $\log_3 3$ **18.** $\log 10$ **19.** $\log_5 5^4$ **20.** $\log_4 4^5$

21. $6^{\log_6 11}$ **22.** $7^{\log_7 2}$ **23.** $\log 10^3$ **24.** $\log_6 6^3$

25. $\log_3 1$ **26.** $\log_8 1$ **27.** $10^{\log 9}$ **28.** $8^{\log_8 5}$

29. $\log_{1/2} 1$ **30.** $\log_{1/3}\left(\dfrac{1}{3}\right)$ **31.** $\log_2 1 + \log_2 2^3$ **32.** $\log 10^4 + \log 10$

33. $\log_4 4 + \log_2 1$ **34.** $\log_7 7 + \log_4 4^2$ **35.** $\log_{1/4}\left(\dfrac{1}{4}\right)^{2x}$ **36.** $\log_{2/3}\left(\dfrac{2}{3}\right)^p$

37. $\log_a a^4$ **38.** $\log_y y^2$ **39.** $\log 10^2 - \log_3 3^2$ **40.** $\log_6 6^4 - \log 10^4$

41. Compare the expressions by approximating their values on a calculator. Which two expressions are equivalent?

   a. $\log (3 \cdot 5)$ **b.** $\log 3 \cdot \log 5$ **c.** $\log 3 + \log 5$

42. Compare the expressions by approximating their values on a calculator. Which two expressions are equivalent?

   a. $\log\left(\dfrac{6}{5}\right)$ **b.** $\dfrac{\log 6}{\log 5}$ **c.** $\log 6 - \log 5$

43. Compare the expressions by approximating their values on a calculator. Which two expressions are equivalent?

   a. $\log 20^2$ **b.** $[\log 20]^2$ **c.** $2 \log 20$

44. Compare the expressions by approximating their values on a calculator. Which two expressions are equivalent?

a. $\log \sqrt{4}$ **b.** $\dfrac{1}{2}\log 4$ **c.** $\sqrt{\log 4}$

Concept 2: Expanded Logarithmic Expressions

For Exercises 45–62, expand into sums and/or differences of logarithms. Assume all variables represent positive real numbers. **(See Examples 2–4.)**

45. $\log_3\left(\dfrac{x}{5}\right)$

46. $\log_2\left(\dfrac{y}{z}\right)$

47. $\log(2x)$

48. $\log_6(xyz)$

49. $\log_5 x^4$

50. $\log_7 z^{1/3}$

51. $\log_4\left(\dfrac{ab}{c}\right)$

52. $\log_2\left(\dfrac{x}{yz}\right)$

53. $\log_b\left(\dfrac{\sqrt{x}\,y}{z^3 w}\right)$

54. $\log\left(\dfrac{a\sqrt[3]{b}}{cd^2}\right)$

55. $\log_2\left(\dfrac{x+1}{y^2\sqrt{z}}\right)$

56. $\log\left(\dfrac{a+1}{b\sqrt[3]{c}}\right)$

57. $\log\left(\sqrt[3]{\dfrac{ab^2}{c}}\right)$

58. $\log_5\left(\sqrt[4]{\dfrac{w^3 z}{x^2}}\right)$

59. $\log\left(\dfrac{1}{w^5}\right)$

60. $\log_3\left(\dfrac{1}{z^4}\right)$

61. $\log_b\left(\dfrac{\sqrt{a}}{b^3 c}\right)$

62. $\log_x\left(\dfrac{x}{y\sqrt{z}}\right)$

Concept 3: Single Logarithmic Expressions

For Exercises 63–78, write the expressions as a single logarithm and simplify if possible. Assume all variable expressions represent positive real numbers. **(See Examples 5 and 6.)**

63. $\log_3 270 - \log_3 2 - \log_3 5$

64. $\log_5 8 + \log_5 50 - \log_5 16$

65. $\log_7 98 - \log_7 2$

66. $\log_6 24 - \log_6 4$

67. $2\log_3 x - 3\log_3 y + \log_3 z$

68. $\log C + \log A + \log B + \log I + \log N$

69. $2\log_3 a - \dfrac{1}{4}\log_3 b + \log_3 c$

70. $\log_5 a - \dfrac{1}{2}\log_5 b - 3\log_5 c$

71. $\log_b x - 3\log_b x + 4\log_b x$

72. $2\log_3 z + \log_3 z - \dfrac{1}{2}\log_3 z$

73. $5\log_8 a - \log_8 1 + \log_8 8$

74. $\log_2 2 + 2\log_2 b - \log_2 1$

75. $2\log(x+6) + \dfrac{1}{3}\log y - 5\log z$

76. $\dfrac{1}{4}\log(a+1) - 2\log b - 4\log c$

77. $\log_b(x+1) - \log_b(x^2-1)$

78. $\log_x(p^2-4) - \log_x(p-2)$

Expanding Your Skills

For Exercises 79–90, find the values of the logarithms given that $\log_b 2 \approx 0.693$, $\log_b 3 \approx 1.099$, and $\log_b 5 \approx 1.609$.

79. $\log_b 6$ **80.** $\log_b 4$ **81.** $\log_b 12$ **82.** $\log_b 25$

83. $\log_b 81$ **84.** $\log_b 30$ **85.** $\log_b \left(\dfrac{5}{2}\right)$ **86.** $\log_b \left(\dfrac{25}{3}\right)$

87. $\log_b 10^6$ **88.** $\log_b 15^3$ **89.** $\log_b 5^{10}$ **90.** $\log_b 2^{12}$

91. The intensity of sound waves is measured in decibels and is calculated by the formula

$$B = 10 \log \left(\frac{I}{I_0}\right)$$

where I_0 is the minimum detectable decibel level.

  a. Expand this formula by using the properties of logarithms.

  b. Let $I_0 = 10^{-16}$ W/cm^2 and simplify.

92. The Richter scale is used to measure the intensity of an earthquake and is calculated by the formula

$$R = \log \left(\frac{I}{I_0}\right)$$

where I_0 is the minimum level detectable by a seismograph.

  a. Expand this formula by using the properties of logarithms.

  b. Let $I_0 = 1$ and simplify.

Graphing Calculator Exercises

93. **a.** Graph $Y_1 = \log x^2$ and state its domain.

  b. Graph $Y_2 = 2 \log x$ and state its domain.

  c. For what values of x are the expressions $\log x^2$ and $2 \log x$ equivalent?

94. **a.** Graph $Y_1 = \log (x - 1)^2$ and state its domain.

  b. Graph $Y_2 = 2 \log (x - 1)$ and state its domain.

  c. For what values of x are the expressions $\log (x - 1)^2$ and $2 \log (x - 1)$ equivalent?

The Irrational Number *e*   Section 9.5

1. The Irrational Number *e*

The exponential function base 10 is particularly easy to work with because integral powers of 10 represent different place positions in the base-10 numbering system. In this section, we introduce another important exponential function whose base is an irrational number called *e*.

Consider the expression $(1 + \frac{1}{x})^x$. The value of the expression for increasingly large values of *x* approaches a constant (Table 9-7).

As *x* approaches infinity, the expression $(1 + \frac{1}{x})^x$ approaches a constant value that we call *e*. From Table 9-7, this value is approximately 2.718281828.

$$e \approx 2.718281828$$

The value of *e* is an irrational number (a nonterminating, nonrepeating decimal) and like the number π, it is a universal constant.

Table 9-7

x	$\left(1 + \dfrac{1}{x}\right)^x$
100	2.70481382942
1000	2.71692393224
10,000	2.71814592683
100,000	2.71826823717
1,000,000	2.71828046932
1,000,000,000	2.71828182710

Concepts

1. The Irrational Number *e*
2. Computing Compound Interest
3. The Natural Logarithmic Function
4. Change-of-Base Formula
5. Applications of the Natural Logarithmic Function

Example 1   Graphing $f(x) = e^x$

Graph the function defined by $f(x) = e^x$.

Solution:

Because the base of the function is greater than 1 ($e \approx 2.718281828$), the graph is an increasing exponential function. We can use a calculator to evaluate $f(x) = e^x$ at several values of x.

Calculator Connections

Topic: Evaluating Exponential Expressions Base e

Practice using your calculator by evaluating e^x for $x = 1$, $x = 2$, and $x = -1$.

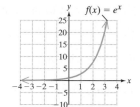

```
e^(1)
           2.718281828
e^(2)
           7.389056099
e^(-1)
           .3678794412
```

TIP: On a scientific calculator, you may need to enter the base and exponent of an exponential expression in reverse order. For example, enter e^2 as:

2 **2nd** **LN**

If you are using your calculator correctly, your answers should match those found in Table 9-8. Values are rounded to three decimal places. The corresponding graph of $f(x) = e^x$ is shown in Figure 9-14.

Table 9-8

x	$f(x) = e^x$
-3	0.050
-2	0.135
-1	0.368
0	1.000
1	2.718
2	7.389
3	20.086

Figure 9-14

---

2. Computing Compound Interest

One particularly interesting application of exponential functions is in computing compound interest.

1. If the number of compounding periods per year is finite, then the amount in an account is given by

$$A(t) = P\left(1 + \frac{r}{n}\right)^{nt}$$

where P is the initial principal, r is the annual interest rate, n is the number of times compounded per year, and t is the time in years that the money is invested.

2. If the number of compound periods per year is infinite, then interest is said to be **compounded continuously**. In such a case, the amount in an account is given by

$$A(t) = Pe^{rt}$$

where P is the initial principal, r is the annual interest rate, and t is the time in years that the money is invested.

Example 2   Computing the Balance on an Account

Suppose $5000 is invested in an account earning 6.5% interest. Find the balance in the account after 10 yr under the following compounding options.

a. Compounded annually

b. Compounded quarterly

c. Compounded monthly

d. Compounded daily

e. Compounded continuously

Solution:

Compounding Option	*n* Value	Formula	Result
Annually	$n = 1$	$A(10) = 5000\left(1 + \dfrac{0.065}{1}\right)^{(1)(10)}$	$9385.69
Quarterly	$n = 4$	$A(10) = 5000\left(1 + \dfrac{0.065}{4}\right)^{(4)(10)}$	$9527.79
Monthly	$n = 12$	$A(10) = 5000\left(1 + \dfrac{0.065}{12}\right)^{(12)(10)}$	$9560.92
Daily	$n = 365$	$A(10) = 5000\left(1 + \dfrac{0.065}{365}\right)^{(365)(10)}$	$9577.15
Continuously	Not applicable	$A(10) = 5000e^{(0.065)(10)}$	$9577.70

Notice that there is a $191.46 difference in the account balance between annual compounding and daily compounding. However, the difference between compounding daily and compounding continuously is small—$0.55. As *n* gets infinitely large, the function defined by

$$A(t) = P\left(1 + \frac{r}{n}\right)^{nt} \qquad \text{converges to } A(t) = Pe^{rt}.$$

Skill Practice

2. Suppose $1000 is invested at 5%. Find the balance after 8 yr under the following options.

a. Compounded annually

b. Compounded quarterly

c. Compounded monthly

d. Compounded daily

e. Compounded continuously

3. The Natural Logarithmic Function

Recall that the common logarithmic function $y = \log x$ has a base of 10. Another important logarithmic function is called the **natural logarithmic function**. The natural logarithmic function has a base of *e* and is written as $y = \ln x$. That is,

$$y = \ln x = \log_e x$$

Example 3   Graphing $y = \ln x$

Graph $y = \ln x$.

Solution:

Because the base of the function $y = \ln x$ is *e* and $e > 1$, the graph is an increasing logarithmic function. We can use a calculator to find specific points on the graph of $y = \ln x$ by pressing the   **LN**   key.

Answer

2. **a.** $1477.46   **b.** $1488.13
   c. $1490.59   **d.** $1491.78
   e. $1491.82

Practice using your calculator by evaluating $\ln x$ for the following values of x. If you are using your calculator correctly, your answers should match those found in Table 9-9. Values are rounded to three decimal places. The corresponding graph of $y = \ln x$ is shown in Figure 9-15.

Calculator Connections

Topic: Evaluating Natural Logarithms

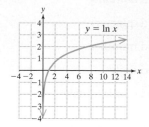

Table 9-9

x	$\ln x$
1	0.000
2	0.693
3	1.099
4	1.386
5	1.609
6	1.792
7	1.946

Figure 9-15

The properties of logarithms stated in Section 9.4 are also true for natural logarithms.

Properties of the Natural Logarithmic Function

Let x and y be positive real numbers, and let p be a real number. Then the following properties are true.

1. $\ln 1 = 0$
2. $\ln e = 1$
3. $\ln e^p = p$
4. $e^{\ln x} = x$

5. $\ln (xy) = \ln x + \ln y$ Product property for logarithms
6. $\ln \left(\dfrac{x}{y}\right) = \ln x - \ln y$ Quotient property for logarithms
7. $\ln x^p = p \ln x$ Power property for logarithms

Example 4 Simplifying Expressions with Natural Logarithms

Simplify the expressions. Assume that all variable expressions within the logarithms represent positive real numbers.

a. $\ln e$ **b.** $\ln 1$ **c.** $\ln (e^{x+1})$ **d.** $e^{\ln (x+1)}$

Solution:

a. $\ln e = 1$ Property 2

b. $\ln 1 = 0$ Property 1

c. $\ln (e^{x+1}) = x + 1$ Property 3

d. $e^{\ln (x+1)} = x + 1$ Property 4

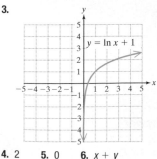

Example 5 Writing a Sum or Difference of Natural Logarithms as a Single Logarithm

Write the expression as a single logarithm. Assume that all variable expressions within the logarithms represent positive real numbers.

$$\ln x - \frac{1}{3}\ln y - \ln z$$

Solution:

$$\ln x - \frac{1}{3}\ln y - \ln z$$

$= \ln x - \ln y^{1/3} - \ln z$ Power property for logarithms (property 7)

$= \ln x - (\ln y^{1/3} + \ln z)$ Factor out -1 from the last two terms.

$= \ln x - \ln (y^{1/3} \cdot z)$ Product property for logarithms (property 5)

$= \ln \left(\dfrac{x}{y^{1/3}z} \right)$ Quotient property for logarithms (property 6)

$= \ln \left(\dfrac{x}{\sqrt[3]{y}\, z} \right)$ Write the rational exponent as a radical.

Skill Practice

Write as a single logarithm.

8. $\dfrac{1}{4}\ln a - \ln b + \ln c$

Example 6 Writing a Logarithmic Expression in Expanded Form

Write the expression as a sum or difference of logarithms of x and y. Assume all variable expressions within the logarithm represent positive real numbers.

$$\ln \left(\frac{e}{x^2\sqrt{y}} \right)$$

Solution:

$$\ln \left(\frac{e}{x^2\sqrt{y}} \right)$$

$= \ln e - \ln (x^2\sqrt{y})$ Quotient property for logarithms (property 6)

$= \ln e - (\ln x^2 + \ln \sqrt{y})$ Product property for logarithms (property 5)

$= 1 - \ln x^2 - \ln y^{1/2}$ Distributive property. Also simplify $\ln e = 1$ (property 2).

$= 1 - 2\ln x - \dfrac{1}{2}\ln y$ Power property for logarithms (property 7)

Skill Practice

Write as a sum or difference of logarithms of x and y.

9. $\ln \left(\dfrac{x^3\sqrt{y}}{e^2} \right)$

4. Change-of-Base Formula

A calculator can be used to approximate the value of a logarithm with a base of 10 or a base of e by using the LOG key or the LN key, respectively. However, to use a calculator to evaluate a logarithmic expression with a base other than 10 or e, we must use the **change-of-base formula**.

Answers

8. $\ln \left(\dfrac{\sqrt[4]{a}\, c}{b} \right)$

9. $3\ln x + \dfrac{1}{2}\ln y - 2$

Change-of-Base Formula

Let a and b be positive real numbers such that $a \neq 1$ and $b \neq 1$. Then for any positive real number x,

$$\log_b x = \frac{\log_a x}{\log_a b}$$

Proof:

Let $M = \log_b x$, which implies that $b^M = x$.

Take the logarithm, base a, on both sides: $\log_a b^M = \log_a x$

Apply the power property for logarithms: $M \cdot \log_a b - \log_a x$

Divide both sides by $\log_a b$: $\dfrac{M \cdot \cancel{\log_a b}}{\cancel{\log_a b}} = \dfrac{\log_a x}{\log_a b}$

$$M = \frac{\log_a x}{\log_a b}$$

Because $M = \log_b x$, we have $\log_b x = \dfrac{\log_a x}{\log_a b}$ ✓

The change-of-base formula converts a logarithm of one base to a ratio of logarithms of a different base. For the sake of using a calculator, we often apply the change-of-base formula with base 10 or base e.

Skill Practice

10. Use the change-of-base formula to evaluate $\log_5 95$ by using base 10. Round to three decimal places.

11. Use the change-of-base formula to evaluate $\log_5 95$ by using base e. Round to three decimal places.

Example 7   Using the Change-of-Base Formula

a. Use the change-of-base formula to evaluate $\log_4 80$ by using base 10. (Round to three decimal places.)

b. Use the change-of-base formula to evaluate $\log_4 80$ by using base e. (Round to three decimal places.)

Solution:

a. $\log_4 80 = \dfrac{\log_{10} 80}{\log_{10} 4} = \dfrac{\log 80}{\log 4} \approx \dfrac{1.903089987}{0.6020599913} \approx 3.161$

b. $\log_4 80 = \dfrac{\log_e 80}{\log_e 4} = \dfrac{\ln 80}{\ln 4} \approx \dfrac{4.382026635}{1.386294361} \approx 3.161$

To check the result, we see that $4^{3.161} \approx 80$.

Calculator Connections

Topic: Using the Change-of-Base Formula to Graph Logarithmic Functions

The change-of-base formula can be used to graph logarithmic functions with bases other than 10 or e. For example, to graph $Y_1 = \log_2 x$, we can enter the function as either

$$Y_1 = (\log x)/(\log 2) \quad \text{or} \quad Y_1 = (\ln x)/(\ln 2)$$

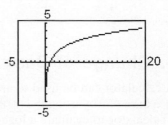

Answers

10. 2.829   11. 2.829

5. Applications of the Natural Logarithmic Function

Plant and animal tissue contains both carbon 12 and carbon 14. Carbon 12 is a stable form of carbon, whereas carbon 14 is a radioactive isotope with a half-life of approximately 5730 yr. While a plant or animal is living, it takes in carbon from the atmosphere either through photosynthesis or through its food. The ratio of carbon 14 to carbon 12 in a living organism is constant and is the same as the ratio found in the atmosphere.

When a plant or animal dies, it no longer ingests carbon from the atmosphere. The amount of stable carbon 12 remains unchanged from the time of death, but the carbon 14 begins to decay. Because the rate of decay is constant, a tissue sample can be dated by comparing the percent of carbon 14 still present to the percentage of carbon 14 assumed to be in its original living state.

The age of a tissue sample is a function of the percent of carbon 14 still present in the organism according to the following model:

$$A(p) = \frac{\ln p}{-0.000121}$$

where $A(p)$ is the age in years and p is the percentage (in decimal form) of carbon 14 still present.

Example 8 **Applying the Natural Logarithmic Function to Radioactive Decay**

Using the formula

$$A(p) = \frac{\ln p}{-0.000121}$$

a. Find the age of a bone that has 72% of its initial carbon 14.

b. Find the age of the Iceman, a body uncovered in the mountains of northern Italy in 1991. Samples of his hair revealed that 51.4% of the original carbon 14 was present after his death.

Solution:

a. $A(p) = \dfrac{\ln p}{-0.000121}$

$A(0.72) = \dfrac{\ln 0.72}{-0.000121}$ Substitute 0.72 for p.

≈ 2715 years

b. $A(p) = \dfrac{\ln p}{-0.000121}$

$A(0.514) = \dfrac{\ln 0.514}{-0.000121}$ Substitute 0.514 for p.

≈ 5500 years   The body of the Iceman is approximately 5500 yr old.

Skill Practice

12. Use the formula
$$A(p) = \frac{\ln p}{-0.000121}$$
(where $A(p)$ is the age in years and p is the percent of carbon 14 still present) to determine the age of a human skull that has 90% of its initial carbon 14.

Answer

12. ≈ 871 yr

Section 9.5   Practice Exercises

Vocabulary and Key Concepts

1. **a.** As x becomes increasingly large, the value of $(1 + \frac{1}{x})^x$ approaches _____ ≈ 2.71828.

   b. The function $f(x) = e^x$ is the exponential function base _____.

   c. The logarithmic function base e is called the _____ logarithmic function and is denoted by $y =$ _____.

   d. If x is a positive real number, then $\ln 1 =$ _____, $\ln e =$ _____, $\ln e^p =$ _____, and $e^{\ln x} =$ _____.

   e. If x and y are positive real numbers, then $\ln(xy) =$ _____ and $\ln\left(\dfrac{x}{y}\right) =$ _____.

   f. If x is a positive real number, then for any real number p, $\ln x^p$ can be written as _____.

   g. The change-of-base formula states that $\log_b x = \dfrac{\log_a x}{\square}$, where a is a positive real number and $a \neq 1$.

Review Exercises

For Exercises 2–5, fill out the tables and graph the functions. For Exercises 4 and 5 round to two decimal places.

2. $f(x) = \left(\dfrac{3}{2}\right)^x$

x	$f(x)$
-3	
-2	
-1	
0	
1	
2	
3	

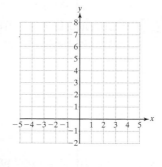

3. $g(x) = \left(\dfrac{1}{5}\right)^x$

x	$g(x)$
-3	
-2	
-1	
0	
1	
2	
3	

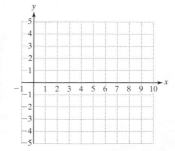

4. $q(x) = \log(x + 1)$

x	$q(x)$
-0.5	
0	
4	
9	

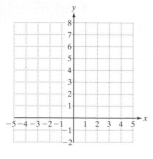

5. $r(x) = \log x$

x	$r(x)$
0.5	
1	
5	
10	

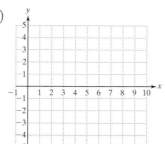

Concept 1: The Irrational Number e

6. From memory, write a decimal approximation of the number e, correct to three decimal places.

For Exercises 7–10, graph the equation by completing the table and plotting points. Identify the domain. Round to two decimal places when necessary. **(See Example 1.)**

7. $y = e^{x+1}$

x	y
-4	
-3	
-2	
-1	
0	
1	

8. $y = e^{x+2}$

x	y
-5	
-4	
-3	
-2	
-1	
0	

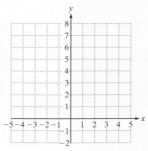

9. $y = e^x + 2$

x	y
−2	
−1	
0	
1	
2	
3	

10. $y = e^x - 1$

x	y
−4	
−3	
−2	
−1	
0	
1	

Concept 2: Computing Compound Interest

In Exercises 11–16, use the model

$$A(t) = P\left(1 + \frac{r}{n}\right)^{nt}$$

for interest compounded *n* times per year. Use the model $A(t) = Pe^{rt}$ for interest compounded continuously.

11. Suppose an investor deposits $10,000 in a certificate of deposit for 5 yr for which the interest is compounded monthly. Find the total amount of money in the account for the following interest rates. Compare your answers and comment on the effect of interest rate on an investment. **(See Example 2.)**

   a. $r = 4.0\%$　　**b.** $r = 6.0\%$　　**c.** $r = 8.0\%$　　**d.** $r = 9.5\%$

12. Suppose an investor deposits $5000 in a certificate of deposit for 8 yr for which the interest is compounded quarterly. Find the total amount of money in the account for the following interest rates. Compare your answers and comment on the effect of interest rate on an investment.

   a. $r = 4.5\%$　　**b.** $r = 5.5\%$　　**c.** $r = 7.0\%$　　**d.** $r = 9.0\%$

13. Suppose an investor deposits $8000 in a savings account for 10 yr at 4.5% interest. Find the total amount of money in the account for the following compounding options. Compare your answers. How does the number of compound periods per year affect the total investment?

   a. Compounded annually　　　　　　　　**b.** Compounded quarterly

   c. Compounded monthly　　　　　　　　　**d.** Compounded daily

   e. Compounded continuously

14. Suppose an investor deposits $15,000 in a savings account for 8 yr at 5.0% interest. Find the total amount of money in the account for the following compounding options. Compare your answers. How does the number of compound periods per year affect the total investment?

   a. Compounded annually　　　　　　　　**b.** Compounded quarterly

   c. Compounded monthly　　　　　　　　　**d.** Compounded daily

   e. Compounded continuously

15. Suppose an investor deposits $5000 in an account earning 6.5% interest compounded continuously. Find the total amount in the account for the following time periods. How does the length of time affect the amount of interest earned?

   a. 5 yr　　　**b.** 10 yr　　　**c.** 15 yr　　　**d.** 20 yr　　　**e.** 30 yr

16. Suppose an investor deposits $10,000 in an account earning 6.0% interest compounded continuously. Find the total amount in the account for the following time periods. How does the length of time affect the amount of interest earned?

   a. 5 yr　　　**b.** 10 yr　　　**c.** 15 yr　　　**d.** 20 yr　　　**e.** 30 yr

Concept 3: The Natural Logarithmic Function

For Exercises 17–20, graph the equation by completing the table and plotting the points. Identify the domain. Round to two decimal places when necessary. **(See Example 3.)**

17. $y = \ln(x - 2)$

x	y
2.25	
2.50	
2.75	
3	
4	
5	
6	

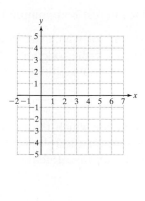

18. $y = \ln(x - 1)$

x	y
1.25	
1.50	
1.75	
2	
3	
4	
5	

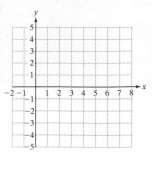

19. $y = \ln x - 1$

x	y
0.25	
0.5	
0.75	
1	
2	
3	
4	

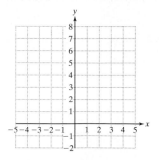

20. $y = \ln x + 2$

x	y
0.25	
0.5	
0.75	
1	
2	
3	
4	

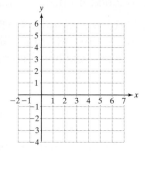

21. a. Graph $f(x) = 10^x$

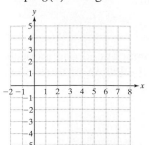

b. Identify the domain and range of f.

c. Graph $g(x) = \log x$.

d. Identify the domain and range of g.

22. a. Graph $f(x) = e^x$.

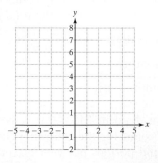

b. Identify the domain and range of f.

c. Graph $g(x) = \ln x$.

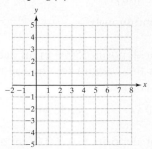

d. Identify the domain and range of g.

For Exercises 23–30, simplify the expressions. Assume all variables represent positive real numbers. **(See Example 4.)**

23. $\ln e$ **24.** $\ln e^2$ **25.** $\ln 1$ **26.** $e^{\ln x}$

27. $\ln e^{-6}$ **28.** $\ln e^{(5-3x)}$ **29.** $e^{\ln (2x+3)}$ **30.** $e^{\ln 4}$

For Exercises 31–38, write the expression as a single logarithm. Assume all variables represent positive real numbers. **(See Example 5.)**

31. $6 \ln p + \dfrac{1}{3} \ln q$ **32.** $2 \ln w + \ln z$ **33.** $\dfrac{1}{2} (\ln x - 3 \ln y)$

34. $\dfrac{1}{3} (4 \ln a - \ln b)$ **35.** $2 \ln a - \ln b - \dfrac{1}{3} \ln c$ **36.** $-\ln x + 3 \ln y - \ln z$

37. $4 \ln x - 3 \ln y - \ln z$ **38.** $\dfrac{1}{2} \ln c + \ln a - 2 \ln b$

For Exercises 39–46, write the expression as a sum and/or difference of $\ln a$, $\ln b$, and $\ln c$. Assume all variables represent positive real numbers. **(See Example 6.)**

39. $\ln \left(\dfrac{a}{b} \right)^2$ **40.** $\ln \sqrt[3]{\dfrac{a}{b}}$ **41.** $\ln (b^2 \cdot e)$ **42.** $\ln (\sqrt{c} \cdot e)$

43. $\ln \left(\dfrac{a^4 \sqrt{b}}{c} \right)$ **44.** $\ln \left(\dfrac{\sqrt{ab}}{c^3} \right)$ **45.** $\ln \left(\dfrac{ab}{c^2} \right)^{1/5}$ **46.** $\ln \sqrt{2ab}$

Concept 4: Change-of-Base Formula

47. a. Evaluate $\log_6 200$ by computing $\dfrac{\log 200}{\log 6}$ to four decimal places.

     b. Evaluate $\log_6 200$ by computing $\dfrac{\ln 200}{\ln 6}$ to four decimal places.

     c. How do your answers to parts (a) and (b) compare?

48. a. Evaluate $\log_8 120$ by computing $\dfrac{\log 120}{\log 8}$ to four decimal places.

     b. Evaluate $\log_8 120$ by computing $\dfrac{\ln 120}{\ln 8}$ to four decimal places.

     c. How do your answers to parts (a) and (b) compare?

For Exercises 49–60, use the change-of-base formula to approximate the logarithms to four decimal places. Check your answers by using the exponential key on your calculator. **(See Example 7.)**

49. $\log_2 7$ **50.** $\log_3 5$ **51.** $\log_8 24$ **52.** $\log_4 17$

53. $\log_8 0.012$ **54.** $\log_7 0.251$ **55.** $\log_9 1$ **56.** $\log_2 \left(\dfrac{1}{5} \right)$

57. $\log_4 \left(\dfrac{1}{100} \right)$ **58.** $\log_5 0.0025$ **59.** $\log_7 0.0006$ **60.** $\log_2 0.24$

Concept 5: Applications of the Natural Logarithmic Function

Under continuous compounding, the amount of time t in years required for an investment to double is a function of the interest rate r according to the formula:

$$t = \frac{\ln 2}{r}$$

 Use the formula for Exercises 61–63. **(See Example 8.)**

61. a. If you invest $5000, how long will it take the investment to reach $10,000 if the interest rate is 4.5%? Round to one decimal place.

b. If you invest $5000, how long will it take the investment to reach $10,000 if the interest rate is 10%? Round to one decimal place.

c. Using the doubling time found in part (b), how long would it take a $5000 investment to reach $20,000 if the interest rate is 10%?

62. a. If you invest $3000, how long will it take the investment to reach $6000 if the interest rate is 5.5%? Round to one decimal place.

b. If you invest $3000, how long will it take the investment to reach $6000 if the interest rate is 8%? Round to one decimal place.

c. Using the doubling time found in part (b), how long would it take a $3000 investment to reach $12,000 if the interest rate is 8%?

63. a. If you invest $4000, how long will it take the investment to reach $8000 if the interest rate is 3.5%? Round to one decimal place.

b. If you invest $4000, how long will it take the investment to reach $8000 if the interest rate is 5%? Round to one decimal place.

c. Using the doubling time found in part (b), how long would it take a $4000 investment to reach $16,000 if the interest rate is 5%?

64. On August 31, 1854, an epidemic of cholera was discovered in London, England, resulting from a contaminated community water pump at Broad Street. By the end of September more than 600 citizens who drank water from the pump had died.

The cumulative number of deaths from cholera in the 1854 London epidemic can be approximated by

$$D(t) = 91 + 160 \ln (t + 1)$$

where t is the number of days after the start of the epidemic ($t = 0$ corresponds to September 1, 1854).

a. Approximate the total number of deaths as of September 1 ($t = 0$).

b. Approximate the total number of deaths as of September 5, September 10, and September 20.

Graphing Calculator Exercises

65. a. Graph the function defined by $f(x) = \log_3 x$ by graphing $Y_1 = \dfrac{\log x}{\log 3}$.

b. Graph the function defined by $f(x) = \log_3 x$ by graphing $Y_2 = \dfrac{\ln x}{\ln 3}$.

c. Does it appear that $Y_1 = Y_2$ on the standard viewing window?

66. a. Graph the function defined by $f(x) = \log_7 x$ by graphing $Y_1 = \dfrac{\log x}{\log 7}$.

b. Graph the function defined by $f(x) = \log_7 x$ by graphing $Y_2 = \dfrac{\ln x}{\ln 7}$.

c. Does it appear that $Y_1 = Y_2$ on the standard viewing window?

For Exercises 67–69, graph the functions on your calculator.

67. Graph $s(x) = \log_{1/2} x$ **68.** Graph $y = e^{x-2}$ **69.** Graph $y = e^{x-1}$

Problem Recognition Exercises

Logarithmic and Exponential Forms

Fill out the table by writing the exponential expressions in logarithmic form, and the logarithmic expressions in exponential form. Use the fact that:

$$y = \log_b x \quad \text{is equivalent to} \quad b^y = x$$

	Exponential Form	**Logarithmic Form**
1.	$2^5 = 32$	
2.		$\log_3 81 = 4$
3.	$z^y = x$	
4.		$\log_b a = c$
5.	$10^3 = 1000$	
6.		$\log 10 = 1$
7.	$e^a = b$	
8.		$\ln p = q$
9.	$\left(\frac{1}{2}\right)^2 = \frac{1}{4}$	
10.		$\log_{1/3} 9 = -2$
11.	$10^{-2} = 0.01$	
12.		$\log 4 = x$
13.	$e^0 = 1$	
14.		$\ln e = 1$
15.	$25^{1/2} = 5$	
16.		$\log_{16} 2 = \frac{1}{4}$
17.	$e^t = s$	
18.		$\ln w = r$
19.	$15^{-2} = \frac{1}{225}$	
20.		$\log_3 p = -1$

Section 9.6   Exponential Equations and Applications

1. Solving Exponential Equations

Suppose that the population $P(t)$ (in thousands) of the District of Columbia t years after 2010 can be modeled by $P(t) = 602e^{0.02623t}$ (based on data from the U.S. Census Bureau, www.census.gov). To find the amount of time for the population to reach 1,000,000, we would solve the equation:

$$1000 = 602e^{0.02623t}$$

This equation is called an **exponential equation** because the equation contains a variable in the exponent. Other examples of exponential equations are:

$$5^x = 5^3 \qquad 4^{2x-9} = 64 \qquad 4^x = 25$$

To solve an exponential equation, first note that all exponential functions are one-to-one. Therefore, $b^x = b^y$ implies that $x = y$. This is called the equivalence property of exponential expressions.

> **Equivalence of Exponential Expressions**
> Let x, y, and b be real numbers such that $b > 0$ and $b \neq 1$. Then
> $$b^x = b^y \qquad \text{implies} \qquad x = y$$

The equivalence of exponential expressions indicates that if two exponential expressions of the same base are equal, their exponents must be equal.

Skill Practice

Solve the equation.

1. $2^{3x+1} = 8$

Example 1   Solving an Exponential Equation

Solve the equation. $4^{2x-9} = 64$

Solution:

$$4^{2x-9} = 64$$

$$4^{2x-9} = 4^3 \qquad \text{Write both sides with a common base.}$$

$$2x - 9 = 3 \qquad \text{If } b^x = b^y, \text{ then } x = y.$$

$$2x = 12 \qquad \text{Solve for } x.$$

$$x = 6$$

To check, substitute $x = 6$ into the original equation.

$$4^{2(6)-9} \stackrel{?}{=} 64$$

$$4^{12-9} \stackrel{?}{=} 64$$

$$4^3 \stackrel{?}{=} 64 \checkmark \text{ True}$$

The solution is 6.

Answer

1. $\dfrac{2}{3}$

Example 2 Solving an Exponential Equation

Solve the equation. $(2^x)^{x+3} = \dfrac{1}{4}$

Solution:

$$(2^x)^{x+3} = \dfrac{1}{4}$$

$2^{x^2+3x} = 2^{-2}$ Apply the multiplication property of exponents. Write both sides of the equation with a common base.

$x^2 + 3x = -2$ If $b^x = b^y$, then $x = y$.

$x^2 + 3x + 2 = 0$ The resulting equation is quadratic.

$(x+2)(x+1) = 0$ Solve for x.

$x = -2$ or $x = -1$

The solutions -2 and -1 both check in the original equation.

Skill Practice

Solve the equation.

2. $(3^x)^{x-5} = \dfrac{1}{81}$

In Examples 1 and 2, we were able to write the left and right sides of the equation with a common base. However, most exponential equations cannot be written in this form by inspection. For example:

$$4^x = 25 \quad\quad \text{25 is not a recognizable}$$
$$4^x = 4^? \quad\quad \text{power of 4.}$$

To solve such an equation, we can use the definition of a logarithm to rewrite the equation, or we can take a logarithm of the same base on each side of the equation and then apply the power property of logarithms. These methods are demonstrated in Example 3.

Example 3 Solving an Exponential Equation

Solve the equation. $4^x = 25$

Solution:

Because 25 cannot be written as an integral power of 4, we cannot use the property that if $b^x = b^y$, then $x = y$. Instead we can rewrite the equation in its corresponding logarithmic form to solve for x.

$$4^x = 25$$

$x = \log_4 25$ Write the equation in logarithmic form

$= \dfrac{\ln 25}{\ln 4} \approx 2.322$ Change-of-base formula

Skill Practice

Solve the equation.
3. $5^x = 32$

Answers

2. $1, 4$ **3.** $\dfrac{\ln 32}{\ln 5} \approx 2.153$

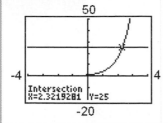

The same result can be reached by taking a logarithm of any base on both sides of the equation. Then by applying the power property of logarithms, the unknown exponent can be written as a factor.

$$4^x = 25$$

$$\log 4^x = \log 25 \qquad \text{Take the common logarithm of both sides.}$$

$$x \log 4 = \log 25 \qquad \text{Apply the power property of logarithms to express the exponent as a factor. This is now a linear equation in } x.$$

$$\frac{x \log 4}{\log 4} = \frac{\log 25}{\log 4} \qquad \text{Solve for } x.$$

$$x = \frac{\log 25}{\log 4}$$

The solution is $\dfrac{\log 25}{\log 4}$ or approximately 2.322.

To solve the equation from Example 3, we can take a logarithm of any base. For example:

$$4^x = 25 \qquad\qquad\qquad\qquad 4^x = 25$$

$$\log_4 4^x = \log_4 25 \quad \text{Take the logarithm} \qquad \ln 4^x = \ln 25 \quad \text{Take the natural}$$
$$\text{base 4 on both} \qquad\qquad\qquad\qquad \text{logarithm on}$$
$$\text{sides.} \qquad\qquad\qquad x \ln 4 = \ln 25 \quad \text{both sides.}$$

$$x = \log_4 25 \text{ (solution)} \qquad\qquad x = \frac{\ln 25}{\ln 4} \text{ (solution)}$$

The values $\log_4 25, \dfrac{\log 25}{\log 4},$ and $\dfrac{\ln 25}{\ln 4}$ are all equivalent. However, common logarithms and natural logarithms are often used to express the solution to an exponential equation so that the solution can be approximated on a calculator.

Solving Exponential Equations

Step 1 Isolate one of the exponential expressions in the equation.

Step 2 Take a logarithm on both sides of the equation. (The natural logarithmic function or the common logarithmic function is often used so that the final answer can be approximated with a calculator.)

Step 3 Use the power property of logarithms (property 7) to write exponents as factors. Recall: $\log_b M^p = p \log_b M$.

Step 4 Solve the resulting equation from step 3.

Example 4 Solving an Exponential Equation by Taking a Logarithm on Both Sides

Solve the equation. $e^{-3.6x} - 2 = 7.74$

Skill Practice

Solve the equation.
4. $e^{-0.2t} + 1 = 8.52$

Solution:

$e^{-3.6x} - 2 = 7.74$ Add 2 to both sides to isolate the exponential expression.

$e^{-3.6x} = 9.74$

$\ln e^{-3.6x} = \ln 9.74$ The exponential expression has a base of e, so it is convenient to take the natural logarithm of both sides.

$(-3.6x)\ln e = \ln 9.74$ Use the power property of logarithms.

$-3.6x = \ln 9.74$ Simplify (recall that $\ln e = 1$).

$x = \dfrac{\ln 9.74}{-3.6}$

The solution is $-\dfrac{\ln 9.74}{3.6}$ or approximately -0.632.

Example 5 Solving an Exponential Equation by Taking a Logarithm on Both Sides

Solve the equation. $2^{x+3} = 7^x$

Skill Practice

Solve the equation.
5. $3^x = 8^{x+2}$

Solution:

$2^{x+3} = 7^x$

$\ln 2^{(x+3)} = \ln 7^x$ Take the natural logarithm of both sides.

$(x+3)\ln 2 = x\ln 7$ Use the power property of logarithms.

$x(\ln 2) + 3(\ln 2) = x\ln 7$ Apply the distributive property.

$x(\ln 2) - x(\ln 7) = -3\ln 2$ Collect x-terms on one side.

$x(\ln 2 - \ln 7) = -3\ln 2$ Factor out x.

$\dfrac{x(\ln 2 - \ln 7)}{(\ln 2 - \ln 7)} = \dfrac{-3\ln 2}{\ln 2 - \ln 7}$ Solve for x.

$x = \dfrac{-3\ln 2}{\ln 2 - \ln 7}$

The solution is $-\dfrac{3\ln 2}{\ln 2 - \ln 7}$ or approximately 1.660.

TIP: The exponential equation $2^{x+3} = 7^x$ could have been solved by taking a logarithm of *any* base on both sides of the equation.

Answers

4. $\dfrac{\ln 7.52}{-0.2} \approx -10.088$

5. $\dfrac{2\ln 8}{\ln 3 - \ln 8} \approx -4.240$

2. Applications of Exponential Equations

Example 6 Applying an Exponential Function to World Population

The population of the world was estimated to have reached 6.5 billion in April 2006. The population growth rate for the world is estimated to be 1.4%. (*Source:* U.S. Census Bureau)

$$P(t) = 6.5(1.014)^t$$

represents the world population in billions as a function of the number of years after April 2006 ($t = 0$ represents April 2006).

a. Use the function to estimate the world population in April 2020.

b. Use the function to estimate the amount of time after April 2006 required for the world population to reach 13 billion.

Solution:

a. $P(t) = 6.5(1.014)^t$

$P(14) = 6.5(1.014)^{14}$ The year 2020 corresponds to $t = 14$.

≈ 7.9 In 2020, the world's population will be approximately 7.9 billion.

b. $P(t) = 6.5(1.014)^t$

$13 = 6.5(1.014)^t$ Substitute $P(t) = 13$ and solve for t.

$\dfrac{13}{6.5} = \dfrac{6.5(1.014)^t}{6.5}$ Isolate the exponential expression on one side of the equation.

$2 = 1.014^t$

$\ln 2 = \ln 1.014^t$ Take the natural logarithm of both sides.

$\ln 2 = t \ln 1.014$ Use the power property of logarithms.

$\dfrac{\ln 2}{\ln 1.014} = \dfrac{t \ln 1.014}{\ln 1.014}$ Solve for t.

$t = \dfrac{\ln 2}{\ln 1.014} \approx 50$ The population will reach 13 billion (double the April 2006 value) approximately 50 yr after 2006.

Note: It has taken thousands of years for the world's population to reach 6.5 billion. However, with a growth rate of 1.4%, it will take only 50 yr to gain an additional 6.5 billion, if this trend continues.

On Friday, April 25, 1986, a nuclear accident occurred at the Chernobyl nuclear reactor, resulting in radioactive contaminates being released into the atmosphere. The most hazardous isotopes released in this accident were ^{137}Cs (cesium 137), ^{131}I (iodine 131), and ^{90}Sr (strontium 90). People living close to Chernobyl (in Ukraine) were at risk of radiation exposure from inhalation, from absorption through the skin, and from food contamination. Years after the incident, scientists have seen an increase in the incidence of thyroid disease among children living in the contaminated areas. Because iodine is readily absorbed in the thyroid gland, scientists suspect that radiation from iodine 131 is the cause.

Example 7 **Applying an Exponential Equation to Radioactive Decay**

The half-life of radioactive iodine ^{131}I is 8.04 days. If 10 g of iodine 131 is initially present, then the amount of radioactive iodine still present after t days is approximated by

$$A(t) = 10e^{-0.0862t}$$

where t is the time in days.

a. Use the model to approximate the amount of ^{131}I still present after 2 weeks. Round to the nearest 0.1 g.

b. How long will it take for the amount of ^{131}I to decay to 0.5 g? Round to the nearest 0.1 day.

Solution:

a. $A(t) = 10e^{-0.0862t}$

$A(14) = 10e^{-0.0862(14)}$ Substitute $t = 14$ (2 weeks).

≈ 3.0 g

b. $A(t) = 10e^{-0.0862t}$

$0.5 = 10e^{-0.0862t}$ Substitute $A = 0.5$.

$\dfrac{0.5}{10} = \dfrac{10e^{-0.0862t}}{10}$ Isolate the exponential expression.

$0.05 = e^{-0.0862t}$

$\ln 0.05 = \ln e^{-0.0862t}$ Take the natural logarithm of both sides.

$\ln 0.05 = -0.0862t$ The resulting equation is linear.

$\dfrac{\ln 0.05}{-0.0862} = \dfrac{-0.0862t}{-0.0862}$ Solve for t.

$t = \dfrac{\ln 0.05}{-0.0862} \approx 34.8$ days

Note: Radioactive iodine (^{131}I) is used in medicine in appropriate dosages to treat patients with hyperactive (overactive) thyroids. Because iodine is readily absorbed in the thyroid gland, the radiation is localized and will reduce the size of the thyroid while minimizing damage to surrounding tissues.

Skill Practice

Radioactive strontium 90 (^{90}Sr) has a half-life of 28 yr. If 100 g of strontium 90 is initially present, the amount left after t years is approximated by

$$A(t) = 100e^{-0.0248t}$$

8. Find the amount of ^{90}Sr present after 85 yr.

9. How long will it take for the amount of ^{90}Sr to decay to 40 g?

Answers

8. 12.1 g   **9.** 36.9 years

Section 9.6 Practice Exercises

Vocabulary and Key Concepts

1. a. An equation such as $5^x = 147$ is called an _____ equation because the equation contains a variable in the exponent.

   b. The equivalence property of exponential expressions states that if $b^x = b^y$ then _____ = _____.

Review Exercises

For Exercises 2–5, write the equivalent logarithmic form.

2. $400 = e^{-0.2t}$ **3.** $8721 = 10^{-0.003k}$ **4.** $e^{6t} = 152$ **5.** $10^{-4x} = 0.00125$

For Exercises 6–9, state whether the equation is linear or quadratic and solve the equation.

6. $4x - 5 = 2x + 77$ **7.** $1 - 7x = -3x - 11$ **8.** $4x^2 - 5x = 6$ **9.** $1 - 7x^2 = -5x - 1$

10. Which expressions are equivalent to $\log_4 18$?

     a. $\dfrac{\log 18}{\log 4}$ **b.** $\dfrac{\ln 18}{\ln 4}$ **c.** $\dfrac{\log_{18} 4}{\log_4 18}$

Concept 1: Solving Exponential Equations

For Exercises 11–14, solve the equation by inspection.

11. a. Given $4^x = 16$, then $x =$ _____.   **12. a.** Given $2^x = 16$, then $x =$ _____.

     b. Given $4^x = 64$, then $x =$ _____.   **b.** Given $2^x = 32$, then $x =$ _____.

     c. If $4^x = 50$, then x is between which two consecutive integers?   **c.** If $2^x = 30$, then x is between which two consecutive integers?

13. Given $\sqrt[3]{6} = 6^x$, then $x =$ _____.   **14.** Given $\sqrt{11} = 11^x$, then $x =$ _____.

For Exercises 15–30, solve the exponential equation by using the property that $b^x = b^y$ implies $x = y$, for $b > 0$ and $b \neq 1$. **(See Examples 1 and 2.)**

15. $5^x = 625$ **16.** $3^x = 81$ **17.** $2^{-x} = 64$ **18.** $6^{-x} = 216$

19. $36^x = 6$ **20.** $343^x = 7$ **21.** $4^{2x-1} = 64$ **22.** $5^{3x-1} = 125$

23. $81^{3x-4} = \dfrac{1}{243}$ **24.** $4^{2x-7} = \dfrac{1}{128}$ **25.** $\left(\dfrac{2}{3}\right)^{-x+4} = \dfrac{8}{27}$ **26.** $\left(\dfrac{1}{4}\right)^{3x+2} = \dfrac{1}{64}$

27. $16^{-x+1} = 8^{5x}$ **28.** $27^{(1/3)x} = \left(\dfrac{1}{9}\right)^{2x-1}$ **29.** $(4^x)^{x+1} = 16$ **30.** $(3^x)^{x+2} = \dfrac{1}{3}$

For Exercises 31–54, solve the exponential equations by taking the logarithm of both sides. (Round the answers to three decimal places.) **(See Examples 3–5.)**

31. $8^a = 21$ **32.** $6^y = 39$ **33.** $e^x = 8.1254$ **34.** $e^x = 0.3151$

35. $10^t = 0.0138$ **36.** $10^p = 16.8125$ **37.** $e^{0.07h} = 15$ **38.** $e^{0.03k} = 4$

39. $e^{1.2t} = 3$ **40.** $e^{1.5t} = 7$ **41.** $3^{x+1} = 5^x$ **42.** $2^{x-1} = 7^x$

43. $2^{x+2} = 6^x$ **44.** $5^{x-2} = 3^x$ **45.** $32e^{0.04m} = 128$ **46.** $8e^{0.05n} = 160$

47. $6e^{x/3} = 125$ **48.** $8e^{x/5} = 155$ **49.** $9^x = 20^{2x}$ **50.** $4^{2x} = 5^x$

51. $e^{1.3x} + 3 = 20.8$ **52.** $e^{0.4x} - 5 = 9.2$ **53.** $5^{x-2} - 4 = 16$ **54.** $2^{x+4} + 17 = 50$

Concept 2: Applications of Exponential Equations

55. The population of China can be modeled by the function

$$P(t) = 1237(1.0095)^t$$

where $P(t)$ is in millions and t is the number of years since 1998. **(See Example 6.)**

a. Using this model, what was the population in the year 2002?

b. Predict the population in the year 2016.

c. If this growth rate continues, in what year will the population reach 2 billion people (2 billion is 2000 million)?

56. The population of Delhi, India, can be modeled by the function

$$P(t) = 9817(1.031)^t$$

where $P(t)$ is in thousands and t is the number of years since 2001.

a. Using this model, predict the population in the year 2018.

b. If this growth rate continues, in what year will the population reach 15 million (15 million is 15,000 thousand)?

57. The growth of a certain bacteria in a culture is given by the model

$$A(t) = 500e^{0.0277t}$$

where $A(t)$ is the number of bacteria and t is time in minutes.

a. What is the initial number of bacteria?

b. What is the population after 10 min?

c. How long will it take for the population to double (that is, reach 1000)?

58. The population of the bacteria *Salmonella typhimurium* is given by the model

$$A(t) = 300e^{0.01733t}$$

where $A(t)$ is the number of bacteria and t is time in minutes.

a. What is the initial number of bacteria?

b. What is the population after 10 min?

c. How long will it take for the population to double?

59. Suppose $5000 is invested at 7% interest compounded continuously. How long will it take for the investment to grow to $10,000? Use the model $A(t) = Pe^{rt}$ and round to the nearest tenth of a year.

60. Suppose $2000 is invested at 10% interest compounded continuously. How long will it take for the investment to triple? Use the model $A(t) = Pe^{rt}$ and round to the nearest year.

61. Phosphorus 32 (^{32}P) has a half-life of approximately 14 days. If 10 g of ^{32}P is present initially, then the amount of phosphorus 32 still present after t days is given by $A(t) = 10(0.5)^{t/14}$. **(See Example 7.)**

a. Find the amount of phosphorus 32 still present after 5 days. Round to the nearest tenth of a gram.

b. Find the amount of time necessary for the amount of ^{32}P to decay to 4 g. Round to the nearest tenth of a day.

62. Polonium 210 (^{210}Po) has a half-life of approximately 138.6 days. If 4 g of ^{210}Po is present initially, then the amount of polonium 210 still present after t days is given by $A(t) = 4e^{-0.005t}$.

a. Find the amount of polonium 210 still present after 50 days. Round to the nearest tenth of a gram.

b. Find the amount of time necessary for the amount of ^{210}P to decay to 0.5 g. Round to the nearest tenth of a day.

63. Suppose you save $10,000 from working an extra job. Rather than spending the money, you decide to save the money for retirement by investing in a mutual fund that averages 12% per year. How long will it take for this money to grow to $1,000,000? Use the model $A(t) = Pe^{rt}$ and round to the nearest tenth of a year.

64. The model $A = Pe^{rt}$ is used to compute the total amount of money in an account after t years at an interest rate r, compounded continuously. The value P is the initial principal. Find the amount of time required for the investment to double as a function of the interest rate. (*Hint:* Substitute $A = 2P$ and solve for t.)

Expanding Your Skills

65. The isotope of plutonium of mass 238 (written ^{238}Pu) is used to make thermoelectric power sources for spacecraft. The heat and electric power derived from such units have made the Voyager, Gallileo, and Cassini missions to the outer reaches of our solar system possible. The half-life of ^{238}Pu is 87.7 yr.

Suppose a space probe was launched in the year 2002 with 2.0 kg of ^{238}Pu. Then the amount of ^{238}Pu available to power the spacecraft decays over time according to

$$P(t) = 2e^{-0.0079t}$$

where $t \geq 0$ is the number of years since 2002 and $P(t)$ is the amount of plutonium still present (in kilograms).

a. Suppose the space probe is due to arrive at Pluto in the year 2045. How much plutonium will remain when the spacecraft reaches Pluto? Round to two decimal places.

b. If 1.5 kg of ^{238}Pu is required to power the spacecraft's data transmitter, will there be enough power in the year 2045 for us to receive close-up images of Pluto?

66. ^{99m}Tc is a radionuclide of technetium that is widely used in nuclear medicine. Although its half-life is only 6 hr, the isotope is continuously produced via the decay of its longer-lived parent ^{99}Mo (molybdenum 99) whose half-life is approximately 3 days. The ^{99}Mo generators (or "cows") are sold to hospitals in which the ^{99m}Tc can be "milked" as needed over a period of a few weeks. Once separated from its parent, the ^{99m}Tc may be chemically incorporated into a variety of imaging agents, each of which is designed to be taken up by a specific target organ within the body. Special cameras, sensitive to the gamma rays emitted by the technetium, are then used to record a "picture" (similar in appearance to an X-ray film) of the selected organ.

Suppose a technician prepares a sample of ^{99m}Tc-pyrophosphate to image the heart of a patient suspected of having had a mild heart attack. If the injection contains 10 millicuries (mCi) of ^{99m}Tc at 1:00 P.M., then the amount of technetium still present is given by

$$T(t) = 10e^{-0.1155t}$$

where $t > 0$ represents the time in hours after 1:00 P.M. and $T(t)$ represents the amount of ^{99m}Tc (in millicuries) still present.

a. How many millicuries of ^{99m}Tc will remain at 4:20 P.M. when the image is recorded? Round to the nearest tenth of a millicurie.

b. How long will it take for the radioactive level of the ^{99m}Tc to reach 2 mCi? Round to the nearest tenth of an hour.

Graphing Calculator Exercises

67. Graph $Y_1 = 8^x$ and $Y_2 = 21$ on a window where $0 \leq x \leq 5$ and $0 \leq y \leq 40$. Use the graph and an *Intersect* feature to support your answer to Exercise 31.

68. Graph $Y_1 = 6^x$ and $Y_2 = 39$ on a window where $0 \leq x \leq 5$ and $0 \leq y \leq 50$. Use the graph and an *Intersect* feature to support your answer to Exercise 32.

Logarithmic Equations and Applications

1. Solving Logarithmic Equations

An equation containing a variable within a logarithmic expression is called a **logarithmic equation**. For example:

$$\log_3 (5x - 1) = \log_3 (x + 11) \quad \text{and} \quad \ln(2x + 5) = 1 \text{ are logarithmic equations.}$$

Given an equation in which two logarithms of the same base are equated, we can apply the equivalence property of logarithms. Since all logarithmic functions are one-to-one, $\log_b x = \log_b y$ implies that $x = y$.

> ### Equivalence Property of Logarithmic Expressions
>
> Let b, x, and y be positive real numbers with $b \neq 1$. Then,
>
> $$\log_b x = \log_b y \quad \text{implies that } x = y.$$

Concepts

1. Solving Logarithmic Equations
2. Applications of Logarithmic Equations

TIP: The equivalence property tells us that if two logarithmic expressions with the same base are equal, then their arguments must be equal.

Example 1 Solving a Logarithmic Equation Using the Equivalence Property

Solve. $\log_3 (5x - 1) = \log_3 (x + 11)$

Solution:

$\log_3 (5x - 1) = \log_3 (x + 11)$	Two logarithms of the same base are set equal to each other.
$5x - 1 = x + 11$	Equate the arguments.
$4x = 12$	Solve for x.
$x = 3$	Because the domain of a logarithm function is restricted, it is mandatory that we check all potential solutions to a logarithmic equation.

$$\underline{\text{Check:}} \quad \log_3 (5x - 1) = \log_3 (x + 11)$$

$$\log_3 [5(3) - 1] \overset{?}{=} \log_3 [(3) + 11]$$

The solution is 3. $\qquad \log_3 (14) = \log_3 (14) \checkmark$

Skill Practice

Solve the equation.

1. $\log_2 (7x - 19) = \log_2 (2x + 1)$

Many logarithmic equations involve logarithmic terms and constant terms. In such a case, we can apply the properties of logarithms to write the equation in the form $\log_b x = k$, where k is a constant. We can solve for x, by writing the equation in its equivalent exponential form $x = b^k$. For example, consider the equation $\ln(2x + 5) = 1$. First write the equation in exponential form:

$$\ln(2x + 5) = 1 \Rightarrow e^1 = 2x + 5. \text{ Now solve for } x.$$

$$e - 5 = 2x$$

$$\frac{e - 5}{2} = x \quad \text{or} \quad x \approx -1.14$$

To solve equations containing more than one logarithm of first degree, use the following guidelines.

Answer

1. 4

Solving Logarithmic Equations

Step 1 Isolate the logarithms on one side of the equation.

Step 2 Write a sum or difference of logarithms as a single logarithm.

Step 3 Rewrite the equation in step 2 in exponential form.

Step 4 Solve the resulting equation from step 3.

Step 5 Check all solutions to verify that they are within the domain of the logarithmic expressions in the original equation.

Example 2 Solving a Logarithmic Equation

Solve the equation.

$$\log_4 x = 1 - \log_4(x - 3)$$

Solution:

$$\log_4 x = 1 - \log_4(x - 3)$$

$$\log_4 x + \log_4(x - 3) = 1 \qquad \text{Isolate the logarithms on one side of the equation.}$$

$$\log_4[x(x - 3)] = 1 \qquad \text{Write as a single logarithm.}$$

$$\log_4(x^2 - 3x) = 1 \qquad \text{Simplify inside the parentheses.}$$

$$x^2 - 3x = 4^1 \qquad \text{Write the equation in exponential form.}$$

$$x^2 - 3x - 4 = 0 \qquad \text{The resulting equation is quadratic.}$$

$$(x - 4)(x + 1) = 0 \qquad \text{Factor.}$$

$$x = 4 \quad \text{or} \quad x = -1 \qquad \text{Apply the zero product rule.}$$

Notice that -1 is *not* a solution because $\log_4 x$ is not defined at $x = -1$. However, $x = 4$ *is* defined in both expressions $\log_4 x$ and $\log_4(x - 3)$. We can substitute $x = 4$ into the original equation to show that it checks.

Check: $x = 4$

$$\log_4 x = 1 - \log_4(x - 3)$$

$$\log_4 4 \stackrel{?}{=} 1 - \log_4(4 - 3)$$

$$1 \stackrel{?}{=} 1 - \log_4 1$$

$$1 \stackrel{?}{=} 1 - 0 \checkmark \text{True}$$

The solution is 4.

Example 3 Solving a Logarithmic Equation

Solve the equation.

$$\log(x + 300) = 3.7$$

Answers

2. 9 (The value -1 does not check.)

3. 2

Solution:

$$\log (x + 300) = 3.7$$ The equation has a single logarithm that is already isolated.

$$10^{3.7} = x + 300$$ Write the equation in exponential form.

$$10^{3.7} - 300 = x$$ Solve for x.

$$x = 10^{3.7} - 300$$

$$\approx 4711.87$$

Check the exact value of x in the original equation.

Check: $x = 10^{3.7} - 300$

$$\log (x + 300) = 3.7$$

$$\log [(10^{3.7} - 300) + 300] \overset{?}{=} 3.7$$

$$\log (10^{3.7} - 300 + 300) \overset{?}{=} 3.7$$

$$\log 10^{3.7} \overset{?}{=} 3.7$$ Property 3 of logarithms: $\log_b b^p = p$

$$3.7 \overset{?}{=} 3.7 \checkmark \text{ True}$$

The solution $10^{3.7} - 300$ checks.

Example 4 **Solving a Logarithmic Equation**

Solve the equation.

$$\ln (x + 2) + \ln (x - 1) = \ln (9x - 17)$$

Solution:

$$\ln (x + 2) + \ln (x - 1) = \ln (9x - 17)$$

$$\ln (x + 2) + \ln (x - 1) - \ln (9x - 17) = 0$$ Isolate the logarithms on one side.

$$\ln \left[\frac{(x + 2)(x - 1)}{9x - 17} \right] = 0$$ Write as a single logarithm.

$$e^0 = \frac{(x + 2)(x - 1)}{9x - 17}$$ Write the equation in exponential form.

$$1 = \frac{(x + 2)(x - 1)}{9x - 17}$$ Simplify.

$$(1) \cdot (9x - 17) = \left[\frac{(x + 2)(x - 1)}{9x - 17} \right] \cdot (9x - 17)$$ Multiply by the LCD.

$$9x - 17 = (x + 2)(x - 1)$$ The equation is quadratic.

$$9x - 17 = x^2 + x - 2$$

$$0 = x^2 - 8x + 15$$

$$0 = (x - 5)(x - 3)$$

$$x = 5 \quad \text{or} \quad x = 3$$

The solutions 5 and 3 are both within the domain of the logarithmic functions in the original equation.

Skill Practice

Solve the equation.

4. $\ln (t - 3) + \ln (t - 1) = \ln (2t - 5)$

Answer

4. 4 (The value 2 does not check.)

Check: $x = 5$

$\ln(x + 2) + \ln(x - 1) = \ln(9x - 17)$

$\ln(5 + 2) + \ln(5 - 1) \overset{?}{=} \ln[9(5) - 17]$

$\ln 7 + \ln 4 \overset{?}{=} \ln(45 - 17)$

$\ln(7 \cdot 4) \overset{?}{=} \ln(28)$

$\ln 28 \overset{?}{=} \ln 28 \checkmark$ True

Check: $x = 3$

$\ln(x + 2) + \ln(x - 1) = \ln(9x - 17)$

$\ln(3 + 2) + \ln(3 - 1) \overset{?}{=} \ln[9(3) - 17]$

$\ln 5 + \ln 2 \overset{?}{=} \ln(27 - 17)$

$\ln(5 \cdot 2) \overset{?}{=} \ln(10)$

$\ln 10 \overset{?}{=} \ln 10 \checkmark$ True

Both solutions check.

2. Applications of Logarithmic Equations

Example 5 Applying a Logarithmic Equation to Earthquake Intensity

The magnitude of an earthquake (the amount of seismic energy released at the hypocenter of the earthquake) is measured on the Richter scale. The Richter scale value R is determined by the formula

$$R = \log\left(\frac{I}{I_0}\right)$$

where I is the intensity of the earthquake and I_0 is the minimum measurable intensity of an earthquake. (I_0 is a "zero-level" quake—one that is barely detected by a seismograph.)

a. Compare the Richter scale values of earthquakes that are (i) 100,000 times (10^5 times) as intense than I_0 and (ii) 1,000,000 times (10^6 times) as intense than I_0.

b. On October 17, 1989, an earthquake measuring 7.1 on the Richter scale occurred in the Loma Prieta area in the Santa Cruz Mountains. The quake devastated parts of San Francisco and Oakland, California, bringing 63 deaths and over 3700 injuries. Determine how many times as intense this earthquake was than a zero-level quake.

Solution:

a. $R = \log\left(\dfrac{I}{I_0}\right)$

i. Earthquake 100,000 times I_0

$R = \log\left(\dfrac{10^5 \cdot I_0}{I_0}\right)$ Substitute $I = 10^5 I_0$.

$= \log 10^5$

$= 5$

ii. Earthquake 1,000,000 times I_0

$R = \log\left(\dfrac{10^6 \cdot I_0}{I_0}\right)$ Substitute $I = 10^6 I_0$.

$= \log 10^6$

$= 6$

b. $R = \log\left(\dfrac{I}{I_0}\right)$

$7.1 = \log\left(\dfrac{I}{I_0}\right)$ Substitute $R = 7.1$.

$\dfrac{I}{I_0} = 10^{7.1}$ Write the equation in exponential form.

$I = 10^{7.1} \cdot I_0$ Solve for I.

The Loma Prieta earthquake in 1989 was $10^{7.1}$ times ($\approx 12{,}590{,}000$ times) as intense than a zero-level earthquake.

Section 9.7 Practice Exercises

Vocabulary and Key Concepts

1. a. An equation containing a variable within a logarithmic expression is called a _____ equation.

b. The equivalence property of logarithmic expressions states that if $\log_b x = \log_b y$, then _____ = _____.

Review Exercises

For Exercises 2–6, write the expression as a single logarithm. Assume all variables represent positive real numbers.

2. $\log_b (x - 1) + \log_b (x + 2)$

3. $\log_b x + \log_b (2x + 3)$

4. $\log_b x - \log_b (1 - x)$

5. $\log_b (x + 2) - \log_b (3x - 5)$

6. $2 \log_b x - \log_b (x + 1)$

Concept 1: Solving Logarithmic Equations

For Exercises 7–46, solve the logarithmic equations. **(See Examples 1–4.)**

7. $\log_3 (x + 1) = \log_3 7$

8. $\log_6 8 = \log_6 (x - 1)$

9. $\log_5 (2x - 4) = \log_5 (x + 3)$

10. $\log_8 (3x + 3) = \log_8 (4x - 1)$

11. $\log (3x + 1) = \log (x - 3)$

12. $\ln (y - 4) = \ln (2y - 1)$

13. $\log_2 (a^2 - a) = \log_2 12$

14. $\log_5 (c^2 - c) = \log_5 20$

15. $\log_3 x = 2$

16. $\log_4 x = 9$

17. $\log p = 42$

18. $\log q = \dfrac{1}{2}$

19. $\ln x = 0.08$

20. $\ln x = 19$

21. $\log (x + 40) = -9.2$

22. $\log (z - 3) = -6.7$

23. $\log_x 25 = 2$ $(x > 0)$

24. $\log_x 100 = 2$ $(x > 0)$

25. $\log_b 10{,}000 = 4$ $(b > 0)$

26. $\log_b e^3 = 3$ $(b > 0)$

27. $\log_y 5 = \dfrac{1}{2}$ $(y > 0)$

28. $\log_b 8 = \dfrac{1}{2}$ $(b > 0)$

29. $\log_4 (c + 5) = 3$

30. $\log_5 (a - 4) = 2$

31. $\log_5 (4y + 1) = 1$

32. $\log_6 (5t - 2) = 1$

33. $\ln (1 - x) = 0$

34. $\log_4 (2 - x) = 1$

35. $\log_3 8 - \log_3 (x + 5) = 2$

36. $\log_2 (x + 3) - \log_2 (x + 2) = 1$

37. $\log_2(h - 1) + \log_2(h + 1) = 3$ **38.** $\log_3 k + \log_3(2k + 3) = 2$ **39.** $\log(x + 2) = \log(3x - 6)$

40. $\log x = \log(1 - x)$ **41.** $\ln x - \ln(4x - 9) = 0$ **42.** $\ln(x + 5) - \ln x = \ln(4x)$

43. $\log_5(3t + 2) - \log_5 t = \log_5 4$ **44.** $\log(6y - 7) + \log y = \log 5$ ▶▣ **45.** $\log(4m) = \log 2 + \log(m - 3)$

46. $\log(-h) + \log 3 = \log(2h - 15)$

Concept 2: Applications of Logarithmic Equations

47. On May 28, 2004, an earthquake that measured 6.3 on the Richter scale occurred in northern Iran. Use the

formula $R = \log\left(\dfrac{I}{I_0}\right)$ to determine how many times as intense this earthquake was than a zero-level quake.

(See Example 5.)

48. Papua, Indonesia, had a 7.0 earthquake on February 4, 2004. Determine how many times as intense this earthquake was than a zero-level quake.

The decibel level of sound can be found by the equation $L = 10 \log\left(\dfrac{I}{I_0}\right)$, where I is the intensity of the sound and I_0 is the intensity of the least audible sound that an average person can hear. Generally I_0 is found as 10^{-12} watt per square meter (W/m^2). Use this information to answer Exercises 49 and 50.

49. Given that heavy traffic has a decibel level of 89.3 and that $I_0 = 10^{-12}$, find the intensity of the sound of heavy traffic.

50. Given that normal conversation has a decibel level of 65 and $I_0 = 10^{-12}$, find the intensity of the sound of normal conversation.

The apparent magnitude of a star refers to the brightness of the star. Objects with lower magnitude appear brighter to the eye than objects with higher magnitude. For example, the full moon has a magnitude of -12.6, whereas the naked eye can barely see an object of magnitude 6. For higher magnitudes, a telescope is required. However, even a telescope has its limitations. The limiting magnitude L of an optical telescope with lens diameter D (in inches) is given by $L(D) = 8.8 + 5.1 \log D$. Use this function for Exercises 51 and 52.

51. a. Find the limiting magnitude for a reflecting telescope 5 in. in diameter. Round to one decimal place.

    b. Find the diameter of a telescope whose limiting magnitude is 14. Round to the nearest tenth of an inch.

52. a. Find the limiting magnitude for a telescope with a lens 7 in. in diameter. Round to one decimal place.

    b. How large a telescope is needed to see a galaxy with magnitude 17.2? Round to the nearest tenth of an inch.

Scientists use the pH scale to represent the level of acidity or alkalinity of a liquid. This is based on the molar concentration of hydrogen ions, $[H^+]$. Since the values of $[H^+]$ vary over a large range, 1×10^0 mole per liter to 1×10^{-14} mole per liter (mol/L), a logarithmic scale is used to compute pH. The formula

$$pH = -\log[H^+]$$

represents the pH of a liquid as a function of its concentration of hydrogen ions, $[H^+]$.

The pH scale ranges from 0 to 14. Pure water is taken as neutral having a pH of 7. A pH less than 7 is acidic. A pH greater than 7 is alkaline (or basic).

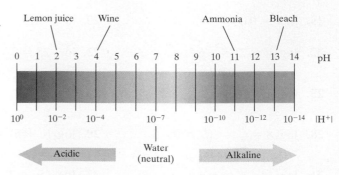

For Exercises 53 and 54, determine the value of H^+ (in mol/L) for the following liquids, given their pH values. Write the answers in scientific notation.

53. a. Seawater pH $= 8.5$

    b. Acid rain pH $= 2.3$

54. a. Milk pH $= 6.2$

    b. Sodium bicarbonate pH $= 8.4$

Mixed Exercises

For Exercises 55–70 solve the equations.

55. $\log_3 (2x + 1) = 4$

56. $\log_5 (3 - x) = 1$

57. $8^{2x+1} = 64$

58. $5^{4-x} = 125$

59. $4^{3x-2} = \dfrac{1}{32}$

60. $27^{3-x} = \dfrac{1}{9}$

61. $\log (x^2 + 7x) = \log(18)$

62. $\ln(x^2) = \ln(x + 2)$

63. $\log_5 (x + 4) + \log_5 x = 1$

64. $\log_2 (17x) - \log_2 (2x + 3) = 3$

65. $\ln(3x - 3) - \ln(6x) = \ln 2$

66. $\log_6 (3x + 1) - \log_6 2 = \log_6 x$

67. $e^{0.2t} = 6$

68. $e^{1.5t} = 20$

69. $2^{x+3} = 5^x$

70. $8^{2x-2} = 3^x$

For Exercises 71–78, write an equation for the inverse function.

71. $f(x) = 3^x - 2$

72. $g(x) = 4^x + 1$

73. $k(x) = 10^{x-2} + 6$

74. $m(x) = 10^{x+1} - 2$

75. $h(x) = \ln(x + 10)$

76. $f(x) = \ln(x - 5)$

77. $n(x) = \log(x + 3) - 1$

78. $c(x) = \log(x - 8) + 3$

Expanding Your Skills

For Exercises 79–82, solve the equations.

79. $(\log x)^2 - 2 \log x - 15 = 0$
(*Hint:* Let $u = \log x$.)

80. $(\log_2 z)^2 - 3 \log_2 z - 4 = 0$

81. $(\log_3 w)^2 + 5 \log_3 w + 6 = 0$

82. $(\ln x)^2 - 2 \ln x = 0$

Graphing Calculator Exercises

83. Compare the graphs of the functions.

    $Y_1 = \ln(3x)$ and $Y_2 = \ln x + \ln 3$

84. Compare the graphs of the functions.

    $Y_1 = \ln\left(\dfrac{x}{3}\right)$ and $Y_2 = \ln x - \ln 3$

Group Activity

Creating a Population Model

Materials: A computer with Internet access and a calculator

Estimated time: 15–20 minutes

Group Size: 3

1. Go to a website such as http://www.census.gov/ and determine the population of your state for the year 2000, and then for another year after that time. For example, the population of Florida in the year 2000 was approximately 16,050,000. Florida's population in 2006 was approximately 18,090,000.

2. Let the variable t represent the number of years since the year 2000. Therefore, $t = 0$ represents the year 2000, $t = 1$ represents the year 2001, and so on. Write your data from step 1 as ordered pairs. For example: $(0, 16050000)$ and $(6, 18090000)$. (_____, _____) (_____, _____)

3. Population growth can be modeled by an exponential function of the form:

   $P(t) = P_0 e^{kt}$ where P_0 is the initial population for $t = 0$

                           $P(t)$ is the population at time, t

                           k is the growth constant

   Substitute the population for the year 2000 into the model for P_0, and write the function here:

   $P(t) = $ _____ e^{kt} [For example: $P(t) = 16,050,000e^{kt}$]

4. Substitute the population value and t value from the second ordered pair into the function for $P(t)$ and t, respectively.   [For example: $18,090,000 = 16,050,000e^{k(6)}$]

5. Solve the equation from step 4 for the value of k. Round k to five decimal places.

6. Substitute the values of P_0 and k in the original model, $P(t) = P_0 e^{kt}$, and write the function here:

   $P(t) = $ _____

   You should now have a function that models the population of your state as a function of the time in years since 2000.

7. Assuming that the population follows this trend, predict the population of your state in the year 2015.

8. Assuming that the population follows this trend, predict the population of your state in the year 2050.

Chapter 9   Summary

Key Concepts

Horizontal Line Test

Consider a function defined by a set of points (x, y) in a rectangular coordinate system. Then the graph defines y as a one-to-one function of x if no horizontal line intersects the graph in more than one point.

Finding an Equation of the Inverse of a Function

For a one-to-one function defined by $y = f(x)$, the equation of the inverse can be found as follows:

1. Replace $f(x)$ by y.
2. Interchange x and y.
3. Solve for y.
4. Replace y by $f^{-1}(x)$.

The graphs defined by $y = f(x)$ and $y = f^{-1}(x)$ are symmetric with respect to the line $y = x$.

Definition of an Inverse Function

If f is a one-to-one function represented by ordered pairs of the form (x, y), then the inverse function, denoted f^{-1}, is the set of ordered pairs (y, x).

Inverse Function Property

If f is a one-to-one function, then g is the inverse of f if and only if $(f \circ g)(x) = x$ for all x in the domain of g, and $(g \circ f)(x) = x$ for all x in the domain of f.

Examples

Example 1

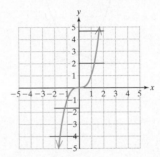

The function is one-to-one because it passes the horizontal line test.

Example 2

Find the inverse of the one-to-one function defined by $f(x) = 3 - x^3$.

1. $y = 3 - x^3$

2. $x = 3 - y^3$

3. $\quad x - 3 = -y^3$

   $\quad -x + 3 = y^3$

   $\quad \sqrt[3]{-x + 3} = y$

4. $f^{-1}(x) = \sqrt[3]{-x + 3}$

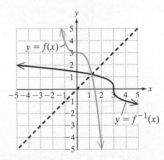

Example 3

Verify that the functions defined by $f(x) = x - 1$ and $g(x) = x + 1$ are inverses.

$(f \circ g)(x) = f(x + 1) = (x + 1) - 1 = x$

$(g \circ f)(x) = g(x - 1) = (x - 1) + 1 = x$

Section 9.2 Exponential Functions

Key Concepts

A function $f(x) = b^x$ $(b > 0, b \neq 1)$ is an **exponential function**.

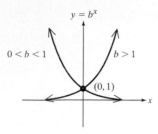

The domain is $(-\infty, \infty)$.

The range is $(0, \infty)$.

The line $y = 0$ (x-axis) is a horizontal asymptote.

The y-intercept is $(0, 1)$.

Example

Example 1

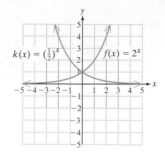

Section 9.3 Logarithmic Functions

Key Concepts

The function $y = \log_b x$ is a **logarithmic function**.

$$y = \log_b x \Leftrightarrow b^y = x \quad (x > 0, b > 0, b \neq 1)$$

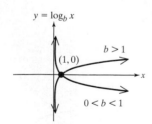

For $y = \log_b x$, the domain is $(0, \infty)$.

The range is $(-\infty, \infty)$.

The line $x = 0$ (y-axis) is a vertical asymptote.

The x-intercept is $(1, 0)$.

The function $y = \log x$ is the **common logarithmic function** (base 10).

Examples

Example 1

$\log_4 64 = 3$ because $4^3 = 64$

Example 2

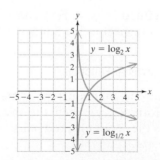

Example 3

$\log 10,000 = 4$ because $10^4 = 10,000$

Section 9.4   Properties of Logarithms

Key Concepts

Let b, x, and y be positive real numbers where $b \neq 1$, and let p be a real number. Then the following properties are true.

1. $\log_b 1 = 0$

2. $\log_b b = 1$

3. $\log_b b^p = p$

4. $b^{\log_b x} = x$

5. $\log_b (xy) = \log_b x + \log_b y$

6. $\log_b \left(\dfrac{x}{y} \right) = \log_b x - \log_b y$

7. $\log_b x^p = p \log_b x$

The **properties of logarithms** can be used to write multiple logarithms as a single logarithm.

The **properties of logarithms** can be used to write a single logarithm as a sum or difference of logarithms.

Examples

Example 1

1. $\log_5 1 = 0$

2. $\log_6 6 = 1$

3. $\log_4 4^7 = 7$

4. $2^{\log_2 5} = 5$

5. $\log (5x) = \log 5 + \log x$

6. $\log_7 \left(\dfrac{z}{10} \right) = \log_7 z - \log_7 10$

7. $\log x^5 = 5 \log x$

Example 2

$$\log x - \frac{1}{2} \log y - 3 \log z$$

$$= \log x - (\log y^{1/2} + \log z^3)$$

$$= \log x - \log (\sqrt{y} z^3)$$

$$= \log \left(\frac{x}{\sqrt{y} z^3} \right)$$

Example 3

$$\log \sqrt[3]{\frac{x}{y^2}}$$

$$= \frac{1}{3} \log \left(\frac{x}{y^2} \right)$$

$$= \frac{1}{3} (\log x - \log y^2)$$

$$= \frac{1}{3} (\log x - 2 \log y)$$

$$= \frac{1}{3} \log x - \frac{2}{3} \log y$$

Section 9.5   The Irrational Number e

Key Concepts

The function $y = e^x$ is the exponential function with base e.

The **natural logarithm function** $y = \ln x$ is the logarithm function with base e.

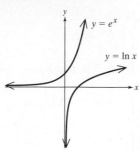

Change-of-Base Formula

$$\log_b x = \frac{\log_a x}{\log_a b} \qquad a > 0, a \neq 1, b > 0, b \neq 1$$

Examples

Example 1

Use a calculator to approximate the value of the expressions.

$e^{7.5} \approx 1808.0424$

$e^{-\pi} \approx 0.0432$

$\ln 107 \approx 4.6728$

$\ln\left(\dfrac{1}{\sqrt{2}}\right) \approx -0.3466$

Example 2

$$\log_3 59 = \frac{\log 59}{\log 3} \approx 3.7115$$

Section 9.6   Exponential Equations and Applications

Key Concepts

The equivalence of exponential expressions can be used to solve **exponential equations**.

If $b^x = b^y$, then $x = y$

Guidelines to Solve Exponential Equations

1. Isolate one of the exponential expressions in the equation.
2. Take a logarithm of both sides of the equation.
3. Use the power property of logarithms to write exponents as factors.
4. Solve the resulting equation from step 3.

Examples

Example 1

$5^{2x} = 125$

$5^{2x} = 5^3$ implies that $2x = 3$

$$x = \frac{3}{2}$$

Example 2

$$4^{x+1} - 2 = 1055$$

Step 1: $4^{x+1} = 1057$

Step 2: $\ln 4^{x+1} = \ln 1057$

Step 3: $(x + 1)\ln 4 = \ln 1057$

Step 4: $x + 1 = \dfrac{\ln 1057}{\ln 4}$

$$x = \frac{\ln 1057}{\ln 4} - 1 \approx 4.023$$

Section 9.7   Logarithmic Equations and Applications

Key Concepts

The equivalence property of logarithmic expressions can be used to solve logarithmic equations.

If $\log_b x = \log_b y$, then $x = y$.

Guidelines to Solve Logarithmic Equations

1. Isolate the logarithms on one side of the equation.
2. Write a sum or difference of logarithms as a single logarithm.
3. Rewrite the equation in step 2 in exponential form.
4. Solve the resulting equation from step 3.
5. Check all solutions to verify that they are within the domain of the logarithmic expressions in the equation.

Examples

Example 1

$\log(4x) = \log(x + 6)$ implies $4x = x + 6$

$$3x = 6$$

$$x = 2$$

Example 2

$\log(3x - 1) + 1 = \log(2x + 1)$

Step 1: $\log(3x - 1) - \log(2x + 1) = -1$

Step 2: $\log\left(\dfrac{3x - 1}{2x + 1}\right) = -1$

Step 3: $10^{-1} = \dfrac{3x - 1}{2x + 1}$

Step 4: $\dfrac{1}{10} = \dfrac{3x - 1}{2x + 1}$

$$2x + 1 = 10(3x - 1)$$

$$2x + 1 = 30x - 10$$

$$-28x = -11$$

$$x = \dfrac{11}{28}$$

Step 5: $x = \dfrac{11}{28}$ checks in original equation

Chapter 9   Review Exercises

Section 9.1

For Exercises 1 and 2, determine if the function is one-to-one by using the horizontal line test.

1.

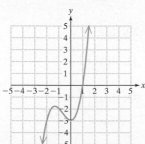

2.

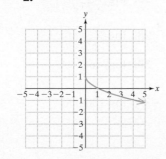

For Exercises 3–7, write the inverse for each one-to-one function.

3. $\{(3, 5), (2, 9), (0, -1), (4, 1)\}$

4. $q(x) = \dfrac{3}{4}x - 2$

5. $g(x) = \sqrt[5]{x} + 3$

6. $f(x) = (x - 1)^3$

7. $n(x) = \dfrac{4}{x - 2}$

8. Verify that the functions defined by $f(x) = 5x - 2$ and $g(x) = \frac{1}{5}x + \frac{2}{5}$ are inverses by showing that $(f \circ g)(x) = x$ and $(g \circ f)(x) = x$.

9. Graph the functions q and q^{-1} from Exercise 8 on the same grid. What can you say about the relationship between these two graphs?

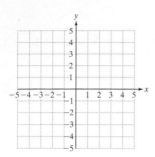

10. a. Find the domain and range of the function defined by $h(x) = \sqrt{x + 1}$.

   b. Find the domain and range of the function defined by $k(x) = x^2 - 1, x \geq 0$.

11. Determine the inverse of the function $p(x) = \sqrt{x} + 2$.

Section 9.2

 For Exercises 12–19, evaluate the exponential expressions. Use a calculator and round to three decimal places, if necessary.

12. 4^5

13. 6^{-2}

14. $8^{1/3}$

15. $\left(\dfrac{1}{100}\right)^{-1/2}$

16. 2^π

17. $5^{\sqrt{3}}$

18. $(\sqrt{7})^{1/2}$

19. $\left(\dfrac{3}{4}\right)^{4/3}$

For Exercises 20–23, graph the functions.

20. $f(x) = 3^x$

21. $g(x) = \left(\dfrac{1}{4}\right)^x$

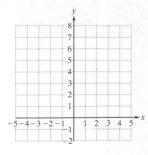　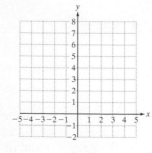

22. $h(x) = 5^{-x}$

23. $k(x) = \left(\dfrac{2}{5}\right)^{-x}$

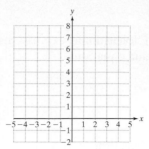　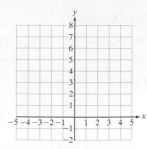

24. a. Does the graph of $y = b^x, b > 0, b \neq 1$, have a vertical or a horizontal asymptote?

   b. Write an equation of the asymptote.

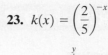 **25.** Background radiation is radiation that we are exposed to from naturally occurring sources including the soil, the foods we eat, and the sun. Background radiation varies depending on where we live. A typical background radiation level is 150 millirems (mrem) per year. (A rem is a measure of energy produced from radiation.) Suppose a substance emits 30,000 mrem per year and has a half-life of 5 yr. The function defined by

$$A(t) = 30{,}000\left(\dfrac{1}{2}\right)^{t/5}$$

gives the radiation level (in millirems) of this substance after t years.

   a. What is the radiation level after 5 yr?

   b. What is the radiation level after 15 yr?

   c. Will the radiation level of this substance be below the background level of 150 mrem after 50 yr?

Section 9.3

For Exercises 26–33, evaluate the logarithms without using a calculator.

26. $\log_3\left(\dfrac{1}{27}\right)$

27. $\log_5 1$

28. $\log_7 7$

29. $\log_2 2^8$

30. $\log_2 16$

31. $\log_3 81$

32. $\log 100{,}000$

33. $\log_8\left(\dfrac{1}{8}\right)$

For Exercises 34 and 35, graph the logarithmic functions.

34. $q(x) = \log_3 x$ **35.** $r(x) = \log_{1/2} x$

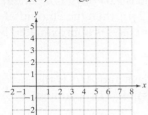

36. a. Does the graph of $y = \log_b x$ have a vertical or a horizontal asymptote?

b. Write an equation of the asymptote.

37. Acidity of a substance is measured by its pH. The pH can be calculated by the formula $pH = -\log[H^+]$, where $[H^+]$ is the hydrogen ion concentration.

a. What is the pH of a fruit with a hydrogen ion concentration of 0.00316 mol/L? Round to one decimal place.

b. What is the pH of an antacid tablet with $[H^+] = 3.16 \times 10^{-10}$? Round to one decimal place.

Section 9.4

For Exercises 38–41, evaluate the logarithms without using a calculator.

38. $\log_8 8$ **39.** $\log_{11} 11^6$

40. $\log_{1/2} 1$ **41.** $12^{\log_{12} 7}$

42. Complete the properties. Assume x, y, and b are positive real numbers such that $b \neq 1$.

a. $\log_b(xy) =$

b. $\log_b x - \log_b y =$

c. $\log_b x^p =$

For Exercises 43–46, write the logarithmic expressions as a single logarithm and simplify if possible.

43. $\dfrac{1}{4}(\log_b y - 4\log_b z + 3\log_b x)$

44. $\dfrac{1}{2}\log_3 a + \dfrac{1}{2}\log_3 b - 2\log_3 c - 4\log_3 d$

45. $\log 540 - 3\log 3 - 2\log 2$

46. $-\log_4 18 + \log_4 6 + \log_4 3 - \log_4 1$

47. Which of the following is equivalent to

$$\frac{2\log 7}{\log 7 + \log 6}?$$

a. $\dfrac{\log 7}{\log 6}$ **b.** $\dfrac{\log 49}{\log 42}$ **c.** $\log\left(\dfrac{7}{6}\right)$

48. Which of the following is equivalent to

$$\frac{\log 8^{-3}}{\log 2 + \log 4}?$$

a. -3 **b.** $-3\log\left(\dfrac{4}{3}\right)$ **c.** $\dfrac{-3\log 4}{\log 3}$

Section 9.5

For Exercises 49–56, use a calculator to approximate the expressions to four decimal places.

49. e^5 **50.** $e^{\sqrt{7}}$

51. $32e^{0.008}$ **52.** $58e^{-0.0125}$

53. $\ln 6$ **54.** $\ln\left(\dfrac{1}{9}\right)$

55. $\log 22$ **56.** $\log e^3$

For Exercises 57–60, use the change-of-base formula to approximate the logarithms to four decimal places.

57. $\log_2 10$ **58.** $\log_9 80$

59. $\log_5 0.26$ **60.** $\log_4 0.0062$

61. An investor wants to deposit \$20,000 in an account for 10 yr at 5.25% interest. Compare the amount she would have if her money were invested with the following different compounding options. Use

$$A(t) = P\left(1 + \frac{r}{n}\right)^{nt}$$

for interest compounded n times per year and $A(t) = Pe^{rt}$ for interest compounded continuously.

a. Compounded annually

b. Compounded quarterly

c. Compounded monthly

d. Compounded continuously

62. To measure a student's retention of material at the end of a course, researchers give the student a test on the material every month for 24 months after the course is over. The student's average score t months after completing the course is given by

$$S(t) = 75e^{-0.5t} + 20$$

where $S(t)$ is the test score.

a. Find $S(0)$ and interpret the result.

b. Find $S(6)$ and interpret the result.

c. Find $S(12)$ and interpret the result.

d. The graph of $y = S(t)$ is shown here. Does it appear that the student's average score is approaching a limiting value?

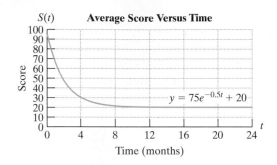

For Exercises 63–70, identify the domain. Write the answer in interval notation.

63. $f(x) = e^x$ **64.** $g(x) = e^{x+6}$

65. $h(x) = e^{x-3}$ **66.** $k(x) = \ln x$

67. $q(x) = \ln(x + 5)$ **68.** $p(x) = \ln(x - 7)$

69. $r(x) = \ln(3x - 4)$ **70.** $w(x) = \ln(5 - x)$

Section 9.6

 Solve the exponential equations in Exercises 71–80. If necessary, round to four decimal places.

71. $4^{3x+5} = 16$ **72.** $5^{7x} = 625$

73. $4^a = 21$ **74.** $5^a = 18$

75. $e^{-x} = 0.1$ **76.** $e^{-2x} = 0.06$

77. $10^{2n} = 1512$ **78.** $10^{-3m} = \dfrac{1}{821}$

79. $2^{x+3} = 7^x$ **80.** $14^{x-5} = 6^x$

81. Radioactive iodine (^{131}I) is used to treat patients with a hyperactive (overactive) thyroid. The half-life of radioactive iodine is 8.04 days. If a patient is given an initial dose of 2 μg, then the amount of iodine in the body after t days is approximated by

$$A(t) = 2e^{-0.0862t}$$

where t is the time in days and $A(t)$ is the amount (in micrograms) of ^{131}I remaining.

a. How much radioactive iodine is present after a week? Round to two decimal places.

b. How much radioactive iodine is present after 30 days? Round to two decimal places.

c. How long will it take for the level of radioactive iodine to reach 0.5 μg?

82. The growth of a certain bacteria in a culture is given by the model $A(t) = 150e^{0.007t}$, where $A(t)$ is the number of bacteria and t is time in minutes.

a. What is the initial number of bacteria?

b. What is the population after $\frac{1}{2}$ hr?

c. How long will it take for the population to double?

 83. The value of a car is depreciated with time according to

$$V(t) = 15,000e^{-0.15t}$$

where $V(t)$ is the value in dollars and t is the time in years.

a. Find $V(0)$ and interpret the result in the context of this problem.

b. Find $V(10)$ and interpret the result in the context of this problem. Round to the nearest dollar.

c. Find the time required for the value of the car to drop to $5000. Round to the nearest tenth of a year.

d. The graph of $y = V(t)$ is shown here. Does it appear that the value of the car is approaching a limiting value?

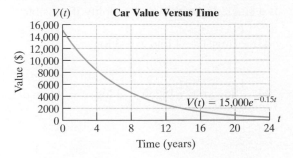

Section 9.7

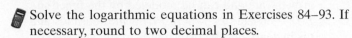 Solve the logarithmic equations in Exercises 84–93. If necessary, round to two decimal places.

84. $\log_3 (10 - t) = \log_3 (2t + 7)$

85. $\ln(4x - 9) = \ln(16 - x)$

86. $\log_5 x = 3$ **87.** $\log_7 x = -2$

88. $\log_6 y = 3$ **89.** $\log_3 y = \dfrac{1}{12}$

90. $\log (2w - 1) = 3$

91. $\log_2 (3w + 5) = 5$

92. $-1 + \log p = -\log (p - 3)$

93. $\log_4 (2 + t) - 3 = \log_4 (3 - 5t)$

94. In 2011, an earthquake that measured 9.0 on the Richter scale occurred near the east coast of Honshu, Japan. Use the formula $R = \log\left(\dfrac{I}{I_0}\right)$ to determine how many times as intense this earthquake was than a zero-level quake.

95. What is the Richter scale value of an earthquake that is 100,000,000 times as intense than I_0?

Chapter 9   Test

1. Explain how to determine graphically if a function is one-to-one.

2. Which of the functions is one-to-one?

a.
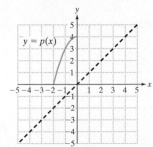

b.
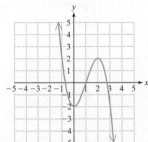

3. Write an equation of the inverse of the one-to-one function defined by $f(x) = \frac{1}{4}x + 3$.

4. Write an equation of the inverse of the function defined by $g(x) = (x - 1)^2, \quad x \geq 1$.

5. Given the graph of the function $y = p(x)$, graph its inverse $p^{-1}(x)$.

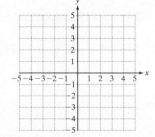

6. Use a calculator to approximate the expression to four decimal places.

a. $10^{2/3}$ **b.** $3^{\sqrt{10}}$ **c.** 8^{π}

7. Graph $f(x) = 4^{x-1}$.

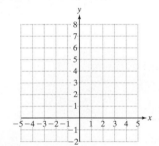

8. a. Write in logarithmic form. $16^{3/4} = 8$

b. Write in exponential form. $\log_x 31 = 5$

9. Graph $g(x) = \log_3 x$.

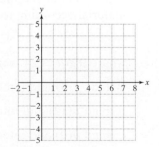

10. Complete the change-of-base formula:
$\log_b n = $ _____

11. Use a calculator to approximate the expression to four decimal places:

   a. $\log 21$ **b.** $\log_4 13$ **c.** $\log_{1/2} 6$

12. Using the properties of logarithms, expand and simplify. Assume all variables represent positive real numbers.

   a. $-\log_3\left(\dfrac{3}{9x}\right)$ **b.** $\log\left(\dfrac{1}{10^5}\right)$

13. Write as a single logarithm. Assume all variables represent positive real numbers.

   a. $\dfrac{1}{2}\log_b x + 3\log_b y$ **b.** $\log a - 4\log a$

 14. Use a calculator to approximate the expression to four decimal places, if necessary.

   a. $e^{1/2}$ **b.** e^{-3}

   c. $\ln\left(\dfrac{1}{3}\right)$ **d.** $\ln e$

15. Identify the graphs as $y = e^x$ or $y = \ln x$.

   a.   **b.**

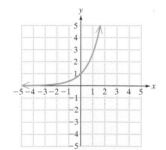

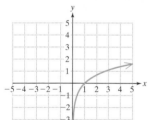 **16.** Researchers found that t months after taking a course, students remembered $p\%$ of the material according to

$$p(t) = 92 - 20\ln(t + 1)$$

where $0 \le t \le 24$ is the time in months.

   a. Find $p(4)$ and interpret the results.

   b. Find $p(12)$ and interpret the results.

   c. Find $p(0)$ and interpret the results.

17. The population of New York City has a 2% growth rate and can be modeled by the function $P(t) = 8008(1.02)^t$, where $P(t)$ is in thousands and t is in years ($t = 0$ corresponds to the year 2000).

   a. Using this model, predict the population in the year 2010.

   b. If this growth rate continues, in what year will the population reach 12 million (12 million is 12,000 thousand)?

 18. A certain bacterial culture grows according to

$$P(t) = \frac{1,500,000}{1 + 5000e^{-0.8t}}$$

where P is the population of the bacteria and t is the time in hours.

   a. Find $P(0)$ and interpret the result. Round to the nearest whole number.

   b. How many bacteria will be present after 6 hr?

   c. How many bacteria will be present after 12 hr?

   d. How many bacteria will be present after 18 hr?

   e. From the graph does it appear that the population of bacteria is reaching a limiting value?

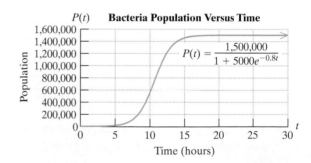

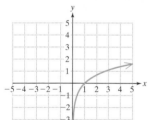 For Exercises 19–26, solve the exponential and logarithmic equations. If necessary, round to three decimal places.

19. $\log(2p - 4) = \log(8 - p)$

20. $\log_4 12 = \log_4 (x^2 - 24)$

21. $\log x + \log(x - 21) = 2$

22. $\log_{1/2} x = -5$

23. $\ln(x + 7) = 2.4$ **24.** $3^{x+4} = \dfrac{1}{27}$

25. $4^x = 50$ **26.** $e^{2.4x} = 250$

27. Atmospheric pressure P decreases exponentially with altitude x according to

$$P(x) = 760e^{-0.000122x}$$

where $P(x)$ is the pressure measured in millimeters of mercury (mm Hg) and x is the altitude measured in meters.

a. Find $P(2500)$ and interpret the result. Round to one decimal place.

b. Find the pressure at sea level.

c. Find the altitude at which the pressure is 633 mm Hg.

28. Use the formula $A(t) = Pe^{rt}$ to compute the value of an investment under continuous compounding.

a. If \$2000 is invested at 7.5% compounded continuously, find the value of the investment after 5 yr.

b. How long will it take the investment to double? Round to two decimal places.

Chapters 1–9 Cumulative Review Exercises

1. Simplify completely.

$$\frac{8 - 4 \cdot 2^2 + 15 \div 5}{|-3 + 7|}$$

2. Divide. $\dfrac{-8p^2 + 4p^3 + 6p^5}{8p^2}$

3. Divide $(t^4 - 13t^2 + 36) \div (t - 2)$. Identify the quotient and remainder.

4. Simplify. $\sqrt{x^2 - 6x + 9}$

5. Simplify. $\dfrac{4}{\sqrt[3]{40}}$

6. Find the length of the missing side.

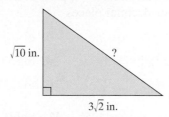

$\sqrt{10}$ in.   ?

$3\sqrt{2}$ in.

7. Simplify. Write the answer with positive exponents only.

$$\frac{2^{2/5}c^{-1/4}d^{1/5}}{2^{-8/5}c^{3/4}d^{1/10}}$$

8. Find the area of the rectangle.

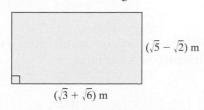

$(\sqrt{5} - \sqrt{2})$ m

$(\sqrt{3} + \sqrt{6})$ m

9. Perform the indicated operation.

$$\frac{4 - 3i}{2 + 5i}$$

10. Solve. $2x(x - 7) = x - 18$

11. How many liters of pure alcohol must be mixed with 8 L of 20% alcohol to bring the concentration up to 50% alcohol?

12. Bank robbers leave the scene of a crime and head north through winding mountain roads to their hideaway. Their average rate of speed is 40 mph. The police leave 6 min later in pursuit. If the police car averages 50 mph traveling the same route, how long will it take the police to catch the bank robbers?

13. Solve the system by using the Gauss-Jordan method.

$$5x + 10y = 25$$
$$-2x + 6y = -20$$

14. Solve. $-2[w - 3(w + 1)] = 4 - 7(w + 3)$

15. Solve for x. $ax - c = bx + d$

16. Solve for t. $s = \frac{1}{2}gt^2$ $t \geq 0$

17. Solve for T. $\sqrt{1 - kT} = \dfrac{V_0}{V}$

18. Find the x-intercepts of the function defined by $f(x) = |x - 5| - 2$.

19. Let $f(t) = 6$, $g(t) = -5t$, and $h(t) = 2t^2$. Find

     a. $(f \cdot g)(t)$ **b.** $(g \circ h)(t)$ **c.** $(h - g)(t)$

20. Solve. $|2q - 5| = |2q + 5|$

21. a. Find an equation of the line parallel to the y-axis and passing through the point $(2, 6)$.

     b. Find an equation of the line perpendicular to the y-axis and passing through the point $(2, 6)$.

     c. Find an equation of the line perpendicular to the line $2x + y = 4$ and passing through the point $(2, 6)$. Write the answer in slope-intercept form.

22. The smallest angle in a triangle measures one-half the largest angle. The smallest angle measures $20°$ less than the middle angle. Find the measures of all three angles.

23. Solve the system.

$$\frac{1}{2}x - \frac{1}{4}y = 1$$

$$-2x + y = -4$$

24. Find the domain. $f(x) = \sqrt{2x - 1}$.

25. Find $f^{-1}(x)$, given $f(x) = 5x - \frac{2}{3}$.

26. The volume of a gas varies directly as its temperature and inversely with pressure. At a temperature of 100 kelvins (K) and a pressure of 10 newtons per square meter (N/m²), the gas occupies a volume of 30 m³. Find the volume at a temperature of 200 K and pressure of 15 N/m².

27. Perform the indicated operations.

$$\frac{5x - 10}{x^2 - 4x + 4} \div \frac{5x^2 - 125}{25 - 5x} \cdot \frac{x^3 + 125}{10x + 5}$$

28. Perform the indicated operations.

$$\frac{x}{x - y} + \frac{y}{y - x} + x$$

29. Given the equation

$$\frac{2}{x - 4} = \frac{5}{x + 2}$$

     a. Are there any restrictions on x for which the rational expressions are undefined?

     b. Solve the equation.

     c. Solve the related inequality.

$$\frac{2}{x - 4} \geq \frac{5}{x + 2}$$

     Write the answer in interval notation.

30. Two more than 3 times the reciprocal of a number is $\frac{5}{4}$ less than the number. Find all such numbers.

31. Solve. $\sqrt{-x} = x + 6$

32. Solve the inequality $2|x - 3| + 1 > -7$. Write the answer in interval notation.

33. Four million *Escherichia coli* (*E. coli*) bacteria are present in a laboratory culture. An antibacterial agent is introduced and the population of bacteria decreases by one-half every 6 hr according to

$$P(t) = 4{,}000{,}000\left(\frac{1}{2}\right)^{t/6} \qquad t \geq 0$$

     where t is the time in hours.

     a. Find the population of bacteria after 6, 12, 18, 24, and 30 hr.

     b. Predict the time required for the population to decrease to 15,625 bacteria.

34. Evaluate the expressions without a calculator.

     a. $\log_7 49$ **b.** $\log_4\left(\frac{1}{64}\right)$

     c. $\log 1{,}000{,}000$ **d.** $\ln e^3$

35. Use a calculator to approximate the expressions to four decimal places.

     a. $\pi^{4.7}$ **b.** e^{π}

     c. $(\sqrt{2})^{-5}$ **d.** $\log 5362$

     e. $\ln 0.67$ **f.** $\log_4 37$

36. Solve. $5^2 = 125^x$ **37.** Solve. $e^x = 100$

38. Solve. $\log_3(x + 6) - 3 = -\log_3 x$

39. Write the following expression as a single logarithm.

$$\frac{1}{2}\log z - 2\log x - 3\log y$$

40. Write the following expression as a sum or difference of logarithms

$$\ln \sqrt[3]{\frac{x^2}{y}}$$

Conic Sections

10

Chapter 10

In this chapter, we present several new types of graphs, called conic sections. These include the parabola, the circle, the ellipse, and the hyperbola. These shapes are found in a variety of applications. For example, a reflecting telescope has a mirror whose cross section is in the shape of a parabola, and planetary orbits are modeled by ellipses.

Review Your Skills

To prepare yourself for this chapter, review the process of completing the square. For Exercises 1–6 below, find the value of k that makes the trinomial a perfect square trinomial. Then factor the trinomial. Find the answer at the right and then record the corresponding letter in the space at the bottom of the page.

Answers

1. $x^2 + 6x + k$ **2.** $x^2 - 8x + k$ **I** $\dfrac{49}{4}; \left(x - \dfrac{7}{2}\right)^2$ **O** $\dfrac{9}{4}; \left(x + \dfrac{3}{2}\right)^2$

3. $x^2 - 7x + k$ **4.** $x^2 + 3x + k$ **C** $\dfrac{1}{25}; \left(x - \dfrac{1}{5}\right)^2$ **N** $\dfrac{1}{64}; \left(x - \dfrac{1}{8}\right)^2$

5. $x^2 - \dfrac{2}{5}x + k$ **6.** $x^2 - \dfrac{1}{4}x + k$ **S** $9; (x + 3)^2$ **T** $16; (x - 4)^2$

7. Plot the points (3, 1), (7, 1), and (7, 6). Use the Pythagorean theorem to determine the distance between the points (3, 1) and (7, 6) and choose the correct answer.

    A. 4   **B.** 5   **C.** $\sqrt{29}$ **D.** 3   **E.** $\sqrt{41}$

The parabola, ellipse, hyperbola, are all

$\overline{}\ \overline{}\ \overline{}\ \overline{}\ \overline{}\quad\ \overline{}\ \overline{}\ \overline{}\ \overline{}\ \overline{}\ \overline{}\ \overline{}\ \overline{}$
 5   4   6   3   5   1   7   5   2   3   4   6   1

Section 10.1 │ Distance Formula, Midpoint Formula, and Circles

1. Distance Formula

Suppose we are given two points (x_1, y_1) and (x_2, y_2) in a rectangular coordinate system. The distance between the two points can be found by using the Pythagorean theorem (Figure 10-1).

First draw a right triangle with the distance d as the hypotenuse. The length of the horizontal leg a is $|x_2 - x_1|$, and the length of the vertical leg b is $|y_2 - y_1|$. From the Pythagorean theorem we have

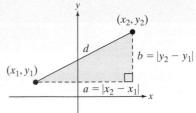

$$d^2 = a^2 + b^2$$
$$= (x_2 - x_1)^2 + (y_2 - y_1)^2$$
$$d = \pm\sqrt{(x_2 - x_1)^2 + (y_2 - y_1)^2}$$
$$= \sqrt{(x_2 - x_1)^2 + (y_2 - y_1)^2}$$

Figure 10-1

Because distance is positive, reject the negative value.

The Distance Formula

The distance d between the points (x_1, y_1) and (x_2, y_2) is

$$d = \sqrt{(x_2 - x_1)^2 + (y_2 - y_1)^2}$$

Example 1 Finding the Distance Between Two Points

Find the distance between the points $(-2, 3)$ and $(4, -1)$ (Figure 10-2).

Solution:

$$\underset{(x_1, y_1)}{(-2, 3)} \quad \text{and} \quad \underset{(x_2, y_2)}{(4, -1)}$$

Label the points.

$$d = \sqrt{(x_2 - x_1)^2 + (y_2 - y_1)^2}$$
$$= \sqrt{[(4) - (-2)]^2 + [(-1) - (3)]^2}$$
$$= \sqrt{(6)^2 + (-4)^2}$$
$$= \sqrt{36 + 16}$$
$$= \sqrt{52}$$
$$= \sqrt{4 \cdot 13}$$
$$= 2\sqrt{13}$$

Apply the distance formula.

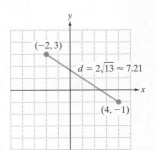

Figure 10-2

TIP: The order in which the points are labeled does not affect the result of the distance formula. For example, if the points in Example 1 had been labeled in reverse, the distance formula would still yield the same result:

$$(-2, 3) \quad \text{and} \quad (4, -1) \quad d = \sqrt{(x_2 - x_1)^2 + (y_2 - y_1)^2}$$

$$(x_2, y_2) \quad \text{and} \quad (x_1, y_1) \qquad = \sqrt{[(-2) - (4)]^2 + [3 - (-1)]^2}$$

$$= \sqrt{(-6)^2 + (4)^2}$$

$$= \sqrt{36 + 16}$$

$$= \sqrt{52}$$

$$= 2\sqrt{13}$$

2. Circles

A **circle** is defined as the set of all points in a plane that are equidistant from a fixed point called the **center**. The fixed distance from the center is called the **radius** and is denoted by r, where $r > 0$.

Suppose a circle is centered at the point (h, k) and has radius, r (Figure 10-3). The distance formula can be used to derive an equation of the circle.

Let (x, y) be any arbitrary point on the circle. Then, by definition, the distance between (h, k) and (x, y) must be r.

$$\sqrt{(x - h)^2 + (y - k)^2} = r$$

$$(x - h)^2 + (y - k)^2 = r^2 \qquad \text{Square both sides.}$$

Figure 10-3

Standard Equation of a Circle

The **standard equation of a circle**, centered at (h, k) with radius r, is given by

$$(x - h)^2 + (y - k)^2 = r^2 \qquad \text{where } r > 0$$

Note: If a circle is centered at the origin $(0, 0)$, then $h = 0$ and $k = 0$, and the equation simplifies to $x^2 + y^2 = r^2$.

Example 2 **Graphing a Circle**

Find the center and radius of the circle. Then graph the circle.

$$(x - 3)^2 + (y + 4)^2 = 36$$

Solution:

$$(x - 3)^2 + (y + 4)^2 = 36$$

$$(x - 3)^2 + [y - (-4)]^2 = (6)^2$$

$$h = 3, k = -4, \text{ and } r = 6$$

The equation is in standard form $(x - h)^2 + (y - k)^2 = r^2$. The center is $(3, -4)$, and the radius is given by $r = 6$. To graph the circle, first locate the center. From the center, mark points 6 units to the right, left, above, and below the center. Then sketch the circle through these four points (Figure 10-4).

Note that the center of the circle is not actually part of the circle. It is drawn as an open dot for reference.

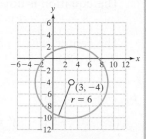

Figure 10-4

Skill Practice

Find the center and radius of the circle. Then graph the circle.

2. $(x + 1)^2 + (y - 2)^2 = 9$

Answer

2. Center: $(-1, 2)$; radius: 3

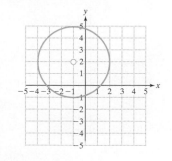

Calculator Connections

Topic: Graphing a Circle

Graphing calculators are designed to graph *functions,* in which y is written in terms of x. A circle is not a function. However, it can be graphed as the union of two functions—one representing the top semicircle and the other representing the bottom semicircle.

Solving for y in Example 2, we have

$$(x - 3)^2 + (y + 4)^2 = 36$$
Graph these functions as Y_1 and Y_2, using a square viewing window.

$$(y + 4)^2 = 36 - (x - 3)^2$$
$$Y_1 = -4 + \sqrt{36 - (x - 3)^2}$$

$$y + 4 = \pm\sqrt{36 - (x - 3)^2}$$
$$Y_2 = -4 - \sqrt{36 - (x - 3)^2}$$

$$y = -4 \pm \sqrt{36 - (x - 3)^2}$$

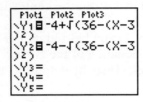

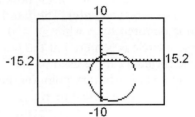

Notice that the image from the calculator does not show the upper and lower semicircles connecting at their endpoints, when in fact the semicircles should "hook up." This is due to the calculator's limited resolution.

Skill Practice

Find the center and radius of the circle. Then graph the circle.

3. $\left(x + \dfrac{7}{2}\right)^2 + y^2 = \dfrac{9}{4}$

Answer

3. Center: $\left(-\dfrac{7}{2}, 0\right)$; radius: $\dfrac{3}{2}$

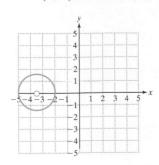

Example 3 **Graphing a Circle**

Find the center and radius of each circle. Then graph the circle.

a. $x^2 + \left(y - \dfrac{10}{3}\right)^2 = \dfrac{25}{9}$

b. $x^2 + y^2 = 10$

Solution:

a. $x^2 + \left(y - \dfrac{10}{3}\right)^2 = \dfrac{25}{9}$

$$(x - 0)^2 + \left(y - \dfrac{10}{3}\right)^2 = \left(\dfrac{5}{3}\right)^2$$

The equation is now in standard form $(x - h)^2 + (y - k)^2 = r^2$.

The center is $(0, \frac{10}{3})$ and the radius is $\frac{5}{3}$ (Figure 10-5).

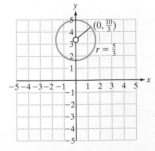

Figure 10-5

b. $x^2 + y^2 = 10$

$$(x - 0)^2 + (y - 0)^2 = (\sqrt{10})^2$$

The equation is now in standard form $(x - h)^2 + (y - k)^2 = r^2$.

The center is $(0, 0)$ and the radius is $\sqrt{10} \approx 3.16$ (Figure 10-6).

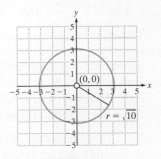

Figure 10-6

Sometimes it is necessary to complete the square to write an equation of a circle in standard form.

Example 4 Writing the Equation of a Circle in the Form $(x - h)^2 + (y - k)^2 = r^2$

Identify the center and radius of the circle given by the equation $x^2 + y^2 + 2x - 16y + 61 = 0$.

Solution:

$x^2 + y^2 + 2x - 16y + 61 = 0$ — To identify the center and radius, write the equation in the form $(x - h)^2 + (y - k)^2 = r^2$.

$(x^2 + 2x \quad) + (y^2 - 16y \quad) = -61$ — Group the x-terms and group the y-terms. Move the constant to the right-hand side.

$(x^2 + 2x + 1) + (y^2 - 16y + 64) = -61 + 1 + 64$
- Complete the square on x. Add $[\frac{1}{2}(2)]^2 = 1$ to both sides of the equation.
- Complete the square on y. Add $[\frac{1}{2}(-16)]^2 = 64$ to both sides of the equation.

$(x + 1)^2 + (y - 8)^2 = 4$ — Factor and simplify.

$[x - (-1)]^2 + (y - 8)^2 = 2^2$ — Standard form: $(x - h)^2 + (y - k)^2 = r^2$.

The center is $(-1, 8)$ and the radius is 2.

3. Writing an Equation of a Circle

Example 5 Writing an Equation of a Circle

Write an equation of the circle shown in Figure 10-7.

Solution:

The center is at $(-3, 2)$; therefore, $h = -3$ and $k = 2$. From the graph, $r = 2$.

$$(x - h)^2 + (y - k)^2 = r^2$$
$$[x - (-3)]^2 + (y - 2)^2 = (2)^2$$
$$(x + 3)^2 + (y - 2)^2 = 4$$

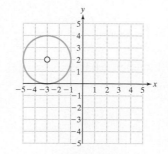

Figure 10-7

4. The Midpoint Formula

Consider two points in the coordinate plane and the line segment determined by the points. It is sometimes necessary to determine the point that is halfway between the endpoints of the segment. This point is called the *midpoint*. If the coordinates of the endpoints are represented by (x_1, y_1) and (x_2, y_2), then the midpoint of the segment is given by the following formula.

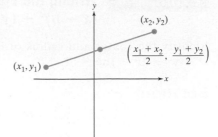

Midpoint formula: $\left(\dfrac{x_1 + x_2}{2}, \dfrac{y_1 + y_2}{2} \right)$

> **TIP:** The midpoint of a line segment is found by taking the *average* of the x-coordinates and the *average* of the y-coordinates of the endpoints.

Skill Practice

Find the midpoint of the line segment with the given endpoints.

6. $(5, 6)$ and $(-10, 4)$

7. $(-2.6, -6.3)$ and $(1.2, 4.1)$

Example 6   **Finding the Midpoint of a Segment**

Find the midpoint of the line segment with the given endpoints.

a. $(-4, 6)$ and $(8, 1)$ **b.** $(-1.2, -3.1)$ and $(-6.6, 1.2)$

Solution:

a. $(-4, 6)$ and $(8, 1)$
    (x_1, y_1) (x_2, y_2)

$\left(\dfrac{-4 + 8}{2}, \dfrac{6 + 1}{2} \right)$ Apply the midpoint formula $\left(\dfrac{x_1 + x_2}{2}, \dfrac{y_1 + y_2}{2} \right)$.

$\left(2, \dfrac{7}{2} \right)$ Simplify.

The midpoint of the segment is $(2, \frac{7}{2})$.

b. $(-1.2, -3.1)$ and $(-6.6, 1.2)$
    (x_1, y_1) (x_2, y_2)

$\left(\dfrac{-1.2 + (-6.6)}{2}, \dfrac{-3.1 + 1.2}{2} \right)$ Apply the midpoint formula.

$(-3.9, -0.95)$ Simplify.

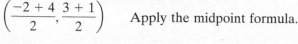

Example 7 **Applying the Midpoint and Distance Formulas**

Suppose that $(-2, 3)$ and $(4, 1)$ are endpoints of a diameter of a circle.

a. Find the center of the circle.　　**b.** Write an equation of the circle.

Solution:

a. Because the midpoint of a diameter of a circle is the center of the circle, apply the midpoint formula. See Figure 10-8.

$$(-2, 3) \quad \text{and} \quad (4, 1)$$
$$(x_1, y_1) \qquad\quad (x_2, y_2)$$

$$\left(\frac{x_1 + x_2}{2}, \frac{y_1 + y_2}{2} \right)$$

$$\left(\frac{-2 + 4}{2}, \frac{3 + 1}{2} \right) \qquad \text{Apply the midpoint formula.}$$

$$(1, 2) \qquad\qquad\qquad \text{Simplify.}$$

The center of the circle is $(1, 2)$.

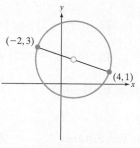

Figure 10-8

b. The radius can be determined by finding the distance between the center and either endpoint of the diameter.

$$(1, 2) \quad \text{and} \quad (4, 1)$$
$$(x_1, y_1) \qquad\quad (x_2, y_2)$$

$$d = \sqrt{(x_2 - x_1)^2 + (y_2 - y_1)^2} \qquad \text{Apply the distance formula.}$$

$$d = \sqrt{(4 - 1)^2 + (1 - 2)^2} \qquad \text{Substitute } (1, 2) \text{ and } (4, 1) \text{ for } (x_1, y_1)$$
$$\text{and } (x_2, y_2).$$

$$= \sqrt{(3)^2 + (-1)^2} \qquad\qquad \text{Simplify.}$$

$$= \sqrt{10}$$

The circle is represented by $(x - 1)^2 + (y - 2)^2 = (\sqrt{10})^2$

or $\quad (x - 1)^2 + (y - 2)^2 = 10$.

Answer
8. a. $(5, 6)$
　b. $(x - 5)^2 + (y - 6)^2 = 20$

Section 10.1 Practice Exercises

Vocabulary and Key Concepts

1. a. The distance between two distinct points (x_1, y_1) and (x_2, y_2) is given by $d = $ _____.

b. A _____ is the set of all points equidistant from a fixed point called the _____.

c. The distance from the center of a circle to any point on the circle is called the _____ and is often denoted by r.

d. The standard equation of a circle with center (h, k) and radius r is given by _____.

e. The midpoint of the line segment with endpoints (x_1, y_1) and (x_2, y_2) is given by _____.

Concept 1: Distance Formula

For Exercises 2–16, use the distance formula to find the distance between the two points. **(See Example 1.)**

2. $(-2, 7)$ and $(4, -5)$

3. $(1, 10)$ and $(-2, 4)$

4. $(0, 5)$ and $(-3, 8)$

5. $(6, 7)$ and $(3, 2)$

6. $\left(\dfrac{2}{3}, \dfrac{1}{5}\right)$ and $\left(-\dfrac{5}{6}, \dfrac{3}{10}\right)$

7. $\left(-\dfrac{1}{2}, \dfrac{5}{8}\right)$ and $\left(-\dfrac{3}{2}, \dfrac{1}{4}\right)$

8. $(4, 13)$ and $(4, -6)$

9. $(-2, 5)$ and $(-2, 9)$

10. $(8, -6)$ and $(-2, -6)$

11. $(7, 2)$ and $(15, 2)$

12. $(-6, -2)$ and $(-3, -5)$

13. $(-1, -5)$ and $(-5, -9)$

14. $(3\sqrt{5}, 2\sqrt{7})$ and $(-\sqrt{5}, -3\sqrt{7})$ **15.** $(4\sqrt{6}, -2\sqrt{2})$ and $(2\sqrt{6}, \sqrt{2})$

16. $(6, 0)$ and $(0, -1)$

17. Explain how to find the distance between 5 and -7 on the y-axis.

18. Explain how to find the distance between 15 and -37 on the x-axis.

19. Find the values of y such that the distance between the points $(4, 7)$ and $(-4, y)$ is 10 units.

20. Find the values of x such that the distance between the points $(-4, -2)$ and $(x, 3)$ is 13 units.

21. Find the values of x such that the distance between the points $(x, 2)$ and $(4, -1)$ is 5 units.

22. Find the values of y such that the distance between the points $(-5, 2)$ and $(3, y)$ is 10 units.

For Exercises 23–26, determine if the three points define the vertices of a right triangle.

23. $(-3, 2), (-2, -4),$ and $(3, 3)$

24. $(1, -2), (-2, 4),$ and $(7, 1)$

25. $(-3, -2), (4, -3),$ and $(1, 5)$

26. $(1, 4), (5, 3),$ and $(2, 0)$

Concept 2: Circles

For Exercises 27–47, identify the center and radius of the circle and then graph the circle. Complete the square, if necessary. **(See Examples 2–4.)**

27. $(x - 4)^2 + (y + 2)^2 = 9$

28. $(x - 3)^2 + (y + 1)^2 = 16$

29. $(x + 1)^2 + (y + 1)^2 = 1$

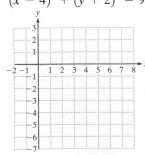

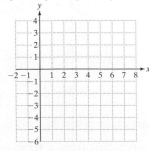

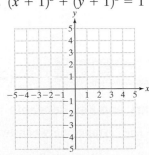

30. $(x - 4)^2 + (y - 4)^2 = 4$

31. $x^2 + (y - 2)^2 = 4$

32. $(x + 1)^2 + y^2 = 1$

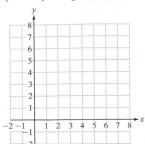

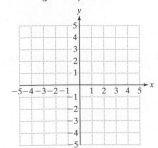

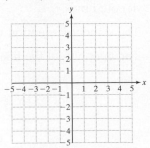

33. $(x - 3)^2 + y^2 = 8$

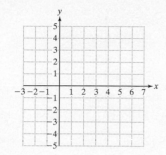

34. $x^2 + (y + 2)^2 = 20$

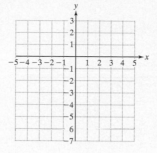

35. $x^2 + y^2 = 6$

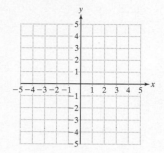

36. $x^2 + y^2 = 15$

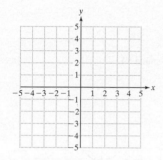

37. $\left(x + \dfrac{4}{5}\right)^2 + y^2 = \dfrac{64}{25}$

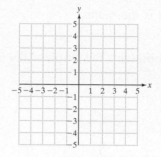

38. $x^2 + \left(y - \dfrac{5}{2}\right)^2 = \dfrac{9}{4}$

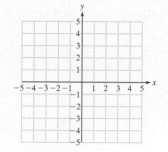

39. $x^2 + y^2 - 2x - 6y - 26 = 0$

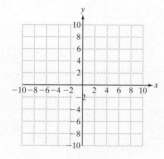

40. $x^2 + y^2 + 4x - 8y + 16 = 0$

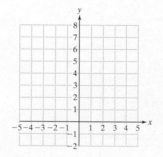

41. $x^2 + y^2 - 6y + 5 = 0$

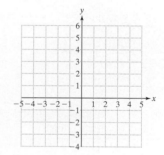

42. $x^2 + 2x + y^2 - 15 = 0$

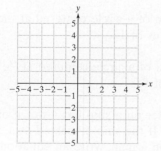

43. $x^2 + y^2 + 6y + \dfrac{65}{9} = 0$

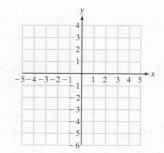

44. $x^2 + y^2 - 12x + 12y + 71 = 0$

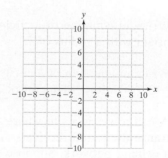

45. $x^2 + y^2 + 2x + 4y - 4 = 0$ **46.** $2x^2 + 2y^2 = 32$ **47.** $3x^2 + 3y^2 = 3$

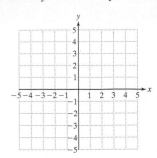
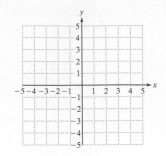
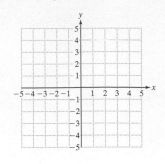

Concept 3: Writing an Equation of a Circle

48. If the diameter of a circle is 10 ft, what is the radius?

For Exercises 49–54, write an equation that represents the graph of the circle. **(See Example 5.)**

49.

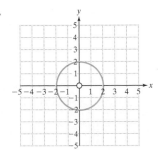

50.

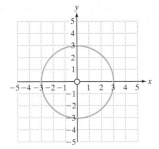

51.

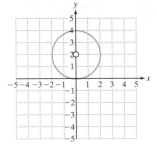

52.

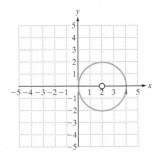

53.

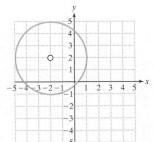

54.

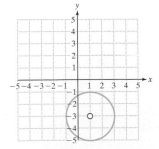

55. Write an equation of a circle centered at the origin with a radius of 7.

56. Write an equation of a circle centered at the origin with a radius of 12.

57. Write an equation of a circle centered at $(-3, -4)$ with a diameter of 12.

58. Write an equation of a circle centered at $(5, -1)$ with a diameter of 8.

59. A cell tower is a site where antennas, transmitters, and receivers are placed to create a cellular network. Suppose that a cell tower is located at a point $A(5, 3)$ on a map and its range is 1.5 mi. Write an equation that represents the circular boundary of the area that can receive a signal from the tower. Assume that all distances are in miles.

60. A radar transmitter on a ship has a range of 20 nautical miles. If the ship is located at a point $(-28, 32)$ on a map, write an equation for the circular boundary of the area within the range of the ship's radar. Assume that all distances on the map are represented in nautical miles.

Concept 4: The Midpoint Formula

For Exercises 61–64, find the midpoint of the line segment. Check your answers by plotting the midpoint on the graph.

61.

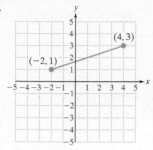

62.

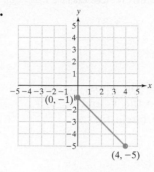

63.

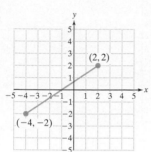

64.

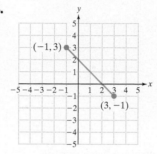

For Exercises 65–72, find the midpoint of the line segment between the two given points. **(See Example 6.)**

65. $(4, 0)$ and $(-6, 12)$

66. $(-7, 2)$ and $(-3, -2)$

67. $(-3, 8)$ and $(3, -2)$

68. $(0, 5)$ and $(4, -5)$

69. $(5, 2)$ and $(-6, 1)$

70. $(-9, 3)$ and $(0, -4)$

71. $(-2.4, -3.1)$ and $(1.6, 1.1)$

72. $(0.8, 5.3)$ and $(-4.2, 7.1)$

73. Two courier trucks leave the warehouse to make deliveries. One travels 20 mi north and 30 mi east. The other truck travels 5 mi south and 50 mi east. If the two drivers want to meet for lunch at a restaurant at a point halfway between them, where should they meet relative to the warehouse? (*Hint:* Label the warehouse as the origin, and find the coordinates of the restaurant. See the figure.)

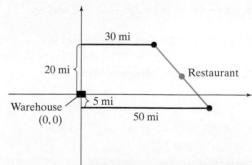

74. A map of a hiking area is drawn so that the Visitor Center is at the origin of a rectangular grid. Two hikers are located at positions $(-1, 1)$ and $(-3, -2)$ with respect to the Visitor Center where all units are in miles. A campground is located exactly halfway between the hikers. What are the coordinates of the campground?

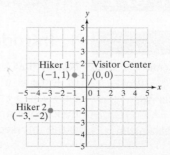

For Exercises 75–78, the two given points are endpoints of a diameter of a circle.

a. Find the center of the circle.

b. Write an equation of the circle. **(See Example 7.)**

75. $(-1, 2)$ and $(3, 4)$　　**76.** $(-3, 3)$ and $(7, -1)$　　**77.** $(-2, 3)$ and $(2, 3)$　　**78.** $(-1, 3)$ and $(-1, -3)$

Expanding Your Skills

79. Write an equation of a circle whose center is $(4, 4)$ and is tangent to the x- and y-axes. (*Hint:* Sketch the circle first.)

80. Write an equation of a circle whose center is $(-3, 3)$ and is tangent to the x- and y-axes. (*Hint:* Sketch the circle first.)

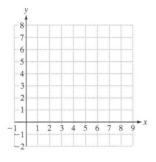

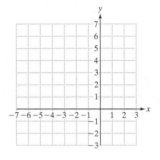

81. Write an equation of a circle whose center is $(1, 1)$ and that passes through the point $(-4, 3)$.

82. Write an equation of a circle whose center is $(-3, -1)$ and that passes through the point $(5, -2)$.

Graphing Calculator Exercises

For Exercises 83–88, graph the circles from the indicated exercise on a square viewing window, and approximate the center and the radius from the graph.

83. $(x - 4)^2 + (y + 2)^2 = 9$　(Exercise 27)

84. $(x - 3)^2 + (y + 1)^2 = 16$　(Exercise 28)

85. $x^2 + (y - 2)^2 = 4$　(Exercise 31)

86. $(x + 1)^2 + y^2 = 1$　(Exercise 32)

87. $x^2 + y^2 = 6$　(Exercise 35)

88. $x^2 + y^2 = 15$　(Exercise 36)

Section 10.2 | More on the Parabola

Concepts

1. **Introduction to Conic Sections**
2. **Parabola—Vertical Axis of Symmetry**
3. **Parabola—Horizontal Axis of Symmetry**
4. **Vertex Formula**

1. Introduction to Conic Sections

Recall that the graph of a second-degree equation of the form $y = ax^2 + bx + c$ ($a \neq 0$) is a parabola. In Section 10.1 we learned that the graph of $(x - h)^2 + (y - k)^2 = r^2$ is a circle. These and two other types of figures called ellipses and hyperbolas are called **conic sections**. Conic sections derive their names because each is the intersection of a plane and a double-napped cone.

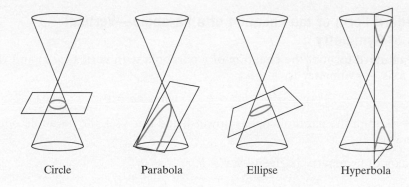

Circle   Parabola   Ellipse   Hyperbola

2. Parabola—Vertical Axis of Symmetry

A **parabola** is defined by a set of points in a plane that are equidistant from a fixed line (called the directrix) and a fixed point (called the focus) not on the directrix (Figure 10-9). Parabolas have numerous real-world applications. For example, a flashlight has a mirror with the cross section in the shape of a parabola. The bulb is located at the focus. The light hits the mirror and is reflected outward parallel to the sides of the cylinder. This forms a beam of light (Figure 10-10).

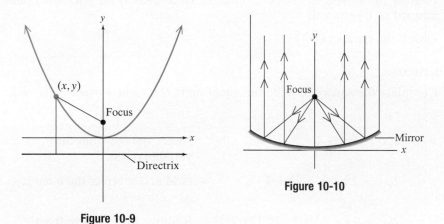

Figure 10-9

Figure 10-10

The graph of the solutions to the quadratic equation $y = ax^2 + bx + c$ is a parabola. We graphed parabolas of this type in Section 8.4. Recall that the **vertex** is the highest or the lowest point of a parabola. The **axis of symmetry** of the parabola is a line that passes through the vertex and is perpendicular to the directrix (Figure 10-11).

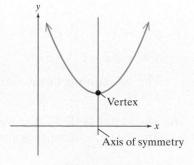

Figure 10-11

> ## Standard Form of the Equation of a Parabola—Vertical Axis of Symmetry
>
> The standard form of the equation of a parabola with vertex (h, k) and vertical axis of symmetry is
>
> $$y = a(x - h)^2 + k \qquad \text{where } a \neq 0$$
>
> If $a > 0$, then the parabola opens upward; and if $a < 0$, the parabola opens downward.
>
> The axis of symmetry is given by $x = h$.

In Section 8.5, we learned that it is sometimes necessary to complete the square to write the equation of a parabola in standard form.

Skill Practice

Given the equation of the parabola $y = 3x^2 + 6x + 4$:

1. Write the equation in standard form.
2. Identify the vertex and axis of symmetry.
3. Graph the parabola.

Example 1   Graphing a Parabola by First Completing the Square

Given the equation of the parabola $y = -2x^2 + 4x + 1$:

a. Write the equation in standard form $y = a(x - h)^2 + k$.

b. Identify the vertex and axis of symmetry. Determine if the parabola opens upward or downward.

c. Graph the parabola.

Solution:

a. Complete the square to write the equation in the form $y = a(x - h)^2 + k$.

$y = -2x^2 + 4x + 1$

$y = -2(x^2 - 2x) + 1$ Factor out -2 from the variable terms.

$y = -2(x^2 - 2x + 1 - 1) + 1$ Add and subtract the quantity $\left[\frac{1}{2}(-2)\right]^2 = 1$.

$y = -2(x^2 - 2x + 1) + (-2)(-1) + 1$ Remove the -1 term from within the parentheses by first applying the distributive property. When -1 is removed from the parentheses, it carries with it the factor of -2 from outside the parentheses.

$y = -2(x - 1)^2 + 2 + 1$

$y = -2(x - 1)^2 + 3$

The equation is in the form $y = a(x - h)^2 + k$ where $a = -2$, $h = 1$, and $k = 3$.

b. The vertex is $(1, 3)$. Because a is negative, the parabola opens downward. The axis of symmetry is $x = 1$.

Answers

1. $y = 3(x + 1)^2 + 1$
2. Vertex: $(-1, 1)$; axis of symmetry: $x = -1$
3.

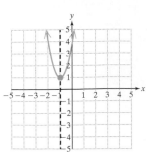

c. To graph the parabola, we know that its orientation is downward because the leading term is negative. Furthermore, we know the vertex is (1, 3). To find other points on the parabola, select several values of x and solve for y. Recall that the y-intercept is found by substituting $x = 0$ and solving for y.

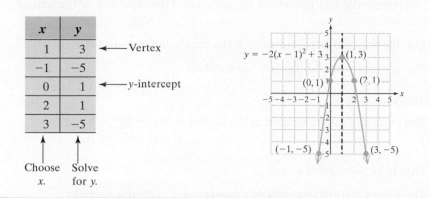

3. Parabola—Horizontal Axis of Symmetry

We have seen that the graph of a parabola $y = ax^2 + bx + c$ opens upward if $a > 0$ and downward if $a < 0$. A parabola can also open to the left or right. In such a case, the "roles" of x and y are essentially interchanged in the equation. Thus, the graph of $x = ay^2 + by + c$ opens to the right if $a > 0$ (Figure 10-12) and to the left if $a < 0$ (Figure 10-13).

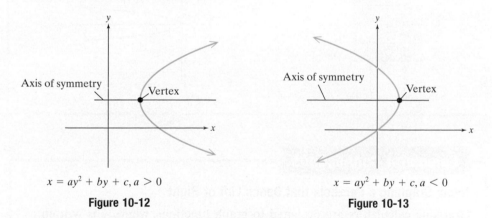

Figure 10-12   **Figure 10-13**

Standard Form of the Equation of a Parabola—Horizontal Axis of Symmetry

The standard form of the equation of a parabola with vertex (h, k) and horizontal axis of symmetry is

$$x = a(y - k)^2 + h \qquad \text{where } a \neq 0$$

If $a > 0$, then the parabola opens to the right and if $a < 0$, the parabola opens to the left.

The axis of symmetry is given by $y = k$.

Skill Practice

Given the equation $x = -y^2 + 1$:

4. Identify the vertex and the axis of symmetry.

5. Determine if the parabola opens to the right or to the left.

6. Graph the parabola.

Example 2 **Graphing a Parabola with a Horizontal Axis of Symmetry**

Given the equation of the parabola $x = 4y^2$:

a. Determine the coordinates of the vertex and the equation of the axis of symmetry.

b. Use the value of a to determine if the parabola opens to the right or left.

c. Plot several points and graph the parabola.

Solution:

a. The equation can be written in the form $x = a(y - k)^2 + h$:
$$x = 4(y - 0)^2 + 0$$

Therefore, $h = 0$ and $k = 0$.

The vertex is $(0, 0)$. The axis of symmetry is $y = 0$ (the x-axis).

b. Because a is positive, the parabola opens to the right.

c. The vertex of the parabola is $(0, 0)$. To find other points on the graph, select values for y and solve for x.

x	y	
0	0	←—Vertex
4	1	
4	−1	
16	2	
16	−2	

Solve for x.   Choose y.

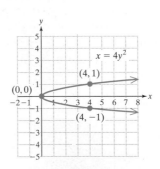

Calculator Connections

Topic: Graphing a Parabola that Opens Left or Right

Graphing calculators are designed to graph functions, where y is written as a function of x. The parabola from Example 2 is not a function. It can be graphed, however, as the union of two functions, one representing the top branch and the other representing the bottom branch.

Solving for y in Example 2, we have: $x = 4y^2$

$$y^2 = \frac{x}{4}$$

$$y = \pm\sqrt{\frac{x}{4}}$$

$$y = \pm\frac{\sqrt{x}}{2}$$

Graph these functions as Y_1 and Y_2.

$$Y_1 = \frac{\sqrt{x}}{2} \quad \text{and} \quad Y_2 = -\frac{\sqrt{x}}{2}$$

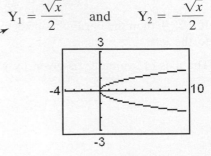

Answers

4. Vertex: $(1, 0)$; axis of symmetry: $y = 0$
5. Parabola opens left.
6.

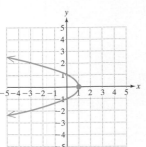

Example 3 Graphing a Parabola by First Completing the Square

Given the equation of the parabola $x = -y^2 + 8y - 14$:

a. Write the equation in standard form $x = a(y - k)^2 + h$.

b. Identify the vertex and axis of symmetry. Determine if the parabola opens to the right or left.

c. Graph the parabola.

Solution:

a. Complete the square to write the equation in the form $x = a(y - k)^2 + h$.

$x = -y^2 + 8y - 14$

$x = -1(y^2 - 8y) - 14$ Factor out -1 from the variable terms.

$x = -1(y^2 - 8y + 16 - 16) - 14$ Add and subtract the quantity $\left[\frac{1}{2}(-8)\right]^2 = 16$.

$x = -1(y^2 - 8y + 16) + (-1)(-16) - 14$ Remove the -16 term from within the parentheses by first applying the distributive property. When -16 is removed from the parentheses, it carries with it the factor of -1 from outside the parentheses.

$x = -1(y - 4)^2 + 16 - 14$

$x = -1(y - 4)^2 + 2$

The equation is in the form $x = a(y - k)^2 + h$, where $a = -1$, $h = 2$, and $k = 4$.

b. The vertex is $(2, 4)$. Because a is negative, the parabola opens to the left. The axis of symmetry is $y = 4$.

c.

x	y
2	4	← Vertex
1	5
1	3
−2	6
−2	2

↑ Solve for x.   ↑ Choose y.

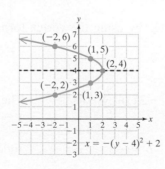

$x = -(y - 4)^2 + 2$

Skill Practice

Given the equation of the parabola $x = y^2 + 4y + 7$:

7. Determine the coordinates of the vertex.

8. Determine if the parabola opens to the right or to the left.

9. Graph the parabola.

Answers

7. Vertex: $(3, -2)$

8. Parabola opens right.

9.

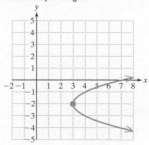

4. Vertex Formula

From Section 8.5, we learned that the vertex formula can also be used to find the vertex of a parabola.

For a parabola defined by $y = ax^2 + bx + c$,

- The x-coordinate of the vertex is given by $x = \frac{-b}{2a}$.
- The y-coordinate of the vertex is found by substituting this value for x into the original equation and solving for y.

For a parabola defined by $x = ay^2 + by + c$,

- The y-coordinate of the vertex is given by $y = \frac{-b}{2a}$.
- The x-coordinate of the vertex is found by substituting this value for y into the original equation and solving for x.

<table>
<tr><td>

Skill Practice

Find the vertex by using the vertex formula.

10. $x = -4y^2 + 12y$

</td></tr>
</table>

Example 4 Finding the Vertex of a Parabola by Using the Vertex Formula

Find the vertex by using the vertex formula. $x = y^2 + 4y + 5$

Solution:

$x = y^2 + 4y + 5$	The parabola is in the form $x = ay^2 + by + c$.
$a = 1 \quad b = 4 \quad c = 5$	Identify a, b, and c.

The y-coordinate of the vertex is given by $y = \dfrac{-b}{2a} = \dfrac{-(4)}{2(1)} = -2$.

The x-coordinate of the vertex is found by substitution:
$x = (-2)^2 + 4(-2) + 5 = 1$.

The vertex is $(1, -2)$.

<table>
<tr><td>

Skill Practice

Find the vertex by using the vertex formula.

11. $y = \frac{3}{4}x^2 + 3x + 5$

</td></tr>
</table>

Example 5 Find a Vertex of a Parabola by Using the Vertex Formula

Find the vertex by using the vertex formula. $y = \dfrac{1}{2}x^2 - 3x + \dfrac{5}{2}$

Solution:

$y = \dfrac{1}{2}x^2 - 3x + \dfrac{5}{2}$	The parabola is in the form $y = ax^2 + bx + c$.
$a = \dfrac{1}{2} \quad b = -3 \quad c = \dfrac{5}{2}$	Identify a, b, and c.

The x-coordinate of the vertex is given by $x = \dfrac{-b}{2a} = \dfrac{-(-3)}{2\left(\frac{1}{2}\right)} = 3$.

Answers

10. Vertex: $\left(9, \dfrac{3}{2}\right)$ **11.** $(-2, 2)$

The y-coordinate of the vertex is found by substitution.

$$y = \frac{1}{2}(3)^2 - 3(3) + \frac{5}{2}$$

$$= \frac{9}{2} - 9 + \frac{5}{2}$$

$$= \frac{14}{2} - 9$$

$$= 7 - 9$$

$$= -2$$

The vertex is $(3, -2)$.

Section 10.2 Practice Exercises

Vocabulary and Key Concepts

1. **a.** A circle, a parabola, an ellipse, and a hyperbola are curves that collectively are called _____ sections. These types of curves derive their names because each is the intersection of a _____ and a double-napped cone.

   b. A _____ is defined by a set of points in a plane that are equidistant from a fixed line (called the _____) and a fixed point (called the _____).

   c. Given a parabola defined by $y = a(x - h)^2 + k$ $(a \neq 0)$, the _____ is (h, k) and is the highest or lowest point on the graph. The axis of _____ is the vertical line through the vertex whose equation is $x = h$.

   d. Given a parabola defined by $x = a(y - k)^2 + h$ $(a \neq 0)$, the axis of symmetry is the horizontal line whose equation is _____.

Review Exercises

2. Determine the distance between the points $(1, 1)$ and $(2, -2)$.

3. Determine the distance from the origin to the point $(4, -3)$.

4. Determine the distance between the points $(11, 3)$ and $(5, 3)$.

For Exercises 5 and 6, identify the center and radius of the circle and then graph the circle.

5. $x^2 + (y + 1)^2 = 16$

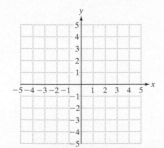

6. $(x - 3)^2 + y^2 = 4$

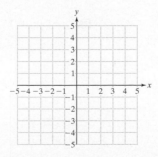

7. Find the midpoint of the line segment between the points $(7, -3)$ and $(4, 5)$.

8. Find the midpoint of the line segment between the points $(-3, 0)$ and $(-1, 9)$.

Concept 2: Parabola—Vertical Axis of Symmetry

For Exercises 9–16, use the equation of the parabola in standard form $y = a(x - h)^2 + k$ to determine the coordinates of the vertex and the equation of the axis of symmetry (complete the square if necessary). Then graph the parabola. **(See Example 1.)**

9. $y = (x + 2)^2 + 1$

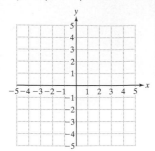

10. $y = (x - 1)^2 + 1$

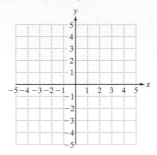

11. $y = x^2 - 4x + 3$

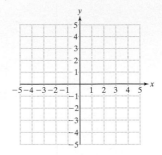

12. $y = x^2 + 6x + 5$

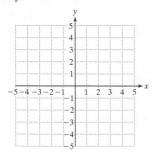

13. $y = -2x^2 + 8x$

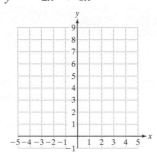

14. $y = -3x^2 - 6x$

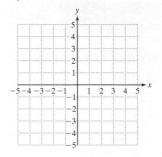

15. $y = -x^2 - 3x + 2$

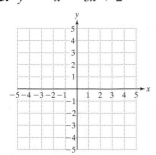

16. $y = -x^2 + x - 4$

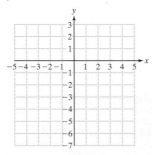

Animation

Concept 3: Parabola—Horizontal Axis of Symmetry

For Exercises 17–24, use the equation of the parabola in standard form $x = a(y - k)^2 + h$ to determine the coordinates of the vertex and the axis of symmetry (complete the square if necessary). Then graph the parabola. **(See Examples 2 and 3.)**

17. $x = y^2 - 3$

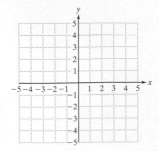

18. $x = y^2 + 1$

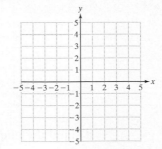

19. $x = -(y - 3)^2 - 3$

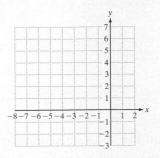

20. $x = -2(y + 2)^2 + 1$

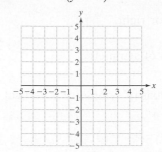

21. $x = -y^2 + 4y - 4$

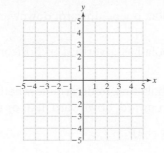

22. $x = -4y^2 - 4y - 2$

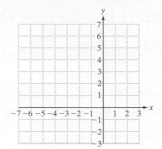

23. $x = y^2 - 2y + 2$

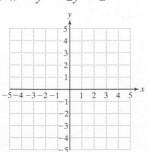

24. $x = y^2 + 4y + 1$

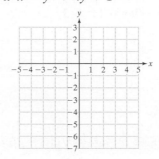

Concept 4: Vertex Formula

For Exercises 25–33, determine the vertex by using the vertex formula. **(See Examples 4 and 5.)**

25. $y = x^2 - 4x + 3$

26. $y = x^2 + 6x - 2$

27. $x = y^2 + 2y + 6$

28. $x = y^2 - 8y + 3$

29. $y = -\dfrac{1}{4}x^2 + x + \dfrac{3}{4}$

30. $y = -\dfrac{1}{2}x^2 - x + \dfrac{1}{2}$

31. $y = x^2 - 3x + 2$

32. $x = -2y^2 + 16y + 1$

33. $x = -3y^2 - 6y + 7$

34. Ricardo has a satellite dish for his television. The cross sections of the satellite dish are parabolic in shape. A cross section through the vertex can be described by the equation

$$d(x) = \frac{2}{25}x^2 - \frac{2}{5}x$$

where $d(x)$ is the depth of the dish (in feet) at a distance of x ft from the edge of the dish. How deep is the satellite dish?

35. Water from an outdoor fountain is projected outward from a jet on the side wall of the fountain. The water follows a parabolic path, and can be modeled by $h(x) = -x^2 + 10x - 3$. In this function, $h(x)$ represents the height of the water, x ft from the jet. Find the maximum height of the water.

Mixed Exercises

36. Explain how to determine whether a parabola opens upward, downward, left, or right.

37. Explain how to determine whether a parabola has a vertical or horizontal axis of symmetry.

For Exercises 38–49, use the equation of the parabola first to determine whether the axis of symmetry is vertical or horizontal. Then determine if the parabola opens upward, downward, left, or right.

38. $y = (x - 2)^2 + 3$

39. $y = (x - 4)^2 + 2$

40. $y = -2(x + 1)^2 - 4$

41. $y = -3(x + 2)^2 - 1$

42. $x = y^2 + 4$

43. $x = y^2 - 2$

44. $x = -(y + 3)^2 + 2$

45. $x = -2(y - 1)^2 - 3$

46. $y = -2x^2 - 5$

47. $y = -x^2 + 3$

48. $x = 2y^2 + 3y - 2$

49. $x = y^2 - 5y + 1$

Section 10.3 The Ellipse and Hyperbola

Concepts

1. Standard Form of an Equation of an Ellipse
2. Standard Forms of an Equation of a Hyperbola

1. Standard Form of an Equation of an Ellipse

In this section, we will study the two remaining conic sections: the ellipse and the hyperbola. An **ellipse** is the set of all points (x, y) such that the sum of the distances between (x, y) and two distinct points is a constant. The fixed points are called the foci (plural of *focus*) of the ellipse.

To visualize an ellipse, consider the following application. Suppose Sonya wants to cut an elliptical rug from a rectangular rug to avoid a stain made by the family dog. She places two tacks along the center horizontal line. Then she ties the ends of a slack piece of rope to each tack. With the rope pulled tight, she traces out a curve. This curve is an ellipse, and the tacks are located at the foci of the ellipse (Figure 10-14).

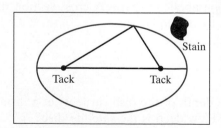

Figure 10-14

We will first graph ellipses that are centered at the origin.

Standard Form of an Equation of an Ellipse Centered at the Origin

An ellipse with the center at the origin has the equation $\dfrac{x^2}{a^2} + \dfrac{y^2}{b^2} = 1$, where a and b are positive real numbers. In the standard form of the equation, the right side must equal 1.

To graph an ellipse centered at the origin, find the x- and y-intercepts.

To find the x-intercepts, let $y = 0$.

To find the y-intercepts, let $x = 0$.

$$\frac{x^2}{a^2} + \frac{0}{b^2} = 1$$

$$\frac{x^2}{a^2} = 1$$

$$x^2 = a^2$$

$$x - \pm\sqrt{a^2}$$

$$x = \pm a$$

$$\frac{0}{a^2} + \frac{y^2}{b^2} = 1$$

$$\frac{y^2}{b^2} = 1$$

$$y^2 = b^2$$

$$y = \pm\sqrt{b^2}$$

$$y = \pm b$$

The x-intercepts are $(a, 0)$ and $(-a, 0)$. See Figure 10-15.

The y-intercepts are $(0, b)$ and $(0, -b)$. See Figure 10-15.

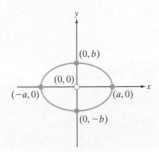

Figure 10-15

Example 1 **Graphing an Ellipse**

Graph the ellipse given by the equation. $\dfrac{x^2}{9} + \dfrac{y^2}{4} = 1$

Solution:

The equation can be written as $\dfrac{x^2}{3^2} + \dfrac{y^2}{2^2} = 1$.

This is in standard form $\dfrac{x^2}{a^2} + \dfrac{y^2}{b^2} = 1$

with $a = 3$ and $b = 2$.

Graph the intercepts $(3, 0)$, $(-3, 0)$, $(0, 2)$, $(0, -2)$ and sketch the ellipse.

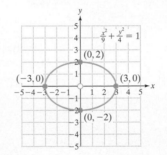

Example 2 **Graphing an Ellipse**

Graph the ellipse given by the equation. $25x^2 + y^2 = 25$

Solution:

$25x^2 + y^2 = 25$ Divide both sides by 25.

$$\frac{25x^2}{25} + \frac{y^2}{25} = \frac{25}{25}$$

$$x^2 + \frac{y^2}{25} = 1$$

The equation can then be written as $\dfrac{x^2}{1^2} + \dfrac{y^2}{5^2} = 1$.

The equation is now in standard form $\dfrac{x^2}{a^2} + \dfrac{y^2}{b^2} = 1$,

with $a = 1$ and $b = 5$.

Graph the intercepts $(1, 0)$, $(-1, 0)$, $(0, 5)$, and $(0, -5)$ and sketch the ellipse.

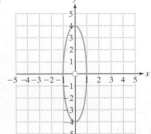

Answers

1.

2.

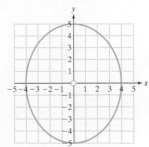

A circle is a special case of an ellipse where $a = b$. Therefore, it is not surprising that we graph an ellipse centered at the point (h, k) in much the same way we graph a circle.

Example 3   **Graphing an Ellipse Whose Center Is Not at the Origin**

Graph the ellipse given by the equation. $\dfrac{(x - 1)^2}{16} + \dfrac{(y + 3)^2}{4} = 1$

Solution:

Just as we would find the center of a circle, we see that the center of the ellipse is $(1, -3)$. Now we can use the values of a and b to help us plot four strategic points to define the curve.

The equation can be written as $\dfrac{(x - 1)^2}{(4)^2} + \dfrac{(y + 3)^2}{(2)^2} = 1$.

From this, we have $a = 4$ and $b = 2$. To sketch the curve, locate the center at $(1, -3)$. Then move $a = 4$ units to the left and right of the center and plot two points. Similarly, move $b = 2$ units up and down from the center and plot two points.

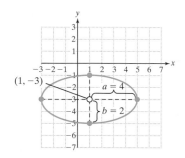

2. Standard Forms of an Equation of a Hyperbola

A **hyperbola** is the set of all points (x, y) such that the absolute value of the *difference* of the distances between (x, y) and two distinct points is a constant. The fixed points are called the foci of the hyperbola. The graph of a hyperbola has two parts, called branches. Each part resembles a parabola but is a slightly different shape. A hyperbola has two vertices that lie on an axis of symmetry called the **transverse axis**. For the hyperbolas studied here, the transverse axis is either horizontal or vertical.

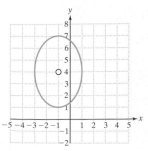
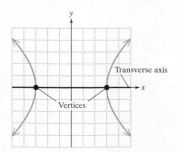

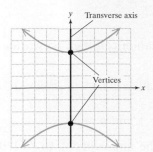

Standard Forms of an Equation of a Hyperbola with Center at the Origin

Let a and b represent positive real numbers.

<u>Horizontal transverse axis:</u>

The standard form of an equation of a hyperbola with a *horizontal transverse axis* and center at the origin is given by $\dfrac{x^2}{a^2} - \dfrac{y^2}{b^2} = 1$.

Note: The x-term is positive. The branches of the hyperbola open left and right.

<u>Vertical transverse axis:</u>

The standard form of an equation of a hyperbola with a *vertical transverse axis* and center at the origin is given by $\dfrac{y^2}{b^2} - \dfrac{x^2}{a^2} = 1$.

Note: The y-term is positive. The branches of the hyperbola open up and down.

In the standard forms of an equation of a hyperbola, the right side must equal 1.

To graph a hyperbola centered at the origin, first construct a reference rectangle. Draw this rectangle by using the points (a, b), $(-a, b)$, $(a, -b)$, and $(-a, -b)$. Asymptotes lie on the diagonals of the rectangle. The branches of the hyperbola are drawn to approach the asymptotes.

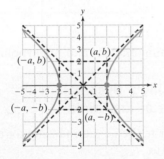
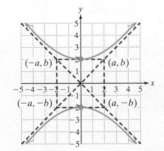

Example 4 **Graphing a Hyperbola**

Graph the hyperbola given by the equation. $\dfrac{x^2}{36} - \dfrac{y^2}{9} = 1$

a. Determine whether the transverse axis is horizontal or vertical.

b. Draw the reference rectangle and asymptotes.

c. Graph the hyperbola and label the vertices.

Solution:

a. Since the x-term is positive, the transverse axis is horizontal.

b. The equation can be written $\dfrac{x^2}{6^2} - \dfrac{y^2}{3^2} = 1$;

 therefore, $a = 6$ and $b = 3$. Graph the reference rectangle from the points $(6, 3), (6, -3), (-6, 3), (-6, -3)$.

c. The vertices are $(-6, 0)$ and $(6, 0)$.

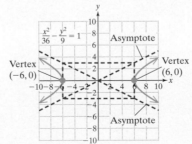

Skill Practice

Graph the hyperbola.

4. $\dfrac{x^2}{1} - \dfrac{y^2}{9} = 1$

Answer

4.

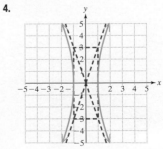

Skill Practice

Graph the hyperbola.

5. $9y^2 - 4x^2 = 36$

Example 5 Graphing a Hyperbola

Graph the hyperbola given by the equation. $y^2 - 4x^2 - 16 = 0$

a. Write the equation in standard form to determine whether the transverse axis is horizontal or vertical.

b. Draw the reference rectangle and asymptotes.

c. Graph the hyperbola and label the vertices.

Solution:

a. Isolate the variable terms and divide by 16.

$$\frac{y^2}{16} - \frac{x^2}{4} = 1$$

Since the y-term is positive, the transverse axis is vertical.

b. The equation can be written $\dfrac{y^2}{4^2} - \dfrac{x^2}{2^2} = 1$;

therefore, $a = 2$ and $b = 4$. Graph the

reference rectangle from the points $(2, 4)$, $(2, -4), (-2, 4), (-2, -4)$.

c. The vertices are $(0, 4)$ and $(0, -4)$.

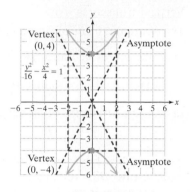

The following summarizes the steps for graphing ellipses and hyperbolas.

Graphing an Ellipse with Center at the Origin

1. Write the equation in standard form:
$$\frac{x^2}{a^2} + \frac{y^2}{b^2} = 1$$

2. Plot the intercepts.
   x-intercepts: $(a, 0)$ and $(-a, 0)$
   y-intercepts: $(0, b)$ and $(0, -b)$

3. Sketch the ellipse through the intercepts.

Graphing a Hyperbola with Center at the Origin

1. Write the equation in standard form:
   - Horizontal transverse axis:
   $$\frac{x^2}{a^2} - \frac{y^2}{b^2} = 1$$
   - Vertical transverse axis:
   $$\frac{y^2}{b^2} - \frac{x^2}{a^2} = 1$$

2. Draw the reference rectangle through the points $(a, b), (a, -b)$, $(-a, b)$, and $(-a, -b)$.

3. Draw the asymptotes.

4. Plot the vertices and sketch the hyperbola.

Answer

5.

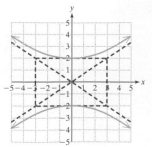

Section 10.3   Practice Exercises

Vocabulary and Key Concepts

1. a. A(n) _____ is the set of all points (x, y) such that the sum of the distances between (x, y) and two distinct points (called _____) is constant.

b. The standard form of an equation of an _____ with center at the origin is $\dfrac{x^2}{a^2} + \dfrac{y^2}{b^2} = 1$.

c. A(n) _____ is the set of all points (x, y) such that the absolute value of the difference in distances between (x, y) and two distinct points (called _____) is a constant.

d. A hyperbola has two vertices that lie on an axis of symmetry called the _____ axis.

e. The standard form of an equation of a hyperbola with a *horizontal* transverse axis and center at $(0, 0)$, is given by _____.

f. The standard form of an equation of a hyperbola with a *vertical* transverse axis and center at $(0, 0)$, is given by _____.

Review Exercises

2. Write the standard form of a circle with center at (h, k) and radius, r.

For Exercises 3 and 4, identify the center and radius of the circle.

3. $x^2 + y^2 - 16x + 12y = 0$

4. $x^2 + y^2 + 4x + 4y - 1 = 0$

For Exercises 5 and 6, identify the vertex and the axis of symmetry.

5. $y = 3(x + 3)^2 - 1$

6. $x = -\dfrac{1}{4}(y - 1)^2 - 6$

7. Write an equation for a circle whose center has coordinates $\left(\frac{1}{2}, \frac{5}{2}\right)$ with radius equal to $\frac{1}{2}$.

8. Write the equation for the circle centered at the origin and with radius equal to $\frac{1}{8}$.

Concept 1: Standard Form of an Equation of an Ellipse

For Exercises 9–16, find the x- and y-intercepts and graph the ellipse. **(See Examples 1 and 2.)**

9. $\dfrac{x^2}{4} + \dfrac{y^2}{9} = 1$

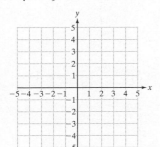

10. $\dfrac{x^2}{16} + \dfrac{y^2}{25} = 1$

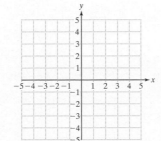

11. $\dfrac{x^2}{16} + \dfrac{y^2}{9} = 1$

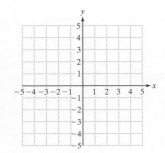

12. $\dfrac{x^2}{36} + \dfrac{y^2}{4} = 1$

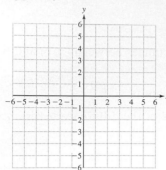

13. $4x^2 + y^2 = 4$

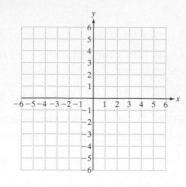

14. $9x^2 + y^2 = 36$

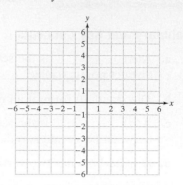

15. $x^2 + 25y^2 - 25 = 0$

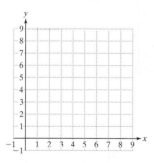

16. $4x^2 + 9y^2 = 144$

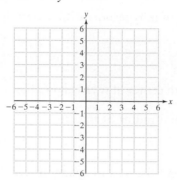

For Exercises 17–24, identify the center (h, k) of the ellipse. Then graph the ellipse. **(See Example 3.)**

17. $\dfrac{(x - 4)^2}{4} + \dfrac{(y - 5)^2}{9} = 1$

18. $\dfrac{(x + 2)^2}{9} + \dfrac{(y - 1)^2}{16} = 1$

19. $\dfrac{(x + 1)^2}{25} + \dfrac{(y - 2)^2}{9} = 1$

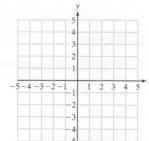

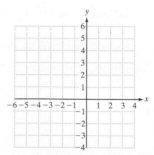

20. $\dfrac{(x - 4)^2}{25} + \dfrac{(y + 1)^2}{16} = 1$

21. $\dfrac{(x - 2)^2}{9} + (y + 3)^2 = 1$

22. $(x - 5)^2 + \dfrac{(y - 3)^2}{4} = 1$

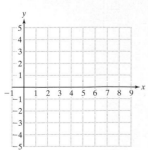

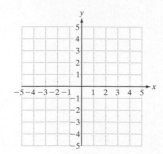

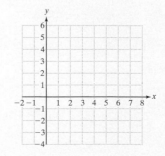

23. $\dfrac{x^2}{36} + \dfrac{(y-1)^2}{25} = 1$

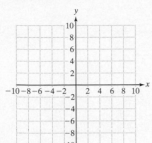

24. $\dfrac{(x+5)^2}{4} + \dfrac{y^2}{16} = 1$

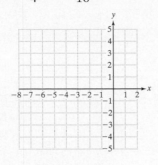

Concept 2: Standard Forms of an Equation of a Hyperbola

For Exercises 25–32, determine whether the transverse axis is horizontal or vertical.

25. $\dfrac{y^2}{6} - \dfrac{x^2}{18} = 1$

26. $\dfrac{y^2}{10} - \dfrac{x^2}{4} = 1$

27. $\dfrac{x^2}{20} - \dfrac{y^2}{15} = 1$

28. $\dfrac{x^2}{12} - \dfrac{y^2}{9} = 1$

29. $x^2 - y^2 = 12$

30. $x^2 - y^2 = 15$

31. $x^2 - 3y^2 = -9$

32. $2x^2 - y^2 = -10$

For Exercises 33–40, use the equation in standard form to graph the hyperbola. Label the vertices of the hyperbola. **(See Examples 4 and 5.)**

33. $\dfrac{x^2}{25} - \dfrac{y^2}{16} = 1$

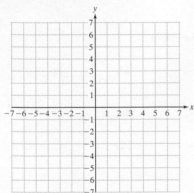

34. $\dfrac{x^2}{9} - \dfrac{y^2}{36} = 1$

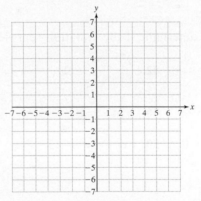

35. $\dfrac{y^2}{4} - \dfrac{x^2}{4} = 1$

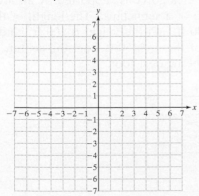

36. $\dfrac{y^2}{9} - \dfrac{x^2}{9} = 1$

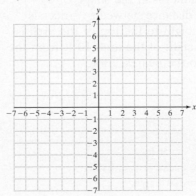

37. $36x^2 - y^2 = 36$

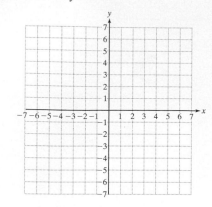

38. $x^2 - 25y^2 = 25$

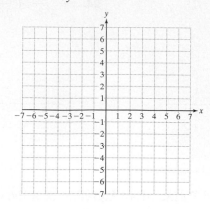

39. $y^2 - 4x^2 - 16 = 0$

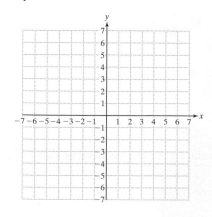

40. $y^2 - 4x^2 - 36 = 0$

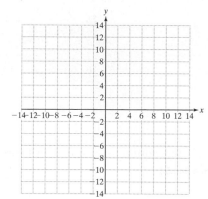

Mixed Exercises

For Exercises 41–52, determine if the equation represents an ellipse or a hyperbola.

41. $\dfrac{x^2}{6} - \dfrac{y^2}{10} = 1$

42. $\dfrac{x^2}{14} + \dfrac{y^2}{2} = 1$

43. $\dfrac{y^2}{4} + \dfrac{x^2}{16} = 1$

44. $\dfrac{x^2}{5} + \dfrac{y^2}{10} = 1$

45. $4x^2 + y^2 = 16$

46. $-3x^2 - 4y^2 = -36$

47. $-y^2 + 2x^2 = -10$

48. $x^2 - y^2 = -1$

49. $5x^2 + y^2 - 10 = 0$

50. $5x^2 - 3y^2 = 15$

51. $y^2 - 6x^2 = 6$

52. $3x^2 + 5y^2 = 15$

Expanding Your Skills

53. An arch for a tunnel is in the shape of a semiellipse. The distance between vertices is 120 ft, and the height to the top of the arch is 50 ft. Find the height of the arch 10 ft from the center. Round to the nearest foot.

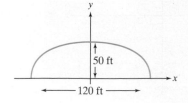

54. A bridge over a gorge is supported by an arch in the shape of a semiellipse. The length of the bridge is 400 ft, and the height is 100 ft. Find the height of the arch 50 ft from the center. Round to the nearest foot.

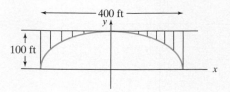

In Exercises 55–58, graph the hyperbola centered at (h, k) by following these guidelines. First locate the center. Then draw the reference rectangle relative to the center (h, k). Using the reference rectangle as a guide, sketch the hyperbola. Label the center and vertices.

55. $\dfrac{(x-1)^2}{9} - \dfrac{(y+2)^2}{4} = 1$

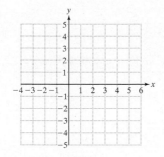

56. $\dfrac{(x+2)^2}{16} - \dfrac{(y-3)^2}{4} = 1$

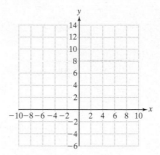

57. $\dfrac{(y-1)^2}{4} - (x+3)^2 = 1$

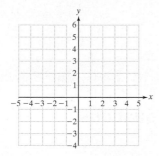

58. $(y+2)^2 - \dfrac{(x+3)^2}{4} = 1$

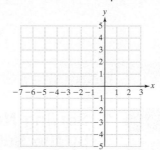

Problem Recognition Exercises

Formulas and Conic Sections

For Exercises 1–8, identify the formula.

1. $(x-h)^2 + (y-k)^2 = r^2$

2. $\dfrac{x^2}{a^2} + \dfrac{y^2}{b^2} = 1$

3. $\sqrt{(x_2 - x_1)^2 + (y_2 - y_1)^2}$

4. $\left(\dfrac{x_1 + x_2}{2}, \dfrac{y_1 + y_2}{2}\right)$

5. $y = a(x-h)^2 + k$

6. $\dfrac{x^2}{a^2} - \dfrac{y^2}{b^2} = 1$

7. $\dfrac{y^2}{b^2} - \dfrac{x^2}{a^2} = 1$

8. $x = a(y-k)^2 + h$

For Exercises 9–30, identify the equation as representing a circle, parabola, ellipse, hyperbola, or none of these.

9. $y = -2(x-3)^2 + 4$

10. $\dfrac{x^2}{4} - \dfrac{y^2}{1} = 1$

11. $(x+3)^2 + (y+2)^2 = 4$

12. $(x-2)^2 + (y-4)^2 = 9$

13. $\dfrac{x^2}{9} - \dfrac{y^2}{9} = 1$

14. $\dfrac{x^2}{9} + \dfrac{y^2}{16} = 1$

15. $\dfrac{x^2}{16} + \dfrac{y^2}{4} = 0$

16. $x^2 + y^2 - 2x + 4y - 4 = 0$

17. $y = \dfrac{1}{2}(x + 2)^2 - 3$

18. $\dfrac{x^2}{4} - \dfrac{y^2}{2} = 1$

19. $x = (y + 2)^2 - 4$

20. $x^2 + y^2 + 6x + 8 = 0$

21. $(x - 1)^2 + (y + 1)^2 = 0$

22. $x = -(y - 2)^2 - 1$

23. $\dfrac{x^2}{25} + \dfrac{y^2}{4} = 1$

24. $x^2 + y^2 = 15$

25. $y = (x - 6)^2 + 4$

26. $\dfrac{(x + 1)^2}{2} + \dfrac{(y + 1)^2}{5} = 1$

27. $\dfrac{y^2}{3} - \dfrac{x^2}{3} = 1$

28. $3x^2 + 3y^2 = 1$

29. $\dfrac{x^2}{9} + \dfrac{y^2}{12} = 1$

30. $x = (y + 2)^2 - 5$

| Section 10.4 | Nonlinear Systems of Equations in Two Variables |

Concepts

1. Solving Nonlinear Systems of Equations by the Substitution Method
2. Solving Nonlinear Systems of Equations by the Addition Method

1. Solving Nonlinear Systems of Equations by the Substitution Method

Recall that a linear equation in two variables x and y is an equation that can be written in the form $Ax + By = C$, where A and B are not both zero. In Sections 4.1–4.3, we solved systems of linear equations in two variables by using the graphing method, the substitution method, and the addition method. In this section, we will solve *nonlinear* systems of equations by using the same methods. A **nonlinear system of equations** is a system in which at least one of the equations is nonlinear.

Graphing the equations in a nonlinear system helps to determine the number of solutions and to approximate the coordinates of the solutions. The substitution method is used most often to solve a nonlinear system of equations analytically.

Example 1 Solving a Nonlinear System of Equations

Given the system

$$x - 7y = -25$$
$$x^2 + y^2 = 25$$

a. Solve the system by graphing.

b. Solve the system by the substitution method.

Solution:

a. $x - 7y = -25$ is a line (the slope-intercept form is $y = \frac{1}{7}x + \frac{25}{7}$).

$x^2 + y^2 = 25$ is a circle centered at the origin with radius 5.

From Figure 10-16, we appear to have two solutions $(-4, 3)$ and $(3, 4)$.

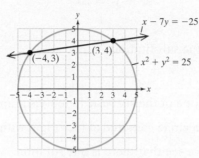

Figure 10-16

Skill Practice

Given the system

$2x + y = 5$

$x^2 + y^2 = 50$

1. Solve the system by graphing.
2. Solve the system by the substitution method.

b. To use the substitution method, isolate one of the variables from one of the equations. We will solve for x in the first equation.

$\boxed{A}$ $x - 7y = -25$ $\xrightarrow{\text{Solve for } x.}$ $x = 7y - 25$

$\boxed{B}$ $x^2 + y^2 = 25$

$\boxed{B}$ $(7y - 25)^2 + y^2 = 25$ Substitute $(7y - 25)$ for x in the second equation.

$49y^2 - 350y + 625 + y^2 = 25$ The resulting equation is quadratic in y.

$50y^2 - 350y + 600 = 0$ Set the equation equal to zero.

$50(y^2 - 7y + 12) = 0$ Factor.

$50(y - 3)(y - 4) = 0$

$y = 3$ or $y = 4$

For each value of y, find the corresponding x value from the equation $x = 7y - 25$.

$y = 3$: $x = 7(3) - 25 = -4$ The solution point is $(-4, 3)$.

$y = 4$: $x = 7(4) - 25 = 3$ The solution point is $(3, 4)$.
                                (See Figure 10-16.)

Answers

1.

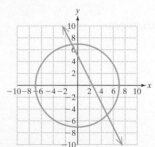

The solutions appear to be $(-1, 7)$ and $(5, -5)$.

2. $(-1, 7)$ and $(5, -5)$

Example 2 Solving a Nonlinear System by the Substitution Method

Given the system

$$y = \sqrt{x}$$
$$x^2 + y^2 = 20$$

a. Sketch the graphs.

b. Solve the system by the substitution method.

Solution:

a. $y = \sqrt{x}$ is one of the six basic functions graphed in Section 3.3.

$x^2 + y^2 = 20$ is a circle centered at the origin with radius $\sqrt{20} \approx 4.5$.

From Figure 10-17, we see that there is one solution.

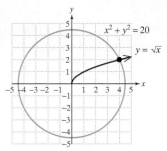

Figure 10-17

b. To use the substitution method, we will substitute $y = \sqrt{x}$ into equation $\boxed{B}$.

$\boxed{A}$ $y = \sqrt{x}$ Substitute $y = \sqrt{x}$ into the second equation.

$\boxed{B}$ $x^2 + y^2 = 20$

$\boxed{B}$ $x^2 + (\sqrt{x})^2 = 20$

        $x^2 + x = 20$

     $x^2 + x - 20 = 0$ Set the second equation equal to zero.

    $(x + 5)(x - 4) = 0$ Factor.

$x = -5$ or $x = 4$ Reject $x = -5$ because it is not in the domain of $y = \sqrt{x}$.

Substitute $x = 4$ into the equation $y = \sqrt{x}$.

If $x = 4$, then $y = \sqrt{4} = 2$.   The solution is $(4, 2)$.

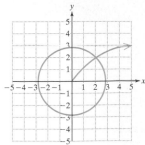

Calculator Connections

Topic: Graphing a Nonlinear System of Equations

Graph the equations from Example 2 to confirm your solution to the system of equations. Use an *Intersect* feature to approximate the point of intersection. Recall that the circle must be entered into the calculator as two functions:

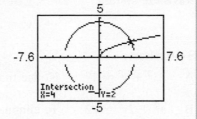

$$Y_1 = \sqrt{20 - x^2}$$

$$Y_2 = -\sqrt{20 - x^2}$$

$$Y_3 = \sqrt{x}$$

Example 3   **Solving a Nonlinear System by Using the Substitution Method**

Solve the system by using the substitution method.

$$y = \sqrt[3]{x}$$

$$y = x$$

Solution:

$\boxed{A}$ $y = \sqrt[3]{x}$

$\boxed{B}$ $y = x$

$\sqrt[3]{x} = x$ Because y is isolated in both equations, we can equate the expressions for y.

$(\sqrt[3]{x})^3 = (x)^3$ To solve the radical equation, raise both sides to the third power.

$x = x^3$ This is a third-degree polynomial equation.

$0 = x^3 - x$ Set the equation equal to zero.

$0 = x(x^2 - 1)$ Factor out the GCF.

$0 = x(x + 1)(x - 1)$ Factor completely.

$x = 0$ or $x = -1$ or $x = 1$

For each value of x, find the corresponding y-value from either original equation. We will use equation $\boxed{B}$: $y = x$.

If $x = 0$, then $y = 0$.   The solution point is $(0, 0)$.

If $x = -1$, then $y = -1$.   The solution point is $(-1, -1)$.

If $x = 1$, then $y = 1$.   The solution point is $(1, 1)$. See Figure 10-18.

Figure 10-18

2. Solving Nonlinear Systems of Equations by the Addition Method

The substitution method is used most often to solve a system of nonlinear equations. In some situations, however, the addition method offers an efficient means of finding a solution. Example 4 demonstrates that we can eliminate a variable from both equations provided the terms containing that variable are *like* terms.

Answer

5. $(0, 0)$; $(3, 3)$; $(-3, -3)$

Example 4   Solving a Nonlinear System of Equations by the Addition Method

Solve the system. $2x^2 + y^2 = 17$
                    $x^2 + 2y^2 = 22$

Solution:

$\boxed{A}$ $2x^2 + y^2 = 17$ Notice that the y^2-terms are *like* in each equation.

$\boxed{B}$ $x^2 + 2y^2 = 22$ To eliminate the y^2-terms, multiply the first equation by -2.

$\boxed{A}$ $2x^2 + y^2 = 17$ $\xrightarrow{\text{Multiply by } -2.}$ $-4x^2 - 2y^2 = -34$
$\boxed{B}$ $x^2 + 2y^2 = 22$ $\longrightarrow$ $\underline{x^2 + 2y^2 = 22}$

$-3x^2 = -12$ Eliminate the y^2 term.

$$\frac{-3x^2}{-3} = \frac{-12}{-3}$$

$$x^2 = 4$$ Use the square root
$$x = \pm 2$$ property.

Substitute each value of x into one of the original equations to solve for y. We will use equation $\boxed{A}$: $2x^2 + y^2 = 17$.

$x = 2$: $\boxed{A}$ $2(2)^2 + y^2 = 17$

$8 + y^2 = 17$

$y^2 = 9$

$y = \pm 3$ The solution points are $(2, 3)$ and $(2, -3)$.

$x = -2$: $\boxed{A}$ $2(-2)^2 + y^2 = 17$

$8 + y^2 = 17$

$y^2 = 9$

$y = \pm 3$ The solution points are $(-2, 3)$ and $(-2, -3)$.

The two equations in this example are ellipses but an exact graph is difficult to render by hand.

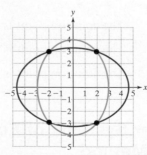

To check the solutions to the system, verify that each ordered pair satisfies both equations.

TIP: It is important to note that the addition method can be used only if two equations share a pair of *like* terms. The substitution method is effective in solving a wider range of systems of equations. The system in Example 4 could also have been solved by using substitution.

A $2x^2 + y^2 = 17$ $\xrightarrow{\text{Solve for } y^2}$ $y^2 = 17 - 2x^2$

B $x^2 + 2y^2 = 22$

B $x^2 + 2(17 - 2x^2) = 22$ $x = 2$: $y^2 = 17 - 2(2)^2$

$x^2 + 34 - 4x^2 = 22$ $y^2 = 9$

$-3x^2 = -12$ $y = \pm 3$ The solutions are
                                                            $(2, 3)$ and $(2, -3)$.

$x^2 = 4$

$x = \pm 2$ $\longrightarrow$ $x = -2$: $y^2 = 17 - 2(-2)^2$

$y^2 = 9$

$y = \pm 3$ The solutions are
$(-2, 3)$ and $(-2, -3)$.

Section 10.4   Practice Exercises

Vocabulary and Key Concepts

1. **a.** A _____ system of equations in two variables is a system in which at least one of the equations is nonlinear.

    b. Graphically, the solution set to a nonlinear system of equations is the set of ordered pairs representing the points of _____ of the graphs of the two equations.

Review Exercises

2. Write the distance formula between two points (x_1, y_1) and (x_2, y_2) from memory.

3. Find the distance between the two points $(8, -1)$ and $(1, -8)$.

4. Write an equation representing the set of all points 2 units from the point $(-1, 1)$.

5. Write an equation representing the set of all points 8 units from the point $(-5, 3)$.

6. Write the formula for the midpoint between the points (x_1, y_1) and (x_2, y_2).

7. Find the midpoint of the segment with the endpoints $(3, -9)$ and $(-4, 2)$.

8. Find the midpoint of the segment with the endpoints $(2.4, 1.8)$ and $(-1.6, -5.2)$.

Concept 1: Solving Nonlinear Systems of Equations by the Substitution Method

For Exercises 9–16, use sketches to explain.

9. How many points of intersection are possible between a line and a parabola?

10. How many points of intersection are possible between a line and a circle?

11. How many points of intersection are possible between two distinct circles?

12. How many points of intersection are possible between two distinct parabolas of the form $y = ax^2 + bx + c$, $a \neq 0$?

13. How many points of intersection are possible between a circle and a parabola?

14. How many points of intersection are possible between two distinct lines?

15. How many points of intersection are possible with an ellipse and a hyperbola?

16. How many points of intersection are possible with an ellipse and a parabola?

For Exercises 17–22, sketch each system of equations. Then solve the system by the substitution method. **(See Example 1.)**

17.
$$y = x + 3$$
$$x^2 + y = 9$$

18.
$$y = x - 2$$
$$x^2 + y = 4$$

19. $x^2 + y^2 = 1$
$$y = x + 1$$

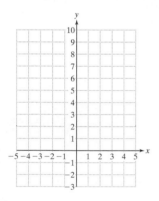

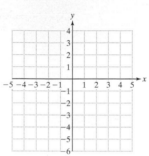

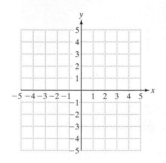

20. $x^2 + y^2 = 25$
$$y = 2x$$

21. $x^2 + y^2 = 6$
$$y = x^2$$

22. $x^2 + y^2 = 12$
$$y = x^2$$

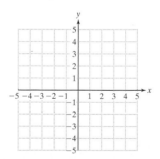

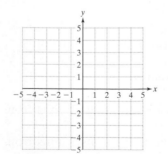

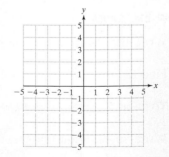

For Exercises 23–31, solve the system by the substitution method. **(See Examples 2 and 3.)**

23. $y = \sqrt{x}$
$$2x^2 - y^2 = 1$$

24. $x^2 + y^2 = 30$
$$y = \sqrt{x}$$

25. $y = x^2$
$$y = -\sqrt{x}$$

26. $y = -x^2$
$$y = -\sqrt{x}$$

27. $y = x^2$
$$y = (x - 3)^2$$

28. $y = (x + 4)^2$
$$y = x^2$$

29. $y = x^2 + 6x$
$$y = 4x$$

30. $y = 3x^2 - 6x$
$$y = 3x$$

31. $x^2 - 5x + y = 0$
$$y = 3x + 1$$

Concept 2: Solving Nonlinear Systems of Equations by the Addition Method

For Exercises 32–46, solve the system of nonlinear equations by the addition method. **(See Example 4.)**

32. $x^2 + y^2 = 13$
$x^2 - y^2 = 5$

33. $4x^2 - y^2 = 4$
$4x^2 + y^2 = 4$

34. $9x^2 + 4y^2 = 36$
$x^2 + y^2 = 9$

35. $x^2 + y^2 = 4$
$2x^2 + y^2 = 8$

36. $3x^2 + 4y^2 = 16$
$2x^2 - 3y^2 = 5$

37. $2x^2 - 5y^2 = -2$
$3x^2 + 2y^2 = 35$

38. $y = x^2 - 2$
$y = -x^2 + 2$

39. $y = x^2$
$y = -x^2 + 8$

40. $\dfrac{x^2}{4} + \dfrac{y^2}{9} = 1$
$x^2 + y^2 = 4$

41. $\dfrac{x^2}{16} + \dfrac{y^2}{4} = 1$
$x^2 + y^2 = 4$

42. $x^2 + 6y^2 = 9$
$\dfrac{x^2}{9} + \dfrac{y^2}{12} = 1$

43. $\dfrac{x^2}{10} + \dfrac{y^2}{10} = 1$
$2x^2 + y^2 = 11$

44. $x^2 - xy = -4$
$2x^2 - xy = 12$

45. $x^2 - xy = 3$
$2x^2 + xy = 6$

46. $y = x^2 + 4$
$y = 2(x^2 + 2)$

Expanding Your Skills

47. The sum of two numbers is 7. The sum of the squares of the numbers is 25. Find the numbers.

48. The sum of the squares of two numbers is 100. The sum of the numbers is 2. Find the numbers.

49. The sum of the squares of two numbers is 32. The difference of the squares of the numbers is 18. Find the numbers.

50. The sum of the squares of two numbers is 24. The difference of the squares of the numbers is 8. Find the numbers.

Graphing Calculator Exercises

For Exercises 51–54, use the *Intersect* feature to approximate the solutions to the system.

51. $y = x + 3$ (Exercise 17)
$x^2 + y = 9$

52. $y = x - 2$ (Exercise 18)
$x^2 + y = 4$

53. $y = x^2$ (Exercise 25)
$y = -\sqrt{x}$

54. $y = -x^2$ (Exercise 26)
$y = -\sqrt{x}$

For Exercises 55 and 56, graph the system on a square viewing window. What can be said about the solution to the system?

55. $x^2 + y^2 = 4$
$y = x^2 + 3$

56. $x^2 + y^2 = 16$
$y = -x^2 - 5$

Section 10.5   Nonlinear Inequalities and Systems of Inequalities

1. Nonlinear Inequalities in Two Variables

In Section 4.5 we graphed the solution sets to linear inequalities in two variables. For example, to graph the solution set to $y \leq 2x + 1$, first graph the related equation $y = 2x + 1$. This is the line shown in Figure 10-19. Then using test points, we see that points on and below the line make up the solution set to the inequality.

Test Point Above: (–2, 2)	Test Point Below: (0, 0)
$y \leq 2x + 1$	$y \leq 2x + 1$
$2 \overset{?}{\leq} 2(-2) + 1$	$0 \overset{?}{\leq} 2(0) + 1$
$2 \overset{?}{\leq} -3$ False	$0 \overset{?}{\leq} 1$ ✓True

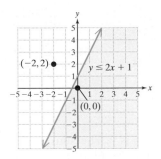

Figure 10-19

Example 1   Graphing a Nonlinear Inequality in Two Variables

Graph the solution set of the inequality $x^2 + y^2 < 16$.

Skill Practice

1. Graph the solution set of the inequality $x^2 + y^2 \geq 9$.

TIP: The first step in solving an inequality in two variables is to graph the related equation.

Solution:

The related equation $x^2 + y^2 = 16$ is a circle of radius 4, centered at the origin. Graph the related equation by using a dashed curve because the points satisfying the equation $x^2 + y^2 = 16$ are not part of the solution to the strict inequality $x^2 + y^2 < 16$. See Figure 10-20.

Notice that the dashed curve separates the xy-plane into two regions, one "inside" the circle, the other "outside" the circle. Select a test point from each region and test the point in the original inequality.

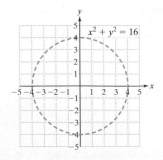

Figure 10-20

Test Point "Inside": (0, 0)	Test Point "Outside": (4, 4)
$x^2 + y^2 < 16$	$x^2 + y^2 < 16$
$(0)^2 + (0)^2 \overset{?}{<} 16$	$(4)^2 + (4)^2 \overset{?}{<} 16$
$0 \overset{?}{<} 16$ True	$32 \overset{?}{<} 16$ False

Answer

1.

The inequality $x^2 + y^2 < 16$ is true at the test point $(0, 0)$. Therefore, the solution set is the region "inside" the circle. See Figure 10-21.

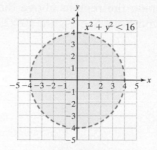

Figure 10-21

Example 2 **Graphing a Nonlinear Inequality in Two Variables**

Graph the solution set of the inequality $9y^2 \geq 36 + 4x^2$.

Solution:

First graph the related equation $9y^2 = 36 + 4x^2$. Notice that the equation can be written in the standard form of a hyperbola.

$$9y^2 = 36 + 4x^2$$

$$9y^2 - 4x^2 = 36 \qquad \text{Subtract } 4x^2 \text{ from both sides.}$$

$$\frac{9y^2}{36} - \frac{4x^2}{36} = \frac{36}{36} \qquad \text{Divide both sides by 36.}$$

$$\frac{y^2}{4} - \frac{x^2}{9} = 1$$

Graph the hyperbola as a solid curve, because the original inequality includes equality. See Figure 10-22.

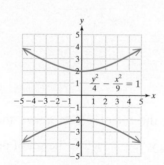

Figure 10-22

The hyperbola divides the xy-plane into three regions: a region above the upper branch, a region between the branches, and a region below the lower branch. Select a test point from each region.

$$9y^2 \geq 36 + 4x^2$$

Test: $(0, 3)$	**Test: $(0, 0)$**	**Test: $(0, -3)$**
$9(3)^2 \overset{?}{\geq} 36 + 4(0)^2$	$9(0)^2 \overset{?}{\geq} 36 + 4(0)^2$	$9(-3)^2 \overset{?}{\geq} 36 + 4(0)^2$
$81 \overset{?}{\geq} 36$ True	$0 \overset{?}{\geq} 36$ False	$81 \overset{?}{\geq} 36$ True

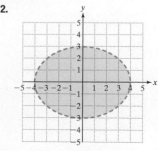

Shade the regions above the top branch and below the bottom branch of the hyperbola. See Figure 10-23.

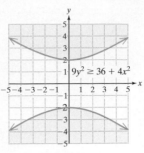

Figure 10-23

2. Systems of Nonlinear Inequalities in Two Variables

In Section 10.4 we solved systems of nonlinear equations in two variables. The solution set for such a system is the set of ordered pairs that satisfy both equations simultaneously. We will now solve systems of nonlinear inequalities in two variables. Similarly, the solution set is the set of all ordered pairs that simultaneously satisfy each inequality. To solve a system of inequalities, we will graph the solution to each individual inequality and then take the intersection of the solution sets.

> **Example 3** Graphing a System of Nonlinear Inequalities in Two Variables

Graph the solution set.
$$y > \frac{1}{3}x^2$$
$$y < -x^2 + 2$$

Solution:

The solution to $y > \dfrac{1}{3}x^2$ is the set of points above the parabola $y = \dfrac{1}{3}x^2$. See Figure 10-24. The solution to $y < -x^2 + 2$ is the set of points below the parabola $y = -x^2 + 2$. See Figure 10-25.

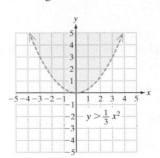

Figure 10-24

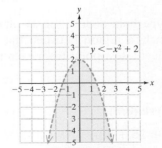

Figure 10-25

The solution to the system of inequalities is the intersection of the solution sets of the individual inequalities. See Figure 10-26.

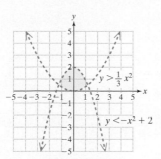

Figure 10-26

Example 4 **Graphing a System of Nonlinear Inequalities in Two Variables**

Graph the solution set.

$$y > e^x$$

$$y < -x^2 + 4$$

Solution:

The solution to $y > e^x$ is the set of points above the curve $y = e^x$. See Figure 10-27. The solution to $y < -x^2 + 4$ is the set of points below the parabola $y = -x^2 + 4$. See Figure 10-28.

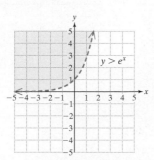

Figure 10-27

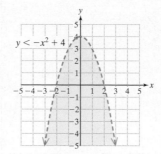

Figure 10-28

The solution to the system of inequalities is the intersection of the solution sets of the individual inequalities. See Figure 10-29.

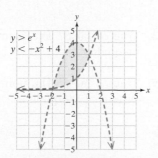

Figure 10-29

Answer

4.

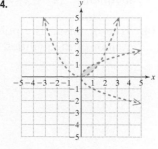

Section 10.5 Practice Exercises

Review Exercises

For Exercises 1–11, match the equation with its graph.

1. $y = \left(\dfrac{1}{3}\right)^x$

2. $y = 4x - 1$

3. $y = -4x^2$

4. $y = e^x$

5. $y = x^3$

6. $\dfrac{x^2}{4} + \dfrac{y^2}{9} = 1$

7. $\dfrac{x^2}{4} - \dfrac{y^2}{9} = 1$

8. $y = \dfrac{1}{x}$

9. $y = \log_2(x)$

10. $y = \sqrt{x}$

11. $(x + 2)^2 + (y - 1)^2 = 4$

a.
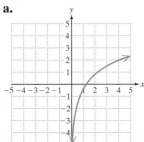

b.
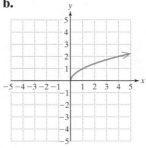

c.
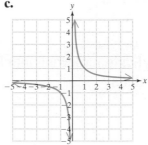

d.
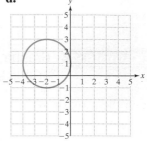

e.
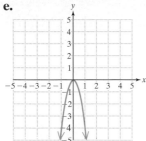

f.
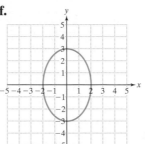

g.
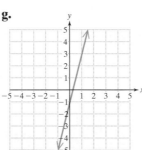

h.
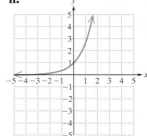

i.
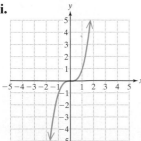

j.
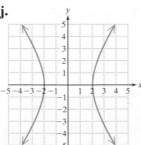

k.
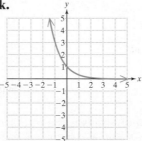

Concept 1: Nonlinear Inequalities in Two Variables

12. True or false? The point $(2, 3)$ satisfies the inequality $-x^2 + y^3 > 1$.

13. True or false? The point $(4, -2)$ satisfies the inequality $4x^2 - 2x + 1 + y^2 < 3$.

14. True or false? The point $(5, 4)$ satisfies the system of inequalities.

$$\frac{x^2}{36} + \frac{y^2}{25} < 1$$

$$x^2 + y^2 \geq 4$$

15. True or false? The point $(1, -2)$ satisfies the system of inequalities.

$$y < x^2$$

$$y > x^2 - 4$$

16. True or false? The point $(-3, 5)$ satisfies the system of inequalities. $(x + 3)^2 + (y - 5)^2 \leq 2$
$$y > x^2$$

17. a. Graph the solution set for $x^2 + y^2 \leq 9$.

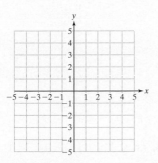

18. a. Graph the solution set for $\dfrac{x^2}{4} + \dfrac{y^2}{9} \geq 1$.

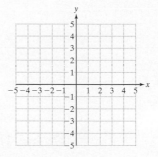

b. Describe the solution set for the inequality $x^2 + y^2 \geq 9$.

c. Describe the solution set for the equation $x^2 + y^2 = 9$

b. Describe the solution set for the inequality $\dfrac{x^2}{4} + \dfrac{y^2}{9} \leq 1$.

c. Describe the solution set for the equation $\dfrac{x^2}{4} + \dfrac{y^2}{9} = 1$

19. a. Graph the solution set for $y \geq x^2 + 1$.

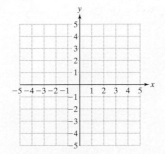

20. a. Graph the solution set for $\dfrac{x^2}{4} - \dfrac{y^2}{9} \leq 1$.

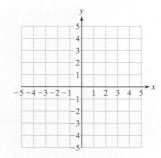

b. How would the solution change for the strict inequality $y > x^2 + 1$?

b. How would the solution change for the strict inequality $\dfrac{x^2}{4} - \dfrac{y^2}{9} < 1$?

21. A weak earthquake occurred in northern California roughly 4 mi south and 3 mi east of Sunol, California. The quake could be felt 25 mi away. Suppose the origin of a map is placed at the center of Sunol with the positive x-axis pointing east and the positive y-axis pointing north. Find an inequality that describes the points on the map for which the earthquake could be felt.

22. A coordinate system is placed at the center of a town with the positive x-axis pointing east, and the positive y-axis pointing north. A cell tower is located 2 mi west and 4 mi north of the center of town. If the tower has a 30-mi range, write an inequality that represents the points on the map serviced by this tower.

For Exercises 23–37, graph the solution set. **(See Examples 1 and 2.)**

23. $2x + y \geq 1$

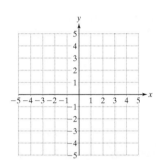

24. $3x + 2y \geq 6$

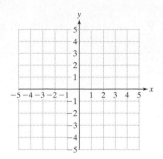

25. $x \leq y^2$

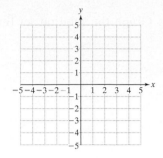

26. $y \leq -x^2$

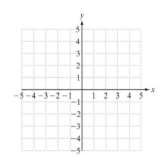

27. $(x - 1)^2 + (y + 2)^2 > 9$

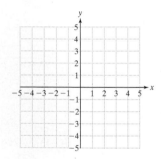

28. $(x + 1)^2 + (y - 4)^2 > 1$

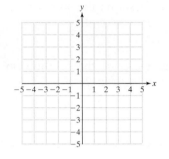

29. $x + y^2 \geq 4$

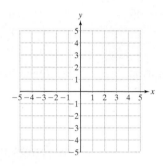

30. $x^2 + 2x + y - 1 \leq 0$

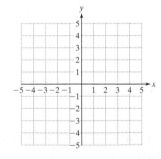

31. $9x^2 - y^2 > 9$

32. $y^2 - 4x^2 \leq 4$

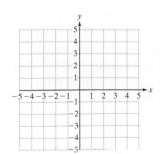

33. $x^2 + 16y^2 \leq 16$

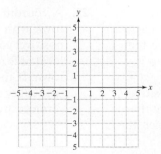

34. $4x^2 + y^2 \leq 4$

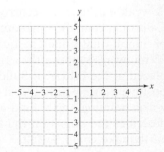

35. $y \leq \ln x$

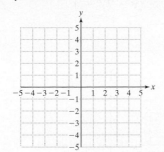

36. $y \leq \log x$

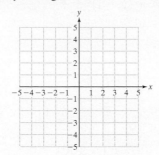

37. $y > 5^x$

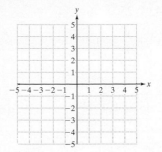

Concept 2: Systems of Nonlinear Inequalities in Two Variables

For Exercises 38–51, graph the solution set to the system of inequalities. **(See Examples 3 and 4.)**

38. $y < \sqrt{x}$
   $x > 1$

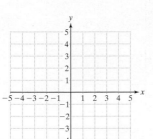

39. $y \geq \sqrt{x}$
   $x \geq 0$

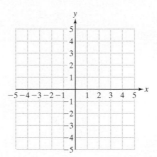

40. $\dfrac{x^2}{36} + \dfrac{y^2}{25} < 1$
   $x^2 + y^2 \geq 4$

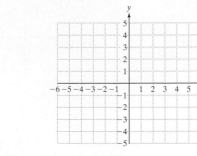

41. $x^2 - y^2 \geq 1$
   $x \leq 0$

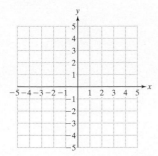

42. $y < x^2$
   $y > x^2 - 4$

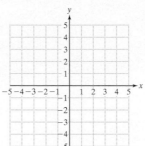

43. $y^2 - x^2 \geq 1$
   $y \geq 0$

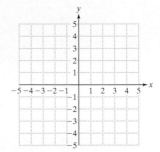

44. $y < \dfrac{1}{x}$
   $y > 0$
   $y < x$

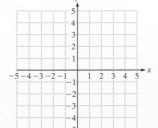

45. $y > x^3$
   $y < 8$
   $x > 0$

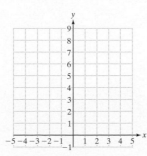

46. $x^2 + y^2 \geq 25$
   $x^2 + y^2 \leq 9$

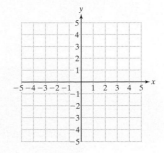

47. $\dfrac{x^2}{4} + \dfrac{y^2}{25} \geq 1$

$x^2 + \dfrac{y^2}{4} \leq 1$

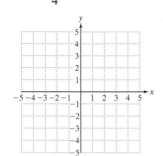

48. $x < -(y - 1)^2 + 3$

$x + y > 2$

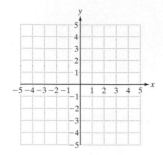

49. $x > (y - 2)^2 + 1$

$x - y < 1$

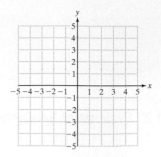

50. $x^2 + y^2 < 25$

$y < \dfrac{4}{3}x$

$y > -\dfrac{4}{3}x$

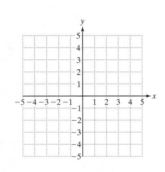

51. $y < e^x$

$y > 1$

$x < 2$

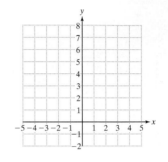

Group Activity

Investigating the Graphs of Conic Sections on a Calculator

Materials: Graphing Calculator

Estimated time: 15 minutes

Group Size: 3

1. a. Given $x^2 + y^2 = 25$, use analytical methods to determine the x- and y-intercepts.

x-intercepts: _____ y-intercepts: _____

   b. Solve the equation for y: $x^2 + y^2 = 25$. (*Hint:* When taking the square root of both sides of the equation, don't forget the $\pm$.)

$Y_1 = $ _____

$Y_2 = $ _____

   c. Graph the two functions from part (b) on a graphing calculator. Use a window defined by $-7.6 \le x \le 7.6$ and $-5 \le y \le 5$.

   d. Does it look like the functions graphed on the calculator intersect with the x-axis? Why do the x-intercepts not show up in the graph? If you relied solely on your calculator to find the x-intercepts, do you think that you would always get reliable results?

2. a. Given $(x + 2)^2 - \dfrac{y^2}{9} = 1$, use analytical methods to find the x- and y-intercepts.

x-intercepts: _____ y-intercepts: _____

   b. Solve the equation for y. $(x + 2)^2 - \dfrac{y^2}{9} = 1$

$Y_1 = $ _____

$Y_2 = $ _____

   c. Graph the two functions from part (b) on a graphing calculator. Use a window defined by $-7.6 \le x \le 7.6$ and $-5 \le y \le 5$.

   d. Does it look like the functions graphed on the calculator intersect with the x-axis? Why do the x-intercepts not show up in the graph? If you relied solely on your calculator to find the x-intercepts, do you think that you would always get reliable results?

3. Solve the equation for y and graph the functions on a graphing calculator. Use a window defined by $-7.6 \le x \le 7.6$ and $-5 \le y \le 5$.

   a. $\dfrac{x^2}{4} + \dfrac{y^2}{16} = 1$ $Y_1 = $ _____

$Y_2 = $ _____

   b. What are the x-intercepts for the equation $\dfrac{x^2}{4} + \dfrac{y^2}{16} = 1$?

x-intercepts: _____

Chapter 10   Summary

Section 10.1   Distance Formula, Midpoint Formula, and Circles

Key Concepts

The **distance between two points** (x_1, y_1) and (x_2, y_2) is

$$d = \sqrt{(x_2 - x_1)^2 + (y_2 - y_1)^2}$$

The standard equation of a **circle** with center (h, k) and radius r is

$$(x - h)^2 + (y - k)^2 = r^2$$

The midpoint between two points is found by using the formula:

$$\left(\frac{x_1 + x_2}{2}, \frac{y_1 + y_2}{2} \right)$$

Examples

Example 1

Find the distance between two points.

$(5, -2)$ and $(-1, -6)$

$$\begin{aligned}
d &= \sqrt{(-1 - 5)^2 + [-6 - (-2)]^2} \\
&= \sqrt{(-6)^2 + (-4)^2} \\
&= \sqrt{36 + 16} \\
&= \sqrt{52} = 2\sqrt{13}
\end{aligned}$$

Example 2

Find the center and radius of the circle.

$$x^2 + y^2 - 8x + 6y = 0$$

$$(x^2 - 8x + 16) + (y^2 + 6y + 9) = 16 + 9$$

$$(x - 4)^2 + (y + 3)^2 = 25$$

The center is $(4, -3)$ and the radius is 5.

Example 3

Find the midpoint between $(-3, 1)$ and $(5, 7)$.

$$\left(\frac{-3 + 5}{2}, \frac{1 + 7}{2} \right) = (1, 4)$$

Section 10.2   More on the Parabola

Key Concepts

A **parabola** is the set of points in a plane that are equidistant from a fixed line (called the directrix) and a fixed point (called the focus) not on the directrix.

The standard form of an equation of a parabola with **vertex** (h, k) and vertical **axis of symmetry** is

$$y = a(x - h)^2 + k \qquad \text{where } a \neq 0$$

- The equation of the axis of symmetry is $x = h$.
- If $a > 0$, the parabola opens upward.
- If $a < 0$, the parabola opens downward.

The standard form of an equation of a parabola with vertex (h, k) and horizontal axis of symmetry is

$$x = a(y - k)^2 + h \qquad \text{where } a \neq 0$$

- The equation of the axis of symmetry is $y = k$.
- If $a > 0$, the parabola opens to the right.
- If $a < 0$, the parabola opens to the left.

Examples

Example 1

Given the parabola, $y = (x - 2)^2 + 1$

The vertex is $(2, 1)$.

The axis of symmetry is $x = 2$.

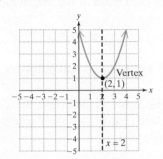

Example 2

Given the parabola $x = -\dfrac{1}{4}y^2 + 1$,

determine the coordinates of the vertex and the equation of the axis of symmetry.

$$x = -\frac{1}{4}(y - 0)^2 + 1$$

The vertex is $(1, 0)$.

The axis of symmetry is $y = 0$.

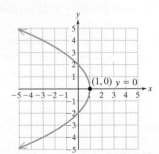

Section 10.3   The Ellipse and Hyperbola

Key Concepts

An **ellipse** is the set of all points (x, y) such that the sum of the distances between (x, y) and two distinct points (called foci) is constant.

Standard Form of an Ellipse with Center at the Origin

An ellipse with the center at the origin has the equation

$$\frac{x^2}{a^2} + \frac{y^2}{b^2} = 1$$

where a and b are positive real numbers.

For an ellipse centered at the origin, the x-intercepts are given by $(a, 0)$ and $(-a, 0)$, and the y-intercepts are given by $(0, b)$ and $(0, -b)$.

A **hyperbola** is the set of all points (x, y) such that the absolute value of the difference of the distances between (x, y) and two distinct points is a constant. The fixed points are called the foci of the hyperbola.

Standard Forms of an Equation of a Hyperbola

Let a and b represent positive real numbers.

Horizontal Transverse Axis. The standard form of an equation of a hyperbola with a horizontal transverse axis and center at the origin is given by

$$\frac{x^2}{a^2} - \frac{y^2}{b^2} = 1$$

Vertical Transverse Axis. The standard form of an equation of a hyperbola with a vertical transverse axis and center at the origin is given by

$$\frac{y^2}{b^2} - \frac{x^2}{a^2} = 1$$

Examples

Example 1

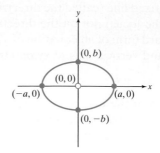

Example 2

$$\frac{x^2}{25} + \frac{y^2}{9} = 1$$

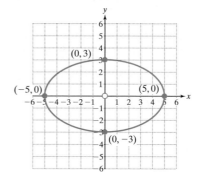

Example 3

$$\frac{x^2}{4} - \frac{y^2}{16} = 1$$

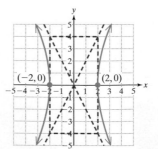

Example 4

$$\frac{y^2}{4} - \frac{x^2}{16} = 1$$

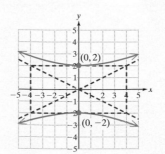

Section 10.4   Nonlinear Systems of Equations in Two Variables

Key Concepts

A **nonlinear system of equations** can be solved by graphing or by the substitution method.

$$2x^2 + y^2 = 15$$
$$x^2 - y = 0$$

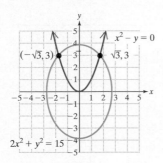

Examples

Example 1

$\boxed{A}$ $2x^2 + y^2 = 15$

$\boxed{B}$ $x^2 - y = 0$ Solve for y: $y = x^2$

$\boxed{A}$ $2x^2 + (x^2)^2 = 15$ Substitute in first equation.

$\qquad 2x^2 + x^4 = 15$

$\quad x^4 + 2x^2 - 15 = 0$

$\quad (x^2 + 5)(x^2 - 3) = 0$

$x^2 + 5 = 0$ or $x^2 - 3 = 0$

$\cancel{x^2 = -5}$ or $x^2 = 3$

$\qquad\qquad\qquad\qquad\qquad x = \pm\sqrt{3}$

If $x = \sqrt{3}$ then $y = (\sqrt{3})^2 = 3$.

If $x = -\sqrt{3}$ then $y = (-\sqrt{3})^2 = 3$.

Points of intersection are $(\sqrt{3}, 3)$ and $(-\sqrt{3}, 3)$.

Section 10.5 Nonlinear Inequalities and Systems of Inequalities

Key Concepts

Graph a nonlinear inequality by using the test point method. That is, graph the related equation. Then choose test points in each region to determine where the inequality is true.

Examples

Example 1

$y \geq x^2$

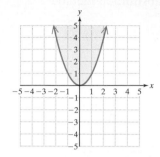

Example 2

$x^2 + y^2 < 4$

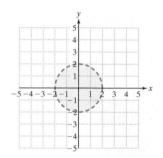

Graph a system of nonlinear inequalities by finding the intersection of the solution sets. That is, graph the solution set for each individual inequality, then take the intersection.

Example 3

$x \geq 0, \qquad y > x^2, \qquad$ and $\qquad x^2 + y^2 < 4$

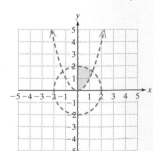

Chapter 10   Review Exercises

Section 10.1

For Exercises 1 and 2, find the distance between the two points by using the distance formula.

1. $(-6, 3)$ and $(0, 1)$ **2.** $(4, 13)$ and $(-1, 5)$

3. Find x such that $(x, 5)$ is 5 units from $(2, 9)$.

4. Find x such that $(-3, 4)$ is 3 units from $(x, 1)$.

For Exercises 5–8, find the center and the radius of the circle.

5. $(x - 12)^2 + (y - 3)^2 = 16$

6. $(x + 7)^2 + (y - 5)^2 = 81$

7. $(x + 3)^2 + (y + 8)^2 = 20$

8. $(x - 1)^2 + (y + 6)^2 = 32$

9. A stained glass window is in the shape of a circle with a 16-in. diameter. Find an equation of the circle relative to the origin for each of the following graphs.

a.   **b.**

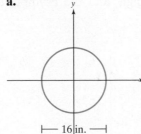   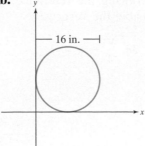

For Exercises 10–13, write the equation of the circle in standard form by completing the square.

10. $x^2 + y^2 + 12x - 10y + 51 = 0$

11. $x^2 + y^2 + 4x + 16y + 60 = 0$

12. $x^2 + y^2 - x - 4y + \dfrac{1}{4} = 0$

13. $x^2 + y^2 - 6x - \dfrac{2}{3}y + \dfrac{1}{9} = 0$

14. Write an equation of a circle with center at the origin and a diameter of 7 m.

15. Write an equation of a circle with center at $(0, 2)$ and a diameter of 6 m.

For Exercises 16 and 17, find the midpoint of the segment with the given endpoints.

16. $(-3, 1)$ and $(-5, -5)$ **17.** $(0, 9)$ and $(-2, 7)$

Section 10.2

For Exercises 18–21, determine whether the axis of symmetry is vertical or horizontal and if the parabola opens upward, downward, left, or right.

18. $y = -2(x - 3)^2 + 2$

19. $x = 3(y - 9)^2 + 1$

20. $x = -(y + 4)^2 - 8$

21. $y = (x + 3)^2 - 10$

For Exercises 22–25, determine the coordinates of the vertex and the equation of the axis of symmetry. Then use this information to graph the parabola.

22. $x = -(y - 1)^2$ **23.** $y = (x + 2)^2$

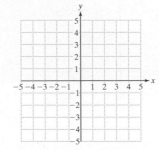   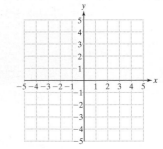

24. $y = -\dfrac{1}{4}x^2$ **25.** $x = 2y^2 - 1$

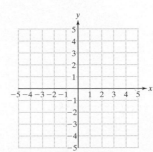   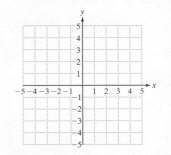

For Exercises 26–29, write the equation in standard form $y = a(x - h)^2 + k$ or $x = a(y - k)^2 + h$. Then identify the vertex and axis of symmetry.

26. $y = x^2 - 6x + 5$

27. $x = y^2 + 4y + 2$

28. $x = -4y^2 + 4y$

29. $y = -2x^2 - 2x$

33. $\dfrac{x^2}{25} + \dfrac{(y - 2)^2}{9} = 1$

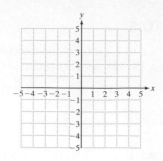

Section 10.3

For Exercises 30 and 31, identify the x- and y-intercepts. Then graph the ellipse.

30. $\dfrac{x^2}{9} + \dfrac{y^2}{25} = 1$

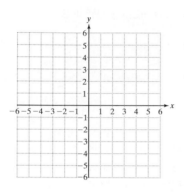

31. $x^2 + 4y^2 = 36$

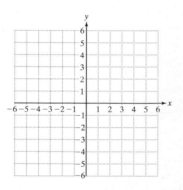

For Exercises 34–37, determine whether the transverse axis is horizontal or vertical.

34. $\dfrac{x^2}{12} - \dfrac{y^2}{16} = 1$

35. $\dfrac{y^2}{9} - \dfrac{x^2}{9} = 1$

36. $y^2 - 8x^2 = 16$

37. $3x^2 - y^2 = 18$

For Exercises 38 and 39, graph the hyperbola by first drawing the reference rectangle and the asymptotes. Label the vertices.

38. $\dfrac{x^2}{4} - y^2 = 1$

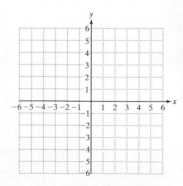

For Exercises 32 and 33, identify the center of the ellipse and graph the ellipse.

32. $\dfrac{(x - 5)^2}{4} + \dfrac{(y + 3)^2}{16} = 1$

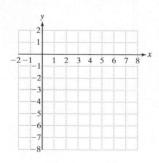

39. $y^2 - x^2 = 16$

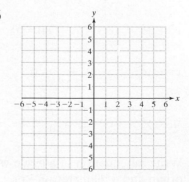

For Exercises 40–43, identify the equations as representing an ellipse or a hyperbola.

40. $\dfrac{x^2}{4} - \dfrac{y^2}{9} = 1$ **41.** $\dfrac{x^2}{16} + \dfrac{y^2}{9} = 1$

42. $\dfrac{x^2}{4} + \dfrac{y^2}{1} = 1$ **43.** $\dfrac{y^2}{1} - \dfrac{x^2}{16} = 1$

Section 10.4

For Exercises 44–47,

    a. Identify each equation as representing a line, a parabola, a circle, an ellipse, or a hyperbola.

    b. Graph both equations on the same coordinate system.

    c. Solve the system analytically and verify the answers from the graph.

44. $3x + 2y = 10$

    $y = x^2 - 5$

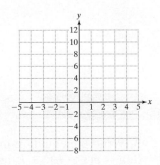

45. $4x + 2y = 10$

    $y = x^2 - 10$

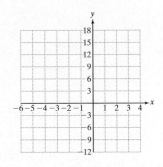

46. $x^2 + y^2 = 9$

    $2x + y = 3$

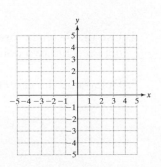

47. $x^2 + y^2 = 16$

    $x - 2y = 8$

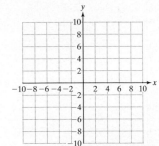

For Exercises 48–53, solve the system of nonlinear equations by using either the substitution method or the addition method.

48. $x^2 + 2y^2 = 8$ **49.** $x^2 + 4y^2 = 29$

    $2x - y = 2$ $x - y = -4$

50. $x - y = 4$ **51.** $y = x^2$

    $y^2 = 2x$ $6x^2 - y^2 = 8$

52. $x^2 + y^2 = 10$ **53.** $x^2 + y^2 = 61$

    $x^2 + 9y^2 = 18$ $x^2 - y^2 = 11$

Section 10.5

For Exercises 54–59, graph the solution set to the inequality.

54. $\dfrac{x^2}{16} + \dfrac{y^2}{81} < 1$ **55.** $\dfrac{x^2}{25} + \dfrac{y^2}{4} > 1$

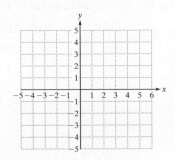
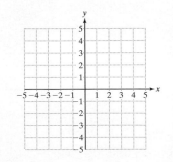

56. $(x - 3)^2 + (y + 1)^2 \geq 9$

57. $(x + 2)^2 + (y + 1)^2 \leq 4$

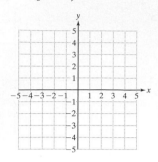

For Exercises 60 and 61, graph the solution set to the system of nonlinear inequalities.

60. $y > 2^x$

$x^2 + y^2 < 4$

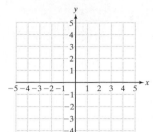

61. $y < x^2$

$x^2 + y^2 < 9$

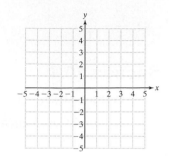

58. $y > (x - 1)^2$

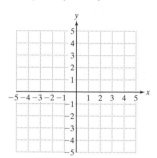

59. $x^2 - \dfrac{y^2}{4} \leq 1$

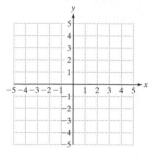

Chapter 10   Test

1. Determine the vertex and the equation of the axis of symmetry. Then graph the parabola.

$$x = -(y - 2)^2 + 3$$

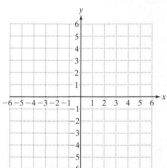

2. Determine the center and radius of the circle.

$$\left(x - \frac{5}{6}\right)^2 + \left(y + \frac{1}{3}\right)^2 = \frac{25}{49}$$

3. Write the equation in standard form $y = a(x - h)^2 + k$, and graph the parabola.

$$y = x^2 + 4x + 5$$

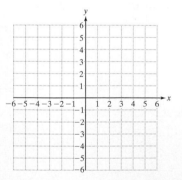

4. Use the distance formula to find the distance between the two points $(-5, 19)$ and $(-3, 13)$.

5. Determine the center and radius of the circle.

$$x^2 + y^2 - 4y - 5 = 0$$

6. Let $(0, 4)$ be the center of a circle that passes through the point $(-2, 5)$.

    a. What is the radius of the circle?

    b. Write the equation of the circle in standard form.

7. Find the center of the circle that has a diameter with endpoints $(7.3, -1.2)$ and $(0.3, 5.1)$.

8. Graph the ellipse.

$$\frac{x^2}{16} + \frac{y^2}{49} = 1$$

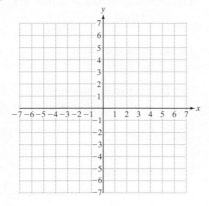

9. Graph the ellipse.

$$\frac{(x + 4)^2}{25} + (y - 3)^2 = 1$$

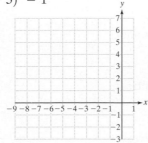

10. Graph the hyperbola.

$$y^2 - \frac{x^2}{4} = 1$$

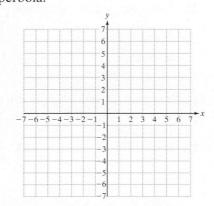

11. Solve each system and identify the correct graph.

    a. $16x^2 + 9y^2 = 144$ **b.** $x^2 + 4y^2 = 4$

        $4x - 3y = -12$ $4x - 3y = -12$

i.   ii.

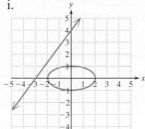   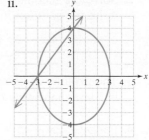

12. Describe the circumstances in which a nonlinear system of equations can be solved by using the addition method.

13. Solve the system by using either the substitution method or the addition method.

$$25x^2 + 4y^2 = 100$$
$$25x^2 - 4y^2 = 100$$

For Exercises 14–17, graph the solution set.

14. $x \le y^2 + 1$ **15.** $y \ge -\frac{1}{3}x + 1$

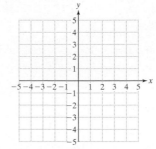   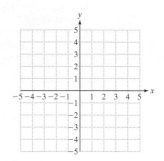

16. $x < y^2 + 1$ **17.** $y < \sqrt{x}$

    $y > -\frac{1}{3}x + 1$ $y > x - 2$

                           $x > 0$

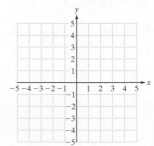   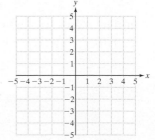

Chapters 1–10   Cumulative Review Exercises

1. Solve. $5(2y - 1) = 2y - 4 + 8y - 1$

2. Solve the inequality. Graph the solution and write the solution in interval notation.

$$4(x - 1) + 2 > 3x + 8 - 2x$$

3. The product of two integers is 150. If one integer is 5 less than twice the other, find the integers.

4. For $5y - 3x - 15 = 0$:

   a. Find the x- and y-intercepts.

   b. Find the slope.

   c. Graph the line.

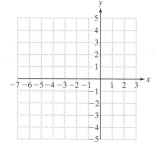

5. Find the slope and y-intercept of $3x - 4y = 6$ by first writing the equation in slope-intercept form.

6. A collection of dimes and quarters has a total value of \$2.45. If there are 17 coins, how many of each type are there?

7. Solve the system.

$$\begin{aligned} x + y \quad\;\; &= -1 \\ 2x \quad\;\; - z &= 3 \\ y + 2z &= -1 \end{aligned}$$

8. Solve the following system.

$$4x - 2y = 7$$
$$-3x + 5y = 0$$

9. Solve by using the Gauss-Jordan method.

$$3x - 4y = 6$$
$$x + 2y = 12$$

10. For $f(x) = 3x - x^2 - 12$, find the function values $f(0), f(-1), f(2)$, and $f(4)$.

11. For $g = \{(2, 5), (8, -1), (3, 0), (-5, 5)\}$ find the function values $g(2), g(8), g(3)$, and $g(-5)$.

12. The quantity z varies jointly as y and as the square of x. If z is 80 when x is 5 and y is 2, find z when $x = 2$ and $y = 5$.

13. For $f(x) = \sqrt{x + 1}$ and $g(x) = x^2 + 6$, find $(g \circ f)(x)$.

14. **a.** Find the value of the expression
$x^3 + x^2 + x + 1$ for $x = -2$.

   b. Factor the expression $x^3 + x^2 + x + 1$ and find the value when x is -2.

   c. Compare the values for parts (a) and (b).

15. Factor completely.

$$x^2 - y^2 - 6x - 6y$$

16. Multiply. $(x - 4)(x^2 + 2x + 1)$

17. Solve. $2x(x - 7) = x - 18$

18. Simplify. $\dfrac{3a^2 - a - 2}{3a^2 + 8a + 4}$

19. Subtract. $\dfrac{2}{x + 3} - \dfrac{x}{x - 2}$

20. Solve. $\dfrac{2}{x + 3} - \dfrac{x}{x - 2} = \dfrac{-4}{x^2 + x - 6}$

21. Solve the radical equations.

   a. $\sqrt{2x - 5} = -3$

   b. $\sqrt[3]{2x - 5} = -3$

For Exercises 22 and 23, perform the indicated operations with complex numbers.

22. $6i(4 + 5i)$

23. $\dfrac{3}{4 - 5i}$

24. Find the length of the missing side.

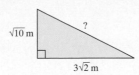

25. An automobile starts from rest and accelerates at a constant rate for 10 sec. The distance, $d(t)$, in feet traveled by the car is given by

$$d(t) = 4.4t^2$$

where $0 \leq t \leq 10$ is the time in seconds.

  a. How far has the car traveled after 2, 3, and 4 sec, respectively?

  b. How long will it take for the car to travel 281.6 ft?

26. Solve the equation $125w^3 + 1 = 0$ by factoring and using the quadratic formula. (*Hint:* You will find one real solution and two imaginary solutions.)

27. Solve. $\dfrac{x}{x + 2} - \dfrac{3}{x - 1} = \dfrac{1}{x^2 + x - 2}$

28. Find the coordinates of the vertex of the parabola defined by $f(x) = x^2 + 10x - 11$ by completing the square.

29. Graph the quadratic function defined by $g(x) = -x^2 - 2x + 3$.

  a. Label the x-intercepts.

  b. Label the y-intercept.

  c. Label the vertex.

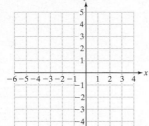

30. Solve the inequality and write the answer in interval notation.

$$|x - 9| - 3 < 7$$

31. Solve the inequality: $|2x - 5| \geq 4$

32. Write the expression in logarithmic form. $8^{5/3} = 32$

33. Solve. $5^2 = 125^x$

34. For $h(x) = x^3 - 1$, find $h^{-1}(x)$.

35. Write an equation representing the set of all points 4 units from the point $(0, 5)$.

36. Graph the ellipse. $\dfrac{x^2}{4} + \dfrac{y^2}{25} = 1$

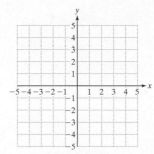

37. Find the center of the circle.

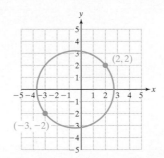

38. Solve the system of nonlinear equations.

$$x^2 + y^2 = 16$$
$$y = -x^2 - 4$$

39. Graph the solution set.

$$y^2 - x^2 < 1$$

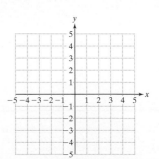

40. Graph the solution set to this system.

$$y > \left(\dfrac{1}{2}\right)^x$$

$$x < 0$$

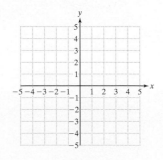

Additional Topics Appendix

Binomial Expansions

1. Binomial Expansions and Pascal's Triangle

In Section 5.3 we learned how to square a binomial.

$$(a + b)^2 = a^2 + 2ab + b^2$$

The expression $a^2 + 2ab + b^2$ is called the **binomial expansion** of $(a + b)^2$. To expand $(a + b)^3$, we can find the product $(a + b)(a + b)^2$.

$$(a + b)(a + b)^2 = (a + b)(a^2 + 2ab + b^2)$$

$$= a^3 + 2a^2b + ab^2 + a^2b + 2ab^2 + b^3$$
$$= a^3 + 3a^2b + 3ab^2 + b^3$$

Similarly, to expand $(a + b)^4$, we can multiply $(a + b)$ by $(a + b)^3$. Using this method, we can expand several powers of $(a + b)$ to find the following pattern:

$$(a + b)^0 = 1$$
$$(a + b)^1 = a + b$$
$$(a + b)^2 = a^2 + 2ab + b^2$$
$$(a + b)^3 = a^3 + 3a^2b + 3ab^2 + b^3$$
$$(a + b)^4 = a^4 + 4a^3b + 6a^2b^2 + 4ab^3 + b^4$$
$$(a + b)^5 = a^5 + 5a^4b + 10a^3b^2 + 10a^2b^3 + 5ab^4 + b^5$$

Notice that the exponents on a decrease from left to right, while the exponents on b increase from left to right. Also observe that for each term, the sum of the exponents is equal to the exponent to which $(a + b)$ is raised. Finally, notice that the number of terms in the expansion is exactly 1 more than the power to which $(a + b)$ is raised. For example, the expansion of $(a + b)^4$ has five terms, and the expansion of $(a + b)^5$ has six terms.

With these guidelines in mind, we know that $(a + b)^6$ will contain seven terms involving

$$a^6 \qquad a^5b \qquad a^4b^2 \qquad a^3b^3 \qquad a^2b^4 \qquad ab^5 \qquad b^6$$

We can complete the expansion of $(a + b)^6$ if we can determine the correct coefficients of each term.

If we write the coefficients for several expansions of $(a + b)^n$, where $n \geq 0$, we have a triangular array of numbers.

Concepts

1. Binomial Expansions and Pascal's Triangle
2. Factorial Notation
3. The Binomial Theorem

$$(a + b)^0 = 1$$
$$(a + b)^1 = 1a + 1b$$
$$(a + b)^2 = 1a^2 + 2ab + 1b^2$$
$$(a + b)^3 = 1a^3 + 3a^2b + 3ab^2 + 1b^3$$
$$(a + b)^4 = 1a^4 + 4a^3b + 6a^2b^2 + 4ab^3 + 1b^4$$
$$(a + b)^5 = 1a^5 + 5a^4b + 10a^3b^2 + 10a^2b^3 + 5ab^4 + 1b^5$$

```
          1
        1   1
      1   2   1
    1   3   3   1
  1   4   6   4   1
1   5   10   10   5   1
```

Each row begins and ends with a 1, and each entry in between is the sum of the two entries from the line above. For example, in the last row, $5 = 1 + 4$, $10 = 4 + 6$, and so on. This triangular array of coefficients for a binomial expansion is called **Pascal's triangle**, named after French mathematician Blaise Pascal (1623–1662).

```
            1
          1   1
        1   2   1
      1   3   3   1
    1   4   6   4   1
  1   5   10   10   5   1
1   6   15   20   15   6   1
```

By using the pattern shown in Pascal's triangle, the coefficients corresponding to $(a + b)^6$ would be 1, 6, 15, 20, 15, 6, 1. By inserting the coefficients, the sum becomes

$$(a + b)^6 = 1a^6 + 6a^5b + 15a^4b^2 + 20a^3b^3 + 15a^2b^4 + 6ab^5 + 1b^6$$

2. Factorial Notation

Although Pascal's triangle provides an easy method to find the coefficients of $(a + b)^n$, it is impractical for large values of n. A more efficient method to find the coefficients of a binomial expansion involves **factorial notation**.

> **Definition of n!**
>
> Let n be a positive integer. Then $n!$ (read as "n factorial") is defined as the product of integers from 1 through n. That is,
>
> $$n! = n(n - 1)(n - 2) \cdots (2)(1)$$
>
> *Note:* We define $0! = 1$.

Example 1   Evaluating Factorial Notation

Evaluate the expressions.

a. 4!   **b.** 10!   **c.** 0!

Solution:

a. $4! = 4 \cdot 3 \cdot 2 \cdot 1 = 24$

b. $10! = 10 \cdot 9 \cdot 8 \cdot 7 \cdot 6 \cdot 5 \cdot 4 \cdot 3 \cdot 2 \cdot 1 = 3,628,800$

c. $0! = 1$ by definition

Calculator Connections

Topic: Evaluating Expressions with Factorial Notation

Most calculators have a [!] function. Try evaluating the expressions from Example 1 on a calculator:

```
4!
                    24
10!
            3628800
0!
                     1
```

Sometimes factorial notation is used with other operations such as multiplication and division. It is also helpful to note that $n! = n(n-1)!$.

From the definition, $n! = n\underbrace{(n-1)(n-2)\cdots(2)(1)}_{(n-1)!}$.

Therefore, $n! = n(n-1)!$.

Example 2 Operations with Factorials

Evaluate the expressions.

a. $\dfrac{4!}{4! \cdot 0!}$ **b.** $\dfrac{4!}{3! \cdot 1!}$ **c.** $\dfrac{4!}{2! \cdot 2!}$ **d.** $\dfrac{4!}{1! \cdot 3!}$ **e.** $\dfrac{4!}{0! \cdot 4!}$

Skill Practice

Evaluate.

4. $\dfrac{5!}{5!0!}$ **5.** $\dfrac{5!}{4!1!}$ **6.** $\dfrac{5!}{3!2!}$

Solution:

a. $\dfrac{4!}{4! \cdot 0!} = \dfrac{4!}{4! \cdot 1} = 1$

b. $\dfrac{4!}{3! \cdot 1!} = \dfrac{4 \cdot 3!}{3! \cdot 1} = 4$

c. $\dfrac{4!}{2! \cdot 2!} = \dfrac{4 \cdot 3 \cdot 2!}{2! \cdot 2 \cdot 1} = \dfrac{12}{2} = 6$

d. $\dfrac{4!}{1! \cdot 3!} = \dfrac{4 \cdot 3!}{1 \cdot 3!} = 4$

e. $\dfrac{4!}{0! \cdot 4!} = \dfrac{4!}{1 \cdot 4!} = 1$

Calculator Connection

Topic: Multiplying and Dividing Factorial Expressions

To evaluate the expressions from Example 2 on a calculator, use parentheses around the factors in the denominator.

```
4!/(4!*0!)
                    1
4!/(3!*1!)
                    4
4!/(2!*2!)
                    6
```

3. The Binomial Theorem

Notice from Example 2 that the values of

$$\dfrac{4!}{4! \cdot 0!} \qquad \dfrac{4!}{3! \cdot 1!} \qquad \dfrac{4!}{2! \cdot 2!} \qquad \dfrac{4!}{1! \cdot 3!} \quad \text{and} \quad \dfrac{4!}{0! \cdot 4!}$$

Answers

4. 1 **5.** 5 **6.** 10

correspond to the values $1, 4, 6, 4, 1$, which are the coefficients for the expansion of $(a + b)^4$. Generalizing this pattern, we see the coefficients for the terms in the expansion of $(a + b)^n$ are given by

$$\frac{n!}{r! \cdot (n - r)!}$$

where r corresponds to the exponent on the factor of a and $(n - r)$ corresponds to the exponent on the factor of b. Using this formula for the coefficients in a binomial expansion results in the **binomial theorem**.

The Binomial Theorem

For any positive integer n,

$$(a + b)^n = \frac{n!}{n! \cdot 0!} a^n + \frac{n!}{(n - 1)! \cdot 1!} a^{(n-1)}b + \frac{n!}{(n - 2)! \cdot 2!} a^{(n-2)}b^2$$
$$+ \cdots + \frac{n!}{0! \cdot n!} b^n$$

Skill Practice

7. Write out the first three terms of $(x + y)^5$.

Example 3   Applying the Binomial Theorem

Write out the first three terms of the expansion of $(a + b)^{10}$.

Solution:

The first three terms of $(a + b)^{10}$ are

$$\frac{10!}{10! \cdot 0!} a^{10} + \frac{10!}{9! \cdot 1!} a^9 b + \frac{10!}{8! \cdot 2!} a^8 b^2$$

$$= \frac{\cancel{10!}}{\cancel{10!} \cdot 1} a^{10} + \frac{10 \cdot \cancel{9!}}{\cancel{9!} \cdot 1!} a^9 b + \frac{10 \cdot 9 \cdot \cancel{8!}}{\cancel{8!} \cdot 2 \cdot 1} a^8 b^2$$

$$= a^{10} + 10a^9 b + 45a^8 b^2$$

Skill Practice

8. Use the binomial theorem to find the expansion of $(2a - 3b^2)^4$.

Example 4   Applying the Binomial Theorem

Use the binomial theorem to find the expansion of $(3x^2 - 5y)^4$.

Solution:

Write $(3x^2 - 5y)^4$ as $[(3x^2) + (-5y)]^4$. In this case, the expressions $3x^2$ and $-5y$ may be substituted for a and b in the expansion of $(a + b)^4$.

TIP: The values of $\frac{n!}{r!(n - r)!}$ can also be found by using Pascal's triangle.

$$1$$
$$1 \quad 1$$
$$1 \quad 2 \quad 1$$
$$1 \quad 3 \quad 3 \quad 1$$
$$1 \quad 4 \quad 6 \quad 4 \quad 1$$

$$(a + b)^4 = \frac{4!}{4! \cdot 0!} a^4 + \frac{4!}{3! \cdot 1!} a^3 b + \frac{4!}{2! \cdot 2!} a^2 b^2 + \frac{4!}{1! \cdot 3!} ab^3 + \frac{4!}{0! \cdot 4!} b^4$$

$$= \frac{4!}{4! \cdot 0!} (3x^2)^4 + \frac{4!}{3! \cdot 1!} (3x^2)^3(-5y) + \frac{4!}{2! \cdot 2!} (3x^2)^2(-5y)^2$$

$$+ \frac{4!}{1! \cdot 3!} (3x^2)(-5y)^3 + \frac{4!}{0! \cdot 4!} (-5y)^4$$

$$= 1 \cdot (81x^8) + 4 \cdot (27x^6)(-5y) + 6 \cdot (9x^4)(25y^2) + 4 \cdot (3x^2)(-125y^3)$$
$$+ 1 \cdot (625y^4)$$

$$= 81x^8 - 540x^6 y + 1350x^4 y^2 - 1500x^2 y^3 + 625y^4$$

Answers

7. $x^5 + 5x^4 y + 10x^3 y^2$
8. $16a^4 - 96a^3 b^2 + 216a^2 b^4$
    $- 216ab^6 + 81b^8$

The binomial theorem may also be used to find a specific term in a binomial expansion.

Example 5   Finding a Specific Term in a Binomial Expansion

Find the fourth term of the expansion $(a + b)^{13}$.

Solution:

There are 14 terms in the expansion of $(a + b)^{13}$. The first term will have variable factors $a^{13}b^0$. The second term will have variable factors $a^{12}b^1$. The third term will have $a^{11}b^2$, and the fourth term will have $a^{10}b^3$. Hence, the fourth term is

$$\frac{13!}{10! \cdot 3!} a^{10}b^3 = 286a^{10}b^3$$

From Example 5, we see that for the kth term in the expansion $(a + b)^n$, where k is an integer greater than zero, the variable factors are $a^{n-(k-1)}$ and b^{k-1}. Therefore, to find the kth term of $(a + b)^n$, we can make the following generalization.

Finding a Specific Term in a Binomial Expansion

Let n and k be positive integers such that $k \le n$. Then the kth term in the expansion of $(a + b)^n$ is given by

$$\frac{n!}{[n - (k - 1)]! \cdot (k - 1)!} \cdot a^{n-(k-1)} \cdot b^{k-1}$$

Example 6   Finding a Specific Term in a Binomial Expansion

Find the sixth term of $(p^3 + 2w)^8$.

Solution:

Apply the formula.

$$\frac{n!}{[n - (k - 1)]! \cdot (k - 1)!} \cdot a^{n-(k-1)} \cdot b^{k-1} \quad \text{with } n = 8, k = 6, a = p^3, \text{ and } b = 2w$$

$$\frac{8!}{[8 - (6 - 1)]! \cdot (6 - 1)!} \cdot (p^3)^{8-(6-1)} \cdot (2w)^{6-1}$$

$$= \frac{8!}{(3)! \cdot (5)!} \cdot (p^3)^3 \cdot (2w)^5$$

$$= 56 \cdot (p^9) \cdot (32w^5)$$

$$= 1792p^9w^5$$

Section A.1   Practice Exercises

Vocabulary and Key Concepts

1. **a.** The expanded form of $(a + b)^2 =$ _____. The expanded form of $(a + b)^3 =$ _____.
    These are both called _____ expansions.

    b. Given a positive integer n, the value of $n!$ is _____. Furthermore, $n!$ is read
    as "n _____."

    c. $3! =$ _____, $2! =$ _____, $1! =$ _____, and $0! =$ _____.

    d. The _____ theorem states that for any positive integer n,
    $$(a + b)^n = \frac{n!}{n! \cdot 0!}a^n + \frac{n!}{(n-1)! \cdot 1!}a^{(n-1)}b + \frac{n!}{(n-2)! \cdot 2!}a^{(n-2)}b^2 + \cdots + \frac{n!}{0! \cdot n!}b^n.$$

    e. The coefficients of the binomial expansion of $(a + b)^n$ can also be found by using _____ triangle.

Concept 1: Binomial Expansions and Pascal's Triangle

For Exercises 2–9, expand the binomials. Use Pascal's triangle to find the coefficients.

2. $(x + y)^4$ 3. $(a + b)^3$ 4. $(4 + p)^3$ 5. $(1 + g)^4$

6. $(a^2 + b)^6$ 7. $(p + q^2)^7$ 8. $(t + 2)^5$ 9. $(s + t)^5$

For Exercises 10–13, rewrite each binomial of the form $(a - b)^n$ as $[a + (-b)]^n$. Then expand the binomials. Use Pascal's triangle to find the coefficients.

10. $(p^2 - w)^3$ 11. $(5 - u^3)^4$ 12. $(a - b^2)^4$ 13. $(x^2 - 4)^3$

14. For $a > 0$ and $b > 0$, what happens to the signs of the terms when expanding the binomial $(a - b)^n$ compared with $(a + b)^n$?

Concept 2: Factorial Notation

For Exercises 15–18, evaluate the expression. **(See Example 1.)**

15. $5!$ 16. $3!$ 17. $0!$ 18. $1!$

19. True or false: $0! \neq 1!$ 20. True or false: $n!$ is defined for negative integers.

21. True or false: $n! = n$ for $n = 1$ and 2.   22. Show that $9! = 9 \cdot 8!$

23. Show that $6! = 6 \cdot 5!$ 24. Show that $8! = 8 \cdot 7!$

For Exercises 25–32, evaluate the expression. **(See Example 2.)**

25. $\dfrac{8!}{4!}$ 26. $\dfrac{7!}{5!}$ 27. $\dfrac{3!}{0!}$ 28. $\dfrac{4!}{0!}$

29. $\dfrac{8!}{3! \, 5!}$ 30. $\dfrac{6!}{2! \, 4!}$ 31. $\dfrac{4!}{0! \, 4!}$ 32. $\dfrac{6!}{0! \, 6!}$

Concept 3: The Binomial Theorem

For Exercises 33–36, find the first three terms of the expansion. **(See Example 3.)**

33. $(m + n)^{11}$ 34. $(p + q)^9$ 35. $(u^2 - v)^{12}$ 36. $(r - s^2)^8$

37. How many terms are in the expansion of $(a + b)^8$?

38. How many terms are in the expansion of $(x + y)^{13}$?

For Exercises 39–50, use the binomial theorem to expand the binomials. **(See Example 4.)**

39. $(s + t)^6$

40. $(h + k)^4$

41. $(b - 3)^3$

42. $(c - 2)^5$

43. $(2x + y)^4$

44. $(p + 3q)^3$

45. $(c^2 - d)^7$

46. $(u - v^3)^6$

47. $\left(\dfrac{a}{2} - b\right)^5$

48. $\left(\dfrac{s}{3} + t\right)^5$

49. $(x + 4y)^4$

50. $(3y - w)^3$

For Exercises 51–56, find the indicated term of the binomial expansion. **(See Examples 5 and 6.)**

51. $(m - n)^{11}$; sixth term

52. $(p - q)^9$; fourth term

53. $(u^2 - v)^{12}$; fifth term

54. $(r - s^2)^8$; sixth term

55. $(5f + g)^9$; 10th term

56. $(4m + n)^{10}$; 11th term

Determinants and Cramer's Rule

1. Introduction to Determinants

Associated with every square matrix is a real number called the **determinant** of the matrix. A determinant of a square matrix $\mathbf{A}$, denoted $\det \mathbf{A}$, is written by enclosing the elements of the matrix within two vertical bars. For example,

$$\text{If} \quad \mathbf{A} = \begin{bmatrix} 2 & -1 \\ 6 & 0 \end{bmatrix} \quad \text{then} \quad \det \mathbf{A} = \begin{vmatrix} 2 & -1 \\ 6 & 0 \end{vmatrix}$$

$$\text{If} \quad \mathbf{B} = \begin{bmatrix} 0 & -5 & 1 \\ 4 & 0 & \frac{1}{2} \\ -2 & 10 & 1 \end{bmatrix} \quad \text{then} \quad \det \mathbf{B} = \begin{vmatrix} 0 & -5 & 1 \\ 4 & 0 & \frac{1}{2} \\ -2 & 10 & 1 \end{vmatrix}$$

Determinants have many applications in mathematics, including solving systems of linear equations, finding the area of a triangle, determining whether three points are collinear, and finding an equation of a line between two points.

The determinant of a 2×2 matrix is defined as follows:

Concepts

1. Introduction to Determinants
2. Determinant of a 3×3 Matrix
3. Cramer's Rule

Determinant of a 2 × 2 Matrix

The determinant of the matrix $\begin{bmatrix} a & b \\ c & d \end{bmatrix}$ is the real number $ad - bc$. It is written as

$$\begin{vmatrix} a & b \\ c & d \end{vmatrix} = ad - bc$$

Animation

Example 1   **Evaluating a 2 × 2 Determinant**

Evaluate the determinants.

a. $\begin{vmatrix} 6 & -2 \\ 5 & \frac{1}{3} \end{vmatrix}$ b. $\begin{vmatrix} 2 & -11 \\ 0 & 0 \end{vmatrix}$

Solution:

a. $\begin{vmatrix} 6 & -2 \\ 5 & \frac{1}{3} \end{vmatrix}$ For this determinant, $a = 6, b = -2, c = 5$, and $d = \frac{1}{3}$.

$$ad - bc = (6)\left(\frac{1}{3}\right) - (-2)(5)$$

$$= 2 + 10$$

$$= 12$$

b. $\begin{vmatrix} 2 & -11 \\ 0 & 0 \end{vmatrix}$ For this determinant, $a = 2, b = -11, c = 0, d = 0$.

$$ad - bc = (2)(0) - (-11)(0)$$

$$= 0 - 0$$

$$= 0$$

TIP: Example 1(b) illustrates that the value of a determinant having a row of all zeros is 0. The same is true for a determinant having a column of all zeros.

2. Determinant of a 3 × 3 Matrix

To find the determinant of a 3 × 3 matrix, we first need to define the **minor** of an element of the matrix. For any element of a 3 × 3 matrix, the minor of that element is the determinant of the 2 × 2 matrix obtained by deleting the row and column in which the element resides. For example, consider the matrix

$$\begin{bmatrix} 5 & -1 & 6 \\ 0 & -7 & 1 \\ 4 & 2 & 6 \end{bmatrix}$$

The minor of the element 5 is found by deleting the first row and first column and then evaluating the determinant of the remaining 2 × 2 matrix:

$$\begin{bmatrix} 5 & -1 & 6 \\ 0 & -7 & 1 \\ 4 & 2 & 6 \end{bmatrix}$$ Now evaluate the determinant: $\begin{vmatrix} -7 & 1 \\ 2 & 6 \end{vmatrix} = (-7)(6) - (1)(2)$

$$= -44$$

For this matrix, the minor for the element 5 is −44.

To find the minor of the element −7, delete the second row and second column, and then evaluate the determinant of the remaining 2 × 2 matrix.

$$\begin{bmatrix} 5 & -1 & 6 \\ 0 & -7 & 1 \\ 4 & 2 & 6 \end{bmatrix}$$ Now evaluate the determinant: $\begin{vmatrix} 5 & 6 \\ 4 & 6 \end{vmatrix} = (5)(6) - (6)(4) = 6$

For this matrix, the minor for the element −7 is 6.

Example 2 Determining the Minor for Elements in a 3 × 3 matrix

Find the minor for each element in the first column of the matrix.

$$\begin{bmatrix} 3 & 4 & -1 \\ 2 & -4 & 5 \\ 0 & 1 & -6 \end{bmatrix}$$

Solution:

For 3: $\begin{bmatrix} 3 & 4 & -1 \\ 2 & -4 & 5 \\ 0 & 1 & -6 \end{bmatrix}$ The minor is: $\begin{vmatrix} -4 & 5 \\ 1 & -6 \end{vmatrix} = (-4)(-6) - (5)(1) = 19$

For 2: $\begin{bmatrix} 3 & 4 & -1 \\ 2 & -4 & 5 \\ 0 & 1 & -6 \end{bmatrix}$ The minor is: $\begin{vmatrix} 4 & -1 \\ 1 & -6 \end{vmatrix} = (4)(-6) - (-1)(1) = -23$

For 0: $\begin{bmatrix} 3 & 4 & -1 \\ 2 & -4 & 5 \\ 0 & 1 & -6 \end{bmatrix}$ The minor is $\begin{vmatrix} 4 & -1 \\ -4 & 5 \end{vmatrix} = (4)(5) - (-1)(-4) = 16$

The determinant of a 3 × 3 matrix is defined as follows.

Determinant of a 3 × 3 Matrix

$$\begin{vmatrix} a_1 & b_1 & c_1 \\ a_2 & b_2 & c_2 \\ a_3 & b_3 & c_3 \end{vmatrix} = a_1 \cdot \begin{vmatrix} b_2 & c_2 \\ b_3 & c_3 \end{vmatrix} - a_2 \cdot \begin{vmatrix} b_1 & c_1 \\ b_3 & c_3 \end{vmatrix} + a_3 \cdot \begin{vmatrix} b_1 & c_1 \\ b_2 & c_2 \end{vmatrix}$$

From this definition, we see that the determinant of a 3 × 3 matrix can be written as

$$a_1 \cdot (\text{minor of } a_1) - a_2 \cdot (\text{minor of } a_2) + a_3 \cdot (\text{minor of } a_3)$$

Evaluating determinants in this way is called *expanding minors*.

Example 3 Evaluating a 3 × 3 Determinant

Evaluate the determinant.

$$\begin{vmatrix} 2 & 4 & 2 \\ 1 & -3 & 0 \\ -5 & 5 & -1 \end{vmatrix}$$

Solution:

$$\begin{vmatrix} 2 & 4 & 2 \\ 1 & -3 & 0 \\ -5 & 5 & -1 \end{vmatrix} = 2 \cdot \begin{vmatrix} -3 & 0 \\ 5 & -1 \end{vmatrix} - (1) \cdot \begin{vmatrix} 4 & 2 \\ 5 & -1 \end{vmatrix} + (-5) \cdot \begin{vmatrix} 4 & 2 \\ -3 & 0 \end{vmatrix}$$

$$= 2[(-3)(-1) - (0)(5)] - 1[(4)(-1) - (2)(5)] - 5[(4)(0) - (2)(-3)]$$

$$= 2(3) - 1(-14) - 5(6)$$

$$= 6 + 14 - 30$$

$$= -10$$

Animation

TIP: There is another method to determine the signs for each term of the expansion. For the a_{ij} element, multiply the term by $(-1)^{i+j}$.

Although we defined the determinant of a matrix by expanding the minors of the elements in the first column, *any row or column can be used.* However, we must choose the correct sign to apply to each term in the expansion. The following array of signs is helpful.

$$\begin{array}{ccc} + & - & + \\ - & + & - \\ + & - & + \end{array}$$

The signs alternate for each row and column, beginning with $+$ in the first row, first column.

Skill Practice

Evaluate the determinant.

5. $\begin{vmatrix} 4 & -1 & 2 \\ 3 & 6 & -8 \\ 0 & \frac{1}{2} & 5 \end{vmatrix}$

Example 4 Evaluating a 3 × 3 Determinant

Evaluate the determinant, by expanding minors about the elements in the second row.

$$\begin{vmatrix} 2 & 4 & 2 \\ 1 & -3 & 0 \\ -5 & 5 & -1 \end{vmatrix}$$

Solution:

Signs obtained from the array of signs

$$\begin{vmatrix} 2 & 4 & 2 \\ 1 & -3 & 0 \\ -5 & 5 & -1 \end{vmatrix} = -(1) \cdot \begin{vmatrix} 4 & 2 \\ 5 & -1 \end{vmatrix} + (-3) \cdot \begin{vmatrix} 2 & 2 \\ -5 & -1 \end{vmatrix} - (0) \cdot \begin{vmatrix} 2 & 4 \\ -5 & 5 \end{vmatrix}$$

$$= -1[(4)(-1) - (2)(5)] - 3[(2)(-1) - (2)(-5)] - 0$$

$$= -1(-14) - 3(8)$$

$$= 14 - 24$$

$$= -10$$

Notice that the value of the determinant obtained in Examples 3 and 4 is the same.

Calculator Connections

Topic: Evaluating a Determinant

The determinant of a matrix can be evaluated on a graphing calculator. First use the matrix editor to enter the elements of the matrix. Then use a *det* function to evaluate the determinant. The determinant from Examples 3 and 4 is evaluated below.

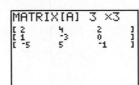
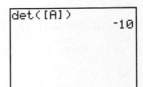

Answer

5. 154

In Example 4, the third term in the expansion of minors was zero because the element 0 when multiplied by its minor is zero. To simplify the arithmetic in evaluating a determinant of a 3×3 matrix, expand about the row or column that has the most 0 elements.

3. Cramer's Rule

In Sections 4.2, 4.3, and 4.6, we learned three methods to solve a system of linear equations: the substitution method, the addition method, and the Gauss-Jordan method. In this section, we will learn another method called **Cramer's rule** to solve a system of linear equations.

Cramer's Rule for a 2 × 2 System of Linear Equations

The solution to the system
$$a_1 x + b_1 y = c_1$$
$$a_2 x + b_2 y = c_2$$

is given by $\quad x = \dfrac{\mathbf{D}_x}{\mathbf{D}} \quad$ and $\quad y = \dfrac{\mathbf{D}_y}{\mathbf{D}}$

where $\mathbf{D} = \begin{vmatrix} a_1 & b_1 \\ a_2 & b_2 \end{vmatrix}$ (and $\mathbf{D} \neq 0$) $\quad \mathbf{D}_x = \begin{vmatrix} c_1 & b_1 \\ c_2 & b_2 \end{vmatrix} \quad \mathbf{D}_y = \begin{vmatrix} a_1 & c_1 \\ a_2 & c_2 \end{vmatrix}$

Animation

Example 5   Using Cramer's Rule to Solve a 2 × 2 System of Linear Equations

Solve the system by using Cramer's rule.

$$3x - 5y = 11$$

$$-x + 3y = -5$$

Solution:

For this system: $\quad a_1 = 3 \qquad b_1 = -5 \qquad c_1 = 11$

$$a_2 = -1 \qquad b_2 = 3 \qquad c_2 = -5$$

$$\mathbf{D} = \begin{vmatrix} 3 & -5 \\ -1 & 3 \end{vmatrix} = (3)(3) - (-5)(-1) = 9 - 5 = 4$$

$$\mathbf{D}_x = \begin{vmatrix} 11 & -5 \\ -5 & 3 \end{vmatrix} = (11)(3) - (-5)(-5) = 33 - 25 = 8$$

$$\mathbf{D}_y = \begin{vmatrix} 3 & 11 \\ -1 & -5 \end{vmatrix} = (3)(-5) - (11)(-1) = -15 + 11 = -4$$

Therefore, $\quad x = \dfrac{\mathbf{D}_x}{\mathbf{D}} = \dfrac{8}{4} = 2 \qquad y = \dfrac{\mathbf{D}_y}{\mathbf{D}} = \dfrac{-4}{4} = -1$

The solution is $(2, -1)$. <u>Check:</u> $3x - 5y = 11 \longrightarrow 3(2) - 5(-1) \overset{?}{=} 11$ ✔

$$-x + 3y = -5 \longrightarrow -(2) + 3(-1) \overset{?}{=} -5 \; ✔$$

Skill Practice

Solve using Cramer's rule.

6. $2x + y = 5$
      $-x - 3y = 5$

Answer

6. $(4, -3)$

TIP: Here are some memory tips to help you remember Cramer's rule.

Coefficients of
x-terms   y-terms

1. The determinant **D** is the determinant of the coefficients of x and y.

$$\mathbf{D} = \begin{vmatrix} a_1 & b_1 \\ a_2 & b_2 \end{vmatrix}$$

x-coefficients
replaced by c_1 and c_2

2. The determinant $\mathbf{D}_x$ has the column of x-term coefficients replaced by c_1 and c_2.

$$\mathbf{D}_x = \begin{vmatrix} c_1 & b_1 \\ c_2 & b_2 \end{vmatrix}$$

y-coefficients
replaced by c_1 and c_2

3. The determinant $\mathbf{D}_y$ has the column of y-term coefficients replaced by c_1 and c_2.

$$\mathbf{D}_y = \begin{vmatrix} a_1 & c_1 \\ a_2 & c_2 \end{vmatrix}$$

It is important to note that the linear equations must be written in standard form to apply Cramer's rule.

Skill Practice

Solve by using Cramer's rule.

7. $9x = 12y - 8$
$30y = -18x - 7$

Example 6 Using Cramer's Rule to Solve a 2 × 2 System of Linear Equations

Solve the system by using Cramer's rule.

$$-16y = -40x - 7$$
$$40y = 24x + 27$$

Solution:

$$\begin{array}{ll} -16y = -40x - 7 & \longrightarrow \quad 40x - 16y = -7 \\ 40y = 24x + 27 & \longrightarrow \quad -24x + 40y = 27 \end{array} \quad \text{Rewrite each equation in standard form.}$$

For this system: $a_1 = 40 \qquad b_1 = -16 \qquad c_1 = -7$
$a_2 = -24 \qquad b_2 = 40 \qquad c_2 = 27$

$$\mathbf{D} = \begin{vmatrix} 40 & -16 \\ -24 & 40 \end{vmatrix} = (40)(40) - (-16)(-24) = 1216$$

$$\mathbf{D}_x = \begin{vmatrix} -7 & -16 \\ 27 & 40 \end{vmatrix} = (-7)(40) - (-16)(27) = 152$$

$$\mathbf{D}_y = \begin{vmatrix} 40 & -7 \\ -24 & 27 \end{vmatrix} = (40)(27) - (-7)(-24) = 912$$

Therefore, $x = \dfrac{\mathbf{D}_x}{\mathbf{D}} = \dfrac{152}{1216} = \dfrac{1}{8} \qquad y = \dfrac{\mathbf{D}_y}{\mathbf{D}} = \dfrac{912}{1216} = \dfrac{3}{4}$

The solution $\left(\frac{1}{8}, \frac{3}{4}\right)$ checks in the original equations.

Answer

7. $\left(-\dfrac{2}{3}, \dfrac{1}{6}\right)$

Cramer's rule can be used to solve a 3×3 system of linear equations by using a similar pattern of determinants.

Cramer's Rule for a 3 × 3 System of Linear Equations

The solution to the system

$$a_1x + b_1y + c_1z = d_1$$

$$a_2x + b_2y + c_2z = d_2$$

$$a_3x + b_3y + c_3z = d_3$$

is given by $\qquad x = \dfrac{\mathbf{D}_x}{\mathbf{D}} \qquad y = \dfrac{\mathbf{D}_y}{\mathbf{D}} \qquad$ and $\qquad z = \dfrac{\mathbf{D}_z}{\mathbf{D}}$

where $\mathbf{D} = \begin{vmatrix} a_1 & b_1 & c_1 \\ a_2 & b_2 & c_2 \\ a_3 & b_3 & c_3 \end{vmatrix}$ (and $\mathbf{D} \neq 0$) $\qquad \mathbf{D}_x = \begin{vmatrix} d_1 & b_1 & c_1 \\ d_2 & b_2 & c_2 \\ d_3 & b_3 & c_3 \end{vmatrix}$

$$\mathbf{D}_y = \begin{vmatrix} a_1 & d_1 & c_1 \\ a_2 & d_2 & c_2 \\ a_3 & d_3 & c_3 \end{vmatrix} \qquad\qquad \mathbf{D}_z = \begin{vmatrix} a_1 & b_1 & d_1 \\ a_2 & b_2 & d_2 \\ a_3 & b_3 & d_3 \end{vmatrix}$$

Example 7   Using Cramer's Rule to Solve a 3 × 3 System of Linear Equations

Solve the system by using Cramer's rule.

$$x - 2y + 4z = 3$$

$$x - 4y + 3z = -5$$

$$x + 3y - 2z = 6$$

Solution:

$$\mathbf{D} = \begin{vmatrix} 1 & -2 & 4 \\ 1 & -4 & 3 \\ 1 & 3 & -2 \end{vmatrix} = 1 \cdot \begin{vmatrix} -4 & 3 \\ 3 & -2 \end{vmatrix} - 1 \cdot \begin{vmatrix} -2 & 4 \\ 3 & -2 \end{vmatrix} + 1 \cdot \begin{vmatrix} -2 & 4 \\ -4 & 3 \end{vmatrix}$$

$$= 1(-1) - 1(-8) + 1(10)$$

$$= 17$$

$$\mathbf{D}_x = \begin{vmatrix} 3 & -2 & 4 \\ -5 & -4 & 3 \\ 6 & 3 & -2 \end{vmatrix} = 3 \cdot \begin{vmatrix} -4 & 3 \\ 3 & -2 \end{vmatrix} - (-5) \cdot \begin{vmatrix} -2 & 4 \\ 3 & -2 \end{vmatrix} + 6 \cdot \begin{vmatrix} -2 & 4 \\ -4 & 3 \end{vmatrix}$$

$$= 3(-1) + 5(-8) + 6(10)$$

$$= 17$$

$$\mathbf{D}_y = \begin{vmatrix} 1 & 3 & 4 \\ 1 & -5 & 3 \\ 1 & 6 & -2 \end{vmatrix} = 1 \cdot \begin{vmatrix} -5 & 3 \\ 6 & -2 \end{vmatrix} - 1 \cdot \begin{vmatrix} 3 & 4 \\ 6 & -2 \end{vmatrix} + 1 \cdot \begin{vmatrix} 3 & 4 \\ -5 & 3 \end{vmatrix}$$

$$= 1(-8) - 1(-30) + 1(29)$$

$$= 51$$

Skill Practice

Solve by using Cramer's rule.

8. $x + 3y - 3z = -14$
$\quad\ x - 4y + \ z = 2$
$\quad\ x + \ \ y + 2z = 6$

TIP: In Example 7, we expanded the determinants about the first column.

Answer

8. $(-2, 0, 4)$

$$D_z = \begin{vmatrix} 1 & -2 & 3 \\ 1 & -4 & -5 \\ 1 & 3 & 6 \end{vmatrix} = 1 \cdot \begin{vmatrix} -4 & -5 \\ 3 & 6 \end{vmatrix} - 1 \cdot \begin{vmatrix} -2 & 3 \\ 3 & 6 \end{vmatrix} + 1 \cdot \begin{vmatrix} -2 & 3 \\ -4 & -5 \end{vmatrix}$$

$$= 1(-9) - 1(-21) + 1(22)$$

$$= 34$$

Hence,

$$x = \frac{D_x}{D} = \frac{17}{17} = 1 \qquad y = \frac{D_y}{D} = \frac{51}{17} = 3 \quad \text{and} \quad z = \frac{D_z}{D} = \frac{34}{17} = 2$$

The solution is $(1, 3, 2)$.

$$\underline{\text{Check:}} \quad \begin{aligned} x - 2y + 4z &= 3 & (1) - 2(3) + 4(2) &\overset{?}{=} 3 \quad \checkmark \\ x - 4y + 3z &= -5 & (1) - 4(3) + 3(2) &\overset{?}{=} -5 \quad \checkmark \\ x + 3y - 2z &= 6 & (1) + 3(3) - 2(2) &\overset{?}{=} 6 \quad \checkmark \end{aligned}$$

Cramer's rule may seem cumbersome for solving a 3×3 system of linear equations. However, it provides convenient formulas that can be programmed into a computer or calculator to solve for x, y, and z. Cramer's rule can also be extended to solve a 4×4 system of linear equations, a 5×5 system of linear equations, and in general an $n \times n$ system of linear equations.

It is important to remember that Cramer's rule does not apply if $D = 0$. In such a case, either the equations are dependent or the system is inconsistent, and another method may be needed to analyze the system.

Example 8   Analyzing a System of Dependent Equations

Solve the system. Use Cramer's rule if possible.

$$2x - 3y = 6$$
$$-6x + 9y = -18$$

Solution:

$$D = \begin{vmatrix} 2 & -3 \\ -6 & 9 \end{vmatrix} = (2)(9) - (-3)(-6) = 18 - 18 = 0$$

Because $D = 0$, Cramer's rule does not apply. Using the addition method to solve the system, we have

$$\begin{aligned} 2x - 3y &= 6 & \xrightarrow{\text{Multiply by 3.}} & & 6x - 9y &= 18 \\ -6x + 9y &= -18 & \longrightarrow & & \underline{-6x + 9y} &= \underline{-18} \\ & & & & 0 &= 0 \end{aligned} \quad \text{The equations are dependent.}$$

The solution is $\{(x, y) \mid 2x - 3y = 6\}$.

Skill Practice

Solve. Use Cramer's rule if possible.

9. $x - 6y = 2$
   $2x - 12y = -2$

TIP: When Cramer's rule does not apply, that is, when $D = 0$, you may also use the substitution method or the Gauss-Jordan method to get a solution.

TIP: In a 2×2 system of equations, if $D = 0$, then the equations are dependent or the system is inconsistent.

- If $D = 0$ and both $D_x = 0$ and $D_y = 0$, then the equations are dependent and the system has infinitely many solutions.
- If $D = 0$ and either $D_x \neq 0$ or $D_y \neq 0$, then the system is inconsistent and has no solution.

Answer

9. No solution (inconsistent system)

Section A.2 Practice Exercises

Vocabulary and Key Concepts

1. **a.** Given the matrix $\mathbf{A} = \begin{bmatrix} a & b \\ c & d \end{bmatrix}$, the _____ of $\mathbf{A}$ is denoted det $\mathbf{A}$ and is written as $\begin{vmatrix} a & b \\ c & d \end{vmatrix}$. The value of det $\mathbf{A}$ is the real number equal to _____.

   b. Given a 3×3 matrix, the _____ of an element in the matrix is the determinant of the 2×2 matrix formed by deleting the row and column in which the element resides.

   c. Complete the expression on the right to represent the value of the 3×3 determinant shown here.

   $$\begin{vmatrix} a_1 & b_1 & c_1 \\ a_2 & b_2 & c_2 \\ a_3 & b_3 & c_3 \end{vmatrix} = a_1 \begin{vmatrix} b_2 & c_2 \\ b_3 & c_3 \end{vmatrix} - \square \begin{vmatrix} b_1 & c_1 \\ b_3 & c_3 \end{vmatrix} + a_3 \begin{vmatrix} \square & \square \\ \square & \square \end{vmatrix}$$

Concept 1: Introduction to Determinants

For Exercises 2–7, evaluate the determinant of the 2×2 matrix. **(See Example 1.)**

2. $\begin{vmatrix} -3 & 1 \\ 5 & 2 \end{vmatrix}$ **3.** $\begin{vmatrix} 5 & 6 \\ 4 & 8 \end{vmatrix}$ **4.** $\begin{vmatrix} -2 & 2 \\ -3 & -5 \end{vmatrix}$

5. $\begin{vmatrix} 5 & -1 \\ 1 & 0 \end{vmatrix}$ **6.** $\begin{vmatrix} \frac{1}{2} & 3 \\ -2 & 4 \end{vmatrix}$ **7.** $\begin{vmatrix} -3 & \frac{1}{4} \\ 8 & -2 \end{vmatrix}$

Concept 2: Determinant of a 3 × 3 Matrix

For Exercises 8–11, evaluate the minor corresponding to the given element from matrix $\mathbf{A}$. **(See Example 2.)**

$$\mathbf{A} = \begin{bmatrix} 4 & -1 & 8 \\ 2 & 6 & 0 \\ -7 & 5 & 3 \end{bmatrix}$$

8. 4   **9.** -1 **10.** 2   **11.** 3

For Exercises 12–15, evaluate the minor corresponding to the given element from matrix $\mathbf{B}$.

$$\mathbf{B} = \begin{bmatrix} -2 & 6 & 0 \\ 4 & -2 & 1 \\ 5 & 9 & -1 \end{bmatrix}$$

12. 6   **13.** 5   **14.** 1   **15.** 0

16. Construct the sign array for a 3×3 matrix.

17. Evaluate the determinant of matrix $\mathbf{B}$, using expansion by minors. **(See Exercises 3 and 4.)**

$$\mathbf{B} = \begin{bmatrix} 0 & 1 & 2 \\ 3 & -1 & 2 \\ 3 & 2 & -2 \end{bmatrix}$$

    a. About the first column

    b. About the second row

18. Evaluate the determinant of matrix $\mathbf{C}$, using expansion by minors.

$$\mathbf{C} = \begin{bmatrix} 4 & 1 & 3 \\ 2 & -2 & 1 \\ 3 & 1 & 2 \end{bmatrix}$$

    a. About the first row

    b. About the second column

19. When evaluating the determinant of a 3×3 matrix, explain the advantage of being able to choose any row or column about which to expand minors.

For Exercises 20–25, evaluate the determinants. **(See Examples 3 and 4.)**

20. $\begin{vmatrix} 8 & 2 & -4 \\ 4 & 0 & 2 \\ 3 & 0 & -1 \end{vmatrix}$

21. $\begin{vmatrix} 5 & 2 & 1 \\ 3 & -6 & 0 \\ -2 & 8 & 0 \end{vmatrix}$

22. $\begin{vmatrix} -2 & 1 & 3 \\ 1 & 4 & 4 \\ 1 & 0 & 2 \end{vmatrix}$

23. $\begin{vmatrix} 3 & 2 & 1 \\ 1 & -1 & 2 \\ 1 & 0 & 4 \end{vmatrix}$

24. $\begin{vmatrix} -5 & 4 & 2 \\ 0 & 0 & 0 \\ 3 & -1 & 5 \end{vmatrix}$

25. $\begin{vmatrix} 0 & 5 & -8 \\ 0 & -4 & 1 \\ 0 & 3 & 6 \end{vmatrix}$

For Exercises 26–31, evaluate the determinants.

26. $\begin{vmatrix} x & 3 \\ y & -2 \end{vmatrix}$

27. $\begin{vmatrix} a & 2 \\ b & 8 \end{vmatrix}$

28. $\begin{vmatrix} a & 5 & -1 \\ b & -3 & 0 \\ c & 3 & 4 \end{vmatrix}$

29. $\begin{vmatrix} x & 0 & 3 \\ y & -2 & 6 \\ z & -1 & 1 \end{vmatrix}$

30. $\begin{vmatrix} p & 0 & q \\ r & 0 & s \\ t & 0 & u \end{vmatrix}$

31. $\begin{vmatrix} f & e & 0 \\ d & c & 0 \\ b & a & 0 \end{vmatrix}$

Concept 3: Cramer's Rule

For Exercises 32–34, evaluate the determinants represented by $\mathbf{D}$, $\mathbf{D}_x$, and $\mathbf{D}_y$.

32. $x - 4y = 2$
$3x + 2y = 1$

33. $4x + 6y = 9$
$-2x + y = 12$

34. $-3x + 8y = -10$
$5x + 5y = -13$

For Exercises 35–40, solve the system by using Cramer's rule. **(See Examples 5 and 6.)**

35. $2x + y = 3$
$x - 4y = 6$

36. $2x - y = -1$
$3x + y = 6$

37. $4y = x - 8$
$3x = -7y + 5$

38. $7x - 4 = -3y$
$5x = 4y + 9$

39. $4x - 3y = 5$
$2x + 5y = 7$

40. $2x + 3y = 4$
$6x - 12y = -5$

41. When does Cramer's rule not apply in solving a system of equations?

42. How can a system be solved if Cramer's rule does not apply?

For Exercises 43–48, solve the system of equations by using Cramer's rule, if possible. If not possible, use another method. **(See Example 8.)**

43. $4x - 2y = 3$
$-2x + y = 1$

44. $6x - 6y = 5$
$x - y = 8$

45. $4x + y = 0$
$x - 7y = 0$

46. $-3x - 2y = 0$
$-x + 5y = 0$

47. $x + 5y = 3$
$2x + 10y = 6$

48. $-2x - 10y = -4$
$x + 5y = 2$

For Exercises 49–54, solve for the indicated variable by using Cramer's rule. **(See Example 7.)**

49. $2x - y + 3z = 9$

    $x + 4y + 4z = 5$ for x

    $3x + 2y + 2z = 5$

50. $x + 2y + 3z = 8$

    $2x - 3y + z = 5$ for y

    $3x - 4y + 2z = 9$

51. $3x - 2y + 2z = 5$

    $6x + 3y - 4z = -1$ for z

    $3x - y + 2z = 4$

52. $4x + 4y - 3z = 3$

    $8x + 2y + 3z = 0$ for x

    $4x - 4y + 6z = -3$

53. $5x + 6z = 5$

    $-2x + y = -6$ for y

    $3y - z = 3$

54. $8x + y = 1$

    $7y + z = 0$ for y

    $x - 3z = -2$

For Exercises 55–58, solve the system by using Cramer's rule, if possible.

55. $x = 3$

    $-x + 3y = 3$

    $y + 2z = 4$

56. $4x + z = 7$

    $y = 2$

    $x + z = 4$

57. $x + y + 8z = 3$

    $2x + y + 11z = 4$

    $x + 3z = 0$

58. $-8x + y + z = 6$

    $2x - y + z = 3$

    $3x - z = 0$

Expanding Your Skills

For Exercises 59–62, solve the equation.

59. $\begin{vmatrix} 6 & x \\ 2 & -4 \end{vmatrix} = 14$

60. $\begin{vmatrix} y & -2 \\ 8 & 7 \end{vmatrix} = 30$

61. $\begin{vmatrix} 3 & 1 & 0 \\ 0 & 4 & -2 \\ 1 & 0 & w \end{vmatrix} = 10$

62. $\begin{vmatrix} -1 & 0 & 2 \\ 4 & t & 0 \\ 0 & -5 & 3 \end{vmatrix} = -4$

For Exercises 63 and 64, evaluate the determinant by using expansion by minors about the first column.

63. $\begin{vmatrix} 1 & 0 & 3 & 0 \\ 0 & 1 & 2 & 4 \\ -2 & 0 & 0 & 1 \\ 4 & -1 & -2 & 0 \end{vmatrix}$

64. $\begin{vmatrix} 5 & 2 & 0 & 0 \\ 0 & 4 & -1 & 1 \\ -1 & 0 & 3 & 0 \\ 0 & -2 & 1 & 0 \end{vmatrix}$

For Exercises 65 and 66, refer to the following system of four variables.

$$x + y + z + w = 0$$
$$2x - z + w = 5$$
$$2x + y - w = 0$$
$$y + z = -1$$

65. a. Evaluate the determinant $\mathbf{D}$.

    b. Evaluate the determinant $\mathbf{D}_x$.

    c. Solve for x by computing $\dfrac{\mathbf{D}_x}{\mathbf{D}}$.

66. a. Evaluate the determinant $\mathbf{D}_y$.

    b. Solve for y by computing $\dfrac{\mathbf{D}_y}{\mathbf{D}}$.

67. Two angles are complementary. The measure of one angle is $\frac{5}{7}$ the measure of the other. Find the measures of the two angles.

68. Two angles are supplementary. The measure of the larger angle is 61° more than $\frac{3}{4}$ the measure of the smaller angle. Find the measures of the angles.

69. An electronics store sells iPods, iPads, and iPhones. In a recent month, twice as many iPods as iPhones were sold. The number of iPods sold was five more than the sum of iPads and iPhones. The total number of devices sold was 75. Find the number iPods, iPads, and iPhones sold.

70. A group was surveyed regarding e-readers. The group represented a total of 100 e-reader users. The number of Kindles used was 44 more than the sum of the number Kobos and iPads. The number of Kobos was 50 less than the number of Kindles. Find the number of each type of e-reader.

71. Suppose 1000 people were surveyed in southern California, and 445 said that they worked-out at least three times a week. If $\frac{1}{2}$ of the women and $\frac{3}{8}$ of the men said that they worked-out at least three times a week, how many men and how many women were in the survey?

72. During a 1-hr television program, there were 22 commercials. Some commercials were 15 sec and some were 30 sec long. Find the number of 15-sec commercials and the number of 30-sec commercials if the total playing time for commercials was 9.5 min.

Section A.3   Sequences and Series

Concepts

1. Finite and Infinite Sequences
2. Series

1. Finite and Infinite Sequences

In everyday life, we think of a sequence as a set of events or items with some order or pattern. In mathematics, a sequence is a list of terms that correspond with the set of positive integers. For example, the sequence

$$1, 4, 9, 16, 25$$

represents the squares of the first five positive integers. This sequence has a finite number of terms and is called a *finite sequence*. The sequence

$$1, 4, 9, 16, 25, \ldots$$

represents the squares of *all* positive integers. This sequence continues indefinitely and is called an *infinite sequence*.

Because the terms in a sequence are related to the set of positive integers, we give a formal definition of finite and infinite sequences, using the language of functions.

> **Definition of Finite and Infinite Sequences**
>
> An **infinite sequence** is a function whose domain is the set of positive integers. A **finite sequence** is a function whose domain is the set of the first n positive integers.

For any positive integer n, the value of the sequence is denoted by a_n (read as "a sub n"). The values $a_1, a_2, a_3, \ldots$ are called the **terms of the sequence**. The expression a_n defines the **nth term** (or general term) **of the sequence**.

Example 1 **Listing the Terms of a Sequence**

List the terms of the following sequences.

a. $a_n = 3n^2 - 4; \quad 1 \leq n \leq 4$ **b.** $a_n = 3 \cdot 2^n$

Solution:

a. The domain is restricted to the first four positive integers, indicating that the sequence is finite.

n	a_n
1	$3(1)^2 - 4 = -1$
2	$3(2)^2 - 4 = \ \ 8$
3	$3(3)^2 - 4 = 23$
4	$3(4)^2 - 4 = 44$

The sequence is $-1, 8, 23, 44$.

b. The sequence $a_n = 3 \cdot 2^n$ has no restrictions on its domain; therefore, it is an infinite sequence.

n	a_n
1	$3 \cdot 2^1 = \ \ 6$
2	$3 \cdot 2^2 = 12$
3	$3 \cdot 2^3 = 24$
4	$3 \cdot 2^4 = 48$

The sequence is $6, 12, 24, 48, \ldots$

$\ldots$

Skill Practice

List the terms of the sequences.
1. $a_n = n^3 - 2 \quad 1 \leq n \leq 3$
2. $a_n = \left(\dfrac{2}{3}\right)^n$

Calculator Connections

Topic: Displaying Terms of a Sequence

If the nth term of a sequence is known, a seq function on a graphing utility can display a list of terms. Finding the first four terms of the sequence from Example 1(b) is outlined here: $a_n = 3 \cdot 2^n$.

- Select the MODE key and highlight SEQ.
- Then access the seq function from the LIST menu and enter the parameters as prompted.
- Paste the seq command on the home screen and select ENTER.

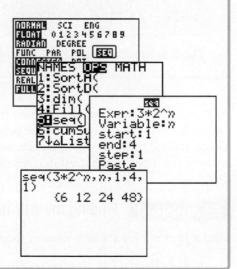

Sometimes the terms of a sequence may have alternating signs. Such a sequence is called an **alternating sequence**.

Answers

1. $-1, 6, 25$ **2.** $\dfrac{2}{3}, \dfrac{4}{9}, \dfrac{8}{27}, \ldots$

Skill Practice

List the first four terms of each alternating sequence.

3. $a_n = (-1)^n \left(\dfrac{1}{2}\right)^n$

4. $a_n = (-1)^{n+1} \left(\dfrac{2}{n}\right)$

Example 2 Listing the Terms of an Alternating Sequence

List the first four terms of each alternating sequence.

a. $a_n = (-1)^n \cdot \dfrac{1}{n}$ **b.** $a_n = (-1)^{n+1} \cdot \left(\dfrac{2}{3}\right)^n$

Solution:

a.

n	a_n
1	$(-1)^1 \cdot \dfrac{1}{1} = -1$
2	$(-1)^2 \cdot \dfrac{1}{2} = \dfrac{1}{2}$
3	$(-1)^3 \cdot \dfrac{1}{3} = -\dfrac{1}{3}$
4	$(-1)^4 \cdot \dfrac{1}{4} = \dfrac{1}{4}$

The first four terms are $-1, \frac{1}{2}, -\frac{1}{3}, \frac{1}{4}$.

TIP: Notice that the factor $(-1)^n$ makes the even-numbered terms positive and the odd-numbered terms negative.

b.

n	a_n
1	$(-1)^{1+1} \cdot \left(\dfrac{2}{3}\right)^1 = (-1)^2 \cdot \left(\dfrac{2}{3}\right) = \dfrac{2}{3}$
2	$(-1)^{2+1} \cdot \left(\dfrac{2}{3}\right)^2 = (-1)^3 \cdot \left(\dfrac{4}{9}\right) = -\dfrac{4}{9}$
3	$(-1)^{3+1} \cdot \left(\dfrac{2}{3}\right)^3 = (-1)^4 \cdot \left(\dfrac{8}{27}\right) = \dfrac{8}{27}$
4	$(-1)^{4+1} \cdot \left(\dfrac{2}{3}\right)^4 = (-1)^5 \cdot \left(\dfrac{16}{81}\right) = -\dfrac{16}{81}$

TIP: Notice that the factor $(-1)^{n+1}$ makes the odd-numbered terms positive and the even-numbered terms negative.

The first four terms are $\frac{2}{3}, -\frac{4}{9}, \frac{8}{27}$, and $-\frac{16}{81}$.

Skill Practice

Find a formula for the nth term of the sequence.

5. $\dfrac{1}{3}, \dfrac{1}{9}, \dfrac{1}{27}, \dfrac{1}{81}, \ldots$

6. $-1, 4, -9, 16, -25, \ldots$

7. $\dfrac{3}{2}, \dfrac{4}{3}, \dfrac{5}{4}, \dfrac{6}{5}, \dfrac{7}{6}, \ldots$

In Examples 1 and 2, we were given the formula for the nth term of a sequence and asked to list several terms of the sequence. We now consider the reverse process. Given several terms of the sequence, we will find a formula for the nth term. To do so, look for a pattern that establishes each term as a function of the term number.

Example 3 Finding the nth Term of a Sequence

Find a formula for the nth term of the sequence.

a. $\dfrac{1}{2}, \dfrac{2}{3}, \dfrac{3}{4}, \dfrac{4}{5}, \ldots$ **b.** $-2, 4, -6, 8, -10, \ldots$ **c.** $\dfrac{1}{2}, \dfrac{1}{4}, \dfrac{1}{8}, \dfrac{1}{16}, \ldots$

Solution:

a. For each term in the sequence, the numerator is equal to the term number, and the denominator is equal to 1 more than the term number. Therefore, the nth term may be given by

$$a_n = \dfrac{n}{n+1}$$

Answers

3. $-\dfrac{1}{2}, \dfrac{1}{4}, -\dfrac{1}{8}, \dfrac{1}{16}$ **4.** $2, -1, \dfrac{2}{3}, -\dfrac{1}{2}$

5. $a_n = \dfrac{1}{3^n}$ **6.** $a_n = (-1)^n n^2$

7. $a_n = \dfrac{n+2}{n+1}$

b. The odd-numbered terms are negative, and the even-numbered terms are positive. The factor $(-1)^n$ will produce the required alternation of signs. The numbers $2, 4, 6, 8, 10, \ldots$ are equal to $2(1), 2(2), 2(3), 2(4), 2(5), \ldots$. Therefore, the nth term may be given by

$$a_n = (-1)^n \cdot 2n$$

c. The denominators are consecutive powers of 2. The sequence can be written as

$$\frac{1}{2^1}, \frac{1}{2^2}, \frac{1}{2^3}, \frac{1}{2^4}, \ldots$$

Therefore, the nth term may be given by

$$a_n = \frac{1}{2^n}$$

TIP: The first few terms of a sequence are not sufficient to define the sequence uniquely. For example, consider the sequence

$$\frac{1}{2}, \frac{1}{4}, \frac{1}{8}, \ldots$$

The following formulas both produce the first three terms, but differ at the fourth term:

$$a_n = \frac{1}{2^n} \quad \text{and} \quad b_n = \frac{1}{n^2 - n + 2}$$

$$a_4 = \frac{1}{16} \quad \text{whereas} \quad b_4 = \frac{1}{14}$$

To define a sequence uniquely, the nth term must be provided.

Example 4   Using a Sequence in an Application

A child drops a ball from a height of 4 ft. With each bounce, the ball rebounds to 50% of its height. Write a sequence whose terms represent the heights from which the ball falls (begin with the initial height from which the ball was dropped).

Skill Practice

8. The first swing of a pendulum measures 30°. If each swing that follows is 25% less, list the first 3 terms of this sequence.

Solution:

The ball first drops 4 ft and then rebounds to a new height of $0.50(4 \text{ ft}) = 2$ ft. Similarly, it falls from 2 ft and rebounds $0.50(2 \text{ ft}) = 1$ ft. Repeating this process, we have

$$4, 2, 1, \frac{1}{2}, \frac{1}{4}, \frac{1}{8}, \ldots$$

The nth term can be represented by $a_n = 4 \cdot (0.50)^{n-1}$.

Answer

8. 30°, 22.5°, 16.875°

2. Series

In many mathematical applications it is important to find the sum of the terms of a sequence. For example, suppose that the yearly interest earned in an account over a 4-yr period is given by the sequence

$$\$250, \quad \$265, \quad \$278.25, \quad \$292.16$$

The sum of the terms gives the total interest earned

$$\$250 + \$265 + \$278.25 + \$292.16 = \$1085.41$$

By adding the terms of a sequence, we obtain a series.

> ### Definition of a Series
> The indicated sum of the terms of a sequence is called a **series**.

As with a sequence, a series may be a finite or an infinite sum of terms.

A convenient notation used to denote the sum of a set of terms is called **summation notation** or **sigma notation**. The Greek letter Σ (sigma) is used to indicate sums. For example, the sum of the first four terms of the sequence defined by $a_n = n^3$ are denoted by

$$\sum_{n=1}^{4} n^3$$

This is read as "the sum from n equals 1 to 4 of n^3" and is simplified as

$$\sum_{n=1}^{4} n^3 = (1)^3 + (2)^3 + (3)^3 + (4)^3$$
$$= 1 + 8 + 27 + 64$$
$$= 100$$

In this example, the letter n is called the **index of summation**. Many times, the letters i, j, and k are also used for the index of summation.

Calculator Connections

Topic: Finding the Sum of a Finite Sequence of Terms

The sum of a finite series may be found on a graphing calculator if the nth term of the corresponding sequence is known. The calculator takes the sum of a sequence by using the *Sum* and *Seq* functions.

```
sum(seq(n^3,n,1,
4,1))
              100
```

Skill Practice

Find the sum.

9. $\displaystyle\sum_{n=1}^{4} n(n+4)$

Answer

9. 70

Example 5   Finding a Sum from Summation Notation

Find the sum. $\displaystyle\sum_{n=1}^{4} 2^n$

Solution:

$$\sum_{n=1}^{4} 2^n = 2^1 + 2^2 + 2^3 + 2^4$$
$$= 2 + 4 + 8 + 16$$
$$= 30$$

Example 6 Finding a Sum from Summation Notation

Find the sum. $\displaystyle\sum_{i=1}^{3}(-1)^{i+1}\cdot(3i+4)$

Solution:

$$\sum_{i=1}^{3}(-1)^{i+1}\cdot(3i+4) = (-1)^{1+1}\cdot[3(1)+4]+(-1)^{2+1}\cdot[3(2)+4]$$
$$+ (-1)^{3+1}\cdot[3(3)+4]$$
$$= (-1)^2\cdot(7)+(-1)^3\cdot(10)+(-1)^4\cdot(13)$$
$$= 7-10+13$$
$$= 10$$

Skill Practice

Find the sum.

10. $\displaystyle\sum_{i=1}^{4}(-1)^{i}(4i-3)$

Avoiding Mistakes

The letter i is used as a variable for the index of summation. In this context, it does not represent an imaginary number.

Example 7 Converting to Summation Notation

Write the series in summation notation.

a. $\dfrac{2}{3}+\dfrac{4}{9}+\dfrac{8}{27}+\dfrac{16}{81}$ Use n as the index of summation.

b. $1-\dfrac{1}{2}+\dfrac{1}{3}-\dfrac{1}{4}+\dfrac{1}{5}-\dfrac{1}{6}$ Use j as the index of summation.

Solution:

a. The sum can be written as

$$\left(\frac{2}{3}\right)^1+\left(\frac{2}{3}\right)^2+\left(\frac{2}{3}\right)^3+\left(\frac{2}{3}\right)^4$$

Taking n from 1 to 4, we have

$$\sum_{n=1}^{4}\left(\frac{2}{3}\right)^n$$

b. The even-numbered terms are negative. The factor $(-1)^{j+1}$ is negative for even values of j. Therefore, the series can be written as

$$\sum_{j=1}^{6}(-1)^{j+1}\cdot\frac{1}{j}$$

Skill Practice

Write the series in summation notation.

11. $\dfrac{1}{4}+\dfrac{1}{16}+\dfrac{1}{64}+\dfrac{1}{256}$

12. $\dfrac{1}{2}-\dfrac{1}{4}+\dfrac{1}{6}-\dfrac{1}{8}+\dfrac{1}{10}$

Answers

10. 8   **11.** $\displaystyle\sum_{i=1}^{4}\frac{1}{4^i}$

12. $\displaystyle\sum_{i=1}^{5}(-1)^{i+1}\frac{1}{2i}$

Section A.3 Practice Exercises

Vocabulary and Key Concepts

1. a. A(n) (finite/infinite) sequence is a function whose domain is the set of positive integers. A(n) (finite/infinite) sequence is a function whose domain is the set of the first n positive integers.

b. Given a sequence, the values a_1, a_2, a_3, and so on are called the _____ of the sequence. The expression a_n defines the _____ term (or general term) of the sequence.

c. In an _____ sequence, consecutive terms change sign.

d. A sum of a sequence is called a _____.

e. To represent a sum of terms, _____ notation is often used and is represented by the Greek letter $\sum$.

f. Given the sum $\sum\limits_{n=1}^{5} n^2$, the letter n is called the _____ of summation. In expanded form, this series is
$(1)^2 + (2)^2 +$ _____ $= 55$.

Concept 1: Finite and Infinite Sequences

For Exercises 2–17, list the terms of each sequence. **(See Examples 1 and 2.)**

2. $a_n = 3n + 1, \quad 1 \le n \le 5$

3. $a_n = -2n + 3, \quad 1 \le n \le 5$

4. $a_n = \sqrt{n + 2}, \quad 1 \le n \le 4$

5. $a_n = \sqrt{n - 1}, \quad 1 \le n \le 4$

6. $a_n = \dfrac{3}{n}, \quad 1 \le n \le 5$

7. $a_n = \dfrac{n}{n + 2}, \quad 1 \le n \le 5$

8. $a_n = (-1)^n \dfrac{n + 1}{n + 2}, \quad 1 \le n \le 4$

9. $a_n = (-1)^n \dfrac{n - 1}{n + 2}, \quad 1 \le n \le 4$

10. $a_n = (-1)^{n+1}(n^2 - 1), \quad 1 \le n \le 3$

11. $a_n = (-1)^{n+1}(n^2), \quad 1 \le n \le 3$

12. $a_n = 3 - \dfrac{1}{n}, \quad 2 \le n \le 5$

13. $a_n = 2 - \dfrac{1}{n + 1}, \quad 2 \le n \le 5$

14. $a_n = n^2 - n, \quad 1 \le n \le 6$

15. $a_n = n(n^2 - 1), \quad 1 \le n \le 6$

16. $a_n = (-1)^n 3^n, \quad 1 \le n \le 4$

17. $a_n = (-1)^n n, \quad 1 \le n \le 4$

18. If the nth term of a sequence is $a_n = (-1)^n \cdot n^2$, which terms are positive and which are negative?

19. If the nth term of a sequence is $a_n = (-1)^{n-1} \cdot \frac{1}{n}$, which terms are positive and which are negative?

For Exercises 20–31, find a formula for the nth term of the sequence. Answers may vary. **(See Example 3.)**

20. $2, 4, 6, 8, \ldots$

21. $3, 6, 9, 12, \ldots$

22. $1, 3, 5, 7, \ldots$

23. $3, 5, 7, 9, \ldots$

24. $1, \dfrac{1}{4}, \dfrac{1}{9}, \dfrac{1}{16}, \ldots$

25. $\dfrac{1}{2}, \dfrac{2}{3}, \dfrac{3}{4}, \dfrac{4}{5}, \ldots$

26. $1, -1, 1, -1, \ldots$

27. $-1, 1 -1, 1, \ldots$

28. $-2, 4, -8, 16, \ldots$

29. $3, -9, 27, -81, \ldots$

30. $\dfrac{3}{5}, \dfrac{3}{25}, \dfrac{3}{125}, \dfrac{3}{625}, \ldots$

31. $\dfrac{1}{4}, \dfrac{1}{16}, \dfrac{1}{64}, \dfrac{1}{256}, \ldots$

32. Edmond borrowed \$500. To pay off the loan, he agreed to pay 2% of the balance plus \$50 each month. Write a sequence representing the amount Edmond will pay each month for the next 4 months. Round each term to the nearest cent.

33. Janice deposited $1000 in a savings account that pays 3% interest compounded annually. Write a sequence representing the amount Janice receives in interest each year for the first 4 yr. Round each term to the nearest cent. **(See Example 4.)**

34. A certain bacteria culture doubles its size each day. If there are 25,000 bacteria on the first day, write a sequence representing the population each day for the first week (7 days).

35. A radioactive chemical decays by one-half of its amount each week. If there is 16 g of the chemical in week 1, write a sequence representing the amount present each week for 2 months (8 weeks).

Concept 2: Series

36. What is the difference between a sequence and a series?

For Exercises 37–52, find the sums. **(See Examples 5 and 6.)**

37. $\displaystyle\sum_{i=1}^{4}(3i^2)$

38. $\displaystyle\sum_{i=1}^{4}(2i^2)$

39. $\displaystyle\sum_{j=0}^{4}\left(\frac{1}{2}\right)^{j}$

40. $\displaystyle\sum_{j=0}^{4}\left(\frac{1}{3}\right)^{j}$

41. $\displaystyle\sum_{i=1}^{6}5$

42. $\displaystyle\sum_{i=1}^{7}3$

43. $\displaystyle\sum_{j=1}^{4}(-1)^{j}(5j)$

44. $\displaystyle\sum_{j=1}^{4}(-1)^{j}(4j)$

45. $\displaystyle\sum_{i=1}^{4}\frac{i+1}{i}$

46. $\displaystyle\sum_{i=2}^{5}\frac{i-1}{i}$

47. $\displaystyle\sum_{j=1}^{3}(j+1)(j+2)$

48. $\displaystyle\sum_{j=1}^{3}j(j+2)$

49. $\displaystyle\sum_{k=1}^{7}(-1)^{k}$

50. $\displaystyle\sum_{k=0}^{5}(-1)^{k+1}$

51. $\displaystyle\sum_{k=1}^{5}k^2$

52. $\displaystyle\sum_{k=1}^{5}2^{k}$

For Exercises 53–62, write the series in summation notation. **(See Example 7.)**

53. $1+2+3+4+5+6$

54. $1-2+3-4+5-6$

55. $4+4+4+4+4$

56. $8+8+8+8+8$

57. $4+8+12+16+20$

58. $3+6+9+12+15$

59. $\dfrac{1}{3}-\dfrac{1}{9}+\dfrac{1}{27}-\dfrac{1}{81}$

60. $\dfrac{1}{2}-\dfrac{1}{4}+\dfrac{1}{8}-\dfrac{1}{16}$

61. $x+x^2+x^3+x^4+x^5$

62. $y^2+y^4+y^6+y^8+y^{10}$

Summation notation is used extensively in statistics. The sample mean (average), $\bar{x}$, of a set of n values $x_1, x_2, x_3, \ldots, x_n$ is given by

$$\bar{x}=\frac{1}{n}\sum_{i=1}^{n}x_i$$

Use this formula for Exercises 63 and 64.

63. Find the mean number of grams of protein for a sample of five energy bars. 10, 15, 12, 18, 22

64. Find the mean age of a sample of viewers who regularly watch CNN on television. 29, 56, 62, 39, 58, 74

The sample standard deviation, s, of a set of n values $x_1, x_2, x_3, \ldots, x_n$ is given by

$$s = \sqrt{\frac{n \cdot \sum_{i=1}^{n} x_i^2 - \left(\sum_{i=1}^{n} x_i\right)^2}{n(n-1)}}$$

Use this formula for Exercises 65 and 66.

65. Find the standard deviation for the number of grams of protein for a sample of five energy bars: 10, 15, 12, 18, 22. Round to one decimal place.

66. Find the standard deviation of a sample of viewers who regularly watch CNN on television: 29, 56, 62, 39, 58, 74. Round to one decimal place.

Expanding Your Skills

Some sequences are defined by a recursion formula, which defines each term of a sequence in terms of one or more of its preceding terms. For example, if $a_1 = 5$ and $a_n = 2a_{n-1} + 1$ for $n > 1$, then the terms of the sequence are 5, 11, 23, 47, ... In this case, each term after the first is one more than twice the term before it.

For Exercises 67–70, list the first five terms of the sequence.

67. $a_1 = -3, a_n = a_{n-1} + 5$ for $n > 1$

68. $a_1 = 10, a_n = a_{n-1} - 3$ for $n > 1$

69. $a_1 = 5, a_n = 4a_{n-1} + 1$ for $n > 1$

70. $a_1 = -2, a_n = -3a_{n-1} + 4$ for $n > 1$

71. A famous sequence in mathematics is called the Fibonacci sequence, named after the Italian mathematician Leonardo Fibonacci of the thirteenth century. The Fibonacci sequence is defined by

$$a_1 = 1$$
$$a_2 = 1$$
$$a_n = a_{n-1} + a_{n-2} \quad \text{for } n > 2$$

This definition implies that beginning with the third term, each term is the sum of the preceding two terms. Write out the first 10 terms of the Fibonacci sequence.

Section A.4 | Arithmetic and Geometric Sequences and Series

Concepts

1. Arithmetic Sequences
2. Arithmetic Series
3. Geometric Sequences
4. Geometric Series

1. Arithmetic Sequences

In this section, we will study two special types of sequences and series. The first is called an arithmetic sequence, for example:

$$4, 7, 10, 13, 16, \ldots$$

This sequence is an arithmetic sequence. Note the characteristic that each successive term after the first is a fixed value more than the previous term (in this case the terms differ by 3).

Definition of Arithmetic Sequence

An **arithmetic sequence** is a sequence in which the difference between consecutive terms is constant.

The fixed difference between a term and its predecessor is called the **common difference** and is denoted by the letter d. The common difference is the difference between a term and its predecessor. That is,

$$d = a_{n+1} - a_n$$

Furthermore, if a_1 is the first term, then

$$a_1 = a_1 + 0d$$

$$a_2 = a_1 + 1d \qquad \text{is the second term}$$

$$a_3 = a_1 + 2d \qquad \text{is the third term}$$

$$a_4 = a_1 + 3d \qquad \text{is the fourth term and so on}$$

$$\cdots$$

In general, $a_n = a_1 + (n-1)d$.

nth Term of an Arithmetic Sequence

The nth term of an arithmetic sequence is given by

$$a_n = a_1 + (n-1)d$$

where a_1 is the first term of the sequence and d is the common difference.

Example 1   **Writing the nth Term of an Arithmetic Sequence**

Write the nth term of the sequence. $9, 2, -5, -12, \ldots$

Solution:

$$9, \quad 2, \quad -5, \quad -12, \ldots$$

$$\underbrace{}_{-7} \; \underbrace{}_{-7} \; \underbrace{}_{-7}$$

The common difference can be found by subtracting a term from its predecessor: $2 - 9 = -7$

With $a_1 = 9$ and $d = -7$, we have

$$a_n = 9 + (n-1)(-7)$$

$$= 9 - 7n + 7$$

$$= -7n + 16$$

In Example 1, the common difference between terms is -7. Accordingly, each term of the sequence *decreases* by 7.

The formula $a_n = a_1 + (n-1)d$ contains four variables: $a_n, a_1, n,$ and d. Consequently, if we know the value of three of the four variables, we can solve for the fourth.

Skill Practice

Write the nth term of the sequence.

1. $-3, -1, 1, 3, 5, \ldots$

Answer

1. $2n - 5$

Skill Practice

2. Find the tenth term of the arithmetic sequence in which $a_1 = 14$ and $a_{15} = -28$. (*Hint:* First find *d*.)

Example 2   Finding a Specified Term of an Arithmetic Sequence

Find the ninth term of the arithmetic sequence in which $a_1 = -4$ and $a_{22} = 164$.

Solution:

To find the value of the ninth term a_9, we need to determine the value of d. To find d, substitute $a_1 = -4$, $n = 22$, and $a_{22} = 164$ into the formula $a_n = a_1 + (n - 1)d$.

$$a_n = a_1 + (n - 1)d$$
$$164 = -4 + (22 - 1)d$$
$$164 = -4 + 21d$$
$$168 = 21d$$
$$d = 8$$

Therefore, $a_n = -4 + (n - 1)(8)$

$$a_9 = -4 + (9 - 1)(8)$$
$$= -4 + (8)(8)$$
$$= 60$$

Skill Practice

Find the number of terms of the sequence.

3. $-15, -11, -7, -3, \ldots 81$

Example 3   Finding the Number of Terms in an Arithmetic Sequence

Find the number of terms of the sequence $7, 3, -1, -5, \ldots, -113$.

Solution:

To find the number of terms n, we can substitute $a_1 = 7$, $d = -4$, and $a_n = -113$ into the formula for the nth term.

$$a_n = a_1 + (n - 1)d$$
$$-113 = 7 + (n - 1)(-4)$$
$$-113 = 7 - 4n + 4$$
$$-113 = 11 - 4n$$
$$-124 = -4n$$
$$n = 31$$

2. Arithmetic Series

The indicated sum of an arithmetic sequence is called an **arithmetic series**. For example, the series

$$3 + 7 + 11 + 15 + 19 + 23$$

is an arithmetic series because the common difference between terms is constant (4). Adding the terms in a lengthy sum is cumbersome, so we offer the following

Answers

2. $d = -3$, $a_{10} = -13$ **3.** 25

"short cut," which is developed here. Let S represent the sum of the terms in the series.

$$
\begin{array}{lll}
S = & 3 + \ 7 + 11 + 15 + 19 + 23 & \text{Add the terms in ascending order.} \\
\underline{S = 23 + 19 + 15 + 11 + \ 7 + \ 3} & \text{Add the terms in descending order.} \\
2S = 26 + 26 + 26 + 26 + 26 + 26 & \text{Adding the two series produces} \\
& & \text{six terms of 26.}
\end{array}
$$

$$2S = 6 \cdot 26$$

$$S = \frac{6 \cdot 26}{2}$$

$$= 78$$

By adding the terms in ascending and descending order, we double the sum but create a pattern that is easily added. This is true in general. To find the sum, S_n, of the first n terms of the arithmetic series $a_1 + a_2 + a_3 + \cdots + a_n$, we have

$$
\begin{array}{lll}
S_n = a_1 & + (a_1 + d) + (a_1 + 2d) + \cdots + a_n & \text{Ascending order} \\
\underline{S_n = a_n \quad + (a_n - d) + (a_n - 2d) + \cdots + a_1} & \text{Descending order} \\
2S_n = (a_1 + a_n) + (a_1 + a_n) + (a_1 + a_n) + \cdots + (a_1 + a_n) \\
2S_n = n(a_1 + a_n) \\
S_n = \dfrac{n}{2}(a_1 + a_n)
\end{array}
$$

Sum of an Arithmetic Series

The sum, S_n, of the first n terms of an arithmetic series is given by

$$S_n = \frac{n}{2}(a_1 + a_n)$$

where a_1 is the first term of the series and a_n is the nth term of the series.

Example 4 **Finding the Sum of an Arithmetic Series**

Find the sum of the series. $\displaystyle\sum_{i=1}^{25}(2i + 3)$

Solution:

In this series, $n = 25$. Furthermore, $a_1 = 2(1) + 3 = 5$ and $a_{25} = 2(25) + 3 = 53$. Therefore,

$$S_{25} = \frac{25}{2}(5 + 53) \qquad \text{Apply the formula } S_n = \frac{n}{2}(a_1 + a_n).$$

$$= \frac{25}{2}(58) \qquad\qquad \text{Simplify.}$$

$$= 725$$

Skill Practice

Find the sum of the series.

4. $\displaystyle\sum_{i=1}^{20}(3i - 2)$

Avoiding Mistakes

When we apply the sum formula $S_n = \dfrac{n}{2}(a_1 + a_n)$ to find the sum $\displaystyle\sum_{i=1}^{n} a_n$, the index of summation, i, must begin at 1.

Answer
4. 590

Skill Practice

Find the sum of the series.
5. $8 + 3 + (-2) + \cdots + (-137)$

Example 5 **Finding the Sum of an Arithmetic Series**

Find the sum of the series.

$$-3 + (-5) + (-7) + \cdots + (-127)$$

Solution:

For this series, $a_1 = -3$ and $a_n = -127$. However, to determine the sum, we also need to find the value of n. The difference between the second term and its predecessor is $-5 - (-3) = -2$. Thus, $d = -2$. We have

$-127 = -3 + (n - 1)(-2)$	Apply the formula $a_n = a_1 + (n - 1)d$.
$-127 = -3 - 2n + 2$	Apply the distributive property.
$-127 = -1 - 2n$	Combine *like* terms.
$-126 = -2n$	Solve for n.
$n = 63$	

Using $n = 63$, $a_1 = -3$, and $a_{63} = -127$, we have

$$S_n = \frac{n}{2}(a_1 + a_n) = \frac{63}{2}[-3 + (-127)] = \frac{63}{2}(-130) = -4095$$

3. Geometric Sequences

The sequence $2, 4, 8, 16, 32, \ldots$ is not an arithmetic sequence because the difference between terms is not constant. However, a different pattern exists. Notice that each term after the first is 2 times the preceding term. This sequence is called a geometric sequence.

> **Definition of a Geometric Sequence**
>
> A **geometric sequence** is a sequence in which each term after the first term is a constant multiple of the preceding term.

The constant multiple between a term and its predecessor is called the **common ratio** and is denoted by r. The common ratio is found by dividing a term by the preceding term. That is,

$$r = \frac{a_{n+1}}{a_n}$$

For the sequence $2, 4, 8, 16, 32, \ldots$ we have $r = \frac{4}{2} = \frac{8}{4} = \frac{16}{8} = \frac{32}{16} = 2$.

If a_1 denotes the first term of a geometric sequence, then

$$a_2 = a_1 r \qquad \text{is the second term.}$$

$$a_3 = a_1 r^2 \qquad \text{is the third term.}$$

$$a_4 = a_1 r^3 \qquad \text{is the fourth term, and so on.}$$

$$\cdots$$

This pattern gives $a_n = a_1 r^{n-1}$.

Answer

5. -1935

---

*n*th Term of a Geometric Sequence

The *n*th term of a geometric sequence is given by

$$a_n = a_1 r^{n-1}$$

where a_1 is the first term and r is the common ratio.

Example 6 Finding the *n*th Term of a Geometric Sequence

Find the *n*th term of the sequence.

a. $-1, 4, -16, 64, \ldots$ **b.** $12, 8, \dfrac{16}{3}, \dfrac{32}{9}, \ldots$

Solution:

a. The common ratio is found by dividing any term (after the first) by its predecessor.

$$r = \frac{4}{-1} = -4$$

With $r = -4$ and $a_1 = -1$, we have $a_n = -1(-4)^{n-1}$.

b. The common ratio is $r = \frac{8}{12} = \frac{2}{3}$. With $a_1 = 12$ and $r = \frac{2}{3}$, we have

$$a_n = 12\left(\frac{2}{3}\right)^{n-1}$$

The formula $a_n = a_1 r^{n-1}$ contains the variables $a_n, a_1, n,$ and r. If we know the value of three of the four variables, we can find the fourth.

Example 7 Finding a Specified Term of a Geometric Sequence

Given $a_n = 6\left(\frac{1}{2}\right)^{n-1}$, find a_5.

Solution:

$$a_n = 6\left(\frac{1}{2}\right)^{n-1}$$

$$a_5 = 6\left(\frac{1}{2}\right)^{5-1} = 6\left(\frac{1}{2}\right)^{4} = 6\left(\frac{1}{16}\right) = \frac{3}{8}$$

Example 8 Finding a Specified Term of a Geometric Sequence

Find the first term of the geometric sequence where $a_5 = -162$ and $r = 3$.

Solution:

$-162 = a_1(3)^{5-1}$ Substitute $a_5 = -162$, $n = 5$, and $r = 3$ into the formula $a_n = a_1 r^{n-1}$.

$-162 = a_1(3)^4$ Simplify and solve for a_1.

$-162 = a_1(81)$

$a_1 = -2$

4. Geometric Series

The indicated sum of a geometric sequence is called a **geometric series**. For example,

$$1 + 3 + 9 + 27 + 81 + 243$$

is a geometric series. To find the sum, consider the following procedure. Let S represent the sum

$$S = 1 + 3 + 9 + 27 + 81 + 243$$

Now multiply S by the common ratio, which in this case is 3.

$$3S = 3 + 9 + 27 + 81 + 243 + 729$$

Then

$$3S - S = (3 + 9 + 27 + 81 + 243 + 729) - (1 + 3 + 9 + 27 + 81 + 243)$$

$$2S = 3 + 9 + 27 + 81 + 243 + 729 - 1 - 3 - 9 - 27 - 81 - 243$$

$$2S = 729 - 1 \qquad \text{The terms in red form a sum of zero.}$$

$$2S = 728$$

$$S = 364$$

A similar procedure can be used to find the sum S_n of the first n terms of any geometric series. Subtract rS_n from S_n.

$$S_n = a_1 \qquad\;\; + a_1 r \qquad\;\; + a_1 r^2 \qquad\;\; + \cdots + a_1 r^{n-1}$$
$$\underline{rS_n = a_1 r \qquad\;\; + a_1 r^2 \qquad\;\; + a_1 r^3 \qquad\;\; + \cdots + a_1 r^n}$$

$$S_n - rS_n = (a_1 - a_1 r) + (a_1 r - a_1 r^2) + (a_1 r^2 - a_1 r^3) + \cdots + (a_1 r^{n-1} - a_1 r^n)$$

$$S_n - rS_n = a_1 - a_1 r^n \qquad \text{The terms in red form a sum of zero.}$$

$$S_n(1 - r) = a_1(1 - r^n) \qquad \text{Factor each side of the equation.}$$

$$S_n = \frac{a_1(1 - r^n)}{1 - r} \qquad \text{Divide by } (1 - r).$$

Sum of a Geometric Series

The sum, S_n, of the first n terms of a geometric series $\displaystyle\sum_{i=1}^{n} a_1 r^{i-1}$ is given by

$$S_n = \frac{a_1(1 - r^n)}{1 - r}$$

where a_1 is the first term of the series, r is the common ratio, and $r \neq 1$.

Skill Practice

Find the sum of the geometric series.

10. $\dfrac{1}{9} + \dfrac{1}{3} + 1 + 3 + 9 + 27$

Answer

10. $\dfrac{364}{9}$

Example 9   Finding the Sum of a Geometric Series

Find the sum of the series. $\displaystyle\sum_{i=1}^{6} 4\left(\frac{1}{2}\right)^{i-1}$

Solution:

By expanding the terms of this series, we see that the series is geometric.

$$\sum_{i=1}^{6} 4\left(\frac{1}{2}\right)^{i-1} = 4 + 2 + 1 + \frac{1}{2} + \frac{1}{4} + \frac{1}{8}$$

$$a_1 = 4 \qquad r = \frac{1}{2} \qquad \text{and} \qquad n = 6$$

$$S_n = \frac{a_1(1 - r^n)}{1 - r} = \frac{4[1 - (\frac{1}{2})^6]}{1 - \frac{1}{2}} = \frac{4(1 - \frac{1}{64})}{\frac{1}{2}} = 8\left(\frac{63}{64}\right) = \frac{63}{8}$$

Example 10 Finding the Sum of a Geometric Series

Find the sum of the series $5 + 10 + 20 + \cdots + 5120$.

Solution:

The common ratio is 2 and $a_1 = 5$. The nth term of the sequence can be written as $a_n = 5(2)^{n-1}$. To find the value of n, substitute 5120 for a_n.

$5120 = 5(2)^{n-1}$

$\dfrac{5120}{5} = \dfrac{5(2)^{n-1}}{5}$ Divide both sides by 5.

$1024 = 2^{n-1}$ To solve the exponential equation, write each side as

$2^{10} = 2^{n-1}$ a power of 2.

$10 = n - 1$ From Section 9.6, we know that if $b^x = b^y$, then $x = y$.

$n = 11$

With $a_1 = 5$, $r = 2$, and $n = 11$, we have

$$S_n = \frac{a_1(1 - r^n)}{1 - r} = \frac{5(1 - 2^{11})}{1 - 2} = \frac{5(1 - 2048)}{-1} = \frac{5(-2047)}{-1} = 10{,}235$$

Skill Practice

Find the sum of the geometric series.

11. $3 + 6 + 12 + \cdots + 768$

Consider a geometric series where $|r| < 1$. For increasing values of n, r^n decreases. For example,

$$\left(\frac{1}{2}\right)^5 = 0.03125 \qquad \left(\frac{1}{2}\right)^{10} \approx 0.00097656 \qquad \left(\frac{1}{2}\right)^{15} \approx 0.00003052$$

For $|r| < 1$, r^n approaches 0 as n gets larger and larger. As n approaches infinity, the sum

$$S = \frac{a_1(1 - r^n)}{1 - r} \qquad \text{approaches} \qquad \frac{a_1(1 - 0)}{1 - r} = \frac{a_1}{1 - r}$$

Sum of an Infinite Geometric Series

Given an infinite geometric series $a_1 + a_1r + a_1r^2 + \cdots$, with $|r| < 1$, the sum S of all terms in the series is given by

$$S = \frac{a_1}{1 - r}$$

Note: If $|r| \geq 1$, then the sum does not exist.

Example 11 Finding the Sum of an Infinite Geometric Series

Find the sum of the series. $\displaystyle\sum_{i=1}^{\infty} \left(\frac{1}{3}\right)^{i-1}$

Solution:

In this example, the upper limit of summation is ∞. This indicates that we have an infinite series where the pattern for the sum of terms continues indefinitely.

$$\sum_{i=1}^{\infty} \left(\frac{1}{3}\right)^{i-1} = 1 + \frac{1}{3} + \frac{1}{9} + \frac{1}{27} + \frac{1}{81} + \cdots$$

Skill Practice

Find the sum of the series.

12. $\displaystyle\sum_{i=1}^{\infty} 4\left(\frac{3}{4}\right)^{i-1}$

Answers

11. 1533   **12.** 16

This is a geometric series with $a_1 = 1$ and $r = \frac{1}{3}$. Because $|r| = \left|\frac{1}{3}\right| < 1$, we have

$$S = \frac{a_1}{1 - r} = \frac{1}{1 - \frac{1}{3}} = \frac{1}{\frac{2}{3}} = \frac{3}{2}$$

The sum is $\frac{3}{2}$.

Skill Practice

13. A child drops a ball from a height of 3 ft. With each bounce, the ball rebounds to $\frac{1}{3}$ of its original height. Determine the total distance traveled by the ball.

Example 12 **Using Geometric Series in a Physics Application**

A child drops a ball from a height of 4 ft. With each bounce, the ball rebounds to 50% of its original height. Determine the total distance traveled by the ball.

Solution:

The heights (in ft) from which the ball drops are given by the sequence

$$4, 2, 1, \frac{1}{2}, \frac{1}{4}, \dots$$

After the ball falls from its initial height of 4 ft, the distance traveled for every bounce thereafter is doubled

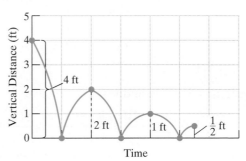

(the ball travels up and down). Therefore, the total distance traveled is given by the series

$$4 + 2 \cdot 2 + 2 \cdot 1 + 2 \cdot \frac{1}{2} + 2 \cdot \frac{1}{4} + \cdots$$

or equivalently

$$4 + 4 + 2 + 1 + \frac{1}{2} + \cdots$$

The series $4 + 2 + 1 + \frac{1}{2} + \cdots$ is an infinite geometric series with $a_1 = 4$ and $r = \frac{1}{2}$.

Initial height from which the ball was dropped   Infinite geometric series

$$4 \quad + \quad 4 + 2 + 1 + \frac{1}{2} + \cdots$$

$$= 4 + \frac{4}{1 - \frac{1}{2}}$$

$$= 4 + \frac{4}{\frac{1}{2}}$$

$$= 4 + 8 = 12$$

The ball traveled a total of 12 ft.

Answer

13. 6 ft.

Section A.4   Practice Exercises

Vocabulary and Key Concepts

1. a. An _____ sequence is a sequence in which the difference between consecutive terms is constant.

    b. The common _____ between a term and its predecessor in an arithmetic sequence is often denoted by the letter d.

    c. The nth term of an arithmetic sequence is given by $a_n =$ _____ , where a_n is the nth term, a_1 is the first term, n is the number of terms, and _____ is the common difference between terms.

    d. An indicated sum of an arithmetic sequence is called an arithmetic _____ .

    e. The sum S_n of the first n terms of an arithmetic series is given by $S_n =$ _____ , where a_1 is the first term and a_n is the nth term.

    f. A _____ sequence is a sequence in which the ratio between each term and its predecessor is constant.

    g. The common _____ between a term and its predecessor in a geometric sequence is often denoted by r.

    h. The nth term of a geometric sequence is given by $a_n =$ _____ , where a_n is the nth term in the sequence, a_1 is the first term, and r is the common ratio.

    i. The sum S_n of the first n terms of a geometric series $\displaystyle\sum_{i=1}^{n} a_1 r^{i-1}$ is given by $S_n =$ _____ , where a_1 is the first term and r is the common ratio.

    j. Given an infinite geometric series $a_1 + a_2 r + a_1 r^2 + \cdots$ with $|r| < 1$, the sum of all terms in the sequence is given by $S =$ _____ . If $|r| \geq$ _____ , then the sum does not exist.

Concept 1: Arithmetic Sequences

2. Explain how to determine if a sequence is arithmetic.

For Exercises 3–8, find the common difference d for each arithmetic sequence.

3. $1, 3, 5, 7, 9, \ldots$

4. $2, 8, 14, 20, 26, \ldots$

5. $6, 3, 0, -3, -6, \ldots$

6. $8, 3, -2, -7, -12, \ldots$

7. $-7, -9, -11, -13, -15, \ldots$

8. $-15, -11, -7, -3, 1, \ldots$

For Exercises 9–14, write the first five terms of the arithmetic sequence.

9. $a_1 = 3, d = 5$

10. $a_1 = -3, d = 2$

11. $a_1 = 2, d = \dfrac{1}{2}$

12. $a_1 = 0, d = \dfrac{1}{3}$

13. $a_1 = 2, d = -4$

14. $a_1 = 10, d = -6$

For Exercises 15–23, write the nth term of the sequence. **(See Example 1.)**

15. $0, 5, 10, 15, 20, \ldots$

16. $7, 12, 17, 22, 27, \ldots$

17. $-2, -4, -6, -8, -10, \ldots$

18. $1, -3, -7, -11, -15, \ldots$

19. $2, \dfrac{5}{2}, 3, \dfrac{7}{2}, 4, \ldots$

20. $1, \dfrac{4}{3}, \dfrac{5}{3}, 2, \dfrac{7}{3}, \ldots$

21. $21, 17, 13, 9, 5, \ldots$

22. $9, 6, 3, 0, -3, \ldots$

23. $-8, -2, 4, 10, 16, \ldots$

For Exercises 24–31, find the indicated term of each arithmetic sequence. **(See Example 2.)**

24. Find the eighth term given $a_1 = -6$ and $d = -3$.

25. Find the sixth term given $a_1 = -3$ and $d = 4$.

26. Find the 12th term given $a_1 = -8$ and $d = -2$.

27. Find the ninth term given $a_1 = -1$ and $d = 6$.

28. Find the tenth term given $a_1 = 1$ and $a_7 = 31$.

29. Find the seventh term given $a_1 = 0$ and $a_{10} = -45$.

30. Find the sixth term given $a_1 = 3$ and $a_{13} = 39$.

31. Find the 11th term given $a_1 = 12$ and $a_6 = -18$.

For Exercises 32–39, find the number of terms, n, of each arithmetic sequence. **(See Example 3.)**

32. $2, 0, -2, \ldots, -56$

33. $8, 13, 18, \ldots, 98$

34. $1, -3, -7, \ldots, -67$

35. $1, 5, 9, \ldots, 85$

36. $1, \dfrac{3}{4}, \dfrac{1}{2}, \ldots, -4$

37. $2, \dfrac{5}{2}, 3, \ldots, 13$

38. $-\dfrac{5}{3}, -1, -\dfrac{1}{3}, \ldots, \dfrac{37}{3}$

39. $\dfrac{13}{3}, \dfrac{19}{3}, \dfrac{25}{3}, \ldots, \dfrac{73}{3}$

40. If the third and fourth terms of an arithmetic sequence are 18 and 21, what are the first and second terms?

41. If the third and fourth terms of an arithmetic sequence are -8 and -11, what are the first and second terms?

Concept 2: Arithmetic Series

42. Explain the difference between an arithmetic sequence and an arithmetic series.

For Exercises 43–56, find the sum of the arithmetic series. **(See Examples 4 and 5.)**

43. $\displaystyle\sum_{i=1}^{20} (3i + 2)$

44. $\displaystyle\sum_{i=1}^{15} (2i - 3)$

45. $\displaystyle\sum_{i=1}^{20} (i + 4)$

46. $\displaystyle\sum_{i=1}^{25} (i - 3)$

47. $\displaystyle\sum_{j=1}^{10} (4 - j)$

48. $\displaystyle\sum_{j=1}^{10} (6 - j)$

49. $\displaystyle\sum_{j=1}^{15} \left(\dfrac{2}{3}j + 1\right)$

50. $\displaystyle\sum_{j=1}^{15} \left(\dfrac{1}{2}j - 2\right)$

51. $4 + 8 + 12 + \cdots + 84$

52. $4 + 9 + 14 + \cdots + 49$

53. $6 + 8 + 10 + \cdots + 34$

54. $-4 + (-3) + (-2) + \cdots + 12$

55. $-3 + (-7) + (-11) + \cdots + (-39)$

56. $2 + 5 + 8 + \cdots + 53$

57. Find the sum of the first 100 positive integers.

58. Find the sum of the first 50 positive even integers.

59. The seating in a certain theater is arranged so that there are 30 seats in row 1, 32 in row 2, 34 in row 3, and so on. If there are 20 rows, how many total seats are there? What is the total revenue if the average ticket price is $15 per seat and the theater is sold out?

60. A triangular array of dominoes has one domino in the first row, two dominoes in the second row, three dominoes in the third row, and so on. If there are 15 rows, how many dominoes are in the array?

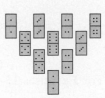

Concept 3: Geometric Sequences

61. Explain how to determine if a sequence is geometric.

For Exercises 62–67, determine the common ratio, r, for the geometric sequence.

62. $5, 10, 20, 40, \ldots$

63. $-2, -1, -\dfrac{1}{2}, -\dfrac{1}{4}, \ldots$

64. $8, -2, \dfrac{1}{2}, -\dfrac{1}{8}, \ldots$

65. $4, -12, 36, -108, \ldots$

66. $3, -6, 12, -24, \ldots$

67. $1, 4, 16, 64, \ldots$

For Exercises 68–73, write the first five terms of the geometric sequence.

68. $a_1 = -3, r = -2$

69. $a_1 = -4, r = -1$

70. $a_1 = 6, r = \dfrac{1}{2}$

71. $a_1 = 8, r = \dfrac{1}{4}$

72. $a_1 = -1, r = 6$

73. $a_1 = 2, r = -3$

For Exercises 74–79, find the nth term of each geometric sequence. **(See Example 6.)**

74. $3, 12, 48, 192, \ldots$

75. $2, 6, 18, 54, \ldots$

76. $-5, 15, -45, 135, \ldots$

77. $-6, 12, -24, 48, \ldots$

78. $\dfrac{1}{2}, 2, 8, 32, \ldots$

79. $\dfrac{16}{3}, 4, 3, \dfrac{9}{4}, \ldots$

For Exercises 80–89, find the indicated term of each geometric sequence. **(See Examples 7 and 8.)**

80. Given $a_n = 2(\frac{1}{2})^{n-1}$, find a_8.

81. Given $a_n = -3(\frac{1}{2})^{n-1}$, find a_8.

82. Given $a_n = 4(-\frac{3}{2})^{n-1}$, find a_6.

83. Given $a_n = 6(-\frac{1}{3})^{n-1}$, find a_6.

84. Given $a_n = -3(2)^{n-1}$, find a_5.

85. Given $a_n = 5(3)^{n-1}$, find a_4.

86. Given $a_5 = -\frac{16}{9}$ and $r = -\frac{2}{3}$, find a_1.

87. Given $a_6 = \frac{5}{16}$ and $r = -\frac{1}{2}$, find a_1.

88. Given $a_7 = 8$ and $r = 2$, find a_1.

89. Given $a_6 = 27$ and $r = 3$, find a_1.

90. If the second and third terms of a geometric sequence are 16 and 64, what is the first term?

91. If the second and third terms of a geometric sequence are $\frac{1}{3}$ and $\frac{1}{9}$, what is the first term?

Concept 4: Geometric Series

92. Explain the difference between a geometric sequence and a geometric series.

For Exercises 93–102, find the sum of the geometric series. **(See Examples 9 and 10.)**

93. $10 + 2 + \dfrac{2}{5} + \dfrac{2}{25} + \dfrac{2}{125}$

94. $1 + 3 + 9 + 27 + 81 + 243$

95. $-2 + 1 + \left(-\dfrac{1}{2}\right) + \dfrac{1}{4} + \left(-\dfrac{1}{8}\right)$

96. $\dfrac{1}{4} + (-1) + 4 + (-16) + 64$

97. $12 + 16 + \dfrac{64}{3} + \dfrac{256}{9} + \dfrac{1024}{27}$

98. $9 + 6 + 4 + \dfrac{8}{3} + \dfrac{16}{9}$

99. $1 + \dfrac{2}{3} + \dfrac{4}{9} + \cdots + \dfrac{64}{729}$

100. $\dfrac{8}{3} + 2 + \dfrac{3}{2} + \cdots + \dfrac{243}{512}$

101. $-4 + 8 + (-16) + \cdots + (-256)$ **102.** $-\dfrac{7}{3} + 7 + (-21) + \cdots + (-1701)$

103. A deposit of $1000 is made in an account that earns 5% interest compounded annually. The balance in the account after n years is given by

$$a_n = 1000(1.05)^n \quad \text{for } n \geq 1$$

a. List the first four terms of the sequence. Round to the nearest cent.

b. Find the balance after 10 yr, 20 yr, and 40 yr by computing a_{10}, a_{20}, and a_{40}. Round to the nearest cent.

104. A home purchased for $125,000 increases by 4% of its value each year. The value of the home after n years is given by

$$a_n = 125{,}000(1.04)^n \quad \text{for } n \geq 1$$

a. List the first four terms of the sequence. Round to the nearest dollar.

b. Find the value of the home after 5 yr, 10 yr, and 20 yr by computing a_5, a_{10}, and a_{20}. Round to the nearest dollar.

For Exercises 105–112, first find the common ratio r. Then determine the sum of the infinite series, if it exists. **(See Example 11.)**

105. $1 + \dfrac{1}{6} + \dfrac{1}{36} + \dfrac{1}{216} + \cdots$

106. $-2 + \left(-\dfrac{1}{2}\right) + \left(-\dfrac{1}{8}\right) + \left(-\dfrac{1}{32}\right) + \cdots$

107. $\displaystyle\sum_{i=1}^{\infty}\left(-\dfrac{1}{4}\right)^{i-1}$

108. $\displaystyle\sum_{i=1}^{\infty}\left(-\dfrac{1}{5}\right)^{i-1}$

109. $-3 + 1 + \left(-\dfrac{1}{3}\right) + \dfrac{1}{9} + \cdots$

110. $\dfrac{1}{2} + \left(-\dfrac{1}{10}\right) + \dfrac{1}{50} + \left(-\dfrac{1}{250}\right) + \cdots$

111. $\dfrac{2}{3} + (-1) + \dfrac{3}{2} + \left(-\dfrac{9}{4}\right) + \cdots$

112. $3 + 5 + \dfrac{25}{3} + \dfrac{125}{9} + \cdots$

113. Suppose \$200 million is spent by tourists at a certain resort town. Further suppose that 75% of the revenue is respent in the community and then respent over and over, each time at a rate of 75%. The series

$$200 + 200(0.75) + 200(0.75)^2 + 200(0.75)^3 + \cdots$$

gives the total amount spent (and respent) in the community. Find the sum of the infinite series. **(See Example 12.)**

114. A bungee jumper jumps off a platform and stretches the cord 80 ft before rebounding upward. Each successive bounce stretches the cord 60% of its previous length. The total vertical distance traveled is given by

$$80 + 2(0.60)(80) + 2(0.60)^2(80) + 2(0.60)^3(80) + \cdots$$

After the first term, the series is an infinite geometric series. Compute the total vertical distance traveled.

115. A ball drops from a height of 4 ft. With each bounce, the ball rebounds to $\frac{3}{4}$ of its height. The total vertical distance traveled is given by

$$4 + 2\left(\dfrac{3}{4}\right)(4) + 2\left(\dfrac{3}{4}\right)^2(4) + 2\left(\dfrac{3}{4}\right)^3(4) + \cdots$$

After the first term, the series is an infinite geometric series. Compute the total vertical distance traveled.

Expanding Your Skills

116. The yearly salary for job A is \$40,000 initially with an annual raise of \$3000 per year. The yearly salary for job B is \$38,000 initially with an annual raise of 6% per year.

   a. Find the total earnings for job A over 20 yr. Is this an arithmetic or geometric series?

   b. For job B, what is the amount of the raise after 1 yr?

   c. Find the total earnings for job B over 20 yr. Round to the nearest dollar. Is this an arithmetic or geometric series?

   d. What is the difference in total salary earned over 20 yr between job A and job B?

117. a. Brook has a job that pays \$48,000 the first year. She receives a 4% raise each year. Find the sum of her yearly salaries over a 20-yr period. Round to the nearest dollar.

   b. Chamille has a job that pays \$48,000 the first year. She receives a 4.5% raise each year. Find the sum of her yearly salaries over a 20-yr period. Round to the nearest dollar.

   c. Chamille's raise each year was 0.5% higher than Brook's raise. How much more total income did Chamille receive than Brook over 20 yr?

Student Answer Appendix

Chapter R

Chapter Opener Puzzle

<u>D</u> <u>I</u> <u>S</u> <u>T</u> <u>R</u> <u>I</u> <u>B</u> <u>U</u> <u>T</u> <u>I</u> <u>V</u> <u>E</u>   <u>P</u> <u>R</u> <u>O</u> <u>P</u> <u>E</u> <u>R</u> <u>T</u> Y
4 3 10 1 5 3 2 12 1 3 11 8   7 5 9 7 8 5 1 6

Section R.2 Practice Exercises, pp. 12–15

1. a. set **b.** inequalities **c.** a is less than b **d.** c is greater than or equal to d **e.** 5 is not equal to 6 **f.** infinity; negative infinity **g.** $\{x \mid x > 5\}$; interval **h.** excludes; includes **i.** parenthesis

3.

5. $\dfrac{-10}{1}$ **7.** $\dfrac{-3}{5}$ **9.** $\dfrac{0}{1}$

11.

	Real Numbers	Irrational Numbers	Rational Numbers	Integers	Whole Numbers	Natural Numbers
5	✓		✓	✓	✓	✓
$-\sqrt{9}$	✓		✓	✓		
-1.7	✓		✓			
$\dfrac{1}{2}$	✓		✓			
$\sqrt{7}$	✓	✓				
$\dfrac{0}{4}$	✓		✓	✓	✓	
$0.\overline{2}$	✓		✓			

13. $<$ **15.** $>$ **17.** $>$ **19.** $<$ **21.** $(2, \infty)$
23. $(-\infty, 0]$ **25.** $(-5, 0]$ **27.** $[-4.7, \infty)$
29. $(-1, \infty)$ **31.** $(-\infty, -2]$

33. $(-\infty, \frac{9}{2})$ **35.** $(-2.5, 4.5]$

37. $(-\infty, -3)$ **39.** $(\frac{5}{2}, \infty)$

41. $[2, \infty)$ **43.** $(-4, 4)$

45. $[-3, 0]$

47. All real numbers less than -4 **49.** All real numbers greater than -2 and less than or equal to 7 **51.** All real numbers between -180 and 90, inclusive **53.** All real numbers

55. $a \geq 18$ **57.** $c \leq 25$ **59.** $s \geq 261$
61. $r \leq 4.5$ **63.** $18 \leq a \leq 25$ **65.** $p < 130$
67. $130 \leq p \leq 139$ **69.** $2.2 \leq \text{pH} \leq 2.4$ acidic
71. $3.0 \leq \text{pH} \leq 3.5$ acidic

Section R.3 Practice Exercises, pp. 26–29

1. a. opposites **b.** $|a|$; 0 **c.** base; n **d.** radical; square **e.** negative; positive **f.** positive; negative **g.** $-b$; positive **h.** $\dfrac{1}{a}$; 1 **i.** 0 **j.** 0; undefined **3.** $(-\infty, 4)$ **5.** $[-3, -1)$

7. Distance can never be negative. **9.** Negative

11.

Number	Opposite	Reciprocal	Absolute Value
6	-6	$\frac{1}{6}$	6
$\frac{1}{11}$	$-\frac{1}{11}$	11	$\frac{1}{11}$
-8	8	$-\frac{1}{8}$	8
$-\frac{13}{10}$	$\frac{13}{10}$	$-\frac{10}{13}$	$\frac{13}{10}$
0	0	Undefined	0
-3	3	$-0.\overline{3}$	3

13. $<$ **15.** $=$ **17.** $<$ **19.** $<$ **21.** -4 **23.** -19
25. -7 **27.** 14 **29.** -22.1 **31.** -8.1
33. $-\dfrac{5}{3}$ or $-1\dfrac{2}{3}$ **35.** $-\dfrac{67}{45}$ **37.** -32 **39.** $\dfrac{8}{21}$
41. $\dfrac{3}{5}$ **43.** $-\dfrac{18}{5}$ **45.** Undefined **47.** 0 **49.** 3.72
51. $\dfrac{5}{11}$ **53.** 64 **55.** -49 **57.** 49 **59.** $\dfrac{125}{27}$
61. 3 **63.** Not a real number **65.** $\dfrac{1}{2}$ **67.** -7
69. 32 **71.** 40 **73.** 25 **75.** 13 **77.** 17 **79.** -11
81. -603 **83.** $\dfrac{109}{150}$ **85.** 5.4375 **87.** $\dfrac{2}{3}$ **89.** -1
91. 21 **93.** Undefined **95.** $\dfrac{9}{10}$ **97.** $-10.1°\text{C}$
99. a. $25°\text{C}$ **b.** $100°\text{C}$ **c.** $0°\text{C}$ **d.** $-40°\text{C}$ **101.** $4\dfrac{1}{6}$ gal
103. 9 in.^{2} **105.** 8.06 cm^2 **107.** 14.1 ft^3 **109.** 26.8 ft^3
111. 141.4 in.^{3}

Section R.3 Graphing Calculator Exercises, p. 29

113. 12/(6 − 2) **114.** (24 − 6)/3

115. $\frac{2}{3}$ or 0.6666667

```
(√(10²-8²))/3²
         .6666666667
```

116. 6

```
(√(16-7)+3²)/(√(
16)-√(4))
              6
```

Section R.4 Practice Exercises, pp. 34–37

1. a. constant **b.** coefficient **c.** 1; 1 **d.** *like* **3.** $-\frac{1}{2}$

5. $(3, \infty)$ **7.** $\left(-\frac{5}{2}, 3\right]$ **9. a.** 3 terms **b.** 6 **c.** $2, -5, 6$

11. a. 5 terms **b.** -7 **c.** $1, -7, 1, -4, 1$ **13.** a **15.** f
17. e **19.** i **21.** b **23.** g **25.** d **27.** h **29.** a
31. $2x - 6y + 16$ **33.** $-40s + 90t + 30$ **35.** $7w - 5z$
37. $\frac{1}{2}a - 2b + \frac{8}{5}$ **39.** $7.8x - 12.3$
41. $14c - 16 - 30d + 5f$ **43.** $14y - 2x$
45. $6p^2 + p - 6$ **47.** $-p^2 - 3p$ **49.** $n^3 + m - 6$
51. $7ab + 8a$ **53.** $16xy^2 - 5y^2$ **55.** $8x - 23$
57. $-4c - 6$ **59.** $-9w + 10$ **61.** $4z - 16$
63. $7s - 26$ **65.** $-12w + 13$ **67.** 0 **69.** $4c + 2$
71. $1.4x + 10.2$ **73.** $-2a^2 + 3a + 38$
75. $2y^2 - 3y - 5$ **77.** $-62.7x + 220$
79. $-4m + 15n + 2$ **81.** 0, for example: $3 + 0 = 3$
83. Reciprocal **85.** No, for example: $6 - 5 \neq 5 - 6$
$1 \neq -1$ **87. a.** $x(y + z)$ **b.** xy **c.** xz **d.** $xy + xz$
e. $x(y + z) = xy + xz$ the distributive property of
multiplication over addition

Chapter R Review Exercises, pp. 39–40

1. 0 **2.** For example: $-\frac{1}{2}, -\frac{3}{4}, \frac{5}{2}$
3. For example: $-2, -1, 0, 1, 2$
4. All real numbers between 7 and 16 **5.** All real
numbers greater than 0 but less than or equal to 2.6
6. All real numbers between -6 and -3, inclusive
7. All real numbers greater than 8
8. All real numbers less than or equal to 13
9. All real numbers
10. $(-\infty, 2)$
11. $[0, \infty)$
12. $(-1, 5)$

13. True **14.** $8, -\frac{1}{8}, 8$ **15.** $-\frac{4}{9}, \frac{9}{4}, \frac{4}{9}$ **16.** $16, 2$

17. $625, 5$ **18.** -2 **19.** 3 **20.** -21.6 **21.** -8.151
22. $-\frac{25}{13}$ or $-1\frac{12}{13}$ **23.** $\frac{4}{11}$ **24.** $\frac{119}{100}$ or $1\frac{19}{100}$

25. $\frac{7}{6}$ or $1\frac{1}{6}$ **26.** 3 **27.** -8 **28.** 6 **29.** 48
30. 11 **31.** 37 **32.** 75 **33.** 4 **34.** 256

35. 756 in.^{2} **36.** $3x + 15y$ **37.** $\frac{1}{2}x + 4y - \frac{5}{2}$

38. $4x - 10y + z$ **39.** $-13a + b + 5c$ **40.** $7q - 14$
41. $9p + 3$ **42.** $-6y - 5$ **43.** $9x - 1$
44. For example: $3 + x = x + 3$
45. For example: $5(2y) = (5 \cdot 2)y$

Chapter R Test, p. 41

1. a. $-5, -4, -3, -2, -1, 0, 1, 2$ **b.** For example: $\frac{3}{2}, \frac{5}{4}, \frac{8}{5}$

2. a. $\frac{1}{2}, -2, \frac{1}{2}$ **b.** $-4, \frac{1}{4}, 4$ **c.** 0, no reciprocal exists, 0

3. The interval $[4, \infty)$ includes all real numbers 4 and
greater, whereas the interval $(4, \infty)$ does not include the
endpoint, 4.
4. True

5. $\left(-\infty, -\frac{4}{3}\right)$ **6.** $[12, \infty)$

7. $x \leq 5$ **8.** $p \geq 7$ **9.** 6 **10.** -17 **11.** $\frac{1}{4}$

12. $-\frac{16}{3}$ **13.** $z = 1.1$ **14.** $-2b - 6$ **15.** $2x - 1$

16. $-2x + 1$ **17.** False **18.** True **19.** True
20. True

Chapter 1

Chapter Opener Puzzle

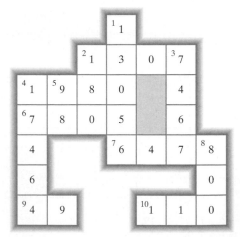

Section 1.1 Practice Exercises, pp. 52–54

1. a. equation **b.** solution **c.** linear **d.** first
e. solutions **f.** conditional **g.** contradiction **h.** identity
3. $3x - 3y + 14xy$ **5.** $5z - 20$ **7.** Linear
9. Nonlinear **11.** Linear **13.** b **15.** 12

17. -2 **19.** $\frac{21}{20}$ **21.** -40 **23.** -1.1

25. 6.7 **27.** 11 **29.** 13 **31.** -7

33. 13 **35.** $\frac{3}{2}$ **37.** 0 **39.** 3 **41.** $\frac{7}{4}$

43. 0 **45.** $\frac{4}{3}$ **47.** -4 **49.** 3 **51.** -6

53. 1 **55.** 2

57. It is an equation that is true for some values of the
variable but false for other values.
59. All real numbers; an identity
61. 0; a conditional equation
63. No solution; a contradiction **65.** 16

67. 2   **69.** 3   **71.** All real numbers

73. No solution   **75.** $\dfrac{8}{5}$ **77.** $-\dfrac{3}{2}$

79. All real numbers   **81.** $\dfrac{33}{5}$ **83.** No solution

85. $\dfrac{3}{7}$ **87.** $\dfrac{3}{4}$ **89.** 60   **91.** 6   **93.** No solution

95. $-\dfrac{9}{2}$ **97.** The family used 1024 kWh.

99. a. $y + 8$ **b.** -8 **c.** To simplify an equation, clear parentheses and combine *like* terms. To solve an equation, isolate the variable to find a solution.

Chapter 1 Problem Recognition Exercises, pp. 54–55

1. Expression; $-4x + 4$ **2.** Expression; $-7y + 5$

3. Equation; 1   **4.** Equation; 0

5. Expression; $-10a - 39$ **6.** Expression; $28x - 10$

7. Equation; $\dfrac{1}{2}$ **8.** Equation; $\dfrac{3}{4}$

9. Equation; $-\dfrac{19}{6}$ **10.** Expression; $-\dfrac{1}{6}v - \dfrac{1}{10}$

11. Expression; $-\dfrac{17}{8}t + \dfrac{1}{2}u$ **12.** Equation; No solution

13. Equation; No solution   **14.** Equation; $\dfrac{39}{8}$

15. Equation; $\dfrac{2}{25}$ **16.** Expression; $0.17c + 4.495$

17. Expression; $-1.006k - 0.78$
18. Equation; All real numbers   **19.** Equation; All real numbers

20. Equation; $\dfrac{25}{22}$

Section 1.2 Practice Exercises, pp. 62–66

1. a. consecutive   **b.** even; odd   **c.** 1; 2; 2   **d.** $x + 1$
e. $x + 2$ **f.** $x + 2$; $x + 4$ **g.** *Prt*; interest   **h.** $1300
i. 0.48 L; $0.12(x + 8)$ **j.** $\dfrac{d}{t}, \dfrac{d}{r}$ **3.** $\dfrac{13}{7}$ **5.** 6   **7.** $\dfrac{18}{5}$

9. $x + 5$ **11.** $2t - 7$ **13.** The numbers are 5 and 13.
15. The number is -3.   **17.** The page numbers are 111 and 112.   **19.** The integers are -75 and -73.   **21.** The integers are -22 and -20.   **23.** The integers are 31, 33, and 35.   **25.** She would pay $5100 for 4 yr at 8.5% and $5812.50 for 5 yr at 7.75%; the 8.5% option for 4 yr requires less interest.   **27.** She must sell $60,000.   **29.** The total for merchandise was $1197.02 and the sales tax was $96.36.
31. The price before markup was $35.90.
33. He invested $8500 in the 2% account and $4000 in the 5% account.   **35.** $12,000 was borrowed at 6% and $6000 was borrowed at 11%.   **37.** She invested $12,000 in the 4% account and $8000 in the 3% account.   **39.** 8 oz should be used.   **41.** 2 L should be used.   **43.** 12.5 L of 18% solution should be added to 7.5 L of 10% solution.
45. 2 oz should be added.   **47.** The plane flies 300 mph from Atlanta to Fort Lauderdale and 240 mph on the return trip.   **49.** The speeds are 46 mph and 50 mph.
51. The integers are 10 and 20.   **53.** The original price was $199.   **55.** The speeds are 20 mph and 40 mph.

57. She deposited $4500 at 5% and $9000 at 6%.
59. Mix 2.5 lb of black tea and 1.5 lb of green tea.
61. The median price the previous year was $215,000.

Section 1.3 Practice Exercises, pp. 70–74

1. a. $2l + 2w$ **b.** 90°   **c.** supplementary   **d.** 180°
3. $\dfrac{40}{13}$ **5.** No solution   **7.** The court is 29.5 ft wide and 59 ft long.   **9.** The sides are 6 m, 8 m, and 10 m.
11. a. The dimensions are $12\frac{1}{2}$ yd by 8 yd.   **b.** Raoul needs 41 yd.   **13.** The width is 9 ft and the length is 11 ft.
15. The angles are 30°, 30°, and 120°.
17. The angles are 15° and 75°.   **19.** $x = 20$; 139°, 41°
21. $x = 27.5$; 60°, 30°   **23.** $x = 18$; 36°, 91°, 53°
25. $x = 23$; 42°, 48°   **27. a.** $r = \dfrac{d}{t}$
b. Franchitti's average speed was 161.6 mph.

29. a. $t = \dfrac{I}{Pr}$ **b.** 7 yr   **31.** $l = \dfrac{A}{w}$ **33.** $P = \dfrac{I}{rt}$

35. $K_1 = K_2 - W$ **37.** $C = \frac{5}{9}(F - 32)$ or $C = \dfrac{5F - 160}{9}$

39. $v^2 = \dfrac{2K}{m}$ **41.** $a = \dfrac{v - v_0}{t}$

43. $v_2 = \dfrac{w}{p} + v_1$ or $v_2 = \dfrac{w + pv_1}{p}$ **45.** $y = \dfrac{c - ax}{b}$

47. $B = \dfrac{3V}{h}$ **49.** $y = -3x + 6$ **51.** $y = \dfrac{5}{4}x - 5$

53. $y = -3x - \dfrac{13}{2}$ **55.** $y = x - 2$

57. $y = -\dfrac{27}{4}x + \dfrac{15}{4}$ **59.** $y = \dfrac{3}{2}x$ **61. a.** $x = z\sigma + \mu$

b. $x = 130$ **63.** a, b, c   **65.** a, b   **67.** $t = \dfrac{12}{6 - r}$

69. $x = \dfrac{-2}{a - 6}$ or $x = \dfrac{2}{6 - a}$ **71.** $P = \dfrac{A}{1 + rt}$

73. $m = \dfrac{T}{g - f}$ **75.** $x = \dfrac{z - by}{a - c}$ or $x = \dfrac{by - z}{c - a}$

Section 1.4 Practice Exercises, pp. 79–81

1. a. linear inequality   **b.** negative   **c.** Both statements are correct.

	Set-Builder Notation	Interval Notation	Graph
3.	$\{x \mid x > 5\}$	$(5, \infty)$	←———→ 5
5.	$\{x \mid -3 < x \le 6\}$	$(-3, 6]$	←——→ -3 6
7.	$\{x \mid x \ge 4\}$	$[4, \infty)$	←———→ 4

9. ←———→ -3 **a.** $\{x \mid x > -3\}$ **b.** $(-3, \infty)$

11. ←———→ -1 **a.** $\{y \mid y \le -1\}$ **b.** $(-\infty, -1]$

13. ←———→ 10   **a.** $\{x \mid x \ge 10\}$ **b.** $[10, \infty)$

15. ←———→ $\frac{13}{6}$ **a.** $\{z \mid z > \frac{13}{6}\}$ **b.** $\left(\dfrac{13}{6}, \infty\right)$

17. ⟵⟶ -12 **a.** $\{t\,|\,t < -12\}$ **b.** $(-\infty, -12)$

19. ⟵⟶ $\frac{13}{8}$ **a.** $\{y\,|\,y < \frac{13}{8}\}$ **b.** $\left(-\infty, \frac{13}{8}\right)$

21. ⟵⟶ -21 **a.** $\{a\,|\,a \le -21\}$ **b.** $(-\infty, -21]$

23. ⟵⟶ -3 **a.** $\{x\,|\,x > -3\}$ **b.** $(-3, \infty)$

25. ⟵⟶ $\frac{9}{10}$ **a.** $\{x\,|\,x \ge \frac{9}{10}\}$ **b.** $\left[\frac{9}{10}, \infty\right)$

27. ⟵⟶ -3 **a.** $\{p\,|\,p < -3\}$ **b.** $(-\infty, -3)$

29. ⟵⟶ 5 **a.** $\{t\,|\,t < 5\}$ **b.** $(-\infty, 5)$

31. ⟵⟶ $-\frac{5}{4}$ **a.** $\{y\,|\,y \ge -\frac{5}{4}\}$ **b.** $\left[-\frac{5}{4}, \infty\right)$

33. ⟵⟶ 1.5 **a.** $\{k\,|\,k \ge 1.5\}$ **b.** $[1.5, \infty)$

35. ⟵⟶ 12 **a.** $\{x\,|\,x \le 12\}$ **b.** $(-\infty, 12]$

37. ⟵⟶ -0.5 **a.** $\{b\,|\,b \le -0.5\}$ **b.** $(-\infty, -0.5]$

39. ⟵⟶ $-\frac{5}{11}$ **a.** $\{c\,|\,c \le -\frac{5}{11}\}$ **b.** $\left(-\infty, -\frac{5}{11}\right]$

41. ⟵⟶ -6 **a.** $\{y\,|\,y < -6\}$ **b.** $(-\infty, -6)$

43. ⟵⟶ -3 **a.** $\{x\,|\,x > -3\}$ **b.** $(-3, \infty)$

45. ⟵⟶ 0 **a.** $\{a\,|\,a \le 0\}$ **b.** $(-\infty, 0]$

47. a. Nadia needs to score at least a 70 but less than 120 to get a B average.   **b.** It would be impossible for Nadia to get an A because she would have to earn 120 on her last quiz and it is impossible to earn more than 100.
49. Boys 8 yr old or older will be on average at least 51 in. tall.   **51.** Boys 6 yr old or younger will be on average no more than 46 in. tall.   **53. a.** She needs to sell in excess of \$375,000.   **b.** She needs to sell in excess of \$1,375,000.
c. The base salary is still the same. The increase comes solely from commission.   **55.** More than 73 jackets must be sold.   **57.** >   **59.** <

Section 1.5 Practice Exercises, pp. 89–93

1. a. union; $A \cup B$ **b.** intersection; $A \cap B$
c. intersection   **d.** $a < x < b$ **e.** union
3. $(-\infty, 1)$ **5.** $[-16, \infty)$
7. a. $\{-3, -1\}$ **b.** $\{-4, -3, -2, -1, 0, 1, 3, 5\}$
9. $[-7, -4)$ **11.** $(-\infty, -4) \cup (2, \infty)$ **13.** No solution
15. $[-7, \infty)$ **17.** $[0, 5)$ **19.** $[-7, \infty)$
21. a. $[-1, 5)$ **b.** $(-2, \infty)$
23. a. $(-1, 3)$ **b.** $\left(-\frac{5}{2}, \frac{9}{2}\right)$ **25. a.** $(0, 2]$ **b.** $(-4, 5]$
27. $[-2, 3]$ ⟵⟶ -2 3
29. $(3, 6)$ ⟵⟶ 3 6

31. $(-\infty, 2]$ ⟵⟶ 2
33. $[8, \infty)$ ⟵⟶ 8
35. No solution   **37.** $-4 \le t$ and $t < \frac{3}{4}$
39. The statement $6 < x < 2$ is equivalent to $6 < x$ and $x < 2$. However, no real number is greater than 6 and also less than 2.
41. The statement $-5 > y > -2$ is equivalent to $-5 > y$ and $y > -2$. However, no real number is less than -5 and also greater than -2.
43. $\left[\frac{5}{2}, 7\right)$ ⟵⟶ $\frac{5}{2}$ 7
45. $(-6, 6]$ ⟵⟶ -6 6
47. $(2, 8)$ ⟵⟶ 2 8
49. $\left[-\frac{10}{3}, -\frac{7}{3}\right]$ ⟵⟶ $-\frac{10}{3}$ $-\frac{7}{3}$
51. $\left[-\frac{1}{2}, \frac{3}{2}\right)$ ⟵⟶ $-\frac{1}{2}$ $\frac{3}{2}$
53. $(-0.4, 0.2)$ ⟵⟶ -0.4 0.2
55. $(-\infty, -2) \cup [2, \infty)$ ⟵⟶ -2 2
57. $(-\infty, 2]$ ⟵⟶ 2
59. $[-6, \infty)$ ⟵⟶ -6
61. $(-\infty, \infty)$ ⟵⟶
63. $(-\infty, -4] \cup (4.5, \infty)$ ⟵⟶ -4 4.5
65. a. $(-10, 8)$ **b.** $(-\infty, \infty)$
67. a. $(-\infty, -8] \cup [1.3, \infty)$ **b.** No solution
69. $\left(-\frac{11}{2}, \frac{7}{2}\right]$ **71.** $\left(-\frac{13}{2}, \infty\right)$ **73.** $(-\infty, 1) \cup (5, \infty)$
75. a. $4800 \le x \le 10{,}800$ **b.** $x < 4800$ or $x > 10{,}800$
77. a. $44\% < x < 48\%$ **b.** $x \le 44\%$ or $x \ge 48\%$
79. All real numbers between $-\frac{3}{2}$ and 6
81. All real numbers greater than 2 or less than -1
83. a. Amy would need 82% or better on her final exam.
b. If Amy scores at least 32% and less than 82% on her final exam she will receive a "B" in the class.
85. $32° \le F \le 42.08°$

Section 1.6 Practice Exercises, pp. 98–99

1. a. absolute; a and $-a$ **b.** Subtract 5 from both sides.
c. y; $-y$ **d.** no; -4
3. $(-\infty, -2] \cup [2, \infty)$ **5.** $[-6, 10)$
7. $7, -7$ **9.** $6, -6$ **11.** No solution
13. $-2, 2$ **15.** $\sqrt{2}, -\sqrt{2}$
17. No solution   **19.** 0 **21.** $4, -\frac{4}{3}$
23. $\frac{9}{2}, -\frac{1}{2}$ **25.** $\frac{10}{7}, -\frac{8}{7}$ **27.** No solution
29. $-12, 28$ **31.** $\frac{5}{2}, -\frac{7}{2}$ **33.** $\frac{7}{3}$

35. No solution   **37.** $\dfrac{3}{2}$ **39.** $-4, \dfrac{16}{5}$

41. $-\dfrac{1}{2}, \dfrac{7}{6}$ **43.** $\dfrac{5}{2}, -\dfrac{3}{2}$ **45.** $-4, \dfrac{1}{3}$

47. $\dfrac{1}{2}$ **49.** $-\dfrac{1}{16}$ **51.** All real numbers

53. $-1.44, -0.4$ **55.** No solution
57. All real numbers

59. No solution   **61.** $|x| = 6$ **63.** $|x| = \dfrac{4}{3}$

Section 1.7 Practice Exercises, pp. 106–108

1. a. $-a; a$ **b.** $-a; >$ **c.** no; $(-\infty, \infty)$
d. includes; excludes

3. $\dfrac{4}{7}, \dfrac{6}{7}$ **5.** $(-3, -1]$

7. $(-\infty, \infty)$ **9. a.** $-5, 5$

b. $(-\infty, -5) \cup (5, \infty)$

c. $(-5, 5)$

11. a. $10, -4$
b. $(-\infty, -4) \cup (10, \infty)$

c. $(-4, 10)$

13. a. No solution   **b.** $(-\infty, \infty)$
c. No solution
15. a. No solution   **b.** $(-\infty, \infty)$
c. No solution
17. a. 0 **b.** $(-\infty, 0) \cup (0, \infty)$

c. No solution   **19. a.** 7 **b.** $(-\infty, 7) \cup (7, \infty)$
c. No solution

21. $(-\infty, -6) \cup (6, \infty)$

23. $[-3, 3]$

25. $(-\infty, \infty)$
27. $(-\infty, -2] \cup [3, \infty)$

29. No solution   **31.** $[-10, 14]$

33. $\left(-\infty, -\dfrac{5}{4}\right] \cup \left[\dfrac{23}{4}, \infty\right)$

35. $\left(-\dfrac{21}{2}, \dfrac{19}{2}\right)$

37. $(-\infty, \infty)$ **39.** No solution
41. -5

43. $(-\infty, 6) \cup (6, \infty)$

45. $(-\infty, 2] \cup [4, \infty)$

47. $(4, 8)$

49. $(-10.5, 4.5)$

51. $|x| > 7$ **53.** $|x - 2| \le 13$ **55.** $|x - 32| \le 0.05$

57. $\left|x - 6\dfrac{3}{4}\right| \le \dfrac{1}{8}$

59. The solution is $\{w \mid 1.99 \le w \le 2.01\}$ or, equivalently in interval notation, $[1.99, 2.01]$. This means that the actual width of the bolt could be between 1.99 cm and 2.01 cm, inclusive.   **61.** b   **63.** a

Chapter 1 Problem Recognition Exercises, p. 109

1. a. 9 **b.** $9, -3$ **c.** $(-3, 9)$ **d.** $(-\infty, -3] \cup [9, \infty)$

2. a. $-\dfrac{22}{5}$ **b.** No solution   **c.** No solution

d. $(-\infty, \infty)$ **3. a.** -7 **b.** $(-\infty, -7)$ **c.** $[-7, \infty)$
4. a. 29 **b.** $[29, \infty)$ **c.** $(-\infty, 29)$

5. a. $\dfrac{1}{2}, -\dfrac{1}{6}$ **b.** $\dfrac{1}{2}$ **6. a.** $(-7, 6]$ **b.** $(-7, 6]$

7. a. $(-5, \infty)$ **b.** $[1, \infty)$

8. a. $(-\infty, -7) \cup (2, \infty)$ **b.** No solution
9. a. Linear equation   **b.** -6
10. a. Linear equation   **b.** 9
11. a. Absolute value inequality   **b.** $[-6, -2]$
12. a. Absolute value inequality   **b.** No solution
13. a. Compound inequality   **b.** $(-6, 9)$
14. a. Compound inequality   **b.** $(-\infty, 2) \cup [7, \infty)$
15. a. Absolute value equation   **b.** $4, -16$
16. a. Absolute value equation   **b.** $1, -6$

17. a. Linear inequality   **b.** $(-\infty, 12]$ **18. a.** Linear inequality   **b.** $[-1, \infty)$ **19. a.** Absolute value inequality   **b.** $(-\infty, -3] \cup [12, \infty)$

20. a. Absolute value inequality   **b.** $(-5, 25)$
21. a. Absolute value equation   **b.** No solution
22. a. Absolute value equation   **b.** $-\dfrac{1}{5}$

23. a. Linear equation   **b.** -16
24. a. Linear equation   **b.** -11
25. a. Compound inequality   **b.** $(8, 12]$
26. a. Compound inequality   **b.** $[-6, -3)$
27. a. Linear equation   **b.** $(-\infty, \infty)$
28. a. Linear equation   **b.** No solution

Chapter 1 Review Exercises, pp. 118–121

1. No solution   **2.** All real numbers

3. -5; conditional equation

4. $\dfrac{1}{8}$; conditional equation

5. -18.075; conditional equation
6. 0.22; conditional equation
7. $\dfrac{31}{6}$; conditional equation

8. 3; conditional equation

9. $-\dfrac{21}{8}$; conditional equation

10. No solution; contradiction
11. No solution; contradiction
12. All real numbers; identity

13. $x, x + 1, x + 2$ **14.** $x, x + 2$ **15.** Distance equals rate times time.   **16.** Simple interest equals principal times rate times time.   **17. a.** $23,856   **b.** $61,344

18. There were 6.7 million men.   **19.** There were 16,600 deaths due to alcohol-related accidents.   **20.** The integers are $-4, -2$, and 0.   **21.** The pieces are 2 ft and $\frac{2}{3}$ ft.

22. She invested $1500 in 6% account and $3500 in 9% account.   **23.** Mix 2 L of 10% solution.

24. Lynn drives 45 mph and Linda drives 30 mph.

25. The width is 9 ft, and the length is 11 ft.

26. $x = 52; 27°$

27. $x = 30; 29°, 61°$ **28.** $y = \dfrac{3x - 4}{2}$ or $y = \dfrac{3}{2}x - 2$

29. $y = 6x + 12$ **30.** $h = \dfrac{S - 2\pi r}{\pi r^2}$ or $h = \dfrac{s}{\pi r^2} - \dfrac{2}{r}$

31. $b = \dfrac{2A}{h}$ **32. a.** $\pi = \dfrac{C}{2r}$ **b.** $\pi \approx 3.14$

33. ← → graph   **a.** $\{x \mid x < -\frac{4}{3}\}$ **b.** $(-\infty, -\frac{4}{3})$

34. ← → graph   **a.** $\{x \mid x \geq -\frac{3}{2}\}$ **b.** $[-\frac{3}{2}, \infty)$

35. ← → graph   **a.** $\{x \mid x < -\frac{8}{13}\}$ **b.** $(-\infty, -\frac{8}{13})$

36. ← → graph   **a.** $\{x \mid x \geq \frac{46}{7}\}$ **b.** $[\frac{46}{7}, \infty)$

37. ← → graph   **a.** $\{x \mid x \leq -\frac{67}{4}\}$ **b.** $(-\infty, -\frac{67}{4}]$

38. ← → graph   **a.** $\{x \mid x \leq \frac{29}{2}\}$ **b.** $(-\infty, \frac{29}{2}]$

39. Dave must earn at least 95 on his fifth test.

40. $C \cap D$ is the set of elements common to both sets C and D. $C \cup D$ is the set of elements in either set C, set D, or both sets.

41. $[-10, 1)$ **42.** $(-\infty, \infty)$ **43.** $(-\infty, \infty)$ **44.** $(-1, 1)$

45. $(-\infty, -3] \cup (-1, \infty)$ **46.** No solution   **47.** $\left(-\dfrac{11}{4}, 4\right]$

48. $[-5, 2)$ **49.** No solution   **50.** No solution

51. $(-\infty, 6] \cup (12, \infty)$ **52.** $(-\infty, -2) \cup [2, \infty)$

53. $\left(-\infty, \dfrac{1}{2}\right)$ **54.** $(-5, \infty)$ **55.** $\left[0, \dfrac{4}{3}\right]$

56. $[-7, -2)$ **57.** All real numbers between -6 and 12

58. a. $140 \leq x \leq 225$ **b.** $x < 140$ or $x > 225$

59. a. $125 \leq x \leq 200$ **b.** $x < 125$ or $x > 200$

60. a. For example, for a 47-year-old person, the maximum recommended heart rate is 173.   **b.** Given a maximum recommended heart rate of 173, the interval is (104, 130).

61. $10, -10$ **62.** $17, -17$ **63.** $1.3, 7.4$

64. $-0.44, 2.54$ **65.** $5, -9$ **66.** $3, 1$

67. No solution   **68.** No solution   **69.** $\dfrac{3}{7}$

70. $-\dfrac{5}{4}$ **71.** $6, \dfrac{4}{5}$ **72.** $-\dfrac{1}{2}$

73. All real numbers   **74.** No solution

75. Both expressions give the distance between 3 and -2.

76. $|x| > 5$ **77.** $|x| < 4$ **78.** $|x| < 6$ **79.** $|x| > \dfrac{2}{3}$

80. $(-\infty, -14] \cup [2, \infty)$ ← → graph

81. $[-11, -5]$ ← → graph

82. $\left(-\infty, \dfrac{1}{7}\right) \cup \left(\dfrac{1}{7}, \infty\right)$ ← → graph

83. $\left(-\infty, -\dfrac{1}{5}\right) \cup \left(-\dfrac{1}{5}, \infty\right)$ ← → graph

84. $\left[-2, -\dfrac{2}{3}\right]$ ← → graph

85. $\left[0, \dfrac{6}{5}\right]$ ← → graph

86. $(2, 22)$ ← → graph

87. $(-12, 0)$ ← → graph

88. $(-\infty, \infty)$ ← → graph
89. $(-\infty, \infty)$ ← → graph
90. $(-\infty, \infty)$ ← → graph
91. $(-\infty, \infty)$ ← → graph

92. No solution   **93.** No solution   **94.** If an absolute value is less than a negative number, there will be no solution.   **95.** If an absolute value is greater than a negative number, then all real numbers are solutions.

96. $0.17 \leq p \leq 0.23$ or, equivalently in interval notation, $[0.17, 0.23]$. This means that the actual percentage of viewers is estimated to be between 17% and 23%, inclusive.

97. $3\frac{1}{8} \leq L \leq 3\frac{5}{8}$ or, equivalently in interval notation, $[3\frac{1}{8}, 3\frac{5}{8}]$. This means that the actual length of the screw may be between $3\frac{1}{8}$ in. and $3\frac{5}{8}$ in., inclusive.

Chapter 1 Test, pp. 121–122

1. 133   **2.** $\dfrac{4}{5}$ **3.** 142,500   **4.** $\dfrac{14}{9}$ **5.** $10, -22$

6. $-8, 2$ **7.** $3, 0$ **8.** No solution   **9.** -1

10. a. Identity   **b.** Contradiction   **c.** Conditional equation

11. The numbers are 18 and 90.

12. a. It took 0.4 hr (24 min).   **b.** The distance is 1.8 mi.

13. Shawnna invested $1000 in the CD.

14. Each side is 27 in.

15. The numbers are $31, 33$, and 35.

16. 24 gal of 20% solution must be used.

17. $y = -2x + 3$

18. $z = \dfrac{x - \mu}{\sigma}$

19. $(34, \infty)$ ← → graph

20. $(-\infty, \frac{18}{5}]$ ← → graph

21. $(-\frac{1}{3}, 2]$ ← → graph

22. $\left[-2, \dfrac{1}{2}\right]$ **23.** $(-2, 13]$ **24.** $(-\infty, -24] \cup [-15, \infty)$

25. $[-3, 0)$ **26.** $(-\infty, \infty)$ **27.** No solution

28. No solution   **29.** $\left(-\infty, -\dfrac{1}{3}\right] \cup \left[\dfrac{17}{3}, \infty\right)$

30. $(-18.75, 17.25)$ **31.** $(-\infty, \infty)$ **32.** $[-3, 8]$

33. It can carry at most seven more passengers.

34. a. $9 \le x \le 33$ **b.** $x < 9$ or $x > 33$

35. $|x - 15.41| \le 0.01$

Chapter 2

Chapter Opener Puzzle

A graph intersects the x-axis at an

$\underset{1}{X} \text{-} \underset{2}{I} \underset{4}{N} \underset{3}{T} \underset{5}{E} \underset{3}{R} \underset{3}{C} \underset{4}{E} \underset{}{P} \underset{}{T}$.

Section 2.1 Practice Exercises, pp. 134–140

1. a. x; y-axis   **b.** ordered   **c.** origin; $(0, 0)$
d. quadrants   **e.** negative   **f.** III   **g.** $Ax + By = C$
h. x-intercept   **i.** y-intercept   **j.** vertical   **k.** horizontal

3. For (x, y), if $x > 0$, $y > 0$ the point is in Quadrant I. If $x < 0$, $y > 0$ the point is in Quadrant II. If $x < 0$, $y < 0$ the point is in Quadrant III. If $x > 0$, $y < 0$ the point is in Quadrant IV.

5.

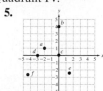

7. 0   **9.** $A(-4, 5)$, II; $B(-2, 0)$, x-axis; $C(1, 1)$, I; $D(4, -2)$, IV; $E(-5, -3)$, III   **11. a.** Yes   **b.** No   **c.** Yes

13. a. No   **b.** Yes   **c.** No

15.

x	y
0	-2
4	4
-1	$-\dfrac{7}{2}$

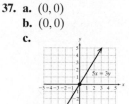

17.

x	y
0	0
5	-1
-5	1

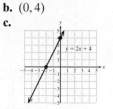

19.

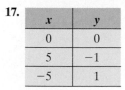

21.

23.

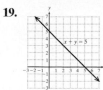

25.

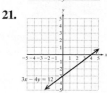

27.

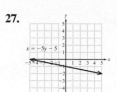

29.

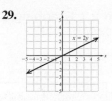

31. To find an x-intercept, substitute $y = 0$ and solve for x. To find a y-intercept, substitute $x = 0$ and solve for y.

33. a. $(9, 0)$ **35. a.** $(4, 0)$
b. $(0, 6)$ **b.** $(0, -2)$
c.   **c.**

     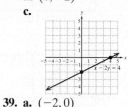

37. a. $(0, 0)$ **39. a.** $(-2, 0)$
b. $(0, 0)$ **b.** $(0, 4)$
c.   **c.**

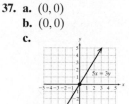 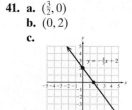   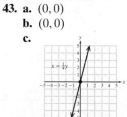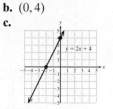

41. a. $\left(\dfrac{3}{2}, 0\right)$ **43. a.** $(0, 0)$
b. $(0, 2)$ **b.** $(0, 0)$
c.   **c.**

     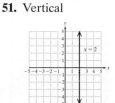

45. a. \$35,000   **b.** \$25,000   **c.** For \$0 in sales the salary is \$10,000.   **d.** Total sales cannot be negative.

47. a. \$1200   **b.** 4 yr   **c.** $(0, 1500)$; The y-intercept represents the initial value.   **d.** $(5, 0)$; The x-intercept indicates that when the computer is 5 yr old, its value is \$0.

49. Horizontal   **51.** Vertical

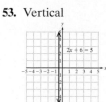

53. Vertical   **55.** Horizontal

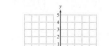   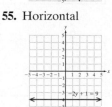

57. x-intercept $(2, 0)$ y-intercept $(0, 3)$

59. x-intercept $(a, 0)$ y-intercept $(0, b)$

Section 2.1 Graphing Calculator Exercises, p. 140

61. $y = \dfrac{2}{3}x - \dfrac{7}{3}$

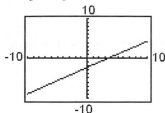

62. $y = -2x - 1$

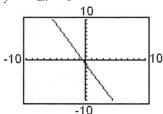

63. $y = 3$

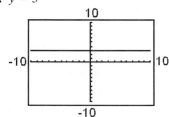

64. $y = -5$

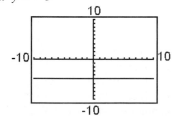

65.

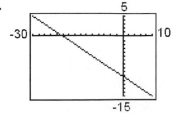

66.

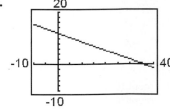

67.

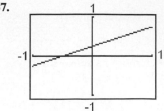

68.

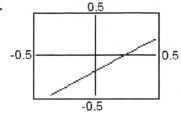

69. The lines look nearly indistinguishable. However, the linear equations are different.

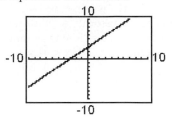

70. The lines look nearly indistinguishable. However, the linear equations are different.

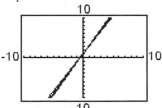

Section 2.2 Practice Exercises, pp. 147–152

1. a. slope; $\dfrac{y_2 - y_1}{x_2 - x_1}$ **b.** parallel, same   **c.** right   **d.** -1

3. a. 4   **b.** 8   **c.** 6

5. No x-intercept; y-intercept: $(0, 2)$

7. $m = -\dfrac{24}{7}$ **9.** $m = \dfrac{1}{9}$ **11.** $m = \dfrac{4}{100}$ or $\dfrac{1}{25}$

13. $m = \dfrac{1}{2}$ **15.** $m = -\dfrac{5}{3}$ **17.** $m = \dfrac{4}{5}$

19. $m = -\dfrac{3}{4}$ **21.** m is undefined.   **23.** $m = 0$

25. $m = \dfrac{1}{2}$ **27.** $m = -\dfrac{1}{6}$ **29.** $m = 0$

31. Line rising to the right: positive slope   Line falling to the right: negative slope   Horizontal line: zero slope   Vertical line: undefined slope   **33.** $m = 0$

35. $m = \dfrac{1}{10}$ **37.** $m = -1$ **39.** No, because the product of slopes of perpendicular lines must be -1. The product of two positive numbers is not negative.   **41. a.** $m = 5$ **b.** $m = -\dfrac{1}{5}$ **43. a.** $m = -\dfrac{4}{7}$ **b.** $m = \dfrac{7}{4}$ **45. a.** $m = 0$ **b.** m is undefined.   **47.** $m_1 = 2; m_2 = -\dfrac{1}{2}$; perpendicular

49. $m_1 = -1; m_2 = 3$; neither   **51.** m_1 is undefined; $m_2 = 0$; perpendicular. One line is horizontal and one is vertical.   **53.** $m_1 = 1; m_2 = 1$; parallel   **55. a.** $m = 20.25$ **b.** The number of cell phone subscriptions increased at a rate of 20.25 million per year during this period.   **57. a.** $m = 6$ **b.** The weight of boys tends to increase at a rate of 6 lb/yr during this period of growth.

59. a. $m = 2$ **b.** $m = 2$ **c.** $m = 2$
61. a. $m = -2$ **b.** $m = 2$ **c.** $m = 5$
63. For example: $(1, 2)$ **65.** For example: $(2, 0)$

    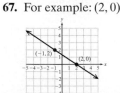

67. For example: $(2, 0)$

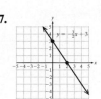

Section 2.3 Practice Exercises, pp. 160–164

1. a. $y = mx + b$ **b.** standard **c.** horizontal **d.** vertical **e.** slope; y-intercept **f.** $y - y_1 = m(x - x_1)$
3. a. 0 **b.** undefined

5. If the slope of one line is the opposite of the reciprocal of the other then the lines are perpendicular.   **7.** Slope: $-\dfrac{2}{3}$; y-intercept: $(0, -4)$ **9.** Slope: 3; y-intercept: $(0, 2)$

11. Slope: -17; y-intercept: $(0, 0)$ **13.** Slope: 0; y-intercept: $(0, 9)$ **15.** Slope: $-\dfrac{2}{3}$; y-intercept: $\left(0, \dfrac{3}{4}\right)$

17. Slope: 0.625; y-intercept: $(0, -1.2)$
19. d   **21.** f   **23.** b
25.   **27.**

29.

31. $y = -\dfrac{A}{B}x + \dfrac{C}{B}$. The slope is given by $m = -\dfrac{A}{B}$. The y-intercept is $\left(0, \dfrac{C}{B}\right)$.   **33.** Perpendicular

35. Parallel   **37.** Neither   **39.** $y = 3x + 5$

41. $y = 2x - 11$ **43.** $y = -\dfrac{4}{5}x + 8$ **45.** $y = 3x - 2$ or $3x - y = 2$ **47.** $y = 2x + 3$ or $2x - y = -3$
49. $y = -3x - 11$ or $3x + y = -11$ **51.** $y = -\dfrac{4}{5}x + \dfrac{9}{5}$ or $4x + 5y = 9$ **53.** $y = -\dfrac{3}{4}x + 4$ or $4x + 3y = 12$
55. $y = x + 6$ or $x - y = -6$ **57.** $y = 2$
59. $y = -\dfrac{3}{4}x + \dfrac{17}{4}$ or $3x + 4y = 17$ **61.** $y = \dfrac{4}{3}x - 2$ or $4x - 3y = 6$ **63.** $y = \dfrac{3}{4}x - \dfrac{13}{2}$ or $3x - 4y = 26$
65. $y = -\dfrac{1}{5}x - \dfrac{13}{5}$ or $x + 5y = -13$ **67.** $y = \dfrac{3}{2}x - 6$ or $3x - 2y = 12$ **69.** $y = \dfrac{3}{2}x + 1$ or $3x - 2y = -2$
71. $y = -\dfrac{1}{2}x + \dfrac{7}{2}$ or $x + 2y = 7$ **73.** $y = -3$
75. $x = 2$ **77.** $y = 5$ **79.** $x = 5$ **81.** $x = -2$ is not in slope-intercept form. No y-intercept, undefined slope.
83. $y = 3$ is in slope-intercept form, $y = 0x + 3$. Slope is 0 and y-intercept is $(0, 3)$.   **85.** $y = -2x + 3$ **87.** $y = 2$

Section 2.3 Graphing Calculator Exercises, p. 164

89. The lines have the same slope but different y-intercepts; they are parallel lines.

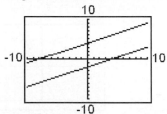

90. The lines have the same slope but different y-intercepts; they are parallel lines.

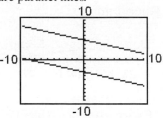

91. The lines have different slopes but the same y-intercept.

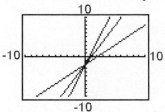

92. The lines have different slopes but the same y-intercept.

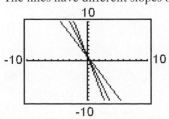

93. The lines are perpendicular.

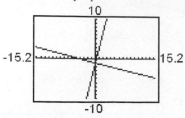

94. The lines are perpendicular.

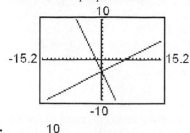

95.

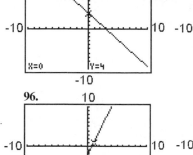

96.

Chapter 2 Problem Recognition Exercises, p. 165

1. b, f   **2.** a, c, d, h   **3.** a   **4.** b, g   **5.** c, e
6. a   **7.** c, h   **8.** b   **9.** e   **10.** g   **11.** c, h
12. a   **13.** g   **14.** e   **15.** h   **16.** f   **17.** e
18. g   **19.** d, h   **20.** b, f

Section 2.4 Practice Exercises, pp. 169–176

1. model
3. a. $m = -\frac{1}{3}$ **b.** $y = -\frac{1}{3}x - 1$ or $x + 3y = -3$
c.

5. a. $m = 0$ **b.** $y = 3$
c.

7. a. $y = 120x + 65$
b.

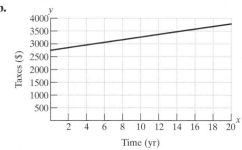

c. $(0, 65)$; The base cost to rent the car is $65.   **d.** A 2-day rental costs $305. A 5-day rental costs $665, and a 7-day rental costs $905.   **e.** Yes; $799 is less expensive than $905.
f. $577.70   **g.** No, the car cannot be driven for a negative number of days.
9. a. $y = 52x + 2742$
b.

c. $m = 52$. Taxes increase at a rate of $52 per year.
d. $(0, 2742)$. Initial year ($x = 0$) taxes were $2742.
e. 10 yr: $3262; 15 yr: $3522
11. a. 0.8 mi, 2.4 mi, and 3.2 mi   **b.** 21 sec

13. a. $73.30   **b.** $54.50; The approximate value differs from the actual value by $1.30.   **c.** $m = 9.4$; The amount spent per person on video games increased by an average rate of $9.40 per year.   **d.** $(0, 35.7)$; The y-intercept means that the average amount spent on video games per person was $35.70 in the year 2006.
15. a. $y = 0.026x + 3.3$ **b.** 3.82 m   **c.** 5.38 m   **d.** Yes. The linear model can only be used to approximate the winning heights.   **e.** $m = 0.026$. Winning heights have increased at a rate of 0.026 m/yr.
17. a. $y = 15x + 155$ **b.** The slope is 15 and means that the number of associate degrees awarded in the United States increased by 15 thousand per year.   **c.** 830 thousand
19. a.

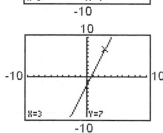

b. $y = -175x + 1087.5$ **c.** Approximately 388 hot dogs would be sold.
21. a.

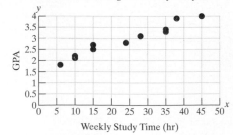

b. yes   **c.** $y = 0.05x + 1.7$ **d.** 3.2   **e.** This model is not reasonable for study times greater than 46 hours per week, because the GPA would exceed 4.0.   **23.** Collinear
25. Not collinear

Section 2.4 Graphing Calculator Exercises, p. 176

27. 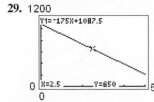   **28.**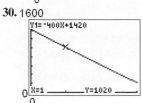

29. 1200   1200

30. 1600   1600

Chapter 2 Review Exercises, p. 182–184

1.   c   **2.** No   **3.** Yes

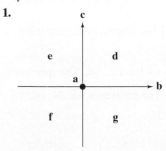

4. $A(0,0); B(2,1); C(0,-4); D(-2,-4); E(-2,0);$
$F(-5,1); G(4,-3)$

5.

x	y
0	3
-2	0
1	$\dfrac{9}{2}$

6.

x	y
0	$\dfrac{13}{2}$
5	$\dfrac{13}{2}$
-4	$\dfrac{13}{2}$

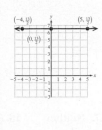

7.

x	y
4	0
4	1
4	-2

8. x-intercept $(-3,0)$;
y-intercept $(0,2)$

9. x- and y-intercept $(0,0)$

10. No x-intercept;
y-intercept $(0,3)$

11. x-intercept $(-2,0)$;
no y-intercept

12. a. $m = \dfrac{1}{2}$ **b.** $m = -3$ **c.** $m = 0$

13. For example: $y = 2x$

14. For example: $y = -\dfrac{3}{4}x$

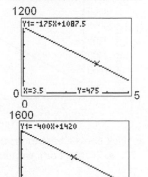

15. $m = 2$ **16.** $m = \dfrac{7}{10}$ **17.** $m = 0$

18. Undefined slope   **19.** Perpendicular
20. Perpendicular   **21.** Neither   **22.** Parallel
23. a. $m = 53$ **b.** There is an increase of 53 students per year.

24. $m = \dfrac{3}{4}$ **25. a.** $y = k$ **b.** $y - y_1 = m(x - x_1)$

c. $Ax + By = C$ **d.** $x = k$ **e.** $y = mx + b$

26. $y = \dfrac{1}{9}x + 6$ or $x - 9y = -54$

27. $y = -\dfrac{2}{3}x + 2$ or $2x + 3y = 6$

28. $y = \dfrac{10}{3}x + \dfrac{77}{3}$ or $10x - 3y = -77$

29. $y = 3x - 20$ or $3x - y = 20$

30. $y = -\dfrac{4}{3}x - 3$ or $4x + 3y = -9$

31. a. $y = -2$ **b.** $x = -3$ **c.** $x = -3$ **d.** $y = -2$
32. Yes, (a) and (d) are the same; (b) and (c) are the same.
33. a. $y = 0.25x + 20$

b.

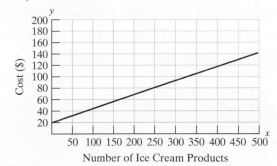

c. The daily fixed cost is $20 if no ice cream is sold.
d. $132.50   **e.** 0.25   **f.** The daily cost increases at a rate of $0.25 per ice cream product.

34. a.

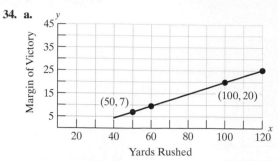

Yards Rushed

b. $y = 0.26x - 6$
c. The margin of victory would be -6 points, which means they would lose by 6 points.

Chapter 2 Test, pp. 185–186

1. $(0, -9)\ (6, 0)\ (4, -3)$

2. a. False, the product is positive in Quadrant III also.
b. False, the quotient is negative in Quadrant II also.   **c.** True
d. True   **3.** Yes   **4.** To find the x-intercept, let $y = 0$ and solve for x. To find the y-intercept, let $x = 0$ and solve for y.
5. x-intercept: $(4, 0)$;   **6.** x-intercept: $(-4, 0)$;
y-intercept: $(0, -3)$ no y-intercept

7. x- and y-intercept: $(0, 0)$ **8.** no x-intercept;
                                         y-intercept: $(0, -3)$

9. a. $m = \dfrac{5}{8}$ **b.** $m = \dfrac{6}{5}$ **10. a.** The slopes are the same.
b. The slope of one line is the opposite of the reciprocal of the slope of the other.   **11. a.** -7 **b.** $\dfrac{1}{7}$ **12.** Neither
13. Parallel   **14. a.** $y = \frac{3}{4}x + 1$ **b.** Slope: $\frac{3}{4}$;
y-intercept: $(0, 1)$
c.

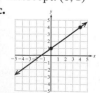

15. a. Perpendicular   **b.** Parallel   **c.** Perpendicular
d. Neither   **16. a.** For example: $y = 3x + 2$
b. For example: $x = 2$ **c.** For example: $y = 3; m = 0$
d. For example: $y = -2x$ **17.** $y = -2x + \dfrac{31}{2}$
18. $y = \frac{3}{2}x - 6$ or $3x - 2y = 12$
19. $y = 2x - 11$ or $2x - y = 11$
20. $y = \dfrac{1}{3}x + \dfrac{1}{3}$ **21. a.** $y = 300x + 800$
b.

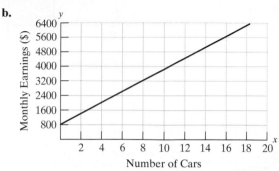

Number of Cars

c. Jack earns a base monthly salary of $800.   **d.** $5900
22. a. $(0, 66)$. For a woman born in 1940, life expectancy was about 66 yr.   **b.** $m = \dfrac{3}{10}$; Life expectancy rises at a rate of 3 yr for every 10 yr.   **c.** $y = \dfrac{3}{10}x + 66$
d. 82.2 yr old

Chapters 1–2 Cumulative Review Exercises, p. 187

1. $-\dfrac{5}{7}$ **2.** 1   **3.** No solution   **4.** The speed of the faster car is 168 mph.   **5.** The measures are $28°$ and $62°$.
6. $y = \dfrac{d - ax}{b}$ or $-\dfrac{ax - d}{b}$ **7.** $\left(-\infty, -\dfrac{7}{10}\right)$
8. $(-3, 1]$ **9.** $(-\infty, -3]\cup\left[-\dfrac{5}{9}, \infty\right)$
10. $36, -12$ **11.** $\left(-\dfrac{3}{4}, \dfrac{7}{4}\right)$ **12.** $-\dfrac{8}{7}, -2$
13.

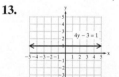

14.

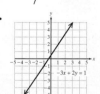

15.

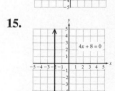

16. Perpendicular   **17.** Neither   **18.** $y = -6x + 9$
19. $y = 2x + 2$ **20.** $x = 7$

Chapter 3

Chapter Opener Puzzle

Section 3.1 Practice Exercises, pp. 194–198

1. a. relation   **b.** domain   **c.** range

3. a. Vertical   **b.** Undefined   **c.** $\left(\dfrac{7}{2}, 0\right)$ **d.** None

5. a. Slanted   **b.** $m = \dfrac{3}{2}$ **c.** $\left(\dfrac{4}{3}, 0\right)$ **d.** $(0, -2)$

7. a. {(Northeast, 54.1), (Midwest, 65.6), (South, 110.7), (West, 70.7)}   **b.** Domain: {Northeast, Midwest, South, West}; Range: {54.1, 65.6, 110.7, 70.7}   **9. a.** {(USSR, 1961), (USA, 1962), (Poland, 1978), (Vietnam, 1980), (Cuba, 1980)}   **b.** Domain: {USSR, USA, Poland, Vietnam, Cuba}; range: {1961, 1962, 1978, 1980}

11. a. {(A, 1), (A, 2), (B, 2), (C, 3), (D, 5), (E, 4)}   **b.** Domain: {A, B, C, D, E}; range: {1, 2, 3, 4, 5}

13. a. {(−4, 4), (1, 1), (2, 1), (3, 1), (4, −2)}   **b.** Domain: {−4, 1, 2, 3, 4}; range: {−2, 1, 4}

15. Domain: [0, 4]; range: [−2, 2]   **17.** Domain: [−5, 3]; range: [−2.1, 2.8]   **19.** Domain: $(-\infty, 2]$; range: $(-\infty, \infty)$ **21.** Domain: $(-\infty, \infty)$; range: $(-\infty, \infty)$ **23.** Domain: {−3}; range: $(-\infty, \infty)$ **25.** Domain: $(-\infty, 2)$; range: $[-1.3, \infty)$ **27.** Domain: {−3, −1, 1, 3}; range: {0, 1, 2, 3}

29. Domain: [−4, 5); range: {−2, 1, 3}   **31. a.** 2.85   **b.** 9.33   **c.** Dec.   **d.** Nov.   **e.** 7.63   **f.** {Jan., Feb., Mar., Apr., May, June, July, Aug., Sept., Oct., Nov., Dec.}

33. a. 31.876 million or 31,876,000   **b.** The year 2012   **35. a.** For example: {(Julie, New York), (Peggy, Florida), (Stephen, Kansas), (Pat, New York)}   **b.** Domain: {Julie, Peggy, Stephen, Pat}; range: {New York, Florida, Kansas}

37. $y = 2x - 1$ **39.** $y = x^2$

Section 3.2 Practice Exercises, pp. 205–210

1. a. function   **b.** vertical   **c.** $2x + 1$ **d.** domain; range   **e.** denominator; negative   **f.** −2   **g.** 3   **h.** (1, 6)

3. a. {(−2, −4), (−1, −1), (0, 0), (1, −1), (2, −4)}   **b.** {−2, −1, 0, 1, 2}   **c.** {−4, −1, 0}   **5.** Function

7. Not a function   **9.** Function   **11.** Not a function   **13.** Function   **15.** Not a function   **17.** −11   **19.** 1

21. 2   **23.** $6t - 2$ **25.** 7   **27.** 4   **29.** 4

31. $6x + 4$ **33.** $-4x^2 - 8x + 1$ **35.** $-\pi^2 + 4\pi + 1$

37. 7   **39.** $-6a - 2$ **41.** $|-c - 2|$ **43.** 1   **45.** 7

47. −18.8   **49.** −7   **51.** 2π **53.** −5   **55.** 4

57. a. 2   **b.** 1   **c.** 1   **d.** $x = -3$ **e.** $x = 1$ **f.** $(-\infty, 3]$ **g.** $(-\infty, 3]$ **59. a.** 3   **b.** $H(4)$ is not defined because 4 is not in the domain of H.   **c.** 4   **d.** $x = -3$ and $x = 2$ **e.** All x on the interval $[-2, 1]$ **f.** $[-4, 4)$ **g.** [2, 5)

61. a. −4   **b.** 0   **c.** −3   **d.** $x = -1$ **e.** There are no such values of x.   **f.** $(-\infty, \infty)$ **g.** $(-\infty, -3] \cup (-2, \infty)$

63. The domain is the set of all real numbers for which the denominator is not zero. Set the denominator equal to zero, and solve the resulting equation. The solution(s) to the

equation must be excluded from the domain. In this case, setting $x - 2 = 0$ indicates that $x = 2$ must be excluded from the domain. The domain is $(-\infty, 2) \cup (2, \infty)$.

65. $(-\infty, -6) \cup (-6, \infty)$ **67.** $(-\infty, 0) \cup (0, \infty)$

69. $(-\infty, \infty)$ **71.** $[-7, \infty)$ **73.** $[3, \infty)$ **75.** $\left(-\infty, \frac{1}{2}\right]$

77. $(-\infty, \infty)$ **79.** $(-\infty, \infty)$ **81. a.** $h(1) = 64$ and $h(1.5) = 44$ **b.** $h(1) = 64$ means that after 1 sec, the height of the ball is 64 ft. $h(1.5) = 44$ means that after 1.5 sec, the height of the ball is 44 ft.   **83. a.** $d(1) = 11.5$ and $d(1.5) = 17.25$ **b.** $d(1) = 11.5$ means that after 1 hr, the distance traveled is 11.5 mi. $d(1.5) = 17.25$ means that after 1.5 hr, the distance traveled is 17.25 mi.   **85. a.** $P(0) = 0\%$

$$P(5) = \frac{100}{3} \approx 33.3\% \quad P(10) = \frac{200}{3} \approx 66.7\%$$

$$P(15) = \frac{900}{11} \approx 81.8\% \quad P(20) = \frac{800}{9} \approx 88.9\%$$

$$P(25) = \frac{2500}{27} \approx 92.6\%.$$ If Brian studies 25 hours he will

get a score of 92.6%.   **b.** $A(0, 0)$ $B\left(5, \dfrac{100}{3}\right)$ $C\left(10, \dfrac{200}{3}\right)$

$D\left(15, \dfrac{900}{11}\right)$ $E\left(20, \dfrac{800}{9}\right)$ $F\left(25, \dfrac{2500}{27}\right)$ **87.** $(4, \infty)$

89. {5, −3, 4}   **91.** {0, 2, 6, 1}   **93.** −7   **95.** 0 and 2   **97.** 6

Section 3.2 Graphing Calculator Exercises, p. 210

98.

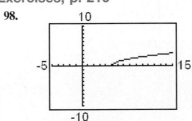

99.

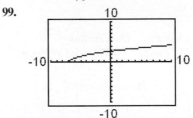

100. a.

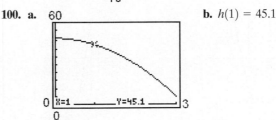

b. $h(1) = 45.1$

101. a.

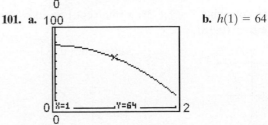

b. $h(1) = 64$

Section 3.3 Practice Exercises, pp. 217–221

1. a. linear   **b.** constant   **c.** quadratic   **d.** parabola   **e.** 0   **f.** y **3. a.** Yes   **b.** $\{7, 2, -5\}$ **c.** $\{3\}$
5. $f(3) = 9$ means that 9 lb of force is required to stretch the spring 3 in. $f(10) = 30$ means that 30 lb of force is required to stretch the spring 10 in.   **7.** horizontal
9. Domain $(-\infty, \infty)$; range $\{2\}$

11.

x	$f(x)$	x	$f(x)$
-2	$-\frac{1}{2}$	$\frac{1}{4}$	4
-1	-1	$\frac{1}{2}$	2
$-\frac{1}{2}$	-2	1	1
$-\frac{1}{4}$	-4	2	$\frac{1}{2}$

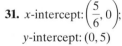

13.

x	$h(x)$
-2	-8
-1	-1
0	0
1	1
2	8

15.

x	$q(x)$
-2	4
-1	1
0	0
1	1
2	4

17. Quadratic   **19.** Linear   **21.** Constant
23. None of these   **25.** Linear   **27.** None of these

29. x-intercept: $(2, 0)$;
    y-intercept: $(0, -10)$

31. x-intercept: $\left(\frac{5}{6}, 0\right)$;
    y-intercept: $(0, 5)$

33. x-intercept: none;
    y-intercept: $(0, 18)$

35. x-intercept: $\left(-\frac{3}{8}, 0\right)$;
    y-intercept: $\left(0, \frac{1}{4}\right)$

37. a. $x = -1$ **b.** $f(0) = 1$ **39. a.** $x = -2$ and $x = 2$
b. $f(0) = -2$ **41. a.** None   **b.** $f(0) = 2$
43. a. $(-\infty, \infty)$ **b.** $(0, 0)$ **c.** vi   **45. a.** $(-\infty, \infty)$
b. $(0, 1)$ **c.** viii   **47. a.** $[-1, \infty)$ **b.** $(0, 1)$ **c.** vii
49. a. $(-\infty, 3) \cup (3, \infty)$ **b.** $\left(0, -\frac{1}{3}\right)$ **c.** ii
51. a. $(-\infty, \infty)$ **b.** $(0, 2)$ **c.** iv   **53. a.** Linear
b. $G(90) = 82.5$. This means that if the student gets a 90% score on her final exam, then her overall course average is 82.5%.   **c.** $G(50) = 72.5$. This means that if the student gets a 50% score on her final exam, then her overall course average is 72.5%.   **55.** $f(x) = x^2 + 2$ **57.** $f(x) = 3$
59. $f(x) = \frac{1}{2}x - 2$

Section 3.3 Graphing Calculator Exercises, p. 221

61.

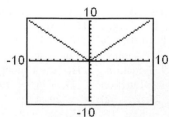

62.

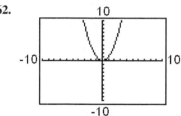

63.

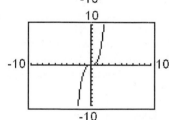

64.

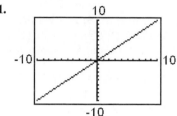

65.

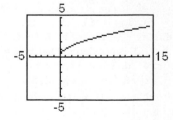

66.

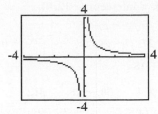

67. x-intercept: (8, 0); y-intercept: (0, 1)   **68.** x-intercept: (−6, 0); y-intercept: (0, −3)   **69.** x-intercept: (−5, 0); y-intercept: (0, 4)   **70.** x-intercept: (2, 0); y-intercept: (0, −7)

Chapter 3 Problem Recognition Exercises, p. 222

1. a, c, d, f, g   **2.** a, b, d, f, h   **3.** 4   **4.** 2
5. $\{0, 1, \frac{1}{2}, -3, 2\}$ **6.** $\{4, 3, 6, 1, 10\}$ **7.** $[-2, 4]$
8. $[0, \infty)$ **9.** 0   **10.** $\left(\frac{9}{5}, 0\right)$ **11.** $(0, -1), (0, 1)$
12. 1   **13.** c   **14.** d   **15.** 3

Section 3.4 Practice Exercises, pp. 226–229

1. a. $f(x); g(x)$ **b.** $g(x)$ **c.** $f(g(x))$
3. $(f + g)(x) = 2x^2 + 5x + 4$
5. $(g - f)(x) = 2x^2 + 3x - 4$
7. $(f \cdot h)(x) = x^3 + 4x^2 + x + 4$
9. $(g \cdot f)(x) = 2x^3 + 12x^2 + 16x$
11. $\left(\dfrac{h}{f}\right)(x) = \dfrac{x^2 + 1}{x + 4}, x \neq -4$
13. $\left(\dfrac{f}{g}\right)(x) = \dfrac{x + 4}{2x^2 + 4x}, x \neq 0, x \neq -2$
15. $(f \circ g)(x) = 2x^2 + 4x + 4$
17. $(g \circ f)(x) = 2x^2 + 20x + 48$ **19.** $(k \circ h)(x) = \dfrac{1}{x^2 + 1}$
21. $(k \circ g)(x) = \dfrac{1}{2x^2 + 4x}, x \neq 0, x \neq -2$ **23.** No
25. $(f \circ g)(x) = 25x^2 - 15x + 1; (g \circ f)(x) = 5x^2 - 15x + 5$
27. $(f \circ g)(x) = |x^3 - 1|; (g \circ f)(x) = |x|^3 - 1$
29. $(h \circ h)(x) = 25x - 24$ **31.** 0   **33.** −64   **35.** 2
37. 1   **39.** $\dfrac{1}{64}$ **41.** 0   **43.** Undefined
45. $\dfrac{4}{9}$ **47.** −2   **49.** 2   **51.** 0   **53.** 1   **55.** 0
57. Undefined   **59.** −1   **61.** 2   **63.** 2   **65.** −1
67. −2   **69.** 3   **71.** −6   **73.** −1   **75.** 4
77. 0   **79.** 2   **81. a.** $P(x) = 3.78x - 1$ **b.** $188
83. a. $F(t) = 0.2t + 6.4$; F represents the amount of child support (in billion dollars) not paid.   **b.** $F(4) = 7.2$ means that in 2004, $7.2 billion of child support was not paid.

Difference Between Child Support Due and Child Support Paid, United States

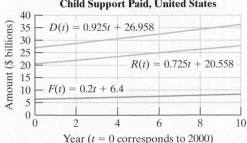

85. a. $(D \circ r)(t) = 560t$; This function represents the total distance Joe travels as a function of time that he rides.
b. 5600 ft

Chapter 3 Review Exercises, pp. 233–236

1. Domain $\left\{\frac{1}{3}, 6, \frac{1}{4}, 7\right\}$; range $\left\{10, -\frac{1}{2}, 4, \frac{2}{5}\right\}$
2. Domain $\{-3, -1, 0, 2, 3\}$; range $\left\{-2, 0, 1, \frac{5}{2}\right\}$
3. Domain $[-3, 9]$; range $[0, 60]$
4. Domain $[-4, 4]$; range $[1, 3]$
5.

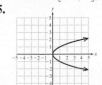

6.

7. a. Not a function   **b.** $[1, 3]$ **c.** $[-4, 4]$
8. a. Function   **b.** $(-\infty, \infty)$ **c.** $(-\infty, 0.35]$
9. a. Function   **b.** $\{1, 2, 3, 4\}$ **c.** $\{3\}$
10. a. Not a function   **b.** $\{0, 4\}$ **c.** $\{2, 3, 4, 5\}$
11. a. Not a function   **b.** $\{4, 9\}$ **c.** $\{2, -2, 3, -3\}$
12. a. Function   **b.** $\{6, 7, 8, 9\}$ **c.** $\{9, 10, 11, 12\}$
13. −4   **14.** 2   **15.** 2   **16.** $6t^2 - 4$ **17.** $6b^2 - 4$
18. $6\pi^2 - 4$ **19.** $6\square^2 - 4$ **20.** 20   **21.** $(-\infty, \infty)$
22. $(-\infty, 11)\cup(11, \infty)$ **23.** $[8, \infty)$ **24.** $[-2, \infty)$
25. a. $98   **b.** $123   **c.** $148
26.

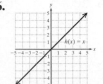

27.

28.

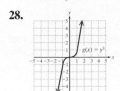

29.

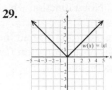

30.

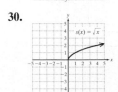

31.

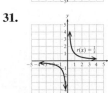

32.

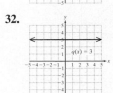

33.

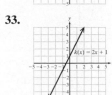

34. a. $s(4) = 4, s(-3) = 25, s(2) = 0, s(1) = 1, s(0) = 4$
b. $(-\infty, \infty)$ **35. a.** $r(2)$ is not a real number, $r(4) = 0, r(5) = 2, r(8) = 4$ **b.** $[4, \infty)$
36. a. $h(-3) = -\dfrac{1}{2}, h(0) = -1, h(2) = -3, h(5) = \dfrac{3}{2}$
b. $(-\infty, 3)\cup(3, \infty)$

37. a. $k(-5) = -2, k(-4) = -1, k(-3) = 0, k(2) = -5$
b. $(-\infty, \infty)$ **38.** x-intercept: $\left(\dfrac{7}{4}, 0\right)$; y-intercept: $(0, -7)$

39. x-intercept: $\left(\dfrac{9}{2}, 0\right)$; y-intercept: $(0, 9)$

40. a. $b(0) = 4.5$. In 1985 consumption was 4.5 gal of bottled water per capita. $b(7) = 9.4$. In 1992 consumption was 9.4 gal of bottled water per capita.   **b.** $m = 0.7$. Consumption increased at a rate of 0.7 gal/year.
41. -1 **42.** 0   **43.** $x = 0$ and $x = 4$
44. There are no values of x for which $g(x) = 4$.
45. $(-4, \infty)$ **46.** $(-\infty, 1]$
47. $(f - g)(x) = 2x^3 + 9x - 7$
48. $(f + g)(x) = -2x^3 - 7x - 7$
49. $(f \cdot n)(x) = \dfrac{x - 7}{x - 2}, x \neq 2$
50. $(f \cdot m)(x) = x^3 - 7x^2$
51. $\left(\dfrac{f}{g}\right)(x) = \dfrac{x - 7}{-2x^3 - 8x}; x \neq 0$
52. $\left(\dfrac{g}{f}\right)(x) = \dfrac{-2x^3 - 8x}{x - 7}, x \neq 7$
53. $(m \circ f)(x) = (x - 7)^2$ or $x^2 - 14x + 49$
54. $(n \circ f)(x) = \dfrac{1}{x - 9}, x \neq 9$

55. 100   **56.** $\dfrac{1}{8}$ **57.** -167 **58.** -10
59. a. $(2x + 1)^2$ or $4x^2 + 4x + 1$ **b.** $2x^2 + 1$
c. No, $f \circ g \neq g \circ f$ **60.** $\dfrac{1}{4}$ **61.** -3 **62.** 0
63. -1 **64.** 4   **65.** 1

Chapter 3 Test, pp. 236–237

1. a. Not a function   **b.** $\{-3, -1, 1, 3\}$
c. $\{-2, -1, 1, 3\}$ **2. a.** Function   **b.** $(-\infty, \infty)$
c. $(-\infty, 0]$ **3.** To find the x-intercept(s), solve for the real solutions of the equation $f(x) = 0$. To find the y-intercept, find $f(0)$.
4.   **5.**

6.   **7.**

8. $(-\infty, -7) \cup (-7, \infty)$ **9.** $[-7, \infty)$ **10.** $(-\infty, \infty)$
11. a. $r(-2) = 9, r(-1) = 4, r(0) = 1, r(2) = 1, r(3) = 4$
b. $(-\infty, \infty)$ **12. a.** $s(0) = 36$. In 1985 the per capita consumption was 36 gal. $s(7) = 47.2$. In 1992 the per capita consumption was 47.2 gal.   **b.** $m = 1.6$. Consumption increases at a rate of 1.6 gal/year.   **13.** 1   **14.** 2
15. $(-1, 7]$ **16.** $[-1, 4)$ **17.** False   **18.** $(6, 0)$
19. $x = 6$ **20.** All x in the interval $[1, 3]$ and $x = 5$
21. Quadratic   **22.** Linear   **23.** Constant

24. None of these   **25.** x-intercept: $(-12, 0)$; y-intercept: $(0, 9)$
26. $\left(\dfrac{f}{g}\right)(x) = \dfrac{x - 4}{x^2 + 2}$ **27.** $(h \cdot g)(x) = \dfrac{x^2 + 2}{x}, x \neq 0$
28. $(g \circ f)(x) = x^2 - 8x + 18$
29. $(h \circ f)(x) = \dfrac{1}{x - 4}, x \neq 4$ **30.** -48 **31.** $-\dfrac{3}{2}$
32. $\dfrac{1}{18}$ **33.** 18   **34.** $\left(\dfrac{g}{f}\right)(x) = \dfrac{x^2 + 2}{x - 4}, x \neq 4$

Chapters 1–3 Cumulative Review Exercises, pp. 237–238

1. -1 **2.** 4   **3.** -7 **4.** -15 **5.** 60 mi
6. a. 1   **b.** 33   **7. a.** $[-2, \infty)$ **b.** $(-\infty, 3)$ **8.** $-\dfrac{1}{5}$
9. No   **10. a.** x-intercept: $\left(\dfrac{10}{3}, 0\right)$ y-intercept: $(0, -2)$
b. $m = \dfrac{3}{5}$
c.

11. a. no x-intercept; y-intercept: $(0, 3)$ **b.** $m = 0$
c.
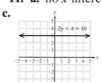

12. $x = 7$ **13.** Domain: $\{3, 4\}$; range: $\{-1, -5, -8\}$; not a function   **14.** $y = -2x - 2$ **15.** $y = -4x$
16. a. 31   **b.** $4(x - 5)^2$ **17.** $(-\infty, 15) \cup (15, \infty)$
18. She used 10 L of the 20% solution.   **19.** Los Angeles gets 16.9 in. per year and Seattle gets 33.1 in. per year.
20. $21x - 62y + 12$

Chapter 4

Chapter Opener Puzzle

4	A 2	3	6	B 5	1
6	1	5	2	C 4	3
5	3	E 4	1	2	F 6
2	6	1	4	3	5
3	4	6	5	1	2
D 1	5	2	3	6	4

Section 4.1 Practice Exercises, pp. 245–249

1. a. system **b.** solution **c.** intersect **d.** consistent
e. no **f.** dependent **g.** independent
3. $(2, 11)$ is a solution. **5.** $(-1, 4)$ is a solution.
7. None **9. a.** Consistent **b.** Independent
c. One solution **11. a.** Inconsistent **b.** Independent
c. Zero solutions **13. a.** Consistent **b.** Dependent
c. Infinitely many solutions
15. $(3, -2)$ **17.** $(1, 1)$

19. $(3, -4)$ **21.** $(4, 5)$

23. No solution; inconsistent system **25.** Infinitely many solutions; $\left\{(x, y) \mid y = \dfrac{2}{3}x - 1\right\}$; dependent equations

 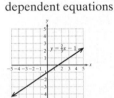

27. $(2, -2)$ **29.** Infinitely many solutions; $\{(x, y) \mid -x + 3y = 6\}$; dependent equations

31. No solution; inconsistent system

33. False **35.** True
37. For example: $x + y = 9$ **39.** $C = 5, D = 4$
$2x + y = 13$

Section 4.1 Graphing Calculator Exercises, p. 249

41. $(-3.5, -4.21)$

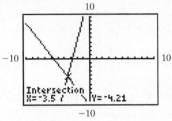

42. $(-4.6, 5.1)$

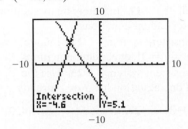

43. $(2.14, 3.02)$

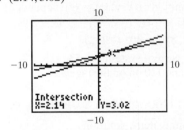

44. $(4.5, 2.2)$

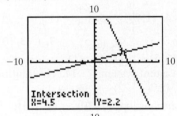

Section 4.2 Practice Exercises, pp. 253–256

1. No **3.** Infinitely many solutions
5. No solution
7. $(0, 3)$

9. $(-1, 2)$ **11.** $(3, 1)$ **13.** $(12, 0)$ **15.** $(5, -3)$
17. $\left(-\dfrac{1}{4}, \dfrac{1}{2}\right)$ **19.** $(2, 0)$ **21.** Infinitely many solutions;
$\{(x, y) \mid x = 3y - 1\}$; dependent equations **23.** No solution;
inconsistent system **25.** No solution; inconsistent system
27. Infinitely many solutions; $\{(x, y) \mid 3x - y = 7\}$;
dependent equations **29.** When solving a system, if you
get an identity, such as $0 = 0$ or $5 = 5$, then the equations are
dependent. **31.** $(8, 5)$ **33.** $(5, 3)$ **35.** No solution;
inconsistent system **37.** $(0, 3)$ **39.** $(-7, 6)$

41. Infinitely many solutions; $\{(x, y) \mid 2x - y = 6\}$; dependent equations   **43.** $(4, 480)$ **45.** $(10, -32.1)$

47. $\left(\dfrac{1}{2}, \dfrac{3}{4}\right)$ **49.** $(-2, 4)$ **51.** $(1, 1)$ **53.** $(6, 1)$

55. a. $y = \dfrac{4}{9}x + \dfrac{25}{9}$ **b.** $y = -\dfrac{4}{7}x + \dfrac{23}{7}$ **c.** $(0.5, 3)$

57. a. At Glendale Lakes: $y = 800x + 250$
    At the Breakers: $y = 750x + 500$ **b.** 5 months

Section 4.3 Practice Exercises, pp. 261–263

1. a. -3 **b.** 5 **3.** One solution   **5.** $(-2, -5)$
7. $(3, -1)$ **9.** $(12, -8)$ **11.** $(1, -1)$
13. $(2, -1)$ **15.** $(0, 0)$ **17.** Infinitely many solutions; $\{(x, y) \mid 3x - 2y = 1\}$; dependent equations
19. No solution; inconsistent system   **21.** Infinitely many solutions; $\{(x, y) \mid 12x - 4y = 2\}$; dependent equations
23. No solution; inconsistent system   **25.** Use the substitution method if one equation has x or y already isolated.   **27.** $(-2, -3)$ **29.** $\left(-\dfrac{1}{2}, 2\right)$ **31.** $(4, 0)$
33. No solution; inconsistent system   **35.** $(12, 30)$
37. Infinitely many solutions; $\{(x, y) \mid x = \frac{3}{2}y\}$; dependent equations   **39.** $(0, 4)$ **41.** $\left(-\frac{1}{2}, -\frac{5}{2}\right)$ **43.** $(9, 9)$
45. $\left(\frac{4}{3}, -1\right)$ **47.** $(1, 5)$ **49.** $(-1, 2)$
51. No solution; inconsistent system   **53.** $(18, -8)$
55. 112 mi in the city and 120 mi on the highway

Chapter 4 Problem Recognition Exercises, p. 263

1. $(2, 4)$ **2.** Infinitely many solutions; $\{(x, y) \mid 3x - 2y = 4\}$; dependent equations   **3.** No solution; inconsistent system   **4.** $(1, 2)$ **5.** $\left(\dfrac{1}{2}, -11\right)$
6. $(-3, 1)$ **7.** $(-2, -4)$ **8.** $(2, 2)$ **9.** $(-3, 10)$
10. $(-9, -1)$ **11.** $(3, 3)$ **12.** $(2, -1)$
13. No solution; inconsistent system   **14.** No solution; inconsistent system   **15.** $(3, -2)$ **16.** $(4, 1)$
17. Infinitely many solutions; $\{(x, y) \mid 10x - 6y = 4\}$; dependent equations   **18.** Infinitely many solutions; $\left\{(x, y) \,\middle|\, y = \dfrac{3}{5}x - 2\right\}$; dependent equations   **19.** $(5, 1)$
20. $(20, 10)$

Section 4.4 Practice Exercises, pp. 269–272

1. a. $\$60; 12x$ **b.** 2 L; $0.10x$ **c.** $\$200; 0.04y$
d. $b - c; b + c$ **e.** $180°$ **f.** $180°$ **g.** $90°$ **h.** $90°$
3. $(5, -1)$ **5.** $(12, -27)$ **7.** There are 102 nonstudent tickets and 84 student tickets.
9. A hamburger costs $\$4.60$ and a fish sandwich costs $\$5.20$.
11. Vanilla has 12 g of fat per scoop and mud pie has 16 g of fat per scoop.   **13.** Combine 4 oz of 18% moisturizer with 8 oz of 24% moisturizer.   **15.** Mix 2 L of 8% nitrogen fertilizer and 6 L of 12% nitrogen fertilizer.   **17.** Mix 1 oz of pure bleach with 11 oz of 4% solution.   **19.** He invested $\$4500$ in the stock fund and $\$1500$ in the bond fund.
21. He borrowed $\$2800$ at 5.5% and $\$2600$ at 3.5%.
23. She invested $\$12,500$ in the 2% account and $\$14,500$ in the 3% account.   **25.** The speed of the boat is 6 mph and the speed of the current is 2 mph.

27. The speed of the plane is 210 mph and the speed of the wind is 30 mph.   **29.** Stephen's walking speed on nonmoving ground is 3.5 ft/sec. The sidewalk moves at 1.5 ft/sec.
31. The measures are $69°$ and $21°$.   **33.** The measures are $134.5°$ and $45.5°$.   **35.** The measures are $29°$ and $61°$.
37. Mix 7.5 g of pure gold.   **39.** The boat's speed in still water is 5.2 mph and the current speed is 1.2 mph.
41. Grandstand tickets cost $\$330$ each and general admission tickets cost $\$175$ each.   **43.** She invested $\$1750$ at 2% and $\$1250$ at 1.3%.   **45.** The length is 11 m and the width is 10 m.   **47.** There are eleven 50-cent pieces and ten $\$1$ coins.   **49. a.** $c(x) = 20 + 0.25x$
b. $m(x) = 30 + 0.20x$ **c.** 200 mi

Section 4.5 Practice Exercises, pp. 279–286

1. a. linear **b.** is not; is **c.** dashed; is not **d.** solid; is
3. $[1, 3)$ **5.** $(-\infty, -4] \cup (-2, \infty)$ **7. a.** No **b.** No
c. Yes **d.** Yes   **9. a.** No **b.** Yes **c.** Yes **d.** No
11. $>$ **13.** $\geq$ **15.** $\geq, \leq$

17. 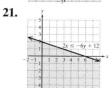   **19.**

21. 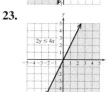   **23.**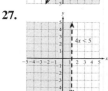

25.    **27.**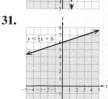

29. 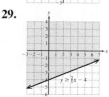   **31.**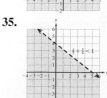

33. 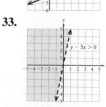   **35.**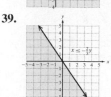

37. 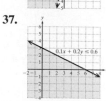   **39.**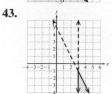

41. 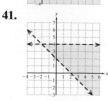   **43.**

45.

47.

49.

51.

53.

55.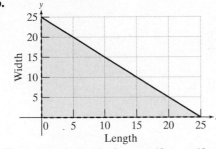

57.

59.

61.

63. a. $2x + 2y \le 50$

b.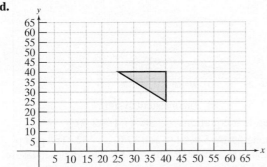

Length

65. a. $x \ge 0, y \ge 0$ **b.** $x \le 40, y \le 40$ **c.** $x + y \ge 65$

d.

e. Yes; The point $(35, 40)$ means that Karen works 35 hours and Todd works 40 hours.   **f.** No; The point $(20, 40)$ means that Karen works 20 hours and Todd works 40 hours. This does not satisfy the constraint that there must be at least 65 hours total.

Section 4.6 Practice Exercises, pp. 292–295

1. a. linear   **b.** ordered triples   **3.** $(1, 1)$
5. Marge's speed in still air is 15 mph and the wind speed is 3 mph.   **7.** $(4, 0, 2)$ is a solution.
9. $(1, 1, 1)$ is a solution.   **11.** $(1, -2, 4)$
13. $(-1, 2, 0)$ **15.** $(1, -4, -2)$ **17.** $(1, 2, 3)$
19. $(-6, 1, 7)$ **21.** $\left(\frac{1}{2}, \frac{2}{3}, -\frac{5}{6}\right)$ **23.** The measures are
$67°, 82°,$ and $31°.$ **25.** The lengths are 10 cm, 18 cm, and 27 cm.   **27.** The fiber supplement has 3 g, the oatmeal has 4 g, and the cereal has 1.5 g.   **29.** There are three par 3s, nine par 4s, and six par 5s.   **31.** The price of the hat is $15, T-shirt is $25, and jacket is $45.   **33.** Walter invested $7000 in small caps, $6000 in the balanced fund, and $12,000 in global markets.   **35.** Dependent equations   **37.** Inconsistent system   **39.** $(1, 3, 1)$ **41.** $(-2, -1, -3)$
43. Dependent equations   **45.** Inconsistent system
47. $(0, 0, 0)$ is the only solution.   **49.** Dependent equations

Section 4.7 Practice Exercises, pp. 301–304

1. a. matrix; rows; columns   **b.** column; one; square
c. coefficient; augmented   **d.** row echelon
3. $\left(12, \frac{1}{2}\right)$ **5.** $(1, 3, -3)$ **7. a.** $3 \times 1,$ **b.** column matrix
9. a. 2×2 **b.** square matrix   **11. a.** 1×4
b. row matrix   **13. a.** 2×3 **b.** none of these
15. $\begin{bmatrix} 1 & -2 & | & -1 \\ 2 & 1 & | & -7 \end{bmatrix}$ **17.** $\begin{bmatrix} 1 & -2 & 1 & | & 5 \\ 2 & 6 & 3 & | & -2 \\ 3 & -1 & -2 & | & 1 \end{bmatrix}$
19. $\begin{aligned} 4x + 3y &= 6 \\ 12x + 5y &= -6 \end{aligned}$ **21.** $x = 4, y = -1, z = 7$
23. a. 7   **b.** -2 **25.** $\begin{bmatrix} 1 & \frac{1}{2} & | & \frac{11}{2} \\ 2 & -1 & | & 1 \end{bmatrix}$
27. $\begin{bmatrix} 1 & -4 & | & 3 \\ 5 & 2 & | & 1 \end{bmatrix}$ **29.** $\begin{bmatrix} 1 & 5 & | & 2 \\ 0 & 11 & | & 5 \end{bmatrix}$
31. a. $\begin{bmatrix} 1 & 3 & 0 & | & -1 \\ 0 & -11 & -5 & | & 10 \\ -2 & 0 & -3 & | & 10 \end{bmatrix}$ **b.** $\begin{bmatrix} 1 & 3 & 0 & | & -1 \\ 0 & -11 & -5 & | & 10 \\ 0 & 6 & -3 & | & 8 \end{bmatrix}$
33. True   **35.** True   **37.** Interchange rows 1 and 2.
39. Multiply row 1 by -3 and add to row 2. Replace row 2 with the result.   **41.** $(-3, -1)$ **43.** $(-21, 9)$
45. Infinitely many solutions; $\{(x, y) \mid x + 3y = 3\}$; dependent equations   **47.** $(3, -1)$ **49.** $(-10, 3)$
51. No solution; inconsistent system
53. $(1, 2, 3)$ **55.** $(1, -2, 0)$

Section 4.7 Graphing Calculator Exercises, p. 304

57.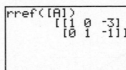

58.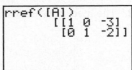

59.
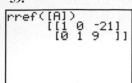
```
rref([A])
     [[1 0 -21]
      [0 1 9  ]]
```

60.
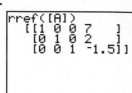
```
rref([A])
          [[1 0 5]
           [0 1 4]]
```

61.
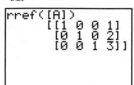
```
rref([A])
     [[1 0 0 1]
      [0 1 0 2]
      [0 0 1 3]]
```

62.
```
rref([A])
     [[1 0 0 7  ]
      [0 1 0 2  ]
      [0 0 1 -1.5]]
```

Chapter 4 Review Exercises, pp. 311–315

1. a. No   **b.** Yes   **2.** False   **3.** True   **4.** True
5. $(3, 2)$ **6.** $(-4, -1)$

7. Infinitely many solutions;   **8.** No solution; inconsistent
$\{(x, y) \mid 6x + 2y = 4\}$;   system
dependent equations

9. $(4, -1)$ **10.** No solution; inconsistent system
11. Infinitely many solutions; $\{(x, y) \mid 4x + y = 7\}$;
dependent equations   **12.** $(2, -7)$ **13.** $(0, 2)$
14. $\left(\dfrac{1}{2}, 2\right)$ **15.** 30 months   **16.** 8 days   **17.** $(1, 1)$
18. $(2, -1)$ **19.** $(2, -1)$ **20.** Infinitely many solutions;
$\{(x, y) \mid 3x + y = 1\}$; dependent equations   **21.** No solution;
inconsistent system   **22.** $(6, -2)$ **23.** $(2, -3)$
24. $(-10, 25)$ **25.** $(0, 6)$ **26.** $(-2, 7)$
27. She invested \$4500 at 5%.   **28.** There were
21 student tickets and 33 adult tickets.   **29.** 10 L of 20%
solution must be mixed with 6 L of 50% solution.
30. The speed of the plane is 150 mph and the speed of the
wind is 10 mph.   **31. a.** $f(x) = 0.1x + 9.95$
b. $g(x) = 0.08x + 12.95$ **c.** 150 min
32. The measures are 76° and 14°.

33.

34.

35.

36.

37.

38.

39.

40.

41.

42.

43.

44.

45.
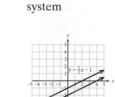

46. a. $x \geq 0, y \geq 0$ **b.** $x + y \leq 100$ **c.** $x \geq 4y$
d.
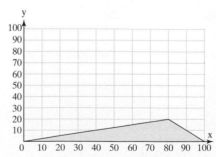

47. $(2, -1, 5)$ **48.** Inconsistent system
49. Dependent equations   **50.** $(-1, -2, 2)$
51. The lengths are 5 ft, 12 ft, and 13 ft.   **52.** The pumps
can drain 250, 300, and 400 gal/hr.   **53.** The angles measure
113°, 38°, and 29°.
54. 3×3 **55.** 3×2 **56.** 1×4 **57.** 3×1
58. $\begin{bmatrix} 1 & 1 & | & 3 \\ 1 & -1 & | & -1 \end{bmatrix}$ **59.** $\begin{bmatrix} 1 & -1 & 1 & | & 4 \\ 2 & -1 & 3 & | & 8 \\ -2 & 2 & -1 & | & -9 \end{bmatrix}$

60. $x = 9$ **61.** $x = -5$
   $y = -3$ $y = 2$
              $z = -8$

62. a. 4   **b.** $\begin{bmatrix} 1 & 3 & | & 1 \\ 0 & -13 & | & 2 \end{bmatrix}$

63. a. $\begin{bmatrix} 1 & 2 & 0 & | & -3 \\ 0 & -9 & 1 & | & 12 \\ -3 & 2 & 2 & | & 5 \end{bmatrix}$ **b.** $\begin{bmatrix} 1 & 2 & 0 & | & -3 \\ 0 & -9 & 1 & | & 12 \\ 0 & 8 & 2 & | & -4 \end{bmatrix}$

64. $(1, 2)$ **65.** $(-3, 6)$
66. $(1, 3, -2)$ **67.** $(6, 1, -1)$

Chapter 4 Test, pp. 315–317

1. Yes   **2.** b   **3.** c   **4.** a
5. $(1, 4)$ **6.** $(-2, 1)$

7. $(-4, 5)$ **8.** $\left(\dfrac{1}{2}, \dfrac{1}{4}\right)$ **9.** $(0, -3)$

10. Infinitely many solutions; $\{(x, y) \mid 3x - 5y = -7\}$; dependent equations   **11.** $(2, -6)$ **12.** No solution; inconsistent system   **13.** $(5, 0)$ **14.** Mix 80 L of 20% solution with 120 L of 60% solution.   **15.** The angles are 80° and 10°.

16. 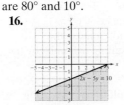   **17.**

18.

19. a. $x \geq 0, y \geq 0$ **b.** $300x + 400y \geq 1200$
c.

20. $(16, -37, 9)$ **21.** $(2, -2, 5)$ **22.** Joanne can process 142 orders, Kent can process 162 orders, and Geoff can process 200 orders.

23. For example: $\begin{bmatrix} 2 & 1 \\ 0 & -4 \\ 2.6 & 7 \end{bmatrix}$

24. a. $\begin{bmatrix} 1 & 2 & 1 & | & -3 \\ 0 & -8 & -3 & | & 10 \\ -5 & -6 & 3 & | & 0 \end{bmatrix}$

b. $\begin{bmatrix} 1 & 2 & 1 & | & -3 \\ 0 & -8 & -3 & | & 10 \\ 0 & 4 & 8 & | & -15 \end{bmatrix}$ **25.** $(6, -1)$ **26.** $(2, -4, 3)$

Chapters 1–4 Cumulative Review Exercises, pp. 317–318

1. -14 **2.** $-56w + 97$ **3.** No solution   **4.** $-\dfrac{19}{12}$

5. $\dfrac{a}{36b^2}$ **6.** $\left(-\dfrac{7}{5}, \infty\right)$ **7.** Slope $\frac{5}{2}$; x-intercept $(3, 0)$; y-intercept $\left(0, -\dfrac{15}{2}\right)$

8.   **9.**

10. 0 **11.** $y = -4x + 4$ **12.** No solution; inconsistent system   **13.** $(1, 2)$ **14. a.** $f(x) = 2.50x + 25$
b. $g(x) = 3x + 10$ **c.** 30 DVDs
15.

16. $(2, 0, -1)$ **17.** 2×3
18. For example: $\begin{bmatrix} 2 & 3 & 1 & 7 \\ -1 & 4 & 2 & 6 \end{bmatrix}$
19. Interchange two rows. Multiply a row by a nonzero constant. Add a multiple of one row to another row.
20. $(1, 1)$

Chapter 5

Chapter Opener Puzzle

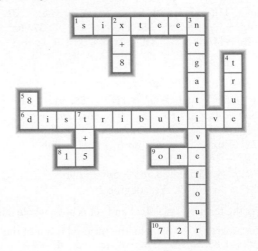

Section 5.1 Practice Exercises, pp. 326–328

1. a. exponent   **b.** 1   **c.** $\left(\dfrac{1}{b}\right)^n$ or $\dfrac{1}{b^n}$ **d.** scientific notation

3. $ab^3 = a \cdot b \cdot b \cdot b$
$(ab)^3 = (ab) \cdot (ab) \cdot (ab) = a \cdot a \cdot a \cdot b \cdot b \cdot b = a^3 b^3$

5. For example: $(xy)^3 = x^3 y^3$ **7.** For example: $\dfrac{x^5}{x^2} = x^3$

9. For example: $x^0 = 1$ $(x \neq 0)$ **11.** 3   **13.** $\dfrac{1}{25}$

15. $-\dfrac{1}{25}$ **17.** $\dfrac{1}{25}$ **19.** -64 **21.** $\dfrac{16}{81}$ **23.** $-\dfrac{125}{8}$

25. 1 **27.** $10a$ **29.** y^8 **31.** 13^2 or 169 **33.** y^8

35. 3^4x^8 or $81x^8$ **37.** $\dfrac{1}{p^3}$ **39.** $\dfrac{1}{7^3}$ or $\dfrac{1}{343}$ **41.** $\dfrac{1}{w^2}$

43. $\dfrac{1}{a^7}$ **45.** r^2 **47.** $\dfrac{1}{z^4}$ **49.** a^3b^2 **51.** 1 **53.** $\dfrac{65}{4}$

55. $\dfrac{26}{25}$ **57.** 3 **59.** $\dfrac{5}{2}$ **61.** $\dfrac{q^2}{p^3}$ **63.** $-\dfrac{3b^7}{2a^3}$

65. $\dfrac{x^{16}}{81y^{20}z^8}$ **67.** $-\dfrac{4m^4}{n^2}$ **69.** $\dfrac{4q^{11}}{p^4}$ **71.** $\dfrac{5x^8}{y^2}$

73. $\dfrac{16a^2}{b^6}$ **75.** $-\dfrac{27y^{27}}{8x^{24}}$ **77.** $\dfrac{4x^{18}}{y^{10}}$ **79.** $27x^3y^9$

81. a. $\$8 \times 10^9$ **b.** 3×10^6 DVDs   **c.** 1.4×10^{13} eV
d. 1.602×10^{-19} J   **83. a.** $200{,}000{,}000{,}000$ **b.** 0.000004 m
c. $108{,}200{,}000{,}000$ m   **85.** 3.5×10^5 **87.** Proper
89. Proper   **91.** 3.38×10^{-4} **93.** 1.608×10^4
95. 3.4×10^{13} **97.** 2.5×10^3 **99.** 1.204×10^{24}
hydrogen atoms and 6.02×10^{23} oxygen atoms
101. There are 2×10^4 or $20{,}000$ people per square mile.
103. $\$5.25 \times 10^{10}$ **105. a.** 540 months   **b.** $\$10{,}800$
c. $\$55{,}395.45$ **107.** y^{2a+2} **109.** x^{2a-4} **111.** $x^{a-6}y^6$

Section 5.2 Practice Exercises, pp. 334–338

1. a. polynomial   **b.** coefficient; n **c.** $1; 1$ **d.** one
e. binomial   **f.** trinomial   **g.** leading; leading coefficient
h. greatest   **i.** zero   **j.** exponents   **k.** polynomial
3. $10c^2$ **5.** 1.7×10^4 **7.** $-6a^3 + a^2 - a$; leading
coefficient -6; degree 3 **9.** $3x^4 + 6x^2 - x - 1$; leading
coefficient 3; degree 4 **11.** $-t^2 + 100$; leading
coefficient -1; degree 2 **13.** For example: $3x^5$
15. For example: $x^2 + 2x + 1$ **17.** For example: $6x^4 - x^2$
19. $m^2 + 10m$ **21.** $3x^4 + 2x^3 - 8x^2 + 2x$
23. $2w^3 + \dfrac{1}{9}w^2 + 0.9w$ **25.** $17x^2y - 4xy - 14$
27. $4a^2 - 11a - 7$ **29.** $9x^3 - 5x^2 + 2x - 8$ **31.** $30y^3$
33. $-4p^3 - 2p + 12$ **35.** $11ab^2 - a^2b$ **37.** $6z^5 - 6z^2$
39. $-2x^3 + 4x^2 + 5$ **41.** $-2xy^3 + 3x^2y + xy + 5$
43. $t^3 - 13t^2 - 9t - 13$ **45.** $\dfrac{1}{2}a^2 - \dfrac{9}{10}ab + \dfrac{3}{5}b^2 + 8$
47. $-x^2 + 6x - 16$ **49.** $3x^5 - x^4 - 4x^3 + 11$
51. $4y^3 + 5y^2$ **53.** $-7r^4 - 11r$ **55.** $9x^2 + 5xy + 11y$
57. $18ab - 42b^2$ **59.** $3p - 9$ **61.** $2m^2 + 6$
63. $7x^3 + 4x - 5$ **65.** $12a^2b + ab - 5ab^2$
67. $12x^3 - 6x^2 + 1$ **69.** $8a^2b - 5ab^2 + 4ab$
71. $-3x^3 - 16x^2 + 10x + 1$
73. $-2.2p^5 - 15.5p^4 - 8.5p^3 - 5p^2 - 7.9p$
75. $12x^3 + 2x$ **77.** Yes; degree 2
79. No; the term $-\dfrac{3}{x} = -3x^{-1}$ and -1 is not a whole number.
81. Yes; degree 0 **83.** No; the term $|x|$ is not of the
form ax^n.   **85. a.** -17 **b.** -8 **c.** -5 **d.** -4
87. a. $\dfrac{1}{4}$ **b.** $\dfrac{9}{4}$ **c.** $-\dfrac{7}{4}$ **d.** $\dfrac{3}{4}$ **89.** $P(x) = 4x + 6$
91. a. $P(x) = 6.6x - 99$ **b.** $\$231$ **93. a.** $D(0) = 1636$
means that in the year 1990, the annual dormitory charge
was $\$1636$. $D(18) = 4048$ means that in the year 2008,
the annual dormitory charge was $\$4048$.   **b.** $\$5896$
95. a. $W(0) = 6580$; $W(5) = 7295$; $W(10) = 8010$
b. $W(10) = 8010$ means that in the year 2010, 8010 thousand
women (8,010,000) were due child support.

97. a. $(0, 0)$; at $t = 0$ sec, the position of the rocket is at the
origin.   **b.** $(25, 27.3)$; after 1 sec, the position of the rocket is
$(25, 27.3)$.   **c.** $(50, 22.6)$

Section 5.3 Practice Exercises, pp. 345–348

1. a. distributive   **b.** $4x - 7$ **c.** squares; $a^2 - b^2$
d. perfect; $a^2 + 2ab + b^2$ **3.** $6x^2 + x - 6$
5. a. -3 **b.** 9 **c.** -3 **7.** $-42x^5y^6$ **9.** $11a^7b^8c^{10}$
11. $\dfrac{2}{5}a - \dfrac{3}{5}$ **13.** $2m^5n^5 - 6m^4n^4 + 8m^3n^3$
15. $3x^2y^2 - 4x^2y^3$ **17.** $x^2 - xy - 2y^2$
19. $12x^2 + 28x - 5$ **21.** $2y^4 - 21y^2 - 36$
23. $25s^2 + 5st - 6t^2$ **25.** $5n^3 + 3n^2 + 50n + 30$
27. $3.25a^2 - 0.9ab - 28b^2$ **29.** $6x^3 + 7x^2y + 4xy^2 + y^3$
31. $x^3 - 343$ **33.** $4a^4 - 17a^3b + 8a^2b^2 - 5ab^3 + b^4$
35. $\dfrac{1}{2}a^2 + ab + \dfrac{1}{2}ac - 12b^2 + 8bc - c^2$
37. $-3x^3 + 11x^2 - 7x - 5$ **39.** $\dfrac{1}{10}y^2 - 8y + 150$
41. $a^2 - 64$ **43.** $9p^2 - 1$ **45.** $x^2 - \dfrac{1}{9}$ **47.** $9h^2 - k^2$
49. $9h^2 - 6hk + k^2$ **51.** $t^2 - 14t + 49$
53. $u^2 + 6uv + 9v^2$ **55.** $h^2 + \dfrac{1}{3}hk + \dfrac{1}{36}k^2$
57. $4z^4 - w^6$ **59.** $25x^4 - 30x^2y + 9y^2$
61. a. Yes   **b.** No   **63. a.** $A^2 - B^2$
b. $x^2 + 2xy + y^2 - B^2$; Both are examples of multiplying
conjugates to get a difference of squares.
65. $w^2 + 2wv + v^2 - 4$ **67.** $4 - x^2 - 2xy - y^2$
69. $9a^2 - 24a + 16 - b^2$ **71.** Write $(x + y)^3$ as
$(x + y)^2(x + y)$. Square the binomial and then use the
distributive property to multiply the resulting trinomial by the
remaining factor of $(x + y)$.   **73.** $8x^3 + 12x^2y + 6xy^2 + y^3$
75. $64a^3 - 48a^2b + 12ab^2 - b^3$ **77.** Multiply and
simplify the first two binomials. Then multiply the resulting
trinomial by the third binomial, using the distributive property.
79. $6a^4 + 32a^3 + 10a^2$ **81.** $x^3 + 5x^2 - 9x - 45$
83. $-16x^2 + 2x - 22$ **85.** $-3y^2 - 10y - 8$
87. $(r + t)^2$ **89.** $x^2 - y^3$ **91.** The sum of p cubed and
q squared   **93.** The product of x and the square of y
95. $A(x) = 4x^2 + 70x + 300$
97. a. $V(x) = 4x^3 - 32x^2 + 64x$ **b.** 36 in.^{3}
99. $x^2 - 4x + 4$ **101.** $x^2 - 4$ **103.** $x^2 - 9$
105. $9x^3 + 30x^2$ **107.** $2x + h - 3$ **109.** Multiply
$(x + 2)^2(x + 2)^2$ by squaring the binomials. Then multiply the
resulting trinomials, using the distributive property.
111. $(5x - 6)$ **113.** $(2y - 1)$

Section 5.4 Practice Exercises, pp. 355–358

1. a. division; quotient; remainder   **b.** Synthetic
3. a. $-4a - 11b$ **b.** $5a^2 - 49ab - 10b^2$
5. a. $7x^2 + 2$ **b.** $6x^4 - 5x^3 + x^2 - 2x$
7. For example: $(5y + 1)^2 = 25y^2 + 10y + 1$
9. $-4t^3 + t - 5$ **11.** $12 + 8y + 2y^2$
13. $x^2 + 3xy - y^2$ **15.** $2y^2 + 3y - 8$
17. $-\dfrac{1}{2}p^3 + p^2 - \dfrac{1}{3}p + \dfrac{1}{6}$ **19.** $a^2 + 5a + 1 - \dfrac{5}{a}$
21. $-3s^2t + 4s - \dfrac{5}{t^2}$ **23.** $8p^2q^6 - 9p^3q^5 - 11p - \dfrac{4}{p^2q}$
25. a. Divisor $(x - 2)$; quotient $(2x^2 - 3x - 1)$;
remainder (-3) **b.** Multiply the quotient and the divisor,
then add the remainder. The result should equal the dividend.

27. $x + 7 + \dfrac{-9}{x + 4}$ **29.** $3y^2 + 2y + 2 + \dfrac{9}{y - 3}$

31. $-4a + 11$ **33.** $6y - 5$ **35.** $6x^2 + 4x + 5 + \dfrac{22}{3x - 2}$

37. $4a^2 - 2a + 1$ **39.** $x^2 - 2x + 2$ **41.** $x^2 + 2x + 5$

43. $x^2 - 1 + \dfrac{8}{x^2 - 2}$ **45.** $n^3 + 2n^2 + 4n + 8$

47. The divisor must be of the form $x - r$.   **49.** No. The divisor is not of the form $x - r$.   **51. a.** $x - 5$

b. $x^2 + 3x + 11$ **c.** 58   **53.** $x + 6$ **55.** $t - 4$

57. $5y + 10 + \dfrac{11}{y - 1}$ **59.** $3y^2 - 2y + 2 + \dfrac{-3}{y + 3}$

61. $x^2 - x - 2$ **63.** $a^4 + 2a^3 + 4a^2 + 8a + 16$

65. $x^2 + 6x + 36$ **67.** $6t^2 + 3t - 3 + \dfrac{5}{t + \frac{2}{3}}$

69. $4w^3 + 2w^2 + 6$ **71.** $-x^2 - 4x + 13 + \dfrac{-54}{x + 4}$

73. $2x - 1 + \dfrac{3}{x}$ **75.** $4y - 3 + \dfrac{4y + 5}{3y^2 - 2y + 5}$

77. $2x^2 + 3x - 1$ **79.** $2k^3 - 4k^2 + 1 - \dfrac{5}{k^4}$

81. $x + \dfrac{9}{5} + \dfrac{2}{x}$ **83. a.** -84 **b.** $4x^2 - 6x + 16 + \dfrac{-84}{x + 4}$

c. The values are the same.   **85.** $P(r)$ equals the remainder of $P(x) \div (x - r)$.   **87. a.** $8x + 5$ **b.** Yes

Chapter 5 Problem Recognition Exercises, pp. 358–359

1. a. $9x^2 + 6x + 1$ **b.** $9x^2 - 1$ **c.** 2   **2. a.** -10

b. $81m^2 - 25$ **c.** $81m^2 - 90m + 25$ **3. a.** $2x + 4 - \dfrac{5}{x}$

b. $2x + 5 + \dfrac{-5}{2x - 1}$ **c.** $4x + 12 + \dfrac{2}{x - 1}$

4. a. $y - 5 + \dfrac{4}{3y}$ **b.** $y - 7 + \dfrac{46}{3y + 6}$ **c.** $3y - 33 + \dfrac{202}{y + 6}$

5. a. -30 **b.** $-10p - 50$ **c.** 0   **6. a.** $-8x - 32$

b. -20 **c.** 0   **7.** $2t^2 + t - 1$ **8.** $-15x^4 - 5x^3 + 10x^2$

9. $36z^2 - 25$ **10.** $9y^3 + 2y^2 - 3y + 1$

11. $6b^2 - 11b + 4$ **12.** $10a^3 + 19a^2 + 11a + 2$

13. $t^3 - 6t^2 + 8t + 3$ **14.** $2b^2 + 4b + 5$

15. $k^2 + 4k + 25$ **16.** $3x^3 + x^2 - 5$

17. $7t^3 - 12t^2 + 2t$ **18.** $-xy^2 + 2y + \dfrac{x}{7y}$

19. $\dfrac{11}{12}p^3 - \dfrac{1}{2}p^2 + \dfrac{1}{5}p + 5$

20. $-30.6w^6 + 15.6w^5 - 7.2w^4$ **21.** $36a^4 - 48a^2b + 16b^2$

22. $\dfrac{1}{4}z^4 - \dfrac{1}{9}$ **23.** $m^2 - 8m - 7$ **24.** $x^2 + 3x - 14$

25. $2m^4 - 8m^3 - 13m^2 + 46m - 21$ **26.** $x^2 + 4x + 16$

27. $25 - 10a - 10b + a^2 + 2ab + b^2$

28. $a^2 - x^2 + 2xy - y^2$ **29.** $4xy$

30. $a^3 - 12a^2 + 48a - 64$ **31.** $-\dfrac{1}{8}x^2 + \dfrac{1}{3}x - \dfrac{1}{6}$

32. $-\dfrac{1}{2}x^6y^4z^5w^3$

Section 5.5 Practice Exercises, pp. 364–367

1. a. product   **b.** greatest common factor   **c.** greatest common factor   **d.** grouping   **3.** $9t^4 + 5t^3 - 6t^2 - 6t$

5. $5y^4 + 5y^3 + 7y^2 - 3y - 6$ **7.** $-3v^2 + 6v - 1$

9. $3(x + 4)$ **11.** $2z(3z + 2)$ **13.** $4p(p^5 - 1)$

15. $12x^2(x^2 - 3)$ **17.** $9t(st + 3)$

19. $9a^2b^3(a^2 + 3ab - 2b^2)$ **21.** $5xy(2x + 3y - 1)$

23. $b(13b - 11a^2 - 12a)$ **25.** $-1(x^2 + 10x - 7)$

27. $-3xy(4x^2 + 2x + 1)$ **29.** $-t(2t^2 - 11t + 3)$

31. $(3z - 2b)(2a - 5)$ **33.** $(2x - 3)(2x^2 + 1)$

35. $(2x + 1)^2(y - 3)$ **37.** $3(x - 2)^2(y + 2)$

39. For example: $3x^3 + 6x^2 + 12x^4$ **41.** For example: $6(c + d) + y(c + d)$ **43. a.** $(2x - y)(a + 3b)$

b. $(2w - 1)(5w - 3b)$ **c.** In part (b), $-3b$ was factored out so that the signs in the last two terms were changed. The resulting binomial factor matches the binomial factor from the first two terms.   **45.** $(y + 4)(y^2 + 3)$

47. $(p - 7)(6 + q)$ **49.** $(m + n)(2x + 3y)$

51. $(2x - 3y)(5a - 4b)$ **53.** $(x^2 - 3)(x - 1)$

55. $6p(p + 3)(q - 5)$ **57.** $100(x - 3)(x^2 + 2)$

59. $(3a + b)(2x - y)$ **61.** $(4 - b)(a + 3)$

63. Cannot be factored   **65.** It is not possible to get a common binomial factor regardless of the order of terms.

67. $A = \dfrac{U}{v + cw}$ **69.** $y = \dfrac{bx}{c - a}$ or $y = \dfrac{-bx}{a - c}$

71. Length $= 2w + 1$ **73.** $(a + 3)^4(6a + 19)$

75. $18(3x + 5)^2(4x + 5)$ **77.** $(t + 4)(t + 3)$

79. $5w^2(2w - 1)^2(7w - 3)$

Section 5.6 Practice Exercises, pp. 377–379

1. a. positive   **b.** opposite   **c.** Both are correct.

d. $2(3x - 5)(x + 1)$ **e.** $(a + b)^2$; $(a - b)^2$

3. $6c^2d^4e^7(6d^3e^4 + 2cde^8 - 1)$ **5.** $(3a - b)(2x - 1)$

7. $(z + 2)(wz - 33a)$ **9.** $(b - 8)(b - 4)$

11. $(y + 12)(y - 2)$ **13.** $(x + 10)(x + 3)$

15. $(c - 8)(c + 2)$ **17.** $(2x + 3)(x - 5)$

19. $(6a - 5)(a + 1)$ **21.** $(s + 3t)(s - 2t)$

23. $3(x - 18)(x - 2)$ **25.** $2(c - 4)(c + 3)$

27. $2(x - y)(x + 5y)$ **29.** Prime

31. $(3x + 5y)(x + 3y)$ **33.** $5uv(u - 3v)^2$

35. $x(x - 7)(x + 2)$ **37.** $(2z - 5)(5z + 1)$ **39.** Prime

41. $-2(t + 4)(t - 10)$ **43.** $(7a - 4)(2a + 3)$

45. $2b(3a + 2)(a + 3)$ **47. a.** $x^2 + 10x + 25$ **b.** $(x + 5)^2$

49. a. $9x^2 - 12xy + 4y^2$ **b.** $(3x - 2y)^2$ **51.** $30x$

53. $16z^2t$ **55.** $(y - 4)^2$ **57.** $(8m + 5)^2$

59. Not a perfect square trinomial   **61.** $(3a - 5b)^2$

63. $4(4t^2 - 20tv + 5v^2)$; Not a perfect square trinomial

65. $5(b^2 - 2)^2$ **67. a.** $(u - 5)^2$ **b.** $(x^2 - 5)^2$ **c.** $(a - 4)^2$

69. a. $(u + 13)(u - 2)$ **b.** $(w^3 + 13)(w^3 - 2)$

c. $(y + 9)(y - 6)$ **71.** $(3x - 4)(3x + 1)$

73. $(2x - 9)(x - 1)$ **75.** $(3y + 11)(y + 6)$

77. $(3y^3 + 2)(y^3 + 3)$ **79.** $(4p^2 + 1)(p^2 + 1)$

81. $(x^2 + 12)(x^2 + 3)$ **83.** The factorization $(2y - 1)(2y - 4)$ is not factored completely because the factor $2y - 4$ has a GCF of 2.

85. $(w^2 + 6)^2$ **87.** $(9w + 5)^2$ **89.** $3(a + b)(x - 2)$

91. $2abc^2(6a + 2b - 3c)$ **93.** $-2x(5x - 6)(2x - 5)$

95. Prime   **97.** $(2w^2 - 15)(w^2 - 2)$

99. $(1 - d)(1 - 3d)$ or $(3d - 1)(d - 1)$

101. $(a + 2b)(x - 5a)$ **103.** $8(z - 4w)(z + 7w)$

105. $(y + x)(a - 5c)$ **107.** $g(x) = (3x + 2)(x + 4)$

109. $n(t) = (t + 10)^2$ **111.** $Q(x) = x^2(x + 4)(x + 2)$

113. $k(a) = (a - 4)(a^2 + 2)$

Section 5.7 Practice Exercises, pp. 386–389

1. a. difference; $(a + b)(a - b)$ **b.** sum **c.** is not
d. square **e.** sum; cubes **f.** difference; cubes
g. $(a - b)(a^2 + ab + b^2)$ **h.** $(a + b)(a^2 - ab + b^2)$
3. $(2x - 5)^2$ **5.** $(5 + 3y)(2x + 1)$
7. $4(8p + 1)(p - 1)$ **9.** Look for a binomial of the
form $a^2 - b^2$. This factors as $(a + b)(a - b)$.
11. $(x - 3)(x + 3)$ **13.** $(4 - 7w)(4 + 7w)$
15. $2(2a - 9b)(2a + 9b)$ **17.** Prime
19. $2(a^2 + 4)(a - 2)(a + 2)$ **21.** $(7 - k^3)(7 + k^3)$
23. $(x - 4)(x + 4)(x - 1)$ **25.** $(2x + 1)(2x - 1)(x + 3)$
27. $(9y + 7)(y - 2)(y + 2)$ **29.** $(7x + 2 - y)(7x + 2 + y)$
31. $(w - 3n + 1)(w + 3n - 1)$
33. $(p^2 - 5 - t^2)(p^2 - 5 + t^2)$
35. $(3u^2 - 2v^2 + 5)(3u^2 + 2v^2 - 5)$
37. Look for a binomial of the form $a^3 + b^3$. This factors as
$(a + b)(a^2 - ab + b^2)$. **39.** $(2x - 1)(4x^2 + 2x + 1)$
41. $(5c + 3)(25c^2 - 15c + 9)$
43. $(x - 10)(x^2 + 10x + 100)$
45. $(4t^2 + 1)(16t^4 - 4t^2 + 1)$
47. $2(10y^2 + x)(100y^4 - 10y^2x + x^2)$
49. $2z(2z - 3)(4z^2 + 6z + 9)$
51. $(p^4 - 5)(p^8 + 5p^4 + 25)$ **53.** $\left(6y - \dfrac{1}{5}\right)\left(6y + \dfrac{1}{5}\right)$
55. $2(3d^6 - 4)(3d^6 + 4)$ **57.** $2(121v^2 + 16)$
59. $4(x - 2)(x + 2)$ **61.** $(5 - 7q)(5 + 7q)$
63. $(t + 2s - 6)(t + 2s + 6)$ **65.** $(3 - t)(9 + 3t + t^2)$
67. $\left(3a + \dfrac{1}{2}\right)\left(9a^2 - \dfrac{3}{2}a + \dfrac{1}{4}\right)$
69. $2(m + 2)(m^2 - 2m + 4)$ **71.** $(x - y)(x + y)(x^2 + y^2)$
73. $(a + b)(a^2 - ab + b^2)(a^6 - a^3b^3 + b^6)$
75. $\left(\dfrac{1}{2}p - \dfrac{1}{5}\right)\left(\dfrac{1}{4}p^2 + \dfrac{1}{10}p + \dfrac{1}{25}\right)$ **77.** Prime
79. $\left(\dfrac{1}{5}x - \dfrac{1}{2}y\right)\left(\dfrac{1}{5}x + \dfrac{1}{2}y\right)$
81. $(a + b)(a^2 - ab + b^2)(a - b)(a^2 + ab + b^2)$
83. $(2 + y)(4 - 2y + y^2)(2 - y)(4 + 2y + y^2)$
85. $(h^2 + k^2)(h^4 - h^2k^2 + k^4)$
87. $(2x^2 + 5)(4x^4 - 10x^2 + 25)$ **89.** $4x^2 - 9$
91. $8a^3 - 27$ **93.** $64x^6 + y^3$ **95. a.** $x^2 - y^2$
b. $(x + y)(x - y)$ **c.** 20 in.^{2} **97.** $(x + y)(x - y + 1)$
99. $(x + y)(x^2 - xy + y^2 + 1)$
101. $(3a - c)(3a + c)(4a - 1)(16a^2 + 4a + 1)$

Chapter 5 Problem Recognition Exercises, pp. 389–391

1. A polynomial whose only factors are 1 and itself
2. Factor out the GCF. **3.** Difference of squares
$a^2 - b^2$, difference of cubes $a^3 - b^3$, or sum of cubes $a^3 + b^3$
4. Look for a perfect square trinomial,
$a^2 + 2ab + b^2$ or $a^2 - 2ab + b^2$.
5. Try factoring by grouping 2 terms by 2 terms or by
grouping 3 terms by 1 term. **6.** Let $u = (4x^2 + 1)$. The
polynomial becomes $3u^2 + 20u + 12$. Factor this simpler
expression and then back substitute.
7. a. Trinomial **b.** $3(2x + 3)(x - 5)$
8. a. Trinomial **b.** $m(4m + 1)(2m - 3)$

9. a. Difference of squares **b.** $2(2a - 5)(2a + 5)$
10. a. Grouping **b.** $(b + y)(a - b)$
11. a. Trinomial **b.** $(2u - v)(7u - 2v)$
12. a. Perfect square trinomial **b.** $(3p - 2q)^2$
13. a. Difference of cubes **b.** $2(2x - 1)(4x^2 + 2x + 1)$
14. a. Sum of squares **b.** Prime
15. a. Sum of cubes **b.** $(3y + 5)(9y^2 - 15y + 25)$
16. a. None of these **b.** Prime
17. a. Sum of cubes **b.** $2(4p^2 + 3q)(16p^4 - 12p^2q + 9q^2)$
18. a. Perfect square trinomial **b.** $5(b - 3)^2$
19. a. Difference of squares **b.** $(2a - 1)(2a + 1)(4a^2 + 1)$
20. a. Perfect square trinomial **b.** $(9u - 5v)^2$
21. a. Grouping **b.** $(p - 6 - c)(p - 6 + c)$
22. a. Sum of squares **b.** $4(x^2 + 4)$
23. a. Grouping **b.** $2(2x - y)(3a + b)$
24. a. Difference of cubes **b.** $(5y - 2)(25y^2 + 10y + 4)$
25. a. Trinomial **b.** $(5y - 1)(y + 3)$
26. a. Difference of squares **b.** $2(m^2 - 8)(m^2 + 8)$
27. a. Difference of squares **b.** $(t - 10)(t + 10)$
28. a. Difference of squares **b.** $(2m - 7n)(2m + 7n)$
29. a. Sum of cubes **b.** $(y + 3)(y^2 - 3y + 9)$
30. a. Sum of cubes **b.** $(x + 1)(x^2 - x + 1)$
31. a. Trinomial **b.** $(d - 4)(d + 7)$
32. a. Trinomial **b.** $(c + 8)(c - 3)$
33. a. Perfect square trinomial **b.** $(x - 6)^2$
34. a. Perfect square trinomial **b.** $(p + 8)^2$
35. a. Grouping **b.** $(ax + b)(2x - 5)$
36. a. Grouping **b.** $(4x + a)(2x - b)$
37. a. Trinomial **b.** $(2y - 1)(5y + 4)$
38. a. Trinomial **b.** $(3z + 2)(4z + 1)$
39. a. Difference of squares **b.** $10(p - 8)(p + 8)$
40. a. Difference of squares **b.** $2(5a - 6)(5a + 6)$
41. a. Difference of cubes **b.** $z(z - 4)(z^2 + 4z + 16)$
42. a. Difference of cubes **b.** $t(t - 2)(t^2 + 2t + 4)$
43. a. Trinomial **b.** $b(b + 5)(b - 9)$
44. a. Trinomial **b.** $y(y - 4)(y - 10)$
45. a. Perfect square trinomial **b.** $(3w + 4x)^2$
46. a. Perfect square trinomial **b.** $(2k - 5p)^2$
47. a. Grouping **b.** $10(2x + a)(3x - 1)$
48. a. Grouping **b.** $10(5x + c)(x - 4)$
49. a. Difference of squares **b.** $(w^2 + 4)(w - 2)(w + 2)$
50. a. Difference of squares **b.** $(k^2 + 9)(k - 3)(k + 3)$
51. a. Difference of cubes **b.** $(t^2 - 2)(t^4 + 2t^2 + 4)$
52. a. Sum of cubes **b.** $(p^2 + 3)(p^4 - 3p^2 + 9)$
53. a. Trinomial **b.** $(4p - 1)(2p - 5)$
54. a. Trinomial **b.** $(3m + 4)(3m - 5)$
55. a. Perfect square trinomial **b.** $(6y - 1)^2$
56. a. Perfect square trinomial **b.** $(3a + 7)^2$
57. a. Sum of squares **b.** $2(x^2 + 25)$
58. a. Sum of squares **b.** $4(y^2 + 16)$
59. a. Trinomial **b.** $s^2(3r - 2)(4r + 5)$
60. a. Trinomial **b.** $w^2(7z + 4)(z - 2)$
61. a. Trinomial **b.** $(x - 3y)(x + 11y)$
62. a. Trinomial **b.** $(s + 3t)(s - 12t)$
63. a. Sum of cubes **b.** $(m^2 + n)(m^4 - m^2n + n^2)$
64. a. Difference of cubes **b.** $(a - b^2)(a^2 + ab^2 + b^4)$
65. a. None of these **b.** $x(x - 4)$
66. a. None of these **b.** $y(y - 9)$ **67.** $(x - y)(x + y)^2$
68. $(u - v)^2(u + v)$ **69.** $(a + 3)^4(6a + 19)$
70. $(4 - b)^3(2 - b)$ **71.** $18(3x + 5)^2(4x + 5)$

72. $5(2y + 3)^2(6y + 11)$ **73.** $\left(\dfrac{1}{10}x + \dfrac{1}{7}\right)^2$

74. $\left(\dfrac{1}{5}a + \dfrac{1}{6}\right)^2$ **75.** $5x^2(5x^2 - 6)$ **76.** $x^3(x^3 - 2)$

77. $(4p^2 + q^2)(2p - q)(2p + q)$

78. $(s^2t^2 + 9)(st - 3)(st + 3)$

79. $\left(y + \dfrac{1}{4}\right)\left(y^2 - \dfrac{1}{4}y + \dfrac{1}{16}\right)$

80. $\left(z + \dfrac{1}{5}\right)\left(z^2 - \dfrac{1}{5}z + \dfrac{1}{25}\right)$

81. $(a + b)(a - b)(6a + b)$

82. $(2p - q)(2p + q)(p + 3q)$ **83.** $\left(\dfrac{1}{3}t + \dfrac{1}{4}\right)^2$

84. $\left(\dfrac{1}{5}y + \dfrac{1}{2}\right)^2$ **85.** $(x + 6 - a)(x + 6 + a)$

86. $(a + 5 + b)(a + 5 - b)$ **87.** $(p + q - 9)(p + q + 9)$
88. $(m - n - 3)(m - n + 3)$ **89.** $(b - x - 2)(b + x + 2)$
90. $(p - y + 3)(p + y - 3)$ **91.** $(2 + u - v)(2 - u + v)$
92. $(5 - a - b)(5 + a + b)$ **93.** $(3a + b)(2x - y)$
94. $(q + 3)(5p - 4)$
95. $(u - 2)(u + 2)(u^2 + 2u + 4)(u^2 - 2u + 4)$
96. $(1 - v)(1 + v)(1 + v + v^2)(1 - v + v^2)$
97. $(x^4 + 1)(x^2 + 1)(x + 1)(x - 1)$
98. $(y^4 + 16)(y^2 + 4)(y + 2)(y - 2)$
99. $(a + b)(a - b + 1)$ **100.** $(5c - 3d)(5c + 3d + 1)$
101. $(x + y)(x^2 - xy + y^2)(5w - 2z)$

Section 5.8 Practice Exercises, pp. 400–404

1. a. quadratic   **b.** $0; 0$ **c.** Pythagorean; c^2 **d.** quadratic
e. $f(x) = 0; y$ **f.** $x + 1; x + 2; x + 2$ **g.** lw **h.** $\dfrac{1}{2}bh$

3. $x(10x + 3)$ **5.** $(p - 5)(2p + 1)$
7. $(t - 1)(t^2 + t + 1)$
9. The equation must be set equal to 0, and the polynomial must be factored.   **11.** Correct form
13. Incorrect form. Polynomial is not factored.
15. Incorrect form. Equation is not set equal to 0.
17. a. $(w + 9)(w - 9)$ **b.** $-9, 9$

19. a. $(3x - 1)(x + 5)$ **b.** $\dfrac{1}{3}, -5$ **21.** $-3, -5$

23. $-\dfrac{9}{2}, \dfrac{1}{5}$ **25.** $0, -4, \dfrac{3}{10}$ **27.** $0.4, -2.1$

29. $-9, 3$ **31.** $-3, \dfrac{1}{2}$ **33.** $0, \dfrac{3}{2}$ **35.** $\dfrac{23}{3}$

37. -3 **39.** $-\dfrac{1}{3}, 2$ **41.** $-1, \dfrac{1}{2}, 3$ **43.** $-5, 4$

45. $\dfrac{5}{2}, -1$ **47.** $-12, 5$ **49.** $-\dfrac{1}{11}$ **51.** $0, 6, -2$

53. $0, 4, -4$ **55.** $3, -3, -\dfrac{5}{2}$ **57.** The numbers are 5 and -5.   **59.** The numbers are 4 and -3.   **61.** The integers are -7 and -6 or 6 and 7.   **63.** The integers are -9 and -7 or 7 and 9.   **65.** The length is 7 ft, and the width is 5 ft.   **67.** The length is 20 yd, and the width is 15 yd.   **69. a.** The base is 5 in., and the height is 6 in.
b. The area is 15 in.^{2} **71.** The base is 10 ft, and the height is 5 ft.   **73.** The integers are 4 and 5.   **75. a.** 14 mi
b. The alternative route using superhighways
77. The lengths are 6 m, 8 m, and 10 m.   **79.** The radius is

2 units.   **81. a.** $0, 3$ **b.** $f(0) = 0$ **83. a.** $7, -1$
b. $f(0) = -7$ **85.** x-intercepts: $(2, 0), (-1, 0), (0, 0)$;
y-intercept: $(0, 0)$ **87.** x-intercept: $(1, 0)$; y-intercept: $(0, 1)$
89. $(3, 0)(-3, 0)$; d   **91.** $(-1, 0)$; a   **93. a.** The function
is in the form $s(t) = at^2 + bt + c$.   **b.** $(0, 0)$ and $(100, 0)$
c. At 0 sec and 100 sec, the rocket is at ground level
(height $= 0$).   **d.** At 1 sec and 99 sec
95. $f(x) = (x - 5)(x - 2); x = 5$ and $x = 2$ represent
x-intercepts.   **97.** $f(x) = (x + 1)^2; x = -1$ represents the
x-intercept.   **99.** $f(x) = -(x + 1)(x + 5); x = -1$ and
$x = -5$ represent the x-intercepts.   **101.** The radius is 6 ft.
103. The length is 8 ft and the width is 6 ft.
105. $(x - 2)(x + 2) = 0$ or $x^2 - 4 = 0$
107. $(x - 0)(x + 3) = 0$ or $x^2 + 3x = 0$

Section 5.8 Graphing Calculator Exercises, p. 404

108. $(2, 0)(-1, 0)$

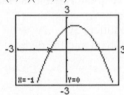
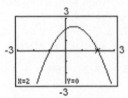

109. $(-5, 0)(4, 0)$

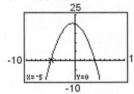
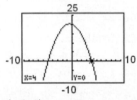

110. $(3, 0)$ **111.** $(-2, 0)$

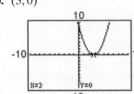
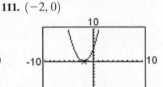

Chapter 5 Review Exercises, pp. 411–414

1. 3^5x^5 or $243x^5$ **2.** $-\dfrac{18}{x^{12}}$ **3.** $-3xy^2$ **4.** $\dfrac{3x^3}{2y^2}$

5. $-\dfrac{b^{15}}{8a^6}$ **6.** $\dfrac{a^4}{16b^6}$ **7.** $\dfrac{5^4y^{24}}{4^4x^{20}}$ **8.** $\dfrac{x^{10}y^5}{5^5}$

9. a. 3.6866×10^9 **b.** 1×10^{-6} **10. a.** 1×10^{-3}
b. 5.1557×10^9 **11. a.** 0.001 **b.** 0.000000001
12. a. $5,230,000,000$ ft^2 **b.** $\$1,091,000,000,000$
13. 6.25×10^9 **14.** 2×10^{-8} **15.** 3.24×10^7
16. 3.64×10^{-8} **17.** Trinomial; degree 4
18. Monomial; degree 0   **19. a.** -7 **b.** -23 **c.** 5
20. a. 12   **b.** 10   **c.** -2 **21. a.** $A(5) \approx 25$ means that
in the year 2005, Americans on average consumed
approximately 25 gal of bottled water each.
b. $A(15) \approx 49$ means that in the year 2015, Americans will
consume approximately 49 gal of bottled water each if this
trend continues.   **22.** $-2x^2 - 3x - xy - 1$
23. $20xy - 18xz - 3yz$ **24.** $3a^3 + 5a^2 + 6a$
25. $-2a^2 + 6a$ **26.** $x^4 - x^2 - 1$ **27.** $x^4 + \dfrac{3}{4}x^2$

28. $6x - 6y$ **29.** $-5x + 9y$ **30.** $-11x + 1$

31. $4x^2 - 11x$ **32.** $-3x - 11$ **33.** $-3x$
34. $2x^3 - 14x^2 - 8x$ **35.** $-18x^3 + 15x^2 - 12x$
36. $x^2 - x - 42$ **37.** $x^2 - 11x + 18$ **38.** $\frac{1}{4}x^2 - 2x - 5$
39. $2y^2 + \frac{1}{5}y - \frac{1}{25}$ **40.** $27x^3 + 125$ **41.** $x^3 - y^3$
42. $4x^2 - 20x + 25$ **43.** $\frac{1}{4}x^2 + 4x + 16$
44. $9y^2 - 121$ **45.** $36w^2 - 1$ **46.** $\frac{4}{9}t^2 - 16$
47. $z^2 - \frac{1}{16}$ **48.** $x^2 + 4x + 4 - b^2$
49. $c^2 - w^2 - 6w - 9$ **50.** $8x^3 + 12x^2 + 6x + 1$
51. $y^6 - 9y^4 + 27y^2 - 27$
52. a. $4x^2 + 12x + 9$ **b.** $16x^2 + 24x + 9$ **c.** $12x^2 + 12x$
53. a. $P(x) = 8x + 4$ **b.** $A(x) = 3x^2 + 2x$
54. a. False; cannot add unlike terms   True   **b.** True
False; when subtracting *like* terms, keep the variable the same.
55. $2x^2 + 4xy - 3y^2$ **56.** $-2x^2 - 3x + 4$
57. a. $3y^3 - 2y^2 + 6y - 4$ **b.** Quotient $3y^3 - 2y^2 + 6y - 4$;
no remainder   **c.** Multiply the quotient and the divisor.
58. $x + 2$ **59.** $x + 4 + \dfrac{-32}{x + 4}$
60. $2x^3 + 2x^2 + 8x + 24 + \dfrac{72x - 4}{x^2 - 3x}$
61. $2x^3 - 2x^2 + 5x - 4 + \dfrac{4x - 4}{x^2 + x}$
62. Synthetic division can be used with a divisor in the form
$x - r$.   **63. a.** $x - 3$ **b.** $2x^3 + 11x^2 + 31x + 99$ **c.** 298
64. $t^2 - t + 6$ **65.** $x + 2 + \dfrac{4}{x + 5}$ **66.** $x + 4 + \dfrac{4}{x + 4}$
67. $w^2 - 3w - 9 + \dfrac{-19}{w - 3}$ **68.** $p^3 + 2p^2 + 4p + 8$
69. $x(-x^2 - 4x + 11)$ or $-x(x^2 + 4x - 11)$
70. $7(3w^3 - w + 2)$ **71.** $(x - 7)(5x - 2)$
72. $(t + 4)(3t + 5)$ **73.** $(m - 8)(m^2 + 1)$
74. $12(2x - 3)(x^2 + 3)$ **75.** $x(2a + b)(2x - 3)$
76. $(y - 6)(y^2 + 1)$ **77.** The trinomial must be of the
form $a^2 + 2ab + b^2$ or $a^2 - 2ab + b^2$.
78. $(3x + 2y)(6x + 5y)$ **79.** $(3m - 5t)(m + 2t)$
80. $5a^2(3 + 4a)(4 - a)$ or $-5a^2(4a + 3)(a - 4)$
81. $k^2(3k + 2)(2k + 1)$ **82.** $(7x - 6)^2$ **83.** $2(5z + 4)^2$
84. $(9w + 7)(9w + 1)$ **85.** $(4x - 3)^2$
86. $3(3a^2 - 1)(2a^2 + 5)$ **87.** $(w^2 + 1)(3w^2 - 5)$
88. $(5 - y)(5 + y)$ **89.** $\left(x - \dfrac{1}{3}\right)\left(x^2 + \dfrac{1}{3}x + \dfrac{1}{9}\right)$
90. Prime   **91.** $h(h^2 + 9)$ **92.** $(a + 4)(a^2 - 4a + 16)$
93. $(k^2 + 4)(k - 2)(k + 2)$ **94.** $y(3y - 2)(3y + 2)$
95. $(x - 4y - 3)(x - 4y + 3)$ **96.** $(a + 6 - b)(a + 6 + b)$
97. $(t + 8 - 5c)(t + 8 + 5c)$ **98.** $(y - 3 - 4x)(y - 3 + 4x)$
99. It can be written in the form $ax^2 + bx + c = 0$ $(a \neq 0)$.
100. It is a parabola.   **101.** Quadratic   **102.** Quadratic
103. Linear   **104.** Quadratic   **105. a.** $(5x - 4)(x + 2)$
b. $\dfrac{4}{5}, -2$ **106. a.** $(3x - 7)(x - 4)$ **b.** $\dfrac{7}{3}, 4$
107. $-3, 5$ **108.** $\dfrac{3}{8}, 7$ **109.** $-\dfrac{4}{3}, -1$ **110.** $1, -5, \dfrac{9}{2}$
111. $(1, 0)(-1, 0)(0, 4)$; b   **112.** $(1, 0)(-1, 0)(0, -2)$; d
113. $(2, 0)(-2, 0)(0, 40)$; c   **114.** $(2, 0)(-2, 0)\left(0, \dfrac{1}{2}\right)$; a
115. The length is 15 ft, the width is 8 ft, and the height is 10 ft.

116. a.

Time t (sec)	Height $h(t)$ (ft)
0	-1280
1	-624
3	592
10	3840
20	5760
30	4480
42	-1280

b. The position of the missile is below sea level.   **c.** The
missile will be at sea level after 2 sec and again after 40 sec.

Chapter 5 Test, pp. 415–416

1. $5a^{13}$ **2.** x^{11} **3.** $\dfrac{9x^{12}}{25y^{14}}$ **4.** $\dfrac{8y^{12}}{x^7}$ **5.** 5.68
6. 2.3×10^9 **7.** $F(-1) = 1$ $F(2) = 40$ $F(0) = 8$
8. $8x^2 - 8x + 8$ **9.** $2a^3 - 13a^2 + 2a + 45$
10. $2x^2 - \dfrac{23}{3}x - 6$ **11.** $25x^2 - 16y^4$
12. The expression $25x^2 + 49$ does not include the middle
term $70x$.   **13.** $49x^2 - 56x + 16$
14. $x^2y^3 + \dfrac{5}{2}xy - 3y^2 - \dfrac{1}{2}$ **15.** $5p^2 - p + 1$
16. $y^3 + 2y^2 + 4y + 6 + \dfrac{17}{y - 2}$ **17.** See page 389.
18. See page 392.   **19.** $3(a + 6b)(a + 3b)$
20. $(c - 1)(c + 1)(c^2 + 1)$ **21.** $(y - 7)(x + 3)$
22. Prime   **23.** $-10(u - 2)(u - 1)$
24. $3(2t - 5)(2t + 5)$ **25.** $5(y - 5)^2$ **26.** $7q(3q + 2)$
27. $(2x + 1)(x - 2)(x + 2)$ **28.** $(y - 5)(y^2 + 5y + 25)$
29. $(x + 4 - y)(x + 4 + y)$ **30.** $r^2(r^2 + 16)(r - 4)(r + 4)$
31. $(x^2 + 3)(x^2 + 2)$ **32.** $(2 - c)(6a + b)$ **33.** $\dfrac{3}{2}, -5$
34. $0, 7$ **35.** $8, -2$ **36.** $\dfrac{1}{5}, -1$ **37.** $0, \dfrac{1}{4}, -\dfrac{1}{4}$
38. $-\dfrac{1}{4}$ **39.** The rocket hit the ground 256 ft from the pad.
40. a. $P(4) = 127.292$. In the year 2004, the population of
Japan was approximately 127.292 million.   **b.** 126.968 million
c. 124.52 million
41. $(4, 0)(2, 0)(0, 8)$; c   **42.** $(-4, 0)(3, 0)(-3, 0)(0, -36)$; b
43. $(-3, 0)(-1, 0)(0, -6)$; d   **44.** $(4, 0)(-3, 0)(0, 0)$; a

Chapters 1–5 Cumulative Review Exercises, pp. 416–417

1.  $[5, 12]$
2. $-x^2 - 5x - 10$
3. a. 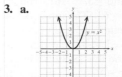   **b.**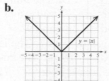

4. $\dfrac{71}{8}$ **5.** 1.97×10^8 people   **6.** Penn State scored
26 points and Florida State scored 23 points.
7. The measures are $37°, 74°,$ and $69°$.
8. $x^2 - 4x + 16$
9. $m = \dfrac{4}{3}$; y-intercept $(0, 3)$

10. $y = -5$ **11.** $\dfrac{a^6}{8b^{30}}$ **12.** $(0, 3, 2)$ **13. a.** Function
b. Not a function   **14.** $m = 7$ **15.** 7   **16.** 8
17. $y = \dfrac{3}{2}x - \dfrac{5}{2}$
18. a. The fourth test would have to be 107; therefore, it is
not possible.   **b.** Student can score between 67 and 100,
inclusive.   **19.** Mix 15 L of the 40% solution with 10 L of
the 15% solution.   **20.** $8b^3 - 6b^2 + 4b - 3$
21. $3a^3 + 5a^2 - 2a + 5$ **22.** $3w - \dfrac{5}{2} - \dfrac{1}{w}$
23. $7, -2$ **24.** $0, -4, -5$ **25.** $(\infty, 1) \cup (4, \infty)$

Chapter 6

Chapter Opener Puzzle

14	−25	C 19	−2
−10	E 30	−7	B −7
D −1	−24	A 24	7
F 3	25	−30	8

Section 6.1 Practice Exercises, pp. 427–430

1. a. rational   **b.** denominator; zero   **c.** $\dfrac{p}{q}$ **d.** $1; -1$
e. rational   **f.** denominator
3. undefined; $-4; -\dfrac{2}{3}; 0$ **5.** $2; 1; 0; -1$
7. $y = 0$ **9.** $v = 8$ **11.** $x = \dfrac{5}{2}$ **13.** $q = -9, q = 3$
15. There are no restricted values.   **17.** $x = 5, x = -5$
19. a. $\dfrac{(x + 4)(x + 2)}{(x + 4)(x - 1)}$ **b.** $x = -4, x = 1$
c. $\dfrac{x + 2}{x - 1}$ provided $x \neq -4, x \neq 1$ **21. a.** $\dfrac{(x - 9)(x - 9)}{(x - 9)(x + 9)}$
b. $x = 9, x = -9$ **c.** $\dfrac{x - 9}{x + 9}$ provided $x \neq 9, x \neq -9$

23. $\dfrac{25x^2}{9y^3}$ **25.** $\dfrac{w^8z^3}{2}$ **27.** $-\dfrac{1}{4m^2n^3}$ **29.** $\dfrac{2}{3}$
31. $\dfrac{1}{x + 5}$ **33.** $-\dfrac{1}{3c - 5}$ **35.** $-\dfrac{t + 4}{t + 3}$ **37.** $\dfrac{(2p - 1)^2}{p + 1}$
39. $\dfrac{3 - z}{2z - 5}$ **41.** $\dfrac{2(z^2 - 2z + 4)}{z - 5}$ **43.** $-\dfrac{2x - 5}{2}$
45. 1   **47.** -1 **49.** -1 **51.** -2
53. $\dfrac{c + 4}{c - 4}$ cannot be simplified   **55.** $-\dfrac{1}{12(x + y)}$
57. $\dfrac{t - 1}{t + 6}$ **59.** $\dfrac{4}{p - 3}$ **61.** $-\dfrac{2a}{b}$ **63.** $-\dfrac{x + y}{8}$
65. $\dfrac{1}{2b - 3}$ **67.** $\dfrac{x - 17}{2x - 3}$ **69.** $\dfrac{(a - 5)^2}{a - 2}$ **71.** $-\dfrac{2x}{5}$
73. $\dfrac{x - 5}{x + 2}$ **75.** $\dfrac{t^2 - 2t + 4}{3t - 5}$
77. $h(0) = 3, h(1)$ undefined, $h(-3) = \dfrac{3}{4}$,
$h(-1) = \dfrac{3}{2}, h\left(\dfrac{1}{2}\right) = 6$ **79.** $(-\infty, -4) \cup (-4, \infty)$; b
81. $(-\infty, 4) \cup (4, \infty)$; d   **83.** $\left(-\infty, \dfrac{1}{3}\right) \cup \left(\dfrac{1}{3}, \infty\right)$
85. $(-\infty, \infty)$ **87.** $(-\infty, 2) \cup (2, \infty)$
89. $\left(-\infty, -\dfrac{5}{6}\right) \cup \left(-\dfrac{5}{6}, 2\right) \cup (2, \infty)$
91. $\left(-\infty, -\dfrac{3}{2}\right) \cup \left(-\dfrac{3}{2}, 0\right) \cup \left(0, \dfrac{5}{3}\right) \cup \left(\dfrac{5}{3}, \infty\right)$
93. For example: $\dfrac{1}{x - 3}$
95. For example: $f(x) = \dfrac{1}{x + 6}$

Section 6.2 Practice Exercises, pp. 433–434

1. a. $\dfrac{pr}{qs}$ **b.** $\dfrac{ps}{qr}$ **3.** $\dfrac{t + 1}{t - 1}$ **5.** $-\dfrac{1}{p + 1}$ **7.** $\dfrac{1}{x - 3}$
9. $\dfrac{bc^3}{2}$ **11.** $\dfrac{2}{z^3}$ **13.** $12r^3$ **15.** $\dfrac{x(2x - 3)}{y^2(x + 1)}$
17. $\dfrac{2(3w - 7)}{5w + 4}$ **19.** 15   **21.** $y^2(y - 2)$ **23.** $\dfrac{r + 3}{4r^2}$
25. $\dfrac{1}{(p + 2)(6p - 7)}$ **27.** $\dfrac{4}{(b + 2)(b + 3)}$ **29.** $\dfrac{2s + 5t}{s + 5t}$
31. $\dfrac{2a + b^2}{a^2 + b^2}$ **33.** $-\dfrac{4x}{3(x + 2)}$ **35.** $\dfrac{9}{4y^2}$ **37.** $\dfrac{24}{y^2}$
39. $\dfrac{2a}{5}$ **41.** $m - n$ **43.** $-\dfrac{x - 3y}{x + 3y}$ or $\dfrac{3y - x}{x + 3y}$
45. 1   **47.** $\dfrac{a - 3}{a + 4}$ **49.** $\dfrac{(x - 6)(3x + 2)}{3x - 2}$
51. $\dfrac{2k}{h^3}$ cm^2 **53.** $\dfrac{5x}{4}$ ft^2

Section 6.3 Practice Exercises, pp. 442–444

1. a. $\dfrac{p + r}{q}; \dfrac{p - r}{q}$ **b.** least common denominator
3. $\dfrac{3(b + 1)}{2(b - 1)}$ **5.** $\dfrac{5a + 1}{2}$ **7.** $\dfrac{2}{x}$ **9.** $\dfrac{1}{x + 1}$
11. $\dfrac{2x + 5}{(2x + 9)(x - 6)}$ **13.** 2   **15.** $40x$ **17.** $30m^4n^7$

19. $(x - 4)(x + 2)(x - 6)$ **21.** $x^2(x - 1)(x + 7)^2$
23. $(x - 6)(x - 2)$ **25.** $a - 4$ or $4 - a$
27. $15xy$ **29.** $2x^3 + 4x^2$ **31.** $y^2 - y$
33. $\dfrac{8p - 15}{6p^2}$ **35.** $\dfrac{-t - s}{st}$ **37.** $\dfrac{1}{3}$
39. $\dfrac{2b + 20}{b(b + 5)}$ **41.** $\dfrac{2x}{x - 6}$ or $\dfrac{-2x}{6 - x}$
43. $\dfrac{6b^2 + 5b + 4}{(b - 4)(b + 1)}$ **45.** $\dfrac{10x}{(2x + 1)(x - 2)}$ **47.** $\dfrac{3y - 1}{y + 4}$
49. $\dfrac{11x + 6}{(x - 6)(x + 6)(x + 3)}$ **51.** $-\dfrac{3}{w}$ **53.** 1
55. $\dfrac{10 - x}{3(x - 5)(x + 5)}$ or $\dfrac{x - 10}{3(5 - x)(5 + x)}$
57. $\dfrac{x^2 + 7x + 6}{2x^2}$ **59.** $\dfrac{w^2 - 3}{w - 2}$ **61.** $\dfrac{-x^2 + 13x - 3}{x(x - 1)^2}$
63. $-\dfrac{3}{t + 3}$ **65.** $\dfrac{x^2 + x + 6}{2(x + 1)}$ **67.** $\dfrac{2z^2 + 15z}{(z - 3)(z + 4)}$
69. $\dfrac{-2y}{(x - y)(x + y)}$ or $\dfrac{2y}{(y - x)(y + x)}$
71. $\dfrac{2p^2 + 2p - 3}{3(p + 4)}$ **73.** $\dfrac{5y - 18}{y(y - 6)}$
75. $\dfrac{3x^2 + 5x + 18}{3x^2}$ cm **77.** $\dfrac{4x^2 - 2x + 50}{(x - 3)(x + 5)}$ m

Section 6.4 Practice Exercises, pp. 449–450

1. complex **3.** $\dfrac{5b^2}{3a}$ **5.** $\dfrac{10 + 3x}{2x^2}$ **7.** $\dfrac{2a + 8}{(a - 5)(a + 1)}$
9. $\dfrac{5x^2}{27}$ **11.** $\dfrac{1}{x}$ **13.** $\dfrac{10}{3}$ **15.** $\dfrac{2y - 1}{4y + 1}$ **17.** $28y$
19. -8 **21.** $\dfrac{3 - p}{p - 1}$ **23.** $\dfrac{2a + 3}{4 - 9a}$ **25.** $-\dfrac{t}{t + 1}$
27. $-\dfrac{4(w - 1)}{w + 2}$ **29.** $\dfrac{1}{y + 4}$ **31.** $\dfrac{x + 2}{x - 1}$
33. $\dfrac{t^2}{(t + 1)^2}$ **35.** $\dfrac{-a + 2}{-a - 3}$ **37.** $-\dfrac{y + 1}{y - 5}$
39. $-\dfrac{2}{x(x + h)}$ **41.** $\dfrac{1}{x(x^2 + 3)}$ **43.** $\dfrac{2b^2 + 3a}{b(b - a)}$
45. $m = \dfrac{y_2 - y_1}{x_2 - x_1}$ **47.** $\dfrac{63}{10}$ **49.** $\dfrac{4}{3}$
51. $(x^{-1} + y^{-1})^{-1} = \dfrac{1}{x^{-1} + y^{-1}} = \dfrac{1}{(1/x) + (1/y)}$ simplifies to $\dfrac{xy}{y + x}$
53. $-x(x - 1)$

Chapter 6 Problem Recognition Exercises, p. 451

1. $\dfrac{4y^2 - 8y + 9}{2y(2y - 3)}$ **2.** $\dfrac{x^2 + x - 13}{x - 4}$ **3.** $\dfrac{4x - 1}{4x - 3}$
4. $\dfrac{a - 5}{3a}$ **5.** $\dfrac{2y - 5}{(y - 1)(y + 1)}$ **6.** $\dfrac{4}{w + 4}$
7. $\dfrac{a + 4}{2}$ **8.** $\dfrac{(t - 3)(t + 2)}{t}$ **9.** $\dfrac{a}{2a - 1}$
10. $\dfrac{2xy^2}{5}$ **11.** $\dfrac{-x^2 + 4xy - y^2}{(x - y)(x + y)}$ or $\dfrac{x^2 - 4xy + y^2}{(y - x)(y + x)}$
12. $3(x - 4)$ **13.** $-\dfrac{1}{3}$ **14.** $\dfrac{1}{9}$ **15.** $y - 1$

16. $\dfrac{5t^2 - 6t - 17}{(t + 2)(t - 3)}$ **17.** $\dfrac{3(x + 4)}{2x - 3}$ **18.** $\dfrac{3z^3}{2x^2y^2}$
19. $\dfrac{y^2 + 11y - 1}{(y + 3)(y - 2)}$ **20.** $a - 10$

Section 6.5 Practice Exercises, pp. 456–458

1. a. rational **b.** denominator **c.** No
3. $\dfrac{y - 5}{(y + 1)(y - 1)^2}$ **5.** $\dfrac{1}{2t - 1}$ **7.** xy **9.** -14
11. -24 **13.** $-\dfrac{15}{22}$ **15.** 6 **17.** 5 **19.** 6
21. -25 **23.** $3, 1$ **25.** $4, -4$
27. $8, -2$ **29.** -2 (The value 2 does not check.)
31. 60 **33.** No solution (The value 5 does not check.)
35. $\dfrac{31}{5}$ **37.** 5 (The value -2 does not check.)
39. No solution (The value 4 does not check.)
41. $m = \dfrac{FK}{a}$ **43.** $E = \dfrac{IR}{K}$
45. $R = \dfrac{E - Ir}{I}$ or $R = \dfrac{E}{I} - r$
47. $B = \dfrac{2A - hb}{h}$ or $B = \dfrac{2A}{h} - b$
49. $t = \dfrac{b}{x - a}$ or $t = \dfrac{-b}{a - x}$
51. $x = \dfrac{y}{1 - yz}$ or $x = \dfrac{-y}{yz - 1}$ **53.** $h = \dfrac{2A}{a + b}$
55. $R = \dfrac{R_1 R_2}{R_2 + R_1}$
57. $t_2 = \dfrac{s_2 - s_1 + vt_1}{v}$ or $t_2 = \dfrac{s_2 - s_1}{v} + t_1$
59. $-\dfrac{8}{5}$ **61.** $-6, 2$ **63.** $\dfrac{11}{2}$ **65.** $\dfrac{11}{2}$ **67.** $-5, 5$
69. All real numbers **71.** $-\dfrac{5}{2}$ **73.** $y = 5$ **75.** $x = 5$

Chapter 6 Problem Recognition Exercises, p. 459

1. a. $\dfrac{2w + 30}{(w - 5)(w + 5)}$ **b.** -15 **c.** The problem in part (a) is an expression, and the problem in part (b) is an equation. **2. a.** $\dfrac{-3x - 8}{6(x + 2)}$ **b.** $-\dfrac{8}{3}$ **c.** The problem in part (a) is an expression, and the problem in part (b) is an equation. **3.** $\dfrac{1}{a + 1}$ **4.** $\dfrac{1}{c + 2}$ **5.** 1
6. 5 **7.** $\dfrac{x^2 - 12}{x(x - 1)}$ **8.** $\dfrac{23}{5(t - 4)}$ **9.** -1
10. $\dfrac{3}{2}, -1$ **11.** $\dfrac{8p - 11}{4(2p - 3)}$ **12.** $\dfrac{-(2x + 1)(x - 1)}{2(x + 2)}$
13. $\dfrac{x + 3}{6x^2}$ **14.** $\dfrac{15a + 2}{12a^2}$ **15.** $-\dfrac{4}{15}$ **16.** -10
17. $\dfrac{c + 7}{(c + 3)^2(c + 1)}$ **18.** $\dfrac{-2y + 13}{(y - 5)^2(y - 2)}$
19. No solution (The value 4 does not check.)
20. No solution (The value -2 does not check.)

Section 6.6 Practice Exercises, pp. 466–470

1. a. proportion **b.** proportional   **3.** $-6, 1$
5. $\dfrac{9}{2(5t-1)}$ **7.** $-\dfrac{6}{7}, 3$ **9.** $\dfrac{12(p+1)}{p+8}$ **11.** 8

13. 6   **15.** -15 **17.** $\dfrac{1}{3}$ **19.** $\dfrac{20}{9}$ **21.** $5, -1$

23. $\dfrac{3}{7}, -\dfrac{3}{7}$ **25.** $8, -5$ **27.** There must be 6 adults.

29. A 14-oz box has 84 g of fat.   **31.** 1000 swordfish were caught.   **33.** Pam needs 11.5 gal of gas.

35. There are approximately 4000 bison in the park.

37. 54 are men and 27 are women.   **39.** 595 are men and 500 are women.   **41.** $a = 8$ ft, $z = 8.4$ ft

43. $x = 12$ in., $y = 13$ in., $a = 4.2$ in.

45. The number is 3.   **47.** The number is $\dfrac{2}{5}$.

49. a. $x + 7$ **b.** $\dfrac{48}{x}$ **c.** $\dfrac{83}{x+7}$ **51.** The motorist drives 40 mph in the rain and 60 mph in sunny weather.
53. The speed is 11 mph.   **55.** He rode 10 mph.
57. Celeste walks 5 ft/sec on the ground and 7 ft/sec on the moving walkway.
59. Joe runs 6 mph and Beatrice runs 8 mph.

61. It takes $3\frac{3}{7}$ hr.   **63.** It would take $\dfrac{20}{3}$ hr or $6\frac{2}{3}$ hr.

65. It would take 8 days.   **67.** Gus would take 6 hr and Sid would take 12 hr.

Section 6.7 Practice Exercises, pp. 475–478

1. a. kx **b.** $\dfrac{k}{x}$ **c.** kxw **d.** increases **e.** decreases

3. $\dfrac{44}{3}$ **5.** $\dfrac{16}{3}$ **7.** $\dfrac{1}{8}$ **9.** Inversely   **11.** $T = kq$

13. $b = \dfrac{k}{c}$ **15.** $Q = \dfrac{kx}{y}$ **17.** $c = kst$

19. $L = kw\sqrt{v}$ **21.** $x = \dfrac{ky^2}{z}$ **23.** $k = \dfrac{9}{2}$

25. $k = 512$ **27.** $k = 1.75$ **29.** $x = 70$ **31.** $b = 6$

33. $Z = 56$ **35.** $Q = 9$ **37.** $L = 9$ **39.** $B = \dfrac{15}{2}$

41. a. 3.6 g **b.** 4.5 g **c.** 2.4 g   **43. a.** \$0.40 **b.** \$0.30 **c.** \$1.00   **45.** 355,000 tons   **47.** 156.25 lumens/m^2
49. 18.5 A   **51.** \$3500   **53.** 42.6 ft   **55.** 150 ft^2
57. 90 W   **59. a.** $A = kl^2$ **b.** The area is 4 times the original.   **c.** The area is 9 times the original.

Chapter 6 Review Exercises, pp. 485–488

1. a. $-\dfrac{2}{9}, -\dfrac{1}{10}, 0, -\dfrac{5}{6}$, undefined **b.** $t = -9$

2. a. $-\dfrac{1}{5}, -\dfrac{1}{2}$, undefined, $0, \dfrac{1}{7}$ **b.** $k = 5$

3. a. $k(2) = \dfrac{2}{3}, k(0) = 0,$

$k(1)$ undefined, $k(-1)$ undefined, $k\left(\dfrac{1}{2}\right) = -\dfrac{2}{3}$
b. $(-\infty, -1) \cup (-1, 1) \cup (1, \infty)$

4. a. $h(1) = \dfrac{1}{2}, h(0) = 0, h(-1) = -\dfrac{1}{2}, h(-3) = -\dfrac{3}{10},$

$h\left(\dfrac{1}{2}\right) = \dfrac{2}{5}$ **b.** $(-\infty, \infty)$ **5.** $2a$ **6.** $\dfrac{xz^2}{5}$ **7.** $x - 1$

8. $\dfrac{k+5}{k-3}$ **9.** $-\dfrac{x^2+3x+9}{3+x}$ **10.** $-(a^2+9)(a+3)$

11. $-\dfrac{2t+5}{t+7}$ **12.** $\dfrac{y(y+2)}{y-3}$ **13.** c; $(-\infty, 3) \cup (3, \infty)$

14. a; $(-\infty, -2) \cup (-2, \infty)$

15. b; $(-\infty, 0) \cup (0, 3) \cup (3, \infty)$ **16.** d; $(-\infty, \infty)$

17. $\dfrac{a}{2}$ **18.** $-\dfrac{3}{2}$ **19.** $-\dfrac{x-y}{5x}$ or $\dfrac{y-x}{5x}$ **20.** $(x+8)^2$

21. $\dfrac{7(k-4)}{2(k-2)}$ **22.** $\dfrac{a+1}{a-1}$ **23.** $\dfrac{x-5}{x-4}$ **24.** $-\dfrac{1}{b}$

25. $\dfrac{8}{9w^2}$ **26.** $\dfrac{5(y^2+1)}{7(y^2-2y+4)}$ **27.** $\dfrac{5}{2}$

28. $(3k+5)(k+5)$ **29.** $\dfrac{x^2+x-1}{x^3}$

30. $\dfrac{2(3x+4)}{(x+2)(x-2)}$ **31.** $\dfrac{y-3}{2y-1}$ or $\dfrac{3-y}{1-2y}$

32. $\dfrac{a-4}{2(a+3)}$ **33.** $\dfrac{4k^2-k+3}{(k+1)^2(k-1)}$ **34.** $\dfrac{4x^2+17x+11}{x+4}$

35. $\dfrac{2(a^2-5)}{(a-5)(a+3)}$ **36.** $\dfrac{13x+19}{(x+1)(x+3)(x+2)}$

37. $\dfrac{6(7-4x)}{3x-5}$ or $\dfrac{-6(4x-7)}{3x-5}$

38. $\dfrac{2(16k-9)}{(4k+3)(k-1)(4k-3)}$ **39.** $\dfrac{9a^2+a+4}{(3a-1)(a-2)}$

40. $\dfrac{5y^2-17y-9}{(y-3)(y+2)}$ **41.** $\dfrac{1}{x+1}$ **42.** $\dfrac{(k-2)(k+2)}{15}$

43. $\dfrac{x(2y+1)}{4y}$ **44.** $-y$ **45.** $\dfrac{1+a}{1-a}$ or $-\dfrac{a+1}{a-1}$

46. $2x$ **47.** $\dfrac{y}{x-y}$ **48.** $\dfrac{a(5b+1)}{3b}$ **49.** $m = \dfrac{1}{18}$

50. $m = \dfrac{64}{23}$ **51.** 3 (The value 1 does not check.)

52. 9 (The value -3 does not check.)   **53.** $0, 17$

54. $-\dfrac{1}{3}$ **55.** $5, 1$ **56.** 1   **57.** $x = \dfrac{b}{c-a}$ or $x = \dfrac{-b}{a-c}$

58. $P = \dfrac{A}{rt+1}$ **59.** $\dfrac{15}{2}$ **60.** $\dfrac{216}{7}$ **61.** $-\dfrac{7}{11}$

62. $-\dfrac{3}{4}$ **63.** Manning would gain 231 yd.

64. Erik can buy \$253.80 Canadian.   **65.** His speed was 60 mph.   **66.** It would take the larger pipe 9 hr and the smaller pipe 18 hr.
67. a. $F = kd$ **b.** $k = 3$ **c.** 12.6 lb
68. $y = 16$ **69.** $y = 12$ **70.** 48 km

Chapter 6 Test, pp. 488–489

1. a. $x = 4, x = -3$ **b.** $\dfrac{2}{x-4}$

2. a. $h(0) = \dfrac{2}{7}, h(5) = \dfrac{1}{6}, h(7)$ is undefined,

$h(-7)$ is undefined **b.** $(-\infty, -7) \cup (-7, 7) \cup (7, \infty)$

$-\infty, \infty)$ **4.** $\dfrac{2m^2}{3n}$ **5.** $\dfrac{9(x+1)}{3x+5}$

6. $m = -\dfrac{23}{9}$ **7.** $-(x-3)$ or $-x+3$ **8.** $x-4$

9. $\dfrac{x^2+5x+2}{x+1}$ **10.** $\dfrac{3}{4}$ **11.** $\dfrac{u^2v^2}{2(v^2-uv+u^2)}$

12. $-a$ **13.** $\dfrac{1}{(x+5)(x+3)}$ **14.** 3

15. $0, 4$ **16.** No solution (The value 4 does not check.)

17. $T = \dfrac{1}{p-v}$ or $T = \dfrac{-1}{v-p}$ **18.** $m_1 = \dfrac{Fr^2}{Gm_2}$

19. The number is $\dfrac{1}{6}$ or 2.   **20.** $a = 14$ m, $y = 15$ m

21. The distance is 1960 mi.   **22.** Lance rides 16 mph against the wind and 20 mph with the wind.

23. It would take $2\frac{6}{7}$ hr.   **24.** $x = \dfrac{ky}{t^2}$ **25.** 3.3 sec

Chapters 1–6 Cumulative Review Exercises, pp. 489–490

1.

Set \ Number	-22	π	6	$-\sqrt{2}$
Real numbers	✓	✓	✓	✓
Irrational numbers		✓		✓
Rational numbers	✓		✓	
Integers	✓		✓	
Whole numbers			✓	
Natural numbers			✓	

2. $\left(-2, \dfrac{1}{2}\right)$ **3.** $x^2 - x - 13$

4. a. $b_1 = \dfrac{2A - hb_2}{h}$ or $b_1 = \dfrac{2A}{h} - b_2$ **b.** 10 cm

5. 50 m by 30 m   **6.** $(-1, 2, -4)$ **7.** $y = -\dfrac{1}{3}x + 4$

8.   **9.** 75 mph

10. $(0,0), \left(\dfrac{3}{4}, 0\right), \left(\dfrac{2}{3}, 0\right)$ **11.** $8(2y - z^2)(4y^2 + 2yz^2 + z^4)$

12. $(5x + 2)(2x - 1)$ **13.** $\dfrac{x+a}{4(x-7)}$

14. $\dfrac{x+3}{(x+10)(x-2)}$ **15.** $\dfrac{c-7}{c}$ **16.** $1, -1, 5$

17. $3, -1$ **18.** The distance is 278 mi.

19. a. Vertical line; slope is undefined   **b.** Horizontal line; $m = 0$ **20.** \$3500

Chapter 7

Chapter Opener Puzzle

<u>I</u> <u>M</u> <u>A</u> <u>G</u> <u>I</u> <u>N</u> <u>A</u> <u>R</u> <u>Y</u>   <u>N</u> <u>U</u> <u>M</u> <u>B</u> <u>E</u> <u>R</u>.
4 8 7 1 4 3 7 6 9   3 5 8 10 2 6

Section 7.1 Practice Exercises, pp. 500–503

1. a. $b; a$ **b.** principal   **c.** $b^n; a$ **d.** index; radicand   **e.** cube   **f.** is not; is   **g.** even; odd   **h.** Pythagorean; c^2 **i.** $[0, \infty); (-\infty, \infty)$ **j.** -5 and -4

3. a. $8, -8$ **b.** 8   **c.** There are two square roots for every positive number. $\sqrt{64}$ identifies the positive square root.

5. a. 9   **b.** -9 **7.** There is no real number b such that $b^2 = -36$.   **9.** 7   **11.** -7 **13.** Not a real number

15. $\dfrac{8}{3}$ **17.** 0.9   **19.** -0.4 **21. a.** 8   **b.** 4

c. -8 **d.** -4 **e.** Not a real number   **f.** -4

23. -3 **25.** $\dfrac{1}{2}$ **27.** 2   **29.** $-\dfrac{5}{4}$ **31.** Not a real number   **33.** 10   **35.** -0.2 **37.** 0.5   **39.** $|a|$

41. a **43.** $|a|$ **45.** $|x+1|$ **47.** $|x|y^2$ **49.** $-\dfrac{x}{y}$

51. $\dfrac{2}{|x|}$ **53.** -92 **55.** 2   **57.** -923

59. y^4 **61.** $\dfrac{a^3}{b}$ **63.** $-\dfrac{5}{q}$ **65.** $3xy^2z$ **67.** $\dfrac{hk^2}{4}$

69. $-\dfrac{t}{3}$ **71.** $2y^2$ **73.** $2p^2q^3$ **75.** 9 cm   **77.** 13 ft

79. They were 5 mi apart.   **81.** They are 25 mi apart.

83. a. Not a real number   **b.** Not a real number   **c.** 0   **d.** 1   **e.** 2; Domain: $[2, \infty)$

85. a. -2 **b.** -1 **c.** 0   **d.** 1; Domain: $(-\infty, \infty)$

87. $\left(-\infty, \dfrac{5}{2}\right]$ **89.** $(-\infty, \infty)$ **91.** $[5, \infty)$ **93.** $(-\infty, \infty)$

95. a. $(-\infty, 1]$ **97. a.** $[-3, \infty)$

b.   **b.**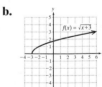

99. a. $[0, \infty)$ **101. a.** $(-\infty, \infty)$
b.   **b.**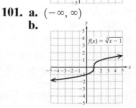

103. $q + p^2$ **105.** $\dfrac{6}{\sqrt[3]{x}}$ **107.** 8 in.

Section 7.1 Graphing Calculator Exercises, pp. 503–504

109. 8.3066   **110.** 76.1446   **111.** 3.7100
112. -0.5566 **113.** 15.6525   **114.** -6.2403
115. -0.1235 **116.** 1.0622

117.

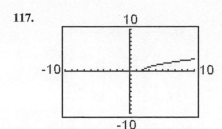

118.

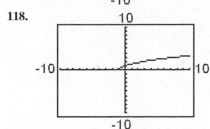

119.

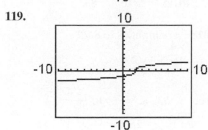

120.

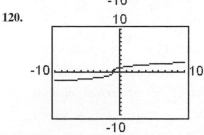

Section 7.2 Practice Exercises, pp. 508–511

1. a. $\sqrt[n]{a}$ **b.** $(\sqrt[n]{a})^m$ or $\sqrt[n]{a^m}$ **c.** $\dfrac{1}{\sqrt{x}}$ **d.** $2; \dfrac{1}{2}$

3. Index: 3, radicand: 27 **5.** 5 **7.** 3

9. 12 **11.** −12 **13.** Not a real number

15. −4 **17.** $\dfrac{1}{5}$ **19.** $-\dfrac{1}{7}$

21. $a^{m/n} = \sqrt[n]{a^m}$; The numerator of the exponent represents the power of the base. The denominator of the exponent represents the index of the radical. **23. a.** 8 **b.** −8

c. Not a real number **d.** $\dfrac{1}{8}$ **e.** $-\dfrac{1}{8}$ **f.** Not a real number

25. a. 125 **b.** −125 **c.** Not a real number **d.** $\dfrac{1}{125}$

e. $-\dfrac{1}{125}$ **f.** Not a real number **27.** $\dfrac{1}{512}$ **29.** 27

31. $-\dfrac{1}{81}$ **33.** $\dfrac{27}{1000}$ **35.** Not a real number

37. −2 **39.** −2 **41.** 6 **43.** 10 **45.** $\dfrac{3}{4}$ **47.** 1

49. $\dfrac{9}{2}$ **51.** $\sqrt[3]{q^2}$ **53.** $6\sqrt[4]{y^3}$ **55.** $\sqrt[3]{x^2y}$ **57.** $\dfrac{6}{\sqrt[5]{r^2}}$

59. $x^{1/3}$ **61.** $10b^{1/2}$ **63.** $y^{2/3}$ **65.** $(a^2b^3)^{1/4}$ **67.** $\dfrac{1}{x}$

69. p **71.** y^2 **73.** $6^{2/5}$ **75.** $\dfrac{4}{t^{5/3}}$ **77.** a^7

79. $\dfrac{25a^4d}{c}$ **81.** $\dfrac{y^9}{x^8}$ **83.** $\dfrac{2z^3}{w}$ **85.** $5xy^2z^3$

87. $x^{13}z^{4/3}$ **89.** $\dfrac{x^3y^2}{z^5}$ **91. a.** 10.9% **b.** 8.8%

c. The account in part (a) has the higher return.
93. 2.7 in. **95.** $\sqrt{y}$ **97.** $\sqrt[4]{z}$ **99.** $\sqrt[3]{x^2}$
101. $x^2\sqrt[4]{y}$ **103.** $\sqrt[6]{x}$ **105.** $\sqrt[15]{w}$ **107.** 3
109. 0.3761 **111.** 2.9240 **113.** 31.6228

Section 7.3 Practice Exercises, pp. 516–518

1. a. $\sqrt[n]{a}; \sqrt[n]{b}$ **b.** The exponent within the radicand is greater than the index. **c.** is not **d.** 3 **e.** t^{12} **f.** No. $\sqrt{2}$ is an irrational number and the decimal form is a nonterminating, nonrepeating decimal.

3. $\dfrac{p}{q^5}$ **5.** $\sqrt[4]{x^4}$ **7.** $y^{9/2}$ **9.** $x^2\sqrt{x}$ **11.** $q^2\sqrt[3]{q}$

13. $a^2b^2\sqrt{a}$ **15.** $-x^2y^3\sqrt[4]{y}$ **17.** $2\sqrt{7}$ **19.** $15\sqrt{2}$
21. $3\sqrt[3]{2}$ **23.** $5b\sqrt{ab}$ **25.** $-2x^2z\sqrt[3]{2y}$

27. $2wz\sqrt[4]{5z^3}$ **29.** x **31.** p^2 **33.** 5 **35.** $\dfrac{1}{2}$

37. $\dfrac{5\sqrt[3]{2}}{3}$ **39.** $\dfrac{5\sqrt[3]{9}}{6}$ **41.** $4\sqrt{5}$ **43.** $-30\sqrt{3}$

45. $5x^2y\sqrt{y}$ **47.** $3yz\sqrt[3]{x^2z}$ **49.** $2w^2$ **51.** $\dfrac{1}{10y^6}$

53. $\dfrac{2}{b}$ **55.** $2a^7b^4c^{15}d^{11}\sqrt{2c}$ **57.** $3a^2b\sqrt[3]{2b}$

59. $-10a^2b^2\sqrt{3ac}$ **61.** $x\sqrt[4]{7xy}$ **63.** $3a^2b\sqrt{6}$

65. $2\sqrt{3}$ **67.** $\dfrac{3\sqrt{5}}{4}$ **69.** $\dfrac{1}{\sqrt[3]{w^6}}$ simplifies to $\dfrac{1}{w^2}$

71. $\sqrt{k^3}$ simplifies to $k\sqrt{k}$ **73.** $2\sqrt{41}$ ft **75.** $6\sqrt{5}$ m
77. The distance is $90\sqrt{2}$ ft ≈ 127.3 ft. **79.** The path from A to B and B to C is faster.

Section 7.4 Practice Exercises, pp. 521–524

1. a. index; radicand **b.** $2\sqrt{3x}$ **c.** cannot; can **d.** $4\sqrt{2}$

3. $-xy\sqrt[4]{x^3}$ **5.** $\dfrac{b^2}{2}$ **7.** $(x^3y)^{1/4}$ **9.** $y^{11/12}$

11. a. Not *like* radicals **b.** *Like* radicals **c.** Not *like* radicals **13. a.** Both expressions can be simplified by using the distributive property. **b.** Neither expression can be simplified because the terms do not contain *like* terms or *like* radicals. **15.** $9\sqrt{5}$ **17.** $2\sqrt[3]{tw}$ **19.** $5\sqrt{10}$
21. $8\sqrt[4]{3} - \sqrt[4]{14}$ **23.** $2\sqrt{x} + 2\sqrt{y}$ **25.** Cannot be simplified further **27.** Cannot be simplified further

29. $\dfrac{29}{18}z\sqrt[3]{6}$ **31.** $0.7x\sqrt{y}$ **33.** $(5x + 6)\sqrt{x}$

35. $(14a - 3)\sqrt{2a}$ **37.** 15 **39.** $8\sqrt{3}$ **41.** $3\sqrt{7}$
43. $-5\sqrt{2} + 3\sqrt{3}$ **45.** $\sqrt[3]{3}$ **47.** $-5\sqrt{2a}$
49. $8s^2t^2\sqrt[3]{s^2}$ **51.** $6x\sqrt[3]{x}$ **53.** $14p^2\sqrt{5}$ **55.** $-\sqrt[3]{a^2b}$
57. $33d\sqrt[3]{2c}$ **59.** $2a^2b\sqrt{6a}$ **61.** $5x\sqrt[3]{2} - 6\sqrt[3]{x}$
63. False, $\sqrt{9} + \sqrt{16} \neq \sqrt{9 + 16}$ **65.** True
67. False, $\sqrt{y} + \sqrt{y} = 2\sqrt{y} \neq \sqrt{2y}$
69. $\sqrt{48} + \sqrt{12}$ simplifies to $6\sqrt{3}$
71. $5\sqrt[3]{x^6} - x^2$ simplifies to $4x^2$ **73.** The difference of the principal square root of 18 and the square of 5

principal fourth root of x and the cube
≈ 22.0 cm   **79.** $x = 2\sqrt{2}$ ft
. 22.36 yd   **c.** $105.95

ctice Exercises, pp. 529–532

1. a. $\sqrt{ab}$ **b.** x **c.** a **d.** conjugates   **e.** $m - n$
f. $c + 8\sqrt{c} + 16$
   3. $-2ab\sqrt{5bc}$ **5.** $x^{1/6}y^{7/12}$ **7.** $2\sqrt[3]{7}$
9. $\sqrt[3]{21}$ **11.** $2\sqrt{5}$ **13.** $4\sqrt[4]{4}$ **15.** $8\sqrt[3]{20}$
17. $-24ab\sqrt{a}$ **19.** $6\sqrt{10}$ **21.** $6x\sqrt{2}$
23. $10a^3b^2\sqrt{b}$ **25.** $-24x^2y^2\sqrt{2y}$ **27.** $6ab\sqrt[3]{2a^2b^2}$
29. $12 - 6\sqrt{3}$ **31.** $2\sqrt{3} - \sqrt{6}$ **33.** $-2x - \dfrac{7}{3}\sqrt{x}$
35. $-8 + 7\sqrt{30}$ **37.** $x - 5\sqrt{x} - 36$
39. $\sqrt[3]{y^2} - \sqrt[3]{y} - 6$ **41.** $9a - 28\sqrt{ab} + 3b$
43. $8\sqrt{p} + 3p + 5\sqrt{pq} + 16\sqrt{q} - 2q$ **45.** 15
47. $3y$ **49.** 6   **51.** 709   **53. a.** $x^2 - y^2$ **b.** $x^2 - 25$
55. $29 + 8\sqrt{13}$ **57.** $p - 2\sqrt{7p} + 7$
59. $2a - 6\sqrt{2ab} + 9b$ **61.** $3 - x^2$
63. 4   **65.** $\dfrac{4}{9}x - \dfrac{1}{4}y$ **67. a.** $3 - x$ **b.** $3 + 2\sqrt{3x} + x$
c. $3 - 2\sqrt{3x} + x$ **69.** True   **71.** False;
$(x - \sqrt{5})^2 = x^2 - 2x\sqrt{5} + 5$.   **73.** False; 5 is multiplied
by 3 only.   **75.** True   **77.** $6x$ **79.** $3x + 1$
81. $x + 19 - 8\sqrt{x + 3}$ **83.** $2t + 10\sqrt{2t - 3} + 22$
85. $12\sqrt{5}$ ft^2 **87.** $18\sqrt{15}$ in.^{2} **89.** $\sqrt[4]{x^3}$
91. $\sqrt[15]{(2z)^8}$ **93.** $p^2\sqrt[6]{p}$ **95.** $u\sqrt[6]{u}$
97. $\sqrt[4]{x^2y}$ **99.** $\sqrt[4]{2^3 \cdot 3^2}$ or $\sqrt[4]{72}$ **101.** $a + b$

Chapter 7 Problem Recognition Exercises, p. 532

1. a. $2\sqrt{6}$ **b.** $2\sqrt[3]{3}$ **2. a.** $3\sqrt{6}$ **b.** $3\sqrt[3]{2}$
3. a. $10y^3\sqrt{2}$ **b.** $2y^2\sqrt[3]{25}$ **4. a.** $4z^7\sqrt{2z}$ **b.** $2z^5\sqrt[3]{4}$
5. a. $4\sqrt{5}$ **b.** $2\sqrt[3]{10}$ **c.** $2\sqrt[4]{5}$ **6. a.** $4\sqrt{3}$ **b.** $2\sqrt[3]{6}$
c. $2\sqrt[4]{3}$ **7. a.** $x^2y^3\sqrt{x}$ **b.** $xy^2\sqrt[3]{x^2}$ **c.** $xy\sqrt[4]{xy^2}$
8. a. $a^5b^4\sqrt{b}$ **b.** $a^3b^3\sqrt[3]{a}$ **c.** $a^2b^2\sqrt[4]{a^2b}$
9. a. $2st^2\sqrt[3]{4s^2}$ **b.** $2st\sqrt[4]{2st^2}$ **c.** $2st\sqrt[5]{t}$
10. a. $2v^2w^6\sqrt[3]{12vw^2}$ **b.** $2vw^5\sqrt[4]{6v^3}$ **c.** $2vw^4\sqrt[5]{3v^2}$
11. a. $2\sqrt{5}$ **b.** 5   **12. a.** $2\sqrt{10}$ **b.** 10
13. a. $-3\sqrt{6}$ **b.** 60   **14. a.** $-7\sqrt{7}$ **b.** 210
15. a. $3\sqrt{2}$ **b.** 4   **16. a.** $3\sqrt{3}$ **b.** 6
17. a. $7\sqrt{2}$ **b.** 240   **18. a.** $-\sqrt{2}$ **b.** 60
19. a. $14\sqrt[3]{3}$ **b.** $48\sqrt[3]{9}$ **20. a.** $\sqrt[3]{2}$ **b.** $30\sqrt[3]{4}$

Section 7.6 Practice Exercises, pp. 539–542

1. a. radical   **b.** $\dfrac{\sqrt[n]{a}}{\sqrt[n]{b}}$ **c.** $\dfrac{4}{x^2}$ **d.** rationalizing
e. is; is not   **f.** denominator
   3. $12y\sqrt{5}$ **5.** $-18y + 3\sqrt{y} + 3$ **7.** $64 - 16\sqrt{t} + t$
9. -5 **11.** $\dfrac{7x^2}{y^3}$ **13.** $\dfrac{2a\sqrt{2}}{x^3}$ **15.** $\dfrac{-2j\sqrt[3]{2}}{k}$
17. $\dfrac{t\sqrt[4]{3s^2}}{10}$ **19.** $3b^2$ **21.** $\dfrac{1}{2b^2}$ **23.** $3x^2\sqrt{2}$
25. $\dfrac{z\sqrt{3y}}{w^2}$ **27.** $\dfrac{\sqrt{5}}{\sqrt{5}}$ **29.** $\dfrac{\sqrt[3]{x^2}}{\sqrt[3]{x^2}}$ **31.** $\dfrac{\sqrt{3z}}{\sqrt{3z}}$

33. $\dfrac{\sqrt[4]{2a^2}}{\sqrt[4]{2a^2}}$ **35.** $\dfrac{\sqrt{3}}{3}$ **37.** $\dfrac{\sqrt{x}}{x}$ **39.** $\dfrac{3\sqrt{2y}}{y}$
41. $\dfrac{a\sqrt{2a}}{2}$ **43.** $\dfrac{3\sqrt{2}}{2}$ **45.** $\dfrac{3\sqrt[3]{4}}{2}$ **47.** $\dfrac{-6\sqrt[4]{x^3}}{x}$
49. $\dfrac{7\sqrt[3]{2}}{2}$ **51.** $\dfrac{\sqrt[3]{4w}}{w}$ **53.** $\dfrac{2\sqrt[4]{27}}{3}$ **55.** $\dfrac{\sqrt[3]{2x}}{x}$
57. $\dfrac{2\sqrt{6}}{21}$ **59.** $\dfrac{\sqrt{x}}{x^4}$ **61.** $\dfrac{\sqrt{2x}}{2x^3}$ **63.** $\sqrt{2} + \sqrt{6}$
65. $\sqrt{x} - 23$ **67.** $\dfrac{4\sqrt{2} - 12}{-7}$ or $\dfrac{-4\sqrt{2} + 12}{7}$
69. $4\sqrt{6} + 8$ **71.** $-\sqrt{21} + 2\sqrt{7}$ **73.** $\dfrac{-\sqrt{p} + \sqrt{q}}{p - q}$
75. $\sqrt{x} - \sqrt{5}$ **77.** $\dfrac{-14\sqrt{a} - 35\sqrt{b}}{4a - 25b}$
79. $\dfrac{4\sqrt{xy} - 3x - y}{y - x}$ **81.** $5 - \sqrt{10}$ **83.** $\dfrac{5 + \sqrt{21}}{4}$
85. $-6\sqrt{5} - 13$ **87.** $\dfrac{16}{\sqrt[3]{4}}$ simplifies to $8\sqrt[3]{2}$
89. $\dfrac{4}{x - \sqrt{2}}$ simplifies to $\dfrac{4x + 4\sqrt{2}}{x^2 - 2}$
91. $\dfrac{\pi\sqrt{2}}{4}$ sec ≈ 1.11 sec   **93. a.** $\dfrac{\sqrt{2}}{2}$ **b.** $\dfrac{\sqrt[3]{4}}{2}$
95. a. $\dfrac{\sqrt{5a}}{5a}$ **b.** $\dfrac{\sqrt{5} - a}{5 - a^2}$ **97.** $\dfrac{2\sqrt{6}}{3}$ **99.** $\dfrac{17\sqrt{15}}{15}$
101. $\dfrac{8\sqrt[3]{25}}{5}$ **103.** $\dfrac{-33}{2\sqrt{3} - 12}$ **105.** $\dfrac{a - b}{a + 2\sqrt{ab} + b}$

Section 7.7 Practice Exercises, pp. 549–551

1. a. radical   **b.** isolate; 7   **c.** extraneous   **d.** third
3. $\dfrac{3w\sqrt{w}}{4}$ **5.** $3c\sqrt[3]{2c}$ **7.** $4x - 6$ **9.** $9p + 7$
11. 4   **13.** 3   **15.** 42   **17.** 29   **19.** 140   **21.** No
solution (The value 25 does not check.)   **23.** 7 (The value
-1 does not check.)   **25.** No solution (The value 3 does
not check.)   **27.** $V = \dfrac{4\pi r^3}{3}$ **29.** $h^2 = \dfrac{r^2 - \pi^2 r^2}{\pi^2}$ or
$h^2 = \dfrac{r^2}{\pi^2} - r^2$ **31.** $a^2 + 10a + 25$ **33.** $5a - 6\sqrt{5a} + 9$
35. $r + 22 + 10\sqrt{r - 3}$ **37.** -3 **39.** No solution
(The value $\dfrac{1}{2}$ does not check.)   **41.** $0, 3$ **43.** 9
45. -4 **47.** $\dfrac{9}{5}$ **49.** 2   **51.** No solution (The value 9
does not check.)   **53.** $-\dfrac{11}{4}$ **55.** No solution (The value
$\dfrac{8}{3}$ does not check.)   **57.** -1 (The value 3 does not check.)
59. $\dfrac{1}{3}, -1$ **61.** No solution (The values 3 and 23 do not
check.)   **63.** 6   **65.** 2 (The value 18 does not check.)
67. a. 30.25 ft   **b.** 34.5 m   **69. a.** $2 million
b. $1.2 million   **c.** 50,000 passengers   **71. a.** 12 lb
b. $t(18) = 5.1$. An 18-lb turkey will take about 5.1 hr to cook.
73. $b = \sqrt{25 - h^2}$ **75.** $a = \sqrt{k^2 - 196}$

Section 7.7 Graphing Calculator Exercises, p. 551

76.

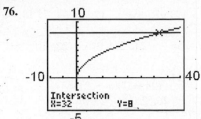

77.

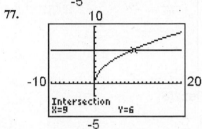

78.

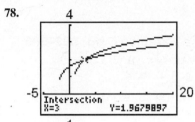

79.

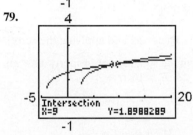

Section 7.8 Practice Exercises, pp. 559–561

1. a. imaginary   **b.** $\sqrt{-1}; -1$ **c.** $i\sqrt{b}$ **d.** $a + bi; \sqrt{-1}$
e. real; b **f.** $a + bi$ **g.** True   **h.** True   **3.** $9 - x$

5. -8 **7.** $\dfrac{1}{25}$ **9.** $\sqrt{-1} = i$ and $-\sqrt{1} = -1$

11. $12i$ **13.** $i\sqrt{3}$ **15.** $-2i\sqrt{5}$ **17.** -60
19. $13i\sqrt{7}$ **21.** -7 **23.** -12 **25.** $-3\sqrt{10}$
27. $i\sqrt{2}$ **29.** $3i$ **31.** $-i$ **33.** 1 **35.** i
37. 1 **39.** $-i$ **41.** -1 **43.** $a - bi$
45. Real: -5; imaginary: 12   **47.** Real: 0; imaginary: -6

49. Real: 35; imaginary: 0   **51.** Real: $\dfrac{3}{5}$; imaginary: 1

53. $7 + 6i$ **55.** $\dfrac{3}{10} + \dfrac{3}{2}i$ **57.** $5i\sqrt{2}$

59. $-1 + 10i$ **61.** -24 **63.** $18 + 6i$ **65.** $26 - 26i$
67. $-29 + 0i$ **69.** $-9 + 40i$ **71.** $35 + 20i$

73. $-20 + 48i$ **75.** $\dfrac{13}{16} + 0i$ **77.** $\dfrac{1}{5} - \dfrac{3}{5}i$

79. $\dfrac{3}{25} - \dfrac{4}{25}i$ **81.** $\dfrac{21}{29} + \dfrac{20}{29}i$ **83.** $-\dfrac{17}{10} - \dfrac{1}{10}i$

85. $-\dfrac{1}{2} - \dfrac{5}{2}i$ **87.** $\dfrac{1}{2} - \dfrac{1}{3}i$ **89.** $1 + 10i$ **91.** $\dfrac{1}{4} + \dfrac{1}{2}i$

93. $-1 + i\sqrt{2}$ **95.** $-2 - i\sqrt{3}$ **97.** $-\dfrac{1}{2} + \dfrac{\sqrt{2}}{2}i$

99. Yes   **101.** Yes

Chapter 7 Review Exercises, pp. 569–571

1. a. False; $\sqrt{0} = 0$ is not positive.   **b.** False; $\sqrt[3]{-8} = -2$

2. $\sqrt{(-3)^2} = \sqrt{9} = 3$ **3. a.** False   **b.** True   **4.** $\dfrac{5}{4}$

5. 5   **6.** 6   **7. a.** 3   **b.** 0   **c.** $\sqrt{7}$ **d.** $[1, \infty)$
8. a. 0   **b.** 1   **c.** 3   **d.** $[-5, \infty)$

9. $\dfrac{\sqrt[3]{2x}}{\sqrt[4]{2x}} + 4$ **10. a.** $|x|$ **b.** x **c.** $|x|$ **d.** $x + 1$

11. a. $2|y|$ **b.** $3y$ **c.** $|y|$ **d.** y **12.** 8 cm

13. Yes, provided the expressions are defined. For example: $x^5 \cdot x^3 = x^8$ and $x^{1/5} \cdot x^{2/5} = x^{3/5}$ **14.** n represents the root.   **15.** Take the reciprocal of the base and change the exponent to positive.   **16.** -5

17. $\dfrac{1}{2}$ **18.** 4   **19.** b^{10} **20.** $\dfrac{16y^{12}}{xz^9}$ **21.** $\dfrac{a^3c}{b^2}$

22. $x^{3/4}$ **23.** $(2y^2)^{1/3}$ **24.** 2.1544   **25.** 6.8173

26. 54.1819   **27.** See page 512.   **28.** $6\sqrt{3}$

29. $xz\sqrt[4]{xy}$ **30.** $-10ab^3\sqrt[3]{2b}$ **31.** $-\dfrac{2a}{b}$

32. a. The principal square root of the quotient of 2 and x.
b. The cube of the sum of x and 1.
33. 31 ft   **34.** they are *like* radicals.
35. Cannot be combined; the indices are different.
36. Cannot be combined; one term has a radical, but the other does not.
37. Can be combined: $3\sqrt[4]{3xy}$
38. Can be added after simplifying: $19\sqrt{2}$
39. $5\sqrt{7}$ **40.** $9\sqrt[3]{2} - 8$ **41.** $10\sqrt{2}$ **42.** $13x\sqrt[3]{2x^2}$
43. False; 5 and $3\sqrt{x}$ are not *like* radicals.
44. False; $\sqrt{y} + \sqrt{y} = 2\sqrt{y}$ (Add the coefficients.)
45. 6   **46.** $2\sqrt[4]{2}$ **47.** $-2\sqrt{21} + 6\sqrt{33}$
48. $-6\sqrt{15} + 15$ **49.** $4x - 9$ **50.** $y - 16$
51. $7y - 2\sqrt{21xy} + 3x$ **52.** $12w + 20\sqrt{3w} + 25$
53. $-2z - 9\sqrt{6z} - 42$ **54.** $3a + 5\sqrt{5a} - 10$

55. $u^2\sqrt[6]{u^5}$ **56.** $\sqrt[4]{4w^3}$ **57.** $\dfrac{y^2\sqrt{3y}}{5x^3}$

58. $\dfrac{-2x^2y^2\sqrt[3]{2x}}{z^3}$ **59.** $9w^2$ **60.** $\dfrac{t^4}{4}$ **61.** $\dfrac{\sqrt{14y}}{2y}$

62. $\dfrac{\sqrt{15w}}{3w}$ **63.** $\dfrac{4\sqrt[3]{3p}}{3p}$ **64.** $-\dfrac{\sqrt[3]{4x^2}}{x}$

65. $-\sqrt{15} - \sqrt{10}$ **66.** $-3\sqrt{7} - 3\sqrt{5}$

67. $\sqrt{t} + \sqrt{3}$ **68.** $\sqrt{w} + \sqrt{7}$ **69.** The quotient of the principal square root of 2 and the square of x

70. $\dfrac{49}{2}$ **71.** 31   **72.** -12 **73.** 7   **74.** 9

75. No solution (The value -2 does not check.)

76. $\dfrac{1}{2}, 4$ **77.** -2 (The value 2 does not check.)

78. $6\sqrt{5}$ m ≈ 13.4 m   **79. a.** $v(20) \approx 25.3$ ft/sec. When the water depth is 20 ft, a wave travels about 25.3 ft/sec.   **b.** 8 ft

80. $a + bi$, where a and b are real numbers and $i = \sqrt{-1}$
81. $a + bi$, where $b \neq 0$
82. Multiply the numerator and denominator by $4 - 6i$, which is the complex conjugate of the denominator.
83. $4i$ **84.** $-i\sqrt{5}$ **85.** -15 **86.** $-2i$ **87.** -1
88. i **89.** $-i$ **90.** 0 **91.** $-5 + 5i$
92. $9 + 17i$ **93.** $25 + 0i$ **94.** $24 - 10i$
95. $-\dfrac{17}{4} + i$; Real part: $-\dfrac{17}{4}$; Imaginary part: 1
96. $-2 - i$; Real part: -2; Imaginary part: -1
97. $\dfrac{4}{13} - \dfrac{7}{13}i$ **98.** $3 + 4i$ **99.** $-\dfrac{3}{2} + \dfrac{5}{2}i$
100. $-\dfrac{4}{17} + \dfrac{16}{17}i$ **101.** $-\dfrac{2}{3} + \dfrac{\sqrt{10}}{6}i$ **102.** $2 - 4i$

Chapter 7 Test, p. 572

1. a. 6 **b.** -6 **2. a.** Real **b.** Not real **c.** Real
d. Real   **3. a.** y **b.** $|y|$ **4.** 3 **5.** $\dfrac{4}{3}$ **6.** $2\sqrt[3]{4}$
7. $a^2bc^2\sqrt{bc}$ **8.** $3x^2yz^2\sqrt{2xy}$ **9.** $4w^2\sqrt{w}$ **10.** $\dfrac{x^2}{5y}$
11. $\dfrac{3\sqrt{2}}{2}$ **12. a.** $f(-8) = 2\sqrt{3}$; $f(-6) = 2\sqrt{2}$;
$f(-4) = 2$; $f(-2) = 0$ **b.** $(-\infty, -2]$ **13.** -0.3080
14. -3 **15.** $\dfrac{1}{t^{3/4}}$ **16.** $\sqrt[6]{7y^3}$ **17.** $\sqrt[12]{10}$ **18.** $3\sqrt{5}$
19. $3\sqrt{2x} - 3\sqrt{5x}$ **20.** $40 - 10\sqrt{5x} - 3x$
21. $\dfrac{-2\sqrt[3]{x^2}}{x}$ **22.** $\dfrac{x + 6 + 5\sqrt{x}}{9 - x}$ **23. a.** $2i\sqrt{2}$ **b.** $8i$
c. $\dfrac{1 + i\sqrt{2}}{2}$ **24.** $1 - 11i$ **25.** $30 + 16i$ **26.** -28
27. $-33 - 56i$ **28.** 104 **29.** $\dfrac{17}{25} + \dfrac{6}{25}i$ **30.** $36 - 11i$
31. $r(10) \approx 1.34$; the radius of a sphere of volume 10 cubic units is approximately 1.34 units.   **32.** 21 ft   **33.** -16
34. $\dfrac{17}{5}$ **35.** 2 (The value 42 does not check.)

Chapters 1–7 Cumulative Review Exercises, pp. 573–574

1. 54 **2.** $15x - 5y - 5$ **3.** -3 **4.** $(-\infty, -5)$
5. $y = -2x + 5$ **6.**

7. $\left(\dfrac{1}{2}, \dfrac{1}{3}\right)$ **8.** $\left(2, -2, \dfrac{1}{2}\right)$ is not a solution.
9. $x = 6$, $y = 3$, $z = 8$ **10. a.** $f(-2) = -10$; $f(0) = -2$;
$f(4) = 14$; $f\left(\dfrac{1}{2}\right) = 0$ **b.** $(-2, -10), (0, -2), (4, 14), \left(\dfrac{1}{2}, 0\right)$
c.

11. Not a function   **12.** $a^4b^8c^2$ **13.** $a^6b^{12}c^4$
14. a. 1.4×10^{-4} **b.** 3.14×10^9 **15.** $2x^2 - x - 15$;
2nd degree   **16.** $\sqrt{15} + 3\sqrt{2} + 3$ **17.** $x - 4$ **18.** $-\dfrac{1}{6}$
19. $\dfrac{3c\sqrt[3]{2}}{d}$ **20.** $32b\sqrt{5b}$ **21.** $2 + 3i$ **22.** 10
23. $\dfrac{x - 15}{x(x + 5)(x - 5)}$ **24.** $\dfrac{-1}{(2a + 1)(a + 2)}$
25. $-14x^2 + 26x - 17$ **26.** $2\sqrt{3} + 2\sqrt{5}$
27. $\dfrac{6}{17} + \dfrac{10}{17}i$ **28.** $\dfrac{7}{6}, -\dfrac{3}{2}$ **29.** $(x + 3 - y)(x + 3 + y)$
30. $(x^2 + 2)(x^4 - 2x^2 + 4)$

Chapter 8

Chapter Opener Puzzle

$$\underline{Q}\ \underline{U}\ \underline{A}\ \underline{D}\ \underline{R}\ \underline{A}\ \underline{T}\ \underline{I}\ \underline{C}$$
$$2\ \ 1\ \ 6\ \ 5\ \ 8\ \ 6\ \ 4\ \ 7\ \ 3$$

Section 8.1 Practice Exercises, pp. 582–585

1. a. $0; 0$ **b.** 0 **c.** $\sqrt{k}; -\sqrt{k}$ **d.** $2; 3$ and -3
e. completing **f.** 100 **g.** $4; 1$ **h.** 8
3. ± 2 **5.** $\pm\dfrac{11}{6}$ **7.** $\pm 5i$ **9.** $-1, -5$
11. $\dfrac{-3 \pm \sqrt{7}}{2}$ **13.** $-5 \pm 3i\sqrt{2}$ **15.** $\pm\dfrac{\sqrt{33}}{3}$
17. $-\dfrac{4}{5} \pm \dfrac{\sqrt{3}}{5}$ **19.** 1. Factoring and applying the zero
product rule. 2. Applying the square root property. $x = \pm 6$
21. $n = 9; (x - 3)^2$ **23.** $n = 16; (t + 4)^2$
25. $n = \dfrac{1}{4}; \left(c - \dfrac{1}{2}\right)^2$ **27.** $n = \dfrac{25}{4}; \left(y + \dfrac{5}{2}\right)^2$
29. $n = \dfrac{1}{25}; \left(b + \dfrac{1}{5}\right)^2$ **31.** $n = \dfrac{1}{9}; \left(p - \dfrac{1}{3}\right)^2$
33. See page 579.   **35.** $-3, -5$ **37.** $-3 \pm i\sqrt{7}$
39. $-2 \pm i\sqrt{2}$ **41.** $5, -2$ **43.** $-1 \pm \dfrac{\sqrt{6}}{2}i$
45. $2 \pm \dfrac{2}{3}i$ **47.** $\dfrac{1}{5} \pm \dfrac{\sqrt{3}}{5}$ **49.** $-\dfrac{3}{4} \pm \dfrac{\sqrt{65}}{4}$
51. $2 \pm \sqrt{11}$ **53.** $1, -7$
55. a. $t = \dfrac{\sqrt{d}}{4}$ **b.** The time required is 8 sec.
57. $r = \sqrt{\dfrac{A}{\pi}}$ or $r = \dfrac{\sqrt{A\pi}}{\pi}$ **59.** $a = \sqrt{d^2 - b^2 - c^2}$
61. $r = \sqrt{\dfrac{3V}{\pi h}}$ or $r = \dfrac{\sqrt{3V\pi h}}{\pi h}$
63. The shelf extends 4.2 ft.   **65.** The sides are 7.1 in.
67. a. The number of books required is approximately
4.5 thousand textbooks or 35.5 thousand textbooks.
b. Profit increases to a point as more books are produced.
Beyond that point, the market is "flooded," and profit
decreases. There are two points at which the profit is $20,000.
Producing 4.5 thousand books makes the same profit using
fewer resources as producing 35.5 thousand books.

Section 8.2 Practice Exercises, pp. 595–599

1. a. quadratic; $\dfrac{-b \pm \sqrt{b^2 - 4ac}}{2a}$ **b.** $ax^2 + bx + c = 0$

c. $8; -42; -27$ **d.** $7; 97$ **e.** $b^2 - 4ac$; discriminant
f. imaginary   **g.** real   **h.** less than   **3.** $2, -12$
5. $1 \pm 3i$ **7.** $4 - 2\sqrt{5}$ **9.** $2 - i\sqrt{3}$

11. linear   **13.** $-12, 1$ **15.** $\dfrac{1}{9} \pm \dfrac{2\sqrt{11}}{9}i$

17. $\dfrac{1}{6} \pm \dfrac{\sqrt{14}}{6}i$ **19.** $7, -5$ **21.** $\dfrac{-3 \pm \sqrt{41}}{2}$

23. $\dfrac{2}{5}$ **25.** $3 \pm i\sqrt{5}$ **27.** $\dfrac{1 \pm \sqrt{29}}{2}$

29. $\dfrac{-1 \pm \sqrt{13}}{4}$ **31.** $-\dfrac{2}{3} \pm \dfrac{2\sqrt{2}}{3}i$

33. $\dfrac{-5 \pm \sqrt{13}}{2}$ **35.** $-2, -4$ **37.** $\dfrac{-7 \pm \sqrt{109}}{6}$

39. a. $(x - 3)(x^2 + 3x + 9)$ **b.** $3, -\dfrac{3}{2} \pm \dfrac{3\sqrt{3}}{2}i$

41. a. $3x(x^2 - 2x + 2)$ **b.** $0, 1 \pm i$

43. The length of each side is 3 ft.   **45.** The lengths are
1.6 in. and 3.6 in.   **47.** The lengths are 8.2 m and 6.1 m.
49. a. Approximately 4.5 fatalities per 100 million miles
driven   **b.** Approximately 1 fatality per 100 million miles
driven   **c.** Approximately 4.2 fatalities per 100 million miles
driven   **d.** For drivers 26 yr old and 71 yr old

51. $\dfrac{3 + \sqrt{5}}{2} \approx 2.62$ sec or $\dfrac{3 - \sqrt{5}}{2} \approx 0.38$ sec

53. a. $x^2 + 2x + 1 = 0$ **b.** 0   **c.** 1 rational solution
55. a. $19m^2 - 8m + 0 = 0$ **b.** 64   **c.** 2 rational solutions
57. a. $5p^2 + 0p - 21 = 0$ **b.** 420   **c.** 2 irrational solutions
59. a. $n^2 + 3n + 4 = 0$ **b.** -7 **c.** 2 imaginary solutions

61. x-intercepts: $\left(\dfrac{5 + \sqrt{13}}{2}, 0\right), \left(\dfrac{5 - \sqrt{13}}{2}, 0\right)$;

y-intercept: $(0, 3)$ **63.** x-intercepts: none; y-intercept: $(0, -1)$

65. x-intercepts: $\left(\dfrac{-5 - \sqrt{41}}{4}, 0\right), \left(\dfrac{-5 + \sqrt{41}}{4}, 0\right)$;

y-intercept: $(0, -2)$ **67.** $-\dfrac{3}{2} \pm \dfrac{\sqrt{7}}{2}i$ **69.** $-\dfrac{1}{3}, 6$

71. $-\dfrac{5}{2}, \dfrac{1}{2}$ **73.** $-\dfrac{1}{2} \pm 2i$ **75.** $\dfrac{3 \pm \sqrt{7}}{2}$

77. $1 \pm 3i\sqrt{2}$ **79.** $2 \pm \sqrt{2}$ **81.** $-\dfrac{1}{8}, \dfrac{3}{4}$ **83.** $3, -3$

85. a. $-3 \pm \sqrt{14}$ **b.** $-3 \pm \sqrt{14}$ **c.** Answers will vary

87. b. $x^2 = \left(\dfrac{18 - x}{2}\right)^2 + \left(\dfrac{18 - x}{2}\right)^2$

c. $x^2 + 36x - 324 = 0$ **d.** 7.5 in., -43.5 in.   **e.** 7.5 in. is
appropriate; the other solution is negative.

Section 8.2 Graphing Calculator Exercises, p. 599

88.

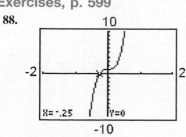

89.

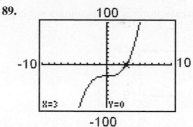

90.

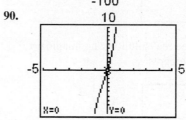

91.

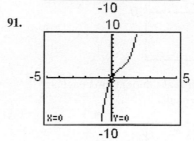

Section 8.3 Practice Exercises, pp. 603–604

1. a. quadratic   **b.** $3x - 1$ **c.** $p^{1/3}$

3. $\dfrac{3}{2} \pm \dfrac{\sqrt{7}}{2}$ **5.** -4 **7.** $2 \pm \sqrt{26}$

9. a. $7, -5$ **b.** $7, -1, 5, 1$ **11. a.** $3, 1$ **b.** $-\dfrac{3}{2}, 1, \dfrac{1}{2}, -1$

13. $-2, 1, -3, 2$ **15.** $\dfrac{27}{8}, -125$ **17.** $-32, 1$ **19.** $25, 9$

21. $\dfrac{1}{9}$ (The value 4 does not check.)   **23.** $-\dfrac{7}{3}$

25. $\pm\sqrt{3}, \pm 2i$ **27.** $\pm\dfrac{\sqrt{2}}{3}, \pm i$ **29.** $4, 6$ **31.** $\dfrac{9 \pm \sqrt{73}}{4}$

33. $2 \pm \sqrt{3}$ **35.** $\pm 2, \pm 2i$ **37.** $-\dfrac{7}{4}, -\dfrac{3}{2}$

39. $\dfrac{1}{2}, -\dfrac{1}{2}, \sqrt{2}, -\sqrt{2}$ **41.** $2, 1, -1 \pm i\sqrt{3}, -\dfrac{1}{2} \pm \dfrac{\sqrt{3}}{2}i$

43. $5, 4$ **45.** $1, \dfrac{17}{2}$ **47.** $64, -125$ **49.** $\pm\sqrt{2}, \pm 2i$

51. $\pm 4i, 1$ **53.** $4, \pm i\sqrt{5}$

Section 8.3 Graphing Calculator Exercises, pp. 604–605

55. a. $\pm i\sqrt{2}$ **b.** Two imaginary solutions; no real solutions
c. No x-intercepts
d.

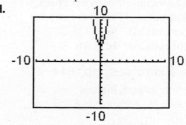

56. a. $1, -1$ **b.** Two real solutions; no imaginary solutions
c. Two x-intercepts
d.

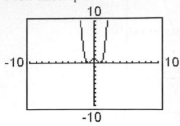

57. a. $0, 3, -2$ **b.** Three real solutions; no imaginary
solutions   **c.** Three x-intercepts
d.

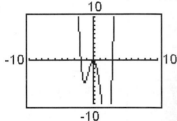

58. a. $\pm 1, \pm 3$ **b.** Four real solutions; no imaginary
solutions   **c.** Four x-intercepts
d.

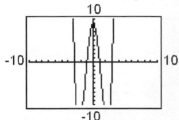

Chapter 8 Problem Recognition Exercises, p. 605

1. $-5 \pm \sqrt{22}$ **2.** $8 \pm \sqrt{59}$ **3.** $-\dfrac{1}{6} \pm \dfrac{\sqrt{47}}{6}i$

4. $-\dfrac{3}{8} \pm \dfrac{\sqrt{71}}{8}i$ **5. a.** Quadratic   **b.** $2, -7$

6. a. Quadratic   **b.** $4, 5$ **7. a.** Quadratic form
b. $-3, -1, 1, 3$ **8. a.** Quadratic form   **b.** $-5, -2, 2, 5$

9. a. Quadratic   **b.** $\dfrac{-7 \pm \sqrt{33}}{2}$ **10. a.** Quadratic

b. $\dfrac{9 \pm \sqrt{89}}{2}$ **11. a.** Linear   **b.** 4 **12. a.** Linear   **b.** -1

13. a. Quadratic   **b.** $\pm i\sqrt{2}$ **14. a.** Quadratic
b. $\pm i\sqrt{6}$ **15. a.** Rational   **b.** $2, -1$ **16. a.** Rational
b. $-2, 4$ **17. a.** Quadratic   **b.** $10 \pm \sqrt{101}$
18. a. Quadratic   **b.** $-9 \pm \sqrt{77}$ **19. a.** Quadratic
b. $\dfrac{1 \pm \sqrt{17}}{2}$ **20. a.** Quadratic   **b.** $\dfrac{1 \pm \sqrt{13}}{2}$

21. a. Radical   **b.** 3; (The value -1 does not check.)
22. a. Radical   **b.** 6; (The value -1 does not check.)
23. a. Quadratic form (or radical)   **b.** $-125, 27$
24. a. Quadratic form (or radical)   **b.** $-64, -1$

Section 8.4 Practice Exercises, pp. 614–618

1. a. parabola   **b.** $>; <$ **c.** lowest; highest
d. (h, k); upward; downward   **e.** vertex; $x = h$

3. $3 \pm 2i$ **5.** $\dfrac{5 \pm \sqrt{10}}{5}$ **7.** $-27; -8$

9. The value of k shifts the graph of $f(x) = x^2$ vertically.
11.

13.

15.

17.

19.

21.

23.

25.

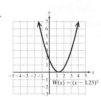

27. The value of a vertically stretches or shrinks the graph
of $f(x) = x^2$.
29.

31.

33.

35.

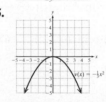

37. d   **39.** g   **41.** a   **43.** e
45.

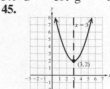

47.

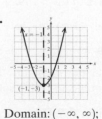

Domain: $(-\infty, \infty)$;
range: $[2, \infty)$

Domain: $(-\infty, \infty)$;
range: $[-3, \infty)$

49.

51.

Domain: $(-\infty, \infty)$;
range: $(-\infty, -2]$

Domain: $(-\infty, \infty)$;
range: $(-\infty, 3]$

53.

Domain: $(-\infty, \infty)$;
range: $[1, \infty)$

55.

Domain: $(-\infty, \infty)$;
range: $[0, \infty)$

57.

Domain: $(-\infty, \infty)$;
range: $(-\infty, 3]$

59.

Domain: $(-\infty, \infty)$;
range: $[-1, \infty)$

61.

Domain: $(-\infty, \infty)$;
range: $(-\infty, 2]$

63.

Domain: $(-\infty, \infty)$;
range: $[1, \infty)$

65. a. $y = x^2 + 3$ is $y = x^2$ shifted up 3 units.
b. $y = (x + 3)^2$ is $y = x^2$ shifted left 3 units.   **c.** $y = 3x^2$ is $y = x^2$ with a vertical stretch by a factor of 3 units.
67. Vertex $(6, -9)$ minimum point; minimum value: -9
69. Vertex $(2, 5)$; maximum point; maximum value: 5
71. Vertex $(-8, 0)$; minimum point; minimum value: 0
73. Vertex $\left(0, \dfrac{21}{4}\right)$; maximum point; maximum value: $\dfrac{21}{4}$
75. Vertex $\left(7, -\dfrac{3}{2}\right)$; minimum point; minimum value: $-\dfrac{3}{2}$
77. Vertex $(0, 0)$; minimum point; minimum value: 0
79. True   **81.** False   **83. a.** $(60, 30)$ **b.** 30 ft   **c.** 70 ft

Section 8.4 Graphing Calculator Exercises, p. 618

85.

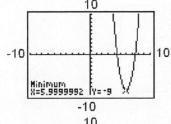

86.

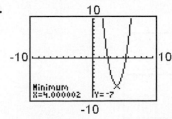

87.

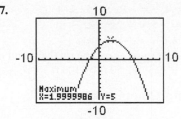

88.

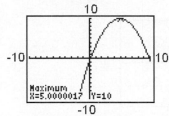

Section 8.5 Practice Exercises, pp. 625–628

1. a. $\dfrac{-b}{2a}; \dfrac{-b}{2a}$ **b.** True   **c.** True   **d.** True   **e.** False
3. The graph of p is the graph of $y = x^2$ shrunk vertically by a factor of $\frac{1}{4}$.
5. The graph of r is the graph of $y = x^2$ shifted up 7 units.
7. The graph of t is the graph of $y = x^2$ shifted to the left 10 units.   **9.** $16; (x - 4)^2$ **11.** $\dfrac{49}{4}; \left(y + \dfrac{7}{2}\right)^2$
13. $\dfrac{1}{81}; \left(b + \dfrac{1}{9}\right)^2$ **15.** $\dfrac{1}{36}; \left(t - \dfrac{1}{6}\right)^2$
17. $g(x) = (x - 4)^2 - 11; (4, -11)$
19. $n(x) = 2(x + 3)^2 - 5; (-3, -5)$
21. $p(x) = -3(x - 1)^2 - 2; (1, -2)$
23. $k(x) = \left(x + \dfrac{7}{2}\right)^2 - \dfrac{89}{4}; \left(-\dfrac{7}{2}, -\dfrac{89}{4}\right)$
25. $F(x) = 5(x + 1)^2 - 4; (-1, -4)$
27. $P(x) = -2\left(x - \dfrac{1}{4}\right)^2 + \dfrac{1}{8}; \left(\dfrac{1}{4}, \dfrac{1}{8}\right)$ **29.** $(2, 3)$
31. $(-1, -2)$ **33.** $(-4, -15)$ **35.** $(-1, 2)$ **37.** $(1, 3)$
39. $\left(\dfrac{3}{2}, \dfrac{3}{4}\right)$ **41.** $(-4, -15)$ **43.** $(-1, 4)$
45. a. $(-1, -4)$ **b.** $(0, -3)$ **47. a.** $(\frac{1}{2}, \frac{7}{2})$ **b.** $(0, 4)$
c. $(1, 0), (-3, 0)$ **c.** No x-intercepts
d.

d.

49. a. $(\frac{3}{2}, 0)$ **b.** $(0, -\frac{9}{4})$ **51. a.** $(-1, 4)$ **b.** $(0, 3)$
c. $(\frac{3}{2}, 0)$ **c.** $(1, 0), (-3, 0)$
d.

d.

53. Mia must package 10 MP3 players.   **55. a.** 33 psi
b. 38 thousand miles   **57.** The speed should be 45 mph.
59. It will take 48 hr.

61. $a = -9, b = 5, c = 4; y = -9x^2 + 5x + 4$
63. $a = 2, b = -1, c = -5; y = 2x^2 - x - 5$
65. $a = -3, b = 4, c = 0; y = -3x^2 + 4x$
67. a. The sum of the sides must equal the total amount of fencing. **b.** $A = x(200 - 2x)$ **c.** 50 ft by 100 ft

Section 8.5 Graphing Calculator Exercises, p. 628

69.

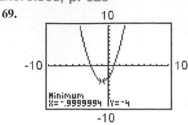

70.

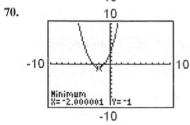

71.

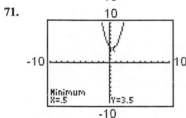

72.

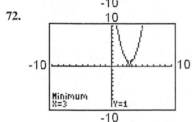

73.

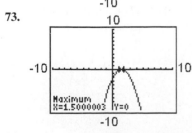

74.

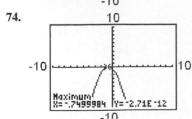

Section 8.6 Practice Exercises, pp. 637–640

1. a. quadratic **b.** solutions; undefined **c.** test point
d. true **e.** rational **f.** denominator **g.** True **h.** True
i. excludes; includes

3. $(-\infty, -4) \cup \left(\dfrac{1}{3}, \infty\right)$ **5.** $\left[3, \dfrac{16}{5}\right]$ **7.** $[-1, 3]$

9. a. $(-\infty, -2) \cup (3, \infty)$ **b.** $(-2, 3)$ **c.** $[-2, 3]$
d. $(-\infty, -2] \cup [3, \infty)$ **11. a.** $(-2, 0) \cup (3, \infty)$
b. $(-\infty, -2) \cup (0, 3)$ **c.** $(-\infty, -2] \cup [0, 3]$
d. $[-2, 0] \cup [3, \infty)$ **13. a.** $4, -\dfrac{1}{2}$ **b.** $(-\infty, -\dfrac{1}{2}) \cup (4, \infty)$
c. $(-\dfrac{1}{2}, 4)$ **15. a.** $-10, 3$ **b.** $(-10, 3)$

c. $(-\infty, -10) \cup (3, \infty)$ **17. a.** $-\dfrac{1}{2}, 3$ **b.** $\left[-\dfrac{1}{2}, 3\right]$

c. $\left(-\infty, -\dfrac{1}{2}\right] \cup [3, \infty)$ **19.** $(-1, 7)$

21. $(-\infty, -2) \cup (5, \infty)$ **23.** $[-5, 0]$ **25.** $[4, 8]$
27. $\left(-\infty, \dfrac{1 - \sqrt{6}}{5}\right) \cup \left(\dfrac{1 + \sqrt{6}}{5}, \infty\right)$

29. $\left[\dfrac{-3 - \sqrt{33}}{2}, \dfrac{-3 + \sqrt{33}}{2}\right]$ **31.** $(-11, 11)$

33. $\left(-\infty, -\dfrac{1}{3}\right] \cup [3, \infty)$ **35.** $(-\infty, -3] \cup [0, 4]$
37. $(-2, -1) \cup (2, \infty)$ **39. a.** 7 **b.** $(-\infty, 5) \cup (7, \infty)$
c. $(5, 7)$ **41. a.** 4 **b.** $[4, 6)$ **c.** $(-\infty, 4] \cup (6, \infty)$
43. $(1, \infty)$ **45.** $(-\infty, -4) \cup (4, \infty)$
47. $\left(2, \dfrac{7}{2}\right)$ **49.** $(5, 7]$ **51.** $(-\infty, 0) \cup \left[\dfrac{1}{2}, \infty\right)$
53. $(0, \infty)$ **55.** $(-\infty, \infty)$ **57.** No solution
59. $\{0\}$ **61.** No solution **63.** No solution
65. $(-\infty, \infty)$ **67.** Quadratic; $[-4, 4]$
69. Quadratic; $(-\infty, \infty)$ **71.** Linear; $(-\dfrac{5}{2}, \infty)$
73. Rational; $(-\infty, -1)$ **75.** Polynomial (degree > 2);
$(0, 5) \cup (5, \infty)$ **77.** Polynomial (degree > 2); $(2, \infty)$
79. Polynomial (degree > 2); $(-\infty, 5]$
81. Quadratic; $(-\sqrt{2}, \sqrt{2})$
83. Quadratic; $\left(-\infty, \dfrac{-5 - \sqrt{33}}{2}\right] \cup \left[\dfrac{-5 + \sqrt{33}}{2}, \infty\right)$
85. Rational; $(-\infty, -2] \cup (5, \infty)$ **87.** Linear; $(-\infty, 5]$
89. Quadratic; $(-\infty, \infty)$

Section 8.6 Graphing Calculator Exercises, p. 640

91. $(-\infty, 0) \cup (2, \infty)$

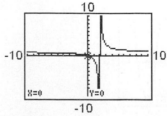

92. $(0, 2)$

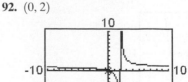

93. $(-1, 1)$

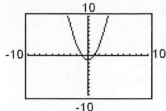

94. $(-\infty, -1) \cup (1, \infty)$

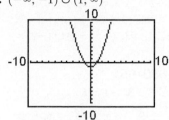

95. $\{-5\}$

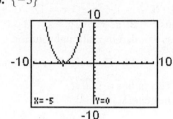

96. $\{5\}$

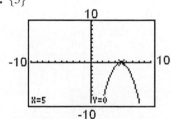

97. No solution

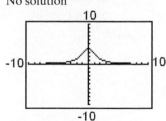

98. No solution

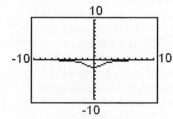

Chapter 8 Problem Recognition Exercises, p. 640–641

1. a. Equation quadratic in form and polynomial equation
b. $\pm 2\sqrt{2}, \pm 1$ **2. a.** Absolute value inequality   **b.** $(-\frac{1}{2}, 1)$
3. a. Polynomial inequality   **b.** $[-\frac{1}{2}, 5]$
4. a. Radical equation   **b.** 1
5. a. Absolute value equation   **b.** $9, -1$
6. a. Rational equation   **b.** $4 \pm 2\sqrt{6}$

7. a. Polynomial inequality   **b.** $[-5, -2] \cup [2, \infty)$
8. a. Compound inequality   **b.** $(-\infty, 1)$
9. a. Linear inequality   **b.** $[-23, \infty)$
10. a. Absolute value equation   **b.** $9, 1$
11. a. Rational inequality   **b.** $(-\infty, 2) \cup [5, \infty)$
12. a. Absolute value inequality   **b.** $(-\infty, -13) \cup (5, \infty)$
13. a. Radical equation   **b.** -4 (The value -7 does not
check.)   **14. a.** Quadratic equation   **b.** $\dfrac{3}{4} \pm \dfrac{\sqrt{10}}{4}i$
15. a. Compound inequality   **b.** $(-\infty, -6) \cup (4, \infty)$
16. a. Linear equation and rational equation   **b.** $-\dfrac{11}{2}$
17. a. Polynomial inequality   **b.** 5
18. a. Rational inequality   **b.** No solution
19. a. Radical equation and equation quadratic in form
b. $16, 81$ **20. a.** Polynomial equation and equation
quadratic in form   **b.** $\pm 2, \pm 3$

Chapter 8 Review Exercises, pp. 647–650

1. $\pm\sqrt{5}$ **2.** $\pm 2i$ **3.** ± 9 **4.** $\pm\dfrac{\sqrt{57}}{3}i$

5. $2 \pm 6\sqrt{2}$ **6.** $\dfrac{5}{2} \pm \dfrac{3}{2}i$ **7.** $\dfrac{1 \pm \sqrt{3}}{3}$ **8.** $4 \pm \sqrt{5}$
9. The height is $5\sqrt{3}$ in. ≈ 8.7 in.
10. The length is 9 in.   **11.** The length is
$5\sqrt{6}$ in. ≈ 12.2 in.   **12.** $n = 64; (x + 8)^2$
13. $n = \dfrac{81}{4}; \left(x - \dfrac{9}{2}\right)^2$ **14.** $n = \dfrac{1}{16}; \left(y + \dfrac{1}{4}\right)^2$

15. $n = \dfrac{1}{25}; \left(z - \dfrac{1}{5}\right)^2$ **16.** $-2 \pm 3i$ **17.** $\dfrac{3}{2} \pm i$

18. $\dfrac{1}{3}, -1$ **19.** $\dfrac{1}{2}, -4$ **20.** $3 \pm 2\sqrt{3}$ **21.** $4 \pm 3i$

22. $r = \sqrt{\dfrac{V}{\pi h}}$ or $r = \dfrac{\sqrt{V\pi h}}{\pi h}$

23. $s = \sqrt{\dfrac{A}{6}}$ or $s = \dfrac{\sqrt{6A}}{6}$ **24.** See page 590.

25. Two rational solutions   **26.** Two rational solutions
27. Two irrational solutions   **28.** Two imaginary solutions
29. One rational solution   **30.** Two imaginary solutions

31. $2 \pm \sqrt{3}$ **32.** $\dfrac{5}{2} \pm \dfrac{5\sqrt{3}}{2}i$ **33.** $2, -\dfrac{5}{6}$ **34.** $2, \dfrac{4}{3}$

35. $\dfrac{4}{5}, -\dfrac{1}{5}$ **36.** $\dfrac{1}{10}, -\dfrac{1}{2}$ **37.** $2 \pm 2i\sqrt{7}$ **38.** $4 \pm \sqrt{6}$

39. $\dfrac{2 \pm \sqrt{22}}{3}$ **40.** $8, -2$ **41.** $-7 \pm \sqrt{3}$ **42.** $-1 \pm \sqrt{11}$

43. a. 1822 ft   **b.** 115 ft/s   **44. a.** $\approx 53,939$ thousand
b. 2021   **45.** 49 (The value 9 does not check.)

46. $4, 16$ **47.** $\pm 3, \pm\sqrt{2}$ **48.** $\pm\dfrac{\sqrt{6}}{2}, \pm i$

49. $-243, 32$ **50.** $32, 1$ **51.** $3 \pm \sqrt{10}$

52. $\dfrac{1 \pm \sqrt{181}}{6}$ **53.** $\pm 3i, \pm i\sqrt{3}$ **54.** $\pm\sqrt{7}, \pm 2$

55.

Domain: $(-\infty, \infty)$;
range: $[-5, \infty)$

56.

Domain: $(-\infty, \infty)$;
range: $[3, \infty)$

57.

Domain: $(-\infty, \infty)$;
range: $[0, \infty)$

58.

Domain: $(-\infty, \infty)$;
range: $[0, \infty)$

59.

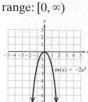

Domain: $(-\infty, \infty)$;
range: $(-\infty, 0]$

60.

Domain: $(-\infty, \infty)$;
range: $(-\infty, 0]$

61.

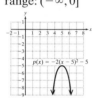

Domain: $(-\infty, \infty)$;
range: $(-\infty, -5]$

62.

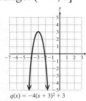

Domain: $(-\infty, \infty)$;
range: $(-\infty, 3]$

63. $\left(4, \dfrac{5}{3}\right)$ is the minimum point. The minimum value is $\dfrac{5}{3}$.

64. $\left(1, -\dfrac{1}{7}\right)$ is the maximum point. The maximum value is $-\dfrac{1}{7}$.

65. $x = -\dfrac{2}{11}$ **66.** $x = \dfrac{3}{16}$

67. $z(x) = (x - 3)^2 - 2; (3, -2)$
68. $b(x) = (x - 2)^2 - 48; (2, -48)$
69. $p(x) = -5(x + 1)^2 - 8; (-1, -8)$
70. $q(x) = -3(x + 4)^2 - 6; (-4, -6)$ **71.** $(1, -15)$

72. $(-1, 7)$ **73.** $\left(\dfrac{1}{2}, \dfrac{41}{4}\right)$ **74.** $\left(-\dfrac{1}{3}, -\dfrac{22}{3}\right)$

75. a. $(2, -3)$ **b.** $(0, 0), (4, 0)$
c.

76. a. $(-2, 4)$ **b.** $(0, 0), (-4, 0)$
c.

77. It will take 3 sec.   **78. a.** 150 meals   **b.** $1200
79. $a = 1, b = 4, c = -1; y = x^2 + 4x - 1$
80. a. $(-2, 0)(2, 0)$ are the x-intercepts.   **b.** On the interval $(-2, 2)$, the graph is below the x-axis.   **c.** On the intervals $(-\infty, -2)$ and $(2, \infty)$, the graph is above the x-axis.

81. a. $x = 0; (0, 0)$ is the x-intercept.   **b.** $x = 2$ is the vertical asymptote.   **c.** On the intervals $(-\infty, 0]$ and $(2, \infty)$ the graph is on or above the x-axis.   **d.** On the interval $[0, 2)$ the graph is on or below the x-axis.
82. $(-2, 6)$ **83.** $(-\infty, \infty)$ **84.** $(-\infty, -2) \cup [0, \infty)$
85. $(-\infty, -1] \cup (1, \infty)$ **86.** $(-2, 0) \cup (5, \infty)$
87. $(-2, 0) \cup \left(\dfrac{5}{2}, \infty\right)$
88. $(-\infty, -2 - \sqrt{3}] \cup [-2 + \sqrt{3}, \infty)$
89. $(-\infty, -5) \cup (1, \infty)$ **90.** $(3, \infty)$
91. $(-3, \infty)$ **92.** $\{-5\}$ **93.** $(-\infty, -2) \cup (-2, \infty)$

Chapter 8 Test, pp. 650–651

1. $2, -8$ **2.** $2 \pm 2\sqrt{3}$ **3.** $-1 \pm i$
4. $n = \dfrac{121}{4}; \left(d + \dfrac{11}{2}\right)^2$ **5.** $-3 \pm 3\sqrt{3}$ **6.** $\dfrac{3}{4} \pm \dfrac{\sqrt{47}}{4}i$
7. a. $x^2 - 3x + 12 = 0$ **b.** $a = 1, b = -3, c = 12$ **c.** -39
d. Two imaginary solutions   **8. a.** $y^2 - 2y + 1 = 0$
b. $a = 1, b = -2, c = 1$ **c.** 0   **d.** One rational solution
9. $1, \dfrac{1}{3}$ **10.** $\dfrac{-7 \pm \sqrt{5}}{2}$ **11.** The height is 4.6 ft and the base if 6.2 ft.   **12.** The radius is 12.0 ft.
13. 9 (The value 4 does not check.)
14. $8, -64$ **15.** $\dfrac{11}{3}, 6$ **16.** $\pm\sqrt{6}, \pm 3$ **17.** $\dfrac{5 \pm \sqrt{57}}{2}$
18. b. The vertex is $(3, -17)$.
19.

Domain: $(-\infty, \infty)$;
range: $[-4, \infty)$

20.

Domain: $(-\infty, \infty)$;
range: $(-\infty, 0]$

21.

Domain: $(-\infty, \infty)$; range: $[-3, \infty)$
22. The rocket will hit the ground in 256 ft.
23. a. 1320 million   **b.** 1999   **24.** The graph of $y = x^2 - 2$ is the graph of $y = x^2$ shifted down 2 units.
25. The graph of $y = (x + 3)^2$ is the graph of $y = x^2$ shifted 3 units to the left.   **26.** The graph of $y = -4x^2$ is the graph of $y = 4x^2$ opening downward instead of upward.
27. a. $(4, 2)$ **b.** Downward   **c.** Maximum point
d. The maximum value is 2.   **e.** $x = 4$
28. a. $g(x) = 2(x - 5)^2 + 1; (5, 1)$ **b.** $(5, 1)$
29. a. $f(x) = (x + 2)^2 - 16$ **b.** Vertex: $(-2, -16)$
c. x-intercepts: $(-6, 0)$ and $(2, 0)$; y-intercept: $(0, -12)$
d. The minimum value is -16 **e.** $x = -2$
30. a. 200 ft   **b.** 20,000 ft^2

31. $\left[\dfrac{1}{2}, 6\right)$ **32.** $(-5, 5)$ **33.** $(-\infty, -3) \cup (-2, 2)$

34. $\left(-3, -\dfrac{3}{2}\right)$ **35.** No solution   **36.** $\{-11\}$

Chapters 1–8 Cumulative Review Exercises, pp. 652–653

1. a. $\{2, 4, 6, 8, 10, 12, 16\}$ **b.** $\{2, 8\}$ **2.** $-3x^2 - 13x + 1$
3. -16 **4.** 1.8×10^{10} **5. a.** $(x + 2)(x + 3)(x - 3)$
b. Quotient: $x^2 + 5x + 6$; remainder: 0 **6.** $x - 2$
7. $\dfrac{2\sqrt{2x}}{x}$ **8.** Jacques invested \$8000 in the 12%
account and \$2000 in the 3% account.   **9.** $(8, 7)$
10. a. 720 ft   **b.** 720 ft   **c.** 12 sec
11. $3 \pm 4i$ **12.** $\dfrac{-5 \pm \sqrt{33}}{4}$ **13.** $n = 25; (x + 5)^2$
14. $2(x + 5)(x^2 - 5x + 25)$
15.

16. a. $\left(\dfrac{5}{2}, 0\right)(2, 0)$ **b.** $(0, 10)$ **17.** He made 565 free
throws, 851 2-pt shots, and 30 3-pt shots.   **18.** The domain
element 3 has more than one corresponding range element.
19.

20. $y = 39$ **21. a.** Linear   **b.** $(0, 300,000)$. If there are
no passengers, the airport runs 300,000 flights per year.
c. $m = 0.008$ or $m = \dfrac{8}{1000}$. There are eight additional flights
per 1000 passengers.   **22. a.** 3
b. $\sqrt{2}$ **c.** Not a real number   **23. a.** $[-4, \infty)$
b. $(-\infty, \infty)$ **24. a.** $(-\infty, 2]$ **b.** $(-\infty, 4]$ **c.** 4   **d.** 3
e. $x = -1$ **25.** $\dfrac{1}{18}, \dfrac{1}{2}$ **26.** $f = \dfrac{pq}{p + q}$ **27.** 7
28. $\dfrac{y + 1}{y - 3}$ **29. a.** $(3, 1)$ **b.** Upward   **c.** $(0, 19)$
d. No x-intercept
e.

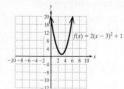

30. $(-\infty, 0] \cup (2, \infty)$

Chapter 9

Chapter Opener Puzzle

5	1	B 4	C 3	6	2
3	F 6	2	1	4	5
G 1	5	3	6	D 2	4
A 2	4	6	5	1	3
6	2	5	E 4	3	1
4	3	H 1	2	5	6

Section 9.1 Practice Exercises, pp. 661–665

1. a. $\{(2, 1), (3, 2), (4, 3)\}$ **b.** one-to-one; y
c. is not   **d.** is   **e.** $y = x$ **f.** $x; x$ **g.** f^{-1} **h.** (b, a)
3. Yes   **5.** No   **7.** Yes
9. $g^{-1} = \{(5, 3), (1, 8), (9, -3), (2, 0)\}$
11. $r^{-1} = \{(3, a), (6, b), (9, c)\}$ **13.** The function is not
one-to-one.   **15.** Yes   **17.** No   **19.** Yes
21. a. $(f \circ g)(x) = 6\left(\dfrac{x - 1}{6}\right) + 1 = x$
b. $(g \circ f)(x) = \dfrac{(6x + 1) - 1}{6} = x$
23. a. $(f \circ g)(x) = \dfrac{\sqrt[3]{8x^3}}{2} = x$ **b.** $(g \circ f)(x) = 8\left(\dfrac{\sqrt[3]{x}}{2}\right)^3 = x$
25. a. $(f \circ g)(x) = (\sqrt{x - 1})^2 + 1 = x$
b. $(g \circ f)(x) = \sqrt{(x^2 + 1) - 1} = x$
27. $h^{-1}(x) = x - 4$ **29.** $m^{-1}(x) = 3(x + 2)$
31. $p^{-1}(x) = -x + 10$ **33.** $n^{-1}(x) = \dfrac{5x - 2}{3}$
35. $h^{-1}(x) = \dfrac{3x + 1}{4}$ **37.** $f^{-1}(x) = \sqrt[3]{x - 1}$
39. $g^{-1}(x) = \dfrac{x^3 + 1}{2}$ **41.** $g^{-1}(x) = \sqrt{x - 9}$
43. $q^{-1}(x) = x^2 - 4, \; x \geq 0$ **45.** $z^{-1}(x) = x^2 - 4, \; x \leq 0$
47. a. 1.2192 m, 15.24 m   **b.** $f^{-1}(x) = \dfrac{x}{0.3048}$ **c.** 4921.3 ft
49. False   **51.** True   **53.** False   **55.** True   **57.** $(b, 0)$
59. a. Domain $[1, \infty)$, range $[0, \infty)$ **b.** Domain $[0, \infty)$,
range $[1, \infty)$
61. a. $[-4, 0]$ **b.** $[0, 2]$ **c.** $[0, 2]$ **d.** $[-4, 0]$

63. a. $[0, 2]$ **b.** $[0, 4]$ **c.** $[0, 4]$ **d.** $[0, 2]$

65. $f^{-1}(x) = \dfrac{x + 1}{1 - x}$ **67.** $t^{-1}(x) = \dfrac{x + 2}{x}$

69. $n^{-1}(x) = -\sqrt{x - 9}$

Section 9.1 Graphing Calculator Exercises, p. 665

71. $f^{-1}(x) = x^3 - 5$

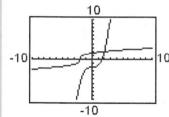

72. $k^{-1}(x) = \sqrt[3]{x + 4}$

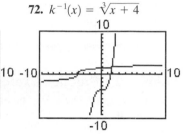

73. $g^{-1}(x) = \sqrt[3]{2x + 4}$

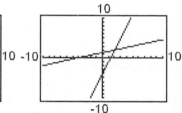

74. $m^{-1}(x) = \dfrac{x + 4}{3}$

Section 9.2 Practice Exercises, pp. 671–674

1. a. b^x **b.** is not; is   **c.** increasing   **d.** decreasing
e. $(-\infty, \infty); (0, \infty)$ **f.** $(0, 1)$ **g.** $y = 0$ **h.** is not

3. $-2x^2 + 2x - 3$ **5.** $\dfrac{3x - 1}{2x^2 + x + 2}$ **7.** $6x^2 + 3x + 5$

9. 25   **11.** $\dfrac{1}{1000}$ **13.** 6   **15.** 8   **17.** 5.8731

19. 1385.4557   **21.** 0.0063   **23.** 0.8950
25. a. $x = 2$ **b.** $x = 3$ **c.** Between 2 and 3
27. a. $x = 4$ **b.** $x = 5$ **c.** Between 4 and 5

29. $f(0) = 1, f(1) = \dfrac{1}{5}, f(2) = \dfrac{1}{25}, f(-1) = 5, f(-2) = 25$

31. $h(0) = 1, h(1) = 3, h(-1) = \dfrac{1}{3}, h(\sqrt{2}) \approx 4.73,$

$h(\pi) \approx 31.54$ **33.** If $b > 1$, the graph is increasing. If $0 < b < 1$, the graph is decreasing.

35.

37.

39.

41.

43. a. 0.25 g   **b.** ≈ 0.16 g   **45. a.** 758,000   **b.** 379,000
c. 144,000   **47. a.** $P(t) = 153,000,000(1.0125)^t$
b. $P(41) \approx 255,000,000$ **49. a.** \$1640.67   **b.** \$2691.80
c. $A(0) = 1000$. The initial amount of the investment is \$1000.
$A(7) = 2000$. The amount of the investment doubles in 7 yr.

Section 9.2 Graphing Calculator Exercises, p. 674

51.

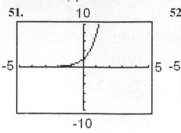

52.

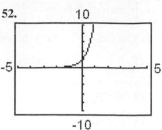

53.

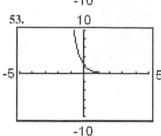

54.

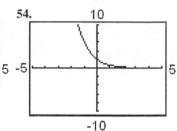

55.

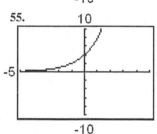

56.

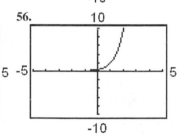

57.

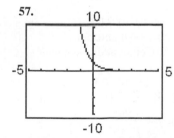

58.

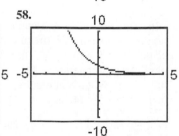

Section 9.3 Practice Exercises, pp. 683–687

1. a. logarithmic; b **b.** logarithm; base; argument
c. $(0, \infty); (-\infty, \infty)$ **d.** exponential   **e.** common
f. increasing; decreasing   **g.** 2, 3, and 4   **h.** $y = x$
3. i

5. a. $g(-2) = \dfrac{1}{9}, g(-1) = \dfrac{1}{3}, g(0) = 1, g(1) = 3, g(2) = 9$

b.

7. a. $s(-2) = \dfrac{25}{4}, s(-1) = \dfrac{5}{2}, s(0) = 1, s(1) = \dfrac{2}{5}, s(2) = \dfrac{4}{25}$

b.

9. $b^y = x$ **11.** $5^4 = 625$ **13.** $10^{-4} = 0.0001$
15. $6^2 = 36$ **17.** $b^x = 15$ **19.** $3^x = 5$

21. $\left(\dfrac{1}{4}\right)^{10} = x$ **23.** $\log_3 81 = x$ **25.** $\log_5 25 = 2$

27. $\log_7\left(\dfrac{1}{7}\right) = -1$ **29.** $\log_b y = x$ **31.** $\log_e y = x$

33. $\log_{1/3} 9 = -2$ **35.** 2   **37.** -1 **39.** $\dfrac{1}{2}$ **41.** 0

43. 5   **45.** 1   **47.** 3   **49.** $\dfrac{1}{2}$ **51.** 1   **53.** 3

55. 6   **57.** -2 **59.** 0.7782   **61.** 0.4971
63. -1.5051 **65.** -2.2676 **67.** 5.5315
69. -7.4202 **71. a.** Slightly less than 2   **b.** Slightly
more than 1   **c.** $\log 93 \approx 1.9685, \log 12 \approx 1.0792$

73. a. $f\left(\dfrac{1}{64}\right) = -3, f\left(\dfrac{1}{16}\right) = -2, f\left(\dfrac{1}{4}\right) = -1, f(1) = 0,$
$f(4) = 1, f(16) = 2, f(64) = 3$

b.

75. $3^y = x$

x	y
$\frac{1}{9}$	-2
$\frac{1}{3}$	-1
1	0
3	1
9	2

77. $\left(\dfrac{1}{2}\right)^y = x$

x	y
4	-2
2	-1
1	0
$\frac{1}{2}$	1
$\frac{1}{4}$	2

79. $(5, \infty)$ **81.** $(-\infty, 2)$ **83.** $\left(\dfrac{1}{3}, \infty\right)$ **85.** $(-1.2, \infty)$

87. $(-\infty, 2)$ **89.** $(-\infty, 0) \cup (0, \infty)$ **91.** ≈ 7.35

93. a.

t (months)	0	1	2	6	12	24
$S_1(t)$	91	82.0	76.7	65.6	57.6	49.1
$S_2(t)$	88	83.5	80.8	75.3	71.3	67.0

b. Group 1: 91; Group 2: 88   **c.** Method II

Section 9.3 Graphing Calculator Exercises, p. 687

95. Domain: $(-6, \infty)$; asymptote: $x = -6$

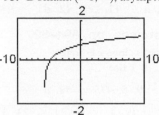

96. Domain: $(-2, \infty)$; asymptote: $x = -2$

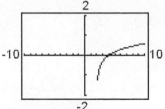

97. Domain: $(2, \infty)$; asymptote: $x = 2$

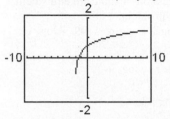

98. Domain: $(-8, \infty)$; asymptote:

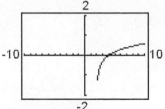

99. Domain: $(-\infty, 2)$; asymptote: $x = 2$

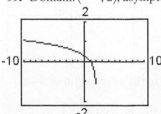

100. Domain: $(-\infty, 3)$; asymptote: $x = 3$

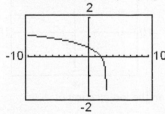

Chapter 9 Problem Recognition Exercises, p. 688

1. e **2.** l **3.** j **4.** g **5.** c **6.** a
7. k **8.** b **9.** i **10.** h **11.** f **12.** d

Section 9.4 Practice Exercises, pp. 694–697

1. a. $1; 0; x; x$ **b.** $\log_b x + \log_b y; \log_b x - \log_b y$
c. $p \log_b x$ **d.** False **e.** False **f.** False

3. 4 **5.** $\dfrac{1}{6}$ **7.** 0.9031 **9.** 5.0475 **11.** a

13. c **15.** a, b, c **17.** 1 **19.** 4 **21.** 11 **23.** 3
25. 0 **27.** 9 **29.** 0 **31.** 3 **33.** 1 **35.** $2x$
37. 4 **39.** 0 **41.** Expressions a and c are equivalent.
43. Expressions a and c are equivalent.
45. $\log_3 x - \log_3 5$ **47.** $\log 2 + \log x$ **49.** $4 \log_5 x$
51. $\log_4 a + \log_4 b - \log_4 c$
53. $\dfrac{1}{2} \log_b x + \log_b y - 3 \log_b z - \log_b w$

55. $\log_2 (x + 1) - 2 \log_2 y - \dfrac{1}{2} \log_2 z$

57. $\dfrac{1}{3} \log a + \dfrac{2}{3} \log b - \dfrac{1}{3} \log c$ **59.** $-5 \log w$

61. $\dfrac{1}{2} \log_b a - 3 - \log_b c$ **63.** 3 **65.** 2

67. $\log_3 \left(\dfrac{x^2 z}{y^3} \right)$ **69.** $\log_3 \left(\dfrac{a^2 c}{\sqrt[4]{b}} \right)$ **71.** $\log_b x^2$

73. $\log_8 (8a^5)$ or $\log_8 a^5 + 1$ **75.** $\log \left[\dfrac{(x + 6)^2 \sqrt[3]{y}}{z^5} \right]$

77. $\log_b \left(\dfrac{1}{x - 1} \right)$ **79.** 1.792 **81.** 2.485 **83.** 4.396
85. 0.916 **87.** 13.812 **89.** 16.09
91. a. $B = 10 \log I - 10 \log I_0$ **b.** $10 \log I + 160$

Section 9.4 Graphing Calculator Exercises, p. 697

93. a. Domain: $(-\infty, 0) \cup (0, \infty)$ **b.** Domain: $(0, \infty)$

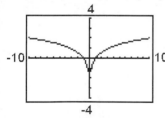
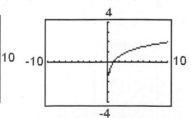

c. They are equivalent for all x in the intersection of their domains, $(0, \infty)$.
94. a. Domain: $(-\infty, 1) \cup (1, \infty)$ **b.** Domain: $(1, \infty)$

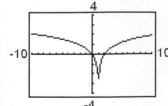
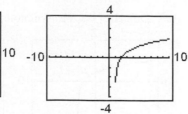

c. They are equivalent for all x in the intersection of their domains, $(1, \infty)$.

Section 9.5 Practice Exercises, pp. 704–708

1. a. e **b.** e **c.** natural; $\ln x$ **d.** $0; 1; p; x$
e. $\ln x + \ln y; \ln x - \ln y$ **f.** $p \ln x$ **g.** $\log_a b$

3.

x	$g(x)$
-3	125
-2	25
-1	5
0	1
1	$\frac{1}{5}$
2	$\frac{1}{25}$
3	$\frac{1}{125}$

5.

x	$r(x)$
0.5	-0.30
1	0
5	0.70
10	1.00

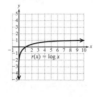

7.

x	y
-4	0.05
-3	0.14
-2	0.37
-1	1
0	2.72
1	7.39

Domain: $(-\infty, \infty)$

9.

x	y
-2	2.14
-1	2.37
0	3
1	4.72
2	9.39
3	22.09

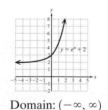

Domain: $(-\infty, \infty)$

11. An investment grows more rapidly at higher interest rates.
a. \$12,209.97 **b.** \$13,488.50 **c.** \$14,898.46 **d.** \$16,050.09
13. More money is earned at a greater number of compound periods per year. **a.** \$12,423.76 **b.** \$12,515.01 **c.** \$12,535.94
d. \$12,546.15 **e.** \$12,546.50 **15.** More money is earned over a longer period of time. **a.** \$6920.15 **b.** \$9577.70
c. \$13,255.84 **d.** \$18,346.48 **e.** \$35,143.44

17.

x	y
2.25	-1.39
2.50	-0.69
2.75	-0.29
3	0
4	0.69
5	1.10
6	1.39

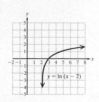

Domain: $(2, \infty)$

19.

x	y
0.25	−2.39
0.5	−1.69
0.75	−1.29
1	−1.00
2	−0.31
3	0.10
4	0.39

Domain: $(0, \infty)$

21. a.

b. Domain: $(-\infty, \infty)$; range: $(0, \infty)$

c.

d. Domain: $(0, \infty)$; range: $(-\infty, \infty)$ **23.** 1   **25.** 0

27. −6   **29.** $2x + 3$ **31.** $\ln\left(p^6 \sqrt[3]{q}\right)$ **33.** $\ln\sqrt{\dfrac{x}{y^3}}$

35. $\ln\left(\dfrac{a^2}{b\sqrt[3]{c}}\right)$ **37.** $\ln\left(\dfrac{x^4}{y^3 z}\right)$ **39.** $2\ln a - 2\ln b$

41. $2\ln b + 1$ **43.** $4\ln a + \dfrac{1}{2}\ln b - \ln c$

45. $\dfrac{1}{5}\ln a + \dfrac{1}{5}\ln b - \dfrac{2}{5}\ln c$ **47. a.** 2.9570   **b.** 2.9570

c. They are the same.   **49.** 2.8074   **51.** 1.5283
53. −2.1269   **55.** 0   **57.** −3.3219   **59.** −3.8124
61. a. 15.4 yr   **b.** 6.9 yr   **c.** 13.8 yr   **63. a.** 19.8 yr
b. 13.9 yr   **c.** 27.8 yr

Section 9.5 Graphing Calculator Exercises, pp. 708–709

65. a., b.

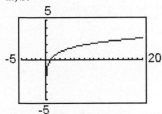

c. They appear to be the same.

66. a., b.

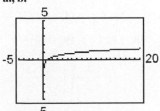

c. They appear to be the same.

67.

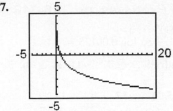

68.

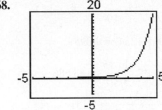

69.

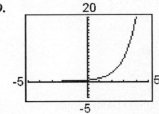

Chapter 9 Problem Recognition Exercises, p. 709

	Exponential Form	**Logarithmic Form**
1.	$2^5 = 32$	$\log_2 32 = 5$
2.	$3^4 = 81$	$\log_3 81 = 4$
3.	$z^y = x$	$\log_z x = y$
4.	$b^c = a$	$\log_b a = c$
5.	$10^3 = 1000$	$\log 1000 = 3$
6.	$10^1 = 10$	$\log 10 = 1$
7.	$e^a = b$	$\ln b = a$
8.	$e^q = p$	$\ln p = q$
9.	$\left(\frac{1}{2}\right)^2 = \frac{1}{4}$	$\log_{1/2}\left(\frac{1}{4}\right) = 2$
10.	$\left(\frac{1}{3}\right)^{-2} = 9$	$\log_{1/3} 9 = -2$
11.	$10^{-2} = 0.01$	$\log 0.01 = -2$
12.	$10^x = 4$	$\log 4 = x$
13.	$e^0 = 1$	$\ln 1 = 0$
14.	$e^1 = e$	$\ln e = 1$
15.	$25^{1/2} = 5$	$\log_{25} 5 = \frac{1}{2}$
16.	$16^{1/4} = 2$	$\log_{16} 2 = \frac{1}{4}$
17.	$e^t = s$	$\ln s = t$
18.	$e^r = w$	$\ln w = r$
19.	$15^{-2} = \frac{1}{225}$	$\log_{15}\left(\frac{1}{225}\right) = -2$
20.	$3^{-1} = p$	$\log_3 p = -1$

Section 9.6 Practice Exercises, pp. 716–718

1. a. exponential   **b.** $x; y$

3. $\log 8721 = -0.003k$ **5.** $\log 0.00125 = -4x$

7. Linear; 3   **9.** Quadratic; $-\dfrac{2}{7}, 1$ **11. a.** 2   **b.** 3

c. 2 and 3   **13.** $\dfrac{1}{3}$ **15.** 4   **17.** -6 **19.** $\dfrac{1}{2}$

21. 2   **23.** $\dfrac{11}{12}$ **25.** 1   **27.** $\dfrac{4}{19}$ **29.** $-2, 1$

31. $\dfrac{\ln 21}{\ln 8} \approx 1.464$ **33.** $\ln 8.1254 \approx 2.095$

35. $\log 0.0138 \approx -1.860$ **37.** $\dfrac{\ln 15}{0.07} \approx 38.686$

39. $\dfrac{\ln 3}{1.2} \approx 0.916$ **41.** $\dfrac{\ln 3}{\ln 5 - \ln 3} \approx 2.151$

43. $\dfrac{2 \ln 2}{\ln 6 - \ln 2} \approx 1.262$ **45.** $\dfrac{\ln 4}{0.04} \approx 34.657$

47. $3 \ln\left(\dfrac{125}{6}\right) \approx 9.110$ **49.** 0

51. $\dfrac{\ln 17.8}{1.3} \approx 2.215$ **53.** $\dfrac{\ln 20}{\ln 5} + 2 \approx 3.861$

55. a. ≈ 1285 million (or 1,285,000,000) people
b. ≈ 1466.5 million (or 1,466,500,000) people
c. The year 2049 ($t \approx 50.8$)
57. a. 500 bacteria   **b.** ≈ 660 bacteria   **c.** ≈ 25 min
59. 9.9 yr   **61. a.** 7.8 g   **b.** 18.5 days   **63.** 38.4 yr
65. a. 1.42 kg   **b.** No

Section 9.6 Graphing Calculator Exercises, p. 718

67.

68.

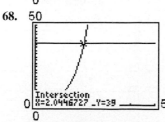

Section 9.7 Practice Exercises, pp. 723–725

1. a. logarithmic   **b.** $x; y$ **3.** $\log_b [x(2x + 3)]$

5. $\log_b\left(\dfrac{x + 2}{3x - 5}\right)$ **7.** 6   **9.** 7   **11.** No solution

13. $4, -3$ **15.** 9   **17.** 10^{42} **19.** $e^{0.08}$

21. $10^{-9.2} - 40$ **23.** 5   **25.** 10   **27.** 25

29. 59   **31.** 1   **33.** 0   **35.** $-\dfrac{37}{9}$

37. 3 (The value -3 does not check.)   **39.** 4   **41.** 3
43. 2   **45.** No solution (The value -3 does not check.)
47. $10^{6.3}$ (or approximately 1,995,262) times as intense
49. $10^{-3.07}$ W/m^2 **51. a.** 12.4   **b.** 10.5 in.
53. a. 3.16×10^{-9} mol/L   **b.** 5.01×10^{-3} mol/L

55. 40   **57.** $\dfrac{1}{2}$ **59.** $-\dfrac{1}{6}$ **61.** $-9, 2$

63. 1 (The value -5 does not check.)
65. No solution (The value $-\frac{1}{3}$ does not check.)

67. $\dfrac{\ln 6}{0.2}$ **69.** $\dfrac{3 \ln 2}{\ln 5 - \ln 2}$ **71.** $f^{-1}(x) = \log_3(x + 2)$

73. $k^{-1}(x) = \log(x - 6) + 2$ **75.** $h^{-1}(x) = e^x - 10$

77. $n^{-1}(x) = 10^{x+1} - 3$ **79.** $10^5, 10^{-3}$ **81.** $\dfrac{1}{27}, \dfrac{1}{9}$

Section 9.7 Graphing Calculator Exercises, p. 725

83. The graphs are the same.

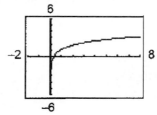

84. The graphs are the same.

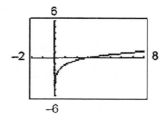

Chapter 9 Review Exercises, pp. 731–735

1. No   **2.** Yes   **3.** $\{(5, 3), (9, 2), (-1, 0), (1, 4)\}$

4. $q^{-1}(x) = \dfrac{4}{3}(x + 2)$ **5.** $g^{-1}(x) = (x - 3)^5$

6. $f^{-1}(x) = \sqrt[3]{x} + 1$ **7.** $n^{-1}(x) = \dfrac{2x + 4}{x}$

8. $(f \circ g)(x) = 5\left(\dfrac{1}{5}x + \dfrac{2}{5}\right) - 2 = x + 2 - 2 = x$

$(g \circ f)(x) = \dfrac{1}{5}(5x - 2) + \dfrac{2}{5} = x - \dfrac{2}{5} + \dfrac{2}{5} = x$

9. The graphs are symmetric about the line $y = x$.

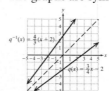

10. a. Domain: $[-1, \infty)$; range: $[0, \infty)$ **b.** Domain: $[0, \infty)$
range: $[-1, \infty)$ **11.** $p^{-1}(x) = (x - 2)^2, x \geq 2$

12. 1024   **13.** $\dfrac{1}{36} \approx 0.028$ **14.** 2   **15.** 10

16. 8.825   **17.** 16.242   **18.** 1.627   **19.** 0.681

20.

21.

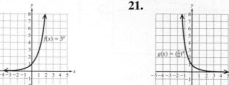

22.

23.

24. a. Horizontal   **b.** $y = 0$ **25. a.** 15,000 mrem
b. 3750 mrem   **c.** Yes   **26.** -3 **27.** 0   **28.** 1
29. 8   **30.** 4   **31.** 4   **32.** 5   **33.** -1
34.

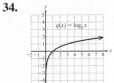

35.

36. a. Vertical asymptote   **b.** $x = 0$
37. a. 2.5   **b.** 9.5   **38.** 1   **39.** 6   **40.** 0   **41.** 7

42. a. $\log_b x + \log_b y$ **b.** $\log_b\left(\dfrac{x}{y}\right)$ **c.** $p\log_b x$

43. $\log_b\left(\dfrac{\sqrt[4]{x^3 y}}{z}\right)$ **44.** $\log_3\left(\dfrac{\sqrt{ab}}{c^2 d^4}\right)$ **45.** $\log 5$ **46.** 0

47. b   **48.** a   **49.** 148.4132   **50.** 14.0940
51. 32.2570   **52.** 57.2795   **53.** 1.7918
54. -2.1972 **55.** 1.3424   **56.** 1.3029
57. 3.3219   **58.** 1.9943   **59.** -0.8370
60. -3.6668 **61. a.** \$33,361.92   **b.** \$33,693.90
c. \$33,770.48   **d.** \$33,809.18
62. a. $S(0) = 95$; the student's score is 95 at the end of the
course.   **b.** $S(6) \approx 23.7$; the student's score is 23.7 after
6 months.   **c.** $S(12) \approx 20.2$; the student's score is 20.2
after 1 year.   **d.** The limiting value is 20.
63. $(-\infty, \infty)$ **64.** $(-\infty, \infty)$ **65.** $(-\infty, \infty)$ **66.** $(0, \infty)$
67. $(-5, \infty)$ **68.** $(7, \infty)$ **69.** $\left(\dfrac{4}{3}, \infty\right)$ **70.** $(-\infty, 5)$

71. -1 **72.** $\dfrac{4}{7}$ **73.** $\dfrac{\ln 21}{\ln 4} \approx 2.1962$

74. $\dfrac{\ln 18}{\ln 5} \approx 1.7959$ **75.** $-\ln 0.1 \approx 2.3026$

76. $-\dfrac{\ln 0.06}{2} \approx 1.4067$ **77.** $\dfrac{\log 1512}{2} \approx 1.5898$

78. $\dfrac{\log\left(\dfrac{1}{821}\right)}{-3} \approx 0.9714$ **79.** $\dfrac{3\ln 2}{\ln 7 - \ln 2} \approx 1.6599$

80. $\dfrac{5\ln 14}{\ln 14 - \ln 6} \approx 15.5734$

81. a. 1.09 μg   **b.** 0.15 μg   **c.** 16.08 days
82. a. 150 bacteria   **b.** ≈ 185 bacteria   **c.** ≈ 99 min
83. a. $V(0) = 15,000$; the initial value of the car is \$15,000.
b. $V(10) = 3347$; the value of the car after 10 years is \$3347.
c. 7.3 yr   **d.** The limiting value is 0.
84. 1   **85.** 5   **86.** 125   **87.** $\dfrac{1}{49} \approx 0.02$ **88.** 216

89. $3^{1/12} \approx 1.10$ **90.** $\dfrac{1001}{2} = 500.5$ **91.** 9

92. 5 (The value -2 does not check.)
93. $\dfrac{190}{321} \approx 0.59$

94. 10^9 (or 1,000,000,000) times as intense
95. 8

Chapter 9 Test, pp. 735–737

1. A function is one-to-one if it passes the horizontal
line test.   **2.** b   **3.** $f^{-1}(x) = 4x - 12$
4. $g^{-1}(x) = \sqrt{x} + 1$
5.

6. a. 4.6416   **b.** 32.2693   **c.** 687.2913
7.

8. a. $\log_{16} 8 = \dfrac{3}{4}$ **b.** $x^5 = 31$
9.

10. $\dfrac{\log_a n}{\log_a b}$ **11. a.** 1.3222   **b.** 1.8502   **c.** -2.5850
12. a. $1 + \log_3 x$ **b.** -5 **13. a.** $\log_b\left(\sqrt{xy^3}\right)$
b. $\log\left(\dfrac{1}{a^3}\right)$ or $-\log a^3$ **14. a.** 1.6487   **b.** 0.0498
c. -1.0986 **d.** 1   **15. a.** $y = \ln x$ **b.** $y = e^x$
16. a. $p(4) \approx 59.8$; 59.8% of the material is retained after
4 months.   **b.** $p(12) \approx 40.7$; 40.7% of the material is retained
after 1 yr.   **c.** $p(0) = 92$; 92% of the material is retained at
the end of the course.   **17. a.** 9762 thousand
(or 9,762,000) people   **b.** The year 2020 ($t \approx 20.4$)
18. a. $P(0) = 300$; there are 300 bacteria initially.
b. 35,588 bacteria   **c.** 1,120,537 bacteria   **d.** 1,495,831
bacteria   **e.** The limiting value appears to be 1,500,000.
19. 4   **20.** 6, -6 **21.** 25 (The value -4 does
not check.)   **22.** 32   **23.** $e^{2.4} - 7 \approx 4.023$
24. -7 **25.** $\dfrac{\ln 50}{\ln 4} \approx 2.822$ **26.** $\dfrac{\ln 250}{2.4} \approx 2.301$
27. a. $P(2500) = 560.2$; at 2500 m the atmospheric pressure
is 560.2 mm Hg.   **b.** 760 mm Hg   **c.** 1498.8 m
28. a. \$2909.98   **b.** 9.24 yr to double

Chapters 1–9 Cumulative Review Exercises, pp. 737–738

1. $-\dfrac{5}{4}$ **2.** $-1 + \dfrac{p}{2} + \dfrac{3p^3}{4}$ **3.** Quotient:
$t^3 + 2t^2 - 9t - 18$; remainder: 0   **4.** $|x - 3|$
5. $\dfrac{2\sqrt[3]{25}}{5}$ **6.** $2\sqrt{7}$ in.   **7.** $\dfrac{4d^{1/10}}{c}$

8. $(\sqrt{15} - \sqrt{6} + \sqrt{30} - 2\sqrt{3})\,\mathrm{m}^2$ **9.** $-\dfrac{7}{29} - \dfrac{26}{29}i$

10. $6, \dfrac{3}{2}$ **11.** 4.8 L   **12.** 24 min   **13.** $(7, -1)$

14. $-\dfrac{23}{11}$ **15.** $x = \dfrac{c+d}{a-b}$ **16.** $t = \dfrac{\sqrt{2sg}}{g}$

17. $T = \dfrac{1 - \left(\dfrac{V_0}{V}\right)^2}{k}$ or $\dfrac{V^2 - V_0^2}{kV^2}$ **18.** $(7, 0), (3, 0)$

19. a. $-30t$ **b.** $-10t^2$ **c.** $2t^2 + 5t$ **20.** 0

21. a. $x = 2$ **b.** $y = 6$ **c.** $y = \dfrac{1}{2}x + 5$

22. $40°, 80°, 60°$ **23.** Infinitely many solutions;
$\{(x, y)\,|\,-2x + y = -4\}$; dependent equations

24. $\left[\dfrac{1}{2}, \infty\right)$ **25.** $f^{-1}(x) = \dfrac{1}{5}x + \dfrac{2}{15}$ **26.** 40 m^3

27. $\dfrac{-x^2 + 5x - 25}{(2x+1)(x-2)}$ **28.** $1 + x$ **29. a.** Yes;

$x \neq 4, x \neq -2$ **b.** 8   **c.** $(-\infty, -2) \cup (4, 8]$ **30.** The

numbers are $-\dfrac{3}{4}, 4$.   **31.** -4 (The value -9 does not check.)

32. $(-\infty, \infty)$ **33. a.** $P(6) = 2{,}000{,}000, P(12) = 1{,}000{,}000,$
$P(18) = 500{,}000, P(24) = 250{,}000, P(30) = 125{,}000$ **b.** 48 hr
34. a. 2   **b.** -3 **c.** 6   **d.** 3   **35. a.** 217.0723
b. 23.1407   **c.** 0.1768   **d.** 3.7293   **e.** -0.4005 **f.** 2.6047

36. $\dfrac{2}{3}$ **37.** $\ln(100) \approx 4.6052$ **38.** 3 (The value -9

does not check.)   **39.** $\log\left(\dfrac{\sqrt{z}}{x^2 y^3}\right)$ **40.** $\dfrac{2}{3}\ln x - \dfrac{1}{3}\ln y$

Chapter 10

Chapter Opener Puzzle

$$\underset{5}{\mathrm{C}}\,\underset{4}{\mathrm{O}}\,\underset{6}{\mathrm{N}}\,\underset{3}{\mathrm{I}}\,\underset{5}{\mathrm{C}} \qquad \underset{1}{\mathrm{S}}\,\underset{7}{\mathrm{E}}\,\underset{5}{\mathrm{C}}\,\underset{2}{\mathrm{T}}\,\underset{3}{\mathrm{I}}\,\underset{4}{\mathrm{O}}\,\underset{6}{\mathrm{N}}\,\underset{1}{\mathrm{S}}$$

Section 10.1 Practice Exercises, pp. 745–750

1. a. $\sqrt{(x_2 - x_1)^2 + (y_2 - y_1)^2}$ **b.** circle; center   **c.** radius

d. $(x - h)^2 + (y - k)^2 = r^2$ **e.** $\left(\dfrac{x_1 + x_2}{2}, \dfrac{y_1 + y_2}{2}\right)$

3. $3\sqrt{5}$ **5.** $\sqrt{34}$ **7.** $\dfrac{\sqrt{73}}{8}$ **9.** 4   **11.** 8

13. $4\sqrt{2}$ **15.** $\sqrt{42}$ **17.** Subtract 5 and -7. This
becomes $5 - (-7) = 12$.   **19.** $y = 13, y = 1$
21. $x = 0, x = 8$ **23.** Yes   **25.** No
27. Center $(4, -2); r = 3$ **29.** Center $(-1, -1); r = 1$

    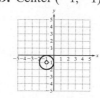

31. Center $(0, 2); r = 2$

33. Center $(3, 0); r = 2\sqrt{2}$

35. Center $(0, 0); r = \sqrt{6}$

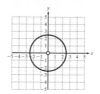

37. Center $\left(-\dfrac{4}{5}, 0\right); r = \dfrac{8}{5}$

39. Center $(1, 3); r = 6$

41. Center: $(0, 3); r = 2$

43. Center $(0, -3); r = \dfrac{4}{3}$

45. Center: $(-1, -2); r = 3$

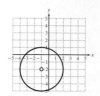

47. Center $(0, 0); r = 1$

49. $x^2 + y^2 = 4$ **51.** $x^2 + (y - 2)^2 = 4$
53. $(x + 2)^2 + (y - 2)^2 = 9$ **55.** $x^2 + y^2 = 49$
57. $(x + 3)^2 + (y + 4)^2 = 36$ **59.** $(x - 5)^2 + (y - 3)^2 = 2.25$
61. $(1, 2)$ **63.** $(-1, 0)$ **65.** $(-1, 6)$ **67.** $(0, 3)$

69. $\left(-\dfrac{1}{2}, \dfrac{3}{2}\right)$ **71.** $(-0.4, -1)$ **73.** $(40, 7\tfrac{1}{2})$; they should

meet 40 mi east, $7\tfrac{1}{2}$ mi north of the warehouse.
75. a. $(1, 3)$ **b.** $(x - 1)^2 + (y - 3)^2 = 5$
77. a. $(0, 3)$ **b.** $x^2 + (y - 3)^2 = 4$
79. $(x - 4)^2 + (y - 4)^2 = 16$

81. $(x - 1)^2 + (y - 1)^2 = 29$

Section 10.1 Graphing Calculator Exercises, p. 750

83.

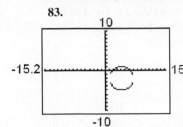

84.

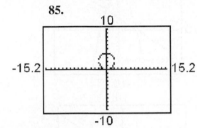

85.

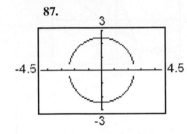

86.

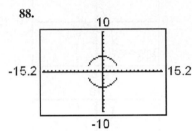

87.

88.

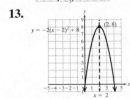

Section 10.2 Practice Exercises, pp. 757–760

1. a. conic; plane   **b.** parabola; directrix; focus
c. vertex; symmetry   **d.** $y = k$ **3.** 5
5. Center: $(0, -1)$; radius: 4

7. $\left(\dfrac{11}{2}, 1\right)$

9.

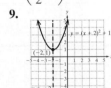

11.

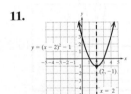

13.

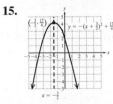

15.

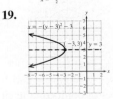

17.

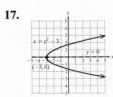

19.

21.

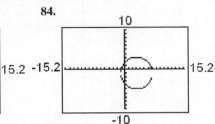

23.

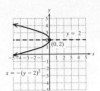

25. $(2, -1)$ **27.** $(5, -1)$ **29.** $\left(2, \dfrac{7}{4}\right)$ **31.** $\left(\dfrac{3}{2}, -\dfrac{1}{4}\right)$

33. $(10, -1)$ **35.** 22 ft   **37.** A parabola whose equation is in the form $y = a(x - h)^2 + k$ has a vertical axis of symmetry. A parabola whose equation is in the form $x = a(y - k)^2 + h$ has a horizontal axis of symmetry.
39. Vertical axis of symmetry; opens upward
41. Vertical axis of symmetry; opens downward
43. Horizontal axis of symmetry; opens right
45. Horizontal axis of symmetry; opens left   **47.** Vertical axis of symmetry; opens downward   **49.** Horizontal axis of symmetry; opens right

Section 10.3 Practice Exercises, pp. 765–769

1. a. ellipse; foci   **b.** ellipse   **c.** hyperbola; foci
d. transverse   **e.** $\dfrac{x^2}{a^2} - \dfrac{y^2}{b^2} = 1$ **f.** $\dfrac{y^2}{b^2} - \dfrac{x^2}{a^2} = 1$
3. Center: $(8, -6)$; radius: 10
5. Vertex: $(-3, -1)$; $x = -3$
7. $\left(x - \dfrac{1}{2}\right)^2 + \left(y - \dfrac{5}{2}\right)^2 = \dfrac{1}{4}$

9.

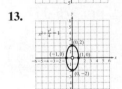

11.

13.

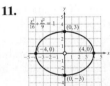

15.

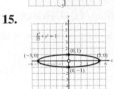

17. Center: $(4, 5)$

19. Center: $(-1, 2)$

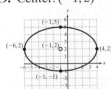

21. Center: $(2, -3)$

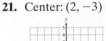

23. Center: $(0, 1)$

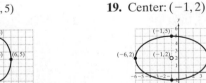

25. Vertical   **27.** Horizontal   **29.** Horizontal
31. Vertical
33.

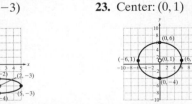

35.

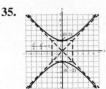

37.   **39.**

37. $(3, 2)(-3, 2)(3, -2)(-3, -2)$ **39.** $(-2, 4)(2, 4)$
41. $(0, -2)(0, 2)$ **43.** $(1, 3)(1, -3)(-1, 3)(-1, -3)$
45. $(\sqrt{3}, 0)(-\sqrt{3}, 0)$ **47.** The numbers are 3 and 4.
49. The numbers are 5 and $\sqrt{7}$, -5 and $\sqrt{7}$, 5 and $-\sqrt{7}$, or -5 and $-\sqrt{7}$.

41. Hyperbola   **43.** Ellipse   **45.** Ellipse
47. Hyperbola   **49.** Ellipse   **51.** Hyperbola   **53.** 49 ft
55.   **57.**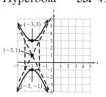

Section 10.3 Problem Recognition Exercises, pp. 769–770

1. Standard equation of a circle
2. Ellipse centered at the origin
3. Distance between two points
4. Midpoint between two points
5. Parabola with vertical axis of symmetry
6. Hyperbola with horizontal transverse axis
7. Hyperbola with vertical transverse axis
8. Parabola with horizontal axis of symmetry
9. Parabola   **10.** Hyperbola   **11.** Circle
12. Circle   **13.** Hyperbola   **14.** Ellipse
15. None of the above   **16.** Circle   **17.** Parabola
18. Hyperbola   **19.** Parabola   **20.** Circle
21. None of the above   **22.** Parabola   **23.** Ellipse
24. Circle   **25.** Parabola   **26.** Ellipse
27. Hyperbola   **28.** Circle   **29.** Ellipse   **30.** Parabola

Section 10.4 Practice Exercises, pp. 775–777

1. **a.** nonlinear **b.** intersection

3. $7\sqrt{2}$ **5.** $(x + 5)^2 + (y - 3)^2 = 64$ **7.** $\left(-\dfrac{1}{2}, -\dfrac{7}{2}\right)$

9. Zero, one, or two   **11.** Zero, one, or two
13. Zero, one, two, three, or four   **15.** Zero, one, two, three, or four

17. $(-3, 0)(2, 5)$ **19.** $(0, 1)(-1, 0)$

    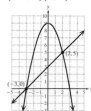

21. $(\sqrt{2}, 2)(-\sqrt{2}, 2)$

23. $(1, 1)$ **25.** $(0, 0)$ **27.** $\left(\dfrac{3}{2}, \dfrac{9}{4}\right)$ **29.** $(0, 0), (-2, -8)$
31. $(1, 4)$ **33.** $(1, 0)(-1, 0)$ **35.** $(2, 0)(-2, 0)$

Section 10.4 Graphing Calculator Exercises, p. 777

51.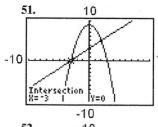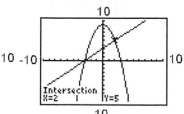

52.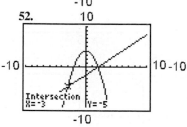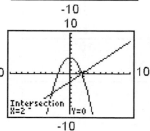

53.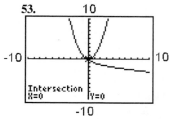

54.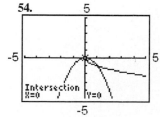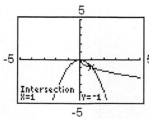

55. 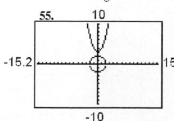   **56.**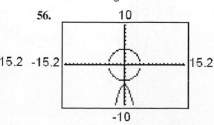

Section 10.5 Practice Exercises, pp. 782–786

1. k   **3.** e   **5.** i   **7.** j   **9.** a   **11.** d
13. False   **15.** True
17. **a.**

b. The set of points on and "outside" the circle $x^2 + y^2 = 9$
c. The set of points on the circle $x^2 + y^2 = 9$

19. a.

b. The parabola $y = x^2 + 1$ would be drawn as a dashed curve.
21. $(x - 3)^2 + (y + 4)^2 \leq 625$

23.

25.

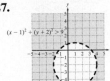

27.

29.

31.

33.

35.

37.

39.

41.

43.

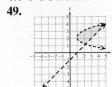

45.

47. No Solution
49.

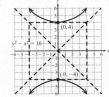

51.

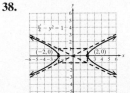

Chapter 10 Review Exercises, pp. 793–796

1. $2\sqrt{10}$ **2.** $\sqrt{89}$ **3.** $x = 5$ or $x = -1$
4. $x = -3$ **5.** Center $(12, 3)$; $r = 4$
6. Center $(-7, 5)$; $r = 9$ **7.** Center $(-3, -8)$; $r = 2\sqrt{5}$
8. Center $(1, -6)$; $r = 4\sqrt{2}$ **9. a.** $x^2 + y^2 = 64$
b. $(x - 8)^2 + (y - 8)^2 = 64$ **10.** $(x + 6)^2 + (y - 5)^2 = 10$

11. $(x + 2)^2 + (y + 8)^2 = 8$ **12.** $\left(x - \dfrac{1}{2}\right)^2 + (y - 2)^2 = 4$

13. $(x - 3)^2 + \left(y - \dfrac{1}{3}\right)^2 = 9$ **14.** $x^2 + y^2 = \dfrac{49}{4}$

15. $x^2 + (y - 2)^2 = 9$ **16.** $(-4, -2)$ **17.** $(-1, 8)$
18. Vertical axis of symmetry; parabola opens downward
19. Horizontal axis of symmetry; parabola opens right
20. Horizontal axis of symmetry; parabola opens left
21. Vertical axis of symmetry; parabola opens upward
22.

23.

24.

25.

26. $y = (x - 3)^2 - 4$; vertex: $(3, -4)$; axis of symmetry: $x = 3$ **27.** $x = (y + 2)^2 - 2$; vertex: $(-2, -2)$; axis of symmetry: $y = -2$

28. $x = -4\left(y - \dfrac{1}{2}\right)^2 + 1$; vertex: $\left(1, \dfrac{1}{2}\right)$;

axis of symmetry: $y = \dfrac{1}{2}$

29. $y = -2\left(x + \dfrac{1}{2}\right)^2 + \dfrac{1}{2}$; vertex: $\left(-\dfrac{1}{2}, \dfrac{1}{2}\right)$;

axis of symmetry: $x = -\dfrac{1}{2}$

30.

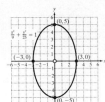

31.

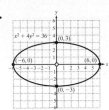

32. Center: $(5, -3)$ **33.** Center: $(0, 2)$

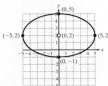

34. Horizontal   **35.** Vertical
36. Vertical   **37.** Horizontal
38.

39.

40. Hyperbola   **41.** Ellipse

42. Ellipse   **43.** Hyperbola

44. a. Line and parabola   **45. a.** Line and parabola

b.

c. $\left(\dfrac{5}{2}, \dfrac{5}{4}\right)$, $(-4, 11)$

b.

c. $(-5, 15)$, $(3, -1)$

46. a. Circle and line   **47. a.** Circle and line

b.

c. $(0, 3)$, $\left(\dfrac{12}{5}, -\dfrac{9}{5}\right)$

b.
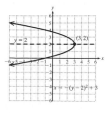

c. $(0, -4)$, $\left(\dfrac{16}{5}, -\dfrac{12}{5}\right)$

48. $(0, -2)$, $\left(\dfrac{16}{9}, \dfrac{14}{9}\right)$ **49.** $\left(-\dfrac{7}{5}, \dfrac{13}{5}\right)$, $(-5, -1)$

50. $(8, 4)$, $(2, -2)$ **51.** $(2, 4)$, $(-2, 4)$, $(\sqrt{2}, 2)$, $(-\sqrt{2}, 2)$

52. $(3, 1)$, $(3, -1)$, $(-3, 1)$, $(-3, -1)$

53. $(6, 5)$, $(-6, 5)$, $(6, -5)$, $(-6, -5)$

54.

55.

56.
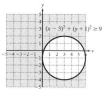

57.

58.

59.

60.

61.

Chapter 10 Test, pp. 796–797

1.

2. Center: $\left(\dfrac{5}{6}, -\dfrac{1}{3}\right)$; $r = \dfrac{5}{7}$

3.

4. $2\sqrt{10}$ **5.** Center: $(0, 2)$; $r = 3$ **6. a.** $\sqrt{5}$

b. $x^2 + (y - 4)^2 = 5$ **7.** The center is the midpoint $(3.8, 1.95)$.

8.
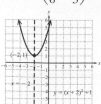

9.
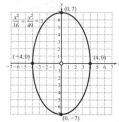

10.
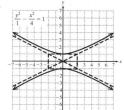

11. a. $(-3, 0)(0, 4)$; ii   **b.** No solution; i   **12.** The addition method can be used if the equations have corresponding *like* terms.   **13.** $(2, 0)(-2, 0)$

14.

15.

16.

17.

Chapters 1–10 Cumulative Review Exercises, pp. 798–799

1. All real numbers

2. $\left(\dfrac{10}{3}, \infty\right)$

3. The integers are 10 and 15.

4. a. $(-5, 0)$, $(0, 3)$ **b.** $m = \dfrac{3}{5}$

c.

5. slope: $\dfrac{3}{4}$; y-intercept: $\left(0, -\dfrac{3}{2}\right)$

6. There are 12 dimes and 5 quarters.

7. $(2, -3, 1)$ **8.** $\left(\dfrac{5}{2}, \dfrac{3}{2}\right)$ **9.** $(6, 3)$

10. $f(0) = -12; f(-1) = -16; f(2) = -10; f(4) = -16$

11. $g(2) = 5; g(8) = -1; g(3) = 0; g(-5) = 5$

12. $z = 32$ **13.** $(g \circ f)(x) = x + 7; x \geq -1$

14. **a.** -5 **b.** $(x + 1)(x^2 + 1); -5$ **c.** They are the same.

15. $(x + y)(x - y - 6)$ **16.** $x^3 - 2x^2 - 7x - 4$

17. $6, \dfrac{3}{2}$ **18.** $\dfrac{a - 1}{a + 2}$ **19.** $\dfrac{-x^2 - x - 4}{(x + 3)(x - 2)}$ **20.** $0, -1$

21. **a.** No solution   **b.** -11 **22.** $-30 + 24i$

23. $\dfrac{12}{41} + \dfrac{15}{41}i$ **24.** $2\sqrt{7}$ m

25. **a.** 17.6 ft; 39.6 ft; 70.4 ft   **b.** 8 sec

26. $-\dfrac{1}{5}, \dfrac{1}{10} \pm \dfrac{\sqrt{3}}{10}i$ **27.** $2 \pm \sqrt{11}$ **28.** $(-5, -36)$

29. **a.** $(-3, 0), (1, 0)$ **b.** $(0, 3)$ **c** $(-1, 4)$

30. $(-1, 19)$ **31.** $\left(-\infty, \dfrac{1}{2}\right] \cup \left[\dfrac{9}{2}, \infty\right)$ **32.** $\log_8 32 = \frac{5}{3}$

33. $\dfrac{2}{3}$ **34.** $h^{-1}(x) = \sqrt[3]{x + 1}$ **35.** $x^2 + (y - 5)^2 = 16$

36.

37. $\left(-\dfrac{1}{2}, 0\right)$ **38.** $(0, -4)$

39.

40.

Additional Topic Appendix

Section A.1 Practice Exercises, pp. A6–A7

1. **a.** $a^2 + 2ab + b^2$; $a^3 + 3a^2b + 3ab^2 + b^3$; binomial
b. $n(n - 1)(n - 2) \cdots (2)(1)$; factorial   **c.** $6; 2; 1; 1$
d. binomial   **e.** Pascal's

3. $a^3 + 3a^2b + 3ab^2 + b^3$ **5.** $1 + 4g + 6g^2 + 4g^3 + g^4$

7. $p^7 + 7p^6q^2 + 21p^5q^4 + 35p^4q^6 + 35p^3q^8 + 21p^2q^{10} + 7pq^{12} + q^{14}$

9. $s^5 + 5s^4t + 10s^3t^2 + 10s^2t^3 + 5st^4 + t^5$

11. $625 - 500u^3 + 150u^6 - 20u^9 + u^{12}$

13. $x^6 - 12x^4 + 48x^2 - 64$ **15.** 120

17. 1   **19.** False   **21.** True

23. $6! = 6 \cdot (5 \cdot 4 \cdot 3 \cdot 2 \cdot 1) = 6 \cdot 5!$ **25.** 1680

27. 6   **29.** 56   **31.** 1   **33.** $m^{11} + 11m^{10}n + 55m^9n^2$

35. $u^{24} - 12u^{22}v + 66u^{20}v^2$ **37.** 9 terms

39. $s^6 + 6s^5t + 15s^4t^2 + 20s^3t^3 + 15s^2t^4 + 6st^5 + t^6$

41. $b^3 - 9b^2 + 27b - 27$

43. $16x^4 + 32x^3y + 24x^2y^2 + 8xy^3 + y^4$

45. $c^{14} - 7c^{12}d + 21c^{10}d^2 - 35c^8d^3 + 35c^6d^4 - 21c^4d^5 + 7c^2d^6 - d^7$

47. $\dfrac{1}{32}a^5 - \dfrac{5}{16}a^4b + \dfrac{5}{4}a^3b^2 - \dfrac{5}{2}a^2b^3 + \dfrac{5}{2}ab^4 - b^5$

49. $x^4 + 16x^3y + 96x^2y^2 + 256xy^3 + 256y^4$

51. $-462m^6n^5$ **53.** $495u^{16}v^4$ **55.** g^9

Section A.2 Practice Exercises, pp. A15–A18

1. **a.** determinant; $ad - bc$ **b.** minor   **c.** $a_2; \begin{vmatrix} b_1 & c_1 \\ b_2 & c_2 \end{vmatrix}$

3. 16   **5.** 1   **7.** 4   **9.** 6   **11.** 26   **13.** 6

15. 46   **17.** **a.** 30   **b.** 30   **19.** Choosing the row or column with the most zero elements simplifies the arithmetic when evaluating a determinant.

21. 12   **23.** -15 **25.** 0   **27.** $8a - 2b$

29. $4x - 3y + 6z$ **31.** 0

33. $\mathbf{D} = 16; \mathbf{D}_x = -63; \mathbf{D}_y = 66$ **35.** $(2, -1)$

37. $(4, -1)$ **39.** $\left(\dfrac{23}{13}, \dfrac{9}{13}\right)$

41. Cramer's rule does not apply when the determinant $\mathbf{D} = 0$.

43. No solution; Inconsistent system   **45.** $(0, 0)$

47. Infinitely many solutions; $\{(x, y) \mid x + 5y = 3\}$; Dependent equations   **49.** $x = 1$

51. $z = \dfrac{1}{2}$ **53.** $y = \dfrac{16}{41}$ **55.** $(3, 2, 1)$

57. Cramer's rule does not apply.   **59.** $x = -19$

61. $w = 1$ **63.** 36   **65.** **a.** 2   **b.** -2 **c.** $x = -1$

67. The measures are 37.5° and 52.5°.

69. 40 iPods, 15 iPads, and 20 iPhones were sold.

71. There were 560 women and 440 men in the survey.

Section A.3 Practice Exercises, pp. A23–A26

1. **a.** infinite; finite   **b.** terms; nth   **c.** alternating
d. series   **e.** summation   **f.** index; $(3)^2 + (4)^2 + (5)^2$

3. $1, -1, -3, -5, -7$ **5.** $0, 1, \sqrt{2}, \sqrt{3}$ **7.** $\dfrac{1}{3}, \dfrac{1}{2}, \dfrac{3}{5}, \dfrac{2}{3}, \dfrac{5}{7}$

9. $0, \dfrac{1}{4}, -\dfrac{2}{5}, \dfrac{1}{2}$ **11.** $1, -4, 9$ **13.** $\dfrac{5}{3}, \dfrac{7}{4}, \dfrac{9}{5}, \dfrac{11}{6}$

15. $0, 6, 24, 60, 120, 210$ **17.** $-1, 2, -3, 4$ **19.** When n is odd, the term is positive. When n is even, the term is negative.

21. $a_n = 3n$ **23.** $a_n = 2n + 1$ **25.** $a_n = \dfrac{n}{n + 1}$

27. $a_n = (-1)^n$ **29.** $a_n = (-1)^{n+1} 3^n$ **31.** $a_n = \dfrac{1}{4^n}$

33. \$30.00, \$30.90, \$31.83, \$32.78

35. $16, 8, 4, 2, 1, \dfrac{1}{2}, \dfrac{1}{4}, \dfrac{1}{8}$ (grams)   **37.** 90   **39.** $\dfrac{31}{16}$

41. 30   **43.** 10   **45.** $\dfrac{73}{12}$ **47.** 38   **49.** -1

51. 55   **53.** $\displaystyle\sum_{n=1}^{6} n$ **55.** $\displaystyle\sum_{i=1}^{5} 4$ **57.** $\displaystyle\sum_{j=1}^{5} 4j$

59. $\displaystyle\sum_{k=1}^{4} (-1)^{k+1}\frac{1}{3^k}$ **61.** $\displaystyle\sum_{n=1}^{5} x^n$ **63.** 15.4 g

65. 4.8 g   **67.** $-3, 2, 7, 12, 17$ **69.** $5, 21, 85, 341, 1365$
71. $1, 1, 2, 3, 5, 8, 13, 21, 34, 55$

Section A.4 Practice Exercises, pp. A35–A38

1. a. arithmetic   **b.** difference   **c.** $a_1 + (n-1)d; d$

d. series   **e.** $\dfrac{n}{2}(a_1 + a_n)$ **f.** geometric   **g.** ratio

h. $a_1 r^{n-1}$ **i.** $\dfrac{a_1(1 - r^n)}{1 - r}$ **j.** $\dfrac{a_1}{1 - r}; 1$

3. 2   **5.** -3 **7.** -2 **9.** $3, 8, 13, 18, 23$

11. $2, \dfrac{5}{2}, 3, \dfrac{7}{2}, 4$ **13.** $2, -2, -6, -10, -14$

15. $a_n = -5 + 5n$ **17.** $a_n = -2n$ **19.** $a_n = \dfrac{3}{2} + \dfrac{1}{2}n$

21. $a_n = 25 - 4n$ **23.** $a_n = -14 + 6n$ **25.** $a_6 = 17$
27. $a_9 = 47$ **29.** $a_7 = -30$ **31.** $a_{11} = -48$ **33.** 19
35. 22   **37.** 23   **39.** 11   **41.** $a_1 = -2, a_2 = -5$
43. 670   **45.** 290   **47.** -15 **49.** 95   **51.** 924
53. 300   **55.** -210 **57.** 5050   **59.** 980 seats; $14,700
61. A sequence is geometric if the ratio between each term and the preceding term is constant.

63. $\dfrac{1}{2}$ **65.** -3 **67.** 4   **69.** $-4, 4, -4, 4, -4$

71. $8, 2, \dfrac{1}{2}, \dfrac{1}{8}, \dfrac{1}{32}$ **73.** $2, -6, 18, -54, 162$

75. $a_n = 2(3)^{n-1}$ **77.** $a_n = -6(-2)^{n-1}$

79. $a_n = \dfrac{16}{3}\left(\dfrac{3}{4}\right)^{n-1}$ **81.** $-\dfrac{3}{128}$ **83.** $-\dfrac{2}{81}$ **85.** 135

87. -10 **89.** $\dfrac{1}{9}$ **91.** 1   **93.** $\dfrac{1562}{125}$ **95.** $-\dfrac{11}{8}$

97. $\dfrac{3124}{27}$ **99.** $\dfrac{2059}{729}$ **101.** -172

103. a. \$1050.00, \$1102.50, \$1157.63, \$1215.51
b. $a_{10} = \$1628.89; a_{20} = \$2653.30; a_{40} = \$7039.99$

105. $r = \dfrac{1}{6}$; sum is $\dfrac{6}{5}$ **107.** $r = -\dfrac{1}{4}$; sum is $\dfrac{4}{5}$

109. $r = -\dfrac{1}{3}$; sum is $-\dfrac{9}{4}$ **111.** $r = -\dfrac{3}{2}$; sum does not exist

113. \$800 million   **115.** 28 ft
117. a. \$1,429,348   **b.** \$1,505,828   **c.** \$76,480

Photo Credits

Application Index

Biology/Health/Life Sciences

Adenosine deaminase levels in humans, 122

Age range for average height of boys, 81

Amount of medicine prescribed by weight of patient, 476

Average weight for boys/girls based on age, 151

Bacteria population as function of time, 627, 674, 717, 736, 738

Bison population in Wyoming, 467

Blood pressure, normal range of, 15

Calcium intake program, 316

Calories in candy, 467

Cholesterol level ranges, 120, 461

Epidemic deaths as function of time, 708

Fat amount in foods, 484

Feasible cage configurations at animal shelter, 278

Femur length vs. height in women, 190, 193–194

Hemoglobin range in humans, 92

Length of human pregnancy, 88

Life expectancy vs. year of birth, 186

Longevity of animals, by species, 191

Maximum heart rate vs. age, 120, 197

Normal level of TSH for adults, 88

Number of manatees in Florida, 467

pH level in ammonia, 682

pH level in antacid tablet, 733

pH level in blood, 686

pH level in shampoo, 682

pH ranges in food substances, 15, 733

Pollution in atmosphere, 477

Radioactive decay, 703

Ratio of cats to dogs in shelter, 461

Sodium content of foods, from total sodium intakes, 290–291

Water level retention in pond, 78–79

Weight vs. age for children, 151

White blood cell range in humans, 92

Business and Economics

Airline flights as function of passenger volume, 653

Amount of each nut in nut mixture, from relative amounts, 294

Cell phone subscription trends, 150

Commission, 64

Cost of airline operation, 550

Cost of cab ride per mile, 138

Cost of running business, 184

Cost per item as function of number produced, 403, 477

Cost to package product, 627

Currency exchange rates, 487

Depreciation, 130–131, 139, 734

Drinks, price of related to number sold, 174

Earnings needed to meet average salary level, 81

Hot dogs, price of related to number sold, 174

Interest
  compound, 584, 699, 705, 717, 733, 737, A–37
  simple, 58, 59, 64–65, 66, 112, 269, 474, 490

Markups/markdowns, 58–59, 64, 66

Median weekly salary of individual with degree, 66

Minimum wage trends, 167

Mixing foods for resale, 66, 294

Mixing money among investments, 59–60

Monthly salary, 78

Number of fish, from weight of total catch, 467

Orders processed per day by each worker, 316

Profit as function of number of items produced, 81, 228, 337, 584

Profit as function of number of items sold, 81, 477, 649

Ratio of men to women in accounting firm, 467

Salary plan, 665–666

Salary with commissions, vs. sales, 138, 171, 186

Salary with tips, vs. number of tables served, 234

Sales as function of advertising expenditures, 687

Sales by each salesperson, from relative totals, 290

Time required for each person working alone to finish job, 469, 470

Time required for two workers to finish job, 285, 469

Tourist revenue, A–38

Construction and Design

Bridge support design, 618

Bulge height in heat-expansion of bridge, 570

Cell tower location, 748, 783

Cost to carpet a room, 229

Dimensions of fenced area, 628

Elevator passenger capacity, 122

Fencing needed to enclose area, 71, 285, 518, 628

Length of roof, from structure dimensions, 572

Length of screw, 121

Length of tower guy wires, 571

Margin of error in measurement, 108, 122

Maximum enclosable area with given amount of fence, 71

Outdoor fountain water projection, 759

Production of two models of desks, 286

Pump performance, from relative performance, 314, 469

Slope of hill, 148

Slope of ladder, 142, 148, 402

Slope of leaning telephone pole, 417

Slope of roof, 142, 148

Slope of treadmill, 148

Stained glass window design, 599, 793

Strength of beam as function of dimensions, 478

Time required for two workers to finish job, 285, 465, 489

Time required to fill tank, from pipe size, 488

Window film needed for project, 25

Consumer Applications

Car rental cost vs. mileage, 272

Cellular phone charges, 167

Cellular phone subscriptions in U.S., 150

City and highway miles driven, 263

Coins by type, from relative number, 272, 798

Cook time vs. weight for turkey, 551

Cost of bottle of laundry detergent by size, 467

Cost of cab ride, 138

Cost of car rental, 170

Cost of items, from money spent, 269–270, 294, 467

Cost of nights in motel, 256

Cost to rent apartment, 256

Dog run dimensions, 71, 285

Gas mileage, 467

Gas mileage as function of speed, 627

Gas required to complete trip, 29

Home value, A–37

Subject Index

Perimeter and Circumference

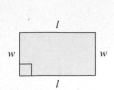

Rectangle

$P = 2l + 2w$

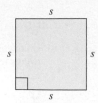

Square

$P = 4s$

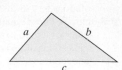

Triangle

$P = a + b + c$

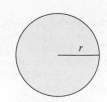

Circle

Circumference: $C = 2\pi r$

Area

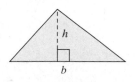

Rectangle

$A = lw$

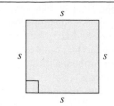

Square

$A = s^2$

Parallelogram

$A = bh$

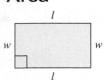

Triangle

$A = \frac{1}{2}bh$

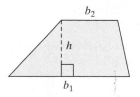

Trapezoid

$A = \frac{1}{2}(b_1 + b_2)h$

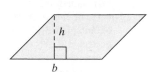

Circle

$A = \pi r^2$

Volume

Rectangular Solid

$V = lwh$

Cube

$V = s^3$

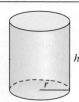

Right Circular Cylinder

$V = \pi r^2 h$

Right Circular Cone

$V = \frac{1}{3}\pi r^2 h$

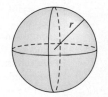

Sphere

$V = \frac{4}{3}\pi r^3$

Angles

- Two angles are **complementary** if the sum of their measures is 90°.

- Two angles are **supplementary** if the sum of their measures is 180°.

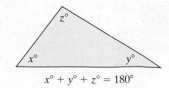

- $\angle a$ and $\angle c$ are vertical angles, and $\angle b$ and $\angle d$ are vertical angles. The measures of vertical angles are equal.

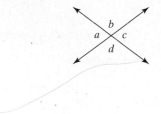

- The sum of the measures of the angles of a triangle is 180°.

$$x° + y° + z° = 180°$$